How to Use This Book Effectively

This book is organized by chapters and sections within chapters. Fundamental concepts and associated equations within each section lay the foundation for applications of engineering thermodynamics provided in solved examples, end-of-chapter problems and exercises, and accompanying discussions. **Boxed material** within sections of the book allows you to explore selected topics in greater depth, as in the boxed discussion of properties and nonproperties.

Contemporary issues related to thermodynamics are introduced throughout the text with three unique features: **Energy & Environment** discussions explore issues related to energy resource use and the environment. **BioConnections** tie topics to applications in bioengineering and biomedicine. **Horizons** link subject matter to emerging technologies and thought-provoking issues.

Other core features of this book that facilitate your study and contribute to your understanding include:

Examples

- Numerous annotated solved examples are provided that feature the **solution methodology** presented in Sec. 1.9 and illustrated in Example 1.1. We encourage you to study these examples, including the accompanying comments.
- Each solved example concludes with a list of the **SKILLS DEVELOPED** in solving the example and a **Quick Quiz** that allows an immediate check of understanding.
- Less formal examples are given throughout the text indicated by **FOR EXAMPLE**. These examples also should be studied.

Exercises

- Each chapter has a set of discussion questions under the heading **Exercises: Things Engineers Think About** that may be done on an individual or small-group basis. They allow you to gain a deeper understanding of the text material and think critically.
- Every chapter has a set of questions in a section called **Checking Understanding** that provide opportunity for individual or small group *self-testing* of the fundamental ideas presented in the chapter. Included are a variety of exercises, such as matching, fill-in-the-blank, short answer, and true-and-false questions.
- A large number of end-of-chapter problems also are provided under the heading **Problems: Developing Engineering Skills**. The problems are sequenced to coordinate with the subject matter. The problems are also classified under headings to expedite the process of selecting review problems to solve. Answers to selected problems are provided on the **student companion website** that accompanies this book at www.wiley.com/college/moran.
- Because one purpose of this book is to help you prepare to use thermodynamics in engineering practice, design considerations related to thermodynamics are included. Every chapter has a set of problems under the heading **Design & Open-Ended Problems: Exploring Engineering Practice** that provide opportunities for practicing creativity, formulating and solving design and open-ended problems, using the Internet and library resources to find relevant information, making engineering judgments, and developing communications skills.

Further Study Aids

- Each chapter opens with an introduction giving the **engineering context**, stating the **chapter objective**, and listing the **learning outcomes**.
- Each chapter concludes with a Chapter Summary and Study Guide that provides a point of departure to study for examinations.
- For easy reference, each chapter also concludes with lists of Key Engineering Concepts and Key Equations.
- Important terms are listed in the margins and coordinated with the text material at those locations.
- Important equations are boxed and numbered throughout.
- **TAKE NOTE...** in the margin provides just-in-time information that illuminates the current discussion or refines our problem-solving methodology.
- **(→) Animation** in the margin identifies an animation that reinforces the text presentation at that point. Animations can be viewed by going to the **student companion website** for this book. See **TAKE NOTE...** in Section 1.2 for further detail about accessing animations.
- **C** in the margin denotes end-of-chapter problems where the use of appropriate computer software is recommended.
- For quick reference, conversion factors and important constants are provided within the text and as an online resource.
- A list of symbols is provided on the inside back cover.

Conversion Factors

Mass and Density

1 kg	= 2.2046 lb
1 g/cm^3	= 10^3 kg/m^3
1 g/cm^3	= 62.428 lb/ft^3
1 lb	= 0.4536 kg
1 lb/ft^3	= 0.016018 g/cm^3
1 lb/ft^3	= 16.018 kg/m^3

Length

1 cm	= 0.3937 in.
1 m	= 3.2808 ft
1 in.	= 2.54 cm
1 ft	= 0.3048 m

Velocity

1 km/h	= 0.62137 mile/h
1 mile/h	= 1.6093 km/h

Volume

1 cm^3	= 0.061024 in.^{3}
1 m^3	= 35.315 ft^3
1 L	= 10^{-3} m^3
1 L	= 0.0353 ft^3
1 in.^{3}	= 16.387 cm^3
1 ft^3	= 0.028317 m^3
1 gal	= 0.13368 ft^3
1 gal	= 3.7854 × 10^{-3} m^3

Force

1 N	= 1 kg · m/s^2
1 N	= 0.22481 lbf
1 lbf	= 32.174 lb · ft/s^2
1 lbf	= 4.4482 N

Pressure

1 Pa	= 1 N/m^2
	= 1.4504 × 10^{-4} lbf/in.^{2}
1 bar	= 10^5 N/m^2
1 atm	= 1.01325 bar
1 lbf/in.^{2}	= 6894.8 Pa
1 lbf/in.^{2}	= 144 lbf/ft^2
1 atm	= 14.696 lbf/in.^{2}

Energy and Specific Energy

1 J	= 1 N · m = 0.73756 ft · lbf
1 kJ	= 737.56 ft · lbf
1 kJ	= 0.9478 Btu
1 kJ/kg	= 0.42992 Btu/lb
1 ft · lbf	= 1.35582 J
1 Btu	= 778.17 ft · lbf
1 Btu	= 1.0551 kJ
1 Btu/lb	= 2.326 kJ/kg
1 kcal	= 4.1868 kJ

Energy Transfer Rate

1 W	= 1 J/s = 3.413 Btu/h
1 kW	= 1.341 hp
1 Btu/h	= 0.293 W
1 hp	= 2545 Btu/h
1 hp	= 550 ft · lbf/s
1 hp	= 0.7457 kW

Specific Heat

1 kJ/kg · K	= 0.238846 Btu/lb · °R
1 kcal/kg · K	= 1 Btu/lb · °R
1 Btu/lb · °R	= 4.1868 kJ/kg · K

Others

1 ton of refrigeration = 200 Btu/min = 211 kJ/min
1 volt = 1 watt per ampere

Constants

Universal Gas Constant

$$\bar{R} = \begin{cases} 8.314 \text{ kJ/kmol} \cdot \text{K} \\ 1545 \text{ ft} \cdot \text{lbf/lbmol} \cdot \text{°R} \\ 1.986 \text{ Btu/lbmol} \cdot \text{°R} \end{cases}$$

Standard Acceleration of Gravity

$$g = \begin{cases} 9.80665 \text{ m/s}^2 \\ 32.174 \text{ ft/s}^2 \end{cases}$$

Standard Atmospheric Pressure

$$1 \text{ atm} = \begin{cases} 1.01325 \text{ bar} \\ 14.696 \text{ lbf/in.}^2 \\ 760 \text{ mm Hg} = 29.92 \text{ in. Hg} \end{cases}$$

Temperature Relations

$T(\text{°R}) = 1.8 \, T(\text{K})$
$T(\text{°C}) = T(\text{K}) - 273.15$
$T(\text{°F}) = T(\text{°R}) - 459.67$

Moran's Principles of Engineering Thermodynamics

SI Version

Global Edition

MICHAEL J. MORAN
The Ohio State University

HOWARD N. SHAPIRO
Iowa State University

DAISIE D. BOETTNER
Brigadier General (Retired), USA

MARGARET B. BAILEY
Rochester Institute of Technology

WILEY

ISBN: 978-1-119-45406-9

Printed and bound by CPI Group (UK) Ltd, Croydon, CR0 4YY

C9781119454069_090523

A Textbook for the 21st Century

In the 21st century, engineering thermodynamics plays a central role in developing improved ways to provide and use energy, while mitigating the serious human health and environmental consequences accompanying energy—including air and water pollution and global climate change. Applications in bioengineering, biomedical systems, and nanotechnology also continue to emerge. This book provides the tools needed by specialists working in all such fields. For non-specialists, this book provides background for making decisions about technology related to thermodynamics—on the job and as informed citizens.

Engineers in the 21st century need a solid set of analytical and problem-solving skills as the foundation for tackling important societal issues relating to engineering thermodynamics. The global edition develops these skills and significantly expands our coverage of their applications to provide

- current context for the study of thermodynamic principles.
- relevant background to make the subject meaningful for meeting the challenges of the decades ahead.
- significant material related to existing technologies in light of new challenges.

In the global edition, we build on the **core features** that have made the text the global leader in engineering thermodynamics education. We are known for our clear and concise explanations grounded in the fundamentals, pioneering pedagogy for effective learning, and relevant, up-to-date applications. Through the creativity and experience of our author team, and based on excellent feedback from instructors and students, we continue to enhance what has become the leading text in the field.

New in the Global Edition

The global edition is aimed at helping students

- better understand and apply the subject matter, and
- fully appreciate the relevance of the topics to engineering practice and to society.

The end-of-chapter problem set is modified and is completely SI.

Other Core Features

This edition also provides, under the heading `How to Use` `This Book Effectively,` an updated roadmap to core features of this text that make it so effective for student learning. To fully understand all of the many features we have built into the book, be sure to see this important element.

In this edition, several enhancements to improve student learning have been introduced or upgraded:

- The p–h diagrams for two refrigerants: CO_2 (R-744) and R-410A are included as Figs. A-10 and A-11, respectively, in the appendix. The ability to locate states on property diagrams is an important skill that is used selectively in end-of-chapter problems.
- **Animations** are offered at key subject matter locations to improve student learning. When viewing the animations, students will develop deeper understanding by visualizing key processes and phenomena.
- Special text elements feature important illustrations of engineering thermodynamics applied to our environment, society, and world:
  - **Energy & Environment** presentations explore topics related to energy resource use and environmental issues in engineering.
  - **BioConnections** discussions tie textbook topics to contemporary applications in biomedicine and bioengineering.
  - **Horizons** features have been included that link subject matter to thought-provoking 21st-century issues and emerging technologies.

Suggestions for additional reading and sources for topical content presented in these elements provided on request.

- End-of-chapter problems in each of the four modes: **conceptual, checking understanding, skill building, and design** have been extensively revised.
- New and revised class-tested material contributes to student learning and instructor effectiveness:
  - Significant content explores how thermodynamics contributes to meet the challenges of the 21st century.
  - Key aspects of fundamentals and applications within the text have been enhanced.
- In response to instructor and student needs, class-tested changes that contribute to a more **just-in-time** presentation have been introduced:
  - TAKE NOTE... entries in the margins are expanded throughout the textbook to improve student learning. For example, see Section 1.2.3.
  - **Boxed material** allows students and instructors to explore topics in greater depth. For example, see Section 3.5.2.
  - **Margin terms** throughout aid in navigating subject matter.

Supplements

The following supplements are available with the text:

- Outstanding *Instructor* and *Student* companion web sites (visit www.wiley.com/college/moran) that greatly enhance teaching and learning:
  - Instructor Companion Site: Assists instructors in delivering an effective course with resources including
    - a Steam Table Process Overview to assist students in mastering the use of the steam tables for retrieving data.
    - animations—with just-in-time labels in the margins.
    - a complete solution manual that is easy to navigate.
    - IT: Interactive Thermodynamics Software.
    - sample syllabi on semester and quarter bases.
    - Lecture PowerPoints
  - Student Companion Site: Helps students learn the subject matter with resources including
    - Steam Table Process Overview.
    - animations.
    - answers to selected problems.
- *Interactive Thermodynamics: IT software* is a highly valuable learning tool that allows students to develop engineering models, perform "what-if" analyses, and examine principles in more detail to enhance their learning. Brief tutorials of *IT* are included within the text and the use of *IT* is illustrated within selected solved examples.
- Skillful use of tables and property diagrams is prerequisite for the effective use of software to retrieve thermodynamic property data. The latest version of *IT* provides data for CO_2 (R-744) and R-410A using as its source Mini REFPROP by permission of the National Institute of Standards and Technology (NIST).

Visit www.wiley.com/college/moran or contact your local Wiley representative for information on the above-mentioned supplements.

Ways to Meet Different Course Needs

In recognition of the evolving nature of engineering curricula, and in particular of the diverse ways engineering thermodynamics is presented, the text is structured to meet a variety of course needs. The following table illustrates several possible uses of the textbook assuming a semester basis (3 credits). Courses could be taught using this textbook to engineering students with appropriate background beginning in their second year of study.

Type of course	Intended audience	Chapter coverage
Survey courses	Non-majors	• <u>Principles</u>. Chaps. 1–6. • <u>Applications</u>. Selected topics from Chaps. 8–10 (omit compressible flow in Chap. 9).
	Majors	• <u>Principles</u>. Chaps. 1–6. • <u>Applications</u>. Same as above plus selected topics from Chaps. 12 and 13.
Two-course sequences	Majors	• <u>First course</u>. Chaps. 1–7. (Chap. 7 may be deferred to second course or omitted.) • <u>Second course</u>. Selected topics from Chaps. 8–14 to meet particular course needs.

Acknowledgments

We thank the many users of our previous editions, located at hundreds of universities and colleges in the United States, Canada, and world-wide, who continue to contribute to the development of our text through their comments and constructive criticism.

The following colleagues have assisted in the development of this text. We greatly appreciate their contributions:

Hisham A. Abdel-Aal, University of North Carolina Charlotte
Alexis Abramson, Case Western Reserve University
Edward Anderson, Texas Tech University
Jason Armstrong, University of Buffalo
Euiwon Bae, Purdue University
H. Ed. Bargar, University of Alaska
Amy Betz, Kansas State University
John Biddle, California Polytechnic State University, Pomona
Jim Braun, Purdue University
Robert Brown, Iowa State University
Marcello Canova, The Ohio State University
Bruce Carroll, University of Florida
Gary L. Catchen, The Pennsylvania State University
Cho Lik Chan, University of Arizona
John Cipolla, Northeastern University
Matthew Clarke, University of Calgary
Stephen Crown, University of Texas Pan American
Ram Devireddy, Louisiana State University
Jon F. Edd, Vanderbilt University
Gloria Elliott, University of North Carolina Charlotte
P. J. Florio, New Jersey Institute of Technology
Stephen Gent, South Dakota State University
Nick Glumac, University of Illinois, Urbana-Champaign
Jay Gore, Purdue University
Nanak S. Grewal, University of North Dakota
John Haglund, University of Texas at Austin
Davyda Hammond, Germanna Community College
Kelly O. Homan, Missouri University of Science and Technology-Rolla
Andrew Kean, California Polytechnic State University, San Luis Obispo
Jan Kleissl, University of California, San Diego
Deify Law, Baylor University
Xiaohua Li, University of North Texas
Randall D. Manteufei, University of Texas at San Antonio
Michael Martin, Louisiana State University
Mason Medizade, California Polytechnic State University
Alex Moutsoglou, South Dakota State University
Sameer Naik, Purdue University
Jay M. Ochterbeck, Clemson University
Jason Olfert, University of Alberta
Juan Ordonez, Florida State University
Tayhas Palmore, Brown University
Arne Pearlstein, University of Illinois, Urbana-Champaign
Laurent Pilon, University of California, Los Angeles
Michele Putko, University of Massachusetts Lowell

Albert Ratner, The University of Iowa
John Reisel, University of Wisconsin-Milwaukee
Michael Renfro, University of Connecticut
Michael Reynolds, University of Arkansas
Donald E. Richards, Rose-Hulman Institute of Technology
Robert Richards, Washington State University
Edward Roberts, University of Calgary
David Salac, University at Buffalo SUNY
Brian Sangeorzan, Oakland University
Alexei V. Saveliev, North Carolina State University
Enrico Sciubba, University of Roma- Sapienza
Dusan P. Sekulic, University of Kentucky
Benjamin D. Shaw, University of California-Davis
Angela Shih, California Polytechnic State University Pomona
Gary L. Solbrekken, University of Missouri
Clement C. Tang, University of North Dakota
Constantine Tarawneh, University of Texas Pan American
Evgeny Timofeev, McGill University
Elisa Toulson, Michigan State University
V. Ismet Ugursal, Dalhousie University
Joseph Wang, University of California—San Diego
Kevin Wanklyn, Kansas State University
K. Max Zhang, Cornell University

The views expressed in this text are those of the authors and do not necessarily reflect those of individual contributors listed, The Ohio State University, Iowa State University, Rochester Institute of Technology, the United States Military Academy, the Department of the Army, or the Department of Defense.

We also acknowledge the efforts of many individuals in the John Wiley and Sons, Inc., organization who have contributed their talents and energy to this edition. We applaud their professionalism and commitment.

We continue to be extremely gratified by the reception this book has enjoyed over the years. With this edition we have made the text more effective for teaching the subject of engineering thermodynamics and have greatly enhanced the relevance of the subject matter for students who will shape the 21st century. As always, we welcome your comments, criticisms, and suggestions.

MICHAEL J. MORAN
moran.4@osu.edu

HOWARD N. SHAPIRO
hshapiro513@gmail.com

DAISIE D. BOETTNER
BoettnerD@aol.com

MARGARET B. BAILEY
Margaret.Bailey@rit.edu

Contents

4 Control Volume Analysis Using Energy 129

5 The Second Law of Thermodynamics 179

Getting Started

Introductory Concepts and Definitions

Engineering Context

Although aspects of thermodynamics have been studied since ancient times, the formal study of thermodynamics began in the early nineteenth century through consideration of the capacity of hot objects to produce work. Today the scope is much larger. Thermodynamics now provides essential concepts and methods for addressing critical twenty-first-century issues, such as using fossil fuels more effectively, fostering renewable energy technologies, and developing more fuel-efficient means of transportation. Also critical are the related issues of greenhouse gas emissions and air and water pollution.

Thermodynamics is both a branch of science and an engineering specialty. The scientist is normally interested in gaining a fundamental understanding of the physical and chemical behavior of fixed quantities of matter at rest and uses the principles of thermodynamics to relate the properties of matter. Engineers are generally interested in studying *systems* and how they interact with their *surroundings*. To facilitate this, thermodynamics has been extended to the study of systems through which matter flows, including bioengineering and biomedical systems.

The **objective** of this chapter is to introduce you to some of the fundamental concepts and definitions that are used in our study of engineering thermodynamics. In most instances this introduction is brief, and further elaboration is provided in subsequent chapters.

LEARNING OUTCOMES

When you complete your study of this chapter, you will be able to...

- Explain several fundamental concepts used throughout the book, including closed system, control volume, boundary and surroundings, property, state, process, the distinction between extensive and intensive properties, and equilibrium.
- Identify SI Engineering units, including units for specific volume, pressure, and temperature.
- Work with Kelvin and Celsius temperature scales.
- Apply the problem-solving methodology used in this book.

1.1 Using Thermodynamics

Engineers use principles drawn from thermodynamics and other engineering sciences, including fluid mechanics and heat and mass transfer, to analyze and design devices intended to meet human needs. Throughout the twentieth century, engineering applications of thermodynamics helped pave the way for significant improvements in our quality of life with advances in major areas such as surface transportation, air travel, space flight, electricity generation and transmission, building heating and cooling, and improved medical practices. The wide realm of these applications is suggested by **Table 1.1**.

In the twenty-first century, engineers will create the technology needed to achieve a sustainable future. Thermodynamics will continue to advance human well-being by addressing looming societal challenges owing to declining supplies of energy resources: oil, natural gas, coal, and fissionable material; effects of global climate change; and burgeoning population. Life in the United States is expected to change in several important respects by mid-century. In the area of power use, for example, electricity will play an even greater role than today. **Table 1.2** provides predictions of other changes experts say will be observed.

If this vision of mid-century life is correct, it will be necessary to evolve quickly from our present energy posture. As was the case in the twentieth century, thermodynamics will contribute significantly to meeting the challenges of the twenty-first century, including using fossil fuels more effectively, advancing renewable energy technologies, and developing more energy-efficient transportation systems, buildings, and industrial practices. Thermodynamics also will play a role in mitigating global climate change, air pollution, and water pollution. Applications will be observed in bioengineering, biomedical systems, and the deployment of nanotechnology. This book provides the tools needed by specialists working in all such fields. For nonspecialists, the book provides background for making decisions about technology related to thermodynamics—on the job, as informed citizens, and as government leaders and policy makers.

1.2 Defining Systems

The key initial step in any engineering analysis is to describe precisely what is being studied. In mechanics, if the motion of a body is to be determined, normally the first step is to define a *free body* and identify all the forces exerted on it by other bodies. Newton's second law of motion is then applied. In thermodynamics the term *system* is used to identify the subject of the analysis. Once the system is defined and the relevant interactions with other systems are identified, one or more physical laws or relations are applied.

system
  The **system** is whatever we want to study. It may be as simple as a free body or as complex as an entire chemical refinery. We may want to study a quantity of matter contained within a closed, rigid-walled tank, or we may want to consider something such as a pipeline through which natural gas flows. The composition of the matter inside the system may be fixed or may be changing through chemical or nuclear reactions. The shape or volume of the system being analyzed is not necessarily constant, as when a gas in a cylinder is compressed by a piston or a balloon is inflated.

surroundings
boundary
  Everything external to the system is considered to be part of the system's **surroundings**. The system is distinguished from its surroundings by a specified **boundary**, which may be at rest or in motion. You will see that the interactions between a system and its surroundings, which take place across the boundary, play an important part in engineering thermodynamics.

Two basic kinds of systems are distinguished in this book. These are referred to, respectively, as *closed systems* and *control volumes*. A closed system refers to a fixed quantity of matter, whereas a control volume is a region of space through which mass may flow. The term *control mass* is sometimes used in place of closed system, and the term *open system* is used interchangeably with control volume. When the terms *control mass* and *control volume* are used, the system boundary is often referred to as a *control surface*.

TABLE 1.1

Selected Areas of Application of Engineering Thermodynamics

Aircraft and rocket propulsion
Alternative energy systems
  Fuel cells
  Geothermal systems
  Magnetohydrodynamic (MHD) converters
  Ocean thermal, wave, and tidal power generation
  Solar-activated heating, cooling, and power generation
  Thermoelectric and thermionic devices
  Wind turbines
Automobile engines
Bioengineering applications
Biomedical applications
Combustion systems
Compressors, pumps
Cooling of electronic equipment
Cryogenic systems, gas separation, and liquefaction
Fossil and nuclear-fueled power stations
Heating, ventilating, and air-conditioning systems
  Absorption refrigeration and heat pumps
  Vapor-compression refrigeration and heat pumps
Steam and gas turbines
  Power production
  Propulsion

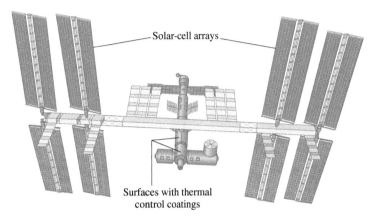

International Space Station

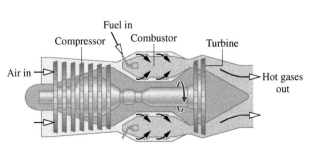

Refrigerator

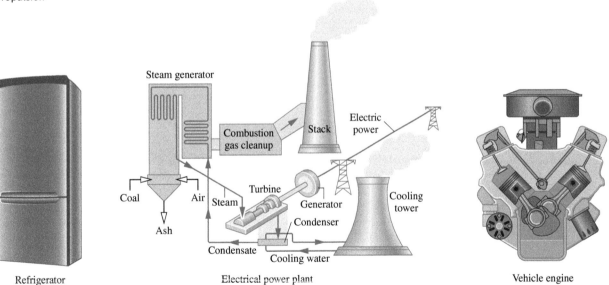

Electrical power plant

Vehicle engine

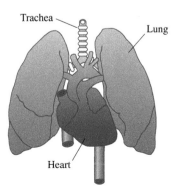

Biomedical applications

Turbojet engine

TABLE 1.2

Predictions of Life in 2050

At home
- ▶ Homes are constructed better to reduce heating and cooling needs.
- ▶ Homes have systems for electronically monitoring and regulating energy use.
- ▶ Appliances and heating and air-conditioning systems are more energy-efficient.
- ▶ Use of solar energy for space and water heating is common.
- ▶ More food is produced locally.

Transportation
- ▶ Plug-in hybrid vehicles and all-electric vehicles dominate.
- ▶ One-quarter of transport fuel is biofuels.
- ▶ Use of public transportation within and between cities is common.
- ▶ An expanded passenger railway system is widely used.

Lifestyle
- ▶ Efficient energy-use practices are utilized throughout society.
- ▶ Recycling is widely practiced, including recycling of water.
- ▶ Distance learning is common at most educational levels.
- ▶ Telecommuting and teleconferencing are the norm.
- ▶ The Internet is predominately used for consumer and business commerce.

Power generation
- ▶ Electricity plays a greater role throughout society.
- ▶ Wind, solar, and other renewable technologies contribute a significant share of the nation's electricity needs.
- ▶ A mix of conventional fossil-fueled and nuclear power plants provides a smaller, but still significant, share of the nation's electricity needs.
- ▶ A smart and secure national power transmission grid is in place.

1.2.1 Closed Systems

closed system

A **closed system** is defined when a particular quantity of matter is under study. A closed system always contains the same matter. There can be no transfer of mass across its boundary. A special type of closed system that does not interact in any way with its surroundings is called an **isolated system**.

isolated system

Figure 1.1 shows a gas in a piston–cylinder assembly. When the valves are closed, we can consider the gas to be a closed system. The boundary lies just inside the piston and cylinder walls, as shown by the dashed lines on the figure. Since the portion of the boundary between the gas and the piston moves with the piston, the system volume varies. No mass would cross this or any other part of the boundary. If combustion occurs, the composition of the system changes as the initial combustible mixture becomes products of combustion.

1.2.2 Control Volumes

In subsequent sections of this book, we perform thermodynamic analyses of devices such as turbines and pumps through which mass flows. These analyses can be conducted in principle by studying a particular quantity of matter, a closed system, as it passes through the device. In most cases it is simpler to think instead in terms of a given region of space through which mass flows. With this approach, a *region* within a prescribed boundary is studied. The region is called a **control volume**. Mass crosses the boundary of a control volume.

A diagram of an engine is shown in **Fig. 1.2a**. The dashed line defines a control volume that surrounds the engine. Observe that air, fuel, and exhaust gases cross the boundary. A schematic such as in **Fig. 1.2b** often suffices for engineering analysis. Control volume applications in biology and botany are illustrated is **Figs. 1.3** and **1.4** respectively.

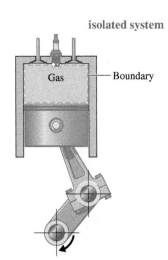

Fig. 1.1 Closed system: A gas in a piston–cylinder assembly.

control volume

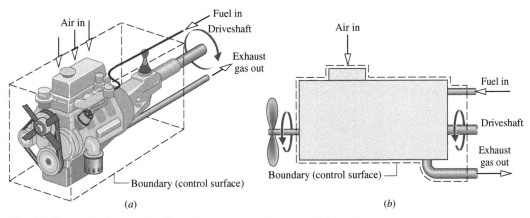

Fig. 1.2 Example of a control volume (open system). An automobile engine.

BioConnections

Living things and their organs can be studied as control volumes. For the pet shown in Fig. 1.3a, air, food, and drink essential to sustain life and for activity enter across the boundary, and waste products exit. A schematic such as Fig. 1.3b can suffice for biological analysis. Particular organs, such as the heart, also can be studied as control volumes. As shown in Fig. 1.4, plants can be studied from a control volume viewpoint. Intercepted solar radiation is used in the production of essential chemical substances within plants by *photosynthesis*. During photosynthesis, plants take in carbon dioxide from the atmosphere and discharge oxygen to the atmosphere. Plants also draw in water and nutrients through their roots.

1.2.3 Selecting the System Boundary

The system boundary should be delineated carefully before proceeding with any thermodynamic analysis. However, the same physical phenomena often can be analyzed in terms of alternative choices of the system, boundary, and surroundings. The choice of a particular boundary defining a particular system depends heavily on the convenience it allows in the subsequent analysis.

In general, the choice of system boundary is governed by two considerations: (1) what is known about a possible system, particularly at its boundaries, and (2) the objective of the analysis.

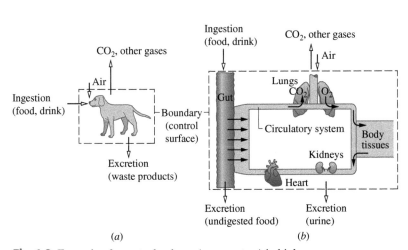

(a)   *(b)*

Fig. 1.3 Example of a control volume (open system) in biology.

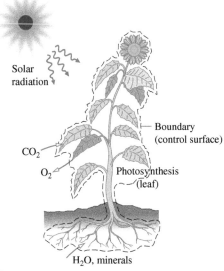

Fig. 1.4 Example of a control volume (open system) in botany.

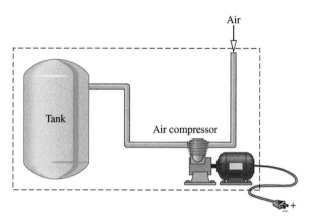

Fig. 1.5 Air compressor and storage tank.

System Types Tabs a, b, and c

> **FOR EXAMPLE**
>
> **Figure 1.5** shows a sketch of an air compressor connected to a storage tank. The system boundary shown on the figure encloses the compressor, tank, and all of the piping. This boundary might be selected if the electrical power input is known, and the objective of the analysis is to determine how long the compressor must operate for the pressure in the tank to rise to a specified value. Since mass crosses the boundary, the system would be a control volume. A control volume enclosing only the compressor might be chosen if the condition of the air entering and exiting the compressor is known, and the objective is to determine the electric power input.

1.3 Describing Systems and Their Behavior

Engineers are interested in studying systems and how they interact with their surroundings. In this section, we introduce several terms and concepts used to describe systems and how they behave.

1.3.1 Macroscopic and Microscopic Views of Thermodynamics

Systems can be studied from a macroscopic or a microscopic point of view. The macroscopic approach to thermodynamics is concerned with the gross or overall behavior. This is sometimes called *classical* thermodynamics. No model of the structure of matter at the molecular, atomic, and subatomic levels is directly used in classical thermodynamics. Although the behavior of systems is affected by molecular structure, classical thermodynamics allows important aspects of system behavior to be evaluated from observations of the overall system.

The microscopic approach to thermodynamics, known as *statistical* thermodynamics, is concerned directly with the structure of matter. The objective of statistical thermodynamics is to characterize by statistical means the average behavior of the particles making up a system of interest and relate this information to the observed macroscopic behavior of the system. For applications involving lasers, plasmas, high-speed gas flows, chemical kinetics, very low temperatures (cryogenics), and others, the methods of statistical thermodynamics are essential. The microscopic approach is used in this text to interpret *internal energy* in Chap. 2 and *entropy* in Chap 6. Moreover, as noted in Chap. 3, the microscopic approach is instrumental in developing certain data, for example *ideal gas specific heats*.

For a wide range of engineering applications, classical thermodynamics not only provides a considerably more direct approach for analysis and design but also requires far fewer mathematical complications. For these reasons the macroscopic viewpoint is the one adopted in this book. Finally, relativity effects are not significant for the systems under consideration in this book.

1.3.2 Property, State, and Process

To describe a system and predict its behavior requires knowledge of its properties and how those properties are related. A **property** is a macroscopic characteristic of a system such as mass, volume, energy, pressure, and temperature to which a numerical value can be assigned at a given time without knowledge of the previous behavior (*history*) of the system.

property

The word **state** refers to the condition of a system as described by its properties. Since there are normally relations among the properties of a system, the state often can be specified by providing the values of a subset of the properties. All other properties can be determined in terms of these few.

state

When any of the properties of a system changes, the state changes and the system is said to undergo a **process**. A process is a transformation from one state to another. If a system exhibits the same values of its properties at two different times, it is in the same state at these times. A system is said to be at **steady state** if none of its properties changes with time.

process

steady state

Many properties are considered during the course of our study of engineering thermodynamics. Thermodynamics also deals with quantities that are not properties, such as mass flow rates and energy transfers by work and heat. Additional examples of quantities that are not properties are provided in subsequent chapters. For a way to distinguish properties from *non*properties, see the following box.

Animation

Property, State and Process Tab a

Distinguishing Properties from Nonproperties

At a given state, each property has a definite value that can be assigned without knowledge of how the system arrived at that state. The change in value of a property as the system is altered from one state to another is determined, therefore, solely by the two end states and is independent of the particular way the change of state occurred. The change is independent of the details of the process. Conversely, if the value of a quantity is independent of the process between two states, then that quantity is the change in a property. This provides a test for determining whether a quantity is a property: *A quantity is a property if, and only if, its change in value between two states is independent of the process.* It follows that if the value of a particular quantity depends on the details of the process, and not solely on the end states, that quantity cannot be a property.

1.3.3 Extensive and Intensive Properties

Thermodynamic properties can be placed in two general classes: extensive and intensive. A property is called **extensive** if its value for an overall system is the sum of its values for the parts into which the system is divided. Mass, volume, energy, and several other properties introduced later are extensive. Extensive properties depend on the size or extent of a system. The extensive properties of a system can change with time, and many thermodynamic analyses consist mainly of carefully accounting for changes in extensive properties such as mass and energy as a system interacts with its surroundings.

extensive property

Intensive properties are not additive in the sense previously considered. Their values are independent of the size or extent of a system and may vary from place to place within the system at any moment. Intensive properties may be functions of both position and time, whereas extensive properties can vary only with time. Specific volume (Sec. 1.5), pressure, and temperature are important intensive properties; several other intensive properties are introduced in subsequent chapters.

intensive property

FOR EXAMPLE

To illustrate the difference between extensive and intensive properties, consider an amount of matter that is uniform in temperature, and imagine that it is composed of several parts, as illustrated in **Fig. 1.6**. The mass of the whole is the sum of the masses of the parts, and the overall volume is the sum of the volumes of the parts. However, the temperature of the whole is not the sum of the temperatures of the parts; it is the same for each part. Mass and volume are extensive, but temperature is intensive.

Animation

Extensive and Intensive Properties Tab a

Fig. 1.6 Figure used to discuss the extensive and intensive property concepts.

(a)   (b)

1.3.4 Equilibrium

equilibrium

Classical thermodynamics places primary emphasis on equilibrium states and changes from one equilibrium state to another. Thus, the concept of **equilibrium** is fundamental. In mechanics, equilibrium means a condition of balance maintained by an equality of opposing forces. In thermodynamics, the concept is more far-reaching, including not only a balance of forces but also a balance of other influences. Each kind of influence refers to a particular aspect of thermodynamic, or complete, equilibrium. Accordingly, several types of equilibrium must exist individually to fulfill the condition of complete equilibrium; among these are mechanical, thermal, phase, and chemical equilibrium.

Criteria for these four types of equilibrium are considered in subsequent discussions. For the present, we may think of testing to see if a system is in thermodynamic equilibrium by the following procedure: Isolate the system from its surroundings and watch for changes in its observable properties. If there are no changes, we conclude that the system was in equilibrium at the moment it was isolated. The system can be said to be at an **equilibrium state**.

equilibrium state

When a system is isolated, it does not interact with its surroundings; however, its state can change as a consequence of spontaneous events occurring internally as its intensive properties, such as temperature and pressure, tend toward uniform values. When all such changes cease, the system is in equilibrium. At equilibrium, temperature is uniform throughout the system. Also, pressure can be regarded as uniform throughout as long as the effect of gravity is not significant; otherwise, a pressure variation can exist, as in a vertical column of liquid.

It is not necessary that a system undergoing a process be in equilibrium *during* the process. Some or all of the intervening states may be nonequilibrium states. For many such processes, we are limited to knowing the state before the process occurs and the state after the process is completed.

1.4 Measuring Mass, Length, Time, and Force

When engineering calculations are performed, it is necessary to be concerned with the *units* of the physical quantities involved. A unit is any specified amount of a quantity by comparison with which any other quantity of the same kind is measured. For example, meters, centimeters, kilometers, feet, inches, and miles are all *units of length*. Seconds, minutes, and hours are alternative *time units*.

Because physical quantities are related by definitions and laws, a relatively small number of physical quantities suffice to conceive of and measure all others. These are called *primary dimensions*. The others are measured in terms of the primary dimensions and are called *secondary*. For example, if length and time were regarded as primary, velocity and area would be secondary.

A set of primary dimensions that suffice for applications in *mechanics* is mass, length, and time. Additional primary dimensions are required when additional physical phenomena come under consideration. Temperature is included for thermodynamics, and electric current is introduced for applications involving electricity.

base unit

Once a set of primary dimensions is adopted, a **base unit** for each primary dimension is specified. Units for all other quantities are then derived in terms of the base units. Let us illustrate these ideas by considering briefly two systems of units: SI units and English Engineering units.

TABLE 1.3

Units for Mass, Length, Time, and Force

| Quantity | SI | |
	Unit	Symbol
mass	kilogram	kg
length	meter	m
time	second	s
force	newton	N
	$(= 1 \text{ kg} \cdot \text{m/s}^2)$	

1.4.1 SI Units

In the present discussion we consider the SI system of units that takes mass, length, and time as primary dimensions and regards force as secondary. SI is the abbreviation for Système International d'Unités (International System of Units), which is the legally accepted system in most countries. The conventions of the SI are published and controlled by an international treaty organization. The **SI base units** for mass, length, and time are listed in **Table 1.3** and discussed in the following paragraphs. The SI base unit for temperature is the kelvin, K.

SI base units

The SI base unit of mass is the kilogram, kg. It is equal to the mass of a particular cylinder of platinum–iridium alloy kept by the International Bureau of Weights and Measures near Paris. The mass standard for the United States is maintained by the National Institute of Standards and Technology (NIST). The kilogram is the only base unit still defined relative to a fabricated object.

The SI base unit of length is the meter (metre), m, defined as the length of the path traveled by light in a vacuum during a specified time interval. The base unit of time is the second, s. The second is defined as the duration of 9,192,631,770 cycles of the radiation associated with a specified transition of the cesium atom.

The SI unit of force, called the newton, is a secondary unit, defined in terms of the base units for mass, length, and time. Newton's second law of motion states that the net force acting on a body is proportional to the product of the mass and the acceleration, written $F \propto ma$. The newton is defined so that the proportionality constant in the expression is equal to unity. That is, Newton's second law is expressed as the equality

$$F = ma \tag{1.1}$$

The newton, N, is the force required to accelerate a mass of 1 kilogram at the rate of 1 meter per second per second. With Eq. 1.1

$$1 \text{ N} = (1 \text{ kg})(1 \text{ m/s}^2) = 1 \text{ kg} \cdot \text{m/s}^2 \tag{1.2}$$

FOR EXAMPLE

To illustrate the use of the SI units introduced thus far, let us determine the weight in newtons of an object whose mass is 1000 kg, at a place on Earth's surface where the acceleration due to gravity equals a *standard* value defined as 9.80665 m/s². Recalling that the weight of an object refers to the force of gravity and is calculated using the mass of the object, m, and the local acceleration of gravity, g, with Eq. 1.1 we get

$$F = mg$$
$$= (1000 \text{ kg})(9.80665 \text{ m/s}^2) = 9806.65 \text{ kg} \cdot \text{m/s}^2$$

This force can be expressed in terms of the newton by using Eq. 1.2 as a *unit conversion factor*. That is,

$$F = \left(9806.65 \frac{\text{kg} \cdot \text{m}}{\text{s}^2} \right) \left| \frac{1 \text{ N}}{1 \text{ kg} \cdot \text{m/s}^2} \right| = 9806.65 \text{ N}$$

TABLE 1.4

SI Unit Prefixes

Factor	Prefix	Symbol
10^{12}	tera	T
10^{9}	giga	G
10^{6}	mega	M
10^{3}	kilo	k
10^{2}	hecto	h
10^{-2}	centi	c
10^{-3}	milli	m
10^{-6}	micro	μ
10^{-9}	nano	n
10^{-12}	pico	p

English base units

Since weight is calculated in terms of the mass and the local acceleration due to gravity, the weight of an object can vary because of the variation of the acceleration of gravity with location, but its mass remains constant.

FOR EXAMPLE

If the object considered previously were on the surface of a planet at a point where the acceleration of gravity is one-tenth of the value used in the above calculation, the mass would remain the same but the weight would be one-tenth of the calculated value.

SI units for other physical quantities are also derived in terms of the SI base units. Some of the derived units occur so frequently that they are given special names and symbols, such as the newton. SI units for quantities pertinent to thermodynamics are given as they are introduced in the text. Since it is frequently necessary to work with extremely large or small values when using the SI unit system, a set of standard prefixes is provided in **Table 1.4** to simplify matters. For example, km denotes kilometer, that is, 10^3 m.

1.4.2 English Engineering Units

Although SI units are the worldwide standard, at the present time many segments of the engineering community in the United States regularly use other units. A large portion of America's stock of tools and industrial machines and much valuable engineering data utilize units other than SI units. For many years to come, engineers in the United States will have to be conversant with a variety of units.

In this section we consider a system of units that is commonly used in the United States, called the English Engineering system. The **English base units** for mass, length, and time are listed in Table 1.3 and discussed in the following paragraphs.

The base unit for length is the foot, ft, defined in terms of the meter as

$$1 \text{ ft} = 0.3048 \text{ m} \tag{1.3}$$

The inch, in., is defined in terms of the foot:

$$12 \text{ in.} = 1 \text{ ft}$$

One inch equals 2.54 cm. Although units such as the minute and the hour are often used in engineering, it is convenient to select the second as the English Engineering base unit for time.

The English Engineering base unit of mass is the pound mass, lb, defined in terms of the kilogram as

$$1 \text{ lb} = 0.45359237 \text{ kg} \tag{1.4}$$

The symbol lbm also may be used to denote the pound mass.

Once base units have been specified for mass, length, and time in the English Engineering system of units, a force unit can be defined, as for the newton, using Newton's second law written as Eq. 1.1. From this viewpoint, the English unit of force, the pound force, lbf, is the force required to accelerate one pound mass at 32.1740 ft/s^2, which is the standard acceleration of gravity. Substituting values into Eq. 1.1,

$$1 \text{ lbf} = (1 \text{ lb})(32.1740 \text{ ft/s}^2) = 32.1740 \text{ lb} \cdot \text{ft/s}^2 \tag{1.5}$$

With this approach force is regarded as _secondary_.

The pound force, lbf, is not equal to the pound mass, lb, introduced previously. Force and mass are fundamentally different, as are their units. The double use of the word "pound" can be confusing, so care must be taken to avoid error.

To show the use of these units in a single calculation, let us determine the weight of an object whose mass is 1000 lb at a location where the local acceleration of gravity is 32.0 ft/s². By inserting values into Eq. 1.1 and using Eq. 1.5 as a unit conversion factor, we get

$$F = mg = (1000 \text{ lb}) \left(32.0 \frac{\text{ft}}{\text{s}^2} \right) \left| \frac{1 \text{ lbf}}{32.1740 \text{ lb} \cdot \text{ft/s}^2} \right| = 994.59 \text{ lbf}$$

This calculation illustrates that the pound force is a unit of force distinct from the pound mass, a unit of mass.

1.5 Specific Volume

Three measurable intensive properties that are particularly important in engineering thermodynamics are specific volume, pressure, and temperature. Specific volume is considered in this section. Pressure and temperature are considered in Secs. 1.6 and 1.7, respectively.

From the macroscopic perspective, the description of matter is simplified by considering it to be distributed continuously throughout a region. The correctness of this idealization, known as the *continuum* hypothesis, is inferred from the fact that for an extremely large class of phenomena of engineering interest the resulting description of the behavior of matter is in agreement with measured data.

When substances can be treated as continua, it is possible to speak of their intensive thermodynamic properties "at a point." Thus, at any instant the density ρ at a point is defined as

$$\rho = \lim_{V \to V'} \left(\frac{m}{V} \right) \tag{1.6}$$

where V' is the smallest volume for which a definite value of the ratio exists. The volume V' contains enough particles for statistical averages to be significant. It is the smallest volume for which the matter can be considered a continuum and is normally small enough that it can be considered a "point." With density defined by Eq. 1.6, density can be described mathematically as a continuous function of position and time.

Animation

Extensive and Intensive Properties Tabs b and c

The density, or local mass per unit volume, is an intensive property that may vary from point to point within a system. Thus, the mass associated with a particular volume V is determined in principle by integration

$$m = \int_V \rho \, dV \tag{1.7}$$

and *not* simply as the product of density and volume.

The **specific volume** v is defined as the reciprocal of the density, $v = 1/\rho$. It is the volume per unit mass. Like density, specific volume is an intensive property and may vary from point to point. SI units for density and specific volume are kg/m³ and m³/kg, respectively. They are also often expressed, respectively, as g/cm³ and cm³/g.

specific volume

In certain applications it is convenient to express properties such as specific volume on a molar basis rather than on a mass basis. A mole is an amount of a given substance numerically equal to its molecular weight. In this book we express the amount of substance on a **molar basis** in terms of the kilomole (kmol). We use

molar basis

$$n = \frac{m}{M} \tag{1.8}$$

The number of kilomoles of a substance, n, is obtained by dividing the mass, m, in kilograms by the molecular weight, M, in kg/kmol. When m is in grams, Eq. 1.8 gives n in gram moles, or *mol* for short. Recall from chemistry that the number of molecules in a gram mole, called Avogadro's number, is 6.022×10^{23}. Appendix Table A-1 provides molecular weights for several substances.

To signal that a property is on a molar basis, a bar is used over its symbol. Thus, $\bar{v}$ signifies the volume per kmol or lbmol, as appropriate. In this text, the units used for $\bar{v}$ are m³/kmol. With Eq. 1.8, the relationship between $\bar{v}$ and v is

$$\bar{v} = Mv \tag{1.9}$$

where M is the molecular weight in kg/kmol.

1.6 Pressure

Next, we introduce the concept of pressure from the continuum viewpoint. Let us begin by considering a small area, A, passing through a point in a fluid at rest. The fluid on one side of the area exerts a compressive force on it that is normal to the area, F_{normal}. An equal but oppositely directed force is exerted on the area by the fluid on the other side. For a fluid at rest, no other forces than these act on the area. The **pressure**, p, at the specified point is defined as the limit

pressure

Animation

Extensive and Intensive Properties Tab d

$$p = \lim_{A \to A'} \left(\frac{F_{normal}}{A} \right) \tag{1.10}$$

where A′ is the area at the "point" in the same limiting sense as used in the definition of density.

If the area A′ was given new orientations by rotating it around the given point, and the pressure determined for each new orientation, it would be found that the pressure at the point is the same in all directions *as long as the fluid is at rest*. This is a consequence of the equilibrium of forces acting on an element of volume surrounding the point. However, the pressure can vary from point to point within a fluid at rest; examples are the variation of atmospheric pressure with elevation and the pressure variation with depth in oceans, lakes, and other bodies of water.

Consider next a fluid in motion. In this case the force exerted on an area passing through a point in the fluid may be resolved into three mutually perpendicular components: one normal to the area and two in the plane of the area. When expressed on a unit area basis, the component normal to the area is called the *normal stress*, and the two components in the plane of the area are termed *shear stresses*. The magnitudes of the stresses generally vary with the orientation of the area. The state of stress in a fluid in motion is a topic that is normally treated thoroughly in *fluid mechanics*. The deviation of a normal stress from the pressure, the normal stress that would exist were the fluid at rest, is typically very small. In this book we assume that the normal stress at a point is equal to the pressure at that point. This assumption yields results of acceptable accuracy for the applications considered. Also, the term *pressure*, unless stated otherwise, refers to **absolute pressure**: pressure with respect to the zero pressure of a complete vacuum. The lowest possible value of absolute pressure is zero.

absolute pressure

1.6.1 Pressure Measurement

Manometers and barometers measure pressure in terms of the length of a column of liquid such as mercury, water, or oil. The manometer shown in **Fig. 1.7** has one end open to the atmosphere and the other attached to a tank containing a gas at a uniform pressure. Since pressures at equal elevations in a *continuous* mass of a liquid or gas *at rest* are equal, the pressures at points *a* and *b* of Fig. 1.7 are equal. Applying an elementary force balance, the gas pressure is

$$p = p_{atm} + \rho g L \tag{1.11}$$

where p_{atm} is the local atmospheric pressure, ρ is the density of the manometer liquid, g is the acceleration of gravity, and L is the difference in the liquid levels.

The barometer shown in **Fig. 1.8** is formed by a closed tube filled with liquid mercury and a small amount of mercury vapor inverted in an open container of liquid mercury. Since the pressures at points *a* and *b* are equal, a force balance gives the atmospheric pressure as

$$p_{atm} = p_{vapor} + \rho_m g L \tag{1.12}$$

where ρ_m is the density of liquid mercury. Because the pressure of the mercury vapor is much less than that of the atmosphere, Eq. 1.12 can be approximated closely as $p_{atm} = \rho_m g L$. For short columns of liquid, ρ and g in Eqs. 1.11 and 1.12 may be taken as constant.

Pressures measured with manometers and barometers are frequently expressed in terms of the length L in millimeters of mercury (mmHg), inches of mercury (inHg), inches of water (inH$_2$O), and so on.

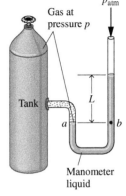

Fig. 1.7 Manometer.

FOR EXAMPLE

A barometer reads 750 mmHg. If $\rho_m = 13.59$ g/cm^3 and $g = 9.81$ m/s^2, the atmospheric pressure, in N/m^2, is calculated as follows:

$$p_{atm} = \rho_m g L$$

$$= \left[\left(13.59 \frac{g}{cm^3} \right) \left| \frac{1 \text{ kg}}{10^3 \text{ g}} \right| \left| \frac{10^2 \text{ cm}}{1 \text{ m}} \right|^3 \right] \left[9.81 \frac{m}{s^2} \right] \left[(750 \text{ mmHg}) \left| \frac{1 \text{ m}}{10^3 \text{ mm}} \right| \right] \left| \frac{1 \text{ N}}{1 \text{ kg} \cdot \text{m/s}^2} \right|$$

$$= 10^5 \text{ N/m}^2$$

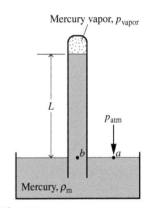

Fig. 1.8 Barometer.

A Bourdon tube gage is shown in **Fig. 1.9**. The figure shows a curved tube having an elliptical cross section with one end attached to the pressure to be measured and the other end connected to a pointer by a mechanism. When fluid under pressure fills the tube, the elliptical section tends to become circular, and the tube straightens. This motion is transmitted by the mechanism to the pointer. By calibrating the deflection of the pointer for known pressures, a graduated scale can be determined from which any applied pressure can be read in suitable units. Because of its construction, the Bourdon tube measures the pressure relative to the pressure of the surroundings existing at the instrument. Accordingly, the dial reads zero when the inside and outside of the tube are at the same pressure.

Pressure can be measured by other means as well. An important class of sensors utilizes the *piezoelectric* effect: A charge is generated within certain solid materials when they are deformed. This mechanical input/electrical output provides the basis for pressure measurement as well as displacement and force measurements. Another important type of sensor employs a diaphragm that deflects when a force is applied, altering an inductance, resistance, or capacitance. **Figure 1.10** shows a piezoelectric pressure sensor together with an automatic data acquisition system.

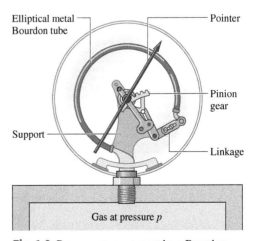

Fig. 1.9 Pressure measurement by a Bourdon tube gage.

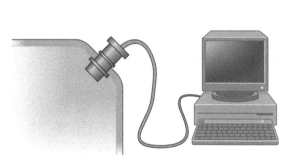

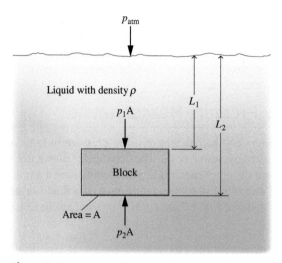

Fig. 1.10 Pressure sensor with automatic data acquisition.

Fig. 1.11 Evaluation of buoyant force for a submerged body.

1.6.2 Buoyancy

buoyant force

When a body is completely or partially submerged in a liquid, the resultant pressure force acting on the body is called the **buoyant force**. Since pressure increases with depth from the liquid surface, pressure forces acting from below are greater than pressure forces acting from above; thus, the buoyant force acts vertically upward. The buoyant force has a magnitude equal to the weight of the displaced liquid (*Archimedes' principle*).

> **FOR EXAMPLE**
>
> Applying Eq. 1.11 to the submerged rectangular block shown in **Fig. 1.11**, the magnitude of the net force of pressure acting upward, the buoyant force, is
>
> $$\begin{aligned} F &= A(p_2 - p_1) = A(p_{atm} + \rho g L_2) - A(p_{atm} + \rho g L_1) \\ &= \rho g A(L_2 - L_1) \\ &= \rho g V \end{aligned}$$
>
> where V is the volume of the block and ρ is the density of the surrounding liquid. Thus, the magnitude of the buoyant force acting on the block is equal to the weight of the displaced liquid.

1.6.3 Pressure Units

The SI unit of pressure and stress is the pascal:

$$1 \text{ pascal} = 1 \text{ N/m}^2$$

Multiples of the pascal, the kPa, the bar, and the MPa, are frequently used.

$$1 \text{ kPa} = 10^3 \text{ N/m}^2$$
$$1 \text{ bar} = 10^5 \text{ N/m}^2$$
$$1 \text{ MPa} = 10^6 \text{ N/m}^2$$

Although atmospheric pressure varies with location on the earth, a standard reference value can be defined and used to express other pressures.

$$1 \text{ standard atmosphere (atm)} = \begin{cases} 1.01325 \times 10^5 \text{ N/m}^2 \\ 760 \text{ mmHg} = 29.92 \text{ inHg} \end{cases} \tag{1.13}$$

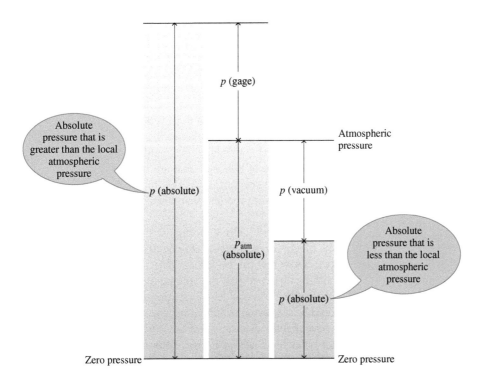

Fig. 1.12 Relationships among the absolute, atmospheric, gage, and vacuum pressures.

Since 1 bar (10^5 N/m^2) closely equals one standard atmosphere, it is a convenient pressure unit despite not being a standard SI unit. When working in SI, the bar, MPa, and kPa are all used in this text.

Although absolute pressures must be used in thermodynamic relations, pressure-measuring devices often indicate the *difference* between the absolute pressure of a system and the absolute pressure of the atmosphere existing outside the measuring device. The magnitude of the difference is called a **gage pressure** or a **vacuum pressure**. The term *gage pressure* is applied when the pressure of the system is greater than the local atmospheric pressure, p_{atm}.

gage pressure

vacuum pressure

$$p(\text{gage}) = p(\text{absolute}) - p_{atm}(\text{absolute}) \qquad (1.14)$$

When the local atmospheric pressure is greater than the pressure of the system, the term *vacuum pressure* is used.

$$p(\text{vacuum}) = p_{atm}(\text{absolute}) - p(\text{absolute}) \qquad (1.15)$$

Engineers frequently use the letters a and g to distinguish between absolute and gage pressures. The relationship among the various ways of expressing pressure measurements is shown in Fig. 1.12.

TAKE NOTE...

In this book, the term *pressure* refers to absolute pressure unless indicated otherwise.

 BioConnections

One in three Americans is said to have high blood pressure. Since this can lead to heart disease, strokes, and other serious medical complications, medical practitioners recommend regular blood pressure checks for everyone. Blood pressure measurement aims to determine the maximum pressure (systolic pressure) in an artery when the heart is pumping blood and the minimum pressure (diastolic pressure) when the heart is resting, each pressure expressed in millimeters of mercury, mmHg. The systolic and diastolic pressures of healthy persons should be less than about 120 mmHg and 80 mmHg, respectively.

The standard blood pressure measurement apparatus in use for decades involving an inflatable cuff, mercury manometer, and stethoscope is gradually being replaced because of concerns over mercury toxicity and in response to special requirements, including monitoring during clinical exercise and during anesthesia. Also, for home use and self-monitoring, many patients prefer easy-to-use automated devices that provide digital displays of blood pressure data. This has prompted biomedical engineers to rethink blood pressure measurement and develop new mercury-free and stethoscope-free approaches. One of these uses a highly sensitive pressure transducer to detect pressure oscillations within an inflated cuff placed around the patient's arm. The monitor's software uses these data to calculate the systolic and diastolic pressures, which are displayed digitally.

1.7 Temperature

Animation

Extensive and Intensive
Properties Tab e

In this section the intensive property temperature is considered along with means for measuring it. A concept of temperature, like our concept of force, originates with our sense perceptions. Temperature is rooted in the notion of the "hotness" or "coldness" of objects. We use our sense of touch to distinguish hot objects from cold objects and to arrange objects in their order of "hotness," deciding that 1 is hotter than 2, 2 hotter than 3, and so on. But however sensitive human touch may be, we are unable to gauge this quality precisely.

A definition of temperature in terms of concepts that are independently defined or accepted as primitive is difficult to give. However, it is possible to arrive at an objective understanding of *equality* of temperature by using the fact that when the temperature of an object changes, other properties also change.

To illustrate this, consider two copper blocks, and suppose that our senses tell us that one is warmer than the other. If the blocks were brought into contact and isolated from their surroundings, they would interact in a way that can be described as a **thermal (heat) interaction**. During this interaction, it would be observed that the volume of the warmer block decreases somewhat with time, while the volume of the colder block increases with time. Eventually, no further changes in volume would be observed, and the blocks would feel equally warm. Similarly, we would be able to observe that the electrical resistance of the warmer block decreases with time and that of the colder block increases with time; eventually the electrical resistances would become constant also. When all changes in such observable properties cease, the interaction is at an end. The two blocks are then in **thermal equilibrium**. Considerations such as these lead us to infer that the blocks have a physical property that determines whether they will be in thermal equilibrium. This property is called **temperature**, and we postulate that when the two blocks are in thermal equilibrium, their temperatures are equal.

It is a matter of experience that when two objects are in thermal equilibrium with a third object, they are in thermal equilibrium with one another. This statement, which is sometimes called the **zeroth law of thermodynamics**, is tacitly assumed in every measurement of temperature. If we want to know if two objects are at the same temperature, it is not necessary to bring them into contact and see whether their observable properties change with time, as described previously. It is necessary only to see if they are individually in thermal equilibrium with a third object. The third object is usually a *thermometer*.

thermal (heat) interaction

thermal equilibrium

temperature

zeroth law of thermodynamics

1.7.1 Thermometers

Any object with at least one measurable property that changes as its temperature changes can be used as a thermometer. Such a property is called a **thermometric property**. The particular substance that exhibits changes in the thermometric property is known as a *thermometric substance*.

A familiar device for temperature measurement is the liquid-in-glass thermometer pictured in **Fig. 1.13a**, which consists of a glass capillary tube connected to a bulb filled with a liquid such as alcohol and sealed at the other end. The space above the liquid is occupied by the vapor of the liquid or an inert gas. As temperature increases, the liquid expands in volume and rises in the capillary. The length L of the liquid in the capillary depends on the temperature. Accordingly, the liquid is the thermometric substance and L is the thermometric property. Although this type of thermometer is commonly used for ordinary temperature measurements, it is not well suited for applications where extreme accuracy is required.

More accurate sensors known as *thermocouples* are based on the principle that when two dissimilar metals are joined, an electromotive force (emf) that is primarily a function of temperature will exist in a circuit. In certain thermocouples, one thermocouple wire is platinum of a specified purity and the other is an alloy of platinum and rhodium. Thermocouples also utilize copper and constantan (an alloy of copper and nickel), iron and constantan, as well as several other pairs of materials. Electrical-resistance sensors are another important class of temperature measurement devices. These sensors are based on the fact that the electrical resistance of various materials changes in a predictable manner with temperature. The materials used for this purpose are normally conductors (such as platinum, nickel, or copper)

thermometric property

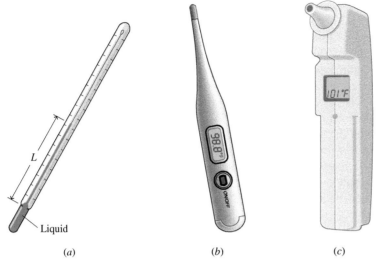

Fig. 1.13 Thermometers. (*a*) Liquid-in-glass. (*b*) Electrical-resistance. (*c*) Infrared-sensing ear thermometer.

or semiconductors. Devices using conductors are known as *resistance temperature detectors*. Semiconductor types are called *thermistors*. A battery-powered electrical-resistance thermometer commonly used today is shown in Fig. 1.13b.

A variety of instruments measure temperature by sensing radiation, such as the ear thermometer shown in Fig. 1.13c. They are known by terms such as *radiation thermometers* and *optical pyrometers*. This type of thermometer differs from those previously considered because it is not required to come in contact with an object to determine its temperature, an advantage when dealing with moving objects or objects at extremely high temperatures.

Energy & Environment

The mercury-in-glass thermometer once prevalent in home medicine cabinets and industrial settings is fast disappearing because of toxicity of mercury and its harmful effects on humans. The *American Academy of Pediatrics* has designated mercury as too toxic to be present in the home. After 110 years of calibration service for mercury thermometers, the *National Institute of Standards and Technology* (NIST) terminated this service in 2011 to encourage industry to seek safer temperature measurement options. Alternative options for home and industrial use include digital electronic thermometers, alcohol-in-glass thermometers, and other patented liquid mixtures-in-glass thermometers.

Proper disposal of millions of obsolete mercury-filled thermometers has emerged as an environmental issue because mercury is toxic and nonbiodegradable. These thermometers must be taken to hazardous-waste collection stations rather than simply thrown in the trash where they can be easily broken, releasing mercury. Mercury can be recycled for use in other products such as fluorescent and compact fluorescent bulbs, household switches, and thermostats.

1.7.2 Kelvin Temperature Scales

Empirical means of measuring temperature such as considered in Sec. 1.7.1 have inherent limitations.

FOR EXAMPLE

The tendency of the liquid in a liquid-in-glass thermometer to freeze at low temperatures imposes a lower limit on the range of temperatures that can be measured. At high temperatures liquids vaporize and, therefore, these temperatures also cannot be determined by a liquid-in-glass thermometer. Accordingly, several *different* thermometers might be required to cover a wide temperature interval.

Kelvin scale

In view of the limitations of empirical means for measuring temperature, it is desirable to have a procedure for assigning temperature values that do not depend on the properties of any particular substance or class of substances. Such a scale is called a *thermodynamic* temperature scale. The **Kelvin scale** is an absolute thermodynamic temperature scale that provides a continuous definition of temperature, valid over all ranges of temperature. The unit of temperature on the Kelvin scale is the kelvin (K). The kelvin is the SI base unit for temperature. The lowest possible value of temperature on an absolute thermodynamic temperature scale is zero.

To develop the Kelvin scale, it is necessary to use the conservation of energy principle and the second law of thermodynamics; therefore, further discussion is deferred to Sec. 5.8 after these principles have been introduced. We note here, however, that the Kelvin scale has a zero of 0 K, and lower temperatures than this are not defined.

In thermodynamic relationships, temperature is always in terms of the Kelvin scale unless specifically stated otherwise. Still, the Celsius scale considered next is commonly encountered.

1.7.3 Celsius Scales

The relationship of the Kelvin and Celsius scales is shown in **Fig. 1.14** together with values for temperature at three fixed points: the triple point, ice point, and steam point.

By international agreement, temperature scales are defined by the numerical value assigned to the easily reproducible **triple point** of water: the state of equilibrium among steam, ice, and liquid water (Sec. 3.2). As a matter of convenience, the temperature at this standard fixed point is defined as 273.16 kelvins, abbreviated as 273.16 K. This makes the temperature interval from the *ice point*[1] (273.15 K) to the *steam point*[2] equal to 100 K and thus in agreement with the Celsius scale, which assigns 100 degrees to the same interval.

The **Celsius temperature scale** uses the unit degree Celsius (°C), which has the same magnitude as the kelvin. Thus, temperature *differences* are identical on both scales. However,

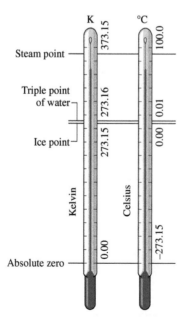

Fig. 1.14 Comparison of temperature scales.

---

[1]The state of equilibrium between ice and air-saturated water at a pressure of 1 atm.

[2]The state of equilibrium between steam and liquid water at a pressure of 1 atm.

the zero point on the Celsius scale is shifted to 273.15 K, as shown by the following relationship between the Celsius temperature and the Kelvin temperature:

$$T(^\circ\text{C}) = T(\text{K}) - 273.15 \qquad (1.16)$$

From this it can be concluded that on the Celsius scale the triple point of water is 0.01°C and that 0 K corresponds to −273.15°C. These values are shown on Fig. 1.14.

BioConnections

Cryobiology, the science of life at low temperatures, comprises the study of biological materials and systems (proteins, cells, tissues, and organs) at temperatures ranging from the cryogenic (below about 120 K) to the hypothermic (low body temperature). Applications include freeze-drying pharmaceuticals, cryosurgery for removing unhealthy tissue, study of cold-adaptation of animals and plants, and long-term storage of cells and tissues (called *cryopreservation*).

Cryobiology has challenging engineering aspects owing to the need for refrigerators capable of achieving the low temperatures required by researchers. Freezers to support research requiring cryogenic temperatures in the low-gravity environment of the International Space Station, shown in Table 1.1, are illustrative. Such freezers must be extremely compact and miserly in power use. Further, they must pose no hazards. On-board research requiring a freezer might include the growth of near-perfect protein crystals, important for understanding the structure and function of proteins and ultimately in the design of new drugs.

1.8 Engineering Design and Analysis

The word *engineer* traces its roots to the Latin *ingeniare*, relating to *invention*. Today invention remains a key engineering function having many aspects ranging from developing new devices to addressing complex social issues using technology. In pursuit of many such activities, engineers are called upon to design and analyze devices intended to meet human needs. Design and analysis are considered in this section.

1.8.1 Design

Engineering design is a decision-making process in which principles drawn from engineering and other fields such as economics and statistics are applied, usually iteratively, to devise a system, system component, or process. Fundamental elements of design include the establishment of objectives, synthesis, analysis, construction, testing, evaluation, and redesign (as necessary). Designs typically are subject to a variety of **constraints** related to economics, safety, environmental impact, and so on.

design constraints

Design projects usually originate from the recognition of a need or an opportunity that is only partially understood initially. Thus, before seeking solutions it is important to define the design objectives. Early steps in engineering design include developing quantitative performance specifications and identifying alternative *workable* designs that meet the specifications. Among the workable designs are generally one or more that are "best" according to some criteria: lowest cost, highest efficiency, smallest size, lightest weight, and so on. Other important factors in the selection of a final design include reliability, manufacturability, maintainability, and marketplace considerations. Accordingly, a compromise must be sought among competing criteria, and there may be alternative design solutions that are feasible.[3]

---

[3]For further discussion, see A. Bejan, G. Tsatsaronis, and M. J. Moran, *Thermal Design and Optimization*, John Wiley & Sons, New York, 1996, Chap. 1.

1.8.2 Analysis

Design requires synthesis: selecting and putting together components to form a coordinated whole. However, as each individual component can vary in size, performance, cost, and so on, it is generally necessary to subject each to considerable study or analysis before a final selection can be made.

FOR EXAMPLE

A proposed design for a fire-protection system might entail an overhead piping network together with numerous sprinkler heads. Once an overall configuration has been determined, detailed engineering analysis is necessary to specify the number and type of the spray heads, the piping material, and the pipe diameters of the various branches of the network. The analysis also must aim to ensure all components form a smoothly working whole while meeting relevant cost constraints and applicable codes and standards.

Engineers frequently do analysis, whether explicitly as part of a design process or for some other purpose. Analyses involving systems of the kind considered in this book use, directly or indirectly, one or more of three basic laws. These laws, which are independent of the particular substance or substances under consideration, are

1. the conservation of mass principle
2. the conservation of energy principle
3. the second law of thermodynamics

In addition, relationships among the properties of the particular substance or substances considered are usually necessary (Chaps. 3, 6, 11–14). Newton's second law of motion (Chaps. 1, 2, 9), relations such as Fourier's conduction model (Chap. 2), and principles of engineering economics (Chap. 7) also may play a part.

    The first steps in a thermodynamic analysis are defining the system and identifying relevant interactions with the surroundings. Attention then turns to the pertinent physical laws and **engineering model** relationships that allow the behavior of the system to be described in terms of an **engineering model**. The objective in modeling is to obtain a simplified representation of system behavior that is sufficiently faithful for the purpose of the analysis, even if many aspects exhibited by the actual system are ignored. For example, idealizations often used in mechanics to simplify an analysis and arrive at a manageable model include the assumptions of point masses, frictionless pulleys, and rigid beams. Satisfactory modeling takes experience and is a part of the *art* of engineering.

    Engineering analysis is most effective when it is done systematically. This is considered next.

1.9 Methodology for Solving Thermodynamics Problems

A major goal of this textbook is to help you learn how to solve engineering problems that involve thermodynamic principles. To this end, numerous solved examples and end-of-chapter problems are provided. It is extremely important for you to study the examples *and* solve problems, for mastery of the fundamentals comes only through practice.

    To maximize the results of your efforts, it is necessary to develop a systematic approach. You must think carefully about your solutions and avoid the temptation of starting problems *in the middle* by selecting some seemingly appropriate equation, substituting in numbers, and quickly "punching up" a result on your calculator. Such a haphazard problem-solving approach can lead to difficulties as problems become more complicated. Accordingly, it is strongly recommended that problem solutions be organized using the following *five steps*, which are employed in the solved examples of this text.

① Known   State briefly in your own words what is known. This requires that you read the problem carefully *and* think about it.

② Find   State concisely in your own words what is to be determined.

③ Schematic and Given Data   Draw a sketch of the system to be considered. Decide whether a closed system or control volume is appropriate for the analysis, and then carefully identify the boundary. Label the diagram with relevant information from the problem statement.

Record all property values you are given or anticipate may be required for subsequent calculations. Sketch appropriate property diagrams (see Sec. 3.2), locating key state points and indicating, if possible, the processes executed by the system.

The importance of good sketches of the system and property diagrams cannot be overemphasized. They are often instrumental in enabling you to think clearly about the problem.

④ Engineering Model   To form a record of how you *model* the problem, list all *simplifying assumptions* and *idealizations* made to reduce it to one that is manageable. Sometimes this information also can be noted on the sketches of the previous step. The development of an appropriate model is a key aspect of successful problem solving.

⑤ Analysis   Using your assumptions and idealizations, reduce the appropriate governing equations and relationships to forms that will produce the desired results.

It is advisable to work with equations as long as possible before substituting numerical data. When the equations are reduced to final forms, consider them to determine what additional data may be required. Identify the tables, charts, or property equations that provide the required values. Additional property diagram sketches may be helpful at this point to clarify states and processes.

When all equations and data are in hand, substitute numerical values into the equations. Carefully check that a consistent and appropriate set of units is being employed. Then perform the needed calculations.

Finally, consider whether the magnitudes of the numerical values are reasonable and the algebraic signs associated with the numerical values are correct.

The problem solution format used in this text is intended to *guide* your thinking, not substitute for it. Accordingly, you are cautioned to avoid the rote application of these five steps, for this alone would provide few benefits. Indeed, as a particular solution evolves you may have to return to an earlier step and revise it in light of a better understanding of the problem. For example, it might be necessary to add or delete an assumption, revise a sketch, determine additional property data, and so on.

The solved examples provided in the book are frequently annotated with various comments intended to assist learning, including commenting on what was learned, identifying key aspects of the solution, and discussing how better results might be obtained by relaxing certain assumptions.

In some of the earlier examples and end-of-chapter problems, the solution format may seem unnecessary or unwieldy. However, as the problems become more complicated you will see that it reduces errors, saves time, and provides a deeper understanding of the problem at hand.

The example to follow illustrates the use of this solution methodology together with important system concepts introduced previously, including identification of interactions occurring at the boundary.

Using the Solution Methodology and System Concepts

A wind turbine–electric generator is mounted atop a tower. As wind blows steadily across the turbine blades, electricity is generated. The electrical output of the generator is fed to a storage battery.

a. Considering only the wind turbine–electric generator as the system, identify locations on the system boundary where the system interacts with the surroundings. Describe changes occurring within the system with time.

b. Repeat for a system that includes only the storage battery.

SOLUTION

Known A wind turbine–electric generator provides electricity to a storage battery.

Find For a system consisting of (a) the wind turbine–electric generator, (b) the storage battery, identify locations where the system interacts with its surroundings, and describe changes occurring within the system with time.

Schematic and Given Data:

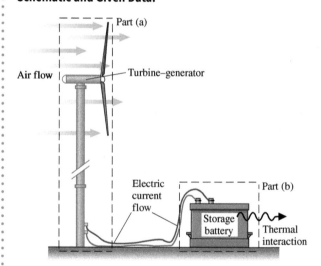

Fig. E1.1

Engineering Model

1. In part (a), the system is the control volume shown by the dashed line on the figure.

2. In part (b), the system is the closed system shown by the dashed line on the figure.

3. The wind is steady.

Analysis

a. In this case, the wind turbine is studied as a control volume with air flowing across the boundary. Another principal interaction between the system and surroundings is the electric current passing through the wires. From the macroscopic perspective, such an interaction is not considered a mass transfer, however. With a steady wind, the turbine–generator is likely to reach steady-state operation, where the rotational speed of the blades is constant and a steady electric current is generated.

❶ b. In this case, the battery is studied as a closed system. The principal interaction between the system and its surroundings is the electric current passing into the battery through the wires. As noted in part (a), this interaction is not considered a mass transfer. As the battery is charged and chemical reactions occur within it, the temperature of the battery surface may become somewhat elevated and a thermal interaction might occur between the battery and its surroundings. This interaction is likely to be of secondary importance. Also, as the battery is charged, the state within changes with time. The battery is not at steady state.

---

❶ Using terms familiar from a previous physics course, the system of part (a) involves the *conversion* of kinetic energy to electricity, whereas the system of part (b) involves energy *storage* within the battery.

SKILLS DEVELOPED

Ability to…

- apply the problem-solving methodology used in this book.
- define a control volume and identify interactions on its boundary.
- define a closed system and identify interactions on its boundary.
- distinguish steady-state operation from nonsteady operation.

Quick Quiz

May an overall system consisting of the turbine–generator and battery be considered as operating at steady state? Explain. Ans. No. A system is at steady state only if *none* of its properties changes with time.

CHAPTER SUMMARY AND STUDY GUIDE

In this chapter, we have introduced some of the fundamental concepts and definitions used in the study of thermodynamics. The principles of thermodynamics are applied by engineers to analyze and design a wide variety of devices intended to meet human needs.

An important aspect of thermodynamic analysis is to identify systems and to describe system behavior in terms of properties and processes. Three important properties discussed in this chapter are specific volume, pressure, and temperature.

In thermodynamics, we consider systems at equilibrium states and systems undergoing processes (changes of state). We study processes during which the intervening states are not equilibrium states and processes during which the departure from equilibrium is negligible.

In this chapter, we have introduced SI units for mass, length, time, force, and temperature.

Chapter 1 concludes with discussions of how thermodynamics is used in engineering design and how to solve thermodynamics problems systematically.

This book has several features that facilitate study and contribute to understanding. For an overview, see *How to Use This Book Effectively* on the book companion site.

The following checklist provides a study guide for this chapter. When your study of the text and the end-of-chapter exercises has been completed you should be able to

- write out the meanings of the terms listed in the margin throughout the chapter and explain each of the related concepts. The subset of key concepts listed is particularly important in subsequent chapters.

- identify an appropriate system boundary and describe the interactions between the system and its surroundings.

- work on a molar basis using Eq. 1.8.

- use SI units for mass, length, time, force, and temperature and apply appropriately Newton's second law and Eqs. 1.16.

- apply the methodology for problem solving discussed in Sec. 1.9.

KEY ENGINEERING CONCEPTS

system	property	equilibrium
surroundings	state	specific volume
boundary	process	pressure
closed system	extensive property	temperature
control volume	intensive property	Kelvin scale

KEY EQUATIONS

$n = m/M$	(1.8)	Relation between amounts of matter on a mass basis, m, and on a molar basis, n.
$T(\degree C) = T(K) - 273.15$	(1.16)	Relation between the Celsius and Kelvin temperatures.

EXERCISES: THINGS ENGINEERS THINK ABOUT

1.1 What components would be required for a high school laboratory project aimed at determining the electricity generated by a hamster running on its exercise wheel?

1.2 Operating rooms in hospitals typically have a *positive pressure* relative to adjacent spaces. What does this mean and why is it done?

1.3 Based on the macroscopic view, a quantity of air at 100 kPa, 20°C is in equilibrium. Yet the atoms and molecules of the air are in constant motion. How do you reconcile this apparent contradiction?

1.4 Laura takes an elevator from the tenth floor of her office building to the lobby. Should she expect the air pressure on the two levels to differ much?

1.5 Why does ocean water temperature vary with depth?

1.6 Are the *systolic* and *diastolic* pressures reported in blood pressure measurements absolute, gage, or vacuum pressures?

1.7 Air at 1 atm, 20°C in a closed tank adheres to the continuum hypothesis. Yet when sufficient air has been drawn from the tank, the hypothesis no longer applies to the remaining air. Why?

1.8 When one walks barefoot from a carpet onto a ceramic tile floor, the tiles feel *colder* than the carpet even though each surface is at the same temperature. Explain.

1.9 What causes changes in atmospheric pressure?

1.10 A data sheet indicates that the pressure at the inlet to a pump is –15 kPa. What does the negative sign denote?

1.11 When the instrument panel of a car provides the outside air temperature, where is the temperature sensor located?

1.12 What is a *nanotube*?

1.13 When buildings have large exhaust fans, exterior doors can be difficult to open due to a pressure difference between the inside and outside. Do you think you could open a 1.2 m by 2.1 m door if the inside pressure were 0.28 kPa (vacuum)?

CHECKING UNDERSTANDING

For problems 1.1–1.6, match the appropriate definition in the right column with each term in the left column.

1.1 Boundary

1.2 Control volume

1.3 Extensive property

1.4 Intensive property

1.5 Property

1.6 Surroundings

A. A region of space through which mass may flow

B. A property whose value for an overall system is the sum of its values for the parts into which the system is divided

C. Everything external to the system

D. A property whose value is independent of the size or extent of a system and may vary from place to place within the system at any moment

E. Distinguishes the system from its surroundings

F. A macroscopic characteristic of a system to which a numerical value can be assigned at a given time without knowledge of the previous behavior of the system

1.7 A special type of closed system that does not interact in any way with its surroundings is an _____.

1.8 Describe the difference between specific volume expressed on a *mass* basis and a *molar* basis _____.

1.9 A control volume is a system that

a. always contains the same matter.

b. allows a transfer of matter across its boundary.

c. does not interact in any way with its surroundings.

d. always has a constant volume.

1.10 _____ is pressure with respect to the zero pressure of a complete vacuum.

1.11 A gas contained within a piston–cylinder assembly undergoes Process 1–2–3 shown on the *pressure–volume* diagram in **Fig. P1.11C**. Process 1–2–3 is

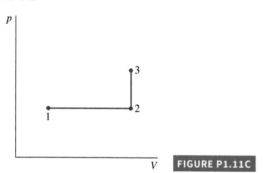

FIGURE P1.11C

a. a constant-volume process followed by constant-pressure compression.

b. constant-pressure compression followed by a constant-volume process.

c. a constant-volume process followed by a constant-pressure expansion.

d. constant-pressure expansion followed by a constant-volume process.

1.12 The statement, "When two objects are in thermal equilibrium with a third object, they are in thermal equilibrium with each other," is called the _____.

1.13 SI base units include

a. kilogram (kg), meter (m), newton (N).

b. kelvin (K), meter (m), second (s).

c. second (s), meter (m), pound mass.

d. kelvin (K), newton (N), second (s).

1.14 Explain why the value for *gage* pressure is always less than the corresponding value for *absolute* pressure.

1.15 When a system is isolated,

a. its mass remains constant.

b. its temperature may change.

c. its pressure may change.

d. all of the above.

1.16 The list consisting only of intensive properties is

a. volume, temperature, pressure.

b. specific volume, mass, volume.

c. pressure, temperature, specific volume.

d. mass, temperature, pressure.

Indicate whether the following statements are true or false. Explain.

1.17 Gage pressure indicates the difference between the absolute pressure of a system and the absolute pressure of the atmosphere existing outside the measuring device.

1.18 Kilogram, second, foot, and newton are all examples of SI units.

1.19 Temperature is an extensive property.

1.20 Mass is an intensive property.

1.21 Intensive properties may be functions of both position and time, whereas extensive properties can vary only with time.

1.22 Devices that measure pressure include barometers, Bourdon tube gages, and manometers.

1.23 If a system is isolated from its surroundings and no changes occur in its observable properties, the system was in equilibrium at the moment it was isolated.

1.24 The specific volume is the reciprocal of the density.

1.25 Volume is an extensive property.

1.26 Pressure is an intensive property.

1.27 A closed system always contains the same matter; there is no transfer of matter across its boundary.

1.28 When a closed system undergoes a process between two specified states, the change in temperature between the end states is independent of details of the process.

1.29 Specific volume, the volume per unit of mass, is an intensive property whereas volume and mass are extensive properties.

1.30 In local surroundings at standard atmospheric pressure, a gage will indicate a pressure of 0.2 atm for a refrigerant whose absolute pressure is 1.2 atm.

1.31 If the value of *any* property of a system changes with time, that system cannot be at steady state.

1.32 The composition of a closed system cannot change.

1.33 Temperature is the property that is the same for each of two systems when they are in thermal equilibrium.

PROBLEMS: DEVELOPING ENGINEERING SKILLS

Working with Force and Mass

1.1 A fully loaded shipping container has a mass of 30,000 kg. If *local* acceleration of gravity is 9.81 m/s², determine the container's weight, in kN.

1.2 The *Phoenix* with a mass of 350 kg was a spacecraft used for exploration of Mars. Determine the weight of the *Phoenix*, in N, (a) on the surface of Mars where the acceleration of gravity is 3.73 m/s² and (b) on Earth where the acceleration of gravity is 9.81 m/s².

1.3 A gas occupying a volume of 0.68 m³ weighs 15.6 N on the moon, where the acceleration of gravity is 1.67 m/s². Determine its weight, in N, and density, in kg/m³, on Mars, where $g = 3.92$ m/s².

1.4 When an object of mass 5 kg is suspended from a spring, the spring is observed to stretch by 8 cm. The deflection of the spring is related linearly to the weight of the suspended mass. What is the proportionality constant, in newtons per cm, if $g = 9.81$ m/s²?

1.5 A spring compresses in length by 3 mm, for every 4.5 N of applied force. Determine the deflection, in inches, of the spring caused by the weight of an object whose mass is 6.8 kg. The local acceleration of gravity is $g = 9.8$ m/s^2.

1.6 A simple instrument for measuring the acceleration of gravity employs a *linear* spring from which a mass is suspended. At a location on Earth where the acceleration of gravity is 9.81 m/s^2, the spring extends 7.391 mm. If the spring extends 2.946 mm when the instrument is on Mars, what is the Martian acceleration of gravity? How much would the spring extend on the moon, where $g = 1.668$ m/s^2?

1.7 An astronaut weighs 700 N on Earth where $g = 9.81$ m/s^2. What is the astronaut's weight, in N, on an orbiting space station where the acceleration of gravity is 6 m/s^2? Express each weight in N.

1.8 If the variation of the acceleration of gravity, in m/s^2, with elevation z, in m, above sea level is $g = 9.81 - (3.3 \times 10^{-6})\, z$, determine the percent change in weight of an airliner landing from a cruising altitude of 10 km on a runway at sea level.

1.9 Using local acceleration of gravity data from the Internet, determine the weight, in N, of a person whose mass is 80 kg living in:

   a. Mexico City, Mexico
   b. Cape Town, South Africa
   c. Tokyo, Japan
   d. Chicago, IL
   e. Copenhagen, Denmark

1.10 As shown in **Fig. P1.10**, a cylinder of compacted scrap metal measuring 2 m in length and 0.5 m in diameter is suspended from a spring scale at a location where the acceleration of gravity is 9.78 m/s^2. If the scrap metal density, in kg/m^3, varies with position z, in m, according to $\rho = 7800 - 360(z/L)^2$, determine the reading of the scale, in N.

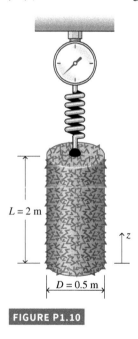

$L = 2$ m

$D = 0.5$ m

FIGURE P1.10

Using Specific Volume, Volume, and Pressure

1.11 A closed system consists of 0.5 kmol of ammonia occupying a volume of 6 m^3. Determine (a) the weight of the system, in N, and (b) the specific volume, in m^3/kmol and m^3/kg. Let $g = 9.81$ m/s^2.

1.12 A spherical balloon holding 16 kg of air has a diameter of 3 m. For the air, determine (a) the specific volume, in m^3/kg and m^3/kmol, and (b) the weight, in N. Let $g = 9.8$ m/s^2.

1.13 A closed vessel having a volume of 1 liter holds 2.5×10^{22} molecules of ammonia vapor. For the ammonia, determine (a) the amount present, in kg and kmol, and (b) the specific volume, in m^3/kg and m^3/kmol.

1.14 The specific volume of 5 kg of water vapor at 1.5 MPa, 440°C is 0.2160 m^3/kg. Determine (a) the volume, in m^3, occupied by the water vapor, (b) the amount of water vapor present, in gram moles, and (c) the number of molecules.

1.15 A closed system consisting of 1 kg of a gas undergoes a process during which the relation between pressure and volume is $pV^n = constant$. The process begins with $p_1 = 140$ kPa, $V_1 = 0.27$ m^3 and ends with $p_2 = 690$ kPa, $V_2 = 0.078$ m^3. Determine (a) the value of n and (b) the specific volume at states 1 and 2, each in m^3/kg. (c) Sketch the process on pressure–volume coordinates.

1.16 As shown in **Fig. P1.16**, a vertical piston–cylinder assembly containing a gas is placed on a hot plate. The piston initially rests on the stops. With the onset of heating, the gas pressure increases. At what pressure, in bar, does the piston start rising? The piston moves smoothly in the cylinder and $g = 9.81$ m/s^2.

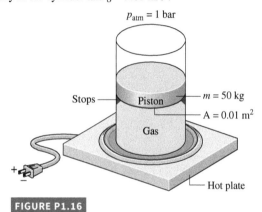

$p_{atm} = 1$ bar

Stops

Piston

$m = 50$ kg

$A = 0.01$ m^2

Gas

Hot plate

FIGURE P1.16

1.17 A closed system consisting of 5 kg of a gas undergoes a process during which the relationship between pressure and specific volume is $pv^{1.3} = constant$. The process begins with $p_1 = 1$ bar, $v_1 = 0.2$ m^3/kg and ends with $p_2 = 0.25$ bar. Determine the final volume, in m^3, and plot the process on a graph of pressure versus specific volume.

1.18 **Figure P1.18** shows a gas contained in a vertical piston–cylinder assembly. A vertical shaft whose cross-sectional area is 0.8 cm^2 is attached to the top of the piston. Determine the magnitude, F, of the force acting on the shaft, in N, required if the gas pressure is 3 bar. The masses of the piston and attached shaft are 24.5 kg and 0.5 kg, respectively. The piston diameter is 10 cm. The local atmospheric pressure is 1 bar. The piston moves smoothly in the cylinder and $g = 9.81$ m/s^2.

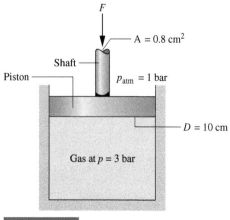

F

$A = 0.8$ cm^2

Shaft

Piston

$p_{atm} = 1$ bar

$D = 10$ cm

Gas at $p = 3$ bar

FIGURE P1.18

1.19 A gas contained within a piston–cylinder assembly undergoes four processes in series:

> **Process 1–2:** Constant-pressure expansion at 1 bar from $V_1 = 0.5$ m^3 to $V_2 = 2$ m^3
>
> **Process 2–3:** Constant volume to 2 bar
>
> **Process 3–4:** Constant-pressure compression to 1 m^3
>
> **Process 4–1:** Compression with $pV^{-1} = constant$

Sketch the processes in series on a p–V diagram labeled with pressure and volume values at each numbered state.

1.20 Referring to Fig. 1.7,

> **a.** if the pressure in the tank is 1.5 bar and atmospheric pressure is 1 bar, determine L, in m, for water with a density of 997 kg/m^3 as the manometer liquid. Let $g = 9.81$ m/s^2.
>
> **b.** determine L, in cm, if the manometer liquid is mercury with a density of 13.59 g/cm^3 and the gas pressure is 1.3 bar. A barometer indicates the local atmospheric pressure is 750 mmHg. Let $g = 9.81$ m/s^2.

1.21 **Figure P1.21** shows a storage tank holding natural gas. In an adjacent instrument room, a U-tube mercury manometer in communication with the storage tank reads $L = 1.0$ m. If the atmospheric pressure is 101 kPa, the density of the mercury is 13.59 g/cm^3, and $g = 9.81$ m/s^2, determine the pressure of the natural gas, in kPa.

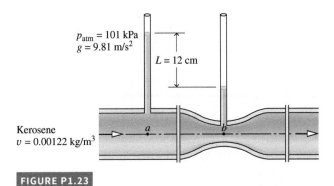

FIGURE P1.21

1.22 The absolute pressure inside a tank is 0.4 bar, and the surrounding atmospheric pressure is 98 kPa. What reading would a Bourdon gage mounted in the tank wall give, in kPa? Is this a *gage* or *vacuum* reading?

1.23 Liquid kerosene flows through a Venturi meter, as shown in **Fig. P1.23**. The pressure of the kerosene in the pipe supports columns of kerosene that differ in height by 12 cm. Determine the difference in pressure between points a and b, in kPa. Does the pressure increase or decrease as the kerosene flows from point a to point b as the pipe diameter decreases? The atmospheric pressure is 101 kPa, the specific volume of kerosene is 0.00122 m^3/kg, and the acceleration of gravity is $g = 9.81$ m/s^2.

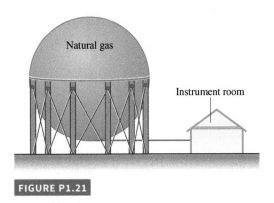

FIGURE P1.23

1.24 **Figure P1.24** shows a tank within a tank, each containing air. The absolute pressure in tank A is 267.7 kPa. Pressure gage A is located inside tank B and reads 140 kPa. The U-tube manometer connected to tank B contains mercury. Using data on the diagram, determine the absolute pressure inside tank B, in kPa, and the column length L, in cm. The atmospheric pressure surrounding tank B is 101 kPa. The acceleration of gravity is $g = 9.81$ m/s^2.

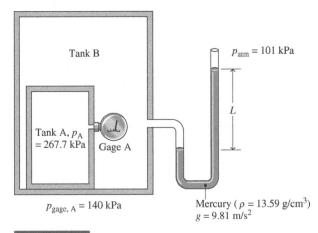

FIGURE P1.24

1.25 As shown in **Fig. P1.25**, an underwater exploration vehicle submerges to a depth of 300 m. If the atmospheric pressure at the surface is 101.3 kPa, the water density is 1000 kg/m^3 and $g = 9.8$ m/s^2, determine the pressure on the vehicle, in atm.

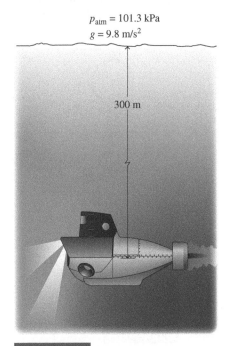

FIGURE P1.25

1.26 Show that a standard atmospheric pressure of 760 mmHg is equivalent to 101.3 kPa. The density of mercury is 13,590 kg/m^3 and $g = 9.81$ m/s^2.

1.27 Refrigerant 22 vapor enters the compressor of a refrigeration system at an absolute pressure of 140 kPa. A pressure gage at the compressor exit indicates a pressure of 1930 kPa (gage). The atmospheric pressure is 101.3 kPa. Determine the change in absolute pressure from inlet to exit, in kPa, and the ratio of exit to inlet pressure.

1.28 As shown in Fig. P1.28, air is contained in a vertical piston-cylinder assembly fitted with an electrical resistor. The atmosphere exerts a pressure of 101.3 kPa on the top of the piston, which has a mass of 45 kg and face area of 0.09 m². As electric current passes through the resistor, the volume of the air increases while the piston moves smoothly in the cylinder. The local acceleration of gravity is $g = 9.8$ m/s² Determine the pressure of the air in the piston-cylinder assembly, in kPa and N/m².

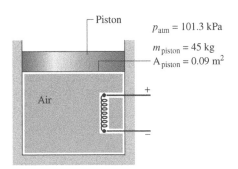

$p_{atm} = 101.3$ kPa

$m_{piston} = 45$ kg

$A_{piston} = 0.09$ m²

FIGURE P1.28

1.29 Air is contained in a vertical piston–cylinder assembly such that the piston is in static equilibrium. The atmosphere exerts a pressure of 101 kPa on top of the 0.5-m-diameter piston. The gage pressure of the air inside the cylinder is 1.2 kPa. The local acceleration of gravity is $g = 9.81$ m/s². Subsequently, a weight is placed on top of the piston causing the piston to fall until reaching a new static equilibrium position. At this position, the gage pressure of the air inside the cylinder is 2.8 kPa. Determine (a) the mass of the piston, in kg, and (b) the mass of the added weight, in kg.

1.30 Figure P1.30 shows a tank used to collect rainwater having a diameter of 4 m. As shown in the figure, the depth of the tank varies linearly from 3.5 m at its center to 3 m along the perimeter. The local atmospheric pressure is 1 bar, the acceleration of gravity is 9.8 m/s², and the density of the water is 987.1 kg/m³. When the tank is filled with water, determine

   a. the pressure, in kPa, at the bottom center of the tank.

   b. the total force, in kN, acting on the bottom of the tank.

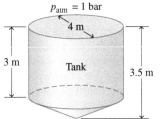

$p_{atm} = 1$ bar

4 m

3 m   Tank   3.5 m

FIGURE P1.30

1.31 The pressure from water mains located at street level may be insufficient for delivering water to the upper floors of tall buildings. In such a case, water may be pumped up to a tank that feeds water to the building by gravity. For an open storage tank atop a 90 m tall building, determine the pressure, in kPa, at the bottom of the tank when filled to a depth of 4 m. The density of water is 1000 kg/m³, $g = 9.8$ m/s², and the local atmospheric pressure is 101.3 kPa.

1.32 As shown in Fig. P1.32, an *inclined* manometer is used to measure the pressure of the gas within the reservoir. (a) Using data on the figure, determine the gas pressure, in kPa. (b) Express the pressure as a gage or a vacuum pressure, as appropriate, in kPa. (c) What advantage does an inclined manometer have over the U-tube manometer shown in Fig. 1.7?

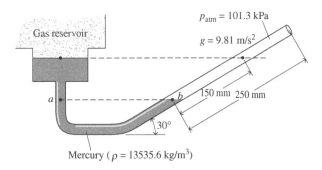

$p_{atm} = 101.3$ kPa

$g = 9.81$ m/s²

Gas reservoir

a b 150 mm   250 mm

30°

Mercury ($\rho = 13535.6$ kg/m³)

FIGURE P1.32

1.33 Figure P1.33 shows a spherical buoy, having a diameter of 1.5 m and weighing 8500 N, anchored to the floor of a lake by a cable. Determine the force exerted by the cable, in N. The density of the lake water is 10³ kg/m³ and $g = 9.81$ m/s².

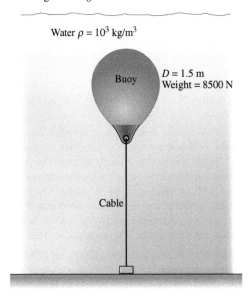

Water $\rho = 10^3$ kg/m³

Buoy $D = 1.5$ m   Weight = 8500 N

Cable

FIGURE P1.33

1.34 Determine the total force, in kN, on the bottom of a 100 × 50 m swimming pool. The depth of the pool varies linearly along its length from 1 m to 4 m. Also, determine the pressure on the floor at the center of the pool, in kPa. The atmospheric pressure is 0.98 bar, the density of the water is 998.2 kg/m³, and the local acceleration of gravity is 9.8 m/s².

Exploring Temperature

1.35 Convert the following temperatures from °C to K: (a) 21°C, (b) −40°C, (c) 500°C, (d) 0°C, (e) 100°C, (f) −273.15°C.

1.36 Convert the following temperatures from K to °C: (a) 293.15 K, (b) 233.15 K, (c) 533.15 K, (d) 255.4 K, (e) 373.15 K, (f) 0 K.

1.37 As shown in Fig. P1.37, a small-diameter water pipe passes through the 150 mm thick exterior wall of a dwelling. Assuming that temperature varies linearly with position x through the wall from 20°C to −7°C, would the water in the pipe freeze? Explain.

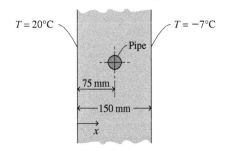

$T = 20°C$ $T = -7°C$

Pipe

75 mm

150 mm

x

FIGURE P1.37

1.38 What is (a) the lowest *naturally* occurring temperature recorded on Earth, (b) the lowest temperature recorded in a laboratory on Earth, (c) the lowest temperature recorded in the Earth's solar system, and (d) the temperature of deep space, each in K?

1.39 Calculate the value of temperature x such that $x°F$ is equal to $x°C$. Also convert the unit of x into Kelvin.

DESIGN & OPEN-ENDED PROBLEMS: EXPLORING ENGINEERING PRACTICE

1.1D *Hyperbaric chamber*s are used by medical professionals to treat several conditions. Research how a hyperbaric chamber functions, and identity at least three medical conditions for which time in a hyperbaric chamber may be part of the prescribed treatment. Describe how this treatment addresses each condition. Present your findings in a memorandum.

1.2D List several aspects of engineering economics relevant to design. What are the important contributors to *cost* that should be considered in engineering design? Discuss what is meant by *annualized costs*. Present your findings in a memorandum.

1.3D One type of prosthetic limb relies on suction to attach to an amputee's residual limb. The engineer must consider the required difference between atmospheric pressure and the pressure inside the socket of the prosthetic limb to develop suction sufficient to maintain attachment. What other considerations are important as engineers design this type of prosthetic device? Write a report of your findings including at least three references.

1.4D Design a low-cost, compact, lightweight, hand-held, human-powered air pump capable of directing a stream of air for cleaning computer keyboards, circuit boards, and hard-to-reach locations in electronic devices. The pump cannot use electricity, including batteries, or employ any chemical propellants. All materials must be recyclable. Owing to existing patent protections, the pump must be a *distinct alternative* to the familiar tube and plunger bicycle pump and to existing products aimed at accomplishing the specified computer and electronic cleaning tasks.

1.5D Design an experiment to determine the specific volume of water. For the experiment develop written procedures that include identification of all equipment needed and specification of all required calculations. Conduct the experiment, compare your result to *steam table* data, and communicate your results in an executive summary.

1.6D Magnetic resonance imaging (MRI) employs a strong magnetic field to produce detailed pictures of internal organs and tissues. As shown in **Fig. P1.6D**, the patient reclines on a table that slides into the cylindrical opening where the field is created. Considering a MRI scanner as a system, identify locations on the system boundary where the system interacts with its surroundings. Also describe events occurring within the system and the measures taken for patient comfort and safety. Write a report including at least three references.

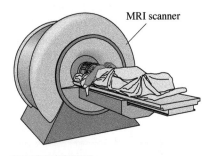

MRI scanner

FIGURE P1.6D

1.7D The *sphygmomanometer*, commonly used to measure blood pressure, is shown in **Fig. P1.7D**. During testing, the cuff is placed around the patient's arm and fully inflated by repeated squeezing of the inflation bulb. Then, as the cuff pressure is gradually reduced,

arterial sounds known as *Korotkoff* sounds are monitored with a stethoscope. Using these sounds as cues, the *systolic* and *diastolic* pressures can be identified. These pressures are reported in terms of the mercury column length, in mmHg. Investigate the physical basis for the Korotkoff sounds, their role in identifying the systolic and diastolic pressures, and why these pressures are significant in medical practice. Write a report including at least three references.

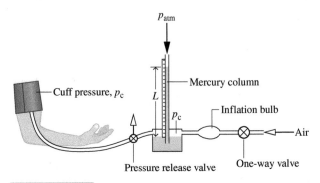

FIGURE P1.7D

1.8D In Bangladesh, unsafe levels of arsenic, which is a tasteless, odorless, and colorless poison, are present in underground wells providing drinking water to millions of people living in rural areas. The task is to identify affordable, easy-to-use treatment technologies for removing arsenic from their drinking water. Technologies considered should include, but not be limited to, applications of *smart materials* and other nanotechnology approaches. Write a report including at least three references.

1.9D Conduct a term-length design project in the realm of bioengineering done on either an independent or a small-group basis. The project might involve a device or technique for minimally invasive surgery, an implantable drug-delivery device, a biosensor, artificial blood, or something of special interest to you or your design group. Take several days to research your project idea and then prepare a brief written proposal, including several references, that provides a general statement of the core concept plus a list of objectives. During the project, observe good design practices such as discussed in Sec. 1.3 of *Thermal Design and Optimization,* John Wiley & Sons Inc., New York, 1996, by A. Bejan, G. Tsatsaronis, and M. J. Moran. Provide a well-documented final report, including several references.

1.10D Conduct a term-length design project involving the International Space Station pictured in Table 1.1 done on either an independent or a small-group basis. The project might involve an experiment that is best conducted in a low-gravity environment, a device for the comfort or use of the astronauts, or something of special interest to you or your design group. Take several days to research your project idea and then prepare a brief written proposal, including several references, that provides a general statement of the core concept plus a list of objectives. During the project, observe good design practices such as discussed in Sec. 1.3 of *Thermal Design and Optimization,* John Wiley & Sons Inc., New York, 1996, by A. Bejan, G. Tsatsaronis, and M. J. Moran. Provide a well-documented final report, including several references.

CHAPTER 2

Energy and the First Law of Thermodynamics

Engineering Context

Energy is a fundamental concept of thermodynamics and one of the most significant aspects of engineering analysis. In this chapter we discuss energy and develop equations for applying the principle of conservation of energy. The current presentation is limited to closed systems. In Chap. 4 the discussion is extended to control volumes.

Energy is a familiar notion, and you already know a great deal about it. In the present chapter several important aspects of the energy concept are developed. Some of these you have encountered before. A basic idea is that energy can be *stored* within systems in various forms. Energy also can be *converted* from one form to another and *transferred* between systems. For closed systems, energy can be transferred by *work* and *heat transfer*. The total amount of energy is *conserved* in all conversions and transfers.

The **objective** of this chapter is to organize these ideas about energy into forms suitable for engineering analysis. The presentation begins with a review of energy concepts from mechanics. The thermodynamic concept of energy is then introduced as an extension of the concept of energy in mechanics.

LEARNING OUTCOMES

When you complete your study of this chapter, you will be able to...

- Explain key concepts related to energy and the first law of thermodynamics . . . including internal, kinetic, and potential energy, work and power, heat transfer and heat transfer modes, heat transfer rate, power cycle, refrigeration cycle, and heat pump cycle.

- Analyze closed systems including applying energy balances, appropriately modeling the case at hand, and correctly observing sign conventions for work and heat transfer.

- Conduct energy analyses of systems undergoing thermodynamic cycles, evaluating as appropriate thermal efficiencies of power cycles and coefficients of performance of refrigeration and heat pump cycles.

2.1 Reviewing Mechanical Concepts of Energy

Building on the contributions of Galileo and others, Newton formulated a general description of the motions of objects under the influence of applied forces. Newton's laws of motion, which provide the basis for classical mechanics, led to the concepts of *work, kinetic energy,* and *potential energy,* and these led eventually to a broadened concept of energy. The present discussion begins with an application of Newton's second law of motion.

2.1.1 Work and Kinetic Energy

The curved line in **Fig. 2.1** represents the path of a body of mass m (a closed system) moving relative to the x–y coordinate frame shown. The velocity of the center of mass of the body is denoted by **V**. The body is acted on by a resultant force **F**, which may vary in magnitude from location to location along the path. The resultant force is resolved into a component $\mathbf{F}_s$ along the path and a component $\mathbf{F}_n$ normal to the path. The effect of the component $\mathbf{F}_s$ is to change the magnitude of the velocity, whereas the effect of the component $\mathbf{F}_n$ is to change the direction of the velocity. As shown in Fig. 2.1, s is the instantaneous position of the body measured along the path from some fixed point denoted by 0. Since the magnitude of **F** can vary from location to location along the path, the magnitudes of $\mathbf{F}_s$ and $\mathbf{F}_n$ are, in general, functions of s.

Let us consider the body as it moves from $s = s_1$, where the magnitude of its velocity is V_1, to $s = s_2$, where its velocity is V_2. Assume for the present discussion that the only interaction between the body and its surroundings involves the force **F**. By Newton's second law of motion, the magnitude of the component $\mathbf{F}_s$ is related to the change in the magnitude of **V** by

$$F_s = m\frac{dV}{dt} \tag{2.1}$$

Using the chain rule, this can be written as

$$F_s = m\frac{dV}{ds}\frac{ds}{dt} = mV\frac{dV}{ds} \tag{2.2}$$

where $V = ds/dt$. Rearranging Eq. 2.2 and integrating from s_1 to s_2 gives

$$\int_{V_1}^{V_2} mV\,dV = \int_{s_1}^{s_2} F_s\,ds \tag{2.3}$$

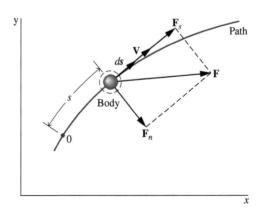

Fig. 2.1 Forces acting on a moving system.

The integral on the left of Eq. 2.3 is evaluated as follows

$$\int_{V_1}^{V_2} mV\,dV = \frac{1}{2}mV^2 \bigg]_{V_1}^{V_2} = \frac{1}{2}m(V_2^2 - V_1^2) \tag{2.4}$$

The quantity $\frac{1}{2}mV^2$ is the **kinetic energy**, KE, of the body. Kinetic energy is a scalar quantity. The *change* in kinetic energy, ΔKE, of the body is

kinetic energy

$$\Delta KE = KE_2 - KE_1 = \frac{1}{2}m(V_2^2 - V_1^2) \tag{2.5}$$

The integral on the right of Eq. 2.3 is the *work* of the force F_s as the body moves from s_1 to s_2 along the path. Work is also a scalar quantity.

With Eq. 2.4, Eq. 2.3 becomes

$$\frac{1}{2}m(V_2^2 - V_1^2) = \int_{s_1}^{s_2} \mathbf{F} \cdot d\mathbf{s} \tag{2.6}$$

TAKE NOTE...

The symbol Δ always means "final value minus initial value."

where the expression for work has been written in terms of the scalar product (dot product) of the force vector $\mathbf{F}$ and the displacement vector $d\mathbf{s}$. Equation 2.6 states that the work of the resultant force on the body equals the change in its kinetic energy. When the body is accelerated by the resultant force, the work done on the body can be considered a *transfer* of energy *to* the body, where it is *stored* as kinetic energy.

Kinetic energy can be assigned a value knowing only the mass of the body and the magnitude of its instantaneous velocity relative to a specified coordinate frame, without regard for how this velocity was attained. Hence, *kinetic energy is a property* of the body. Since kinetic energy is associated with the body as a whole, it is an *extensive* property.

Energy & Environment

Did you ever wonder what happens to the kinetic energy when you step on the brakes of your moving car? Automotive engineers have, and the result is the *hybrid vehicle* combining *regenerative* braking, batteries, an electric motor, and a conventional engine. When brakes are applied in a hybrid vehicle, some of the vehicle's kinetic energy is harvested and stored on board electrically for use when needed. Through regenerative braking and other innovative features, hybrids get much better mileage than comparably sized conventional vehicles.

Hybrid vehicle technology is quickly evolving. Today's hybrids use electricity to supplement conventional engine power, while future *plug-in* hybrids will use the power of a smaller engine to supplement electricity. The hybrids we now see on the road have enough battery power on board for acceleration to about 20 miles per hour and after that assist the engine when necessary. This improves fuel mileage, but the batteries are recharged by the engine—they are never *plugged in*.

Plug-in hybrids achieve even better fuel economy. Instead of relying on the engine to recharge batteries, most recharging will be received from an electrical outlet while the car is idle—overnight, for example. This will allow cars to get the energy they need mainly from the electrical grid, not the fuel pump. Widespread deployment of plug-ins awaits development of a new generation of batteries and ultra-capacitors (see Sec. 2.7).

Better fuel economy not only allows our society to be less reliant on oil to meet transportation needs but also reduces release of CO_2 into the atmosphere from vehicles. Each gallon of gasoline burned by a car's engine produces about 9 kg (20 lb) of CO_2. A conventional vehicle produces several tons of CO_2 annually; fuel-thrifty hybrids produce much less. Still, since hybrids use electricity from the grid, we will have to make a greater effort to reduce power plant emissions by including more wind power, solar power, and other renewables in the national mix.

2.1.2 Potential Energy

Equation 2.6 is a principal result of the previous section. Derived from Newton's second law, the equation gives a relationship between two *defined* concepts: kinetic energy and work. In this section it is used as a point of departure to extend the concept of energy. To begin, refer

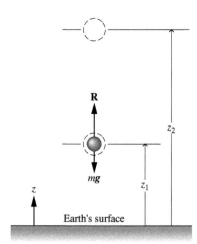

Fig. 2.2 Illustration used to introduce the potential energy concept.

to **Fig. 2.2**, which shows a body of mass m that moves vertically from an elevation z_1 to an elevation z_2 relative to the surface of Earth. Two forces are shown acting on the system: a downward force due to gravity with magnitude mg and a vertical force with magnitude R representing the resultant of all *other* forces acting on the system.

The work of each force acting on the body shown in Fig. 2.2 can be determined by using the definition previously given. The total work is the algebraic sum of these individual values. In accordance with Eq. 2.6, the total work equals the change in kinetic energy. That is,

$$\frac{1}{2}m(V_2^2 - V_1^2) = \int_{z_1}^{z_2} R\, dz - \int_{z_1}^{z_2} mg\, dz \tag{2.7}$$

A minus sign is introduced before the second term on the right because the gravitational force is directed downward and z is taken as positive upward.

The first integral on the right of Eq. 2.7 represents the work done by the force **R** on the body as it moves vertically from z_1 to z_2. The second integral can be evaluated as follows:

$$\int_{z_1}^{z_2} mg\, dz = mg(z_2 - z_1) \tag{2.8}$$

where the acceleration of gravity has been assumed to be constant with elevation. By incorporating Eq. 2.8 into Eq. 2.7 and rearranging

$$\frac{1}{2}m(V_2^2 - V_1^2) + mg(z_2 - z_1) = \int_{z_1}^{z_2} R\, dz \tag{2.9}$$

TAKE NOTE...

Throughout this book it is assumed that the acceleration of gravity, g, can be assumed constant unless otherwise noted.

gravitational potential energy

The quantity mgz is the **gravitational potential energy**, PE. The *change* in gravitational potential energy, ΔPE, is

$$\boxed{\Delta PE = PE_2 - PE_1 = mg(z_2 - z_1)} \tag{2.10}$$

Potential energy is associated with the force of gravity and is therefore an attribute of a system consisting of the body and Earth together. However, evaluating the force of gravity as mg enables the gravitational potential energy to be determined for a specified value of g knowing only the mass of the body and its elevation. With this view, potential energy is regarded as an *extensive property* of the body. Throughout this book it is assumed that elevation differences are small enough that the acceleration of gravity, g, can be considered constant. The concept of gravitational potential energy can be formulated to account for the variation of the acceleration of gravity with elevation, however.

To assign a value to the kinetic energy or the potential energy of a system, it is necessary to assume a datum and specify a value for the quantity at the datum. Values of kinetic and potential energy are then determined relative to this arbitrary choice of datum and reference value. However, since only *changes* in kinetic and potential energy between two states are required, these arbitrary reference specifications cancel.

2.1.3 Units for Energy

Work has units of force times distance. The units of kinetic energy and potential energy are the same as for work. In SI, the energy unit is the newton-meter, N · m, called the joule, J. In this book it is convenient to use the kilojoule, kJ.

When a system undergoes a process where there are changes in kinetic and potential energy, special care is required to obtain a consistent set of units.

FOR EXAMPLE

To illustrate the proper use of units in the calculation of such terms, consider a system having a mass of 1 kg whose velocity increases from 15 m/s to 30 m/s while its elevation decreases by 10 m at a location where $g = 9.7$ m/s^2. Then

$$\Delta KE = \frac{1}{2}m(V_2^2 - V_1^2)$$

$$= \frac{1}{2}(1\,kg)\left[\left(30\frac{m}{s}\right)^2 - \left(15\frac{m}{s}\right)^2\right]\left|\frac{1\,N}{1\,kg \cdot m/s^2}\right|\left|\frac{1\,kJ}{10^3\,N \cdot m}\right|$$

$$= 0.34\,kJ$$

$$\Delta PE = mg(z_2 - z_1)$$

$$= (1\,kg)\left(9.7\frac{m}{s^2}\right)(-10\,m)\left|\frac{1\,N}{1\,kg \cdot m/s^2}\right|\left|\frac{1\,kJ}{10^3\,N \cdot m}\right|$$

$$= -0.10\,kJ$$

2.1.4 Conservation of Energy in Mechanics

Equation 2.9 states that the total work of all forces acting on the body from the surroundings, with the exception of the gravitational force, equals the sum of the changes in the kinetic and potential energies of the body. When the resultant force causes the elevation to be increased, the body to be accelerated, or both, the work done by the force can be considered a *transfer* of energy *to* the body, where it is stored as gravitational potential energy and/or kinetic energy. The notion that *energy is conserved* underlies this interpretation.

The interpretation of Eq. 2.9 as an expression of the conservation of energy principle can be reinforced by considering the special case of a body on which the only force acting is that due to gravity, for then the right side of the equation vanishes and the equation reduces to

$$\frac{1}{2}m(V_2^2 - V_1^2) + mg(z_2 - z_1) = 0 \qquad (2.11)$$

or

$$\frac{1}{2}mV_2^2 + mgz_2 = \frac{1}{2}mV_1^2 + mgz_1$$

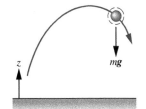

Under these conditions, the *sum* of the kinetic and gravitational potential energies *remains constant*. Equation 2.11 also illustrates that energy can be *converted* from one form to another: For an object falling under the influence of gravity *only*, the potential energy would decrease as the kinetic energy increases by an equal amount.

2.1.5 Closing Comment

The presentation thus far has centered on systems for which applied forces affect only their overall velocity and position. However, systems of engineering interest normally interact with their surroundings in more complicated ways, with changes in other properties as well. To analyze such systems, the concepts of kinetic and potential energy alone do not suffice, nor does the rudimentary conservation of energy principle introduced in this section. In thermodynamics the concept of energy is broadened to account for other observed changes, and the principle of *conservation of energy* is extended to include a wide variety of ways in which systems interact with their surroundings. The basis for such generalizations is experimental evidence. These extensions of the concept of energy are developed in the remainder of the chapter, beginning in the next section with a fuller discussion of work.

2.2 Broadening Our Understanding of Work

The work W done by, or on, a system evaluated in terms of macroscopically observable forces and displacements is

$$W = \int_{s_1}^{s_2} \mathbf{F} \cdot d\mathbf{s}$$

(2.12)

This relationship is important in thermodynamics and is used later in the present section to evaluate the work done in the compression or expansion of gas (or liquid), the extension of a solid bar, and the stretching of a liquid film. However, thermodynamics also deals with phenomena not included within the scope of mechanics, so it is necessary to adopt a broader interpretation of work, as follows.

A particular interaction is categorized as a work interaction if it satisfies the following criterion, which can be considered the **thermodynamic definition of work**: *Work is done by a system on its surroundings if the sole effect on everything external to the system could have been the raising of a weight.* Notice that the raising of a weight is, in effect, a force acting through a distance, so the concept of work in thermodynamics is a natural extension of the concept of work in mechanics. However, the test of whether a work interaction has taken place is not that the elevation of a weight has actually taken place, or that a force has actually acted through a distance, but that the sole effect *could have been* an increase in the elevation of a weight.

thermodynamic definition of work

FOR EXAMPLE

Consider Fig. 2.3 showing two systems labeled A and B. In system A, a gas is stirred by a paddle wheel: The paddle wheel does work on the gas. In principle, the work could be evaluated in terms of the forces and motions at the boundary between the paddle wheel and the gas. Such an evaluation of work is consistent with Eq. 2.12, where work is the product of force and displacement. By contrast, consider system B, which includes only the battery. At the boundary of system B, forces and motions are not evident. Rather, there is an electric current i driven by an electrical potential difference existing across the terminals a and b. That this type of interaction at the boundary can be classified as work follows from the thermodynamic definition of work given previously: We can imagine the current is supplied to a *hypothetical* electric motor that lifts a weight in the surroundings.

TAKE NOTE...

The term work does not refer to what is being transferred between systems or to what is stored within systems. Energy is transferred and stored when work is done.

Work is a means for transferring energy. Accordingly, the term *work* does not refer to what is being transferred between systems or to what is stored within systems. Energy is transferred and stored when work is done.

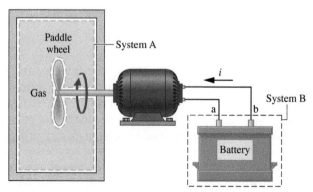

Fig. 2.3 Two examples of work.

2.2.1 Sign Convention and Notation

Engineering thermodynamics is frequently concerned with devices such as internal combustion engines and turbines whose purpose is to do work. Hence, in contrast to the approach generally taken in mechanics, it is often convenient to consider such work as positive. That is,

$$W > 0: \text{work done } by \text{ the system}$$

$$W < 0: \text{work done } on \text{ the system}$$

This **sign convention** is used throughout the book. In certain instances, however, it is convenient to regard the work done *on* the system to be positive, as has been done in the discussion of Sec. 2.1. To reduce the possibility of misunderstanding in any such case, the direction of energy transfer is shown by an arrow on a sketch of the system, and work is regarded as positive in the direction of the arrow.

  To evaluate the integral in Eq. 2.12, it is necessary to know how the force varies with the displacement. This brings out an important idea about work: The value of W depends on the details of the interactions taking place between the system and surroundings during a process and not just the initial and final states of the system. It follows that **work is not a property** of the system or the surroundings. In addition, the limits on the integral of Eq. 2.12 mean "from state 1 to state 2" and cannot be interpreted as the *values* of work at these states. The notion of work at a state *has no meaning,* so the value of this integral should never be indicated as $W_2 - W_1$.

sign convention for work

work is not a property

Horizons

Nanoscale Machines on the Move

Engineers working in the field of nanotechnology, the engineering of molecular-sized devices, look forward to the time when practical nanoscale machines can be fabricated that are capable of movement, sensing and responding to stimuli such as light and sound, delivering medication within the body, performing computations, and numerous other functions that promote human well-being. For inspiration, engineers study biological nanoscale *machines* in living things that perform functions such as creating and repairing cells, circulating oxygen, and digesting food. These studies have yielded positive results. Molecules mimicking the function of mechanical devices have been fabricated, including gears, rotors, ratchets, brakes, switches, and abacus-like structures. A particular success is the development of molecular motors that convert light to rotary or linear motion. Although devices produced thus far are rudimentary, they do demonstrate the feasibility of constructing nanomachines, researchers say.

  The differential of work, δW, is said to be *inexact* because, in general, the following integral cannot be evaluated without specifying the details of the process

$$\int_1^2 \delta W = W$$

On the other hand, the differential of a property is said to be *exact* because the change in a property between two particular states depends in no way on the details of the process linking the two states. For example, the change in volume between two states can be determined by integrating the differential dV, without regard for the details of the process, as follows

$$\int_{V_1}^{V_2} dV = V_2 - V_1$$

where V_1 is the volume *at* state 1 and V_2 is the volume *at* state 2. The differential of every property is exact. Exact differentials are written, as above, using the symbol d. To stress the difference between exact and inexact differentials, the differential of work is written as δW. The symbol δ is also used to identify other inexact differentials encountered later.

2.2.2 Power

power

Many thermodynamic analyses are concerned with the time rate at which energy transfer occurs. The rate of energy transfer by work is called **power** and is denoted by $\dot{W}$. When a work interaction involves a macroscopically observable force, the rate of energy transfer by work is equal to the product of the force and the velocity at the point of application of the force

$$\dot{W} = \mathbf{F} \cdot \mathbf{V}$$

(2.13)

A dot appearing over a symbol, as in $\dot{W}$, is used throughout this book to indicate a time rate. In principle, Eq. 2.13 can be integrated from time t_1 to time t_2 to get the total energy transfer by work during the time interval

$$W = \int_{t_1}^{t_2} \dot{W}\, dt = \int_{t_1}^{t_2} \mathbf{F} \cdot \mathbf{V}\, dt$$

(2.14)

units for power

The same sign convention applies for $\dot{W}$ as for W. Since power is a time rate of doing work, it can be expressed in terms of any units for energy and time. In SI, the **unit for power** is J/s, called the watt. In this book the kilowatt, kW, is generally used.

> **FOR EXAMPLE**
>
> To illustrate the use of Eq. 2.13, let us evaluate the power required for a bicyclist traveling at 20 miles per hour to overcome the drag force imposed by the surrounding air. This *aerodynamic drag* force is given by
>
> $$F_d = \tfrac{1}{2} C_d A \rho V^2$$
>
> where C_d is a constant called the *drag coefficient*, A is the frontal area of the bicycle and rider, and ρ is the air density. By Eq. 2.13 the required power is $F_d \cdot V$ or
>
> $$\dot{W} = \left(\tfrac{1}{2} C_d A \rho V^2\right) V$$
> $$= \tfrac{1}{2} C_d A \rho V^3$$
>
> Using typical values: $C_d = 0.88$, A = 0.362 m², and $\rho = 1.2$ kg/m³, together with V = 8.94 m/s, the power required is
>
> $$\dot{W} = \frac{1}{2}(0.88)(0.362\ \text{m}^2)(1.2\ \text{kg/m}^3)(8.94\ \text{m/s})$$
> $$= 136.6\ \text{W}$$

Drag can be reduced by *streamlining* the shape of a moving object and using the strategy known as *drafting* (see box).

Drafting

Drafting occurs when two or more moving vehicles or individuals align closely to reduce the overall effect of drag. Drafting is seen in competitive events such as auto racing, bicycle racing, speed-skating, and running.

Studies show that air flow over a single vehicle or individual in motion is characterized by a high-pressure region in front and a low-pressure region behind. The difference between these pressures creates a force, called drag, impeding motion. During drafting, as seen in the sketch below, a second vehicle or individual is closely aligned with another, and air flows over the pair nearly as if they were a single entity, thereby altering the pressure between them and reducing the drag each experiences. While race-car drivers use drafting to increase speed, non–motor sport competitors usually aim to reduce demands on their bodies while maintaining the same speed.

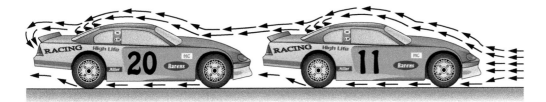

2.2.3 Modeling Expansion or Compression Work

There are many ways in which work can be done by or on a system. The remainder of this section is devoted to considering several examples, beginning with the important case of the work done when the volume of a quantity of a gas (or liquid) changes by expansion or compression.

Let us evaluate the work done by the closed system shown in **Fig. 2.4** consisting of a gas (or liquid) contained in a piston–cylinder assembly as the gas expands. During the process, the gas pressure exerts a normal force on the piston. Let p denote the pressure acting at the interface between the gas and the piston. The force exerted by the gas on the piston is simply the product $p\text{A}$, where A is the area of the piston face. The work done by the system as the piston is displaced a distance dx is

$$\delta W = p\text{A}\, dx \qquad (2.15)$$

The product A dx in Eq. 2.15 equals the change in volume of the system, dV. Thus, the work expression can be written as

$$\delta W = p\, dV \qquad (2.16)$$

Since dV is positive when volume increases, the work at the moving boundary is positive when the gas expands. For a compression, dV is negative, and so is work found from Eq. 2.16. These signs are in agreement with the previously stated sign convention for work.

For a change in volume from V_1 to V_2, the work is obtained by integrating Eq. 2.16

$$\boxed{W = \int_{V_1}^{V_2} p\, dV} \qquad (2.17)$$

Although Eq. 2.17 is derived for the case of a gas (or liquid) in a piston–cylinder assembly, it is applicable to systems of *any* shape provided the pressure is uniform with position over the moving boundary.

2.2.4 Expansion or Compression Work in Actual Processes

There is no requirement that a system undergoing a process be in equilibrium *during* the process. Some or all of the intervening states may be nonequilibrium states. For many such processes we are limited to knowing the state before the process occurs and the state after the process is completed.

Typically, at a nonequilibrium state intensive properties vary with position at a given time. Also, at a specified position intensive properties may vary with time, sometimes chaotically. In certain cases, spatial and temporal variations in properties such as temperature, pressure, and velocity can be measured or obtained by solving appropriate governing equations, which are generally differential equations.

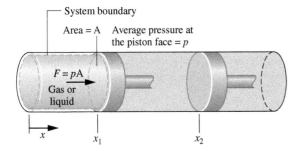

Fig. 2.4 Expansion or compression of a gas or liquid.

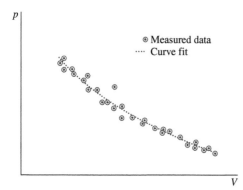

Fig. 2.5 Pressure at the piston face versus cylinder volume.

To perform the integral of Eq. 2.17 requires a relationship between the gas pressure *at the moving boundary* and the system volume. However, due to nonequilibrium effects during an *actual* expansion or compression process, this relationship may be difficult, or even impossible, to obtain. In the cylinder of an automobile engine, for example, combustion and other nonequilibrium effects give rise to non-uniformities throughout the cylinder. Accordingly, if a pressure transducer were mounted on the cylinder head, the recorded output might provide only an approximation for the pressure at the piston face required by Eq. 2.17. Moreover, even when the measured pressure is essentially equal to that at the piston face, scatter might exist in the pressure–volume data, as illustrated in **Fig. 2.5**. Still, performing the integral of Eq. 2.17 based on a curve fitted to the data could give a *plausible estimate* of the work. We will see later that in some cases where lack of the required pressure–volume relationship keeps us from evaluating the work from Eq. 2.17, the work can be determined alternatively from an *energy balance* (Sec. 2.5).

2.2.5 Expansion or Compression Work in Quasiequilibrium Processes

quasiequilibrium process

Processes are sometime modeled as an idealized type of process called a **quasiequilibrium (or quasistatic) process**. A quasiequilibrium process is one in which the departure from thermodynamic equilibrium is at most infinitesimal. All states through which the system passes in a quasiequilibrium process may be considered equilibrium states. Because nonequilibrium effects are inevitably present during actual processes, systems of engineering interest can at best approach, but never realize, a quasiequilibrium process. Still the quasiequilibrium process plays a role in our study of engineering thermodynamics. For details, see the box.

Incremental masses removed during an expansion of the gas or liquid

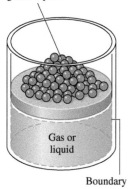

Gas or liquid

Boundary

Fig. 2.6 Illustration of a quasiequilibrium expansion or compression.

To consider how a gas (or liquid) might be expanded or compressed in a quasiequilibrium fashion, refer to **Fig. 2.6**, which shows a system consisting of a gas initially at an equilibrium state. As shown in the figure, the gas pressure is maintained uniform throughout by a number of small masses resting on the freely moving piston. Imagine that one of the masses is removed, allowing the piston to move upward as the gas expands slightly. During such an expansion, the state of the gas would depart only slightly from equilibrium. The system would eventually come to a new equilibrium state, where the pressure and all other intensive properties would again be uniform in value. Moreover, were the mass replaced, the gas would be restored to its initial state, while again the departure from equilibrium would be slight. If several of the masses were removed one after another, the gas would pass through a sequence of equilibrium states without ever being far from equilibrium. In the limit as the increments of mass are made vanishingly small, the gas would undergo a quasiequilibrium expansion process. A quasiequilibrium compression can be visualized with similar considerations.

Using the Quasiequilibrium Process Concept

Our interest in the quasiequilibrium process concept stems mainly from two considerations:

- Simple thermodynamic models giving at least *qualitative* information about the behavior of actual systems of interest often can be developed using the quasiequilibrium process concept. This is akin to the use of idealizations such as the point mass or the frictionless pulley in mechanics for the purpose of simplifying an analysis.

- The quasiequilibrium process concept is instrumental in deducing relationships that exist among the properties of systems at equilibrium (Chaps. 3, 6, and 11).

Equation 2.17 can be applied to evaluate the work in quasiequilibrium expansion or compression processes. For such idealized processes the pressure p in the equation is the pressure of the entire quantity of gas (or liquid) undergoing the process, and not just the pressure at the moving boundary. The relationship between the pressure and volume may be graphical or analytical. Let us first consider a graphical relationship.

A graphical relationship is shown in the pressure–volume diagram (*p–V* diagram) of **Fig. 2.7**. Initially, the piston face is at position x_1, and the gas pressure is p_1; at the conclusion of a quasiequilibrium expansion process the piston face is at position x_2, and the pressure is reduced

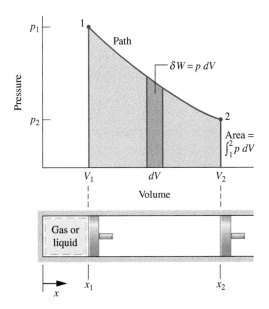

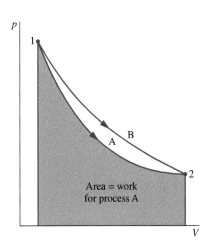

Fig. 2.7 Work of a quasiequilibrium expansion or compression process.

Fig. 2.8 Illustration that work depends on the process.

to p_2. At *each* intervening piston position, the uniform pressure throughout the gas is shown as a point on the diagram. The curve, or *path,* connecting states 1 and 2 on the diagram represents the equilibrium states through which the system has passed during the process. The work done by the gas on the piston during the expansion is given by $\int p\,dV$, which can be interpreted as the area under the curve of pressure versus volume. Thus, the shaded area on Fig. 2.7 is equal to the work for the process. Had the gas been *compressed* from 2 to 1 along the same path on the p–V diagram, the *magnitude* of the work would be the same, but the sign would be negative, indicating that for the compression the energy transfer was from the piston to the gas.

The area interpretation of work in a quasiequilibrium expansion or compression process allows a simple demonstration of the idea that work depends on the process. This can be brought out by referring to Fig. 2.8. Suppose the gas in a piston–cylinder assembly goes from an initial equilibrium state 1 to a final equilibrium state 2 along two different paths, labeled A and B on Fig. 2.8. Since the area beneath each path represents the work for that process, the work depends on the details of the process as defined by the particular curve and not just on the end states. Using the test for a property given in Sec. 1.3.3, we can conclude again (Sec. 2.2.1) that *work is not a property.* The value of work depends on the nature of the process between the end states.

The relation between pressure and volume, or pressure and specific volume, also can be described analytically. A quasiequilibrium process described by $pV^n = constant$, or $pv^n = constant,$ where n is a constant, is called a **polytropic process**. Additional analytical forms for the pressure–volume relationship also may be considered.

The example to follow illustrates the application of Eq. 2.17 when the relationship between pressure and volume during an expansion is described analytically as $pV^n = constant.$

Compression Work All Tabs
Expansion Work All Tabs

polytropic process

▶ **EXAMPLE 2.1** ▶

Evaluating Expansion Work

A gas in a piston–cylinder assembly undergoes an expansion process for which the relationship between pressure and volume is given by

$$pV^n = constant$$

The initial pressure is 3 bar, the initial volume is 0.1 m³, and the final volume is 0.2 m³. Determine the work for the process, in kJ, if **(a)** $n = 1.5$, **(b)** $n = 1.0$, and **(c)** $n = 0$.

SOLUTION

Known A gas in a piston–cylinder assembly undergoes an expansion for which $pV^n = constant.$

Find Evaluate the work if (a) $n = 1.5$, (b) $n = 1.0$, (c) $n = 0$.

Schematic and Given Data:

The given p–V relationship and the given data for pressure and volume can be used to construct the accompanying pressure–volume diagram of the process.

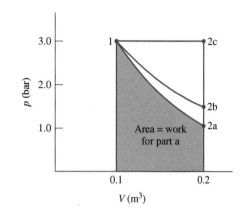

Fig. E2.1

Engineering Model

1. The gas is a closed system.

2. The moving boundary is the only work mode.

❷ 3. The expansion is a polytropic process.

Analysis The required values for the work are obtained by integration of Eq. 2.17 using the given pressure–volume relation.

a. Introducing the relationship $p = constant/V^n$ into Eq. 2.17 and performing the integration

$$W = \int_{V_1}^{V_2} p\, dV = \int_{V_1}^{V_2} \frac{constant}{V^n}\, dV$$

$$= \frac{(constant)V_2^{1-n} - (constant)V_1^{1-n}}{1 - n}$$

The constant in this expression can be evaluated at either end state: $constant = p_1V_1^n = p_2V_2^n$. The work expression then becomes

$$W = \frac{(p_2V_2^n)V_2^{1-n} - (p_1V_1^n)V_1^{1-n}}{1 - n} = \frac{p_2V_2 - p_1V_1}{1 - n} \qquad (a)$$

This expression is valid for all values of n except $n = 1.0$. The case $n = 1.0$ is taken up in part (b).

To evaluate W, the pressure at state 2 is required. This can be found by using $p_1V_1^n = p_2V_2^n$, which on rearrangement yields

$$p_2 = p_1\left(\frac{V_1}{V_2}\right)^n = (3\ \text{bar})\left(\frac{0.1}{0.2}\right)^{1.5} = 1.06\ \text{bar}$$

Accordingly,

❸ $$W = \left(\frac{(1.06\ \text{bar})(0.2\ \text{m}^3) - (3)(0.1)}{1 - 1.5}\right)\left|\frac{10^5\ \text{N/m}^2}{1\ \text{bar}}\right|\left|\frac{1\ \text{kJ}}{10^3\ \text{N} \cdot \text{m}}\right|$$

$$= +17.6\ \text{kJ}$$

b. For $n = 1.0$, the pressure–volume relationship is $pV = constant$ or $p = constant/V$. The work is

$$W = constant\int_{V_1}^{V_2} \frac{dV}{V} = (constant)\ln\frac{V_2}{V_1} = (p_1V_1)\ln\frac{V_2}{V_1} \qquad (b)$$

Substituting values

$$W = (3\ \text{bar})(0.1\ \text{m}^3)\left|\frac{10^5\ \text{N/m}^2}{1\ \text{bar}}\right|\left|\frac{1\ \text{kJ}}{10^3\ \text{N} \cdot \text{m}}\right|\ln\left(\frac{0.2}{0.1}\right) = +20.79\ \text{kJ}$$

c. For $n = 0$, the pressure–volume relation reduces to $p = constant$, and ❹ the integral becomes $W = p(V_2 - V_1)$, which is a special case of the expression found in part (a). Substituting values and converting units as above, $W = +30$ kJ.

---

❶ In each case, the work for the process can be interpreted as the area under the curve representing the process on the accompanying p–V diagram. Note that the relative areas are in agreement with the numerical results.

❷ The assumption of a polytropic process is significant. If the given pressure–volume relationship were obtained as a fit to experimental pressure–volume data, the value of $\int p\, dV$ would provide a plausible estimate of the work only when the measured pressure is essentially equal to that exerted at the piston face.

❸ Observe the use of unit conversion factors here and in part (b).

❹ In each of the cases considered, it is not necessary to identify the gas (or liquid) contained within the piston–cylinder assembly. The calculated values for W are determined by the process path and the end states. However, if it is desired to evaluate a property such as temperature, both the nature and amount of the substance must be provided because appropriate relations among the properties of the particular substance would then be required.

SKILLS DEVELOPED

Ability to...

- apply the problem-solving methodology.
- define a closed system and identify interactions on its boundary.
- evaluate work using Eq. 2.17.
- apply the pressure–volume relation $pV^n = constant$.

Quick Quiz

Evaluate the work, in kJ, for a two-step process consisting of an expansion with $n = 1.0$ from $p_1 = 3$ bar, $V_1 = 0.1\ \text{m}^3$ to $V = 0.15\ \text{m}^3$, followed by an expansion with $n = 0$ from $V = 0.15\ \text{m}^3$ to $V_2 = 0.2\ \text{m}^3$. Ans. 22.16 kJ.

2.2.6 Further Examples of Work

To broaden our understanding of the work concept, we now briefly consider several other examples of work.

Extension of a Solid Bar   Consider a system consisting of a solid bar under tension, as shown in **Fig. 2.9**. The bar is fixed at $x = 0$, and a force F is applied at the other end. Let the force be represented as $F = \sigma A$, where A is the cross-sectional area of the bar and σ the *normal stress acting at the end* of the bar. The work done as the end of the bar moves a distance dx is given by $\delta W = -\sigma A\, dx$. The minus sign is required because work is done *on* the bar when dx is positive. The work for a change in length from x_1 to x_2 is found by integration

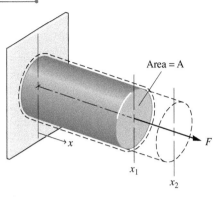

Fig. 2.9 Elongation of a solid bar.

$$W = -\int_{x_1}^{x_2} \sigma A\, dx \qquad (2.18)$$

Equation 2.18 for a solid is the counterpart of Eq. 2.17 for a gas undergoing an expansion or compression.

Stretching of a Liquid Film   Figure 2.10 shows a system consisting of a liquid film suspended on a wire frame. The two surfaces of the film support the thin liquid layer inside by the effect of *surface tension,* owing to microscopic forces between molecules near the liquid–air interfaces. These forces give rise to a macroscopically measurable force perpendicular to any line in the surface. The force per unit length across such a line is the surface tension. Denoting the surface tension *acting at the movable wire* by τ, the force F indicated on the figure can be expressed as $F = 2l\tau$, where the factor 2 is introduced because two film surfaces act at the wire. If the movable wire is displaced by dx, the work is given by $\delta W = -2l\tau\, dx$. The minus sign is required because work is done *on* the system when dx is positive. Corresponding to a displacement dx is a change in the total area of the surfaces in contact with the wire of $dA = 2l\, dx$, so the expression for work can be written alternatively as $\delta W = -\tau\, dA$. The work for an increase in surface area from A_1 to A_2 is found by integrating this expression

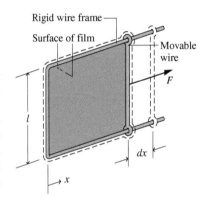

Fig. 2.10 Stretching of a liquid film.

$$W = -\int_{A_1}^{A_2} \tau\, dA \qquad (2.19)$$

Power Transmitted by a Shaft   A rotating shaft is a commonly encountered machine element. Consider a shaft rotating with angular velocity ω and exerting a torque $\mathcal{T}$ on its surroundings. Let the torque be expressed in terms of a tangential force F_t and radius R: $\mathcal{T} = F_t R$. The velocity at the point of application of the force is $V = R\omega$, where ω is in radians per unit time. Using these relations with Eq. 2.13, we obtain an expression for the *power* transmitted from the shaft to the surroundings

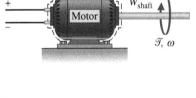

$$\dot{W} = F_t V = (\mathcal{T}/R)(R\omega) = \mathcal{T}\omega \qquad (2.20)$$

A related case involving a gas stirred by a paddle wheel is considered in the discussion of Fig. 2.3.

Electric Power   Shown in **Fig. 2.11** is a system consisting of an electrolytic cell. The cell is connected to an external circuit through which an electric current, i, is flowing. The current is driven by the electrical potential difference $\mathcal{E}$ existing across the terminals labeled a and b. That this type of interaction can be classed as work is considered in the discussion of Fig. 2.3.

The rate of energy transfer by work, or the power, is

$$\dot{W} = -\mathcal{E}i \qquad (2.21)$$

Since the current i equals dZ/dt, the work can be expressed in differential form as

$$\delta W = -\mathcal{E}\, dZ \qquad (2.22)$$

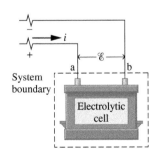

Fig. 2.11 Electrolytic cell used to discuss electric power.

where dZ is the amount of electrical charge that flows into the system. The minus signs appearing in Eqs. 2.21 and 2.22 are required to be in accord with our previously stated sign convention for work.

Work Due to Polarization or Magnetization   Let us next refer briefly to the types of work that can be done on systems residing in electric or magnetic fields, known as the work of polarization and magnetization, respectively. From the microscopic viewpoint, electrical dipoles within dielectrics resist turning, so work is done when they are aligned by an electric field. Similarly, magnetic dipoles resist turning, so work is done on certain other materials when their magnetization is changed. Polarization and magnetization give rise to *macroscopically* detectable changes in the total dipole moment as the particles making up the material are given new alignments. In these cases the work is associated with forces imposed on the overall system by fields in the surroundings. Forces acting on the material in the system interior are called *body forces*. For such forces the appropriate displacement in evaluating work is the displacement of the matter on which the body force acts.

2.2.7 Further Examples of Work in Quasiequilibrium Processes

Systems other than a gas or liquid in a piston–cylinder assembly also can be envisioned as undergoing processes in a quasiequilibrium fashion. To apply the quasiequilibrium process concept in any such case, it is necessary to conceive of an *ideal situation* in which the external forces acting on the system can be varied so slightly that the resulting imbalance is infinitesimal. As a consequence, the system undergoes a process without ever departing significantly from thermodynamic equilibrium.

The extension of a solid bar and the stretching of a liquid surface can readily be envisioned to occur in a quasiequilibrium manner by direct analogy to the piston–cylinder case. For the bar in Fig. 2.9 the external force can be applied in such a way that it differs only slightly from the opposing force within. The normal stress is then essentially uniform throughout and can be determined as a function of the instantaneous length: $\sigma = \sigma(x)$. Similarly, for the liquid film shown in Fig. 2.10 the external force can be applied to the movable wire in such a way that the force differs only slightly from the opposing force within the film. During such a process, the surface tension is essentially uniform throughout the film and is functionally related to the instantaneous area: $\tau = \tau(A)$. In each of these cases, once the required functional relationship is known, the work can be evaluated using Eq. 2.18 or 2.19, respectively, in terms of properties of the system as a whole as it passes through equilibrium states.

Other systems also can be imagined as undergoing quasiequilibrium processes. For example, it is possible to envision an electrolytic cell being charged or discharged in a quasiequilibrium manner by adjusting the potential difference across the terminals to be slightly greater, or slightly less, than an ideal potential called the cell *electromotive force* (emf). The energy transfer by work for passage of a differential quantity of charge *to* the cell, dZ, is given by the relation

$$\delta W = -\mathscr{E}\, dZ \tag{2.23}$$

In this equation $\mathscr{E}$ denotes the cell emf, an intensive property of the cell, and not just the potential difference across the terminals as in Eq. 2.22.

Consider next a dielectric material residing in a *uniform electric field*. The energy transferred by work from the field when the polarization is increased slightly is

$$\delta W = -\mathbf{E} \cdot d(V\mathbf{P}) \tag{2.24}$$

where the vector **E** is the electric field strength within the system, the vector **P** is the electric dipole moment per unit volume, and V is the volume of the system. A similar equation for energy transfer by work from a *uniform magnetic field* when the magnetization is increased slightly is

$$\delta W = -\mu_0 \mathbf{H} \cdot d(V\mathbf{M}) \tag{2.25}$$

where the vector **H** is the magnetic field strength within the system, the vector **M** is the magnetic dipole moment per unit volume, and μ_0 is a constant, the permeability of free space. The minus signs appearing in the last three equations are in accord with our previously stated sign convention for work: W takes on a negative value when the energy transfer is *into* the system.

2.2.8 Generalized Forces and Displacements

The similarity between the expressions for work in the quasiequilibrium processes considered thus far should be noted. In each case, the work expression is written in the form of an intensive property and the differential of an extensive property. This is brought out by the following expression, which allows for one or more of these work modes to be involved in a process

$$\delta W = p \, dV - \sigma d(\mathrm{A}x) - \tau \, d\mathrm{A} - \mathcal{E} \, dZ - \mathbf{E} \cdot d(V\mathbf{P}) - \mu_0\mathbf{H} \cdot d(V\mathbf{M}) + \cdots \tag{2.26}$$

where the last three dots represent other products of an intensive property and the differential of a related extensive property that account for work. Because of the notion of work being a product of force and displacement, the intensive property in these relations is sometimes referred to as a "generalized" force and the extensive property as a "generalized" displacement, even though the quantities making up the work expressions may not bring to mind actual forces and displacements.

Owing to the underlying quasiequilibrium restriction, Eq. 2.26 does not represent every type of work of practical interest. An example is provided by a paddle wheel that stirs a gas or liquid taken as the system. Whenever any shearing action takes place, the system necessarily passes through nonequilibrium states. To appreciate more fully the implications of the quasiequilibrium process concept requires consideration of the second law of thermodynamics, so this concept is discussed again in Chap. 5 after the second law has been introduced.

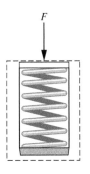

2.3 Broadening Our Understanding of Energy

The objective in this section is to use our deeper understanding of work developed in Sec. 2.2 to broaden our understanding of the energy of a system. In particular, we consider the *total* energy of a system, which includes kinetic energy, gravitational potential energy, and other forms of energy. The examples to follow illustrate some of these forms of energy. Many other examples could be provided that enlarge on the same idea.

When work is done to compress a spring, energy is stored within the spring. When a battery is charged, the energy stored within it is increased. And when a gas (or liquid) initially at an equilibrium state in a closed, insulated vessel is stirred vigorously and allowed to come to a final equilibrium state, the energy of the gas is increased in the process. In keeping with the discussion of work in Sec. 2.2, we can also think of other ways in which work done on systems increases energy stored within those systems—work related to magnetization, for example. In each of these examples the change in system energy cannot be attributed to changes in the system's *overall* kinetic or gravitational potential energy as given by Eqs. 2.5

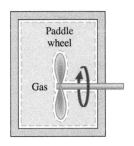

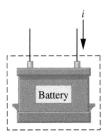

and 2.10, respectively. The change in energy can be accounted for in terms of *internal energy,* as considered next.

In engineering thermodynamics the change in the total energy of a system is considered to be made up of three *macroscopic* contributions. One is the change in kinetic energy, associated with the motion of the system *as a whole* relative to an external coordinate frame. Another is the change in gravitational potential energy, associated with the position of the system *as a whole* in Earth's gravitational field. All other energy changes are lumped together in the **internal energy** of the system. Like kinetic energy and gravitational potential energy, *internal energy is an extensive property* of the system, as is the total energy.

internal energy

Internal energy is represented by the symbol U, and the change in internal energy in a process is $U_2 - U_1$. The specific internal energy is symbolized by u or $\bar{u}$ respectively, depending on whether it is expressed on a unit mass or per mole basis.

The change in the total energy of a system is

$$\boxed{E_2 - E_1 = (U_2 - U_1) + (KE_2 - KE_1) + (PE_2 - PE_1)} \qquad (2.27a)$$

Total Energy Tab a

or

$$\boxed{\Delta E = \Delta U + \Delta KE + \Delta PE} \qquad (2.27b)$$

All quantities in Eq. 2.27 are expressed in terms of the energy units previously introduced.

The identification of internal energy as a macroscopic form of energy is a significant step in the present development, for it sets the concept of energy in thermodynamics apart from that of mechanics. In Chap. 3 we will learn how to evaluate changes in internal energy for practically important cases involving gases, liquids, and solids by using empirical data.

To further our understanding of internal energy, consider a system we will often encounter in subsequent sections of the book, a system consisting of a gas contained in a tank. Let us develop a **microscopic interpretation of internal energy** by thinking of the energy attributed to the motions and configurations of the individual molecules, atoms, and subatomic particles making up the matter in the system. Gas molecules move about, encountering other molecules or the walls of the container. Part of the internal energy of the gas is the *translational* kinetic energy of the molecules. Other contributions to the internal energy include the kinetic energy due to *rotation* of the molecules relative to their centers of mass and the kinetic energy associated with *vibrational* motions within the molecules. In addition, energy is stored in the chemical bonds between the atoms that make up the molecules. Energy storage on the atomic level includes energy associated with electron orbital states, nuclear spin, and binding forces in the nucleus. In dense gases, liquids, and solids, intermolecular forces play an important role in affecting the internal energy.

microscopic interpretation of internal energy for a gas

2.4 Energy Transfer by Heat

Thus far, we have considered quantitatively only those interactions between a system and its surroundings that can be classed as work. However, closed systems also can interact with their surroundings in a way that cannot be categorized as work.

FOR EXAMPLE

When a gas in a rigid container interacts with a hot plate, the energy of the gas is increased even though no work is done.

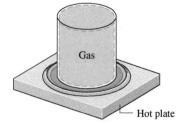

energy transfer by heat

This type of interaction is called an **energy transfer by heat.**

On the basis of experiment, beginning with the work of Joule in the early part of the nineteenth century, we know that energy transfers by heat are induced only as a result of a temperature difference between the system and its surroundings and occur only in the direction of decreasing temperature. Because the underlying concept is so important in thermodynamics, this section is devoted to a further consideration of energy transfer by heat.

2.4.1 Sign Convention, Notation, and Heat Transfer Rate

The symbol Q denotes an amount of energy transferred across the boundary of a system in a heat interaction with the system's surroundings. Heat transfer *into* a system is taken to be *positive,* and heat transfer *from* a system is taken as *negative*.

$Q > 0$: heat transfer *to* the system

$Q < 0$: heat transfer *from* the system

This **sign convention** is used throughout the book. However, as was indicated for work, it is sometimes convenient to show the direction of energy transfer by an arrow on a sketch of the system. Then the heat transfer is regarded as positive in the direction of the arrow.

    The sign convention for heat transfer is just the *reverse* of the one adopted for work, where a positive value for W signifies an energy transfer *from* the system to the surroundings. These signs for heat and work are a legacy from engineers and scientists who were concerned mainly with steam engines and other devices that develop a work output from an energy input by heat transfer. For such applications, it was convenient to regard both the work developed and the energy input by heat transfer as positive quantities.

    The value of a heat transfer depends on the details of a process and not just the end states. Thus, like work, **heat is not a property**, and its differential is written as δQ. The amount of energy transfer by heat for a process is given by the integral

$$Q = \int_1^2 \delta Q \tag{2.28}$$

where the limits mean "from state 1 to state 2" and do not refer to the values of heat at those states. As for work, the notion of "heat" at a state has no meaning, and the integral should *never* be evaluated as $Q_2 - Q_1$.

    The net **rate of heat transfer** is denoted by $\dot{Q}$. In principle, the amount of energy transfer by heat during a period of time can be found by integrating from time t_1 to time t_2

$$Q = \int_{t_1}^{t_2} \dot{Q}\, dt \tag{2.29}$$

To perform the integration, it is necessary to know how the rate of heat transfer varies with time.

    In some cases it is convenient to use the *heat flux, $\dot{q}$,* which is the heat transfer rate per unit of system surface area. The net rate of heat transfer, $\dot{Q}$, is related to the heat flux $\dot{q}$ by the integral

$$\dot{Q} = \int_A \dot{q}\, dA \tag{2.30}$$

where A represents the area on the boundary of the system where heat transfer occurs.

    The units for heat transfer Q and heat transfer rate $\dot{Q}$ are the same as those introduced previously for W and $\dot{W}$, respectively. The units for the heat flux are those of the heat transfer rate per unit area: kW/m^2.

    The word **adiabatic** means *without heat transfer.* Thus, if a system undergoes a process involving no heat transfer with its surroundings, that process is called an *adiabatic process.*

sign convention for heat transfer

Animation

Heat Transfer Modes Tab a

heat is not a property

rate of heat transfer

adiabatic

BioConnections

Medical researchers have found that by gradually increasing the temperature of cancerous tissue to 41–45°C the effectiveness of chemotherapy and radiation therapy is enhanced for some patients. Different approaches can be used, including raising the temperature of the entire body with heating devices and, more selectively, by beaming microwaves or ultrasound onto the tumor or affected organ. Speculation about why a temperature increase may be beneficial varies. Some say it helps chemotherapy penetrate certain tumors more readily by dilating blood vessels. Others think it helps radiation therapy by increasing the amount of oxygen in tumor cells, making them more receptive to radiation. Researchers report that further study is needed before the efficacy of this approach is established and the mechanisms whereby it achieves positive results are known.

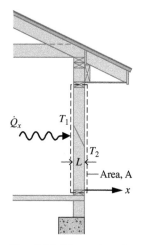

Fig. 2.12 Illustration of Fourier's conduction law.

2.4.2 Heat Transfer Modes

Methods based on experiment are available for evaluating energy transfer by heat. These methods recognize two basic transfer mechanisms: *conduction* and *thermal radiation*. In addition, empirical relationships are available for evaluating energy transfer involving a *combined* mode called *convection*. A brief description of each of these is given next. A detailed consideration is left to a course in engineering heat transfer, where these topics are studied in depth.

Conduction Energy transfer by *conduction* can take place in solids, liquids, and gases. Conduction can be thought of as the transfer of energy from the more energetic particles of a substance to adjacent particles that are less energetic due to interactions between particles. The time rate of energy transfer by conduction is quantified macroscopically by *Fourier's law*. As an elementary application, consider **Fig. 2.12** showing a plane wall of thickness L at steady state, where the temperature $T(x)$ varies linearly with position x. By **Fourier's law**, the rate of heat transfer across any plane normal to the x direction, $\dot{Q}_x$, is proportional to the wall area, A, and the temperature gradient in the x direction, dT/dx:

Fourier's law

$$\dot{Q}_x = -\kappa A \frac{dT}{dx} \tag{2.31}$$

where the proportionality constant κ is a property called the *thermal conductivity*. The minus sign is a consequence of energy transfer in the direction of *decreasing* temperature.

> **FOR EXAMPLE**
>
> In the case of Fig. 2.12 the temperature varies linearly; thus, the temperature gradient is
>
> $$\frac{dT}{dx} = \frac{T_2 - T_1}{L}(< 0)$$
>
> and the rate of heat transfer in the *x* direction is then
>
> $$\dot{Q}_x = -\kappa A \left[\frac{T_2 - T_1}{L} \right]$$

Heat Transfer Modes Tab b

Values of thermal conductivity are given in Table A-19 for common materials. Substances with large values of thermal conductivity such as copper are good conductors, and those with small conductivities (cork and polystyrene foam) are good insulators.

Radiation *Thermal radiation* is emitted by matter as a result of changes in the electronic configurations of the atoms or molecules within it. The energy is transported by electromagnetic waves (or photons). Unlike conduction, thermal radiation requires no intervening medium to propagate and can even take place in a vacuum. Solid surfaces, gases, and liquids all emit, absorb, and transmit thermal radiation to varying degrees. The rate at which energy is emitted, $\dot{Q}_e$, *from* a surface of area A is quantified macroscopically by a modified form of the **Stefan–Boltzmann law**

Stefan–Boltzmann law

$$\dot{Q}_e = \varepsilon \sigma A T_b^4 \tag{2.32}$$

which shows that thermal radiation is associated with the fourth power of the absolute temperature of the surface, T_b. The emissivity, ε, is a property of the surface that indicates how effectively the surface radiates ($0 \le \varepsilon \le 1.0$), and σ is the Stefan–Boltzmann constant:

$$\sigma = 5.67 \times 10^{-8}\, W/m^2 \cdot K^4$$

In general, the *net* rate of energy transfer by thermal radiation between two surfaces involves relationships among the properties of the surfaces, their orientations with respect to each other, the extent to which the intervening medium scatters, emits, and absorbs thermal radiation, and other factors. A special case that occurs frequently is radiation exchange between a surface at temperature T_b and a much larger surrounding surface at T_s, as shown in **Fig. 2.13**.

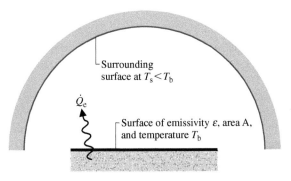

Fig. 2.13 Net radiation exchange.

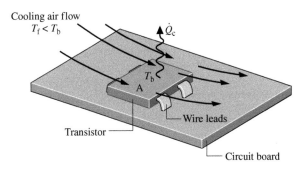

Fig. 2.14 Illustration of Newton's law of cooling.

The *net* rate of radiant exchange between the smaller surface, whose area is A and emissivity is ε, and the larger surroundings is

$$\dot{Q}_e = \varepsilon \sigma A [T_b^4 - T_s^4] \tag{2.33}$$

Convection   Energy transfer between a solid surface at a temperature T_b and an adjacent gas or liquid at another temperature T_f plays a prominent role in the performance of many devices of practical interest. This is commonly referred to as *convection*. As an illustration, consider Fig. 2.14, where $T_b > T_f$. In this case, energy is transferred *in the direction indicated by the arrow* due to the *combined* effects of conduction within the air and the bulk motion of the air. The rate of energy transfer *from* the surface *to* the air can be quantified by the following *empirical* expression:

$$\dot{Q}_c = hA(T_b - T_f) \tag{2.34}$$

known as **Newton's law of cooling**. In Eq. 2.34, A is the surface area and the proportionality factor h is called the *heat transfer coefficient*. In subsequent applications of Eq. 2.34, a minus sign may be introduced on the right side to conform to the sign convention for heat transfer introduced in Sec. 2.4.1.

The heat transfer coefficient is *not* a thermodynamic property. It is an empirical parameter that incorporates into the heat transfer relationship the nature of the flow pattern near the surface, the fluid properties, and the geometry. When fans or pumps cause the fluid to move, the value of the heat transfer coefficient is generally greater than when relatively slow buoyancy-induced motions occur. These two general categories are called *forced* and *free* (or natural) convection, respectively. Table 2.1 provides typical values of the convection heat transfer coefficient for forced and free convection.

Heat Transfer Modes Tab d

Newton's law of cooling

Heat Transfer Modes Tab c

TABLE 2.1

Typical Values of the Convection Heat Transfer Coefficient

Applications	h (W/m² · K)
Free convection	
Gases	2–25
Liquids	50–1000
Forced convection	
Gases	25–250
Liquids	50–20,000

2.4.3 Closing Comments

The first step in a thermodynamic analysis is to define the system. It is only after the system boundary has been specified that possible heat interactions with the surroundings are considered, for these are *always* evaluated at the system boundary.

In ordinary conversation, the term *heat* is often used when the word *energy* would be more correct thermodynamically. For example, one might hear, "Please close the door or 'heat' will be lost." In *thermodynamics,* heat refers only to a particular means whereby energy is transferred. It does not refer to what is being transferred between systems or to what is stored within systems. Energy is transferred and stored, not heat.

Sometimes the heat transfer of energy to, or from, a system can be neglected. This might occur for several reasons related to the mechanisms for heat transfer discussed above. One might be that the materials surrounding the system are good insulators, or heat transfer might not be significant because there is a small temperature difference between the system and its surroundings. A third reason is that there might not be enough surface area to allow significant heat transfer to occur. When heat transfer is neglected, it is because one or more of these considerations apply.

In the discussions to follow, the value of Q is provided or it is an unknown in the analysis. When Q is provided, it can be assumed that the value has been determined by the methods introduced above. When Q is the unknown, its value is usually found by using the *energy balance,* discussed next.

TAKE NOTE...

The term *heat* does not refer to what is being transferred or to what is stored within systems. Energy is transferred and stored when heat transfer occurs.

2.5 Energy Accounting: Energy Balance for Closed Systems

As our previous discussions indicate, the *only ways* the energy of a closed system can be changed are through transfer of energy by work or by heat. Further, based on the experiments of Joule and others, a fundamental aspect of the energy concept is that *energy is conserved;* we call this the **first law of thermodynamics**. For further discussion of the first law, see the box.

first law of thermodynamics

Joule's Experiments and the First Law

In classic experiments conducted in the early part of the nineteenth century, Joule studied processes by which a closed system can be taken from one equilibrium state to another. In particular, he considered processes that involve work interactions but no heat interactions between the system and its surroundings. Any such process is an *adiabatic process*, in keeping with the discussion of Sec. 2.4.1.

Based on his experiments Joule deduced that the value of the net work is the same for *all* adiabatic processes between two equilibrium states. In other words, the value of the net work done by or on a closed system undergoing an adiabatic process between two given states *depends solely on the end states* and not on the details of the adiabatic process.

If the net work is the same for all adiabatic processes of a closed system between a given pair of end states, it follows from the definition of property (Sec. 1.3) that the net work for such a process is the change in some property of the system. This property is called *energy*.

Following Joule's reasoning, the *change in energy* between the two states is *defined* by

$$E_2 - E_1 = -W_{ad} \qquad \text{(a)}$$

where the symbol E denotes the energy of a system and W_{ad} represents the net work for *any* adiabatic process between the two states. The minus sign before the work term is in accord with the previously stated sign convention for work. Finally, note that since any arbitrary value E_1 can be assigned to the energy of a system at a given state 1, no particular significance can be attached to the value of the energy at state 1 or at *any* other state. Only *changes* in the energy of a system have significance.

The foregoing discussion is based on experimental evidence beginning with the experiments of Joule. Because of inevitable experimental uncertainties, it is not possible to prove by measurements that the net work is *exactly* the same for all adiabatic processes between the same end states. However, the preponderance of experimental findings supports this conclusion, so it is adopted as a fundamental principle that the work actually is the same. This principle is an alternative formulation of the *first law* and has been used by subsequent scientists and engineers as a springboard for developing the *conservation of energy* concept and the *energy balance* as we know them today.

Summarizing Energy Concepts   All energy aspects introduced in this book thus far are summarized in words as follows:

$$
\begin{bmatrix}
\textit{change} \text{ in the amount} \\
\text{of energy contained} \\
\text{within a system} \\
\text{during some time} \\
\text{interval}
\end{bmatrix}
=
\begin{bmatrix}
\textit{net} \text{ amount of energy} \\
\text{transferred } \textit{in} \text{ across} \\
\text{the system boundary by} \\
\textit{heat} \text{ transfer during} \\
\text{the time interval}
\end{bmatrix}
-
\begin{bmatrix}
\textit{net} \text{ amount of energy} \\
\text{transferred } \textit{out} \text{ across} \\
\text{the system boundary} \\
\text{by } \textit{work} \text{ during the} \\
\text{time interval}
\end{bmatrix}
$$

This word statement is just an accounting balance for energy, an energy balance. It requires that in any process of a closed system the energy of the system increases or decreases by an amount equal to the net amount of energy transferred across its boundary.

The phrase *net amount* used in the word statement of the energy balance must be carefully interpreted, for there may be heat or work transfers of energy at many different places on the boundary of a system. At some locations the energy transfers may be into the system, whereas at others they are out of the system. The two terms on the right side account for the *net* results of all the energy transfers by heat and work, respectively, taking place during the time interval under consideration.

The **energy balance** can be expressed in symbols as

$$E_2 - E_1 = Q - W \tag{2.35a}$$

energy balance

Introducing Eq. 2.27 an alternative form is

$$\Delta KE + \Delta PE + \Delta U = Q - W \tag{2.35b}$$

which shows that an energy transfer across the system boundary results in a change in one or more of the macroscopic energy forms: kinetic energy, gravitational potential energy, and internal energy. All previous references to energy as a conserved quantity are included as special cases of Eqs. 2.35.

Note that the algebraic signs before the heat and work terms of Eqs. 2.35 are different. This follows from the sign conventions previously adopted. A minus sign appears before W because energy transfer by work *from* the system *to* the surroundings is taken to be positive. A plus sign appears before Q because it is regarded to be positive when the heat transfer of energy is *into* the system *from* the surroundings.

Energy Balance Closed System All Tabs

 BioConnections

The energy required by animals to sustain life is derived from oxidation of ingested food. We often speak of food being *burned* in the body. This is an appropriate expression because experiments show that when food is burned with oxygen, approximately the same energy is released as when the food is oxidized in the body. Such an experimental device is the well-insulated, constant-volume *calorimeter* shown in **Fig. 2.15**.

A carefully weighed food sample is placed in the chamber of the calorimeter together with oxygen (O_2). The entire chamber is submerged in the calorimeter's water bath. The chamber contents are then electrically ignited, fully oxidizing the food sample. The energy released during the reaction within the chamber results in an increase in calorimeter temperature. Using the measured temperature rise, the energy released can be calculated from an energy balance for the calorimeter as the system. This is reported as the calorie value of the food sample, usually in terms of kilocalorie (kcal), which is the "Calorie" seen on food labels.

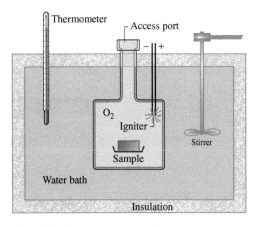

Fig. 2.15 Constant-volume calorimeter.

2.5.1 Important Aspects of the Energy Balance

Various special forms of the energy balance can be written. For example, the energy balance in differential form is

$$dE = \delta Q - \delta W \tag{2.36}$$

where dE is the differential of energy, a property. Since Q and W are not properties, their differentials are written as δQ and δW, respectively.

time rate form of the energy balance

The instantaneous **time rate form of the energy balance** is

$$\boxed{\frac{dE}{dt} = \dot{Q} - \dot{W}} \tag{2.37}$$

The rate form of the energy balance expressed in words is

$$
\begin{bmatrix} \text{time } \textit{rate of change} \\ \text{of the energy} \\ \text{contained within} \\ \text{the system } \textit{at} \\ \textit{time t} \end{bmatrix}
=
\begin{bmatrix} \text{net } \textit{rate} \text{ at which} \\ \text{energy is being} \\ \text{transferred in} \\ \text{by heat transfer} \\ \textit{at time t} \end{bmatrix}
-
\begin{bmatrix} \text{net } \textit{rate} \text{ at which} \\ \text{energy is being} \\ \text{transferred out} \\ \text{by work } \textit{at} \\ \textit{time t} \end{bmatrix}
$$

Since the time rate of change of energy is given by

$$\frac{dE}{dt} = \frac{d\,\text{KE}}{dt} + \frac{d\,\text{PE}}{dt} + \frac{dU}{dt}$$

Equation 2.37 can be expressed alternatively as

$$\frac{d\,\text{KE}}{dt} + \frac{d\,\text{PE}}{dt} + \frac{dU}{dt} = \dot{Q} - \dot{W} \tag{2.38}$$

Equations 2.35 through 2.38 provide alternative forms of the energy balance that are convenient starting points when applying the principle of conservation of energy to closed systems. In Chap. 4 the conservation of energy principle is expressed in forms suitable for the analysis of control volumes. When applying the energy balance in any of its forms, it is important to be careful about signs and units and to distinguish carefully between rates and amounts. In addition, it is important to recognize that the location of the system boundary can be relevant in determining whether a particular energy transfer is regarded as heat or work.

FOR EXAMPLE

Consider **Fig. 2.16**, in which three alternative systems are shown that include a quantity of a gas (or liquid) in a rigid, well-insulated container. In Fig. 2.16*a*, the gas itself is the system. As current flows through the copper plate, there is an energy transfer from the copper plate to the gas. Since this energy transfer occurs as a result of the temperature difference between the plate and the gas, it is classified as a heat transfer. Next, refer to Fig. 2.16*b*, where the boundary is drawn to include the copper plate. It follows from the thermodynamic definition of work that the energy transfer that occurs as current crosses the boundary of this system must be regarded as work. Finally, in Fig. 2.16*c*, the boundary is located so that no energy is transferred across it by heat or work.

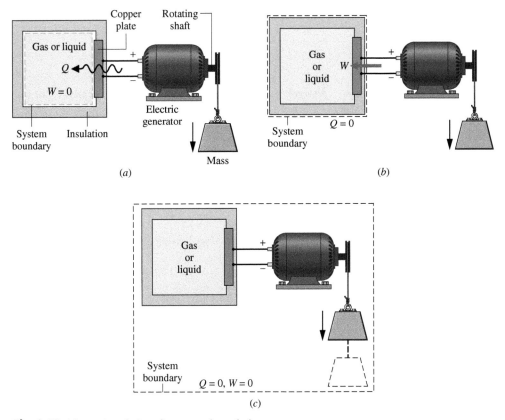

Fig. 2.16 Alternative choices for system boundaries.

Closing Comments   Thus far, we have been careful to emphasize that the quantities symbolized by W and Q in the foregoing equations account for transfers of *energy* and not transfers of work and heat, respectively. The terms *work* and *heat* denote different *means* whereby energy is transferred and not *what* is transferred. However, to achieve economy of expression in subsequent discussions, W and Q are often referred to simply as work and heat transfer, respectively. This less formal manner of speaking is commonly used in engineering practice.

The five solved examples provided in Secs. 2.5.2–2.5.4 bring out important ideas about energy and the energy balance. They should be studied carefully, and similar approaches should be used when solving the end-of-chapter problems. In this text, most applications of the energy balance will not involve significant kinetic or potential energy changes. Thus, to expedite the solutions of many subsequent examples and end-of-chapter problems, we indicate in the problem statement that such changes can be neglected. If this is not made explicit in a problem statement, you should decide on the basis of the problem at hand how best to handle the kinetic and potential energy terms of the energy balance.

> **TAKE NOTE...**
>
> The terms *heat* and *work* denote *means* whereby energy is transferred. However, W and Q are often referred to informally as work and heat transfer, respectively.

2.5.2 Using the Energy Balance: Processes of Closed Systems

The next two examples illustrate the use of the energy balance for processes of closed systems. In these examples, internal energy data are provided. In Chap. 3, we learn how to obtain internal energy and other thermodynamic property data using tables, graphs, equations, and computer software.

▶▶ **EXAMPLE 2.2** ▶

Cooling a Gas in a Piston–Cylinder

Four-tenths kilogram of a certain gas is contained within a piston–cylinder assembly. The gas undergoes a process for which the pressure–volume relationship is

$$pV^{1.5} = constant$$

The initial pressure is 3 bar, the initial volume is 0.1 m^3, and the final volume is 0.2 m^3. The change in specific internal energy of the gas in the process is $u_2 - u_1 = -55$ kJ/kg. There are no significant changes in kinetic or potential energy. Determine the net heat transfer for the process, in kJ.

SOLUTION

Known A gas within a piston–cylinder assembly undergoes an expansion process for which the pressure–volume relation and the change in specific internal energy are specified.

Find Determine the net heat transfer for the process.

Schematic and Given Data:

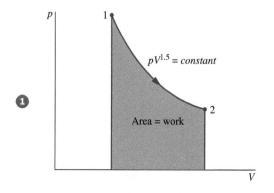

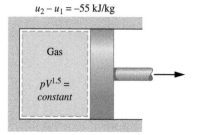

Fig. E2.2

Engineering Model

1. The gas is a closed system.
2. The process is described by $pV^{1.5} = constant$.
3. There is no change in the kinetic or potential energy of the system.

Analysis An energy balance for the closed system takes the form

$$\Delta\cancel{KE}^0 + \Delta\cancel{PE}^0 + \Delta U = Q - W$$

where the kinetic and potential energy terms drop out by assumption 3. Then, writing ΔU in terms of specific internal energies, the energy balance becomes

$$m(u_2 - u_1) = Q - W$$

where m is the system mass. Solving for Q

$$Q = m(u_2 - u_1) + W$$

The value of the work for this process is determined in the solution to part (a) of Example 2.1: $W = +17.6$ kJ. The change in internal energy is obtained using given data as

$$m(u_2 - u_1) = 0.4 \text{ kg}\left(-55\frac{\text{kJ}}{\text{kg}}\right) = -22 \text{ kJ}$$

Substituting values

❷ $$Q = -22 + 17.6 = -4.4 \text{ kJ}$$

---

❶ The given relationship between pressure and volume allows the process to be represented by the path shown on the accompanying diagram. The area under the curve represents the work. Since they are not properties, the values of the work and heat transfer depend on the details of the process and cannot be determined from the end states only.

❷ The minus sign for the value of Q means that a net amount of energy has been transferred from the system to its surroundings by heat transfer.

SKILLS DEVELOPED

Ability to...

• define a closed system and identify interactions on its boundary.

• apply the closed-system energy balance.

Quick Quiz

If the gas undergoes a process for which $pV = constant$ and $\Delta u = 0$, determine the heat transfer, in kJ, keeping the initial pressure and given volumes fixed. Ans. 20.79 kJ.

In the next example, we follow up the discussion of Fig. 2.16 by considering two alternative systems. This example highlights the need to account correctly for the heat and work interactions occurring on the boundary as well as the energy change.

▶ ▶ **EXAMPLE 2.3** ▶

Considering Alternative Systems

Air is contained in a vertical piston–cylinder assembly fitted with an electrical resistor. The atmosphere exerts a pressure of 1 bar on the top of the piston, which has a mass of 45 kg and a face area of 0.09 m². Electric current passes through the resistor, and the volume of the air slowly increases by 0.045 m³ while its pressure remains constant. The mass of the air is 0.27 kg, and its specific internal energy increases by 42 kJ/kg. The air and piston are at rest initially and finally. The piston–cylinder material is a ceramic composite and thus a good insulator. Friction between the piston and cylinder wall can be ignored, and the local acceleration of gravity is $g = 9.81$ m/s². Determine the heat transfer from the resistor to the air, in kJ, for a system consisting of **(a)** the air alone, **(b)** the air and the piston.

SOLUTION

Known   Data are provided for air contained in a vertical piston-cylinder fitted with an electrical resistor.

Find   Considering each of two alternative systems, determine the heat transfer from the resistor to the air.

Schematic and Given Data:

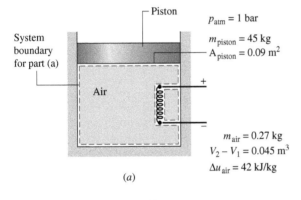

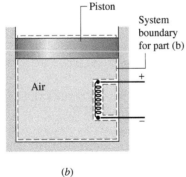

Fig. E2.3

Engineering Model

1. Two closed systems are under consideration, as shown in the schematic.

2. The only significant heat transfer is from the resistor to the air, during which the air expands slowly and its pressure remains constant.

3. There is no net change in kinetic energy; the change in potential energy of the air is negligible; and since the piston material is a good insulator, the internal energy of the piston is not affected by the heat transfer.

4. Friction between the piston and cylinder wall is negligible.

5. The acceleration of gravity is constant; $g = 9.81$ m/s².

Analysis

a. Taking the air as the system, the energy balance, Eq. 2.35, reduces with assumption 3 to

$$(\cancel{\Delta KE}^{0} + \cancel{\Delta PE}^{0} + \Delta U)_{\text{air}} = Q - W$$

Or, solving for Q

$$Q = W + \Delta U_{\text{air}}$$

For this system, work is done by the force of the pressure p acting on the *bottom* of the piston as the air expands. With Eq. 2.17 and the assumption of constant pressure

$$W = \int_{V_1}^{V_2} p \, dV = p(V_2 - V_1)$$

To determine the pressure p, we use a force balance on the slowly moving, frictionless piston. The upward force exerted by the air on the *bottom* of the piston equals the weight of the piston plus the downward force of the atmosphere acting on the *top* of the piston. In symbols

$$p \, A_{\text{piston}} = m_{\text{piston}} \, g + p_{\text{atm}} A_{\text{piston}}$$

Solving for p and inserting values

$$p = \frac{m_{\text{piston}} g}{A_{\text{piston}}} + p_{\text{atm}}$$

$$= \frac{(45 \text{ kg})(9.81 \text{ m/s}^2)}{0.09 \text{ m}^2} \left| \frac{1 \text{ bar}}{10^5 \text{ N/m}^2} \right| + 1 \text{ bar} = 1.049 \text{ bar}$$

Thus, the work is

$$W = p(V_2 - V_1)$$

$$= (1.049 \text{ bar})(0.045 \text{ m}^2) \left| \frac{10^5 \text{ N/m}^2}{1 \text{ bar}} \right| \left| \frac{1 \text{ kJ}}{10^3 \text{ N} \cdot \text{m}} \right| = 4.72 \text{ kJ}$$

With $\Delta U_{\text{air}} = m_{\text{air}}(\Delta u_{\text{air}})$, the heat transfer is

$$Q = W + m_{\text{air}}(\Delta u_{\text{air}})$$

$$= 4.72 \text{ kJ} + 11.07 \text{ kJ} = 15.8 \text{ kJ}$$

b. Consider next a system consisting of the air and the piston. The energy change of the overall system is the sum of the energy changes of the air and the piston. Thus, the energy balance, Eq. 2.35, reads

$$(\cancel{\Delta KE}^{0} + \cancel{\Delta PE}^{0} + \Delta U)_{\text{air}} + (\cancel{\Delta KE}^{0} + \Delta PE + \cancel{\Delta U}^{0})_{\text{piston}} = Q - W$$

where the indicated terms drop out by assumption 3. Solving for Q

$$Q = W + (\Delta PE)_{\text{piston}} + (\Delta U)_{\text{air}}$$

For this system, work is done at the *top* of the piston as it pushes aside the surrounding atmosphere. Applying Eq. 2.17

$$W = \int_{V_1}^{V_2} p\, dV = p_{atm}(V_2 - V_1)$$

$$= (1 \text{ bar}) (0.045 \text{ m}^2) \left| \frac{10^5 \text{N/m}^2}{1 \text{ bar}} \right| \left| \frac{1 \text{ kJ}}{10^3 \text{ N} \cdot \text{m}} \right| = 4.5 \text{ kJ}$$

The elevation change, Δz, required to evaluate the potential energy change of the piston can be found from the volume change of the air and the area of the piston face as

$$\Delta z = \frac{V_2 - V_1}{A_{piston}} = \frac{0.045 \text{ m}^3}{0.09 \text{ m}^2} = 0.5 \text{ m}$$

Thus, the potential energy change of the piston is

$$(\Delta PE)_{piston} = m_{piston} g \Delta z$$

$$= (45 \text{ kg})(9.81 \text{ m/s}^2)(0.5 \text{ m}) = 0.22 \text{ kJ}$$

Finally,

$$Q = W + (\Delta PE)_{piston} + m_{air}\Delta u_{air}$$

$$= 4.5 \text{ kJ} + 0.22 \text{ kJ} + 11.07 \text{ kJ} = 15.8 \text{ kJ}$$

❶ which agrees with the result of part (a).

---

❶ Although the value of Q is the same for each system, observe that the values for W differ. Also, observe that the energy changes differ, depending on whether the air alone or the air and the piston is the system.

SKILLS DEVELOPED

Ability to...

• define alternative closed systems and identify interactions on their boundaries.

• evaluate work using Eq. 2.17.

• apply the closed-system energy balance.

Quick Quiz

What is the change in potential energy of the air, in kJ?
Ans. $\approx 1.055 \times 10^{-3}$ kJ

2.5.3 Using the Energy Rate Balance: Steady-State Operation

A system is at steady state if none of its properties change with time (Sec. 1.3). Many devices operate at steady state or nearly at steady state, meaning that property variations with time are small enough to ignore. The two examples to follow illustrate the application of the energy rate equation to closed systems at steady state.

> ▶▶ **EXAMPLE 2.4** ▶

Evaluating Energy Transfer Rates of a Gearbox at Steady State

During steady-state operation, a gearbox receives 60 kW through the input shaft and delivers power through the output shaft. For the gearbox as the system, the rate of energy transfer by convection is

$$\dot{Q} = -hA(T_b - T_f)$$

where h = 0.171 kW/m² · K is the heat transfer coefficient, A = 1.0 m² is the outer surface area of the gearbox, T_b = 300 K (27°C) is the temperature at the outer surface, and T_f = 293 K (20°C) is the temperature of the surrounding air away from the immediate vicinity of the gearbox. For the gearbox, evaluate the heat transfer rate and the power delivered through the output shaft, each in kW.

SOLUTION

Known A gearbox operates at steady state with a known power input. An expression for the heat transfer rate from the outer surface is also known.

Find Determine the heat transfer rate and the power delivered through the output shaft, each in kW.

Schematic and Given Data:

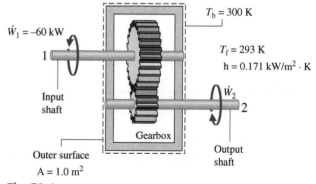

Fig. E2.4

Engineering Model

1. The gearbox is a closed system at steady state.

2. For the gearbox, convection is the dominant heat transfer mode.

Analysis Using the given expression for $\dot{Q}$ together with known data, the rate of energy transfer by heat is

1
$$\dot{Q} = -hA(T_b - T_f)$$
$$= -\left(0.171\frac{kW}{m^2 \cdot K}\right)(1.0\ m^2)(300 - 293)\ K$$
$$= -1.2\ kW$$

The minus sign for $\dot{Q}$ signals that energy is carried *out* of the gearbox by heat transfer.

The energy rate balance, Eq. 2.37, reduces at steady state to

2
$$\frac{d\cancel{E}^{\,0}}{dt} = \dot{Q} - \dot{W} \quad \text{or} \quad \dot{W} = \dot{Q}$$

The symbol $\dot{W}$ represents the *net* power from the system. The net power is the sum of $\dot{W}_1$ and the output power $\dot{W}_2$

$$\dot{W} = \dot{W}_1 + \dot{W}_2$$

With this expression for $\dot{W}$, the energy rate balance becomes

$$\dot{W}_1 + \dot{W}_2 = \dot{Q}$$

Solving for $\dot{W}_2$, inserting $\dot{Q} = -1.2$ kW, and $\dot{W}_1 = -60$ kW, where the minus sign is required because the input shaft brings energy *into* the system, we have

3
$$\dot{W}_2 = \dot{Q} - \dot{W}_1$$
$$= (-1.2\ kW) - (-60\ kW)$$
$$= +58.8\ kW$$

4 The positive sign for $\dot{W}_2$ indicates that energy is transferred from the system through the output shaft, as expected.

1 In accord with the sign convention for the heat transfer rate in the energy rate balance (Eq. 2.37), Eq. 2.34 is written with a minus sign: $\dot{Q}$ is negative since T_b is greater than T_f.

2 Properties of a system at steady state do not change with time. Energy E is a property, but heat transfer and work are not properties.

3 For this system, energy transfer by work occurs at two different locations, and the signs associated with their values differ.

4 At steady state, the rate of heat transfer from the gearbox accounts for the difference between the input and output power. This can be summarized by the following energy rate "balance sheet" in terms of *magnitudes*:

Input	Output
60 kW (input shaft)	58.8 kW (output shaft)
	1.2 kW (heat transfer)
Total: 60 kW	60 kW

SKILLS DEVELOPED

Ability to...

- define a closed system and identify interactions on its boundary.
- evaluate the rate of energy transfer by convection.
- apply the energy rate balance for *steady-state* operation.
- develop an energy rate *balance sheet*.

Quick Quiz

For an emissivity of 0.8 and taking $T_s = T_f$, use Eq. 2.33 to determine the net rate at which energy is radiated from the outer surface of the gearbox, in kW. Ans. 0.03 kW.

EXAMPLE 2.5

Determining Surface Temperature of a Silicon Chip at Steady State

A silicon chip measuring 5 mm on a side and 1 mm in thickness is embedded in a ceramic substrate. At steady state, the chip has an electrical power input of 0.225 W. The top surface of the chip is exposed to a coolant whose temperature is 20°C. The heat transfer coefficient for convection between the chip and the coolant is 150 W/m² · K. If heat transfer by conduction between the chip and the substrate is negligible, determine the surface temperature of the chip, in °C.

Schematic and Given Data:

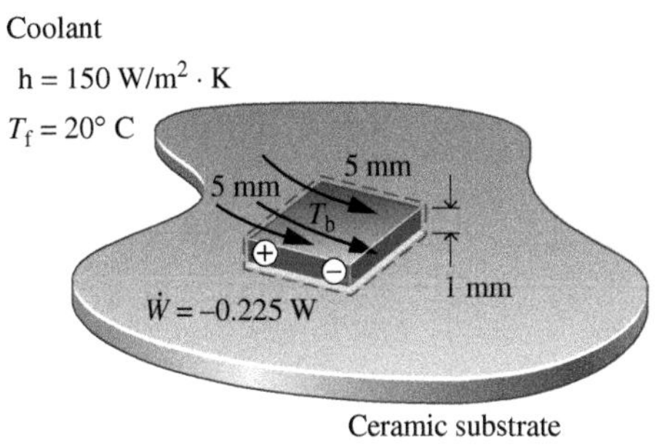

Fig. E2.5

SOLUTION

Known A silicon chip of known dimensions is exposed on its top surface to a coolant. The electrical power input and convective heat transfer coefficient are known.

Find Determine the surface temperature of the chip at steady state.

Engineering Model

1. The chip is a closed system at steady state.

2. There is no heat transfer between the chip and the substrate.

Analysis The surface temperature of the chip, T_b, can be determined using the energy rate balance, Eq. 2.37, which at steady state reduces as follows

❶
$$\frac{dE}{dt}^{0} = \dot{Q} - \dot{W}$$

With assumption 2, the only heat transfer is by convection to the coolant. In this application, Newton's law of cooling, Eq. 2.34, takes the form

❷
$$\dot{Q} = -hA(T_b - T_f)$$

Collecting results

$$0 = -hA(T_b - T_f) - \dot{W}$$

Solving for T_b

$$T_b = \frac{-\dot{W}}{hA} + T_f$$

In this expression, $\dot{W} = -0.225$ W, $A = 25 \times 10^{-6}$ m^2, h = 150 W/m$^2 \cdot$ K, and $T_f = 293$ K, giving

$$T_b = \frac{-(-0.225 \text{ W})}{(150 \text{ W/m}^2 \cdot \text{K})(25 \times 10^{-6} \text{ m}^2)} + 293 \text{ K}$$

$$= 353 \text{ K } (80°\text{C})$$

❶ Properties of a system at steady state do not change with time. Energy E is a property, but heat transfer and work are not properties.

❷ In accord with the sign convention for heat transfer in the energy rate balance (Eq. 2.37), Eq. 2.34 is written with a minus sign: $\dot{Q}$ is negative since T_b is greater than T_f.

SKILLS DEVELOPED

Ability to...

• define a closed system and identify interactions on its boundary.

• evaluate the rate of energy transfer by convection.

• apply the energy rate balance for *steady-state* operation.

Quick Quiz

If the surface temperature of the chip must be no greater than 60°C, what is the corresponding range of values required for the convective heat transfer coefficient, assuming all other quantities remain unchanged? Ans. h ≥ 225 W/m$^2 \cdot$ K.

2.5.4 Using the Energy Rate Balance: Transient Operation

Many devices undergo periods of transient operation where the state changes with time. This is observed during startup and shutdown periods. The next example illustrates the application of the energy rate balance to an electric motor during startup. The example also involves both electrical work and power transmitted by a shaft.

EXAMPLE 2.6

Investigating Transient Operation of a Motor

The rate of heat transfer between a certain electric motor and its surroundings varies with time as

$$\dot{Q} = -0.2[1 - e^{(-0.05t)}]$$

where t is in seconds and $\dot{Q}$ is in kW. The shaft of the motor rotates at a constant speed of $\omega = 100$ rad/s (about 955 revolutions per minute, or RPM) and applies a constant torque of $\mathcal{T} = 18$ N $\cdot$ m to an external load. The motor draws a constant electric power input equal to 2.0 kW. For the motor, plot $\dot{Q}$ and $\dot{W}$, each in kW, and the change in energy ΔE, in kJ, as functions of time from $t = 0$ to $t = 120$ s. Discuss.

SOLUTION

Known A motor operates with constant electric power input, shaft speed, and applied torque. The time-varying rate of heat transfer between the motor and its surroundings is given.

Find Plot $\dot{Q}$, $\dot{W}$, and ΔE versus time. Discuss.

Schematic and Given Data:

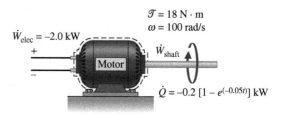

$$\dot{W}_{elec} = -2.0 \text{ kW}$$

$$\mathcal{T} = 18 \text{ N} \cdot \text{m}$$
$$\omega = 100 \text{ rad/s}$$

$$\dot{W}_{shaft}$$

$$\dot{Q} = -0.2 [1 - e^{(-0.05t)}] \text{ kW}$$

Fig. E2.6a

Engineering Model The system shown in the accompanying sketch is a closed system.

Analysis The time rate of change of system energy is

$$\frac{dE}{dt} = \dot{Q} - \dot{W}$$

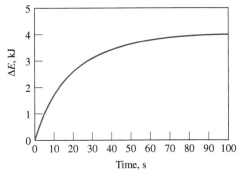
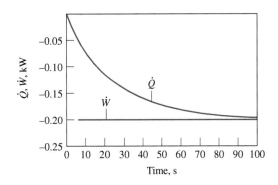

Fig. E2.6*b* and *c*

$\dot{W}$ represents the *net* power *from* the system: the sum of the power associated with the rotating shaft, $\dot{W}_{shaft}$, and the power associated with the electricity flow, $\dot{W}_{elec}$:

$$\dot{W} = \dot{W}_{shaft} + \dot{W}_{elec}$$

The rate $\dot{W}_{elec}$ is known from the problem statement: $\dot{W}_{elec} = -2.0$ kW, where the negative sign is required because energy is carried into the system by electrical work. The term $\dot{W}_{shaft}$ can be evaluated with Eq. 2.20 as

$$\dot{W}_{shaft} = \mathcal{T}\omega = (18 \text{ N} \cdot \text{m})(100 \text{ rad/s}) = 1800 \text{ W} = +1.8 \text{ kW}$$

Because energy exits the system along the rotating shaft, this energy transfer rate is positive.

In summary,

$$\dot{W} = \dot{W}_{elec} + \dot{W}_{shaft} = (-2.0 \text{ kW}) + (+1.8 \text{ kW}) = -0.2 \text{ kW}$$

where the minus sign means that the electrical power input is greater than the power transferred out along the shaft.

With the foregoing result for $\dot{W}$ and the given expression for $\dot{Q}$, the energy rate balance becomes

$$\frac{dE}{dt} = -0.2[1 - e^{(-0.05t)}] - (-0.2) = 0.2e^{(-0.05t)}$$

Integrating

$$\Delta E = \int_0^t 0.2e^{(-0.05t)} dt$$

$$= \frac{0.2}{(-0.05)} e^{(-0.05t)} \Big]_0^t = 4[1 - e^{(-0.05t)}]$$

❶ The accompanying plots, Figs. E2.6*b* and *c*, are developed using the given expression for $\dot{Q}$ and the expressions for $\dot{W}$ and ΔE obtained in the analysis. Because of our sign conventions for

heat and work, the values of $\dot{Q}$ and $\dot{W}$ are negative. In the first few seconds, the *net* rate that energy is carried in by work greatly exceeds the rate that energy is carried out by heat transfer. Consequently, the energy stored in the motor increases rapidly as the motor "warms up." As time elapses, the value of $\dot{Q}$ approaches $\dot{W}$, and the rate of energy storage diminishes. After about 100 s, this *transient* operating mode is nearly over, and there is little further change in the amount of energy stored, or in any other **❷** property. We may say that the motor is then at steady state.

❶ Figures E.2.6*b* and *c* can be developed using appropriate software or can be drawn by hand.

❷ At steady state, the value of $\dot{Q}$ is constant at −0.2 kW. This constant value for the heat transfer rate can be thought of as the portion of the electrical power input that is not obtained as a mechanical power output because of effects within the motor such as electrical resistance and friction.

SKILLS DEVELOPED

Ability to…

- define a closed system and identify interactions on its boundary.
- apply the energy rate balance for *transient* operation.
- develop and interpret graphical data.

Quick Quiz

If the dominant mode of heat transfer from the outer surface of the motor is convection, determine at *steady state* the temperature T_b on the outer surface, in K, for h = 0.17 kW/m^2 · K, A = 0.3 m^2, and T_f = 293 K. Ans. 297 K.

2.6 Energy Analysis of Cycles

In this section the energy concepts developed thus far are illustrated further by application to systems undergoing thermodynamic cycles. A **thermodynamic cycle** is a sequence of processes that begins and ends at the same state. At the conclusion of a cycle all properties have the same values they had at the beginning. Consequently, over the cycle the system experiences no *net* change of state. Cycles that are repeated periodically play prominent roles in many areas of application. For example, steam circulating through an electrical power plant executes a cycle.

thermodynamic cycle

The study of systems undergoing cycles has played an important role in the development of the subject of engineering thermodynamics. Both the first and second laws of thermodynamics have roots in the study of cycles. Additionally, there are many important practical applications involving power generation, vehicle propulsion, and refrigeration for which an understanding of thermodynamic cycles is essential. In this section, cycles are considered from the perspective of the conservation of energy principle. Cycles are studied in greater detail in subsequent chapters, using both the conservation of energy principle and the second law of thermodynamics.

2.6.1 Cycle Energy Balance

The energy balance for any system undergoing a thermodynamic cycle takes the form

$$\Delta E_{\text{cycle}} = Q_{\text{cycle}} - W_{\text{cycle}} \tag{2.39}$$

where Q_{cycle} and W_{cycle} represent *net* amounts of energy transfer by heat and work, respectively, for the cycle. Since the system is returned to its initial state after the cycle, there is no *net* change in its energy. Therefore, the left side of Eq. 2.39 equals zero, and the equation reduces to

$$\boxed{W_{\text{cycle}} = Q_{\text{cycle}}} \tag{2.40}$$

Equation 2.40 is an expression of the conservation of energy principle that must be satisfied by *every* thermodynamic cycle, regardless of the sequence of processes followed by the system undergoing the cycle or the nature of the substances making up the system.

Figure 2.17 provides simplified schematics of two general classes of cycles considered in this book: power cycles and refrigeration and heat pump cycles. In each case pictured, a system undergoes a cycle while communicating thermally with two bodies, one hot and the other cold. These bodies are systems located in the surroundings of the system undergoing the cycle. During each cycle there is also a net amount of energy exchanged with the surroundings by work. Carefully observe that in using the symbols Q_{in} and Q_{out} on Fig. 2.17 we have departed from the previously stated sign convention for heat transfer. In this section it is advantageous to regard Q_{in} and Q_{out} as transfers of energy in the *directions indicated by the arrows*. The direction of the net work of the cycle, W_{cycle}, is *also indicated by an arrow*. Finally, note that the directions of the energy transfers shown in Fig. 2.17*b* are opposite to those of Fig. 2.17*a*.

TAKE NOTE...

When analyzing cycles, we normally take energy transfers as positive in the directions of arrows on a sketch of the system and write the energy balance accordingly.

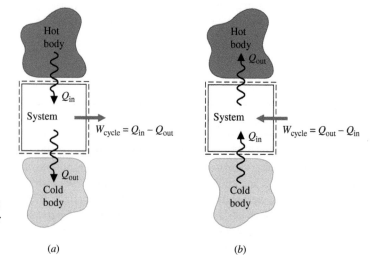

Fig. 2.17 Schematic diagrams of two important classes of cycles. (*a*) Power cycles. (*b*) Refrigeration and heat pump cycles.

2.6.2 Power Cycles

Systems undergoing cycles of the type shown in Fig. 2.17*a* deliver a net work transfer of energy to their surroundings during each cycle. Any such cycle is called a **power cycle**. From Eq. 2.40, the net work output equals the net heat transfer to the cycle, or

power cycle

$$\boxed{W_{\text{cycle}} = Q_{\text{in}} - Q_{\text{out}} \qquad \text{(power cycle)}}$$ (2.41)

where Q_{in} represents the heat transfer of energy *into* the system from the hot body, and Q_{out} represents heat transfer *out* of the system to the cold body. From Eq. 2.41 it is clear that Q_{in} must be greater than Q_{out} for a *power* cycle. The energy supplied by heat transfer to a system undergoing a power cycle is normally derived from the combustion of fuel or a moderated nuclear reaction; it can also be obtained from solar radiation. The energy Q_{out} is generally discharged to the surrounding atmosphere or a nearby body of water.

The performance of a system undergoing a *power cycle* can be described in terms of the extent to which the energy added by heat, Q_{in}, is *converted* to a net work output, W_{cycle}. The extent of the energy conversion from heat to work is expressed by the following ratio, commonly called the **thermal efficiency**:

thermal efficiency

$$\boxed{\eta = \frac{W_{\text{cycle}}}{Q_{\text{in}}} \qquad \text{(power cycle)}}$$ (2.42)

Introducing Eq. 2.41, an alternative form is obtained as

$$\eta = \frac{Q_{\text{in}} - Q_{\text{out}}}{Q_{\text{in}}} = 1 - \frac{Q_{\text{out}}}{Q_{\text{in}}} \qquad \text{(power cycle)}$$ (2.43)

Since energy is conserved, it follows that the thermal efficiency can never be greater than unity (100%). However, experience with *actual* power cycles shows that the value of thermal efficiency is invariably *less* than unity. That is, not all the energy added to the system by heat transfer is converted to work; a portion is discharged to the cold body by heat transfer. Using the second law of thermodynamics, we will show in Chap. 5 that the conversion from heat to work cannot be fully accomplished by any power cycle. The thermal efficiency of *every* power cycle must be less than unity: $\eta < 1$ (100%).

Power Cycles Tabs a & b

Energy & Environment

Today fossil-fueled power plants can have thermal efficiencies of 40%, or more. This means that up to 60% of the energy added by heat transfer during the power plant cycle is discharged from the plant other than by work, principally by heat transfer. One way power plant cooling is achieved is to use water drawn from a nearby river or lake. The water is eventually returned to the river or lake but at a higher temperature, which is a practice having several possible environmental consequences.

The return of large quantities of warm water to a river or lake can affect its ability to hold dissolved gases, including the oxygen required for aquatic life. If the return water temperature is greater than about 35°C, the dissolved oxygen may be too low to support some species of fish. If the return water temperature is too great, some species also can be stressed. As rivers and lakes become warmer, nonnative species that thrive in the warmth can take over. Warmer water also fosters bacterial populations and algae growth.

Regulatory agencies have acted to limit warm-water discharges from power plants, which has made cooling towers (Sec. 12.9) adjacent to power plants a common sight.

2.6.3 Refrigeration and Heat Pump Cycles

Next, consider the **refrigeration and heat pump cycles** shown in Fig. 2.17*b*. For cycles of this type, Q_{in} is the energy transferred by heat *into* the system undergoing the cycle *from the* cold body, and Q_{out} is the energy discharged by heat transfer *from* the system *to* the hot body. To accomplish these energy transfers requires a net work *input*, W_{cycle}. The quantities Q_{in}, Q_{out},

refrigeration and heat pump cycles

and W_{cycle} are related by the energy balance, which for refrigeration and heat pump cycles takes the form

$$\boxed{W_{cycle} = Q_{out} - Q_{in}} \quad \text{(refrigeration and heat pump cycles)} \tag{2.44}$$

Since W_{cycle} is positive in this equation, it follows that Q_{out} is greater than Q_{in}.

Although we have treated them as the same to this point, refrigeration and heat pump cycles actually have different objectives. The objective of a refrigeration cycle is to cool a refrigerated space or to maintain the temperature within a dwelling or other building *below* that of the surroundings. The objective of a heat pump is to maintain the temperature within a dwelling or other building *above* that of the surroundings or to provide heating for certain industrial processes that occur at elevated temperatures.

Since refrigeration and heat pump cycles have different objectives, their performance parameters, called *coefficients of performance,* are defined differently. These coefficients of performance are considered next.

Refrigeration Cycles

coefficient of performance: refrigeration

The performance of *refrigeration cycles* can be described as the ratio of the amount of energy received by the system undergoing the cycle from the cold body, Q_{in}, to the net work into the system to accomplish this effect, W_{cycle}. Thus, the **coefficient of performance**, β, is

$$\boxed{\beta = \frac{Q_{in}}{W_{cycle}}} \quad \text{(refrigeration cycle)} \tag{2.45}$$

Introducing Eq. 2.44, an alternative expression for β is obtained as

$$\beta = \frac{Q_{in}}{Q_{out} - Q_{in}} \quad \text{(refrigeration cycles)} \tag{2.46}$$

For a household refrigerator, Q_{out} is discharged to the space in which the refrigerator is located. W_{cycle} is usually provided in the form of electricity to run the motor that drives the refrigerator.

> **FOR EXAMPLE**
>
> In a refrigerator the inside compartment acts as the cold body and the ambient air surrounding the refrigerator is the hot body. Energy Q_{in} passes to the circulating refrigerant *from* the food and other contents of the inside compartment. For this heat transfer to occur, the refrigerant temperature is necessarily below that of the refrigerator contents. Energy Q_{out} passes *from* the refrigerant *to* the surrounding air. For this heat transfer to occur, the temperature of the circulating refrigerant must necessarily be above that of the surrounding air. To achieve these effects, a work *input* is required. For a refrigerator, W_{cycle} is provided in the form of electricity.

Heat Pump Cycles

coefficient of performance: heat pump

The performance of *heat pumps* can be described as the ratio of the amount of energy discharged from the system undergoing the cycle to the hot body, Q_{out}, to the net work into the system to accomplish this effect, W_{cycle}. Thus, the **coefficient of performance**, γ, is

$$\boxed{\gamma = \frac{Q_{out}}{W_{cycle}}} \quad \text{(heat pump cycle)} \tag{2.47}$$

Introducing Eq. 2.44, an alternative expression for this coefficient of performance is obtained as

$$\gamma = \frac{Q_{out}}{Q_{out} - Q_{in}} \quad \text{(heat pump cycle)} \tag{2.48}$$

Refrigeration Cycles Tabs a & b
Heat Pump Cycles Tabs a & b

From this equation it can be seen that the value of γ is never less than unity. For residential heat pumps, the energy quantity Q_{in} is normally drawn from the surrounding atmosphere, the ground, or a nearby body of water. W_{cycle} is usually provided by electricity.

The coefficients of performance β and γ are defined as ratios of the desired heat transfer effect to the cost in terms of work to accomplish that effect. Based on the definitions, it is desirable thermodynamically that these coefficients have values that are as large as possible. However, as discussed in Chap. 5, coefficients of performance must satisfy restrictions imposed by the second law of thermodynamics.

2.7 Energy Storage

In this section we consider energy storage, which is deemed a critical national need today and likely will continue to be so in years ahead. The need is widespread, including use with conventional fossil- and nuclear-fueled power plants, power plants using renewable sources like solar and wind, and countless applications in transportation, industry, business, and the home.

2.7.1 Overview

While aspects of the present discussion of energy storage are broadly relevant, we are mainly concerned here with storage and recapture of electricity. Electricity can be stored as internal energy, kinetic energy, and gravitational potential energy and converted back to electricity when needed. Owing to thermodynamic limitations associated with such conversions, the effects of friction and electrical resistance for instance, an overall input-to-output *loss* of electricity is *always* observed, however.

Among technically feasible storage options, economics usually determines if, when, and how storage is implemented. For power companies, consumer demand for electricity is a key issue in storage decisions. Consumer demand varies over the day and typically is greatest in the 8 a.m. to 8 p.m. period, with demand *spikes* during that interval. Demand is least in nighttime hours, on weekends, and on major holidays. Accordingly, power companies must decide which option makes the greatest economic sense: marketing electricity as generated, storing it for later use, or a combination—and if stored, how to store it.

2.7.2 Storage Technologies

The focus in this section is on five storage technologies: batteries, ultra-capacitors, superconducting magnets, flywheels, and hydrogen production. Thermal storage is considered in Sec. 3.8. Pumped-hydro and compressed-air storage are considered in Sec. 4.8.3.

Batteries are a widely deployed means of electricity storage appearing in cell phones, laptop computers, automobiles, power-generating systems, and numerous other applications. Yet battery makers struggle to keep up with demands for lighter-weight, greater-capacity, longer-lasting, and more quickly recharged units. For years batteries have been the subject of vigorous research and development programs. Through these efforts, batteries have been developed providing significant improvements over the *lead-acid* batteries used for decades. These include utility-scale *sodium-sulfur* batteries and the *lithium-ion* and *nickel-metal hydride* types seen in consumer products and hybrid vehicles. Novel nanotechnology-based batteries promise even better performance: greater capacity, longer service life, and a quicker recharge time, all of which are essential for use in hybrid vehicles.

Ultra-capacitors are energy storage devices that work like large versions of common electrical capacitors. When an ultra-capacitor is charged electrically, energy is stored as a charge on the surface of a material. In contrast to batteries, ultra-capacitors require no chemical reactions and consequently enjoy a much longer service life. This storage type is also capable of very rapid charging and discharging. Applications today include starting railroad locomotives and diesel trucks. Ultra-capacitors are also used in hybrid vehicles, where they work in tandem with batteries. In hybrids, ultra-capacitors are best suited for performing short-duration tasks, such as storing electricity from regenerative braking and delivering power for acceleration during start–stop driving, while batteries provide energy needed for sustained vehicle motion, all with less total mass and longer service life than with batteries alone.

Superconducting magnetic systems store an electrical input in the magnetic field created by flow of electric current in a coil of cryogenically cooled, superconducting material. This storage type provides power nearly instantaneously, and with very low input-to-output loss of electricity. Superconducting magnetic systems are used by high-speed magnetic-levitation trains, by utilities for power-quality control, and by industry for special applications such as microchip fabrication.

Flywheels provide another way to store an electrical input—as kinetic energy. When electricity is required, kinetic energy is drained from the spinning flywheel and provided to a generator. Flywheels typically exhibit low input-to-output loss of electricity. Flywheel storage is used, for instance, by Internet providers to protect equipment from power outages.

Hydrogen has also been proposed as an energy storage medium for electricity. With this approach, electricity is used to *dissociate* water to hydrogen via the *electrolysis* reaction, $H_2O \rightarrow H_2 + \frac{1}{2} O_2$. Hydrogen produced this way can be stored to meet various needs, including generating electricity by fuel cells via the *inverse* reaction: $H_2 + \frac{1}{2} O_2 \rightarrow H_2O$. A shortcoming of this type of storage is its characteristically significant input-to-output loss of electricity.

Energy & Environment

Batteries provide convenient and portable power for many household uses as well as many commercial and industrial applications. They play an ever-increasing role in our lives and their use will expand significantly in coming years. Batteries are great while they work, but when their useful life is over, we must find ways to dispose of them that don't harm the environment. Already a big concern, this challenge becomes more concerning as battery use proliferates.

Most batteries rely on chemical reactions involving heavy metals such as lead, cadmium, and nickel with electrolytes to produce power. Finding ways to deal with these chemicals is a difficult yet critical environmental issue. Federal, state, and local regulations are in place to govern the collection and management of used batteries and other hazardous wastes. Many municipal and commercial programs attempt to reduce the quantities of these materials that end up in solid waste landfills or incinerators. The success of such programs depends greatly on consumer behavior to separate used batteries from other domestic waste and to make sure that the batteries are disposed of properly.

A related approach to the problem of disposal is to greatly expand the use of rechargeable batteries and battery recycling. Widespread use of rechargeable batteries can greatly reduce the waste stream. Despite their higher initial cost, rechargeable batteries have been shown to be lower in total cost than disposable types. Recycling is also very promising since the heavy metals and other materials that make up the batteries can often be reclaimed for reuse. Many local communities have established programs for battery recycling and such programs can also reduce the waste stream significantly.

CHAPTER SUMMARY AND STUDY GUIDE

In this chapter, we have considered the concept of energy from an engineering perspective and have introduced energy balances for applying the conservation of energy principle to closed systems. A basic idea is that energy can be stored within systems in three macroscopic forms: internal energy, kinetic energy, and gravitational potential energy. Energy also can be transferred to and from systems.

Energy can be transferred to and from closed systems by two means only: work and heat transfer. Work and heat transfer are identified at the system boundary and are not properties. In mechanics, work is energy transfer associated with macroscopic forces and displacements. The thermodynamic definition of work introduced in this chapter extends the notion of work from mechanics to include other types of work. Energy transfer by heat to or from a system is due to a temperature difference between the system and its surroundings and occurs in the direction of decreasing temperature. Heat transfer modes include conduction, radiation, and convection. These sign conventions are used for work and heat transfer:

- $W, \dot{W}$ $\begin{cases} > 0 \text{: work done by the system} \\ < 0 \text{: work done on the system} \end{cases}$

- $Q, \dot{Q}$ $\begin{cases} > 0 \text{: heat transfer to the system} \\ < 0 \text{: heat transfer from the system} \end{cases}$

Energy is an extensive property of a system. Only changes in the energy of a system have significance. Energy changes are accounted for by the energy balance. The energy balance for a process of a closed system is Eq. 2.35 and an accompanying time rate form is Eq. 2.37. Equation 2.40 is a special form of the energy balance for a system undergoing a thermodynamic cycle.

The following checklist provides a study guide for this chapter. When your study of the text and end-of-chapter exercises has been completed, you should be able to

- write out the meanings of the terms listed in the margins throughout the chapter and understand each of the related concepts. The subset of key concepts listed below is particularly important in subsequent chapters.
- evaluate these energy quantities
  –kinetic and potential energy changes using Eqs. 2.5 and 2.10, respectively.

  –work and power using Eqs. 2.12 and 2.13, respectively.
  –expansion or compression work using Eq. 2.17
- apply closed system energy balances in each of several alternative forms, appropriately modeling the case at hand, correctly observing sign conventions for work and heat transfer, and carefully applying SI units.
- conduct energy analyses for systems undergoing thermodynamic cycles using Eq. 2.40, and evaluating, as appropriate, the thermal efficiencies of power cycles and coefficients of performance of refrigeration and heat pump cycles.

KEY ENGINEERING CONCEPTS

kinetic energy	internal energy	energy balance
gravitational potential energy	heat transfer	thermodynamic cycle
work	sign convention for heat transfer	power cycle
sign convention for work	adiabatic	refrigeration cycle
power	first law of thermodynamics	heat pump cycle

KEY EQUATIONS

$\Delta E = \Delta U + \Delta KE + \Delta PE$	(2.27)	Change in total energy of a system.
$\Delta KE = KE_2 - KE_1 = \frac{1}{2}m(V_2^2 - V_1^2)$	(2.5)	Change in kinetic energy of a mass m.
$\Delta PE = PE_2 - PE_1 = mg(z_2 - z_1)$	(2.10)	Change in gravitational potential energy of a mass m at constant g.
$E_2 - E_1 = Q - W$	(2.35a)	Energy balance for closed systems.
$\dfrac{dE}{dt} = \dot{Q} - \dot{W}$	(2.37)	Energy rate balance for closed systems.
$W = \displaystyle\int_{s_1}^{s_2} \mathbf{F} \cdot \mathbf{ds}$	(2.12)	Work due to action of a force F.
$\dot{W} = \mathbf{F} \cdot \mathbf{V}$	(2.13)	Power due to action of a force F.
$W = \displaystyle\int_{V_1}^{V_2} p\, dV$	(2.17)	Expansion or compression work related to fluid pressure. See Fig. 2.4.

Thermodynamic Cycles

$W_{cycle} = Q_{in} - Q_{out}$	(2.41)	Energy balance for a *power cycle*. As in Fig. 2.17a, all quantities are regarded as positive.
$\eta = \dfrac{W_{cycle}}{Q_{in}}$	(2.42)	Thermal efficiency of a power cycle.
$W_{cycle} = Q_{out} - Q_{in}$	(2.44)	Energy balance for a *refrigeration* or *heat pump cycle*. As in Fig. 2.17b, all quantities are regarded as positive.
$\beta = \dfrac{Q_{in}}{W_{cycle}}$	(2.45)	Coefficient of performance of a refrigeration cycle.
$\gamma = \dfrac{Q_{out}}{W_{cycle}}$	(2.47)	Coefficient of performance of a heat pump cycle.

2.1 What are several things you as an individual can do to reduce energy use in your home? While meeting your transportation needs?

2.2 How does the kilowatt-hour meter in your house measure electric energy usage?

2.3 Why is it incorrect to say that a system *contains* heat?

2.4 What examples of heat transfer by conduction, radiation, and convection do you encounter when using a charcoal grill?

2.5 After running 8 km on a treadmill at her campus rec center, Ashley observes that the treadmill belt is warm to the touch. Why is the belt warm?

2.6 When microwaves are beamed onto a tumor during cancer therapy to increase the tumor's temperature, this interaction is considered work and not heat transfer. Why?

2.7 For good acceleration, what is more important for an automobile engine, horsepower or torque?

2.8 When cooking ingredients are mixed in a blender, what happens to the energy transferred to the ingredients? Is the energy transfer by work, by heat transfer, or by work and heat transfer?

2.9 For polytropic expansion or compression, what causes the value of n to vary from process to process?

2.10 In the *differential* form of the closed system energy balance, $dE = \delta Q - \delta W$, why is d and not δ used for the differential on the left?

2.11 When two amusement park bumper cars collide head-on and come to a stop, how do you account for the kinetic energy the pair had just before the collision?

2.12 If the change in energy of a closed system is known for a process between two end states, can you determine if the energy change was due to work, to heat transfer, or to some combination of work and heat transfer?

2.13 What forms of energy and energy transfer are present in the life cycle of a thunderstorm?

2.14 How would you define an *efficiency* for the motor of Example 2.6?

2.15 How many tons of CO_2 are produced annually by a conventional automobile?

Match the appropriate definition or expression in the right column with the corresponding term in the left column.

2.1 Refrigeration cycle

2.2 Adiabatic

2.3 Energy balance

2.4 Thermodynamic cycle

2.5 Energy transfer by heat

2.6 Energy transfer by work

A. Energy transfer for which the *sole* effect on everything external to the system *could have been* the raising of a weight

B. A sequence of processes that begins and ends at the same state

C. Energy transfer induced only as a result of a temperature difference between a system and its surroundings

D. A cycle where energy is transferred by heat into the system undergoing the cycle *from* the cold body and energy is transferred by heat from the system *to* the hot body

E. A process involving no energy transfer by heat

F. $\Delta E = Q - W$

2.7 Why does evaluating work using Eq. 2.17 for expansion of a gas require knowing the pressure at the interface between the gas and the moving piston during the process?

2.8 Each of the cycle performance parameters defined in this chapter is in the form of the desired energy transfer divided by an energy input quantity. For each of the three types of cycles considered, identify the energy transfers that play the respective roles.

2.9 In mechanics, the work of a resultant force acting on a body equals the change in its _____.

2.10 What direction is the *net* energy transfer by work for a power cycle: in or out? The *net* energy transfer by heat?

2.11 The differential of work, δW, is said to be an _____ differential.

2.12 Kinetic and gravitational potential energies are *extensive properties* of a closed system. Explain.

2.13 What direction is the *net* energy transfer by work for a refrigeration or heat pump cycle: in or out? The *net* energy transfer by heat?

2.14 An object of known mass and initially at rest falls from a specified elevation. It hits the ground and comes to rest at zero elevation. Is energy conserved in this process? Discuss.

2.15 List the three modes of energy transfer by heat and discuss the differences among them.

2.16 In order to evaluate work using $W = \int_{V_1}^{V_2} p \, dV$, we must specify how p varies with V during the process. It follows that work is not a _____.

2.17 What are the three modes of energy storage for individual atoms and molecules making up the matter within a system?

2.18 When a system undergoes a process, the terms *work* and *heat* do not refer to what is being transferred. _____ is transferred when work and/or heat transfer occurs.

2.19 The change in total energy of a closed system other than changes in kinetic and gravitational potential energy are accounted for by the change in _____.

Indicate whether the following statements are true or false. Explain.

2.20 A spring is compressed adiabatically. Its internal energy increases.

2.21 If a system's temperature increases, it must have experienced heat transfer.

2.22 If a closed system undergoes a thermodynamic cycle, there can be no net work or heat transfer.

2.23 In principle, expansion or compression work can be evaluated using $\int p \, dV$ for both actual and quasiequilibrium expansion processes.

2.24 For heat pumps, the coefficient of performance γ is always greater than or equal to one.

2.25 The heat transfer coefficient, h, in *Newton's law of cooling* is not a thermodynamic property. It is an empirical parameter that incorporates into the heat transfer relationship the nature of the flow pattern near the surface, the fluid properties, and the geometry.

2.26 Only *changes* in the internal energy of a system between two states have significance: No significance can be attached to the internal energy *at* a state.

2.27 Thermal radiation can occur in vacuum.

2.28 Current passes through an electrical resistor inside a tank of gas. Depending on where the system boundary is located, the energy transfer can be considered work or heat.

2.29 For any cycle, the net amounts of energy transfer by heat and work are equal.

2.30 A rotating flywheel stores energy in the form of kinetic energy.

2.31 If a closed system undergoes a process for which the change in total energy is positive, the heat transfer must be positive.

2.32 If a closed system undergoes a process for which the work is negative and the heat transfer is positive, the total energy of the system must increase.

2.33 According to the *Stefan–Boltzmann law*, all objects emit thermal radiation at temperatures higher than 0 K.

2.34 Power is related mathematically to the amount of energy transfer by work by integrating over time.

PROBLEMS: DEVELOPING ENGINEERING SKILLS

Exploring Energy Concepts

2.1 A baseball has a mass of 0.14 kg. What is the kinetic energy relative to home plate of a 151 km/hr fastball, in kJ?

FIGURE P2.1

2.2 Determine the gravitational potential energy, in kJ, of 2 m³ of liquid water at an elevation of 30 m above the surface of Earth. The acceleration of gravity is constant at 9.7 m/s² and the density of the water is uniform at 1000 kg/m³. Determine the change in gravitational potential energy as the elevation decreases by 15 m.

2.3 An object whose mass is 45 kg experiences a decrease in kinetic energy of 0.68 kJ and an increase in potential energy of 2kJ. The initial velocity and elevation of the object, each relative to the surface of the earth, are 12 m/s and 9 m, respectively. If $g = 9.8$ m/s², determine

   a. the final velocity, in m/s
   b. the final elevation, in m

2.4 A 30-seat turboprop airliner whose mass is 14,000 kg takes off from an airport and eventually achieves its cruising speed of 620 km/h at an altitude of 10,000 m. For $g = 9.78$ m/s², determine the change in kinetic energy and the change in gravitational potential energy of the airliner, each in kJ.

2.5 An object whose mass is 25 kg is projected upward from the surface of the earth with an initial velocity of 60 m/s. The only force acting on the object is the force of gravity. Plot the velocity of the object versus elevation. Determine the elevation of the object, in m, when its velocity reaches zero. The acceleration of gravity is $g = 9.8$ m/s².

2.6 An object of mass 15 kg is at an elevation of 100 m relative to the surface of the Earth. What is the potential energy of the object, in kJ? If the object were initially at rest, to what velocity, in m/s, would you have to accelerate it for the kinetic energy to have the same value as the potential energy you calculated above? The acceleration of gravity is 9.8 m/s².

2.7 An automobile having a mass of 900 kg initially moves along a level highway at 100 km/h relative to the highway. It then climbs a hill whose crest is 50 m above the level highway and parks at a rest area located there. For the automobile, determine its changes in kinetic and potential energy, each in kJ. For each quantity, kinetic energy and potential energy, specify your choice of datum and reference value at that datum. Let $g = 9.81$ m/s².

2.8 An object whose mass is 136 kg experiences changes in its kinetic and potential energies owing to the action of a resultant force **R**. The work done on the object by the resultant force is 148 kJ. There are no other interactions between the object and its surroundings. If the object's elevation increases by 30.5 m and its final velocity is 61 m/s, what is its initial velocity, in m/s? Let $g = 9.81$ m/s².

2.9 Beginning from rest, an object of mass 200 kg slides down a 10-m-long ramp. The ramp is inclined at an angle of 40° from the horizontal. If air resistance and friction between the object and the ramp are negligible, determine the velocity of the object, in m/s, at the bottom of the ramp. Let $g = 9.81$ m/s².

2.10 Using KE = $I\omega^2/2$, how fast would a flywheel whose moment of inertia is 8.4 kg · m² have to spin, in RPM, to store an amount of kinetic energy equivalent to the potential energy of a 45 kg mass raised to an elevation of 9 m above the surface of the earth? Let $g = 9.81$ m/s².

2.11 A block of mass 10 kg moves along a surface inclined 30° relative to the horizontal. The center of gravity of the block is elevated by 3.0 m and the kinetic energy of the block *decreases* by 50 J. The block is acted upon by a constant force **R** parallel to the incline, and by the force of gravity. Assume frictionless surfaces and let $g = 9.81$ m/s². Determine the magnitude and direction of the constant force **R**, in N.

2.12 During the packaging process, a can of soda of mass 0.4 kg moves down a surface inclined 20° relative to the horizontal, as shown in Fig. P2.12. The can is acted upon by a constant force **R** parallel to the incline and by the force of gravity. The magnitude of the constant force **R** is 0.05 N. Ignoring friction between the can and the inclined surface, determine the can's change in kinetic energy, in J, and whether it is *increasing* or *decreasing*. If friction between the can and the inclined surface were significant, what effect would that have on the value of the change in kinetic energy? Let $g = 9.8$ m/s².

2.13 Jack, whose mass is 68 kg, runs 8 km in 43 minutes on a treadmill set at a one-degree incline. The treadmill display shows he has

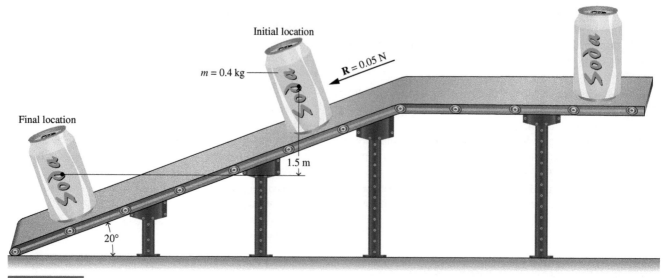

Initial location

$m = 0.4$ kg

$\mathbf{R} = 0.05$ N

Final location

1.5 m

20°

FIGURE P2.12

burned 620 kcal. For Jack to break even caloriewise, how much vanilla ice cream, in cups, may he have after his workout?

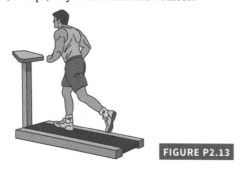

FIGURE P2.13

Evaluating Work

2.14 An object initially at an elevation of 5 m relative to Earth's surface with a velocity of 50 m/s is acted on by an applied force $\mathbf{R}$ and moves along a path. Its final elevation is 20 m and its velocity is 100 m/s. The acceleration of gravity is 9.81 m/s^2. Determine the work done on the object by the applied force, in kJ.

2.15 An object of mass 10 kg, initially at rest, experiences a constant horizontal acceleration of 4 m/s^2 due to the action of a resultant force applied for 20 s. Determine the total amount of energy transfer by work, in kJ.

2.16 A system with a mass of 5 kg, initially moving horizontally with a velocity of 40 m/s, experiences a constant horizontal *deceleration* of 2 m/s^2 due to the action of a resultant force. As a result, the system comes to rest. Determine the length of time, in s, the force is applied and the amount of energy transfer by work, in kJ.

2.17 A gas in a piston–cylinder assembly undergoes a process for which the relationship between pressure and volume is $pV^2 = constant$. The initial pressure is 1 bar, the initial volume is 0.1 m^3, and the final pressure is 9 bar. Determine (a) the final volume, in m^3, and (b) the work for the process, in kJ.

2.18 Carbon dioxide (CO_2) gas within a piston–cylinder assembly undergoes an expansion from a state where $p_1 = 138$ kPa, $V_1 = 0.014$ m^3 to a state where $p_2 = 34.5$ kPa, $V_2 = 0.07$ m^3. The relationship between pressure and volume during the process is $p = A + BV$, where A and B are constants. (a) For the CO_2, evaluate the work, in N $\cdot$ m and kJ. (b) Evaluate A, in kPa, and B, in (kPa)/m^3.

2.19 A gas in a piston–cylinder assembly undergoes a compression process for which the relation between pressure and volume is given by

$pV^n = constant$. The initial volume is 0.1 m^3, the final volume is 0.04 m^3, and the final pressure is 2 bar. Determine the initial pressure, in bar, and the work for the process, in kJ, if (a) $n = 0$, (b) $n = 1$, (c) $n = 1.3$.

2.20 Oxygen (O_2) gas within a piston–cylinder assembly undergoes an expansion from a volume $V_1 = 0.01$ m^3 to a volume $V_2 = 0.03$ m^3. The relationship between pressure and volume during the process is $p = AV^{-1} + B$, where $A = 0.06$ bar $\cdot$ m^3 and $B = 3.0$ bar. For the O_2, determine (a) the initial and final pressures, each in bar, and (b) the work, in kJ.

2.21 Air is compressed slowly in a piston–cylinder assembly from an initial state where $p_1 = 1.4$ bar, $V_1 = 4.25$ m^3, to a final state where $p_2 = 6.8$ bar. During the process, the relation between pressure and volume follows $pV = constant$. For the air as the closed system, determine the work, in kJ.

2.22 Air contained within a piston–cylinder assembly is slowly heated. As shown in Fig. P2.22, during this process the pressure first varies linearly with volume and then remains constant. Determine the total work, in kJ.

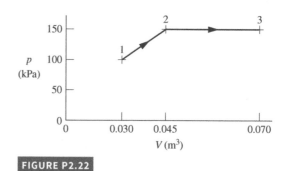

FIGURE P2.22

2.23 A gas contained within a piston–cylinder assembly undergoes three processes in series:

    Process 1–2: Constant volume from $p_1 = 1$ bar, $V_1 = 4$ m^3 to state 2, where $p_2 = 2$ bar.

    Process 2–3: Compression to $V_3 = 2$ m^3, during which the pressure–volume relationship is $pV = constant$.

    Process 3–4: Constant pressure to state 4, where $V_4 = 1$ m^3.

Sketch the processes in series on p–V coordinates and evaluate the work for each process, in kJ.

2.24 Carbon dioxide (CO_2) gas in a piston-cylinder assembly undergoes three processes in series that begin and end at the same state (a cycle).

Process 1-2: Expansion from state 1 where $p_1 = 10$ bar, $V_1 = 1$ m^3, to state 2 where $V_2 = 4$ m^3. During the process, pressure and volume are related by $pV^{1.5} = constant$.

Process 2-3: Constant volume heating to state 3 where $p_3 = 10$ bar.

Process 3-1: Constant pressure compression to state 1.

Sketch the processes on p–V coordinates and evaluate the work for each process, in kJ. What is the *net* work for the cycle, in kJ?

2.25 Air contained within a piston–cylinder assembly undergoes three processes in series:

Process 1–2: Compression at constant pressure from $p_1 = 69$ kPa, $V_1 = 0.11$ m^3 to state 2.

Process 2–3: Constant-volume heating to state 3, where $p_3 = 345$ kPa.

Process 3–1: Expansion to the initial state, during which the pressure–volume relationship is $pV = constant$.

Sketch the processes in series on p-V coordinates. Evaluate (a) the volume at state 2, in m^3, and (b) the work for each process, in kJ.

2.26 A 0.15-m-diameter pulley turns a belt rotating the driveshaft of a power plant pump. The torque applied by the belt on the pulley is 200 N · m, and the power transmitted is 7 kW. Determine the net force applied by the belt on the pulley, in kN, and the rotational speed of the driveshaft, in RPM.

2.27 A 10-V battery supplies a constant current of 0.5 amp to a resistance for 30 min. (a) Determine the resistance, in ohms. (b) For the battery, determine the amount of energy transfer by work, in kJ.

2.28 An electric heater draws a constant current of 6 amp, with an applied voltage of 220 V, for 24 h. Determine the instantaneous electric power provided to the heater, in kW, and the total amount of energy supplied to the heater by electrical work, in kW · h. If electric power is valued at $0.08/kW · h, determine the cost of operation for one day.

2.29 The driveshaft of a building's air-handling fan is turned at 300 RPM by a belt running on a 0.3-m-diameter pulley. The net force applied by the belt on the pulley is 2,000 N. Determine the torque applied by the belt on the pulley, in N · m, and the power transmitted, in kW.

2.30 The belt sander shown in Fig. P2.30 has a belt speed of 7.6 m/s. The coefficient of friction between the sander and a plywood surface being finished is 0.2. If the downward (normal) force on the sander is 66.7 N, determine (a) the power transmitted by the belt, in kJ/s and kW, and (b) the work done in one minute of sanding, in kJ.

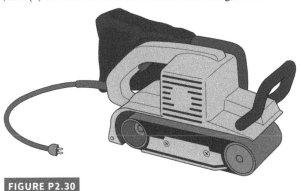

FIGURE P2.30

2.31 A soap film is suspended on a wire frame, as shown in Fig. 2.10. The movable wire is displaced by an applied force, F. If the surface tension remains constant

a. obtain an expression for the work done in stretching the film in terms of the surface tension τ, length ℓ, and displacement Δx.

b. evaluate the work done, in J, if $\ell = 5$ cm, $\Delta x = 0.5$ cm, and $\tau = 25 \times 10^{-5}$ N/cm.

2.32 As shown in **Fig. P2.32**, a steel wire suspended vertically having a cross-section area A and an initial length x_0 is stretched by a downward force F applied to the end of the wire. The normal stress in the wire varies linearly according to $\sigma = C\varepsilon$, where ε is the strain, given by $\varepsilon = (x - x_0)/x_0$, and x is the stretched length of the wire. C is a material constant (Young's modulus). Assuming the cross-sectional area remains constant

a. obtain an expression for the work done on the wire.

b. evaluate the work done on the wire, in N · m, and the magnitude of the downward force, in N, if $x_0 = 3.048$ m, $x = 3.051$ m, A = 64.5 mm^2, and $C = 17.2 \times 10^7$ kPa.

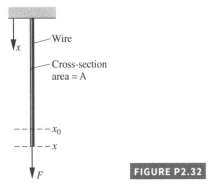

FIGURE P2.32

2.33 As shown in **Fig. P2.33**, a spring having an initial unstretched length of ℓ_0 is stretched by a force F applied at its end. The stretched length is ℓ. By *Hooke's law,* the force is linearly related to the spring extension by $F = k(\ell - \ell_0)$ where k is the *stiffness*. If stiffness is constant,

a. obtain an expression for the work done in changing the spring's length from ℓ_1 to ℓ_2.

b. evaluate the work done, in J, if $\ell_0 = 3$ cm, $\ell_1 = 6$ cm, $\ell_2 = 10$ cm, and the stiffness is $k = 10^4$ N/m.

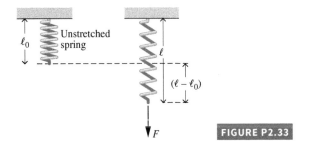

FIGURE P2.33

Evaluating Heat Transfer

2.34 A flat surface having an area of 2 m^2 and a temperature of 350 K is cooled convectively by a gas at 300 K. Using data from Table 2.1, determine the largest and smallest heat transfer rates, in kW, that might be encountered for (a) free convection, (b) forced convection.

2.35 A fan forces air over a computer circuit board with surface area of 70 cm^2 to avoid overheating. The air temperature is 300 K while the circuit board surface temperature is 340 K. Using data from Table 2.1, determine the largest and smallest heat transfer rates, in W, that might be encountered for this forced convection.

2.36 As shown in **Fig. P2.36**, an oven wall consists of a 6.35 mm-thick layer of steel ($\kappa_s = 15.05 \times 10^{-3}$ kW/m · K) and a layer of brick ($\kappa_b = 0.73 \times 10^{-3}$ kW/m · K). At steady state, a temperature decrease of 0.67°C occurs over the steel layer. The inner temperature of the steel layer is 282°C. If the temperature of the outer surface of the brick must be no greater than 40°C, determine the minimum thickness of brick, in in., that ensures this limit is met.

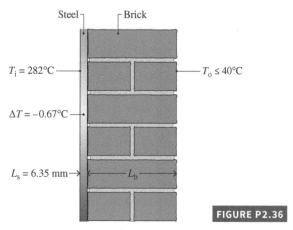

FIGURE P2.36

2.37 A composite plane wall consists of a 75-mm-thick layer of insulation ($\kappa_i = 0.05$ W/m · K) and a 25-mm-thick layer of siding ($\kappa = 0.10$ W/m · K). The inner temperature of the insulation is 20°C. The outer temperature of the siding is −13°C. Determine at steady state (a) the temperature at the interface of the two layers, in °C, and (b) the rate of heat transfer through the wall, in W per m² of surface area.

2.38 Complete the following exercise using heat transfer relations:

 a. Referring to Fig. 2.12, determine the rate of conduction heat transfer, in W, for $\kappa = 0.07$ W/m · K, A = 0.125 m², $T_1 = 298$ K, $T_2 = 273$ K.

 b. Referring to Fig. 2.14, determine the rate of convection heat transfer from the surface to the air, in W, for h = 10 W/m² · K, A = 0.125 m², $T_b = 305$ K, $T_f = 298$ K.

2.39 At steady state, a spherical interplanetary electronics-laden probe having a diameter of 0.6 m transfers energy by radiation from its outer surface at a rate of 160 W. If the probe does not receive radiation from the sun or deep space, what is the surface temperature, in K?

2.40 A body whose surface area is 0.25 m², emissivity is 0.85, and temperature is 175°C is placed in a large, evacuated chamber whose walls are at 27°C. What is the rate at which radiation is *emitted* by the surface, in W? What is the *net* rate at which radiation is *exchanged* between the surface and the chamber walls, in W?

2.41 The outer surface of the grill hood shown in Fig. P2.41 is at 47°C and the emissivity is 0.93. The heat transfer coefficient for convection between the hood and the surroundings at 27°C is 10 W/m² · K. Determine the net rate of heat transfer between the grill hood and the surroundings by convection and radiation, in kW per m² of surface area.

$$T_0 = 27°C$$
$$h = 10 \text{ W/m}^2 \cdot k$$

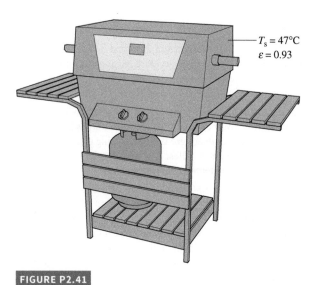

$T_s = 47°C$
$\varepsilon = 0.93$

FIGURE P2.41

Using the Energy Balance

2.42 Each line in the following table gives information about a process of a closed system. Each entry has the same energy units. Fill in the blank spaces in the table.

Process	Q	W	E_1	E_2	ΔE
a		−20		+50	+70
b	+50		+20	+50	
c		−60		+60	+20
d		−90		+50	0
e	+50	+150	+20		

2.43 A mass of 10 kg undergoes a process during which there is heat transfer from the mass at a rate of 5 kJ per kg, an elevation decrease of 50 m, and an increase in velocity from 15 m/s to 30 m/s. The specific internal energy decreases by 5 kJ/kg and the acceleration of gravity is constant at 9.7 m/s². Determine the work for the process, in kJ.

2.44 As shown in Fig. P2.44, a gas contained within a piston–cylinder assembly, initially at a volume of 0.1 m³, undergoes a constant-pressure expansion at 2 bar to a final volume of 0.12 m³, while being slowly heated through the base. The change in internal energy of the gas is 0.25 kJ. The piston and cylinder walls are fabricated from heat-resistant material, and the piston moves smoothly in the cylinder. The local atmospheric pressure is 1 bar.

 a. For the gas as the system, evaluate work and heat transfer, each in kJ.

 b. For the piston as the system, evaluate work and change in potential energy, each in kJ.

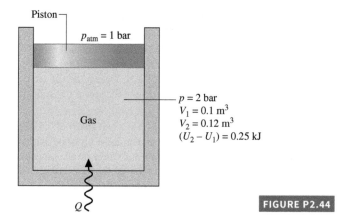

FIGURE P2.44

2.45 A gas contained within a piston–cylinder assembly undergoes two processes, A and B, between the *same end states*, 1 and 2, where $p_1 = 10$ bar, $V_1 = 0.1$ m³, $U_1 = 400$ kJ and $p_2 = 1$ bar, $V_2 = 1.0$ m³, $U_2 = 200$ kJ:

 Process A: Process from 1 to 2 during which the pressure-volume relation is $pV = constant$.

 Process B: Constant-volume process from state 1 to a pressure of 2 bar, followed by a linear pressure-volume process to state 2.

Kinetic and potential energy effects can be ignored. For each of the processes A and B, (a) sketch the process on $p–V$ coordinates, (b) evaluate the work, in kJ, and (c) evaluate the heat transfer, in kJ.

2.46 An electric motor draws a current of 10 amp with a voltage of 110 V. The output shaft develops a torque of 10.2 N · m and a rotational speed of 1,000 RPM. For operation at steady state, determine for the motor, each in kW

a. the electric power required.

b. the power developed by the output shaft.

c. the rate of heat transfer.

2.47 An electric motor draws a current of 10 amp with a voltage of 110 V, as shown in **Fig. P2.47**. The output shaft develops a torque of 9.7 N $\cdot$ m and a rotational speed of 1000 RPM. For operation at steady state, determine for the motor

a. the electric power required, in kW.

b. the power developed by the output shaft, in kW.

c. the average surface temperature, T_s, in °C, if heat transfer occurs by convection to the surroundings at $T_f = 21$°C.

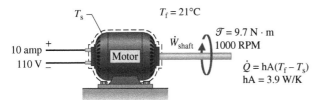

T_s

$T_f = 21$°C

10 amp

110 V

Motor

$\dot{W}_{shaft}$

$\mathcal{T} = 9.7$ N $\cdot$ m

1000 RPM

$\dot{Q} = hA(T_f - T_s)$

$hA = 3.9$ W/K

FIGURE P2.47

2.48 In a rigid insulated container of volume 0.8 m³, 2.5 kg of air is filled. A paddle wheel is fitted in the container and it transfers energy to the contained air at a constant rate of 12 W for a period of 1 h. There is no change in the potential or kinetic energy of the system. Determine the energy transfer by the wheel to the air per kg of air.

2.49 A gas expands in a piston–cylinder assembly from $p_1 = 8$ bar, $V_1 = 0.02$ m³ to $p_2 = 2$ bar in a process during which the relation between pressure and volume is $pV^{1.2} = constant$. The mass of the gas is 0.25 kg. If the specific internal energy of the gas *decreases* by 55 kJ/kg during the process, determine the heat transfer, in kJ. Kinetic and potential energy effects are negligible.

2.50 Four kilograms of carbon monoxide (CO) is contained in a rigid tank with a volume of 1 m³. The tank is fitted with a paddle wheel that transfers energy to the CO at a constant rate of 14 W for 1 h. During the process, the specific internal energy of the carbon monoxide increases by 10 kJ/kg. If no overall changes in kinetic and potential energy occur, determine

a. the specific volume at the final state, in m³/kg.

b. the energy transfer by work, in kJ.

c. the energy transfer by heat transfer, in kJ, and the direction of the heat transfer.

2.51 Steam in a piston–cylinder assembly undergoes a polytropic process, with $n = 2$, from an initial state where $p_1 = 3.45$ MPa, $v_1 = 0.106$ m³/kg, $u_1 = 3171.1$ kJ/kg, to a final state where $u_2 = 2303.9$ kJ/kg. During the process, there is a heat transfer from the steam of magnitude 361.76 kJ. The mass of steam is 0.54 kg. Neglecting changes in kinetic and potential energy, determine the work, in kJ, and the final specific volume, in m³/kg.

2.52 Gaseous CO_2 is contained in a vertical piston–cylinder assembly by a piston of mass 50 kg and having a face area of 0.01 m². The mass of CO_2 is 4 g. The CO_2 initially occupies a volume of 0.005 m³ and has a specific internal energy of 657 kJ/kg. The atmosphere exerts a pressure of 100 kPa on the top of the piston. Heat transfer in the amount of 1.95 kJ occurs slowly from the CO_2 to the surroundings, and the volume of the CO_2 decreases to 0.0025 m³. Friction between the piston and the cylinder wall can be neglected. The local acceleration of gravity is 9.81 m/s². For the CO_2, determine (a) the pressure, in kPa, and (b) the final specific internal energy, in kJ/kg.

2.53 **Figure P2.53** shows a gas contained in a vertical piston–cylinder assembly. A vertical shaft whose cross-sectional area is 0.8 cm² is attached to the top of the piston. The total mass of the piston and shaft is 25 kg. While the gas is slowly heated, the internal energy of

the gas increases by 0.1 kJ, the potential energy of the piston–shaft combination increases by 0.2 kJ, and a force of 1334 N is exerted on the shaft as shown in the figure. The piston and cylinder are poor conductors, and friction between them is negligible. The local atmospheric pressure is 1 bar and $g = 9.81$ m/s². Determine, (a) the work done by the shaft, (b) the work done in displacing the atmosphere, and (c) the heat transfer to the gas, all in kJ. (d) Using calculated and given data, develop a detailed *accounting* of the heat transfer of energy to the gas.

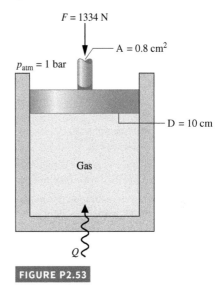

$F = 1334$ N

$A = 0.8$ cm²

$p_{atm} = 1$ bar

$D = 10$ cm

Gas

Q

FIGURE P2.53

Analyzing Thermodynamic Cycles

2.54 The following table gives data, in kJ, for a system undergoing a power cycle consisting of four processes in series. Determine (a) the missing table entries, each in kJ, and (b) the thermal efficiency.

Process	ΔU	ΔKE	ΔPE	ΔE	Q	W
1–2	1002	53	0		1005	
2–3		0	53	−475		475
3–4	−686		0	−633		0
4–1	211	−105.5	−53		0	

2.55 A gas within a piston–cylinder assembly undergoes a thermodynamic cycle consisting of three processes:

Process 1–2: Constant volume $V_1 = 1$ m³, $p_1 = 1$ bar, to $p_2 = 3$ bar, $U_2 - U_1 = 400$ kJ.

Process 2–3: Constant pressure expansion to $V_3 = 2$ m³.

Process 3–1: Adiabatic compression, with $W_{31} = -1120$ kJ.

There are no significant changes in kinetic or potential energy. Determine the net work for the cycle, in kJ, and the heat transfers for Processes 1–2 and 2–3, in kJ. Is this a power cycle or refrigeration cycle? Explain.

2.56 A gas undergoes a cycle in a piston–cylinder assembly consisting of the following three processes:

Process 1–2: Constant pressure, $p = 1.4$ bar, $V_1 = 0.028$ m³, $W_{12} = 10.5$ kJ

Process 2–3: Compression with $pV = constant$, $U_3 = U_2$

Process 3–1: Constant volume, $U_1 - U_3 = -26.4$ kJ

There are no significant changes in kinetic or potential energy.

a. Sketch the cycle on a p–V diagram.

b. Calculate the net work for the cycle, in kJ.

c. Calculate the heat transfer for process 1–2, in kJ.

2.57 The net work of a power cycle operating as in Fig. 2.17a is 10,000 kJ, and the thermal efficiency is 0.4. Determine the heat transfers Q_{in} and Q_{out}, each in kJ.

2.58 For a power cycle operating as shown in Fig. 2.17a, the energy transfer by heat into the cycle, Q_{in}, is 500 MJ. What is the net work developed, in MJ, if the cycle thermal efficiency is 30%? What is the value of Q_{out}, in MJ?

2.59 For a power cycle operating as in Fig. 2.17a, W_{cycle} = 844 kJ and Q_{out} = 1,900 kJ. What is the thermal efficiency?

2.60 A power cycle receives energy by heat transfer from the combustion of fuel and develops power at a net rate of 150 MW. The thermal efficiency of the cycle is 40%.

    a. Determine the net rate at which the cycle receives energy by heat transfer, in MW.
    b. For 8,000 hours of operation annually, determine the net work output, in kW · h per year.
    c. Evaluating the net work output at $0.08 per kW · h determine the value of net work, in $ per year.

2.61 In what ways do automobile engines operate analogously to the power cycle shown in Fig. 2.17a? How are they different? Discuss.

2.62 Shown in Fig. P2.62 is a *cogeneration* power plant operating in a thermodynamic cycle at steady state. The plant provides electricity to a community at a rate of 80 MW. The energy discharged from the power plant by heat transfer is denoted on the figure by $\dot{Q}_{out}$. Of this, 70 MW is provided to the community for water heating and the remainder is discarded to the environment without use. The electricity is valued at $0.08 per kW · h. If the cycle thermal efficiency is 40%, determine the (a) rate energy is added by heat transfer, $\dot{Q}_{in}$, in MW, (b) rate energy is discarded to the environment, in MW, and (c) value of the electricity generated, in $ per year.

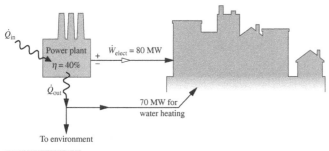

FIGURE P2.62

2.63 Figure P2.63 shows two power cycles, A and B, operating in series, with the energy transfer by heat *into* cycle B equal in magnitude to the energy transfer by heat *from* cycle A. All energy transfers are positive in the directions of the arrows. Determine an expression for the thermal efficiency of an *overall* cycle consisting of cycles A and B together in terms of their individual thermal efficiencies.

2.64 A refrigeration cycle operating as shown in Fig. 2.17b has heat transfer Q_{out} = 2,530 kJ and net work of W_{cycle} = 844 kJ. Determine the coefficient of performance for the cycle.

2.65 A refrigeration cycle operating as shown in Fig. 2.17b has a coefficient of performance β = 1.8. For the cycle, Q_{out} = 250 kJ. Determine Q_{in} and W_{cycle}, each in kJ.

2.66 The refrigerator shown in Fig. P2.66 steadily receives a power input of 0.15 kW while rejecting energy by heat transfer to the

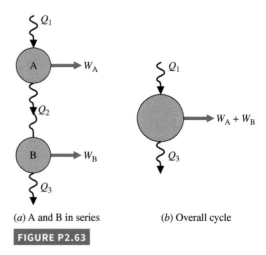

(a) A and B in series   (b) Overall cycle

FIGURE P2.63

surroundings at a rate of 0.6 kW. Determine the rate at which energy is removed by heat transfer from the refrigerated space, in kW, and the refrigerator's coefficient of performance.

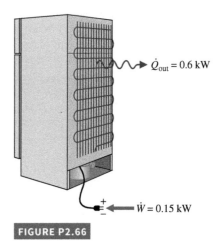

FIGURE P2.66

2.67 A household refrigerator operating steadily and with a coefficient of performance of 2.4 removes energy from a refrigerated space at a rate of 633 kJ/h. Evaluating electricity at $0.08 per kW · h, determine the cost of electricity in a month when the refrigerator operates for 360 hours.

2.68 A window-mounted room air conditioner removes energy by heat transfer from a room and rejects energy by heat transfer to the outside air. For steady-state operation, the air conditioner cycle requires a power input of 0.434 kW and has a coefficient of performance of 6.22. Determine the rate that energy is removed from the room air, in kW. If electricity is valued at $0.1/kW · h, determine the cost of operation for 24 hours of operation.

2.69 A heat pump cycle delivers energy by heat transfer to a dwelling at a rate of 63,300 kJ/h. The power input to the cycle is 5.82 kW.

    a. Determine the coefficient of performance of the cycle.
    b. Evaluating electricity at $0.08 per kW · h, determine the cost of electricity in a month when the heat pump operates for 200 hours.

2.70 A heat pump cycle delivers energy by heat transfer to a dwelling at a rate of 11.7 kW. The coefficient of performance of the cycle is 2.8.

    a. Determine the power input to the cycle, in kW.
    b. Evaluating electricity at $0.10 per kW · h, determine the cost of electricity during the heating season when the heat pump operates for 1800 hours.

DESIGN & OPEN-ENDED PROBLEMS: EXPLORING ENGINEERING PRACTICE

2.1D Design a go-anywhere, use-anywhere wind screen for outdoor recreational and casual-living activities, including sunbathing, reading, cooking, and picnicking. The wind screen must be lightweight, portable, easy to deploy, and low cost. A key constraint is that the wind screen can be set up anywhere, including hard surfaces such as parking lots for tailgating, wood decks, brick and concrete patios, and at the beach. A cost analysis should accompany the design.

2.2D In living things, energy is stored in the molecule *adenosine triphosphate,* called ATP for short. ATP is said to *act like a battery,* storing energy when it is not needed and instantly releasing energy when it is required. Investigate how energy is stored and the role of ATP in biological processes. Write a report including at least three references.

2.3D Visit a local appliance store and collect data on energy requirements for different models within various classes of appliances, including but not limited to refrigerators with and without ice makers, dishwashers, and clothes washers and driers. Prepare a memorandum ranking the different models in each class on an energy-use basis together with an accompanying discussion considering retail cost and other pertinent issues.

2.4D For a 2-week period, keep a diary of your calorie intake for food and drink and the calories *burned* due to your full range of activities over the period, each in kcal. Interpret your results using concepts introduced in this chapter. In a memorandum, summarize results and interpretations.

2.5D The global reach of the Internet supports a rapid increase in consumer and business *e-commerce.* Some say e-commerce will result in net reductions in both energy use and global climate change. Using the Internet, interviews with experts, and design-group *brainstorming*, identify several major ways e-commerce can lead to such reductions. Report your findings in a memorandum having at least three references.

2.6D Develop a list of the most common home-heating options in your locale. For a 225 m² dwelling, what is the annual fuel cost or electricity cost of each option? Also, what is the installed cost of each option? For a 15-year life, which option is the most economical?

2.7D Using data from your state utility regulatory body, determine the breakdown of sources of energy for electric generation. What fraction of your state's needs is met by renewable resources such as wind, geothermal, hydroelectric, and solar energy? Present your findings in a report that summarizes current electric power sources in your state and projections in place to meet needs within the next 10 years.

2.8D Fossil-fuel power plants produce most of the electricity generated annually in the United States. The cost of electricity is determined by several factors, including the power plant thermal efficiency, the unit cost of the fuel, in $ per kW · h, and the plant capital cost, in $ per kW of power generated. Prepare a memorandum comparing typical ranges of these three factors for coal-fired steam power plants and natural gas-fired gas turbine power plants. Which type of plant is most prevalent in the United States?

2.9D Battery disposal presents significant concerns for the environment (see *Energy and Environment*, Sec. 2.7). Research the current federal regulations and those in your state and local area that govern the collection and management of used batteries. Prepare a PowerPoint presentation that summarizes the regulations and the programs and services in place to assist consumers in complying with those regulations. Present data on the effectiveness of those efforts in your area based on compliance and environmental benefits.

2.10D An advertisement describes a portable heater claimed to cut home-heating bills by up to 50%. The heater is said to be able to heat large rooms in minutes without having a high outer-surface temperature, reducing humidity and oxygen levels, or producing carbon monoxide.

A typical deployment is shown in **Fig. P2.10D**. The heater is an enclosure containing electrically powered quartz infrared lamps that shine on copper tubes. Air drawn into the enclosure by a fan flows over the tubes and then is directed back into the living space. Heaters requiring 500 watts of power cost about $400 while the 1,000-W model costs about $500. Critically evaluate the technical and economic merit of such heaters. Write a report including at least three references.

FIGURE P2.10D

2.11D An inventor proposes *borrowing* water from municipal water mains and storing it *temporarily* in a tank on the premises of a dwelling equipped with a heat pump. As shown in **Fig. P2.11D**, the stored water serves as the cold body for the heat pump and the dwelling itself serves as the hot body. To maintain the cold body temperature within a proper operating range, water is drawn from the mains periodically and an equal amount of water is returned to the mains. As the invention requires no *net* water from the mains, the inventor maintains that nothing should be paid for water usage. The inventor also maintains that this approach not only gives a coefficient of performance superior to those of *air-source* heat pumps but also avoids the installation costs associated with *ground-source* heat pumps. In all, significant cost savings result, the inventor says. Critically evaluate the inventor's claims. Write a report including at least three references.

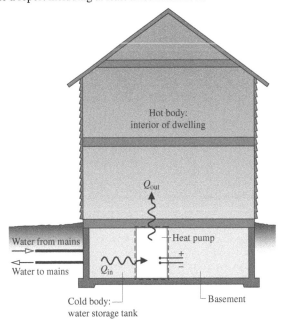

FIGURE P2.11D

CHAPTER 3

Evaluating Properties

Engineering Context

To apply the energy balance to a system of interest requires knowledge of the properties of the system and how the properties are related. The **objectives** of this chapter are to introduce property relations relevant to engineering thermodynamics and provide several examples illustrating the use of the closed system energy balance together with the property relations considered in this chapter.

LEARNING OUTCOMES

When you complete your study of this chapter, you will be able to...

- Explain key concepts, including phase and pure substance, state principle for simple compressible systems, p–v–T surface, saturation temperature and saturation pressure, two-phase liquid–vapor mixture, quality, enthalpy, and specific heats.
- Analyze closed systems, including applying the energy balance with property data.
- Sketch T–v, p–v, and phase diagrams, and locate states on these diagrams.
- Retrieve property data from Tables A-1 through A-23.
- Apply the ideal gas model for thermodynamic analysis, including determining when use of the model is warranted.

3.1 Getting Started

In this section, we introduce concepts that support our study of property relations, including phase, pure substance, and the state principle for simple systems.

3.1.1 Phase and Pure Substance

phase

The term **phase** refers to a quantity of matter that is homogeneous throughout in both chemical composition and physical structure. Homogeneity in physical structure means that the matter is all *solid,* or all *liquid,* or all *vapor* (or equivalently all *gas*). A system can contain one or more phases.

Phase and Pure Substance Tabs a and b

> **FOR EXAMPLE**
>
> A system of liquid water and water vapor (steam) contains *two* phases. A system of liquid water and ice, including the case of *slush,* also contains *two* phases. Gases, oxygen and nitrogen for instance, can be mixed in any proportion to form a *single* gas phase. Certain liquids, such as alcohol and water, can be mixed to form a *single* liquid phase. But liquids such as oil and water, which are not miscible, form *two* liquid phases.

Two phases coexist during the *changes in phase* called *vaporization, melting,* and *sublimation.*

pure substance

A **pure substance** is one that is uniform and invariable in chemical composition. A pure substance can exist in more than one phase, but its chemical composition must be the same in each phase.

> **FOR EXAMPLE**
>
> If liquid water and water vapor form a system with two phases, the system can be regarded as a pure substance because each phase has the same composition. A uniform mixture of gases can be regarded as a pure substance provided it remains a gas and doesn't react chemically. Air can be regarded as a pure substance as long as it is a mixture of gases, but if a liquid phase should form on cooling, the liquid would have a different composition than the gas phase, and the system would no longer be considered a pure substance.

Changes in composition due to chemical reaction are considered in Chap. 13.

3.1.2 Fixing the State

TAKE NOTE...

Temperature *T*, pressure *p*, specific volume v, specific internal energy *u*, and specific enthalpy *h* are all *intensive* properties. See Secs. 1.3.3, 1.5–1.7, and 3.6.1.

The *intensive* state of a closed system *at equilibrium* is its condition as described by the values of its intensive thermodynamic properties. From observation of many thermodynamic systems, we know that not all of these properties are independent of one another, and the state can be uniquely determined by giving the values of a subset of *the independent* intensive properties. Values for all other intensive thermodynamic properties are determined once this independent subset is specified. A general rule known as the **state principle** has been developed as a guide in determining the number of independent properties required to fix the state of a system.

state principle

For the applications considered in this book, we are interested in what the state principle says about the intensive states of systems of commonly encountered pure substances, such as water or a uniform mixture of nonreacting gases. These systems are called **simple compressible systems**. Experience shows simple compressible systems occur in a wide range of engineering applications. For such systems, the state principle indicates that specification of the values for *any two independent* intensive thermodynamic properties will fix the values of all other intensive thermodynamic properties.

simple compressible systems

FOR EXAMPLE

In the case of a gas, temperature and another intensive property such as a specific volume might be selected as the two independent properties. The state principle then affirms that pressure, specific internal energy, and all other pertinent *intensive* properties are functions of T and v: $p = p(T, v)$, $u = u(T, v)$, and so on. The functional relations would be developed using experimental data and would depend explicitly on the particular chemical identity of the substances making up the system. The development of such functions is discussed in Chap. 11.

Intensive properties such as velocity and elevation that are assigned values relative to datums *outside* the system are excluded from present considerations. Also, as suggested by the name, changes in volume can have a significant influence on the energy of *simple compressible systems*. The only mode of energy transfer by work that can occur as a simple compressible system undergoes *quasiequilibrium* processes (Sec. 2.2.5) is associated with volume change and is given by $\int p\, dV$. For further discussion of simple systems and the state principle, see the box.

TAKE NOTE...

For a *simple compressible system*, specification of the values for *any two independent* intensive thermodynamic properties will fix the values of all other intensive thermodynamic properties.

State Principle for Simple Systems

Based on empirical evidence, there is one independent property for each way a system's energy can be varied independently. We saw in Chap. 2 that the energy of a closed system can be altered independently by heat or by work. Accordingly, an independent property can be associated with heat transfer as one way of varying the energy, and another independent property can be counted for each relevant way the energy can be changed through work. On the basis of experimental evidence, therefore, the *state principle* asserts that the number of independent properties is one plus the number of *relevant work* interactions. When counting the number of relevant work interactions, only those that would be significant in *quasiequilibrium* processes of the system need to be considered.

The term *simple system* is applied when there is only *one* way the system energy can be significantly altered by work as the system undergoes quasiequilibrium processes. Therefore, counting one independent property for heat transfer and another for the single work mode gives a total of two independent properties needed to fix the state of a simple system. *This is the state principle for simple systems.* Although no system is ever truly simple, many systems can be modeled as simple systems for the purpose of thermodynamic analysis. The most important of these models for the applications considered in this book is the *simple compressible system*. Other types of simple systems are simple *elastic* systems and simple *magnetic* systems.

Evaluating Properties: General Considerations

The first part of the chapter is concerned generally with the thermodynamic properties of simple compressible systems consisting of *pure* substances. A pure substance is one of uniform and invariable chemical composition. In the second part of this chapter, we consider property evaluation for a special case: the *ideal gas model.* Property relations for systems in which composition changes by chemical reaction are considered in Chap. 13.

3.2 *p–v–T* Relation

We begin our study of the properties of pure, simple compressible substances and the relations among these properties with pressure, specific volume, and temperature. From experiment it is known that temperature and specific volume can be regarded as independent and pressure determined as a function of these two: $p = p(T, v)$. The graph of such a function is a *surface,* the *p–v–T* surface.

p–v–T surface

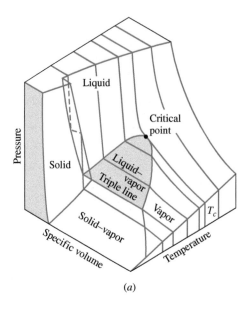

(a)

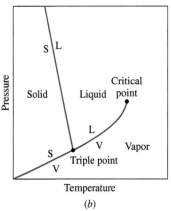

(b)

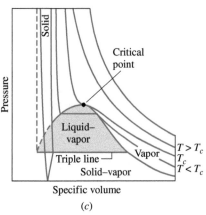

(c)

Fig. 3.1 p–υ–T surface and projections for a substance that expands on freezing. (a) Three-dimensional view. (b) Phase diagram. (c) p–υ diagram.

3.2.1 p–υ–T Surface

Figure 3.1 is the p–υ–T surface of a substance such as water that expands on freezing. **Figure 3.2** is for a substance that contracts on freezing, and most substances exhibit this characteristic. The coordinates of a point on the p–υ–T surfaces represent the values that pressure, specific volume, and temperature would assume when the substance is at equilibrium.

There are regions on the p–υ–T surfaces of Figs. 3.1 and 3.2 labeled *solid, liquid,* and *vapor*. In these *single-phase* regions, the state is fixed by *any* two of the properties: pressure, specific volume, and temperature, since all of these are independent when there is a single phase present. Located between the single-phase regions are **two-phase regions** where two phases exist in equilibrium: liquid–vapor, solid–liquid, and solid–vapor. Two phases can coexist during changes in phase such as vaporization, melting, and sublimation. Within the two-phase regions pressure and temperature are not independent; one cannot be changed without changing the other. In these regions the state cannot be fixed by temperature and pressure alone; however, the state can be fixed by specific volume and either pressure or temperature. Three phases can exist in equilibrium along the line labeled **triple line**.

A state at which a phase change begins or ends is called a **saturation state**. The dome-shaped region composed of the two-phase liquid–vapor states is called the **vapor dome**. The lines bordering the vapor dome are called saturated liquid and saturated vapor lines. At the top of the dome, where the saturated liquid and saturated vapor lines meet, is the **critical point**.

two-phase regions

triple line
saturation state
vapor dome
critical point

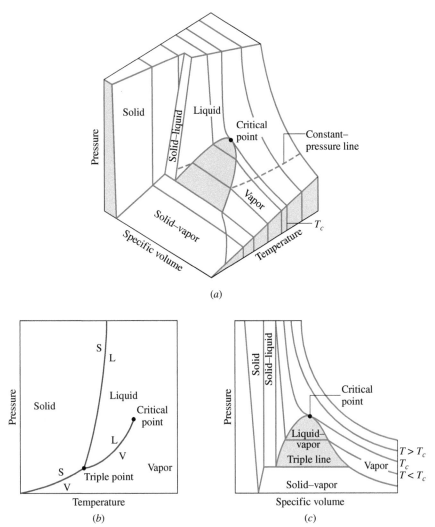

Fig. 3.2 p–v–T surface and projections for a substance that contracts on freezing.
(*a*) Three-dimensional view. (*b*) Phase diagram. (*c*) p–v diagram.

The *critical temperature T_c* of a pure substance is the maximum temperature at which liquid and vapor phases can coexist in equilibrium. The pressure at the critical point is called the *critical pressure, p_c*. The specific volume at this state is the *critical specific volume*. Values of the critical point properties for a number of substances are given in Table A-1 located in the Appendix.

The three-dimensional p–v–T surface is useful for bringing out the general relationships among the three phases of matter normally under consideration. However, it is often more convenient to work with two-dimensional projections of the surface. These projections are considered next.

3.2.2 Projections of the p–v–T Surface

The Phase Diagram   If the p–v–T surface is projected onto the pressure–temperature plane, a property diagram known as a **phase diagram** results. As illustrated by Figs. 3.1*b* and 3.2*b*, when the surface is projected in this way, the two-phase *regions* reduce to *lines*. A point on any of these lines represents all two-phase mixtures at that particular temperature and pressure.

The term **saturation temperature** designates the temperature at which a phase change takes place at a given pressure, and this pressure is called the **saturation pressure** for the given temperature. It is apparent from the phase diagrams that for each saturation pressure there is a unique saturation temperature, and conversely.

phase diagram

saturation temperature

saturation pressure

triple point

The triple *line* of the three-dimensional $p–v–T$ surface projects onto a *point* on the phase diagram. This is called the **triple point**. Recall that the triple point of water is used as a reference in defining temperature scales (Sec. 1.7.3). By agreement, the temperature *assigned* to the triple point of water is 273.16 K. The *measured* pressure at the triple point of water is 0.6113 kPa.

The line representing the two-phase solid–liquid region on the phase diagram slopes to the left for substances that expand on freezing and to the right for those that contract. Although a single solid phase region is shown on the phase diagrams of Figs. 3.1 and 3.2, solids can exist in different solid phases. For example, seventeen different crystalline forms have been identified for water as a solid (ice).

Phases of Solids

In addition to the three phases of solid, liquid, and vapor, it is also possible that distinct phases can exist *within* solids or liquids. One example is from metallurgy, where scientists and engineers study *crystalline* structures within solids. Also, diamond and graphite are two forms of carbon called *allotropes*. Iron and its alloys can exist in several solid phases, and these structures are often formed purposefully to provide desired strength and reliability characteristics by metallurgists. Even ice can exist in as many as 17 different crystalline phases. Although solids are not a focus of this text, the thermodynamics of solids plays important roles in many engineering applications.

p–v diagram

$p–v$ Diagram   Projecting the $p–v–T$ surface onto the pressure–specific volume plane results in a $p–v$ **diagram**, as shown by Figs. 3.1c and 3.2c. The figures are labeled with terms that have already been introduced.

When solving problems, a sketch of the $p–v$ diagram is frequently convenient. To facilitate the use of such a sketch, note the appearance of constant-temperature lines (isotherms). By inspection of Figs. 3.1c and 3.2c, it can be seen that for any specified temperature *less than* the critical temperature, pressure remains constant as the two-phase liquid–vapor region is traversed, but in the single-phase liquid and vapor regions the pressure decreases at fixed temperature as specific volume increases. For temperatures greater than or equal to the critical temperature, pressure decreases continuously at fixed temperature as specific volume increases. There is no passage across the two-phase liquid–vapor region. The critical isotherm passes through a point of inflection at the critical point and the slope is zero there.

T–v diagram

$T–v$ Diagram   Projecting the liquid, two-phase liquid–vapor, and vapor regions of the $p–v–T$ surface onto the temperature–specific volume plane results in a $T–v$ **diagram** as in Fig. 3.3. Since consistent patterns are revealed in the $p–v–T$ behavior of all pure substances, Fig. 3.3 showing a $T–v$ diagram for water can be regarded as representative.

As for the $p–v$ diagram, a sketch of the $T–v$ diagram is often convenient for problem solving. To facilitate the use of such a sketch, note the appearance of constant-pressure lines (isobars). For pressures *less than* the critical pressure, such as the 10 MPa isobar on Fig. 3.3,

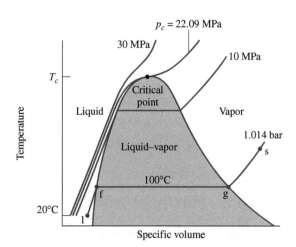

Fig. 3.3 Sketch of a temperature–specific volume diagram for water showing the liquid, two-phase liquid–vapor, and vapor regions (not to scale).

the pressure remains constant with temperature as the two-phase region is traversed. In the single-phase liquid and vapor regions the temperature increases at fixed pressure as the specific volume increases. For pressures greater than or equal to the critical pressure, such as the one marked 30 MPa on Fig. 3.3, temperature increases continuously at fixed pressure as the specific volume increases. There is no passage across the two-phase liquid–vapor region.

The projections of the p–v–T surface used in this book to illustrate processes are not generally drawn to scale. A similar comment applies to other property diagrams introduced later.

3.3 Studying Phase Change

It is instructive to study the events that occur as a pure substance undergoes a phase change. To begin, consider a closed system consisting of a unit mass (1 kg) of liquid water at 20°C contained within a piston–cylinder assembly, as illustrated in Fig. 3.4a. This state is represented by point 1 on Fig. 3.3. Suppose the water is slowly heated while its pressure is kept constant and uniform throughout at 1.014 bar.

Liquid States   As the system is heated at constant pressure, the temperature increases considerably while the specific volume increases slightly. Eventually, the system is brought to the state represented by f on Fig. 3.3. This is the saturated liquid state corresponding to the specified pressure. For water at 1.014 bar the saturation temperature is 100°C. The liquid states along the line segment 1–f of Fig. 3.3 are sometimes referred to as **subcooled liquid** states because the temperature at these states is less than the saturation temperature at the given pressure. These states are also referred to as **compressed liquid** states because the pressure at each state is higher than the saturation pressure corresponding to the temperature at the state. The names *liquid*, *subcooled liquid*, and *compressed liquid* are used interchangeably.

subcooled liquid

compressed liquid

Two-Phase Liquid–Vapor Mixture   When the system is at the saturated liquid state (state f of Fig. 3.3), additional heat transfer at fixed pressure results in the formation of vapor without any change in temperature but with a considerable increase in specific volume. As shown in Fig. 3.4b, the system would now consist of a two-phase liquid–vapor mixture. When a mixture of liquid and vapor exists in equilibrium, the liquid phase is a saturated liquid and the vapor phase is a saturated vapor. If the system is heated further until the last bit of liquid has vaporized, it is brought to point g on Fig. 3.3, the saturated vapor state. The intervening **two-phase liquid–vapor mixture** states can be distinguished from one another by the *quality,* an intensive property.

two-phase liquid–vapor mixture

For a two-phase liquid–vapor mixture, the ratio of the mass of vapor present to the total mass of the mixture is its **quality**, x. In symbols,

quality

$$x = \frac{m_{vapor}}{m_{liquid} + m_{vapor}} \qquad (3.1)$$

Fig. 3.4 Illustration of constant-pressure change from liquid to vapor for water.

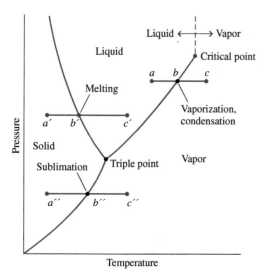

Fig. 3.5 Phase diagram for water (not to scale).

The value of the quality ranges from zero to unity: At saturated liquid states, $x = 0$, and at saturated vapor states, $x = 1.0$. Although defined as a ratio, the quality is frequently given as a percentage. Examples illustrating the use of quality are provided in Sec. 3.5. Similar parameters can be defined for two-phase solid–vapor and two-phase solid–liquid mixtures.

Vapor States   Let us return to Figs. 3.3 and 3.4. When the system is at the saturated vapor state (state g on Fig. 3.3), further heating at fixed pressure results in increases in both temperature and specific volume. The condition of the system would now be as shown in Fig. 3.4c. The state labeled s on Fig. 3.3 is representative of the states that would be attained by further heating while keeping the pressure constant. A state such as s is often referred to as a **superheated vapor** state because the system would be at a temperature greater than the saturation temperature corresponding to the given pressure.

Consider next the same thought experiment at the other constant pressures labeled on Fig. 3.3, 10 MPa, 22.09 MPa, and 30 MPa. The first of these pressures is less than the critical pressure of water, the second is the critical pressure, and the third is greater than the critical pressure. As before, let the system initially contain a liquid at 20°C. First, let us study the system if it were heated slowly at 10 MPa. At this pressure, vapor would form at a higher temperature than in the previous example because the saturation pressure is higher (refer to Fig. 3.3). In addition, there would be somewhat less of an increase in specific volume from saturated liquid to saturated vapor, as evidenced by the narrowing of the vapor dome. Apart from this, the general behavior would be the same as before.

superheated vapor

Consider next the behavior of the system if it were heated at the critical pressure or higher. As seen by following the critical isobar on Fig. 3.3, there would be no change in phase from liquid to vapor. At all states there would be only one phase. As shown by line *a–b–c* of the phase diagram sketched in Fig. 3.5, *vaporization* and the inverse process of *condensation* can occur only when the pressure is less than the critical pressure. Thus, at states where pressure is greater than the critical pressure, the terms *liquid* and *vapor* tend to lose their significance. Still, for ease of reference to such states, we use the term *liquid* when the temperature is less than the critical temperature and *vapor* when the temperature is greater than the critical temperature. This convention is labeled on Fig. 3.5.

Liquid to Vapor
Vapor to Liquid

While condensation of water vapor to liquid and further cooling to lower-temperature liquid are easily imagined and even a part of our everyday experience, liquefying gases other than water vapor may not be so familiar. Still, there are important applications for liquefied gases. See the box for applications of nitrogen in liquid *and* gas forms.

Nitrogen, Unsung Workhorse

Nitrogen is obtained using commercial air-separation technology that extracts oxygen and nitrogen from air. While applications for oxygen are widely recognized, uses for nitrogen tend to be less heralded but still touch on things people use every day.

Liquid nitrogen is used to fast-freeze foods. Tunnel freezers employ a conveyer belt to pass food through a liquid-nitrogen spray, while batch freezers immerse food in a liquid-nitrogen bath. Each freezer type operates at temperatures less than about −185°C. Liquid nitrogen is also used to preserve specimens employed in medical research and by dermatologists to remove lesions.

As a gas, nitrogen, with other gases, is inserted into food packaging to replace oxygen-bearing air, thereby prolonging shelf life—examples include gas-inflated bags of potato chips, salad greens, and shredded cheese. For improved tire performance, nitrogen is used to inflate the tires of airplanes and race cars. Nitrogen is among several alternative substances injected into underground rock formations to stimulate flow of trapped oil and natural gas to the surface—a procedure known as hydraulic fracturing. Chemical plants and refineries use nitrogen gas as a blanketing agent to prevent explosion. Laser-cutting machines also use nitrogen and other specialty gases.

Melting and Sublimation Although the phase changes from liquid to vapor (vaporization) and vapor to liquid (condensation) are of principal interest in this book, it is also instructive to consider the phase changes from solid to liquid (melting) and from solid to vapor (sublimation). To study these transitions, consider a system consisting of a unit mass of ice at a temperature below the triple point temperature. Let us begin with the case where the pressure is greater than the triple point pressure and the system is at state a' of Fig. 3.5. Suppose the system is slowly heated while maintaining the pressure constant and uniform throughout. The temperature increases with heating until point b' on Fig. 3.5 is attained. At this state the ice is a saturated solid. Additional heat transfer at fixed pressure results in the formation of liquid without any change in temperature. As the system is heated further, the ice continues to melt until eventually the last bit melts, and the system contains only saturated liquid. During the melting process the temperature and pressure remain constant. For most substances, the specific volume increases during melting, but for water the specific volume of the liquid is less than the specific volume of the solid. Further heating at fixed pressure results in an increase in temperature as the system is brought to point c' on Fig. 3.5. Next, consider the case where the pressure is less than the triple point pressure and the system is at state a'' of Fig. 3.5. In this case, if the system is heated at constant pressure it passes through the two-phase solid–vapor region into the vapor region along the line a''–b''–c'' shown on Fig. 3.5. That is, sublimation occurs.

BioConnections

As discussed in the box devoted to nitrogen in this section, nitrogen has many applications, including medical applications. One medical application is the practice of *cryosurgery* by dermatologists. Cryosurgery entails the localized freezing of skin tissue for the removal of unwanted lesions, including precancerous lesions. For this type of surgery, liquid nitrogen is applied as a spray or with a probe. Cryosurgery is quickly performed and generally without anesthetic. Dermatologists store liquid nitrogen required for up to several months in containers called *Dewar* flasks that are similar to vacuum bottles.

3.4 Retrieving Thermodynamic Properties

Thermodynamic property data can be retrieved in various ways, including tables, graphs, equations, and computer software. The emphasis of Secs. 3.5 and 3.6 to follow is on the use of *tables* of thermodynamic properties, which are commonly available for pure, simple compressible substances of engineering interest. The use of these tables is an important skill. The ability to locate states on property diagrams is an important related skill. The software available with this text, *Interactive Thermodynamics: IT,* is introduced in Sec. 3.7. *IT* is used selectively in examples and end-of-chapter problems throughout the book. Skillful use of tables and property diagrams is prerequisite for the effective use of software to retrieve thermodynamic property data.

Since tables for different substances are frequently set up in the same general format, the present discussion centers mainly on Tables A-2 through A-6 giving the properties of water; these are commonly referred to as the **steam tables**. Tables A-7 through A-9 for Refrigerant 22, Tables A-10 through A-12 for Refrigerant 134a, Tables A-13 through A-15 for ammonia, and Tables A-16 through A-18 for propane are used similarly, as are tables for other substances found in the engineering literature. Tables are provided in the Appendix in SI units.

steam tables

The substances for which tabulated data are provided in this book have been selected because of their importance in current practice. Still, they are merely representative of a wide range of industrially important substances. To meet changing requirements and address special needs, new substances are frequently introduced while others become obsolete.

Development in the twentieth century of chlorine-containing refrigerants such as Refrigerant 12 helped pave the way for the refrigerators and air conditioners we enjoy today. Still, concern over the effect of chlorine on Earth's protective ozone layer led to international agreements to phase out these refrigerants. Substitutes for them also have come under criticism as being environmentally harmful. Accordingly, a search is on for alternatives, and *natural refrigerants* are getting a close look. Natural refrigerants include ammonia, certain hydrocarbons—propane, for example—carbon dioxide, water, and air.

Ammonia, once widely used as a refrigerant for domestic applications but dropped owing to its toxicity, is receiving renewed interest because it is both effective as a refrigerant and chlorine free. Refrigerators using propane are available on the global market despite lingering worries over flammability. Carbon dioxide is well suited for small, lightweight systems such as automotive and portable air-conditioning units. Although CO_2 released to the environment contributes to global climate change, only a tiny amount is present in a typical unit, and ideally even this would be contained under proper maintenance and refrigeration unit disposal protocols.

3.5 Evaluating Pressure, Specific Volume, and Temperature

3.5.1 Vapor and Liquid Tables

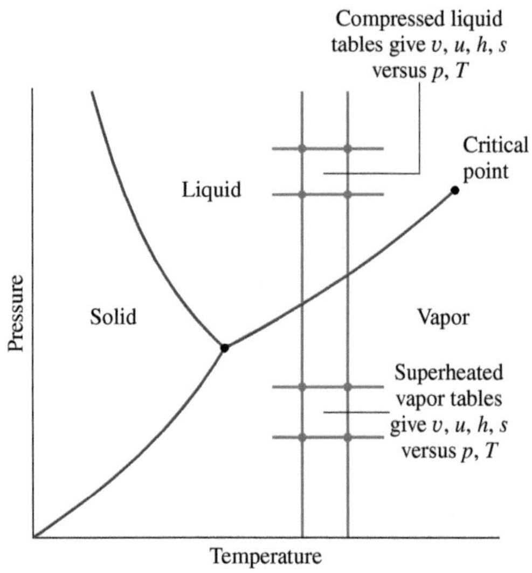

The properties of water vapor are listed in Table A-4 and of liquid water in Table A-5. These are often referred to as the *superheated* vapor tables and *compressed* liquid tables, respectively. The sketch of the phase diagram shown in **Fig. 3.6** brings out the structure of these tables. Since pressure and temperature are independent properties in the single-phase liquid and vapor regions, they can be used to fix the state in these regions. Accordingly, Tables A-4 and A-5 are set up to give values of several properties as functions of pressure and temperature. The first property listed is specific volume. The remaining properties are discussed in subsequent sections.

For each pressure listed, the values given in the superheated vapor table (Table A-4) *begin* with the saturated vapor state and then proceed to higher temperatures. The data in the compressed liquid table (Table A-5) *end* with saturated liquid states. That is, for a given pressure the property values are given as the temperature increases to the saturation temperature. In these tables, the value shown in parentheses after the pressure in the table heading is the corresponding saturation temperature.

Compressed liquid tables give v, u, h, s versus p, T

Critical point

Liquid

Solid

Vapor

Superheated vapor tables give v, u, h, s versus p, T

Pressure

Temperature

Fig. 3.6 Sketch of the phase diagram for water used to discuss the structure of the superheated vapor and compressed liquid tables (not to scale).

FOR EXAMPLE

In Tables A-4 and A-5, at a pressure of 10.0 MPa, the saturation temperature is listed as 311.06°C.

FOR EXAMPLE

To gain more experience with Tables A-4 and A-5 verify the following: Table A-4 gives the specific volume of water vapor at 10.0 MPa and 600°C as 0.03837 m^3/kg. At 10.0 MPa and 100°C, Table A-5 gives the specific volume of liquid water as 1.0385×10^{-3} m^3/kg.

The states encountered when solving problems often do not fall exactly on the grid of values provided by property tables. *Interpolation* between adjacent table entries then becomes necessary. Care always must be exercised when interpolating table values. The tables provided

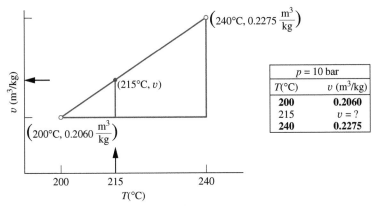

Fig. 3.7 Illustration of linear interpolation.

in the Appendix are extracted from more extensive tables that are set up so that **linear inter-** **linear interpolation**
polation, illustrated in the following example, can be used with acceptable accuracy. Linear
interpolation is assumed to remain valid when using the abridged tables of the text for the
solved examples and end-of-chapter problems.

FOR EXAMPLE

Let us determine the specific volume of water vapor at a state where $p = 10$ bar and $T = 215°C$.
Shown in **Fig. 3.7** is a sampling of data from Table A-4. At a pressure of 10 bar, the specified
temperature of 215°C falls between the table values of 200 and 240°C, which are shown in
boldface. The corresponding specific volume values are also shown in boldface. To determine
the specific volume v corresponding to 215°C, we think of the *slope* of a straight line joining
the adjacent table entries, as follows

$$slope = \frac{(0.2275 - 0.2060)\ m^3/kg}{(240 - 200)°C} = \frac{(v - 0.2060)\ m^3/kg}{(215 - 200)°C}$$

Solving for v, the result is $v = 0.2141\ m^3/kg$.

The following example features the use of sketches of p–v and T–v diagrams in conjunction with tabular data to fix the end states of a process. In accord with the state principle,
two independent intensive properties must be known to fix the states of the system under
consideration.

► EXAMPLE 3.1 ► . . . . ▪▪

Heating Ammonia at Constant Pressure

A vertical piston–cylinder assembly containing 0.05 kg of ammonia, initially a saturated vapor, is placed on a hot plate. Due to the
weight of the piston and the surrounding atmospheric pressure, the
pressure of the ammonia is 1.5 bars. Heating occurs slowly, and the
ammonia expands at constant pressure until the final temperature is
25°C. Show the initial and final states on T–v and p–v diagrams, and
determine

  a. the volume occupied by the ammonia at each end state, in m^3.

  b. the work for the process. in kJ.

SOLUTION

Known Ammonia is heated at constant pressure in a vertical piston–
cylinder assembly from the saturated vapor state to a known final
temperature.

Find Show the initial and final states on T–v and p–v diagrams,
and determine the volume at each end state and the work for the
process.

Schematic and Given Data:

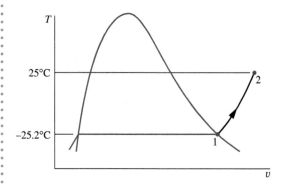

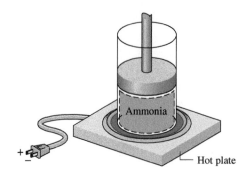

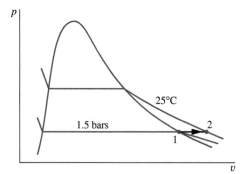

Fig. E3.1

Engineering Model

1. The ammonia is a closed system.

2. States 1 and 2 are equilibrium states.

3. The process occurs at constant pressure.

Analysis   The initial state is a saturated vapor condition at 1.5 bars. Since the process occurs at constant pressure, the final state is in the superheated vapor region and is fixed by 1.5 bars and $T_2 = 25°C$. The initial and final states are shown on the T–v and p–v diagrams above.

a. The volumes occupied by the ammonia at states 1 and 2 are obtained using the given mass and the respective specific volumes. From Table A-14 at $P_2 = 1.5$ bars and corresponding to *Sat.* in the temperature column, we get $v_1 = v_g = 0.7787$ m³/kg. Thus

$$V_1 = mv_1 = (0.05 \text{ kg})(0.7787 \text{ m}^3/\text{kg})$$
$$= 0.0389 \text{ m}^3$$

Interpolating in Table A-14 at $p_2 = 1.5$ bar and $T_2 = 25°C$, we get $v_2 = 0.9553$ m³/kg. Thus

$$V_2 = mv_2 = (0.05 \text{ kg})(0.9553 \text{ m}^3/\text{kg}) = 0.0478 \text{ m}^3$$

b. In this case, the work can be evaluated using Eq. 2.17. Since the pressure is constant

$$W = \int_{V_1}^{V_2} p \, dV = p(V_2 - V_1)$$

Inserting values

❶ $W = (1.5 \text{ bars})(0.0478 - 0.0389)\text{m}^3 \left| \frac{10^5 \text{ N/m}^2}{1 \text{ bar}} \right| \left| \frac{\text{kJ}}{10^3 \text{ N} \cdot \text{m}} \right|$

$= 1.335$ kJ

❶ Note the use of conversion factors in this calculation.

SKILLS DEVELOPED

Ability to...

• define a closed system and identify interactions on its boundary.

• sketch T-v and p-v diagrams and locate states on them.

• evaluate work using Eq. 2.17.

• retrieve property data for ammonia at vapor states.

Quick Quiz

If heating continues at 1.5 bars from $T_2 = 25°C$ to $T_3 = 32°C$, determine the work for Process 2-3, in kJ. Ans. 0.16 kJ

3.5.2 Saturation Tables

Tables A-2, A-3, and A-6 provide property data for water at saturated liquid, saturated vapor, and saturated solid states. Tables A-2 and A-3 are the focus of the present discussion. Each of these tables gives saturated liquid and saturated vapor data. Property values at saturated liquid and saturated vapor states are denoted by the subscripts f and g, respectively. Table A-2 is called the *temperature table* because temperatures are listed in the first column in convenient increments. The second column gives the corresponding saturation pressures. The next two columns give, respectively, the specific volume of saturated liquid, v_f, and the specific volume of saturated vapor, v_g. Table A-3 is called the *pressure table*, because pressures are listed in the first column in convenient increments. The corresponding saturation temperatures are given in the second column. The next two columns give v_f and v_g, respectively.

The specific volume of a two-phase liquid–vapor mixture can be determined by using the saturation tables and the definition of quality given by Eq. 3.1 as follows. The total volume of the mixture is the sum of the volumes of the liquid and vapor phases

$$V = V_{\text{liq}} + V_{\text{vap}}$$

Dividing by the total mass of the mixture, m, an *average* specific volume for the mixture is obtained

$$v = \frac{V}{m} = \frac{V_{\text{liq}}}{m} + \frac{V_{\text{vap}}}{m}$$

Since the liquid phase is a saturated liquid and the vapor phase is a saturated vapor, $V_{\text{liq}} = m_{\text{liq}} v_f$ and $V_{\text{vap}} = m_{\text{vap}} v_g$, so

$$v = \left(\frac{m_{\text{liq}}}{m}\right) v_f + \left(\frac{m_{\text{vap}}}{m}\right) v_g$$

Introducing the definition of quality, $x = m_{\text{vap}}/m$, and noting that $m_{\text{liq}}/m = 1 - x$, the above expression becomes

$$\boxed{v = (1 - x)v_f + xv_g = v_f + x(v_g - v_f)} \qquad (3.2)$$

The increase in specific volume on vaporization $(v_g - v_f)$ is also denoted by v_{fg}.

FOR EXAMPLE

Consider a system consisting of a two-phase liquid–vapor mixture of water at 100°C and a quality of 0.9. From Table A-2 at 100°C, $v_f = 1.0435 \times 10^{-3}$ m³/kg and $v_g = 1.673$ m³/kg. The specific volume of the mixture is

$$v = v_f + x(v_g - v_f) = 1.0435 \times 10^{-3} + (0.9)(1.673 - 1.0435 \times 10^{-3}) = 1.506 \text{ m}^3/\text{kg}$$

where the v_f and v_g values are obtained from Table A-2.

To facilitate locating states in the tables, it is often convenient to use values from the saturation tables together with a sketch of a T–v or p–v diagram. For example, if the specific volume v and temperature T are known, refer to the appropriate temperature table, Table A-2, and determine the values of v_f and v_g. A T–v diagram illustrating these data is given in **Fig. 3.8**. If the

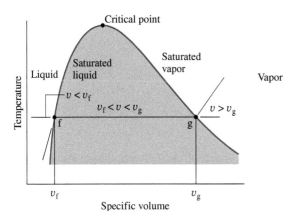

Fig. 3.8 Sketch of a T–v diagram for water used to discuss locating states in the tables.

given specific volume falls between v_f and v_g, the system consists of a two-phase liquid–vapor mixture, and the pressure is the saturation pressure corresponding to the given temperature. The quality can be found by solving Eq. 3.2. If the given specific volume is greater than v_g, the state is in the superheated vapor region. Then, by interpolating in Table A-4, the pressure and other properties listed can be determined. If the given specific volume is less than v_f, Table A-5 would be used to determine the pressure and other properties.

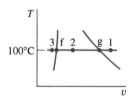

FOR EXAMPLE

Let us determine the pressure of water at each of three states defined by a temperature of 100°C and specific volumes, respectively, of $v_1 = 2.434$ m³/kg, $v_2 = 1.0$ m³/kg, and $v_3 = 1.0423 \times 10^{-3}$ m³/kg. Using the known temperature, Table A-2 provides the values of v_f and v_g: $v_f = 1.0435 \times 10^{-3}$ m³/kg, $v_g = 1.673$ m³/kg. Since v_1 is greater than v_g, state 1 is in the vapor region. Table A-4 gives the pressure as 0.70 bar. Next, since v_2 falls between v_f and v_g, the pressure is the saturation pressure corresponding to 100°C, which is 1.014 bar. Finally, since v_3 is less than v_f, state 3 is in the liquid region. Table A-5 gives the pressure as 25 bar.

Finding States in the Steam Tables: Overview Using p–v Diagram

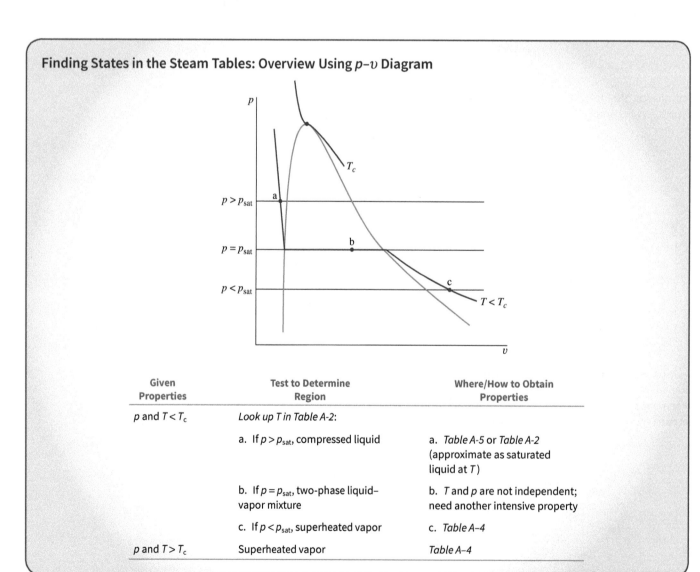

Given Properties	Test to Determine Region	Where/How to Obtain Properties
p and $T < T_c$	Look up T in Table A-2:	
	a. If $p > p_{sat}$, compressed liquid	a. Table A-5 or Table A-2 (approximate as saturated liquid at T)
	b. If $p = p_{sat}$, two-phase liquid–vapor mixture	b. T and p are not independent; need another intensive property
	c. If $p < p_{sat}$, superheated vapor	c. Table A-4
p and $T > T_c$	Superheated vapor	Table A-4

The following example features the use of a sketch of the T–v diagram in conjunction with tabular data to fix the end states of processes. In accord with the state principle, two independent intensive properties must be known to fix the states of the system under consideration.

▸▸▸ EXAMPLE 3.2 ▸ . . . .

Heating Water at Constant Volume

A closed, rigid container of volume 0.5 m³ is placed on a hot plate. Initially, the container holds a two-phase mixture of saturated liquid water and saturated water vapor at $p_1 = 1$ bar with a quality of 0.5. After heating, the pressure in the container is $p_2 = 1.5$ bar. Indicate the initial and final states on a T–v diagram, and determine

a. the temperature, in °C, at states 1 and 2.

b. the mass of vapor present at states 1 and 2, in kg.

c. If heating continues, determine the pressure, in bar, when the container holds only saturated vapor.

SOLUTION

Known   A two-phase liquid–vapor mixture of water in a closed, rigid container is heated on a hot plate. The initial pressure and quality and the final pressure are known.

Find   Indicate the initial and final states on a T–v diagram and determine at each state the temperature and the mass of water vapor present. Also, if heating continues, determine the pressure when the container holds only saturated vapor.

Schematic and Given Data:

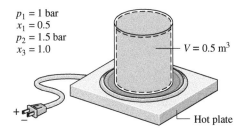

$p_1 = 1$ bar
$x_1 = 0.5$
$p_2 = 1.5$ bar
$x_3 = 1.0$

— $V = 0.5$ m³

— Hot plate

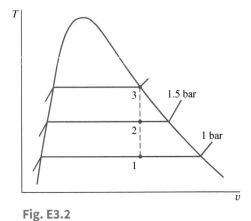

Fig. E3.2

Engineering Model

1. The water in the container is a closed system.

2. States 1, 2, and 3 are equilibrium states.

3. The volume of the container remains constant.

Analysis   Two independent properties are required to fix states 1 and 2. At the initial state, the pressure and quality are known. As these are independent, the state is fixed. State 1 is shown on the T–v diagram in the two-phase region. The specific volume at state 1 is found using the given quality and Eq. 3.2. That is,

$$v_1 = v_{f1} + x_1(v_{g1} - v_{f1})$$

From Table A-3 at $p_1 = 1$ bar, $v_{f1} = 1.0432 \times 10^{-3}$ m³/kg and $v_{g1} = 1.694$ m³/kg. Thus,

$$v_1 = 1.0432 \times 10^{-3} + 0.5(1.694 - 1.0432 \times 10^{-3}) = 0.8475 \text{ m}^3/\text{kg}$$

At state 2, the pressure is known. The other property required to fix the state is the specific volume v_2. Volume and mass are each constant, so $v_2 = v_1 = 0.8475$ m³/kg. For $p_2 = 1.5$ bar, Table A-3 gives $v_{f2} = 1.0582 \times 10^{-3}$ m³/kg and $v_{g2} = 1.59$ m³/kg. Since

①
$$v_f < v_2 < v_{g2}$$

② state 2 must be in the two-phase region as well. State 2 is also shown on the T–v diagram above.

a. Since states 1 and 2 are in the two-phase liquid–vapor region, the temperatures correspond to the saturation temperatures for the given pressures. Table A-3 gives

$$T_1 = 99.63°C \qquad \text{and} \qquad T_2 = 111.4°C$$

b. To find the mass of water vapor present, we first use the volume and the specific volume to find the *total* mass, m. That is,

$$m = \frac{V}{v} = \frac{0.5 \text{ m}^3}{0.8475 \text{ m}^3/\text{kg}} = 0.59 \text{ kg}$$

Then, with Eq. 3.1 and the given value of quality, the mass of vapor at state 1 is

$$m_{g1} = x_1 m = 0.5(0.59 \text{ kg}) = 0.295 \text{ kg}$$

The mass of vapor at state 2 is found similarly using the quality x_2. To determine x_2, solve Eq. 3.2 for quality and insert specific volume data from Table A-3 at a pressure of 1.5 bar, along with the known value of v, as follows

$$x_2 = \frac{v - v_{f2}}{v_{g2} - v_{f2}}$$

$$= \frac{0.8475 - 1.0528 \times 10^{-3}}{1.159 - 1.0528 \times 10^{-3}} = 0.731$$

Then, with Eq. 3.1

$$m_{g2} = 0.731(0.59 \text{ kg}) = 0.431 \text{ kg}$$

c. If heating continued, state 3 would be on the saturated vapor line, as shown on the T–v diagram of Fig. E3.2. Thus, the pressure would be the corresponding saturation pressure. Interpolating in Table A-3 at $v_g = 0.8475$ m³/kg, we get $p_3 = 2.11$ bar.

---

① The procedure for fixing state 2 is the same as illustrated in the discussion of Fig. 3.8.

2 Since the process occurs at constant specific volume, the states lie along a vertical line.

SKILLS DEVELOPED

Ability to...

• define a closed system and identify interactions on its boundary.

• sketch a T–v diagram and locate states on it.
• retrieve property data for water at liquid–vapor states, using quality.

Quick Quiz

If heating continues at constant specific volume from state 3 to a state where pressure is 3 bar, determine the temperature at that state, in °C. Ans. 282°C

3.6 Evaluating Specific Internal Energy and Enthalpy

3.6.1 Introducing Enthalpy

enthalpy

In many thermodynamic analyses the sum of the internal energy U and the product of pressure p and volume V appears. Because the sum $U + pV$ occurs so frequently in subsequent discussions, it is convenient to give the combination a name, **enthalpy**, and a distinct symbol, H. By definition

$$H = U + pV \tag{3.3}$$

Since U, p, and V are all properties, this combination is also a property. Enthalpy can be expressed on a unit mass basis

$$h = u + pv \tag{3.4}$$

and per mole

$$\bar{h} = \bar{u} + p\bar{v} \tag{3.5}$$

Units for enthalpy are the same as those for internal energy.

3.6.2 Retrieving u and h Data

The property tables introduced in Sec. 3.5 giving pressure, specific volume, and temperature also provide values of specific internal energy u, enthalpy h, and entropy s. Use of these tables to evaluate u and h is described in the present section; the consideration of entropy is deferred until it is introduced in Chap. 6.

Data for specific internal energy u and enthalpy h are retrieved from the property tables in the same way as for specific volume. For saturation states, the values of u_f and u_g, as well as h_f and h_g, are tabulated versus both saturation pressure and saturation temperature. The specific internal energy for a two-phase liquid–vapor mixture is calculated for a given quality in the same way the specific volume is calculated:

$$u = (1 - x)u_f + xu_g = u_f + x(u_g - u_f) \tag{3.6}$$

The increase in specific internal energy on vaporization $(u_g - u_f)$ is often denoted by u_{fg}. Similarly, the specific enthalpy for a two-phase liquid–vapor mixture is given in terms of the quality by

$$h = (1 - x)h_f + xh_g = h_f + x(h_g - h_f) \tag{3.7}$$

The increase in enthalpy during vaporization $(h_g - h_f)$ is often tabulated for convenience under the heading h_{fg}.

FOR EXAMPLE

To illustrate the use of Eqs. 3.6 and 3.7, we determine the specific enthalpy of Refrigerant 22 when its temperature is 12°C and its specific internal energy is 144.58 kJ/kg. Referring to Table A-7, the given internal energy value falls between u_f and u_g at 12°C, so the state is a two-phase liquid–vapor mixture. The quality of the mixture is found by using Eq. 3.6 and data from Table A-7 as follows:

$$x = \frac{u - u_f}{u_g - u_f} = \frac{144.58 - 58.77}{230.38 - 58.77} = 0.5$$

Then, with the values from Table A-7, Eq. 3.7 gives

$$h = (1 - x)h_f + xh_g$$
$$= (1 - 0.5)(59.35) + 0.5(253.99) = 156.67 \text{ kJ/kg}$$

In the superheated vapor tables, u and h are tabulated along with v as functions of temperature and pressure.

FOR EXAMPLE

Let us evaluate T, v, and h for water at 0.10 MPa and a specific internal energy of 2537.3 kJ/kg. Turning to Table A-3, note that the given value of u is greater than u_g at 0.1 MPa ($u_g = 2506.1$ kJ/kg). This suggests that the state lies in the superheated vapor region. By inspection of Table A-4 we get $T = 120°C$, $v = 1.793$ m³/kg, and $h = 2716.6$ kJ/kg. Alternatively, h and u are related by the definition of h:

$$h = u + pv$$
$$= 2537.3\frac{\text{kJ}}{\text{kg}} + \left(10^5 \frac{\text{N}}{\text{m}^2}\right)\left(1.793\frac{\text{m}^3}{\text{kg}}\right)\left|\frac{1 \text{ kJ}}{10^3 \text{ N} \cdot \text{m}}\right|$$
$$= 2537.3 + 179.3 = 2716.6 \text{ kJ/kg}$$

Specific internal energy and enthalpy data for liquid states of water are presented in Table A-5. The format of these tables is the same as that of the superheated vapor tables considered previously. Accordingly, property values for liquid states are retrieved in the same manner as those of vapor states.

For water, Table A-6 give the equilibrium properties of saturated solid and saturated vapor. The first column lists the temperature, and the second column gives the corresponding saturation pressure. These states are at pressures and temperatures *below* those at the triple point. The next two columns give the specific volume of saturated solid, v_i, and saturated vapor, v_g, respectively. The table also provides the specific internal energy, enthalpy, and entropy values for the saturated solid and the saturated vapor at each of the temperatures listed.

3.6.3 Reference States and Reference Values

The values of u, h, and s given in the property tables are not obtained by direct measurement but are calculated from other data that can be more readily determined experimentally. The computational procedures require use of the second law of thermodynamics, so consideration of these procedures is deferred to Chap. 11 after the second law has been introduced. However, because u, h, and s are calculated, the matter of **reference states** and **reference values** becomes important and is considered briefly in the following paragraphs.

When applying the energy balance, it is *differences* in internal, kinetic, and potential energy between two states that are important, and *not* the values of these energy quantities at each of the two states.

reference states

reference values

> **FOR EXAMPLE**
>
> Consider the case of potential energy. The numerical value of potential energy determined relative to the surface of Earth is not the same as the value relative to the top of a tall building at the same location. However, the difference in potential energy between any two elevations is precisely the same regardless of the datum selected, because the datum cancels in the calculation.

Similarly, values can be assigned to specific internal energy and enthalpy relative to arbitrary reference values at arbitrary reference states. As for the case of potential energy considered above, the use of values of a particular property determined relative to an arbitrary reference is unambiguous as long as the calculations being performed involve only differences in that property, for then the reference value cancels. When chemical reactions take place among the substances under consideration, special attention must be given to the matter of reference states and values, however. A discussion of how property values are assigned when analyzing reactive systems is given in Chap. 13.

The tabular values of u and h for water, ammonia, propane, and Refrigerants 22 and 134a provided in the Appendix are relative to the following reference states and values. For water, the reference state is saturated liquid at 0.01°C. At this state, the specific internal energy is set to zero. Values of the specific enthalpy are calculated from $h = u + pv$, using the tabulated values for p, v, and u. For ammonia, propane, and the refrigerants, the reference state is saturated liquid at −40°C. At this reference state the specific enthalpy is set to zero. Values of specific internal energy are calculated from $u = h − pv$ by using the tabulated values for p, v, and h. Notice in Table A-7 that this leads to a negative value for internal energy at the reference state, which emphasizes that it is not the numerical values assigned to u and h at a given state that are important but their *differences* between states. The values assigned to particular states change if the reference state or reference values change, but the differences remain the same.

Evaluating Properties Using Computer Software

The use of computer software for evaluating thermodynamic properties is becoming prevalent in engineering. Computer software falls into two general categories: those that provide data only at individual states and those that provide property data as part of a more general simulation package. The software available with this text, *Interactive Thermodynamics: IT,* is a tool that can be used not only for routine problem solving by providing data at individual state points but also for simulation and analysis. Software other than *IT* also can be used for these purposes. See the box for discussion of software use in engineering thermodynamics.

Using Software In Thermodynamics

The computer software tool *Interactive Thermodynamics: IT* is available for use with this text. Used properly, *IT* provides an important adjunct to learning engineering thermodynamics and solving engineering problems. The program is built around an equation solver enhanced with thermodynamic property data and other valuable features. With *IT* you can obtain a single numerical solution or vary parameters to investigate their effects. You also can obtain graphical output, and the Windows-based format allows you to use any Windows word-processing software or spreadsheet to generate reports. Additionally, functions in *IT* can be called from *Excel* through use of the *Excel Add-in Manager,* allowing you to use these thermodynamic functions while working within *Excel*. Other features of *IT* include

- a guided series of help screens and a number of sample solved examples to help you learn how to use the program.

- drag-and-drop templates for many of the standard problem types, including a list of assumptions that you can customize to the problem at hand.

- predetermined scenarios for power plants and other important applications.

- thermodynamic property data for water, Refrigerants 22, 134a, and 410a, ammonia, carbon dioxide, air–water vapor mixtures, and a number of ideal gases.*

- the capability to input user-supplied data.

- the capability to interface with user-supplied routines.

Many features of *IT* are found in the popular *Engineering Equation Solver (EES)*. Readers already proficient with *EES* may prefer its use for solving problems in this text.

The use of computer software for engineering analysis is a powerful approach. Still, there are some rules to observe:

- Software *complements* and *extends* careful analysis but does not substitute for it.

- Computer-generated values should be checked selectively against hand-calculated or otherwise independently determined values.

- Computer-generated plots should be studied to see if the curves appear reasonable and exhibit expected trends.

*In the latest version of *IT*, some property data are evaluated using Mini REFPROP by permission from the National Institute of Standards and Technology (NIST).

IT provides data for substances represented in the Appendix tables. Generally, data are retrieved by simple call statements that are placed in the workspace of the program.

FOR EXAMPLE

Consider the two-phase liquid–vapor mixture at state 1 of Example 3.2 for which $p = 1$ bar, $v = 0.8475$ m³/kg. The following illustrates how data for saturation temperature, quality, and specific internal energy are retrieved using *IT*. The functions for T, v, and u are obtained by selecting Water/Steam from the Properties menu. Choosing SI units from the Units menu, with p in bar, T in °C, and amount of substance in kg, the *IT* program is

```
p = 1//bar
v = 0.8475//m³/kg
T = Tsat_P("Water/Steam",p)
v = vsat_Px("Water/Steam",p,x)
u = usat_Px("Water/Steam",p,x)
```

Clicking the Solve button, the software returns values of $T = 99.63$°C, $x = 0.5$, and $u = 1462$ kJ/kg. These values can be verified using data from Table A-3. Note that text inserted between the symbol // and a line return is treated as a comment.

The previous example illustrates an important feature of *IT*. Although the quality, x, is implicit in the list of arguments in the expression for specific volume, there is no need to solve the expression algebraically for x. Rather, the program can solve for x as long as the number of equations equals the number of unknowns.

IT also retrieves property values in the superheated vapor region.

FOR EXAMPLE

Consider the superheated ammonia vapor at state 2 in Example 3.1, for which $p = 1.5$ bar and $T = 25$°C. Selecting Ammonia from the Properties menu and choosing SI units from the Units menu, data for specific volume, internal energy, and enthalpy are obtained from *IT* as follows:

```
p = 1.5//bar
T = 25//°C
v = v_PT("Ammonia",p,T)
u = u_PT("Ammonia",p,T)
h = h_PT("Ammonia",p,T)
```

Clicking the Solve button, the software returns values of $v = 0.9553$ m³/kg, $u = 1380.9$ kJ/kg, and $h = 1524.21$ kJ/kg, respectively. These values agree closely with the respective values obtained by interpolation in Table A-15.

3.8 Applying the Energy Balance Using Property Tables and Software

The energy balance for closed systems is introduced in Sec. 2.5. Alternative expressions are given by Eqs. 2.35a and 2.35b, which are forms applicable to processes between end states denoted 1 and 2, and by Eq. 2.37, the time rate form. In applications where changes in kinetic energy and gravitational potential energy between the end states can be ignored, Eq. 2.35b reduces to

$$U_2 - U_1 = Q - W \qquad \text{(a)}$$

where Q and W account, respectively, for the transfer of energy by heat and work between the system and its surroundings during the process. The term $U_2 - U_1$ accounts for change in internal energy between the end states.

Taking water for simplicity, let's consider how the internal energy term is evaluated in three representative cases of systems involving a *single* substance.

Case 1 Consider a system consisting initially and finally of a single phase of water, vapor or liquid. Then Eq. (a) takes the form

$$m(u_2 - u_1) = Q - W \qquad \text{(b)}$$

where m is the system mass and u_1 and u_2 denote, respectively, the initial and final specific internal energies. When the initial and final temperatures T_1, T_2 and pressures p_1, p_2 are known, for instance, the internal energies u_1 and u_2 can be readily obtained from the *steam tables* or using computer software.

Case 2 Consider a system consisting initially of water vapor and finally as a two-phase mixture of liquid water and water vapor. As in Case 1, we write $U_1 = mu_1$ in Eq. (a), but now

$$
\begin{aligned}
U_2 &= (U_{\text{liq}} + U_{\text{vap}}) \\
&= m_{\text{liq}} u_{\text{f}} + m_{\text{vap}} u_{\text{g}}
\end{aligned}
\qquad \text{(c)}
$$

where m_{liq} and m_{vap} account, respectively, for the masses of saturated liquid and saturated vapor present finally, and u_{f} and u_{g} are the corresponding specific internal energies determined by the final temperature T_2 (or final pressure p_2).

If quality x_2 is known, Eq. 3.6 can be invoked to evaluate the specific internal energy of the two-phase liquid–vapor mixture, u_2. Then, $U_2 = mu_2$, thereby preserving the form of the energy balance expressed by Eq. (b).

Case 3 Consider a system consisting initially of two separate masses of water vapor that mix to form a total mass of water vapor. In this case

$$
\begin{aligned}
U_1 &= m'u(T', p') + m''u(T'', p'') \qquad \text{(d)} \\
U_2 &= (m' + m'')u(T_2, p_2) \\
&= mu(T_2, p_2) \qquad \text{(e)}
\end{aligned}
$$

where m' and m'' are masses of water vapor initially separate at T', p' and T'', p'', respectively, that mix to form a total mass, $m = m' + m''$, at a final state where temperature is T_2 and pressure is p_2. When temperatures and pressures at the respective states are known, for instance, the specific internal energies of Eqs. (d) and (e) can be readily obtained from the *steam tables* or using computer software.

Thermal Energy Storage

Energy is often available at one time but is needed at other times. For example, solar energy collected during daylight hours is often needed to heat buildings overnight. These conditions give rise to the need to store energy through methods introduced in Sec. 2.7 and by means discussed here. *Thermal energy storage* systems have been developed to meet solar and other similar energy storage needs. The term *thermal energy* here should be understood as *internal energy*.

Various mediums are used in thermal energy storage systems that change temperature and/or change phase. Some simply store energy by heating or cooling liquids like water or mineral oil, or solids like concrete, held in insulated storage tanks until the stored energy is needed. Systems that use *phase-change materials* (PCM) store energy by melting or freezing a substance, often water, paraffin, or a *molten (eutectic) salt.* The choice of storage medium is determined by the temperature requirements of the storage application together with capital and operating costs related to the storage system. When substances change phase, significant energy is stored at nearly constant temperature. This gives PCM systems an advantage over systems that store energy through temperature change alone because the high energy storage per unit volume tends to make PCM systems smaller and more cost-effective.

The availability of relatively inexpensive electricity generated in low-demand periods, usually overnight or during weekends, impacts storage strategies. In one approach, low-cost electricity is provided to a refrigeration system that chills water and/or produces ice during nighttime hours when air-conditioning needs are low. The chilled water and/or ice can then be used to satisfy building-cooling needs during the day when electricity is in highest demand and is more costly.

These cases show that when applying the energy balance, an important consideration is whether the system has one or two phases. A pertinent application is that of *thermal energy storage*, considered in the box.

3.8.1 Using Property Tables

In Examples 3.3 and 3.4, closed systems undergoing processes are analyzed using the energy balance. In each case, sketches of *p–υ* and/or *T–υ* diagrams are used in conjunction with appropriate tables to obtain the required property data. Using property diagrams and table data introduces an additional level of complexity compared to similar problems in Chap. 2.

TAKE NOTE...

On property diagrams, solid lines are reserved for processes that pass through equilibrium states: quasiequilibrium processes (Sec. 2.2.5). A dashed line on a property diagram signals only that a process has occurred between initial and final equilibrium states and does not define a path for the process.

▸▸ **EXAMPLE 3.3** ▸

Stirring Water at Constant Volume

A well-insulated rigid tank having a volume of 0.25 m^3 contains saturated water vapor at 100°C. The water is rapidly stirred until the pressure is 1.5 bars. Determine the temperature at the final state, in °C, and the work during the process, in kJ.

SOLUTION

Known   By rapid stirring, water vapor in a well-insulated rigid tank is brought from the saturated vapor state at 100°C to a pressure of 1.5 bars.

Find   Determine the temperature at the final state and the work.

Schematic and Given Data:

Engineering Model

1. The water is a closed system.

2. The initial and final states are at equilibrium. There is no net change in kinetic or potential energy.

3. There is no heat transfer with the surroundings.

4. The tank volume remains constant.

Analysis   To determine the final equilibrium state, the values of two independent intensive properties are required. One of these is

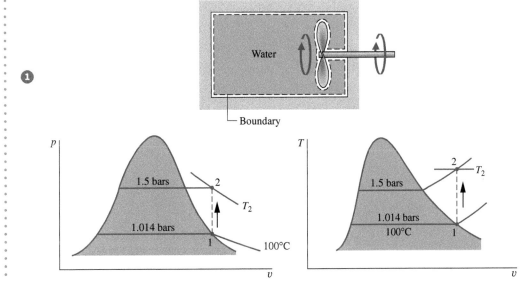

Fig. E3.3

pressure, $p_2 = 1.5$ bars, and the other is the specific volume: $v_2 = v_1$. The initial and final specific volumes are equal because the total mass and total volume are unchanged in the process. The initial and final states are located on the accompanying T–v and p–v diagrams.

From Table A-2, $v_1 = v_g(100°C) = 1.673$ m³/kg, $u_1 = u_g(100°C) = 2,506.5$ kJ/kg. By using $v_2 = v_1$ and interpolating in Table A-4 at $p_2 = 1.5$ bar,

$$T_2 = 273°C, \qquad u_2 = 2,767.8 \text{ kJ/kg}$$

Next, with assumptions 2 and 3 an energy balance for the system reduces to

$$\Delta U + \Delta \cancel{KE}^{\,0} + \Delta \cancel{PE}^{\,0} = \cancel{Q}^{\,0} - W$$

On rearrangement

$$W = -(U_2 - U_1) = -m(u_2 - u_1)$$

To evaluate W requires the system mass. This can be determined from the volume and specific volume

$$m = \frac{V}{v_1} = \left(\frac{0.25 \text{ m}^3}{1.673 \text{ m}^3/\text{kg}} \right) = 0.149 \text{ kg}$$

Finally, by inserting values into the expression for W

$$W = -(1.49 \text{ kg})(2,767.8 - 2,506.5) \text{ kJ/kg} = -38.9 \text{ kJ}$$

where the minus sign signifies that the energy transfer by work is to the system.

① Although the initial and final states are equilibrium states, the intervening states are not at equilibrium. To emphasize this, the process has been indicated on the T–v and p–v diagrams by a dashed line. Solid lines on property diagrams are reserved for processes that pass through equilibrium states only (quasiequilibrium processes). The analysis illustrates the importance of carefully sketched property diagrams as an adjunct to problem solving.

SKILLS DEVELOPED

Ability to...

• define a closed system and identify interactions on its boundary.

• apply the energy balance with steam table data.

• sketch T–v and p–v diagrams and locate states on them.

Quick Quiz

If insulation were removed from the tank and the water cooled at constant volume from $T_2 = 229.4°C$ to $T_3 = 148.89°C$ while no stirring occurs, determine the heat transfer, in kJ. Ans. –20.57 kJ.

▶ ▶ ▶ **EXAMPLE 3.4** ▶

Analyzing Two Processes in Series

Water contained in a piston–cylinder assembly undergoes two processes in series from an initial state where the pressure is 10 bar and the temperature is 400°C.

Process 1–2   The water is cooled as it is compressed at a constant pressure of 10 bar to the saturated vapor state.

Process 2–3   The water is cooled at constant volume to 150°C.

a. Sketch both processes on T–v and p–v diagrams.

b. For the overall process determine the work, in kJ/kg.

c. For the overall process determine the heat transfer, in kJ/kg.

SOLUTION

Known   Water contained in a piston–cylinder assembly undergoes two processes: It is cooled and compressed, while keeping the pressure constant, and then cooled at constant volume.

Find   Sketch both processes on T–v and p–v diagrams. Determine the net work and the net heat transfer for the overall process per unit of mass contained within the piston–cylinder assembly.

Schematic and Given Data:

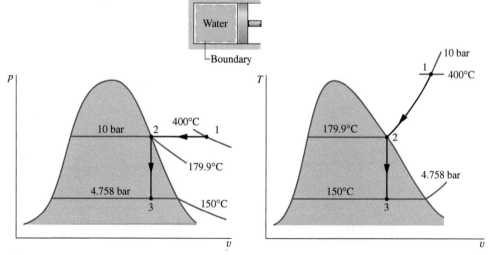

Fig. E3.4

Engineering Model

1. The water is a closed system.

2. The piston is the only work mode.

3. There are no changes in kinetic or potential energy.

Analysis

a. The accompanying T–v and p–v diagrams show the two processes. Since the temperature at state 1, $T_1 = 400°C$, is greater than the saturation temperature corresponding to $p_1 = 10$ bar: $179.9°C$, state 1 is located in the superheat region.

b. Since the piston is the only work mechanism

$$W = \int_1^3 p\, dV = \int_1^2 p\, dV + \int_2^3 p_2 \overset{0}{\cancel{dV}}$$

The second integral vanishes because the volume is constant in Process 2–3. Dividing by the mass and noting that the pressure is constant for Process 1–2

$$\frac{W}{m} = p(v_2 - v_1)$$

The specific volume at state 1 is found from Table A-4 using $p_1 = 10$ bar and $T_1 = 400°C$: $v_1 = 0.3066$ m³/kg. Also, $u_1 = 2957.3$ kJ/kg. The specific volume at state 2 is the saturated vapor value at 10 bar: $v_2 = 0.1944$ m³/kg, from Table A-3. Hence,

$$\frac{W}{m} = (10\text{ bar})(0.1944 - 0.3066)\left(\frac{\text{m}^3}{\text{kg}}\right)\left|\frac{10^5\text{ N/m}^2}{1\text{ bar}}\right|\left|\frac{1\text{ kJ}}{10^3\text{ N}\cdot\text{m}}\right|$$

$$= -112.2\text{ kJ/kg}$$

The minus sign indicates that work is done *on* the water vapor by the piston.

c. An energy balance for the *overall* process reduces to

$$m(u_3 - u_1) = Q - W$$

By rearranging

$$\frac{Q}{m} = (u_3 - u_1) + \frac{W}{m}$$

To evaluate the heat transfer requires u_3, the specific internal energy at state 3. Since T_3 is given and $v_3 = v_2$, two independent intensive properties are known that together fix state 3. To find u_3, first solve for the quality

$$x_3 = \frac{v_3 - v_{f3}}{v_{g3} - v_{f3}} = \frac{0.1944 - 1.0905 \times 10^{-3}}{0.3928 - 1.0905 \times 10^{-3}} = 0.494$$

where v_{f3} and v_{g3} are from Table A-2 at $150°C$. Then

$$u_3 = u_{f3} + x_3(u_{g3} - u_{f3}) = 631.68 + 0.494(2559.5 - 631.68)$$

$$= 1584.0\text{ kJ/kg}$$

where u_{f3} and u_{g3} are from Table A-2 at $150°C$.
Substituting values into the energy balance

$$\frac{Q}{m} = 1584.0 - 2957.3 + (-112.2) = -1485.5\text{ kJ/kg}$$

The minus sign shows that energy is transferred *out* by heat transfer.

SKILLS DEVELOPED

Ability to...

- define a closed system and identify interactions on its boundary.
- evaluate work using Eq. 2.17.
- apply the energy balance with steam table data.
- sketch T–v and p–v diagrams and locate states on them.

Quick Quiz

If the two specified processes were followed by Process 3-4, during which the water expands at a constant temperature of $150°C$ to saturated vapor, determine the work, in kJ/kg, for the *overall* process from 1 to 4. Ans. $W/m = -17.8$ kJ/kg

3.8.2 Using Software

Example 3.5 illustrates the use of *Interactive Thermodynamics: IT* for solving problems. In this case, the software evaluates the property data, calculates the results, and displays the results graphically.

▶▶ **EXAMPLE 3.5** ▶

Plotting Thermodynamic Data Using Software

For the system of Example 3.2, plot the heat transfer, in kJ, and the mass of saturated vapor present, in kg, each versus pressure at state 2 ranging from 1 to 2 bar. Discuss the results.

Find   Plot the heat transfer and the mass of saturated vapor present, each versus pressure at the final state. Discuss.

SOLUTION

Known   A two-phase liquid–vapor mixture of water in a closed, rigid container is heated on a hot plate. The initial pressure and quality are known. The pressure at the final state ranges from 1 to 2 bar.

Schematic and Given Data: See Fig. E3.2.

Engineering Model

1. There is no work.

2. Kinetic and potential energy effects are negligible.

3. See Example 3.2 for other assumptions.

Analysis The heat transfer is obtained from the energy balance. With assumptions 1 and 2, the energy balance reduces to

$$\Delta U + \Delta \cancel{KE}^0 + \Delta \cancel{PE}^0 = Q - \cancel{W}^0$$

or

$$Q = m(u_2 - u_1)$$

Selecting Water/Steam from the **Properties** menu and choosing SI Units from the **Units** menu, the *IT* program for obtaining the required data and making the plots is

```
// Given data—State 1
  p1 = 1//bar
  x1 = 0.5
  V = 0.5//m³
// Evaluate property data—State 1
  v1 = vsat_Px("Water/Steam", p1,x1)
  u1 = usat_Px("Water/Steam", p1,x1)
// Calculate the mass
  m = V/v1
// Fix state 2
  v2 = v1
  p2 = 1.5//bar
// Evaluate property data—State 2
  v2 = vsat_Px ("Water/Steam", p2,x2)
  u2 = usat_Px("Water/Steam", p2,x2)
// Calculate the mass of saturated vapor present
  mg2 = x2 * m
// Determine the pressure for which the quality
  is unity
① v3 = v1
  v3 = vsat_Px( "Water/Steam",p3,1)
// Energy balance to determine the heat transfer
  m * (u2 - u1) = Q - W
  W = 0
```

Click the **Solve** button to obtain a solution for p_2 = 1.5 bar. The program returns values of v_1 = 0.8475 m³/kg and m = 0.59 kg. Also, at p_2 = 1.5 bar, the program gives m_{g2} = 0.4311 kg. These values agree with the values determined in Example 3.2.

Now that the computer program has been verified, use the **Explore** button to vary pressure from 1 to 2 bar in steps of 0.1 bar. Then use the **Graph** button to construct the required plots. The results are

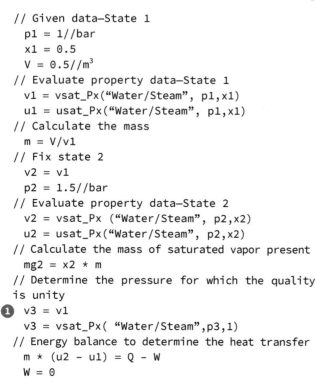

Fig. E3.5

We conclude from the first of these graphs that the heat transfer to the water varies directly with the pressure. The plot of m_g shows that the mass of saturated vapor present also increases as the pressure increases. Both of these results are in accord with expectations for the process.

① Using the **Browse** button, the computer solution indicates that the pressure for which the quality becomes unity is 2.096 bar. Thus, for pressures ranging from 1 to 2 bar, all of the states are in the two-phase liquid–vapor region.

SKILLS DEVELOPED

Ability to…

• apply the closed system energy balance.

• use *IT* to retrieve property data for water and plot calculated data.

Quick Quiz

If heating continues at constant specific volume to a state where the pressure is 3 bar, modify the *IT* program to give the temperature at that state, in °C.

Ans. v4 = v1
  p4 = 3//bar
  v4 = v_PT ("Water/Steam", p4, T4)
  T4 = 282.4°C

BioConnections

What do first responders, military flight crews, costumed characters at theme parks, and athletes have in common? They share a need to avoid heat stress while performing their duty, job, and pastime, respectively. To meet this need, *wearable coolers* have been developed such as cooling vests and cooling collars. Wearable coolers may feature ice pack inserts, channels through which a cool liquid is circulated, encapsulated *phase-change materials*, or a combination. A familiar example of a phase-change material (PCM) is ice, which on melting at 0°C absorbs energy of about 334 kJ/kg.

When worn close to the body, PCM-laced apparel absorbs energy from persons working or exercising in hot environments, keeping them cool. When specifying a PCM for a wearable cooler, the material must change phase at the desired cooler operating temperature. Hydrocarbons known as *paraffins* are frequently used for such duty. Many coolers available today employ PCM beads with diameters as small as 0.5 micron, encapsulated in a durable polymer shell. Encapsulated phase-change materials also are found in other products.

3.9 Introducing Specific Heats c_v and c_p

Several properties related to internal energy are important in thermodynamics. One of these is the property enthalpy introduced in Sec. 3.6.1. Two others, known as **specific heats**, are considered in this section. The specific heats, denoted c_v and c_p, are particularly useful for thermodynamic calculations involving the *ideal gas model* introduced in Sec. 3.12.

specific heats

The intensive properties c_v and c_p are defined for pure, simple compressible substances as partial derivatives of the functions $u(T, v)$ and $h(T, p)$, respectively,

$$c_v = \left(\frac{\partial u}{\partial T} \right)_v \tag{3.8}$$

$$c_p = \left(\frac{\partial h}{\partial T} \right)_p \tag{3.9}$$

where the subscripts v and p denote, respectively, the variables held fixed during differentiation. Values for c_v and c_p can be obtained via statistical mechanics using *spectroscopic* measurements. They also can be determined macroscopically through exacting property measurements. Since u and h can be expressed either on a unit mass basis or per mole, values of the specific heats can be similarly expressed. SI units are kJ/kg · K or kJ/kmol · K.

The property k, called the *specific heat ratio*, is simply the ratio

$$k = \frac{c_p}{c_v} \tag{3.10}$$

The properties c_v and c_p are referred to as *specific heats* (or *heat capacities*) because under certain *special conditions* they relate the temperature change of a system to the amount of energy added by heat transfer. However, it is generally preferable to think of c_v and c_p in terms of their definitions, Eqs. 3.8 and 3.9, and not with reference to this limited interpretation involving heat transfer.

In general, c_v is a function of v and T (or p and T), and c_p depends on both p and T (or v and T). **Figure 3.9** shows how c_p for water vapor varies as a function of temperature and

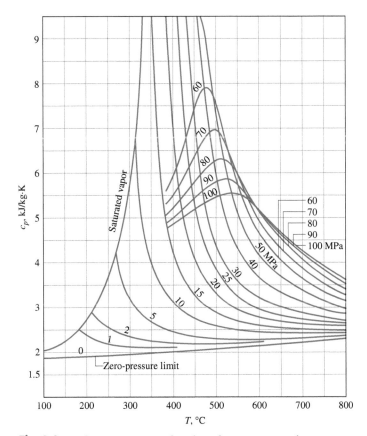

Fig. 3.9 c_p of water vapor as a function of temperature and pressure.

pressure. The vapor phases of other substances exhibit similar behavior. Note that the figure gives the variation of c_p with temperature in the limit as pressure tends to zero. In this limit, c_p increases with increasing temperature, which is a characteristic exhibited by other gases as well. We will refer again to such *zero-pressure* values for c_v and c_p in Sec. 3.13.2.

Specific heat data are available for common gases, liquids, and solids. Data for gases are introduced in Sec. 3.13.2 as a part of the discussion of the ideal gas model. Specific heat values for some common liquids and solids are introduced in Sec. 3.10.2 as a part of the discussion of the incompressible substance model.

3.10 Evaluating Properties of Liquids and Solids

Special methods often can be used to evaluate properties of liquids and solids. These methods provide simple, yet accurate, approximations that do not require exact compilations like the compressed liquid tables for water, Table A-5. Two such special methods are discussed next: approximations using saturated liquid data and the incompressible substance model.

3.10.1 Approximations for Liquids Using Saturated Liquid Data

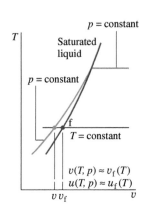

Approximate values for v, u, and h at liquid states can be obtained using saturated liquid data. To illustrate, refer to the compressed liquid tables, Table A-5. These tables show that the specific volume and specific internal energy change very little with pressure *at a fixed temperature*. Because the values of v and u vary only gradually as pressure changes at fixed temperature, the following approximations are reasonable for many engineering calculations:

$$v(T, p) \approx v_f(T) \tag{3.11}$$

$$u(T, p) \approx u_f(T) \tag{3.12}$$

That is, for liquids v and u may be evaluated at the saturated liquid state corresponding to the temperature at the given state.

An approximate value of h at liquid states can be obtained by using Eqs. 3.11 and 3.12 in the definition $h = u + pv$; thus,

$$h(T, p) \approx u_f(T) + pv_f(T)$$

This can be expressed alternatively as

$$h(T, p) \approx h_f(T) + \underline{v_f(T)[p - p_{sat}(T)]} \tag{3.13}$$

where p_{sat} denotes the saturation pressure at the given temperature. The derivation is left as an exercise. When the contribution of the underlined term of Eq. 3.13 is small, the specific enthalpy can be approximated by the saturated liquid value, as for v and u. That is,

$$h(T, p) \approx h_f(T) \tag{3.14}$$

Although the approximations given here have been presented with reference to liquid water, they also provide plausible approximations for other substances *when the only liquid data available are for saturated liquid states*. In this text, compressed liquid data are presented only for water (Table A-5). Also note that *Interactive Thermodynamics: IT* does not provide compressed liquid data for *any* substance but uses Eqs. 3.11, 3.12, and 3.14 to return liquid values for v, u, and h, respectively. When greater accuracy is required than provided by these approximations, other data sources should be consulted for more complete property compilations for the substance under consideration.

3.10.2 Incompressible Substance Model

As noted above, there are regions where the specific volume of liquid water varies little and the specific internal energy varies mainly with temperature. The same general behavior is exhibited by the liquid phases of other substances and by solids. The approximations of Eqs. 3.11–3.14 are based on these observations, as is the **incompressible substance model** under present consideration.

To simplify evaluations involving liquids or solids, the specific volume (density) is often assumed to be constant and the specific internal energy assumed to vary only with temperature. A substance idealized in this way is called *incompressible*.

Since the specific internal energy of a substance modeled as incompressible depends only on temperature, the specific heat c_v is also a function of temperature alone:

$$c_v(T) = \frac{du}{dT} \quad \text{(incompressible)} \tag{3.15}$$

This is expressed as an ordinary derivative because u depends only on T.

Although the specific volume is constant and internal energy depends on temperature only, enthalpy varies with both pressure and temperature according to

$$h(T, p) = u(T) + pv \quad \text{(incompressible)} \tag{3.16}$$

For a substance modeled as incompressible, the specific heats c_v and c_p are equal. This is seen by differentiating Eq. 3.16 with respect to temperature while holding pressure fixed to obtain

$$\left(\frac{\partial h}{\partial T} \right)_p = \frac{du}{dT}$$

The left side of this expression is c_p by definition (Eq. 3.9), so using Eq. 3.15 on the right side gives

$$c_p = c_v \quad \text{(incompressible)} \tag{3.17}$$

Thus, for an incompressible substance it is unnecessary to distinguish between c_p and c_v, and both can be represented by the same symbol, c. Specific heats of some common liquids and solids are given in Table A-19. Over limited temperature intervals the variation of c with temperature can be small. In such instances, the specific heat c can be treated as constant without a serious loss of accuracy.

Using Eqs. 3.15 and 3.16, the changes in specific internal energy and specific enthalpy between two states are given, respectively, by

$$u_2 - u_1 = \int_{T_1}^{T_2} c(T)\, dT \quad \text{(incompressible)} \tag{3.18}$$

$$h_2 - h_1 = u_2 - u_1 + v(p_2 - p_1)$$
$$= \int_{T_1}^{T_2} c(T)\, dT + v(p_2 - p_1) \quad \text{(incompressible)} \tag{3.19}$$

If the specific heat c is taken as constant, Eqs. 3.18 and 3.19 become, respectively,

$$u_2 - u_1 = c(T_2 - T_1) \tag{3.20a}$$

$$h_2 - h_1 = c(T_2 - T_1) + \underline{v(p_2 - p_1)} \quad \text{(incompressible, constant } c) \tag{3.20b}$$

In Eq. 3.20b, the underlined term is often small relative to the first term on the right side and then may be dropped.

The next example illustrates use of the incompressible substance model in an application involving the constant-volume calorimeter.

TAKE NOTE...

For a substance modeled as incompressible,
$v = \text{constant}$
$u = u(T)$

▶ ▶ ▶ **EXAMPLE 3.6** ▶ •

Measuring the Calorie Value of Cooking Oil

One-tenth milliliter of cooking oil is placed in the chamber of a constant-volume calorimeter filled with sufficient oxygen for the oil to be completely burned. The chamber is immersed in a water bath. The mass of the water bath is 2.15 kg. For the purpose of this analysis, the metal parts of the apparatus are modeled as equivalent to an additional 0.5 kg of water. The calorimeter is well-insulated, and initially the temperature throughout is 25°C. The oil is ignited by a spark. When equilibrium is again attained, the temperature throughout is 25.3°C. Determine the change in internal energy of the chamber contents, in kcal per mL of cooking oil and in kcal per tablespoon of cooking oil.

Known Data are provided for a constant-volume calorimeter testing cooking oil for calorie value.

Find Determine the change in internal energy of the contents of the calorimeter chamber.

Schematic and Given Data:

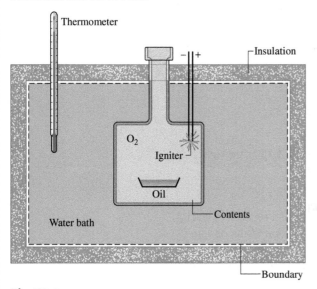

Fig. E3.6

Engineering Model

1. The closed system is shown by the dashed line in the accompanying figure.

2. The total volume remains constant, including the chamber, water bath, and the amount of water modeling the metal parts.

3. Water is modeled as incompressible with constant specific heat c.

4. Heat transfer with the surroundings is negligible, and there is no change in kinetic or potential energy.

Analysis With the assumptions listed, the closed system energy balance reads

$$\Delta U + \Delta KE^0 + \Delta PE^0 = Q^0 - W^0$$

or

$$(\Delta U)_{contents} + (\Delta U)_{water} = 0$$

thus

$$(\Delta U)_{contents} = -(\Delta U)_{water} \qquad (a)$$

The change in internal energy of the contents is equal and opposite to the change in internal energy of the water.

Since water is modeled as incompressible, Eq. 3.20a is used to evaluate the right side of Eq. (a), giving

❶ ❷ $$(\Delta U)_{contents} = -m_w c_w (T_2 - T_1) \qquad (b)$$

With $m_w = 2.15$ kg + 0.5 kg = 2.65 kg, $(T_2 - T_1) = 0.3$ K, and $c_w = 4.18$ kJ/kg · K from Table A-19, Eq. (b) gives

$$(\Delta U)_{contents} = -(2.65 \text{ kg})(4.18 \text{ kJ/kg} \cdot \text{K})(0.3 \text{ K}) = -3.32 \text{ kJ}$$

Converting to kcal, and expressing the result on a per milliliter of oil basis using the oil volume, 0.1 mL, we get

$$\frac{(\Delta U)_{contents}}{V_{oil}} = \frac{-3.32 \text{ kJ}}{0.1 \text{ mL}} \left| \frac{1 \text{ kcal}}{4.1868 \text{ kJ}} \right|$$

$$= -7.9 \text{ kcal/mL}$$

The calorie value of the cooking oil is the magnitude—that is, 7.9 kcal/mL. Labels on cooking oil containers usually give calorie value for a serving size of 1 tablespoon (15 mL). Using the calculated value, we get 119 kcal per tablespoon.

---

❶ The change in internal energy for water can be found alternatively using Eq. 3.12 together with saturated liquid internal energy data from Table A-2.

❷ The change in internal energy of the chamber contents cannot be evaluated using a specific heat because specific heats are defined (Sec. 3.9) only for *pure* substances—that is, substances that are unchanging in composition.

SKILLS DEVELOPED

Ability to...

• define a closed system and identify interactions within it and on its boundary.

• apply the energy balance using the incompressible substance model.

Quick Quiz

Using Eq. 3.12 together with saturated liquid internal energy data from Table A-2, find the change in internal energy of the water, in kJ, and compare with the value obtained assuming water is incompressible. **Ans. 3.32 kJ**

BioConnections

Is your diet bad for the environment? It could be. The fruits, vegetables, and animal products found in grocery stores require a lot of fossil fuel just to get there. While study of the linkage of the human diet to the environment is in its infancy, some preliminary findings are interesting.

One study of U.S. dietary patterns evaluated the amount of fossil fuel—and, implicitly, the level of greenhouse gas production—required to support several different diets. Diets rich in meat and fish were found to require the most fossil fuel, owing to the significant energy resources required to produce these products and bring them to market. But for those who enjoy meat

and fish, the news is not all bad. Only a fraction of the fossil fuel needed to get food to stores is used to grow the food; most is spent on processing and distribution. Accordingly, eating favorite foods originating close to home can be a good choice environmentally.

Still, the connection between the food we eat, energy resource use, and accompanying environmental impact requires further study, including the vast amounts of agricultural land needed, huge water requirements, emissions related to fertilizer production and use, methane emitted from waste produced by billions of animals raised for food annually, and fuel for transporting food to market.

3.11 Generalized Compressibility Chart

The object of the present section is to gain a better understanding of the relationship among pressure, specific volume, and temperature of gases. This is important not only as a basis for analyses involving gases but also for the discussions of the second part of the chapter, where the *ideal gas model* is introduced. The current presentation is conducted in terms of the *compressibility factor* and begins with the introduction of the *universal gas constant*.

3.11.1 Universal Gas Constant, $\bar{R}$

Let a gas be confined in a cylinder by a piston and the entire assembly held at a constant temperature. The piston can be moved to various positions so that a series of equilibrium states at constant temperature can be visited. Suppose the pressure and specific volume are measured at each state and the value of the ratio $p\bar{v}/T$ ($\bar{v}$ is volume per mole) determined. These ratios can then be plotted versus pressure at constant temperature. The results for several temperatures are sketched in **Fig. 3.10**. When the ratios are extrapolated to zero pressure, *precisely the same limiting value is obtained* for each curve. That is,

$$\lim_{p \to 0} \frac{p\bar{v}}{T} = \bar{R} \qquad (3.21)$$

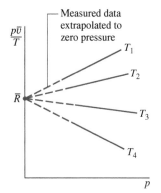

Fig. 3.10 Sketch of $p\bar{v}/T$ versus pressure for a gas at several specified values of temperature.

where $\bar{R}$ denotes the common limit for all temperatures. If this procedure were repeated for other gases, it would be found in every instance that the limit of the ratio $p\bar{v}/T$ as p tends to zero at fixed temperature is the same, namely, $\bar{R}$. Since the same limiting value is exhibited by all gases, $\bar{R}$ is called the **universal gas constant**. Its value as determined experimentally is

$$\bar{R} = 8.314 \text{ kJ/kmol} \cdot \text{K} \qquad (3.22)$$

Having introduced the universal gas constant, we turn next to the compressibility factor.

universal gas constant

3.11.2 Compressibility Factor, Z

The dimensionless ratio $p\bar{v}/\bar{R}T$ is called the **compressibility factor** and is denoted by Z. That is,

compressibility factor

$$\boxed{Z = \frac{p\bar{v}}{\bar{R}T}} \qquad (3.23)$$

As illustrated by subsequent calculations, when values for p, $\bar{v}$, $\bar{R}$, and T are used in consistent units, Z is unitless.

TABLE 3.1

Values of the Gas Constant R of Selected Elements and Compounds

Substance	Chemical Formula	R (kJ/kg · K)
Air	—	0.2870
Ammonia	NH_3	0.4882
Argon	Ar	0.2082
Carbon dioxide	CO_2	0.1889
Carbon monoxide	CO	0.2968
Helium	He	2.0769
Hydrogen	H_2	4.1240
Methane	CH_4	0.5183
Nitrogen	N_2	0.2968
Oxygen	O_2	0.2598
Water	H_2O	0.4614

Source: R values are calculated in terms of the universal gas constant
$\bar{R} = 8.314$ kJ/kmol · K and the molecular weight M provided in Table A-1 using
$R = \bar{R}/M$ (Eq. 3.25).

With $\bar{v} = Mv$ (Eq. 1.9), where M is the atomic or molecular weight, the compressibility factor can be expressed alternatively as

$$Z = \frac{pv}{RT} \qquad (3.24)$$

where

$$R = \frac{\bar{R}}{M} \qquad (3.25)$$

R is a constant for the particular gas whose molecular weight is M. Alternative units for R are kJ/kg · K. **Table 3.1** provides a sampling of values for the gas constant R calculated from Eq. 3.25.

Equation 3.21 can be expressed in terms of the compressibility factor as

$$\lim_{p \to 0} Z = 1 \qquad (3.26)$$

That is, the compressibility factor Z tends to unity as pressure tends to zero at fixed temperature. This can be illustrated by reference to **Fig. 3.11**, which shows Z for hydrogen plotted versus pressure at a number of different temperatures. In general, at states of a gas where pressure is small relative to the critical pressure, Z is approximately 1.

3.11.3 Generalized Compressibility Data, Z Chart

Figure 3.11 gives the compressibility factor for hydrogen versus pressure at specified values of temperature. Similar charts have been prepared for other gases. When these charts are studied, they are found to be *qualitatively* similar. Further study shows that when the coordinates are suitably modified, the curves for several different gases coincide closely when plotted together on the same coordinate axes, and so *quantitative* similarity also can be achieved. This is referred to as the *principle of corresponding states*. In one such approach, the compressibility factor Z is plotted versus a dimensionless **reduced pressure** p_R and **reduced temperature** T_R, defined as

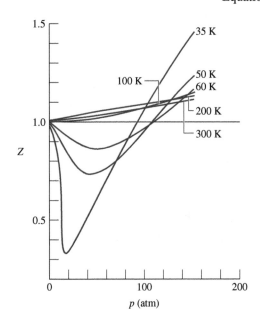

Fig. 3.11 Variation of the compressibility factor of hydrogen with pressure at constant temperature.

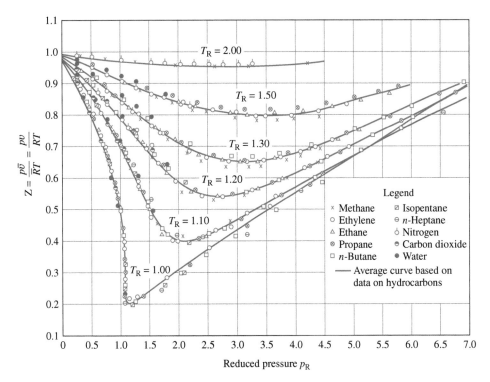

Fig. 3.12 Generalized compressibility chart for various gases.

$$p_R = p/p_c \qquad (3.27)$$

$$T_R = T/T_c \qquad (3.28)$$

reduced pressure
and temperature

where p_c and T_c denote the critical pressure and temperature, respectively. This results in a **generalized compressibility chart** of the form $Z = f(p_R, T_R)$. **Figure 3.12** shows experimental data for 10 different gases on a chart of this type. The solid lines denoting reduced isotherms represent the best curves fitted to the data. Observe that Table A-1 provide the critical temperature and critical pressure for a sampling of substances.

generalized
compressibility chart

A generalized chart more suitable for problem solving than Fig. 3.12 is given in the Appendix as Figs. A-1, A-2, and A-3. In Fig. A-1, p_R ranges from 0 to 1.0; in Fig. A-2, p_R ranges from 0 to 10.0; and in Fig. A-3, p_R ranges from 10.0 to 40.0. At any one temperature, the deviation of observed values from those of the generalized chart increases with pressure. However, for the 30 gases used in developing the chart, the deviation is *at most* on the order of 5% and for most ranges is much less.[1]

Values of specific volume are included on the generalized chart through the variable v_R', called the **pseudoreduced specific volume**, defined by

pseudoreduced specific volume

$$v_R' = \frac{\bar{v}}{\bar{R}T_c/p_c} \qquad (3.29)$$

TAKE NOTE...

Study of Fig. A-2 shows that the value of Z tends to unity at fixed reduced temperature T_R as reduced pressure p_R tends to zero. That is, $Z \to 1$ as $p_R \to 0$ at fixed T_R. Figure A-2 also shows that Z tends to unity at fixed reduced pressure as reduced temperature becomes large.

The pseudoreduced specific volume gives a better correlation of the data than does the *reduced* specific volume $v_R = \bar{v}/\bar{v}_c$, where $\bar{v}_c$ is the critical specific volume.

Using the critical pressure and critical temperature of a substance of interest, the generalized chart can be entered with various pairs of the variables T_R, p_R, and v_R': (T_R, p_R), (p_R, v_R'), or (T_R, v_R'). The merit of the generalized chart for relating p, v, and T data for gases is simplicity coupled with accuracy. However, the generalized compressibility chart should not be used as a substitute for p–v–T data for a given substance as provided by tables or computer software. The chart is mainly useful for obtaining reasonable estimates in the absence of more accurate data.

---

[1]To determine Z for hydrogen, helium, and neon above a T_R of 5, the reduced temperature and pressure should be calculated using $T_R = T/(T_c + 8)$ and $p_R = p/(p_c + 8)$, where temperatures are in K and pressures are in atm.

Animation

Ideal Gas Tab a

The next example provides an illustration of the use of the generalized compressibility chart.

EXAMPLE 3.7

Using the Generalized Compressibility Chart

A closed, rigid tank filled with water vapor, initially at 20 MPa, 520°C, is cooled until its temperature reaches 400°C. Using the compressibility chart, determine

a. the specific volume of the water vapor in m³/kg at the initial state.

b. the pressure in MPa at the final state.

Compare the results of parts (a) and (b) with the values obtained from the superheated vapor table, Table A-4.

SOLUTION

Known Water vapor is cooled at constant volume from 20 MPa, 520°C to 400°C.

Find Use the compressibility chart and the superheated vapor table to determine the specific volume and final pressure and compare the results.

Schematic and Given Data:

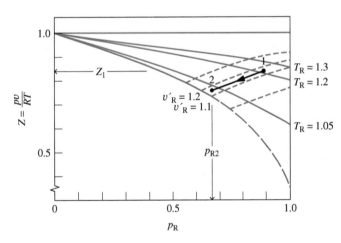

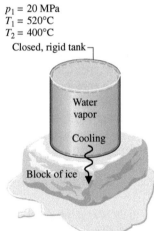

$p_1 = 20$ MPa
$T_1 = 520°C$
$T_2 = 400°C$

Closed, rigid tank

Water vapor

Cooling

Block of ice

Fig. E3.7

Engineering Model

1. The water vapor is a closed system.

2. The initial and final states are at equilibrium.

3. The volume is constant.

Analysis

a. From Table A-1, $T_c = 647.3$ K and $p_c = 22.09$ MPa for water. Thus,

❶
$$T_{R1} = \frac{793}{647.3} = 1.23, \qquad p_{R1} = \frac{20}{22.09} = 0.91$$

With these values for the reduced temperature and reduced pressure, the value of Z obtained from Fig. A-1 is approximately

0.83. Since $Z = pv/RT$, the specific volume at state 1 can be determined as follows:

$$v_1 = Z_1 \frac{RT_1}{p_1} = 0.83 \frac{\bar{R}T_1}{Mp_1}$$

❷
$$= 0.83 \frac{\left(8314 \dfrac{\text{N} \cdot \text{m}}{\text{kmol} \cdot \text{k}}\right)\left(793 \text{ K}\right)}{\left(18.02 \dfrac{\text{kg}}{\text{kmol}}\right)\left(20 \times 10^6 \dfrac{\text{N}}{\text{m}^2}\right)} = 0.0152 \text{ m}^3/\text{kg}$$

The molecular weight of water is from Table A-1.

Turning to Table A-4, the specific volume at the initial state is 0.01551 m³/kg. This is in good agreement with the compressibility chart value, as expected.

b. Since both mass and volume remain constant, the water vapor cools at constant specific volume and, thus, at constant v_R'. Using the value for specific volume determined in part (a), the constant v_R' value is

$$v_R' = \frac{vp_c}{RT_c} = \frac{\left(0.0152 \dfrac{\text{m}^3}{\text{kg}}\right)\left(22.09 \times 10^6 \dfrac{\text{N}}{\text{m}^2}\right)}{\left(\dfrac{8314 \text{ N} \cdot \text{m}}{18.02 \text{ kg} \cdot \text{K}}\right)(647.3 \text{ K})} = 1.12$$

At state 2

$$T_{R2} = \frac{673}{647.3} = 1.04$$

Locating the point on the compressibility chart where $v_R' = 1.12$ and $T_R = 1.04$, the corresponding value for p_R is about 0.69. Accordingly

$$p_2 = p_c(p_{R2}) = (22.09 \text{ MPa})(0.69) = 15.24 \text{ MPa}$$

Interpolating in the superheated vapor tables gives $p_2 = 15.16$ MPa. As before, the compressibility chart value is in good agreement with the table value.

---

❶ *Absolute* temperature and *absolute* pressure must be used in evaluating the compressibility factor Z, the reduced temperature T_R, and reduced pressure p_R.

❷ Since Z is unitless, values for p, v, R, and T must be used in consistent units.

3.11.4 Equations of State

Considering the curves of Figs. 3.11 and 3.12, it is reasonable to think that the variation with pressure and temperature of the compressibility factor for gases might be expressible as an equation, at least for certain intervals of p and T. Two expressions can be written that enjoy a theoretical basis. One gives the compressibility factor as an infinite series expansion in pressure:

$$Z = 1 + \hat{B}(T)p + \hat{C}(T)p^2 + \hat{D}(T)p^3 + \cdots \qquad (3.30)$$

where the coefficients $\hat{B}, \hat{C}, \hat{D}, \ldots$ depend on temperature only. The dots in Eq. 3.30 represent higher-order terms. The other is a series form entirely analogous to Eq. 3.30 but expressed in terms of $1/\bar{v}$ instead of p

$$Z = 1 + \frac{B(T)}{\bar{v}} + \frac{C(T)}{\bar{v}^2} + \frac{D(T)}{\bar{v}^3} + \cdots \qquad (3.31)$$

Equations 3.30 and 3.31 are known as **virial equations of state**, and the coefficients $\hat{B}, \hat{C}, \hat{D}, \ldots$ and $B, C, D, \ldots$ are called *virial coefficients*. The word *virial* stems from the Latin word for force. In the present usage it is force interactions among molecules that are intended.

    The virial expansions can be derived by the methods of statistical mechanics, and physical significance can be attributed to the coefficients: $B/\bar{v}$ accounts for two-molecule interactions, $C/\bar{v}^2$ accounts for three-molecule interactions, and so on. In principle, the virial coefficients can be calculated by using expressions from statistical mechanics derived from consideration of the force fields around the molecules of a gas. The virial coefficients also can be determined from experimental $p–v–T$ data. The virial expansions are used in Sec. 11.1 as a point of departure for the further study of analytical representations of the $p–v–T$ relationship of gases known generically as *equations of state*.

    The virial expansions and the physical significance attributed to the terms making up the expansions can be used to clarify the nature of gas behavior in the limit as pressure tends to zero at fixed temperature. From Eq. 3.30 it is seen that if pressure decreases at fixed temperature, the terms $\hat{B}p, \hat{C}p^2$, and so on, accounting for various molecular interactions tend to decrease, suggesting that the force interactions become weaker under these circumstances. In the limit as pressure approaches zero, these terms vanish, and the equation reduces to $Z = 1$ in accordance with Eq. 3.26. Similarly, since specific volume increases when the pressure decreases at fixed temperature, the terms $B/\bar{v}, C/\bar{v}^2$, etc. of Eq. 3.31 also vanish in the limit, giving $Z = 1$ when the force interactions between molecules are no longer significant.

virial equations of state

Evaluating Properties Using the Ideal Gas Model

3.12 Introducing the Ideal Gas Model

In this section the ideal gas model is introduced. The ideal gas model has many applications in engineering practice and is frequently used in subsequent sections of this text.

3.12.1 Ideal Gas Equation of State

As observed in Sec. 3.11.3, study of the generalized compressibility chart Fig. A-2 shows that at states where the pressure p is small relative to the critical pressure p_c (low p_R) and/or the temperature T is large relative to the critical temperature T_c (high T_R), the compressibility factor, $Z = pv/RT$, is approximately 1. At such states, we can assume with reasonable accuracy that $Z = 1$, or

$$pv = RT \tag{3.32}$$

ideal gas equation of state

Known as the **ideal gas equation of state**, Eq. 3.32 underlies the second part of this chapter dealing with the ideal gas model.

Alternative forms of the same basic relationship among pressure, specific volume, and temperature are obtained as follows. With $v = V/m$, Eq. 3.32 can be expressed as

$$pV = mRT \tag{3.33}$$

In addition, using $v = \bar{v}/M$ and $R = \bar{R}/M$, which are Eqs. 1.9 and 3.25, respectively, where M is the molecular weight, Eq. 3.32 can be expressed as

$$p\bar{v} = \bar{R}T \tag{3.34}$$

Ideal Gas Tab b

or, with $\bar{v} = V/n$, as

$$pV = n\bar{R}T \tag{3.35}$$

3.12.2 Ideal Gas Model

For any gas whose equation of state is given *exactly* by $pv = RT$, the specific internal energy depends on temperature *only*. This conclusion is demonstrated formally in Sec. 11.4. It is also supported by experimental observations, beginning with the work of Joule, who showed in 1843 that the internal energy of air at low density (large specific volume) depends primarily on temperature. Further motivation from the microscopic viewpoint is provided shortly. The specific enthalpy of a gas described by $pv = RT$ also depends on temperature only, as can be shown by combining the definition of enthalpy, $h = u + pv$, with $u = u(T)$ and the ideal gas equation of state to obtain $h = u(T) + RT$. Taken together, these specifications

ideal gas model

constitute the **ideal gas model**, summarized as follows

$$pv = RT \tag{3.32}$$
$$u = u(T) \tag{3.36}$$
$$h = h(T) = u(T) + RT \tag{3.37}$$

TAKE NOTE...

To expedite the solutions of many subsequent examples and end-of-chapter problems involving air, oxygen (O_2), nitrogen (N_2), carbon dioxide (CO_2), carbon monoxide (CO), hydrogen (H_2), and other common gases, we indicate in the problem statements that the ideal gas model should be used. If not indicated explicitly, the suitability of the ideal gas model should be checked using the Z chart or other data.

The specific internal energy and enthalpy of gases generally depend on two independent properties, not just temperature as presumed by the ideal gas model. Moreover, the ideal gas equation of state does not provide an acceptable approximation at all states. Accordingly, whether the ideal gas model is used depends on the error acceptable in a given calculation. Still, gases often do *approach* ideal gas behavior, and a particularly simplified description is obtained with the ideal gas model.

To verify that a gas can be modeled as an ideal gas, the states of interest can be located on a compressibility chart to determine how well $Z = 1$ is satisfied. As shown in subsequent discussions, other tabular or graphical property data can also be used to determine the suitability of the ideal gas model.

The next example illustrates the use of the ideal gas equation of state and reinforces the use of property diagrams to locate principal states during processes.

▶ ▶ EXAMPLE 3.8 ▶ ·

Air as an Ideal Gas Undergoing a Cycle

One pound of air in a piston–cylinder assembly undergoes a thermodynamic cycle consisting of three processes.

Process 1–2 Constant specific volume

Process 2–3 Constant-temperature expansion

Process 3–1 Constant-pressure compression

At state 1, the temperature is 300 K, and the pressure is 1 atm. At state 2, the pressure is 2 bars. Employing the ideal gas equation of state,

a. sketch the cycle on p–v coordinates.

b. determine the temperature at state 2, in K.

c. determine the specific volume at state 3, in m³/kg.

SOLUTION

Known Air executes a thermodynamic cycle consisting of three processes: Process 1–2, $v = constant$; Process 2–3, $T = constant$; Process 3–1, $p = constant$. Values are given for T_1, p_1, and p_2.

Find Sketch the cycle on p–v coordinates and determine T_2 and v_3.

Schematic and Given Data:

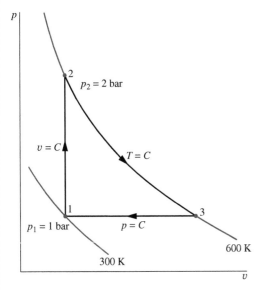

Fig. E3.8

Engineering Model

    1. The air is a closed system.

❶ **2.** The air behaves as an ideal gas.

    3. Volume change is the only work mode.

Analysis

a. The cycle is shown on p–v coordinates in the accompanying figure. Note that since $p = RT/v$ and temperature is constant, the variation of p with v for the process from 2 to 3 is nonlinear.

b. Using $pv = RT$, the temperature at state 2 is

$$T_2 = p_2 v_2 / R$$

To obtain the specific volume v_2 required by this relationship, note that $v_2 = v_1$, so

❷ $$v_2 = RT_1/p_1$$

Combining these two results gives

$$T_2 = \frac{p_2}{p_1} T_1 = \left(\frac{2 \text{ bars}}{1 \text{ bar}} \right) (300 \text{ K}) = 600 \text{ K}$$

c. Since $pv = RT$, the specific volume at state 3 is

$$v_3 = RT_3 / p_3$$

Noting that $T_3 = T_2$, $p_3 = p_1$, and $R = \bar{R}/M$

$$v_3 = \frac{\bar{R} T_2}{M p_1}$$

$$= \left(\frac{8.314 \dfrac{\text{kJ}}{\text{kmol} \cdot \text{J}}}{28.97 \dfrac{\text{kg}}{\text{mol}}} \right) \left(\frac{600 \text{ k}}{1 \text{ bar}} \right) \left(\frac{1 \text{ bar}}{10^5 \text{ N/m}^2} \right) \left(\frac{10^3 \text{ N} \cdot \text{m}}{1 \text{ kJ}} \right)$$

$$= 1.72 \text{ m}^3/\text{kg}$$

where the molecular weight of air is from Table A-1.

---

❶ Table A-1 gives $p_c = 37.3$ bar, $T_c = 133$ K for air. Therefore, $p_{R2} = 0.054$, $T_{R2} = 4.52$. Referring to Fig. A-1, the value of the compressibility factor at this state is $Z \approx 1$. The same conclusion results when states 1 and 3 are checked. Accordingly, $pv = RT$ adequately describes the p–v–T relation for the air at these states.

❷ Carefully note that the equation of state $pv = RT$ requires the use of *absolute* temperature T and *absolute* pressure p.

SKILLS DEVELOPED

Ability to...

• evaluate p–v–T data using the ideal gas equation of state.

• sketch processes on a p–v diagram

Quick Quiz

Is the cycle sketched in Fig. E3.8 a power cycle or a refrigeration cycle? Explain. Ans. A power cycle. As represented by enclosed area 1-2-3-1, the net work is positive.

3.12.3 Microscopic Interpretation

A picture of the dependence of the internal energy of gases on temperature at low density (large specific volume) can be obtained with reference to the discussion of the virial equations: Eqs. 3.30 and 3.31. As $p \rightarrow 0$ ($\bar{v} \rightarrow \infty$), the force interactions between molecules of a gas become weaker, and the virial expansions approach $Z = 1$ in the limit. The study of gases from the microscopic point of view shows that the dependence of the internal energy of a gas on pressure, or specific volume, at a specified temperature arises primarily because of molecular interactions. Accordingly, as the density of a gas decreases (specific volume increases) at fixed temperature, there comes a point where the effects of intermolecular forces are minimal. The internal energy is then determined principally by the temperature.

From the microscopic point of view, the ideal gas model adheres to several idealizations: The gas consists of molecules that are in random motion and obey the laws of mechanics; the total number of molecules is large, but the volume of the molecules is a negligibly small fraction of the volume occupied by the gas; and no appreciable forces act on the molecules except during collisions. Further discussion of the ideal gas using the microscopic approach is provided in Sec. 3.13.2.

3.13 Internal Energy, Enthalpy, and Specific Heats of Ideal Gases

3.13.1 Δu, Δh, c_v, and c_p Relations

For a gas obeying the ideal gas model, specific internal energy depends only on temperature. Hence, the specific heat c_v, defined by Eq. 3.8, is also a function of temperature alone. That is,

$$c_v(T) = \frac{du}{dT} \qquad \text{(ideal gas)} \tag{3.38}$$

This is expressed as an ordinary derivative because u depends only on T.

By separating variables in Eq. 3.38

$$du = c_v(T)\, dT \tag{3.39}$$

On integration, the change in specific internal energy is

$$\boxed{u(T_2) - u(T_1) = \int_{T_1}^{T_2} c_v(T)\, dT \qquad \text{(ideal gas)}} \tag{3.40}$$

Similarly, for a gas obeying the ideal gas model, the specific enthalpy depends only on temperature, so the specific heat c_p, defined by Eq. 3.9, is also a function of temperature alone. That is,

$$c_p(T) = \frac{dh}{dT} \qquad \text{(ideal gas)} \tag{3.41}$$

Separating variables in Eq. 3.41

$$dh = c_p(T)\, dT \tag{3.42}$$

On integration, the change in specific enthalpy is

$$\boxed{h(T_2) - h(T_1) = \int_{T_1}^{T_2} c_p(T)\, dT \qquad \text{(ideal gas)}} \tag{3.43}$$

An important relationship between the ideal gas specific heats can be developed by differentiating Eq. 3.37 with respect to temperature

$$\frac{dh}{dT} = \frac{du}{dT} + R$$

and introducing Eqs. 3.38 and 3.41 to obtain

$$c_p(T) = c_v(T) + R \quad \text{(ideal gas)} \tag{3.44}$$

On a molar basis, this is written as

$$\overline{c}_p(T) = \overline{c}_v(T) + \overline{R} \quad \text{(ideal gas)} \tag{3.45}$$

Although each of the two ideal gas specific heats is a function of temperature, Eqs. 3.44 and 3.45 show that the specific heats differ by just a constant: the gas constant. Knowledge of either specific heat for a particular gas allows the other to be calculated by using only the gas constant. The above equations also show that $c_p > c_v$ and $\overline{c}_p > \overline{c}_v$, respectively.

For an ideal gas, the specific heat ratio, k, is also a function of temperature only

$$k = \frac{c_p(T)}{c_v(T)} \quad \text{(ideal gas)} \tag{3.46}$$

Since $c_p > c_v$, it follows that $k > 1$. Combining Eqs. 3.44 and 3.46 results in

$$c_p(T) = \frac{kR}{k-1} \tag{3.47a}$$
$$\text{(ideal gas)}$$
$$c_v(T) = \frac{R}{k-1} \tag{3.47b}$$

Similar expressions can be written for the specific heats on a molar basis, with R being replaced by $\overline{R}$.

3.13.2 Using Specific Heat Functions

The foregoing expressions require the ideal gas specific heats as functions of temperature. These functions are available for gases of practical interest in various forms, including graphs, tables, and equations. **Figure 3.13** illustrates the variation of $\overline{c}_p$ (molar basis) with temperature for a number of common gases. In the range of temperatures shown, $\overline{c}_p$ increases with

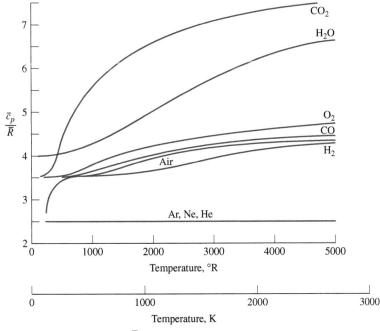

Fig. 3.13 Variation of $\overline{c}_p/\overline{R}$ with temperature for a number of gases modeled as ideal gases.

temperature for all gases, except for the monatonic gases Ar, Ne, and He. For these, $\bar{c}_p$ is constant at the value predicted by kinetic theory: $\bar{c}_p = \frac{5}{2}\bar{R}$. Tabular specific heat data for selected gases are presented versus temperature in Table A-20. Specific heats are also available in equation form. Several alternative forms of such equations are found in the engineering literature. An equation that is relatively easy to integrate is the polynomial form

$$\frac{\bar{c}_p}{\bar{R}} = \alpha + \beta T + \gamma T^2 + \delta T^3 + \varepsilon T^4 \tag{3.48}$$

Values of the constants α, β, γ, δ, and ε are listed in Table A-21 for several gases in the temperature range 300 to 1000 K.

FOR EXAMPLE

To illustrate the use of Eq. 3.48, let us evaluate the change in specific enthalpy, in kJ/kg, of air modeled as an ideal gas from a state where $T_1 = 400$ K to a state where $T_2 = 900$ K. Inserting the expression for $\bar{c}_p(T)$ given by Eq. 3.48 into Eq. 3.43 and integrating with respect to temperature

$$h_2 - h_1 = \frac{\bar{R}}{M}\int_{T_1}^{T_2}(\alpha + \beta T + \gamma T^2 + \delta T^3 + \varepsilon T^4)\,dT$$

$$= \frac{\bar{R}}{M}\left[\alpha(T_2 - T_1) + \frac{\beta}{2}(T_2^2 - T_1^2) + \frac{\gamma}{3}(T_2^3 - T_1^3) + \frac{\delta}{4}(T_2^4 - T_1^4) + \frac{\varepsilon}{5}(T_2^5 - T_1^5)\right]$$

where the molecular weight M has been introduced to obtain the result on a unit mass basis. With values for the constants from Table A-21

$$h_2 - h_1 = \frac{8.314}{28.97}\left\{3.653(900 - 400) - \frac{1.337}{2(10)^3}[(900)^2 - (400)^2]\right.$$

$$+ \frac{3.294}{3(10)^6}[(900)^3 - (400)^3] - \frac{1.913}{4(10)^9}[(900)^4 - (400)^4]$$

$$\left. + \frac{0.2763}{5(10)^{12}}[(900)^5 - (400)^5]\right\} = 531.69 \text{ kJ/kg}$$

Specific heat functions $c_v(T)$ and $c_p(T)$ are also available in *IT: Interactive Thermodynamics* in the PROPERTIES menu. These functions can be integrated using the integral function of the program to calculate Δu and Δh, respectively.

FOR EXAMPLE

Let us repeat the immediately preceding example using *IT*. For air, the *IT* code is

```
cp = cp_T ("Air",T)
delh = Integral(cp,T)
```

Pushing SOLVE and sweeping T from 400 K to 900 K, the change in specific enthalpy is delh = 531.7 kJ/kg, which agrees closely with the value obtained by integrating the specific heat function from Table A-21, as illustrated above.

The source of ideal gas specific heat data is experiment. Specific heats can be determined macroscopically from painstaking property measurements. In the limit as pressure tends to zero, the properties of a gas tend to merge into those of its ideal gas model, so macroscopically determined specific heats of a gas extrapolated to very low pressures may be called either *zero-pressure* specific heats or *ideal gas* specific heats. Although zero-pressure specific heats can be obtained by extrapolating macroscopically determined experimental data, this is rarely done nowadays because ideal gas specific heats can be readily calculated with expressions from statistical mechanics by using *spectral* data, which can be obtained experimentally with precision. The determination of ideal gas specific heats is one of the important areas where the *microscopic approach* contributes significantly to the application of engineering thermodynamics.

3.14 Applying the Energy Balance Using Ideal Gas Tables, Constant Specific Heats, and Software

Although changes in specific enthalpy and specific internal energy can be obtained by integrating specific heat expressions, as illustrated in Sec. 3.13.2, such evaluations are more easily conducted using ideal gas tables, the assumption of constant specific heats, and computer software, all introduced in the present section. These procedures are also illustrated in the present section via solved examples using the closed system energy balance.

3.14.1 Using Ideal Gas Tables

For a number of common gases, evaluations of specific internal energy and enthalpy changes are facilitated by the use of the *ideal gas tables*, Tables A-22 and A-23, which give u and h (or $\overline{u}$ and $\overline{h}$) versus temperature.

To obtain enthalpy versus temperature, write Eq. 3.43 as

$$h(T) = \int_{T_{\text{ref}}}^{T} c_p(T)\, dT + h(T_{\text{ref}})$$

where T_{ref} is an arbitrary reference temperature and $h(T_{\text{ref}})$ is an arbitrary value for enthalpy at the reference temperature. Tables A-22 and A-23 are based on the selection $h = 0$ at $T_{\text{ref}} = 0$ K. Accordingly, a tabulation of enthalpy versus temperature is developed through the integral[2]

$$h(T) = \int_{0}^{T} c_p(T)\, dT \qquad (3.49)$$

Tabulations of internal energy versus temperature are obtained from the tabulated enthalpy values by using $u = h - RT$.

For air as an ideal gas, h and u are given in Table A-22 with units of kJ/kg. Values of molar specific enthalpy $\overline{h}$ and internal energy $\overline{u}$ for several other common gases modeled as ideal gases are given in Table A-23 with units of kJ/kmol. Quantities other than specific internal energy and enthalpy appearing in these tables are introduced in Chap. 6 and should be ignored at present. Tables A-22 and A-23 are convenient for evaluations involving ideal gases, not only because the variation of the specific heats with temperature is accounted for automatically but also because the tables are easy to use.

The next example illustrates the use of the ideal gas tables, together with the closed system energy balance.

----

[2]The simple specific heat variation given by Eq. 3.48 is valid only for a limited temperature range, so tabular enthalpy values are calculated from Eq. 3.49 using other expressions that enable the integral to be evaluated accurately over wider ranges of temperature.

▶▶ **EXAMPLE 3.9** ▶

Using the Energy Balance and Ideal Gas Tables

A piston–cylinder assembly contains 0.9 kg of air at a temperature of 300 K and a pressure of 1 bar. The air is compressed to a state where the temperature is 470 K and the pressure is 6 bars. During the compression, there is a heat transfer from the air to the surroundings equal to 20 kJ. Using the ideal gas model for air, determine the work during the process, in kJ.

SOLUTION

Known   0.9 kg of air are compressed between two specified states while there is heat transfer from the air of a known amount.

Find   Determine the work, in kJ.

Schematic and Given Data:

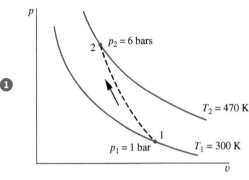

Fig. E3.9

Engineering Model

1. The air is a closed system.

2. The initial and final states are equilibrium states. There is no change in kinetic or potential energy.

❷ 3. The air is modeled as an ideal gas.

Analysis An energy balance for the closed system is

$$\Delta \cancel{KE}^0 + \Delta \cancel{PE}^0 + \Delta U = Q - W$$

where the kinetic and potential energy terms vanish by assumption 2. Solving for W

❸ $$W = Q - \Delta U = Q - m(u_2 - u_1)$$

From the problem statement, $Q = -20$ kJ. Also, from Table A-22, at $T_1 = 300$ K, $u_1 = 214.07$ kJ/kg, and at $T_2 = 470$ K, $u_2 = 337.32$ kJ/kg. Accordingly

$$W = -20 \text{ kJ} - (0.9 \text{ kg})(337.32 - 214.07) = -130.9 \text{ kJ}$$

The minus sign indicates that work is done on the system in the process.

❶ Although the initial and final states are assumed to be equilibrium states, the intervening states are not necessarily equilibrium states, so the process has been indicated on the accompanying p–v diagram by a dashed line. This dashed line does not define a "path" for the process.

❷ Table A-1 gives $p_c = 37.7$ bars, $T_c = 133$ K for air. Therefore, at state 1, $p_{R1} = 0.03$, $T_{R1} = 2.26$, and at state 2, $p_{R2} = 0.16$, $T_{R2} = 3.51$. Referring to Fig. A-1, we conclude that at these states $Z \approx 1$, as assumed in the solution.

❸ In principle, the work could be evaluated through $\int p \, dV$, but because the variation of pressure at the piston face with volume is not known, the integration cannot be performed without more information.

SKILLS DEVELOPED

Ability to…

- define a closed system and identify interactions on its boundary.
- apply the energy balance using the ideal gas model.

Quick Quiz

Replacing air by carbon dioxide, but keeping all other problem statement details the same, evaluate work, in kJ. Ans. −131.99 kJ

3.14.2 Using Constant Specific Heats

When the specific heats are taken as constants, Eqs. 3.40 and 3.43 reduce, respectively, to

$$u(T_2) - u(T_1) = c_v(T_2 - T_1) \tag{3.50}$$
$$h(T_2) - h(T_1) = c_p(T_2 - T_1) \tag{3.51}$$

Equations 3.50 and 3.51 are often used for thermodynamic analyses involving ideal gases because they enable simple closed-form equations to be developed for many processes.

The constant values of c_v and c_p in Eqs. 3.50 and 3.51 are, strictly speaking, mean values calculated as follows:

$$c_v = \frac{\int_{T_1}^{T_2} c_v(T) \, dT}{T_2 - T_1}, \qquad c_p = \frac{\int_{T_1}^{T_2} c_p(T) \, dT}{T_2 - T_1}$$

However, when the variation of c_v or c_p over a given temperature interval is slight, little error is normally introduced by taking the specific heat required by Eq. 3.50 or 3.51 as the arithmetic average of the specific heat values at the two end temperatures. Alternatively, the specific heat at the average temperature over the interval can be used. These methods are particularly convenient when tabular specific heat data are available, as in Table A-20, for then the *constant* specific heat values often can be determined by inspection.

FOR EXAMPLE

Assuming the specific heat c_v is a constant and using Eq. 3.50, the expression for work in the solution of Example 3.9 reads

$$W = Q - mc_v(T_2 - T_1)$$

Evaluating c_v at the average temperature, 383 K (110°C), Table A-20 gives $c_v = 0.721$ kJ/kg · K. Inserting this value for c_v together with other data from Example 3.9

$$W = -20 \text{ kJ} - (0.9 \text{ kg})(0.721 \text{ kJ/kg} \cdot \text{K})(470 - 300) \text{ K}$$

$$= -130.31 \text{ kJ}$$

which agrees closely with the answer obtained in Example 3.9 using Table A-22 data.

The following example illustrates the use of the closed system energy balance, together with the ideal gas model and the assumption of constant specific heats.

▶▶ EXAMPLE 3.10 ▶

Using the Energy Balance and Constant Specific Heats

Two tanks are connected by a valve. One tank contains 2 kg of carbon monoxide gas at 77°C and 0.7 bar. The other tank holds 8 kg of the same gas at 27°C and 1.2 bar. The valve is opened and the gases are allowed to mix while receiving energy by heat transfer from the surroundings. The final equilibrium temperature is 42°C. Using the ideal gas model with constant c_v, determine (a) the final equilibrium pressure, in bar, (b) the heat transfer for the process, in kJ.

SOLUTION

Known Two tanks containing different amounts of carbon monoxide gas at initially different states are connected by a valve. The valve is opened and the gas allowed to mix while receiving energy by heat transfer. The final equilibrium temperature is known.

Find Determine the final pressure and the heat transfer for the process.

Schematic and Given Data:

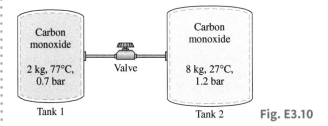

Tank 1 — Carbon monoxide — 2 kg, 77°C, 0.7 bar

Valve

Tank 2 — Carbon monoxide — 8 kg, 27°C, 1.2 bar

Fig. E3.10

Engineering Model

1. The total amount of carbon monoxide gas is a closed system.
❶ 2. The gas is modeled as an ideal gas with constant c_v.
3. The gas initially in each tank is in equilibrium. The final state is an equilibrium state.
4. No energy is transferred to, or from, the gas by work.
5. There is no change in kinetic or potential energy.

Analysis

a. The final equilibrium pressure p_f can be determined from the ideal gas equation of state

$$p_f = \frac{mRT_f}{V}$$

where m is the sum of the initial amounts of mass present in the two tanks, V is the total volume of the two tanks, and T_f is the final equilibrium temperature. Thus,

$$p_f = \frac{(m_1 + m_2)RT_f}{V_1 + V_2}$$

Denoting the initial temperature and pressure in tank 1 as T_1 and p_1, respectively, $V_1 = m_1RT_1/p_1$. Similarly, if the initial temperature and pressure in tank 2 are T_2 and p_2, $V_2 = m_2RT_2/p_2$. Thus, the final pressure is

$$p_f = \frac{(m_1 + m_2)RT_f}{\left(\dfrac{m_1RT_1}{p_1}\right) + \left(\dfrac{m_2RT_2}{p_2}\right)} = \frac{(m_1 + m_2)T_f}{\left(\dfrac{m_1T_1}{p_1}\right) + \left(\dfrac{m_2T_2}{p_2}\right)}$$

Inserting values

$$p_f = \frac{(10 \text{ kg})(315 \text{ K})}{\dfrac{(2 \text{ kg})(350 \text{ K})}{0.7 \text{ bar}} + \dfrac{(8 \text{ kg})(300 \text{ K})}{1.2 \text{ bar}}} = 1.05 \text{ bar}$$

b. The heat transfer can be found from an energy balance, which reduces with assumptions 4 and 5 to give

$$\Delta U = Q - \cancel{W}^0$$

or

$$Q = U_f - U_i$$

U_i is the initial internal energy, given by

$$U_i = m_1 u(T_1) + m_2 u(T_2)$$

where T_1 and T_2 are the initial temperatures of the CO in tanks 1 and 2, respectively. The final internal energy is U_f

$$U_f = (m_1 + m_2)u(T_f)$$

Introducing these expressions for internal energy, the energy balance becomes

$$Q = m_1[u(T_f) - u(T_1)] + m_2[u(T_f) - u(T_2)]$$

Since the specific heat c_v is constant (assumption 2)

$$Q = m_1 c_v (T_f - T_1) + m_2 c_v (T_f - T_2)$$

Evaluating c_v as the average of the values listed in Table A-20 at 300 K and 350 K, $c_v = 0.745$ kJ/kg · K. Hence,

$$Q = (2 \text{ kg})\left(0.745 \frac{\text{kJ}}{\text{kg} \cdot \text{K}}\right)(315 \text{ K} - 350 \text{ K})$$

$$+ (8 \text{ kg})\left(0.745 \frac{\text{kJ}}{\text{kg} \cdot \text{K}}\right)(315 \text{ K} - 300 \text{ K})$$

$$= +37.25 \text{ kJ}$$

The plus sign indicates that the heat transfer is into the system.

❶ By referring to a generalized compressibility chart, it can be verified that the ideal gas equation of state is appropriate for CO in this range of temperature and pressure. Since the specific heat c_v of CO varies little over the temperature interval from 300 to 350 K (Table A-20), it can be treated as constant with acceptable accuracy.

SKILLS DEVELOPED

Ability to...

• define a closed system and identify interactions on its boundary.

• apply the energy balance using the ideal gas model when the specific heat c_v is constant.

Quick Quiz

Evaluate Q using specific internal energy values for CO from Table A-23. Compare with the result using constant c_v. Ans. 36.99 kJ

3.14.3 Using Computer Software

Interactive Thermodynamics: IT also provides values of the specific internal energy and enthalpy for a wide range of gases modeled as ideal gases. Let us consider the use of *IT*, first for air, and then for other gases.

Air For air, *IT* uses the same reference state and reference value as in Table A-22, and the values computed by *IT* agree closely with table data.

FOR EXAMPLE

Let us use *IT* to evaluate the change in specific enthalpy of air from a state where $T_1 = 400$ K to a state where $T_2 = 900$ K. Selecting Air from the Properties menu, the following code would be used by *IT* to determine Δh (delh), in kJ/kg

```
h1 = h_T("Air",T1)
h2 = h_T("Air",T2)
T1 = 400//K
T2 = 900//K
delh = h2 - h1
```

Choosing K for the temperature unit and kg for the amount under the Units menu, the results returned by *IT* are $h_1 = 400.8$, $h_2 = 932.5$, and $\Delta h = 531.7$ kJ/kg, respectively. These values agree closely with those obtained from Table A-22: $h_1 = 400.98$, $h_2 = 932.93$, and $\Delta h = 531.95$ kJ/kg.

Other Gases *IT* also provides data for each of the gases included in Table A-23. For these gases, the values of specific internal energy $\bar{u}$ and enthalpy $\bar{h}$ returned by *IT* are determined relative to a *standard reference state* that differs from that used in Table A-23. This equips *IT* for use in combustion applications; see Sec. 13.2.1 for further discussion. Consequently, the values of $\bar{u}$ and $\bar{h}$ returned by *IT* for the gases of Table A-23 differ from those obtained directly from the table. Still, the property differences between two states remain the same, for datums cancel when differences are calculated.

FOR EXAMPLE

Let us use *IT* to evaluate the change in specific enthalpy, in kJ/kmol, for carbon dioxide (CO_2) as an ideal gas from a state where $T_1 = 300$ K to a state where $T_2 = 500$ K. Selecting CO_2 from the Properties menu, the following code would be used by *IT*:

```
h1 = h_T("CO₂",T1)
h2 = h_T("CO₂",T2)
T1 = 300//K
T2 = 500//K
delh = h2 - h1
```

Choosing K for the temperature unit and moles for the amount under the Units menu, the results returned by *IT* are $\bar{h}_1 = -3.935 \times 10^5$, $\bar{h}_2 = -3.852 \times 10^5$, and $\Delta\bar{h} = 8238$ kJ/mol, respectively. The large negative values for $\bar{h}_1$ and $\bar{h}_2$ are a consequence of the reference state and reference value used by *IT* for CO_2. Although these values for specific enthalpy at states 1 and 2 differ from the corresponding values read from Table A-23: $\bar{h}_1 = 9,431$ and $\bar{h}_2 = 17,678$, which give $\Delta\bar{h} = 8247$ kJ/kmol, the *differences* in specific enthalpy determined with each set of data agree closely.

The next example illustrates the use of software for problem solving with the ideal gas model. The results obtained are compared with those determined assuming the specific heat $\bar{c}_v$ is constant.

EXAMPLE 3.11

Using the Energy Balance and Software

One kmol of carbon dioxide gas (CO_2) in a piston–cylinder assembly undergoes a constant-pressure process at 1 bar from $T_1 = 300$ K to T_2. Plot the heat transfer to the gas, in kJ, versus T_2 ranging from 300 to 1500 K. Assume the ideal gas model, and determine the specific internal energy change of the gas using

   a. $\bar{u}$ data from *IT*.

   b. a constant $\bar{c}_v$ evaluated at T_1 from *IT*.

SOLUTION

Known   One kmol of CO_2 undergoes a constant-pressure process in a piston–cylinder assembly. The initial temperature, T_1, and the pressure are known.

Find   Plot the heat transfer versus the final temperature, T_2. Use the ideal gas model and evaluate $\Delta\bar{u}$ using (**a**) $\bar{u}$ data from *IT*, (**b**) constant $\bar{c}_v$ evaluated at T_1 from *IT*.

Schematic and Given Data:

$T_1 = 300$ K
$n = 1$ kmol
$p = 1$ bar

Carbon dioxide

Fig. E3.11a

Engineering Model

1. The carbon dioxide is a closed system.

2. The piston is the only work mode, and the process occurs at constant pressure.

3. The carbon dioxide behaves as an ideal gas.

4. Kinetic and potential energy effects are negligible.

Analysis   The heat transfer is found using the closed system energy balance, which reduces to

$$U_2 - U_1 = Q - W$$

Using Eq. 2.17 at constant pressure (assumption 2)

$$W = p(V_2 - V_1) = pn(\bar{v}_2 - \bar{v}_1)$$

Then, with $\Delta U = n(\bar{u}_2 - \bar{u}_1)$, the energy balance becomes

$$n(\bar{u}_2 - \bar{u}_1) = Q - pn(\bar{v}_2 - \bar{v}_1)$$

Solving for Q

①

$$Q = n[(\bar{u}_2 - \bar{u}_1) + p(\bar{v}_2 - \bar{v}_1)]$$

With $p\bar{v} = \bar{R}T$, this becomes

$$Q = n[(\bar{u}_2 - \bar{u}_1) + \bar{R}(T_2 - T_1)]$$

The object is to plot Q versus T_2 for each of the following cases: (**a**) values for $\bar{u}_1$ and $\bar{u}_2$ at T_1 and T_2, respectively, are provided by *IT*, (**b**) Eq. 3.50 is used on a molar basis, namely,

$$\bar{u}_2 - \bar{u}_1 = \bar{c}_v(T_2 - T_1)$$

where the value of $\bar{c}_v$ is evaluated at T_1 using *IT*.

The *IT* program follows, where Rbar denotes $\bar{R}$, cvb denotes $\bar{c}_v$, and ubar1 and ubar2 denote $\bar{u}_1$ and $\bar{u}_2$, respectively.

```
//Using the Units menu, select "mole" for the
substance amount.
//Given Data
T1 = 300//K
T2 = 1500//K
n = 1//kmol
Rbar = 8.314//kJ/kmol·K
```

```
// (a) Obtain molar specific internal energy data
using IT.
ubar1 = u_T ("CO2", T1)
ubar2 = u_T ("CO2", T2)
Qa = n*(ubar2 - ubar1) + n*Rbar*(T2 - T1)

// (b) Use Eq. 3.50 with cv evaluated at T1.
cvb = cv_T ("CO2", T1)
Qb = n*cvb*(T2 - T1) + n*Rbar*(T2 - T1)
```

Use the **Solve** button to obtain the solution for the sample case of $T_2 =$ 1500 K. For part (a), the program returns $Q_a = 6.16 \times 10^4$ kJ. The solution can be checked using CO_2 data from Table A-23, as follows:

$$Q_a = n[(\bar{u}_2 - \bar{u}_1) + \bar{R}(T_2 - T_1)]$$

$$= (1 \text{ kmol})[(58,606 - 6939)\text{kJ/kmol}$$
$$+ (8.314 \text{ kJ/kmol} \cdot \text{K})(1500 - 300)\text{K}]$$

$$= 61,644 \text{ kJ}$$

Thus, the result obtained using CO_2 data from Table A-23 is in close agreement with the computer solution for the sample case. For part (b), *IT* returns $\bar{c}_v = 28.95$ kJ/kmol · K at T_1, giving $Q_b = 4.472 \times 10^4$ kJ when $T_2 = 1500$ K. This value agrees with the result obtained using the specific heat c_v at 300 K from Table A-20, as can be verified.

Now that the computer program has been verified, use the **Explore** button to vary T_2 from 300 to 1500 K in steps of 10. Construct the following graph using the **Graph** button:

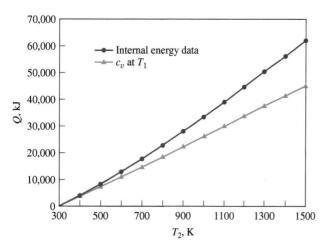

Fig. E3.11b

As expected, the heat transfer is seen to increase as the final temperature increases. From the plots, we also see that using constant $\bar{c}_v$ evaluated at T_1 for calculating $\Delta \bar{u}$, and hence Q, can lead to considerable error when compared to using $\bar{u}$ data. The two solutions compare favorably up to about 500 K, but differ by approximately 27% when heating to a temperature of 1500 K.

① Alternatively, this expression for Q can be written as

$$Q = n[(\bar{u}_2 + p\bar{v}_2) - (\bar{u}_1 + p\bar{v}_1)]$$

Introducing $\bar{h} = \bar{u} + p\bar{v}$, the expression for Q becomes

$$Q = n(\bar{h}_2 - \bar{h}_1)$$

SKILLS DEVELOPED

Ability to...

- define a closed system and identify interactions on its boundary.
- apply the energy balance using the ideal gas model.
- use *IT* to retrieve property data for CO_2 as an ideal gas and plot calculated data.

Quick Quiz

Repeat part (b) using $\bar{c}_v$ evaluated at $T_{average} = (T_1 + T_2)/2$. Which approach gives better agreement with the results of part (a): evaluating $\bar{c}_v$ at T_1 or at $T_{average}$? Ans. At $T_{average}$

3.15 Polytropic Process Relations

A *polytropic process* is a quasiequilibrium process (Sec. 2.2.5) described by

$$pV^n = constant \tag{3.52}$$

or, in terms of specific volumes, by $pv^n = constant$. In these expressions, n is a constant.
For a polytropic process between two states

$$p_1 V_1^n = p_2 V_2^n$$

or

$$\frac{p_2}{p_1} = \left(\frac{V_1}{V_2}\right)^n \tag{3.53}$$

The exponent n may take on any value from $-\infty$ to $+\infty$ depending on the particular process. When $n = 0$, the process is an isobaric (constant-pressure) process, and when $n = \pm\infty$ the process is an isometric (constant-volume) process.

For a polytropic process

$$\int_1^2 p\, dV = \frac{p_2 V_2 - p_1 V_1}{1 - n} \qquad (n \neq 1) \tag{3.54}$$

for any exponent n except $n = 1$. When $n = 1$,

$$\int_1^2 p\, dV = p_1 V_1 \ln\frac{V_2}{V_1} \qquad (n = 1) \tag{3.55}$$

Example 2.1 provides the details of these integrations.

Equations 3.52 through 3.55 apply to *any* gas (or liquid) undergoing a polytropic process. When the *additional* idealization of ideal gas behavior is appropriate, further relations can be derived. Thus, when the ideal gas equation of state is introduced into Eqs. 3.53, 3.54, and 3.55, the following expressions are obtained, respectively,

$$\frac{T_2}{T_1} = \left(\frac{p_2}{p_1}\right)^{(n-1)/n} = \left(\frac{V_1}{V_2}\right)^{n-1} \qquad \text{(ideal gas)} \tag{3.56}$$

$$\int_1^2 p\, dV = \frac{mR(T_2 - T_1)}{1 - n} \qquad \text{(ideal gas, } n \neq 1\text{)} \tag{3.57}$$

$$\int_1^2 p\, dV = mRT \ln\frac{V_2}{V_1} \qquad \text{(ideal gas, } n = 1\text{)} \tag{3.58}$$

For an ideal gas, the case $n = 1$ corresponds to an isothermal (constant-temperature) process, as can readily be verified.

Example 3.12 illustrates the use of the closed system energy balance for a system consisting of an ideal gas undergoing a polytropic process.

► EXAMPLE 3.12 ►

Polytropic Process of Air as an Ideal Gas

Air undergoes a polytropic compression in a piston–cylinder assembly from $p_1 = 1$ bar, $T_1 = 22°C$ to $p_2 = 5$ bars. Employing the ideal gas model with constant specific heat ratio k, determine the work and heat transfer per unit mass, in kJ/kg, if (a) $n = 1.3$, (b) $n = k$. Evaluate k at T_1.

SOLUTION

Known Air undergoes a polytropic compression process from a given initial state to a specified final pressure.

Find Determine the work and heat transfer, each in kJ/kg.

Schematic and Given Data:

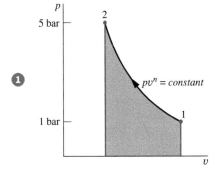

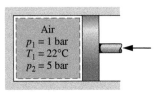

Engineering Model

1. The air is a closed system.

2. The air behaves as an ideal gas with constant specific heat ratio k evaluated at the initial temperature.

3. The compression is polytropic.

4. There is no change in kinetic or potential energy.

Analysis The work can be evaluated in this case from the expression

$$W = \int_1^2 p\, dV$$

Fig. E3.12

With Eq. 3.57

$$\frac{W}{m} = \frac{R(T_2 - T_1)}{1 - n} \quad \text{(a)}$$

The heat transfer can be evaluated from an energy balance. Thus

$$\frac{Q}{m} = \frac{W}{m} + (u_2 - u_1)$$

Inspection of Eq. 3.47b shows that when the specific heat ratio k is constant, c_v is constant. Thus

$$\frac{Q}{m} = \frac{W}{m} + c_v(T_2 - T_1) \quad \text{(b)}$$

a. For $n = 1.3$, the temperature at the final state, T_2, can be evaluated from Eq. 3.56 as follows:

$$T_2 = T_1 \left(\frac{p_2}{p_1} \right)^{(n-1)/n} = 295 \text{ K} \left(\frac{5}{1} \right)^{(1.3-1)/1.3} = 427.68 \text{ K } (154.68°\text{C})$$

Using Eq. (a), the work is then

$$\frac{W}{m} = \frac{R(T_2 - T_1)}{1 - n} = \left(287.21 \frac{\text{J}}{\text{kg} \cdot \text{K}} \right) \left(\frac{427.68 \text{ K} - 295 \text{ K}}{1 - 1.3} \right)$$
$$= -127.02 \text{ kJ/kg}$$

At 22°C, Table A-20 gives $k = 1.401$ and $c_v = 0.716$ kJ/kg · K. Alternatively, c_v can be found using Eq. 3.47b, as follows:

$$c_v = \frac{R}{k - 1} \quad \text{(c)}$$
$$= \frac{287.21 \text{ J/kg} \cdot \text{K}}{(1.401 - 1)} = 0.716 \text{ kJ/kg} \cdot \text{K}$$

Substituting values into Eq. (b), we get

$$\frac{Q}{m} = -127.02 \frac{\text{kJ}}{\text{kg}} + \left(0.716 \frac{\text{kJ}}{\text{kg} \cdot \text{K}} \right)(427.68 - 295) \text{ K}$$
$$= -32.02 \frac{\text{kJ}}{\text{kg}}$$

b. For $n = k$, substituting Eqs. (a) and (c) into Eq. (b) gives

$$\frac{Q}{m} = \frac{R(T_2 - T_1)}{1 - k} + \frac{R(T_2 - T_1)}{k - 1} = 0$$

That is, no heat transfer occurs in the polytropic process of an ideal gas for which $n = k$.

---

1 The states visited in a polytropic compression process are shown by the curve on the accompanying p–v diagram. The magnitude of the work per unit of mass is represented by the shaded area *below* the curve.

SKILLS DEVELOPED

Ability to...

- evaluate work using Eq. 2.17.
- apply the energy balance using the ideal gas model.
- apply the polytropic process concept.

Quick Quiz

For $n = k$, evaluate the temperature at the final state, *in* K and °C. Ans. 466.67 K (193.67°C)

CHAPTER SUMMARY AND STUDY GUIDE

In this chapter, we have considered property relations for a broad range of substances in tabular, graphical, and equation form. Primary emphasis has been placed on the use of tabular data, but computer retrieval also has been considered.

A key aspect of thermodynamic analysis is fixing states. This is guided by the state principle for pure, simple compressible systems, which indicates that the intensive state is fixed by the values of *any two* independent, intensive properties.

Another important aspect of thermodynamic analysis is locating principal states of processes on appropriate diagrams: p–v, T–v, and p–T diagrams. The skills of fixing states and using property diagrams are particularly important when solving problems involving the energy balance.

The ideal gas model is introduced in the second part of this chapter, using the compressibility factor as a point of departure. This arrangement emphasizes the limitations of the ideal gas model. When it is appropriate to use the ideal gas model, we stress that specific heats generally vary with temperature and feature the use of the ideal gas tables in problem solving.

The following checklist provides a study guide for this chapter. When your study of the text and end-of-chapter exercises has been completed you should be able to

- write out the meanings of the terms listed in the margins throughout the chapter and understand each of the related concepts. The subset of key concepts is particularly important in subsequent chapters.
- retrieve property data from Tables A-1 through A-23, using the state principle to fix states and linear interpolation when required.
- sketch T–v, p–v, and p–T diagrams and locate principal states on such diagrams.
- apply the closed system energy balance with property data.
- evaluate the properties of two-phase liquid–vapor mixtures using Eqs. 3.1, 3.2, 3.6, and 3.7.
- estimate the properties of liquids using Eqs. 3.11–3.14.
- apply the incompressible substance model.
- use the generalized compressibility chart to relate p–v–T data of gases.
- apply the ideal gas model for thermodynamic analysis, including determining when use of the ideal gas model is warranted, and appropriately using ideal gas table data or constant specific heat data to determine Δu and Δh.
- apply polytropic process relations.

KEY ENGINEERING CONCEPTS

phase
pure substance
state principle
simple compressible system
p–v–T surface
phase diagram
saturation temperature

saturation pressure
p–v diagram
T–v diagram
compressed liquid
two-phase liquid–vapor mixture
quality
superheated vapor

enthalpy
specific heats
incompressible substance model
universal gas constant
compressibility factor
ideal gas model

KEY EQUATIONS

$x = \dfrac{m_{\text{vapor}}}{m_{\text{liquid}} + m_{\text{vapor}}}$	(3.1)	Quality, x, of a two-phase, liquid–vapor mixture.
$v = (1 - x)v_\text{f} + xv_\text{g} = v_\text{f} + x(v_\text{g} - v_\text{f})$	(3.2)	
$u = (1 - x)u_\text{f} + xu_\text{g} = u_\text{f} + x(u_\text{g} - u_\text{f})$	(3.6)	Specific volume, internal energy and enthalpy of a
$h = (1 - x)h_\text{f} + xh_\text{g} = h_\text{f} + x(h_\text{g} - h_\text{f})$	(3.7)	two-phase liquid–vapor mixture.
$v(T, p) \approx v_\text{f}(T)$	(3.11)	Specific volume, internal energy, and enthalpy of
$u(T, p) \approx u_\text{f}(T)$	(3.12)	liquids, approximated by saturated liquid values,
$h(T, p) \approx h_\text{f}(T)$	(3.14)	respectively.

Ideal Gas Model Relations

$pv = RT$	(3.32)	
$u = u(T)$	(3.36)	Ideal gas model.
$h = h(T) = u(T) + RT$	(3.37)	
$u(T_2) - u(T_1) = \int_{T_1}^{T_2} c_v(T)\, dT$	(3.40)	Change in specific internal energy.
$u(T_2) - u(T_1) = c_v(T_2 - T_1)$	(3.50)	For constant c_v.
$h(T_2) - h(T_1) = \int_{T_1}^{T_2} c_p(T)\, dT$	(3.43)	Change in specific enthalpy.
$h(T_2) - h(T_1) = c_p(T_2 - T_1)$	(3.51)	For constant c_p.

EXERCISES: THINGS ENGINEERS THINK ABOUT

3.1 If water contracted on freezing, what implications might this have for aquatic life?

3.2 A plastic milk jug filled with water and stored within a freezer ruptures. Why?

3.3 Why do many foods have high-altitude cooking instructions?

3.4 What is the *standard* composition of atmospheric air?

3.5 Can a substance that contracts on freezing exist as a solid when at a temperature greater than its triple-point temperature? Repeat for a substance that expands on freezing.

3.6 When should Table A-5 be used for liquid water v, u, and h values? When should Eqs. 3.11–3.14 be used?

3.7 The specific internal energy is arbitrarily set to zero in Table A-2 for saturated liquid water at 0.01°C. If the reference value for u at this reference state were specified differently, would there be any significant effect on thermodynamic analyses using u and h?

3.8 How does a pressure cooker work to cook food faster than an ordinary pan with a lid?

3.9 If a block of iron and a block of tin having equal volumes each received the same energy input by heat transfer, which block would experience the greater temperature increase?

3.10 Why are the tires of airplanes and race cars inflated with nitrogen instead of air?

3.11 How many minutes do you have to exercise to *burn* the calories in a helping of your favorite dessert?

3.12 What is a *molten salt*?

3.13 Do specific volume and specific internal energy fix the state of a simple compressible system? If so, how can you use the *steam tables* to find the state for H_2O?

CHECKING UNDERSTANDING

3.1 The quality of a two-phase liquid–vapor mixture of H_2O at 40°C with a specific volume of 10 m^3/kg is

    a. 0   c. 0.512

    b. 0.486   d. 1

3.2 The quality of a two-phase liquid–vapor mixture of propane at 20 bar with a specific internal energy of 300 kJ/kg is

    a. 0.166   c. 0.575

    b. 0.214   d. 0.627

3.3 A system contains a two-phase liquid–vapor mixture at equilibrium. What does it mean to say that the pressure and temperature are not independently variable for this system?

3.4 A substance that is uniform and invariable in chemical composition is called a _____ substance.

3.5 Is the expression for work of a polytropic process

$$W = \frac{(p_2V_2 - p_1V_1)}{1 - n}$$

restricted to processes of an ideal gas? Explain.

3.6 Show that the specific enthalpy of an ideal gas is a function of temperature only.

3.7 The specific heat ratio, k, must be greater than one. Why?

3.8 Given temperature and specific volume of a two-phase liquid–vapor mixture, how would you determine the specific internal energy?

3.9 The energy of simple compressible systems can be changed by heat transfer and by work associated with _____.

3.10 What is the *state principle* for simple systems?

3.11 The quality of saturated vapor is _____.

3.12 The quality of saturated liquid is _____.

3.13 A system consists of a two-phase liquid–vapor mixture of 5 kg of Refrigerant 134a. One kg is saturated liquid. What is the quality?

3.14 For H_2O at 10 bar and 220°C, what table would you use to find v and u?

3.15 What is the *principle of corresponding states*?

3.16 List three key aspects of the ideal gas model.

3.17 A generalized form of the equation of state for gases that can be derived from the methods of statistical mechanics is called the _____ equation of state.

3.18 Why is subcooled liquid alternatively referred to as compressed liquid?

3.19 Are the data in the *pressure table* and the *temperature table* for saturated two-phase liquid–vapor mixtures compatible?

3.20 The specific volume of saturated liquid water at 0.01°C is $v_f = 1.0002 \times 10^{-3}$ m^3/kg. For saturated solid at 0.01°C, the value of specific volume is $v_i = 1.0908 \times 10^{-3}$ m^3/kg. Why is $v_i > v_f$?

Indicate whether the following statements are true or false. Explain.

3.21 Air can always be regarded as a *pure substance*.

3.22 For liquid water, the approximation $v(T, p) \approx v_f(T)$ is reasonable for many engineering applications.

3.23 A polytropic process with $n = k$ is adiabatic.

3.24 For simple compressible systems, any two intensive thermodynamic properties fix the state.

3.25 For gases modeled as ideal gases, the ratio c_v/c_p must be greater than one.

3.26 A two-phase liquid–vapor mixture with equal volumes of saturated liquid and saturated vapor has a quality of 0.5.

3.27 Carbon dioxide (CO_2) at 320 K and 55 bar can be modeled as an ideal gas.

3.28 When an ideal gas undergoes a *polytropic* process with $n = 1$, the gas temperature remains constant.

3.29 If a closed system consisting of a simple compressible substance is at equilibrium, only one phase can be present.

3.30 A two-phase liquid–vapor mixture has 0.2 kg of saturated water vapor and 0.6 kg of saturated liquid. The quality is 0.25 (25%).

PROBLEMS: DEVELOPING ENGINEERING SKILLS

C Problem may require use of appropriate computer software in order to complete.

Exploring Concepts: Phase and Pure Substance

3.1 A system consists of liquid water in equilibrium with a gaseous mixture of air and water vapor. How many phases are present? Does the system consist of a pure substance? Explain. Repeat for a system consisting of ice and liquid water in equilibrium with a gaseous mixture of air and water vapor.

3.2 A system consists of liquid oxygen in equilibrium with oxygen vapor. How many phases are present? The system undergoes a process during which some of the liquid is vaporized. Can the system be viewed as being a pure substance during the process? Explain.

3.3 A system consisting of liquid water undergoes a process. At the end of the process, some of the liquid water has frozen, and the system contains liquid water and ice. Can the system be viewed as being a pure substance during the process? Explain.

3.4 A dish of liquid water is placed on a table in a room. After a while, all of the water evaporates. Taking the water and the air in the room to be a closed system, can the system be regarded as a pure substance *during* the process? *After* the process is completed? Discuss.

Using p–v–T Data

3.5 For H_2O, plot the following on a p–v diagram drawn to scale on log–log coordinates:

    a. the saturated liquid and saturated vapor lines from the triple point to the critical point, with pressure in MPa and specific volume in m^3/kg.

    b. lines of constant temperature at 100 and 300°C.

3.6 For H_2O, determine the specified property at the indicated state. Locate the state on a sketch of the T–v diagram.

a. $p = 300$ kPa, $v = 0.5$ m³/kg. Find T, in °C.

b. $p = 28$ MPa, $T = 200$°C. Find v, in m³/kg.

c. $p = 1$ MPa, $T = 405$°C. Find v, in m³/kg.

d. $T = 100$°C, $x = 60\%$. Find v, in m³/kg.

3.7 For H₂O, determine the specific volume at the indicated state, in m³/kg. Locate the states on a sketch of the T–v diagram.

a. $T = 400$°C, $p = 20$ MPa.

b. $T = 40$°C, $p = 20$ MPa.

c. $T = 40$°C, $p = 2$ MPa.

3.8 Determine the phase or phases in a system consisting of H₂O at the following conditions and sketch p–v and T–v diagrams showing the location of each state.

a. $p = 5$ bar, $T = 151.9$°C.

b. $p = 5$ bar, $T = 200$°C.

c. $T = 200$°C, $p = 2.5$ MPa.

d. $T = 160$°C, $p = 4.8$ bar.

e. $T = -12$°C, $p = 1$ bar.

3.9 Four kg of water at 100°C fills a closed container having a volume of 1 m³. If the water at this state is a vapor, determine the pressure, in bar. If the water is a two-phase liquid–vapor mixture, determine the quality. Locate the state on sketches of the T–v and p–v diagrams.

3.10 A 1-m³ tank holds a two-phase liquid–vapor mixture of carbon dioxide at -17°C. The quality of the mixture is 70%. For saturated carbon dioxide at -17°C, $v_f = 0.9827 \times 10^{-3}$ m³/kg and $v_g = 1.756 \times 10^{-2}$ m³/kg. Determine the masses of saturated liquid and saturated vapor, each in kg. What is the percent of the total volume occupied by saturated liquid?

3.11 Determine the specific volume of a two-phase liquid vapor mixture of a substance. The pressure is 200 bar, and the mixture occupies a volume of 0.3 m³. The masses of saturated vapor and liquid are 3.5 kg and 4.5 kg, respectively. Also, determine the quality of the mixture.

3.12 A tank contains a two-phase liquid–vapor mixture of Refrigerant 22 at 10 bar. The mass of saturated liquid in the tank is 25 kg and the quality is 60%. Determine the volume of the tank, in m³, and the fraction of the total volume occupied by saturated vapor.

3.13 As shown in Fig. P3.13, 0.1 kg of water is contained within a piston–cylinder assembly at 100°C. The piston is free to move smoothly in the cylinder. The local atmospheric pressure and acceleration of gravity are 100 kPa and 9.81 m/s², respectively. For the water, determine the pressure, in kPa, and volume, in cm³.

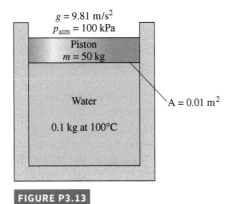

FIGURE P3.13

3.14 Ammonia, initially saturated vapor at -4°C, undergoes a constant-specific volume process to 200 kPa. At the final state, determine the temperature, in °C, and the quality. Locate each state on a sketch of the T–v diagram.

3.15 Water is contained in a closed, rigid, 0.2 m³ tank at an initial pressure of 5 bar and a quality of 50%. Heat transfer occurs until the tank contains only saturated vapor. Determine the final mass of vapor in the tank, in kg, and the final pressure, in bar.

3.16 A rigid tank contains 2.27 kg of a two-phase, liquid–vapor mixture of H₂O, initially at 127°C with a quality of 0.6. Heat transfer to the contents of the tank occurs until the temperature is 160°C. Show the process on a p–v diagram. Determine the mass of vapor, in kg, initially present in the tank and the final pressure, in kPa.

3.17 **C** Ammonia contained in a closed, rigid tank is heated from an initial saturated vapor state at temperature $T_1 = 20$°C to the final temperature, $T_2 = 40$°C. Using IT, determine the final pressure, in bar.

Compare the pressure values determined using IT with those obtained using the appropriate Appendix tables for ammonia.

3.18 A closed, rigid tank contains a two-phase liquid–vapor mixture of Refrigerant 22 initially at -20°C with a quality of 50.36%. Energy transfer by heat into the tank occurs until the refrigerant is at a final pressure of 6 bar. Determine the final temperature, in °C. If the final state is in the superheated vapor region, at what temperature, in °C, does the tank contain only saturated vapor?

3.19 Two thousand kg of water, initially a saturated liquid at 150°C, is heated in a closed, rigid tank to a final state where the pressure is 2.5 MPa. Determine the final temperature, in °C, the volume of the tank, in m³, and sketch the process on T–v and p–v diagrams.

3.20 Ammonia undergoes an isothermal process from an initial state at $T_1 = 27$°C and $v_1 = 0.6$ m²/kg to saturated vapor. Determine the initial and final pressures, in kPa, and sketch the process on T–v and p–v diagrams.

3.21 Steam is contained in a closed rigid container with a volume of 1 m³. Initially, the pressure and temperature of the steam are 7 bar and 500°C, respectively. The temperature drops as a result of heat transfer to the surroundings. Determine the temperature at which condensation first occurs, in °C, and the fraction of the total mass that has condensed when the pressure reaches 0.5 bar. What is the volume, in m³, occupied by saturated liquid at the final state?

3.22 Refrigerant 134a in a piston–cylinder assembly undergoes a process for which the pressure–volume relation is $pv^{1.058} = constant$. At the initial state, $p_1 = 200$ kPa, $T_1 = -10$°C. The final temperature is $T_2 = 50$°C. Determine the final pressure, in kPa, and the work for the process, in kJ/kg of refrigerant.

3.23 Two kilograms of Refrigerant 22 undergo a process for which the pressure–volume relation is $pv^{1.05} = constant$. The initial state of the refrigerant is fixed by $p_1 = 2$ bar, $T_1 = -20$°C, and the final pressure is $p_2 = 10$ bar. Calculate the work for the process, in kJ.

3.24 From an initial state where the pressure is p_1, the temperature is T_1, and the volume is V_1, water vapor contained in a piston–cylinder assembly undergoes each of the following processes:

Process 1–2: Constant-temperature to $p_2 = 2p_1$.

Process 1–3: Constant volume to $p_3 = 2p_1$.

Process 1–4: Constant pressure to $V_4 = 2V_1$.

Process 1–5: Constant temperature to $V_5 = 2V_1$.

On a p–V diagram, sketch each process, identify the work by an area on the diagram, and indicate whether the work is done by, or on, the water vapor.

3.25 As shown in **Fig. P3.25**, Refrigerant 134a is contained in a piston–cylinder assembly, initially as saturated vapor. The refrigerant is slowly heated until its temperature is 160°C. During the process, the piston moves smoothly in the cylinder. For the refrigerant, evaluate the work, in kJ/kg.

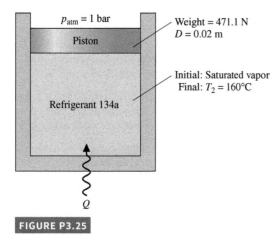

FIGURE P3.25

3.26 A piston–cylinder assembly contains 0.02 kg of Refrigerant 134a. The refrigerant is compressed from an initial state where $p_1 = 68.9$ kPa and $T_1 = -6.7$°C to a final state where $p_2 = 1.1$ MPa. During the process, the pressure and specific volume are related by $p\upsilon = constant$. Determine the work, in kJ, for the refrigerant.

Using u–h Data

3.27 C For each of the following cases, determine the specified properties using tables and using *IT*. Show the states on a sketch of the T-υ diagram.

    a. For water at $p = 80$ bar and $\upsilon = 0.04034$ m³/kg, determine T in °C, u in kJ/kg, and h in kJ/kg.

    b. For Refrigerant 134a at $T = -20$°C and $h = 235.31$ kJ/kg, determine p in bar, υ in m³/kg, and u in kJ/kg.

3.28 Using the tables for water, determine the specified property data at the indicated states. In each case, locate the state by hand on sketches of the p–υ and T–υ diagrams.

    a. At $p = 3$ bar, $T = 240$°C, find υ in m³/kg and u in kJ/kg.

    b. At $p = 3$ bar, $\upsilon = 0.5$ m³/kg, find T in °C and u in kJ/kg.

    c. At $T = 400$°C, $p = 10$ bar, find υ in m³/kg and h in kJ/kg.

    d. At $T = 320$°C, $\upsilon = 0.03$ m³/kg, find p in MPa and u in kJ/kg.

    e. At $p = 28$ MPa, $T = 520$°C, find υ in m³/kg and h in kJ/kg.

    f. At $T = 100$°C, $x = 60\%$, find p in bar and υ in m³/kg.

    g. At $T = 10$°C, $\upsilon = 100$ m³/kg, find p in kPa and h in kJ/kg.

    h. At $p = 4$ MPa, $T = 160$°C, find υ in m³/kg and u in kJ/kg.

3.29 Using the tables for water, determine the specified property data at the indicated states. In each case, locate the state by hand on sketches of the p–υ and T–υ diagrams.

    a. At $p = 1.38$ bar, $T = 204$°C, find υ in m³/kg and u in kJ/kg.

    b. At $p = 1.38$ bar, $\upsilon = 1$ m³/kg, find T in °C and u in kJ/kg.

    c. At $T = 482$°C, $p = 11.7$ bar, find υ in m³/kg and h in kJ/kg.

    d. At $T = 315.6$°C, $\upsilon = 0.04$ m³/kg, find p in kPa and u in kJ/kg.

    e. At $p = 48$ bars, $T = 343$°C, find υ in m³/kg and h in kJ/kg.

    f. At $T = 204$°C, $x = 90\%$, find p in kPa and υ in m³/kg.

    g. At $T = 4$°C, $\upsilon = 121.7$ m³/kg, find p in kPa and h in kJ/kg.

    h. At $p = 41.37$ bars, $T = 160$°C, find υ in m³/kg and u in kJ/kg.

3.30 Evaluate the specific volume, in m³/kg, and the specific enthalpy, in kJ/kg, of water at 204°C and a pressure of 207 bars.

3.31 Evaluate the specific volume, in m³/kg, and the specific enthalpy, in kJ/kg, of Refrigerant 134a at 35°C and 10.3 bar.

Applying the Energy Balance

3.32 Water, initially saturated vapor at 4 bar, fills a closed, rigid container. The water is heated until its temperature is 400°C. For the water, determine the heat transfer, in kJ per kg of water. Kinetic and potential energy effects can be ignored.

3.33 Refrigerant 134a is compressed with no heat transfer in a piston–cylinder assembly from 207 kPa, −6.7°C to 1.1 MPa. The mass of refrigerant is 0.02 kg. For the refrigerant as the system, $W = -0.59$ kJ. Kinetic and potential energy effects are negligible. Determine the final temperature, in °C.

3.34 Propane within a piston-cylinder assembly undergoes a constant-pressure process from saturated vapor at 400 kPa to a temperature of 40°C. Kinetic and potential energy effects are negligible. For the propane, (a) show the process on a p–υ diagram, (b) evaluate the work, in kJ/kg, and (c) evaluate the heat transfer, in kJ/kg.

3.35 A piston-cylinder assembly contains 1 kg of water, initially occupying a volume of 0.5 m³ at 1 bar. Energy transfer by heat to the water results in an expansion at constant temperature to a final volume of 1.694 m³. Kinetic and potential energy effects are negligible. For the water, (a) show the process on a T–υ diagram, (b) evaluate the work, in kJ, and (c) evaluate the heat transfer, in kJ.

3.36 Refrigerant 22 undergoes a constant-pressure process within a piston–cylinder assembly from saturated vapor at 4 bar to a final temperature of 30°C. Kinetic and potential energy effects are negligible. For the refrigerant, show the process on a p–υ diagram. Evaluate the work and the heat transfer, each in kJ per kg of refrigerant.

3.37 For the system of Problem 3.18, determine the amount of energy transfer by heat, in kJ per kg of refrigerant.

3.38 Water in a piston-cylinder assembly, initially at a temperature of 99.63°C and a quality of 65%, is heated at constant pressure to a temperature of 200°C. If the work during the process is +300 kJ, determine (a) the mass of water, in kg, and (b) the heat transfer, in kJ. Changes in kinetic and potential energy are negligible.

3.39 Ammonia vapor in a piston-cylinder assembly undergoes a constant-pressure process from saturated vapor at 10 bar. The work is +16.5 kJ/kg. Changes in kinetic and potential energy are negligible. Determine (a) the final temperature of the ammonia, in °C, and (b) the heat transfer, in kJ/kg.

3.40 A piston–cylinder assembly contains water, initially a saturated vapor at 200°C. The water is cooled at constant temperature to saturated liquid. Kinetic and potential energy effects are negligible.

    a. For the water as a closed system, determine the work per unit mass of water, in kJ/kg.

    b. If the energy transfer by heat for the process is −1200 kJ, determine the mass of the water, in kg.

3.41 A well-insulated, rigid tank contains 1.5 kg of Refrigerant 134a, initially a two-phase liquid–vapor mixture with a quality of 60% and a temperature of 0°C. An electrical resistor transfers energy *to* the contents of the tank at a rate of 2 kW until the tank contains only saturated vapor. For the refrigerant, locate the initial and final states on a T–υ diagram and determine the time it takes, in s, for the process.

3.42 If the two-phase mixture in Example 3.2 requires 50 minutes to undergo process 1-2, determine the rate at which the hot plate transfers energy, in kW, to the mixture. If the rate of energy transfer remains the same, determine the time, in min, required to bring the mixture from state 2 to state 3.

3.43 A closed, rigid tank filled with water, initially at 20 bar, a quality of 80%, and a volume of 0.5 m^3, is cooled until the pressure is 4 bar. Show the process of the water on a sketch of the T–v diagram and evaluate the heat transfer, in kJ.

3.44 As shown in **Fig. P3.44**, a closed, rigid tank fitted with a fine-wire electric resistor is filled with Refrigerant 22, initially at –10°C, a quality of 80%, and a volume of 0.01 m^3. A 12-volt battery provides a 5-amp current to the resistor for 5 minutes. If the final temperature of the refrigerant is 40°C, determine the heat transfer, in kJ, from the refrigerant.

3.45 A piston-cylinder assembly containing water, initially a liquid at 10°C, undergoes a process at a constant pressure of 138 kPa to a final state where the water is a vapor at 149°C. Kinetic and potential energy effects are negligible. Determine the work and heat transfer, in kJ/kg, for each of three parts of the overall process: (a) from the initial liquid state to the saturated liquid state, (b) from saturated liquid to saturated vapor, and (c) from saturated vapor to the final vapor state, all at 138 kPa.

3.46 A closed, rigid tank contains 2 kg of water initially at 80°C and a quality of 0.6. Heat transfer occurs until the tank contains only saturated vapor at a higher pressure. Kinetic and potential energy effects are negligible. For the water as the system, determine the amount of energy transfer by heat, in kJ.

3.47 A piston–cylinder assembly contains ammonia, initially at a temperature of –20°C and a quality of 50%. The ammonia is slowly heated to a final state where the pressure is 6 bar and the temperature is 180°C. While the ammonia is heated, its pressure varies linearly with specific volume. Show the process of the ammonia on a sketch of the p–v diagram. For the ammonia, determine the work and heat transfer, each in kJ/kg.

3.48 🅒 A two-phase liquid–vapor mixture of H$_2$O, initially at 1.0 MPa with a quality of 90%, is contained in a rigid, well-insulated tank. The mass of H$_2$O is 2 kg. An electric resistance heater in the tank transfers energy to the water at a constant rate of 60 W for 1.95 h. Determine the final temperature of the water in the tank, in °C, and the final pressure, in bar.

3.49 A rigid tank is filled with 2.27 kg of water, initially at $T_1 = 127°C$ and $v_1 = 0.44$ m^3/kg. The tank contents are heated until the temperature is $T_2 = 177°C$. Kinetic and potential energy effects are negligible. For the water, determine (a) the initial and final pressure, each in kPa, and (b) the heat transfer, in kJ.

3.50 Two kg of water is contained in a piston–cylinder assembly, initially at 10 bar and 200°C. The water is slowly heated at constant pressure to a final state. If the heat transfer for the process is 1740 kJ, determine the temperature at the final state, in °C, and the work, in kJ. Kinetic and potential energy effects are negligible.

3.51 Referring to **Fig. P3.51**, water contained in a piston–cylinder assembly, initially at 1.5 bar and a quality of 20%, is heated at constant pressure until the piston hits the stops. Heating then continues until the water is saturated vapor. Show the processes of the water in series on a sketch of the T–v diagram. For the overall process of the water, evaluate the work and heat transfer, each in kJ/kg. Kinetic and potential effects are negligible.

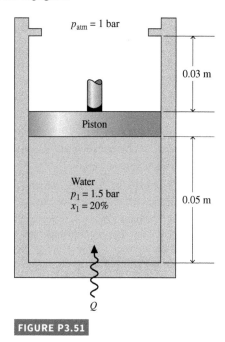

p_{atm} = 1 bar

0.03 m

Piston

Water
p_1 = 1.5 bar
x_1 = 20%

0.05 m

Q

FIGURE P3.51

3.52 A piston-cylinder assembly contains 0.9 kg of water, initially at 149°C. The water undergoes two processes in series: constant-volume heating followed by a constant-pressure process. At the end of the constant-volume process, the pressure is 690 kPa and the water is a two-phase, liquid–vapor mixture with a quality of 80%. At the end of the constant-pressure process, the temperature is 204°C. Neglect kinetic and potential energy effects.

   a. Sketch T–v and p–v diagrams showing the key states and the processes.

   b. Determine the work and heat transfer for each of the two processes, all in kJ.

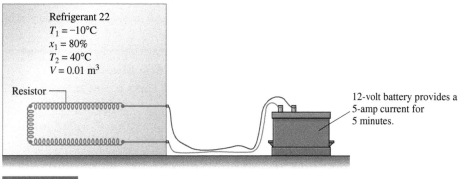

Refrigerant 22
$T_1 = -10°C$
$x_1 = 80\%$
$T_2 = 40°C$
$V = 0.01$ m^3

Resistor

12-volt battery provides a 5-amp current for 5 minutes.

FIGURE P3.44

3.53 A system consisting of 0.9 kg of water vapor, initially at 149°C and occupying a volume of 0.54 m³, is compressed isothermally to a volume of 0.24 m³. The system is then heated at constant volume to a final pressure of 827 kPa. During the isothermal compression there is energy transfer by work of magnitude 95.8 kJ *into* the system. Kinetic and potential energy effects are negligible. Determine the heat transfer, in kJ, for each process.

3.54 Ammonia in a piston-cylinder assembly undergoes two processes in series. At the initial state, $p_1 = 827$ kPa and the quality is 100%. Process 1–2 occurs at constant volume until the temperature is 38°C. The second process, from state 2 to state 3, occurs at constant temperature, with $Q_{23} = 104$ kJ, until the quality is again 100%. Kinetic and potential energy effects are negligible. For 1 kg of ammonia, determine **(a)** the heat transfer for Process 1–2 and **(b)** the work for Process 2–3, each in kJ.

3.55 A system consisting of 1 kg of H₂O undergoes a power cycle composed of the following processes:

> **Process 1–2:** Constant-pressure heating at 10 bar from saturated vapor.
>
> **Process 2–3:** Constant-volume cooling to $p_3 = 5$ bar, $T_3 = 160°C$.
>
> **Process 3–4:** Isothermal compression with $Q_{34} = -815.8$ kJ.
>
> **Process 4–1:** Constant-volume heating.

Sketch the cycle on $T–\upsilon$ and $p–\upsilon$ diagrams. Neglecting kinetic and potential energy effects, determine the thermal efficiency.

3.56 A piston–cylinder assembly contains propane, initially at 27°C, 1 bar, and a volume of 0.2 m³. The propane undergoes a process to a final pressure of 4 bar, during which the pressure–volume relationship is $pV^{1.1} = constant$. For the propane, evaluate the work and heat transfer, each in kJ. Kinetic and potential energy effects can be ignored.

3.57 Four liters of milk at 20°C is placed in a refrigerator. If energy is removed from the milk by heat transfer at a constant rate of 0.084 kJ/s, how long would it take, in minutes, for the milk to cool to 4°C? The specific heat and density of the milk are 4 kJ/kg · K and 1025 kg/m³, respectively.

3.58 Shown in Fig. P3.58 is an insulated copper block that receives energy at a rate of 100 W from an embedded resistor. If the block has a volume of 10^{-3} m³ and an initial temperature of 20°C, how long would it take, in minutes, for the temperature to reach 60°C? Data for copper are provided in Table A-19.

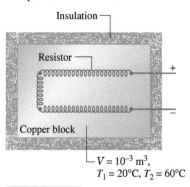

FIGURE P3.58

3.59 In a heat-treating process, a 1-kg metal part, initially at 1075 K, is quenched in a closed tank containing 100 kg of water, initially at 295 K. There is negligible heat transfer between the contents of the tank and their surroundings. Modeling the metal part and water as incompressible with constant specific heats 0.5 kJ/kg · K and 4.4 kJ/kg · K, respectively, determine the final equilibrium temperature after quenching, in K.

3.60 As shown in Fig. P3.60, a closed, insulated tank contains 0.15 kg of liquid water and has a 0.25-kg copper base. The thin walls of the container have negligible mass. Initially, the tank and its contents are all at 30°C. A heating element embedded in the copper base is energized with an electrical current of 10 amps at 12 volts for 100 seconds. Determine the final temperature, in °C, of the tank and its contents. Data for copper and liquid water are provided in Table A-19.

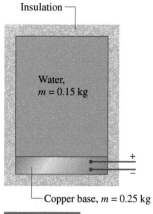

FIGURE P3.60

3.61 As shown in Fig. P3.61, a system consists of a copper tank whose mass is 13 kg, 4 kg of liquid water, and an electrical resistor of negligible mass. The system is insulated on its outer surface. Initially, the temperature of the copper is 27°C and the temperature of the water is 50°C. The electrical resistor transfers 100 kJ of energy to the system. Eventually the system comes to equilibrium. Determine the final equilibrium temperature, in °C.

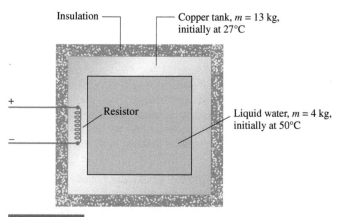

FIGURE P3.61

Using Generalized Compressibility Data

3.62 Determine the volume, in m³, occupied by 2 kg of H₂O at 100 bar, 400°C, using

> **a.** data from the compressibility chart.
>
> **b.** data from the steam tables.

Compare the results of parts (a) and (b) and discuss.

3.63 Five kmol of oxygen (O₂) gas undergoes a process in a closed system from $p_1 = 50$ bar, $T_1 = 170$ K to $p_2 = 25$ bar, $T_2 = 200$ K. Determine the change in volume, in m³.

3.64 2.27 kmol of carbon dioxide (CO₂), initially at 2.2 MPa, 367 K, is compressed at constant pressure in a piston–cylinder assembly. For the gas, $W = -2,110$ kJ. Determine the final temperature, in K.

3.65 Determine the temperature, in °C, of air at 30 bar and a specific volume of 0.013 m³/kg.

3.66 Five kg of butane (C_4H_{10}) in a piston-cylinder assembly undergo a process from $p_1 = 5$ MPa, $T_1 = 500$ K to $p_2 = 3$ MPa, $T_2 = 450$ K during which the relationship between pressure and specific volume is $pv^n = constant$. Determine the work, in kJ.

Working with the Ideal Gas Model

3.67 A tank contains 0.05 m³ of nitrogen (N_2) at −21°C and 10 MPa. Determine the mass of nitrogen, in kg, using

a. the ideal gas model.

b. data from the compressibility chart.

Comment on the applicability of the ideal gas model for nitrogen at this state.

3.68 Determine the specific volume, in m³/kg, of ammonia at 50°C, 10 bar, using

a. Table A-15.

b. Figure A-1.

c. the ideal gas equation of state.

Compare the values obtained in parts (b) and (c) with that of part (a).

3.69 Check the applicability of the ideal gas model

a. for water at 570°C and pressures of 11.03 MPa and 1.1 MPa.

b. for carbon dioxide at 865 K and pressures of 75 bar and 3 bar.

3.70 A closed, rigid tank is filled with a gas modeled as an ideal gas, initially at 27°C and a gage pressure of 300 kPa. The gas is heated, and the gage pressure at the final state is 367 kPa. Determine the final temperature, in °C. The local atmospheric pressure is 1 atm.

3.71 A balloon filled with helium, initially at 27°C, 1 bar, is released and rises in the atmosphere until the helium is at 17°C, 0.9 bar. Determine, as a percent, the change in volume of the helium from its initial volume.

3.72 Determine the total mass of nitrogen (N_2), in kg, required to inflate all four tires of a vehicle, each to a gage pressure of 180 kPa at a temperature of 25°C. The volume of each tire is 0.6 m³, and the atmospheric pressure is 1 atm.

3.73 A tank contains 4.5 kg of air at 21°C with a pressure of 207 kPa. Determine the volume of the air, in m³. Verify that ideal gas behavior can be assumed for air under these conditions.

Using Energy Concepts and the Ideal Gas Model

3.74 As shown in Fig. P3.74, a piston-cylinder assembly fitted with a paddle wheel contains air, initially at $p_1 = 207$ kPa, $T_1 = 282$°C, and $V_1 = 0.108$ m³. The air undergoes a process to a final state where $p_2 = 138$ kPa, $V_2 = 0.12$ m³. During the process, the paddle wheel transfers energy to the air by work in the amount 1.06 kJ, while the air transfers energy by work to the piston in the amount 3.5 kJ. Assuming ideal gas behavior, determine for the air (a) the temperature at state 2, in K and (b) the heat transfer, in kJ.

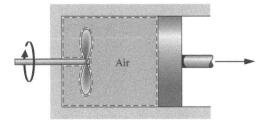

Initially, $p_1 = 207$ kPa, $T_1 = 282$°C, $V_1 = 0.108$ m³.
Finally, $p_2 = 138$ kPa, $V_2 = 0.12$ m³.   **FIGURE P3.74**

3.75 A piston-cylinder assembly contains nitrogen (N_2), initially at 138 kPa, 27°C and occupying a volume of 0.054 m³. The nitrogen is compressed to a final state where the pressure is 1.1 MPa and the temperature is 149°C. During compression, heat transfer of magnitude 1.69 kJ occurs *from* the nitrogen *to* its surroundings. Assuming ideal gas behavior, determine for the nitrogen, (a) the mass, in kg, and (b) the work, in kJ.

3.76 Air contained in a piston–cylinder assembly, initially at 2 bar, 200 K, and a volume of 1 L, undergoes a process to a final state where the pressure is 8 bar and the volume is 2 L. During the process, the pressure–volume relationship is linear. Assuming the ideal gas model for the air, determine the work and heat transfer, each in kJ.

3.77 Carbon dioxide (CO_2) contained in a piston–cylinder arrangement, initially at 6 bar and 400 K, undergoes an expansion to a final temperature of 298 K, during which the pressure–volume relationship is $pV^{1.2} = constant$. Assuming the ideal gas model for the CO_2, determine the final pressure, in bar, and the work and heat transfer, each in kJ/kg.

3.78 **C** Water vapor contained in a piston–cylinder assembly undergoes an isothermal expansion at 240°C from a pressure of 7 bar to a pressure of 3 bar. Evaluate the work, in kJ/kg. Solve two ways: using (a) the ideal gas model, (b) *IT* with *water/steam* data. Comment.

3.79 One kilogram of nitrogen fills the cylinder of a piston-cylinder assembly, as shown in Fig. P3.79. There is no friction between the piston and the cylinder walls, and the surroundings are at 1 atm. The initial volume and pressure in the cylinder are 1 m³ and 1 atm, respectively. Heat transfer to the nitrogen occurs until the volume is doubled. Determine the heat transfer for the process, in kJ, assuming the specific heat ratio is constant, $k = 1.4$.

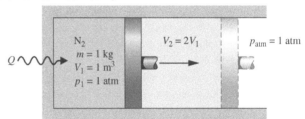

FIGURE P3.79

3.80 As shown in Fig. P3.80, a fan drawing electricity at a rate of 1.5 kW is located within a rigid enclosure, measuring 3 m × 4 m × 5 m. The enclosure is filled with air, initially at 27°C, 0.1 MPa. The fan operates steadily for 30 minutes. Assuming the ideal gas model, determine for the air (a) the mass, in kg, (b) the final temperature, in °C, and (c) the final pressure, in MPa. There is no heat transfer between the enclosure and the surroundings. Ignore the volume occupied by the fan itself and assume the fan stores no energy.

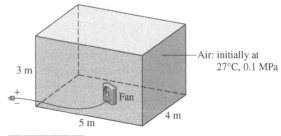

FIGURE P3.80

3.81 **C** Ammonia is contained in a rigid, well-insulated container. The initial pressure is 138 kPa, the mass is 0.05 kg, and the volume is 0.054 m³. The gas is stirred by a paddle wheel, resulting in an energy transfer to the ammonia of magnitude 21 kJ. Assuming the ideal gas model, determine the final temperature of the ammonia, in K. Neglect kinetic and potential energy effects.

3.82 As shown in Fig. P3.82, a piston-cylinder assembly whose piston is resting on a set of stops contains 0.5 kg of helium gas, initially at 100 kPa and 25°C. The mass of the piston and the effect of the atmospheric pressure acting on the piston are such that a pressure of 500 kPa is required to raise it. How much energy must be transferred by heat to the helium, in kJ, before the piston starts rising? Assume ideal gas behavior for the helium.

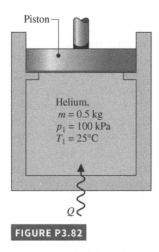

Piston

Helium,
$m = 0.5$ kg
$p_1 = 100$ kPa
$T_1 = 25°C$

Q

FIGURE P3.82

3.83 Argon contained in a closed, rigid tank, initially at 50°C, 2 bar, and a volume of 2 m³, is heated to a final pressure of 8 bar. Assuming the ideal gas model with $k = 1.67$ for the argon, determine the final temperature, in °C, and the heat transfer, in kJ.

3.84 Five kg of oxygen (O_2), initially at 430°C, fills a closed, rigid tank. Heat transfer from the oxygen occurs at the rate 425 W for 30 minutes. Assuming the ideal gas model with $k = 1.350$ for the oxygen, determine its final temperature, in °C.

3.85 A piston–cylinder assembly fitted with a slowly rotating paddle wheel contains 0.13 kg of air, initially at 300 K. The air undergoes a constant-pressure process to a final temperature of 400 K. During the process, energy is gradually transferred to the air by heat transfer in the amount 12 kJ. Assuming the ideal gas model with $k = 1.4$ and negligible changes in kinetic and potential energy for the air, determine the work done (a) by the paddle wheel on the air and (b) by the air to displace the piston, each in kJ.

3.86 A piston-cylinder assembly contains air at a pressure of 207 kPa and a volume of 0.02 m³. The air is heated at constant pressure until its volume is doubled. Assuming the ideal gas model with constant specific heat ratio, $k = 1.4$, for the air, determine the work and heat transfer, each in kJ.

3.87 As shown in Fig. P3.87, a tank fitted with an electrical resistor of negligible mass holds 2 kg of nitrogen (N_2), initially at 27°C, 0.1 MPa. Over a period of 10 minutes, electricity is provided to the resistor at a rate of 0.12 kW. During this same period, a heat transfer of magnitude 12.59 kJ occurs *from* the nitrogen *to* its surroundings. Assuming ideal gas behavior, determine the nitrogen's final temperature, in °C, and final pressure, in MPa.

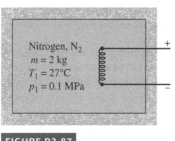

Nitrogen, N_2
$m = 2$ kg
$T_1 = 27°C$
$p_1 = 0.1$ MPa

FIGURE P3.87

3.88 As shown in Fig. P3.88, a piston–cylinder assembly contains 5 g of air holding the piston against the stops. The air, initially at 3 bar, 600 K, is slowly cooled until the piston just begins to move downward in the cylinder. The air behaves as an ideal gas, $g = 9.81$ m/s², and friction is negligible. Sketch the process of the air on a p–V diagram labeled with the temperature and pressure at the end states. Also determine the heat transfer, in kJ, between the air and its surroundings.

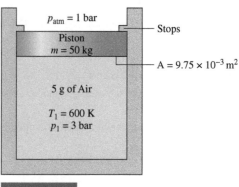

$p_{atm} = 1$ bar

Stops

Piston
$m = 50$ kg

$A = 9.75 \times 10^{-3}$ m²

5 g of Air

$T_1 = 600$ K
$p_1 = 3$ bar

FIGURE P3.88

3.89 A closed, rigid tank fitted with a paddle wheel contains 0.1 kg of air, initially at 300 K, 0.1 MPa. The paddle wheel stirs the air for 20 minutes, with the power input varying with time according to $\dot{W} = -10t$, where $\dot{W}$ is in watts and t is time, in minutes. The final temperature of the air is 1060 K. Assuming ideal gas behavior and no change in kinetic or potential energy, determine for the air (a) the final pressure, in MPa, (b) the work, in kJ, and (c) the heat transfer, in kJ.

3.90 Air is confined to one side of a rigid container divided by a partition, as shown in Fig. P3.90. The other side is initially evacuated. The air is initially at $p_1 = 5$ bar, $T_1 = 500$ K, and $V_1 = 0.2$ m³. When the partition is removed, the air expands to fill the entire chamber. Measurements show that $V_2 = 2 V_1$ and $p_2 = p_1/4$. Assuming the air behaves as an ideal gas, determine (a) the final temperature, in K, and (b) the heat transfer, kJ.

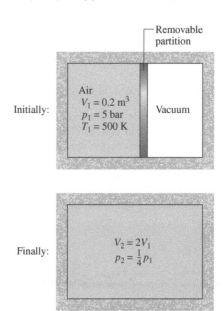

Removable partition

Initially:

Air
$V_1 = 0.2$ m³
$p_1 = 5$ bar
$T_1 = 500$ K

Vacuum

Finally:

$V_2 = 2V_1$
$p_2 = \frac{1}{4}p_1$

FIGURE P3.90

3.91 Carbon dioxide (CO_2) is compressed in a piston–cylinder assembly from $p_1 = 0.7$ bar, $T_1 = 280$ K to $p_2 = 11$ bar. The initial volume is 0.262 m³. The process is described by $pV^{1.25} = constant$. Assuming ideal gas behavior and neglecting kinetic and potential energy effects, determine the work and heat transfer for the process, each in kJ, using (a) constant specific heats evaluated at 300 K, and (b) data from Table A-23. Compare the results and discuss.

3.92 **C** A rigid tank, with a volume of 0.054 m^3, contains air initially at 138 kPa, 278 K. If the air receives a heat transfer of magnitude 6.3 kJ, determine the final temperature, in K and the final pressure, in kPa. Assume ideal gas behavior, and use

  a. a constant specific heat value from Table A-20.

  b. a specific heat function from Table A-21.

  c. data from Table A-22.

3.93 Air contained in a piston–cylinder assembly undergoes two processes in series, as shown in **Fig. P3.93**. Assuming ideal gas behavior for the air, determine the work and heat transfer for the overall process, each in kJ/kg.

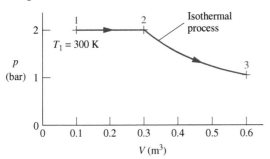

FIGURE P3.93

3.94 Helium (He) gas initially at 2 bar, 200 K undergoes a polytropic process, with $n = k$, to a final pressure of 14 bar. Determine the work and heat transfer for the process, each in kJ per kg of helium. Assume ideal gas behavior.

3.95 Two-tenths kmol of nitrogen (N_2) in a piston–cylinder assembly undergoes two processes in series as follows:

  Process 1–2: Constant pressure at 5 bar from $V_1 = 1.33$ m^3 to $V_2 = 1$ m^3.

  Process 2–3: Constant volume to $p_3 = 4$ bar.

Assuming ideal gas behavior and neglecting kinetic and potential energy effects, determine the work and heat transfer for each process, in kJ.

3.96 One kilogram of air undergoes a power cycle consisting of the following processes:

  Process 1–2: Constant volume from $p_1 = 138$ kPa, $T_1 = 278$ K to $T_2 = 455.6$ K

  Process 2–3: Adiabatic expansion to $v_3 = 1.4v_2$

  Process 3–1: Constant-pressure compression

Sketch the cycle on a p–v diagram. Assuming ideal gas behavior, determine

  a. the pressure at state 2, in kPa.

  b. the temperature at state 3, in K.

  c. the thermal efficiency of the cycle.

3.97 A system consists of 2 kg of carbon dioxide gas initially at state 1, where $p_1 = 1$ bar, $T_1 = 300$ K. The system undergoes a power cycle consisting of the following processes:

  Process 1–2: Constant volume to $p_2 = 4$ bar.

  Process 2–3: Expansion with $pv^{1.28} = constant$.

  Process 3–1: Constant-pressure compression.

Assuming the ideal gas model and neglecting kinetic and potential energy effects,

  a. sketch the cycle on a p–v diagram and calculate thermal efficiency.

  b. plot the thermal efficiency versus p_2/p_1 ranging from 1.05 to 4.

3.98 **C** Air undergoes a polytropic process in a piston–cylinder assembly from $p_1 = 101.3$ kPa, $T_1 = 21°C$ to $p_2 = 690$ kPa. Using *IT*, plot the work and heat transfer, each in kJ per kg of air, for polytropic exponents ranging from 1.0 to 1.6. Investigate the error in the heat transfer introduced by assuming constant c_v evaluated at $21°C$. Discuss.

3.99 **C** Steam, initially at 5 MPa, 280°C undergoes a polytropic process in a piston–cylinder assembly to a final pressure of 20 MPa. Using *IT*, plot the heat transfer, in kJ per kg of steam, for polytropic exponents ranging from 1.0 to 1.6. Investigate the error in the heat transfer introduced by assuming ideal gas behavior for the steam. Discuss.

DESIGN & OPEN-ENDED PROBLEMS: EXPLORING ENGINEERING PRACTICE

3.1D Dermatologists remove skin blemishes from patients by applying sprays from canisters filled with liquid nitrogen (N_2). Investigate how liquid nitrogen is produced and delivered to physicians, and how physicians manage liquid nitrogen in their practices. Also investigate the advantages and disadvantages of this approach for removing blemishes compared to alternative approaches used today. Write a report including at least three references.

3.2D The EPA (Environmental Protection Agency) has developed an online *Personal Emissions Calculator* that helps individuals and families reduce greenhouse emissions. Use the EPA calculator to estimate, in the home and on the road, your personal greenhouse emissions or your family's greenhouse emissions. Also use the calculator to explore steps you as an individual or your family can take to cut emissions by at least 20%. In a memorandum, summarize your findings and present your program for lowering emissions.

3.3D The pressure at about 400 m of depth in an ocean or lake is about 40 atm. A Canadian company is studying a system that will submerge large, empty concrete tanks into which water will flow through turbines to fill the tanks. The turbines will generate electricity that will be used

to run compressors at the surface to store compressed air for later use to produce power on land. The turbines will then be reversed and used as pumps to empty the tanks so the cycle can be repeated. Write a technical paper that summarizes the technical concepts underlying this technology and its potential in a form suitable for briefing a member of Congress.

3.4D Metallurgists use phase diagrams to study *allotropic transformations*, which are phase transitions within the solid region. What features of the phase behavior of solids are important in the fields of metallurgy and materials processing? Write a report including at least three references.

3.5D Design a laboratory flask for containing up to 10 kmol of mercury vapor at pressures up to 3 MPa, and temperatures from 900 to 1000 K. Consider the health and safety of the technicians who would be working with such a mercury vapor-filled container. The p–v–T relation for mercury vapor can be expressed as

$$p = RT/v - T/v^2 \exp(10.3338 - 0.0312095/T - 2.07950\ln T)$$

where T is in K, v is in m^3/kg, and p is in Pa.

3.6D Due to their zero ozone depletion and low global warming potential *natural refrigerants* are actively under consideration for commercial refrigeration applications (see box in Sec. 3.4). Investigate the viability of natural refrigerants in systems to improve human comfort and safeguard food. Consider performance benefits, safety, and cost. On the basis of your study, recommend especially promising natural refrigerants and areas of application where each is particularly well suited. Report your findings in a PowerPoint presentation.

3.7D One method of modeling gas behavior from the microscopic viewpoint is the *kinetic theory of gases*. Using kinetic theory, derive the ideal gas equation of state and explain the variation of the ideal gas specific heat c_v with temperature. Is the use of kinetic theory limited to ideal gas behavior? Summarize your results as a memorandum.

3.8D Some oil and gas companies use *hydraulic fracturing* to access oil and natural gas trapped in deep rock formations. Investigate the process of hydraulic fracturing, its benefits, and environmental impacts. On this basis, write a three-page *brief* for submission to a congressional committee considering whether hydraulic fracturing should continue to be exempt from regulation under the *Safe Drinking Water Act*. The brief can provide either objective technical background to inform committee members or take a position in favor of, or against, continuing the exemption.

3.9D Water is one of our most important resources, yet one of the most poorly managed—it is too often wasted and polluted. Investigate ways we can improve water use as a society in industry, businesses, and households. Record your daily water use for at least three days and compare it to that of those living in the poorest parts of the globe: about one gallon per day. Write a report including at least three references.

3.10D Carbon-dioxide capture and storage is under consideration today for lessening the effects on global warming and climate change traceable to burning fossil fuels. One frequently mentioned approach, shown in **Fig. 3.10D**, is to remove CO_2 from the exhaust gas of a power plant and inject it into an underground geological formation. Critically evaluate this approach for large-scale management of CO_2 generated by human activity, including technical aspects and related costs. Consider the leading methods for CO_2 capture from gas streams, issues related to injecting the CO_2 at great depths, and the consequences of CO_2 migration from storage. Formulate a position in favor of, or in opposition to, the deployment of such technology. Write a report including at least three references.

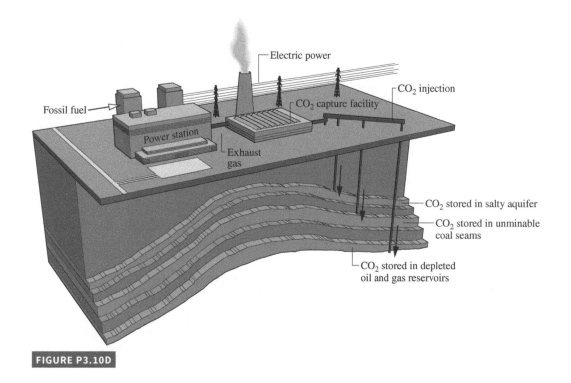

FIGURE P3.10D

CHAPTER 4

Control Volume Analysis Using Energy

Engineering Context

The **objective** of this chapter is to develop and illustrate the use of the control volume forms of the conservation of mass and conservation of energy principles. Mass and energy balances for control volumes are introduced in Secs. 4.1 and 4.4, respectively. These balances are applied in Secs. 4.5–4.11 to control volumes at steady state and in Sec. 4.12 for time-dependent (transient) applications.

Although devices such as turbines, pumps, and compressors through which mass flows can be analyzed in principle by studying a particular quantity of matter (a closed system) as it passes through the device, it is normally preferable to think of a region of space through which mass flows (a control volume). As in the case of a closed system, energy transfer across the boundary of a control volume can occur by means of work and heat. In addition, another type of energy transfer must be accounted for—the energy accompanying mass as it enters or exits.

LEARNING OUTCOMES

When you complete your study of this chapter, you will be able to...

- Describe key concepts related to control volume analysis, including distinguishing between steady-state and transient analysis, distinguishing between mass flow rate and volumetric flow rate, and explaining the meanings of one-dimensional flow and flow work.

- Apply mass and energy balances to control volumes.

- Develop appropriate engineering models for control volumes, with particular attention to analyzing components commonly encountered in engineering practice such as nozzles, diffusers, turbines, compressors, heat exchangers, throttling devices, and integrated systems that incorporate two or more components.

- Obtain and apply appropriate property data for control volume analyses.

4.1 Conservation of Mass for a Control Volume

In this section an expression of the conservation of mass principle for control volumes is developed and illustrated. As a part of the presentation, the one-dimensional flow model is introduced.

4.1.1 Developing the Mass Rate Balance

The mass rate balance for control volumes is introduced by reference to Fig. 4.1, which shows a control volume with mass flowing in at i and flowing out at e, respectively. When applied to such a control volume, the **conservation of mass** principle states

conservation of mass

$$
\begin{bmatrix} \text{time } \textit{rate of change} \text{ of} \\ \text{mass contained within the} \\ \text{control volume } \textit{at time } t \end{bmatrix} = \begin{bmatrix} \text{time } \textit{rate} \text{ of flow of} \\ \text{mass } \textit{in} \text{ across} \\ \text{inlet } i \textit{ at time } t \end{bmatrix} - \begin{bmatrix} \text{time } \textit{rate} \text{ of flow} \\ \text{of mass } \textit{out} \text{ across} \\ \text{exit } e \textit{ at time } t \end{bmatrix}
$$

Denoting the mass contained within the control volume at time t by $m_{cv}(t)$, this statement of the conservation of mass principle can be expressed in symbols as

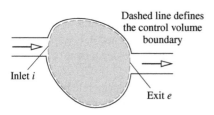

Fig. 4.1 One-inlet, one-exit control volume.

$$
\frac{dm_{cv}}{dt} = \dot{m}_i - \dot{m}_e \tag{4.1}
$$

where dm_{cv}/dt is the time rate of change of mass within the control volume, and $\dot{m}_i$ and $\dot{m}_e$ are the instantaneous **mass flow rates** at the inlet and exit, respectively. As for the symbols $\dot{W}$ and $\dot{Q}$, the dots in the quantities $\dot{m}_i$ and $\dot{m}_e$ denote time rates of transfer. In SI, all terms in Eq. 4.1 are expressed in kg/s. For a discussion of the development of Eq. 4.1, see the box.

mass flow rates

In general, there may be several locations on the boundary through which mass enters or exits. This can be accounted for by summing, as follows

$$
\boxed{\frac{dm_{cv}}{dt} = \sum_i \dot{m}_i - \sum_e \dot{m}_e} \tag{4.2}
$$

mass rate balance

Equation 4.2 is the **mass rate balance** for control volumes with several inlets and exits. It is a form of the conservation of mass principle commonly employed in engineering. Other forms of the mass rate balance are considered in discussions to follow.

Developing the Control Volume Mass Balance

For each of the extensive properties mass, energy, entropy (Chap. 6), and exergy (Chap. 7) the control volume form of the property balance can be obtained by transforming the corresponding closed system form. Let us consider this for mass, recalling that the mass of a closed system is constant.

The figures show a system consisting of a fixed quantity of matter m that occupies different regions at time t and a later time $t + \Delta t$. The mass under consideration is shown in color on the figures. At time t, the mass is the sum $m = m_{cv}(t) + m_i$, where $m_{cv}(t)$ is the mass contained within the control volume, and m_i is the mass within the small region labeled i adjacent to the control volume. Let us study the fixed quantity of matter m as time elapses.

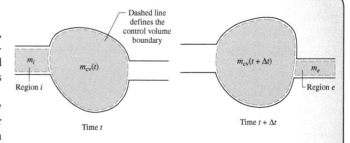

In a time interval Δt all the mass in region i crosses the control volume boundary, while some of the mass, call it m_e, initially contained within the control volume exits to fill the region labeled e adjacent to the control volume. Although the mass in regions i

and e as well as in the control volume differs from time t to $t + \Delta t$, the *total* amount of mass is constant. Accordingly,

$$m_{cv}(t) + m_i = m_{cv}(t + \Delta t) + m_e \qquad \text{(a)}$$

or on rearrangement

$$m_{cv}(t + \Delta t) - m_{cv}(t) = m_i - m_e \qquad \text{(b)}$$

Equation (b) is an *accounting* balance for mass. It states that the change in mass of the control volume during time interval Δt equals the amount of mass that enters less the amount of mass that exits.

Equation (b) can be expressed on a time rate basis. First, divide by Δt to obtain

$$\frac{m_{cv}(t + \Delta t) - m_{cv}(t)}{\Delta t} = \frac{m_i}{\Delta t} - \frac{m_e}{\Delta t} \qquad \text{(c)}$$

Then, in the limit as Δt goes to zero, Eq. (c) becomes Eq. 4.1, the instantaneous *control volume rate equation* for mass

$$\frac{dm_{cv}}{dt} = \dot{m}_i - \dot{m}_e \qquad \text{(4.1)}$$

where dm_{cv}/dt denotes the time rate of change of mass within the control volume, and $\dot{m}_i$ and $\dot{m}_e$ are the inlet and exit mass flow rates, respectively, all at time t.

4.1.2 Evaluating the Mass Flow Rate

An expression for the mass flow rate $\dot{m}$ of the matter entering or exiting a control volume can be obtained in terms of local properties by considering a small quantity of matter flowing with velocity V across an incremental area $d\mathrm{A}$ in a time interval Δt, as shown in **Fig. 4.2**. Since the portion of the control volume boundary through which mass flows is not necessarily at rest, the velocity shown in the figure is understood to be the velocity *relative* to the area $d\mathrm{A}$. The velocity can be resolved into components normal and tangent to the plane containing $d\mathrm{A}$. In the following development $\mathrm{V_n}$ denotes the component of the relative velocity normal to $d\mathrm{A}$ in the direction of flow.

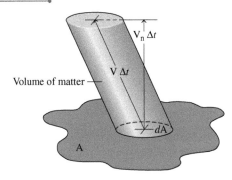

Fig. 4.2 Illustration used to develop an expression for mass flow rate in terms of local fluid properties.

The *volume* of the matter crossing $d\mathrm{A}$ during the time interval Δt shown in Fig. 4.2 is an oblique cylinder with a volume equal to the product of the area of its base $d\mathrm{A}$ and its altitude $\mathrm{V_n}\,\Delta t$. Multiplying by the density ρ gives the amount of mass that crosses $d\mathrm{A}$ in time Δt

$$\begin{bmatrix} \text{amount of mass} \\ \text{crossing } d\mathrm{A} \text{ during} \\ \text{the time interval } \Delta t \end{bmatrix} = \rho(\mathrm{V_n}\,\Delta t)\,d\mathrm{A}$$

Dividing both sides of this equation by Δt and taking the limit as Δt goes to zero, the instantaneous mass flow rate across incremental area $d\mathrm{A}$ is

$$\begin{bmatrix} \text{instantaneous rate} \\ \text{of mass flow} \\ \text{across } d\mathrm{A} \end{bmatrix} = \rho\mathrm{V_n}\,d\mathrm{A}$$

When this is integrated over the area A through which mass passes, an expression for the mass flow rate is obtained

$$\boxed{\dot{m} = \int_A \rho\mathrm{V_n}\,d\mathrm{A}} \qquad \text{(4.3)}$$

Equation 4.3 can be applied at the inlets and exits to account for the rates of mass flow into and out of the control volume.

4.2 Forms of the Mass Rate Balance

The mass rate balance, Eq. 4.2, is a form that is important for control volume analysis. In many cases, however, it is convenient to apply the mass balance in forms suited to particular objectives. Some alternative forms are considered in this section.

4.2.1 One-Dimensional Flow Form of the Mass Rate Balance

one-dimensional flow

When a flowing stream of matter entering or exiting a control volume adheres to the following idealizations, the flow is said to be **one-dimensional**:

TAKE NOTE...

In subsequent control volume analyses, we routinely assume that the idealizations of one-dimensional flow are appropriate. Accordingly, the assumption of one-dimensional flow is not listed explicitly in solved examples.

- The flow is normal to the boundary at locations where mass enters or exits the control volume.
- *All* intensive properties, including velocity and density, are *uniform with position* (bulk average values) over each inlet or exit area through which matter flows.

> **FOR EXAMPLE**
>
> Figure 4.3 illustrates the meaning of one-dimensional flow. The area through which mass flows is denoted by A. The symbol V denotes a single value that represents the velocity of the flowing air. Similarly T and v are single values that represent the temperature and specific volume, respectively, of the flowing air.

When the flow is one-dimensional, Eq. 4.3 for the mass flow rate becomes

$$\dot{m} = \rho A V \quad \text{(one-dimensional flow)} \tag{4.4a}$$

or in terms of specific volume

$$\dot{m} = \frac{A V}{v} \quad \text{(one-dimensional flow)} \tag{4.4b}$$

volumetric flow rate

When area is in m^2, velocity is in m/s, and specific volume is in m^3/kg, the mass flow rate found from Eq. 4.4b is in kg/s, as can be verified. The product AV in Eqs. 4.4 is the **volumetric flow rate**. The volumetric flow rate is expressed in units of m^3/s.

Substituting Eq. 4.4b into Eq. 4.2 results in an expression for the conservation of mass principle for control volumes limited to the case of one-dimensional flow at the inlets and exits

$$\frac{dm_{cv}}{dt} = \sum_i \frac{A_i V_i}{v_i} - \sum_e \frac{A_e V_e}{v_e} \quad \text{(one-dimensional flow)} \tag{4.5}$$

Note that Eq. 4.5 involves summations over the inlets and exits of the control volume. Each individual term in these sums applies to a particular inlet or exit. The area, velocity, and specific volume appearing in each term refer only to the corresponding inlet or exit.

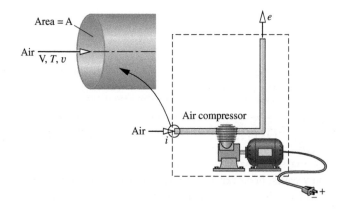

Fig. 4.3 Figure illustrating the one-dimensional flow model.

4.2.2 Steady-State Form of the Mass Rate Balance

Many engineering systems can be idealized as being at **steady state**, meaning that *all* properties are unchanging in time. For a control volume at steady state, the identity of the matter within the control volume changes continuously, but the total amount present at any instant remains constant, so $dm_{cv}/dt = 0$ and Eq. 4.2 reduces to

steady state

$$\boxed{\underset{\text{(mass rate in)}}{\sum_i \dot{m}_i} = \underset{\text{(mass rate out)}}{\sum_e \dot{m}_e}} \tag{4.6}$$

That is, the total incoming and outgoing rates of mass flow are equal.

Note that equality of total incoming and outgoing rates of mass flow does not necessarily imply that a control volume is at steady state. Although the total amount of mass within the control volume at any instant would be constant, other properties such as temperature and pressure might be varying with time. When a control volume is at steady state, *every* property is independent of time. Also note that the steady-state assumption and the one-dimensional flow assumption are independent idealizations. One does not imply the other.

4.2.3 Integral Form of the Mass Rate Balance

We consider next the mass rate balance expressed in terms of local properties. The total mass contained within the control volume at an instant t can be related to the local density as follows

$$m_{cv}(t) = \int_V \rho \, dV \tag{4.7}$$

where the integration is over the volume at time t.

With Eqs. 4.3 and 4.7, the mass rate balance Eq. 4.2 can be written as

$$\frac{d}{dt} \int_V \rho \, dV = \sum_i \left(\int_A \rho V_n \, dA \right)_i - \sum_e \left(\int_A \rho V_n \, dA \right)_e \tag{4.8}$$

where the area integrals are over the areas through which mass enters and exits the control volume, respectively. The product ρV_n appearing in this equation, known as the **mass flux**, gives the time rate of mass flow per unit of area. To evaluate the terms of the right side of Eq. 4.8 requires information about the variation of the mass flux over the flow areas. The form of the conservation of mass principle given by Eq. 4.8 is usually considered in detail in fluid mechanics.

mass flux

4.3 Applications of the Mass Rate Balance

4.3.1 Steady-State Application

For a control volume at steady state, the conditions of the mass within the control volume and at the boundary do not vary with time. The mass flow rates also are constant with time.

Example 4.1 illustrates an application of the steady-state form of the mass rate balance to a control volume enclosing a mixing chamber called a *feedwater heater.* Feedwater heaters are components of the vapor power systems considered in Chap. 8.

> **EXAMPLE 4.1** ▸ •
>
> ## Applying the Mass Rate Balance to a Feedwater Heater at Steady State
>
> A feedwater heater operating at steady state has two inlets and one exit. At inlet 1, water vapor enters at $p_1 = 7$ bar, $T_1 = 200°C$ with a mass flow rate of 40 kg/s. At inlet 2, liquid water at $p_2 = 7$ bar, $T_2 = 40°C$ enters through an area $A_2 = 25$ cm². Saturated liquid at 7 bar exits at 3 with a volumetric flow rate of 0.06 m³/s. Determine the mass flow rates at inlet 2 and at the exit, in kg/s, and the velocity at inlet 2, in m/s.

SOLUTION

Known A stream of water vapor mixes with a liquid water stream to produce a saturated liquid stream at the exit. The states at the inlets and exit are specified. Mass flow rate and volumetric flow rate data are given at one inlet and at the exit, respectively.

Find Determine the mass flow rates at inlet 2 and at the exit, and the velocity V_2.

Schematic and Given Data:

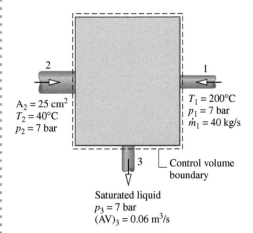

$A_2 = 25\ cm^2$
$T_2 = 40°C$
$p_2 = 7\ bar$

$T_1 = 200°C$
$p_1 = 7\ bar$
$\dot{m}_1 = 40\ kg/s$

3 ⌐ Control volume
boundary

Saturated liquid
$p_3 = 7\ bar$
$(AV)_3 = 0.06\ m^3/s$

Fig. E4.1

Engineering Model The control volume shown on the accompanying figure is at steady state.

Analysis The principal relations to be employed are the mass rate balance (Eq. 4.2) and the expression $\dot{m} = AV/v$ (Eq. 4.4b). At steady state the mass rate balance becomes

$$\frac{dm\!\!\!/^{0}}{dt} = \dot{m}_1 + \dot{m}_2 - \dot{m}_3$$

Solving for $\dot{m}_2$,

$$\dot{m}_2 = \dot{m}_3 - \dot{m}_1$$

The mass flow rate $\dot{m}_1$ is given. The mass flow rate at the exit can be evaluated from the given volumetric flow rate

$$\dot{m}_3 = \frac{(AV)_3}{v_3}$$

where v_3 is the specific volume at the exit. In writing this expression, one-dimensional flow is assumed. From Table A-3, $v_3 = 1.108 \times 10^{-3}\ m^3/kg$. Hence,

$$\dot{m}_3 = \frac{0.06\ m^3/s}{(1.108 \times 10^{-3}\ m^3/kg)} = 54.15\ kg/s$$

The mass flow rate at inlet 2 is then

$$\dot{m}_2 = \dot{m}_3 - \dot{m}_1 = 54.15 - 40 = 14.15\ kg/s$$

For one-dimensional flow at 2, $\dot{m}_2 = A_2 V_2/v_2$, so

$$V_2 = \dot{m}_2 v_2/A_2$$

State 2 is a compressed liquid. The specific volume at this state can be approximated by $v_2 \approx v_f(T_2)$ (Eq. 3.11). From Table A-2 at 40°C, $v_2 = 1.0078 \times 10^{-3}\ m^3/kg$. So,

$$V_2 = \frac{(14.15\ kg/s)(1.0078 \times 10^{-3}\ m^3/kg)}{25\ cm^2}\left|\frac{10^4\ cm^2}{1\ m^2}\right| = 5.7\ m/s$$

❶ In accord with Eq. 4.6, the mass flow rate at the exit equals the sum of the mass flow rates at the inlets. It is left as an exercise to show that the volumetric flow rate at the exit *does not equal* the sum of the volumetric flow rates at the inlets.

SKILLS DEVELOPED

Ability to...

● apply the steady-state mass rate balance.

● apply the mass flow rate expression, Eq. 4.4b.

● retrieve property data for water.

Quick Quiz

Evaluate the volumetric flow rate, in m^3/s, at each inlet. Ans. $(AV)_1 = 12\ m^3/s$, $(AV)_2 = 0.01\ m^3/s$

4.3.2 Time-Dependent (Transient) Application

Many devices undergo periods of operation during which the state changes with time—for example, the startup and shutdown of motors. Other examples include containers being filled or emptied and applications to biological systems. The steady-state model is not appropriate when analyzing time-dependent (transient) cases.

Example 4.2 illustrates a time-dependent, or transient, application of the mass rate balance. In this case, a barrel is filled with water.

▸ **EXAMPLE 4.2** ▸

Filling a Barrel with Water

Water flows into the top of an open barrel at a constant mass flow rate of 7 kg/s. Water exits through a pipe near the base with a mass flow rate proportional to the height of liquid inside: $\dot{m}_e = 1.4\ L$, where L is the instantaneous liquid height, in m. The area of the

base is 0.2 m^2, and the density of water is 1000 kg/m^3. If the barrel is initially empty, plot the variation of liquid height with time and comment on the result.

SOLUTION

Known Water enters and exits an initially empty barrel. The mass flow rate at the inlet is constant. At the exit, the mass flow rate is proportional to the height of the liquid in the barrel.

Find Plot the variation of liquid height with time and comment.

Schematic and Given Data:

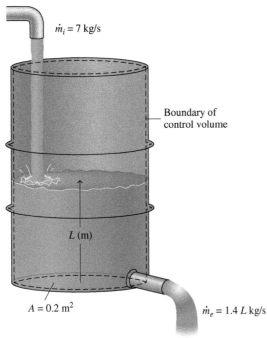

$\dot{m}_i = 7$ kg/s

Boundary of control volume

L (m)

$A = 0.2$ m^2

$\dot{m}_e = 1.4\, L$ kg/s

Fig. E4.2a

Engineering Model

1. The control volume is defined by the dashed line on the accompanying diagram.

2. The water density is constant.

Analysis For the one-inlet, one-exit control volume, Eq. 4.2 reduces to

$$\frac{dm_{cv}}{dt} = \dot{m}_i - \dot{m}_e$$

The mass of water contained within the barrel at time t is given by

$$m_{cv}(t) = \rho A L(t)$$

where ρ is density, A is the area of the base, and $L(t)$ is the instantaneous liquid height. Substituting this into the mass rate balance together with the given mass flow rates

$$\frac{d(\rho A L)}{dt} = 7 - 1.4L$$

Since density and area are constant, this equation can be written as

$$\frac{dL}{dt} + \left(\frac{1.4}{\rho A}\right)L = \frac{7}{\rho A}$$

which is a first-order, ordinary differential equation with constant coefficients. The solution is

❶
$$L = 5 + C \exp\left(-\frac{1.4t}{\rho A}\right)$$

where C is a constant of integration. The solution can be verified by substitution into the differential equation.

To evaluate C, use the initial condition: at $t = 0$, $L = 0$. Thus, $C = -5.0$, and the solution can be written as

$$L = 5.0[1 - \exp(-1.4t/\rho A)]$$

Substituting $\rho = 1000$ kg/m^3 and A $= 0.2$ m^2 results in

$$L = 5[1 - \exp(-0.007t)]$$

This relation can be plotted by hand or using appropriate software. The result is

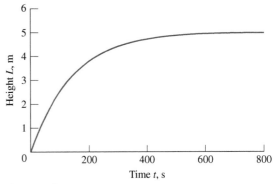

Fig. E4.2b

From the graph, we see that initially the liquid height increases rapidly and then levels out as steady-state operation is approached. After about 100 s, the height stays constant with time. At this point, the rate of water flow into the barrel equals the rate of flow out of the barrel. From the graph, the limiting value of L is 5 m, which also can be verified by taking the limit of the analytical solution as $t \to \infty$.

---

❶ Alternatively, this differential equation can be solved using *Interactive Thermodynamics: IT*. The differential equation can be expressed as

```
der(L, t) + (1.4 * L)/(rho * A) = 7/(rho * A)
rho = 1000 // kg/m³
A = 0.2 // m²
```

where der (L,t) is dL/dt, rho is density ρ, and A is area. Using the **Explore** button, set the initial condition at $L = 0$, and sweep t from 0 to 200 in steps of 0.5. Then, the plot can be constructed using the **Graph** button.

SKILLS DEVELOPED

Ability to…

- apply the time-dependent mass rate balance.
- solve an ordinary differential equation and plot the solution.

Quick Quiz

If the mass flow rate of the water flowing into the barrel were 12.25 kg/s while all other data remained the same, what would be the limiting value of the liquid height, L, in m? Ans. 0.9 m

BioConnections

The human heart provides a good example of how biological systems can be modeled as control volumes. **Figure 4.4** shows the cross section of a human heart. The flow is controlled by valves that intermittently allow blood to enter from veins and exit through arteries as the heart muscles pump. Work is done to increase the pressure of the blood leaving the heart to a level that will propel it through the cardiovascular system of the body. Observe that the boundary of the control volume enclosing the heart is not fixed but moves with time as the heart pulses.

Understanding the medical condition known as *arrhythmia* requires consideration of the time-dependent behavior of the heart. An arrhythmia is a change in the regular beat of the heart. This can take several forms. The heart may beat irregularly, skip a beat, or beat very fast or slowly. An arrhythmia may be detectable by listening to the heart with a stethoscope, but an electrocardiogram offers a more precise approach. Although arrhythmia does occur in people without underlying heart disease, patients having serious symptoms may require treatment to keep their heartbeats regular. Many patients with arrhythmia may require no medical intervention at all.

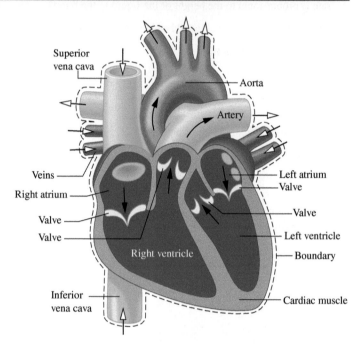

Fig. 4.4 Control volume enclosing the heart.

4.4 Conservation of Energy for a Control Volume

In this section, the rate form of the energy balance for control volumes is obtained. The energy rate balance plays an important role in subsequent sections of this book.

4.4.1 Developing the Energy Rate Balance for a Control Volume

We begin by noting that the control volume form of the energy rate balance can be derived by an approach closely paralleling that considered in the box of Sec. 4.1.1, where the control volume mass rate balance is obtained by transforming the closed system form. The present development proceeds less formally by arguing that, like mass, energy is an extensive property, so it too can be transferred into or out of a control volume as a result of mass crossing the boundary. Since this is the principal difference between the closed system and control volume forms, the control volume energy rate balance can be obtained by modifying the closed system energy rate balance to account for these energy transfers.

Accordingly, the *conservation of energy* principle applied to a control volume states:

$$
\begin{bmatrix}
\text{time } \textit{rate of change} \\
\text{of the energy} \\
\text{contained within} \\
\text{the control volume} \\
\textit{at time t}
\end{bmatrix}
=
\begin{bmatrix}
\textit{net rate} \text{ at which} \\
\text{energy is being} \\
\text{transferred in} \\
\text{by heat } \textit{at} \\
\textit{time t}
\end{bmatrix}
-
\begin{bmatrix}
\textit{net rate} \text{ at which} \\
\text{energy is being} \\
\text{transferred out} \\
\text{by work } \textit{at} \\
\textit{time t}
\end{bmatrix}
+
\begin{bmatrix}
\textit{net rate} \text{ of energy} \\
\text{transfer } \textit{into} \text{ the} \\
\text{control volume} \\
\text{accompanying} \\
\text{mass flow}
\end{bmatrix}
$$

For the one-inlet, one-exit control volume with one-dimensional flow shown in Fig. 4.5 the energy rate balance is

$$\frac{dE_{cv}}{dt} = \dot{Q} - \dot{W} + \dot{m}_i\left(u_i + \frac{V_i^2}{2} + gz_i\right) - \dot{m}_e\left(u_e + \frac{V_e^2}{2} + gz_e\right) \qquad (4.9)$$

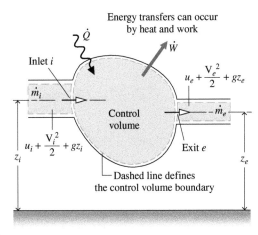

Fig. 4.5 Figure used to develop Eq. 4.9.

where E_{cv} denotes the energy of the control volume at time t. The terms $\dot{Q}$ and $\dot{W}$ account, respectively, for the net rate of energy transfer by heat and work across the boundary of the control volume at t. The underlined terms account for the rates of transfer of internal, kinetic, and potential energy of the entering and exiting streams. If there is no mass flow in or out, the respective mass flow rates vanish and the underlined terms of Eq. 4.9 drop out. The equation then reduces to the rate form of the energy balance for closed systems: Eq. 2.37.

Next, we will place Eq. 4.9 in an alternative form that is more convenient for subsequent applications. This will be accomplished primarily by recasting the work term $\dot{W}$, which represents the net rate of energy transfer by work across *all* portions of the boundary of the control volume.

Energy Balance Control Volume Tab A

4.4.2 Evaluating Work for a Control Volume

Because work is always done on or by a control volume where matter flows across the boundary, it is convenient to separate the work term $\dot{W}$ of Eq. 4.9 into *two contributions:* One contribution is the work associated with the fluid pressure as mass is introduced at inlets and removed at exits. The other contribution, denoted by $\dot{W}_{cv}$, includes *all other* work effects, such as those associated with rotating shafts, displacement of the boundary, and electrical effects.

Consider the work at an exit e associated with the pressure of the flowing matter. Recall from Eq. 2.13 that the rate of energy transfer by work can be expressed as the product of a force and the velocity at the point of application of the force. Accordingly, the *rate* at which work is done at the exit by the normal force (normal to the exit area in the direction of flow) due to pressure is the product of the normal force, $p_e A_e$, and the fluid velocity, V_e. That is

$$\begin{bmatrix} \text{time rate of energy transfer} \\ \text{by work } from \text{ the control} \\ \text{volume at exit } e \end{bmatrix} = (p_e A_e)V_e \qquad (4.10)$$

where p_e is the pressure, A_e is the area, and V_e is the velocity at exit e, respectively. A similar expression can be written for the rate of energy transfer by work into the control volume at inlet i.

With these considerations, the work term $\dot{W}$ of the energy rate equation, Eq. 4.9, can be written as

$$\dot{W} = \dot{W}_{cv} + (p_e A_e)V_e - (p_i A_i)V_i \qquad (4.11)$$

where, in accordance with the sign convention for work, the term at the inlet has a negative sign because energy is transferred into the control volume there. A positive sign precedes the work term at the exit because energy is transferred out of the control volume there. With $AV = \dot{m}v$ from Eq. 4.4b, the above expression for work can be written as

$$\dot{W} = \dot{W}_{cv} + \dot{m}_e(p_e v_e) - \dot{m}_i(p_i v_i) \qquad (4.12)$$

where $\dot{m}_i$ and $\dot{m}_e$ are the mass flow rates and v_i and v_e are the specific volumes evaluated at the inlet and exit, respectively. In Eq. 4.12, the terms $\dot{m}_i(p_i v_i)$ and $\dot{m}_e(p_e v_e)$ account for the work associated with the pressure at the inlet and exit, respectively. They are commonly referred to as **flow work**. The term $\dot{W}_{cv}$ accounts for *all other* energy transfers by work across the boundary of the control volume.

flow work

4.4.3 One-Dimensional Flow Form of the Control Volume Energy Rate Balance

Substituting Eq. 4.12 in Eq. 4.9 and collecting all terms referring to the inlet and the exit into separate expressions, the following form of the control volume energy rate balance results:

$$\frac{dE_{cv}}{dt} = \dot{Q}_{cv} - \dot{W}_{cv} + \dot{m}_i \left(u_i + p_i v_i + \frac{V_i^2}{2} + gz_i \right) - \dot{m}_e \left(u_e + p_e v_e + \frac{V_e^2}{2} + gz_e \right) \quad (4.13)$$

The subscript "cv" has been added to $\dot{Q}$ to emphasize that this is the heat transfer rate over the boundary (control surface) of the *control volume*.

The last two terms of Eq. 4.13 can be rewritten using the specific enthalpy h introduced in Sec. 3.6.1. With $h = u + pv$, the energy rate balance becomes

$$\frac{dE_{cv}}{dt} = \dot{Q}_{cv} - \dot{W}_{cv} + \dot{m}_i \left(h_i + \frac{V_i^2}{2} + gz_i \right) - \dot{m}_e \left(h_e + \frac{V_e^2}{2} + gz_e \right) \quad (4.14)$$

Energy Balance Control Volume Tab B

energy rate balance

TAKE NOTE...

Equation 4.15 is the most general form of the conservation of energy principle for control volumes used in this book. It serves as the starting point for applying the conservation of energy principle to control volumes in problem solving.

The appearance of the sum $u + pv$ in the control volume energy equation is the principal reason for introducing enthalpy previously. It is brought in solely as a *convenience:* The algebraic form of the energy rate balance is simplified by the use of enthalpy and, as we have seen, enthalpy is normally tabulated along with other properties.

In practice there may be several locations on the boundary through which mass enters or exits. This can be accounted for by introducing summations as in the mass balance. Accordingly, the **energy rate balance** is

$$\boxed{\frac{dE_{cv}}{dt} = \dot{Q}_{cv} - \dot{W}_{cv} + \sum_i \dot{m}_i \left(h_i + \frac{V_i^2}{2} + gz_i \right) - \sum_e \dot{m}_e \left(h_e + \frac{V_e^2}{2} + gz_e \right)} \quad (4.15)$$

In writing Eq. 4.15, the one-dimensional flow model is assumed where mass enters and exits the control volume.

Equation 4.15 is an *accounting* balance for the energy of the control volume. It states that the rate of energy increase or decrease within the control volume equals the difference between the rates of energy transfer in and out across the boundary. The mechanisms of energy transfer are heat and work, as for closed systems, and the energy that accompanies the mass entering and exiting.

4.4.4 Integral Form of the Control Volume Energy Rate Balance

As for the case of the mass rate balance, the energy rate balance can be expressed in terms of local properties to obtain forms that are more generally applicable. Thus, the term $E_{cv}(t)$, representing the total energy associated with the control volume at time t, can be written as a volume integral

$$E_{cv}(t) = \int_V \rho e \, dV = \int_V \rho \left(u + \frac{V^2}{2} + gz \right) dV \quad (4.16)$$

Similarly, the terms accounting for the energy transfers accompanying mass flow and flow work at inlets and exits can be expressed as shown in the following form of the energy rate balance:

$$\frac{d}{dt} \int_V \rho e \, dV = \dot{Q}_{cv} - \dot{W}_{cv} + \sum_i \left[\int_A \left(h + \frac{V^2}{2} + gz \right) \rho V_n \, dA \right]_i$$

$$- \sum_e \left[\int_A \left(h + \frac{V^2}{2} + gz \right) \rho V_n \, dA \right]_e \quad (4.17)$$

Additional forms of the energy rate balance can be obtained by expressing the heat transfer rate $\dot{Q}_{cv}$ as the integral of the *heat flux* over the boundary of the control volume, and the work $\dot{W}_{cv}$ in terms of normal and shear stresses at the moving portions of the boundary.

In principle, the change in the energy of a control volume over a time period can be obtained by integration of the energy rate balance with respect to time. Such integrations require information about the time dependences of the work and heat transfer rates, the various mass flow rates, and the states at which mass enters and leaves the control volume. Examples of this type of analysis are presented in Sec. 4.12.

4.5 Analyzing Control Volumes at Steady State

In this section we consider steady-state forms of the mass and energy rate balances. These balances are applied to a variety of devices of engineering interest in Secs. 4.6–4.11. The steady-state forms considered here do not apply to the transient startup or shutdown periods of operation of such devices but only to periods of steady operation. This situation is commonly encountered in engineering.

Animation

System Types
Tab E

4.5.1 Steady-State Forms of the Mass and Energy Rate Balances

For a control volume at steady state, the conditions of the mass within the control volume and at the boundary do not vary with time. The mass flow rates and the rates of energy transfer by heat and work are also constant with time. There can be no accumulation of mass within the control volume, so $dm_{cv}/dt = 0$ and the mass rate balance, Eq. 4.2, takes the form

$$\sum_i \dot{m}_i = \sum_e \dot{m}_e \qquad (4.6)$$
$$\text{(mass rate in)} \qquad \text{(mass rate out)}$$

Furthermore, at steady state $dE_{cv}/dt = 0$, so Eq. 4.15 can be written as

$$0 = \dot{Q}_{cv} - \dot{W}_{cv} + \sum_i \dot{m}_i \left(h_i + \frac{\mathrm{V}_i^2}{2} + gz_i \right) - \sum_e \dot{m}_e \left(h_e + \frac{\mathrm{V}_e^2}{2} + gz_e \right) \qquad (4.18)$$

Alternatively

$$\dot{Q}_{cv} + \sum_i \dot{m}_i \left(h_i + \frac{\mathrm{V}_i^2}{2} + gz_i \right) = \dot{W}_{cv} + \sum_e \dot{m}_e \left(h_e + \frac{\mathrm{V}_e^2}{2} + gz_e \right) \qquad (4.19)$$
$$\text{(energy rate in)} \qquad \text{(energy rate out)}$$

Equation 4.6 asserts that at steady state the total rate at which mass enters the control volume equals the total rate at which mass exits. Similarly, Eq. 4.19 asserts that the total rate at which energy is transferred into the control volume equals the total rate at which energy is transferred out.

Many important applications involve one-inlet, one-exit control volumes at steady state. It is instructive to apply the mass and energy rate balances to this special case. The mass rate balance reduces simply to $\dot{m}_1 = \dot{m}_2$. That is, the mass flow rate must be the same at the exit, 2, as it is at the inlet, 1. The common mass flow rate is designated simply by $\dot{m}$. Next, applying the energy rate balance and factoring the mass flow rate gives

$$0 = \dot{Q}_{cv} - \dot{W}_{cv} + \dot{m} \left[(h_1 - h_2) + \frac{(\mathrm{V}_1^2 - \mathrm{V}_2^2)}{2} + g(z_1 - z_2) \right] \qquad (4.20a)$$

Animation

Energy Balance Control Volume Tab C

Or, dividing by the mass flow rate,

$$0 = \frac{\dot{Q}_{cv}}{\dot{m}} - \frac{\dot{W}_{cv}}{\dot{m}} + (h_1 - h_2) + \frac{(V_1^2 - V_2^2)}{2} + g(z_1 - z_2)$$

(4.20b)

The enthalpy, kinetic energy, and potential energy terms all appear in Eqs. 4.20 as *differences* between their values at the inlet and exit. This illustrates that the datums used to assign values to specific enthalpy, velocity, and elevation cancel. In Eq. 4.20b, the ratios $\dot{Q}_{cv}/\dot{m}$ and $\dot{W}_{cv}/\dot{m}$ are energy transfers *per unit mass flowing through the control volume*.

The foregoing steady-state forms of the energy rate balance relate only energy transfer quantities evaluated at the *boundary* of the control volume. No details concerning properties *within* the control volume are required by, or can be determined with, these equations. When applying the energy rate balance in any of its forms, it is necessary to use the same units for all terms in the equation. For instance, *every* term in Eq. 4.20b must have a unit such as kJ/kg. Appropriate unit conversions are emphasized in examples to follow.

4.5.2 Modeling Considerations for Control Volumes at Steady State

In this section, we provide the basis for subsequent applications by considering the modeling of control volumes *at steady state*. In particular, several applications are given in Secs. 4.6–4.11 showing the use of the principles of conservation of mass and energy, together with relationships among properties for the analysis of control volumes at steady state. The examples are drawn from applications of general interest to engineers and are chosen to illustrate points common to all such analyses. Before studying them, it is recommended that you review the methodology for problem solving outlined in Sec. 1.9. As problems become more complicated, the use of a systematic problem-solving approach becomes increasingly important.

When the mass and energy rate balances are applied to a control volume, simplifications are normally needed to make the analysis manageable. That is, the control volume of interest is *modeled* by making assumptions. The *careful* and *conscious* step of listing assumptions is necessary in every engineering analysis. Therefore, an important part of this section is devoted to considering various assumptions that are commonly made when applying the conservation principles to different types of devices. As you study the examples presented in Secs. 4.6–4.11, it is important to recognize the role played by careful assumption making in arriving at solutions. In each case considered, steady-state operation is assumed. The flow is regarded as one-dimensional at places where mass enters and exits the control volume. Also, at each of these locations equilibrium property relations are assumed to apply.

Horizons

Smaller Can Be Better

Engineers are developing miniature systems for use where weight, portability, and/or compactness are critically important. Some of these applications involve tiny *micro systems* with dimensions in the micrometer to millimeter range. Other somewhat larger *meso-scale systems* can measure up to a few centimeters.

Microelectromechanical systems (MEMS) combining electrical and mechanical features are now widely used for sensing and control. Medical applications of MEMS include pressure sensors that monitor pressure within the balloon inserted into a blood vessel during angioplasty. Air bags are triggered in an automobile crash by tiny acceleration sensors. MEMS are also found in computer hard drives and printers.

Miniature versions of other technologies are being investigated. One study aims at developing an entire gas turbine power plant the size of a shirt button. Another involves micromotors with shafts the diameter of a human hair. Emergency workers wearing fire-, chemical-, or biological-protection suits might in the future be kept cool by tiny heat pumps imbedded in the suit material.

As designers aim at smaller sizes, frictional effects and heat transfers pose special challenges. Fabrication of miniature systems is also demanding. Taking a design from the concept stage to high-volume production can be both expensive and risky, industry representatives say.

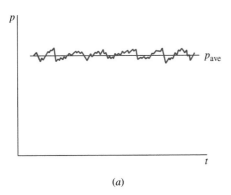
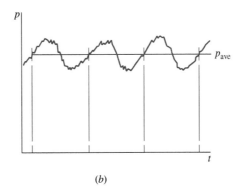

(a)   (b)

Fig. 4.6 Pressure variations about an average. (a) Fluctuation. (b) Periodic.

In several of the examples to follow, the heat transfer term $\dot{Q}_{cv}$ is set to zero in the energy rate balance because it is small relative to other energy transfers across the boundary. This may be the result of one or more of the following factors:

- The outer surface of the control volume is well insulated.
- The outer surface area is too small for there to be effective heat transfer.
- The temperature difference between the control volume and its surroundings is so small that the heat transfer can be ignored.
- The gas or liquid passes through the control volume so quickly that there is not enough time for significant heat transfer to occur.

The work term $\dot{W}_{cv}$ drops out of the energy rate balance when there are no rotating shafts, displacements of the boundary, electrical effects, or other work mechanisms associated with the control volume being considered. The kinetic and potential energies of the matter entering and exiting the control volume are neglected when they are small relative to other energy transfers.

In practice, the properties of control volumes considered to be at steady state *do* vary with time. The steady-state assumption would still apply, however, when properties fluctuate only slightly about their averages, as for pressure in **Fig. 4.6a**. Steady state also might be assumed in cases where *periodic* time variations are observed, as in Fig. 4.6b. For example, in recip-rocating engines and compressors, the entering and exiting flows pulsate as valves open and close. Other parameters also might be time varying. However, the steady-state assumption can apply to control volumes enclosing these devices if the following are satisfied for each suc-cessive period of operation: (1) There is no *net* change in the total energy and the total mass within the control volume. (2) The *time-averaged* mass flow rates, heat transfer rates, work rates, and properties of the substances crossing the control surface all remain constant.

Next, we present brief discussions and examples illustrating the analysis of several devices of interest in engineering, including nozzles and diffusers, turbines, compressors and pumps, heat exchangers, and throttling devices. The discussions highlight some common applications of each device and the modeling typically used in thermodynamic analysis.

4.6 Nozzles and Diffusers

nozzle
diffuser

A **nozzle** is a flow passage of varying cross-sectional area in which the velocity of a gas or liquid increases in the direction of flow. In a **diffuser**, the gas or liq-uid decelerates in the direction of flow. **Figure 4.7** shows a nozzle in which the cross-sectional area decreases in the direction of flow and a diffuser in which the walls of the flow passage diverge. Observe that as velocity increases pressure decreases, and conversely.

For many readers the most familiar application of a nozzle is its use with a garden hose. But nozzles and diffusers have several important engineering applica-tions. In **Fig. 4.8**, a nozzle and diffuser are combined in a wind-tunnel test facility. Ducts with converging and diverging passages are commonly used in distributing cool and warm air in building air-conditioning systems. Nozzles and diffusers also are key components of turbojet engines (Chap. 9).

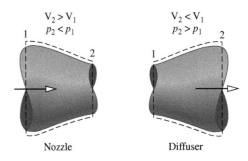

Fig. 4.7 Illustration of a nozzle and a diffuser.

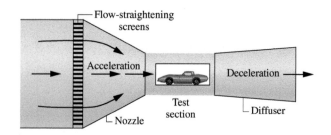

Fig. 4.8 Wind-tunnel test facility.

4.6.1 Nozzle and Diffuser Modeling Considerations

For a control volume enclosing a nozzle or diffuser, the only work is *flow work* at locations where mass enters and exits the control volume, so the term $\dot{W}_{cv}$ drops out of the energy rate balance. The change in potential energy from inlet to exit is negligible under most conditions. Thus, the underlined terms of Eq. 4.20a (repeated below) drop out, leaving the enthalpy, kinetic energy, and heat transfer terms, as shown by Eq. (a)

$$0 = \dot{Q}_{cv} - \dot{W}_{cv} + \dot{m}\left[(h_1 - h_2) + \frac{(V_1^2 - V_2^2)}{2} + \underline{g(z_1 - z_2)}\right]$$

$$0 = \dot{Q}_{cv} + \dot{m}\left[(h_1 - h_2) + \frac{(V_1^2 - V_2^2)}{2}\right] \qquad \text{(a)}$$

where $\dot{m}$ is the mass flow rate. The term $\dot{Q}_{cv}$ representing heat transfer with the surroundings normally would be unavoidable (or stray) heat transfer, and this is often small enough relative to the enthalpy and kinetic energy terms that it also can be neglected, giving simply

$$0 = (h_1 - h_2) + \left(\frac{V_1^2 - V_2^2}{2}\right) \qquad \text{(4.21)}$$

> **Animation**
>
> **Nozzles**
> **Diffusers**

4.6.2 Application to a Steam Nozzle

The modeling introduced in Sec. 4.6.1 is illustrated in the next example involving a steam nozzle. Particularly note the use of unit conversion factors in this application.

▶▶▶ **EXAMPLE 4.3** ▶

Calculating Exit Area of a Steam Nozzle

Steam enters a converging–diverging nozzle operating at steady state with $p_1 = 40$ bar, $T_1 = 400°C$, and a velocity of 10 m/s. The steam flows through the nozzle with negligible heat transfer and no significant change in potential energy. At the exit, $p_2 = 15$ bar, and the velocity is 665 m/s. The mass flow rate is 2 kg/s. Determine the exit area of the nozzle, in m².

SOLUTION

Known Steam flows through a nozzle at steady state with known properties at the inlet and exit, a known mass flow rate, and negligible effects of heat transfer and potential energy.

Find Determine the exit area.

Schematic and Given Data:

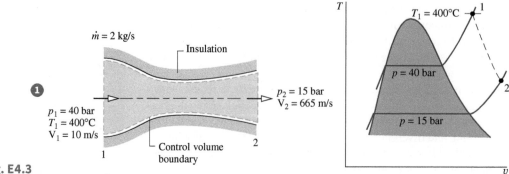

Fig. E4.3

Engineering Model

1. The control volume shown on the accompanying figure is at steady state.

2. Heat transfer is negligible and $\dot{W}_{cv} = 0$.

3. The change in potential energy from inlet to exit can be neglected.

Analysis The exit area can be determined from the mass flow rate $\dot{m}$ and Eq. 4.4b, which can be arranged to read

$$A_2 = \frac{\dot{m}v_2}{V_2}$$

To evaluate A_2 from this equation requires the specific volume v_2 at the exit, and this requires that the exit state be fixed.

The state at the exit is fixed by the values of two independent intensive properties. One is the pressure p_2, which is known. The other is the specific enthalpy h_2, determined from the steady-state energy rate balance Eq. 4.20a, as follows

$$0 = \overset{0}{\cancel{\dot{Q}}}_{cv} - \overset{0}{\cancel{\dot{W}}}_{cv} + \dot{m}\left[(h_1 - h_2) + \frac{(V_1^2 - V_2^2)}{2} + g(z_1 - z_2)\right]$$

The terms $\dot{Q}_{cv}$ and $\dot{W}_{cv}$ are deleted by assumption 2. The change in specific potential energy drops out in accordance with assumption 3 and $\dot{m}$ cancels, leaving

$$0 = (h_1 - h_2) + \left(\frac{V_1^2 - V_2^2}{2}\right)$$

Solving for h_2

$$h_2 = h_1 + \left(\frac{V_1^2 - V_2^2}{2}\right)$$

From Table A-4, $h_1 = 3213.6$ kJ/kg. The velocities V_1 and V_2 are given. Inserting values and converting the units of the kinetic

energy terms to kJ/kg results in

❷ $h_2 = 3213.6 \text{ kJ/kg} + \left[\dfrac{(10)^2 - (665)^2}{2}\right]\left(\dfrac{m^2}{s^2}\right)\left|\dfrac{1 \text{ N}}{1 \text{ kg} \cdot m/s^2}\right|\left|\dfrac{1 \text{ kJ}}{10^3 \text{ N} \cdot m}\right|$

$= 3213.6 - 221.1 = 2992.5 \text{ kJ/kg}$

Finally, referring to Table A-4 at $p_2 = 15$ bar with $h_2 = 2992.5$ kJ/kg, the specific volume at the exit is $v_2 = 0.1627$ m³/kg. The exit area is then

$$A_2 = \frac{(2 \text{ kg/s})(0.1627 \text{ m}^3/\text{kg})}{665 \text{ m/s}} = 4.89 \times 10^{-4} \text{ m}^2$$

---

❶ Although equilibrium property relations apply at the inlet and exit of the control volume, the intervening states of the steam are not necessarily equilibrium states. Accordingly, the expansion through the nozzle is represented on the T–v diagram as a dashed line.

❷ Care must be taken in converting the units for specific kinetic energy to kJ/kg.

SKILLS DEVELOPED

Ability to…

- apply the steady-state energy rate balance to a control volume.
- apply the mass flow rate expression, Eq. 4.4b.
- develop an engineering model.
- retrieve property data for water.

Quick Quiz

Evaluate the nozzle inlet area, in m². Ans. 1.47×10^{-2} m².

4.7 Turbines

A **turbine** is a device in which power is developed as a result of a gas or liquid passing through a set of blades attached to a shaft free to rotate. A schematic of an axial-flow steam or gas turbine is shown in **Fig. 4.9**. Such turbines are widely used for power generation in vapor power plants, gas turbine power plants, and aircraft engines (see Chaps. 8 and 9). In these applications, superheated steam or a gas enters the turbine and expands to a lower pressure as power is generated.

turbine

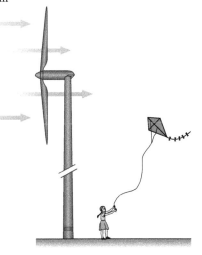

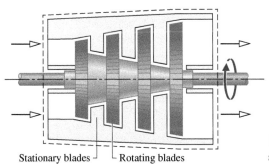

Stationary blades — Rotating blades

Fig. 4.9 Schematic of an axial-flow steam or gas turbine.

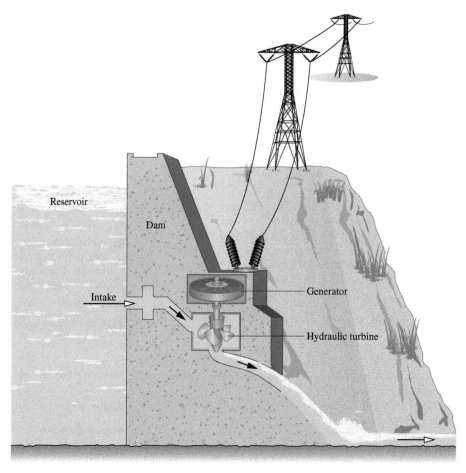

Fig. 4.10 Hydraulic turbine installed in a dam.

A *hydraulic* turbine coupled to a generator installed in a dam is shown in **Fig. 4.10**. As water flows from higher to lower elevation through the turbine, the turbine provides shaft power to the generator. The generator converts shaft power to electricity. This type of generation is called *hydropower*. Today, hydropower is a leading *renewable* means for producing electricity, and it is one of the least expensive ways to do so. Electricity can also be produced from flowing water by using turbines to tap into currents in oceans and rivers.

Turbines are also key components of wind-turbine power plants that, like hydropower plants, are renewable means for generating electricity.

Energy & Environment

Industrial-scale wind turbines can stand as tall as a 30-story building and produce electricity at a rate meeting the needs of hundreds of typical U.S. homes. The three-bladed rotors of these wind turbines have a diameter nearly the length of a football field and can operate in winds up to 90 km per hour. They feature microprocessor control of all functions, ensuring each blade is pitched at the correct angle for current wind conditions. *Wind farms* consisting of dozens of such turbines increasingly dot the landscape over the globe.

In the United States, wind farms at favorable sites in several Great Plains states alone could supply much of the nation's electricity needs—provided the electrical grid is upgraded and expanded (see *Horizons* "Our Electricity Superhighway" in Chap. 8). Offshore wind farms along the U.S. coastline could also contribute significantly to meeting national needs. Experts say wind variability can be managed by producing maximum power when winds are strong and storing some, or all, of the power by various means, including

pumped-hydro storage and *compressed-air* storage, for distribution when consumer demand is highest and electricity has its greatest economic value (see box in Sec. 4.8.3).

Wind power can produce electricity today at costs competitive with all alternative means and within a few years is expected to be among the least costly ways to do it. Wind energy plants take less time to build than conventional power plants and are modular, allowing additional units to be added as warranted. While generating electricity, wind-turbine plants produce no global warming gases or other emissions.

The industrial-scale wind turbines considered thus far are not the only ones available. Companies manufacture smaller, relatively inexpensive wind turbines that can generate electricity with wind speeds as low as 5 or 6 km per hour. These *low-wind* turbines are suitable for small businesses, farms, groups of neighbors, or individual users.

4.7.1 Steam and Gas Turbine Modeling Considerations

With a proper selection of the control volume enclosing a steam or gas turbine, the net kinetic energy of the matter flowing across the boundary is usually small enough to be neglected. The net potential energy of the flowing matter also is typically negligible. Thus, the underlined terms of Eq. 4.20a (repeated below) drop out, leaving the power, enthalpy, and heat transfer terms, as shown by Eq. (a)

$$0 = \dot{Q}_{cv} - \dot{W}_{cv} + \dot{m}\left[(h_1 - h_2) + \frac{(\mathrm{V}_1^2 - \mathrm{V}_2^2)}{2} + g(z_1 - z_2)\right]$$

$$0 = \dot{Q}_{cv} - \dot{W}_{cv} + \dot{m}(h_1 - h_2) \qquad \text{(a)}$$

where $\dot{m}$ is the mass flow rate. The only heat transfer between the turbine and surroundings normally would be unavoidable (or stray) heat transfer, and this is often small enough relative to the power and enthalpy terms that it also can be neglected, giving simply

$$\dot{W}_{cv} = \dot{m}(h_1 - h_2) \qquad \text{(b)}$$

Animation

Turbine
Tabs A, B, and C

4.7.2 Application to a Steam Turbine

In this section, modeling considerations for turbines are illustrated by application to a case involving the practically important steam turbine. Objectives in this example include assessing the significance of the heat transfer and kinetic energy terms of the energy balance and illustrating the appropriate use of unit conversion factors.

▶ EXAMPLE 4.4 ▶

Calculating Heat Transfer from a Steam Turbine

Steam enters a turbine operating at steady state with a mass flow rate of 4600 kg/h. The turbine develops a power output of 1000 kW. At the inlet, the pressure is 60 bar, the temperature is 400°C, and the velocity is 10 m/s. At the exit, the pressure is 0.1 bar, the quality is 0.9 (90%), and the velocity is 30 m/s. Calculate the rate of heat transfer between the turbine and surroundings, in kW.

SOLUTION

Known   A steam turbine operates at steady state. The mass flow rate, power output, and states of the steam at the inlet and exit are known.

Find   Calculate the rate of heat transfer.

Schematic and Given Data:

Engineering Model

1. The control volume shown on the accompanying figure is at steady state.

2. The change in potential energy from inlet to exit can be neglected.

Analysis   To calculate the heat transfer rate, begin with the one-inlet, one-exit form of the energy rate balance for a control volume at steady state, Eq. 4.20a. That is,

$$0 = \dot{Q}_{cv} - \dot{W}_{cv} + \dot{m}\left[(h_1 - h_2) + \frac{(\mathrm{V}_1^2 - \mathrm{V}_2^2)}{2} + g(z_1 - z_2)\right]$$

where $\dot{m}$ is the mass flow rate. Solving for $\dot{Q}_{cv}$ and dropping the potential energy change from inlet to exit

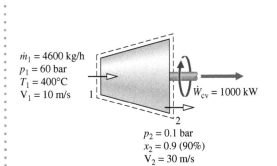

$\dot{m}_1 = 4600$ kg/h
$p_1 = 60$ bar
$T_1 = 400$°C
$V_1 = 10$ m/s

$\dot{W}_{cv} = 1000$ kW

$p_2 = 0.1$ bar
$x_2 = 0.9$ (90%)
$V_2 = 30$ m/s

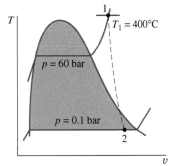

$T_1 = 400$°C

$p = 60$ bar

$p = 0.1$ bar

Fig. E4.4

$$\dot{Q}_{cv} = \dot{W}_{cv} + \dot{m}\left[(h_2 - h_1) + \left(\frac{V_2^2 - V_1^2}{2}\right)\right] \quad \text{(a)}$$

To compare the magnitudes of the enthalpy and kinetic energy terms, and stress the unit conversions needed, each of these terms is evaluated separately.

First, the specific *enthalpy difference* $h_2 - h_1$ is found. Using Table A-4, $h_1 = 3177.2$ kJ/kg. State 2 is a two-phase liquid–vapor mixture, so with data from Table A-3 and the given quality

$$h_2 = h_{f2} + x_2(h_{g2} - h_{f2})$$
$$= 191.83 + (0.9)(2392.8) = 2345.4 \text{ kJ/kg}$$

Hence,

$$h_2 - h_1 = 2345.4 - 3177.2 = -831.8 \text{ kJ/kg}$$

Consider next the specific *kinetic energy difference*. Using the given values for the velocities,

❶ $$\left(\frac{V_2^2 - V_1^2}{2}\right) = \left[\frac{(30)^2 - (10)^2}{2}\right]\left(\frac{m^2}{s^2}\right)\left|\frac{1 \text{ N}}{1 \text{ kg} \cdot \text{m/s}^2}\right|\left|\frac{1 \text{ kJ}}{10^3 \text{ N} \cdot \text{m}}\right|$$
$$= 0.4 \text{ kJ/kg}$$

Calculating $\dot{Q}_{cv}$ from Eq. (a),

❷ $$\dot{Q}_{cv} = (1000 \text{ kW}) + \left(4600\frac{kg}{h}\right)(-831.8 + 0.4)\left(\frac{kJ}{kg}\right)\left|\frac{1 \text{ h}}{3600 \text{ s}}\right|\left|\frac{1 \text{ kW}}{1 \text{ kJ/s}}\right|$$
$$= -62.3 \text{ kW}$$

❶ The magnitude of the change in specific kinetic energy from inlet to exit is much smaller than the specific enthalpy change. Note the use of unit conversion factors here and in the calculation of $\dot{Q}_{cv}$ to follow.

❷ The negative value of $\dot{Q}_{cv}$ means that there is heat transfer from the turbine to its surroundings, as would be expected. The magnitude of $\dot{Q}_{cv}$ is small relative to the power developed.

SKILLS DEVELOPED

Ability to…

- apply the steady-state energy rate balance to a control volume.
- develop an engineering model.
- retrieve property data for water.

Quick Quiz

If the change in kinetic energy from inlet to exit were neglected, evaluate the heat transfer rate, in kW, keeping all other data unchanged. Comment. Ans. −62.9 kW.

4.8 Compressors and Pumps

Compressors, pumps

Compressors and **pumps** are devices in which work is done on the substance flowing through them in order to change the state of the substance, typically to increase the pressure and/or elevation. The term *compressor* is used when the substance is a gas (vapor) and the term *pump* is used when the substance is a liquid. Four compressor types are shown in **Fig. 4.11**. The reciprocating compressor of Fig. 4.11*a* features reciprocating motion while the others have rotating motion.

The axial-flow compressor of Fig. 4.11*b* is a key component of turbojet engines (Chap. 9). Compressors also are essential components of refrigeration and heat pump systems (Chap. 10). In the study of Chap. 8, we find that pumps are important in vapor power systems. Pumps also are commonly used to fill water towers, remove water from flooded basements, and for numerous other domestic and industrial applications.

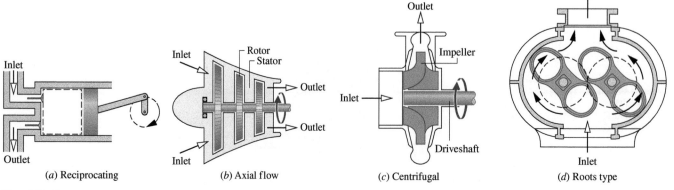

(*a*) Reciprocating　　(*b*) Axial flow　　(*c*) Centrifugal　　(*d*) Roots type

Fig. 4.11 Compressor types.

4.8.1 Compressor and Pump Modeling Considerations

For a control volume enclosing a compressor, the mass and energy rate balances reduce at steady state as for the case of turbines considered in Sec. 4.7.1. Thus, Eq. 4.20a reduces to read

$$0 = \dot{Q}_{cv} - \dot{W}_{cv} + \dot{m}(h_1 - h_2) \tag{a}$$

Heat transfer with the surroundings is frequently a secondary effect that can be neglected, giving as for turbines

$$\dot{W}_{cv} = \dot{m}(h_1 - h_2) \tag{b}$$

For pumps, heat transfer is generally a secondary effect, but the kinetic and potential energy terms of Eq. 4.20a may be significant depending on the application. Be sure to note that for compressors and pumps, the value of $\dot{W}_{cv}$ is *negative* because a power *input* is required.

4.8.2 Applications to an Air Compressor and a Pump System

In this section, modeling considerations for compressors and pumps are illustrated in Examples 4.5 and 4.6, respectively. Applications of compressors and pumps in energy storage systems are described in Sec. 4.8.3.

In Example 4.5 the objectives include assessing the significance of the heat transfer and kinetic energy terms of the energy balance and illustrating the appropriate use of unit conversion factors.

> ▶ ▶ ▶ **EXAMPLE 4.5** ▶

Calculating Compressor Power

Air enters a compressor operating at steady state at a pressure of 1 bar, a temperature of 290 K, and a velocity of 6 m/s through an inlet with an area of 0.1 m^2. At the exit, the pressure is 7 bar, the temperature is 450 K, and the velocity is 2 m/s. Heat transfer from the compressor to its surroundings occurs at a rate of 180 kJ/min. Employing the ideal gas model, calculate the power input to the compressor, in kW.

SOLUTION

Known An air compressor operates at steady state with known inlet and exit states and a known heat transfer rate.

Find Calculate the power required by the compressor.

Schematic and Given Data:

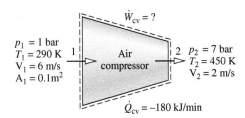

$p_1 = 1$ bar
$T_1 = 290$ K
$V_1 = 6$ m/s
$A_1 = 0.1$m^2

Air compressor

$p_2 = 7$ bar
$T_2 = 450$ K
$V_2 = 2$ m/s

$\dot{W}_{cv} = ?$

$\dot{Q}_{cv} = -180$ kJ/min

Fig. E4.5

Engineering Model

1. The control volume shown on the accompanying figure is at steady state.

2. The change in potential energy from inlet to exit can be neglected.

❶ 3. The ideal gas model applies for the air.

Analysis To calculate the power input to the compressor, begin with the one-inlet, one-exit form of the energy rate balance for a control volume at steady state, Eq. 4.20a. That is,

$$0 = \dot{Q}_{cv} - \dot{W}_{cv} + \dot{m}\left[(h_1 - h_2) + \frac{(V_1^2 - V_2^2)}{2} + g(z_1 - z_2)\right]$$

Solving

$$\dot{W}_{cv} = \dot{Q}_{cv} + \dot{m}\left[(h_1 - h_2) + \left(\frac{V_1^2 - V_2^2}{2}\right)\right]$$

The change in potential energy from inlet to exit drops out by assumption 2.

The mass flow rate $\dot{m}$ can be evaluated with given data at the inlet and the ideal gas equation of state.

$$\dot{m} = \frac{A_1 V_1}{v_1} = \frac{A_1 V_1 p_1}{(\bar{R}/M)T_1} = \frac{(0.1 \text{ m}^2)(6 \text{ m/s})(10^5 \text{ N/m}^2)}{\left(\dfrac{8314 \text{ N} \cdot \text{m}}{28.97 \text{ kg} \cdot \text{K}}\right)(290 \text{ K})} = 0.72 \text{ kg/s}$$

The specific enthalpies h_1 and h_2 can be found from Table A-22. At 290 K, $h_1 = 290.16$ kJ/kg. At 450 K, $h_2 = 451.8$ kJ/kg. Substituting values into the expression for $\dot{W}_{cv}$, and applying appropriate unit conversion factors, we get

$$
\dot{W}_{cv} = \left(-180\,\frac{kJ}{min}\right)\left|\frac{1\ min}{60\ s}\right| + 0.72\,\frac{kg}{s}\left[(290.16 - 451.8)\,\frac{kJ}{kg}\right.
$$

$$
\left. + \left(\frac{(6)^2 - (2)^2}{2}\right)\left(\frac{m^2}{s^2}\right)\left|\frac{1\ N}{1\ kg \cdot m/s^2}\right|\left|\frac{1\ kJ}{10^3\ N \cdot m}\right|\right]
$$

$$
= -3\,\frac{kJ}{s} + 0.72\,\frac{kg}{s}(-161.64 + 0.02)\,\frac{kJ}{kg}
$$

② $\quad = -119.4\,\dfrac{kJ}{s}\left|\dfrac{1\ kW}{1\ kJ/s}\right| = -119.4\ kW$

① The applicability of the ideal gas model can be checked by reference to the generalized compressibility chart.

② In this example $\dot{Q}_{cv}$ and $\dot{W}_{cv}$ have negative values, indicating that the direction of the heat transfer is *from* the compressor and work

is done *on* the air passing through the compressor. The magnitude of the power *input* to the compressor is 119.4 kW. The change in kinetic energy does not contribute significantly.

SKILLS DEVELOPED

Ability to...

- apply the steady-state energy rate balance to a control volume.
- apply the mass flow rate expression, Eq. 4.4b.
- develop an engineering model.
- retrieve property data of air modeled as an ideal gas.

Quick Quiz

If the change in kinetic energy from inlet to exit were neglected, evaluate the compressor power, in kW, keeping all other data unchanged. Comment. Ans. −119.4 kW.

 Animation
Compressor Tabs A, B, and C

    In Example 4.6, a pump is a component of an overall system that delivers a high-velocity stream of water at an elevation greater than at the inlet. Note the modeling considerations in this case, particularly the roles of kinetic and potential energy, and the use of appropriate unit conversion factors.

▶ ▶ EXAMPLE 4.6 ▶

Analyzing a Pump System

A pump steadily draws water from a pond at a volumetric flow rate of 0.83 m³/min through a pipe having a 12-cm diameter inlet. The water is delivered through a hose terminated by a converging nozzle. The nozzle exit has a diameter of 3 cm and is located 10 m above the pipe inlet. Water enters at 20°C, 1 atm and exits with no significant change in temperature or pressure. The magnitude of the rate of heat transfer *from* the pump *to* the surroundings is 5% of the power input. The acceleration of gravity is 9.81 m/s². Determine (**a**) the velocity of the water at the inlet and exit, each in m/s, and (**b**) the power required by the pump, in kW.

SOLUTION

Known A pump system operates at steady state with known inlet and exit conditions. The rate of heat transfer from the pump is specified as a percentage of the power input.

Find Determine the velocities of the water at the inlet and exit of the pump system and the power required.

Schematic and Given Data:

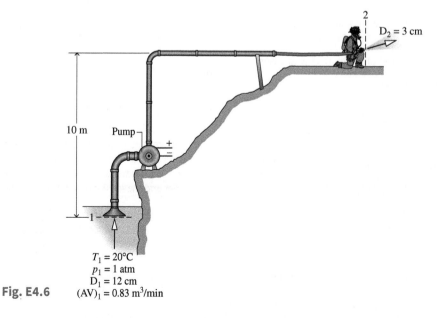

Fig. E4.6
$T_1 = 20°C$
$p_1 = 1\ atm$
$D_1 = 12\ cm$
$(AV)_1 = 0.83\ m^3/min$

Engineering Model

1. A control volume encloses the pump, inlet pipe, and delivery hose.

2. The control volume is at steady state.

3. The magnitude of the heat transfer from the control volume is 5% of the power input.

4. There is no significant change in temperature or pressure.

5. For liquid water, $v \approx v_f(T)$ (Eq. 3.11) and Eq. 3.13 is used to evaluate specific enthalpy.

6. $g = 9.81 \text{ m/s}^2$.

Analysis

a. A mass rate balance reduces at steady state to read $\dot{m}_2 = \dot{m}_1$. The common mass flow rate at the inlet and exit, $\dot{m}$, can be evaluated using Eq. 4.4b together with $v \approx v_f(20°C) = 1.0018 \times 10^{-3} \text{ m}^3/\text{kg}$ from Table A-2. That is,

$$\dot{m} = \frac{AV}{v} = \left(\frac{0.83 \text{ m}^3/\text{min}}{1.0018 \times 10^{-3} \text{ m}^3/\text{kg}}\right)\left|\frac{1 \text{ min}}{60 \text{ s}}\right|$$

$$= 13.8 \frac{\text{kg}}{\text{s}}$$

Thus, the inlet and exit velocities are, respectively,

❶ $V_1 = \frac{\dot{m}v}{A_1} = \frac{(13.8 \text{ kg/s})(1.0018 \times 10^{-3} \text{ m}^3/\text{kg})}{\pi(0.12 \text{ m})^2/4} = 1.22 \text{ m/s}$

$V_2 = \frac{\dot{m}v}{A_2} = \frac{(13.8 \text{ kg/s})(1.0018 \times 10^{-3} \text{ m}^3/\text{kg})}{\pi(0.03 \text{ m})^2/4} = 19.56 \text{ m/s}$

b. To calculate the power input, begin with the one-inlet, one-exit form of the energy rate balance for a control volume at steady state, Eq. 4.20a. That is

$$0 = \dot{Q}_{cv} - \dot{W}_{cv} + \dot{m}\left[(h_1 - h_2) + \left(\frac{V_1^2 - V_2^2}{2}\right) + g(z_1 - z_2)\right]$$

❷ Introducing $\dot{Q}_{cv} = (0.05)\dot{W}_{cv}$, and solving for $\dot{W}_{cv}$,

$$\dot{W}_{cv} = \frac{\dot{m}}{0.95}\left[(h_1 - h_2) + \left(\frac{V_1^2 - V_2^2}{2}\right) + g(z_1 - z_2)\right] \quad \text{(a)}$$

Using Eq. 3.13, the enthalpy term is expressed as

$$h_1 - h_2 = [h_f(T_1) + v_f(T_1)[p_1 - p_{sat}(T_1)]]$$
$$\quad - [h_f(T_2) + v_f(T_2)[p_2 - p_{sat}(T_2)]] \quad \text{(b)}$$

Since there is no significant change in temperature, Eq. (b) reduces to

$$h_1 - h_2 = v_f(T)(p_1 - p_2)$$

As there is also no significant change in pressure, the enthalpy term drops out of the present analysis. Next, evaluating the kinetic energy term

$$\frac{V_1^2 - V_2^2}{2} = \frac{[(1.22)^2 - (19.56)^2]\left(\dfrac{\text{m}}{\text{s}}\right)^2}{2}\left|\frac{1 \text{ N}}{1 \text{ kg} \cdot \text{m/s}^2}\right|\left|\frac{1 \text{ kJ}}{10^3 \text{ N} \cdot \text{m}}\right|$$

$$= -0.191 \text{ kJ/kg}$$

Finally, the potential energy term is

$$g(z_1 - z_2) = (9.81 \text{ m/s}^2)(0 - 10)\text{m}\left|\frac{1 \text{ N}}{1 \text{ kg} \cdot \text{m/s}^2}\right|\left|\frac{1 \text{ kJ}}{10^3 \text{ N} \cdot \text{m}}\right|$$

$$= -0.098 \text{ kJ/kg}$$

Inserting values into Eq. (a)

$$\dot{W}_{cv} = \left(\frac{13.8 \text{ kg/s}}{0.95}\right)[0 - 0.191 - 0.098]\left(\frac{\text{kJ}}{\text{kg}}\right)\left|\frac{1 \text{ kW}}{1 \text{ kJ/s}}\right|$$

$$= -4.2 \text{ kW}$$

where the minus sign indicates that power is provided to the pump.

---

❶ Alternatively, V_1 can be evaluated from the volumetric flow rate at 1. This is left as an exercise.

❷ Since power is required to operate the pump, $\dot{W}_{cv}$ is negative in accord with our sign convention. The energy transfer by heat is from the control volume to the surroundings, and thus $\dot{Q}_{cv}$ is negative as well. Using the value of $\dot{W}_{cv}$ found in part (b), $\dot{Q}_{cv} = (0.05)\dot{W}_{cv} = -0.21 \text{ kW}$.

SKILLS DEVELOPED

Ability to…

- apply the steady-state energy rate balance to a control volume.
- apply the mass flow rate expression, Eq. 4.4b.
- develop an engineering model.
- retrieve properties of liquid water.

Quick Quiz

If the nozzle were removed and water exited directly from the hose, whose diameter is 5 cm, determine the velocity at the exit, in m/s, and the power required, in kW, keeping all other data unchanged. Ans. 7.04 m/s, 1.77 kW.

4.8.3 Pumped-Hydro and Compressed-Air Energy Storage

Owing to the dictates of supply and demand and other economic factors, the value of electricity varies with time. Both the cost to generate electricity and increasingly the price paid by consumers for electricity depend on whether the demand for it is *on-peak* or *off-peak*. The on-peak period is typically weekdays—for example, from 8 a.m. to 8 p.m., while off-peak includes nighttime hours, weekends, and major holidays. Consumers can expect to pay more for

TAKE NOTE...

Cost refers to the amount paid to produce a good or service. Price refers to what consumers pay to acquire that good or service.

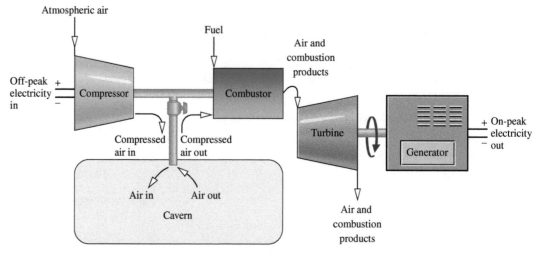

Fig. 4.12 Compressed-air storage.

on-peak electricity. Energy storage methods benefiting from variable electricity rates include thermal storage (see Sec. 3.8) and pumped-hydro and compressed-air storage introduced in the following box.

Economic Aspects of Pumped-Hydro and Compressed-Air Energy Storage

Despite the significant costs of owning and operating utility-scale energy storage systems, various economic strategies, including taking advantage of differing on- and off-peak electricity rates, can make pumped-hydro and compressed-air storage good choices for power generators. In this discussion, we focus on the role of variable electricity rates.

In pumped-hydro storage, water is pumped from a lower reservoir to an upper reservoir, thereby storing energy in the form of gravitational potential energy. (For simplicity, think of the hydro-power plant of Fig. 4.10 operating in the reverse direction.) Off-peak electricity is used to drive the pumps that deliver water to the upper reservoir. Later, during an on-peak period, stored water is released from the upper reservoir to generate electricity as the water flows

through turbines to the lower reservoir. For instance, in the summer water is released from the upper reservoir to generate power to meet a high daytime demand for air conditioning; while at night, when demand is low, water is pumped back to the upper reservoir for use the next day. Owing to friction and other nonidealities, an overall input-to-output loss of electricity occurs with pumped-hydro storage and this adds to operating costs. Still, differing daytime and nighttime electricity rates help make this technology viable.

In compressed-air energy storage, compressors powered with off-peak electricity fill an underground salt cavern, hard-rock mine, or aquifer with pressurized air drawn from the atmosphere. See Fig. 4.12. When electricity demand peaks, high-pressure compressed air is released to the surface, heated by natural gas in combustors, and expanded through a turbine generator, generating electricity for distribution at on-peak rates.

4.9 Heat Exchangers

heat exchanger   **Heat exchangers** have innumerable domestic and industrial applications, including use in home heating and cooling systems, automotive systems, electrical power generation, and chemical processing. Indeed, nearly every area of application listed in Table 1.1 involves heat exchangers.

One common type of heat exchanger is a mixing chamber in which hot and cold streams are mixed directly as shown in **Fig. 4.13a**. The open feedwater heater, which is a component of the vapor power systems considered in Chap. 8, is an example of this type of device.

Another common type of exchanger is one in which a gas or liquid is *separated* from another gas or liquid by a wall through which energy is conducted. These heat exchangers, known as recuperators, take many different forms. Counterflow and parallel tube-within-a-tube configurations are shown in Figs. 4.13b and 4.13c, respectively. Other configurations include cross-flow, as in automobile radiators, and multiple-pass shell-and-tube condensers and evaporators. Figure 4.13d illustrates a cross-flow heat exchanger.

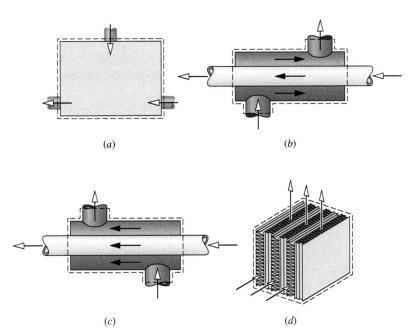

Fig. 4.13 Common heat exchanger types. (*a*) Direct-contact heat exchanger. (*b*) Tube-within-a-tube counterflow heat exchanger. (*c*) Tube-within-a-tube parallel-flow heat exchanger. (*d*) Cross-flow heat exchanger.

 BioConnections

Inflatable blankets such as shown in Fig. 4.14 are used to prevent subnormal body temperatures (hypothermia) during and after surgery. Typically, a heater and blower direct a stream of warm air into the blanket. Air exits the blanket through perforations in its surface. Such *thermal blankets* have been used safely and without incident in millions of surgical procedures. Still, there are obvious risks to patients if temperature controls fail and overheating occurs. Such risks can be anticipated and minimized with good engineering practices.

Warming patients is not always the issue at hospitals; sometimes it is cooling, as in cases involving cardiac arrest, stroke, heart attack, and overheating of the body (hyperthermia). Cardiac arrest, for example, deprives the heart muscle of oxygen and blood, causing part of it to die. This often induces brain damage

among survivors, including irreversible cognitive disability. Studies show when the core body temperature of cardiac patients is reduced to about 33°C, damage is limited because vital organs function more slowly and require less oxygen. To achieve good outcomes, medical specialists say cooling should be done in about 20 minutes or less. A system approved for cooling cardiac arrest victims includes a disposable plastic body suit, pump, and chiller. The pump provides rapidly flowing cold water around the body, in direct contact with the skin of the patient wearing the suit, then recycles coolant to the chiller and back to the patient.

These biomedical applications provide examples of how engineers well versed in thermodynamics principles can bring into the design process their knowledge of heat exchangers, temperature sensing and control, and safety and reliability requirements.

4.9.1 Heat Exchanger Modeling Considerations

As shown by Fig. 4.13, heat exchangers can involve multiple inlets and exits. For a control volume enclosing a heat exchanger, the only work is flow work at the places where matter enters and exits, so the term $\dot{W}_{cv}$ drops out of the energy rate balance. In addition, the kinetic and potential energies of the flowing streams usually can be ignored at the inlets and exits. Thus, the underlined terms of Eq. 4.18 (repeated below) drop out, leaving the enthalpy and heat transfer terms, as shown by Eq. (a). That is,

$$0 = \dot{Q}_{cv} - \dot{W}_{cv} + \sum_i \dot{m}_i \left(h_i + \frac{V_i^2}{2} + gz_i \right) - \sum_e \dot{m}_e \left(h_e + \frac{V_e^2}{2} + gz_e \right)$$

$$0 = \dot{Q}_{cv} + \sum_i \dot{m}_i h_i - \sum_e \dot{m}_e h_e \qquad \text{(a)}$$

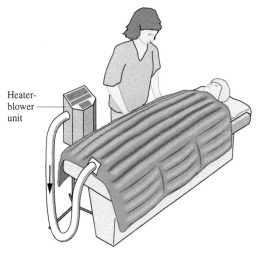

Fig. 4.14 Inflatable thermal blanket.

**Heat Exchanger
Tabs A, B, and C**

Although high rates of energy transfer *within* the heat exchanger occur, heat transfer with the surroundings is often small enough to be neglected. Thus, the $\dot{Q}_{cv}$ term of Eq. (a) would drop out, leaving just the enthalpy terms. The final form of the energy rate balance must be solved together with an appropriate expression of the mass rate balance, recognizing both the number and type of inlets and exits for the case at hand.

4.9.2 Applications to a Power Plant Condenser and Computer Cooling

The next example illustrates how the mass and energy rate balances are applied to a condenser at steady state. Condensers are commonly found in power plants and refrigeration systems.

▶ EXAMPLE 4.7 ▶

Evaluating Performance of a Power Plant Condenser

Steam enters the condenser of a vapor power plant at 0.1 bar with a quality of 0.95 and condensate exits at 0.1 bar and 45°C. Cooling water enters the condenser in a separate stream as a liquid at 20°C and exits as a liquid at 35°C with no change in pressure. Heat transfer from the outside of the condenser and changes in the kinetic and potential energies of the flowing streams can be ignored. For steady-state operation, determine

a. the ratio of the mass flow rate of the cooling water to the mass flow rate of the condensing steam.

b. the energy transfer from the condensing steam to the cooling water, in kJ per kg of steam passing through the condenser.

SOLUTION

Known   Steam is condensed at steady state by interacting with a separate liquid water stream.

Find   Determine the ratio of the mass flow rate of the cooling water to the mass flow rate of the steam and the energy transfer from the steam to the cooling water per kg of steam passing through the condenser.

Schematic and Given Data:

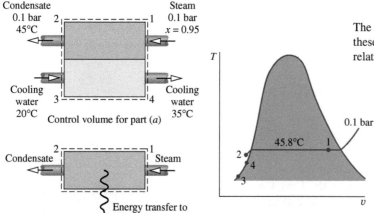

Fig. E4.7

Engineering Model

1. Each of the two control volumes shown on the accompanying sketch is at steady state.

2. There is no significant heat transfer between the overall condenser and its surroundings. $\dot{W}_{cv} = 0$.

3. Changes in the kinetic and potential energies of the flowing streams from inlet to exit can be ignored.

4. At states 2, 3, and 4, $h \approx h_f(T)$ (see Eq. 3.14).

Analysis   The steam and the cooling water streams do not mix. Thus, the mass rate balances for each of the two streams reduce at steady state to give

$$\dot{m}_1 = \dot{m}_2 \quad \text{and} \quad \dot{m}_3 = \dot{m}_4$$

a. The ratio of the mass flow rate of the cooling water to the mass flow rate of the condensing steam, $\dot{m}_3/\dot{m}_1$, can be found from the steady-state form of the energy rate balance, Eq. 4.18, applied to the overall condenser as follows:

$$0 = \underline{\dot{Q}_{cv}} - \underline{\dot{W}_{cv}} + \dot{m}_1\left(h_1 + \underline{\frac{V_1^2}{2} + gz_1}\right) + \dot{m}_3\left(h_3 + \underline{\frac{V_3^2}{2} + gz_3}\right)$$
$$- \dot{m}_2\left(h_2 + \underline{\frac{V_2^2}{2} + gz_2}\right) - \dot{m}_4\left(h_4 + \underline{\frac{V_4^2}{2} + gz_4}\right)$$

The underlined terms drop out by assumptions 2 and 3. With these simplifications, together with the above mass flow rate relations, the energy rate balance becomes simply

$$0 = \dot{m}_1(h_1 - h_2) + \dot{m}_3(h_3 - h_4)$$

Solving, we get

$$\frac{\dot{m}_3}{\dot{m}_1} = \frac{h_1 - h_2}{h_4 - h_3}$$

The specific enthalpy h_1 can be determined using the given quality and data from Table A-3. From Table A-3 at 0.1 bar, $h_{f1} = 191.83$ kJ/kg and $h_{g1} = 2584.7$ kJ/kg, so

$$h_1 = 191.83 + 0.95(2584.7 - 191.83) = 2465.1 \text{ kJ/kg}$$

❶ Using assumption 4, the specific enthalpy at 2 is given by $h_2 \approx h_f(T_2) = 188.45$ kJ/kg. Similarly, $h_3 \approx h_f(T_3)$ and $h_4 \approx h_f(T_4)$, giving $h_4 - h_3 = 62.7$ kJ/kg. Thus,

$$\frac{\dot{m}_3}{\dot{m}_1} = \frac{2465.1 - 188.45}{62.7} = 36.3$$

b. For a control volume enclosing the steam side of the condenser only, begin with the steady-state form of energy rate balance, Eq. 4.20a.

② $$0 = \dot{Q}_{cv} - \underline{\dot{W}_{cv}} + \dot{m}_1\left[(h_1 - h_2) + \underline{\frac{(V_1^2 - V_2^2)}{2}} + \underline{g(z_1 - z_2)}\right]$$

The underlined terms drop out by assumptions 2 and 3. The following expression for the rate of energy transfer between the condensing steam and the cooling water results:

$$\dot{Q}_{cv} = \dot{m}_1(h_2 - h_1)$$

Dividing by the mass flow rate of the steam, $\dot{m}_1$, and inserting values

$$\frac{\dot{Q}_{cv}}{\dot{m}_1} = h_2 - h_1 = 188.45 - 2465.1 = -2276.7 \text{ kJ/kg}$$

where the minus sign signifies that energy is transferred *from* the condensing steam *to* the cooling water.

① Alternatively, $(h_4 - h_3)$ can be evaluated using the incompressible liquid model via Eq. 3.20b.

② Depending on where the boundary of the control volume is located, two different formulations of the energy rate balance are obtained. In part (a), both streams are included in the control volume. Energy transfer between them occurs internally and not across the boundary of the control volume, so the term $\dot{Q}_{cv}$ drops out of the energy rate balance. With the control volume of part (b), however, the term $\dot{Q}_{cv}$ must be included.

SKILLS DEVELOPED

Ability to…

- apply the steady-state mass and energy rate balances to a control volume.
- develop an engineering model.
- retrieve property data for water.

Quick Quiz

If the mass flow rate of the condensing steam is 125 kg/s, determine the mass flow rate of the cooling water, in kg/s. Ans. 4538 kg/s.

Excessive temperatures in electronic components are avoided by providing appropriate cooling. In the next example, we analyze the cooling of computer components, illustrating the use of the control volume form of energy rate balance together with property data for air.

> > > **EXAMPLE 4.8** ►

Cooling Computer Components

The electronic components of a computer are cooled by air flowing through a fan mounted at the inlet of the electronics enclosure. At steady state, air enters at 20°C, 1 atm. For noise control, the velocity of the entering air cannot exceed 1.3 m/s. For temperature control, the temperature of the air at the exit cannot exceed 32°C. The electronic components and fan receive, respectively, 80 W and 18 W of electric power. Determine the smallest fan inlet area, in cm², for which the limits on the entering air velocity and exit air temperature are met.

SOLUTION

Known The electronic components of a computer are cooled by air flowing through a fan mounted at the inlet of the electronics enclosure. Conditions are specified for the air at the inlet and exit. The power required by the electronics and the fan is also specified.

Find Determine the smallest fan area for which the specified limits are met.

Schematic and Given Data:

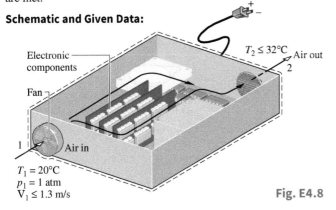

$T_1 = 20°C$
$p_1 = 1$ atm
$V_1 \leq 1.3$ m/s

Fig. E4.8

Engineering Model

1. The control volume shown on the accompanying figure is at steady state.

2. Heat transfer from the *outer* surface of the electronics enclosure to the surroundings is negligible. Thus, $\dot{Q}_{cv} = 0$.

① 3. Changes in kinetic and potential energies can be ignored.

② 4. Air is modeled as an ideal gas with $c_p = 1.005$ kJ/kg · K.

Analysis The inlet area A_1 can be determined from the mass flow rate $\dot{m}$ and Eq. 4.4b, which can be rearranged to read

$$A_1 = \frac{\dot{m}v_1}{V_1} \qquad \text{(a)}$$

The mass flow rate can be evaluated, in turn, from the steady-state energy rate balance, Eq. 4.20a.

$$0 = \underline{\dot{Q}_{cv}} - \dot{W}_{cv} + \dot{m}\left[(h_1 - h_2) + \underline{\left(\frac{V_1^2 - V_2^2}{2}\right)} + \underline{g(z_1 - z_2)}\right]$$

The underlined terms drop out by assumptions 2 and 3, leaving

$$0 = -\dot{W}_{cv} + \dot{m}(h_1 - h_2)$$

where $\dot{W}_{cv}$ accounts for the *total* electric power provided to the electronic components and the fan: $\dot{W}_{cv} = (-80 \text{ W}) + (-18 \text{ W}) = -98$ W. Solving for $\dot{m}$, and using assumption 4 with Eq. 3.51 to evaluate $(h_1 - h_2)$

$$\dot{m} = \frac{(-\dot{W}_{cv})}{c_p(T_2 - T_1)}$$

Introducing this into the expression for A_1, Eq. (a), and using the ideal gas model to evaluate the specific volume v_1

$$A_1 = \frac{1}{V_1}\left[\frac{(-\dot{W}_{cv})}{c_p(T_2 - T_1)}\right]\left(\frac{RT_1}{p_1}\right)$$

From this expression we see that A_1 *increases* when V_1 and/or T_2 *decrease*. Accordingly, since $V_1 \leq 1.3$ m/s and $T_2 \leq 305$ K (32°C), the inlet area must satisfy

$$A_1 \geq \frac{1}{1.3 \text{ m/s}}\left[\frac{98 \text{ W}}{\left(1.005\dfrac{\text{kJ}}{\text{kg} \cdot \text{K}}\right)(305 - 293)\text{K}}\left|\frac{1 \text{ kJ}}{10^3 \text{ J}}\right|\left|\frac{1 \text{ J/s}}{1 \text{ W}}\right|\right]$$

$$\times\left(\frac{\left(\dfrac{8314 \text{ N} \cdot \text{m}}{28.97 \text{ kg} \cdot \text{K}}\right)293 \text{ K}}{1.01325 \times 10^5 \text{ N/m}^2}\right)\left|\frac{10^4 \text{ cm}^2}{1 \text{ m}^2}\right|$$

$$\geq 52 \text{ cm}^2$$

For the specified conditions, the smallest fan area is 52 cm².

① Cooling air typically enters and exits electronic enclosures at low velocities, and thus kinetic energy effects are insignificant.

② The applicability of the ideal gas model can be checked by reference to the generalized compressibility chart. Since the temperature of the air increases by no more than 12°C, the specific heat c_p is nearly constant (Table A-20).

SKILLS DEVELOPED

Ability to...

- apply the steady-state energy rate balance to a control volume.
- apply the mass flow rate expression, Eq. 4.4b.
- develop an engineering model.
- retrieve property data of air modeled as an ideal gas.

Quick Quiz

If heat transfer occurs at a rate of 11 W from the outer surface of the computer case to the surroundings, determine the smallest fan inlet area for which the limits on entering air velocity and exit air temperature are met if the total power input remains at 98 W. Ans. 46 cm².

4.10 Throttling Devices

A significant reduction in pressure can be achieved simply by introducing a restriction into a line through which a gas or liquid flows. This is commonly done by means of a partially opened valve or a porous plug. These *throttling* devices are illustrated in **Fig. 4.15**.

An application of throttling occurs in vapor-compression refrigeration systems, where a valve is used to reduce the pressure of the refrigerant from the pressure at the exit of the *condenser* to the lower pressure existing in the *evaporator*. We consider this further in Chap. 10. Throttling also plays a role in the *Joule–Thomson* expansion considered in Chap. 11. Another **throttling calorimeter**, which is a device for determining the quality of a two-phase liquid–vapor mixture. The throttling calorimeter is considered in Example 4.9.

throttling calorimeter

4.10.1 Throttling Device Modeling Considerations

For a control volume enclosing a throttling device, the only work is flow work at locations where mass enters and exits the control volume, so the term $\dot{W}_{cv}$ drops out of the energy rate balance. There is usually no significant heat transfer with the surroundings, and the change in potential energy from inlet to exit is negligible. Thus, the underlined terms of Eq. 4.20a

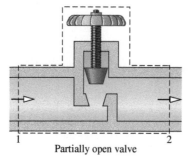
Partially open valve

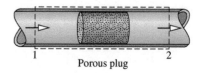
Porous plug

Fig. 4.15 Examples of throttling devices.

(repeated below) drop out, leaving the enthalpy and kinetic energy terms, as shown by Eq. (a). That is,

$$0 = \underline{\dot{Q}_{cv}} - \underline{\dot{W}_{cv}} + \dot{m}\left[(h_1 - h_2) + \frac{(V_1^2 - V_2^2)}{2} + \underline{g(z_1 - z_2)}\right]$$

$$0 = (h_1 - h_2) + \frac{V_1^2 - V_2^2}{2} \qquad \text{(a)}$$

Although velocities may be relatively high in the vicinity of the restriction imposed by the throttling device on the flow through it, measurements made upstream and downstream of the reduced flow area show in most cases that the change in the specific kinetic energy of the flowing substance between these locations can be neglected. With this further simplification, Eq. (a) reduces to

→ **Animation**

Throttling Devices

$$\boxed{h_2 = h_1 \qquad (p_2 < p_1)} \qquad \text{(4.22)}$$

When the flow through the valve or other restriction is idealized in this way, the process is called a **throttling process**.

throttling process

4.10.2 Using a Throttling Calorimeter to Determine Quality

The next example illustrates use of a throttling calorimeter to determine steam quality.

▶▶ **EXAMPLE 4.9** ▶ ·

Measuring Steam Quality

A supply line carries a two-phase liquid–vapor mixture of steam at 20 bar. A small fraction of the flow in the line is diverted through a throttling calorimeter and exhausted to the atmosphere at 1 bar. The temperature of the exhaust steam is measured as 120°C. Determine the quality of the steam in the supply line.

SOLUTION

Known Steam is diverted from a supply line through a throttling calorimeter and exhausted to the atmosphere.

Find Determine the quality of the steam in the supply line.

Schematic and Given Data:

Engineering Model

1. The control volume shown on the accompanying figure is at steady state.

2. The diverted steam undergoes a throttling process.

Analysis For a throttling process, the energy and mass balances reduce to give $h_2 = h_1$, which agrees with Eq. 4.22. Thus, with state 2 fixed, the specific enthalpy in the supply line is known, and state 1 is fixed by the known values of p_1 and h_1.

❶ As shown on the accompanying p–v diagram, state 1 is in the two-phase liquid–vapor region and state 2 is in the superheated vapor region. Thus,

$$h_2 = h_1 = h_{f1} + x_1(h_{g1} - h_{f1})$$

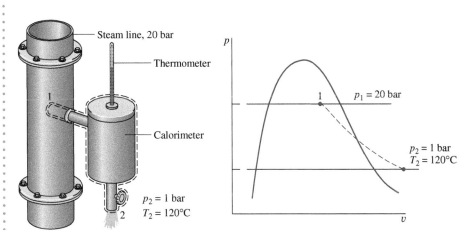

Fig. E4.9

Solving for x_1,

$$x_1 = \frac{h_2 - h_{f1}}{h_{g1} - h_{f1}}$$

From Table A-3 at 20 bar, $h_{f1} = 908.79$ kJ/kg and $h_{g1} = 2,799.5$ kJ/kg. At 1 bar and 120°C, $h_2 = 2,766.6$ kJ/kg from Table A-4. Inserting values into the above expression, the quality of the steam in the line is $x_1 = 0.956$ (95.6%).

---

❶ For throttling calorimeters exhausting to the atmosphere, the quality of the steam in the line must be greater than about 94% to ensure that the steam leaving the calorimeter is superheated.

SKILLS DEVELOPED

Ability to...

- apply Eq. 4.22 for a throttling process.
- retrieve property data for water.

Quick Quiz

If the supply line carried saturated vapor at 20.5 bar, determine the temperature at the calorimeter exit, in °C, for the same exit pressure, 1 bar. Ans. 162.2°C.

4.11 System Integration

Thus far, we have studied several types of components selected from those commonly seen in practice. These components are usually encountered in combination rather than individually. Engineers often must creatively combine components to achieve some overall objective, subject to constraints such as minimum total cost. This important engineering activity is called **system integration**.

system integration

In engineering practice and everyday life, integrated systems are regularly encountered. Many readers are already familiar with a particularly successful system integration: the simple power plant shown in **Fig. 4.16**. This system consists of four components in series: a turbine-generator, condenser, pump, and boiler. We consider such power plants in detail in subsequent sections of the book.

BioConnections

Living things also can be considered integrated systems. **Figure 4.17** shows a control volume enclosing a tree receiving solar radiation. As indicated on the figure, a portion of the incident radiation is reflected to the surroundings. Of the net solar energy received by the tree, about 21% is returned to the surroundings by heat transfer, principally convection. Water management accounts for most of the remaining solar input.

Trees *sweat* as do people; this is called *evapotranspiration*. As shown in Fig. 4.17, about 78% of the net solar energy received by the tree is used to pump liquid water from

the surroundings, primarily the ground, convert it to a vapor, and discharge it to the surroundings through tiny pores (called *stomata*) in the leaves. Nearly all the water taken up is lost in this manner and only a small fraction is used within the tree. Applying an energy balance to the control volume enclosing the tree, just 1% of the net solar energy received by the tree is left for use in the production of biomass (wood and leaves). Evapotranspiration benefits trees but also contributes significantly to water loss from watersheds, illustrating that in nature as in engineering there are *trade-offs*.

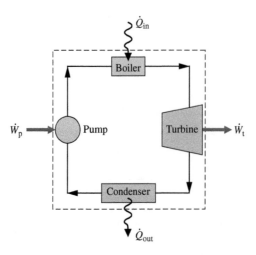

Fig. 4.16 Simple vapor power plant.

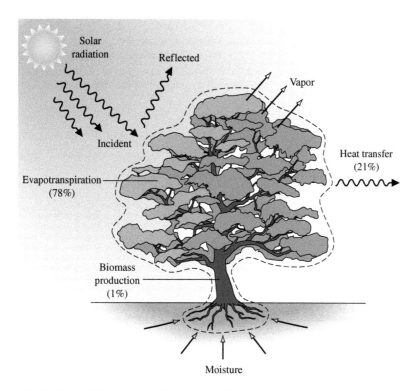

Fig. 4.17 Control volume enclosing a tree.

Example 4.10 provides another illustration of an integrated system. This case involves a *waste heat* recovery system.

▶▶▶ **EXAMPLE 4.10** ▶ ∙

Waste Heat Recovery System

An industrial process discharges gaseous combustion products at 478°K, 1 bar with mass flow rate of 69.78 kg/s. As shown in Fig. E4.10, a proposed system for utilizing the combustion products combines a heat-recovery steam generator with a turbine. At steady state, combustion products exit the steam generator at 470°K, 1 bar and a separate stream of water enters at 0.275 MPa, 38.9°C with a mass flow rate of 2.079 kg/s. At the exit of the turbine, the pressure is 0.07 bar and the quality is 93%. Heat transfer from the outer surfaces of the steam generator and turbine can be ignored, as can the changes in kinetic and potential energies of the flowing streams. There is no significant pressure drop for the water flowing through the steam generator. The combustion products can be modeled as air as an ideal gas.

a. Determine the power developed by the turbine, in kJ/s.

b. Determine the turbine inlet temperature, in °C.

c. Evaluating the power developed at $0.08 per kW · h which is a typical rate for electricity, determine the value of the power, in $/year, for 8000 hours of operation annually.

SOLUTION

Known Steady-state operating data are provided for a system consisting of a heat-recovery steam generator and a turbine.

Find Determine the power developed by the turbine and the turbine inlet temperature. Evaluate the annual value of the power developed.

Schematic and Given Data:

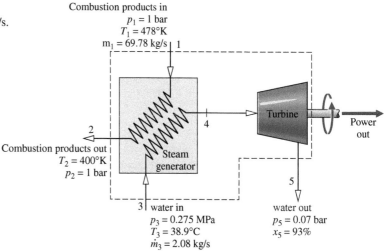

Fig. E4.10

Engineering Model

1. The control volume shown on the accompanying figure is at steady state.

2. Heat transfer is negligible, and changes in kinetic and potential energy can be ignored.

3. There is no pressure drop for water flowing through the steam generator.

4. The combustion products are modeled as air as an ideal gas.

Analysis

a. The power developed by the turbine is determined from a control volume enclosing both the steam generator and the turbine. Since the gas and water streams do not mix, mass rate balances for each of the streams reduce, respectively, to give

$$\dot{m}_1 = \dot{m}_2, \qquad \dot{m}_3 = \dot{m}_5$$

For this control volume, the appropriate form of the steady-state energy rate balance is Eq. 4.18, which reads

$$0 = \dot{Q}_{cv} - \dot{W}_{cv} + \dot{m}_1\left(h_1 + \frac{V_1^2}{2} + gz_1\right) + \dot{m}_3\left(h_3 + \frac{V_3^2}{2} + gz_3\right)$$
$$- \dot{m}_2\left(h_2 + \frac{V_2^2}{2} + gz_2\right) - \dot{m}_5\left(h_5 + \frac{V_5^2}{2} + gz_5\right)$$

The underlined terms drop out by assumption 2. With these simplifications, together with the above mass flow rate relations, the energy rate balance becomes

$$\dot{W}_{cv} = \dot{m}_1(h_1 - h_2) + \dot{m}_3(h_3 - h_5)$$

where $\dot{m}_1 = 69.78$ kg/s, $\dot{m}_3 = 2.08$ kg/s.

The specific enthalpies, h_1 and h_2, can be found from Table A-22: At 478 K, $h_1 = 480.35$ kJ/kg, and at 400 K, $h_2 = 400.98$ kJ/kg. At state 3, water is a liquid. Using Eq. 3.14 and saturated liquid data from Table A-2, $h_3 \approx h_f(T_3) = 162.9$ kJ/kg. State 5 is a two-phase liquid–vapor mixture. With data from Table A-3 and the given quality

$$h_5 = h_{f5} + x_5(h_{g5} - h_{f5})$$
$$= 161 + 0.93(2751.72 - 161) = 2403 \text{ kJ/kg}$$

Substituting values into the expression for $\dot{W}_{cv}$

$$\dot{W}_{cv} = (69.78 \text{ kg/s})(480.3 - 400.98) \text{ kJ/kg}$$
$$+ (2.079 \text{ kg/s})(162.9 - 2403) \text{ kJ/kg}$$
$$= 876.8 \text{ kJ/s}$$
$$= 876.8 \text{ kW}$$

b. To determine T_4, it is necessary to fix the state at 4. This requires two independent property values. With assumption 3, one of these properties is pressure, $p_4 = 0.275$ MPa. The other is the specific enthalpy h_4, which can be found from an energy rate balance for a control volume enclosing just the steam generator. Mass rate balances for each of the two streams give $\dot{m}_1 = \dot{m}_2$ and $\dot{m}_3 = \dot{m}_4$. With assumption 2 and these mass flow rate relations, the steady-state form of the energy rate balance reduces to

❶ $$0 = \dot{m}_1(h_1 - h_2) + \dot{m}_3(h_3 - h_4)$$

Solving for h_4

$$h_4 = h_3 + \frac{\dot{m}_1}{\dot{m}_3}(h_1 - h_2)$$
$$= 162.9 \frac{\text{kJ}}{\text{kg}} + \left(\frac{69.78}{2.079}\right)(480.3 - 400.98)\frac{\text{kJ}}{\text{kg}}$$
$$= 2825 \text{ kJ/kg}$$

Interpolating in Table A-4 at $p_4 = 0.275$ MPa with h_4, we get $T_4 = 180°C$.

c. Using the result of part (a), together with the given economic data and appropriate conversion factors, the value of the power developed for 8000 hours of operation annually is

$$\text{Annual value} = (876.8 \text{ KW})\left(8000\frac{\text{h}}{\text{year}}\right)\left(0.08\frac{\$}{\text{kW}\cdot\text{h}}\right)$$

❷ $$= 561{,}152\frac{\$}{\text{year}}$$

---

❶ Alternatively, to determine h_4 a control volume enclosing just the turbine can be considered.

❷ The decision about implementing this solution to the problem of utilizing the hot combustion products discharged from an industrial process would necessarily rest on the outcome of a detailed economic evaluation, including the cost of purchasing and operating the steam generator, turbine, and auxiliary equipment.

SKILLS DEVELOPED

Ability to...

- apply the steady-state mass and energy rate balances to a control volume.
- apply the mass flow rate expression, Eq. 4.4b.
- develop an engineering model.
- retrieve property data for water and for air modeled as an ideal gas.
- conduct an elementary economic evaluation.

Quick Quiz

Taking a control volume enclosing just the turbine, evaluate the turbine inlet temperature, in °C. Ans. 179°C.

4.12 Transient Analysis

Many devices undergo periods of **transient operation** during which the state changes with time. Examples include the startup or shutdown of turbines, compressors, and motors. Additional examples include vessels being filled or emptied, as considered in Example 4.2 and in the discussion of Fig. 1.5. Because property values, work and heat transfer rates, and mass flow rates may vary with time during transient operation, the steady-state assumption is not appropriate when analyzing such cases. Special care must be exercised when applying the mass and energy rate balances, as discussed next.

**System Types
Tab D**

4.12.1 The Mass Balance in Transient Analysis

First, we place the control volume mass balance in a form that is suitable for transient analysis. We begin by integrating the mass rate balance, Eq. 4.2, from time 0 to a final time t. That is,

$$\int_0^t \left(\frac{dm_{cv}}{dt} \right) dt = \int_0^t \left(\sum_i \dot{m}_i \right) dt - \int_0^t \left(\sum_e \dot{m}_e \right) dt$$

This takes the form

$$m_{cv}(t) - m_{cv}(0) = \sum_i \left(\underline{\int_0^t \dot{m}_i \, dt} \right) - \sum_e \left(\underline{\int_0^t \dot{m}_e \, dt} \right)$$

Introducing the following symbols for the underlined terms

$$m_i = \int_0^t \dot{m}_i \, dt \quad \left\{ \begin{array}{l} \text{amount of mass} \\ \text{entering the control} \\ \text{volume through inlet } i, \\ \text{from time 0 to } t \end{array} \right.$$

$$m_e = \int_0^t \dot{m}_e \, dt \quad \left\{ \begin{array}{l} \text{amount of mass} \\ \text{exiting the control} \\ \text{volume through exit } e, \\ \text{from time 0 to } t \end{array} \right.$$

the mass balance becomes

$$\boxed{m_{cv}(t) - m_{cv}(0) = \sum_i m_i - \sum_e m_e} \tag{4.23}$$

In words, Eq. 4.23 states that the change in the amount of mass contained in the control volume equals the difference between the total incoming and outgoing amounts of mass.

4.12.2 The Energy Balance in Transient Analysis

Next, we integrate the energy rate balance, Eq. 4.15, ignoring the effects of kinetic and potential energy. The result is

$$U_{cv}(t) - U_{cv}(0) = Q_{cv} - W_{cv} + \sum_i \left(\underline{\int_0^t \dot{m}_i h_i \, dt} \right) - \sum_e \left(\underline{\int_0^t \dot{m}_e h_e \, dt} \right) \tag{4.24}$$

where Q_{cv} accounts for the net amount of energy transferred by heat into the control volume and W_{cv} accounts for the net amount of energy transferred by work, except for flow work. The integrals shown underlined in Eq. 4.24 account for the energy carried in at the inlets and out at the exits.

For the *special case* where the states at the inlets and exits are *constant with time*, the respective specific enthalpies, h_i and h_e, are constant, and the underlined terms of Eq. 4.24 become

$$\int_0^t \dot{m}_i h_i \, dt = h_i \int_0^t \dot{m}_i \, dt = h_i m_i$$

$$\int_0^t \dot{m}_e h_e \, dt = h_e \int_0^t \dot{m}_e \, dt = h_e m_e$$

Equation 4.24 then takes the following *special* form

$$U_{cv}(t) - U_{cv}(0) = Q_{cv} - W_{cv} + \sum_i m_i h_i - \sum_e m_e h_e \qquad (4.25)$$

where m_i and m_e account, respectively, for the *amount* of mass entering the control volume through inlet i and exiting the control volume through exit e, each from time 0 to t.

Whether in the general form, Eq. 4.24, or the special form, Eq. 4.25, these equations account for the change in the amount of energy contained within the control volume as the difference between the total incoming and outgoing amounts of energy.

Another *special case* is when the intensive properties within the control volume are *uniform with position* at a particular time t. Accordingly, the specific volume and the specific internal energy are uniform throughout and can depend only on time—that is, $v(t)$ and $u(t)$, respectively. Then

$$m_{cv}(t) = V_{cv}(t)/v(t) \qquad (4.26)$$
$$U_{cv}(t) = m_{cv}(t)u(t) \qquad (4.27)$$

If the control volume is comprised of different phases at time t, the state of each phase is assumed uniform throughout.

Equations 4.23 and 4.25–4.27 are applicable to a wide range of transient cases where inlet and exit states are constant with time *and* intensive properties within the control volume are uniform with position initially and finally.

FOR EXAMPLE

In cases involving the filling of containers having a single inlet and no exit, Eqs. 4.23, 4.25, and 4.27 combine to give

$$m_{cv}(t)u(t) - m_{cv}(0)u(0) = Q_{cv} - W_{cv} + h_i(m_{cv}(t) - m_{cv}(0)) \qquad (4.28)$$

The details are left as an exercise. See Examples 4.12 and 4.13 for this type of transient application.

4.12.3 Transient Analysis Applications

The following examples provide illustrations of the transient analysis of control volumes using the conservation of mass and energy principles. In each case considered, to emphasize fundamentals we begin with general forms of the mass and energy balances and reduce them to forms suited for the case at hand, invoking the idealizations discussed in this section as warranted.

The first example considers a vessel that is partially emptied as mass exits through a valve.

▶▶ **EXAMPLE 4.11** ▶

Evaluating Heat Transfer for a Partially Emptying Tank

A tank having a volume of 0.85 m³ initially contains water as a two-phase liquid–vapor mixture at 260°C and a quality of 0.7. Saturated water vapor at 260°C is slowly withdrawn through a pressure-regulating valve at the top of the tank as energy is transferred by heat to maintain constant pressure in the tank. This continues until the tank is filled with saturated vapor at 260°C. Determine the amount of heat transfer, in kJ. Neglect all kinetic and potential energy effects.

SOLUTION

Known A tank initially holding a two-phase liquid–vapor mixture is heated while saturated water vapor is slowly removed. This continues at constant pressure until the tank is filled only with saturated vapor.

Find Determine the amount of heat transfer.

Schematic and Given Data:

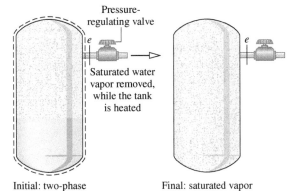

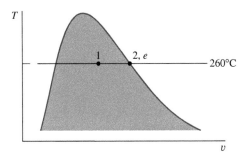

Initial: two-phase   Final: saturated vapor
liquid–vapor mixture

Fig. E4.11

Engineering Model

1. The control volume is defined by the dashed line on the accompanying diagram.

2. For the control volume, $\dot{W}_{cv} = 0$ and kinetic and potential energy effects can be neglected.

3. At the exit the state remains constant.

❶ 4. The initial and final states of the mass within the vessel are equilibrium states.

Analysis Since there is a single exit and no inlet, the mass rate balance Eq. 4.2 takes the form

$$\frac{dm_{cv}}{dt} = -\dot{m}_e$$

With assumption 2, the energy rate balance Eq. 4.15 reduces to

$$\frac{dU_{cv}}{dt} = \dot{Q}_{cv} - \dot{m}_e h_e$$

Combining the mass and energy rate balances results in

$$\frac{dU_{cv}}{dt} = \dot{Q}_{cv} + h_e \frac{dm_{cv}}{dt}$$

By assumption 3, the specific enthalpy at the exit is constant. Accordingly, integration of the last equation gives

$$\Delta U_{cv} = Q_{cv} + h_e \Delta m_{cv}$$

Solving for the heat transfer Q_{cv},

$$Q_{cv} = \Delta U_{cv} - h_e \Delta m_{cv}$$

or

❷ $$Q_{cv} = (m_2 u_2 - m_1 u_1) - h_e(m_2 - m_1)$$

where m_1 and m_2 denote, respectively, the initial and final amounts of mass within the tank.

The terms u_1 and m_1 of the foregoing equation can be evaluated with property values from Table A-2 at 260°C and the given value for quality. Thus,

$$u_1 = u_{f1} + x_1(u_{g1} - u_{f1})$$
$$= 1128.4 + (0.7)(2599.0 - 1128.4) = 2157.8 \text{ kJ/kg}$$

Also,

$$v_1 = v_{f1} + x_1(v_{g1} - v_{f1})$$
$$= 1.2755 \times 10^{-3} + (0.7)(0.04221 - 1.2755 \times 10^{-3})$$
$$= 29.93 \times 10^{-3} \text{ m}^3/\text{kg}$$

Using the specific volume v_1, the mass initially contained in the tank is

$$m_1 = \frac{V}{v_1} = \frac{0.85 \text{ m}^3}{(29.93 \times 10^{-3} \text{ m}^3/\text{kg})} = 28.4 \text{ kg}$$

The final state of the mass in the tank is saturated vapor at 260°C so Table A-2 gives

$$u_2 = u_g(260°C) = 2599.0 \text{ kJ/kg},$$
$$v_2 = v_g(260°C) = 42.21 \times 10^{-3} \text{ m}^3/\text{kg}$$

The mass contained within the tank at the end of the process is

$$m_2 = \frac{V}{v_2} = \frac{0.85 \text{ m}^3}{(42.21 \times 10^{-3} \text{ m}^3/\text{kg})} = 20.14 \text{ kg}$$

Table A-2 also gives $h_e = h_g(260°C) = 2796.6 \text{ kJ/kg}$.

Substituting values into the expression for the heat transfer yields

$$Q_{cv} = (20.14)(2599.0) - (28.4)(2157.8) - 2796.6(20.14 - 28.4)$$
$$= 14,162 \text{ kJ}$$

---

❶ In this case, idealizations are made about the state of the vapor exiting *and* the initial and final states of the mass contained within the tank.

❷ This expression for Q_{cv} can be obtained by applying Eqs. 4.23, 4.25, and 4.27. The details are left as an exercise.

SKILLS DEVELOPED

Ability to...

- apply the time-dependent mass and energy rate balances to a control volume.

- develop an engineering model.

- retrieve property data for water.

Quick Quiz

If the initial quality were 90%, determine the heat transfer, in kJ, keeping all other data unchanged. Ans. 3707 kJ.

In the next two examples we consider cases where tanks are filled. In Example 4.12, an initially evacuated tank is filled with steam as power is developed. In Example 4.13, a compressor is used to store air in a tank.

►►► EXAMPLE 4.12 ►

Using Steam for Emergency Power Generation

Steam at a pressure of 15 bar and a temperature of 320°C is contained in a large vessel. Connected to the vessel through a valve is a turbine followed by a small initially evacuated tank with a volume of 0.6 m³. When emergency power is required, the valve is opened and the tank fills with steam until the pressure is 15 bar. The temperature in the tank is then 400°C. The filling process takes place adiabatically and kinetic and potential energy effects are negligible. Determine the amount of work developed by the turbine, in kJ.

SOLUTION

Known Steam contained in a large vessel at a known state flows from the vessel through a turbine into a small tank of known volume until a specified final condition is attained in the tank.

Find Determine the work developed by the turbine.

Schematic and Given Data:

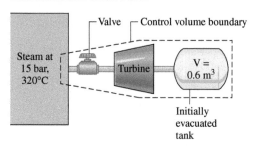

Fig. E4.12

Engineering Model

1. The control volume is defined by the dashed line on the accompanying diagram.

2. For the control volume, $\dot{Q}_{cv} = 0$ and kinetic and potential energy effects are negligible.

❶ 3. The state of the steam within the large vessel remains constant. The final state of the steam in the smaller tank is an equilibrium state.

4. The amount of mass stored within the turbine and the interconnecting piping at the end of the filling process is negligible.

Analysis Since the control volume has a single inlet and no exits, the mass rate balance, Eq. 4.2, reduces to

$$\frac{dm_{cv}}{dt} = \dot{m}_i$$

The energy rate balance, Eq. 4.15, reduces with assumption 2 to

$$\frac{dU_{cv}}{dt} = -\dot{W}_{cv} + \dot{m}_i h_i$$

Combining the mass and energy rate balances gives

$$\frac{dU_{cv}}{dt} = -\dot{W}_{cv} + h_i \frac{dm_{cv}}{dt}$$

Integrating

$$\Delta U_{cv} = -W_{cv} + h_i \Delta m_{cv}$$

In accordance with assumption 3, the specific enthalpy of the steam entering the control volume is constant at the value corresponding to the state in the large vessel.

Solving for W_{cv}

$$W_{cv} = h_i \Delta m_{cv} - \Delta U_{cv}$$

ΔU_{cv} and Δm_{cv} denote, respectively, the changes in internal energy and mass of the control volume. With assumption 4, these terms can be identified with the small tank only.

Since the tank is initially evacuated, the terms ΔU_{cv} and Δm_{cv} reduce to the internal energy and mass within the tank at the end of the process. That is,

$$\Delta U_{cv} = (m_2 u_2) - (m_1 u_1)^{0}, \quad \Delta m_{cv} = m_2 - m_1^{0}$$

where 1 and 2 denote the initial and final states within the tank, respectively.

Collecting results yields

❷❸ $$W_{cv} = m_2(h_i - u_2) \qquad \text{(a)}$$

The mass within the tank at the end of the process can be evaluated from the known volume and the specific volume of steam at 15 bar and 400°C from Table A-4

$$m_2 = \frac{V}{v_2} = \frac{0.6 \text{ m}^3}{(0.203 \text{ m}^3/\text{kg})} = 2.96 \text{ kg}$$

The specific internal energy of steam at 15 bar and 400°C from Table A-4 is 2951.3 kJ/kg. Also, at 15 bar and 320°C, $h_i = 3081.9$ kJ/kg.

Substituting values into Eq. (a)

$$W_{cv} = 2.96 \text{ kg} (3081.9 - 2951.3) \text{kJ/kg} = 386.6 \text{ kJ}$$

❶ In this case idealizations are made about the state of the steam entering the tank and the final state of the steam in the tank. These idealizations make the transient analysis manageable.

❷ A significant aspect of this example is the energy transfer into the control volume by flow work, incorporated in the pv term of the specific enthalpy at the inlet.

❸ This result can also be obtained by reducing Eq. 4.28. The details are left as an exercise.

SKILLS DEVELOPED

Ability to...

• apply the time-dependent mass and energy rate balances to a control volume.

• develop an engineering model.

• retrieve property data for water.

Quick Quiz

If the turbine were removed, and the steam allowed to flow adiabatically into the small tank until the pressure in the tank is 15 bar, determine the final steam temperature in the tank, in °C. Ans. 477°C.

▸ ▸ ▸ **EXAMPLE 4.13** ▸

Storing Compressed Air in a Tank

An air compressor rapidly fills a 0.28 m^3 tank, initially containing air at 21°C, 1 bar, with air drawn from the atmosphere at 21°C, 1 bar. During filling, the relationship between the pressure and specific volume of the air in the tank is $pv^{1.4} = constant$. The ideal gas model applies for the air, and kinetic and potential energy effects are negligible. Plot the pressure, in atm, and the temperature, in °C of the air within the tank, each versus the ratio m/m_1, where m_1 is the initial mass in the tank and m is the mass in the tank at time $t > 0$. Also, plot the compressor work input, in kJ, versus m/m_1. Let m/m_1 vary from 1 to 3.

SOLUTION

Known An air compressor rapidly fills a tank having a known volume. The initial state of the air in the tank and the state of the entering air are known.

Find Plot the pressure and temperature of the air within the tank, and plot the air compressor work input, each versus m/m_1 ranging from 1 to 3.

Schematic and Given Data:

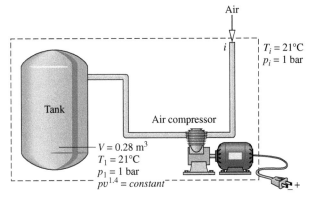

Fig. E4.13a

Engineering Model

1. The control volume is defined by the dashed line on the accompanying diagram.

2. Because the tank is filled rapidly, $\dot{Q}_{cv}$ is ignored.

3. Kinetic and potential energy effects are negligible.

4. The state of the air entering the control volume remains constant.

5. The air stored within the air compressor and interconnecting pipes can be ignored.

① 6. The relationship between pressure and specific volume for the air in the tank is $pv^{1.4} = constant$.

7. The ideal gas model applies for the air.

Analysis The required plots are developed using *Interactive Thermodynamics: IT*. The *IT* program is based on the following analysis. The pressure p in the tank at time $t > 0$ is determined from

$$pv^{1.4} = p_1 v_1^{1.4}$$

where the corresponding specific volume v is obtained using the known tank volume V and the mass m in the tank at that time. That is, $v = V/m$. The specific volume of the air in the tank initially, v_1, is calculated from the ideal gas equation of state and the known initial temperature, T_1, and pressure, p_1. That is

$$v_1 = \frac{RT_1}{p_1} = \frac{\left(\dfrac{8314 \text{ N} \cdot \text{m}}{28.97 \text{ kg} \cdot °\text{K}}\right)(294°\text{K})}{(1 \text{ bar})}\left|\frac{1 \text{ bar}}{10^5 \text{ N/m}^2}\right| = 0.8437 \text{ m}^3/\text{kg}$$

Once the pressure p is known, the corresponding temperature T can be found from the ideal gas equation of state, $T = pv/R$.

To determine the work, begin with the mass rate balance Eq. 4.2, which reduces for the single-inlet control volume to

$$\frac{dm_{cv}}{dt} = \dot{m}_i$$

Then, with assumptions 2 and 3, the energy rate balance Eq. 4.15 reduces to

$$\frac{dU_{cv}}{dt} = -\dot{W}_{cv} + \dot{m}_i h_i$$

Combining the mass and energy rate balances and integrating using assumption 4 gives

$$\Delta U_{cv} = -W_{cv} + h_i \Delta m_{cv}$$

Denoting the work *input* to the compressor by $W_{in} = -W_{cv}$ and using assumption 5, this becomes

$$W_{in} = mu - m_1 u_1 - (m - m_1)h_i \qquad \text{(a)}$$

where m_1 is the initial amount of air in the tank, determined from

$$m_1 = \frac{V}{v_1} = \frac{0.28 \text{ m}^3}{0.8437 \text{ m}^3/\text{kg}} = 0.332 \text{ kg}$$

As a *sample* calculation to validate the *IT* program below, consider the case $m = 0.664$ kg, which corresponds to $m/m_1 = 2$. The specific volume of the air in the tank at that time is

$$v = \frac{V}{m} = \frac{0.28 \text{ m}^3}{0.664 \text{ kg}} = 0.422 \text{ m}^3/\text{kg}$$

The corresponding pressure of the air is

$$p = p_1 \left(\frac{v_1}{v}\right)^{1.4} = (1 \text{ bar})\left(\frac{0.8437 \text{ m}^3/\text{kg}}{0.422 \text{ m}^3/\text{kg}}\right)^{1.4}$$

$$= 2.64 \text{ atm}$$

and the corresponding temperature of the air is

$$T = \frac{pv}{R} = \frac{(2.64 \text{ bar})(0.422 \text{ m}^3/\text{kg})}{\left(\dfrac{8314}{28.97} \dfrac{\text{J}}{\text{kg} \cdot °\text{K}}\right)}\left|\frac{10^5 \text{ N/m}^2}{1 \text{ bar}}\right|$$

$$= 388°\text{K} \ (114.9°\text{C})$$

Evaluating u_1, u, and h_i at the appropriate temperatures from Table A-22, $u_1 = 209.8$ kJ/kg, $u = 27.75$ kJ/kg, $h_i = 294.2$ kJ/kg. Using Eq. (a), the required work input is

$$W_{in} = mu - m_1 u_1 - (m - m_1)h_i$$

$$= (0.664 \text{ kg})\left(277.5\frac{kJ}{kg}\right) - (0.332 \text{ kg})\left(209.8\frac{kJ}{kg}\right)$$

$$- (0.332 \text{ kg})\left(294.2\frac{kJ}{kg}\right)$$

$$= 16.9 \text{ kJ}$$

IT Program Choosing SI units from the **Units** menu, and selecting Air from the **Properties** menu, the *IT* program for solving the problem is

```
//Given Data
p1 = 1//bar
T1 = 21//°C
Ti = 21//°C
V = 0.28//m³
n = 1.4
```

```
// Determine the pressure and temperature for
t > 0

v1 = v_TP("Air", T1, p1)
v = V/m
p * v ^n = p1 * v1 ^n
v = v_TP("Air", T, p)

// Specify the mass and mass ratio r
v1 = V/m1
r = m/m1
r = 2

// Calculate the work using Eq. (a)
Win = m * u - m1 * u1 - hi * (m - m1)
u1 = u_T("Air", T1)
u = u_T("Air", T)
hi = h_T("Air", Ti)
```

Using the **Solve** button, obtain a solution for the sample case $r = m/m_1 = 2$ considered above to validate the program. Good agreement is obtained, as can be verified. Once the program is validated, use the **Explore** button to vary the ratio m/m_1 from 1 to 3 in steps of 0.01. Then, use the **Graph** button to construct the required plots. The results are:

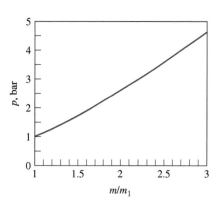

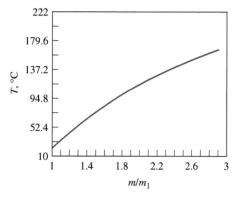

We conclude from the first two plots that the pressure and temperature each increase as the tank fills. The work required to fill the tank increases as well. These results are as expected.

---

① This pressure-specific volume relationship is in accord with what might be measured. The relationship is also consistent with the uniform state idealization, embodied by Eq. 4.25.

SKILLS DEVELOPED

Ability to...

- apply the time-dependent mass and energy rate balances to a control volume.

- develop an engineering model.

- retrieve property data for air modeled as an ideal gas.

- solve iteratively and plot the results using IT.

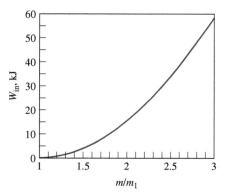

Fig. E4.13b

Quick Quiz

As a *sample* calculation, for the case $m = 1$ kg, evaluate p, in bar. Compare with the value read from the plot of Fig. E4.13b. **Ans. 4.73 bar.**

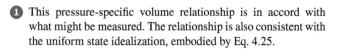

The final example of transient analysis is an application with a *well-stirred* tank. Such process equipment is commonly employed in the chemical and food processing industries.

> ▶ **EXAMPLE 4.14** ▶ ·

Determining Temperature-Time Variation in a Well-Stirred Tank

A tank containing 45 kg of liquid water initially at 45°C has one inlet and one exit with equal mass flow rates. Liquid water enters at 45°C and a mass flow rate of 270 kg/h. A cooling coil immersed in the water removes energy at a rate of 7.6 kW. The water is well mixed by a paddle wheel so that the water temperature is uniform throughout. The power input to the water from the paddle wheel is 0.6 kW. The pressures at the inlet and exit are equal and all kinetic and potential energy effects can be ignored. Plot the variation of water temperature with time.

SOLUTION

Known Liquid water flows into and out of a well-stirred tank with equal mass flow rates as the water in the tank is cooled by a cooling coil.

Find Plot the variation of water temperature with time.

Schematic and Given Data:

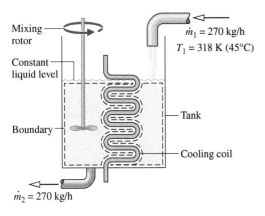

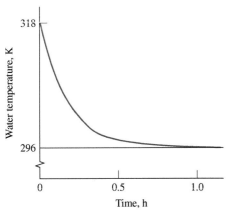

Fig. E4.14

Engineering Model

1. The control volume is defined by the dashed line on the accompanying diagram.

2. For the control volume, the only significant heat transfer is with the cooling coil. Kinetic and potential energy effects can be neglected.

❶ 3. The water temperature is uniform with position throughout and varies only with time: $T = T(t)$.

4. The water in the tank is incompressible, and there is no change in pressure between inlet and exit.

Analysis The energy rate balance, Eq. 4.15, reduces with assumption 2 to

$$\frac{dU_{cv}}{dt} = \dot{Q}_{cv} - \dot{W}_{cv} + \dot{m}(h_1 - h_2)$$

where $\dot{m}$ denotes the mass flow rate.

The mass contained within the control volume remains constant with time, so the term on the left side of the energy rate balance can be expressed as

$$\frac{dU_{cv}}{dt} = \frac{d(m_{cv}u)}{dt} = m_{cv}\frac{du}{dt}$$

Since the water is assumed incompressible, the specific internal energy depends on temperature only. Hence, the chain rule can be used to write

$$\frac{du}{dt} = \frac{du}{dT}\frac{dT}{dt} = c\frac{dT}{dt}$$

where c is the specific heat. Collecting results

$$\frac{dU_{cv}}{dt} = m_{cv}c\frac{dT}{dt}$$

With Eq. 3.20b the enthalpy term of the energy rate balance can be expressed as

$$h_1 - h_2 = c(T_1 - T_2) + v(p_1 - p_2^{\;0})$$

where the pressure term is dropped by assumption 4. Since the water is well mixed, the temperature at the exit equals the temperature of the overall quantity of liquid in the tank, so

$$h_1 - h_2 = c(T_1 - T)$$

where T represents the uniform water temperature at time t.

With the foregoing considerations the energy rate balance becomes

$$m_{cv}c\frac{dT}{dt} = \dot{Q}_{cv} - \dot{W}_{cv} + \dot{m}c(T_1 - T)$$

As can be verified by direct substitution, the solution of this first-order, ordinary differential equation is

$$T = C_1 \exp\left(-\frac{\dot{m}}{m_{cv}}t\right) + \left(\frac{\dot{Q}_{cv} - \dot{W}_{cv}}{\dot{m}c}\right) + T_1$$

The constant C_1 is evaluated using the initial condition: at $t = 0$, $T = T_1$. Finally,

$$T = T_1 + \left(\frac{\dot{Q}_{cv} - \dot{W}_{cv}}{\dot{m}c} \right) \left[1 - \exp\left(-\frac{\dot{m}}{m_{cv}}t \right) \right]$$

Substituting given numerical values together with the specific heat c for liquid water from Table A-19

$$T = 318\ \text{K} + \left[\frac{[-7.6 - (-0.6)]\,\text{kJ/s}}{\left(\dfrac{270}{3600}\ \dfrac{\text{kg}}{\text{s}} \right)\left(4.2\ \dfrac{\text{kJ}}{\text{kg} \cdot \text{K}} \right)} \right] \left[1 - \exp\left(-\frac{270\ \text{kg/h}}{45\ \text{kg}}t \right) \right]$$

$$= 318 - 22[1 - \exp(-6t)]$$

where t is in hours. Using this expression, we construct the accompanying plot showing the variation of temperature with time.

① In this case idealizations are made about the state of the mass contained within the system and the states of the liquid entering and exiting. These idealizations make the transient analysis manageable.

SKILLS DEVELOPED

Ability to...

• apply the time-dependent mass and energy rate balances to a control volume.
• develop an engineering model.
• apply the incompressible substance model for water.
• solve an ordinary differential equation and plot the solution.

Quick Quiz

What is the water temperature, in °C, when _steady state_ is achieved? Ans. 23°C.

CHAPTER SUMMARY AND STUDY GUIDE

The conservation of mass and energy principles for control volumes are embodied in the mass and energy rate balances developed in this chapter. Although the primary emphasis is on cases in which one-dimensional flow is assumed, mass and energy balances are also presented in integral forms that provide a link to subsequent fluid mechanics and heat transfer courses. Control volumes at steady state are featured, but discussions of transient cases are also provided.

The use of mass and energy balances for control volumes at steady state is illustrated for nozzles and diffusers, turbines, compressors and pumps, heat exchangers, throttling devices, and integrated systems. An essential aspect of all such applications is the careful and explicit listing of appropriate assumptions. Such model-building skills are stressed throughout the chapter.

The following checklist provides a study guide for this chapter. When your study of the text and end-of-chapter exercises has been completed you should be able to

• write out the meanings of the terms listed in the margins throughout the chapter and explain each of the related concepts. The subset of key concepts listed below is particularly important in subsequent chapters.

• list the typical modeling assumptions for nozzles and diffusers, turbines, compressors and pumps, heat exchangers, and throttling devices.

• apply Eqs. 4.6, 4.18, and 4.20 to control volumes at steady state, using appropriate assumptions and property data for the case at hand.

• apply mass and energy balances for the transient analysis of control volumes, using appropriate assumptions and property data for the case at hand.

KEY ENGINEERING CONCEPTS

conservation of mass	flow work	heat exchanger
mass flow rates	energy rate balance	throttling calorimeter
mass rate balance	nozzle	throttling process
one-dimensional flow	diffuser	system integration
volumetric flow rate	turbine	transient operation
steady state	compressor	
mass flux	pump	

KEY EQUATIONS

$\dot{m} = \dfrac{AV}{v}$	(4.4b)	Mass flow rate, one-dimensional flow. (See Fig. 4.3.)
$\dfrac{dm_{cv}}{dt} = \sum_i \dot{m}_i - \sum_e \dot{m}_e$	(4.2)	Mass rate balance.
$\underset{\text{(mass rate in)}}{\sum_i \dot{m}_i} = \underset{\text{(mass rate out)}}{\sum_e \dot{m}_e}$	(4.6)	Mass rate balance at steady state.

$\dfrac{dE_{cv}}{dt} = \dot{Q}_{cv} - \dot{W}_{cv} + \sum_i \dot{m}_i \left(h_i + \dfrac{V_i^2}{2} + gz_i \right) - \sum_e \dot{m}_e \left(h_e + \dfrac{V_e^2}{2} + gz_e \right)$	(4.15)	Energy rate balance.
$0 = \dot{Q}_{cv} - \dot{W}_{cv} + \sum_i \dot{m}_i \left(h_i + \dfrac{V_i^2}{2} + gz_i \right) - \sum_e \dot{m}_e \left(h_e + \dfrac{V_e^2}{2} + gz_e \right)$	(4.18)	Energy rate balance at steady state.
$0 = \dot{Q}_{cv} - \dot{W}_{cv} + \dot{m} \left[(h_1 - h_2) + \dfrac{(V_1^2 - V_2^2)}{2} + g(z_1 - z_2) \right]$	(4.20a)	Energy rate balance for one-inlet, one-exit control volumes at steady state.
$0 = \dfrac{\dot{Q}_{cv}}{\dot{m}} - \dfrac{\dot{W}_{cv}}{\dot{m}} + (h_1 - h_2) + \dfrac{(V_1^2 - V_2^2)}{2} + g(z_1 - z_2)$	(4.20b)	
$h_2 = h_1 \quad (p_2 < p_1)$	(4.22)	Throttling process. (See Fig. 4.15.)

EXERCISES: THINGS ENGINEERS THINK ABOUT

4.1 Why might a computer cooled by a *constant-speed* fan operate satisfactorily at sea level but overheat at high altitude?

4.2 When a drip coffeemaker on-off switch is turned to the *on* position, how is cold water in the water reservoir converted to hot water and sent upward to drip down into the coffee grounds in the filter?

4.3 When a slice of bread is placed in a toaster and the toaster is activated, is the toaster in steady-state operation, transient operation, both?

4.4 As a tree grows, its mass increases. Does this violate the conservation of mass principle? Explain.

4.5 Wind turbines and hydraulic turbines develop mechanical power from moving streams of air and water, respectively. In each case, what aspect of the stream is tapped for power?

4.6 For air flowing through a converging-diverging channel, sketch the variation of the air pressure as air accelerates in the converging section and decelerates in the diverging section.

4.7 How does a heart-lung machine maintain blood circulation and oxygen content during surgery?

4.8 Even though their outer surfaces would seem hot to the touch, large steam turbines in power plants might not be covered with much insulation. Why not?

4.9 Where are compressors found within households?

4.10 One stream of a counterflow heat exchanger operating at steady state has saturated water vapor entering and saturated liquid exiting, each at 1 atm. The other stream has ambient air entering at 20°C and exiting at a higher temperature with no significant change in pressure. What is the variation of temperature with position of each stream as it passes through the heat exchanger? Sketch them.

4.11 When selecting a pump to remove water from a flooded basement, how does one size the pump to ensure that it is suitable?

4.12 In what subsystems are pumps found in automobiles?

4.13 Why is it that when air at 1 atm is *throttled* to a pressure of 0.5 atm, its temperature at the valve exit is close to the temperature at the valve inlet, yet when air at 1 atm *leaks* into an insulated, rigid, initially evacuated tank until the tank pressure is 0.5 atm, the temperature of the air in the tank is greater than the air temperature outside the tank?

CHECKING UNDERSTANDING

For problems 4.1–4.5, match the appropriate definition in the right column with each term in the left column.

4.1 Compressor

4.2 Diffuser

4.3 Nozzle

4.4 Pump

4.5 Turbine

A. A device in which power is developed as a result of a gas or liquid passing through a set of blades attached to a shaft free to rotate

B. A device in which work is done on a gas to increase the pressure and/or elevation

C. A device in which work is done on a liquid to increase the pressure and/or elevation

D. A flow passage of varying cross-sectional area in which the velocity of a gas or liquid increases in the direction of flow

E. A flow passage of varying cross-sectional area in which the velocity of a gas or liquid decreases in the direction of flow

4.6 A flow idealized as a throttling process through a device has

    a. $h_2 > h_1$ and $p_2 > p_1$

    b. $h_2 = h_1$ and $p_2 > p_1$

    c. $h_2 > h_1$ and $p_2 < p_1$

    d. $h_2 = h_1$ and $p_2 < p_1$

4.7 _____ is the work associated with the fluid pressure as mass is introduced at inlets and removed at exits.

4.8 Steam enters a horizontal pipe operating at steady state with a specific enthalpy of 3000 kJ/kg and a mass flow rate of 0.5 kg/s. At the exit, the specific enthalpy is 1700 kJ/kg. If there is no significant change in kinetic energy from inlet to exit, the rate of heat transfer between the pipe and its surroundings is

    a. 650 kW from the pipe to the surroundings

    b. 650 kW from the surroundings to the pipe

    c. 2600 kW from the pipe to the surroundings

    d. 2600 kW from the surroundings to the pipe

4.9 A _____ is a device that introduces a restriction into a line to reduce the pressure of a gas or liquid.

4.10 The time rate of mass flow per unit area is called _____.

4.11 _____ means all properties are unchanging in time.

4.12 Air enters a compressor operating at steady state at 1 atm with a specific enthalpy of 290 kJ/kg and exits at a higher pressure with a specific enthalpy of 1023 kJ/kg. The mass flow rate is 0.1 kg/s. Kinetic and potential energy effects are negligible and the air can be modeled as an ideal gas. If the compressor power input is 77 kW, the rate of heat transfer between the air and its surroundings is

    a. 150.3 kW from the surroundings to the air

    b. 150.3 kW from the air to the surroundings

    c. 3.7 kW from the surroundings to the air

    d. 3.7 kW from the air to the surroundings

4.13 Water vapor enters an insulated nozzle operating at steady state with a velocity of 100 m/s and specific enthalpy of 3445.3 kJ/kg, and exits with specific enthalpy of 3051.1 kJ/kg. The velocity at the exit is most closely

    a. 104 m/s

    b. 636 m/s

    c. 888 m/s

    d. 894 m/s

4.14 Mass flow rate for a flow modeled as one-dimensional depends on all except

    a. Density of working fluid

    b. Cross-sectional area through which flow passes

    c. Velocity of working fluid

    d. Total volume of working fluid

4.15 As velocity increases in a nozzle, pressure _____.

Indicate whether the following statements are true or false. Explain.

4.16 For one-dimensional flow, mass flow rate is the product of density, area, and velocity.

4.17 At steady state, conservation of energy asserts the total rate at which energy is transferred into the control volume equals the total rate at which energy is transferred out.

4.18 As velocity decreases in a diffuser, pressure decreases.

4.19 Common heat exchanger types include direct-contact, counter-flow, parallel-flow, and cross-flow heat exchangers.

4.20 A significant increase in pressure can be achieved by introducing a restriction into a line through which a gas or liquid flows.

4.21 System integration is the practice of combining components to achieve an overall objective.

4.22 For a control volume at steady state, mass can accumulate within the control volume.

4.23 *Flow work* is the work done on a flowing stream by a paddle wheel or piston.

4.24 *Transient* operation denotes a change in state with time.

4.25 Where mass crosses the boundary of a control volume, the accompanying energy transfer is accounted for by the internal energy of the mass only.

4.26 When a substance undergoes a *throttling process* through a valve, the specific enthalpies of the substance at the valve inlet and valve exit are equal.

4.27 The thermodynamic performance of a device such as a turbine through which mass flows is best analyzed by studying the flowing mass alone.

4.28 An *open feedwater* heater is a special type of a counterflow heat exchanger.

4.29 A key step in thermodynamic analysis is the careful listing of modeling assumptions.

4.30 An automobile's radiator is an example of a cross-flow heat exchanger.

PROBLEMS: DEVELOPING ENGINEERING SKILLS

C Problem may require use of appropriate computer software in order to complete.

Evaluating Mass Flow Rate

4.1 A *laser Doppler velocimeter* measures a velocity of 8 m/s as water flows in an open channel. The channel has a rectangular cross section of 0.5 m by 0.2 m in the flow direction. If the water density is a constant 998 kg/m³, determine the mass flow rate, in kg/s.

4.2 Air exits a turbine at 200 kPa and 150°C with a mass flow rate of 11.5 kg/s. Modeling air as an ideal gas, determine the volumetric flow rate, in kg/s.

Applying Conservation of Mass

4.3 Figure P4.3 provides data for water entering and exiting a tank. At the inlet and exit of the tank, determine the mass flow rate, each in kg/s. Also find the time rate of change of mass contained within the tank, in kg/s.

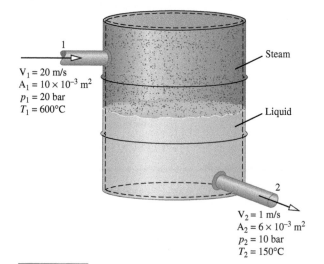

$V_1 = 20$ m/s
$A_1 = 10 \times 10^{-3}$ m²
$p_1 = 20$ bar
$T_1 = 600$°C

$V_2 = 1$ m/s
$A_2 = 6 \times 10^{-3}$ m²
$p_2 = 10$ bar
$T_2 = 150$°C

FIGURE P4.3

4.4 **Figure P4.4** shows a mixing tank initially containing 1,360 kg of liquid water. The tank is fitted with two inlet pipes, one delivering hot water at a mass flow rate of 0.4 kg/s and the other delivering cold water at a mass flow rate of 0.6 kg/s.Water exits through a single exit pipe at a mass flow rate of 1.12 kg/s. Determine the amount of water, in kg, in the tank after one hour.

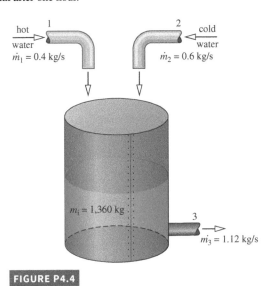

hot water $\dot{m}_1 = 0.4$ kg/s

cold water $\dot{m}_2 = 0.6$ kg/s

$m_i = 1,360$ kg

$\dot{m}_3 = 1.12$ kg/s

FIGURE P4.4

4.5 A 380-L tank contains steam, initially at 400°C, 3 bar. A valve is opened, and steam flows out of the tank at a constant mass flow rate of 0.005 kg/s. During steam removal, a heater maintains the temperature within the tank constant. Determine the time, in s, at which 75% of the initial mass remains in the tank; also determine the specific volume, in m³/kg, and pressure, in bar, in the tank at that time.

4.6 A cylindrical tank contains 1500 kg of liquid water. It has one inlet pipe through which water is entering at a mass flow rate of 1.2 kg/s. The tank is fitted with two outlet pipes and the water is flowing through these exit pipes at the mass flow rates of 0.5 kg/s and 0.8 kg/s. Determine the amount of water that will be left in the tank after thirty minutes.

4.7 The small two-story office building shown in **Fig. P4.7** has 1,020 m³ of occupied space. Due to cracks around windows and outside doors, air leaks in on the windward side of the building and leaks out on the leeward side of the building. Outside air also enters the building when outer doors are opened. On a particular day, tests

were conducted. The outdoor temperature was measured to be −9.5°C. The inside temperature was controlled at 20°C. Keeping the doors closed, the infiltration rate through the cracks was determined to be 0.03 m³/s. The infiltration rate associated with door openings, averaged over the work day, was 0.02 m³/s. The pressure difference was negligible between the inside and outside of the building. (a) Assuming ideal gas behavior, determine at steady state the volumetric flow rate of air exiting the building, in m³/s. (b) When expressed in terms of the volume of the occupied space, determine the number of building air changes per hour.

4.8 Liquid water flows isothermally at 20°C through a one-inlet, one-exit duct operating at steady state. The duct's inlet and exit diameters are 0.02 m and 0.04 m, respectively. At the inlet, the velocity is 40 m/s and pressure is 1 bar. At the exit, determine the mass flow rate, in kg/s, and velocity, in m/s.

4.9 Steam enters a one-inlet, two-exit control volume at location (1) at 360°C, 100 bar, with a mass flow rate of 2 kg/s. The inlet pipe is round with a diameter of 5.2 cm. Fifteen percent of the flow leaves through location (2) and the remainder leaves at (3). For steady-state operation, determine the inlet velocity, in m/s, and the mass flow rate at each exit, in kg/s.

4.10 Air enters a household electric furnace at 24°C, 1 bar, with a volumetric flow rate of 0.38 m³/s. The furnace delivers air at 50°C, 1 bar to a duct system with three branches consisting of two 150 mm diameter ducts and a 300 mm duct. If the velocity in each 150 mm duct is 3 m/s, determine for steady-state operation

   a. the mass flow rate of air entering the furnace, in kg/s.

   b. the volumetric flow rate in each 150 mm duct, in m³/s.

   c. the velocity in the 300 mm duct, in m/s.

4.11 Refrigerant 134a enters the evaporator of a refrigeration system operating at steady state at −4°C and quality of 20% at a velocity of 7 m/s. At the exit, the refrigerant is a saturated vapor at a temperature of −4°C. The evaporator flow channel has constant diameter. If the mass flow rate of the entering refrigerant is 0.1 kg/s, determine

   a. the diameter of the evaporator flow channel, in cm.

   b. the velocity at the exit, in m/s.

4.12 **Figure P4.12** provides steady-state data for water vapor flowing through a piping configuration. At each exit, the volumetric flow rate, pressure, and temperature are equal. Determine the mass flow rate at the inlet and exits, each in kg/s.

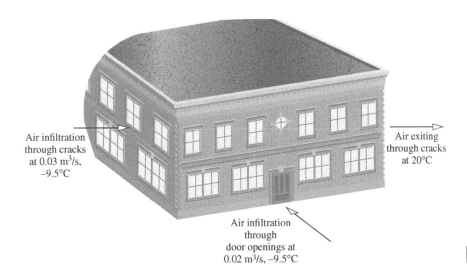

Air infiltration through cracks at 0.03 m³/s, −9.5°C

Air exiting through cracks at 20°C

Air infiltration through door openings at 0.02 m³/s, −9.5°C

FIGURE P4.7

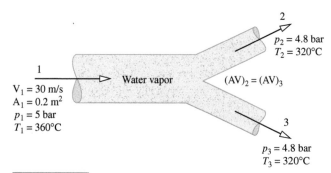

FIGURE P4.12

4.13 Air enters a one-inlet, one-exit control volume at 8 bar, 600 K, and 40 m/s through a flow area of 20 cm². At the exit, the pressure is 2 bar, the temperature is 400 K, and the velocity is 350 m/s. The air behaves as an ideal gas. For steady-state operation, determine

a. the mass flow rate, in kg/s.

b. the exit flow area, in cm².

Energy Analysis of Control Volumes at Steady State

4.14 Refrigerant 134a enters a horizontal pipe operating at steady state at 40°C, 300 kPa, and a mass flow rate of 0.6 kg/s. At the exit, the temperature is 50°C and the pressure is 240 kPa. The pipe diameter is 0.04 m. Determine (a) the velocities at the inlet and exit, in m/s, and (b) the rate of heat transfer between the pipe and its surroundings, in kW.

4.15 Refrigerant 134a enters a horizontal pipe operating at steady state at 40°C, 300 kPa and a velocity of 40 m/s. At the exit, the temperature is 50°C and the pressure is 240 kPa. The pipe diameter is 0.04 m. Determine (a) the mass flow rate of the refrigerant, in kg/s, (b) the velocity at the exit, in m/s, and (c) the rate of heat transfer between the pipe and its surroundings, in kW.

4.16 Air at 600 kPa, 330 K enters a well-insulated, horizontal pipe having a diameter of 1.2 cm and exits at 120 kPa, 300 K. Applying the ideal gas model for air, determine at steady state (a) the inlet and exit velocities, each in m/s, and (b) the mass flow rate, in kg/s.

4.17 At steady state, air at 200 kPa, 52°C and a mass flow rate of 0.5 kg/s enters an insulated duct having differing inlet and exit cross-sectional areas. At the duct exit, the pressure of the air is 100 kPa, the velocity is 255 m/s, and the cross-sectional area is 2×10^{-3} m². Assuming the ideal gas model, determine

a. the temperature of the air at the exit, in °C.

b. the velocity of the air at the inlet, in m/s.

c. the inlet cross-sectional area, in m².

4.18 Refrigerant 134a flows at steady state through a horizontal tube having an inside diameter of 0.05 m. The refrigerant enters the tube with a quality of 0.1, temperature of 36°C, and velocity of 10 m/s. The refrigerant exits the tube at 9 bar as a saturated liquid. Determine

a. the mass flow rate of the refrigerant, in kg/s.

b. the velocity of the refrigerant at the exit, in m/s.

c. the rate of heat transfer, in kW, and its associated direction with respect to the refrigerant.

4.19 Steam enters a well-insulated nozzle at 2068 kPa, 316°C, with a velocity of 30 m/s and exits at 276 kPa with a velocity of 550 m/s. For steady-state operation, and neglecting potential energy effects, determine the exit temperature, in °C.

4.20 Air with a mass flow rate of 2.3 kg/s enters a horizontal nozzle operating at steady state at 450 K, 350 kPa, and velocity of 3 m/s. At

the exit, the temperature is 300 K and the velocity is 460 m/s. Using the ideal gas model for air with constant $c_p = 1.011$ kJ/kg · K, determine

a. the area at the inlet, in m².

b. the heat transfer between the nozzle at its surroundings, in kW. Specify whether the heat transfer is to or from the air.

4.21 Helium gas flows through a well-insulated nozzle at steady state. The temperature and velocity at the inlet are 333 K and 53 m/s, respectively. At the exit, the temperature is 256 K and the pressure is 345 kPa. The mass flow rate is 0.5 kg/s. Using the ideal gas model, and neglecting potential energy effects, determine the exit area, in m².

4.22 Nitrogen, modeled as an ideal gas, flows at a rate of 3 kg/s through a well-insulated horizontal nozzle operating at steady state. The nitrogen enters the nozzle with a velocity of 20 m/s at 340 K, 400 kPa and exits the nozzle at 100 kPa. To achieve an exit velocity of 478.8 m/s, determine

a. the exit temperature, in K.

b. the exit area, in m².

4.23 As shown in Fig. P4.23, air enters the diffuser of a jet engine operating at steady state at 18 kPa, 216 K and a velocity of 265 m/s, all data corresponding to high-altitude flight. The air flows adiabatically through the diffuser and achieves a temperature of 250 K at the diffuser exit. Using the ideal gas model for air, determine the velocity of the air at the diffuser exit, in m/s.

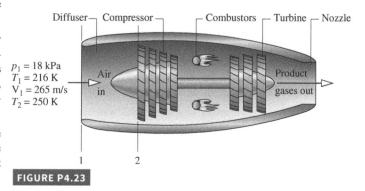

FIGURE P4.23

4.24 Air enters an insulated diffuser operating at steady state with a pressure of 1 bar, a temperature of 300 K, and a velocity of 250 m/s. At the exit, the pressure is 1.13 bar and the velocity is 140 m/s. Potential energy effects can be neglected. Using the ideal gas model, determine

a. the ratio of the exit flow area to the inlet flow area.

b. the exit temperature, in K.

4.25 Refrigerant 134a enters an insulated diffuser as a saturated vapor at 7 bar with a velocity of 370 m/s. At the exit, the pressure is 16 bar and the velocity is negligible. The diffuser operates at steady state and potential energy effects can be neglected. Determine the exit temperature, in °C.

4.26 Steam enters a well-insulated turbine operating at steady state at 4 MPa with a specific enthalpy of 3015.4 kJ/kg and a velocity of 10 m/s. The steam expands to the turbine exit where the pressure is 0.07 MPa, specific enthalpy is 2431.7 kJ/kg, and the velocity is 90 m/s. The mass flow rate is 11.95 kg/s. Neglecting potential energy effects, determine the power developed by the turbine, in kW.

4.27 Hot combustion gases, modeled as air behaving as an ideal gas, enter a turbine at 1000 kPa, 1500 K with a mass flow rate of 0.10 kg/s and exit at 200 kPa and 900 K. If heat transfer from the turbine to its surroundings occurs at a rate of 15 kW, determine the power output of the turbine, in kW.

4.28 A well-insulated turbine operating at steady state develops 23 MW of power for a steam flow rate of 40 kg/s. The steam enters at 360°C with a velocity of 35 m/s and exits as saturated vapor at 0.06 bar with a velocity of 120 m/s. Neglecting potential energy effects, determine the inlet pressure, in bar.

4.29 Steam enters a turbine operating at steady state with a mass flow of 10 kg/min, a specific enthalpy of 3100 kJ/kg, and a velocity of 30 m/s. At the exit, the specific enthalpy is 2300 kJ/kg and the velocity is 45 m/s. The elevation of the inlet is 3 m higher than at the exit. Heat transfer from the turbine to its surroundings occurs at a rate of 1.1 kJ per kg of steam flowing. Let g = 9.81 m/s^2. Determine the power developed by the turbine, in kW.

4.30 Steam enters a turbine operating at steady state at 2 MPa, 360°C with a velocity of 100 m/s. Saturated vapor exits at 0.1 MPa and a velocity of 50 m/s. The elevation of the inlet is 3 m higher than at the exit. The mass flow rate of the steam is 15 kg/s, and the power developed is 7 MW. Let g = 9.81 m/s^2. Determine (a) the area at the inlet, in m^2, and (b) the rate of heat transfer between the turbine and its surroundings, in kW.

4.31 The intake to a hydraulic turbine installed in a flood control dam is located at an elevation of 10 m above the turbine exit. Water enters at 20°C with negligible velocity and exits from the turbine at 10 m/s. The water passes through the turbine with no significant changes in temperature or pressure between the inlet and exit, and heat transfer is negligible. The acceleration of gravity is constant at g = 9.81 m/s^2. If the power output at steady state is 500 kW, what is the mass flow rate of water, in kg/s?

4.32 Steam enters the first-stage turbine shown in **Fig. P4.32** at 40 bar and 500°C with a volumetric flow rate of 90 m^3/min. Steam exits the turbine at 20 bar and 400°C. The steam is then reheated at constant pressure to 500°C before entering the second-stage turbine. Steam leaves the second stage as saturated vapor at 0.6 bar. For operation at steady state, and ignoring stray heat transfer and kinetic and potential energy effects, determine the

  a. mass flow rate of the steam, in kg/h.

  b. total power produced by the two stages of the turbine, in kW.

  c. rate of heat transfer to the steam flowing through the reheater, in kW.

4.33 Air enters a compressor operating at steady state at 1 atm with a specific enthalpy of 290 kJ/kg and exits at a higher pressure with a specific enthalpy of 1023 kJ/kg. The mass flow rate is 0.1 kg/s. If the compressor power input is 77 kW, determine the rate of heat transfer between the compressor and its surroundings, in kW. Neglect kinetic and potential energy effects.

4.34 Hot combustion gases, modeled as air behaving as an ideal gas, enter a turbine at 10 bar, 1500 K and exit at 1.97 bar and 900 K. If

the power output of the turbine is 55.4 kW, determine the rate of heat transfer from the turbine to its surroundings, in kW.

4.35 Air enters a compressor operating at steady state at 1.05 bar, 300 K, with a volumetric flow rate of 12 m^3/min and exits at 12 bar, 400 K. Heat transfer occurs at a rate of 2 kW from the compressor to its surroundings. Assuming the ideal gas model for air and neglecting kinetic and potential energy effects, determine the power input, in kW.

4.36 Nitrogen is compressed in an axial-flow compressor operating at steady state from a pressure of 100 kPa and a temperature of 10°C to a pressure 400 kPa. The gas enters the compressor through a 150 mm-diameter duct with a velocity of 9 m/s and exits at 92°C with a velocity of 24 m/s. Using the ideal gas model, and neglecting stray heat transfer and potential energy effects, determine the compressor power input, in kW.

4.37 Air enters a compressor operating at steady state with a pressure of 100 kPa, a temperature of 27°C, and a volumetric flow rate of 0.57 m^3/s. Air exits the compressor at 345 kPa. Heat transfer from the compressor to its surroundings occurs at a rate of 48 kJ/kg of air flowing. If the compressor power *input* is 78 kW, determine the exit temperature, in °C.

4.38 Refrigerant 134a enters an air conditioner compressor at 3.2 bar, 10°C, and is compressed at steady state to 10 bar, 70°C. The volumetric flow rate of refrigerant entering is 3.0 m^3/min. The power *input* to the compressor is 55.2 kJ per kg of refrigerant flowing. Neglecting kinetic and potential energy effects, determine the heat transfer rate, in kW.

4.39 Refrigerant 134a enters an insulated compressor operating at steady state as saturated vapor at −20°C with a mass flow rate of 1.2 kg/s. Refrigerant exits at 7 bar, 70°C. Changes in kinetic and potential energy from inlet to exit can be ignored. Determine (a) the volumetric flow rates at the inlet and exit, each in m^3/s, and (b) the power input to the compressor, in kW.

4.40 Air enters a compressor operating at steady state with a pressure of 101.4 kPa and a temperature of 21°C. The volumetric flow rate at the inlet is 0.5 m^3/s, and the flow area is 0.02 m^2. At the exit, the pressure is 240 kPa, the temperature is 138°C, and the velocity is 15 m/s. Heat transfer from the compressor to its surroundings occurs at a rate of 2.3 kJ/kg of air flowing. Potential energy effects are negligible, and the ideal gas model can be assumed for the air. Determine (a) the velocity of the air at the inlet, in m/s, (b) the mass flow rate, in kg/s, and (c) the compressor power, in kJ/s and kW.

4.41 At steady state, a well-insulated compressor takes in air at 15°C, 98 kPa, with a volumetric flow rate of 0.57 m^3/s, and compresses it to 260°C, 827 kPa. Kinetic and potential energy changes from inlet to exit can be neglected. Determine the compressor power, in kW, and the volumetric flow rate at the exit, in m^3/s.

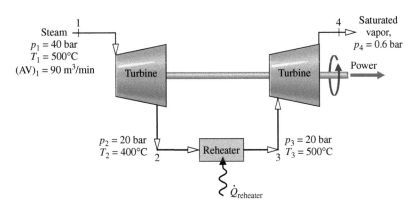

Steam —
p_1 = 40 bar
T_1 = 500°C
$(AV)_1$ = 90 m^3/min

1 Turbine

4 Saturated vapor,
p_4 = 0.6 bar

Turbine   Power

p_2 = 20 bar
T_2 = 400°C   2   Reheater   3 p_3 = 20 bar
T_3 = 500°C

$\dot{Q}_{reheater}$

FIGURE P4.32

4.42 Air is compressed at steady state from 1 bar, 300 K, to 6 bar with a mass flow rate of 4 kg/s. Each unit of mass passing from inlet to exit undergoes a process described by $pv^{1.27} = constant$. Heat transfer occurs at a rate of 46.95 kJ per kg of air flowing to cooling water circulating in a water jacket enclosing the compressor. If kinetic and potential energy changes of the air from inlet to exit are negligible, calculate the compressor power, in kW.

4.43 Air enters a compressor operating at steady state at 101.4 kPa and 15°C and is compressed to a pressure of 1034 kPa. As the air passes through the compressor, it is cooled at a rate of 23 kJ/kg of air flowing by water circulated through the compressor casing. The volumetric flow rate of the air at the inlet is 2 m³/s, and the power input to the compressor is 520 kW. The air behaves as an ideal gas, there is no stray heat transfer, and kinetic and potential effects are negligible. Determine (a) the mass flow rate of the air, kg/s, and (b) the temperature of the air at the compressor exit, in °C.

4.44 Water is drawn steadily by a pump at a volumetric flow rate of 0.8 m³/min through a pipe with a 10 cm diameter inlet. The water is delivered to a tank placed at a height of 10 m above the pipe inlet. The water exits through a nozzle having a diameter of 2.5 cm. Water enters at 25°C and 1 atm pressure but there is no significant change in pressure and temperature at exit. Six percent of the power input to the pump is dissipated in the form of heat into the surroundings. Take the value of acceleration due to gravity (g) equal to 9.8 m/s². Determine (a) the velocity of the water at the inlet and exit, and (b) the power required by the pump.

4.45 **Figure P4.45** provides steady-state operating data for a pump drawing water from a reservoir and delivering it at a pressure of 3 bar to a storage tank perched 15 m above the reservoir. The power input to the pump is 0.52 W. The water temperature remains nearly constant at 15°C, there is no significant change in kinetic energy from inlet to exit, and heat transfer between the pump and its surroundings is negligible. Determine the mass flow rate of water, in kg/s. Let $g = 9.81$ m/s².

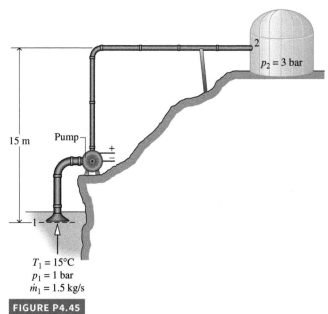

$T_1 = 15°C$
$p_1 = 1$ bar
$\dot{m}_1 = 1.5$ kg/s

FIGURE P4.45

4.46 During cardiac surgery, a heart-lung machine achieves *extracorporeal circulation* of the patient's blood using a pump operating at steady state. Blood enters the well-insulated pump at a rate of 5 liters/min. The temperature change of the blood is negligible as it flows through the pump. The pump requires 20 W of power input. Modeling the blood as an incompressible substance with negligible

kinetic and potential energy effects, determine the pressure change, in kPa, of the blood as it flows through the pump.

4.47 A pump steadily draws water through a pipe from a reservoir at a volumetric flow rate of 1.3 L/s. At the pipe inlet, the pressure is 101.4 kPa, the temperature is 18°C, and the velocity is 3 m/s. At the pump exit, the pressure is 240 kPa, the temperature is 18°C, and the velocity is 12 m/s. The pump exit is located 12 m above the pipe inlet. Ignoring heat transfer, determine the power required by the pump, in kJ/s and kW. The local acceleration of gravity is 9.8 m/s².

4.48 Refrigerant 134a enters a heat exchanger in a refrigeration system operating at steady state as saturated liquid at −18°C and exits at −6°C at a pressure of 1.4 bar. A separate air stream passes in counterflow to the Refrigerant 134a stream, entering at 50°C and exiting at 25°C. The outside of the heat exchanger is well insulated. Neglecting kinetic and potential energy effects and modeling the air as an ideal gas with constant $c_p = 1.05$ kJ/kg · K, determine the mass flow rate of air, in kg/s.

4.49 Oil enters a counterflow heat exchanger at 450 K with a mass flow rate of 10 kg/s and exits at 350 K. A separate stream of liquid water enters at 20°C, 5 bar. Each stream experiences no significant change in pressure. Stray heat transfer with the surroundings of the heat exchanger and kinetic and potential energy effects can be ignored. The specific heat of the oil is constant, $c = 2$ kJ/kg · K. If the designer wants to ensure no water vapor is present in the exiting water stream, what is the allowed range of mass flow rates for the water, in kg/s?

4.50 Air flows steadily through an air compressor at a rate of 0.6 kg/s. Air enters the compressor at 8 m/s velocity, 100 kPa pressure and 0.96 m³/kg volume and leaves at 6 m/s velocity, 700 kPa pressure and 0.2 m³/kg volume. There is an increase in internal energy of 95 kJ/kg. Heat is absorbed by the cooling water in the compressor jackets at a rate of 60 kW.

a. Calculate the rate of shaft work input to the air in kW.

b. Determine the ratio of inlet pipe diameter to outlet pipe diameter.

4.51 Refrigerant 134a enters a heat exchanger in a refrigeration system operating at steady state as saturated vapor at −17°C and exits at −7°C with no change in pressure. A separate liquid stream of Refrigerant 134a passes in counterflow to the vapor stream, entering at 41°C, 11 bar, and exiting at a lower temperature while experiencing no pressure drop. The outside of the heat exchanger is well insulated, and the streams have equal mass flow rates. Neglecting kinetic and potential energy effects, determine the exit temperature of the liquid stream, in °C.

4.52 An air-conditioning system is shown in Fig. P4.52 in which air flows over tubes carrying Refrigerant 134a. Air enters with a volumetric flow rate of 50 m³/min at 32°C, 1 bar, and exits at 22°C, 0.95 bar.

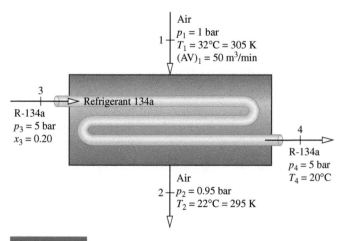

Air
$p_1 = 1$ bar
$T_1 = 32°C = 305$ K
$(AV)_1 = 50$ m³/min

R-134a
$p_3 = 5$ bar
$x_3 = 0.20$

Refrigerant 134a

R-134a
$p_4 = 5$ bar
$T_4 = 20°C$

Air
$p_2 = 0.95$ bar
$T_2 = 22°C = 295$ K

FIGURE P4.52

Refrigerant enters the tubes at 5 bar with a quality of 20% and exits at 5 bar, 20°C. Ignoring heat transfer at the outer surface of the air conditioner, and neglecting kinetic and potential energy effects, determine at steady state

    a. the mass flow rate of the refrigerant, in kg/min.

    b. the rate of heat transfer, in kJ/min, between the air and refrigerant.

4.53 **Figure P4.53** shows a solar collector panel embedded in a roof. The panel, which has a surface area of 2.16 m², receives energy from the sun at a rate of 2344 kJ/h per m² of collector surface. Twenty-five percent of the incoming energy is lost to the surroundings. The remaining energy is used to heat domestic hot water from 32°C to 49°C. The water passes through the solar collector with a negligible pressure drop. Neglecting kinetic and potential effects, determine at steady state how many gallons of water at 49°C the collector generates per hour.

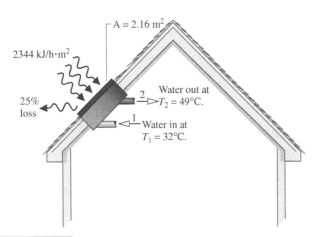

Water out at $T_2 = 49°C$.

Water in at $T_1 = 32°C$.

A = 2.16 m²

2344 kJ/h·m²

25% loss

FIGURE P4.53

4.54 Steam at a pressure of 0.08 bar and a quality of 93.2% enters a shell-and-tube heat exchanger where it condenses on the outside of tubes through which cooling water flows, exiting as saturated liquid at 0.08 bar. The mass flow rate of the condensing steam is 3.4×10^5 kg/h. Cooling water enters the tubes at 15°C and exits at 35°C with negligible change in pressure. Neglecting stray heat transfer and ignoring kinetic and potential energy effects, determine the mass flow rate of the cooling water, in kg/h, for steady-state operation.

4.55 **C** **Figure P4.55** provides steady-state operating data for a *parallel flow* heat exchanger in which there are separate streams of air and water. Each stream experiences no significant change in pressure.

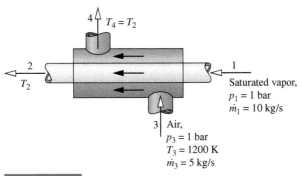

$4\ T_4 = T_2$

$2\ T_2$

1 Saturated vapor, $p_1 = 1$ bar $\dot{m}_1 = 10$ kg/s

3 Air, $p_3 = 1$ bar $T_3 = 1200$ K $\dot{m}_3 = 5$ kg/s

FIGURE P4.55

Stray heat transfer with the surroundings of the heat exchanger and kinetic and potential energy effects can be ignored. The ideal gas model applies to the air. If each stream exits at the same temperature, determine the value of that temperature, in K.

4.56 A 0.75 m³ capacity tank contains a two-phase liquid-vapor mixture at 240°C and a quality of 0.6. Heat is transferred to the mixture and saturated vapor from the tank are withdrawn at 240°C to maintain the constant pressure inside the tank by means of a pressure regulating value. This process continues until the tank has only saturated vapor left in it. Neglect the effect of kinetic and potential energies. Determine the amount of heat transfer in kJ.

4.57 An open feedwater heater operates at steady state with liquid water entering inlet 1 at 10 bar, 50°C. A separate stream of steam enters inlet 2 at 10 bar and 200°C with a mass flow rate of 16 kg/s. Saturated liquid at 10 bar exits the feedwater heater at exit 3. Ignoring heat transfer with the surroundings and neglecting kinetic and potential energy effects, determine the mass flow rate, in kg/s, of the steam at inlet 1.

4.58 Three return steam lines in a chemical processing plant enter a collection tank operating at steady state at 1 bar. Steam enters inlet 1 with flow rate of 0.8 kg/s and quality of 0.9. Steam enters inlet 2 with flow rate of 2 kg/s at 200°C. Steam enters inlet 3 with flow rate of 1.2 kg/s at 95°C. Steam exits the tank at 1 bar. The rate of heat transfer from the collection tank is 40 kW. Neglecting kinetic and potential energy effects, determine for the steam exiting the tank

    a. the mass flow rate, in kg/s.

    b. the temperature, in °C.

4.59 A water heater operating under steady flow conditions receives water at the rate of 5 kg/s at 80°C temperature with specific enthalpy of 320.5 kJ/kg. Water is heated by mixing steam at temperature 100.5°C and specific enthalpy of 2650 kJ/kg. The mixture of water and steam leaves the heater in the form of liquid water at temperature 100°C with specific enthalpy of 421 kJ/kg. Calculate the required steam flow rate to the heater per hour.

4.60 Steam with a quality of 0.7, pressure of 1.5 bar, and flow rate of 10 kg/s enters a steam separator operating at steady state. Saturated vapor at 1.5 bar exits the separator at state 2 at a rate of 6.9 kg/s while saturated liquid at 1.5 bar exits the separator at state 3. Neglecting kinetic and potential energy effects, determine the rate of heat transfer, in kW, and its associated direction.

4.61 Ammonia enters the expansion valve of a refrigeration system at a pressure of 1.4 MPa and a temperature of 32°C and exits at 0.08 MPa. If the refrigerant undergoes a throttling process, what is the quality of the refrigerant exiting the expansion valve?

4.62 **C** As shown in **Fig. P4.62**, 15 kg/s of steam enters a *desuperheater* operating at steady state at 30 bar, 320°C, where it is mixed with liquid water at 25 bar and temperature T_2 to produce saturated vapor at 20 bar. Heat transfer between the device and its surroundings and kinetic and potential energy effects can be neglected.

    a. If $T_2 = 200°C$, determine the mass flow rate of liquid, $\dot{m}_2$, in kg/s.

    b. Plot $\dot{m}_2$, in kg/s, versus T_2 ranging from 20 to 220°C.

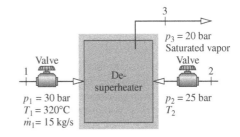

3 $p_3 = 20$ bar Saturated vapor

Valve

De-superheater

Valve 2 $p_2 = 25$ bar T_2

1 $p_1 = 30$ bar $T_1 = 320°C$ $\dot{m}_1 = 15$ kg/s

FIGURE P4.62

4.63 As shown in Fig. P4.63, electronic components mounted on a flat plate are cooled by convection to the surroundings and by liquid water circulating through a U-tube bonded to the plate. At steady state, water enters the tube at 20°C and a velocity of 0.4 m/s and exits at 24°C with a negligible change in pressure. The electrical components receive 0.5 kW of electrical power. The rate of energy transfer by convection from the plate-mounted electronics is estimated to be 0.08 kW. Kinetic and potential energy effects can be ignored. Determine the tube diameter, in cm.

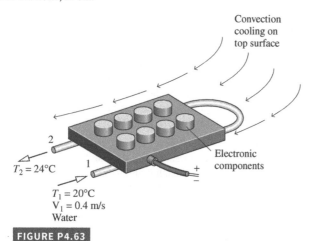

FIGURE P4.63

4.64 Liquid water enters a valve at 300 kPa and exits at 275 kPa. As water flows through the valve, the change in its temperature, stray heat transfer with the surroundings, and potential energy effects are negligible. Operation is at steady state. Modeling the water as incompressible with constant $\rho = 1000$ kg/m³, determine the change in kinetic energy per unit mass of water flowing through the valve, in kJ/kg.

4.65 Figure P4.65 provides steady-state data for a throttling valve in series with a heat exchanger. Saturated liquid Refrigerant 134a enters the valve at a pressure of 9 bar and is throttled to a pressure of 2 bar. The refrigerant then enters the heat exchanger, exiting at a temperature of 10°C with no significant decrease in pressure. In a separate stream, liquid water at 1 bar enters the heat exchanger at a temperature of 25°C with a mass flow rate of 2 kg/s and exits at 1 bar as liquid at a temperature of 15°C. Stray heat transfer and kinetic and potential energy effects can be ignored. Determine

a. the temperature, in °C, of the refrigerant at the exit of the valve.

b. the mass flow rate of the refrigerant, in kg/s.

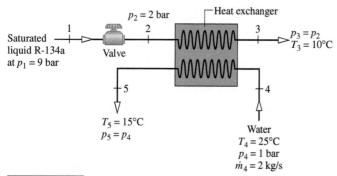

FIGURE P4.65

4.66 Steam enters a turbine in a vapor power plant operating at steady state at 560°C, 80 bar, and exits as a saturated vapor at 8 kPa. The turbine operates adiabatically, and the power developed is 9.43 kW. The steam leaving the turbine enters a condenser heat exchanger, where it is condensed to saturated liquid at 8 kPa through heat transfer

to cooling water passing through the condenser as a separate stream. The cooling water enters at 18°C and exits at 36°C with negligible change in pressure. Ignoring kinetic and potential energy effects and stray heat transfer at the outer surface of the condenser, determine the mass flow rate of cooling water required, in kg/s.

4.67 A large pipe carries steam as a two-phase liquid–vapor mixture at 1.0 MPa. A small quantity is withdrawn through a throttling calorimeter, where it undergoes a throttling process to an exit pressure of 0.1 MPa. For what range of exit temperatures, in °C, can the calorimeter be used to determine the quality of the steam in the pipe? What is the corresponding range of steam quality values?

4.68 As shown in Fig. P4.68, Refrigerant 22 enters the compressor of an air-conditioning unit operating at steady state at 4°C, 5.5 bar and is compressed to 60°C, 14 bar. The refrigerant exiting the compressor enters a condenser where energy transfer to air as a separate stream occurs and the refrigerant exits as a liquid at 14 bar, 32°C. Air enters the condenser at 27°C, 1 bar with a volumetric flow rate of 20.25 m³/min and exits at 43°C. Neglecting stray heat transfer and kinetic and potential energy effects, and assuming ideal gas behavior for the air, determine (a) the mass flow rate of refrigerant, in kg/min, and (b) the compressor power, in kilowatt.

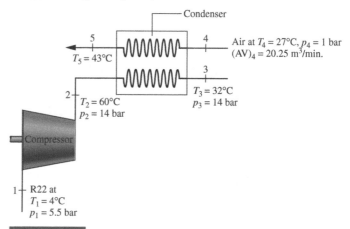

FIGURE P4.68

4.69 Separate streams of air and water flow through the compressor and heat exchanger arrangement shown in Fig. P4.69. Steady-state

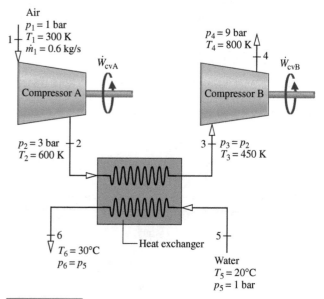

FIGURE P4.69

operating data are provided on the figure. Heat transfer with the surroundings can be neglected, as can all kinetic and potential energy effects. The air is modeled as an ideal gas. Determine

    a. the total power required by both compressors, in kW.

    b. the mass flow rate of the water, in kg/s.

4.70 **C** Refrigerant 134a enters the flash chamber operating at steady state shown in **Fig. P4.70** at 10 bar, 36°C, with a mass flow rate of 482 kg/h. Saturated liquid and saturated vapor exit as separate streams, each at pressure p. Heat transfer to the surroundings and kinetic and potential energy effects can be ignored.

    a. Determine the mass flow rates of the exiting streams, each in kg/h, if $p = 4$ bar.

    b. Plot the mass flow rates of the exiting streams, each in kg/h, versus p ranging from 1 to 9 bar.

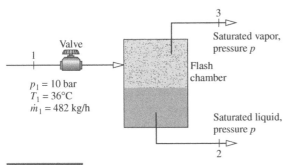

FIGURE P4.70

4.71 Carbon dioxide (CO_2) modeled as an ideal gas flows through the compressor and heat exchanger shown in **Fig. P4.71**. The power input to the compressor is 100 kW. A separate liquid cooling water stream flows through the heat exchanger. All data are for operation at steady state. Stray heat transfer with the surroundings can be neglected, as can all kinetic and potential energy changes. Determine (a) the mass flow rate of the CO_2, in kg/s, and (b) the mass flow rate of the cooling water, in kg/s.

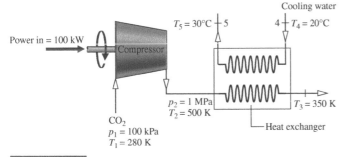

FIGURE P4.71

4.72 **Figure P4.72** provides steady-state operating data for a *cogeneration system* with water vapor at 20 bar, 360°C entering at location 1. Power is developed by the system at a rate of 2.2 MW. Process steam leaves at location 2, and hot water for other process uses leaves at location 3. Evaluate the rate of heat transfer, in MW, between the system and its surroundings. Let $g = 9.81$ m/s^2.

4.73 A simple steam power plant operates at steady state with water circulating through the components with a mass flow rate of 60 kg/s. **Figure P4.73** shows additional data at key points in the cycle. Stray heat transfer and kinetic and potential effects are negligible. Determine

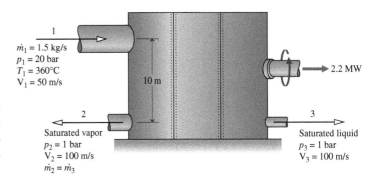

FIGURE P4.72

(a) the thermal efficiency and (b) the mass flow rate of cooling water through the condenser, in kg/s.

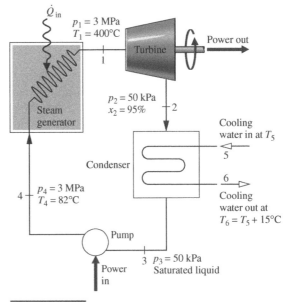

FIGURE P4.73

4.74 A residential air-conditioning system operates at steady state, as shown in **Fig. P4.74**. Refrigerant 22 circulates through the components of the system. Property data at key locations are given on the figure. If the evaporator removes energy by heat transfer from the room air at a rate of 10 kW, determine (a) the rate of heat transfer between the compressor and the surroundings, in kW, and (b) the coefficient of performance.

Transient Analysis

4.75 A rigid tank of volume 0.75 m^3 is initially evacuated. A hole develops in the wall, and air from the surroundings at 1 bar, 25°C flows in until the pressure in the tank reaches 1 bar. Heat transfer between the contents of the tank and the surroundings is negligible. Determine the final temperature in the tank, in °C.

4.76 A rigid, well-insulated tank of volume 0.5 m^3 is initially evacuated. At time $t = 0$, air from the surroundings at 1 bar, 21°C begins to flow into the tank. An electric resistor transfers energy to the air in the tank at a constant rate of 100 W for 500 s, after which time the pressure in the tank is 1 bar. What is the temperature of the air in the tank, in °C, at the final time?

4.77 A rigid tank whose volume is 2 m^3, initially containing air at 1 bar, 295 K, is connected by a valve to a large vessel holding air at 6 bar,

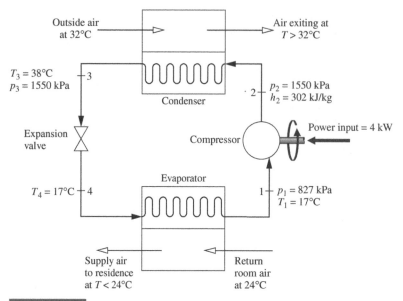

295 K. The valve is opened only as long as required to fill the tank with air to a pressure of 6 bar and a temperature of 350 K. Assuming the ideal gas model for the air, determine the heat transfer between the tank contents and the surroundings, in kJ.

4.78 An insulated, rigid tank whose volume is 0.5 m³ is connected by a valve to a large vessel holding steam at 40 bar, 500°C. The tank is initially evacuated. The valve is opened only as long as required to fill the tank with steam to a pressure of 20 bar. Determine the final temperature of the steam in the tank, in °C, and the final mass of the steam in the tank, in kg.

4.79 **C** The rigid tank illustrated in Fig. P4.79 has a volume of 0.06 m³ and initially contains a two-phase liquid–vapor mixture of H_2O at a pressure of 15 bar and a quality of 20%. As the tank contents are heated, a pressure-regulating valve keeps the pressure constant in the tank by allowing saturated vapor to escape. Neglecting kinetic and potential energy effects

    a. determine the total mass in the tank, in kg, and the amount of heat transfer, in kJ, if heating continues until the final quality is $x = 0.5$.

    b. plot the total mass in the tank, in kg, and the amount of heat transfer, in kJ, versus the final quality x ranging from 0.2 to 1.0.

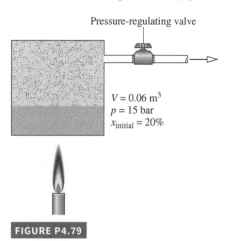

Pressure-regulating valve

$V = 0.06$ m³
$p = 15$ bar
$x_{initial} = 20\%$

4.80 A rigid tank whose volume is 0.5 m³, initially containing ammonia at 20°C, 1.5 bar, is connected by a valve to a large supply line carrying

ammonia at 12 bar, 60°C. The valve is opened only as long as required to fill the tank with additional ammonia, bringing the total mass of ammonia in the tank to 143.36 kg. Finally, the tank holds a two-phase liquid–vapor mixture at 20°C. Determine the heat transfer between the tank contents and the surroundings, in kJ, ignoring kinetic and potential energy effects.

4.81 As shown in **Fig. P4.81**, a 8 m³ tank contains H_2O initially at 207 kPa and a quality of 80%. The tank is connected to a large steam line carrying steam at 1380 kPa, 232°C. Steam flows into the tank through a valve until the tank pressure reaches 690 kPa and the temperature is 204°C, at which time the valve is closed. Determine the amount of mass, in kg, that enters the tank and the heat transfer between the tank and its surroundings, in kJ.

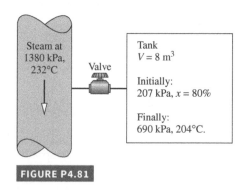

Steam at
1380 kPa,
232°C

Valve

Tank
$V = 8$ m³

Initially:
207 kPa, $x = 80\%$

Finally:
690 kPa, 204°C.

4.82 A two-phase liquid–vapor mixture of Refrigerant 134a is contained in a 0.06 m³, cylindrical storage tank at 670 kPa. Initially, saturated liquid occupies 0.04 m³. The valve at the top of the tank develops a leak, allowing saturated vapor to escape slowly. Eventually, the volume of the liquid drops to 0.02 m³. If the pressure in the tank remains constant, determine the mass of refrigerant that has escaped, in kg, and the heat transfer, in kJ.

4.83 A rigid, well-insulated tank of volume 0.9 m³ is initially evacuated. At time $t = 0$, air from the surroundings at 1 bar, 27°C begins to flow into the tank. An electric resistor transfers energy to the air in the tank at a constant rate for 5 minutes, after which time the pressure in the tank is 1 bar and the temperature is 457°C. Modeling air as an ideal gas, determine the power input to the tank, in kW.

4.84 A rigid tank having a volume of 0.1 m³ initially contains water as a two-phase liquid–vapor mixture at 1 bar and a quality of 1%. The water is heated in two stages:

Stage 1: Constant-volume heating until the pressure is 20 bar.
Stage 2: Continued heating while saturated water vapor is slowly withdrawn from the tank at a constant pressure of 20 bar. Heating ceases when all the water remaining in the tank is saturated vapor at 20 bar.

For the water, evaluate the heat transfer, in kJ, for each stage of heating. Ignore kinetic and potential energy effects.

4.85 A well-insulated rigid tank of volume 0.14 m³ contains oxygen (O_2), initially at 275 kPa and 30°C. The tank is connected to a large supply line carrying oxygen at 690 kPa 38°C. A valve between the line and the tank is opened and gas flows into the tank until the pressure reaches 690 kPa, at which time the valve closes. The contents of the tank eventually cool back to 30°C. Determine

    a. the temperature in the tank at the time when the valve closes, in °C.

    b. the final pressure in the tank, in kPa.

4.86 A well-insulated chamber of volume 0.028 m³ is shown in Fig. P4.86. Initially, the chamber contains air at 101.4 kPa and 38°C. Connected to the chamber are supply and discharge pipes equipped with valves that control the flow rates into and out of the chamber. The supply air is at 207 kPa, 93°C. Both valves are opened simultaneously, allowing air to flow with a mass flow rate $\dot{m}$ through each valve. The air within the chamber is well mixed, so the temperature and pressure at any time can be taken as uniform throughout. Neglecting kinetic and potential energy effects, and using the ideal gas model with constant specific heats for the air, plot the temperature, in °C, and the pressure, in kPa, of the air in the chamber versus time for $\dot{m}$ = 0.01, 0.02, and 0.04 kg/s.

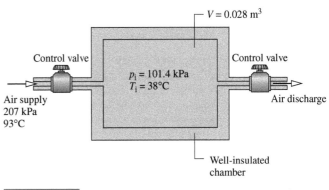

FIGURE P4.86

DESIGN & OPEN-ENDED PROBLEMS: EXPLORING ENGINEERING PRACTICE

4.1D Using the Internet, identify at least five medical applications of *MEMS* technology. In each case, explain the scientific and technological basis for the application, discuss the state of current research, and determine how close the technology is in terms of commercialization. Write a report of your findings, including at least three references.

4.2D Conduct a term-length project centered on using a *low-wind* turbine to meet the electricity needs of a small business, farm, or neighborhood selected by, or assigned to, your project group. Take several days to research the project and then prepare a brief written plan having a statement of purpose, a list of objectives, and several references. As part of your plan, schedule on-site wind-speed measurements for at least three different days to achieve a good match between the requirements of candidate low-wind turbines and local conditions. Your plan also should recognize the need for compliance with applicable zoning codes. During the project, observe good practices such as discussed in Sec. 1.3 of *Thermal Design and Optimization,* John Wiley & Sons Inc., New York, 1996, by A. Bejan, G. Tsatsaronis, and M.J. Moran. Provide a well-documented report, including an assessment of the economic viability of the selected turbine for the application considered.

4.3D Identify sites in your state where wind turbines for utility-scale electrical generation are feasible but do not yet exist. Prepare a memorandum to an appropriate governing or corporate entity with your recommendations as to whether wind-turbine electrical generation should be developed at the most promising sites. Consider engineering, economic, and societal aspects.

4.4D Generating power by harnessing ocean tides and waves is being studied across the globe. Underwater turbines develop power from tidal *currents*. Wave-power devices develop power from the *undulating motion* of ocean waves. Although tides and waves long have been used to meet modest power generation needs, investigators today are aiming at large-scale power generation systems. Some see the oceans as providing a potentially unlimited renewable source of power. Critically evaluate the viability of tidal and wave power, considering both technical and economic issues. Write a report, including at least three references.

4.5D Owing to their relatively compact size, simple construction, and modest power requirement, centrifugal-type blood pumps are under consideration for several medical applications. Still, centrifugal pumps have met with limited success thus far for blood flow because they can cause damage to blood cells and are subject to mechanical failure. The goal of current development efforts is a device having sufficient long-term biocompatibility, performance, and reliability for widespread deployment. Investigate the status of centrifugal blood pump development, including identifying key technical challenges and prospects for overcoming them. Summarize your findings in a report, including at least three references.

4.6D Design an experiment to determine the energy, in kW-h, required to completely evaporate a fixed quantity of water. For the experiment develop written procedures that include identification of all equipment needed and specification of all required calculations. Conduct the experiment, and communicate your results in an executive summary.

4.7D Prepare a memorandum providing guidelines for selecting fans for cooling electronic components. Consider the advantages and disadvantages of locating the fan at the inlet of the enclosure versus at the exit. Consider the relative merits of alternative fan types and of fixed versus variable-speed fans. Explain how *characteristic curves* assist in fan selection.

4.8D The technical literature contains discussions of ways for using tethered kite-mounted wind turbine systems to harvest power from high-altitude winds, including jet streams at elevations from 6 to 15 kilometers (4 to 9 miles). Analysts estimate that if such systems were deployed in sufficient numbers, they could meet a significant share of total U.S. demand for electricity. Critically evaluate the feasibility of such a kite system, selected from the existing literature, to be fully operational by 2025. Consider means for deploying the system to the proper altitude, how the power developed is transferred to earth, infrastructure requirements, environmental impact, cost, and other pertinent issues. Write a report including at least three references.

4.9D *Reverse engineer* a handheld hair dryer by disassembling the dryer into its individual parts. Mount each part onto a presentation board to illustrate how the parts are connected when assembled. Label each part with its name. Next to each part identify its purpose and describe its fundamental operating principle (if applicable). Include a visual trace of the mass and energy flows through the hair dryer during operation. Display the presentation board where others can learn from it.

4.10D Figure P4.10D provides the schematic of a device for producing a combustible fuel gas for transportation from biomass. While several types of solid biomass can be employed in current gasifier designs, wood chips are commonly used. Wood chips are introduced at the top of the gasifier unit. Just below this level, the chips react with oxygen in the combustion air to produce charcoal. At the next depth, the charcoal reacts with hot combustion gases from the charcoal-formation stage to produce a fuel gas consisting mainly of hydrogen, carbon monoxide, and nitrogen from the combustion air. The fuel gas is then cooled, filtered, and ducted to the internal combustion engine served by the gasifier. Critically evaluate the suitability of this technology for transportation use today in the event of a prolonged petroleum shortage in your locale. Document your conclusions in a memorandum.

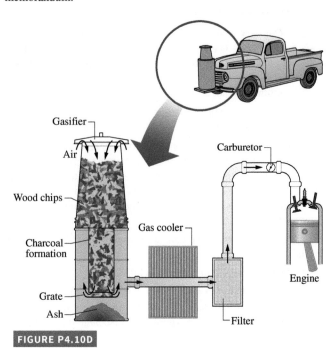

FIGURE P4.10D

CHAPTER 5

The Second Law of Thermodynamics

Engineering Context

The presentation to this point has considered thermodynamic analysis using the conservation of mass and conservation of energy principles together with property relations. In Chaps. 2 through 4 these fundamentals are applied to increasingly complex situations. The conservation principles do not always suffice, however, and the second law of thermodynamics is also often required for thermodynamic analysis. The **objective** of this chapter is to introduce the second law of thermodynamics. A number of deductions that may be called corollaries of the second law are also considered, including performance limits for thermodynamic cycles. The current presentation provides the basis for subsequent developments involving the second law in Chaps. 6 and 7.

LEARNING OUTCOMES

When you complete your study of this chapter, you will be able to...

- Explain key concepts related to the second law of thermodynamics, including alternative statements of the second law, the internally reversible process, and the Kelvin temperature scale.

- List several important irreversibilities.

- Evaluate the performance of power cycles and refrigeration and heat pump cycles using, as appropriate, the corollaries of Secs. 5.6.2 and 5.7.2, together with Eqs. 5.9–5.11.

- Describe the Carnot cycle.

- Apply the Clausius inequality as expressed by Eq. 5.13.

5.1 Introducing the Second Law

The objectives of the present section are to

1. motivate the need for and the usefulness of the second law.
2. introduce statements of the second law that serve as the point of departure for its application.

5.1.1 Motivating the Second Law

It is a matter of everyday experience that there is a definite direction for *spontaneous* processes. This can be brought out by considering the three systems pictured in **Fig. 5.1**.

- System a. An object at an elevated temperature T_i placed in contact with atmospheric air at temperature T_0 eventually cools to the temperature of its much larger surroundings, as illustrated in Fig. 5.1*a*. In conformity with the conservation of energy principle, the decrease in internal energy of the body appears as an increase in the internal energy of the surroundings. The *inverse* process would not take place *spontaneously*, even though energy could be conserved: The internal energy of the surroundings would not decrease spontaneously while the body warmed from T_0 to its initial temperature.

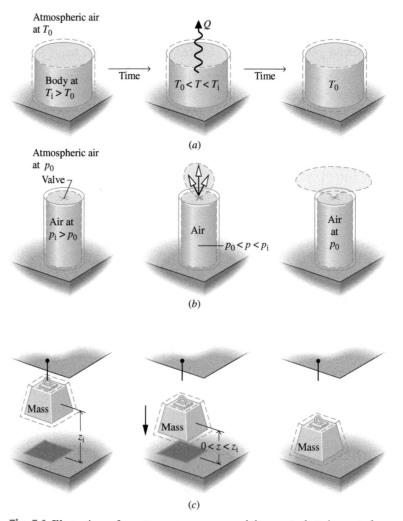

Fig. 5.1 Illustrations of spontaneous processes and the eventual attainment of equilibrium with the surroundings. (*a*) Spontaneous heat transfer. (*b*) Spontaneous expansion. (*c*) Falling mass.

- System b. Air held at a high pressure p_i in a closed tank flows spontaneously to the lower-pressure surroundings at p_0 when the interconnecting valve is opened, as illustrated in Fig. 5.1b. Eventually fluid motions cease and all of the air is at the same pressure as the surroundings. Drawing on experience, it should be clear that the *inverse* process would not take place *spontaneously*, even though energy could be conserved: Air would not flow spontaneously from the surroundings at p_0 into the tank, returning the pressure to its initial value.

- System c. A mass suspended by a cable at elevation z_i falls when released, as illustrated in Fig. 5.1c. When it comes to rest, the potential energy of the mass in its initial condition appears as an increase in the internal energy of the mass and its surroundings, in accordance with the conservation of energy principle. Eventually, the mass also comes to the temperature of its much larger surroundings. The *inverse* process would not take place *spontaneously*, even though energy could be conserved: The mass would not return spontaneously to its initial elevation while its internal energy and/or that of its surroundings decreases.

In each case considered, the initial condition of the system can be restored but not in a spontaneous process. Some auxiliary devices would be required. By such auxiliary means the object could be reheated to its initial temperature, the air could be returned to the tank and restored to its initial pressure, and the mass could be lifted to its initial height. Also in each case, a fuel or electrical input normally would be required for the auxiliary devices to function, so a permanent change in the condition of the surroundings would result.

Further Conclusions   The foregoing discussion indicates that not every process consistent with the principle of energy conservation can occur. Generally, an energy balance alone neither enables the preferred direction to be predicted nor permits the processes that can occur to be distinguished from those that cannot. In elementary cases, such as the ones considered in Fig. 5.1, experience can be drawn upon to deduce whether particular spontaneous processes occur and to deduce their directions. For more complex cases, where experience is lacking or uncertain, a guiding principle is necessary. This is provided by the *second law*.

The foregoing discussion also indicates that when left alone systems tend to undergo spontaneous changes until a condition of equilibrium is achieved, both internally and with their surroundings. In some cases equilibrium is reached quickly, whereas in others it is achieved slowly. For example, some chemical reactions reach equilibrium in fractions of seconds; an ice cube requires a few minutes to melt; and it may take years for an iron bar to rust away. Whether the process is rapid or slow, it must of course satisfy conservation of energy. However, that alone would be insufficient for determining the final equilibrium state. Another general principle is required. This is provided by the *second law*.

 BioConnections

Did you ever wonder why a banana placed in a closed bag or in the refrigerator quickly ripens? The answer is in the ethylene, C_2H_4, naturally produced by bananas, tomatoes, and other fruits and vegetables. Ethylene is a plant hormone that affects growth and development. When a banana is placed in a closed container, ethylene accumulates and stimulates the production of more ethylene. This positive feedback results in more and more ethylene, hastening ripening, aging, and eventually spoilage. In thermodynamic terms, if left alone, the banana tends to undergo spontaneous changes until equilibrium is achieved. Growers have learned to use this natural process to their advantage. Tomatoes picked while still green and shipped to distant markets may be red by the time they arrive; if not, they can be induced to ripen by means of an ethylene spray.

5.1.2 Opportunities for Developing Work

By exploiting the spontaneous processes shown in Fig. 5.1, it is possible, in principle, for work to be developed as equilibrium is attained.

Instead of permitting the body of Fig. 5.1*a* to cool spontaneously with no other result, energy could be delivered by heat transfer to a system undergoing a power cycle that would develop a net amount of work (Sec. 2.6). Once the object attained equilibrium with the surroundings, the process would cease. Although there is an *opportunity* for developing work in this case, the opportunity would be wasted if the body were permitted to cool without developing any work. In the case of Fig. 5.1*b*, instead of permitting the air to expand aimlessly into the lower-pressure surroundings, the stream could be passed through a turbine and work could be developed. Accordingly, in this case there is also a possibility for developing work that would not be exploited in an uncontrolled process. In the case of Fig. 5.1*c*, instead of permitting the mass to fall in an uncontrolled way, it could be lowered gradually while turning a wheel, lifting another mass, and so on.

These considerations can be summarized by noting that when an imbalance exists between two systems, there is an opportunity for developing work that would be irrevocably lost if the systems were allowed to come into equilibrium in an uncontrolled way. Recognizing this possibility for work, we can pose two questions:

1. What is the theoretical maximum value for the work that could be obtained?
2. What are the factors that would preclude the realization of the maximum value?

That there should be a maximum value is fully in accord with experience, for if it were possible to develop unlimited work, few concerns would be voiced over our dwindling fossil fuel supplies. Also in accord with experience is the idea that even the best devices would be subject to factors such as friction that would preclude the attainment of the theoretical maximum work. The second law of thermodynamics provides the means for determining the theoretical maximum and evaluating quantitatively the factors that preclude attaining the maximum.

5.1.3 Aspects of the Second Law

We conclude our introduction to the second law by observing that the second law and deductions from it have many important uses, including means for

1. predicting the direction of processes.
2. establishing conditions for equilibrium.
3. determining the best *theoretical* performance of cycles, engines, and other devices.
4. evaluating quantitatively the factors that preclude the attainment of the best theoretical performance level.

Other uses of the second law include:

5. defining a temperature scale independent of the properties of any thermometric substance.
6. developing means for evaluating properties such as *u* and *h* in terms of properties that are more readily obtained experimentally.

Scientists and engineers have found additional uses of the second law and deductions from it. It also has been used in philosophy, economics, and other disciplines far removed from engineering thermodynamics.

TAKE NOTE...

No single statement of the second law brings out each of its many aspects.

The six points listed can be thought of as aspects of the second law of thermodynamics and not as independent and unrelated ideas. Nonetheless, given the variety of these topic areas, it is easy to understand why there is no single statement of the second law that brings out each one clearly. There are several alternative, yet equivalent, formulations of the second law.

In the next section, three statements of the second law are introduced as *points of departure* for our study of the second law and its consequences. Although the exact relationship of these particular formulations to each of the second law aspects listed above may not be immediately apparent, all aspects listed can be obtained by deduction from these formulations or their corollaries. It is important to add that in every instance where a consequence of the second law has been tested directly or indirectly by experiment, it has been unfailingly verified. Accordingly, the basis of the second law of thermodynamics, like every other physical law, is experimental evidence.

5.2 Statements of the Second Law

Three alternative statements of the second law of thermodynamics are given in this section. They are the (1) Clausius, (2) Kelvin–Planck, and (3) entropy statements. The Clausius and Kelvin–Planck statements are traditional formulations of the second law. You have likely encountered them before in an introductory physics course.

Although the Clausius statement is more in accord with experience and thus easier to accept, the Kelvin–Planck statement provides a more effective means for bringing out second law deductions related to thermodynamic cycles that are the focus of the current chapter. The Kelvin–Planck statement also underlies the entropy statement, which is the most effective form of the second law for an extremely wide range of engineering applications. The entropy statement is the focus of Chap. 6.

5.2.1 Clausius Statement of the Second Law

The **Clausius statement** of the second law asserts that:

It is impossible for any system to operate in such a way that the sole result would be an energy transfer by heat from a cooler to a hotter body.

The Clausius statement does not rule out the possibility of transferring energy by heat from a cooler body to a hotter body, for this is exactly what refrigerators and heat pumps accomplish. However, as the words "sole result" in the statement suggest, when a heat transfer from a cooler body to a hotter body occurs, there must be *other effects* within the system accomplishing the heat transfer, its surroundings, or both. If the system operates in a thermodynamic cycle, its initial state is restored after each cycle, so the only place that must be examined for such *other* effects is its surroundings.

Clausius statement

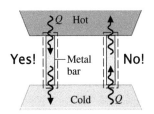

> **FOR EXAMPLE**
>
> Cooling of food is most commonly accomplished by refrigerators driven by electric motors requiring power from their surroundings to operate. The Clausius statement implies it is impossible to construct a refrigeration cycle that operates without a power input.

5.2.2 Kelvin–Planck Statement of the Second Law

Before giving the Kelvin–Planck statement of the second law, the concept of a **thermal reservoir** is introduced. A thermal reservoir, or simply a reservoir, is a special kind of system that always remains at constant temperature even though energy is added or removed by heat transfer. A reservoir is an idealization, of course, but such a system can be approximated in a number of ways—by Earth's atmosphere, large bodies of water (lakes, oceans), a large block of copper, and a system consisting of two phases at a specified pressure (while the ratio of the masses of the two phases changes as the system is heated or cooled at constant pressure, the temperature remains constant as long as both phases coexist). Extensive properties of a thermal reservoir such as internal energy can change in interactions with other systems even though the reservoir temperature remains constant.

Having introduced the thermal reservoir concept, we give the **Kelvin–Planck statement** of the second law:

It is impossible for any system to operate in a thermodynamic cycle and deliver a net amount of energy by work to its surroundings while receiving energy by heat transfer from a single thermal reservoir.

thermal reservoir

Kelvin–Planck statement

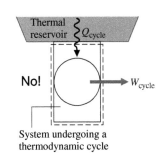

System undergoing a thermodynamic cycle

The Kelvin–Planck statement does not rule out the possibility of a system developing a net amount of work from a heat transfer drawn from a single reservoir. It only denies this possibility if the system undergoes a thermodynamic cycle.

The Kelvin–Planck statement can be expressed analytically. To develop this, let us study a system undergoing a cycle while exchanging energy by heat transfer with a *single* reservoir, as shown by the adjacent figure. The first and second laws each impose constraints:

- A constraint is imposed by the first law on the net work and heat transfer between the system and its surroundings. According to the cycle energy balance (see Eq. 2.40 in Sec. 2.6),

$$W_{cycle} = Q_{cycle}$$

In words, the net work done by (or on) the system undergoing a cycle equals the net heat transfer to (or from) the system. Although the cycle energy balance allows the net work W_{cycle} to be positive *or* negative, the second law imposes a constraint, as considered next.

- According to the Kelvin–Planck statement, a system undergoing a cycle while communicating thermally with a single reservoir *cannot* deliver a net amount of work to its surroundings: The net work of the cycle *cannot be positive*. However, the Kelvin–Planck statement does not rule out the possibility that there is a net work transfer of energy *to* the system during the cycle *or* that the net work is zero. Thus, the **analytical form of the Kelvin–Planck statement** is

analytical form of the Kelvin–Planck statement

$$\boxed{W_{cycle} \leq 0 \quad \text{(single reservoir)}} \qquad (5.1)$$

where the words *single reservoir* are added to emphasize that the system communicates thermally only with a single reservoir as it executes the cycle. In Sec. 5.4, we associate the "less than" and "equal to" signs of Eq. 5.1 with the presence and absence of *internal irreversibilities*, respectively. The concept of irreversibilities is considered in Sec. 5.3.

The equivalence of the Clausius and Kelvin–Planck statements can be demonstrated by showing that the violation of each statement implies the violation of the other. For details, see the box.

Demonstrating the Equivalence of the Clausius and Kelvin–Planck Statements

The equivalence of the Clausius and Kelvin–Planck statements is demonstrated by showing that the violation of each statement implies the violation of the other. That a violation of the Clausius statement implies a violation of the Kelvin–Planck statement is readily shown using Fig. 5.2, which pictures a hot reservoir, a cold reservoir, and two systems. The system on the left transfers energy Q_C from the cold reservoir to the hot reservoir by heat transfer without other effects occurring and thus *violates the Clausius statement*. The system on the right operates in a cycle while receiving Q_H (greater than Q_C) from the hot reservoir, rejecting Q_C to the cold reservoir, and delivering work W_{cycle} to the surroundings. The energy flows labeled on Fig. 5.2 are in the directions indicated by the arrows.

Consider the *combined* system shown by a dotted line on Fig. 5.2, which consists of the cold reservoir and the two devices. The combined system can be regarded as executing a cycle because one part undergoes a cycle and the other two parts experience no net change in their conditions. Moreover, the combined system receives energy $(Q_H - Q_C)$ by heat transfer from a single reservoir, the hot reservoir, and produces an equivalent amount of work. Accordingly, the combined system violates the Kelvin–Planck statement. Thus, a

violation of the Clausius statement implies a violation of the Kelvin–Planck statement. The equivalence of the two second law statements is demonstrated completely when it is also shown that a violation of the Kelvin–Planck statement implies a violation of the Clausius statement. This is left as an exercise (see end-of-chapter Prob. 5.1).

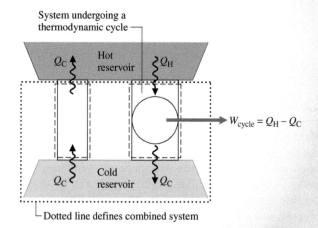

Fig. 5.2 Illustration used to demonstrate the equivalence of the Clausius and Kelvin–Planck statements of the second law.

5.2.3 Entropy Statement of the Second Law

Mass and energy are familiar examples of extensive properties of systems. Entropy is another important extensive property. We show how entropy is evaluated and applied for engineering analysis in Chap. 6. Here we introduce several important aspects.

Just as mass and energy are *accounted for* by mass and energy balances, respectively, entropy is accounted for by an *entropy balance*. In words, the entropy balance states

$$
\begin{bmatrix}
\textit{change} \text{ in the amount} \\
\text{of entropy contained} \\
\text{within the system} \\
\text{during some time} \\
\text{interval}
\end{bmatrix}
=
\begin{bmatrix}
\text{net amount of} \\
\text{entropy } \textit{transferred} \\
\textit{in} \text{ across the system} \\
\text{boundary during the} \\
\text{time interval}
\end{bmatrix}
+
\begin{bmatrix}
\text{amount of entropy} \\
\textit{produced within} \text{ the} \\
\text{system during the} \\
\text{time interval}
\end{bmatrix}
\quad (5.2)
$$

Like mass and energy, *entropy can be transferred* across the system boundary. For closed systems, there is a single means of entropy transfer—namely, entropy transfer accompanying heat transfer. For control volumes entropy also is transferred in and out by streams of matter. These entropy transfers are considered further in Chap. 6.

Unlike mass and energy, which are conserved, *entropy is produced* (or *generated*) within systems whenever *nonidealities* (called *irreversibilities*) such as friction are present. The **entropy statement of the second law** states:

> *It is impossible for any system to operate in a way that entropy is destroyed.*

It follows that the entropy production term of Eq. 5.2 may be positive or zero but *never* negative. Thus, entropy production is an indicator of whether a process is possible or impossible.

entropy statement
of the second law

5.2.4 Second Law Summary

In the remainder of this chapter, we apply the Kelvin–Planck statement of the second law to draw conclusions about systems undergoing thermodynamic cycles. The chapter concludes with a discussion of the *Clausius inequality* (Sec. 5.11), which provides the basis for developing the entropy concept in Chap. 6. This is a traditional approach to the second law in engineering thermodynamics. However, the order can be reversed—namely, the entropy statement can be adopted as the starting point for study of the second law aspects of systems. The box to follow provides such an alternative second law pathway for instructors and for self-study by students:

Alternative Second Law Pathway

- Survey Sec. 5.3, omitting Sec. 5.3.2.
- Survey the discussion in Sec. 6.7 through Sec. 6.7.2 of the closed system entropy balance Eq. 6.24. Omit the box following Eq. 6.25.
- Survey Sec. 6.1, beginning with Eq. 6.2a. Note: The entropy data required to apply the entropy balance are obtained in principle using Eq. 6.2a, a special case of Eq. 6.24.
- Survey Secs. 6.2 through 6.5.
- Survey Sec. 6.6, omitting Sec. 6.6.2.

- Survey Secs. 6.7.3 and 6.7.4.
- Survey Secs. 6.9–6.10.
- Survey Secs. 6.11–6.12.

When taking the entropy balance as the preferred statement of the second law of thermodynamics, it is straightforward to derive the Kelvin–Planck statement as expressed in Sec. 5.4. See the student companion site for "Demonstrating the Equivalence of the Kelvin–Planck and Entropy Statements."

Finally, survey Secs. 5.5 through 5.10, Sec. 6.6.2, and Sec. 6.13. This content is required primarily for study of thermodynamic cycles in Chaps. 8 through 10.

5.3 Irreversible and Reversible Processes

One of the important uses of the second law of thermodynamics in engineering is to determine the best theoretical performance of systems. By comparing actual performance with the best theoretical performance, insights often can be gained into the potential for improvement. As might be surmised, the best performance is evaluated in terms of idealized processes. In this section such idealized processes are introduced and distinguished from actual processes that invariably involve *irreversibilities*.

5.3.1 Irreversible Processes

irreversible process

reversible processes

A process is called **irreversible** if the system and all parts of its surroundings cannot be exactly restored to their respective initial states after the process has occurred. A process is **reversible** if both the system and surroundings can be returned to their initial states. Irreversible processes are the subject of the present discussion. Reversible processes are considered again in Sec. 5.3.3.

A system that has undergone an irreversible process is not necessarily precluded from being restored to its initial state. However, were the system restored to its initial state, it would not be possible also to return the surroundings to the state they were in initially. As demonstrated in Sec. 5.3.2, the second law can be used to determine whether both the system and surroundings can be returned to their initial states after a process has occurred: The second law can be used to determine whether a given process is reversible or irreversible.

It might be apparent from the discussion of the Clausius statement of the second law that any process involving spontaneous heat transfer from a hotter body to a cooler body is irreversible. Otherwise, it would be possible to return this energy from the cooler body to the hotter body with no other effects within the two bodies or their surroundings. However, this possibility is denied by the Clausius statement.

Processes involving other kinds of spontaneous events, such as an unrestrained expansion of a gas or liquid, are also irreversible. Friction, electrical resistance, hysteresis, and inelastic deformation are examples of additional effects whose presence during a process renders it irreversible.

irreversibilities

In summary, irreversible processes normally include one or more of the following **irreversibilities:**

1. Heat transfer through a finite temperature difference
2. Unrestrained expansion of a gas or liquid to a lower pressure
3. Spontaneous chemical reaction
4. Spontaneous mixing of matter at different compositions or states
5. Friction—sliding friction as well as friction in the flow of fluids
6. Electric current flow through a resistance
7. Magnetization or polarization with hysteresis
8. Inelastic deformation

Although the foregoing list is not exhaustive, it does suggest that *all actual processes are irreversible.* That is, every process involves effects such as those listed, whether it is a naturally occurring process or one involving a device of our construction, from the simplest mechanism to the largest industrial plant. The term *irreversibility* is used to identify any of these effects. The above list comprises a few of the irreversibilities that are commonly encountered.

As a system undergoes a process, irreversibilities may be found within the system and its surroundings, although they may be located predominately in one place or the other. For many analyses it is convenient to divide the irreversibilities present into two classes. **Internal irreversibilities** are those that occur within the system. **External irreversibilities** are those that occur within the surroundings, often the immediate surroundings. As this distinction depends solely on the location of the boundary, there is some arbitrariness in the classification, for by extending the boundary to take in a portion of the surroundings, all irreversibilities become "internal." Nonetheless, as shown by subsequent developments, this distinction between irreversibilities is often useful.

internal and external irreversibilities

Engineers should be able to recognize irreversibilities, evaluate their influence, and develop practical means for reducing them. However, certain systems, such as brakes, rely on the effect of friction or other irreversibilities in their operation. The need to achieve profitable rates of production, high heat transfer rates, rapid accelerations, and so on invariably dictates the presence of significant irreversibilities.

Furthermore, irreversibilities are tolerated to some degree in every type of system because the changes in design and operation required to reduce them would be too costly.

Accordingly, although improved thermodynamic performance can accompany the reduction of irreversibilities, steps taken in this direction are constrained by a number of practical factors often related to costs.

> **FOR EXAMPLE**
>
> Consider two bodies at different temperatures that are able to communicate thermally. With a *finite* temperature difference between them, a spontaneous heat transfer would take place and, as discussed previously, this would be a source of irreversibility. It might be expected that the importance of this irreversibility diminishes as the temperature difference between the bodies diminishes, and while this *is* the case, there are practical consequences: From the study of heat transfer (Sec. 2.4), we know that the transfer of a specified amount of energy by heat transfer between bodies whose temperatures differ only slightly requires a considerable amount of time, a large (costly) heat transfer surface area, or both. In the limit as the temperature difference between the bodies vanishes, the amount of time and/or surface area required approach infinity. Such options are clearly impractical; still, they must be imagined when thinking of heat transfer approaching reversibility.

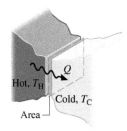

5.3.2 Demonstrating Irreversibility

Whenever an irreversibility is present during a process, that process must necessarily be irreversible. However, the irreversibility of a process can be *demonstrated* rigorously using the Kelvin–Planck statement of the second law and the following procedure: (1) Assume there is a way to return the system and surroundings to their respective initial states. (2) Show that as a consequence of this assumption, it is possible to devise a cycle that violates the Kelvin–Planck statement—namely, a cycle that produces work while interacting thermally with only a single reservoir. Since the existence of such a cycle is denied by the Kelvin–Planck statement, the assumption must be in error and it follows that the process is irreversible.

This procedure can be used to demonstrate that processes involving friction, heat transfer through a finite temperature difference, the unrestrained expansion of a gas or liquid to a lower pressure, and other effects from the list given previously are irreversible. A case involving friction is discussed in the box.

While use of the Kelvin–Planck statement to demonstrate irreversibility is part of a traditional presentation of thermodynamics, such demonstrations can be unwieldy. It is normally easier to use the *entropy production* concept (Sec. 6.7).

Demonstrating Irreversibility: Friction

Let us use the Kelvin–Planck statement to demonstrate the irreversibility of a process involving friction. Consider a system consisting of a block of mass m and an inclined plane. To begin, the block is at rest at the top of the incline. The block then slides down the plane, eventually coming to rest at a lower elevation. There is no significant work or heat transfer between the block–plane system and its surroundings during the process.

Applying the closed system energy balance to the system, we get

$$(U_f - U_i) + mg(z_f - z_i) + \cancelto{0}{(KE_f - KE_i)} = \cancelto{0}{Q} - \cancelto{0}{W}$$

or

$$U_f - U_i = mg(z_i - z_f) \qquad \text{(a)}$$

where U denotes the internal energy of the block–plane system and z is the elevation of the block. Thus, friction between the block and plane during the process acts to convert the potential energy decrease of the block to internal energy of the overall system.

Since no work or heat interactions occur between the block–plane system and its surroundings, the condition of the surroundings remains unchanged during the process. This allows attention to be centered on the system only in demonstrating that the process is irreversible, as follows:

When the block is at rest after sliding down the plane, its elevation is z_f and the internal energy of the block–plane system is U_f. To demonstrate that the process is irreversible using the Kelvin–Planck statement, let us take this condition of the system, shown in **Fig. 5.3a**, as the initial state of a cycle consisting of three processes. We imagine that a pulley–cable arrangement and a thermal reservoir are available to assist in the demonstration.

Process 1 Assume the inverse process occurs with no change in the surroundings: As shown in **Fig. 5.3b**, the block returns *spontaneously* to the top of the plane while the internal energy of the system decreases to its initial value, U_i. (This is the process we want to demonstrate is impossible.)

Process 2 As shown in Fig. 5.3c, we use the pulley–cable arrangement provided to lower the block from z_i to z_f, while allowing the block–plane system to do work by lifting another mass located in the surroundings. The work done equals the decrease in potential energy of the block. This is the only work for the cycle. Thus, $W_{cycle} = mg(z_i - z_f)$.

Process 3 The internal energy of the system is increased from U_i to U_f by bringing it into communication with the reservoir, as shown in Fig. 5.3d. The heat transfer equals $(U_f - U_i)$. This is the only heat transfer for the cycle. Thus, $Q_{cycle} = (U_f - U_i)$, which with Eq. (a) becomes $Q_{cycle} = mg(z_i - z_f)$. At the conclusion of this process the block is again at elevation z_f and the internal energy of the block–plane system is restored to U_f.

The net result of this cycle is to draw energy from a single reservoir by heat transfer, Q_{cycle}, and produce an equivalent amount of work, W_{cycle}. There are no other effects. However, such a cycle is denied by the Kelvin–Planck statement. Since both the heating of the system by the reservoir (Process 3) and the lowering of the mass by the pulley–cable while work is done (Process 2) are possible, we conclude it is Process 1 that is impossible. Since Process 1 is the inverse of the original process where the block slides down the plane, it follows that the original process is irreversible.

To summarize, the effect of friction in this case is an *irreversible* conversion to internal energy of potential energy, a form of *mechanical energy* (Sec. 2.1).

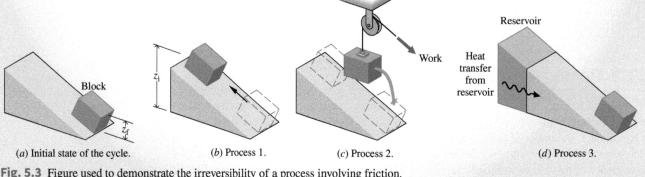

(a) Initial state of the cycle.   (b) Process 1.   (c) Process 2.   (d) Process 3.

Fig. 5.3 Figure used to demonstrate the irreversibility of a process involving friction.

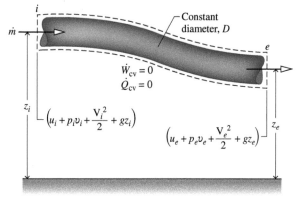

Pipe Friction Friction between solid surfaces is something everyone has experienced, and fluid friction is similar in its effects. Such friction plays an important role in gases expanding through turbines, liquids flowing through pumps and pipelines, and in a wide range of other applications.

To provide an introduction, we now build on the discussion of friction in the box above by considering a control volume at steady state enclosing a constant-diameter pipe carrying a liquid. For the control volume, $\dot{W}_{cv} = 0$ and heat transfer between the control volume and its surroundings is negligible. As before, the present case also exhibits an irreversible conversion of mechanical energy to internal energy owing to friction.

With the specified assumptions, the energy rate balance given by Eq. 4.13 reduces as follows

$$\frac{dE_{cv}}{dt} = \cancel{\dot{Q}_{cv}} - \cancel{\dot{W}_{cv}} + \dot{m}\left(u_i + p_i v_i + \frac{V_i^2}{2} + gz_i\right) - \dot{m}\left(u_e + p_e v_e + \frac{V_e^2}{2} + gz_e\right) \quad \text{(a)}$$

where $\dot{m}$ denotes the common mass flow rate at inlet i and exit e.

Canceling the mass flow rate, Eq. (a) can be rearranged to read

$$\underbrace{\left(p_i v_i + \frac{V_i^2}{2} + gz_i\right) - \left(p_e v_e + \frac{V_e^2}{2} + gz_e\right)}_{\text{decrease in mechanical energy}} = \underbrace{(u_e - u_i)}_{\text{increase in internal energy}} \quad \text{(b)}$$

Each term of Eq. (b) is on a per unit of mass basis. The pv terms account for the transfer of energy by work at i and e associated with the pressure of the flowing matter at these locations. This form of work is called *flow work* in Sec. 4.4.2. The kinetic and potential energy terms,

$V^2/2$ and gz, represent mechanical forms of energy associated with the flowing matter at i and e. For simplicity of expression, all three quantities are referred to here as *mechanical energy*. Terms denoted by u represent internal energy associated with the flowing matter at i and e.

Experience indicates that mechanical energy is more valuable thermodynamically than internal energy and the effect of friction as matter flows from inlet to exit is an *irreversible* conversion of mechanical energy to internal energy. These findings are aspects of the second law. Furthermore, as shown by Eq. (b), the decrease in mechanical energy is matched by an increase in internal energy and, thus, energy is conserved on an overall basis.

Assuming that the specific volume, v of the liquid remains constant, the mass rate balance requires that velocity, V, is also constant throughout the constant-diameter pipe. Thus, Eq. (b) becomes

$$\underbrace{v(p_i - p_e) + g(z_i - z_e)}_{\substack{\text{decrease in} \\ \text{mechanical energy}}} = \underbrace{(u_e - u_i)}_{\substack{\text{increase in} \\ \text{internal energy}}} \qquad \text{(c)}$$

Finally, for the simple pipe under consideration the role of friction is made explicit by expressing the decrease in mechanical energy in terms of the specific kinetic energy of the flowing substance, $V^2/2$, and pipe size. That is

$$v(p_i - p_e) + g(z_i - z_e) = f \frac{L}{D} \frac{V^2}{2} \qquad \text{(d)}$$

where D is the pipe interior diameter, L is the pipe length, and f is an experimentally-determined dimensionless **friction factor**.

friction factor

Equation (d) is the point of departure for applications involving friction in constant-diameter pipes through which an incompressible substance flows.

5.3.3 Reversible Processes

A process of a system is *reversible* if the system and all parts of its surroundings can be exactly restored to their respective initial states after the process has taken place. It should be evident from the discussion of irreversible processes that reversible processes are purely hypothetical. Clearly, no process can be reversible that involves spontaneous heat transfer through a finite temperature difference, an unrestrained expansion of a gas or liquid, friction, or any of the other irreversibilities listed previously. In a strict sense of the word, a reversible process is one that is *perfectly executed*.

All actual processes are irreversible. Reversible processes do not occur. Even so, certain processes that do occur are approximately reversible. The passage of a gas through a properly designed nozzle or diffuser is an example (Sec. 6.12). Many other devices also can be made to approach reversible operation by taking measures to reduce the significance of irreversibilities, such as lubricating surfaces to reduce friction. A reversible process is the *limiting case* as irreversibilities, both internal and external, are reduced further and further.

Although reversible processes cannot actually occur, they can be imagined. In Sec. 5.3.1, we considered how heat transfer would approach reversibility as the temperature difference approaches zero. Let us consider two additional examples:

- A particularly elementary example is a pendulum oscillating in an evacuated space. The pendulum motion approaches reversibility as friction at the pivot point is reduced. In the limit as friction is eliminated, the states of both the pendulum and its surroundings would be completely restored at the end of each period of motion. By definition, such a process is reversible.

- A system consisting of a gas adiabatically compressed and expanded in a frictionless piston–cylinder assembly provides another example. With a very small increase in the external pressure, the piston would compress the gas slightly. At each intermediate volume during the compression, the intensive properties T, p, v, and so on would be uniform throughout: The gas would pass through a series of equilibrium states. With a small decrease in the external pressure, the piston would slowly move out as the gas expands. At

each intermediate volume of the expansion, the intensive properties of the gas would be at the same uniform values they had at the corresponding step during the compression. When the gas volume returned to its initial value, all properties would be restored to their initial values as well. The work done *on* the gas during the compression would equal the work done *by* the gas during the expansion. If the work between the system and its surroundings were delivered to, and received from, a frictionless pulley–mass assembly, or the equivalent, there also would be no net change in the surroundings. This process would be reversible.

Horizons

Second Law Takes Big Bite from Hydrogen Fuel Cells

Hydrogen is not naturally occurring and thus must be produced. Hydrogen can be produced today from water by *electrolysis* and from natural gas by chemical processing called *reforming*. Hydrogen produced by these means and its subsequent utilization is burdened by the second law.

In electrolysis, an electrical input is employed to dissociate water to hydrogen according to $H_2O \rightarrow H_2 + {}^1/_2 O_2$. When the hydrogen is subsequently used by a fuel cell to generate electricity, the cell reaction is $H_2 + {}^1/_2 O_2 \rightarrow H_2O$. Although the cell reaction is the inverse of that occurring in electrolysis, the overall loop from electrical input–to hydrogen–to fuel cell–generated electricity is *not* reversible. Irreversibilities in the electrolyzer and the fuel cell

conspire to ensure that the fuel cell–generated electricity is less than the initial electrical input. This is wasteful because the electricity provided for electrolysis could instead be *fully* directed to most applications envisioned for hydrogen, including transportation. Further, when fossil fuel is burned in a power plant to generate electricity for electrolysis, the greenhouse gases produced can be associated with fuel cells by virtue of the hydrogen they consume. Although technical details differ, similar findings apply to the reforming of natural gas to hydrogen.

While hydrogen and fuel cells are expected to play a role in our energy future, second law barriers and other technical and economic issues stand in the way.

5.3.4 Internally Reversible Processes

internally reversible process

A reversible process is one for which no irreversibilities are present within the system *or* its surroundings. An **internally reversible process** is one for which *there are no irreversibilities within the system*. Irreversibilities may be located within the surroundings, however.

> **FOR EXAMPLE**
>
> Think of water condensing from saturated vapor to saturated liquid at 100°C while flowing through a copper tube whose outer surface is exposed to the ambient at 20°C. The water undergoes an internally reversible process, but there is heat transfer from the water to the ambient through the tube. For a control volume enclosing the water within the tube, such heat transfer is an *external* irreversibility.

TAKE NOTE...

The terms *internally reversible* process and *quasiequilibrium* process can be used interchangeably. However, to avoid having two terms that refer to the same thing, in subsequent sections we will refer to *any* such process as an internally reversible process.

At every intermediate state of an internally reversible process of a closed system, all intensive properties are uniform throughout each phase present. That is, the temperature, pressure, specific volume, and other intensive properties do not vary with position. If there were a spatial variation in temperature, say, there would be a tendency for a spontaneous energy transfer by conduction to occur *within* the system in the direction of decreasing temperature. For reversibility, however, no spontaneous processes can be present. From these considerations it can be concluded that the internally reversible process consists of a series of equilibrium states: It is a quasiequilibrium process.

The use of the internally reversible process concept in thermodynamics is comparable to idealizations made in mechanics: point masses, frictionless pulleys, rigid beams, and so on. In much the same way as idealizations are used in mechanics to simplify an analysis and arrive at a manageable model, simple thermodynamic models of complex situations can be obtained through the use of internally reversible processes. Calculations based on internally reversible processes often can be adjusted with efficiencies or correction factors to obtain reasonable estimates of actual performance under various operating conditions. Internally reversible processes are also useful for investigating the best thermodynamic performance of systems.

Finally, using the internally reversible process concept, we refine the definition of the thermal reservoir introduced in Sec. 5.2.2 as follows: In subsequent discussions we assume no internal irreversibilities are present within a thermal reservoir. That is, every process of a thermal reservoir is *internally reversible*.

5.4 Interpreting the Kelvin–Planck Statement

In this section, we recast Eq. 5.1, the analytical form of the Kelvin–Planck statement, into a more explicit expression, Eq. 5.3. This expression is applied in subsequent sections to obtain a number of significant deductions. In these applications, the following idealizations are assumed: The thermal reservoir and the portion of the surroundings with which work interactions occur are free of irreversibilities. This allows the "less than" sign to be associated with irreversibilities *within* the system of interest and the "equal to" sign to apply when no internal irreversibilites are present.

Accordingly, the **analytical form of the Kelvin–Planck statement** now takes the form

analytical form: Kelvin–Planck statement

$$W_{\text{cycle}} \leq 0 \begin{cases} < 0: & \text{Internal irreversibilities present.} \\ = 0: & \text{No internal irreversibilities.} \end{cases} \quad \text{(single reservoir)} \qquad (5.3)$$

For details, see the *Kelvin–Planck* box below.

Associating Signs with the Kelvin–Planck Statement

Consider a system that undergoes a cycle while exchanging energy by heat transfer with a single reservoir, as shown in **Fig. 5.4**. Work is delivered to, or received from, the pulley–mass assembly located in the surroundings. A flywheel, spring, or some other device also can perform the same function. The pulley–mass assembly, flywheel, or other device to which work is delivered, or from which it is received, is idealized as free of irreversibilities. The thermal reservoir is also assumed free of irreversibilities.

To demonstrate the correspondence of the "equal to" sign of Eq. 5.3 with the absence of irreversibilities, consider a cycle operating as shown in Fig. 5.4 for which the equality applies. At the conclusion of one cycle,

- The system would necessarily be returned to its initial state.
- Since $W_{\text{cycle}} = 0$, there would be no *net* change in the elevation of the mass used to store energy in the surroundings.
- Since $W_{\text{cycle}} = Q_{\text{cycle}}$, it follows that $Q_{\text{cycle}} = 0$, so there also would be no *net* change in the condition of the reservoir.

Thus, the system and all elements of its surroundings would be exactly restored to their respective initial conditions. By definition, such a cycle is reversible. Accordingly, there can be no irreversibilities present within the system or its surroundings. It is left as

an exercise to show the converse: If the cycle occurs reversibly, the equality applies.

Since a cycle is reversible *or* irreversible and we have linked the equality with reversible cycles, we conclude the inequality corresponds to the presence of internal irreversibilities. Moreover, the inequality can be interpreted as follows: Net work done *on* the system per cycle is converted by action of internal irreversibilities to internal energy that is discharged by heat transfer *to* the thermal reservoir in an amount equal to net work.

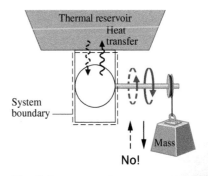

Fig. 5.4 System undergoing a cycle while exchanging energy by heat transfer with a single thermal reservoir.

Concluding Comment   The Kelvin–Planck statement considers systems undergoing *thermodynamic* cycles while exchanging energy by heat transfer with *one* thermal reservoir. These restrictions must be strictly observed—see the *thermal glider* box.

Does the *Thermal Glider* Challenge the Kelvin–Planck Statement?

A *Woods Hole Oceanographic Institute* news release, "Researchers Give New Hybrid Vehicle Its First Test-Drive in the Ocean," announced the successful testing of an underwater *thermal glider* that "harvests . . . energy from the ocean (thermally) to propel itself." Does this submersible vehicle challenge the Kelvin–Planck statement of the second law?

Study of the thermal glider shows it is capable of sustaining forward motion underwater for weeks while interacting thermally only with the ocean and undergoing a *mechanical* cycle. Still, the glider does not mount a challenge to the Kelvin–Planck statement because it does not exchange energy by heat transfer with a *single* thermal reservoir and does not execute a *thermodynamic* cycle.

The glider propels itself by interacting thermally with warmer surface waters and colder, deep-ocean layers to change its buoyancy to dive, rise toward the surface, and dive again, as shown on the accompanying figure. Accordingly, the glider does not interact thermally with a single reservoir as required by the Kelvin–Planck statement. The glider also does not satisfy all energy needs by interacting with the ocean: Batteries are required to power on-board electronics. Although these power needs are relatively minor, the batteries lose charge with use, and so the glider does not execute a thermodynamic cycle as required by the Kelvin–Planck statement.

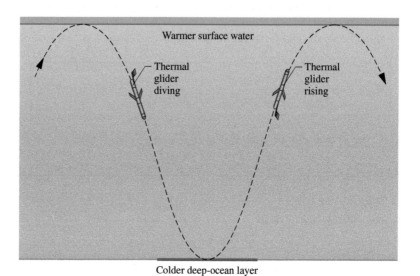

Colder deep-ocean layer

5.5 Applying the Second Law to Thermodynamic Cycles

While the Kelvin–Planck statement of the second law (Eq. 5.3) provides the foundation for the rest of this chapter, application of the second law to thermodynamic cycles is by no means limited to the case of heat transfer with a *single* reservoir or even with *any* reservoirs. Systems undergoing cycles while interacting thermally with *two* thermal reservoirs are considered from a second law viewpoint in Secs. 5.6 and 5.7, providing results having important applications. Moreover, the one- and two-reservoir discussions pave the way for Sec. 5.11, where the *general* case is considered—namely, what the second law says about *any* thermodynamic cycle without regard to the nature of the body or bodies with which energy is exchanged by heat transfer.

In the sections to follow, applications of the second law to power cycles and refrigeration and heat pump cycles are considered. For this content, familiarity with rudimentary thermodynamic cycle principles is required. We recommend you review Sec. 2.6, where cycles are considered from an energy perspective and the thermal efficiency of power cycles and coefficients of performance for refrigeration and heat pump systems are introduced. In particular, Eqs. 2.40–2.48 and the accompanying discussions should be reviewed.

5.6 Second Law Aspects of Power Cycles Interacting with Two Reservoirs

5.6.1 Limit on Thermal Efficiency

A significant limitation on the performance of systems undergoing power cycles can be brought out using the Kelvin–Planck statement of the second law. Consider Fig. 5.5, which shows a system that executes a cycle while communicating thermally with *two* thermal reservoirs, a hot reservoir and a cold reservoir, and developing net work W_{cycle}. The thermal efficiency of the cycle is

$$\eta = \frac{W_{cycle}}{Q_H} = 1 - \frac{Q_C}{Q_H} \qquad (5.4)$$

where Q_H is the amount of energy received by the system from the hot reservoir by heat transfer and Q_C is the amount of energy discharged from the system to the cold reservoir by heat transfer.

If the value of Q_C were zero, the system of Fig. 5.5 would withdraw energy Q_H from the hot reservoir and produce an equal amount of work, while undergoing a cycle. The thermal efficiency of such a cycle would be unity (100%). However, this method of operation violates the Kelvin–Planck statement and thus is not allowed.

It follows that for *any* system executing a power cycle while operating between two reservoirs, only a portion of the heat transfer Q_H can be obtained as work, and the remainder, Q_C, must be discharged by heat transfer to the cold reservoir. That is, the thermal efficiency must be less than 100%.

In arriving at this conclusion it was *not* necessary to

- identify the nature of the substance contained within the system,
- specify the exact series of processes making up the cycle,
- indicate whether the processes are actual processes or somehow idealized.

The conclusion that the thermal efficiency must be less than 100% applies to *all* power cycles whatever their details of operation. This may be regarded as a corollary of the second law. Other corollaries follow.

TAKE NOTE...

The energy transfers labeled on Fig 5.5 are positive in the directions indicated by the arrows.

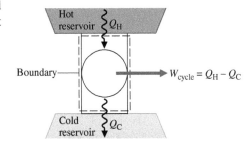

Fig. 5.5 System undergoing a power cycle while exchanging energy by heat transfer with two reservoirs.

5.6.2 Corollaries of the Second Law for Power Cycles

Since no power cycle can have a thermal efficiency of 100%, it is of interest to investigate the maximum theoretical efficiency. The maximum theoretical efficiency for systems undergoing power cycles while communicating thermally with two thermal reservoirs at different temperatures is evaluated in Sec. 5.9 with reference to the following two corollaries of the second law, called the **Carnot corollaries**.

Carnot corollaries

1. The thermal efficiency of an irreversible power cycle is always less than the thermal efficiency of a reversible power cycle when each operates between the same two thermal reservoirs.

2. All reversible power cycles operating between the same two thermal reservoirs have the same thermal efficiency.

A cycle is considered *reversible* when there are no irreversibilities within the system as it undergoes the cycle and heat transfers between the system and reservoirs occur reversibly.

The idea underlying the first Carnot corollary is in agreement with expectations stemming from the discussion of the second law thus far. Namely, the presence of irreversibilities during the execution of a cycle is expected to exact a penalty: If two systems operating between the same reservoirs each receive the same amount of energy Q_H and one executes a reversible cycle while the other executes an irreversible cycle, it is in accord with intuition that the net work developed by the irreversible cycle will be less, and thus the irreversible cycle has the smaller thermal efficiency.

The second Carnot corollary refers only to reversible cycles. All processes of a reversible cycle are perfectly executed. Accordingly, if two reversible cycles operating between the same reservoirs each receive the same amount of energy Q_H but one could produce more work than the other, it could only be as a result of more advantageous selections for the substance making up the system (it is conceivable that, say, air might be better than water vapor) *or* the series of processes making up the cycle (nonflow processes might be preferable to flow processes). This corollary denies both possibilities and indicates that the cycles must have the same efficiency whatever the choices for the working substance or the series of processes.

The two Carnot corollaries can be demonstrated using the Kelvin–Planck statement of the second law. For details, see the box.

Demonstrating the Carnot Corollaries

The first Carnot corollary can be demonstrated using the arrangement of Fig. 5.6. A reversible power cycle R and an irreversible power cycle I operate between the same two reservoirs and each receives the same amount of energy Q_H from the hot reservoir. The reversible cycle produces work W_R while the irreversible cycle produces work W_I. In accord with the conservation of energy principle,

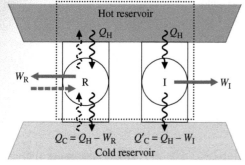

Dotted line defines combined system

Fig. 5.6 Sketch for demonstrating that a reversible cycle R is more efficient than an irreversible cycle I when they operate between the same two reservoirs.

each cycle discharges energy to the cold reservoir equal to the difference between Q_H and the work produced. Let R now operate in the opposite direction as a refrigeration (or heat pump) cycle. Since R is reversible, the magnitudes of the energy transfers W_R, Q_H, and Q_C remain the same, but the energy transfers are oppositely directed, as shown by the dashed lines on Fig. 5.6. Moreover,

with R operating in the opposite direction, the hot reservoir would experience *no net change* in its condition since it would receive Q_H *from* R while passing Q_H *to* I.

The demonstration of the first Carnot corollary is completed by considering the *combined system* shown by the dotted line on Fig. 5.6, which consists of the two cycles and the hot reservoir. Since its parts execute cycles or experience no net change, the combined system operates in a cycle. Moreover, the combined system exchanges energy by heat transfer with a single reservoir: the cold reservoir. Accordingly, the combined system must satisfy Eq. 5.3 expressed as

$$W_{\text{cycle}} < 0 \qquad \text{(single reservoir)}$$

where the inequality is used because the combined system is irreversible in its operation since irreversible cycle I is one of its parts. Evaluating W_{cycle} for the combined system in terms of the work amounts W_I and W_R, the above inequality becomes

$$W_I - W_R < 0$$

which shows that W_I must be less than W_R. Since each cycle receives the same energy input, Q_H, it follows that $\eta_I < \eta_R$ and this completes the demonstration.

The second Carnot corollary can be demonstrated in a parallel way by considering any two reversible cycles R_1 and R_2 operating between the same two reservoirs. Then, letting R_1 play the role of R and R_2 the role of I in the previous development, a combined system consisting of the two cycles and the hot reservoir may be formed that must obey Eq. 5.3. However, in applying Eq. 5.3 to this combined system, the equality is used because the system is reversible in operation. Thus, it can be concluded that $W_{R1} = W_{R2}$ and, therefore, $\eta_{R1} = \eta_{R2}$. The details are left as an exercise.

5.7 Second Law Aspects of Refrigeration and Heat Pump Cycles Interacting with Two Reservoirs

5.7.1 Limits on Coefficients of Performance

The second law of thermodynamics places limits on the performance of refrigeration and heat pump cycles as it does for power cycles. Consider Fig. 5.7, which shows a system undergoing a cycle while communicating thermally with two thermal reservoirs, a hot and a cold reservoir.

The energy transfers labeled on the figure are in the directions indicated by the arrows. In accord with the conservation of energy principle, the cycle discharges energy Q_H by heat transfer to the hot reservoir equal to the sum of the energy Q_C received by heat transfer from the cold reservoir and the net work input. This cycle might be a refrigeration cycle or a heat pump cycle, depending on whether its function is to remove energy Q_C from the cold reservoir or deliver energy Q_H to the hot reservoir.

For a refrigeration cycle the coefficient of performance is

$$\beta = \frac{Q_C}{W_{cycle}} = \frac{Q_C}{Q_H - Q_C} \tag{5.5}$$

The coefficient of performance for a heat pump cycle is

$$\gamma = \frac{Q_H}{W_{cycle}} = \frac{Q_H}{Q_H - Q_C} \tag{5.6}$$

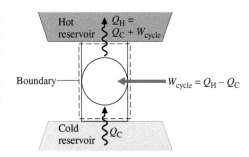

Fig. 5.7 System undergoing a refrigeration or heat pump cycle while exchanging energy by heat transfer with two reservoirs.

As the net work input to the cycle W_{cycle} tends to zero, the coefficients of performance given by Eqs. 5.5 and 5.6 approach a value of infinity. If W_{cycle} were identically zero, the system of Fig. 5.7 would withdraw energy Q_C from the cold reservoir and deliver that energy to the hot reservoir, while undergoing a cycle. However, this method of operation violates the Clausius statement of the second law and thus is not allowed. It follows that the coefficients of performance β and γ must invariably be finite in value. This may be regarded as another corollary of the second law. Further corollaries follow.

5.7.2 Corollaries of the Second Law for Refrigeration and Heat Pump Cycles

The maximum theoretical coefficients of performance for systems undergoing refrigeration and heat pump cycles while communicating thermally with two reservoirs at different temperatures are evaluated in Sec. 5.9 with reference to the following corollaries of the second law:

1. The coefficient of performance of an irreversible refrigeration cycle is always less than the coefficient of performance of a reversible refrigeration cycle when each operates between the same two thermal reservoirs.

2. All reversible refrigeration cycles operating between the same two thermal reservoirs have the same coefficient of performance.

By replacing the term *refrigeration* with *heat pump*, we obtain counterpart corollaries for heat pump cycles.

The first of these corollaries agrees with expectations stemming from the discussion of the second law thus far. To explore this, consider Fig. 5.8, which shows a reversible refrigeration cycle R and an irreversible refrigeration cycle I operating between the same two reservoirs. Each cycle removes the same energy Q_C from the cold reservoir. The net work input required to operate R is W_R, while the net work input for I is W_I. Each cycle discharges energy by heat transfer to the hot reservoir equal to the sum of Q_C and the net work input. The directions of the energy transfers are shown by arrows on Fig. 5.8. The presence of irreversibilities during the operation of a refrigeration cycle is expected to exact a penalty: If two refrigerators working between the same reservoirs each receive an identical energy transfer from the cold reservoir, Q_C, and one executes a reversible cycle while the other executes an irreversible cycle, we expect the irreversible cycle to require a greater net work input and thus have the smaller coefficient of performance. By a simple extension it follows that all reversible refrigeration cycles operating between the same two reservoirs have the same coefficient of performance. Similar arguments apply to the counterpart heat pump cycle statements.

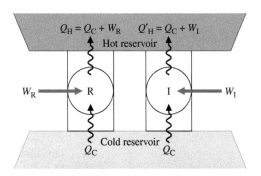

Fig. 5.8 Sketch for demonstrating that a reversible refrigeration cycle R has a greater coefficient of performance than an irreversible cycle I when they operate between the same two reservoirs.

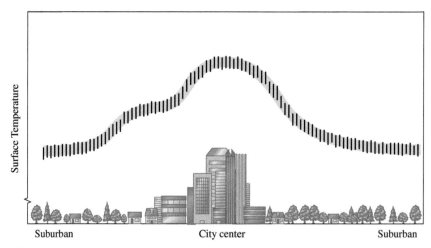

Fig. 5.9 Surface temperature variation in an urban area.

These corollaries can be demonstrated formally using the Kelvin–Planck statement of the second law and a procedure similar to that employed for the Carnot corollaries. The details are left as an exercise (see end-of-chapter Problem 5.8).

Energy & Environment

Warm blankets of pollution-laden air surround major cities. Sunlight-absorbing rooftops and expanses of pavement, together with little greenery, conspire with other features of city living to raise urban temperatures several degrees above adjacent suburban areas. **Figure 5.9** shows the variation of surface temperature in the vicinity of a city as measured by infrared measurements made from low-level flights over the area. Health-care professionals worry about the impact of these "heat islands," especially on the elderly. Paradoxically, the hot exhaust from the air conditioners city dwellers use to keep cool also makes sweltering neighborhoods even hotter. Irreversibilities within air conditioners contribute to the warming effect. Air conditioners may account for as much as 20% of the urban temperature rise. Vehicles and commercial activity also are contributors. Urban planners are combating heat islands in many ways, including the use of highly reflective colored roofing products and the installation of roof top gardens. The shrubs and trees of roof top gardens absorb solar energy, leading to summer roof temperatures significantly below those of nearby buildings without roof top gardens, reducing the need for air conditioning.

5.8 The Kelvin and International Temperature Scales

The results of Secs. 5.6 and 5.7 establish theoretical upper limits on the performance of power, refrigeration, and heat pump cycles communicating thermally with two reservoirs. Expressions for the *maximum* theoretical thermal efficiency of power cycles and the *maximum* theoretical coefficients of performance of refrigeration and heat pump cycles are developed in Sec. 5.9 using the Kelvin temperature scale considered next.

5.8.1 The Kelvin Scale

From the second Carnot corollary we know that all reversible power cycles operating between the same two thermal reservoirs have the same thermal efficiency, regardless of the nature of the substance making up the system executing the cycle or the series of processes. Since the thermal efficiency is independent of these factors, its value can be related only to the nature of the reservoirs themselves. Noting that it is the difference in *temperature* between the two reservoirs that provides the impetus for heat transfer between them, and thereby for the production of work during the cycle, we reason that the thermal efficiency depends *only* on the temperatures of the two reservoirs.

From Eq. 5.4 it also follows that for such reversible power cycles the ratio of the heat transfers Q_C/Q_H depends *only* on the temperatures of the two reservoirs. That is,

$$\left(\frac{Q_C}{Q_H}\right)_{\substack{\text{rev} \\ \text{cycle}}} = \psi(\theta_C, Q_H). \tag{a}$$

where θ_H and θ_C denote the temperatures of the reservoirs and the function ψ is for the present unspecified. Note that the words "rev cycle" are added to this expression to emphasize that it applies only to systems undergoing reversible cycles while operating between two thermal reservoirs.

Equation (a) provides a basis for defining a *thermodynamic* temperature scale: a scale independent of the properties of any substance. There are alternative choices for the function ψ that lead to this end. The **Kelvin scale** is obtained by making a particularly simple choice, namely, $\psi = T_C/T_H$, where T is the symbol used by international agreement to denote temperatures on the Kelvin scale. With this, we get

Kelvin scale

$$\left(\frac{Q_C}{Q_H}\right)_{\substack{\text{rev} \\ \text{cycle}}} = \frac{T_C}{T_H} \tag{5.7}$$

Thus, two temperatures on the Kelvin scale are in the same ratio as the values of the heat transfers absorbed and rejected, respectively, by a system undergoing a reversible cycle while communicating thermally with reservoirs at these temperatures.

If a reversible power cycle were operated in the opposite direction as a refrigeration or heat pump cycle, the magnitudes of the energy transfers Q_C and Q_H would remain the same, but the energy transfers would be oppositely directed. Accordingly, Eq. 5.7 applies to each type of cycle considered thus far, provided the system undergoing the cycle operates between two thermal reservoirs and the cycle is reversible.

TAKE NOTE...

Some readers may prefer to proceed directly to Sec. 5.9, where Eq. 5.7 is applied.

More on the Kelvin Scale

Equation 5.7 gives only a ratio of temperatures. To complete the definition of the Kelvin scale, it is necessary to proceed as in Sec. 1.7.3 by assigning the value 273.16 K to the temperature at the triple point of water. Then, if a reversible cycle is operated between a reservoir at 273.16 K and another reservoir at temperature T, the two temperatures are related according to

$$T = 273.16 \left(\frac{Q}{Q_{tp}}\right)_{\substack{\text{rev} \\ \text{cycle}}} \tag{5.8}$$

where Q_{tp} and Q are the heat transfers between the cycle and reservoirs at 273.16 K and temperature T, respectively. In the present case, the heat transfer Q plays the role of the *thermometric property*. However, since the performance of a reversible cycle is independent of the makeup of the system executing the cycle, the definition of temperature given by Eq. 5.8 depends in no way on the properties of any substance or class of substances.

In Sec. 1.7.2 we noted that the Kelvin scale has a zero of 0 K, and lower temperatures than this are not defined. Let us take up these points by considering a reversible power cycle operating between reservoirs at 273.16 K and a lower temperature T. Referring to Eq. 5.8, we know that the energy rejected from the cycle by heat transfer Q would not be negative, so T must be nonnegative. Equation 5.8 also shows that the smaller the value of Q, the lower the value of T, and conversely. Accordingly, as Q approaches zero the temperature T approaches zero. It can be concluded that a temperature of zero is the lowest temperature on the Kelvin scale. This temperature is called the *absolute* zero, and the Kelvin scale is called an *absolute temperature scale*.

When numerical values of the thermodynamic temperature are to be determined, it is not possible to use reversible cycles, for these exist only in our imaginations. However, temperatures evaluated using the constant-volume gas thermometer discussed in Sec. 5.8.2 to follow are identical to those of the Kelvin scale in the range of temperatures where the gas thermometer can be used. Other empirical approaches can be employed for temperatures above and below the range accessible to gas thermometry. The Kelvin scale provides a continuous definition of temperature valid over all ranges and provides an essential connection between the several empirical measures of temperature.

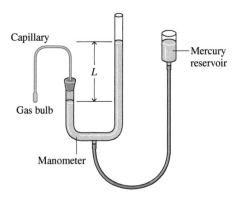

Capillary

Mercury reservoir

L

Gas bulb

Manometer

Fig. 5.10 Constant-volume gas thermometer.

5.8.2 The Gas Thermometer

The constant-volume gas thermometer shown in **Fig. 5.10** is so exceptional in terms of precision and accuracy that it has been adopted internationally as the standard instrument for calibrating other thermometers. The *thermometric substance* is the gas (normally hydrogen or helium), and the *thermometric property* is the pressure exerted by the gas. As shown in the figure, the gas is contained in a bulb, and the pressure exerted by it is measured by an open-tube mercury manometer. As temperature increases, the gas expands, forcing mercury up in the open tube. The gas is kept at constant volume by raising or lowering the reservoir. The gas thermometer is used as a standard worldwide by bureaus of standards and research laboratories. However, because gas thermometers require elaborate apparatus and are large, slowly responding devices that demand painstaking experimental procedures, smaller, more rapidly responding thermometers are used for most temperature measurements and they are calibrated (directly or indirectly) against gas thermometers. For further discussion of gas thermometry, see the box.

Measuring Temperature with the Gas Thermometer—The Gas Scale

It is instructive to consider how numerical values are associated with levels of temperature by the gas thermometer shown in Fig. 5.10. Let p stand for the pressure in the bulb of a constant-volume gas thermometer in thermal equilibrium with a bath. A value can be assigned to the bath temperature by a linear relation

$$T = \alpha p \qquad \text{(a)}$$

where α is an arbitrary constant.

The value of α is determined by inserting the thermometer into another bath maintained at the triple point of water and measuring the pressure, call it p_{tp}, of the confined gas at the triple point temperature, 273.16 K. Substituting values into Eq. (a) and solving for α,

$$\alpha = \frac{273.16}{p_{tp}}$$

Inserting this in Eq. (a), the temperature of the original bath, at which the pressure of the confined gas is p, is then

$$T = 273.16 \left(\frac{p}{p_{tp}} \right) \qquad \text{(b)}$$

However, since the values of both pressures, p and p_{tp}, depend *in part* on the amount of gas in the bulb, the value assigned by Eq. (b) to the bath temperature varies with the amount of gas in the thermometer. This difficulty is overcome in precision thermometry by repeating the measurements (in the original bath and the reference bath) several times with less gas in the bulb in each successive attempt. For each trial the ratio p/p_{tp} is plotted versus the corresponding reference pressure p_{tp} of the gas at the triple point temperature. When several such points have been plotted, the resulting curve is extrapolated to the ordinate where $p_{tp} = 0$. This is illustrated in **Fig. 5.11** for constant-volume thermometers with a number of different gases.

Inspection of Fig. 5.11 shows that at each nonzero value of the reference pressure, the p/p_{tp} values differ with the gas employed in the thermometer. However, as pressure decreases, the p/p_{tp} values from thermometers with different gases approach one another, and

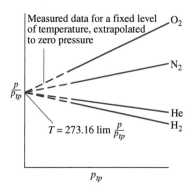

Measured data for a fixed level of temperature, extrapolated to zero pressure

O_2

N_2

$\dfrac{p}{p_{tp}}$

He

H_2

$T = 273.16 \lim \dfrac{p}{p_{tp}}$

p_{tp}

Fig. 5.11 Readings of constant-volume gas thermometers, when several gases are used.

in the limit as pressure tends to zero, *the same value for p/p_{tp} is obtained for each gas.* Based on these general results, the *gas temperature scale* is defined by the relationship

$$T = 273.16 \lim \frac{p}{p_{tp}} \qquad \text{(c)}$$

where "lim" means that both p and p_{tp} tend to zero. It should be evident that the determination of temperatures by this means requires extraordinarily careful and elaborate experimental procedures.

Although the temperature scale of Eq. (c) is independent of the properties of any one gas, it still depends on the properties of gases in general. Accordingly, the measurement of low temperatures requires a gas that does not condense at these temperatures, and this imposes a limit on the range of temperatures that can be measured by a gas thermometer. The lowest temperature that can be measured with such an instrument is about 1 K, obtained with helium. At high temperatures gases dissociate, and therefore these temperatures also cannot be determined by a gas thermometer. Other empirical means, utilizing the properties of other substances, must be employed to measure temperature in ranges where the gas thermometer is inadequate. For further discussion see Sec. 5.8.3.

TABLE 5.1

Defining Fixed Points of the International Temperature Scale of 1990

$T(K)$	Substance[a]	State[b]
3 to 5	He	Vapor pressure point
13.8033	e-H_2	Triple point
≈ 17	e-H_2	Vapor pressure point
≈ 20.3	e-H_2	Vapor pressure point
24.5561	Ne	Triple point
54.3584	O_2	Triple point
83.8058	Ar	Triple point
234.3156	Hg	Triple point
273.16	H_2O	Triple point
302.9146	Ga	Melting point
429.7485	In	Freezing point
505.078	Sn	Freezing point
692.677	Zn	Freezing point
933.473	Al	Freezing point
1234.93	Ag	Freezing point
1337.33	Au	Freezing point
1357.77	Cu	Freezing point

[a]He denotes ^{3}He or ^{4}He; e-H_2 is hydrogen at the equilibrium concentration of the ortho- and para-molecular forms.

[b]Triple point: temperature at which the solid, liquid, and vapor phases are in equilibrium. Melting point, freezing point: temperature, at a pressure of 101.325 kPa, at which the solid and liquid phases are in equilibrium.

Source: H. Preston-Thomas, "The International Temperature Scale of 1990 (ITS-90)," *Metrologia* 27, 3–10 (1990). See also www.ITS-90.com.

5.8.3 International Temperature Scale

To provide a standard for temperature measurement taking into account both theoretical and practical considerations, the International Temperature Scale (ITS) was adopted in 1927. This scale has been refined and extended in several revisions, most recently in 1990. *The International Temperature Scale of 1990 (ITS-90)* is defined in such a way that the temperature measured on it conforms with the thermodynamic temperature, the unit of which is the kelvin, to within the limits of accuracy of measurement obtainable in 1990. The ITS-90 is based on the assigned values of temperature of a number of reproducible *fixed points* (Table 5.1). Interpolation between the fixed-point temperatures is accomplished by formulas that give the relation between readings of standard instruments and values of the ITS. In the range from 0.65 to 5.0 K, ITS-90 is defined by equations giving the temperature as functions of the vapor pressures of particular helium isotopes. The range from 3.0 to 24.5561 K is based on measurements using a helium constant-volume gas thermometer. In the range from 13.8033 to 1234.93 K, ITS-90 is defined by means of certain platinum resistance thermometers. Above 1234.93 K the temperature is defined using *Planck's equation for blackbody radiation* and measurements of the intensity of visible-spectrum radiation.

5.9 Maximum Performance Measures for Cycles Operating Between Two Reservoirs

The discussion continues in this section with the development of expressions for the maximum thermal efficiency of power cycles and the maximum coefficients of performance of refrigeration and heat pump cycles in terms of reservoir temperatures evaluated on the Kelvin scale. These expressions can be used as standards of comparison for actual power, refrigeration, and heat pump cycles.

5.9.1 Power Cycles

The use of Eq. 5.7 in Eq. 5.4 results in an expression for the thermal efficiency of a system undergoing a reversible *power cycle* while operating between thermal reservoirs at temperatures T_H and T_C. That is,

$$\eta_{max} = 1 - \frac{T_C}{T_H} \tag{5.9}$$

Carnot efficiency

which is known as the **Carnot efficiency**. As temperatures on the Rankine scale differ from Kelvin temperatures only by the factor 1.8, the T's in Eq. 5.9 may be on either scale of temperature.

Recalling the two Carnot corollaries, it should be evident that the efficiency given by Eq. 5.9 is the thermal efficiency of *all* reversible power cycles operating between two reservoirs at temperatures T_H and T_C, and the *maximum* efficiency *any* power cycle can have while operating between the two reservoirs. By inspection, the value of the Carnot efficiency increases as T_H increases and/or T_C decreases.

Equation 5.9 is presented graphically in **Fig. 5.12**. The temperature T_C used in constructing the figure is 298 K in recognition that actual power cycles ultimately discharge energy by heat transfer at about the temperature of the local atmosphere or cooling water drawn from a nearby river or lake. Observe that increasing the thermal efficiency of a power cycle by reducing T_C below the ambient temperature is not viable. For instance, reducing T_C below the ambient using an *actual* refrigeration cycle requires a work input to the refrigeration cycle that will exceed the increase in work of the power cycle, giving a lower *net* work output.

Figure 5.12 shows that the thermal efficiency increases with T_H. Referring to segment a–b of the curve, where T_H and η are relatively low, we see that η increases rapidly as T_H increases, showing that in this range even a small increase in T_H can have a large effect on efficiency. Though these conclusions, drawn as they are from Eq. 5.9, apply strictly only to systems undergoing reversible cycles, they are qualitatively correct for actual power cycles. The thermal efficiencies of actual cycles are observed to increase as the *average* temperature at which energy is added by heat transfer increases and/or the *average* temperature at which energy is discharged by heat transfer decreases. However, maximizing the thermal efficiency of a power cycle may not be the only objective. In practice, other considerations such as cost may be overriding.

Power Cycles Tab c

Conventional power-producing cycles have thermal efficiencies ranging up to about 40%. This value may seem low, but the comparison should be made with an appropriate limiting value and not 100%.

FOR EXAMPLE

Consider a system executing a power cycle for which the average temperature of heat addition is 745 K and the average temperature at which heat is discharged is 298 K. For a reversible cycle receiving and discharging energy by heat transfer at these temperatures, the thermal efficiency given by Eq. 5.9 is 60%. When compared to this value, an actual thermal efficiency of 40% does not appear to be so low. The cycle would be operating at two-thirds of the theoretical maximum.

In the next example, we evaluate power cycle performance using the Carnot corollaries together with Eqs. 5.4 and 5.9.

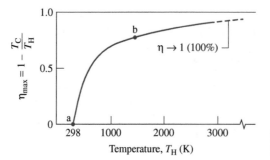

Fig. 5.12 Carnot efficiency versus T_H, for $T_C = 298$ K.

Evaluating Power Cycle Performance

A power cycle operating between two thermal reservoirs receives energy Q_H by heat transfer from a hot reservoir at $T_H = 2000$ K and rejects energy Q_C by heat transfer to a cold reservoir at $T_C = 400$ K. For each of the following cases determine whether the cycle operates reversibly, operates irreversibly, or is impossible.

  a. $Q_H = 1000$ kJ, $\eta = 60\%$

  b. $Q_H = 1000$ kJ, $W_{cycle} = 850$ kJ

  c. $Q_H = 1000$ kJ, $Q_C = 200$ kJ

SOLUTION

Known A system operates in a power cycle while receiving energy by heat transfer from a reservoir at 2000 K and discharging energy by heat transfer to a reservoir at 400 K.

Find For each of three cases determine whether the cycle operates reversibly, operates irreversibly, or is impossible.

Schematic and Given Data:

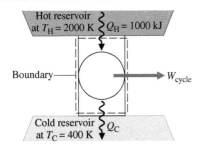

Fig. E5.1

Engineering Model

  1. The system shown in the accompanying figure executes a power cycle.

  2. Each energy transfer is positive in the direction of the arrow.

Analysis The maximum thermal efficiency for *any* power cycle operating between the two thermal reservoirs is given by Eq. 5.9. With the specified temperatures

  ❶
$$\eta_{max} = 1 - \frac{T_C}{T_H} = 1 - \frac{400 \text{ K}}{2000 \text{ K}}$$
$$= 0.8 \, (80\%)$$

a. The given thermal efficiency is $\eta = 60\%$. Since $\eta < \eta_{max}$, the cycle operates irreversibly.

b. Using given the data, $Q_H = 1000$ kJ and $W_{cycle} = 850$ kJ, the thermal efficiency is

$$\eta = \frac{W_{cycle}}{Q_H} = \frac{850 \text{ kJ}}{1000 \text{ kJ}}$$
$$= 0.85 \, (85\%)$$

Since $\eta > \eta_{max}$, the power cycle is impossible.

c. Applying an energy balance together with the given data,

$$W_{cycle} = Q_H - Q_C$$
$$= 1000 \text{ kJ} - 200 \text{ kJ} = 800 \text{ kJ}$$

The thermal efficiency is then

$$\eta = \frac{W_{cycle}}{Q_H} = \frac{800 \text{ kJ}}{1000 \text{ kJ}}$$
$$= 0.80 \, (80\%)$$

Since $\eta = \eta_{max}$, the cycle operates reversibly.

---

❶ The temperatures T_C and T_H used in evaluating η_{max} *must* be in K or °R.

SKILLS DEVELOPED

Ability to...

• apply the Carnot corollaries, using Eqs. 5.4 and 5.9 appropriately.

Quick Quiz

If $Q_C = 300$ kJ and $W_{cycle} = 2700$ kJ, determine whether the power cycle operates reversibly, operates irreversibly, or is impossible.
Ans. Impossible.

5.9.2 Refrigeration and Heat Pump Cycles

Equation 5.7 is also applicable to reversible refrigeration and heat pump cycles operating between two thermal reservoirs, but for these Q_C represents the heat added to the cycle from the cold reservoir at temperature T_C on the Kelvin scale and Q_H is the heat discharged to the hot reservoir at temperature T_H. Introducing Eq. 5.7 in Eq. 5.5 results in the following expression for the coefficient of performance of any system undergoing a reversible refrigeration cycle while operating between the two reservoirs:

$$\boxed{\beta_{max} = \frac{T_C}{T_H - T_C}}$$
(5.10)

Similarly, substituting Eq. 5.7 into Eq. 5.6 gives the following expression for the coefficient of performance of any system undergoing a reversible heat pump cycle while operating between the two reservoirs:

$$\boxed{\gamma_{max} = \frac{T_H}{T_H - T_C}}$$ (5.11)

Note that the temperatures used to evaluate β_{max} and γ_{max} must be absolute temperatures on the Kelvin or Rankine scale.

From the discussion of Sec. 5.7.2, it follows that Eqs. 5.10 and 5.11 are the maximum coefficients of performance that any refrigeration and heat pump cycles can have while operating between reservoirs at temperatures T_H and T_C. As for the case of the Carnot efficiency, these expressions can be used as standards of comparison for actual refrigerators and heat pumps.

In the next example, we evaluate the coefficient of performance of a refrigerator and compare it with the maximum theoretical value, illustrating the use of the second law corollaries of Sec. 5.7.2 together with Eq. 5.10.

Refrigeration Cycles
Tab c

Heat Pump Cycles
Tab c

▶▶▶ **EXAMPLE 5.2** ▶ ·

Evaluating Refrigerator Performance

By steadily circulating a refrigerant at low temperature through passages in the walls of the freezer compartment, a refrigerator maintains the freezer compartment at −5°C when the air surrounding the refrigerator is at 22°C. The rate of heat transfer from the freezer compartment to the refrigerant is 8000 kJ/h and the power input required to operate the refrigerator is 3200 kJ/h. Determine the coefficient of performance of the refrigerator and compare with the coefficient of performance of a reversible refrigeration cycle operating between reservoirs at the same two temperatures.

SOLUTION

Known A refrigerator maintains a freezer compartment at a specified temperature. The rate of heat transfer from the refrigerated space, the power input to operate the refrigerator, and the ambient temperature are known.

Find Determine the coefficient of performance and compare with that of a reversible refrigerator operating between reservoirs at the same two temperatures.

Schematic and Given Data:

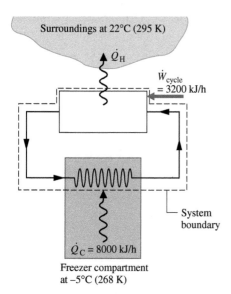

Surroundings at 22°C (295 K)

$\dot{Q}_H$

$\dot{W}_{cycle}$
= 3200 kJ/h

System
boundary

$\dot{Q}_C = 8000$ kJ/h

Freezer compartment
at −5°C (268 K)

Fig. E5.2

Engineering Model

1. The system shown on the accompanying figure is at steady state.

2. The freezer compartment and the surrounding air play the roles of cold and hot reservoirs, respectively.

3. The energy transfers are positive in the directions of the arrows on the schematic.

Analysis Inserting the given operating data into Eq. 5.5 expressed on a *time-rate* basis, the coefficient of performance of the refrigerator is

$$\beta = \frac{\dot{Q}_C}{\dot{W}_{cycle}} = \frac{8000 \text{ kJ/h}}{3200 \text{ kJ/h}} = 2.5$$

Substituting values into Eq. 5.10 gives the coefficient of performance of a reversible refrigeration cycle operating between reservoirs at $T_C = 268$ K and $T_H = 295$ K as

❶ $$\beta_{max} = \frac{T_C}{T_H - T_C} = \frac{268 \text{ K}}{295 \text{ K} - 268 \text{ K}} = 9.9$$

❷ In accord with the corollaries of Sec. 5.7.2, the coefficient of performance of the refrigerator is less than for a reversible refrigeration cycle operating between reservoirs at the same two temperatures. That is, irreversibilities are present within the system.

---

❶ The temperatures T_C and T_H used in evaluating β_{max} *must* be in K.

❷ The difference between the actual and maximum coefficients of performance suggests that there may be some potential for improving the thermodynamic performance. This objective should be approached judiciously, however, for improved performance may require increases in size, complexity, and cost.

In Example 5.3, we determine the minimum theoretical work input and cost for one day of operation of an electric heat pump, illustrating the use of the second law corollaries of Sec. 5.7.2 together with Eq. 5.11.

▶▶ EXAMPLE 5.3 ▶

Evaluating Heat Pump Performance

A dwelling requires 633 MJ per day to maintain its temperature at 21°C when the outside temperature is 0°C. **(a)** If an electric heat pump is used to supply this energy, determine the minimum theoretical work input for one day of operation, in kJ/day. **(b)** Evaluating electricity at 8 cents per kW · h, determine the minimum theoretical cost to operate the heat pump, in $/day.

SOLUTION

Known   A heat pump maintains a dwelling at a specified temperature. The energy supplied to the dwelling, the ambient temperature, and the unit cost of electricity are known.

Find   Determine the *minimum* theoretical work required by the heat pump and the corresponding electricity cost.

Schematic and Given Data:

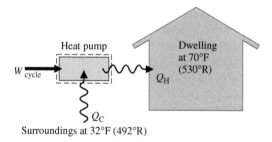

Fig. E5.3

Engineering Model

1. The system shown on the accompanying figure executes a heat pump cycle.

2. The dwelling and the outside air play the roles of hot and cold reservoirs, respectively.

3. The value of electricity is 8 cents per kW · h.

Analysis

a. Using Eq. 5.6, the work for any heat pump cycle can be expressed as $W_{cycle} = Q_H/\gamma$. The coefficient of performance γ of an actual heat pump is less than, or equal to, the coefficient of performance γ_{max} of a reversible heat pump cycle when each operates between the same two thermal reservoirs: $\gamma \leq \gamma_{max}$. Accordingly, for a given value of Q_H, and using Eq. 5.11 to evaluate γ_{max}, we get

$$W_{cycle} \geq \frac{Q_H}{\gamma_{max}}$$

$$\geq \left(1 - \frac{T_C}{T_H}\right) Q_H$$

Inserting values

❶ $W_{cycle} \geq \left(1 - \dfrac{273\ \text{K}}{294\ \text{K}}\right)\left(6.33 \times 10^5\ \dfrac{\text{kJ}}{\text{day}}\right) = 4.5 \times 10^4\ \dfrac{\text{kJ}}{\text{day}}$

The *minimum* theoretical work input is 4.5×10^4 kJ/day.

b. Using the result of part (a) together with the given cost data and an appropriate conversion factor

❷ $\begin{bmatrix} \text{minimum} \\ \text{theoretical} \\ \text{cost per day} \end{bmatrix} = \left(4.5 \times 10^4\ \dfrac{\text{kJ}}{\text{day}}\left|\dfrac{1\ \text{kW} \cdot \text{h}}{3600\ \text{kJ}}\right|\right)\left(0.08\ \dfrac{\$}{\text{kW} \cdot \text{h}}\right)$

$= 1\ \dfrac{\$}{\text{day}}$

---

❶ Note that the temperatures T_C and T_H *must* be in °R or K.

❷ Because of irreversibilities, an actual heat pump must be supplied more work than the minimum to provide the same heating effect. The actual daily cost could be substantially greater than the minimum theoretical cost.

SKILLS DEVELOPED

Ability to...

• apply the second law corollaries of Sec. 5.7.2, using Eqs. 5.6 and 5.11 appropriately.

• conduct an elementary economic evaluation.

5.10 Carnot Cycle

The Carnot cycles introduced in this section provide specific examples of reversible cycles operating between two thermal reservoirs. Other examples are provided in Chap. 9: the Ericsson and Stirling cycles. In a **Carnot cycle**, the system executing the cycle undergoes a series of four internally reversible processes: two adiabatic processes alternated with two isothermal processes.

Carnot cycle

5.10.1 Carnot Power Cycle

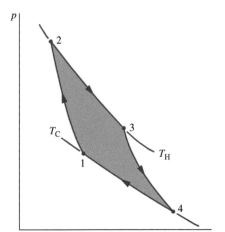

Fig. 5.13 p–v diagram for a Carnot gas power cycle.

Figure 5.13 shows the p–v diagram of a Carnot power cycle in which the system is a gas in a piston–cylinder assembly. **Figure 5.14** provides details of how the cycle is executed. The piston and cylinder walls are nonconducting. The heat transfers are in the directions of the arrows. Also note that there are two reservoirs at temperatures T_H and T_C, respectively, and an insulating stand. Initially, the piston–cylinder assembly is on the insulating stand and the system is at state 1, where the temperature is T_C. The four processes of the cycle are

Process 1–2 The gas is compressed *adiabatically* to state 2, where the temperature is T_H.

Process 2–3 The assembly is placed in contact with the reservoir at T_H. The gas expands *isothermally* while receiving energy Q_H from the hot reservoir by heat transfer.

Process 3–4 The assembly is again placed on the insulating stand and the gas is allowed to continue to expand *adiabatically* until the temperature drops to T_C.

Process 4–1 The assembly is placed in contact with the reservoir at T_C. The gas is compressed *isothermally* to its initial state while it discharges energy Q_C to the cold reservoir by heat transfer.

For the heat transfer during Process 2–3 to be reversible, the difference between the gas temperature and the temperature of the hot reservoir must be vanishingly small. Since the reservoir temperature remains constant, this implies that the temperature of the gas also remains constant during Process 2–3. The same can be concluded for the gas temperature during Process 4–1.

For each of the four internally reversible processes of the Carnot cycle, the work can be represented as an area on Fig. 5.13. The area under the adiabatic process line 1–2 represents the work done per unit of mass to compress the gas in this process. The areas under process lines 2–3 and 3–4 represent the work done per unit of mass by the gas as it expands in these processes. The area under process line 4–1 is the work done per unit of mass to compress the gas in this process. The enclosed area on the p–v diagram, shown shaded, is the net work developed by the cycle per unit of mass. The thermal efficiency of this cycle is given by Eq. 5.9.

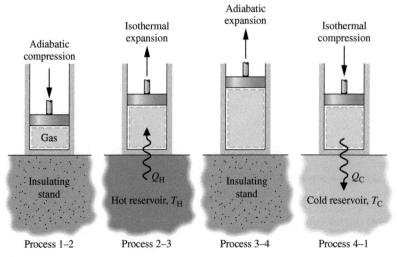

Fig. 5.14 Carnot power cycle executed by a gas in a piston–cylinder assembly.

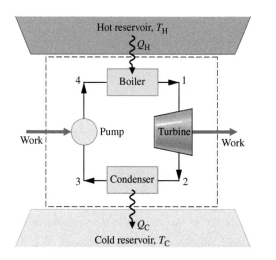

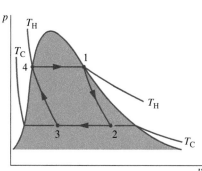

Fig. 5.15 Carnot vapor power cycle.

The Carnot cycle is not limited to processes of a closed system taking place in a piston–cylinder assembly. **Figure 5.15** shows the schematic and accompanying p–v diagram of a Carnot cycle executed by water steadily circulating through a series of four interconnected components that has features in common with the simple vapor power plant shown in Fig. 4.16. As the water flows through the boiler, a *change of phase* from liquid to vapor at constant temperature T_H occurs as a result of heat transfer from the hot reservoir. Since temperature remains constant, pressure also remains constant during the phase change. The steam exiting the boiler expands adiabatically through the turbine and work is developed. In this process the temperature decreases to the temperature of the cold reservoir, T_C, and there is an accompanying decrease in pressure. As the steam passes through the condenser, a heat transfer to the cold reservoir occurs and some of the vapor condenses at constant temperature T_C. Since temperature remains constant, pressure also remains constant as the water passes through the condenser. The fourth component is a pump (or compressor) that receives a two-phase liquid–vapor mixture from the condenser and returns it adiabatically to the state at the boiler entrance. During this process, which requires a work input to increase the pressure, the temperature increases from T_C to T_H. The thermal efficiency of this cycle also is given by Eq. 5.9.

5.10.2 Carnot Refrigeration and Heat Pump Cycles

If a Carnot power cycle is operated in the opposite direction, the magnitudes of all energy transfers remain the same but the energy transfers are oppositely directed. Such a cycle may be regarded as a reversible refrigeration or heat pump cycle, for which the coefficients of performance are given by Eqs. 5.10 and 5.11, respectively. A Carnot refrigeration or heat pump cycle executed by a gas in a piston–cylinder assembly is shown in **Fig. 5.16**. The cycle consists of the following four processes in series:

Process 1–2 The gas expands *isothermally* at T_C while *receiving* energy Q_C from the cold reservoir by heat transfer.

Process 2–3 The gas is compressed *adiabatically* until its temperature is T_H.

Process 3–4 The gas is compressed *isothermally* at T_H while it *discharges* energy Q_H to the hot reservoir by heat transfer.

Process 4–1 The gas expands *adiabatically* until its temperature decreases to T_C.

A refrigeration or heat pump effect can be accomplished in a cycle only if a net work input is supplied to the system executing the cycle. In the case of the cycle shown in Fig. 5.16, the shaded area represents the net work input per unit of mass.

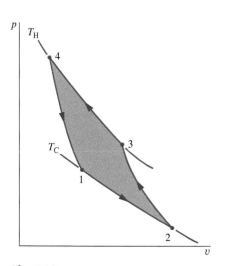

Fig. 5.16 p–v diagram for a Carnot gas refrigeration or heat pump cycle.

5.10.3 Carnot Cycle Summary

In addition to the configurations discussed previously, Carnot cycles also can be devised that are composed of processes in which a capacitor is charged and discharged, a paramagnetic substance is magnetized and demagnetized, and so on. However, regardless of the type of device or the working substance used,

1. the Carnot cycle *always* has the same four internally reversible processes: two adiabatic processes alternated with two isothermal processes.
2. the thermal efficiency of the Carnot power cycle is *always* given by Eq. 5.9 in terms of the temperatures evaluated on the Kelvin or Rankine scale.
3. the coefficients of performance of the Carnot refrigeration and heat pump cycles are *always* given by Eqs. 5.10 and 5.11, respectively, in terms of temperatures evaluated on the Kelvin or Rankine scale.

5.11 Clausius Inequality

Corollaries of the second law developed thus far in this chapter are for systems undergoing cycles while communicating thermally with *one* or *two* thermal energy reservoirs. In the present section a corollary of the second law known as the *Clausius inequality* is introduced that is applicable to *any* cycle without regard for the body, or bodies, from which the cycle receives energy by heat transfer or to which the cycle rejects energy by heat transfer. The Clausius inequality provides the basis for further development in Chap. 6 of the entropy, entropy production, and entropy balance concepts introduced in Sec. 5.2.3.

The *Clausius inequality* states that for any thermodynamic cycle

$$\oint \left(\frac{\delta Q}{T} \right)_b \leq 0 \tag{5.12}$$

where δQ represents the heat transfer at a part of the system boundary during a portion of the cycle, and T is the absolute temperature at that part of the boundary. The subscript "b" serves as a reminder that the integrand is evaluated at the boundary of the system executing the cycle. The symbol $\oint$ indicates that the integral is to be performed over all parts of the boundary and over the entire cycle. The equality and inequality have the same interpretation as in the Kelvin–Planck statement: The equality applies when there are no internal irreversibilities as the system executes the cycle, and the inequality applies when internal irreversibilities are present. The Clausius inequality can be demonstrated using the Kelvin–Planck statement of the second law. See the box for details.

Clausius inequality   The **Clausius inequality** can be expressed equivalently as

$$\boxed{\oint \left(\frac{\delta Q}{T} \right)_b = -\sigma_{\text{cycle}}} \tag{5.13}$$

where σ_{cycle} can be interpreted as representing the "strength" of the inequality. The value of σ_{cycle} is positive when internal irreversibilities are present, zero when no internal irreversibilities are present, and can never be negative.

In summary, the nature of a cycle executed by a system is indicated by the value for σ_{cycle} as follows:

$$\sigma_{\text{cycle}} = 0 \qquad \text{no irreversibilities present within the system}$$
$$\sigma_{\text{cycle}} > 0 \qquad \text{irreversibilities present within the system} \tag{5.14}$$
$$\sigma_{\text{cycle}} < 0 \qquad \text{impossible}$$

FOR EXAMPLE

Applying Eq. 5.13 to the cycle of Example 5.1(c), we get

$$\oint \left(\frac{\delta Q}{T}\right)_b = \frac{Q_H}{T_H} - \frac{Q_C}{T_C} = -\sigma_{cycle}$$

$$= \frac{1000 \text{ kJ}}{2000 \text{ K}} - \frac{200 \text{ kJ}}{400 \text{ K}} = 0 \text{ kJ/K}$$

giving $\sigma_{cycle} = 0$ kJ/K, which indicates no irreversibilities are present within the system undergoing the cycle. This is in keeping with the conclusion of Example 5.1(c). Applying Eq. 5.13 *on a time-rate basis* to the cycle of Example 5.2, we get $\dot{\sigma}_{cycle} = 8.12$ kJ/h · K. The positive value indicates irreversibilities are present within the system undergoing the cycle, which is in keeping with the conclusions of Example 5.2.

In Sec. 6.7, Eq. 5.13 is used to develop the closed system entropy balance. From that development, the term σ_{cycle} of Eq. 5.13 can be interpreted as the entropy *produced (generated)* by internal irreversibilities during the cycle.

Developing the Clausius Inequality

The Clausius inequality can be demonstrated using the arrangement of Fig. 5.17. A system receives energy δQ at a location on its boundary where the absolute temperature is T while the system develops work δW. In keeping with our sign convention for heat transfer, the phrase *receives energy δQ* includes the possibility of heat transfer *from* the system. The energy δQ is received from a thermal reservoir at T_{res}. To ensure that no irreversibility is introduced as a result of heat transfer between the reservoir and the system, let it be accomplished through an intermediary system that undergoes a cycle without irreversibilities of any kind. The cycle receives energy $\delta Q'$ from the reservoir and supplies δQ to the system while producing work $\delta W'$. From the definition of the Kelvin scale (Eq. 5.7), we have the following relationship between the heat transfers and temperatures:

$$\frac{\delta Q'}{T_{res}} = \left(\frac{\delta Q}{T}\right)_b \tag{a}$$

As temperature T may vary, a multiplicity of such reversible cycles may be required.

Consider next the combined system shown by the dotted line on Fig. 5.17. An energy balance for the combined system is

$$dE_C = \delta Q' - \delta W_C$$

where δW_C is the total work of the combined system, the sum of δW and $\delta W'$, and dE_C denotes the change in energy of the combined

system. Solving the energy balance for δW_C and using Eq. (a) to eliminate $\delta Q'$ from the resulting expression yields

$$\delta W_C = T_{res} \left(\frac{\delta Q}{T}\right)_b - dE_C$$

Now, let the system undergo a single cycle while the intermediary system undergoes one or more cycles. The total work of the combined system is

$$W_C = \oint T_{res} \left(\frac{\delta Q}{T}\right)_b - \oint dE_C^{\circ} = T_{res} \oint \left(\frac{\delta Q}{T}\right)_b \tag{b}$$

Since the reservoir temperature is constant, T_{res} can be brought outside the integral. The term involving the energy of the combined system vanishes because the energy change for any cycle is zero. The combined system operates in a cycle because its parts execute cycles. Since the combined system undergoes a cycle and exchanges energy by heat transfer with a single reservoir, Eq. 5.3 expressing the Kelvin–Planck statement of the second law must be satisfied. Using this, Eq. (b) reduces to give Eq. 5.12, where the equality applies when there are *no irreversibilities within the system* as it executes the cycle and the inequality applies when *internal irreversibilities are present*. This interpretation actually refers to the combination of system plus intermediary cycle. However, the intermediary cycle is free of irreversibilities, so the only possible site of irreversibilities is the system alone.

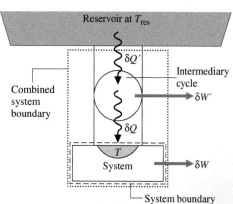

Fig. 5.17 Illustration used to develop the Clausius inequality.

CHAPTER SUMMARY AND STUDY GUIDE

In this chapter, we motivate the need for and usefulness of the second law of thermodynamics and provide the basis for subsequent applications involving the second law in Chaps. 6 and 7. Three statements of the second law, the Clausius, Kelvin–Planck, and entropy statements, are introduced together with several corollaries that establish the best theoretical performance for systems undergoing cycles while interacting with thermal reservoirs. The irreversibility concept is introduced and the related notions of irreversible, reversible, and internally reversible processes are discussed. The Kelvin temperature scale is defined and used to obtain expressions for maximum performance measures of power, refrigeration, and heat pump cycles operating between two thermal reservoirs. The Carnot cycle is introduced to provide a specific example of a reversible cycle operating between two thermal reservoirs. Finally, the Clausius inequality providing a bridge from Chap. 5 to Chap. 6 is presented and discussed.

The following checklist provides a study guide for this chapter. When your study of the text and end-of-chapter exercises has been completed you should be able to

- explain the meanings of the terms listed in the margins throughout the chapter and understand each of the related concepts. The subset of key concepts listed below is particularly important in subsequent chapters.
- give the Kelvin–Planck statement of the second law, correctly interpreting the "less than" and "equal to" signs in Eq. 5.3.
- list several important irreversibilities.
- apply the corollaries of Secs. 5.6.2 and 5.7.2 together with Eqs. 5.9, 5.10, and 5.11 to assess the performance of power cycles and refrigeration and heat pump cycles.
- describe the Carnot cycle.
- apply the Clausius inequality.

KEY ENGINEERING CONCEPTS

second law statements	irreversibilities	Kelvin scale
thermal reservoir	internal and external irreversibilities	Carnot efficiency
irreversible process	internally reversible process	Carnot cycle
reversible process	Carnot corollaries	Clausius inequality

KEY EQUATIONS

$W_{cycle} \leq 0 \begin{cases} < 0: & \text{Internal irreversibilities present.} \\ = 0: & \text{No internal irreversibilities.} \end{cases}$ (single reservoir)	(5.3)	Analytical form of the Kelvin–Planck statement.
$\eta_{max} = 1 - \dfrac{T_C}{T_H}$	(5.9)	Maximum thermal efficiency: power cycle operating between two reservoirs.
$\beta_{max} = \dfrac{T_C}{T_H - T_C}$	(5.10)	Maximum coefficient of performance: refrigeration cycle operating between two reservoirs.
$\gamma_{max} = \dfrac{T_H}{T_H - T_C}$	(5.11)	Maximum coefficient of performance: heat pump cycle operating between two reservoirs.
$\oint \left(\dfrac{\delta Q}{T} \right)_b = -\sigma_{cycle}$	(5.13)	Clausius inequality.

EXERCISES: THINGS ENGINEERS THINK ABOUT

5.1 A system consists of an ice cube in a cup of tap water. The ice cube melts and eventually equilibrium is attained. How might work be developed as the ice and water come to equilibrium?

5.2 Are health risks associated with consuming tomatoes induced to ripen by an ethylene spray? Explain.

5.3 Describe a process that would satisfy the conservation of energy principle, but does not actually occur in nature.

5.4 Are irreversibilities found in living things? Explain.

5.5 Is the power generated by fuel cells limited by the Carnot efficiency? Explain.

5.6 Does the second law impose performance limits on elite athletes seeking world records in events such as track and field and swimming? Explain.

5.7 In years ahead, are we more likely to be driving plug-in hybrid electrical vehicles or fuel cell–powered vehicles operating on hydrogen?

5.8 What is delaying the appearance in new car showrooms of automobiles powered by hydrogen fuel cells?

5.9 If a window air conditioner were placed on a table in a room and operated, would the room temperature increase, decrease, or remain the same? Explain.

5.10 How significant is the roughness at a pipe's inner surface in determining the friction factor? Explain.

5.11 A hot combustion gas enters a turbine operating at steady state and expands adiabatically to a lower pressure. Would you expect the power *output* to be greater in an internally reversible expansion or an actual expansion?

5.12 What factors influence the *actual* coefficient of performance achieved by refrigerators in family residences?

5.13 Refrigerant 22 enters a compressor operating at steady state and is compressed adiabatically to a higher pressure. Would you expect the power *input* to the compressor to be greater in an internally reversible compression or an actual compression?

CHECKING UNDERSTANDING

5.1 A reversible heat pump cycle operates between cold and hot thermal reservoirs at 300°C and 500°C, respectively. The coefficient of performance is closely (a) 1.5, (b) 3.87, (c) 2.87, (d) 2.5.

5.2 Referring to the list of Sec. 5.3.1, irreversibilities present during operation of an internal combustion automobile engine include (a) friction, (b) heat transfer, (c) chemical reaction, (d) all of the above.

5.3 Uses of the second law of thermodynamics include (a) defining the Kelvin scale, (b) predicting the direction of processes, (c) developing means for evaluating internal energy in terms of more readily measured properties, (d) all of the above.

5.4 When placed outside and exposed to the atmosphere, an ice cube melts, forming a thin film of liquid on the ground. Overnight, the liquid freezes, returning to the initial temperature of the ice cube. The water making up the cube undergoes (a) a thermodynamic cycle, (b) a reversible process, (c) an irreversible process, (d) none of the above.

5.5 Extending the discussion of Fig. 5.1a, how might work be developed when T_i is less than T_0?

5.6 A *throttling process* is (a) reversible, (b) internally reversible, (c) irreversible, (d) isobaric.

5.7 The energy of an isolated system remains constant, but change in entropy must satisfy (a) $\Delta S \leq 0$, (b) $\Delta S > 0$, (c) $\Delta S \geq 0$, (d) $\Delta S < 0$.

5.8 The maximum thermal efficiency of *any* power cycle operating between hot and cold reservoirs at 1000°C and 500°C, respectively, is _____.

5.9 A power cycle operating between hot and cold reservoirs at 500 K and 300 K, respectively, receives 1000 kJ by heat transfer from the hot reservoir. The magnitude of the energy discharged by heat transfer to the cold reservoir must satisfy (a) $Q_C > 600$ kJ, (b) $Q_C \geq 600$ kJ, (c) $Q_C = 600$ kJ, (d) $Q_C \leq 600$ kJ.

5.10 An internal irreversibility within a gearbox is (a) chemical reaction, (b) unrestrained expansion of a gas, (c) mixing, (d) friction.

5.11 When hot and cold gas streams pass in counterflow through a heat exchanger, each at constant pressure, the principal internal irreversibility for the heat exchanger is _____.

5.12 Referring to Fig. 5.12, if the temperature corresponding to point b is 1225°C, the Carnot efficiency is _____ %.

5.13 The thermal efficiency of a system that undergoes a power cycle while receiving 1000 kJ of energy by heat transfer from a hot reservoir at 1000 K and discharging 500 kJ of energy by heat transfer to a cold reservoir at 400 K is _____.

5.14 For a closed system, entropy (a) may be produced within the system, (b) may be transferred across its boundary, (c) may remain constant throughout the system, (d) all of the above.

5.15 As shown in **Fig. P5.15C**, a rigid, insulated tank is divided into halves by a partition that has gas on one side and an evacuated space on the other side. When the valve is opened, the gas expands to all fill the entire volume. The principal source of irreversibility is _____.

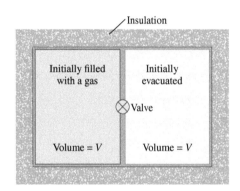

Insulation

Initially filled with a gas | Initially evacuated

Valve

Volume = V | Volume = V

FIGURE P5.15C

Indicate whether the following statements are true or false. Explain.

5.16 The change in entropy of a closed system is the same for every process between two specified end states.

5.17 A process of a closed system that violates the second law of thermodynamics necessarily violates the first law of thermodynamics.

5.18 In principle, the *Clausius inequality* applies to any cycle.

5.19 Friction associated with flow of fluids through pipes and around objects is one type of *irreversibility*.

5.20 The *second* Carnot corollary states that all power cycles operating between the same two thermal reservoirs have the same thermal efficiency.

5.21 Internally reversible processes do not actually occur but serve as hypothetical limiting cases as internal irreversibilities are reduced further and further.

5.22 For reversible refrigeration and heat pump cycles operating between the same hot and cold reservoirs, the relation between their coefficients of performance is $\gamma_{max} = \beta_{max} + 1$.

5.23 Mass, energy, entropy, and temperature are examples of extensive properties.

5.24 The Clausius statement of the second law denies the possibility of transferring energy by heat from a cooler to a hotter body.

5.25 The Kelvin–Planck and Clausius statements of the second law of thermodynamics are equivalent because a violation of one statement implies the violation of the other.

PROBLEMS: DEVELOPING ENGINEERING SKILLS

Exploring the Second Law

5.1 Complete the demonstration of the equivalence of the Clausius and Kelvin–Planck statements of the second law given in Sec. 5.2.2 by showing that a violation of the Kelvin–Planck statement implies a violation of the Clausius statement.

5.2 An inventor claims to have developed a device that undergoes a thermodynamic cycle while communicating thermally with two reservoirs. The system receives energy Q_C from the cold reservoir and discharges energy Q_H to the hot reservoir while delivering a net amount of work to its surroundings. There are no other energy transfers between the device and its surroundings. Evaluate the inventor's claim using (a) the Clausius statement of the second law, and (b) the Kelvin–Planck statement of the second law.

5.3 Classify the following processes of a closed system as *possible*, *impossible*, or *indeterminate*.

	Entropy Change	Entropy Transfer	Entropy Production
(a)	>0	0	
(b)	<0		>0
(c)	0	>0	
(d)	>0	>0	
(e)	0	<0	
(f)	>0		<0
(g)	<0	<0	

5.4 As shown in Fig. P5.4, a hot thermal reservoir is separated from a cold thermal reservoir by a cylindrical rod insulated on its lateral surface. Energy transfer by conduction between the two reservoirs takes place through the rod, which remains at steady state. Using the Kelvin–Planck statement of the second law, demonstrate that such a process is irreversible.

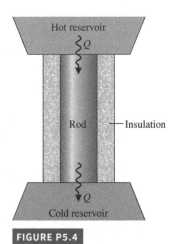

FIGURE P5.4

5.5 Shown in Fig. P5.5 is a proposed system that undergoes a cycle while operating between cold and hot reservoirs. The system receives 500 kJ from the cold reservoir and discharges 400 kJ to the hot reservoir while delivering net work to its surroundings in the amount of 100 kJ. There are no other energy transfers between the system and its surroundings. Evaluate the performance of the system using

a. the Clausius statement of the second law.

b. the Kelvin-Planck statement of the second law.

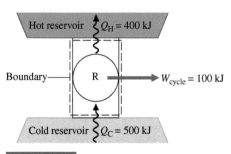

FIGURE P5.5

5.6 A power cycle I and a reversible power cycle R operate between the same two reservoirs, as shown in Fig. 5.6. Cycle I has a thermal efficiency equal to two-thirds of that for cycle R. Using the Kelvin–Planck statement of the second law, prove that cycle I must be irreversible.

5.7 As shown in Fig. P5.7, a rigid insulated tank is divided into halves by a partition. On one side of the partition is a gas. The other side is initially evacuated. A valve in the partition is opened and the gas expands to fill the entire volume. Using the Kelvin–Planck statement of the second law, demonstrate that this process is irreversible.

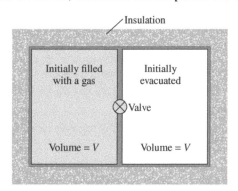

FIGURE P5.7

5.8 Using the Kelvin–Planck statement of the second law of thermodynamics, demonstrate the following corollaries:

a. The coefficient of performance of an irreversible refrigeration cycle is always less than the coefficient of performance of a reversible refrigeration cycle when both exchange energy by heat transfer with the same two reservoirs.

b. All reversible refrigeration cycles operating between the same two reservoirs have the same coefficient of performance.

c. The coefficient of performance of an irreversible heat pump cycle is always less than the coefficient of performance of a reversible heat pump cycle when both exchange energy by heat transfer with the same two reservoirs.

d. All reversible heat pump cycles operating between the same two reservoirs have the same coefficient of performance.

5.9. Complete the discussion of the Kelvin–Planck statement of the second law in the box of Sec. 5.4 by showing that if a system undergoes a thermodynamic cycle reversibly while communicating thermally with a single reservoir, the equality in Eq. 5.3 applies.

5.10 Figure P5.10 shows two power cycles, denoted 1 and 2, operating in series, together with three thermal reservoirs. The energy transfer by heat into cycle 2 is equal in magnitude to the energy transfer by heat from cycle 1. All energy transfers are positive in the directions of the arrows.

a. Determine an expression for the thermal efficiency of an overall cycle consisting of cycles 1 and 2 expressed in terms of their individual thermal efficiencies.

b. If cycles 1 and 2 are each reversible, use the result of part (a) to obtain an expression for the thermal efficiency of the overall cycle in terms of the temperatures of the three reservoirs, T_H, T, and T_C, as required. Comment.

c. If cycles 1 and 2 are each reversible and have the same thermal efficiency, obtain an expression for the intermediate temperature T in terms of T_H and T_C.

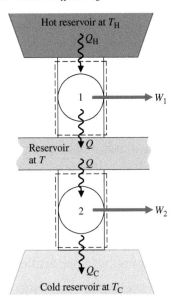

Hot reservoir at T_H

Reservoir at T

Cold reservoir at T_C **FIGURE P5.10**

5.11 The relation between resistance R and temperature T for a *thermistor* closely follows:

$$R = R_0 \exp\left[\beta\left(\frac{1}{T} - \frac{1}{T_0}\right)\right]$$

where R_0 is the resistance, in ohms (Ω), measured at temperature T_0 (K) and β is a material constant with units of K. For a particular thermistor $R_0 = 2.2\ \Omega$ at $T_0 = 310$ K. From a calibration test, it is found that $R = 0.31\ \Omega$ at $T = 422$ K. Determine the value of β for the thermistor and make a plot of resistance versus temperature.

5.12 Two reversible cycles operate between hot and cold reservoirs at temperature T_H and T_C, respectively.

a. If one is a power cycle and the other is a heat pump cycle, what is the relation between the coefficient of performance of the heat pump cycle and the thermal efficiency of the power cycle?

b. If one is a refrigeration cycle and the other is a heat pump cycle, what is the relation between their coefficients of performance?

5.13 Over a limited temperature range, the relation between electrical resistance R and temperature T for a *resistance temperature detector* is

$$R = R_0[1 + \alpha(T - T_0)]$$

where R_0 is the resistance, in ohms (Ω), measured at reference temperature T_0 (in °C) and α is a material constant with units of (°C)$^{-1}$. The following data are obtained for a particular resistance thermometer:

	T (°C)	R (Ω)
Test 1	0	51.39
Test 2	91	51.72

What temperature would correspond to a resistance of 51.47 Ω on this thermometer?

Power Cycle Applications

5.14 A power cycle receives energy Q_H by heat transfer from a hot reservoir at $T_H = 833$ K and rejects energy Q_C by heat transfer to a cold reservoir at $T_C = 278$ K. For each of the following cases, determine whether the cycle operates *reversibly*, operates *irreversibly*, or is *impossible*.

a. $Q_H = 950$ kJ, $W_{cycle} = 475$ kJ

b. $Q_H = 950$ kJ, $Q_C = 315$ kJ

c. $W_{cycle} = 633$ kJ, $Q_C = 422$ kJ

d. $\eta = 70\%$

5.15 The data listed below are claimed for a power cycle operating between hot and cold reservoirs at 1000 K and 300 K, respectively. For each case, determine whether the cycle operates *reversibly*, operates *irreversibly*, or is *impossible*.

a. $Q_H = 600$ kJ, $W_{cycle} = 300$ kJ, $Q_C = 300$ kJ

b. $Q_H = 400$ kJ, $W_{cycle} = 280$ kJ, $Q_C = 120$ kJ

c. $Q_H = 700$ kJ, $W_{cycle} = 300$ kJ, $Q_C = 500$ kJ

d. $Q_H = 800$ kJ, $W_{cycle} = 600$ kJ, $Q_C = 200$ kJ

5.16 A power cycle operating at steady state receives energy by heat transfer at a rate $\dot{Q}_H$ at $T_H = 1800$ K and rejects energy by heat transfer to a cold reservoir at a rate $\dot{Q}_C$ at $T_C = 600$ K. For each of the following cases, determine whether the cycle operates *reversibly*, operates *irreversibly*, or is *impossible*.

a. $\dot{Q}_H = 500$ kW, $\dot{Q}_C = 100$ kW

b. $\dot{Q}_H = 500$ kW, $\dot{W}_{cycle} = 250$ kW, $\dot{Q}_C = 200$ kW

c. $\dot{W}_{cycle} = 350$ kW, $\dot{Q}_C = 150$ kW

d. $\dot{Q}_H = 500$ kW, $\dot{Q}_C = 200$ kW

5.17 As shown in **Fig. P5.17**, a reversible power cycle receives energy Q_H by heat transfer from a hot reservoir at T_H and rejects energy Q_C by heat transfer to a cold reservoir at T_C.

a. If $T_H = 1600$ K and $T_C = 400$ K, what is the thermal efficiency?

b. If $T_H = 500$°C, $T_C = 20$°C, and $W_{cycle} = 1000$ kJ, what are Q_H and Q_C, each in kJ?

c. If $\eta = 40\%$ and $T_H = 727$°C, what is T_C, in °C?

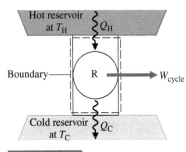

Hot reservoir at T_H

Boundary

Cold reservoir at T_C

FIGURE P5.17

5.18 A reversible power cycle receives 100 kJ by heat transfer from a hot reservoir at 327°C and rejects 40 kJ by heat transfer to a cold reservoir at T_C. Determine (a) the thermal efficiency and (b) the temperature T_C of the cold reservoir, in °C.

5.19 At a particular location, magma exists several kilometers below the Earth's surface at a temperature of 1100°C, while the average temperature of the atmosphere at the surface is 15°C. An inventor claims to have devised a power cycle operating between these temperatures having a thermal efficiency of 79%. Investigate this claim.

5.20 A reversible power cycle operating as in Fig. 5.5 receives energy Q_H by heat transfer from a hot reservoir at T_H and rejects energy Q_C by heat transfer to a cold reservoir at 4°C. If $W_{cycle} = 3\ Q_C$, determine (a) the thermal efficiency and (b) T_H, in °C.

5.21 During January, at a location in Alaska winds at −30°C can be observed. Several meters below ground the temperature remains at 13°C, however. An inventor claims to have devised a power cycle exploiting this situation that has a thermal efficiency of 14%. Evaluate this claim.

5.22 As shown in Fig. **P5.22**, two reversible cycles arranged in series each produce the same net work, W_{cycle}. The first cycle receives energy Q_H by heat transfer from a hot reservoir at 555 K and rejects energy Q by heat transfer to a reservoir at an intermediate temperature, T. The second cycle receives energy Q by heat transfer from the reservoir at temperature T and rejects energy Q_C by heat transfer to a reservoir at 222 K. All energy transfers are positive in the directions of the arrows. Determine

    a. the intermediate temperature T, in K, and the thermal efficiency for each of the two power cycles.

    b. the thermal efficiency of a *single* reversible power cycle operating between hot and cold reservoirs at 555 K and 222 K, respectively. Also, determine the net work developed by the single cycle, expressed in terms of the net work developed by each of the two cycles, W_{cycle}.

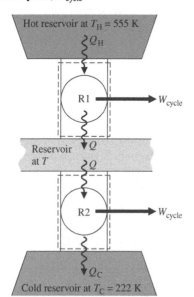

FIGURE P5.22

5.23 The data listed below are claimed for power cycles operating between hot and cold reservoirs at 1000 K and 400 K, respectively. For each case determine whether such a cycle is in keeping with the first and second laws of thermodynamics.

    a. $Q_H = 300$ kJ, $W_{cycle} = 160$ kJ, $Q_C = 140$ kJ.

    b. $Q_H = 300$ kJ, $W_{cycle} = 180$ kJ, $Q_C = 120$ kJ.

    c. $Q_H = 300$ kJ, $W_{cycle} = 170$ kJ, $Q_C = 140$ kJ.

    d. $Q_H = 300$ kJ, $W_{cycle} = 200$ kJ, $Q_C = 100$ kJ.

5.24 Data for two reversible refrigeration is given below:

  Cycle 1: $T_H = 25°C$, $T_C = −10°C$

  Cycle 2: $T_H = 25°C$, $T_C = −25°C$

where T_H and T_C are the temperature of hot and cold reservoirs, respectively. Determine the ratio of net work input values of the two cycles if

same amount of heat energy is removed from cold reservoirs by each refrigerator.

5.25 At steady state, a new power cycle is claimed by its inventor to develop power at a rate of 74.6 kW for a heat addition rate of 5.3×10^5 kJ/h, while operating between hot and cold reservoirs at 1000 and 500 K, respectively. Evaluate this claim.

5.26 A power cycle operates between hot and cold reservoirs at 500 K and 310 K, respectively. At steady state, the cycle rejects energy by heat transfer to the cold reservoir at a rate of 16 MW. Determine the maximum theoretical power that might be developed by such a cycle, in MW.

5.27 An inventor claims to have developed a power cycle operating between hot and cold reservoirs at 1000 K and 250 K, respectively, that develops net work equal to a multiple of the amount of energy, Q_C, rejected to the cold reservoir—that is $W_{cycle} = NQ_C$, where all quantities are positive. What is the maximum theoretical value of the number N for any such cycle?

5.28 At steady state, a power cycle having a thermal efficiency of 38% generates 100 MW of electricity while discharging energy by heat transfer to cooling water at an average temperature of 21°C. The average temperature of the steam passing through the boiler is 480°C. Determine

    a. the rate at which energy is discharged to the cooling water, in kJ/h.

    b. the *minimum* theoretical rate at which energy could be discharged to the cooling water, in kJ/h. Compare with the actual rate and discuss.

Refrigeration and Heat Pump Cycle Applications

5.29 A refrigeration cycle operating between two reservoirs receives energy Q_C from a cold reservoir at $T_C = 275$ K and rejects energy Q_H to a hot reservoir at $T_H = 315$ K. For each of the following cases, determine whether the cycle operates *reversibly*, operates *irreversibly*, or is *impossible*:

    a. $Q_C = 1000$ kJ, $W_{cycle} = 80$ kJ.

    b. $Q_C = 1200$ kJ, $Q_H = 2000$ kJ.

    c. $Q_H = 1575$ kJ, $W_{cycle} = 200$ kJ.

    d. $\beta = 6$.

5.30 A reversible refrigeration cycle operates between cold and hot reservoirs at temperatures T_C and T_H, respectively.

    a. If the coefficient of performance is 3.5 and $T_H = 27°C$, determine T_C, in °C.

    b. If $T_C = −30°C$ and $T_H = 30°C$, determine the coefficient of performance.

    c. If $Q_C = 528$ kJ, $Q_H = 844$ kJ, and $T_C = −7°C$, determine T_H, in °C.

    d. If $T_C = −1°C$ and $T_H = 38°C$, determine the coefficient of performance.

    e. If the coefficient of performance is 8.9 and $T_C = −5°C$, find T_H, in °C.

5.31 At steady state, a reversible heat pump cycle discharges energy at the rate $\dot{Q}_H$ to a hot reservoir at temperature T_H, while receiving energy at the rate $\dot{Q}_C$ from a cold reservoir at T_C.

    a. If $T_H = 21°C$ and $T_C = 7°C$, determine the coefficient performance.

    b. If $\dot{Q}_H = 10.5$ kW, $\dot{Q}_C = 8.75$ kW, and $T_C = 0°C$, determine T_H, in °C.

    c. If the coefficient of performance is 10 and $T_H = 27°C$, determine T_C, in °C.

5.32 A heating system must maintain the interior of a building at 20°C during a period when the outside air temperature is 5°C and the heat transfer from the building through its roof and walls is 3×10^6 kJ. For this duty heat pumps are under consideration that would operate between the dwelling and

  a. the ground at 15°C.

  b. a pond at 10°C.

  c. the outside air at 5°C.

For each case, evaluate the minimum theoretical net work input required by *any* such heat pump, in kJ.

5.33 A refrigeration cycle rejects $Q_H = 528$ kJ per cycle to a hot reservoir at $T_H = 300$ K, while receiving $Q_C = 395$ kJ per cycle from a cold reservoir at temperature T_C. For 10 cycles of operation, determine (a) the net work input, in kJ, and (b) the minimum theoretical temperature T_C, in K.

5.34 A reversible power cycle and a reversible heat pump cycle operate between hot and cold reservoirs at temperature $T_H = 555$ K and T_C, respectively. If the thermal efficiency of the power cycle is 60%, determine (a) T_C, in K, and (b) the coefficient of performance of the heat pump.

5.35 An inventor has developed a refrigerator capable of maintaining its freezer compartment at −7°C while operating in a kitchen at 21°C, and claims the device has a coefficient of performance of (a) 10, (b) 9.5, (c) 4. Evaluate the claim in each of the three cases.

5.36 An inventor claims to have developed a refrigerator that at steady state requires a net power input of 0.52 kW to remove 12,660 kJ/h of energy by heat transfer from the freezer compartment at −17.8°C and discharge energy by heat transfer to a kitchen at 21°C. Evaluate this claim.

5.37 According to an inventor of a refrigerator, the refrigerator can remove heat from the freezer compartment at the rate of 13,000 kJ/h by net input power consumption of 0.65 kW. Heat is discharged into the room at 23°C. The temperature of freezer compartment is −15°C. Evaluate this claim.

5.38 An inventor claims to have devised a refrigeration cycle operating between hot and cold reservoirs at 300 K and 250 K, respectively, that removes an amount of energy Q_C by heat transfer from the cold reservoir that is a multiple of the net work input—that is, $Q_C = NW_{cycle}$, where all quantities are positive. Determine the maximum theoretical value of the number N for any such cycle.

5.39 By removing energy by heat transfer from its freezer compartment at a rate of 1.5 kW, a refrigerator maintains the freezer at −22°C on a day when the temperature of the surroundings is 28°C. Determine the minimum theoretical power, in kW, required by the refrigerator at steady state.

5.40 At steady state, a refrigeration cycle operating between hot and cold reservoirs at 320 K and 272 K, respectively, removes energy by heat transfer from the cold reservoir at a rate of 650 kW.

  a. If the cycle's coefficient of performance is 5, determine the power input required, in kW.

  b. Determine the minimum theoretical power required, in kW, for *any* such cycle.

5.41 An air conditioner operating at steady state maintains a dwelling at 20°C on a day when the outside temperature is 35°C. Energy is removed by heat transfer from the dwelling at a rate of 3 kW while the air conditioner's power input is 1 kW. Determine (a) the coefficient of performance of the air conditioner and (b) the power input required by a reversible refrigeration cycle providing the same cooling effect while operating between hot and cold reservoirs at 35°C and 20°C, respectively.

5.42 At steady state, a reversible refrigeration cycle operates between hot and cold reservoirs at 300 K and 270 K, respectively. Determine the minimum theoretical net power input required, in kW per kW of heat transfer from the cold reservoir.

5.43 By removing energy by heat transfer from a room, a window air conditioner maintains the room at 22°C on a day when the outside temperature is 32°C.

  a. Determine, in kW per kW of cooling, the *minimum* theoretical power required by the air conditioner.

  b. To achieve required rates of heat transfer with practical-sized units, air conditioners typically receive energy by heat transfer at a temperature *below* that of the room being cooled and discharge energy by heat transfer at a temperature *above* that of the surroundings. Consider the effect of this by determining the *minimum* theoretical power, in kW per kW of cooling, required when $T_C = 18$°C and $T_H = 36$°C, and compare with the value found in part (a).

5.44 A heat pump cycle is used to maintain the interior of a building at 20°C. At steady state, the heat pump receives energy by heat transfer from well water at 10°C and discharges energy by heat transfer to the building at a rate of 120,000 kJ/h. Over a period of 14 days, an electric meter records that 1490 kW · h of electricity is provided to the heat pump. Determine

  a. the amount of energy that the heat pump receives over the 14-day period from the well water by heat transfer, in kJ.

  b. the heat pump's coefficient of performance.

  c. the coefficient of performance of a reversible heat pump cycle operating between hot and cold reservoirs at 20°C and 10°C.

5.45 As shown in **Fig. P5.45**, an air conditioner operating at steady state maintains a dwelling at 21°C on a day when the outside temperature is 32°C. If the rate of heat transfer into the dwelling through the walls and roof is 31,650 kJ/h, might a net power input to the air conditioner compressor of 2.2 kW be sufficient? If yes, determine the coefficient of performance. If no, determine the minimum theoretical power input, in kW.

5.46 By supplying energy at an average rate of 24,000 kJ/h, a heat pump maintains the temperature of a dwelling at 20°C. If electricity costs 8.5 cents per kW · h, determine the minimum theoretical operating cost for each day of operation if the heat pump receives energy by heat transfer from

  a. the outdoor air at −7°C.

  b. the ground at 5°C.

5.47 Two reversible refrigeration cycles operate in series. The first cycle receives energy by heat transfer from a cold reservoir at 300 K and rejects energy by heat transfer to a reservoir at an intermediate temperature T greater than 300 K. The second cycle receives energy by heat transfer from the reservoir at temperature T and rejects energy by heat transfer to a higher-temperature reservoir at 883 K. If the refrigeration cycles have the same coefficient of performance, determine (a) T, in K, and (b) the value of each coefficient of performance.

5.48 A heating system must maintain the interior of a building at $T_H = 20$°C when the outside temperature is $T_C = 2$°C. If the rate of heat transfer from the building through its walls and roof is 16.4 kW, determine the electrical power required, in kW, to heat the building using (a) electrical-resistance heating, (b) a heat pump whose coefficient of performance is 3.0, (c) a reversible heat pump operating between hot and cold reservoirs at 20°C and 2°C, respectively.

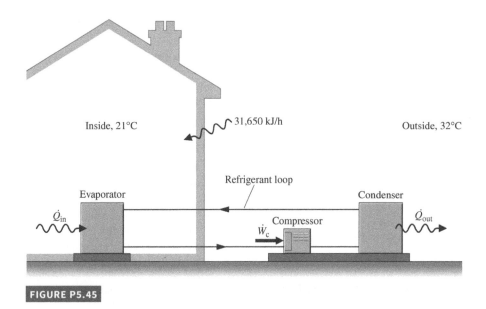

Carnot Cycle Applications

5.49 One-half kg of water executes a Carnot power cycle. During the isothermal expansion, the water is heated at 315°C from a saturated liquid to a saturated vapor. The vapor then expands adiabatically to a temperature of 32°C and a quality of 64.3%.

   a. Sketch the cycle on p–v coordinates.

   b. Evaluate the heat and work for each process, in kJ.

   c. Evaluate the thermal efficiency.

5.50 A gas within a piston-cylinder assembly executes a Carnot power cycle during which the isothermal expansion occurs at $T_H = 600$ K and the isothermal compression occurs at $T_C = 300$ K. Determine

   a. the thermal efficiency.

   b. the percent change in thermal efficiency if T_H increases by 15% while T_C remains the same.

   c. the percent change in thermal efficiency if T_C decreases by 15% while T_H remains the same.

   d. the percent change in thermal efficiency if T_H increases by 15% and T_C decreases by 15%.

5.51 One kilogram of air as an ideal gas executes a Carnot power cycle having a thermal efficiency of 50%. The heat transfer to the air during the isothermal expansion is 50 kJ. At the end of the isothermal expansion, the pressure is 574 kPa and the volume is 0.3 m³. Determine

   a. the maximum and minimum temperatures for the cycle, in K.

   b. the pressure and volume at the beginning of the isothermal expansion in bar and m³, respectively.

   c. the work and heat transfer for each of the four processes, in kJ.

   d. Sketch the cycle on p–v coordinates.

5.52 An ideal gas within a piston–cylinder assembly undergoes a Carnot refrigeration cycle, as shown in Fig. 5.16. The isothermal compression occurs at 325 K from 2 bar to 4 bar. The isothermal expansion occurs at 250 K. Determine (a) the coefficient of performance, (b) the heat transfer to the gas during the isothermal expansion, in kJ per kmol of gas, (c) the magnitude of the net work input, in kJ per kmol of gas.

5.53 Air within a piston-cylinder assembly executes a Carnot heat pump cycle, as shown in Fig. 5.16. For the cycle, $T_H = 600$ K and $T_C = 300$ K. The energy rejected by heat transfer at 600 K has a magnitude of 250 kJ per kg of air. The pressure at the start of the isothermal expansion is 325 kPa. Assuming the ideal gas model for the air, determine (a) the magnitude of the net work input, in kJ per kg of air, and (b) the pressure at the end of the isothermal expansion, in kPa.

5.54 The pressure–volume diagram of a Carnot power cycle executed by an ideal gas with constant specific heat ratio k is shown in Fig. P5.54. Demonstrate that

   a. $V_4 V_2 = V_1 V_3$.

   b. $T_2/T_3 = (p_2/p_3)^{(k-1)/k}$.

   c. $T_2/T_3 = (V_3/V_2)^{k-1}$.

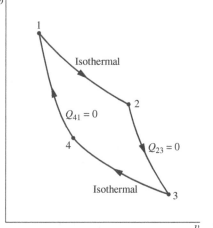

5.55 Carbon Dioxide (CO_2) as an ideal gas executes a Carnot cycle while operating between thermal reservoirs at 232°C and 38°C. The pressures at the initial and final states of the isothermal expansion are 2.8 MPa and 1.4 MPa, respectively. The specific heat ratio is $k = 1.24$. Using the results of Problem 5.54 as needed, determine

   a. the work and heat transfer for each of the four processes, in kJ/kg.

   b. the thermal efficiency.

c. the pressures at the initial and final states of the isothermal compression, in kPa.

5.56 One-tenth kilogram of air as an ideal gas with $k = 1.4$ executes a Carnot refrigeration cycle, as shown in Fig. 5.13. The isothermal expansion occurs at $-23°C$ with a heat transfer to the air of 3.4 kJ. The isothermal compression occurs at $27°C$ to a final volume of 0.01 m³. Using the results of Prob. 5.54 as needed, determine

   a. the pressure, in kPa, at each of the four principal states.

   b. the work, in kJ, for each of the four processes.

   c. the coefficient of performance.

Clausius Inequality Applications

5.57 A system executes a power cycle while receiving 1000 kJ by heat transfer at a temperature of 500 K and discharging energy by heat transfer at a temperature of 300 K. There are no other heat transfers. Applying Eq. 5.13, determine σ_{cycle} if the thermal efficiency is (a) 60%, (b) 40%, (c) 20%. Identify the cases (if any) that are internally reversible or impossible.

5.58 A system executes a power cycle while receiving 1690 kJ by heat transfer at a temperature of 1390 K and discharging 211 kJ by heat transfer at 278 K. A heat transfer from the system also occurs at a temperature of 833 K. There are no other heat transfers. If no internal irreversibilites are present, determine the thermal efficiency.

5.59 The steady-state data listed below are claimed for a power cycle operating between hot and cold reservoirs at 1200 K and 400 K, respectively. For each case, evaluate the net power developed by the cycle, in kW, and the thermal efficiency. Also in each case apply Eq. 5.13 on a time-rate basis to determine whether the cycle operates reversibly, operates irreversibly, or is impossible.

   a. $\dot{Q}_H = 600$ kW, $\dot{Q}_C = 400$ kW

   b. $\dot{Q}_H = 600$ kW, $\dot{Q}_C = 0$ kW

   c. $\dot{Q}_H = 600$ kW, $\dot{Q}_C = 200$ kW

5.60 At steady state, a thermodynamic cycle operating between hot and cold reservoirs at 1000 K and 500 K, respectively, receives energy by heat transfer from the hot reservoir at a rate of 1500 kW, discharges energy by heat transfer to the cold reservoir, and develops power at a rate of (a) 1000 kW, (b) 750 kW, (c) 0 kW. For each case, apply Eq. 5.13 on a time-rate basis to determine whether the cycle operates reversibly, operates irreversibly, or is impossible.

5.61 Figure P5.61 gives the schematic of a vapor power plant in which water steadily circulates through the four components shown. The water flows through the boiler and condenser at constant pressure and through the turbine and pump adiabatically. Kinetic and potential energy effects can be ignored. Process data follow:

   Process 4–1: constant pressure at 8 MPa from saturated liquid to saturated vapor

   Process 2–3: constant pressure at 8 kPa from $x_2 = 67.5\%$ to $x_3 = 34.2\%$

   a. Using Eq. 5.13 expressed on a time-rate basis, determine if the cycle is *internally reversible*, *irreversible*, or *impossible*.

   b. Determine the thermal efficiency using Eq. 5.4 expressed on a time-rate basis and steam table data.

   c. Compare the result of part (b) with the *Carnot efficiency* calculated using Eq. 5.9 with the boiler and condenser temperatures and comment.

5.62 Repeat Problem 5.61 for the following case:

   Process 4–1: constant pressure at 0.15 MPa from saturated liquid to saturated vapor

   Process 2–3: constant pressure at 20 kPa from $x_2 = 90\%$ to $x_3 = 30\%$

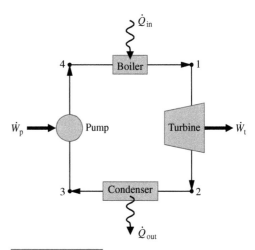

FIGURE P5.61–62

5.63 As shown in Fig. P5.63, a system executes a power cycle while receiving 750 kJ by heat transfer at a temperature of 1500 K and discharging 100 kJ by heat transfer at a temperature of 500 K. Another heat transfer from the system occurs at a temperature of 1000 K. Using Eq. 5.13, plot the thermal efficiency of the cycle versus σ_{cycle}, in kJ/K.

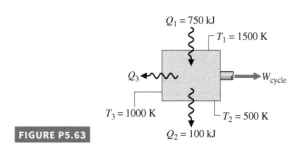

FIGURE P5.63

5.64 Figure P5.64 shows a system consisting of a power cycle driving a heat pump. At steady state, the power cycle receives $\dot{Q}_s$ by heat transfer at T_s from the high-temperature source and delivers $\dot{Q}_1$ to a dwelling at T_d. The heat pump receives $\dot{Q}_0$ from the outdoors at T_0, and delivers $\dot{Q}_2$ to the dwelling. Using Eq. 5.13, obtain an expression for the maximum theoretical value of the performance parameter $(\dot{Q}_1 + \dot{Q}_2)/\dot{Q}_s$ in terms of the temperature ratios T_s/T_d and T_0/T_d.

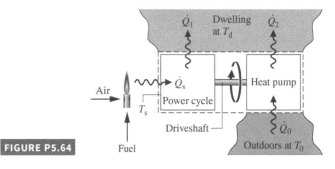

FIGURE P5.64

DESIGN & OPEN-ENDED PROBLEMS: EXPLORING ENGINEERING PRACTICE

5.1D The second law of thermodynamics is sometimes cited in publications of disciplines far removed from engineering and science, including but not limited to philosophy, economics, and sociology. Investigate use of the second law in peer-reviewed nontechnical publications. For three such publications, each in different disciplines, write a three-page critique. For each publication, identify and comment on the key objectives and conclusions. Clearly explain how the second law is used to inform the reader and propel the presentation. Score each publication on a 10-point scale, with 10 denoting a highly effective use of the second law and 1 denoting an ineffective use. Provide a rationale for each score.

5.2D Experiments by Australian researchers suggest that the second law can be violated at the micron scale over time intervals of up to 2 seconds. If validated, some say these findings could place a limit on the engineering of nanomachines because such devices may not behave simply like miniaturized versions of their larger counterparts. Investigate the implications, if any, for nanotechnology of these experimental findings. Write a report including at least three references.

5.3D For each of three comparably sized spaces in your locale, a preschool, an office suite with cubicles, and an assisted-living facility, investigate the suitability of a heat pump/air-conditioning system employing a *natural* refrigerant. Consider factors including, but not limited to, health and safety requirements, applicable codes, performance in meeting occupant comfort needs, annual electricity cost, and environmental impact, each in comparison to systems using conventional refrigerants for the same duty. Summarize your findings in a report, including at least three references.

5.4D In an *Internet* paper titled, "How Lightning Strikes are Produced and Why the Second Law of Thermodynamics Is Invalid," the author claims that electricity can be produced economically by extracting "constant-temperature ambient heat from the atmosphere or the oceans and converting it fully into work." Critically evaluate this claim, and report your findings in a memorandum.

5.5D The objective of this project is to identify a commercially available heat pump system that will meet annual heating and cooling needs of an existing dwelling in a locale of your choice. Consider each of two types of heat pump: air source and ground source. Estimate installation costs, operating costs, and other pertinent costs for each type of heat pump. Assuming a 10-year life, specify the more economical heat pump system. What if electricity were to cost twice its current cost? Prepare a poster presentation of your findings.

5.6D **Figure P5.6D** shows one of those bobbing toy birds that seemingly takes an endless series of sips from a cup filled with water. Prepare a 30-min presentation suitable for a middle school science class explaining the operating principles of this device and whether or not its behavior is at odds with the second law.

5.7D Insulin and several other pharmaceuticals required daily by those suffering from diabetes and other medical conditions have relatively low thermal stability. Those living and traveling in hot climates are especially at risk by heat-induced loss of potency of their pharmaceuticals. Design a wearable, lightweight, and reliable cooler for transporting temperature-sensitive pharmaceuticals. The cooler also must be solely powered by human motion. While the long-term goal is a moderately priced consumer product, the final project report need only provide the costing of a single prototype.

5.8D Four hundred feet below a city in southern Illinois sits an abandoned lead mine filled with an estimated 70 billion gallons of water that remains at a constant temperature of about 58°F. The city engineer has proposed using the impounded water as a resource for heating and cooling the city's central administration building, a two-story, brick building constructed in 1975 having 8500 ft² of office space. You have been asked to develop a preliminary proposal, including a cost estimate. The proposal will specify commercially available systems that utilize the impounded water to meet heating and cooling needs. The cost estimate will include project development, hardware, and annual operating cost. Report your findings in the form of a PowerPoint presentation suitable for the city council.

5.9D The heat transfer rate through the walls and roof of a building is 3570 kJ/h per degree temperature difference between the inside and outside. For outdoor temperatures ranging from 20 to −20°C, make a comparison of the daily cost of maintaining the building at 20°C by means of an electric heat pump, direct electric resistance heating, and a conventional gas-fired furnace. Report your findings in a memorandum.

5.10D **Figure P5.10D** shows that the typical thermal efficiency of U.S. power plants increased rapidly in the early and mid-1900s, but has increased only gradually since then. Investigate the most important factors contributing to this plateauing of thermal efficiency and the most promising near-term and long-term technologies that might lead to appreciable thermal efficiency gains. Report your findings in a memorandum.

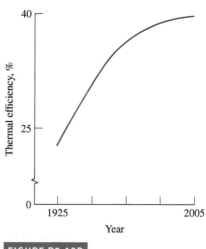

FIGURE P5.10D

FIGURE P5.6D

CHAPTER 6

Using Entropy

Engineering Context

Up to this point, our study of the second law has been concerned primarily with what it says about systems undergoing thermodynamic cycles. In this chapter means are introduced for analyzing systems from the second law perspective as they undergo processes that are not necessarily cycles. The property *entropy* and the *entropy production* concept introduced in Chap. 5 play prominent roles in these considerations.

The **objective** of this chapter is to develop an understanding of entropy concepts, including the use of entropy balances for closed systems and control volumes in forms effective for the analysis of engineering systems. The Clausius inequality developed in Sec. 5.11, expressed as Eq. 5.13, provides the basis.

LEARNING OUTCOMES

When you complete your study of this chapter, you will be able to...

- Explain key concepts related to entropy and the second law, including entropy transfer, entropy production, and the increase in entropy principle.
- Evaluate entropy, evaluate entropy change between two states, and analyze isentropic processes, using appropriate property data.
- Represent heat transfer in an internally reversible process as an area on a temperature–entropy diagram.
- Analyze closed systems and control volumes, including applying entropy balances.
- Use isentropic efficiencies for turbines, nozzles, compressors, and pumps for second law analysis.

6.1 Entropy–A System Property

The word *energy* is so much a part of the language that you were undoubtedly familiar with the term before encountering it in early science courses. This familiarity probably facilitated the study of energy in these courses and in the current course in engineering thermodynamics. In the present chapter you will see that the analysis of systems from a second law perspective is effectively accomplished in terms of the property *entropy*. Energy and entropy are both abstract concepts. However, unlike energy, the word *entropy* is seldom heard in everyday conversation, and you may never have dealt with it quantitatively before. Energy and entropy play important roles in the remaining chapters of this book.

6.1.1 Defining Entropy Change

A quantity is a property if, and only if, its change in value between two states is independent of the process (Sec. 1.3.3). This aspect of the property concept is used in the present section together with the Clausius inequality to introduce entropy change as follows:

Two cycles executed by a closed system are represented in **Fig. 6.1**. One cycle consists of an internally reversible process A from state 1 to state 2, followed by internally reversible process C from state 2 to state 1. The other cycle consists of an internally reversible process B from state 1 to state 2, followed by the same process C from state 2 to state 1 as in the first cycle. For the first cycle, Eq. 5.13 (the Clausius inequality) takes the form

$$\left(\int_1^2 \frac{\delta Q}{T} \right)_A + \left(\int_2^1 \frac{\delta Q}{T} \right)_C = -\sigma_{\text{cycle}}^{\;0} \tag{6.1a}$$

For the second cycle, Eq. 5.13 takes the form

$$\left(\int_1^2 \frac{\delta Q}{T} \right)_B + \left(\int_2^1 \frac{\delta Q}{T} \right)_C = -\sigma_{\text{cycle}}^{\;0} \tag{6.1b}$$

In writing Eqs. 6.1, the term σ_{cycle} has been set to zero since the cycles are composed of internally reversible processes.

When Eq. 6.1b is subtracted from Eq. 6.1a, we get

$$\left(\int_1^2 \frac{\delta Q}{T} \right)_A = \left(\int_1^2 \frac{\delta Q}{T} \right)_B$$

This shows that the integral of $\delta Q/T$ is the same for both processes. Since A and B are arbitrary, it follows that the integral of $\delta Q/T$ has the same value for *any* internally reversible process between the two states. In other words, the value of the integral depends on the end states only. It can be concluded, therefore, that the integral represents the change in some property of the system.

Selecting the symbol S to denote this property, which is called *entropy*, the **change in entropy** is given by

$$\boxed{ S_2 - S_1 = \left(\int_1^2 \frac{\delta Q}{T} \right)_{\substack{\text{int} \\ \text{rev}}} } \tag{6.2a}$$

where the subscript "int rev" is added as a reminder that the integration is carried out for any internally reversible process linking the two states. On a differential basis, the defining equation for entropy change takes the form

$$dS = \left(\frac{\delta Q}{T} \right)_{\substack{\text{int} \\ \text{rev}}} \tag{6.2b}$$

Entropy is an extensive property.

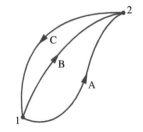

Fig. 6.1 Two internally reversible cycles.

$$\oint \left(\frac{\delta Q}{T} \right)_b = -\sigma_{\text{cycle}}$$

(Eq. 5.13)

definition of entropy change

The **SI unit for entropy** is J/K. However, in this book it is convenient to work in terms of kJ/K. Units in SI for *specific* entropy are kJ/kg · K for s and kJ/kmol · K for $\bar{s}$.

It should be clear that entropy is defined and evaluated in terms of a particular expression (Eq. 6.2a) for which *no accompanying physical picture is given*. We encountered this previously with the property enthalpy. Enthalpy is introduced without physical motivation in Sec. 3.6.1. Then, in Chap. 4, we learned how enthalpy is used for thermodynamic analysis of control volumes. As for the case of enthalpy, to gain an appreciation for entropy you need to understand *how* it is used and *what* it is used for. This is the aim of the rest of this chapter.

6.1.2 Evaluating Entropy

Since entropy is a property, the change in entropy of a system in going from one state to another is the same for *all* processes, both internally reversible and irreversible, between these two states. Thus, Eq. 6.2a allows the determination of the change in entropy, and once it has been evaluated, this is the magnitude of the entropy change for all processes of the system between the two states.

The defining equation for entropy change, Eq. 6.2a, serves as the basis for evaluating entropy relative to a reference value at a reference state. Both the reference value and the reference state can be selected arbitrarily. The value of entropy at any state y relative to the value at the reference state x is obtained in principle from

$$S_y = S_x + \left(\int_x^y \frac{\delta Q}{T} \right)_{\substack{\text{int} \\ \text{rev}}} \tag{6.3}$$

where S_x is the reference value for entropy at the specified reference state.

The use of entropy values determined relative to an arbitrary reference state is satisfactory as long as they are used in calculations involving entropy differences, for then the reference value cancels. This approach suffices for applications where composition remains constant. When chemical reactions occur, it is necessary to work in terms of *absolute* values of entropy determined using the *third law of thermodynamics* (Chap. 13).

6.1.3 Entropy and Probability

The presentation of engineering thermodynamics provided in this book takes a *macroscopic* view as it deals mainly with the gross, or overall, behavior of matter. The macroscopic concepts of engineering thermodynamics introduced thus far, including energy and entropy, rest on operational definitions whose validity is shown directly or indirectly through experimentation. Still, insights concerning energy and entropy can result from considering the microstructure of matter. This brings in the use of *probability* and the notion of *disorder*. Further discussion of entropy, probability, and disorder is provided in Sec. 6.8.2.

6.2 Retrieving Entropy Data

In Chap. 3, we introduced means for retrieving property data, including tables, graphs, equations, and the software available with this text. The emphasis there is on evaluating the properties p, v, T, u, and h required for application of the conservation of mass and energy principles. For application of the second law, entropy values are usually required. In this section, means for retrieving entropy data for water and several refrigerants are considered.

Tables of thermodynamic data are introduced in Secs. 3.5 and 3.6 (Tables A-2 through A-18). Specific entropy is tabulated in the same way as considered there for the properties v, u, and h, and entropy values are retrieved similarly. The specific entropy values given in Tables A-2 through A-18 are relative to the following *reference states and values*. For water, the entropy of saturated liquid at 0.01°C is set to zero. For the refrigerants, the entropy of the saturated liquid at −40°C is assigned a value of zero.

6.2.1 Vapor Data

In the superheat regions of the tables for water and the refrigerants, specific entropy is tabulated along with v, u, and h versus temperature and pressure.

> **FOR EXAMPLE**
>
> Consider two states of water. At state 1 the pressure is 3 MPa and the temperature is 500°C. At state 2, the pressure is 0.3 MPa and the specific entropy is the same as at state 1, $s_2 = s_1$. The object is to determine the temperature at state 2. Using T_1 and p_1, we find the specific entropy at state 1 from Table A-4 as $s_1 = 7.2338$ kJ/kg · K. State 2 is fixed by the pressure, $p_2 = 0.3$ MPa, and the specific entropy, $s_2 = 7.2338$ kJ/kg · K. Returing to Table A-4 at 0.3 MPa and interpolating with s_2 between 160 and 200°C results in $T_2 = 183$°C.

6.2.2 Saturation Data

For saturation states, the values of s_f and s_g are tabulated as a function of either saturation pressure or saturation temperature. The specific entropy of a two-phase liquid–vapor mixture is calculated using the quality

$$\boxed{\begin{aligned} s &= (1 - x)s_f + xs_g \\ &= s_f + x(s_g - s_f) \end{aligned}}$$
(6.4)

These relations are identical in form to those for v, u, and h (Secs. 3.5 and 3.6).

> **FOR EXAMPLE**
>
> Let us determine the specific entropy of Refrigerant 134a at a state where the temperature is 0°C and the specific internal energy is 138.43 kJ/kg. Referring to Table A-10, we see that the given value for u falls between u_f and u_g at 0°C, so the system is a two-phase liquid–vapor mixture. The quality of the mixture can be determined from the known specific internal energy
>
> $$x = \frac{u - u_f}{u_g - u_f} = \frac{138.43 - 49.79}{227.06 - 49.79} = 0.5$$
>
> Then with values from Table A-10, Eq. 6.4 gives
>
> $$\begin{aligned} s &= (1 - x)s_f + xs_g \\ &= (0.5)(0.1970) + (0.5)(0.9190) = 0.5580 \text{ kJ/kg} \cdot \text{K} \end{aligned}$$

6.2.3 Liquid Data

Compressed liquid data are presented for water in Table A-5. In these tables s, v, u, and h are tabulated versus temperature and pressure as in the superheat tables, and the tables are used similarly. In the absence of compressed liquid data, the value of the specific entropy can be estimated in the same way as estimates for v and u are obtained for liquid states (Sec. 3.10.1), by using the saturated liquid value at the given temperature

$$\boxed{s(T, p) \approx s_f(T)}$$
(6.5)

> **FOR EXAMPLE**
>
> Suppose the value of specific entropy is required for water at 25 bar, 200°C. The specific entropy is obtained directly from Table A-5 as $s = 2.3294$ kJ/kg · K. Using the saturated liquid value for specific entropy at 200°C from Table A-2, the specific entropy is approximated with Eq. 6.5 as $s = 2.3309$ kJ/kg · K, which agrees closely with the previous value.

6.2.4 Computer Retrieval

The software available with this text, *Interactive Thermodynamics: IT*, provides data for the substances considered in this section. Entropy data are retrieved by simple call statements placed in the workspace of the program.

> **FOR EXAMPLE**
>
> Consider a two-phase liquid–vapor mixture of H_2O at $p = 1$ bar, $v = 0.8475$ m³/kg. The following illustrates how specific entropy and quality x are obtained using *IT*:
>
> ```
> p = 1 // bar
> v = 0.8475 // m³/kg
> v = vsat_Px("Water/Steam",p,x)
> s = ssat_Px("Water/Steam",p,x)
> ```
>
> The software returns values of $x = 0.5$ and $s = 4.331$ kJ/kg · K, which can be checked using data from Table A-3. Note that quality x is implicit in the expression for specific volume, and it is not necessary to solve explicitly for x. As another example, consider superheated ammonia vapor at $p = 1.5$ bar, $T = 8°C$. Specific entropy is obtained from *IT* as follows:
>
> ```
> p = 1.5 // bar
> T = 8 // °C
> s = s_PT ("Ammonia", p,T)
> ```
>
> The software returns $s = 5.981$ kJ/kg · K, which agrees closely with the value obtained by interpolation in Table A-15.

6.2.5 Using Graphical Entropy Data

The use of property diagrams as an adjunct to problem solving is emphasized throughout this book. When applying the second law, it is frequently helpful to locate states and plot processes on diagrams having entropy as a coordinate. Two commonly used figures having entropy as one of the coordinates are the temperature–entropy diagram and the enthalpy–entropy diagram.

Temperature–Entropy Diagram   The main features of a **temperature–entropy diagram** are shown in **Fig. 6.2**. For detailed figures for water in SI units, see Fig. A-7. Observe that lines of constant enthalpy are shown on these figures. Also note that in the superheated vapor region constant specific volume lines have a steeper slope than constant-pressure lines. Lines of constant quality are shown in the two-phase liquid–vapor region. On some figures, lines of constant quality are marked as *percent moisture* lines. The percent moisture is defined as the ratio of the mass of liquid to the total mass.

T–s diagram

In the superheated vapor region of the *T–s* diagram, constant specific enthalpy lines become nearly horizontal as pressure is reduced. These superheated vapor states are shown as the shaded area on Fig. 6.2. For states in this region of the diagram, the enthalpy is determined primarily by the temperature: $h(T, p) \approx h(T)$. This is the region of the diagram where the ideal gas model provides a reasonable approximation. For superheated vapor states outside the shaded area, both temperature and pressure are required to evaluate enthalpy, and the ideal gas model is not suitable.

Enthalpy–Entropy Diagram   The essential features of an enthalpy–entropy diagram, commonly known as a **Mollier diagram**, are shown in **Fig. 6.3**. For detailed figure for water in SI units, see Fig. A-8. Note the location of the critical point and the appearance of lines of constant temperature and constant pressure. Lines of constant quality are shown in the two-phase liquid–vapor region (some figures give lines of constant percent moisture). The figure is intended for evaluating properties at superheated vapor states and for two-phase liquid–vapor mixtures. Liquid data are seldom shown. In the superheated vapor region, constant-temperature lines become nearly horizontal as pressure is reduced. These superheated vapor states are shown, approximately, as the shaded area on Fig. 6.3. This area corresponds to the shaded area on the temperature–entropy diagram of Fig. 6.2, where the ideal gas model provides a reasonable approximation.

Mollier diagram

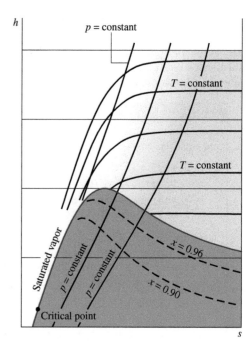

Fig. 6.2 Temperature–entropy diagram. **Fig. 6.3** Enthalpy–entropy diagram.

FOR EXAMPLE

To illustrate the use of the Mollier diagram in SI units, consider two states of water. At state 1, $T_1 = 240°C$, $p_1 = 0.10$ MPa. The specific enthalpy and quality are required at state 2, where $p_2 = 0.01$ MPa and $s_2 = s_1$. Turning to Fig. A-8, state 1 is located in the superheated vapor region. Dropping a vertical line into the two-phase liquid–vapor region, state 2 is located. The quality and specific enthalpy at state 2 read from the figure agree closely with values obtained using Tables A-3 and A-4: $x_2 = 0.98$ and $h_2 = 2537$ kJ/kg.

6.3 Introducing the *T dS* Equations

Although the change in entropy between two states can be determined in principle by using Eq. 6.2a, such evaluations are generally conducted using the *T dS* equations developed in this section. The *T dS* equations allow entropy changes to be evaluated from other more readily determined property data. The use of the *T dS* equations to evaluate entropy changes for an incompressible substance is illustrated in Sec. 6.4 and for ideal gases in Sec. 6.5. The importance of the *T dS* equations is greater than their role in assigning entropy values, however. In Chap. 11 they are used as a point of departure for deriving many important property relations for pure, simple compressible systems, including means for constructing the property tables giving u, h, and s.

The *T dS* equations are developed by considering a pure, simple compressible system undergoing an internally reversible process. In the absence of overall system motion and the effects of gravity, an energy balance in differential form is

$$(\delta Q)_{\substack{\text{int} \\ \text{rev}}} = dU + (\delta W)_{\substack{\text{int} \\ \text{rev}}} \tag{6.6}$$

By definition of simple compressible system (Sec. 3.1.2), the work is

$$(\delta W)_{\substack{\text{int} \\ \text{rev}}} = p \, dV \tag{6.7a}$$

On rearrangement of Eq. 6.2b, the heat transfer is

$$(\delta Q)_{\substack{\text{int} \\ \text{rev}}} = T \, dS \tag{6.7b}$$

first *T dS* equation Substituting Eqs. 6.7 into Eq. 6.6, the **first *T dS* equation** results

$$\boxed{T \, dS = dU + p \, dV} \tag{6.8}$$

The *second T dS* equation is obtained from Eq. 6.8 using $H = U + pV$. Forming the differential

$$dH = dU + d(pV) = dU + p\,dV + V\,dp$$

On rearrangement

$$dU + p\,dV = dH - V\,dp$$

Substituting this into Eq. 6.8 gives the **second *T dS* equation**

$$\boxed{T\,dS = dH - V\,dp} \tag{6.9}$$

second *T dS* equation

The *T dS* equations can be written on a unit mass basis as

$$\boxed{\begin{aligned} T\,ds &= du + p\,dv \\ T\,ds &= dh - v\,dp \end{aligned}}$$

$$(6.10a)$$
$$(6.10b)$$

or on a per mole basis as

$$\boxed{\begin{aligned} T\,d\bar{s} &= d\bar{u} + p\,d\bar{v} \\ T\,d\bar{s} &= d\bar{h} - \bar{v}dp \end{aligned}}$$

$$(6.11a)$$
$$(6.11b)$$

Although the *T dS* equations are derived by considering an internally reversible process, an entropy change obtained by integrating these equations is the change for *any* process, reversible or irreversible, between two equilibrium states of a system. Because entropy is a property, the change in entropy between two states is independent of the details of the process linking the states.

To show the use of the *T dS* equations, consider a change in phase from saturated liquid to saturated vapor at constant temperature and pressure. Since pressure is constant, Eq. 6.10b reduces to give

$$ds = \frac{dh}{T}$$

Then, because temperature is also constant during the phase change,

$$s_g - s_f = \frac{h_g - h_f}{T} \tag{6.12}$$

This relationship shows how $s_g - s_f$ is calculated for tabulation in property tables.

FOR EXAMPLE

Consider Refrigerant 134a at 0°C. From Table A-10, $h_g - h_f = 197.21$ kJ/kg, so with Eq. 6.12

$$s_g - s_f = \frac{197.21 \text{ kJ/kg}}{273.15 \text{ K}} = 0.7220 \frac{\text{kJ}}{\text{kg} \cdot \text{K}}$$

which is the value calculated using s_f and s_g from the table.

6.4 Entropy Change of an Incompressible Substance

In this section, Eq. 6.10a of Sec. 6.3 is used to evaluate the entropy change between two states of an incompressible substance. The incompressible substance model introduced in Sec. 3.10.2 assumes that the specific volume (density) is constant and the specific internal energy depends solely on temperature. Thus, $du = c(T)dT$, where c denotes the specific heat of the substance, and Eq. 6.10a reduces to give

$$ds = \frac{c(T)dT}{T} + \frac{p\,dv}{T}^{0} = \frac{c(T)dT}{T}$$

On integration, the change in specific entropy is

$$s_2 - s_1 = \int_{T_1}^{T_2} \frac{c(T)}{T} dT$$

When the specific heat is constant, this becomes

$$s_2 - s_1 = c \ln \frac{T_2}{T_1} \quad \text{(incompressible, constant } c) \tag{6.13}$$

Equation 6.13, along with Eqs. 3.20 giving Δu and Δh, respectively, are applicable to liquids and solids modeled as incompressible. Specific heats of some common liquids and solids are given in Table A-19.

FOR EXAMPLE

Consider a system consisting of liquid water initially at $T_1 = 300$ K, $p_1 = 2$ bar undergoing a process to a final state at $T_2 = 323$ K, $p_2 = 1$ bar. There are two ways to evaluate the change in specific entropy in this case. The first approach is to use Eq. 6.5 together with saturated liquid data from Table A-2. That is, $s_1 \approx s_f(T_1) = 0.3954$ KJ/kg $\cdot$ K and $s_2 \approx s_f(T_2) = 0.7038$ KJ/kg $\cdot$ K, giving $s_2 - s_1 = 0.308$ KJ/kg $\cdot$ K. The second approach is to use the incompressible model. That is, with Eq. 6.13 and $c = 4.18$ KJ/kg $\cdot$ K from Table A-19, we get

$$s_2 - s_1 = c \ln \frac{T_2}{T_1}$$

$$= \left(4.18 \frac{\text{kJ}}{\text{kg} \cdot \text{K}}\right) \ln \left(\frac{323 \text{ K}}{300 \text{ K}}\right) = 0.309 \text{ KJ/kg} \cdot \text{K}$$

Comparing the values obtained for the change in specific entropy using the two approaches considered here, we see they are in agreement.

6.5 Entropy Change of an Ideal Gas

In this section, the $T\,dS$ equations of Sec. 6.3, Eqs. 6.10, are used to evaluate the entropy change between two states of an ideal gas. For a quick review of ideal gas model relations, see Table 6.1.

TABLE 6.1

Ideal Gas Model Review

Equations of state:

$$pv = RT \tag{3.32}$$
$$pV = mRT \tag{3.33}$$

Changes in u and h:

$$u(T_2) - u(T_1) = \int_{T_1}^{T_2} c_v(T)\,dT \tag{3.40}$$

$$h(T_2) - h(T_1) = \int_{T_1}^{T_2} c_p(T)\,dT \tag{3.43}$$

Constant Specific Heats		Variable Specific Heats
$u(T_2) - u(T_1) = c_v(T_2 - T_1)$	(3.50)	$u(T)$ and $h(T)$ are evaluated from Table A-22
$h(T_2) - h(T_1) = c_p(T_2 - T_1)$	(3.51)	for air (mass basis) and Table A-23 for several other gases (molar basis).
See Table A-20 for c_v and c_p data.		

It is convenient to begin with Eqs. 6.10 expressed as

$$ds = \frac{du}{T} + \frac{p}{T}dv \tag{6.14}$$

$$ds = \frac{dh}{T} - \frac{v}{T}dp \tag{6.15}$$

For an ideal gas, $du = c_v(T)\,dT$, $dh = c_p(T)\,dT$, and $pv = RT$. With these relations, Eqs. 6.14 and 6.15 become, respectively,

$$ds = c_v(T)\frac{dT}{T} + R\frac{dv}{v} \quad \text{and} \quad ds = c_p(T)\frac{dT}{T} - R\frac{dp}{p} \tag{6.16}$$

On integration, Eqs. 6.16 give, respectively,

$$s(T_2, v_2) - s(T_1, v_1) = \int_{T_1}^{T_2} c_v(T)\frac{dT}{T} + R\ln\frac{v_2}{v_1} \tag{6.17}$$

$$s(T_2, p_2) - s(T_1, p_1) = \int_{T_1}^{T_2} c_p(T)\frac{dT}{T} - R\ln\frac{p_2}{p_1} \tag{6.18}$$

Since R is a constant, the last terms of Eqs. 6.16 can be integrated directly. However, because c_v and c_p are functions of temperature for ideal gases, it is necessary to have information about the functional relationships before the integration of the first term in these equations can be performed. Since the two specific heats are related by

$$c_p(T) = c_v(T) + R \tag{3.44}$$

where R is the gas constant, knowledge of either specific function suffices.

6.5.1 Using Ideal Gas Tables

As for internal energy and enthalpy changes of ideal gases, the evaluation of entropy changes for ideal gases can be reduced to a convenient tabular approach. We begin by introducing a new variable $s^\circ(T)$ as

$$s^\circ(T) = \int_{T'}^{T} \frac{c_p(T)}{T}\,dT \tag{6.19}$$

where T' is an arbitrary reference temperature.

The integral of Eq. 6.18 can be expressed in terms of s° as follows:

$$\int_{T_1}^{T_2} c_p\frac{dT}{T} = \int_{T'}^{T_2} c_p\frac{dT}{T} - \int_{T'}^{T_1} c_p\frac{dT}{T}$$
$$= s^\circ(T_2) - s^\circ(T_1)$$

Thus, Eq. 6.18 can be written as

$$s(T_2, p_2) - s(T_1, p_1) = s^\circ(T_2) - s^\circ(T_1) - R\ln\frac{p_2}{p_1} \tag{6.20a}$$

or on a per mole basis as

$$\bar{s}(T_2, p_2) - \bar{s}(T_1, p_1) = \bar{s}^\circ(T_2) - \bar{s}^\circ(T_1) - \bar{R}\ln\frac{p_2}{p_1} \tag{6.20b}$$

Because s° depends only on temperature, it can be tabulated versus temperature, like h and u. For air as an ideal gas, s° with units of kJ/kg · K is given in Table A-22. Values of $\bar{s}^\circ$ for several other common gases are given in Table A-23 with units of kJ/kmol · K. In passing, we note the arbitrary reference temperature T' of Eq. 6.19 is specified differently in Table A-22 than in Table A-23. As discussed in Sec. 13.5.1, Table A-23 give *absolute entropy* values.

Using Eqs. 6.20 and the tabulated values for $s°$ or $\bar{s}°$, as appropriate, entropy changes can be determined that account explicitly for the variation of specific heat with temperature.

FOR EXAMPLE

Let us evaluate the change in specific entropy, in kJ/kg · K, of air modeled as an ideal gas from a state where $T_1 = 300$ K and $p_1 = 1$ bar to a state where $T_2 = 1000$ K and $p_2 = 3$ bar. Using Eq. 6.20a and data from Table A-22,

$$s_2 - s_1 = s°(T_2) - s°(T_1) - R \ln \frac{p_2}{p_1}$$

$$= (2.96770 - 1.70203)\frac{kJ}{kg \cdot K} - \frac{8.314}{28.97}\frac{kJ}{kg \cdot K} \ln \frac{3 \text{ bar}}{1 \text{ bar}}$$

$$= 0.9504 \text{ kJ/kg} \cdot K$$

If a table giving $s°$ (or $\bar{s}°$) is not available for a particular gas of interest, the integrals of Eqs. 6.17 and 6.18 can be performed analytically or numerically using specific heat data such as provided in Tables A-20 and A-21.

6.5.2 Assuming Constant Specific Heats

When the specific heats c_v and c_p are taken as constants, Eqs. 6.17 and 6.18 reduce, respectively, to

$$s(T_2, v_2) - s(T_1, v_1) = c_v \ln \frac{T_2}{T_1} + R \ln \frac{v_2}{v_1} \tag{6.21}$$

$$s(T_2, p_2) - s(T_1, p_1) = c_p \ln \frac{T_2}{T_1} - R \ln \frac{p_2}{p_1} \tag{6.22}$$

These equations, along with Eqs. 3.50 and 3.51 giving Δu and Δh, respectively, are applicable when assuming the ideal gas model with constant specific heats.

FOR EXAMPLE

Let us determine the change in specific entropy, in KJ/kg · K, of air as an ideal gas undergoing a process from $T_1 = 300$ K, $p_1 = 1$ bar to $T_2 = 400$ K, $p_2 = 5$ bar. Because of the relatively small temperature range, we assume a constant value of c_p evaluated at 350 K. Using Eq. 6.22 and $c_p = 1.008$ KJ/kg · K from Table A-20,

$$\Delta s = c_p \ln \frac{T_2}{T_1} - R \ln \frac{p_2}{p_1}$$

$$= \left(1.008 \frac{kJ}{kg \cdot K}\right) \ln \left(\frac{400 \text{ K}}{300 \text{ K}}\right) - \left(\frac{8.314}{28.97}\frac{kJ}{kg \cdot K}\right) \ln \left(\frac{5 \text{ bar}}{1 \text{ bar}}\right)$$

$$= -0.1719 \text{ kJ/kg} \cdot K$$

6.5.3 Computer Retrieval

For gases modeled as ideal gases, *IT directly* returns $s(T, p)$ using the following special form of Eq. 6.18:

$$s(T, p) - s(T_{ref}, p_{ref}) = \int_{T_{ref}}^{T} \frac{c_p(T)}{T} dT - R \ln \frac{p}{p_{ref}}$$

and the following choice of reference state and reference value: $T_{ref} = 0$ K, $p_{ref} = 1$ atm, and $s(T_{ref}, p_{ref}) = 0$, giving

$$s(T, p) = \int_{0}^{T} \frac{c_p(T)}{T} dT - R \ln \frac{p}{p_{ref}} \tag{a}$$

Such reference state and reference value choices equip *IT* for use in combustion applications. See the discussion of *absolute entropy* in Sec. 13.5.1.

Changes in specific entropy evaluated using *IT* agree with entropy *changes* evaluated using ideal gas tables.

> **FOR EXAMPLE**
>
> Consider a process of air as an ideal gas from T_1 = 300 K, p_1 = 1 bar to T_2 = 1000 K, p_2 = 3 bar. The change in specific entropy, denoted as dels, is determined in SI units using *IT* as follows:
>
> ```
> p1 = 1//bar
> T1 = 300//K
> p2 = 3
> T2 = 1000
> s1 = s_TP("Air",T1,p1)
> s2 = s_TP("Air",T2,p2)
> dels = s2 - s1
> ```
>
> The software returns values of s_1 = 1.706, s_2 = 2.656, and dels = 0.9501, all in units of kJ/kg · K. This value for Δs agrees with the value obtained using Table A-22: 0.9504 kJ/kg · K, as shown in the concluding example of Sec. 6.5.1.

Note again that *IT* returns specific entropy directly using Eq. (a) above. *IT* does not use the special function $s°$.

6.6 Entropy Change in Internally Reversible Processes of Closed Systems

In this section the relationship between entropy change and heat transfer for internally reversible processes is considered. The concepts introduced have important applications in subsequent sections of the book. The present discussion is limited to the case of closed systems. Similar considerations for control volumes are presented in Sec. 6.13.

As a closed system undergoes an internally reversible process, its entropy can increase, decrease, or remain constant. This can be brought out using

$$dS = \left(\frac{\delta Q}{T} \right)_{\substack{\text{int} \\ \text{rev}}} \qquad (6.2b)$$

which indicates that when a closed system undergoing an internally reversible process receives energy by heat transfer, the system experiences an increase in entropy. Conversely, when energy is removed from the system by heat transfer, the entropy of the system decreases. This can be interpreted to mean that an **entropy transfer** *accompanies* heat transfer. The direction of the entropy transfer is the same as that of the heat transfer. In an *adiabatic* internally reversible process, entropy remains constant. A constant-entropy process is called an **isentropic process**.

entropy transfer

isentropic process

6.6.1 Area Representation of Heat Transfer

On rearrangement, Eq. 6.2b gives

$$(\delta Q)_{\substack{\text{int} \\ \text{rev}}} = T \, dS$$

Integrating from an initial state 1 to a final state 2

$$\boxed{Q_{\substack{\text{int} \\ \text{rev}}} = \int_1^2 T \, dS} \qquad (6.23)$$

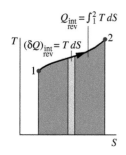

Fig. 6.4 Area representation of heat transfer for an internally reversible process of a closed system.

Carnot cycle

From Eq. 6.23 it can be concluded that an energy transfer by heat to a closed system during an internally reversible process can be represented as an area on a temperature–entropy diagram. **Figure 6.4** illustrates the area representation of heat transfer for an arbitrary internally reversible process in which temperature varies. Carefully note that temperature must be in kelvins or degrees Rankine, and the area is the entire area under the curve (shown shaded). Also note that the area representation of heat transfer is not valid for irreversible processes, as will be demonstrated later.

6.6.2 Carnot Cycle Application

To provide an example illustrating both the entropy change that accompanies heat transfer and the area representation of heat transfer, consider Fig. 6.5a, which shows a **Carnot power cycle** (Sec. 5.10.1). The cycle consists of four internally reversible processes in series: two isothermal processes alternated with two adiabatic processes. In Process 2–3, heat transfer to the system occurs while the temperature of the system remains constant at T_H. The system entropy increases due to the accompanying entropy transfer. For this process, Eq. 6.23 gives $Q_{23} = T_H(S_3 - S_2)$, so area 2–3–a–b–2 on **Fig. 6.5a** represents the heat transfer during the process. Process 3–4 is an adiabatic and internally reversible process and thus is an isentropic (constant-entropy) process. Process 4–1 is an isothermal process at T_C during which heat is transferred *from* the system. Since entropy transfer accompanies the heat transfer, system entropy decreases. For this process, Eq. 6.23 gives $Q_{41} = T_C(S_1 - S_4)$, which is negative in value. Area 4–1–b–a–4 on Fig. 6.5a represents the *magnitude* of the heat transfer Q_{41}. Process 1–2, which completes the cycle, is adiabatic and internally reversible (isentropic).

The net work of any cycle is equal to the net heat transfer, so *enclosed* area 1–2–3–4–1 represents the net work of the cycle. The thermal efficiency of the cycle also can be expressed in terms of areas:

$$\eta = \frac{W_{\text{cycle}}}{Q_{23}} = \frac{\text{area } 1-2-3-4-1}{\text{area } 2-3-a-b-2}$$

The numerator of this expression is $(T_H - T_C)(S_3 - S_2)$ and the denominator is $T_H(S_3 - S_2)$, so the thermal efficiency can be given in terms of temperatures only as $\eta = 1 - T_C/T_H$. This, of course, agrees with Eq. 5.9.

If the cycle were executed as shown in Fig. 6.5b, the result would be a Carnot refrigeration or heat pump cycle. In such a cycle, heat is transferred to the system while its temperature remains at T_C, so entropy increases during Process 1–2. In Process 3–4 heat is transferred from the system while the temperature remains constant at T_H and entropy decreases.

6.6.3 Work and Heat Transfer in an Internally Reversible Process of Water

To further illustrate concepts introduced in this section, Example 6.1 considers water undergoing an internally reversible process while contained in a piston–cylinder assembly.

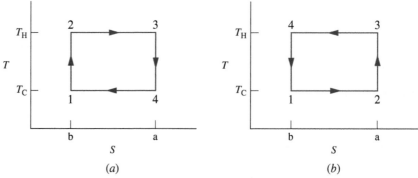

Fig. 6.5 Carnot cycles on the temperature–entropy diagram. (*a*) Power cycle. (*b*) Refrigeration or heat pump cycle.

▶ ▶ ▶ **EXAMPLE 6.1** ▶ ············

Evaluating Work and Heat Transfer for an Internally Reversible Process of Water

Water, initially a saturated liquid at 150°C (423.15 K), is contained in a piston–cylinder assembly. The water undergoes a process to the corresponding saturated vapor state, during which the piston moves freely in the cylinder. If the change of state is brought about by heating the water as it undergoes an internally reversible process at constant pressure and temperature, determine the work and heat transfer per unit of mass, each in kJ/kg.

SOLUTION

Known Water contained in a piston–cylinder assembly undergoes an internally reversible process at 150°C from saturated liquid to saturated vapor.

Find Determine the work and heat transfer per unit mass.

Schematic and Given Data:

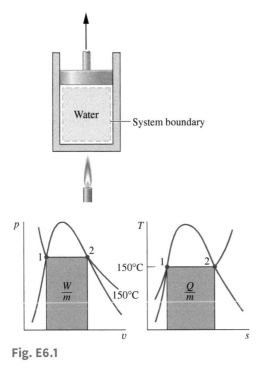

Fig. E6.1

Engineering Model

1. The water in the piston–cylinder assembly is a closed system.

2. The process is internally reversible.

3. Temperature and pressure are constant during the process.

4. There is no change in kinetic or potential energy between the two end states.

Analysis At constant pressure the work is

$$\frac{W}{m} = \int_1^2 p \, dv = p(v_2 - v_1)$$

With values from Table A-2 at 150°C

$$\frac{W}{m} = (4.758 \text{ bar})(0.3928 - 1.0905 \times 10^{-3}) \left(\frac{m^3}{kg}\right) \left|\frac{10^5 \text{ N/m}^2}{1 \text{ bar}}\right| \left|\frac{1 \text{ kJ}}{10^3 \text{ N} \cdot \text{m}}\right|$$

$$= 186.38 \text{ kJ/kg}$$

Since the process is internally reversible and at constant temperature, Eq. 6.23 gives

$$Q = \int_1^2 T \, dS = m \int_1^2 T \, dS$$

or

$$\frac{Q}{m} = T(s_2 - s_1)$$

With values from Table A-2

❶ $\dfrac{Q}{m} = (423.15 \text{ K})(6.8379 - 1.8418) \text{ kJ/kg} \cdot \text{K} = 2114.1 \text{ kJ/kg}$

As shown in the accompanying figure, the work and heat transfer can be represented as areas on p–v and T–s diagrams, respectively.

---

❶ The heat transfer can be evaluated alternatively from an energy balance written on a unit mass basis as

$$u_2 - u_1 = \frac{Q}{m} - \frac{W}{m}$$

Introducing $W/m = p(v_2 - v_1)$ and solving

$$\frac{Q}{m} = (u_2 - u_1) + p(v_2 - v_1)$$
$$= (u_2 + pv_2) - (u_1 + pv_1)$$
$$= h_2 - h_1$$

From Table A-2 at 150°C, $h_2 - h_1 = 2114.3$ kJ/kg, which agrees with the value for Q/m obtained in the solution.

SKILLS DEVELOPED

Ability to...

• evaluate work and heat transfer for an internally reversible process and represent them as areas on p–v and T–s diagrams, respectively.

• retrieve entropy data for water.

Quick Quiz

If the initial and final states were saturation states at 100°C (373.15 K), determine the work and heat transfer per unit of mass, each in kJ/kg. Ans. 170 kJ/kg, 2257 kJ/kg.

6.7 Entropy Balance for Closed Systems

entropy balance

In this section, we begin our study of the **entropy balance**. The entropy balance is an expression of the second law that is particularly effective for thermodynamic analysis. The current presentation is limited to closed systems. The entropy balance is extended to control volumes in Sec. 6.9.

Just as mass and energy are accounted for by mass and energy balances, respectively, entropy is accounted for by an entropy balance. In Eq. 5.2, the entropy balance is introduced in words as

$$
\begin{bmatrix} change \text{ in the amount of} \\ \text{entropy contained} \\ \text{within the system during} \\ \text{some time interval} \end{bmatrix} = \begin{bmatrix} \text{net amount of} \\ \text{entropy } transferred \text{ } in \\ \text{across the system} \\ \text{boundary during the} \\ \text{time interval} \end{bmatrix} + \begin{bmatrix} \text{amount of } entropy \\ produced \text{ within the} \\ \text{system during the time} \\ \text{interval} \end{bmatrix}
$$

closed system entropy balance

In symbols, the **closed system entropy balance** takes the form

$$
\underbrace{S_2 - S_1}_{\substack{\text{entropy} \\ \text{change}}} = \underbrace{\int_1^2 \left(\frac{\delta Q}{T}\right)_b}_{\substack{\text{entropy} \\ \text{transfer}}} + \underbrace{\sigma}_{\substack{\text{entropy} \\ \text{production}}} \tag{6.24}
$$

where subscript b signals that the integrand is evaluated at the system boundary. For the development of Eq. 6.24, see the box.

It is sometimes convenient to use the entropy balance expressed in differential form

$$
dS = \left(\frac{\delta Q}{T}\right)_b + \delta \sigma \tag{6.25}
$$

Note that the differentials of the nonproperties Q and σ are shown, respectively, as δQ and $\delta \sigma$. When there are no internal irreversibilities, $\delta \sigma$ vanishes and Eq. 6.25 reduces to Eq. 6.2b.

In each of its alternative forms the entropy balance can be regarded as a statement of the second law of thermodynamics. For the analysis of engineering systems, the entropy balance is a more effective means for applying the second law than the Clausius and Kelvin–Planck statements given in Chap. 5.

Developing the Closed System Entropy Balance

The entropy balance for closed systems can be developed using the *Clausius inequality* expressed by Eq. 5.13 (Sec. 5.11) and the defining equation for entropy change, Eq. 6.2a, as follows:

Shown in Fig. 6.6 is a cycle executed by a closed system. The cycle consists of process I, during which internal irreversibilities are present, followed by internally reversible process R. For this cycle, Eq. 5.13 takes the form

$$
\int_1^2 \left(\frac{\delta Q}{T}\right)_b + \int_2^1 \left(\frac{\delta Q}{T}\right)_{\substack{\text{int} \\ \text{rev}}} = -\sigma \tag{a}
$$

where the first integral is for process I and the second is for process R. The subscript b in the first integral serves as a reminder that the integrand is evaluated at the system boundary. The subscript is not required in the second integral because temperature is uniform throughout the system at each intermediate state of an internally

reversible process. Since no irreversibilities are associated with process R, the term σ_{cycle} of Eq. 5.13, which accounts for the effect of irreversibilities during the cycle, refers only to process I and is shown in Eq. (a) simply as σ.

Applying the definition of entropy change, Eq. 6.2a, we can express the second integral of Eq. (a) as

$$
\int_2^1 \left(\frac{\delta Q}{T}\right)_{\substack{\text{int} \\ \text{rev}}} = S_1 - S_2 \tag{b}
$$

With this, Eq. (a) becomes

$$
\int_1^2 \left(\frac{\delta Q}{T}\right)_b + (S_1 - S_2) = -\sigma \tag{c}
$$

On rearrangement, Eq. (c) gives Eq. 6.24, the closed system entropy balance.

6.7.1 Interpreting the Closed System Entropy Balance

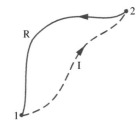

Fig. 6.6 Cycle used to develop the entropy balance.

If the end states are fixed, the entropy change on the left side of Eq. 6.24 can be evaluated independently of the details of the process. However, the two terms on the right side depend explicitly on the nature of the process and cannot be determined solely from knowledge of the end states. The first term on the right side of Eq. 6.24 is associated with heat transfer to or from the system during the process. This term can be interpreted as the **entropy transfer accompanying heat transfer**. The direction of entropy transfer is the same as the direction of the heat transfer, and the same sign convention applies as for heat transfer: A positive value means that entropy is transferred into the system, and a negative value means that entropy is transferred out. When there is no heat transfer, there is no entropy transfer.

The entropy change of a system is not accounted for solely by the entropy transfer but also is due in part to the second term on the right side of Eq. 6.24 denoted by σ. The term σ is positive when internal irreversibilities are present during the process and vanishes when no internal irreversibilities are present. This can be described by saying that **entropy is produced** (or *generated*) within the system by the action of irreversibilities.

The second law of thermodynamics can be interpreted as requiring that entropy is produced by irreversibilities and conserved only in the limit as irreversibilities are reduced to zero. Since σ measures the effect of irreversibilities present within the system during a process, its value depends on the nature of the process and not solely on the end states. Entropy production is *not* a property.

When applying the entropy balance to a closed system, it is essential to remember the requirements imposed by the second law on entropy production: The second law requires that entropy *production* be positive, or zero, in value:

$$\sigma: \begin{cases} > 0 & \text{irreversibilities present within the system} \\ = 0 & \text{no irreversibilities present within the system} \end{cases} \qquad (6.26)$$

The value of the entropy production cannot be negative. In contrast, the *change* in entropy of the system may be positive, negative, or zero:

$$S_2 - S_1: \begin{cases} > 0 \\ = 0 \\ < 0 \end{cases} \qquad (6.27)$$

Like other properties, entropy change for a process between two specified states can be determined without knowledge of the details of the process.

entropy transfer accompanying heat transfer

entropy production

 Animation

Entropy Balance for Closed Systems

6.7.2 Evaluating Entropy Production and Transfer

The objective in many applications of the entropy balance is to evaluate the entropy production term. However, the value of the entropy production for a given process of a system often does not have much significance *by itself*. The significance is normally determined through comparison. For example, the entropy production within a given component might be compared to the entropy production values of the other components included in an overall system formed by these components. By comparing entropy production values, the components where appreciable irreversibilities occur can be identified and rank ordered. This allows attention to be focused on the components that contribute most to inefficient operation of the overall system.

To evaluate the entropy transfer term of the entropy balance requires information regarding both the heat transfer and the temperature on the boundary where the heat transfer occurs. The entropy transfer term is not always subject to direct evaluation, however, because the required information is either unknown or not defined, such as when the system passes through states sufficiently far from equilibrium. In such applications, it may be convenient, therefore, to enlarge the system to include enough of the immediate surroundings that the temperature on the boundary of the *enlarged system* corresponds to the temperature of the surroundings away from the immediate vicinity of the system, T_f. The entropy

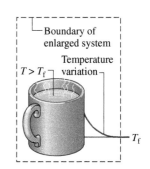

transfer term is then simply Q/T_f. However, as the irreversibilities present would not be just for the system of interest but for the enlarged system, the entropy production term would account for the effects of internal irreversibilities within the original system and external irreversibilities present within that portion of the surroundings included within the enlarged system.

TAKE NOTE...

On property diagrams, solid lines are used for *internally reversible* processes. A dashed line signals only that a process has occurred between initial and final equilibrium states and does not define a *path* for the process.

6.7.3 Applications of the Closed System Entropy Balance

The following examples illustrate the use of the energy and entropy balances for the analysis of closed systems. Property relations and property diagrams also contribute significantly in developing solutions. Example 6.2 reconsiders the system and end states of Example 6.1 to demonstrate that entropy is produced when internal irreversibilities are present and that the amount of entropy production is not a property. In Example 6.3, the entropy balance is used to determine the minimum theoretical compression work.

▶ ▶ ▶ **EXAMPLE 6.2** ▶

Determining Work and Entropy Production for an Irreversible Process of Water

Water, initially a saturated liquid at 150°C, is contained within a piston–cylinder assembly. The water undergoes a process to the corresponding saturated vapor state, during which the piston moves freely in the cylinder. There is no heat transfer with the surroundings. If the change of state is brought about by the action of a paddle wheel, determine the net work per unit mass, in kJ/kg, and the amount of entropy produced per unit mass, in kJ/kg · K.

SOLUTION

Known   Water contained in a piston–cylinder assembly undergoes an adiabatic process from saturated liquid to saturated vapor at 150°C. During the process, the piston moves freely, and the water is rapidly stirred by a paddle wheel.

Find   Determine the net work per unit mass and the entropy produced per unit mass.

Schematic and Given Data:

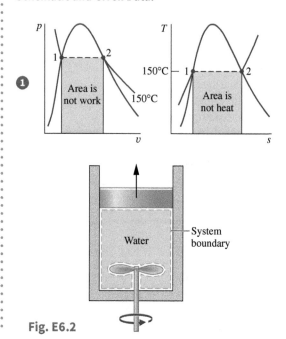

Fig. E6.2

Engineering Model

1. The water in the piston–cylinder assembly is a closed system.

2. There is no heat transfer with the surroundings.

3. The system is at an equilibrium state initially and finally. There is no change in kinetic or potential energy between these two states.

Analysis   As the volume of the system increases during the process, there is an energy transfer by work from the system during the expansion, as well as an energy transfer by work to the system via the paddle wheel. The *net* work can be evaluated from an energy balance, which reduces with assumptions 2 and 3 to

$$\Delta U + \overset{0}{\Delta KE} + \overset{0}{\Delta PE} = \overset{0}{Q} - W$$

On a unit mass basis, the energy balance is then

$$\frac{W}{m} = -(u_2 - u_1)$$

With specific internal energy values from Table A-2 at 150°C, $u_1 = 631.68$ kJ/kg, $u_2 = 2559.5$ kJ/kg, we get

$$\frac{W}{m} = -1927.82 \frac{\text{kJ}}{\text{kg}}$$

The minus sign indicates that the work input by stirring is greater in magnitude than the work done by the water as it expands.

The amount of entropy produced is evaluated by applying the entropy balance Eq. 6.24. Since there is no heat transfer, the term accounting for entropy transfer vanishes:

$$\Delta S = \overset{0}{\int_1^2 \left(\frac{\delta Q}{T}\right)_b} + \sigma$$

On a unit mass basis, this becomes on rearrangement

$$\frac{\sigma}{m} = s_2 - s_1$$

With specific entropy values from Table A-2 at 150°C, $s_1 = 1.8418$ kJ/kg · K, $s_2 = 6.8379$ kJ/kg · K, we get

②
$$\frac{\sigma}{m} = 4.9961\frac{\text{kJ}}{\text{kg} \cdot \text{K}}$$

---

① Although each end state is an equilibrium state at the same pressure and temperature, the pressure and temperature are not necessarily uniform throughout the system at *intervening* states, nor are they necessarily constant in value during the process. Accordingly, there is no well-defined "path" for the process. This is emphasized by the use of dashed lines to represent the process on these *p–v* and *T–s* diagrams. The dashed lines indicate only that a process has taken place, and no "area" should be associated with them. In particular, note that the process is adiabatic, so the "area" below the dashed line on the *T–s* diagram can have no significance as heat transfer. Similarly, the work cannot be associated with an area on the *p–v* diagram.

② The change of state is the same in the present example as in Example 6.1. However, in Example 6.1 the change of state is brought about by heat transfer while the system undergoes an internally reversible process. Accordingly, the value of entropy production for the process of Example 6.1 is zero. Here, fluid friction is present during the process and the entropy production is positive in value. Accordingly, different values of entropy production are obtained for two processes between the *same* end states. This demonstrates that entropy production is not a property.

SKILLS DEVELOPED

Ability to…

• apply the closed system energy and entropy balances.

• retrieve property data for water.

Quick Quiz

If the initial and final states were saturation states at 100°C, determine the net work, in kJ/kg, and the amount of entropy produced, in kJ/kg · K. Ans. −2087.56 kJ/kg, 6.048 kJ/kg · K.

As an illustration of second law reasoning, minimum theoretical compression work is evaluated in Example 6.3 using the fact that the entropy production term of the entropy balance cannot be negative.

▶ EXAMPLE 6.3 ▶

Evaluating Minimum Theoretical Compression Work

Refrigerant 134a is compressed adiabatically in a piston–cylinder assembly from saturated vapor at 0°C to a final pressure of 0.7 MPa. Determine the minimum theoretical work input required per unit mass of refrigerant, in kJ/kg.

SOLUTION

Known Refrigerant 134a is compressed without heat transfer from a specified initial state to a specified final pressure.

Find Determine the minimum theoretical work input required per unit of mass.

Schematic and Given Data:

Engineering Model

1. The Refrigerant 134a is a closed system.

2. There is no heat transfer with the surroundings.

3. The initial and final states are equilibrium states. There is no change in kinetic or potential energy between these states.

Analysis An expression for the work can be obtained from an energy balance. By applying assumptions 2 and 3, we get

$$\Delta U + \Delta KE^{0} + \Delta PE^{0} = Q^{0} - W$$

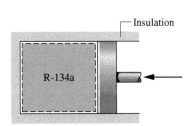

Fig. E6.3

When written on a unit mass basis, the work *input* is then

$$\left(-\frac{W}{m}\right) = u_2 - u_1$$

The specific internal energy u_1 can be obtained from Table A-10E as $u_1 = 227.06$ kJ/kg. Since u_1 is known, the value for the work input depends on the specific internal energy u_2. The minimum work input corresponds to the smallest allowed value for u_2, determined using the second law as follows.

Applying the entropy balance, Eq. 6.24, we get

$$\Delta S = \int_1^2 \cancelto{0}{\left(\frac{\delta Q}{T}\right)_b} + \sigma$$

where the entropy transfer term is set equal to zero because the process is adiabatic. Thus, the *allowed* final states must satisfy

$$s_2 - s_1 = \frac{\sigma}{m} \geq 0$$

The restriction indicated by the foregoing equation can be interpreted using the accompanying T–s diagram. Since σ cannot be negative, states with $s_2 < s_1$ are not accessible adiabatically. When irreversibilities are present during the compression, entropy is produced, so $s_2 > s_1$. The state labeled 2s on the diagram would be attained in the limit as irreversibilities are reduced to zero. This state corresponds to an *isentropic* compression.

By inspection of Table A-12, we see that when pressure is fixed, the specific internal energy decreases as specific entropy decreases. Thus, the smallest allowed value for u_2 corresponds to state 2s. Interpolating in Table A-12 at 0.7 MPa, with $s_{2s} = s_1 = 0.9190$ kJ/kg · K, we find that $u_{2s} = 244.32$ kJ/kg, which corresponds to a temperature at state 2s of about 38.3°C. Finally, the *minimum* work input is

① $\left(-\dfrac{W}{m}\right)_{min} = u_{2s} - u_1 = 244.32 - 227.06 = 17.26$ kJ/kg

---

① The effect of irreversibilities exacts a penalty on the work input required: A greater work input is needed for the actual adiabatic compression process than for an internally reversible adiabatic process between the same initial state and the same final pressure. See the Quick Quiz to follow.

SKILLS DEVELOPED

Ability to…

• apply the closed system energy and entropy balances.

• retrieve property data for Refrigerant 134a.

Quick Quiz

If the refrigerant were compressed adiabatically to a final state where $p_2 = 0.7$ MPa, $T_2 = 49$°C, determine the work input, in kJ/kg, and the amount of entropy produced, in kJ/kg · K. Ans. 39.9 kJ/kg, 36.43 kJ/kg · K.

6.7.4 Closed System Entropy Rate Balance

If the temperature T_b is constant, Eq. 6.24 reads

$$S_2 - S_1 = \frac{Q}{T_b} + \sigma$$

where Q/T_b represents the *amount* of entropy transferred through the portion of the boundary at temperature T_b. Similarly, the quantity $\dot{Q}/T_j$ represents the *time rate* of entropy transfer through the portion of the boundary whose instantaneous temperature is T_j. This quantity appears in the closed system entropy rate balance considered next.

closed system entropy rate balance

On a time rate basis, the **closed system entropy rate balance** is

$$\boxed{\frac{dS}{dt} = \sum_j \frac{\dot{Q}_j}{T_j} + \dot{\sigma}} \tag{6.28}$$

where dS/dt is the time rate of change of entropy of the system. The term $\dot{Q}_j/T_j$ represents the time rate of entropy transfer through the portion of the boundary whose instantaneous temperature is T_j. The term $\dot{\sigma}$ accounts for the time rate of entropy production due to irreversibilities within the system.

To pinpoint the relative significance of the internal and external irreversibilities, Example 6.4 illustrates the application of the entropy rate balance to a system and to an enlarged system consisting of the system and a portion of its immediate surroundings.

▶ ▶ ▶ **EXAMPLE 6.4** ▶

Pinpointing Irreversibilities

Referring to Example 2.4, evaluate the rate of entropy production $\dot{\sigma}$, in kW/K, for **(a)** the gearbox as the system and **(b)** an enlarged system consisting of the gearbox and enough of its surroundings that heat transfer occurs at the temperature of the surroundings away from the immediate vicinity of the gearbox, $T_f = 293$ K (20°C).

SOLUTION

Known   A gearbox operates at steady state with known values for the power input through the high-speed shaft, power output through the low-speed shaft, and heat transfer rate. The temperature on the outer surface of the gearbox and the temperature of the surroundings away from the gearbox are also known.

Find   Evaluate the entropy production rate $\dot{\sigma}$ for each of the two specified systems shown in the schematic.

Schematic and Given Data:

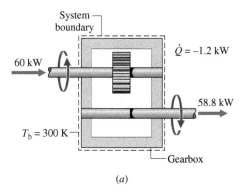

(a)

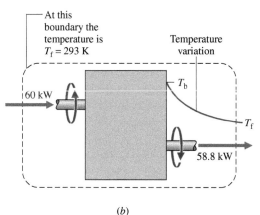

(b)   **Fig. E6.4**

Engineering Model

1. In part (a), the gearbox is taken as a closed system operating at steady state, as shown on the accompanying sketch labeled with data from Example 2.4.

2. In part (b) the gearbox and a portion of its surroundings are taken as a closed system, as shown on the accompanying sketch labeled with data from Example 2.4.

3. The temperature of the outer surface of the gearbox and the temperature of the surroundings do not vary.

Analysis

a. To obtain an expression for the entropy production rate, begin with the entropy balance for a closed system on a time rate basis: Eq. 6.28. Since heat transfer takes place only at temperature T_b, the entropy rate balance reduces at steady state to

$$\frac{dS^{\,0}}{dt} = \frac{\dot{Q}}{T_b} + \dot{\sigma}$$

Solving

$$\dot{\sigma} = -\frac{\dot{Q}}{T_b}$$

Introducing the known values for the heat transfer rate $\dot{Q}$ and the surface temperature T_b

$$\dot{\sigma} = -\frac{(-1.2 \text{ kW})}{(300 \text{ K})} = 4 \times 10^{-3} \text{ kW/K}$$

b. Since heat transfer takes place at temperature T_f for the enlarged system, the entropy rate balance reduces at steady state to

$$\frac{dS^{\,0}}{dt} = \frac{\dot{Q}}{T_f} + \dot{\sigma}$$

Solving

$$\dot{\sigma} = -\frac{\dot{Q}}{T_f}$$

Introducing the known values for the heat transfer rate $\dot{Q}$ and the temperature T_f

❶
$$\dot{\sigma} = -\frac{(-1.2 \text{ kW})}{(293 \text{ K})} = 4.1 \times 10^{-3} \text{ kW/K}$$

---

❶ The value of the entropy production rate calculated in part (a) gauges the significance of irreversibilities associated with friction and heat transfer *within* the gearbox. In part (b), an additional source of irreversibility is included in the enlarged system, namely, the irreversibility associated with the heat transfer from the outer surface of the gearbox at T_b to the surroundings at T_f. In this case, the irreversibilities within the gearbox are dominant, accounting for about 98% of the total rate of entropy production.

SKILLS DEVELOPED

Ability to...

• apply the closed system entropy rate balance.
• develop an engineering model.

Quick Quiz

If the power delivered were 59.32 kW, evaluate the outer surface temperature, in K, and the rate of entropy production, in kW/K, for the gearbox as the system, keeping input power, h, and A from Example 2.4 the same. Ans. 297 K, 2.3×10^{-3} kW/K.

6.8 Directionality of Processes

Our study of the second law of thermodynamics began in Sec. 5.1 with a discussion of the *directionality* of processes. In this section we consider two related aspects for which there are significant applications: the increase in entropy principle and a statistical interpretation of entropy.

6.8.1 Increase of Entropy Principle

In the present discussion, we use the closed system energy and entropy balances to introduce the increase of entropy principle. Discussion centers on an enlarged system consisting of a system and that portion of the surroundings affected by the system as it undergoes a process. Since all energy and mass transfers taking place are included within the boundary of the enlarged system, the enlarged system is an *isolated* system.

An energy balance for the isolated system reduces to

$$\Delta E]_{\text{isol}} = 0 \qquad (6.29a)$$

because no energy transfers take place across its boundary. Thus, the energy of the isolated system remains constant. Since energy is an extensive property, its value for the isolated system is the sum of its values for the system and surroundings, respectively, so Eq. 6.29a can be written as

$$\Delta E]_{\text{system}} + \Delta E]_{\text{surr}} = 0 \qquad (6.29b)$$

In either of these forms, the conservation of energy principle places a constraint on the processes that can occur. For a process to take place, it is necessary for the energy of the system plus the surroundings to remain constant. However, not all processes for which this constraint is satisfied can actually occur. Processes also must satisfy the second law, as discussed next.

An entropy balance for the isolated system reduces to

$$\Delta S]_{\text{isol}} = \int_{1}^{2} \left(\frac{\delta Q}{T} \right)_{\text{b}}^{0} + \sigma_{\text{isol}}$$

or

$$\Delta S]_{\text{isol}} = \sigma_{\text{isol}} \qquad (6.30a)$$

increase of entropy principle

where σ_{isol} is the total amount of entropy produced within the system and its surroundings. Since entropy is produced in all actual processes, the only processes that can occur are those for which the entropy of the isolated system increases. This is known as the **increase of entropy principle**. The increase of entropy principle is sometimes considered an alternative statement of the second law.

Since entropy is an extensive property, its value for the isolated system is the sum of its values for the system and surroundings, respectively, so Eq. 6.30a can be written as

$$\Delta S]_{\text{system}} + \Delta S]_{\text{surr}} = \sigma_{\text{isol}} \qquad (6.30b)$$

Notice that this equation does not require the entropy change to be positive for both the system and surroundings but only that the *sum* of the changes is positive. In either of these forms, the increase of entropy principle dictates the direction in which any process can proceed: Processes occur only in such a direction that the total entropy of the system *plus* surroundings increases.

We observed previously the tendency of systems left to themselves to undergo processes until a condition of equilibrium is attained (Sec. 5.1). The increase of entropy principle suggests that the entropy of an isolated system increases as the state of equilibrium is approached, with the equilibrium state being attained when the entropy reaches a maximum. This interpretation is considered again in Sec. 14.1, which deals with equilibrium criteria.

Example 6.5 illustrates the increase of entropy principle.

▶ ▶ **EXAMPLE 6.5** ▶

Quenching a Hot Metal Bar

A 0.3 kg metal bar initially at 1200 K is removed from an oven and quenched by immersing it in a closed tank containing 9 kg of water initially at 300 K. Each substance can be modeled as incompressible. An appropriate constant specific heat value for the water is $c_w = 4.2$ kJ/kg · K, and an appropriate value for the metal is $c_m = 0.42$ kJ/kg · K. Heat transfer from the tank contents can be neglected. Determine (a) the final equilibrium temperature of the metal bar and the water, in K, and (b) the amount of entropy produced, in kJ/K.

SOLUTION

Known A hot metal bar is quenched by immersing it in a tank containing water.

Find Determine the final equilibrium temperature of the metal bar and the water, and the amount of entropy produced.

Schematic and Given Data:

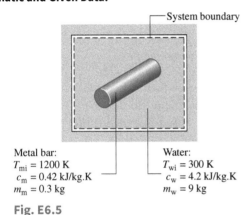

Metal bar:
$T_{mi} = 1200$ K
$c_m = 0.42$ kJ/kg.K
$m_m = 0.3$ kg

Water:
$T_{wi} = 300$ K
$c_w = 4.2$ kJ/kg.K
$m_w = 9$ kg

Fig. E6.5

Engineering Model

1. The metal bar and the water within the tank form a system, as shown on the accompanying sketch.

2. There is no energy transfer by heat or work: The system is isolated.

3. There is no change in kinetic or potential energy.

4. The water and metal bar are each modeled as incompressible with known specific heats.

Analysis

a. The final equilibrium temperature can be evaluated from an energy balance for the isolated system

$$\Delta KE^0 + \Delta PE^0 + \Delta U = Q^0 - W^0$$

where the indicated terms vanish by assumptions 2 and 3. Since internal energy is an extensive property, its value for the isolated system is the sum of the values for the water and metal, respectively. Thus, the energy balance becomes

$$\Delta U]_{water} + \Delta U]_{metal} = 0$$

Using Eq. 3.20a to evaluate the internal energy changes of the water and metal in terms of the constant specific heats

$$m_w c_w (T_f - T_{wi}) + m_m c_m (T_f - T_{mi}) = 0$$

where T_f is the final equilibrium temperature, and T_{wi} and T_{mi} are the initial temperatures of the water and metal, respectively. Solving for T_f and inserting values

$$T_f = \frac{m_w (c_w/c_m) T_{wi} + m_m T_{mi}}{m_w (c_w/c_m) + m_m}$$

$$= \frac{(9 \text{ kg})(10)(300 \text{ K}) + (0.3 \text{ kg})(1200 \text{ K})}{(9 \text{ kg})(10) + (0.3 \text{ kg})}$$

$$= 303 \text{ K}$$

b. The amount of entropy production can be evaluated from an entropy balance. Since no heat transfer occurs between the isolated system and its surroundings, there is no accompanying entropy transfer, and an entropy balance for the isolated system reduces to

$$\Delta S = \int_1^2 \left(\frac{\delta Q}{T} \right)_b^{0} + \sigma$$

Entropy is an extensive property, so its value for the isolated system is the sum of its values for the water and the metal, respectively, and the entropy balance becomes

$$\Delta S]_{water} + \Delta S]_{metal} = \sigma$$

Evaluating the entropy changes using Eq. 6.13 for incompressible substances, the foregoing equation can be written as

$$\sigma = m_w c_w \ln\frac{T_f}{T_{wi}} + m_m c_m \ln\frac{T_f}{T_{mi}}$$

Inserting values

❶ $\sigma = (9 \text{ kg})\left(4.2\frac{\text{kJ}}{\text{kg} \cdot \text{K}}\right)\ln\frac{303}{300} + (0.3 \text{ kg})\left(0.42\frac{\text{kJ}}{\text{kg} \cdot \text{K}}\right)\ln\frac{300}{1200}$

❷ $= \left(0.3761\frac{\text{kJ}}{\text{K}}\right) + \left(-0.1734\frac{\text{kJ}}{\text{K}}\right) = 0.2027$ kJ/K

---

❶ The metal bar experiences a *decrease* in entropy. The entropy of the water *increases*. In accord with the increase of entropy principle, the entropy of the isolated system *increases*.

❷ The value of σ is sensitive to roundoff in the value of T_f.

SKILLS DEVELOPED

Ability to...

• apply the closed system energy and entropy balances.
• apply the incompressible substance model.

Quick Quiz

If the mass of the metal bar were 0.2 kg, determine the final equilibrium temperature, in K, and the amount of entropy produced, in kJ/K, keeping all other given data the same. Ans. 296 K, 0.1049 kJ/K.

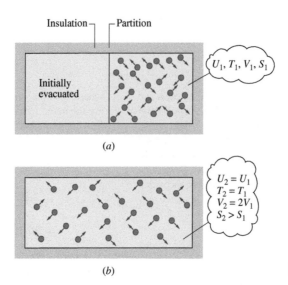

Fig. 6.7 N molecules in a box.

6.8.2 Statistical Interpretation of Entropy

Building on the increase of entropy principle, in this section we introduce an interpretation of entropy from a microscopic perspective based on *probability*.

In *statistical thermodynamics*, entropy is associated with the notion of microscopic *disorder*. From previous considerations we know that in a spontaneous process of an isolated system, the system moves toward equilibrium and the entropy increases. From the microscopic viewpoint, this is equivalent to saying that as an isolated system moves toward equilibrium our knowledge of the condition of individual particles making up the system decreases, which corresponds to an increase in microscopic disorder and a related increase in entropy.

We use an elementary thought experiment to bring out some basic ideas needed to understand this view of entropy. Actual microscopic analysis of systems is more complicated than the discussion given here, but the essential concepts are the same.

Consider N molecules initially contained in one half of the box shown in Fig. 6.7a. The entire box is considered an isolated system. We assume that the ideal gas model applies. In the initial condition, the gas appears to be at equilibrium in terms of temperature, pressure, and other properties. But, on the microscopic level the molecules are moving about randomly. We do know *for sure*, though, that initially all molecules are on the right side of the vessel.

Suppose we remove the partition and wait until equilibrium is reached again, as in Fig. 6.7b. Because the system is isolated, the internal energy U does not change: $U_2 = U_1$. Further, because the internal energy of an ideal gas depends on temperature alone, the temperature is unchanged: $T_2 = T_1$. Still, at the final state a given molecule has twice the volume in which to move: $V_2 = 2V_1$. Just like a coin toss, the probability that the molecule is on one side or the other is now ½, which is the same as the volume ratio V_1/V_2. In the final condition, we have *less knowledge* about where each molecule is than we did originally.

We can evaluate the change in entropy for the process of Fig. 6.7 by applying Eq. 6.17, expressed in terms of volumes and on a molar basis. The entropy change for the constant-temperature process is

$$(S_2 - S_1)/n = \bar{R} \ln(V_2/V_1) \tag{6.31}$$

where n is the amount of substance on a molar basis (Eq. 1.8). Next, we consider how the entropy change would be evaluated from a microscopic point of view.

Through more complete molecular modeling and statistical analysis, the total number of positions and velocities—**microstates**—available to a single molecule can be calculated. This total is called the **thermodynamic probability**, w. For a system of N molecules, the

microstates

thermodynamic probability

"Breaking" the Second Law Has Implications for Nanotechnology

Some 135 years ago, renowned nineteenth-century physicist J. C. Maxwell wrote, ". . . the second law is . . . a statistical . . . truth, for it depends on the fact that the bodies we deal with consist of millions of molecules . . . [Still] the second law is continually being violated . . . in any sufficiently small group of molecules belonging to a real body." Although Maxwell's view was bolstered by theorists over the years, experimental confirmation proved elusive. Then, in 2002, experimenters reported they had demonstrated violations of the second law: At the micron scale over time intervals of up to 2 seconds, entropy was consumed, not produced [see Phys. Rev. Lett. **89**, 050601 (2002)].

While few were surprised that experimental confirmation had at last been achieved, some *were* surprised by implications of the research for the twenty-first-century field of nanotechnology: The experimental results suggest inherent limitations on nanomachines. These tiny devices—only a few molecules in size—may not behave simply as miniaturized versions of their larger counterparts; and the smaller the device, the more likely its motion and operation could be disrupted unpredictably. Occasionally and uncontrollably, nanomachines may not perform as designed, perhaps even capriciously running backward. Still, designers of these machines will applaud the experimental results if they lead to deeper understanding of behavior at the nanoscale.

thermodynamic probability is w^N. In statistical thermodynamics, entropy is considered to be proportional to $\ln(w)^N$. That is, $S \propto N \ln(w)$. This gives the **Boltzmann relation**

$$S/N = k \ln w \qquad (6.32)$$

where the proportionality factor, k, is called *Boltzmann's constant.*

Applying Eq. 6.32 to the process of Fig. 6.7, we get

$$(S_2 - S_1)/N = k \ln(w_2) - k \ln(w_1)$$
$$= k \ln(w_2/w_1) \qquad (6.33)$$

Comparing Eqs. 6.31 and 6.33, the expressions for entropy change coincide when $k = n\overline{R}/N$ and $w_2/w_1 = V_2/V_1$. The first of these expressions allows Boltzmann's constant to be evaluated, giving $k = 1.3806 \times 10^{-23}$ J/K. Also, since $V_2 > V_1$ and $w_2 > w_1$, Eqs. 6.31 and 6.33 each predict an increase of entropy owing to entropy production during the irreversible adiabatic expansion in this example.

From Eq. 6.33, we see that any process that increases the number of possible microstates of a system increases its entropy and conversely. Hence, for an *isolated* system, processes occur only in such a direction that the number of microstates available to the system increases, resulting in our having less knowledge about the condition of individual particles. Because of this concept of decreased knowledge, entropy reflects the microscopic **disorder** of the system. We can then say that the only processes an isolated system can undergo are those that increase the disorder of the system. This interpretation is consistent with the idea of directionality of processes discussed previously.

The notion of entropy as a measure of disorder is sometimes used in fields other than thermodynamics. The concept is employed in information theory, statistics, biology, and even in some economic and social modeling. In these applications, the term *entropy* is used as a measure of disorder without the physical aspects of the thought experiment used here necessarily being implied.

Boltzmann relation

disorder

BioConnections

Do living things violate the second law of thermodynamics because they seem to create order from disorder? Living things are not isolated systems as considered in the previous discussion of entropy and disorder. Living things interact with their surroundings and are influenced by their surroundings. For instance, plants grow into highly ordered cellular structures synthesized from atoms and molecules originating in the earth and its atmosphere.

Through interactions with their surroundings, plants exist in highly organized states and are able to produce within themselves even more organized, lower entropy states. In keeping with the second law, states of lower entropy can be realized within a system as long as the *total* entropy of the system and its surroundings increases. The self-organizing tendency of living things is widely observed and fully in accord with the second law.

6.9 Entropy Rate Balance for Control Volumes

Thus far the discussion of the entropy balance concept has been restricted to the case of closed systems. In the present section the entropy balance is extended to control volumes.

Like mass and energy, entropy is an extensive property, so it too can be transferred into or out of a control volume by streams of matter. Since this is the principal difference between the closed system and control volume forms, the **control volume entropy rate balance** can be obtained by modifying Eq. 6.28 to account for these entropy transfers. The result is

$$
\boxed{
\underbrace{\frac{dS_{cv}}{dt}}_{\substack{\text{rate of} \\ \text{entropy} \\ \text{change}}} = \underbrace{\sum_j \frac{\dot{Q}_j}{T_j} + \sum_i \dot{m}_i s_i - \sum_e \dot{m}_e s_e}_{\substack{\text{rates of} \\ \text{entropy} \\ \text{transfer}}} + \underbrace{\dot{\sigma}_{cv}}_{\substack{\text{rate of} \\ \text{entropy} \\ \text{production}}}
}
\tag{6.34}
$$

control volume entropy rate balance

where dS_{cv}/dt represents the time rate of change of entropy within the control volume. The terms $\dot{m}_i s_i$ and $\dot{m}_e s_e$ account, respectively, for rates of **entropy transfer accompanying mass flow** into and out of the control volume. The term $\dot{Q}_j$ represents the time rate of heat transfer at the location on the boundary where the instantaneous temperature is T_j. The ratio $\dot{Q}_j/T_j$ accounts for the accompanying rate of entropy *transfer*. The term $\dot{\sigma}_{cv}$ denotes the time rate of entropy *production* due to irreversibilities *within* the control volume.

entropy transfer accompanying mass flow

Entropy Rate Balance
Tabs a and b

Integral Form of the Entropy Rate Balance

As for the cases of the control volume mass and energy rate balances, the entropy rate balance can be expressed in terms of local properties to obtain forms that are more generally applicable. Thus, the term $S_{cv}(t)$, representing the total entropy associated with the control volume at time t, can be written as a volume integral

$$
S_{cv}(t) = \int_V \rho s \, dV
$$

where ρ and s denote, respectively, the local density and specific entropy. The rate of entropy transfer accompanying heat transfer can be expressed more generally as an integral over the surface of the control volume

$$
\begin{bmatrix} \text{time rate of entropy} \\ \text{transfer accompanying} \\ \text{heat transfer} \end{bmatrix} = \int_A \left(\frac{\dot{q}}{T}\right)_b dA
$$

where $\dot{q}$ is the *heat flux*, the time rate of heat transfer per unit of surface area, through the location on the boundary where the instantaneous temperature is T. The subscript "b" is added as a reminder that the integrand is evaluated on the boundary of the control volume. In addition, the terms accounting for entropy transfer accompanying mass flow can be expressed as integrals over the inlet and exit flow areas, resulting in the following form of the entropy rate balance

$$
\frac{d}{dt}\int_V \rho s \, dV = \int_A \left(\frac{\dot{q}}{T}\right)_b dA + \sum_i \left(\int_A s\rho V_n dA\right)_i - \sum_e \left(\int_A s\rho V_n dA\right)_e + \dot{\sigma}_{cv}
\tag{6.35}
$$

where V_n denotes the normal component in the direction of flow of the velocity relative to the flow area. In some cases, it is also convenient to express the entropy production rate as a volume integral of the local volumetric rate of entropy production within the control volume. The study of Eq. 6.35 brings out the assumptions underlying Eq. 6.34. Finally, note that for a closed system the sums accounting for entropy transfer at inlets and exits drop out, and Eq. 6.35 reduces to give a more general form of Eq. 6.28.

6.10 Rate Balances for Control Volumes at Steady State

Since a great many engineering analyses involve control volumes at steady state, it is instructive to list steady-state forms of the balances developed for mass, energy, and entropy. At steady state, the conservation of mass principle takes the form

$$\sum_i \dot{m}_i = \sum_e \dot{m}_e \tag{4.6}$$

The energy rate balance at steady state is

$$0 = \dot{Q}_{cv} - \dot{W}_{cv} + \sum_i \dot{m}_i \left(h_i + \frac{V_i^2}{2} + gz_i \right) - \sum_e \dot{m}_e \left(h_e + \frac{V_e^2}{2} + gz_e \right) \tag{4.18}$$

Finally, the **steady-state form of the entropy rate balance** is obtained by reducing Eq. 6.34 to give

$$\boxed{0 = \sum_j \frac{\dot{Q}_j}{T_j} + \sum_i \dot{m}_i s_i - \sum_e \dot{m}_e s_e + \dot{\sigma}_{cv}} \tag{6.36}$$

steady-state entropy rate balance

These equations often must be solved simultaneously, together with appropriate property relations.

Mass and energy are conserved quantities, but entropy is not conserved. Equation 4.6 indicates that at steady state the total rate of mass flow into the control volume equals the total rate of mass flow out of the control volume. Similarly, Eq. 4.18 indicates that the total rate of energy transfer into the control volume equals the total rate of energy transfer out of the control volume. However, Eq. 6.36 requires that the rate at which entropy is transferred out must *exceed* the rate at which entropy enters, the difference being the rate of entropy production within the control volume owing to irreversibilities.

6.10.1 One-Inlet, One-Exit Control Volumes at Steady State

Since many applications involve one-inlet, one-exit control volumes at steady state, let us also list the form of the entropy rate balance for this important case. Thus, Eq. 6.36 reduces to read

$$\boxed{0 = \sum_j \frac{\dot{Q}_j}{T_j} + \dot{m}(s_1 - s_2) + \dot{\sigma}_{cv}} \tag{6.37}$$

Entropy Rate Balance Tab c

Or, on dividing by the mass flow rate $\dot{m}$ and rearranging

$$s_2 - s_1 = \frac{1}{\dot{m}} \left(\sum_j \frac{\dot{Q}_j}{T_j} \right) + \frac{\dot{\sigma}_{cv}}{\dot{m}} \tag{6.38}$$

The two terms on the right side of Eq. 6.38 denote, respectively, the rate of entropy transfer accompanying heat transfer and the rate of entropy production within the control volume, each *per unit of mass flowing through the control volume*. From Eq. 6.38 it can be concluded that the entropy of a unit of mass passing from inlet to exit can increase, decrease, or remain the same. Furthermore, because the value of the second term on the right can never be negative, a decrease in the specific entropy from inlet to exit can be realized only when more entropy is transferred out of the control volume accompanying heat transfer than is produced by irreversibilities within the control volume. When the value of this entropy transfer term is positive, the specific entropy at the exit is greater than the specific entropy at the inlet whether internal irreversibilities are present or not. In the special case where there is no entropy transfer accompanying heat transfer, Eq. 6.38 reduces to

$$s_2 - s_1 = \frac{\dot{\sigma}_{cv}}{\dot{m}} \tag{6.39}$$

Accordingly, when irreversibilities are present within the control volume, the entropy of a unit of mass increases as it passes from inlet to exit. In the limiting case in which no irreversibilities are present, the unit mass passes through the control volume with no change in its entropy—that is, isentropically.

6.10.2 Applications of the Rate Balances to Control Volumes at Steady State

Animation

Turbine Tab d

The following examples illustrate the use of the mass, energy, and entropy balances for the analysis of control volumes at steady state. Carefully note that property relations and property diagrams also play important roles in arriving at solutions.

In Example 6.6, we evaluate the rate of entropy production within a turbine operating at steady state when there is heat transfer from the turbine.

► **EXAMPLE 6.6** ►

Determining Entropy Production in a Steam Turbine

Steam enters a turbine with a pressure of 30 bar, a temperature of 400°C, and a velocity of 160 m/s. Saturated vapor at 100°C exits with a velocity of 100 m/s. At steady state, the turbine develops work equal to 540 kJ per kg of steam flowing through the turbine. Heat transfer between the turbine and its surroundings occurs at an average outer surface temperature of 350 K. Determine the rate at which entropy is produced within the turbine per kg of steam flowing, in kJ/kg · K. Neglect the change in potential energy between inlet and exit.

SOLUTION

Known Steam expands through a turbine at steady state for which data are provided.

Find Determine the rate of entropy production per kg of steam flowing.

Schematic and Given Data:

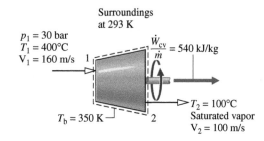

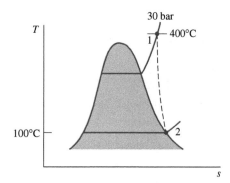

Fig. E6.6

Engineering Model

1. The control volume shown on the accompanying sketch is at steady state.

2. Heat transfer from the turbine to the surroundings occurs at a specified average outer surface temperature.

3. The change in potential energy between inlet and exit can be neglected.

Analysis To determine the entropy production per unit mass flowing through the turbine, begin with mass and entropy rate balances for the one-inlet, one-exit control volume at steady state:

$$\dot{m}_1 = \dot{m}_2$$

$$0 = \sum_j \frac{\dot{Q}_j}{T_j} + \dot{m}_1 s_1 - \dot{m}_2 s_2 + \dot{\sigma}_{cv}$$

Since heat transfer occurs only at $T_b = 350$ K, the first term on the right side of the entropy rate balance reduces to $\dot{Q}_{cv}/T_b$. Combining the mass and entropy rate balances

$$0 = \frac{\dot{Q}_{cv}}{T_b} + \dot{m}(s_1 - s_2) + \dot{\sigma}_{cv}$$

where $\dot{m}$ is the mass flow rate. Solving for $\dot{\sigma}_{cv}/\dot{m}$

$$\frac{\dot{\sigma}_{cv}}{\dot{m}} = -\frac{\dot{Q}_{cv}/\dot{m}}{T_b} + (s_2 - s_1)$$

The heat transfer rate, $\dot{Q}_{cv}/\dot{m}$, required by this expression is evaluated next.

Reduction of the mass and energy rate balances results in

$$\frac{\dot{Q}_{cv}}{\dot{m}} = \frac{\dot{W}_{cv}}{\dot{m}} + (h_2 - h_1) + \left(\frac{V_2^2 - V_1^2}{2} \right)$$

where the potential energy change from inlet to exit is dropped by assumption 3. From Table A-4 at 30 bar, 400°C, $h_1 = 3230.9$ kJ/kg, and from Table A-2, $h_2 = h_g(100°C) = 2676.1$ kJ/kg. Thus,

$$\frac{\dot{Q}_{cv}}{\dot{m}} = 540\,\frac{kJ}{kg} + (2676.1 - 3230.9)\left(\frac{kJ}{kg}\right)$$

$$+ \left[\frac{(100)^2 - (160)^2}{2}\right]\left(\frac{m^2}{s^2}\right)\left|\frac{1\,N}{1\,kg \cdot m/s^2}\right|\left|\frac{1\,kJ}{10^3\,N \cdot m}\right|$$

$$= 540 - 554.8 - 7.8 = -22.6\ kJ/kg$$

From Table A-2, $s_2 = 7.3549$ kJ/kg · K, and from Table A-4, $s_1 = 6.9212$ kJ/kg · K. Inserting values into the expression for entropy production

$$\frac{\dot{\sigma}_{cv}}{\dot{m}} = -\frac{(-22.6\ kJ/kg)}{350\ K} + (7.3549 - 6.9212)\left(\frac{kJ}{kg \cdot K}\right)$$

$$= 0.0646 + 0.4337 = 0.498\ kJ/kg \cdot K$$

SKILLS DEVELOPED

Ability to...

- apply the control volume, mass, energy, and entropy rate balances.
- retrieve property data for water.

Quick Quiz

If the boundary were located to include the turbine and a portion of the immediate surroundings so heat transfer occurs at the temperature of the surroundings, 293 K, determine the rate at which entropy is produced within the enlarged control volume, in kJ/K per kg of steam flowing, keeping all other given data the same. Ans. 0.511 kJ/kg · K.

In Example 6.7, the mass, energy, and entropy rate balances are used to evaluate a performance claim for a device producing hot and cold streams of air from a single stream of air at an intermediate temperature.

Animation

**Heat Exchangers
Tab d**

EXAMPLE 6.7 ▶

Evaluating a Performance Claim

An inventor claims to have developed a device requiring no energy transfer by work or heat transfer, yet able to produce hot and cold streams of air from a single stream of air at an intermediate temperature. The inventor provides steady-state test data indicating that when air enters at a temperature of 21°C and a pressure of 5 bar, separate streams of air exit at temperatures of −18°C and 79°C, respectively, and each at a pressure of 1 bar. Sixty percent of the mass entering the device exits at the lower temperature. Evaluate the inventor's claim, employing the ideal gas model for air and ignoring changes in the kinetic and potential energies of the streams from inlet to exit.

SOLUTION

Known   Data are provided for a device that at steady state produces hot and cold streams of air from a single stream of air at an intermediate temperature without energy transfers by work or heat.

Find   Evaluate whether the device can operate as claimed.

Schematic and Given Data:

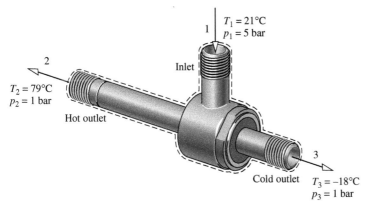

$T_1 = 21°C$
$p_1 = 5$ bar
1 Inlet

2
$T_2 = 79°C$
$p_2 = 1$ bar
Hot outlet

3
Cold outlet $T_3 = -18°C$
$p_3 = 1$ bar

Fig. E6.7

Engineering Model

1. The control volume shown on the accompanying sketch is at steady state.

2. For the control volume, $\dot{W}_{cv} = 0$ and $\dot{Q}_{cv} = 0$.

3. Changes in the kinetic and potential energies from inlet to exit can be ignored.

❶ 4. The air is modeled as an ideal gas with constant $c_p = 1.0$ kJ/kg · K.

Analysis   For the device to operate as claimed, the conservation of mass and energy principles must be satisfied. The second law of thermodynamics also must be satisfied; and in particular the rate of entropy production cannot be negative. Accordingly, the mass, energy and entropy rate balances are considered in turn.

With assumptions 1–3, the mass and energy rate balances reduce, respectively, to

$$\dot{m}_1 = \dot{m}_2 + \dot{m}_3$$

$$0 = \dot{m}_1 h_1 - \dot{m}_2 h_2 - \dot{m}_3 h_3$$

Since $\dot{m}_3 = 0.6\dot{m}_1$, it follows from the mass rate balance that $\dot{m}_2 = 0.4\dot{m}_1$. By combining the mass and energy rate balances and evaluating changes in specific enthalpy using constant c_p, the energy rate balance is also satisfied. That is

❷

$$0 = (\dot{m}_2 + \dot{m}_3)h_1 - \dot{m}_2 h_2 - \dot{m}_3 h_3$$

$$= \dot{m}_2(h_1 - h_2) + \dot{m}_3(h_1 - h_3)$$

$$= 0.4\dot{m}_1[c_p(T_1 - T_2)] + 0.6\dot{m}_1[c_p(T_1 - T_3)]$$

$$= 0.4(T_1 - T_2) + 0.6(T_1 - T_3)$$

$$= 0.4(-105) + 0.6(70)$$

$$= 0$$

Accordingly, with the given data the conservation of mass and energy principles are satisfied.

Since no significant heat transfer occurs, the entropy rate balance at steady state reads

$$0 = \sum_j \frac{\dot{Q}_j}{T_j}^{\,0} + \dot{m}_1 s_1 - \dot{m}_2 s_2 - \dot{m}_3 s_3 + \dot{\sigma}_{cv}$$

Combining the mass and entropy rate balances

$$0 = (\dot{m}_2 + \dot{m}_3)s_1 - \dot{m}_2 s_2 - \dot{m}_3 s_3 + \dot{\sigma}_{cv}$$
$$= \dot{m}_2(s_1 - s_2) + \dot{m}_3(s_1 - s_3) + \dot{\sigma}_{cv}$$
$$= 0.4\dot{m}_1(s_1 - s_2) + 0.6\dot{m}_1(s_1 - s_3) + \dot{\sigma}_{cv}$$

Solving for $\dot{\sigma}_{cv}/\dot{m}_1$ and using Eq. 6.22 to evaluate changes in specific entropy

❸
$$\frac{\dot{\sigma}_{cv}}{\dot{m}_1} = 0.4\left[c_p \ln\frac{T_2}{T_1} - R\ln\frac{p_2}{p_1}\right] + 0.6\left[c_p \ln\frac{T_3}{T_1} - R\ln\frac{p_3}{p_1}\right]$$
$$= 0.4\left[\left(1.0\frac{kJ}{kg\cdot K}\right)\ln\frac{352}{294} - \left(\frac{8.314}{28.97}\frac{kJ}{kg\cdot K}\right)\ln\frac{1}{5.0}\right]$$
$$+ 0.6\left[\left(1.0\frac{kJ}{kg\cdot K}\right)\ln\frac{255}{294} - \left(\frac{8.314}{28.97}\frac{kJ}{kg\cdot K}\right)\ln\frac{1}{5.0}\right]$$

❹
$$= 0.454 \text{ kJ/kg}\cdot K$$

Thus, the second law of thermodynamics is also satisfied.

❺ On the basis of this evaluation, the inventor's claim does not violate principles of thermodynamics.

❶ Since the specific heat c_p of air varies little over the temperature interval from 0 to 79°C, c_p can be taken as constant. From Table A-20, $c_p = 1.0$ kJ/kg · K.

❷ Since temperature *differences* are involved in this calculation, the temperatures can be either in K or °C.

❸ In this calculation involving temperature *ratios*, the temperatures are in K. Temperatures in °C should not be used.

❹ If the value of the rate of entropy production had been negative or zero, the claim would be rejected. A negative value is impossible by the second law and a zero value would indicate operation without irreversibilities.

❺ Such devices *do* exist. They are known as *vortex tubes* and are used in industry for *spot cooling*.

SKILLS DEVELOPED

Ability to...

- apply the control volume mass, energy, and entropy rate balances.
- apply the ideal gas model with constant c_p.

Quick Quiz

If the inventor would claim that the hot and cold streams exit the device at 5 bar, evaluate the revised claim, keeping all other given data the same. Ans. Claim invalid.

Animation
Compressors Tab d

In Example 6.8, we evaluate and compare the rates of entropy production for three components of a heat pump system. Heat pumps are considered in detail in Chap. 10.

▸▸▸ **EXAMPLE 6.8** ▸

Determining Entropy Production in Heat Pump Components

Components of a heat pump for supplying heated air to a dwelling are shown in the schematic below. At steady state, Refrigerant 22 enters the compressor at −5°C, 3.5 bar and is compressed adiabatically to 75°C, 14 bar. From the compressor, the refrigerant passes through the condenser, where it condenses to liquid at 28°C, 14 bar. The refrigerant then expands through a throttling valve to 3.5 bar. The states of the refrigerant are shown on the accompanying T–s diagram. Return air from the dwelling enters the condenser at 20°C, 1 bar with a volumetric flow rate of 0.42 m³/s and exits at 50°C with a negligible change in pressure. Using the ideal gas model for the air and neglecting kinetic and potential energy effects, **(a)** determine the rates of entropy production, in kW/K, for control volumes enclosing the condenser, compressor, and expansion valve, respectively.

(b) Discuss the sources of irreversibility in the components considered in part (a).

SOLUTION

Known Refrigerant 22 is compressed adiabatically, condensed by heat transfer to air passing through a heat exchanger, and then expanded through a throttling valve. Steady-state operating data are known.

Find Determine the entropy production rates for control volumes enclosing the condenser, compressor, and expansion valve, respectively, and discuss the sources of irreversibility in these components.

Schematic and Given Data:

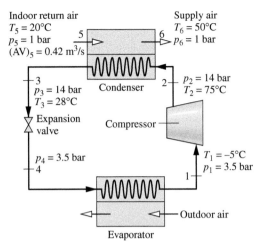

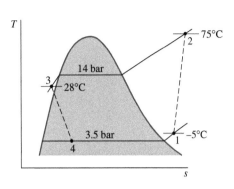

Fig. E6.8

Engineering Model

1. Each component is analyzed as a control volume at steady state.

2. The compressor operates adiabatically, and the expansion across the valve is a *throttling process.*

3. For the control volume enclosing the condenser, $\dot{W}_{cv} = 0$ and $\dot{Q}_{cv} = 0$.

4. Kinetic and potential energy effects can be neglected.

❶ 5. The air is modeled as an ideal gas with constant $c_p = 1.005$ kJ/kg · K.

Analysis

a. Let us begin by obtaining property data at each of the principal refrigerant states located on the accompanying schematic and T–s diagram. At the inlet to the compressor, the refrigerant is a superheated vapor at −5°C, 3.5 bar, so from Table A-9, $s_1 = 0.9572$ kJ/kg · K. Similarly, at state 2, the refrigerant is a superheated vapor at 75°C, 14 bar, so interpolating in Table A-9 gives $s_2 = 0.98225$ kJ/kg · K and $h_2 = 294.17$ kJ/kg.

State 3 is compressed liquid at 28°C, 14 bar. From Table A-7, $s_3 \approx s_f(28°C) = 0.2936$ kJ/kg · K and $h_3 \approx h_f(28°C) = 79.05$ kJ/kg. The expansion through the valve is a *throttling process,* so $h_3 = h_4$. Using data from Table A-8, the quality at state 4 is

$$x_4 = \frac{(h_4 - h_{f4})}{(h_{fg})_4} = \frac{(79.05 - 33.09)}{(212.91)} = 0.216$$

and the specific entropy is

$$s_4 = s_{f4} + x_4(s_{g4} - s_{f4}) = 0.1328 + 0.216(0.9431 - 0.1328)$$
$$= 0.3078 \text{ kJ/kg · K}$$

Condenser

Consider the control volume enclosing the condenser. With assumptions 1 and 3, the entropy rate balance reduces to

$$0 = \dot{m}_{ref}(s_2 - s_3) + \dot{m}_{air}(s_5 - s_6) + \dot{\sigma}_{cond}$$

To evaluate $\dot{\sigma}_{cond}$ requires the two mass flow rates, $\dot{m}_{air}$ and $\dot{m}_{ref}$, and the change in specific entropy for the air. These are obtained next.

Evaluating the mass flow rate of air using the ideal gas model and the known volumetric flow rate

$$\dot{m}_{air} = \frac{(AV)_5}{v_5} = (AV)_5 \frac{p_5}{RT_5}$$

$$= \left(0.42 \frac{\text{m}^3}{\text{s}}\right) \frac{(1 \text{ bar})}{\left(\frac{8.314}{28.97} \frac{\text{kJ}}{\text{kg · K}}\right)(293 \text{ K})} \left|\frac{10^5 \text{ N/m}^2}{1 \text{ bar}}\right| \left|\frac{1 \text{ kJ}}{10^3 \text{ N · m}}\right|$$

$$= 0.5 \text{ kg/s}$$

The refrigerant mass flow rate is determined using an energy balance for the control volume enclosing the condenser together with assumptions 1, 3, and 4 to obtain

$$\dot{m}_{ref} = \frac{\dot{m}_{air}(h_6 - h_5)}{(h_2 - h_3)}$$

With assumption 5, $h_6 - h_5 = c_p(T_6 - T_5)$. Inserting values

❷
$$\dot{m}_{ref} = \frac{\left(0.5 \frac{\text{kg}}{\text{s}}\right)\left(1.005 \frac{\text{kJ}}{\text{kg · K}}\right)(323 - 293)\text{K}}{(294.17 - 79.05) \text{ kJ/kg}} = 0.07 \text{ kg/s}$$

Using Eq. 6.22, the change in specific entropy of the air is

$$s_6 - s_5 = c_p \ln \frac{T_6}{T_5} - R \ln \frac{p_6}{p_5}$$

$$= \left(1.005 \frac{\text{kJ}}{\text{kg · K}}\right) \ln \left(\frac{323}{293}\right) - R \ln \left(\frac{1.0}{1.0}\right)^0 = 0.098 \text{ kJ/kg · K}$$

Finally, solving the entropy balance for $\dot{\sigma}_{cond}$ and inserting values

$$\dot{\sigma}_{cond} = \dot{m}_{ref}(s_3 - s_2) + \dot{m}_{air}(s_6 - s_5)$$

$$= \left[\left(0.07 \frac{\text{kg}}{\text{s}}\right)(0.2936 - 0.98225)\frac{\text{kJ}}{\text{kg · K}} + (0.5)(0.098)\right] \left|\frac{1 \text{ kW}}{1 \text{ kJ/s}}\right|$$

$$= 7.95 \times 10^{-4} \frac{\text{kW}}{\text{K}}$$

Compressor

For the control volume enclosing the compressor, the entropy rate balance reduces with assumptions 1 and 3 to

$$0 = \dot{m}_{ref}(s_1 - s_2) + \dot{\sigma}_{comp}$$

or

$$\dot{\sigma}_{comp} = \dot{m}_{ref}(s_2 - s_1)$$
$$= \left(0.07\frac{kg}{s}\right)(0.98225 - 0.9572)\left(\frac{kJ}{kg \cdot K}\right)\left|\frac{1\ kW}{1\ kJ/s}\right|$$
$$= 17.5 \times 10^{-4}\ kW/K$$

Valve

Finally, for the control volume enclosing the throttling valve, the entropy rate balance reduces to

$$0 = \dot{m}_{ref}(s_3 - s_4) + \dot{\sigma}_{valve}$$

Solving for $\dot{\sigma}_{valve}$ and inserting values

$$\dot{\sigma}_{valve} = \dot{m}_{ref}(s_4 - s_3) = \left(0.07\frac{kg}{s}\right)(0.3078 - 0.2936)\left(\frac{kJ}{kg \cdot K}\right)\left|\frac{1\ kW}{1\ kJ/s}\right|$$
$$= 9.94 \times 10^{-4}\ kW/K$$

b. The following table summarizes, in rank order, the calculated entropy production rates:

Component	$\dot{\sigma}_{cv}(kW/K)$
compressor	17.5×10^{-4}
valve	9.94×10^{-4}
condenser	7.95×10^{-4}

❸ Entropy production in the compressor is due to fluid friction, mechanical friction of the moving parts, and internal heat transfer. For the valve, the irreversibility is primarily due to fluid friction accompanying the expansion across the valve. The principal source of irreversibility in the condenser is the temperature difference between the air and refrigerant streams. In this example, there are no pressure drops for either stream passing through the condenser, but slight pressure drops due to fluid friction would normally contribute to the irreversibility of condensers. The evaporator shown in Fig. E6.8 has not been analyzed.

❶ Due to the relatively small temperature change of the air, the specific heat c_p can be taken as constant at the average of the inlet and exit air temperatures.

❷ Temperatures in K are used to evaluate $\dot{m}_{ref}$, but since a temperature *difference* is involved the same result would be obtained if temperatures in °C were used. Temperatures in K, and not °C, are required when a temperature *ratio* is involved, as in Eq. 6.22 used to evaluate $s_6 - s_5$.

❸ By focusing attention on reducing irreversibilities at the sites with the highest entropy production rates, *thermodynamic improvements* may be possible. However, costs and other constraints must be considered and can be overriding.

SKILLS DEVELOPED

Ability to...

• apply the control volume, mass, energy, and entropy rate balances.
• develop an engineering model.
• retrieve property data for Refrigerant 22.
• apply the ideal gas model with constant c_p.

Quick Quiz

If the compressor operated adiabatically *and* without internal irreversibilities, determine the temperature of the refrigerant at the compressor exit, in °C, keeping the compressor inlet state and exit pressure the same. Ans. 65°C.

6.11 Isentropic Processes

The term *isentropic* means constant entropy. Isentropic processes are encountered in many subsequent discussions. The object of the present section is to show how properties are related at any two states of a process in which there is no change in specific entropy.

6.11.1 General Considerations

The properties at states having the same specific entropy can be related using the graphical and tabular property data discussed in Sec. 6.2. For example, as illustrated by **Fig. 6.8**, temperature–entropy and enthalpy–entropy diagrams are particularly convenient for determining properties at states having the same value of specific entropy. All states on a vertical line passing through a given state have the same entropy. If state 1 on Fig. 6.8 is fixed by pressure p_1 and temperature T_1, states 2 and 3 are readily located once one additional property, such as pressure or temperature, is specified. The values of several other properties at states 2 and 3 can then be read directly from the figures.

Tabular data also can be used to relate two states having the same specific entropy. For the case shown in Fig. 6.8, the specific entropy at state 1 could be determined from the superheated

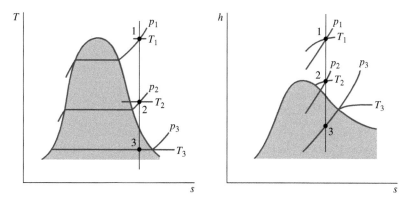

Fig. 6.8 *T*–s and *h*–s diagrams showing states having the same value of specific entropy.

vapor table. Then, with $s_2 = s_1$ and one other property value, such as p_2 or T_2, state 2 could be located in the superheated vapor table. The values of the properties v, u, and h at state 2 can then be read from the table. (An illustration of this procedure is given in Sec. 6.2.1.) Note that state 3 falls in the two-phase liquid–vapor regions of Fig. 6.8. Since $s_3 = s_1$, the quality at state 3 could be determined using Eq. 6.4. With the quality known, other properties such as v, u, and h could then be evaluated. Computer retrieval of entropy data provides an alternative to tabular data.

6.11.2 Using the Ideal Gas Model

Figure 6.9 shows two states of an ideal gas having the same value of specific entropy. Let us consider relations among pressure, specific volume, and temperature at these states, first using the ideal gas tables and then assuming specific heats are constant.

Ideal Gas Tables   For two states having the same specific entropy, Eq. 6.20a reduces to

$$0 = s°(T_2) - s°(T_1) - R \ln \frac{p_2}{p_1} \tag{6.40a}$$

Equation 6.40a involves four property values: p_1, T_1, p_2, and T_2. If any three are known, the fourth can be determined. If, for example, the temperature at state 1 and the pressure ratio p_2/p_1 are known, the temperature at state 2 can be determined from

$$s°(T_2) = s°(T_1) + R \ln \frac{p_2}{p_1} \tag{6.40b}$$

Since T_1 is known, $s°(T_1)$ would be obtained from the appropriate table, the value of $s°(T_2)$ would be calculated, and temperature T_2 would then be determined by interpolation. If p_1, T_1, and T_2 are specified and the pressure at state 2 is the unknown, Eq. 6.40a would be solved to obtain

$$p_2 = p_1 \exp \left[\frac{s°(T_2) - s°(T_1)}{R} \right] \tag{6.40c}$$

Equations 6.40 can be used when $s°$ (or $\bar{s}°$) data are known, as for the gases of Tables A-22 and A-23.

Air.   For the special case of *air* modeled as an ideal gas, Eq. 6.40c provides the basis for an alternative tabular approach for relating the temperatures and pressures at two states having the same specific entropy. To introduce this, rewrite the equation as

$$\frac{p_2}{p_1} = \frac{\exp[s°(T_2)/R]}{\exp[s°(T_1)/R]}$$

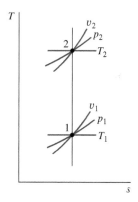

Fig. 6.9 Two states of an ideal gas where $s_2 = s_1$.

The quantity $\exp[s°(T)/R]$ appearing in this expression is solely a function of temperature and is given the symbol $p_r(T)$. A tabulation of p_r versus temperature for *air* is provided in Table A-22.[1] In terms of the function p_r, the last equation becomes

$$\boxed{\frac{p_2}{p_1} = \frac{p_{r2}}{p_{r1}} \qquad (s_1 = s_2, \text{ air only})}$$

(6.41)

where $p_{r1} = p_r(T_1)$ and $p_{r2} = p_r(T_2)$. The function p_r is sometimes called the *relative pressure*. Observe that p_r is not truly a pressure, so the name *relative pressure* has no physical significance. Also, be careful not to confuse p_r with the reduced pressure of the compressibility diagram.

A relation between specific volumes and temperatures for two states of air having the same specific entropy can also be developed. With the ideal gas equation of state, $v = RT/p$, the ratio of the specific volumes is

$$\frac{v_2}{v_1} = \left(\frac{RT_2}{p_2}\right)\left(\frac{p_1}{RT_1}\right)$$

Then, since the two states have the same specific entropy, Eq. 6.41 can be introduced to give

$$\frac{v_2}{v_1} = \left[\frac{RT_2}{p_r(T_2)}\right]\left[\frac{p_r(T_1)}{RT_1}\right]$$

TAKE NOTE...

When applying the software *IT* to relate two states of an ideal gas having the same value of specific entropy, *IT* returns specific entropy *directly* and does not employ the special functions $s°$, p_r, and v_r.

The ratio $RT/p_r(T)$ appearing on the right side of the last equation is solely a function of temperature, and is given the symbol $v_r(T)$. Values of v_r for *air* are tabulated versus temperature in Table A-22. In terms of the function v_r, the last equation becomes

$$\boxed{\frac{v_2}{v_1} = \frac{v_{r2}}{v_{r1}} \qquad (s_1 = s_2, \text{ air only})}$$

(6.42)

where $v_{r1} = v_r(T_1)$ and $v_{r2} = v_r(T_2)$. The function v_r is sometimes called the *relative volume*. Despite the name given to it, $v_r(T)$ is not truly a volume. Also, be careful not to confuse it with the pseudoreduced specific volume of the compressibility diagram.

Assuming Constant Specific Heats Let us consider next how properties are related for isentropic processes of an ideal gas when the specific heats are constants. For any such case, Eqs. 6.21 and 6.22 reduce to the equations

$$0 = c_p \ln\frac{T_2}{T_1} - R \ln\frac{p_2}{p_1}$$

$$0 = c_v \ln\frac{T_2}{T_1} + R \ln\frac{v_2}{v_1}$$

Introducing the ideal gas relations

$$c_p = \frac{kR}{k-1}, \qquad c_v = \frac{R}{k-1}$$

(3.47)

where k is the specific heat ratio and R is the gas constant, these equations can be solved, respectively, to give

$$\boxed{\frac{T_2}{T_1} = \left(\frac{p_2}{p_1}\right)^{(k-1)/k} \qquad (s_1 = s_2, \text{ constant } k)}$$

(6.43)

$$\boxed{\frac{T_2}{T_1} = \left(\frac{v_1}{v_2}\right)^{k-1} \qquad (s_1 = s_2, \text{ constant } k)}$$

(6.44)

[1]The values of p_r determined with this definition are inconveniently large, so they are divided by a scale factor before tabulating to give a convenient range of numbers.

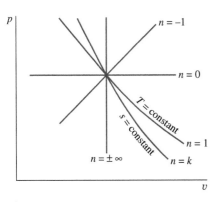
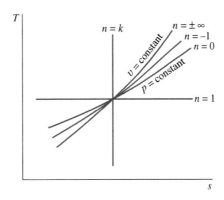

Fig. 6.10 Polytropic processes on p–v and T–s diagrams.

The following relation can be obtained by eliminating the temperature ratio from Eqs. 6.43 and 6.44:

$$\frac{p_2}{p_1} = \left(\frac{v_1}{v_2}\right)^k \qquad (s_1 = s_2, \text{ constant } k) \qquad (6.45)$$

Previously, we have identified an internally reversible process described by $pv^n = constant$, where n is a constant, as a *polytropic process*. From the form of Eq. 6.45, it can be concluded that the polytropic process $pv^k = constant$ of an ideal gas with constant specific heat ratio k is an isentropic process. We observed in Sec. 3.15 that a polytropic process of an ideal gas for which $n = 1$ is an isothermal (constant-temperature) process. For *any* fluid, $n = 0$ corresponds to an isobaric (constant-pressure) process and $n = \pm\infty$ corresponds to an isometric (constant-volume) process. Polytropic processes corresponding to these values of n are shown in Fig. 6.10 on p–v and T–s diagrams.

6.11.3 Illustrations: Isentropic Processes of Air

Means for evaluating data for isentropic processes of air modeled as an ideal gas are illustrated in the next two examples. In Example 6.9, we consider three alternative methods.

▸▸▸ **EXAMPLE 6.9** ▸

Isentropic Process of Air

Air undergoes an isentropic process from $p_1 = 1$ bar, $T_1 = 300$ K to a final state where the temperature is $T_2 = 650$ K. Employing the ideal gas model, determine the final pressure p_2, in bar. Solve using (a) p_r data from Table A-22, (b) *Interactive Thermodynamics: IT,* and (c) a constant specific heat ratio k evaluated at the mean temperature, 475 K, from Table A-20.

SOLUTION

Known Air undergoes an isentropic process from a state where pressure and temperature are known to a state where the temperature is specified.

Find Determine the final pressure using (a) p_r data, (b) *IT*, and (c) a constant value for the specific heat ratio k.

Schematic and Given Data:

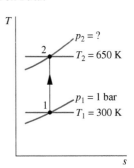

Fig. E6.9

Engineering Model

1. A quantity of air as the system undergoes an isentropic process.

2. The air can be modeled as an ideal gas.

3. In part (c), the specific heat ratio is constant.

Analysis

a. The pressures and temperatures at two states of an ideal gas having the same specific entropy are related by Eq. 6.41

$$\frac{p_2}{p_1} = \frac{p_{r2}}{p_{r1}}$$

Solving

$$p_2 = p_1 \frac{p_{r2}}{p_{r1}}$$

With p_r values from Table A-22

$$p = (1 \text{ bar}) \frac{21.18}{1.3860} = 15.77 \text{ bar}$$

b. The *IT* solution follows:

```
T1 = 300 // K
p1 = 1 // bar
T2 = 650 // K
s_TP("Air", T1,p1) = s_TP("Air",T2,p2)
// Result: p2 = 15.77 bar
```

❶

c. When the specific heat ratio k is assumed constant, the temperatures and pressures at two states of an ideal gas having the same specific entropy are related by Eq. 6.43. Thus

$$p_2 = p_1 \left(\frac{T_2}{T_1} \right)^{k/(k-1)}$$

From Table A-20 at the mean temperature, 202°C (475 K), $k = 1.39$. Inserting values into the above expression

❷
$$p_2 = (1 \text{ bar}) \left(\frac{650}{300} \right)^{1.39/0.39} = 15.81 \text{ bar}$$

---

❶ *IT* returns a value for p_2 even though it is an implicit variable in the specific entropy function. Also note that *IT* returns values for specific entropy *directly* and does not employ special functions such as $s°$, p_r, and v_r.

❷ The close agreement between the answer obtained in part (c) and that of parts (a), (b) can be attributed to the use of an appropriate value for the specific heat ratio k.

SKILLS DEVELOPED

Ability to…

• analyze an isentropic process using Table A-22 data,

• *Interactive Thermodynamics*, and

• a constant specific heat ratio k.

Quick Quiz

Determine the final pressure, in bar, using a constant specific heat ratio k evaluated at $T_1 = 300$ K. Expressed as a percent, how much does this pressure value differ from that of part (c)? Ans. 14.72 bar, −5%

Another illustration of an isentropic process of an ideal gas is provided in Example 6.10 dealing with air leaking from a tank.

▶▶▶ **EXAMPLE 6.10** ▶ •

Considering Air Leaking from a Tank

A rigid, well-insulated tank is filled initially with 5 kg of air at a pressure of 5 bar and a temperature of 500 K. A leak develops, and air slowly escapes until the pressure of the air remaining in the tank is 1 bar. Employing the ideal gas model, determine the amount of mass remaining in the tank and its temperature.

Schematic and Given Data:

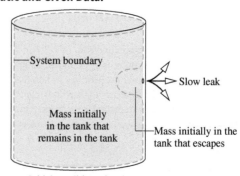

Initial condition of tank

Fig. E6.10

SOLUTION

Known A leak develops in a rigid, insulated tank initially containing air at a known state. Air slowly escapes until the pressure in the tank is reduced to a specified value.

Find Determine the amount of mass remaining in the tank and its temperature.

Engineering Model

1. As shown on the accompanying sketch, the closed system is the mass initially in the tank that remains in the tank.

2. There is no significant heat transfer between the system and its surroundings.

3. Irreversibilities within the tank can be ignored as the air slowly escapes.

4. The air is modeled as an ideal gas.

Analysis With the ideal gas equation of state, the mass initially in the tank that *remains* in the tank at the end of the process is

$$m_2 = \frac{p_2 V}{(\bar{R}/M)T_2}$$

where p_2 and T_2 are the final pressure and temperature, respectively. Similarly, the initial amount of mass within the tank, m_1 is

$$m_1 = \frac{p_1 V}{(\bar{R}/M)T_1}$$

where p_1 and T_1 are the initial pressure and temperature, respectively. Eliminating volume between these two expressions, the mass of the system is

$$m_2 = \left(\frac{p_2}{p_1}\right)\left(\frac{T_1}{T_2}\right)m_1$$

Except for the final temperature of the air remaining in the tank, T_2, all required values are known. The remainder of the solution mainly concerns the evaluation of T_2.

For the closed system under consideration, there are no significant irreversibilities (assumption 3), and no heat transfer occurs

(assumption 2). Accordingly, the entropy balance reduces to

$$\Delta S = \int_1^2 \left(\frac{\delta Q}{T}\right)_b^{\;0} + \cancel{\sigma}^{\;0} = 0$$

Since the system mass remains constant, $\Delta S = m_2 \, \Delta s$, so

$$\Delta s = 0$$

That is, the initial and final states of the system have the same value of *specific* entropy.

Using Eq. 6.41

$$p_{r2} = \left(\frac{p_2}{p_1}\right)p_{r1}$$

where $p_1 = 5$ bar and $p_2 = 1$ bar. With $p_{r1} = 8.411$ from Table A-22 at 500 K, the previous equation gives $p_{r2} = 1.6822$. Using this to interpolate in Table A-22, $T_2 = 317$ K.

Finally, inserting values into the expression for system mass

$$m_2 = \left(\frac{1 \text{ bar}}{5 \text{ bar}}\right)\left(\frac{500 \text{ K}}{317 \text{ K}}\right)(5 \text{ kg}) = 1.58 \text{ kg}$$

SKILLS DEVELOPED

Ability to...

- develop an engineering model.
- apply the closed system entropy balance.
- analyze an isentropic process.

Quick Quiz

Evaluate the tank volume, in m^3. Ans. 1.43 m^3.

6.12 Isentropic Efficiencies of Turbines, Nozzles, Compressors, and Pumps

Engineers make frequent use of efficiencies and many different efficiency definitions are employed. In the present section, *isentropic* efficiencies for turbines, nozzles, compressors, and pumps are introduced. Isentropic efficiencies involve a comparison between the actual performance of a device and the performance that would be achieved under idealized circumstances for the same inlet state and the same exit pressure. These efficiencies are frequently used in subsequent sections of the book.

6.12.1 Isentropic Turbine Efficiency

To introduce the isentropic turbine efficiency, refer to **Fig. 6.11**, which shows a turbine expansion on a Mollier diagram. The state of the matter entering the turbine and the exit pressure are fixed. Heat transfer between the turbine and its surroundings is ignored, as are kinetic and potential energy effects. With these assumptions, the mass and energy

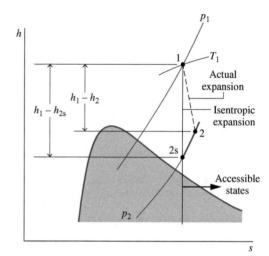

Fig. 6.11 Comparison of actual and isentropic expansions through a turbine.

rate balances reduce, at steady state, to give the work developed per unit of mass flowing through the turbine

$$\frac{\dot{W}_{cv}}{\dot{m}} = h_1 - h_2$$

Since state 1 is fixed, the specific enthalpy h_1 is known. Accordingly, the value of the work depends on the specific enthalpy h_2 only, and increases as h_2 is reduced. The *maximum* value for the turbine work corresponds to the smallest *allowed* value for the specific enthalpy at the turbine exit. This can be determined using the second law as follows.

Since there is no heat transfer, the allowed exit states are constrained by Eq. 6.39:

$$\frac{\dot{\sigma}_{cv}}{\dot{m}} = s_2 - s_1 \geq 0$$

Because the entropy production $\dot{\sigma}_{cv}/\dot{m}$ cannot be negative, states with $s_2 < s_1$ are not accessible in an adiabatic expansion. The only states that actually can be attained *adiabatically* are those with $s_2 > s_1$. The state labeled "2s" on Fig. 6.11 would be attained only in the limit of no internal irreversibilities. This corresponds to an isentropic expansion through the turbine. For fixed exit pressure, the specific enthalpy h_2 decreases as the specific entropy s_2 decreases. Therefore, the *smallest allowed* value for h_2 corresponds to state 2s, and the *maximum* value for the turbine work is

TAKE NOTE...

The subscript s denotes a quantity evaluated for an isentropic process from a specified inlet state to a specified exit pressure.

$$\left(\frac{\dot{W}_{cv}}{\dot{m}}\right)_s = h_1 - h_{2s}$$

In an actual expansion through the turbine $h_2 > h_{2s}$, and thus less work than the maximum would be developed. This difference can be gauged by the **isentropic turbine efficiency** defined by

isentropic turbine efficiency

$$\boxed{\eta_t = \frac{\dot{W}_{cv}/\dot{m}}{(\dot{W}_{cv}/\dot{m})_s} = \frac{h_1 - h_2}{h_1 - h_{2s}}} \tag{6.46}$$

Both the numerator and denominator of this expression are evaluated for the same inlet state and the same exit pressure. The value of η_t is typically 0.7 to 0.9 (70–90%).

The two examples to follow illustrate the isentropic turbine efficiency concept. In Example 6.11 the isentropic efficiency of a steam turbine is known and the objective is to determine the turbine work.

Turbine Tab e

Determining Turbine Work Using the Isentropic Efficiency

A steam turbine operates at steady state with inlet conditions of $p_1 = 5$ bar, $T_1 = 320°C$. Steam leaves the turbine at a pressure of 1 bar. There is no significant heat transfer between the turbine and its surroundings, and kinetic and potential energy changes between inlet and exit are negligible. If the isentropic turbine efficiency is 75%, determine the work developed per unit mass of steam flowing through the turbine, in kJ/kg.

SOLUTION

Known   Steam expands through a turbine operating at steady state from a specified inlet state to a specified exit pressure. The turbine efficiency is known.

Find   Determine the work developed per unit mass of steam flowing through the turbine.

Schematic and Given Data:

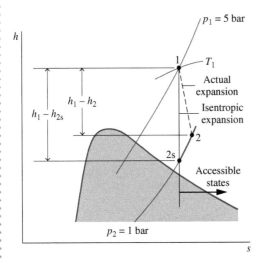

Fig. E6.11

Engineering Model

1. A control volume enclosing the turbine is at steady state.

2. The expansion is adiabatic and changes in kinetic and potential energy between the inlet and exit can be neglected.

Analysis   The work developed can be determined using the isentropic turbine efficiency, Eq. 6.46, which on rearrangement gives

$$\frac{\dot{W}_{cv}}{\dot{m}} = \eta_t \left(\frac{\dot{W}_{cv}}{\dot{m}} \right)_s = \eta_t (h_1 - h_{2s})$$

From Table A-4, $h_1 = 3105.6$ kJ/kg and $s_1 = 7.5308$ kJ/kg · K. The exit state for an isentropic expansion is fixed by $p_2 = 1$ and

❶ $s_{2s} = s_1$. Interpolating with specific entropy in Table A-4 at 1 bar gives $h_{2s} = 2743.0$ kJ/kg. Substituting values

❷ $$\frac{\dot{W}_{cv}}{\dot{m}} = 0.75(3105.6 - 2743.0) = 271.95 \text{ kJ/kg}$$

---

❶ At 2s, the temperature is about 133°C.

❷ The effect of irreversibilities is to exact a penalty on the work output of the turbine. The work is only 75% of what it would be for an isentropic expansion between the given inlet state and the turbine exhaust pressure. This is clearly illustrated in terms of enthalpy differences on the accompanying h–s diagram.

SKILLS DEVELOPED

Ability to...

• apply the isentropic turbine efficiency, Eq. 6.46.

• retrieve *steam table* data.

Quick Quiz

Determine the temperature of the steam at the turbine exit, in °C. Ans. 179°C.

Example 6.12 is similar to Example 6.11, but here the working substance is air as an ideal gas. Moreover, in this case the turbine work is known and the objective is to determine the isentropic turbine efficiency.

Evaluating Isentropic Turbine Efficiency

A turbine operating at steady state receives air at a pressure of $p_1 = 3.0$ bar and a temperature of $T_1 = 390$ K. Air exits the turbine at a pressure of $p_2 = 1.0$ bar. The work developed is measured as 74 kJ per kg of air flowing through the turbine. The turbine operates adiabatically, and changes in kinetic and potential energy between inlet and exit can be neglected. Using the ideal gas model for air, determine the isentropic turbine efficiency.

SOLUTION

Known   Air expands adiabatically through a turbine at steady state from a specified inlet state to a specified exit pressure. The work developed per kg of air flowing through the turbine is known.

Find   Determine the turbine efficiency.

Schematic and Given Data:

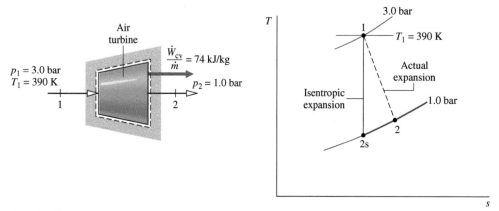

Fig. E6.12

Engineering Model

1. The control volume shown on the accompanying sketch is at steady state.

2. The expansion is adiabatic and changes in kinetic and potential energy between inlet and exit can be neglected.

3. The air is modeled as an ideal gas.

Analysis   The numerator of the isentropic turbine efficiency, Eq. 6.46, is known. The denominator is evaluated as follows.

The work developed in an isentropic expansion from the given inlet state to the specified exit pressure is

$$\left(\frac{\dot{W}_{cv}}{\dot{m}}\right)_s = h_1 - h_{2s}$$

From Table A-22 at 390 K, $h_1 = 390.88$ kJ/kg. To determine h_{2s}, use Eq. 6.41:

$$p_r(T_{2s}) = \left(\frac{p_2}{p_1}\right) p_r(T_1)$$

With $p_1 = 3.0$ bar, $p_2 = 1.0$ bar, and $p_{r1} = 3.481$ from Table A-22 at 390 K

$$p_r(T_{2s}) = \left(\frac{1.0}{3.0}\right)(3.481) = 1.1603$$

Interpolation in Table A-22 gives $h_{2s} = 285.27$ kJ/kg. Thus,

$$\left(\frac{\dot{W}_{cv}}{\dot{m}}\right)_s = 390.88 - 285.27 = 105.6 \text{ kJ/kg}$$

Substituting values into Eq. 6.46

$$\eta_t = \frac{\dot{W}_{cv}/\dot{m}}{(\dot{W}_{cv}/\dot{m})_s} = \frac{74 \text{ kJ/kg}}{105.6 \text{ kJ/kg}} = 0.70(70\%)$$

SKILLS DEVELOPED

Ability to…

- apply the isentropic turbine efficiency, Eq. 6.46.
- retrieve data for air as an ideal gas.

Quick Quiz

Determine the rate of entropy production, in kJ/K per kg of air flowing through the turbine. Ans. 0.105 kJ/kg · K.

6.12.2 Isentropic Nozzle Efficiency

isentropic nozzle efficiency

A similar approach to that for turbines can be used to introduce the isentropic efficiency of nozzles operating at steady state. The **isentropic nozzle efficiency** is defined as the ratio of the actual specific kinetic energy of the gas leaving the nozzle, $V_2^2/2$, to the kinetic energy at the exit that would be achieved in an isentropic expansion between the same inlet state and the same exit pressure, $(V_2^2/2)_s$. That is,

$$\eta_{nozzle} = \frac{V_2^2/2}{(V_2^2/2)_s} \tag{6.47}$$

Nozzles Tab e

Nozzle efficiencies of 95% or more are common, indicating that well-designed nozzles are nearly free of internal irreversibilities.

In Example 6.13, the objective is to determine the isentropic efficiency of a steam nozzle.

▶▶▶ EXAMPLE 6.13 ▶ •

Evaluating the Isentropic Nozzle Efficiency

Steam enters a nozzle operating at steady state at p_1 = 9.7 bar and T_1 = 316°C with a velocity of 30 m/s. The pressure and temperature at the exit are p_2 = 2.8 bar and T_2 = 177°C. There is no significant heat transfer between the nozzle and its surroundings, and changes in potential energy between inlet and exit can be neglected. Determine the nozzle efficiency.

SOLUTION

Known Steam expands through a nozzle at steady state from a specified inlet state to a specified exit state. The velocity at the inlet is known.

Find Determine the nozzle efficiency.

Schematic and Given Data:

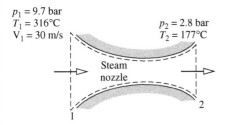

p_1 = 9.7 bar
T_1 = 316°C
V_1 = 30 m/s

p_2 = 2.8 bar
T_2 = 177°C

Steam
nozzle

1
2

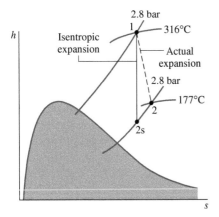

2.8 bar
1 — 316°C
h
Isentropic
expansion
Actual
expansion
2.8 bar
177°C
2
2s

s

Fig. E6.13

Engineering Model

1. The control volume shown on the accompanying sketch operates adiabatically at steady state.

2. For the control volume, $\dot{W}_{cv}$ = 0 and the change in potential energy between inlet and exit can be neglected.

Analysis The nozzle efficiency given by Eq. 6.47 requires the actual specific kinetic energy at the nozzle exit and the specific kinetic energy that would be achieved at the exit in an isentropic expansion from the given inlet state to the given exit pressure.

The energy rate balance for a one-inlet, one-exit control volume at steady state (Eq. 4.20b) reduces to give

$$\frac{V_2^2}{2} = h_1 - h_2 + \frac{V_1^2}{2}$$

This equation applies for both the actual expansion and the isentropic expansion.

From Table A-4 at T_1 = 320°C and p_1 = 1 MPa, h_1 = 3093.9 kJ/kg, s_1 = 7.1962 kJ/kg · K. Also, with T_2 = 180°C and p_2 = 0.3 MPa, h_2 = 2823.9 kJ/kg. Thus, the actual specific kinetic energy at the exit in kJ/kg is

$$\frac{V_2^2}{2} = (3093.9 - 2823.9)\frac{kJ}{kg} + \frac{(30)^2}{2}\left(\frac{m}{s}\right)^2 \left(\frac{IN}{1\,kg \cdot m/s^2}\right)\left(\frac{1\,kJ}{10^3\,N \cdot m}\right)$$

$$= 270.45 \frac{kJ}{kg}$$

Interpolating in Table A-4 at 0.3 MPa, with s_{2s} = s_1 = 7.1962 kJ/kg · K, results in h_{2s} = 2813.3 kJ/kg. Accordingly, the specific kinetic energy at the exit for an isentropic expansion is

$$\left(\frac{V_2^2}{2}\right)_s = 3093.9 - 2813.3 + \frac{(30)^2}{2(10^3)} = 281.05 \text{ kJ/kg}$$

Substituting values into Eq. 6.47

❶ $$\eta_{nozzle} = \frac{(V_2^2/2)}{(V_2^2/2)_s} = \frac{270.45}{281.05} = 0.962 \; (96.2\%)$$

─────────────────────────

❶ The principal irreversibility in nozzles is friction between the flowing gas or liquid and the nozzle wall. The effect of friction is that a smaller exit kinetic energy, and thus a smaller exit velocity, is realized than would have been obtained in an isentropic expansion to the same pressure.

SKILLS DEVELOPED

Ability to...

• apply the control volume energy rate balance.

• apply the isentropic nozzle efficiency, Eq. 6.47.

• retrieve *steam table* data.

Quick Quiz

Determine the temperature, in °C, corresponding to state 2s in Fig. E6.13. Ans. 166°C.

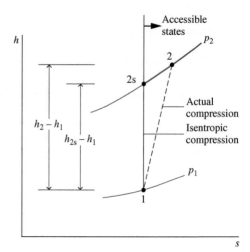

Fig. 6.12 Comparison of actual and isentropic compressions.

6.12.3 Isentropic Compressor and Pump Efficiencies

The form of the isentropic efficiency for compressors and pumps is taken up next. Refer to Fig. 6.12, which shows a compression process on a Mollier diagram. The state of the matter entering the compressor and the exit pressure are fixed. For negligible heat transfer with the surroundings and no appreciable kinetic and potential energy effects, the work *input* per unit of mass flowing through the compressor is

$$\left(-\frac{\dot{W}_{cv}}{\dot{m}} \right) = h_2 - h_1$$

Since state 1 is fixed, the specific enthalpy h_1 is known. Accordingly, the value of the work input depends on the specific enthalpy at the exit, h_2. The above expression shows that the magnitude of the work input decreases as h_2 decreases. The *minimum* work input corresponds to the smallest *allowed* value for the specific enthalpy at the compressor exit. With similar reasoning as for the turbine, the smallest allowed enthalpy at the exit state would be achieved in an isentropic compression from the specified inlet state to the specified exit pressure. The minimum work *input* is given, therefore, by

$$\left(-\frac{\dot{W}_{cv}}{\dot{m}} \right)_s = h_{2s} - h_1$$

**Compressors Tab e
Pumps Tab e**

isentropic compressor efficiency

In an actual compression, $h_2 > h_{2s}$, and thus more work than the minimum would be required. This difference can be gauged by the **isentropic compressor efficiency** defined by

$$\eta_c = \frac{(-\dot{W}_{cv}/\dot{m})_s}{(-\dot{W}_{cv}/\dot{m})} = \frac{h_{2s} - h_1}{h_2 - h_1} \tag{6.48}$$

isentropic pump efficiency

Both the numerator and denominator of this expression are evaluated for the same inlet state and the same exit pressure. The value of η_c is typically 75 to 85% for compressors. An **isentropic pump efficiency**, η_p, is defined similarly.

In Example 6.14, the isentropic efficiency of a refrigerant compressor is evaluated, first using data from property tables and then using *IT*.

> ▶ ▶ ▶ **EXAMPLE 6.14** ▶

Evaluating Isentropic Compressor Efficiency

For the compressor of the heat pump system in Example 6.8, determine the power, in kW, and the isentropic efficiency using **(a)** data from property tables, **(b)** *Interactive Thermodynamics: IT*.

SOLUTION

Known Refrigerant 22 is compressed adiabatically at steady state from a specified inlet state to a specified exit state. The mass flow rate is known.

Find Determine the compressor power and the isentropic efficiency using (a) property tables, (b) *IT*.

Schematic and Given Data:

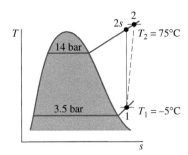

Fig. E6.14

Engineering Model

1. A control volume enclosing the compressor is at steady state.

2. The compression is adiabatic, and changes in kinetic and potential energy between the inlet and the exit can be neglected.

Analysis

a. By assumptions 1 and 2, the mass and energy rate balances reduce to give

$$\dot{W}_{cv} = \dot{m}(h_1 - h_2)$$

From Table A-9, $h_1 = 249.75$ kJ/kg and $h_2 = 294.17$ kJ/kg. Thus, with the mass flow rate determined in Example 6.8

$$\dot{W}_{cv} = (0.07 \text{ kg/s})(249.75 - 294.17) \text{ kJ/kg} \left| \frac{1 \text{ kW}}{1 \text{ kJ/s}} \right| = -3.11 \text{ kW}$$

The isentropic compressor efficiency is determined using Eq. 6.48:

$$\eta_c = \frac{(-\dot{W}_{cv}/\dot{m})_s}{(-\dot{W}_{cv}/\dot{m})} = \frac{(h_{2s} - h_1)}{(h_2 - h_1)}$$

In this expression, the denominator represents the work input per unit mass of refrigerant flowing for the actual compression process, as considered above. The numerator is the work input for an isentropic compression between the initial state and the

same exit pressure. The isentropic exit state is denoted as state 2s on the accompanying *T–s* diagram.

From Table A-9, $s_1 = 0.9572$ kJ/kg · K. With $s_{2s} = s_1$, interpolation in Table A-9 at 14 bar gives $h_{2s} = 285.58$ kJ/kg. Substituting values

$$\eta_c = \frac{(285.58 - 249.75)}{(294.17 - 249.75)} = 0.81 (81\%)$$

b. The *IT* program follows. In the program, $\dot{W}_{cv}$ is denoted as Wdot, $\dot{m}$ as mdot, and η_c as eta_c.

```
// Given Data:
T1 = -5 // °C
p1 = 3.5 // bar
T2 = 75 // °C
p2 = 14 // bar
mdot = 0.07 // kg/s

// Determine the specific enthalpies.
h1 = h_PT("R22",p1,T1)
h2 = h_PT("R22",p2,T2)

// Calculate the power.
Wdot = mdot * (h1 - h2)
// Find h2s:
s1 = s_PT("R22",p1,T1)
s2s = s_Ph("R22",p2,h2s)
s2s = s1

// Determine the isentropic compressor effi-
ciency.
eta_c = (h2s - h1)/(h2 - h1)
```

❶

Use the **Solve** button to obtain $\dot{W}_{cv} = -3.111$ kW and $\eta_c = 80.58\%$, which, as expected, agree closely with the values obtained above.

---

❶ Note that *IT* solves for the value of h_{2s} even though it is an implicit variable in the specific entropy function.

SKILLS DEVELOPED

Ability to...

- apply the control volume energy rate balance.
- apply the isentropic compressor efficiency, Eq. 6.48.
- retrieve data for Refrigerant 22.

Quick Quiz

Determine the minimum theoretical work input, in kJ per kg flowing, for an adiabatic compression from state 1 to the exit pressure of 14 bar. Ans. 35.83 kJ/kg.

6.13 Heat Transfer and Work in Internally Reversible, Steady-State Flow Processes

This section concerns one-inlet, one-exit control volumes at steady state. The objective is to introduce expressions for the heat transfer and the work in the absence of internal irreversibilities. The resulting expressions have several important applications.

6.13.1 Heat Transfer

For a control volume at steady state in which the flow is both *isothermal* at temperature T and *internally reversible*, the appropriate form of the entropy rate balance is

$$0 = \frac{\dot{Q}_{cv}}{T} + \dot{m}(s_1 - s_2) + \dot{\sigma}_{cv}^{0}$$

where 1 and 2 denote the inlet and exit, respectively, and $\dot{m}$ is the mass flow rate. Solving this equation, the heat transfer per unit of mass passing through the control volume is

$$\frac{\dot{Q}_{cv}}{\dot{m}} = T(s_2 - s_1)$$

More generally, temperature varies as the gas or liquid flows through the control volume. We can consider such a temperature variation to consist of a series of infinitesimal steps. Then the heat transfer per unit of mass is given as

$$\left(\frac{\dot{Q}_{cv}}{\dot{m}}\right)_{\substack{int \\ rev}} = \int_1^2 T \, ds \qquad (6.49)$$

The subscript "int rev" serves to remind us that the expression applies only to control volumes in which there are no internal irreversibilities. The integral of Eq. 6.49 is performed from inlet to exit. When the states visited by a unit mass as it passes reversibly from inlet to exit are described by a curve on a T–s diagram, the magnitude of the heat transfer per unit of mass flowing can be represented as the area *under* the curve, as shown in Fig. 6.13.

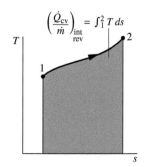

$$\left(\frac{\dot{Q}_{cv}}{\dot{m}}\right)_{\substack{int \\ rev}} = \int_1^2 T \, ds$$

Fig. 6.13 Area representation of heat transfer for an internally reversible flow process.

6.13.2 Work

The work per unit of mass passing through a one-inlet, one-exit control volume can be found from an energy rate balance, which reduces at steady state to give

$$\frac{\dot{W}_{cv}}{\dot{m}} = \frac{\dot{Q}_{cv}}{\dot{m}} + (h_1 - h_2) + \left(\frac{V_1^2 - V_2^2}{2}\right) + g(z_1 - z_2)$$

This equation is a statement of the conservation of energy principle that applies when irreversibilities are present within the control volume as well as when they are absent. However, if consideration is restricted to the internally reversible case, Eq. 6.49 can be introduced to obtain

$$\left(\frac{\dot{W}_{cv}}{\dot{m}}\right)_{\substack{int \\ rev}} = \int_1^2 T \, ds + (h_1 - h_2) + \left(\frac{V_1^2 - V_2^2}{2}\right) + g(z_1 - z_2) \qquad (6.50)$$

where the subscript "int rev" has the same significance as before.

Since internal irreversibilities are absent, a unit of mass traverses a sequence of equilibrium states as it passes from inlet to exit. Entropy, enthalpy, and pressure changes are therefore related by Eq. 6.10b

$$T \, ds = dh - v \, dp$$

which on integration gives

$$\int_1^2 T \, ds = (h_2 - h_1) - \int_1^2 v \, dp$$

Introducing this relation, Eq. 6.50 becomes

$$\left(\frac{\dot{W}_{cv}}{\dot{m}} \right)_{\substack{int \\ rev}} = -\int_1^2 v \, dp + \left(\frac{V_1^2 - V_2^2}{2} \right) + g(z_1 - z_2) \qquad (6.51a)$$

When the states visited by a unit of mass as it passes reversibly from inlet to exit are described by a curve on a p–v diagram as shown in **Fig. 6.14**, the magnitude of the integral $\int v \, dp$ is represented by the shaded area *behind* the curve.

Equation 6.51a is applicable to devices such as turbines, compressors, and pumps. In many of these cases, there is no significant change in kinetic or potential energy from inlet to exit, so

$$\left(\frac{\dot{W}_{cv}}{\dot{m}} \right)_{\substack{int \\ rev}} = -\int_1^2 v \, dp \qquad (\Delta ke = \Delta pe = 0) \qquad (6.51b)$$

This expression shows that the work value is related to the magnitude of the specific volume of the gas or liquid as it flows from inlet to exit.

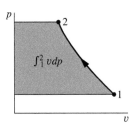

Fig. 6.14 Area representation of $\int_1^2 v \, dp$.

> **FOR EXAMPLE**
>
> Consider two devices: a pump through which liquid water passes and a compressor through which water vapor passes. For the *same pressure rise*, the pump requires a much smaller work *input* per unit of mass flowing than the compressor because the liquid specific volume is much smaller than that of vapor. This conclusion is also qualitatively correct for actual pumps and compressors, where irreversibilities are present during operation.

If the specific volume remains approximately constant, as in many applications with liquids, Eq. 6.51b becomes

$$\left(\frac{\dot{W}_{cv}}{\dot{m}} \right)_{\substack{int \\ rev}} = -v(p_2 - p_1) \qquad (v = \text{constant}, \Delta ke = \Delta pe = 0) \qquad (6.51c)$$

Equation 6.51a also can be applied to study the performance of control volumes at steady state in which $\dot{W}_{cv}$ is zero, as in the case of nozzles and diffusers. For any such case, the equation becomes

$$\int_1^2 v \, dp + \left(\frac{V_2^2 - V_1^2}{2} \right) + g(z_2 - z_1) = 0 \qquad (6.52)$$

which is a form of the **Bernoulli equation** frequently used in fluid mechanics.   **Bernoulli equation**

 BioConnections

Bats, the only mammals that can fly, play several important ecological roles, including feeding on crop-damaging insects. Currently, nearly one-quarter of bat species is listed as endangered or threatened. For unknown reasons, bats are attracted to large wind turbines, where some perish by impact and others from *hemorrhaging*. Near rapidly moving turbine blades there is a drop in air pressure that expands the lungs of bats, causing fine capillaries to burst and their lungs to fill with fluid, killing them.

The relationship between air velocity and pressure in these instances is captured by the following differential form of Eq. 6.52,

the *Bernoulli equation:*

$$v \, dp = -V \, dV$$

which shows that as the *local* velocity V increases, the *local* pressure p decreases. The pressure reduction near the moving turbine blades is the source of peril to bats.

Some say *migrating* bats experience most of the fatalities, so the harm may be decreased at existing wind farms by reducing turbine operation during peak migration periods. New wind farms should be located away from known migratory routes.

6.13.3 Work in Polytropic Processes

We have identified an internally reversible process described by $pv^n = constant$, where n is a constant, as a *polytropic process* (see Sec. 3.15 and the discussion of Fig. 6.10). If each unit of mass passing through a one-inlet, one-exit control volume undergoes a polytropic process, then introducing $pv^n = constant$ in Eq. 6.51b, and performing the integration, gives the work per unit of mass in the absence of internal irreversibilities and significant changes in kinetic and potential energy. That is,

$$\left(\frac{\dot{W}_{cv}}{\dot{m}}\right)_{\substack{int \\ rev}} = -\int_1^2 v \, dp = -(constant)^{1/n} \int_1^2 \frac{dp}{p^{1/n}}$$

$$= -\frac{n}{n-1}(p_2 v_2 - p_1 v_1) \qquad \text{(polytropic, } n \neq 1\text{)} \qquad (6.53)$$

Equation 6.53 applies for any value of n except $n = 1$. When $n = 1$, $pv = constant$, and the work is

$$\left(\frac{\dot{W}_{cv}}{\dot{m}}\right)_{\substack{int \\ rev}} = -\int_1^2 v \, dp = -constant \int_1^2 \frac{dp}{p}$$

$$= -(p_1 v_1) \ln(p_2/p_1) \qquad \text{(polytropic, } n = 1\text{)} \qquad (6.54)$$

Equations 6.53 and 6.54 apply generally to polytropic processes of *any* gas (or liquid).

Ideal Gas Case.   For the special case of an ideal gas, Eq. 6.53 becomes

$$\left(\frac{\dot{W}_{cv}}{\dot{m}}\right)_{\substack{int \\ rev}} = -\frac{nR}{n-1}(T_2 - T_1) \qquad \text{(ideal gas, } n \neq 1\text{)} \qquad (6.55a)$$

For a polytropic process of an ideal gas, Eq. 3.56 applies:

$$\frac{T_2}{T_1} = \left(\frac{p_2}{p_1}\right)^{(n-1)/n}.$$

Thus, Eq. 6.55a can be expressed alternatively as

$$\left(\frac{\dot{W}_{cv}}{\dot{m}}\right)_{\substack{int \\ rev}} = -\frac{nRT_1}{n-1}\left[\left(\frac{p_2}{p_1}\right)^{(n-1)/n} - 1\right] \qquad \text{(ideal gas, } n \neq 1\text{)} \qquad (6.55b)$$

For the case of an ideal gas, Eq. 6.54 becomes

$$\left(\frac{\dot{W}_{cv}}{\dot{m}}\right)_{\substack{int \\ rev}} = -RT \ln(p_2/p_1) \qquad \text{(ideal gas, } n = 1\text{)} \qquad (6.56)$$

In Example 6.15, we consider air modeled as an ideal gas undergoing a polytropic compression process at steady state.

▶ ▶ ▶ **EXAMPLE 6.15** ▶ ·

Determining Work and Heat Transfer for a Polytropic Compression of Air

An air compressor operates at steady state with air entering at $p_1 = 1$ bar, $T_1 = 20°C$, and exiting at $p_2 = 5$ bar. Determine the work and heat transfer per unit of mass passing through the device, in kJ/kg, if the air undergoes a polytropic process with $n = 1.3$. Neglect changes in kinetic and potential energy between the inlet and the exit. Use the ideal gas model for air.

SOLUTION

Known   Air is compressed in a polytropic process from a specified inlet state to a specified exit pressure.

Find   Determine the work and heat transfer per unit of mass passing through the device.

Schematic and Given Data:

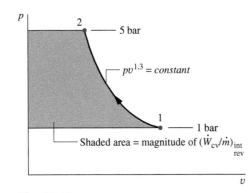

Fig. E6.15

Engineering Model

1. A control volume enclosing the compressor is at steady state.

2. The air undergoes a polytropic process with $n = 1.3$.

3. The air behaves as an ideal gas.

4. Changes in kinetic and potential energy from inlet to exit can be neglected.

Analysis The work is obtained using Eq. 6.55a, which requires the temperature at the exit, T_2. The temperature T_2 can be found using Eq. 3.56:

$$T_2 = T_1 \left(\frac{p_2}{p_1} \right)^{(n-1)/n} = 293 \left(\frac{5}{1} \right)^{(1.3-1)/1.3} = 425 \text{ K}$$

Substituting known values into Eq. 6.55a then gives

$$\frac{\dot{W}_{cv}}{\dot{m}} = -\frac{nR}{n-1}(T_2 - T_1) = -\frac{1.3}{1.3-1} \left(\frac{8.314}{28.97} \frac{\text{kJ}}{\text{kg} \cdot \text{K}} \right) (425 - 293) \text{ K}$$
$$= -164.2 \text{ kJ/kg}$$

The heat transfer is evaluated by reducing the mass and energy rate balances with the appropriate assumptions to obtain

$$\frac{\dot{Q}_{cv}}{\dot{m}} = \frac{\dot{W}_{cv}}{\dot{m}} + h_2 - h_1$$

Using the temperatures T_1 and T_2, the required specific enthalpy values are obtained from Table A-22 as $h_1 = 293.17$ kJ/kg and $h_2 = 426.35$ kJ/kg. Thus,

$$\frac{\dot{Q}_{cv}}{\dot{m}} = -164.15 + (426.35 - 293.17) = -31 \text{ kJ/kg}$$

① The states visited in the polytropic compression process are shown by the curve on the accompanying p–v diagram. The magnitude of the work per unit of mass passing through the compressor is represented by the shaded area *behind* the curve.

SKILLS DEVELOPED

Ability to...

• analyze a polytropic process of an ideal gas.

• apply the control volume energy rate balance.

Quick Quiz

If the air were to undergo a polytropic process with $n = 1.0$, determine the work and heat transfer, each in kJ per kg of air flowing, keeping all other given data the same. Ans. −135.3 kJ/kg.

CHAPTER SUMMARY AND STUDY GUIDE

In this chapter, we have introduced the property entropy and illustrated its use for thermodynamic analysis. Like mass and energy, entropy is an extensive property that can be transferred across system boundaries. Entropy transfer accompanies both heat transfer and mass flow. Unlike mass and energy, entropy is not conserved but is *produced* within systems whenever internal irreversibilities are present.

The use of entropy balances is featured in this chapter. Entropy balances are expressions of the second law that account for the entropy of systems in terms of entropy transfers and entropy production. For processes of closed systems, the entropy balance is Eq. 6.24, and a corresponding rate form is Eq. 6.28. For control volumes, rate forms include Eq. 6.34 and the companion steady-state expression given by Eq. 6.36.

The following checklist provides a study guide for this chapter. When your study of the text and end-of-chapter exercises has been completed you should be able to

• write out meanings of the terms listed in the margins throughout the chapter and understand each of the related concepts. The subset of key concepts listed below is particularly important in subsequent chapters.

• apply entropy balances in each of several alternative forms, appropriately modeling the case at hand, correctly observing sign conventions, and carefully applying SI and English units.

• use entropy data appropriately, to include

• retrieving data from Tables A-2 through A-18, using Eq. 6.4 to evaluate the specific entropy of two-phase liquid–vapor mixtures, sketching T–s and h–s diagrams and locating states on such diagrams, and appropriately using Eqs. 6.5 and 6.13.

• determining Δs of ideal gases using Eq. 6.20 for variable specific heats together with Tables A-22 and A-23, and using Eqs. 6.21 and 6.22 for constant specific heats.

• evaluating isentropic efficiencies for turbines, nozzles, compressors, and pumps from Eqs. 6.46, 6.47, and 6.48, respectively, including for ideal gases the appropriate use of Eqs. 6.41–6.42 for variable specific heats and Eqs. 6.43–6.45 for constant specific heats.

• apply Eq. 6.23 for closed systems and Eqs. 6.49 and 6.51 for one-inlet, one-exit control volumes at steady state, correctly observing the restriction to internally reversible processes.

KEY ENGINEERING CONCEPTS

entropy change	isentropic process	entropy rate balance
T–s diagram	entropy transfer	increase in entropy principle
Mollier diagram	entropy balance	isentropic efficiencies
$T\,dS$ equations	entropy production	

KEY EQUATIONS

$$S_2 - S_1 = \int_1^2 \left(\frac{\delta Q}{T}\right)_b + \sigma \tag{6.24}$$

Closed system entropy balance

$$\frac{dS}{dt} = \sum_j \frac{\dot{Q}_j}{T_j} + \dot{\sigma} \tag{6.28}$$

Closed system entropy rate balance

$$\frac{dS_{cv}}{dt} = \sum_j \frac{\dot{Q}_j}{T_j} + \sum_i \dot{m}_i s_i - \sum_e \dot{m}_e s_e + \dot{\sigma}_{cv} \tag{6.34}$$

Control volume entropy rate balance

$$0 = \sum_j \frac{\dot{Q}_j}{T_j} + \sum_i \dot{m}_i s_i - \sum_e \dot{m}_e s_e + \dot{\sigma}_{cv} \tag{6.36}$$

Steady-state control volume entropy rate balance

$$\eta_t = \frac{\dot{W}_{cv}/\dot{m}}{(\dot{W}_{cv}/\dot{m})_s} = \frac{h_1 - h_2}{h_1 - h_{2s}} \tag{6.46}$$

Isentropic turbine efficiency

$$\eta_{nozzle} = \frac{V_2^2/2}{(V_2^2/2)_s} \tag{6.47}$$

Isentropic nozzle efficiency

$$\eta_c = \frac{(-\dot{W}_{cv}/\dot{m})_s}{(-\dot{W}_{cv}/\dot{m})} = \frac{h_{2s} - h_1}{h_2 - h_1} \tag{6.48}$$

Isentropic compressor (and pump) efficiency

Ideal Gas Model Relations

$$s(T_2, v_2) - s(T_1, v_1) = \int_{T_1}^{T_2} c_v(T)\frac{dT}{T} + R\ln\frac{v_2}{v_1} \tag{6.17}$$

Change in specific entropy; general form for T and v as independent properties.

$$s(T_2, v_2) - s(T_1, v_1) = c_v \ln\frac{T_2}{T_1} + R\ln\frac{v_2}{v_1} \tag{6.21}$$

Constant specific heat, c_v

$$s(T_2, p_2) - s(T_1, p_1) = \int_{T_1}^{T_2} c_p(T)\frac{dT}{T} - R\ln\frac{p_2}{p_1} \tag{6.18}$$

Change in specific entropy; general form for T and p as independent properties

$$s(T_2, p_2) - s(T_1, p_1) = s°(T_2) - s°(T_1) - R\ln\frac{p_2}{p_1} \tag{6.20a}$$

$s°$ for air from Table A-22 ($\overline{s}°$ for other gases from Table A-23)

$$s(T_2, p_2) - s(T_1, p_1) = c_p \ln\frac{T_2}{T_1} - R\ln\frac{p_2}{p_1} \tag{6.22}$$

Constant specific heat, c_v

$$\frac{p_2}{p_1} = \frac{p_{r2}}{p_{r1}} \tag{6.41}$$

$s_1 = s_2$ (air only), p_r and v_r from Table A-22

$$\frac{v_2}{v_1} = \frac{v_{r2}}{v_{r1}} \tag{6.42}$$

$$\frac{T_2}{T_1} = \left(\frac{p_2}{p_1}\right)^{(k-1)/k} \tag{6.43}$$

$$\frac{T_2}{T_1} = \left(\frac{v_1}{v_2}\right)^{k-1} \tag{6.44}$$

$s_1 = s_2$, constant specific heat ratio k

$$\frac{p_2}{p_1} = \left(\frac{v_1}{v_2}\right)^k \tag{6.45}$$

EXERCISES: THINGS ENGINEERS THINK ABOUT

6.1 When applying the entropy balance to a system, which irreversibilities are included in the entropy production term: internal and/or external?

6.2 Is it possible for entropy to be negative? For entropy *change* to be negative? For entropy *production* to be negative?

6.3 Can the entropy balance for a closed system be used to *prove* the Clausius and Kelvin–Planck statements of the second law? Explain.

6.4 Is it possible for the entropy of *both* a closed system and its surroundings to *decrease* during a process? Both to *increase* during a process?

6.5 Can adiabatic mixing of two substances result in decreased entropy? Explain.

6.6 What happens to the entropy produced within an insulated, one-inlet, one-exit control volume operating at steady state?

6.7 When a mixture of olive oil and vinegar *spontaneously* separates into two liquid phases, is the second law violated? Explain.

6.8 If a closed system would undergo an internally reversible process and an irreversible process between the same end states, how would the changes in entropy for the two processes compare? How would the amounts of entropy produced compare?

6.9 The two power cycles shown to the same scale in the figure are composed of internally reversible processes. Compare the net work developed by these cycles. Which cycle has the greater thermal efficiency?

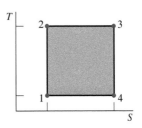 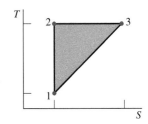

6.10 Is Eq. 6.51a restricted to adiabatic processes and thus to isentropic processes? Explain.

6.11 Could friction be less for a slurry of pulverized coal and water flowing through a pipeline than for water alone? Explain.

6.12 Reducing irreversibilities within a system can improve its *thermodynamic* performance, but steps taken in this direction are usually constrained by other considerations. What are some of these?

CHECKING UNDERSTANDING

For Problems 6.1–6.4, a gas flows through a one-inlet, one-exit control volume operating at steady state. Heat transfer at a rate $\dot{Q}_{cv}$ takes place only at a location on the boundary where the temperature is T_b. For each of the following cases, determine whether the specific entropy of the gas at the exit is greater than, equal to, or less than the specific entropy of the gas at the inlet.

6.1 No internal irreversibilities, $\dot{Q}_{cv} = 0$.

6.2 No internal irreversiblities, $\dot{Q}_{cv} < 0$.

6.3 No internal irreversibilities, $\dot{Q}_{cv} > 0$.

6.4 Internal irreversibilities present, $\dot{Q}_{cv} \geq 0$.

6.5 At steady state, an insulated mixing chamber receives two liquid streams of the same substance at temperatures T_1 and T_2 and mass flow rates $\dot{m}_1$ and $\dot{m}_2$, respectively. A single stream exits at T_3 and $\dot{m}_3$. Assuming the incompressible substance model with constant specific heat c, the exit temperature is

    a. $T_3 = (T_1 + T_2)/2$

    b. $T_3 = (\dot{m}_1 T_1 + \dot{m}_2 T_2)/\dot{m}_3$

    c. $T_3 = c(T_1 - T_2)$

    d. $T_3 = (\dot{m}_1 T_1 - \dot{m}_2 T_2)/\dot{m}_3$

    e. None of the above.

6.6 The *h–s* diagram is commonly known as the _____ diagram.

6.7 Show that for phase change of water from saturated liquid to saturated vapor at constant pressure in a closed system, $(h_g - h_f) = T(s_g - s_f)$.

6.8 The specific internal energy of an ideal gas depends on temperature alone. Is the same statement true for specific entropy of an ideal gas? Explain.

6.9 In the limit, it is possible for a process of a closed system to be internally reversible, but there could be irreversibilities within the surroundings. Give an example of such a process.

6.10 What is the thermodynamic probability w?

6.11 Using the *Bernoulli* equation, show that for an incompressible substance, the pressure must decrease in the direction of flow if the velocity increases, assuming no change in elevation.

6.12 The expression $p_2/p_1 = (T_2/T_1)^{k/(k-1)}$ applies only for _____ _____.

6.13 Ammonia undergoes an isentropic process from an initial state at 10 bar, 40°C to a final pressure of 3.5 bar. What phase or phases are present at the final state?

6.14 Briefly explain the notion of microscopic disorder as it applies to a process of an isolated system.

6.15 Saturated water vapor at 5 bar undergoes a process in a closed system to a final state where the pressure is 10 bar and the temperature is 200°C. Can the process occur adiabatically? Explain.

Indicate whether the following statements are true or false. Explain.

6.16 The change in entropy of a closed system is the same for every process between two specified states.

6.17 A process that violates the second law of thermodynamics violates the first law of thermodynamics.

6.18 One corollary of the second law of thermodynamics states that the change in entropy of a closed system must be greater than or equal to zero.

6.19 Entropy is produced in every internally reversible process of a closed system.

6.20 The entropy of a fixed amount of an ideal gas increases in every isothermal process.

6.21 The energy of an isolated system must remain constant, but the entropy can only decrease.

6.22 The *Carnot* cycle is represented on a T–s diagram as a rectangle.

6.23 For a specified inlet state, exit pressure, and mass flow rate, the power *input* to a compressor operating adiabatically and at steady state is less than what would be required if the compression occurred isentropically.

6.24 The $T\,dS$ equations are fundamentally important in thermodynamics because of their use in deriving important property relations for pure, simple compressible systems.

6.25 At liquid states, the following approximation is reasonable for many engineering applications: $s(T, p) \approx s_g(T)$.

6.26 In *statistical thermodynamics*, entropy is associated with the notion of microscopic *disorder*.

6.27 The only entropy transfers to or from control volumes are those accompanying heat transfer.

PROBLEMS: DEVELOPING ENGINEERING SKILLS

C Problem may require use of appropriate computer software in order to complete.

Using Entropy Data and Concepts

6.1 Using the tables for water, determine the specific entropy at the indicated states, in kJ/kg · K. In each case, locate the state by hand on a sketch of the T–s diagram.

    a. $p = 6.9$ MPa, $T = 400°C$

    b. $p = 6.9$ MPa, $T = 150°C$

    c. $p = 6.9$ MPa, $h = 2170$ kJ/kg

    d. $p = 6.9$ MPa, saturated vapor

6.2 Using the appropriate table, determine the change in specific entropy between the specified states, in kJ/kg · K.

    a. water, $p_1 = 6.9$ MPa, $T_1 = 427°C$, $p_2 = 6.9$ MPa, $T_2 = 38°C$.

    b. Refrigerant 134a, $h_1 = 111.43$ kJ/kg, $T_1 = -40°C$, saturated vapor at $p_2 = 276$ kPa.

    c. air as an ideal gas, $T_1 = 4°C$, $p_1 = 2$ bar, $T_2 = 216°C$, $p_2 = 1$ bar.

    d. carbon dioxide as an ideal gas, $T_1 = 438°C$, $p_1 = 1$ bar, $T_2 = 25°C$, $p_2 = 3.04$ bar.

6.3 Using the appropriate table, determine the indicated property. In each case, locate the state on sketches of the T–v and T–s diagrams.

    a. water at $p = 0.40$ bar, $h = 1477.14$ kJ/kg · K. Find s, in kJ/kg · K.

    b. water at $p = 10$ bar, $u = 3124.4$ kJ/kg. Find s, in kJ/kg · K.

    c. Refrigerant 134a at $T = -22°C$, $x = 0.60$. Find s, in kJ/kg · K.

    d. ammonia at $T = 28°C$, $s = 4.9948$ kJ/kg · K. Find u, in kJ/kg.

6.4 Calculate the specific entropy of water with specific enthalpy of 2300 kJ/kg at 6 MPa pressure.

6.5 Using *steam table* data, determine the indicated property data for a process in which there is no change in specific entropy between state 1 and state 2. In each case, locate the states on a sketch of the T–s diagram.

    a. $T_1 = 40°C$, $x_1 = 100\%$, $p_2 = 150$ kPa. Find T_2, in °C, and Δh, in kJ/kg.

    b. $T_1 = 10°C$, $x_1 = 75\%$, $p_2 = 1$ MPa. Find T_2, in °C, and Δu, in kJ/kg.

6.6 Using the appropriate table, determine the indicated property for a process in which there is no change in specific entropy between state 1 and state 2.

    a. water, $p_1 = 101.3$ kPa, $T_1 = 260°C$, $p_2 = 0.7$ MPa. Find T_2 in °C.

    b. water, $T_1 = 10°C$, $x_1 = 0.75$, saturated vapor at state 2. Find p_2 in bar.

    c. air as an ideal gas, $T_1 = 27°C$, $p_1 = 1.5$ bar, $T_2 = 127°C$. Find p_2 in bar.

    d. air as an ideal gas, $T_1 = 38°C$, $p_1 = 3$ bar, $p_2 = 2$ bar. Find T_2 in °C.

    e. Refrigerant 134a, $T_1 = 20°C$, $p_1 = 5$ bar, $p_2 = 1$ bar. Find v_2 in m³/kg.

6.7 Air in a piston–cylinder assembly undergoes a process from state 1, where $T_1 = 400$ K, $p_1 = 200$ kPa, to state 2, where $T_2 = 600$ K, $p_2 = 750$ kPa. Using the ideal gas model for air, determine the change in specific entropy between these states, in kJ/kg · K, if the process occurs (a) without internal irreversibilities, (b) with internal irreversibilities.

6.8 Propane undergoes a process from state 1, where $p_1 = 1.4$ MPa, $T_1 = 60°C$, to state 2, where $p_1 = 1.0$ MPa, during which the change in specific entropy is $s_2 - s_1 = -0.35$ kJ/kg · K. At state 2, determine the temperature, in °C, and the specific enthalpy, in kJ/kg.

6.9 One-quarter kmol of nitrogen gas (N_2) undergoes a process from $p_1 = 138$ kPa, $T_1 = 278$ K to $p_2 = 1$ MPa. For the process $W = -528$ kJ and $Q = -132.8$ kJ. Employing the ideal gas model, determine

    a. T_2, in K.

    b. the change in entropy, in kJ/K.

Show the initial and final states on a T–s diagram.

6.10 Five kg of nitrogen (N_2) undergoes a process from $p_1 = 5$ bar, $T_1 = 400$ K to $p_2 = 2$ bar, $T_2 = 500$ K. Assuming ideal gas behavior, determine the change in entropy, in kJ/K, with

    a. constant specific heats evaluated at 450 K.

    b. variable specific heats.

Compare the results and discuss.

6.11 One-fifth kmol of carbon monoxide (CO) in a piston–cylinder assembly undergoes a process from $p_1 = 100$ kPa, $T_1 = 298$ K to $p_2 = 400$ kPa, $T_2 = 360$ K. For the process, $W = -250$ kJ. Employing the ideal gas model, determine

    a. the heat transfer, in kJ.

    b. the change in entropy, in kJ/K.

Show the process on a sketch of the T–s diagram.

6.12 Steam enters a turbine operating at steady state at 1.5 MPa, 240°C and exits at 45°C with a quality of 85%. Stray heat transfer and kinetic and potential energy effects are negligible. Determine (a) the power developed by the turbine, in kJ per kg of steam flowing, (b) the change in specific entropy from inlet to exit, in kJ/K per kg of steam flowing.

6.13 Argon in a piston–cylinder assembly is compressed from state 1, where $T_1 = 300$ K, $V_1 = 1$ m³, to state 2, where $T_2 = 200$ K. If the change in specific entropy is $s_2 - s_1 = -0.27$ kJ/kg · K, determine the final volume, in m³. Assume the ideal gas model with $k = 1.67$.

Analyzing Internally Reversible Processes

6.14 One kilogram of water in a piston–cylinder assembly undergoes the two internally reversible processes in series shown in **Fig. P6.14.** For each process, determine, in kJ, the heat transfer and the work.

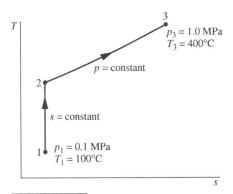

FIGURE P6.14

6.15 A system consisting of 2 kg of water initially at 160°C, 10 bar undergoes an internally reversible, isothermal expansion during which there is energy transfer by heat *into* the system of 2700 kJ. Determine the final pressure, in bar, and the work, in kJ.

6.16 One kilogram of water initially at 160°C, 1.5 bar, undergoes an isothermal, internally reversible compression process to the saturated liquid state. Determine the work and heat transfer, each in kJ. Sketch the process on $p–v$ and $T–s$ coordinates. Associate the work and heat transfer with areas on these diagrams.

6.17 A gas within a piston–cylinder assembly undergoes an isothermal process at 400 K during which the change in entropy is −0.3 kJ/K. Assuming the ideal gas model for the gas and negligible kinetic and potential energy effects, evaluate the work, in kJ.

6.18 One kg mass of water initially a saturated liquid at 100 kPa undergoes a constant-pressure, internally reversible expansion to x = 80%. Determine the work and heat transfer, each in kJ. Sketch the process on $p–v$ and $T–s$ coordinates. Associate the work and heat transfer with areas on these diagrams.

6.19 A gas initially at 2.8 bar and 60°C is compressed to a final pressure of 14 bar in an isothermal internally reversible process. Determine the work

and heat transfer, each in kJ per kg of gas, if the gas is (a) Refrigerant 134a, (b) air as an ideal gas. Sketch the process on $p–v$ and $T–s$ coordinates.

6.20 Nitrogen (N_2) initially occupying 0.5 m^3 at 1.0 bar, 20°C undergoes an internally reversible compression during which $pV^{1.30} = constant$ to a final state where the temperature is 200°C. Assuming the ideal gas model, determine

  a. the pressure at the final state, in bar.

  b. the work and heat transfer, each in kJ.

  c. the entropy change, in kJ/K.

6.21 Air in a piston–cylinder assembly and modeled as an ideal gas undergoes two internally reversible processes in series from state 1, where T_1 = 290 K, p_1 = 1 bar.

  Process 1–2: Compression to p_2 = 5 bar during which $pV^{1.19}$ = *constant.*

  Process 2–3: Isentropic expansion to p_3 = 1 bar.

  a. Sketch the two processes in series on $T–s$ coordinates.

  b. Determine the temperature at state 2, in K.

  c. Determine the net work, in kJ.

6.22 Air in a piston–cylinder assembly undergoes a Carnot power cycle. The isothermal expansion and compression processes occur at 1400 K and 350 K, respectively. The pressures at the beginning and end of the isothermal compression are 100 kPa and 500 kPa, respectively. Assuming the ideal gas model with c_p = 1.005 kJ/kg · K, determine

  a. the pressures at the beginning and end of the isothermal expansion, each in kPa.

  b. the heat transfer and work, in kJ/kg, for each process.

  c. the thermal efficiency.

6.23 **Figure P6.23** shows a Carnot heat pump cycle operating at steady state with ammonia as the working fluid. The condenser temperature is 49°C, with saturated vapor entering and saturated liquid exiting. The evaporator temperature −12°C.

  a. Determine the heat transfer and work for each process, in kJ/kg of ammonia flowing.

  b. Evaluate the coefficient of performance for the heat pump.

  c. Evaluate the coefficient of performance for a Carnot refrigeration cycle operating as shown in the figure.

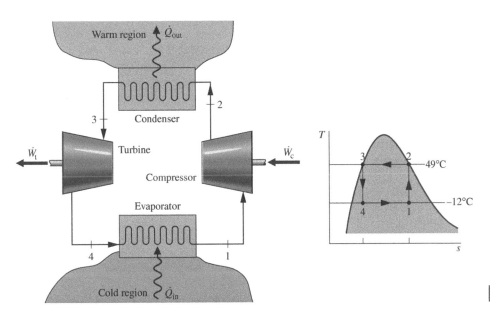

FIGURE P6.23

Applying the Entropy Balance: Closed Systems

6.24 A fixed mass of water m, initially a saturated liquid, is brought to a saturated vapor condition while its pressure and temperature remain constant. Volume change is the only work mode.

 a. Derive expressions for the work and heat transfer in terms of the mass m and properties that can be obtained directly from the *steam tables*.

 b. Demonstrate that this process is internally reversible.

6.25 Two m^3 of air in a rigid, insulated container fitted with a paddle wheel is initially at 293 K, 200 kPa. The air receives 710 kJ by work from the paddle wheel. Assuming the ideal gas model with $c_v = 0.72$ kJ/kg $\cdot$ K, determine for the air (a) the mass, in kg, (b) final temperature, in K, and (c) the amount of entropy produced, in kJ/K.

6.26 Five kg of water contained in a piston-cylinder assembly expand from an initial state where $T_1 = 400°C$, $p_1 = 700$ kPa to a final state where $T_2 = 200°C$, $p_2 = 300$ kPa, with no significant effects of kinetic and potential energy. The accompanying table provides additional data at the two states. It is claimed that the water undergoes an adiabatic process between these states, while developing work. Evaluate this claim.

State	$T(°C)$	$p(kPa)$	$v(m^3/kg)$	$u(kJ/kg)$	$h(kJ/kg)$	$s(kJ/kg \cdot K)$
1	400	700	0.4397	2960.9	3268.7	7.6350
2	200	300	0.7160	2650.7	2865.5	7.3115

6.27 A rigid, insulated container fitted with a paddle wheel contains 2.27 kg of water, initially at 127°C and a quality of 60%. The water is stirred until the temperature is 177°C. For the water, determine (a) the work, in kJ, and (b) the amount of entropy produced, in kJ/K.

6.28 Air contained in a rigid, insulated tank fitted with a paddle wheel, initially at 300 K, 2 bar, and a volume of 2 m^3, is stirred until its temperature is 500 K. Assuming the ideal gas model for the air, and ignoring kinetic and potential energy, determine (a) the final pressure, in bar, (b) the work, in kJ, and (c) the amount of entropy produced, in kJ/K. Solve using

 a. data from Table A-22.

 b. constant c_v read from Table A-20 at 400 K.

Compare the results of parts (a) and (b).

6.29 Three m^3 of air in a rigid, insulated container fitted with a paddle wheel is initially at 295 K, 200 kPa. The air receives 1546 kJ of work from the paddle wheel. Assuming the ideal gas model, determine for the air (a) the mass, in kg, (b) final temperature, in K, and (c) the amount of entropy produced, in KJ/K.

6.30 Air is compressed adiabatically in a piston–cylinder assembly from 1 bar, 300 K to 10 bar, 600 K. The air can be modeled as an ideal gas and kinetic and potential energy effects are negligible. Determine the amount of entropy produced, in kJ/K per kg of air, for the compression. What is the minimum theoretical work input, in kJ per kg of air, for an adiabatic compression from the given initial state to a final pressure of 10 bar?

6.31 Refrigerant 134a contained in a piston–cylinder assembly rapidly expands from an initial state where $T_1 = 60°C$, $p_1 = 1.4$ MPa to a final state where $p_2 = 34.5$ kPa and the quality, x_2, is (a) 99%, (b) 95%. In each case, determine if the process can occur adiabatically. If yes, determine the work, in kJ/kg, for an adiabatic expansion between these states. If no, determine the direction of the heat transfer.

6.32 Steam undergoes an adiabatic expansion in a piston–cylinder assembly from 100 bar, 360°C to 1 bar, 160°C. What is work in kJ per kg

of steam for the process? Calculate the amount of entropy produced, in kJ/K per kg of steam. What is the maximum theoretical work that could be obtained from the given initial state to the same final pressure? Show both processes on a properly labeled sketch of the T–s diagram.

6.33 One kg mass of Refrigerant 134a contained within a piston–cylinder assembly undergoes a process from a state where the pressure is 0.8 MPa and the quality is 40% to a state where the temperature is 10°C and the refrigerant is saturated liquid. Determine the change in specific entropy of the refrigerant, in kJ/kg $\cdot$ K. Can this process be accomplished adiabatically?

6.34 Two kg of air contained in a piston–cylinder assembly undergoes a process from an initial state where $T_1 = 320$ K, $v_1 = 1.2$ m^3/kg to a final state where $T_2 = 440$ K, $v_2 = 0.4$ m^3/kg. Can this process occur adiabatically? If yes, determine the work, in kJ, for an adiabatic process between these states. If no, determine the direction of the heat transfer. Assume the ideal gas model for air.

6.35 One kg of propane initially at 8 bar and 50°C undergoes a process to 3 bar, 20°C while being rapidly expanded in a piston–cylinder assembly. Heat transfer between the propane and its surroundings occurs at an average temperature of 35°C. The work done by the propane is measured as 42.4 kJ. Kinetic and potential energy effects can be ignored. Determine whether it is possible for the work measurement to be correct.

6.36 An inventor claims that the electricity-generating unit shown in Fig. P6.36 receives a heat transfer at the rate of 264 kJ/s at a temperature of 278 K, a second heat transfer at the rate of 360 kJ/s at 390 K, and a third at the rate of 528 kJ at 555 K. For operation at steady state, evaluate this claim.

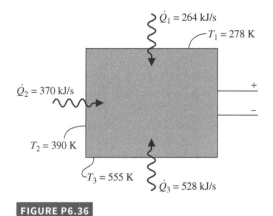

$\dot{Q}_1 = 264$ kJ/s

$T_1 = 278$ K

$\dot{Q}_2 = 370$ kJ/s

$T_2 = 390$ K

$T_3 = 555$ K

$\dot{Q}_3 = 528$ kJ/s

FIGURE P6.36

6.37 Two kg of Refrigerant 134a initially at 1.4 bar, 60°C, are compressed to saturated vapor at 60°C. During this process, the temperature of the refrigerant departs by no more than 0.01°C from 60°C. Determine the minimum theoretical heat transfer from the refrigerant during the process, in kJ.

6.38 One kg mass of carbon dioxide (CO_2) in a piston–cylinder assembly, initially at 21°C and 1.24 MPa, expands isothermally to a final pressure of 0.1 MPa while receiving energy by heat transfer through a wall separating the carbon dioxide from a thermal energy reservoir at 49°C.

 a. For the carbon dioxide as the system, evaluate the work and heat transfer, each in kJ, and the amount of entropy produced, in kJ/K. Model the carbon dioxide as an ideal gas.

 b. Evaluate the entropy production for an enlarged system that includes the carbon dioxide and the wall, assuming the state of the wall remains unchanged. Compare with the entropy production of part (a) and comment on the difference.

6.39 An electric motor at steady state draws a current of 10 amp with a voltage of 110 V. The output shaft develops a torque of 10.2 N · m and a rotational speed of 1000 RPM.

    a. If the outer surface of the motor is at 42°C, determine the rate of entropy production within the motor, in kW/K.

    b. Evaluate the rate of entropy production, in kW/K, for an enlarged system that includes the motor and enough of the nearby surroundings that heat transfer occurs at the ambient temperature, 22°C.

6.40 A power plant has a turbogenerator, shown in **Fig. P6.40**, operating at steady state with an input shaft rotating at 1800 RPM with a torque of 16,700 N · m. The turbogenerator produces current at 230 amp with a voltage of 13,000 V. The rate of heat transfer between the turbogenerator and its surroundings is related to the surface temperature T_b and the lower ambient temperature T_0 and is given by $\dot{Q} = -hA(T_b - T_0)$, where h = 110 W/m² · K, A = 32 m², and T_0 = 298 K.

    a. Determine the temperature T_b, in K.

    b. For the turbogenerator as the system, determine the rate of entropy production, in kW/K.

    c. If the system boundary is located to take in enough of the nearby surroundings for heat transfer to take place at temperature T_0, determine the rate of entropy production, in kW/K, for the enlarged system.

6.41 An isolated system consists of a closed aluminum vessel of mass 0.1 kg containing 1 kg of used engine oil, each initially at 55°C, immersed in a 10-kg bath of liquid water, initially at 20°C. The system is allowed to come to equilibrium. Determine

    a. the final temperature, in degrees centigrade.

    b. the entropy changes, each in kJ/K, for the aluminum vessel, the oil, and the water.

    c. the amount of entropy produced, in kJ/K.

6.42 In a heat-treating process, a 2-kg aluminum rod, initially at 1075 K, is quenched in a tank containing 150 kg of water, initially at 295 K. There is negligible heat transfer between the contents of the tank and their surroundings. Taking the specific heat of the metal rod and water as constant at 0.903 kJ/kg · K and 4.2 kJ/kg · K, respectively, determine (a) the final equilibrium temperature after

quenching, in K, and (b) the amount of entropy produced within the tank, in kJ/K.

6.43 **C** An insulated, rigid tank is divided into two compartments by a frictionless, thermally conducting piston. One compartment initially contains 1 m³ of saturated water vapor at 4 MPa and the other compartment contains 1 m³ of water vapor at 20 MPa, 800°C. The piston is released and equilibrium is attained, with the piston experiencing no change of state. For the water as the system, determine

    a. the final pressure, in MPa.

    b. the final temperature, in °C.

    c. the amount of entropy produced, in kJ/K.

6.44 A 2.64-kg copper part, initially at 400 K, is plunged into a tank containing 4 kg of liquid water, initially at 300 K. The copper part and water can be modeled as incompressible with specific heats 0.385 kJ/kg · K and 4.2 kJ/kg · K, respectively. For the copper part and water as the system, determine (a) the final equilibrium temperature, in K, and (b) the amount of entropy produced within the tank, in kJ/K. Ignore heat transfer between the system and its surroundings.

6.45 In a piston–cylinder assembly water is contained initially at 200°C as a saturated liquid. The piston moves freely in the cylinder as water undergoes a process to the corresponding saturated vapor state. There is no heat transfer with the surroundings. This change of state is brought by the action of paddle wheel. Determine the amount of entropy produced per unit mass, in kJ/kg · K.

6.46 A system consisting of air initially at 300 K and 1 bar experiences the two different types of interactions described below. In each case, the system is brought from the initial state to a state where the temperature is 500 K, while volume remains constant.

    a. The temperature rise is brought about adiabatically by stirring the air with a paddle wheel. Determine the amount of entropy produced, in kJ/kg · K.

    b. The temperature rise is brought about by heat transfer from a reservoir at temperature T. The temperature at the system boundary where heat transfer occurs is also T. Plot the amount of entropy produced, in kJ/kg · K, versus T for T ≥ 500 K. Compare with the result of (a) and discuss.

6.47 As shown in **Fig. P6.47**, an insulated box is initially divided into halves by a frictionless, thermally conducting piston. On one side of the piston is 1.0 m³ of air at 400 K, 3 bar. On the other side is 1.0 m³ of air at 400 K, 1.5 bar. The piston is released and equilibrium is attained,

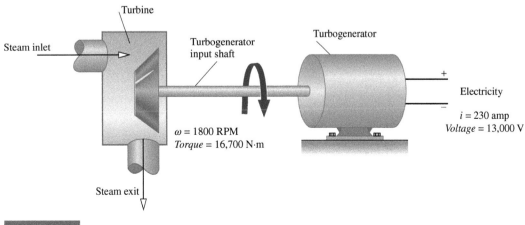

Steam inlet

Turbine

Turbogenerator input shaft

Turbogenerator

Electricity

+

−

ω = 1800 RPM
Torque = 16,700 N·m

i = 230 amp
Voltage = 13,000 V

Steam exit

FIGURE P6.40

with the piston experiencing no change of state. Employing the ideal gas model for the air, determine

a. the final temperature of the air, in K.

b. the final pressure of the air, in bar.

c. the amount of entropy produced, in kJ/K.

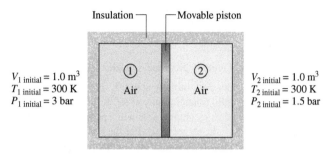

FIGURE P6.47

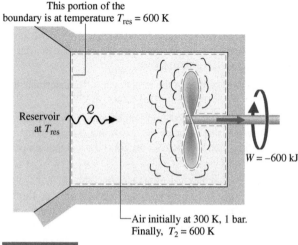

FIGURE P6.50

6.48 **C** A rigid, insulated vessel is divided into two equal-volume compartments connected by a valve. Initially, one compartment contains 1 m³ of water at 20°C, $x = 50\%$, and the other is evacuated. The valve is opened and the water is allowed to fill the entire volume. For the water, determine the final temperature, in °C, and the amount of entropy produced, in kJ/K.

6.49 The heat pump cycle shown in Fig. P6.49 operates at steady state and provides energy by heat transfer at a rate of 15 kW to maintain a dwelling at 22°C when the outside temperature is −22°C. The manufacturer claims that the power input required for this operating condition is 3.2 kW. Applying energy and entropy rate balances evaluate this claim.

6.50 A closed, rigid tank contains 5 kg of air initially at 300 K, 1 bar. As illustrated in Fig. P6.50, the tank is in contact with a thermal reservoir at 600 K and heat transfer occurs at the boundary where the temperature is 600 K. A stirring rod transfers 600 kJ of energy to the air. The final temperature is 600 K. The air can be modeled as an ideal gas with $c_v = 0.733$ kJ/kg · K and kinetic and potential energy effects are negligible. Determine the amount of entropy transferred into the air and the amount of entropy produced, each in kJ/K.

6.51 A system undergoing a thermodynamic cycle receives Q_H at temperature T'_H and discharges Q_C at temperature T'_C. There are no other heat transfers.

a. Show that the net work developed per cycle is given by

$$W_{cycle} = Q_H\left(1 - \frac{T'_C}{T'_H}\right) - T'_C\sigma$$

where σ is the amount of entropy produced per cycle owing to irreversibilities within the system.

b. If the heat transfers Q_H and Q_C are with hot and cold reservoirs, respectively, what is the relationship of T'_H to the temperature of the hot reservoir T_H and the relationship of T'_C to the temperature of the cold reservoir T_C?

c. Obtain an expression for W_{cycle} if there are (i) no internal irreversibilities, (ii) no internal *or* external irreversibilities.

6.52 At steady state, an insulated mixing chamber receives two liquid streams of the same substance at temperatures T_1 and T_2, and mass flow rates $\dot{m}_1$ and $\dot{m}_2$, respectively. A single stream exits at T_3 and $\dot{m}_3$. Using the incompressible substance model with constant specific heat c, obtain an expression for

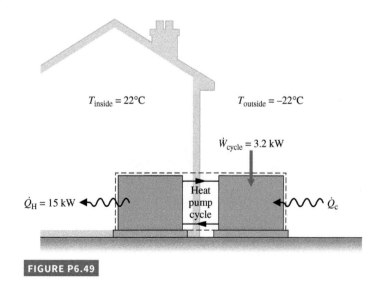

FIGURE P6.49

a. T_3 in terms of T_1, T_2, and the ratio of mass flow rates $\dot{m}_1/\dot{m}_3$.

b. the rate of entropy production per unit of mass exiting the chamber in terms of c, T_1/T_2, and $\dot{m}_1/\dot{m}_3$.

c. For fixed values of c and T_1/T_2, determine the value of $\dot{m}_1/\dot{m}_3$ for which the rate of entropy production is a maximum.

6.53 **C** As shown in Fig. P6.53, a turbine is located between two tanks. Initially, the smaller tank contains steam at 3.0 MPa, 280°C and the larger tank is evacuated. Steam is allowed to flow from the smaller tank, through the turbine, and into the larger tank until equilibrium is attained. If heat transfer with the surroundings is negligible, determine the maximum theoretical work that can be developed, in kJ.

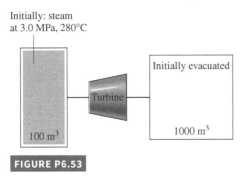

Initially: steam at 3.0 MPa, 280°C

Turbine

Initially evacuated

100 m³

1000 m³

FIGURE P6.53

Applying the Entropy Balance: Control Volumes

6.54 Air enters a turbine operating at steady state at 8 bar, 1400 K and expands to 0.8 bar. The turbine is well insulated, and kinetic and potential energy effects can be neglected. Assuming ideal gas behavior for the air, what is the maximum theoretical work that could be developed by the turbine in kJ per kg of air flow?

6.55 Water at 20 bar, 400°C enters a turbine operating at steady state and exits at 1.5 bar. Stray heat transfer and kinetic and potential energy effects are negligible. A hard-to-read data sheet indicates that the quality at the turbine exit is 98%. Can this quality value be correct? If no, explain. If yes, determine the power developed by the turbine, in kJ per kg of water flowing.

6.56 Propane at 0.1 MPa, 20°C enters an insulated compressor operating at steady state and exits at 0.4 MPa, 90°C. Neglecting kinetic and potential energy effects, determine

a. the power required by the compressor, in kJ per kg of propane flowing.

b. the rate of entropy production within the compressor, in kJ/K per kg of propane flowing.

6.57 An inventor claims that at steady state the device shown in Fig. P6.57 develops power from entering and exiting streams of water

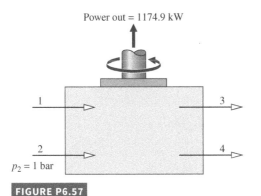

Power out = 1174.9 kW

1

3

2

4

$p_2 = 1$ bar

FIGURE P6.57

at a rate of 1174.9 kW. The accompanying table provides data for inlet 1 and exits 3 and 4. The pressure at inlet 2 is 1 bar. Stray heat transfer and kinetic and potential energy effects are negligible. Evaluate the inventor's claim.

State	$\dot{m}$ (kg/s)	p (bar)	T (°C)	v (m³/kg)	u (kJ/kg)	h (kJ/kg)	s (kJ/kg·K)
1	4	1	450	3.334	3049.0	3382.4	8.6926
3	5	2	200	1.080	2654.4	2870.5	7.5066
4	3	4	400	0.733	2964.4	3273.4	7.8985

6.58 By injecting liquid water into superheated steam, the *desuperheater* shown in Fig. P6.58 has a saturated vapor stream at its exit. Steady-state operating data are provided in the accompanying table. Stray heat transfer and all kinetic and potential energy effects are negligible. (a) Locate states 1, 2, and 3 on a sketch of the *T–s* diagram. (b) Determine the rate of entropy production within the desuperheater, in kW/K.

State	p (MPa)	T (°C)	$v \times 10^3$ (m³/kg)	u (kJ/kg)	h (kJ/kg)	s (kJ/kg·K)
1	2.7	40	1.0066	167.2	169.9	0.5714
2	2.7	300	91.01	2757.0	3002.8	6.6001
3	2.5	sat. vap.	79.98	2603.1	2803.1	6.2575

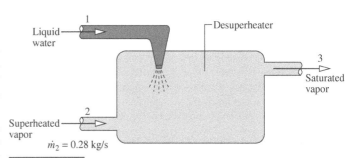

Liquid water

Desuperheater

1

3

Saturated vapor

2

Superheated vapor

$\dot{m}_2 = 0.28$ kg/s

FIGURE P6.58

6.59 Steam enters a well-insulated nozzle operating at steady state at 538°C, 3.4 MPa and a velocity of 3 m/s. At the nozzle exit, the pressure is 101.3 kPa and the velocity is 317 m/s. Determine the rate of entropy production, in kJ/K per kg of steam flowing.

6.60 Air at 400 kPa, 970 K enters a turbine operating at steady state and exits at 100 kPa, 670 K. Heat transfer from the turbine occurs at an average outer surface temperature of 315 K at the rate of 30 kJ per kg of air flowing. Kinetic and potential energy effects are negligible. For air as an ideal gas with $c_p = 1.1$ kJ/kg · K, determine (a) the rate power is developed, in kJ per kg of air flowing, and (b) the rate of entropy production within the turbine, in kJ/K per kg of air flowing.

6.61 By injecting liquid water into superheated vapor, the *desuperheater* shown in Fig. P6.61 has a saturated vapor stream at its exit. Steady-state operating data are shown on the figure. Ignoring stray heat transfer and kinetic and potential energy effects, determine (a) the mass flow rate of the superheated vapor stream, in kg/min, and (b) the rate of entropy production within the desuperheater, in kW/K.

6.62 Air at 200 kPa, 52°C, and a velocity of 355 m/s enters an insulated duct of varying cross-sectional area. The air exits at 100 kPa, 82°C. At the inlet, the cross-sectional area is 6.57 cm². Assuming the ideal gas model for the air, determine

a. the exit velocity, in m/s.

b. the rate of entropy production within the duct, in kW/K.

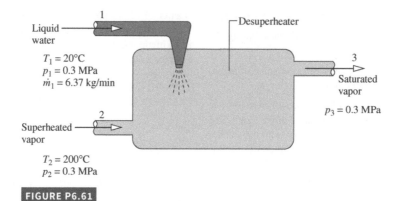

FIGURE P6.61

6.63 For the computer of Example 4.8, determine the rate of entropy production, in W/K, when air exits at 40°C. Ignore the change in pressure between the inlet and exit.

6.64 Students in a laboratory are studying air flowing at steady state through a horizontal insulated duct. One student group reports the measured pressure, temperature, and velocity at one location in the duct as 0.95 bar, 67°C, and 75 m/s, respectively. The group reports the following values at another location in the duct: 0.8 bar, 22°C, and 310 m/s. The group neglected to note the direction of flow on the data sheet, however. Using the data provided, determine the direction of flow.

6.65 An inventor has provided the steady-state operating data shown in **Fig. P6.65** for a *cogeneration* system producing power and increasing the temperature of a stream of air. The system receives and discharges energy by heat transfer at the rates and temperatures indicated on the figure. All heat transfers are in the directions of the accompanying arrows. The ideal gas model applies to the air. Kinetic and potential energy effects are negligible. Using energy and entropy rate balances, evaluate the thermodynamic performance of the system.

6.66 Electronic components are mounted on the inner surface of a horizontal cylindrical duct whose inner diameter is 0.2 m, as shown in **Fig. P6.66**. To prevent overheating of the electronics, the cylinder is cooled by a stream of air flowing through it and by convection from its outer surface. Air enters the duct at 25°C, 1 bar and a velocity of 0.3 m/s and exits at 40°C with negligible changes in kinetic energy and pressure. Convective cooling occurs on the outer surface to the surroundings, which are at 25°C, in accord with hA = 3.4 W/K, where h is the heat transfer coefficient and A is the surface area. The electronic components require 0.20 kW of electric power. For a control volume enclosing the cylinder, determine at steady state (a) the mass flow rate of the air, in kg/s, (b) the temperature on the outer surface of the duct, in °C, and (c) the rate of entropy production, in W/K. Assume the ideal gas model for air.

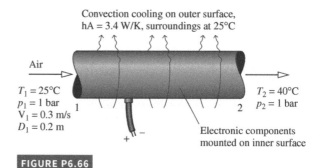

FIGURE P6.66

6.67 Ammonia enters the compressor of an industrial refrigeration plant at 2 bar, −10°C with a mass flow rate of 15 kg/min and is compressed to 12 bar, 140°C. Heat transfer occurs from the compressor to its surroundings at a rate of 6 kW. For steady-state operation with negligible kinetic and potential energy effects, determine (a) the power input to the compressor, in kW, and (b) the rate of entropy production, in kW/K, for a control volume enclosing the compressor and its immediate surroundings such that the heat transfer occurs at 300 K.

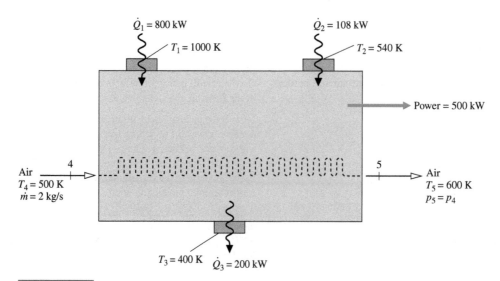

FIGURE P6.65

6.68 Steam enters a horizontal 15-cm-diameter pipe as a saturated vapor at 5 bar with a velocity of 10 m/s and exits at 4.5 bar with a quality of 95%. Heat transfer from the pipe to the surroundings at 300 K takes place at an average outer surface temperature of 400 K. For operation at steady state, determine

    a. the velocity at the exit, in m/s.

    b. the rate of heat transfer from the pipe, in kW.

    c. the rate of entropy production, in kW/K, for a control volume comprising only the pipe and its contents.

    d. the rate of entropy production, in kW/K, for an enlarged control volume that includes the pipe and enough of its immediate surroundings so that heat transfer from the control volume occurs at 300 K.

Why do the answers of parts (c) and (d) differ?

6.69 Refrigerant 134a is compressed from 2 bar, saturated vapor, to 10 bar, 90°C in a compressor operating at steady state. The mass flow rate of refrigerant entering the compressor is 7 kg/min, and the power *input* is 10.85 kW. Kinetic and potential energy effects can be neglected.

    a. Determine the rate of heat transfer, in kW.

    b. If the heat transfer occurs at an average surface temperature of 50°C, determine the rate of entropy production, in kW/K.

    c. Determine the rate of entropy production, in kW/K, for an enlarged control volume that includes the compressor and its immediate surroundings such that the heat transfer occurs at 300 K.

Compare the results of parts (b) and (c) and discuss.

6.70 Nitrogen (N_2) enters a well-insulated diffuser operating at steady state at 0.656 bar, 300 K with a velocity of 282 m/s. The inlet area is 4.8×10^{-3} m^2. At the diffuser exit, the pressure is 0.9 bar and the velocity is 130 m/s. The nitrogen behaves as an ideal gas with $k = 1.4$. Determine the exit temperature, in K, and the exit area, in m^2. For a control volume enclosing the diffuser, determine the rate of entropy production, in kJ/K per kg of nitrogen flowing.

6.71 A counterflow heat exchanger operates at steady state with negligible kinetic and potential energy effects. In one stream, liquid water enters at 15°C and exits at 23°C with a negligible change in pressure. In the other stream, Refrigerant 22 enters at 12 bar, 90°C with a mass flow rate of 150 kg/h and exits at 12 bar, 28°C. Heat transfer from the outer surface of the heat exchanger can be ignored. Determine

    a. the mass flow rate of the liquid water stream, in kg/h.

    b. the rate of entropy production within the heat exchanger, in kW/K.

6.72 Figure P6.72 shows data for a portion of the ducting in a ventilation system operating at steady state. The ducts are well insulated and the pressure is very nearly 1 bar throughout. Assuming the ideal gas model for air with $c_p = 1$ kJ/kg · K, and ignoring kinetic and potential energy effects, determine (a) the temperature of the air at the exit, in °C, (b) the exit diameter, in m, and (c) the rate of entropy production within the duct, in kJ/min.

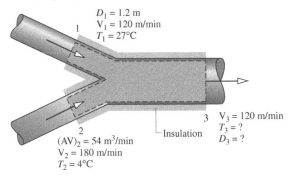

$D_1 = 1.2$ m
$V_1 = 120$ m/min
$T_1 = 27$°C

1

3   $V_3 = 120$ m/min
$T_3 = ?$
Insulation $D_3 = ?$

2
$(AV)_2 = 54$ m^3/min
$V_2 = 180$ m/min
$T_2 = 4$°C

FIGURE P6.72

6.73 Air flows through an insulated circular duct having a diameter of 2 cm. Steady-state pressure and temperature data obtained by measurements at two locations, denoted as 1 and 2, are given in the accompanying table. Modeling air as an ideal gas with $c_p = 1.005$ kJ/kg · K, determine (a) the direction of the flow, (b) the velocity of the air, in m/s, at each of the two locations, and (c) the mass flow rate of the air, in kg/s.

Measurement location	1	2
Pressure (kPa)	100	500
Temperature (°C)	20	50

6.74 Figure P6.74 shows an air compressor and regenerative heat exchanger in a gas turbine system operating at steady state. Air flows from the compressor through the regenerator, and a separate stream of air passes though the regenerator in counterflow. Operating data are provided on the figure. Stray heat transfer to the surroundings and kinetic and potential energy effects can be neglected. The compressor power *input* is 6700 kW. Determine the mass flow rate of air entering the compressor, in kg/s, the temperature of the air exiting the regenerator at state 5, in K, and the rates of entropy production in the compressor and regenerator, in kW/K.

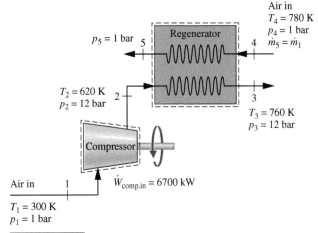

Air in
$T_4 = 780$ K
$p_4 = 1$ bar
$\dot{m}_5 = \dot{m}_1$

$p_5 = 1$ bar   5   Regenerator   4

$T_2 = 620$ K   2
$p_2 = 12$ bar

3
$T_3 = 760$ K
$p_3 = 12$ bar

Compressor

Air in   1 $\dot{W}_{comp.in} = 6700$ kW

$T_1 = 300$ K
$p_1 = 1$ bar

FIGURE P6.74

6.75 A rigid, insulated tank whose volume is 10 L is initially evacuated. A pinhole leak develops and air from the surroundings at 1 bar, 25°C enters the tank until the pressure in the tank becomes 1 bar. Assuming the ideal gas model with $k = 1.4$ for the air, determine (a) the final temperature in the tank, in °C, (b) the amount of air that leaks into the tank, in g, and (c) the amount entropy produced, in J/K.

6.76 **C** A well-insulated, rigid tank of volume 10 m^3 is connected by a valve to a large-diameter supply line carrying air at 227°C and 10 bar. The tank is initially evacuated. Air is allowed to flow into the tank until the tank pressure is p. Using the ideal gas model with constant specific heat ratio k, plot tank temperature, in K, the mass of air in the tank, in kg, and the amount of entropy produced, in kJ/K, versus p in bar.

6.77 A tank of volume 1 m^3 initially contains steam at 60 bar, 320°C. Steam is withdrawn slowly from the tank until the pressure drops to 15 bar. An electric resistor in the tank transfers energy to the steam maintaining the temperature constant at 320°C during the process. Neglecting kinetic and potential energy effects, determine the amount of entropy produced, in kJ/K.

6.78 Air as an ideal gas flows through the turbine and heat exchanger arrangement shown in Fig. P6.78. Steady-state data are given on the figure. Stray heat transfer and kinetic and potential energy effects can be ignored. Determine

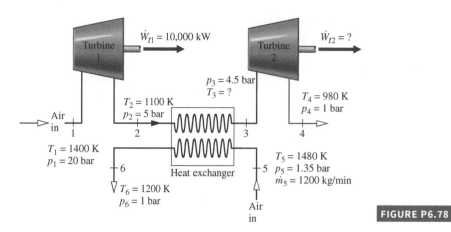

FIGURE P6.78

a. temperature T_3, in K.

b. the power output of the second turbine, in kW.

c. the rates of entropy production, each in kW/K, for the turbines and heat exchanger.

d. Using the result of part (c), place the components in rank order, beginning with the component contributing most to inefficient operation of the overall system.

6.79 A 5 m³ tank initially filled with air at 1 bar and 21°C is evacuated by a device known as a *vacuum pump*, while the tank contents are maintained at 21°C by heat transfer through the tank walls. The vacuum pump discharges air to the surroundings at the temperature and pressure of the surroundings, which are 1 bar and 21°C, respectively. Determine the *minimum* theoretical work required, in kJ.

Using Isentropic Processes/Efficiencies

6.80 Air in a piston–cylinder assembly is compressed isentropically from $T_1 = 16°C$, $p_1 = 138$ kPa to $p_2 = 13.8$ MPa. Assuming the ideal gas model, determine the temperature at state 2, in K, using (a) data from Table A-22, and (b) a constant specific heat ratio, $k = 1.4$. Compare the values obtained in parts (a) and (b) and comment.

6.81 Air in a piston–cylinder assembly is compressed isentropically from state 1, where $T_1 = 45°C$, to state 2, where the specific volume is one-twentieth of the specific volume at state 1. Applying the ideal gas model with $k = 1.4$, determine (a) T_2, in °C and (b) the work, in kJ/kg.

6.82 Steam undergoes an isentropic compression in an insulated piston–cylinder assembly from an initial state where $T_1 = 120°C$, $p_1 = 1$ bar to a final state where the pressure $p_2 = 100$ bar. Determine the final temperature, in °C, and the work, in kJ per kg of steam.

6.83 Propane undergoes an isentropic expansion from an initial state where $T_1 = 40°C$, $p_1 = 1$ MPa to a final state where the temperature and pressure are T_2, p_2, respectively. Determine

a. p_2 in kPa, when $T_2 = -40°C$.

b. T_2, in °C, when $p_2 = 0.8$ MPa.

6.84 Argon in a piston–cylinder assembly is compressed isentropically from state 1, where $p_1 = 150$ kPa, $T_1 = 35°C$, to state 2, where $p_2 = 300$ kPa. Assuming the ideal gas model with $k = 1.67$, determine (a) T_2, in °C, and (b) the work, in kJ per kg of argon.

6.85 Air in a piston–cylinder assembly is compressed isentropically from an initial state where $T_1 = 340$ K to a final state where the pressure is 90% greater than at state 1. Assuming the ideal gas model, determine (a) T_2, in K, and (b) the work, in kJ/kg.

6.86 Water vapor at 6 MPa, 600°C enters a turbine operating at steady state and expands to 10 kPa. The mass flow rate is 2 kg/s, and the power developed is 2626 kW. Stray heat transfer and kinetic and potential energy effects are negligible. Determine (a) the isentropic turbine efficiency and (b) the rate of entropy production within the turbine, in kW/K.

6.87 Air at 25°C, 100 kPa enters a compressor operating at steady state and exits at 260°C, 650 kPa. Stray heat transfer and kinetic and potential energy effects are negligible. Modeling air as an ideal gas with $k = 1.4$, determine the isentropic compressor efficiency.

6.88 The accompanying table provides steady-state data for an isentropic expansion of steam through a turbine. For a mass flow rate of 2.55 kg/s, determine the power developed by the turbine, in MW. Ignore the effects of potential energy.

	p (bar)	T (°C)	V (m/s)	h (kJ/kg)	s (kJ/kg · K)
Inlet	10	300	25	3051.1	7.1214
Exit	1.5	—	100		7.1214

6.89 Refrigerant 22 enters a compressor operating at steady state as saturated vapor at 10 bar and is compressed adiabatically in an internally reversible process to 16 bar. Ignoring kinetic and potential energy effects, determine the required mass flow rate of refrigerant, in kg/s, if the compressor power *input* is 6 kW.

6.90 Figure P6.90 provides the schematic of a heat pump using Refrigerant 134a as the working fluid, together with steady-state data at key points. The mass flow rate of the refrigerant is 7 kg/min, and the power input to the compressor is 5.17 kW. (a) Determine the coefficient of performance for the heat pump. (b) If the valve were replaced by a turbine, power could be produced, thereby reducing the power requirement of the heat pump system. Would you recommend this *power-saving* measure? Explain.

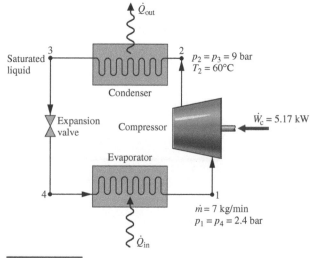

FIGURE P6.90

6.91 Air enters an insulated compressor operating at steady state at 0.95 bar, 27°C with a mass flow rate of 4000 kg/h and exits at 8.7 bar. Kinetic and potential energy effects are negligible.

    a. Determine the minimum theoretical power input required, in kW, and the corresponding exit temperature, in °C.

    b. If the exit temperature is 347°C, determine the power input, in kW, and the isentropic compressor efficiency.

6.92 Water vapor at 10 MPa, 600°C enters a turbine operating at steady state with a volumetric flow rate of 0.36 m^3/s and exits at 0.1 bar and a quality of 92%. Stray heat transfer and kinetic and potential energy effects are negligible. Determine for the turbine (a) the mass flow rate, in kg/s, (b) the power developed by the turbine, in MW, (c) the rate at which entropy is produced, in kW/K, and (d) the isentropic turbine efficiency.

6.93 Water vapor at 5 bar, 320°C enters a turbine operating at steady state with a volumetric flow rate of 0.65 m^3/s and expands adiabatically to an exit state of 1 bar, 160°C. Kinetic and potential energy effects are negligible. Determine for the turbine (a) the power developed, in kW, (b) the rate of entropy production, in kW/K, and (c) the isentropic turbine efficiency.

6.94 Air modeled as an ideal gas enters a turbine operating at steady state at 1040 K, 278 kPa and exits at 120 kPa. The mass flow rate is 5.5 kg/s, and the power developed is 1120 kW. Stray heat transfer and kinetic and potential energy effects are negligible. Assuming $k = 1.4$, determine (a) the temperature of the air at the turbine exit, in K, and (b) the isentropic turbine efficiency.

6.95 Air enters the compressor of a gas turbine power plant operating at steady state at 290 K, 100 kPa and exits at 420 kPa, 330 kPa. Stray heat transfer and kinetic and potential energy effects are negligible. Using the ideal gas model for air, determine the isentropic compressor efficiency.

6.96 Water vapor at 6 MPa, 500°C enters a turbine operating at steady state and expands to 20 kPa. The mass flow rate is 3 kg/s, and the power developed is 2626 kW. Stray heat transfer and kinetic and potential energy effects are negligible. Determine: (a) the isentropic turbine efficiency and (b) the rate of entropy production within the turbine, in kW/K.

6.97 Water vapor at 5 MPa, 320°C enters a turbine operating at steady state and expands to 0.1 bar. The mass flow rate is 2.52 kg/s, and the isentropic turbine efficiency is 92%. Stray heat transfer and kinetic and potential energy effects are negligible. Determine the power developed by the turbine, in kW.

6.98 An ideal gas with constant specific heat ratio k enters a nozzle operating at steady state at pressure p_1, temperature T_1, and velocity V_1. The air expands isentropically to a pressure of p_2.

    a. Develop an expression for the velocity at the exit, V_2, in terms of k, R, V_1, T_1, p_1, and p_2 only.

    b. For $V_1 = 0$, $T_1 = 1000$ K, $p_2/p_1 = 0.1$ and $k = 1.4$, find V_2, in m/s.

6.99 Oxygen (O_2) at 25°C, 100 kPa enters a compressor operating at steady state and exits at 260°C, 650 kPa. Stray heat transfer and kinetic and potential energy effects are negligible. Modeling the oxygen as an ideal gas with $k = 1.379$, determine the isentropic compressor efficiency and the work in kJ per kg of oxygen flowing.

6.100 Carbon dioxide (CO_2) at 1 bar, 300 K enters a compressor operating at steady state and is compressed adiabatically to an exit state of 10 bar, 520 K. The CO_2 is modeled as an ideal gas, and kinetic and potential energy effects are negligible. For the compressor, determine (a) the work input, in kJ per kg of CO_2 flowing, (b) the rate of entropy production, in kJ/K per kg of CO_2 flowing, and (c) the isentropic compressor efficiency.

6.101 Air enters an insulated compressor operating at steady state at 1 bar, 350 K with a mass flow rate of 1 kg/s and exits at 4 bar. The isentropic compressor efficiency is 82%. Determine the power input, in kW, and the rate of entropy production, in kW/K, using the ideal gas model with data from Table A-22.

6.102 Refrigerant 134a enters a compressor operating at steady state as saturated vapor at −6.7°C and exits at a pressure of 0.8 MPa. There is no significant heat transfer with the surroundings, and kinetic and potential energy effects can be ignored.

    a. Determine the minimum theoretical work input required, in kJ per kg of refrigerant flowing through the compressor, and the corresponding exit temperature, in °C.

    b. If the refrigerant exits at a temperature of 49°C, determine the isentropic compressor efficiency.

6.103 Air at 1.3 bar, 423 K and a velocity of 40 m/s enters a nozzle operating at steady state and expands adiabatically to the exit, where the pressure is 0.85 bar and velocity is 307 m/s. For air modeled as an ideal gas with $k = 1.4$, determine for the nozzle (a) the temperature at the exit, in K, and (b) the isentropic nozzle efficiency.

6.104 Water vapor enters an insulated nozzle operating at steady state at 0.7 MPa, 320°C, 35 m/s and expands to 0.15 MPa. If the isentropic nozzle efficiency is 94%, determine the velocity at the exit, in m/s.

6.105 Air modeled as an ideal gas enters a one-inlet, one-exit control volume operating at steady state at 690 kPa, 500 K and expands adiabatically to 0.17 MPa. Kinetic and potential energy effects are negligible. Applying the ideal gas model with $k = 1.4$ and $c_p = 0.241$ kJ/kg · K, determine the rate of entropy production, in kJ/K · kg of air flowing,

    a. if the control volume encloses a turbine having an isentropic turbine efficiency of 89.1%.

    b. if the control volume encloses a throttling valve.

6.106 As part of an industrial process, air as an ideal gas at 10 bar, 400 K expands at steady state through a valve to a pressure of 4 bar. The mass flow rate of air is 0.5 kg/s. The air then passes through a heat exchanger where it is cooled to a temperature of 295 K with negligible change in pressure. The valve can be modeled as a throttling process, and kinetic and potential energy effects can be neglected.

    a. For a control volume enclosing the valve and heat exchanger and enough of the local surroundings that the heat transfer occurs at the ambient temperature of 295 K, determine the rate of entropy production, in kW/K.

    b. If the expansion valve were replaced by an adiabatic turbine operating isentropically, what would be the entropy production, in kW/K?

Compare the results of parts (a) and (b) and discuss.

6.107 Figure P6.107 provides the schematic of a heat pump using Refrigerant 134a as the working fluid, together with steady-state data at key points. The mass flow rate of the refrigerant is 7 kg/min, and the power input to the compressor is 5.17 kW. (a) Determine the coefficient of performance for the heat pump. (b) If the valve were replaced by a turbine, power could be produced, thereby reducing the power requirement of the heat pump system. Would you recommend this *power-saving* measure? Explain.

6.108 Air enters an insulated diffuser operating at steady state at 1 bar, −3°C, and 260 m/s and exits with a velocity of 130 m/s. Employing the ideal gas model and ignoring potential energy, determine

    a. the temperature of the air at the exit, in °C.

    b. The maximum attainable exit pressure, in bar.

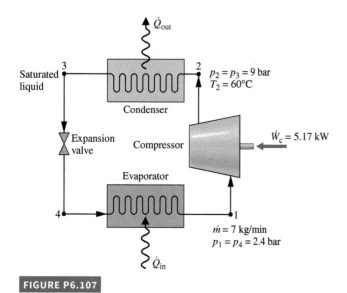

FIGURE P6.107

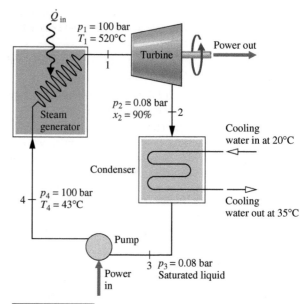

FIGURE P6.110

6.109 As shown in Fig. P6.109, air enters the diffuser of a jet engine at 18 kPa, 216 K with a velocity of 265 m/s, all data corresponding to high-altitude flight. The air flows adiabatically through the diffuser, decelerating to a velocity of 50 m/s at the diffuser exit. Assume steady-state operation, the ideal gas model for air, and negligible potential energy effects.

    a. Determine the temperature of the air at the exit of the diffuser, in K.

    b. If the air would undergo an isentropic process as it flows through the diffuser, determine the pressure of the air at the diffuser exit, in kPa.

    c. If friction were present, would the pressure of the air at the diffuser exit be greater than, less than, or equal to the value found in part (b)? Explain.

6.111 Figure P6.111 shows a power system operating at steady state consisting of three components in series: an air compressor having an isentropic compressor efficiency of 80%, a heat exchanger, and a turbine having an isentropic turbine efficiency of 90%. Air enters the compressor at 1 bar, 300 K with a mass flow rate of 5.8 kg/s and exits at a pressure of 10 bar. Air enters the turbine at 10 bar, 1400 K and exits at a pressure of 1 bar. Air can be modeled as an ideal gas. Stray heat transfer and kinetic and potential energy effects are negligible. Determine, in kW, (a) the power required by the compressor, (b) the power developed by the turbine, and (c) the *net* power output of the overall power system.

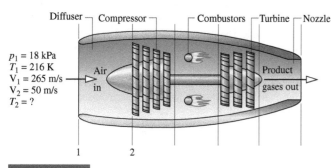

FIGURE P6.109

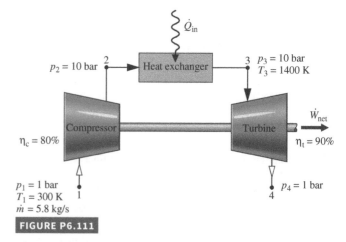

FIGURE P6.111

6.110 Figure P6.110 shows a simple vapor power plant operating at steady state with water as the working fluid. Data at key locations are given on the figure. The mass flow rate of the water circulating through the components is 109 kg/s. Stray heat transfer and kinetic and potential energy effects can be ignored. Determine

    a. the net power developed, in MW.

    b. the thermal efficiency.

    c. the isentropic turbine efficiency.

    d. the isentropic pump efficiency.

    e. the mass flow rate of the cooling water, in kg/s.

    f. the rates of entropy production, each in kW/K, for the turbine, condenser, and pump.

6.112 A rigid tank is filled initially with 5.0 kg of air at a pressure of 0.5 MPa and a temperature of 500 K. The air is allowed to discharge through a turbine into the atmosphere, developing work until the pressure in the tank has fallen to the atmospheric level of 0.1 MPa. Employing the ideal gas model for the air, determine the *maximum* theoretical amount of work that could be developed, in kJ. Ignore heat transfer with the atmosphere and changes in kinetic and potential energy.

6.113 A tank initially containing air at 30.4 bar and 282°C is connected to a small turbine. Air discharges from the tank through the turbine, which produces work in the amount of 105 kJ. The pressure in the tank falls to 3.1 bar during the process and the turbine exhausts to the atmosphere at 1 bar. Employing the ideal gas model for the air and ignoring irreversibilities within the tank and the turbine, determine the volume of the tank, in m^3. Heat transfer with the atmosphere and changes in kinetic and potential energy are negligible.

6.114 Air enters the turbine of a jet engine at 1190 K, 10.8 bar and expands to 5.2 bar. The air then flows through a nozzle and exits at 0.8 bar. Operation is at steady state, and the flow is adiabatic. The nozzle operates with no internal irreversibilities, and the isentropic turbine efficiency is 85%. The air velocities at the turbine inlet and exit are negligible. Assuming the ideal gas model for the air, determine the velocity of the air exiting the nozzle, in m/s.

Analyzing Internally Reversible Flow Processes

6.115 Carbon dioxide (CO_2) expands isothermally at steady state with no irreversibilities through a turbine from 10 bar, 500 K to 2 bar. Assuming the ideal gas model and neglecting kinetic and potential energy effects, determine the heat transfer and work, each in kJ per kg of carbon dioxide flowing.

6.116 Refrigerant 134a enters a compressor operating at steady state at 1.8 bar, −10°C with a volumetric flow rate of 2.4×10^{-2} m^3/s. The refrigerant is compressed to a pressure of 9 bar in an internally reversible process according to $pv^{1.04} = constant$. Neglecting kinetic and potential energy effects, determine

   a. the power required, in kW.

   b. the rate of heat transfer, in kW.

6.117 Water as saturated liquid at 1 bar enters a pump operating at steady state and is pumped isentropically to a pressure of 50 bar. Kinetic and potential energy effects are negligible. Determine the pump work input, in kJ per kg of water flowing, using (a) Eq. 6.51c, (b) an energy balance. Obtain data from Tables A-3 and A-5, as appropriate. Compare the results of parts (a) and (b), and comment.

6.118 An air compressor operates at steady state with air entering at $p_1 = 0.1$ MPa, $T_1 = 16$°C. The air undergoes a polytropic process, and exits at $p_2 = 0.5$ MPa, $T_2 = 146$°C. (a) Evaluate the work and heat transfer, each in kJ/kg of air flowing. (b) Sketch the process on p–v and T–s diagrams and associate areas on the diagrams with work and heat transfer, respectively. Assume the ideal gas model for air and neglect changes in kinetic and potential energy.

6.119 A pump operating at steady state receives saturated liquid water at 50°C with a mass flow rate of 30 kg/s. The pressure of the water at the pump exit is 1.5 MPa. If the pump operates with negligible internal irreversibilities and negligible changes in kinetic and potential energy, determine power required in kW.

6.120 Compare the work required at steady state to compress *water vapor* isentropically to 3 MPa from the saturated vapor state at 0.1 MPa to the work required to pump *liquid water* isentropically to 3 MPa from the saturated liquid state at 0.1 MPa, each in kJ per kg of water flowing through the device. Kinetic and potential energy effects can be ignored.

6.121 A 5-kilowatt pump operating at steady state draws in liquid water at 1 bar, 15°C and delivers it at 5 bar at an elevation 6 m above the inlet. There is no significant change in velocity between the inlet and exit, and the local acceleration of gravity is 9.8 m/s^2. Would it be possible to pump 7.5 m^3 in 10 min or less? Explain.

6.122 An electrically driven pump operating at steady state draws water from a pond at a pressure of 1 bar and a rate of 50 kg/s and delivers the water at a pressure of 4 bar. There is no significant heat transfer with the surroundings, and changes in kinetic and potential energy can be neglected. The isentropic pump efficiency is 75%. Evaluating electricity at 8.5 cents per kW · h, estimate the hourly cost of running the pump.

6.123 As shown in **Fig. P6.123**, water flows from an elevated reservoir through a hydraulic turbine. The pipe diameter is constant, and operation is at steady state. Estimate the minimum mass flow rate, in kg/s, that would be required for a turbine power output of 1 MW. The local acceleration of gravity is 9.8 m/s^2.

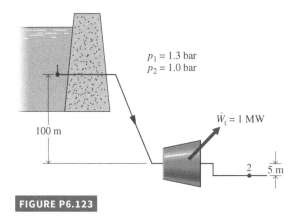

$p_1 = 1.3$ bar
$p_2 = 1.0$ bar

$\dot{W}_t = 1$ MW

100 m

2   5 m

FIGURE P6.123

6.124 Nitrogen (N_2) enters a nozzle operating at steady state at 0.2 MPa, 550K with a velocity of 1 m/s and undergoes a polytropic expansion with $n = 1.3$ to 0.15 MPa. Using the ideal gas model with $k = 1.4$, and ignoring potential energy effects, determine (a) the exit velocity, in m/s, and (b) the rate of heat transfer, in kJ per kg of gas flowing.

DESIGN & OPEN-ENDED PROBLEMS: EXPLORING ENGINEERING PRACTICE

6.1D A thermodynamic diagram used in the refrigeration engineering field has the specific enthalpy and the natural logarithm of pressure as coordinates. Inspection of such a diagram suggests that in the vapor region constant-entropy lines are nearly linear, and thus the relation between s, $\ln p$ and h might be expressible in the vapor region as.

$$h(s, p) = (A\,s + B) \ln p + (C\,s + D)$$

where A, B, C, and D are constants. Investigate the viability of this expression for pressure ranging up to 10 bar, using h, s, p, T, and v data for Refrigerant 134a, together with Eqs. 11.26 and 11.27 deduced from the second Tds equation, Eq. 6.10b. Summarize your conclusions in a memorandum.

6.2D Both electricity and heat transfer are needed for processes within manufacturing settings. *Combined heat and power* (CHP) systems are designed to provide both from a single fuel source such as natural gas (see Sec. 8.5.2). Investigate CHP system designs and prepare a report explaining the types of technology in use in the U.S. manufacturing sector. Discuss the potential for increased use of such technologies and the associated economic considerations. Include at least three references.

6.3D Ocean thermal energy conversion (OTEC) power plants generate electricity on ships or platforms at sea by exploiting the naturally occurring decrease of the temperature of ocean water with depth. One proposal for the use of OTEC-generated electricity is to produce and commercialize ammonia in three steps: Hydrogen (H_2) would first be obtained by electrolysis of desalted seawater. The hydrogen would then be reacted with nitrogen (N_2) from the atmosphere to obtain ammonia (NH_3). Finally, liquid ammonia would be shipped to shore, where it would be reprocessed into hydrogen or used as a feedstock. Some say a major drawback with the proposal is whether *current technology* can be integrated to provide cost-competitive end products. Investigate this issue and summarize your findings in a report with at least three references.

6.4D For a compressor or pump located at your campus or workplace, take data sufficient for evaluating the isentropic compressor or pump efficiency. Compare the experimentally determined isentropic efficiency with data provided by the manufacturer. Rationalize any significant discrepancy between experimental and manufacturer values. Prepare a technical report including a full description of instrumentation, recorded data, results and conclusions, and at least three references.

6.5D Natural gas is currently playing a significant role in meeting our energy needs and hydrogen may be just as important in years ahead. For natural gas *and* hydrogen, energy is required at every stage of distribution between production and end use: for storage, transportation by pipelines, trucks, trains and ships, and liquefaction, if needed. According to some observers, distribution energy requirements will weigh more heavily on hydrogen not only because it has special attributes but also because means for distributing it are less developed than for natural gas. Investigate the energy requirements for distributing hydrogen relative to that for natural gas. Write a report with at least three references.

6.6D Design and execute an experiment to obtain measured property data required to evaluate the change in entropy of a common gas, liquid, or solid undergoing a process of your choice. Compare the experimentally determined entropy change with a value obtained from published engineering data, including property software. Rationalize any significant discrepancy between values. Prepare a technical report including a full description of the experimental set-up and instrumentation, recorded data, sample calculations, results and conclusions, and at least three references.

6.7D The *maximum entropy method* is widely used in the field of astronomical data analysis. Over the last three decades, considerable work has been done using the method for data filtering and removing features in an image that are caused by the telescope itself rather than from light coming from the sky (called deconvolution). To further such aims, refinements of the method have evolved over the years. Investigate the maximum entropy method as it is used today in astronomy, and summarize the state-of-the-art in a memorandum.

6.8D The performance of turbines, compressors, and pumps decreases with use, reducing isentropic efficiency. Select one of these three types of components and develop a detailed understanding of how the component functions. Contact a manufacturer's representative to learn what measurements are typically recorded during operation, causes of degraded performance with use, and maintenance actions that can be taken to extend service life. Visit an industrial site where the selected component can be observed in operation and discuss the same points with personnel there. Prepare a poster presentation of your findings suitable for classroom use.

6.9D Elementary thermodynamic modeling, including the use of the temperature–entropy diagram for water and a form of the *Bernoulli equation* has been employed to study certain types of volcanic eruptions. (See L. G. Mastin, "Thermodynamics of Gas and Steam-Blast Eruptions," *Bull. Volcanol.*, 57, 85–98, 1995.) Write a report *critically evaluating* the underlying assumptions and application of thermodynamic principles, as reported in the article. Include at least three references.

CHAPTER 7

Exergy Analysis

Engineering Context

The **objective** of this chapter is to introduce *exergy analysis*, which uses the conservation of mass and conservation of energy principles together with the second law of thermodynamics for the design and analysis of thermal systems.

The importance of developing thermal systems that make effective use of nonrenewable resources such as oil, natural gas, and coal is apparent. Exergy analysis is particularly suited for furthering the goal of more efficient resource use, since it enables the locations, types, and true magnitudes of waste and loss to be determined. This information can be used to design thermal systems, guide efforts to reduce sources of inefficiency in existing systems, and evaluate system economics.

LEARNING OUTCOMES

When you complete your study of this chapter, you will be able to...

- Demonstrate understanding of key concepts related to exergy analysis, including the exergy reference environment, the dead state, exergy transfer, and exergy destruction.
- Evaluate exergy at a state and exergy change between two states, using appropriate property data.
- Apply exergy balances to closed systems and to control volumes at steady state.
- Define and evaluate exergetic efficiencies.
- Apply exergy costing to heat loss and simple cogeneration systems.

7.1 Introducing Exergy

Energy is conserved in every device or process. It cannot be destroyed. Energy entering a system with fuel, electricity, flowing streams of matter, and so on can be accounted for in the products and by-products. However, the energy conservation idea alone is inadequate for depicting some important aspects of resource utilization.

FOR EXAMPLE

Figure 7.1a shows an *isolated system* consisting initially of a small container of fuel surrounded by air in abundance. Suppose the fuel burns (Fig. 7.1b) so that finally there is a slightly warm mixture of combustion products and air as shown in Fig. 7.1c. The total *quantity* of energy associated with the system is constant because no energy transfers take place across the boundary of an isolated system. Still, the initial fuel–air combination is intrinsically more useful than the final warm mixture. For instance, the fuel might be used in some device to generate electricity or produce superheated steam, whereas the uses of the final slightly warm mixture are far more limited in scope. We can say that the system has a greater *potential for use* initially than it has finally. Since nothing but a final warm mixture is achieved in the process, this potential is largely wasted. More precisely, the initial potential is largely *destroyed* because of the irreversible nature of the process.

Anticipating the main results of this chapter, *exergy* is the property that quantifies *potential for use*. The foregoing example illustrates that, unlike energy, exergy is not conserved but is destroyed by irreversibilities.

Subsequent discussion shows that exergy not only can be destroyed by irreversibilities but also can be transferred *to* and *from* systems. Exergy transferred from a system to its surroundings without use typically represents a *loss*. Improved energy resource utilization can be realized by reducing exergy destruction within a system and/or reducing losses. An objective in exergy analysis is to identify sites where exergy destructions and losses occur and rank order them for significance. This allows attention to be centered on aspects of system operation that offer the greatest opportunities for cost-effective improvements.

Returning to Fig. 7.1, note that the fuel present initially has economic value while the final slightly warm mixture has little value. Accordingly, economic value decreases in this process. From such considerations we might infer there is a link between exergy and economic value, and this is the case as we will see in subsequent discussions.

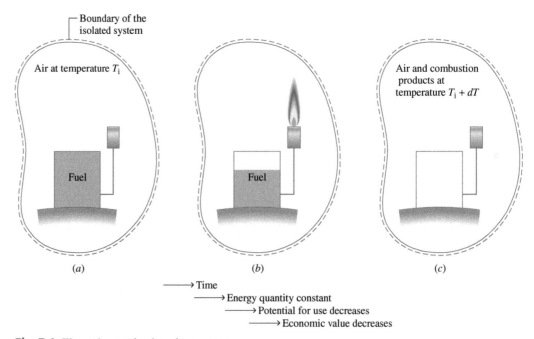

Fig. 7.1 Illustration used to introduce exergy.

7.2 Conceptualizing Exergy

The introduction to the second law in Chap. 5 provides a basis for the exergy concept, as considered next.

Principal conclusions of the discussion of Fig. 5.1 given in Section 5.1 are that

- a potential for developing work exists whenever two systems at different states are brought into communication, and
- work can be developed as the two systems are allowed to come into equilibrium.

In Fig. 5.1a, for example, a body initially at an elevated temperature T_i placed in contact with the atmosphere at temperature T_0 cools spontaneously. To conceptualize how work might be developed in this case, see **Fig. 7.2**. The figure shows an *overall* system with three elements: the body, a power cycle, and the atmosphere at T_0 and p_0. The atmosphere is presumed to be large enough that its temperature and pressure remain constant. W_c denotes the work of the overall system.

Instead of the body cooling spontaneously as considered in Fig. 5.1a, Fig. 7.2 shows that if the heat transfer Q during cooling is passed to the power cycle, work W_c can be developed, while Q_0 is discharged to the atmosphere. These are the only energy transfers. The work W_c is *fully available* for lifting a weight or, equivalently, as shaft work or electrical work. Ultimately the body cools to T_0, and no more work would be developed. At equilibrium, the body and atmosphere each possess energy, but there no longer is any potential for developing work from the two because no further interaction can occur between them.

Note that work W_c also could be developed by the system of Fig. 7.2 if the initial temperature of the body were *less* than that of the atmosphere: $T_i < T_0$. In such a case, the directions of the heat transfers Q and Q_0 shown on Fig. 7.2 would each reverse. Work could be developed as the body *warms* to equilibrium with the atmosphere.

Since there is no net change of state for the power cycle of Fig. 7.2, we conclude that the work W_c is realized solely because the initial state of the body differs from that of the atmosphere. *Exergy is the maximum theoretical value of such work.*

7.2.1 Environment and Dead State

For thermodynamic analysis involving the exergy concept, it is necessary to model the atmosphere used in the foregoing discussion. The resulting model is called the **exergy reference environment**, or simply the **environment**.

    environment

In this book the environment is regarded to be a simple compressible system that is *large* in extent and *uniform* in temperature, T_0, and pressure, p_0. In keeping with the idea that the

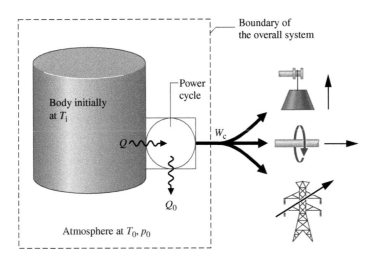

Fig. 7.2 Overall system of body, power cycle, and atmosphere used to conceptualize exergy.

environment represents a portion of the physical world, the values for both p_0 and T_0 used throughout a particular analysis are normally taken as typical ambient conditions, such as 1 atm and 25°C. Additionally, the intensive properties of the environment do not change significantly as a result of any process under consideration, and the environment is free of irreversibilities.

dead state

When a system of interest is at T_0 and p_0 and *at rest* relative to the environment, we say the system is at the **dead state**. At the dead state there can be no interaction between system and environment and, thus, no potential for developing work.

7.2.2 Defining Exergy

definition of exergy

The discussion to this point of the current section can be summarized by the following **definition of exergy:**

> *Exergy is the maximum theoretical work obtainable from an overall system consisting of a system and the environment as the system comes into equilibrium with the environment (passes to the dead state).*

Interactions between the system and the environment may involve auxiliary devices, such as the power cycle of Fig. 7.2, that at least in principle allow the realization of the work. The work developed is fully available for lifting a weight or, equivalently, as shaft work or electrical work. We might expect that the maximum theoretical work would be obtained when there are no irreversibilities. As considered in the next section, this *is* the case.

7.3 Exergy of a System

exergy of a system

The **exergy of a system**, E, at a specified state is given by the expression

$$E = (U - U_0) + p_0(V - V_0) - T_0(S - S_0) + KE + PE \qquad (7.1)$$

where U, KE, PE, V, and S denote, respectively, internal energy, kinetic energy, potential energy, volume, and entropy of the system at the specified state. U_0, V_0, and S_0 denote internal energy, volume, and entropy, respectively, of the system when at the dead state. In this chapter kinetic energy and potential energy are each evaluated relative to the environment. Thus, when the system is at the dead state, it is at rest relative to the environment and the values of its kinetic and potential energies are zero: $KE_0 = PE_0 = 0$. By inspection of Eq. 7.1, the units of exergy are seen to be the same as those of energy.

Equation 7.1 can be derived by applying energy and entropy balances to the overall system shown in **Fig. 7.3** consisting of a closed system and an environment. See the box for the derivation of Eq. 7.1.

Evaluating Exergy of a System

Referring to Fig. 7.3, exergy is the maximum theoretical value of the work W_c obtainable from the overall system as the closed system comes into equilibrium with the environment—that is, as the closed system passes to the dead state.

In keeping with the discussion of Fig. 7.2, the closed system plus the environment is referred to as the *overall* system. The boundary of the overall system is located so there is no energy transfer across it by heat transfer: $Q_c = 0$. Moreover, the boundary of the overall system is located so that the total volume remains constant, even though the volumes of the system and environment can vary.

Accordingly, the work W_c shown on the figure is the *only* energy transfer across the boundary of the overall system and is *fully available* for lifting a weight, turning a shaft, or producing electricity in the surroundings. Next, we apply the energy and entropy balances to determine the maximum theoretical value for W_c.

Energy Balance

Consider a process where the system and the environment come to equilibrium. The energy balance for the overall system is

$$\Delta E_c = \cancel{Q_c}^{0} - W_c \qquad (a)$$

where W_c is the work developed by the overall system and ΔE_c is the change in energy of the overall system: the sum of the energy changes of the system and the environment. The energy of the system initially is denoted by E, which includes the kinetic, potential, and internal energies of the system. Since the kinetic and potential energies are evaluated relative to the environment, the energy of the system at the dead state is just its internal energy, U_0. Accordingly, ΔE_c can be expressed as

$$\Delta E_c = (U_0 - E) + \Delta U_e \qquad \text{(b)}$$

where ΔU_e is the change in internal energy of the environment.

Since T_0 and p_0 are constant, changes in internal energy U_e, entropy S_e, and volume V_e of the environment are related through Eq. 6.8, the *first $T\,dS$ equation*, as

$$\Delta U_e = T_0 \Delta S_e - p_0 \Delta V_e \qquad \text{(c)}$$

Introducing Eq. (c) into Eq. (b) gives

$$\Delta E_c = (U_0 - E) + (T_0 \Delta S_e - p_0 \Delta V_e) \qquad \text{(d)}$$

Substituting Eq. (d) into Eq. (a) and solving for W_c gives

$$W_c = (E - U_0) - (T_0 \Delta S_e - p_0 \Delta V_e)$$

The total volume is constant. Hence, the change in volume of the environment is equal in magnitude and opposite in sign to the volume change of the system: $\Delta V_e = -(V_0 - V)$. With this substitution, the above expression for work becomes

$$W_c = (E - U_0) + p_0 (V - V_0) - (T_0 \Delta S_e) \qquad \text{(e)}$$

This equation gives the work for the overall system as the system passes to the dead state. The maximum theoretical work is determined using the entropy balance as follows.

Entropy Balance

The entropy balance for the overall system reduces to

$$\Delta S_c = \sigma_c$$

where the entropy transfer term is omitted because no heat transfer takes place across the boundary of the overall system. The term σ_c accounts for entropy production due to irreversibilities as the system comes into equilibrium with the environment. The entropy change ΔS_c is the sum of the entropy changes for the system and environment, respectively. That is,

$$\Delta S_c = (S_0 - S) + \Delta S_e$$

where S and S_0 denote the entropy of the system at the given state and the dead state, respectively. Combining the last two equations

$$(S_0 - S) + \Delta S_e = \sigma_c \qquad \text{(f)}$$

Eliminating ΔS_e between Eqs. (e) and (f) results in

$$W_c = (E - U_0) + p_0 (V - V_0) - T_0 (S - S_0) - T_0 \sigma_c \qquad \text{(g)}$$

With $E = U + \text{KE} + \text{PE}$, Eq. (g) becomes

$$\underline{W_c = (U - U_0) + p_0 (V - V_0) - T_0 (S - S_0) + \text{KE} + \text{PE}} - T_0 \sigma_c \qquad \text{(h)}$$

The value of the underlined term in Eq. (h) is determined by the two end states of the system—the given state and the dead state—and is independent of the details of the process linking these states. However, the value of the term $T_0 \sigma_c$ depends on the nature of the process as the system passes to the dead state. In accordance with the second law, $T_0 \sigma_c$ is positive when irreversibilities are present and vanishes in the limiting case where there are no irreversibilities. The value of $T_0 \sigma_c$ cannot be negative. Hence, the *maximum* theoretical value for the work of the overall system W_c is obtained by setting $T_0 \sigma_c$ to zero in Eq. (h). By definition, this maximum value is the exergy, E. Accordingly, Eq. 7.1 is seen to be the appropriate expression for evaluating exergy.

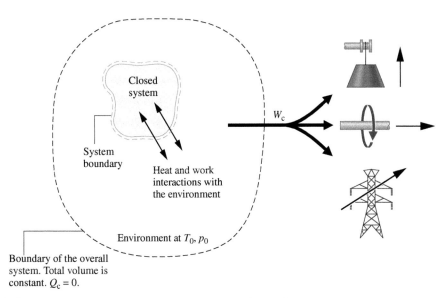

Closed system

System boundary

Heat and work interactions with the environment

W_c

Environment at T_0, p_0

Boundary of the overall system. Total volume is constant. $Q_c = 0$.

Fig. 7.3 Overall system of system and environment used to evaluate exergy.

7.3.1 Exergy Aspects

In this section, we list five important aspects of the exergy concept:

1. Exergy is a measure of the departure of the state of a system from that of the environment. It is therefore an attribute of the system and environment together. However, once the environment is specified, a value can be assigned to exergy in terms of property values for the system only, so exergy can be regarded as a property of the system. Exergy is an extensive property.

2. The value of exergy cannot be negative. If a system were at any state other than the dead state, the system would be able to change its condition *spontaneously* toward the dead state; this tendency would cease when the dead state was reached. No work must be done to effect such a spontaneous change. Accordingly, any change in state of the system to the dead state can be accomplished with *at least zero* work being developed, and thus the *maximum* work (exergy) cannot be negative.

3. Exergy is not conserved but is destroyed by irreversibilities. A limiting case is when exergy is completely destroyed, as would occur if a system were permitted to undergo a spontaneous change to the dead state with no provision to obtain work. The potential to develop work that existed originally would be completely wasted in such a spontaneous process.

4. Exergy has been viewed thus far as the *maximum* theoretical work obtainable from an *overall* system of system plus environment as the system passes *from* a given state *to* the dead state. Alternatively, exergy can be regarded as the magnitude of the *minimum* theoretical work *input* required to bring the system *from* the dead state *to* the given state. Using energy and entropy balances as above, we can readily develop Eq. 7.1 from this viewpoint. This is left as an exercise.

5. When a system is at the dead state, it is in *thermal* and *mechanical* equilibrium with the environment, and the value of exergy is zero. More precisely, the *thermo-mechanical* contribution to exergy is zero. This modifying term distinguishes the exergy concept of the present chapter from another contribution to exergy introduced in Sec. 13.6, where the contents of a system at the dead state are permitted to enter into chemical reaction with environmental components and in so doing develop additional work. This contribution to exergy is called *chemical exergy*. The chemical exergy concept is important in the second law analysis of many types of systems, in particular systems involving combustion. Still, as shown in this chapter, the thermomechanical exergy concept suffices for a wide range of thermodynamic evaluations.

 BioConnections

The U.S. poultry industry produces 50 billion pounds of meat annually, with chicken production accounting for over 80% of the total. The annual waste produced by these birds also reaches into the billions of pounds. A fraction of the waste is spread over cropland as fertilizer. Some is used to manufacture fertilizer pellets for commercial and domestic use. Despite its relatively low chemical exergy, poultry waste also can be used to produce methane through *anaerobic* digestion. The methane can be burned in power plants to make electricity or process steam. Digester systems are now available for use on the farm with potential benefits of lower energy costs and reduced odors. Digester systems produce waste or *effluent* that is nutrient-rich and may still require management of some kind. As costs decline and the need to manage manure increases, systems on larger farms will become more common. Overall, these are positive developments for an industry that has received adverse publicity for concerns over such issues as the arsenic content of poultry waste, runoff of waste into streams and rivers, and excessive odor and fly infestation in the vicinity of huge farming operations.

7.3.2 Specific Exergy

specific exergy

Although exergy is an extensive property, it is often convenient to work with it on a unit mass or molar basis. Expressing Eq. 7.1 on a unit mass basis, the **specific exergy**, e, is

$$\text{e} = (u - u_0) + p_0(v - v_0) - T_0(s - s_0) + V^2/2 + gz \tag{7.2}$$

where u, v, s, $V^2/2$, and gz are the specific internal energy, volume, entropy, kinetic energy, and potential energy, respectively, at the state of interest; u_0, v_0, and s_0 are specific properties at the dead state at T_0, p_0. In Eq. 7.2, the kinetic and potential energies are measured relative to the environment and thus contribute their full values to the exergy magnitude because, in principle, each could be fully converted to work were the system brought to rest at zero elevation relative to the environment. Finally, by inspection of Eq. 7.2, note that the units of specific exergy are the same as for specific energy, kJ/kg.

The specific exergy at a specified state requires properties at that state and at the dead state.

> **TAKE NOTE...**
>
> Kinetic energy and potential energy are each rightfully considered as exergy. But for simplicity of expression in the present chapter, we refer to these terms—whether viewed as energy or exergy—as *accounting for the effects of motion and gravity*. The meaning will be clear in context.

FOR EXAMPLE

Let us use Eq. 7.2 to determine the specific exergy of saturated water vapor at 120°C, having a velocity of 30 m/s and an elevation of 6 m, each relative to an exergy reference environment where $T_0 = 298$ K (25°C), $p_0 = 1$ atm, and $g = 9.8$ m/s². For water as saturated vapor at 120°C, Table A-2 gives $v = 0.8919$ m³/kg, $u = 2529.3$ kJ/kg, $s = 7.1296$ kJ/kg·K. At the dead state, where $T_0 = 298$ K (25°C) and $p_0 = 1$ atm, water is a liquid. Thus, with Eqs. 3.11, 3.12, and 6.5 and values from Table A-2, we get $v_0 = 1.0029 \times 10^{-3}$ m³/kg, $u_0 = 104.88$ kJ/kg, $s_0 = 0.3674$ kJ/kg·K. Substituting values

$$e = (u - u_0) + p_0(v - v_0) - T_0(s - s_0) + \frac{V^2}{2} + gz$$

$$= \left[(2529.3 - 104.88)\frac{kJ}{kg} \right]$$

$$+ \left[\left(1.01325 \times 10^5 \frac{N}{m^2} \right)(0.8919 - 1.0029 \times 10^{-3})\frac{m^3}{kg} \left| \frac{1\ kJ}{10^3\ N \cdot m} \right| \right]$$

$$- \left[(298\ K)(7.1296 - 0.3674)\frac{kJ}{kg \cdot K} \right]$$

$$+ \left[\frac{(30\ m/s)^2}{2} + \left(9.8\frac{m}{s^2} \right)(6\ m) \left| \frac{1\ N}{1\ kg \cdot m/s^2} \right| \left| \frac{1\ kJ}{10^3\ N \cdot m} \right| \right]$$

$$= (2424.42 + 90.27 - 2015.14 + 0.45 + 0.06)\frac{kJ}{kg} = 500\frac{kJ}{kg}$$

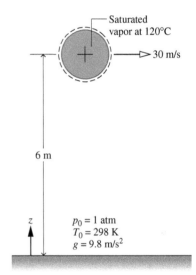

Saturated vapor at 120°C

30 m/s

6 m

z

$p_0 = 1$ atm
$T_0 = 298$ K
$g = 9.8$ m/s²

The following example illustrates the use of Eq. 7.2 together with ideal gas property data.

▸ ▸ ▸ **EXAMPLE 7.1** ▸ ·

Evaluating the Exergy of Exhaust Gas

A cylinder of an internal combustion engine contains 2450 cm³ of gaseous combustion products at a pressure of 7 bar and a temperature of 867°C just before the exhaust valve opens. Determine the specific exergy of the gas, in kJ/kg. Ignore the effects of motion and gravity, and model the combustion products as air behaving as an ideal gas. Take $T_0 = 300$ K (27°C) and $p_0 = 1.013$ bar.

Schematic and Given Data:

2450 cm³ of air at 7 bar, 867°C

Fig. E7.1

SOLUTION

Known Gaseous combustion products at a specified state are contained in the cylinder of an internal combustion engine.

Find Determine the specific exergy.

Engineering Model

1. The gaseous combustion products are a closed system.

2. The combustion products are modeled as air behaving as an ideal gas.

3. The effects of motion and gravity can be ignored.

4. $T_0 = 300$ K (27°C) and $p_0 = 1.013$ bar.

Analysis With assumption 3, Eq. 7.2 becomes

$$e = u - u_0 + p_0(v - v_0) - T_0(s - s_0)$$

The internal energy and entropy terms are evaluated using data from Table A-22, as follows:

$$u - u_0 = (880.35 - 214.07) \text{ kJ/kg} = 666.28 \text{ kJ/kg}$$

$$s - s_0 = s°(T) - s°(T_0) - \frac{\overline{R}}{M} \ln \frac{p}{p_0}$$

$$= \left(3.11883 - 1.70203 - \left(\frac{8.314}{28.97}\right) \ln \left(\frac{7}{1.013}\right)\right) \frac{\text{kJ}}{\text{kg} \cdot \text{K}}$$

$$= 0.8621 \text{ kJ/kg} \cdot \text{K}$$

$$T_0(s - s_0) = (300 \text{ K})(0.8621 \text{ kJ/kg} \cdot \text{K}) = 258.62 \text{ kJ/kg}$$

The $p_0(v - v_0)$ term is evaluated using the ideal gas equation of state: $v = (\overline{R}/M)T/p$ and $v_0 = (\overline{R}/M)T_0/p_0$, so

$$p_0(v - v_0) = \frac{\overline{R}}{M}\left(\frac{p_0 T}{p} - T_0\right)$$

$$= \frac{8.314}{28.97}\left(\frac{(1.013)(1140)}{7} - 300\right)\frac{\text{kJ}}{\text{kg}}$$

$$= -38.75 \text{ kJ/kg}$$

Substituting values into the above expression for the specific exergy

❶ $e = (666.28 + (-38.75) - 258.62) \text{ kJ/kg} = 368.91 \text{ kJ/kg}$

❶ If the gases are discharged directly to the surroundings, the potential for developing work quantified by the exergy value determined in the solution is wasted. However, by venting the gases through a turbine, some work could be developed. This principle is utilized by the *turbochargers* added to some internal combustion engines.

SKILLS DEVELOPED

Ability to...

- evaluate specific exergy.
- apply the ideal gas model.

Quick Quiz

To what elevation, in m, would a 1-kg mass have to be raised from zero elevation with respect to the reference environment for its exergy to equal that of the gas in the cylinder? Assume $g = 9.81$ m/s². Ans. 197 m.

7.3.3 Exergy Change

A closed system at a given state can attain new states by various means, including work and heat interactions with its surroundings. The exergy value associated with a new state generally differs from the exergy value at the initial state. Using Eq. 7.1, we can determine the change in exergy between the two states. At the initial state

$$E_1 = (U_1 - U_0) + p_0(V_1 - V_0) - T_0(S_1 - S_0) + KE_1 + PE_1$$

At the final state

$$E_2 = (U_2 - U_0) + p_0(V_2 - V_0) - T_0(S_2 - S_0) + KE_2 + PE_2$$

Subtracting these we get the **exergy change**

exergy change
$$\boxed{E_2 - E_1 = (U_2 - U_1) + p_0(V_2 - V_1) - T_0(S_2 - S_1) + (KE_2 - KE_1) + (PE_2 - PE_1)} \quad (7.3)$$

Note that the dead state values U_0, V_0, S_0 cancel when we subtract the expressions for E_1 and E_2.

Exergy *change* can be illustrated using **Fig. 7.4**, which shows an exergy-temperature-pressure surface for a gas together with constant-exergy contours projected on temperature-pressure coordinates. For a system undergoing Process A, exergy increases as its state *moves away* from the dead state (from 1 to 2). In Process B, exergy decreases as the state *moves toward* the dead state (from 1' to 2'.)

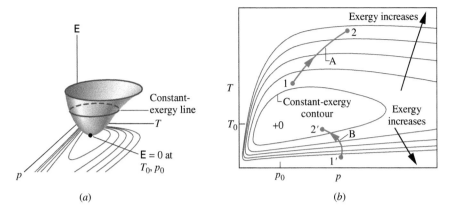

Fig. 7.4 Exergy-temperature-pressure surface for a gas. (*a*) Three-dimensional view (*b*) Constant exergy contours on a *T–p* diagram.

7.4 Closed System Exergy Balance

Like energy, exergy can be transferred across the boundary of a closed system. The change in exergy of a system during a process would not necessarily equal the net exergy transferred because exergy would be destroyed if irreversibilities were present within the system during the process. The concepts of exergy change, exergy transfer, and exergy destruction are related by the closed system exergy balance introduced in this section. The exergy balance concept is extended to control volumes in Sec. 7.5. Exergy balances are expressions of the second law of thermodynamics and provide the basis for exergy analysis.

7.4.1 Introducing the Closed System Exergy Balance

The **closed system exergy balance** is given by Eq. 7.4a. See the box for its development.

$$\underbrace{E_2 - E_1}_{\substack{\text{exergy} \\ \text{change}}} = \underbrace{\int_1^2 \left(1 - \frac{T_0}{T_b}\right)\delta Q - [W - p_0(V_2 - V_1)]}_{\substack{\text{exergy} \\ \text{transfers}}} - \underbrace{T_0\sigma}_{\substack{\text{exergy} \\ \text{destruction}}}$$

(7.4a)   **closed system exergy balance**

For specified end states and given values of p_0 and T_0, the exergy change $E_2 - E_1$ on the left side of Eq. 7.4a can be evaluated from Eq. 7.3. The underlined terms on the right depend explicitly on the nature of the process, however, and cannot be determined by knowing only the end states and the values of p_0 and T_0. These terms are interpreted in the discussions of Eqs. 7.5–7.7, respectively.

Developing the Exergy Balance

The exergy balance for a closed system is developed by combining the closed system energy and entropy balances. The forms of the energy and entropy balances used are, respectively,

$$\Delta U + \Delta KE + \Delta PE = \left(\int_1^2 \delta Q\right) - W$$

$$\Delta S = \int_1^2 \left(\frac{\delta Q}{T}\right)_b + \sigma$$

where W and Q represent, respectively, work and heat transfer between the system and its surroundings. In the entropy balance, T_b denotes the temperature on the system boundary where δQ occurs. The term σ accounts for entropy produced within the system by internal irreversibilities.

As the first step in deriving the exergy balance, multiply the entropy balance by the temperature T_0 and subtract the resulting expression from the energy balance to obtain

$$(\Delta U + \Delta KE + \Delta PE) - T_0\Delta S = \left(\int_1^2 \delta Q\right) - T_0\int_1^2 \left(\frac{\delta Q}{T}\right)_b - W - T_0\sigma$$

Collecting the terms involving δQ on the right side and introducing Eq. 7.3 on the left side, we get

$$(E_2 - E_1) - p_0(V_2 - V_1) = \int_1^2 \left(1 - \frac{T_0}{T_b}\right)\delta Q - W - T_0\sigma$$

On rearrangement, this expression gives Eq. 7.4a, the closed system exergy balance.

Since Eq. 7.4a is obtained by deduction from the energy and entropy balances, it is not an independent result but can be used in place of the entropy balance as an expression of the second law.

The first underlined term on the right side of Eq. 7.4a is associated with heat transfer to or from the system during the process. It is interpreted as the **exergy transfer accompanying heat transfer**. That is,

exergy transfer accompanying heat transfer

$$\mathsf{E}_q = \begin{bmatrix} \textit{exergy transfer} \\ \text{accompanying heat} \\ \text{transfer} \end{bmatrix} = \int_1^2 \left(1 - \frac{T_0}{T_b}\right)\delta Q \qquad (7.5)$$

where T_b denotes the temperature on the boundary where heat transfer occurs.

The second underlined term on the right side of Eq. 7.4a is associated with work. It is interpreted as the **exergy transfer accompanying work**. That is,

exergy transfer accompanying work

$$\mathsf{E}_w = \begin{bmatrix} \textit{exergy transfer} \\ \text{accompanying work} \end{bmatrix} = [W - p_0(V_2 - V_1)] \qquad (7.6)$$

The third underlined term on the right side of Eq. 7.4a accounts for the **destruction of exergy** due to irreversibilities within the system. It is symbolized by E_d. That is

exergy destruction

$$\mathsf{E}_d = T_0\sigma \qquad (7.7)$$

With Eqs. 7.5, 7.6, and 7.7, Eq. 7.4a is expressed alternatively as

$$E_2 - E_1 = \mathsf{E}_q - \mathsf{E}_w - \mathsf{E}_d \qquad (7.4b)$$

Although not required for the practical application of the exergy balance in *any* of its forms, exergy transfer terms can be conceptualized in terms of work, as for the exergy concept itself. See the box for discussion.

Conceptualizing Exergy Transfer

In exergy analysis, heat transfer and work are expressed in terms of a *common measure:* work *fully available* for lifting a weight or, equivalently, as shaft or electrical work. This is the significance of the exergy transfer expressions given by Eqs. 7.5 and 7.6, respectively.

Without regard for the nature of the surroundings with which the system is *actually* interacting, we interpret the *magnitudes* of these exergy transfers as the maximum theoretical work that *could* be developed *were* the system interacting with the environment, as follows:

- On recognizing the term $(1 - T_0/T_b)$ as the Carnot efficiency (Eq. 5.9), the quantity $(1 - T_0/T_b)\delta Q$ appearing in Eq. 7.5

is interpreted as the work developed by a reversible power cycle receiving energy δQ by heat transfer at temperature T_b and discharging energy by heat transfer to the environment at temperature $T_0 < T_b$. When T_b is less than T_0, we also think of the work of a reversible cycle. But in this instance, E_q takes on a negative value signaling that heat transfer and the accompanying exergy transfer are *oppositely* directed.

- The exergy transfer given by Eq. 7.6 is the work W of the system less the work required to displace the environment whose pressure is p_0, namely, $p_0(V_2 - V_1)$.

See Example 7.2 for an illustration of these interpretations.

To summarize, in each of its forms Eq. 7.4 states that the change in exergy of a closed system can be accounted for in terms of exergy transfers and the destruction of exergy due to irreversibilities within the system.

When applying the exergy balance, it is essential to observe the requirements imposed by the second law on the exergy destruction: In accordance with the second law, the exergy destruction is positive when irreversibilities are present within the system during the process and vanishes in the limiting case where there are no irreversibilities. That is,

$$E_d : \begin{cases} > 0 & \text{irreversibilities present within the system} \\ = 0 & \text{no irreversibilities present within the system} \end{cases}$$ (7.8)

The value of the exergy destruction cannot be negative. Moreover, exergy destruction is *not* a property. On the other hand, exergy *is* a property, and like other properties, the *change* in exergy of a system can be positive, negative, or zero:

$$E_2 - E_1 : \begin{cases} > 0 \\ = 0 \\ < 0 \end{cases}$$

For an *isolated* system, no heat or work interactions with the surroundings occur, and thus there are no transfers of exergy between the system and its surroundings. Accordingly, the exergy balance reduces to give

$$\Delta E]_{isol} = -E_d]_{isol}$$ (7.9)

Since the exergy destruction must be positive in any actual process, the only processes of an isolated system that occur are those for which the exergy of the isolated system *decreases*. For exergy, this conclusion is the counterpart of the increase of entropy principle (Sec. 6.8.1) and, like the increase of entropy principle, can be regarded as an alternative statement of the second law.

In Example 7.2, we consider exergy change, exergy transfer, and exergy destruction for the process of water considered in Example 6.1, which should be quickly reviewed before studying the current example.

▶▶ EXAMPLE 7.2 ▶

Exploring Exergy Change, Transfer, and Destruction

Water, initially a saturated liquid at 150°C (423.15 K), is contained in a piston–cylinder assembly. The water is heated to the corresponding saturated vapor state in an internally reversible process at constant temperature and pressure. For $T_0 = 20°C$ (293.15 K), $p_0 = 1$ bar, and ignoring the effects of motion and gravity, determine per unit of mass, each in kJ/kg, **(a)** the change in exergy, **(b)** the exergy transfer accompanying heat transfer, **(c)** the exergy transfer accompanying work, and **(d)** the exergy destruction.

Schematic and Given Data:

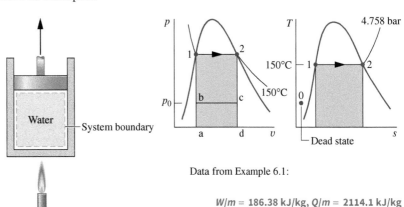

SOLUTION

Known Water contained in a piston–cylinder assembly undergoes an internally reversible process at 150°C from saturated liquid to saturated vapor.

Find Determine the change in exergy, the exergy transfers accompanying heat transfer and work, and the exergy destruction.

Fig. E7.2

Data from Example 6.1:

$W/m = 186.38$ kJ/kg, $Q/m = 2114.1$ kJ/kg

State	v (m³/kg)	u (kJ/kg)	s (kJ/kg · K)
1	1.0905×10^{-3}	631.68	1.8418
2	0.3928	2559.5	6.8379

Engineering Model

1. The water in the piston–cylinder assembly is a closed system.

2. The process is internally reversible.

3. Temperature and pressure are constant during the process.

4. Ignore the effects of motion and gravity.

5. $T_0 = 293.15$ K, $p_0 = 1$ bar.

Analysis

a. Using Eq. 7.3 together with assumption 4, we have per unit of mass

$$e_2 - e_1 = u_2 - u_1 + p_0(v_2 - v_1) - T_0(s_2 - s_1) \qquad \text{(a)}$$

With data from Fig. E7.2

$$
\begin{aligned}
e_2 - e_1 &= (2559.5 - 631.68)\frac{\text{kJ}}{\text{kg}} + \left(1.0 \times 10^5 \frac{\text{N}}{\text{m}^2}\right) \\
&\quad \times (0.3928 - (1.0905 \times 10^{-3}))\frac{\text{m}^3}{\text{kg}}\left|\frac{1\ \text{kJ}}{10^3\ \text{N} \cdot \text{m}}\right| \\
&\quad - 293.15\ \text{K}\,(6.8379 - 1.8418)\frac{\text{kJ}}{\text{kg} \cdot \text{K}} \\
&= (1927.82 + 39.17 - 1464.61)\frac{\text{kJ}}{\text{kg}} = 502.38\,\frac{\text{kJ}}{\text{kg}}
\end{aligned}
$$

b. Noting that temperature remains constant, Eq. 7.5, on a per unit of mass basis, reads

❶
$$\frac{E_q}{m} = \left(1 - \frac{T_0}{T}\right)\frac{Q}{m} \qquad \text{(b)}$$

With $Q/m = 2114.1$ kJ/kg from Fig. E7.2

$$\frac{E_q}{m} = \left(1 - \frac{293.15\ \text{K}}{423.15\ \text{K}}\right)\left(2114.1\,\frac{\text{kJ}}{\text{kg}}\right) = 649.49\,\frac{\text{kJ}}{\text{kg}}$$

c. With $W/m = 186.38$ kJ/kg from Fig. E7.2 and $p_0(v_2 - v_1) = 39.17$ kJ/kg from part (a), Eq. 7.6 gives, per unit of mass,

❷
$$\frac{E_w}{m} = \frac{W}{m} - p_0(v_2 - v_1) \qquad \text{(c)}$$

$$= (186.38 - 39.17)\frac{\text{kJ}}{\text{kg}} = 147.21\,\frac{\text{kJ}}{\text{kg}}$$

d. Since the process is internally reversible, the exergy destruction is necessarily zero. This can be checked by inserting the results of parts (a)–(c) into an exergy balance. Thus, solving Eq. 7.4b for the exergy destruction per unit of mass, evaluating terms, and allowing for roundoff, we get

$$
\begin{aligned}
\frac{E_d}{m} &= -(e_2 - e_1) + \frac{E_q}{m} - \frac{E_w}{m} \\
&= (-502.38 + 649.49 - 147.21)\frac{\text{kJ}}{\text{kg}} = 0
\end{aligned}
$$

Alternatively, the exergy destruction can be evaluated using Eq. 7.7 together with the entropy production obtained from an entropy balance. This is left as an exercise.

---

❶ Recognizing the term $(1 - T_0/T)$ as the Carnot efficiency (Eq. 5.9), the right side of Eq. (b) can be interpreted as the work that *could be* developed by a reversible power cycle *were* it to receive energy Q/m at temperature T and discharge energy to the environment by heat transfer at T_0.

❷ The right side of Eq. (c) shows that *if* the system were interacting with the environment, all of the work, W/m, represented by area 1-2-d-a-1 on the p–v diagram of Fig. E7.2, would not be fully available for lifting a weight. A portion would be spent in pushing aside the environment at pressure p_0. This portion is given by $p_0(v_2 - v_1)$, and is represented by area a-b-c-d-a on the p–v diagram of Fig. E7.2.

SKILLS DEVELOPED

Ability to…

- evaluate exergy change.
- evaluate exergy transfer accompanying heat transfer and work.
- evaluate exergy destruction.

Quick Quiz

If heating from saturated liquid to saturated vapor would occur at 100°C (373.15 K), evaluate the exergy transfers accompanying heat transfer and work, each in kJ/kg. Ans. 484, 0.

7.4.2 Closed System Exergy Rate Balance

As in the case of the mass, energy, and entropy balances, the exergy balance can be expressed in various forms that may be more suitable for particular analyses. A convenient form is the *closed system exergy rate balance* given by

$$\boxed{\frac{d\mathsf{E}}{dt} = \sum_j \left(1 - \frac{T_0}{T_j}\right)\dot{Q}_j - \left(\dot{W} - p_0\frac{dV}{dt}\right) - \dot{E}_d} \qquad (7.10)$$

where dE/dt is the time rate of change of exergy. The term $(1 - T_0/T_j)\dot{Q}_j$ represents the time rate of exergy transfer accompanying heat transfer at the rate $\dot{Q}_j$ occurring where the instantaneous temperature on the boundary is T_j. The term $\dot{W}$ represents the time rate of energy transfer by work. The accompanying rate of exergy transfer is given by $(\dot{W} - p_0 dV/dt)$ where dV/dt is the time rate of change of system volume. The term $\dot{E}_d$ accounts for the time rate of exergy destruction due to irreversibilities within the system.

At steady state, $dE/dt = dV/dt = 0$ and Eq. 7.10 reduces to give the **steady-state form of the exergy rate balance**.

$$0 = \sum_j \left(1 - \frac{T_0}{T_j}\right)\dot{Q}_j - \dot{W} - \dot{E}_d$$
(7.11a)

steady-state form of the closed system exergy rate balance

Note that for a system at steady state, the rate of exergy transfer accompanying the power $\dot{W}$ is simply the power.

The rate of exergy transfer accompanying heat transfer at the rate $\dot{Q}_j$ occurring where the temperature is T_j is compactly expressed as

$$\dot{E}_{qj} = \left(1 - \frac{T_0}{T_j}\right)\dot{Q}_j$$
(7.12)

As shown in the adjacent figure, heat transfer and the accompanying exergy transfer are in the same direction when $T_j > T_0$.

Using Eq. 7.12, Eq. 7.11a reads

$$0 = \sum_j \dot{E}_{qj} - \dot{W} - \dot{E}_d$$
(7.11b)

In Eqs. 7.11, the rate of exergy destruction within the system, $\dot{E}_d$, is related to the rate of entropy production within the system by $\dot{E}_d = T_0 \dot{\sigma}$.

7.4.3 Exergy Destruction and Loss

Most thermal systems are supplied with exergy inputs derived directly or indirectly from the consumption of fossil fuels. Accordingly, *avoidable* destructions and losses of exergy represent the waste of these resources. By devising ways to reduce such inefficiencies, better use can be made of fuels. The exergy balance can be applied to determine the locations, types, and true magnitudes of energy resource waste and thus can play an important part in developing strategies for more effective fuel use.

In Example 7.3, the steady-state form of the closed system energy and exergy rate balances are applied to an oven wall to evaluate exergy destruction and exergy loss, which are interpreted in terms of fossil fuel use.

▶▶ EXAMPLE 7.3 ▶

Evaluating Exergy Destruction in an Oven Wall

The wall of an industrial drying oven is constructed by sandwiching 0.066 m-thick insulation, having a thermal conductivity $\kappa = 0.05 \times 10^{-3}$ kW/m · K, between thin metal sheets. At steady state, the inner metal sheet is at $T_1 = 575$ K and the outer sheet is at $T_2 = 310$ K. Temperature varies linearly through the wall. The temperature of the surroundings away from the oven is 293 K. Determine, in kW per m² of wall surface area, **(a)** the rate of heat transfer through the wall, **(b)** the rates of exergy transfer accompanying heat transfer at the inner and outer wall surfaces, and **(c)** the rate of exergy destruction within the wall. Let $T_0 = 293$ K.

SOLUTION

Known Temperature, thermal conductivity, and wall-thickness data are provided for a plane wall at steady state.

Find For the wall, determine (a) the rate of heat transfer through the wall, (b) the rates of exergy transfer accompanying heat transfer at the inner and outer surfaces, and (c) the rate of exergy destruction, each per m^2 of wall surface area.

Schematic and Given Data:

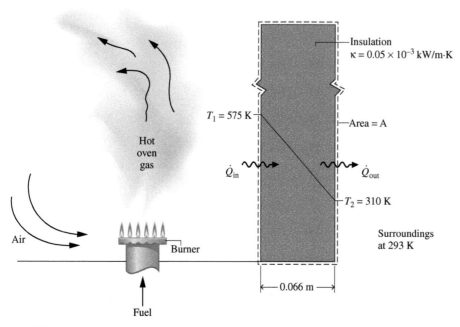

Fig. E7.3

Engineering Model

1. The closed system shown in the accompanying sketch is at steady state.

2. Temperature varies linearly through the wall.

3. $T_0 = 293$ K.

Analysis

a. At steady state, an energy rate balance for the system reduces to give $\dot{Q}_{in} = \dot{Q}_{out}$—namely, the rates of heat transfer into and out of the wall are equal. Let $\dot{Q}$ denote the common heat transfer rate. Using Eq. 2.3.1 with assumption 2, the heat transfer rate is given by

$$(\dot{Q}/A) = -\kappa \left[\frac{T_2 - T_1}{L} \right]$$

$$= -\left(0.05 \times 10^{-3} \frac{kW}{m \cdot K} \right) \left[\frac{(310 - 575)\ K}{0.066\ m} \right] = 0.2 \frac{kW}{m^2}$$

b. The rates of exergy transfer accompanying heat transfer are evaluated using Eq. 7.12. At the inner surface

$$(\dot{E}_{q1}/A) = \left[1 - \frac{T_0}{T_1} \right](\dot{Q}/A)$$

$$= \left[1 - \frac{293}{575} \right]\left(0.2 \frac{kW}{m^2} \right) = 0.1 \frac{kW}{m^2}$$

At the outer surface

❶
$$(\dot{E}_{q2}/A) = \left[1 - \frac{T_0}{T_2} \right](\dot{Q}/A)$$

$$= \left[1 - \frac{293}{310} \right]\left(0.2 \frac{kW}{m^2} \right) = 0.01 \frac{kW}{m^2}$$

c. The rate of exergy destruction within the wall is evaluated using the exergy rate balance. Since $\dot{W} = 0$, Eq. 7.11b gives

❷
$$(\dot{E}_d/A) = (\dot{E}_{q1}/A) - (\dot{E}_{q2}/A)$$

❸
$$= (0.1 - 0.01)\frac{kW}{m^2} = 0.09 \frac{kW}{m^2}$$

---

❶ The rates of heat transfer are the same at the inner and outer walls, but the rates of exergy transfer at these locations are much different. The rate of exergy transfer at the high-temperature inner wall is 10 times the rate of exergy transfer at the low-temperature outer wall. At each of these locations the exergy transfers provide a truer measure of thermodynamic value than the heat transfer rate. This is clearly seen at the outer wall, where the small exergy transfer indicates minimal potential for use and, thus, minimal thermodynamic value.

❷ The exergy transferred into the wall at $T_1 = 575$ K is either destroyed within the wall owing to spontaneous heat transfer or transferred out of the wall at $T_2 = 310$ K where it is *lost* to the surroundings. Exergy transferred to the surroundings accompanying stray heat transfer, as in the present case, is ultimately destroyed in the surroundings. Thicker insulation and/or insulation having a lower thermal conductivity value would reduce the heat transfer rate and thus lower the exergy destruction and loss.

❸ In this example, the exergy destroyed and lost has its origin in the fuel supplied. Thus, cost-effective measures to reduce exergy destruction and loss have benefits in terms of better fuel use.

SKILLS DEVELOPED

Ability to...

- apply the energy and exergy rate balances.
- evaluate exergy transfer accompanying heat transfer.
- evaluate exergy destruction.

Quick Quiz

If the thermal conductivity were reduced to 0.04×10^{-3} kW/m · K, owing to a different selection of insulation material, while the insulation thickness were increased to 0.076 m, determine the rate of exergy destruction in the wall, in kW per m^2 of wall surface area, keeping the same inner and outer wall temperatures and the same temperature of the surroundings. Ans. 0.06 kW/m².

Horizons

Superconducting Power Cable Overcoming All Barriers?

According to industry sources, more than 7% of the electric power conducted through present-day transmission and distribution lines is forfeited en route owing to irreversibility associated with electrical resistance. *Superconducting* cable can nearly eliminate resistance to electricity flow and thereby the accompanying irreversibility. As electricity generation shifts toward technologies that use renewable energy (wind turbines, solar panels, etc.) with locations typically far from end use, these improvements are becoming increasingly important.

For superconducting cable to be effective, however, it must be cooled to about −200°C. Cooling is achieved by refrigeration

systems using liquid nitrogen. Since the refrigeration systems require power to operate, such cooling reduces the power *saved* in superconducting electrical transmission. Moreover, the cost of superconducting cable is much greater today than conventional cable. Factors such as these impose barriers to rapid deployment of superconducting technology.

Still, electrical utilities have partnered with the government to develop and demonstrate superconducting technology that someday may increase U.S. power system efficiency and reliability. The cables are slowly being introduced to the U.S. power system, as companies have begun manufacturing superconducting cables for this purpose.

7.4.4 Exergy Accounting

In the next example, we reconsider the gearbox of Examples 2.4 and 6.4 from an exergy perspective to introduce **exergy accounting**, in which the various terms of an exergy balance for a system are systematically evaluated and compared.

exergy accounting

> ▶ **EXAMPLE 7.4** ▶

Exergy Accounting of a Gearbox

For the gearbox of Examples 2.4 and 6.4(a), develop a full exergy accounting of the power input. Let $T_0 = 293$ K.

SOLUTION

Known A gearbox operates at steady state with known values for the power input, power output, and heat transfer rate. The temperature on the outer surface of the gearbox is also known.

Find Develop a full exergy accounting of the input power.

Schematic and Given Data:

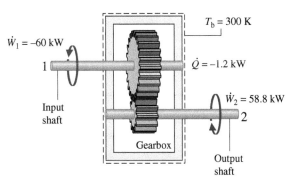

$\dot{W}_1 = -60$ kW

1

Input shaft

$T_b = 300$ K

$\dot{Q} = -1.2$ kW

$\dot{W}_2 = 58.8$ kW

2

Gearbox

Output shaft

Fig. E7.4

Engineering Model

1. The gearbox is taken as a closed system operating at steady state.

2. The temperature at the outer surface does not vary.

3. $T_0 = 293$ K.

Analysis Since the gearbox is at steady state, the rate of exergy transfer accompanying power is simply the power. Accordingly, exergy is transferred *into* the gearbox via the high-speed shaft at a rate equal to the power *input*, 60 kW, and exergy is transferred *out* via the low-speed shaft at a rate equal to the power *output*, 58.8 kW. Additionally, exergy is transferred out accompanying stray heat transfer and destroyed by irreversibilities within the gearbox.

The rate of exergy transfer accompanying heat transfer is evaluated using Eq. 7.12. That is,

$$\dot{E}_q = \left(1 - \frac{T_0}{T_b}\right)\dot{Q}$$

With $\dot{Q} = -1.2$ kW and $T_b = 300$ K from Fig. E7.4, we get

$$\dot{E}_q = \left(1 - \frac{293}{300}\right)(-1.2 \text{ kW})$$

$$= -0.03 \text{ kW}$$

where the minus sign denotes exergy transfer *from* the system.

The rate of exergy destruction is evaluated using the exergy rate balance. On rearrangement, and noting that $\dot{W} = \dot{W}_1 + \dot{W}_2 = -1.2$ kW, Eq. 7.11b gives

① $\dot{E}_d = \dot{E}_q - \dot{W} = -0.03$ kW $- (-1.2$ kW$) = 1.17$ kW

The analysis is summarized by the following exergy *balance sheet* in terms of exergy magnitudes on a rate basis:

Rate of exergy in:
  high-speed shaft   60.00 kW (100%)
Disposition of the exergy:
• Rate of exergy out
  low-speed shaft   58.80 kW (98%)
②   stray heat transfer   0.03 kW (0.05%)
• Rate of exergy destruction   <u>1.17 kW (1.95%)</u>
       60.00 kW (100%)

① Alternatively, the rate of exergy destruction is calculated from $\dot{E}_d = T_0\dot{\sigma}$, where $\dot{\sigma}$ is the rate of entropy production. From the solution to Example 6.4(a), $\dot{\sigma} = 4 \times 10^{-3}$ kW/K. Then

$$\dot{E}_d = T_0\dot{\sigma}$$
$$= (293 \text{ K})(4 \times 10^{-3} \text{ kW/K})$$
$$= 1.17 \text{ kW}$$

② The difference between the input and output power is accounted for primarily by the rate of exergy destruction and only secondarily by the exergy transfer accompanying heat transfer, which is small by comparison. The exergy balance sheet provides a sharper picture of performance than the energy balance sheet of Example 2.4, which does not explicitly consider the effect of irreversibilities within the system.

SKILLS DEVELOPED

Ability to...

• apply the exergy rate balance.

• develop an exergy accounting.

Quick Quiz

By inspection of the exergy balance sheet, specify an exergy-based efficiency for the gearbox. Ans. 98%.

7.5 Exergy Rate Balance for Control Volumes at Steady State

In this section, the exergy balance is extended to a form applicable to control volumes at steady state. The control volume form is generally the most useful for engineering analysis.

The exergy rate balance for a control volume can be derived using an approach like that employed in the box of Sec. 4.1, where the control volume form of the mass rate balance is obtained by transforming the closed system form. However, as in the developments of the energy and entropy rate balances for control volumes (Secs. 4.4.1 and 6.9, respectively), the present derivation is conducted less formally by modifying the closed system rate form, Eq. 7.10, to account for the exergy transfers at the inlets and exits. The result is

$$\frac{dE_{cv}}{dt} = \sum_j \left(1 - \frac{T_0}{T_j}\right)\dot{Q}_j - \left(\dot{W}_{cv} - p_0\frac{dV_{cv}}{dt}\right) + \underline{\sum_i \dot{m}_i e_{fi} - \sum_e \dot{m}_e e_{fe}} - \dot{E}_d$$

where the underlined terms account for exergy transfer where mass enters and exits the control volume, respectively.

At steady state, $dE_{cv}/dt = dV_{cv}/dt = 0$, giving the **steady-state exergy rate balance**

steady-state exergy rate balance: control volumes

$$0 = \sum_j \left(1 - \frac{T_0}{T_j}\right)\dot{Q}_j - \dot{W}_{cv} + \sum_i \dot{m}_i e_{fi} - \sum_e \dot{m}_e e_{fe} - \dot{E}_d \qquad (7.13a)$$

where e_{fi} accounts for the exergy per unit of mass entering at inlet i and e_{fe} accounts for the exergy per unit of mass exiting at exit e. These terms, known as the **specific flow exergy**, are expressed as

specific flow exergy

$$e_f = h - h_0 - T_0(s - s_0) + \frac{V^2}{2} + gz \qquad (7.14)$$

where h and s represent the specific enthalpy and entropy, respectively, at the inlet or exit under consideration; h_0 and s_0 represent the respective values of these properties when evaluated at T_0, p_0. See the box for a derivation of Eq. 7.14 and discussion of the flow exergy concept.

Conceptualizing Specific Flow Exergy

To evaluate the exergy associated with a flowing stream of matter at a state given by h, s, V, and z, let us think of the stream being fed to the control volume operating at steady state shown in Fig. 7.5. At the exit of the control volume, the respective properties are those corresponding to the dead state: h_0, s_0, $V_0 = 0$, $z_0 = 0$. Heat transfer occurs only with the environment at $T_b = T_0$.

For the control volume of Fig. 7.5, energy and entropy balances read, respectively,

$$0 = \dot{Q}_{cv} - \dot{W}_{cv} + \dot{m}\left[(h - h_0) + \frac{V^2 - V_0^2}{2} + g(z - z_0)\right] \quad \text{(a)}$$

$$0 = \frac{\dot{Q}_{cv}}{T_0} + \dot{m}(s - s_0) + \dot{\sigma}_{cv} \quad \text{(b)}$$

Eliminating $\dot{Q}_{cv}$ between Eqs. (a) and (b), the work developed per unit of mass flowing is

$$\frac{\dot{W}_{cv}}{\dot{m}} = \underline{\left[(h - h_0) - T_0(s - s_0) + \frac{V^2}{2} + gz\right]} - T_0\left(\frac{\dot{\sigma}_{cv}}{\dot{m}}\right) \quad \text{(c)}$$

The value of the underlined term in Eq. (c) is determined by two states: the given state and the dead state. However, the value of the entropy production term, which cannot be negative, depends on the nature of the flow. Hence, the maximum theoretical work that could be developed, per unit of mass flowing, corresponds to a zero value for the entropy production—that is,

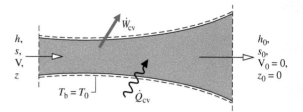

Fig. 7.5 Control volume used to evaluate the specific flow exergy of a stream.

when the flow through the control volume of Fig. 7.5 is internally reversible. The specific flow exergy, e_f, is this work value, and thus Eq. 7.14 is seen to be the appropriate expression for the specific flow exergy.

Subtracting Eq. 7.2 from Eq. 7.14 gives the following relationship between specific flow exergy e_f and specific exergy e,

$$e_f = e + \upsilon(p - p_0) \quad \text{(d)}$$

The underlined term of Eq. (d) has the significance of exergy transfer accompanying *flow work*. Thus, at a control volume inlet or exit, flow exergy e_f accounts for the sum of the exergy accompanying mass flow e and the exergy accompanying flow work. When pressure p at a control volume inlet or exit is less than the dead state pressure p_0, the flow work contribution of Eq. (d) is negative, signaling that the exergy transfer accompanying flow work is opposite to the direction of exergy transfer accompanying mass flow. Flow exergy aspects are also explored in end-of-chapter Problem 7.8.

The steady-state exergy rate balance, Eq. 7.13a, can be expressed more compactly as Eq. 7.13b

$$0 = \sum_j \dot{E}_{qj} - \dot{W}_{cv} + \sum_i \dot{E}_{fi} - \sum_e \dot{E}_{fe} - \dot{E}_d \quad (7.13b)$$

where

$$\dot{E}_{qj} = \left(1 - \frac{T_0}{T_j}\right)\dot{Q}_j \quad (7.15)$$

$$\dot{E}_{fi} = \dot{m}_i e_{fi} \quad (7.16a)$$

$$\dot{E}_{fe} = \dot{m}_e e_{fe} \quad (7.16b)$$

are exergy transfer rates. Equation 7.15 has the same interpretation as given for Eq 7.5 in the box in Sec. 7.4.1, only on a time rate basis. Also note that at steady state the rate of exergy transfer accompanying the power $\dot{W}_{cv}$ is simply the power. Finally, the rate of exergy destruction within the control volume $\dot{E}_d$ is related to the rate of entropy production by $T_0\dot{\sigma}_{cv}$.

If there is a single inlet and a single exit, denoted by 1 and 2, respectively, the steady-state exergy rate balance, Eq. 7.13a, reduces to

$$0 = \sum_j \left(1 - \frac{T_0}{T_j}\right)\dot{Q}_j - \dot{W}_{cv} + \dot{m}(e_{f1} - e_{f2}) - \dot{E}_d \quad (7.17)$$

TAKE NOTE...

Observe that the approach used here to evaluate flow exergy parallels that used in Sec. 7.3 to evaluate exergy of a system. In each case, energy and entropy balances are applied to evaluate maximum theoretical work in the limit as entropy production tends to zero. This approach is also used in Sec. 13.6 to evaluate chemical exergy.

TAKE NOTE...

When the rate of exergy destruction $\dot{E}_d$ is the objective, it can be determined either from an exergy rate balance or from $\dot{E}_d = T_0\dot{\sigma}_{cv}$, where $\dot{\sigma}_{cv}$ is the rate of entropy production evaluated from an entropy rate balance. The second of these procedures normally requires fewer property evaluations and less computation.

where $\dot{m}$ is the mass flow rate. The term $(e_{f1} - e_{f2})$ is evaluated using Eq. 7.14 as

$$e_{f1} - e_{f2} = (h_1 - h_2) - T_0(s_1 - s_2) + \frac{V_1^2 - V_2^2}{2} + g(z_1 - z_2) \qquad (7.18)$$

7.5.1 Comparing Energy and Exergy for Control Volumes at Steady State

Although energy and exergy share common units and exergy transfer accompanies energy transfer, energy and exergy are *fundamentally different* concepts. Energy and exergy relate, respectively, to the first and second laws of thermodynamics:

- Energy is *conserved*. Exergy is *destroyed* by irreversibilities.
- Exergy expresses energy transfer by work, heat, and mass flow in terms of a *common measure*—namely, work that is *fully available* for lifting a weight or, equivalently, as shaft or electrical work.

FOR EXAMPLE

Figure 7.6a shows energy transfer rates for a one-inlet, one-exit control volume at steady state. This includes energy transfers by work and heat and the energy transfers in and out where mass flows across the boundary. **Figure 7.6b** shows the same control volume but now labeled with exergy transfer rates. Be sure to note that the *magnitudes* of the exergy transfers accompanying heat transfer and mass flow *differ* from the corresponding energy transfer magnitudes. These exergy transfer rates are calculated using Eqs. 7.15 and 7.16, respectively. At steady state, the rate of exergy transfer accompanying the power $\dot{W}_{cv}$ is simply the power. In accord with the conservation of energy principle, the total rate energy enters the control volume *equals* the total rate energy exits. However, the total rate exergy enters the control volume *exceeds* the total rate exergy exits. The difference between these exergy values is the rate at which exergy is destroyed by irreversibilities, in accord with the second law.

To summarize, exergy gives a sharper picture of performance than energy because exergy expresses all energy transfers on a common basis and accounts explicitly for the effect of irreversibilities through the exergy destruction concept.

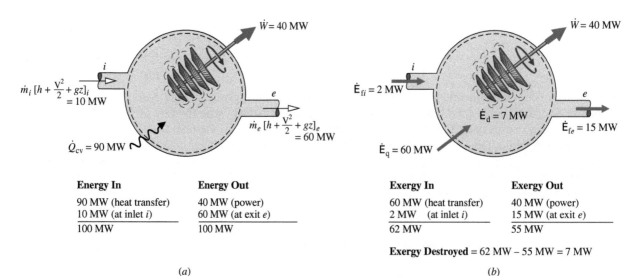

Energy In	Energy Out
90 MW (heat transfer)	40 MW (power)
10 MW (at inlet i)	60 MW (at exit e)
100 MW	100 MW

Exergy In	Exergy Out
60 MW (heat transfer)	40 MW (power)
2 MW (at inlet i)	15 MW (at exit e)
62 MW	55 MW

Exergy Destroyed = 62 MW – 55 MW = 7 MW

(a)   (b)

Fig. 7.6 Comparing energy and exergy for a control volume at steady state. (a) Energy analysis. (b) Exergy analysis.

7.5.2 Evaluating Exergy Destruction in Control Volumes at Steady State

The following examples illustrate the use of mass, energy, and exergy rate balances for the evaluation of exergy destruction in control volumes at steady state. Property data also play an important role in arriving at solutions. The first example involves the expansion of steam through a valve (a throttling process, Sec. 4.10). From an energy perspective, the expansion occurs without loss. Yet, as shown in Example 7.5, such a valve is a site of inefficiency quantified thermodynamically in terms of exergy destruction.

▸▸▸ **EXAMPLE 7.5** ▸ ·

Exergy Destruction in a Throttling Valve

Superheated water vapor enters a valve at 3.0 MPa, 320°C and exits at a pressure of 0.5 MPa. The expansion is a throttling process. Determine the exergy destruction per unit of mass flowing, in kJ/kg. Let $T_0 = 25$°C, $p_0 = 1$ bar.

SOLUTION

Known Water vapor expands in a throttling process through a valve from a specified inlet state to a specified exit pressure.

Find Determine the exergy destruction per unit of mass flowing.

Schematic and Given Data:

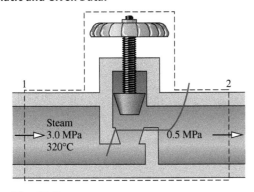

Fig. E7.5

Engineering Model

1. The control volume shown in the accompanying figure is at steady state.

2. For the throttling process, $\dot{Q}_{cv} = \dot{W}_{cv} = 0$, and the effects of motion and gravity can be ignored.

3. $T_0 = 25$°C, $p_0 = 1$ bar.

Analysis The state at the inlet is specified. The state at the exit can be fixed by reducing the steady-state mass and energy rate balances to obtain

$$h_2 = h_1 \qquad (a)$$

Thus, the exit state is fixed by p_2 and h_2. From Table A-4, $h_1 = 3043.4$ kJ/s, $s_1 = 6.6245$ kJ/kg · K. Interpolating at a pressure of 0.5 MPa with $h_2 = h_1$, the specific entropy at the exit is $s_2 = 7.4223$ kJ/kg · K.

With assumptions listed, the steady-state form of the exergy rate balance, Eq. 7.17, reduces to

$$0 = \sum_j \left(1 - \frac{T_0}{T_j}\right)^{\!0} \dot{Q}_j - \dot{W}_{cv}^{\,0} + \dot{m}(e_{f1} - e_{f2}) - \dot{E}_d$$

Dividing by the mass flow rate $\dot{m}$ and solving, the exergy destruction per unit of mass flowing is

$$\frac{\dot{E}_d}{\dot{m}} = (e_{f1} - e_{f2}) \qquad (b)$$

Introducing Eq. 7.18, using Eq. (a), and ignoring the effects of motion and gravity

$$e_{f1} - e_{f2} = (h_1 \overset{0}{\cancel{-} h_2}) - T_0(s_1 - s_2) + \frac{V_1^2 - V_2^{2\,0}}{2} + g(z_1 \overset{0}{\cancel{-} z_2})$$

Eq. (b) becomes

❶
$$\frac{\dot{E}_d}{\dot{m}} = T_0(s_2 - s_1)$$

Inserting values,

❷
$$\frac{\dot{E}_d}{\dot{m}} = 25°C\ (7.4223 - 6.6245)\frac{kJ}{kg \cdot K} = 237.7\ kJ/kg$$

❶ Equation (c) can be obtained alternatively beginning with the relationship $\dot{E}_d = T_0\dot{\sigma}_{cv}$ and then evaluating the rate of entropy production $\dot{\sigma}_{cv}$ from an entropy balance. The details are left as an exercise.

❷ Energy is conserved in the throttling process, but exergy is destroyed. The source of the exergy destruction is the uncontrolled expansion that occurs.

SKILLS DEVELOPED

Ability to...

• apply the energy and exergy rate balances.

• evaluate exergy destruction.

Quick Quiz

If air modeled as an ideal gas were to undergo a throttling process, evaluate the exergy destruction, in kJ per kg of air flowing, for the same inlet conditions and exit pressure as in this example. Ans. 157.5 kJ/kg.

Although heat exchangers appear from an energy perspective to operate without loss when stray heat transfer is ignored, they are a site of thermodynamic inefficiency quantified by exergy destruction. This is illustrated in Example 7.6.

EXAMPLE 7.6

Evaluating Exergy Destruction in a Heat Exchanger

① Compressed air enters a counterflow heat exchanger operating at steady state at 610 K, 10 bar and exits at 860 K, 9.7 bar. Hot combustion gas enters as a separate stream at 1020 K, 1.1 bar and exits at 1 bar. Each stream has a mass flow rate of 90 kg/s. Heat transfer between the outer surface of the heat exchanger and the surroundings can be ignored. The effects of motion and gravity are negligible. Assuming the combustion gas stream has the properties of air, and using the ideal gas model for both streams, determine for the heat exchanger

a. the exit temperature of the combustion gas, in K.

b. the net change in the flow exergy rate from inlet to exit of each stream, in MW.

c. the rate exergy is destroyed, in MW.

Let $T_0 = 300$ K, $p_0 = 1$ bar.

SOLUTION

Known Steady-state operating data are provided for a counterflow heat exchanger.

Find For the heat exchanger, determine the exit temperature of the combustion gas, the change in the flow exergy rate from inlet to exit of each stream, and the rate exergy is destroyed.

Schematic and Given Data:

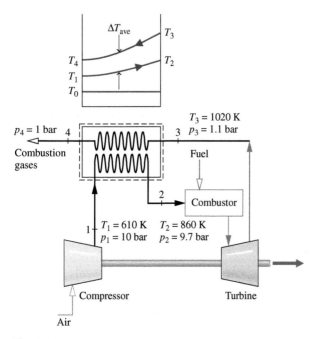

Fig. E7.6

Engineering Model

1. The control volume shown in the accompanying figure is at steady state.

2. For the control volume, $\dot{Q}_{cv} = 0$, $\dot{W}_{cv} = 0$, and the effects of motion and gravity are negligible.

3. Each stream has the properties of air modeled as an ideal gas.

4. $T_0 = 300$ K, $p_0 = 1$ bar.

Analysis

a. The temperature T_4 of the exiting combustion gases can be found by reducing the mass and energy rate balances for the control volume at steady state to obtain

$$0 = \underline{\dot{Q}_{cv}} - \underline{\dot{W}_{cv}} + \dot{m}\left[(h_1 - h_2) + \left(\frac{V_1^2 - V_2^2}{2}\right) + \underline{g(z_1 - z_2)}\right]$$
$$+ \dot{m}\left[(h_3 - h_4) + \left(\frac{V_3^2 - V_4^2}{2}\right) + \underline{g(z_3 - z_4)}\right]$$

where $\dot{m}$ is the common mass flow rate of the two streams. The underlined terms drop out by listed assumptions, leaving

$$0 = \dot{m}(h_1 - h_2) + \dot{m}(h_3 - h_4)$$

Dividing by $\dot{m}$ and solving for h_4

$$h_4 = h_3 + h_1 - h_2$$

From Table A-22, $h_1 = 617.53$ kJ/kg, $h_2 = 888.27$ kJ/kg, $h_3 = 1068.89$ kJ/kg. Inserting values

$$h_4 = 1068.89 + 617.53 - 888.27 = 798.15 \text{ kJ/kg}$$

Interpolating in Table A-22 gives $T_4 = 778$ K (505°C).

b. The net change in the flow exergy rate from inlet to exit for the air stream flowing from 1 to 2 can be evaluated using Eq. 7.18, neglecting the effects of motion and gravity. With Eq. 6.20a and data from Table A-22

$$\dot{m}(e_{f2} - e_{f1}) = \dot{m}[(h_2 - h_1) - T_0(s_2 - s_1)]$$
$$= \dot{m}\left[(h_2 - h_1) - T_0\left(s_2^\circ - s_1^\circ - R\ln\frac{p_2}{p_1}\right)\right]$$
$$= 90\frac{\text{kg}}{\text{s}}\left[(888.27 - 617.53)\frac{\text{kJ}}{\text{kg}} - 300 \text{ K}\right.$$
$$\left. \times\left(2.79783 - 2.42644 - \frac{8.314}{28.97}\ln\frac{9.7}{10}\right)\frac{\text{kJ}}{\text{kg}\cdot\text{K}}\right]$$
$$= 14{,}103\frac{\text{kJ}}{\text{s}}\left|\frac{1 \text{ MW}}{10^3 \text{ kJ/s}}\right| = 14.1 \text{ MW}$$

② As the air flows from 1 to 2, its temperature *increases* relative to T_0 and the flow exergy *increases*.

Similarly, the change in the flow exergy rate from inlet to exit for the combustion gas is

$$\dot{m}(e_{f4} - e_{f3}) = \dot{m}\left[(h_4 - h_3) - T_0\left(s_4^\circ - s_3^\circ - R\ln\frac{p_4}{p_3}\right)\right]$$

$$= 90\left[(798.15 - 1068.89)\right.$$

$$\left. - 300\left(2.68769 - 2.99034 - \frac{8.314}{28.97}\ln\frac{1}{1.1}\right)\right]$$

$$= -16{,}934\,\frac{kJ}{s}\left|\frac{1\,MW}{10^3\ kJ/s}\right| = -16.93\ MW$$

As the combustion gas flows from 3 to 4, its temperature *decreases* relative to T_0 and the flow exergy *decreases*.

❸ **c.** The rate of exergy destruction within the control volume can be determined from an exergy rate balance, Eq. 7.13a,

$$0 = \sum_j\left(1 - \frac{T_0}{T_j}\right)\cancel{\dot{Q}_j}^{\,0} - \cancel{\dot{W}_{cv}}^{\,0} + \dot{m}(e_{f1} - e_{f2}) + \dot{m}(e_{f3} - e_{f4}) - \dot{E}_d$$

Solving for $\dot{E}_d$ and inserting known values

$$\dot{E}_d = \dot{m}(e_{f1} - e_{f2}) + \dot{m}(e_{f3} - e_{f4})$$

❹
$$= (-14.1\ MW) + (16.93\ MW) = 2.83\ MW$$

Comparing results, the exergy increase of the compressed air: 14.1 MW is less than the magnitude of the exergy decrease of the combustion gas: 16.93 MW, even though the energy changes of the two streams are equal in magnitude. The difference in these exergy values is the exergy destroyed: 2.83 MW. Thus, energy is conserved but exergy is not.

---

❶ Heat exchangers of this type are known as *regenerators* (see Sec. 9.7).

❷ The variation in temperature of each stream passing through the heat exchanger is sketched in the schematic. The dead state temperature T_0 also is shown on the schematic for reference.

❸ Alternatively, the rate of exergy destruction can be determined using $\dot{E}_d = T_0\dot{\sigma}_{cv}$, where $\dot{\sigma}_{cv}$ is the rate of entropy production evaluated from an entropy rate balance. This is left as an exercise.

❹ Exergy is destroyed by irreversibilities associated with fluid friction and stream-to-stream heat transfer. The pressure drops for the streams are indicators of frictional irreversibility. The average temperature difference between the streams, ΔT_{ave}, is an indicator of heat transfer irreversibility.

SKILLS DEVELOPED

Ability to…

- apply the energy and exergy rate balances.
- evaluate exergy destruction.

Quick Quiz

If the mass flow rate of each stream were 105 kg/s, what would be the rate of exergy destruction, in MW? Ans. 3.3 MW.

In previous discussions we have noted the effect of irreversibilities on *thermodynamic* performance. Some *economic* consequences of irreversibilities are considered in the next example.

> ▶▶ **EXAMPLE 7.7** ▶ . . . .

Determining Cost of Exergy Destruction

For the heat pump of Examples 6.8 and 6.14, determine the exergy destruction rates, each in kW, for the compressor, condenser, and throttling valve. If exergy is valued at \$0.08 per kW · h, determine the daily cost of electricity to operate the compressor and the daily cost of exergy destruction in each component. Let $T_0 = 273$ K (0°C), which corresponds to the temperature of the outside air.

SOLUTION

Known Refrigerant 22 is compressed adiabatically, condensed by heat transfer to air passing through a heat exchanger, and then expanded through a throttling valve. Data for the refrigerant and air are known.

Find Determine the daily cost to operate the compressor. Also determine the exergy destruction rates and associated daily costs for the compressor, condenser, and throttling valve.

Schematic and Given Data: See Examples 6.8 and 6.14.

Engineering Model

1. See Examples 6.8 and 6.14.

2. $T_0 = 273$ K (0°C).

Analysis The rates of exergy destruction can be calculated using

$$\dot{E}_d = T_0\dot{\sigma}$$

together with data for the entropy production rates from Example 6.8. That is,

$$(\dot{E}_d)_{comp} = (273\ K)(17.5 \times 10^{-4})\left(\frac{kW}{K}\right) = 0.478\ kW$$

$$(\dot{E}_d)_{valve} = (273)(9.94 \times 10^{-4}) = 0.271\ kW$$

$$(\dot{E}_d)_{cond} = (273)(7.95 \times 10^{-4}) = 0.217\ kW$$

The costs of exergy destruction are, respectively,

$$\left(\begin{array}{c}\text{daily cost of exergy destruction due}\\\text{to compressor irreversibilities}\end{array}\right) = (0.478\ kW)$$

$$\times\left(\frac{\$0.08}{kW\cdot h}\right)\left|\frac{24\ h}{day}\right|$$

$$= \$0.92$$

① $\begin{pmatrix} \text{daily cost of exergy destruction due to} \\ \text{irreversibilities in the throttling value} \end{pmatrix} = (0.271)(0.08)|24|$

$$= \$0.52$$

$\begin{pmatrix} \text{daily cost of exergy destruction due to} \\ \text{irreversibilities in the condenser} \end{pmatrix} = (0.217)(0.08)|24|$

$$= \$0.42$$

From the solution to Example 6.14, the magnitude of the compressor power is 3.11 kW. Thus, the daily cost is

$$\begin{pmatrix} \text{daily cost of electricity} \\ \text{to operate compressor} \end{pmatrix} = (3.11 \text{ kW}) \left(\frac{\$0.08}{\text{kW} \cdot \text{h}}\right) \left|\frac{24 \text{ h}}{\text{day}}\right| = \$5.97$$

① Associating exergy destruction with operating costs provides a rational basis for seeking cost-effective design improvements.

Although it may be possible to select components that would destroy less exergy, the trade-off between any resulting reduction in operating cost and the potential increase in equipment cost must be carefully considered.

SKILLS DEVELOPED

Ability to...

- evaluate exergy destruction.
- conduct an elementary economic evaluation using exergy.

Quick Quiz

Expressed as a percent, how much of the cost of electricity to operate the compressor is attributable to exergy destruction in the three components? Ans. 31%.

7.5.3 Exergy Accounting in Control Volumes at Steady State

For a control volume, the location, types, and true magnitudes of inefficiency and loss can be pinpointed by systematically evaluating and comparing the various terms of the exergy balance for the control volume. This is an extension of *exergy accounting*, introduced in Sec. 7.4.4.

The next two examples provide illustrations of exergy accounting in control volumes. The first involves the steam turbine with stray heat transfer considered previously in Example 6.6, which should be quickly reviewed before studying the current example.

▶▶▶ EXAMPLE 7.8 ▶

Exergy Accounting of a Steam Turbine

Steam enters a turbine with a pressure of 30 bar, a temperature of 400°C, and a velocity of 160 m/s. Steam exits as saturated vapor at 100°C with a velocity of 100 m/s. At steady state, the turbine develops work at a rate of 540 kJ per kg of steam flowing through the turbine. Heat transfer between the turbine and its surroundings occurs at an average outer surface temperature of 350 K. Develop a full accounting of the *net exergy carried in* by the steam, in kJ per unit mass of steam flowing. Let $T_0 = 25°C$, $p_0 = 1$ atm.

SOLUTION

Known Steam expands through a turbine for which steady-state data are provided.

Find Develop a full exergy accounting of the *net exergy carried in* by the steam, in kJ per unit mass of steam flowing.

Schematic and Given Data: See Fig. E6.6. From Example 6.6, $\dot{W}_{cv}/\dot{m} = 540$ kJ/kg, $\dot{Q}_{cv}/\dot{m} = -22.6$ kJ/kg.

Engineering Model

1. See the solution to Example 6.6.

2. $T_0 = 25°C$, $p_0 = 1$ atm.

Analysis The *net exergy carried in* per unit mass of steam flowing is obtained using Eq. 7.18:

$$e_{f1} - e_{f2} = (h_1 - h_2) - T_0(s_1 - s_2) + \left(\frac{V_1^2 - V_2^2}{2}\right) + g(z_1 - z_2)^0$$

From Table A-4, $h_1 = 3230.9$ kJ/kg, $s_1 = 6.9212$ kJ/kg · K. From Table A-2, $h_2 = 2676.1$ kJ/kg, $s_2 = 7.3549$ kJ/kg · K. Hence,

$$e_{f1} - e_{f2} = \left[(3230.9 - 2676.1)\frac{\text{kJ}}{\text{kg}} - 298(6.9212 - 7.3549)\frac{\text{kJ}}{\text{kg}}\right.$$
$$\left. + \left[\frac{(160)^2 - (100)^2}{2}\right]\left(\frac{\text{m}}{\text{s}}\right)^2 \left|\frac{1 \text{ N}}{1 \text{ kg} \cdot \text{m/s}^2}\right| \left|\frac{1 \text{ kJ}}{10^3 \text{ N} \cdot \text{m}}\right|\right]$$

$$= 691.84 \text{ kJ/kg}$$

The net exergy carried in can be accounted for in terms of the exergy transfers accompanying work and heat transfer and the exergy destruction within the control volume. At steady state, the exergy transfer accompanying work is simply the work, or $\dot{W}_{cv}/\dot{m} = 540$ kJ/kg. The quantity $\dot{Q}_{cv}/\dot{m}$ is evaluated in the solution to Example 6.6 using the steady-state forms of the mass and energy rate balances: $\dot{Q}_{cv}/\dot{m} = -22.6$ kJ/kg. The accompanying exergy transfer is

$$\frac{\dot{E}_q}{\dot{m}} = \left(1 - \frac{T_0}{T_b}\right)\left(\frac{\dot{Q}_{cv}}{\dot{m}}\right)$$

$$= \left(1 - \frac{298}{350}\right)\left(-22.6\frac{\text{kJ}}{\text{kg}}\right)$$

$$= -3.36\frac{\text{kJ}}{\text{kg}}$$

where T_b denotes the temperature on the boundary where heat transfer occurs.

The exergy destruction can be determined by rearranging the steady-state form of the exergy rate balance, Eq. 7.17, to give

❶
$$\frac{\dot{E}_d}{\dot{m}} = \left(1 - \frac{T_0}{T_b}\right)\left(\frac{\dot{Q}_{cv}}{\dot{m}}\right) - \frac{\dot{W}_{cv}}{\dot{m}} + (e_{f1} - e_{f2})$$

Substituting values

$$\frac{\dot{E}_d}{\dot{m}} = -3.36 - 540 + 691.84 = 148.48 \text{ kJ/kg}$$

The analysis is summarized by the following exergy *balance sheet* in terms of exergy magnitudes on a rate basis:

Net rate of exergy in:	691.84 kJ/kg (100%)
Disposition of the exergy:	
• Rate of exergy out	
work	540.00 kJ/kg (78.05%)
heat transfer	3.36 kJ/kg (0.49%)
• Rate of exergy destruction	148.48 kJ/kg (21.46%)
	691.84 kJ/kg (100%)

Note that the exergy transfer accompanying heat transfer is small relative to the other terms.

---

❶ The exergy destruction can be determined alternatively using $\dot{E}_d = T_0\dot{\sigma}_{cv}$, where $\dot{\sigma}_{cv}$ is the rate of entropy production from an entropy balance. The solution to Example 6.6 provides $\dot{\sigma}_{cv}/\dot{m} = 0.4983$ kJ/kg · K.

SKILLS DEVELOPED

Ability to...

• evaluate exergy quantities for an exergy accounting.

• develop an exergy accounting.

Quick Quiz

By inspection of the exergy balance sheet, specify an exergy-based efficiency for the turbine. Ans. 78.05%.

The next example illustrates the use of exergy accounting to identify opportunities for improving thermodynamic performance of the waste heat recovery system considered in Example 4.10, which should be quickly reviewed before studying the current example.

▶▶ EXAMPLE 7.9 ▶

Exergy Accounting of a Waste Heat Recovery System

Suppose the system of Example 4.10 is one option under consideration for utilizing the combustion products discharged from an industrial process.

a. Develop a full accounting of the *net* exergy carried in by the combustion products.

b. Discuss the design implications of the results.

SOLUTION

Known Steady-state operating data are provided for a heat-recovery steam generator and a turbine.

Find Develop a full accounting of the *net* rate exergy is carried in by the combustion products and discuss the implications for design.

Schematic and Given Data:

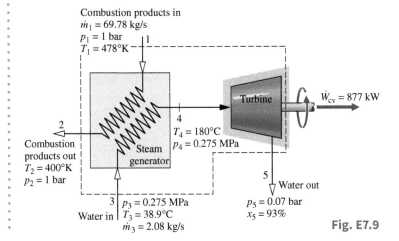

Combustion products in
$\dot{m}_1 = 69.78$ kg/s
$p_1 = 1$ bar
$T_1 = 478°K$

Combustion products out
$T_2 = 400°K$
$p_2 = 1$ bar

$T_4 = 180°C$
$p_4 = 0.275$ MPa

Turbine $\dot{W}_{cv} = 877$ kW

Steam generator

3 $p_3 = 0.275$ MPa
Water in $T_3 = 38.9°C$
$\dot{m}_3 = 2.08$ kg/s

5 Water out
$p_5 = 0.07$ bar
$x_5 = 93\%$

Fig. E7.9

Engineering Model

1. See solution to Example 4.10.

2. $T_0 = 298°K$.

Analysis

a. We begin by determining the *net rate exergy is carried into* the control volume. Modeling the combustion products as an ideal gas, the net rate is determined using Eq. 7.18 together with Eq. 6.20a as

$$\dot{m}_1[e_{f1} - e_{f2}] = \dot{m}_1[h_1 - h_2 - T_0(s_1 - s_2)]$$

$$= \dot{m}_1\left[h_1 - h_2 - T_0\left(s_1^\circ - s_2^\circ - R\ln\frac{p_1}{p_2}\right)\right]$$

With data from Table A-22, $h_1 = 480.35$ kJ/kg, $h_2 = 400.97$ kJ/kg, $s_1^\circ = 2.173$ kJ/kg · °K, $s_2^\circ = 1.992$ kJ/kg · °K, and $p_2 = p_1$, we have

$$\dot{m}_1[e_{f1} - e_{f2}] = 69.8 \text{ kg/s}\left[(480.35 - 400.97)\frac{\text{kJ}}{\text{kg}}\right.$$

$$\left. - 298°K(2.173 - 1.992)\frac{\text{kJ}}{\text{kg} \cdot °C}\right]\frac{\text{kJ}}{\text{s}}$$

$$= 1775.78 \text{ kJ/s}$$

Next, we determine the rate exergy is carried *out* of the control volume. Exergy is carried out of the control volume by work at a rate of 876.8 kJ/s, as shown on the schematic. Additionally, the *net* rate exergy is carried *out* by the water stream is

$$\dot{m}_3[e_{f5} - e_{f3}] = \dot{m}_3[h_5 - h_3 - T_0(s_5 - s_3)]$$

From Table A-2, $h_3 \approx h_f(39°C) = 162.82$ Kg/s, $s_3 \approx s_f(39°C) = 0.5598$ kJ/kg · °K . Using saturation data at 0.07 bar from Table A-3 with $x_5 = 0.93$ gives $h_5 = 2403.27$ kJ/kg and $s_5 = 7.739$ kJ/kg · °K. Substituting values

$$\dot{m}_3[e_{f5} - e_{f3}] = 2.08 \frac{\text{kg}}{\text{s}}\left[(2403.27 - 162.82)\frac{\text{kJ}}{\text{kg}}\right.$$
$$\left. - 298(7.739 - 0.5598)\frac{\text{kJ}}{\text{kg} \cdot °\text{K}}\right]$$
$$= 209.66 \text{ kJ/s}$$

Next, the rate exergy is destroyed in the heat-recovery steam generator can be obtained from an exergy rate balance applied to a control volume enclosing the steam generator. That is, Eq. 7.13a takes the form

$$0 = \sum_j \left(1 - \frac{\cancel{T_0}^{\,0}}{T_j}\right)\dot{Q}_j - \cancel{\dot{W}_{cv}}^{\,0} + \dot{m}_1(e_{f1} - e_{f2}) + \dot{m}_3(e_{f3} - e_{f4}) - \dot{E}_d$$

Evaluating $(e_{f3} - e_{f4})$ with Eq. 7.18 and solving for $\dot{E}_d$

$$\dot{E}_d = \dot{m}_1(e_{f1} - e_{f2}) + \dot{m}_3[h_3 - h_4 - T_0(s_3 - s_4)]$$

The first term on the right is evaluated above. Then, with $h_4 = 2825$ kJ/kg, $s_4 = 7.2196$ kJ/kg · °K at 180°C, 0.275 MPa from Table A-4, and previously determined values for h_3 and s_3

$$\dot{E}_d = 1775.78 \frac{\text{kJ}}{\text{s}} + 2.08 \frac{\text{kg}}{\text{s}}\left[(162 - 2825)\frac{\text{kJ}}{\text{kg}}\right.$$
$$\left. - 298(0.559 - 7.2196)\frac{\text{kJ}}{\text{kg} \cdot °\text{K}}\right]$$
$$= 366.1 \text{ kJ/s}$$

Finally, the rate exergy is destroyed in the turbine can be obtained from an exergy rate balance applied to a control volume enclosing the turbine. That is, Eq. 7.17 takes the form

$$0 = \sum_j \left(1 - \frac{\cancel{T_0}^{\,0}}{T_j}\right)\dot{Q}_j - \dot{W}_{cv} + \dot{m}_4(e_{f4} - e_{f5}) - \dot{E}_d$$

Solving for $\dot{E}_d$, evaluating $(e_{f4} - e_{f5})$ with Eq. 7.18, and using previously determined values

$$\dot{E}_d = -\dot{W}_{cv} + \dot{m}_4[h_4 - h_5 - T_0(s_4 - s_5)]$$

❶
$$= -876.8 \frac{\text{kJ}}{\text{s}} + 2.08 \frac{\text{kg}}{\text{s}}\left[(2885 - 2403)\frac{\text{kJ}}{\text{kg}}\right.$$
$$\left. - 298°\text{K}(7.2196 - 7.739)\frac{\text{kJ}}{\text{kg} \cdot °\text{C}}\right]$$
$$= 320.2 \text{ kJ/s}$$

The analysis is summarized by the following exergy *balance sheet* in terms of exergy magnitudes on a rate basis:

Net rate of exergy in:	1772.8	kJ/s (100%)
Disposition of the exergy:		
• Rate of exergy out		
power developed	876.8	kJ/s (49.5%)
water stream	209.66	kJ/s (11.8%)
• Rate of exergy destruction		
heat-recovery steam		
generator	366.12	kJ/s (22.6%)
turbine	320.2	kJ/s (18%)
	1772.8	kJ/s (100%)

b. The exergy balance sheet suggests an opportunity for improved *thermodynamic* performance because over 50% of the net exergy carried in is either destroyed by irreversibilities or carried out by the water stream. Better thermodynamic performance might be achieved by modifying the design. For example, we might reduce the heat transfer irreversibility by specifying a heat-recovery steam generator with a smaller stream-to-stream temperature difference, and/or reduce friction by specifying a turbine with a higher isentropic efficiency. Thermodynamic performance alone would not determine the *preferred* option, however, for other factors such as cost must be considered, and can be overriding. Further discussion of the use of exergy analysis in design is provided in Sec. 7.7.1.

---

❶ Alternatively, the rates of exergy destruction in control volumes enclosing the heat-recovery steam generator and turbine can be determined using $\dot{E}_d = T_0\dot{\sigma}_{cv}$, where $\dot{\sigma}_{cv}$ is the rate of entropy production for the respective control volume evaluated from an entropy rate balance. This is left as an exercise.

SKILLS DEVELOPED

Ability to…

• evaluate exergy quantities for an exergy accounting.
• develop an exergy accounting.

Quick Quiz

For the turbine of the waste heat recovery system, evaluate the isentropic turbine efficiency and comment. Ans. 74%. This isentropic turbine efficiency value is at the low end of the range for steam turbines today, indicating scope for improved performance of the heat recovery system.

7.6 Exergetic (Second Law) Efficiency

The objective of this section is to show the use of the exergy concept in assessing the effectiveness of energy resource utilization. As part of the presentation, the **exergetic efficiency** concept is introduced and illustrated. Such efficiencies are also known as *second law* efficiencies.

exergetic efficiency

7.6.1 Matching End Use to Source

Tasks such as space heating, heating in industrial furnaces, and process steam generation commonly involve the combustion of coal, oil, or natural gas. When the products of combustion are at a temperature significantly greater than required by a given task, the end use is not well matched to the source and the result is inefficient use of the fuel burned. To illustrate this simply, refer to **Fig. 7.7**, which shows a closed system receiving a heat transfer at the rate $\dot{Q}_s$ at a *source* temperature T_s and delivering $\dot{Q}_u$ at a *use* temperature T_u. Energy is lost to the surroundings by heat transfer at a rate $\dot{Q}_l$ across a portion of the surface at T_l. All energy transfers shown on the figure are in the directions indicated by the arrows.

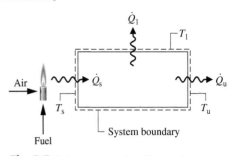

Fig. 7.7 Schematic used to discuss the efficient use of fuel.

Assuming that the system of Fig. 7.7 operates at steady state and there is no work, the closed system energy and exergy rate balances Eqs. 2.37 and 7.10 reduce, respectively, to

$$\frac{dE}{dt}^{\cancel{0}} = (\dot{Q}_s - \dot{Q}_u - \dot{Q}_l) - \cancel{\dot{W}}^0$$

$$\frac{dE}{dt}^{\cancel{0}} = \left[\left(1 - \frac{T_0}{T_s}\right)\dot{Q}_s - \left(1 - \frac{T_0}{T_u}\right)\dot{Q}_u - \left(1 - \frac{T_0}{T_l}\right)\dot{Q}_l\right] - \left[\cancel{\dot{W}}^0 - p_0\frac{dV}{dt}^{\cancel{0}}\right] - \dot{E}_d$$

These equations can be rewritten as follows

$$\dot{Q}_s = \dot{Q}_u + \dot{Q}_l \tag{7.19a}$$

$$\left(1 - \frac{T_0}{T_s}\right)\dot{Q}_s = \left(1 - \frac{T_0}{T_u}\right)\dot{Q}_u + \left(1 - \frac{T_0}{T_l}\right)\dot{Q}_l + \dot{E}_d \tag{7.19b}$$

Equation 7.19a indicates that the energy carried in by heat transfer, $\dot{Q}_s$, is either used, $\dot{Q}_u$, or lost to the surroundings, $\dot{Q}_l$. This can be described by an efficiency in terms of energy rates in the form product/input as

$$\eta = \frac{\dot{Q}_u}{\dot{Q}_s} \tag{7.20}$$

In principle, the value of η can be increased by applying insulation to reduce the loss. The limiting value, when $\dot{Q}_l = 0$, is $\eta = 1$ (100%).

Equation 7.19b shows that the exergy carried into the system accompanying the heat transfer $\dot{Q}_s$ is either transferred from the system accompanying the heat transfers $\dot{Q}_u$ and $\dot{Q}_l$ or destroyed by irreversibilities within the system. This can be described by an efficiency in terms of exergy rates in the form product/input as

$$\varepsilon = \frac{(1 - T_0/T_u)\dot{Q}_u}{(1 - T_0/T_s)\dot{Q}_s} \tag{7.21a}$$

Introducing Eq. 7.20 into Eq. 7.21a results in

$$\boxed{\varepsilon = \eta\left(\frac{1 - T_0/T_u}{1 - T_0/T_s}\right)} \tag{7.21b}$$

The parameter ε, defined with reference to the exergy concept, may be called an *exergetic* efficiency. Note that η and ε each gauge how effectively the input is converted to the product. The parameter η does this on an energy basis, whereas ε does it on an exergy basis. As discussed next, the value of ε is generally less than unity even when $\eta = 1$.

Equation 7.21b indicates that a value for η as close to unity as practical is important for proper utilization of the exergy transferred from the hot combustion gas to the system. However, this alone would not ensure effective utilization. The temperatures T_s and T_u are

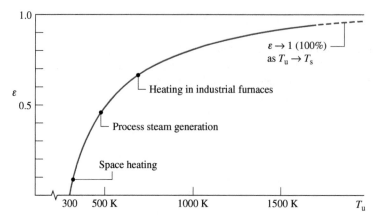

Fig. 7.8 Effect of use temperature T_u on the exergetic efficiency ε ($T_s = 2200$ K, $\eta = 100\%$).

also important, with exergy utilization improving as the use temperature T_u approaches the source temperature T_s. For proper utilization of exergy, therefore, it is desirable to have a value for η as close to unity as practical and also a good *match* between the source and use temperatures.

To emphasize further the central role of the use temperature, a graph of Eq. 7.21b is provided in **Fig. 7.8**. The figure gives the exergetic efficiency ε versus the use temperature T_u for an assumed source temperature $T_s = 2200$ K. Figure 7.8 shows that ε tends to unity (100%) as the use temperature approaches T_s. In most cases, however, the use temperature is substantially below T_s. Indicated on the graph are efficiencies for three applications: space heating at $T_u = 320$ K, process steam generation at $T_u = 480$ K, and heating in industrial furnaces at $T_u = 700$ K. These efficiency values suggest that fuel is used far more effectively in higher-temperature industrial applications than in lower-temperature space heating. The especially low exergetic efficiency for space heating reflects the fact that fuel is consumed to produce only slightly warm air, which from an exergy perspective has little utility. The efficiencies given on Fig. 7.8 are actually on the *high* side, for in constructing the figure we have assumed η to be unity (100%). Moreover, as additional destruction and loss of exergy are associated with combustion, the overall efficiency from fuel input to end use would be much less than indicated by the values shown on the figure.

Costing Heat Loss   For the system in Fig. 7.7, it is instructive to consider further the rate of exergy loss accompanying the heat loss $\dot{Q}_1$; that is, $(1 - T_0/T_1)\dot{Q}_1$. This expression measures the *true* thermodynamic value of the heat loss and is graphed in **Fig. 7.9**. The figure shows that the value of the heat loss in terms of exergy depends *significantly* on the temperature at which the heat loss occurs. We might expect that the *economic* value of such a loss varies similarly with temperature, and this is the case.

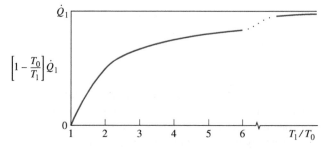

Fig. 7.9 Effect of the temperature ratio T_1/T_0 on the exergy loss associated with heat transfer.

FOR EXAMPLE

Since the source of the exergy loss by heat transfer is the fuel input (see Fig. 7.7), the economic value of the loss can be accounted for in terms of the *unit cost* of fuel based on exergy, c_F (in $\$/kW \cdot h$, for example), as follows

$$\begin{bmatrix} \text{cost rate of heat loss} \\ \dot{Q}_1 \text{ at temperature } T_1 \end{bmatrix} = c_F (1 - T_0/T_1)\dot{Q}_1 \tag{7.22}$$

Equation 7.22 shows that the cost of such a loss is less at lower temperatures than at higher temperatures.

The previous example illustrates what we would expect of a rational costing method. It would not be rational to assign the same economic value for a heat transfer occurring near ambient temperature, where the thermodynamic value is negligible, as for an equal heat transfer occurring at a higher temperature, where the thermodynamic value is significant. Indeed, it would be incorrect to assign the *same cost* to heat loss independent of the temperature at which the loss is occurring. For further discussion of exergy costing, see Sec. 7.7.3.

Horizons

Oil from Shale and Sand Deposits—The Jury Is Still Out

Traditional oil reserves are widely anticipated to decline in years ahead. But the impact could be lessened if cost-effective and environmentally benign technologies can be developed to recover oil-like substances from abundant oil shale and oil sand deposits in the United States and Canada.

Production means available today are both costly and inefficient in terms of exergy demands for the blasting, digging, transporting, crushing, and heating of the materials rendered into oil. Current production means not only use natural gas and large amounts of water but also may cause wide-scale environmental devastation, including air and water pollution and significant

amounts of toxic waste. Moreover, pipeline delivery of Canadian oil sands to the United States has been controversial. Some say that oil sands corrode or damage pipelines more than conventional crude oil, making pipeline failure more likely. Others disagree that oil sands necessarily present significant environmental danger.

Although significant rewards await developers of improved technologies, the challenges also are significant. Some say efforts are better directed to using traditional oil reserves more efficiently and to developing alternatives to oil-based fuels such as *cellulosic* ethanol produced with relatively low-cost biomass from urban, agricultural, and forestry sources.

7.6.2 Exergetic Efficiencies of Common Components

Exergetic efficiency expressions can take many different forms. Several examples are given in the current section for thermal system components of practical interest. In every instance, the efficiency is derived by the use of the exergy rate balance. The approach used here serves as a model for the development of exergetic efficiency expressions for other components. Each of the cases considered involves a control volume at steady state, and we assume no heat transfer between the control volume and its surroundings. The current presentation is not exhaustive. Many other exergetic efficiency expressions can be written.

Turbines For a turbine operating at steady state with no heat transfer with its surroundings, the steady-state form of the exergy rate balance, Eq. 7.17, reduces as follows:

$$0 = \sum_j \left(1 - \frac{T_0}{T_j}\right)\overset{0}{\cancel{\dot{Q}_j}} - \dot{W}_{cv} + \dot{m}(e_{f1} - e_{f2}) - \dot{E}_d$$

This equation can be rearranged to read

$$e_{f1} - e_{f2} = \frac{\dot{W}_{cv}}{\dot{m}} + \frac{\dot{E}_d}{\dot{m}} \tag{7.23}$$

The term on the left of Eq. 7.23 is the decrease in flow exergy from turbine inlet to exit. The equation shows that the flow exergy decrease is accounted for by the turbine work developed, $\dot{W}_{cv}/\dot{m}$, and the exergy destroyed, $\dot{E}_d/\dot{m}$. A parameter that gauges how effectively the flow exergy decrease is converted to the desired product is the *exergetic turbine efficiency*

$$\varepsilon = \frac{\dot{W}_{cv}/\dot{m}}{e_{f1} - e_{f2}} \qquad (7.24)$$

This particular exergetic efficiency is sometimes referred to as the *turbine effectiveness*. Carefully note that the exergetic turbine efficiency is defined differently from the isentropic turbine efficiency introduced in Sec. 6.12.

> **FOR EXAMPLE**
>
> The exergetic efficiency of the turbine considered in Example 6.11 is 81.2% when $T_0 = 298$ K. It is left as an exercise to verify this value.

Compressors and Pumps   For a compressor or pump operating at steady state with no heat transfer with its surroundings, the exergy rate balance, Eq. 7.17, can be placed in the form

$$\left(-\frac{\dot{W}_{cv}}{\dot{m}}\right) = e_{f2} - e_{f1} + \frac{\dot{E}_d}{\dot{m}}$$

Thus, the exergy *input* to the device, $-\dot{W}_{cv}/\dot{m}$, is accounted for by the increase in the flow exergy between inlet and exit and the exergy destroyed. The effectiveness of the conversion from work input to flow exergy increase is gauged by the *exergetic compressor* (or pump) *efficiency*

$$\varepsilon = \frac{e_{f2} - e_{f1}}{(-\dot{W}_{cv}/\dot{m})} \qquad (7.25)$$

> **FOR EXAMPLE**
>
> The exergetic efficiency of the compressor considered in Example 6.14 is 84.6% when $T_0 = 273$ K. It is left as an exercise to verify this value.

Heat Exchanger Without Mixing   The heat exchanger shown in **Fig. 7.10** operates at steady state with no heat transfer with its surroundings and both streams at temperatures above T_0. The exergy rate balance, Eq. 7.13a, reduces to

$$0 = \sum_j \left(1 - \frac{T_0}{T_j}\right) \dot{Q}_j^{\,0} - \dot{W}_{cv}^{\,0} + (\dot{m}_h e_{f1} + \dot{m}_c e_{f3}) - (\dot{m}_h e_{f2} + \dot{m}_c e_{f4}) - \dot{E}_d$$

where $\dot{m}_h$ is the mass flow rate of the hot stream and $\dot{m}_c$ is the mass flow rate of the cold stream. This can be rearranged to read

$$\dot{m}_h(e_{f1} - e_{f2}) = \dot{m}_c(e_{f4} - e_{f3}) + \dot{E}_d \qquad (7.26)$$

The term on the left of Eq. 7.26 accounts for the decrease in the exergy of the hot stream. The first term on the right accounts for the increase in exergy of the cold stream. Regarding the hot stream as supplying the exergy increase of the cold stream as well as the exergy destroyed, we can write an *exergetic heat exchanger efficiency* as

$$\varepsilon = \frac{\dot{m}_c(e_{f4} - e_{f3})}{\dot{m}_h(e_{f1} - e_{f2})} \qquad (7.27)$$

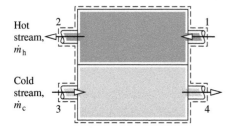

Hot stream, $\dot{m}_h$

Cold stream, $\dot{m}_c$

Fig. 7.10 Counterflow heat exchanger.

> **FOR EXAMPLE**
>
> The exergetic efficiency of the heat exchanger of Example 7.6 is 83.3%. It is left as an exercise to verify this value.

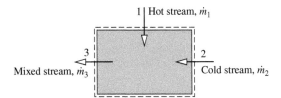

Fig. 7.11 Direct contact heat exchanger.

Direct Contact Heat Exchanger   The direct contact heat exchanger shown in Fig. 7.11 operates at steady state with no heat transfer with its surroundings. The exergy rate balance, Eq. 7.13a, reduces to

$$0 = \sum_j \left(1 - \frac{T_0}{T_j}\right)\overset{0}{\dot{Q}_j} - \overset{0}{\dot{W}_{cv}} + \dot{m}_1 e_{f1} + \dot{m}_2 e_{f2} - \dot{m}_3 e_{f3} - \dot{E}_d$$

With $\dot{m}_3 = \dot{m}_1 + \dot{m}_2$ from a mass rate balance, this can be written as

$$\dot{m}_1(e_{f1} - e_{f3}) = \dot{m}_2(e_{f3} - e_{f2}) + \dot{E}_d \tag{7.28}$$

The term on the left of Eq. 7.28 accounts for the decrease in the exergy of the hot stream between inlet and exit. The first term on the right accounts for the increase in the exergy of the cold stream between inlet and exit. Regarding the hot stream as supplying the exergy increase of the cold stream as well as the exergy destroyed by irreversibilities, we can write an *exergetic efficiency* for a direct contact heat exchanger as

$$\boxed{\varepsilon = \frac{\dot{m}_2(e_{f3} - e_{f2})}{\dot{m}_1(e_{f1} - e_{f3})}} \tag{7.29}$$

7.6.3 Using Exergetic Efficiencies

Exergetic efficiencies are useful for distinguishing means for utilizing fossil fuels that are thermodynamically effective from those that are less so. Exergetic efficiencies also can be used to evaluate the effectiveness of engineering measures taken to improve the performance of systems. This is done by comparing the efficiency values determined before and after modifications have been made to show how much improvement has been achieved. Moreover, exergetic efficiencies can be used to gauge the potential for improvement in the performance of a given system by comparing the efficiency of the system to the efficiency of like systems. A significant difference between these values signals that improved performance is possible.

It is important to recognize that the limit of 100% exergetic efficiency should not be regarded as a practical objective. This theoretical limit could be attained only if there were no exergy destructions or losses. To achieve such idealized processes might require extremely long times to execute processes and/or complex devices, both of which are at odds with the objective of cost-effective operation. In practice, decisions are chiefly made on the basis of *total* costs. An increase in efficiency to reduce fuel consumption, or otherwise utilize fuels better, often requires additional expenditures for facilities and operations. Accordingly, an improvement might not be implemented if an increase in total cost would result. The trade-off between fuel savings and additional investment invariably dictates a lower efficiency than might be achieved *theoretically* and even a lower efficiency than could be achieved using the *best available* technology.

Energy & Environment

A type of exergetic efficiency known as the *well-to-wheel efficiency* is used to compare different options for powering vehicles. The calculation of this efficiency begins at the well where the oil or natural gas feedstock is extracted from the ground and ends with the power delivered to a vehicle's wheels. The well-to-wheel efficiency accounts for how effectively the vehicle's fuel is produced from feedstock, called the *well-to-fuel tank efficiency*, and how effectively the vehicle's power plant converts its fuel to power, called the *fuel tank-to-wheel efficiency*. The product of these gives the *overall* well-to-wheel efficiency.

The table below gives sample well-to-wheel efficiency values for three automotive power options as reported by a manufacturer:

	Well-to-Tank (Fuel Production Efficiency) (%)	×	Tank-to-Wheel (Vehicle Efficiency) (%)	=	Well-to-Wheel (Overall Efficiency) (%)
Conventional gasoline-fueled engine	88	×	16	=	14
Hydrogen-fueled fuel cell[a]	58	×	38	=	22
Gasoline-fueled hybrid electric	88	×	32	=	28

[a]Hydrogen produced from natural gas.

These data show that vehicles using conventional internal combustion engines do not fare well in terms of the well-to-wheel efficiency. The data also show that fuel-cell vehicles operating on hydrogen have the best tank-to-wheel efficiency of the three options but lose out on an overall basis to hybrid vehicles, which enjoy a higher well-to-tank efficiency. Still, the well-to-wheel efficiency is just one consideration when making policy decisions concerning different options for powering vehicles. With increasing concern over global atmospheric CO_2 concentrations, another consideration is the well-to-wheel *total* production of CO_2 in kg per km driven (lb per mile driven). Studies have shown that conventional gasoline-fueled engines typically produce significantly higher levels of CO_2 as compared with hydrogen-fueled fuel cell and hybrid engines.

7.7 Thermoeconomics

Thermal systems typically experience significant work and/or heat interactions with their surroundings, and they can exchange mass with their surroundings in the form of hot and cold streams, including chemically reactive mixtures. Thermal systems appear in almost every industry, and numerous examples are found in our everyday lives. Their design and operation involve the application of principles from thermodynamics, fluid mechanics, and heat transfer, as well as such fields as materials, manufacturing, and mechanical design. The design and operation of thermal systems also require explicit consideration of engineering economics, for cost is always a consideration. The term **thermoeconomics** may be applied to this general area of application, although it is often applied more narrowly to methodologies combining exergy and economics for optimization studies during design of new systems and process improvement of existing systems.

thermoeconomics

7.7.1 Costing

Is costing an art or a science? The answer is a little of both. *Cost engineering* is an important engineering subdiscipline aimed at objectively applying real-world costing experience in engineering design and project management. Costing services are provided by practitioners skilled in the use of specialized methodologies, cost models, and databases, together with costing expertise and judgment garnered from years of professional practice. Depending on need, cost engineers provide services ranging from rough and rapid estimates to in-depth analyses. Ideally, cost engineers are involved with projects from the formative stages, for the *output* of cost engineering is an essential *input* to decision making. Such input can be instrumental in identifying feasible options from a set of alternatives and even pinpointing the best option.

Costing of thermal systems considers costs of owning and operating them. Some observers voice concerns that costs related to the environment often are only weakly taken into consideration in such evaluations. They say companies pay for the right to extract natural resources used in the production of goods and services but rarely pay fully for depleting nonrenewable resources and mitigating accompanying environmental degradation and loss of wildlife habitat, in many cases leaving the cost burden to future generations. Another concern is who pays for the costs of controlling air and water pollution, cleaning up hazardous wastes, and the impacts of pollution and waste on human health—industry, government, the public, or some combination of all three. Yet when agreement about environmental costs is achieved among interested business, governmental, and advocacy groups, such costs are readily integrated in costing of thermal systems, including costing on an exergy basis, which is the present focus.

7.7.2 Using Exergy in Design

To illustrate the use of exergy reasoning in design, consider **Fig. 7.12** showing a boiler at steady state. Fuel and air enter the boiler and react to form hot combustion gases. Feedwater also enters as saturated liquid, receives exergy by heat transfer from the combustion gases, and exits without temperature change as saturated vapor at a specified condition for use elsewhere. Temperatures of the hot gas and water streams are also shown on the figure.

There are two main sources of exergy destruction in the boiler: (1) irreversible heat transfer occurring between the hot combustion gases and the water flowing through the boiler tubes, and (2) the combustion process itself. To simplify the present discussion, the boiler is considered to consist of a combustor unit in which fuel and air are burned to produce hot combustion gases, followed by a heat exchanger unit where water is vaporized as the hot gases cool.

The present discussion centers on the heat exchanger unit. Let us think about its total cost as the sum of fuel-related and capital costs. We will also take the average temperature difference between the two streams, ΔT_{ave}, as the *design variable*. From our study of the second law of thermodynamics, we know that the average temperature difference between the two streams is a measure of exergy destruction associated with heat transfer between them. The exergy destroyed owing to heat transfer originates in the fuel entering the boiler. Accordingly, a cost related to fuel consumption can be attributed to this source of irreversibility. Since exergy destruction increases with temperature difference between the streams, the fuel-related cost increases with *increasing* ΔT_{ave}. This variation is shown in **Fig. 7.13** on an *annualized* basis, in dollars per year.

From our study of heat transfer, we know an inverse relation exists between ΔT_{ave} and the boiler tube surface area required for a desired heat transfer rate between the streams. For

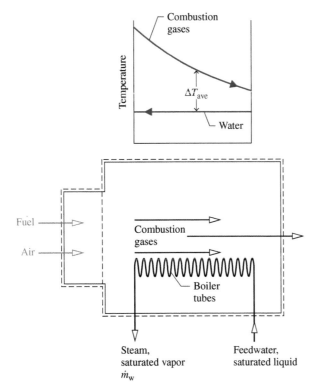

Fig. 7.12 Boiler used to discuss exergy in design.

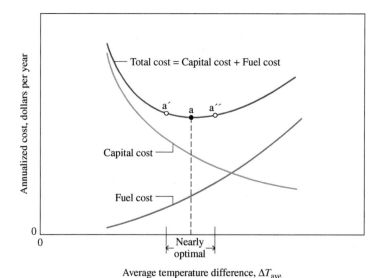

Fig. 7.13 Cost curves for the heat exchanger unit of the boiler of Fig. 7.12.

example, if we design for a small average temperature difference to reduce exergy destruction within the heat exchanger, this dictates a large surface area and typically a more costly boiler. From such considerations, we infer that boiler capital cost increases with *decreasing* ΔT_{ave}. This variation is shown in Fig. 7.13, again on an annualized basis.

The *total cost* is the sum of the capital cost and the fuel cost. The total cost curve shown in Fig. 7.13 exhibits a minimum at the point labeled a. Notice, however, that the curve is relatively flat in the neighborhood of the minimum, so there is a range of ΔT_{ave} values that could be considered *nearly optimal* from the standpoint of minimum total cost. If reducing the fuel cost were deemed more important than minimizing the capital cost, we might choose a design that would operate at point a′. Point a″ would be a more desirable operating point if capital cost were of greater concern. Such trade-offs are common in design situations.

The actual design process differs significantly from the simple case considered here. For one thing, costs cannot be determined as precisely as implied by the curves in Fig. 7.13. Fuel prices vary widely over time, and equipment costs may be difficult to predict as they often depend on a bidding procedure. Equipment is manufactured in discrete sizes, so the cost also would not vary continuously as shown in the figure. Furthermore, thermal systems usually consist of several components that interact with one another. Optimization of components individually, as considered for the heat exchanger unit of the boiler, does not guarantee an optimum for the overall system. Finally, the example involves only ΔT_{ave} as a design variable. Often, several design variables must be considered and optimized simultaneously.

7.7.3 Exergy Costing of a Cogeneration System

Another important aspect of thermoeconomics is the use of exergy for *allocating* costs to the products of a thermal system. This involves assigning to each product the total cost to produce it, namely, the cost of fuel and other inputs plus the cost of owning and operating the system (e.g., capital cost, operating and maintenance costs). Such costing is a common problem in plants where utilities such as electrical power, chilled water, compressed air, and steam are generated in one department and used in others. The plant operator needs to know the cost of generating each utility to ensure that the other departments are charged properly according to the type and amount of each utility used. Common to all such considerations are fundamentals from engineering economics, including procedures for annualizing costs, appropriate means for allocating costs, and reliable cost data.

To explore further the costing of thermal systems, consider the simple *cogeneration system* operating at steady state shown in **Fig. 7.14**. The system consists of a boiler and a turbine,

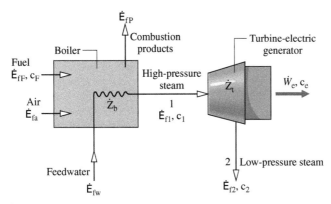

Fig. 7.14 Simple cogeneration system.

with each having no significant heat transfer to its surroundings. The figure is labeled with exergy transfer rates associated with the flowing streams, where the subscripts F, a, P, and w denote fuel, combustion air, combustion products, and feedwater, respectively. The subscripts 1 and 2 denote high- and low-pressure steam, respectively. Means for evaluating the exergies of the fuel and combustion products are introduced in Chap. 13. The cogeneration system has two principal products: electricity, denoted by $\dot{W}_e$, and low-pressure steam for use in some process. The objective is to determine the cost at which each product is generated.

Boiler Analysis Let us begin by evaluating the cost of the high-pressure steam produced by the boiler. For this, we consider a control volume enclosing the boiler. Fuel and air enter the boiler separately and combustion products exit. Feedwater enters and high-pressure steam exits. The total cost to produce the exiting high-pressure steam equals the total cost of the entering streams plus the cost of owning and operating the boiler. This is expressed by the following **cost rate balance** for the boiler

$$\dot{C}_1 = \dot{C}_F + \dot{C}_a + \dot{C}_w + \dot{Z}_b \qquad (7.30)$$

cost rate balance

where $\dot{C}$ is the cost rate of the respective stream (in $ per hour, for instance). $\dot{Z}_b$ accounts for the cost rate associated with owning and operating the boiler, including expenses related to proper disposal of the combustion products. In the present discussion, the cost rate $\dot{Z}_b$ is presumed known from a previous economic analysis.

Although the cost rates denoted by $\dot{C}$ in Eq. 7.30 are evaluated by various means in practice, the present discussion features the use of exergy for this purpose. Since exergy measures the true thermodynamic values of the work, heat, and other interactions between a system and its surroundings as well as the effect of irreversibilities within the system, exergy is a rational basis for assigning costs. With exergy costing, each of the cost rates is evaluated in terms of the associated rate of exergy transfer and a *unit cost*. Thus, for an entering or exiting stream, we write

$$\dot{C} = c\dot{E}_f \qquad (7.31)$$

where c denotes the **cost per unit of exergy** (in $ or cents per kW · h, for example) and $\dot{E}_f$ is the associated exergy transfer rate.

exergy unit cost

For simplicity, we assume the feedwater and combustion air enter the boiler with negligible exergy and cost. Thus, Eq. 7.30 reduces as follows:

$$\dot{C}_1 = \dot{C}_F + \cancel{\dot{C}_a}^{0} + \cancel{\dot{C}_w}^{0} + \dot{Z}_b$$

Then, with Eq. 7.31 we get

$$c_1\dot{E}_{f1} = c_F\dot{E}_{fF} + \dot{Z}_b \qquad (7.32a)$$

Solving for c_1, the unit cost of the high-pressure steam is

$$c_1 = c_F \left(\frac{\dot{E}_{fF}}{\dot{E}_{f1}} \right) + \frac{\dot{Z}_b}{\dot{E}_{f1}} \qquad (7.32b)$$

This equation shows that the unit cost of the high-pressure steam is determined by two contributions related, respectively, to the cost of the fuel and the cost of owning and operating the boiler. Due to exergy destruction and loss, less exergy exits the boiler with the high-pressure steam than enters with the fuel. Thus, $\dot{E}_{fF}/\dot{E}_{f1}$ is invariably greater than one, and the unit cost of the high-pressure steam is invariably greater than the unit cost of the fuel.

Turbine Analysis   Next, consider a control volume enclosing the turbine. The total cost to produce the electricity and low-pressure steam equals the cost of the entering high-pressure steam plus the cost of owning and operating the device. This is expressed by the *cost rate balance* for the turbine

$$\dot{C}_e + \dot{C}_2 = \dot{C}_1 + \dot{Z}_t \qquad (7.33)$$

where $\dot{C}_e$ is the cost rate associated with the electricity, $\dot{C}_1$ and $\dot{C}_2$ are the cost rates associated with the entering and exiting steam, respectively, and $\dot{Z}_t$ accounts for the cost rate associated with owning and operating the turbine. With exergy costing, each of the cost rates $\dot{C}_e$, $\dot{C}_1$, and $\dot{C}_2$ is evaluated in terms of the associated rate of exergy transfer and a unit cost. Equation 7.33 then appears as

$$c_e \dot{W}_e + c_2 \dot{E}_{f2} = c_1 \dot{E}_{f1} + \dot{Z}_t \qquad (7.34a)$$

The unit cost c_1 in Eq. 7.34a is given by Eq. 7.32b. In the present discussion, the same unit cost is assigned to the low-pressure steam; that is, $c_2 = c_1$. This is done on the basis that the purpose of the turbine is to generate electricity, and thus all costs associated with owning and operating the turbine should be charged to the power generated. We can regard this decision as a part of the *cost accounting* considerations that accompany the thermoeconomic analysis of thermal systems. With $c_2 = c_1$, Eq. 7.34a becomes

$$c_e \dot{W}_e = c_1 (\dot{E}_{f1} - \dot{E}_{f2}) + \dot{Z}_t \qquad (7.34b)$$

The first term on the right side accounts for the cost of the exergy used and the second term accounts for the cost of owning and operating the system.

Solving Eq. 7.34b for c_e, and introducing the exergetic turbine efficiency ε from Eq. 7.24

$$c_e = \frac{c_1}{\varepsilon} + \frac{\dot{Z}_t}{\dot{W}_e} \qquad (7.34c)$$

This equation shows that the unit cost of the electricity is determined by the cost of the high-pressure steam and the cost of owning and operating the turbine. Because of exergy destruction within the turbine, the exergetic efficiency is invariably less than one; therefore, the unit cost of electricity is invariably greater than the unit cost of the high-pressure steam.

Summary   By applying cost rate balances to the boiler and the turbine, we are able to determine the cost of each product of the cogeneration system. The unit cost of the electricity is determined by Eq. 7.34c and the unit cost of the low-pressure steam is determined by the expression $c_2 = c_1$ together with Eq. 7.32b. The example to follow provides a detailed illustration. The same general approach is applicable for costing the products of a wide-ranging class of thermal systems.[1]

---

[1] See A. Bejan, G. Tsatsaronis, and M. J. Moran, *Thermal Design and Optimization*, John Wiley & Sons, New York, 1996.

Exergy Costing of a Cogeneration System

A cogeneration system consists of a natural gas–fueled boiler and a steam turbine that develops power and provides steam for an industrial process. At steady state, fuel enters the boiler with an exergy rate of 100 MW. Steam exits the boiler at 50 bar, 466°C with an exergy rate of 35 MW. Steam exits the turbine at 5 bar, 205°C and a mass flow rate of 26.15 kg/s. The unit cost of the fuel is 1.44 cents per kW · h of exergy. The costs of owning and operating the boiler and turbine are, respectively, $1080/h and $92/h. The feedwater and combustion air enter with negligible exergy and cost. Expenses related to proper disposal of the combustion products are included with the cost of owning and operating the boiler. Heat transfer with the surroundings and the effects of motion and gravity are negligible. Let $T_0 = 298$ K.

a. For the turbine, determine the power and the rate exergy exits with the steam, each in MW.

b. Determine the unit costs of the steam exiting the boiler, the steam exiting the turbine, and the power, each in cents per kW · h of exergy.

c. Determine the cost rates of the steam exiting the turbine and the power, each in $/h.

SOLUTION

Known Steady-state operating data are known for a cogeneration system that produces both electricity and low-pressure steam for an industrial process.

Find For the turbine, determine the power and the rate exergy exits with the steam. Determine the unit costs of the steam exiting the boiler, the steam exiting the turbine, and the power developed. Also determine the cost rates of the low-pressure steam and power.

Schematic and Given Data:

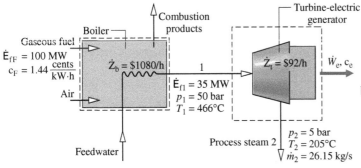

Fig. E7.10

Engineering Model

1. Each control volume shown in the accompanying figure is at steady state.

2. For each control volume, $\dot{Q}_{cv} = 0$ and the effects of motion and gravity are negligible.

3. The feedwater and combustion air enter the boiler with negligible exergy and cost.

4. Expenses related to proper disposal of the combustion products are included with the cost of owning and operating the boiler.

5. The unit costs based on exergy of the high- and low-pressure steam are equal: $c_1 = c_2$.

6. For the environment, $T_0 = 298$ K.

Analysis

a. With assumption 2, the mass and energy rate balances for a control volume enclosing the turbine reduce at steady state to give

$$\dot{W}_e = \dot{m}(h_1 - h_2)$$

From Table A-4, $h_1 = 3353.54$ kJ/kg and $h_2 = 2865.96$ kJ/kg. Thus,

$$\dot{W}_e = \left(26.15\frac{kg}{s}\right)(3353.54 - 2865.96)\left(\frac{kJ}{kg}\right)\left|\frac{1\ MW}{10^3\ kJ/s}\right|$$
$$= 12.75\ MW$$

Using Eq. 7.18, the difference in the rates at which exergy enters and exits the turbine with the steam is

$$\dot{E}_{f2} - \dot{E}_{f1} = \dot{m}(e_{f2} - e_{f1})$$
$$= \dot{m}[h_2 - h_1 - T_0(s_2 - s_1)]$$

Solving for $\dot{E}_{f2}$

$$\dot{E}_{f2} = \dot{E}_{f1} + \dot{m}[h_2 - h_1 - T_0(s_2 - s_1)]$$

With known values for $\dot{E}_{f1}$ and $\dot{m}$, and data from Table A-4: $s_1 = 6.8773$ kJ/kg · K and $s_2 = 7.0806$ kJ/kg · K, the rate exergy exits with the steam is

$$\dot{E}_{f2} = 35\ MW + \left(26.15\frac{kg}{s}\right)\Bigg[(2865.96 - 3353.54)\frac{kJ}{kg}$$
$$- 298\ K(7.0806 - 6.8773)\frac{kJ}{kg \cdot K}\Bigg]\left|\frac{1\ MW}{10^3\ kJ/s}\right|$$
$$= 20.67\ MW$$

b. For a control volume enclosing the boiler, the cost rate balance reduces with assumptions 3 and 4 to give

$$c_1\dot{E}_{f1} = c_F\dot{E}_{fF} + \dot{Z}_b$$

where $\dot{E}_{fF}$ is the exergy rate of the entering fuel, c_F and c_1 are the unit costs of the fuel and exiting steam, respectively, and $\dot{Z}_b$ is the cost rate associated with owning and operating the boiler. Solving for c_1 we get Eq. 7.32b; then, inserting known values, c_1 is determined:

$$c_1 = c_F\left(\frac{\dot{E}_{fF}}{\dot{E}_{f1}}\right) + \frac{\dot{Z}_b}{\dot{E}_{f1}}$$

$$= \left(1.44\frac{cents}{kW \cdot h}\right)\left(\frac{100\ MW}{35\ MW}\right) + \left(\frac{1080\ \$/h}{35\ MW}\right)\left|\frac{1\ MW}{10^3\ kW}\right|\left|\frac{100\ cents}{1\$}\right|$$

$$= (4.11 + 3.09)\frac{cents}{kW \cdot h} = 7.2\frac{cents}{kW \cdot h}$$

The cost rate balance for the control volume enclosing the turbine is given by Eq. 7.34a

$$c_e \dot{W}_e + c_2 \dot{E}_{f2} = c_1 \dot{E}_{f1} + \dot{Z}_t$$

where c_e and c_2 are the unit costs of the power and the exiting steam, respectively, and $\dot{Z}_t$ is the cost rate associated with own-❶ ing and operating the turbine. Assigning the same unit cost to the steam entering and exiting the turbine, $c_2 = c_1 = 7.2$ cents/ kW · h, and solving for c_e

$$c_e = c_1 \left[\frac{\dot{E}_{f1} - \dot{E}_{f2}}{\dot{W}_e} \right] + \frac{\dot{Z}_t}{\dot{W}_e}$$

Inserting known values

$$c_e = \left(7.2 \frac{\text{cents}}{\text{kW} \cdot \text{h}} \right) \left[\frac{(35 - 20.67)\ \text{MW}}{12.75\ \text{MW}} \right]$$
$$+ \left(\frac{92\$/\text{h}}{12.75\ \text{MW}} \right) \left| \frac{1\ \text{MW}}{10^3\ \text{kW}} \right| \left| \frac{100\ \text{cents}}{1\$} \right|$$

❷
$$= (8.09 + 0.72) \frac{\text{cents}}{\text{kW} \cdot \text{h}} = 8.81 \frac{\text{cents}}{\text{kW} \cdot \text{h}}$$

c. For the low-pressure steam and power, the cost rates are, respectively,

$$\dot{C}_2 = c_2 \dot{E}_{f2}$$
$$= \left(7.2 \frac{\text{cents}}{\text{kW} \cdot \text{h}} \right) (20.67\ \text{MW}) \left| \frac{10^3\ \text{kW}}{1\ \text{MW}} \right| \left| \frac{\$1}{100\ \text{cents}} \right|$$
$$= \$1488/\text{h}$$

❸
$$\dot{C}_e = c_e \dot{W}_e$$
$$= \left(8.81 \frac{\text{cents}}{\text{kW} \cdot \text{h}} \right) (12.75\ \text{MW}) \left| \frac{10^3\ \text{kW}}{1\ \text{MW}} \right| \left| \frac{\$1}{100\ \text{cents}} \right|$$
$$= \$1123/\text{h}$$

❶ The purpose of the turbine is to generate power, and thus all costs associated with owning and operating the turbine are charged to the power generated.

❷ Observe that the unit costs c_1 and c_e are significantly greater than the unit cost of the fuel.

❸ Although the unit cost of the steam is less than the unit cost of the power, the steam *cost* rate is greater because the associated exergy rate is much greater.

SKILLS DEVELOPED

Ability to…

- evaluate exergy quantities required for exergy costing.
- apply exergy costing.

Quick Quiz

If the unit cost of the fuel were to double to 2.88 cents/kW · h, what would be the change in the unit cost of power, expressed as a percent, keeping all other given data the same? Ans. +53%.

CHAPTER SUMMARY AND STUDY GUIDE

In this chapter, we have introduced the property exergy and illustrated its use for thermodynamic analysis. Like mass, energy, and entropy, exergy is an extensive property that can be transferred across system boundaries. Exergy transfer accompanies heat transfer, work, and mass flow. Like entropy, exergy is not conserved. Exergy is destroyed within systems whenever internal irreversibilities are present. Entropy production corresponds to exergy destruction.

The use of exergy balances is featured in this chapter. Exergy balances are expressions of the second law that account for exergy in terms of exergy transfers and exergy destruction. For processes of closed systems, the exergy balance is given by Eqs. 7.4 and the companion steady-state forms are Eqs. 7.11. For control volumes, the steady-state expressions are given by Eqs. 7.13. Control volume analyses account for exergy transfer at inlets and exits in terms of flow exergy.

The following checklist provides a study guide for this chapter. When your study of the text and end-of-chapter exercises has been completed, you should be able to

- write out meanings of the terms listed in the margins throughout the chapter and understand each of the related concepts. The subset of key concepts listed below is particularly important.

- evaluate specific exergy at a given state using Eq. 7.2 and exergy change between two states using Eq. 7.3, each relative to a specified reference environment.

- apply exergy balances in each of several alternative forms, appropriately modeling the case at hand, correctly observing sign conventions, and carefully applying SI and English units.

- evaluate the specific flow exergy relative to a specified reference environment using Eq. 7.14.

- define and evaluate exergetic efficiencies for thermal system components of practical interest.

- apply exergy costing to heat loss and simple cogeneration systems.

KEY ENGINEERING CONCEPTS

exergy	closed system exergy balance	exergy accounting
exergy reference environment	exergy transfer	exergetic efficiency
dead state	exergy destruction	thermoeconomics
specific exergy	flow exergy	cost rate balance
exergy change	control volume exergy rate balance	exergy unit cost

KEY EQUATIONS

$E = (U - U_0) + p_0(V - V_0) - T_0(S - S_0) + KE + PE$	(7.1)	Exergy of a system.
$e = (u - u_0) + p_0(v - v_0) - T_0(s - s_0) + V^2/2 + gz$	(7.2)	Specific exergy.
$E_2 - E_1 = (U_2 - U_1) + p_0(V_2 - V_1) - T_0(S_2 - S_1)$ $+ (KE_2 - KE_1) + (PE_2 - PE_1)$	(7.3)	Exergy change.
$E_2 - E_1 = E_q - E_w - E_d$	(7.4b)	Closed system exergy balance. See Eqs. 7.5–7.7 for E_q, E_w, E_d, respectively.
$0 = \sum_j \left(1 - \dfrac{T_0}{T_j}\right)\dot{Q}_j - \dot{W} - \dot{E}_d$	(7.11a)	Steady-state closed system exergy rate balance.
$0 = \sum_j \left(1 - \dfrac{T_0}{T_j}\right)\dot{Q}_j - \dot{W}_{cv} + \sum_i \dot{m}_i e_{fi} - \sum_e \dot{m}_e e_{fe} - \dot{E}_d$	(7.13a)	Steady-state control volume exergy rate balance.
$e_f = h - h_0 - T_0(s - s_0) + \dfrac{V^2}{2} + gz$	(7.14)	Specific flow exergy.

EXERCISES: THINGS ENGINEERS THINK ABOUT

7.1 When you hear the term "energy crisis" used by the news media, do the media really mean *exergy* crisis? Explain.

7.2 A plane is on its final approach to an airport. It lands and parks at the terminal. During this process, what happens to the plane's exergy?

7.3 A pail of water initially at 20°C freezes when left outside on a cold winter's day. Does its exergy increase or decrease? Explain.

7.4 Is it possible for exergy to be negative? For exergy *change* to be negative? For exergy *destruction* to be negative?

7.5 When an automobile brakes to rest, what happens to the exergy associated with its motion?

7.6 After a vehicle receives an oil change and lube job, does the exergy destruction within a control volume enclosing the idling vehicle change? Explain.

7.7 Can the exergetic efficiency of a power cycle ever be greater than the thermal efficiency of the same cycle? Explain.

7.8 Is there a difference between practicing exergy *conservation* and exergy *efficiency*? Explain.

7.9 When installed on the engine of an automobile, which accessory, *supercharger* or *turbocharger*, will result in an engine with the higher exergetic efficiency? Explain.

7.10 Apart from the well-to-wheel efficiency, what other considerations are used to compare different options for powering vehicles?

7.11 In terms of exergy, how does the flight of a bird compare with the flight of a baseball going over the centerfield fence?

7.12 A gasoline-fueled generator is claimed by its inventor to produce electricity at a lower unit cost than the unit cost of the fuel used, where each cost is based on exergy. Comment.

CHECKING UNDERSTANDING

7.1 Which of the following statements is false when describing the exergy associated with an *isolated system* undergoing an actual process?

    a. The exergy of the system decreases.

    b. There are no transfers of exergy between the system and its surroundings.

    c. The exergy of the surroundings increases.

    d. Exergy destruction within the system is greater than zero.

7.2 Steam contained in a piston–cylinder assembly is compressed from an initial volume of 50 m³ to 30 m³. If exergy transfer accompa-

nying work is 500 kJ and the dead state pressure is 1 bar, determine the energy transfer by work, in kJ, for the process.

    a. 150   **c.** 500

    b. −1500   **d.** −500

7.3 Which of the following terms is not included in a cost rate balance for a control volume enclosing a boiler?

    a. Cost rate associated with feedwater

    b. Cost rate associated with owning and operating the boiler

    c. Cost rate associated with electricity

    d. Cost rate associated with combustion air

7.4 The following term reduces to zero within the steady-state form of the closed system exergy rate balance.

    a. The time rate of the change of exergy within the closed system

    b. The time rate of exergy transfer by work

    c. The time rate of exergy transfer by heat

    d. The time rate of exergy destruction

7.5 Air within a closed cylindrical container receives 20 kJ of heat transfer per kg of air from an external source. The boundary and environment temperatures are 320 K and 27°C, respectively. Determine the exergy transfer per unit mass (kJ/kg) accompanying heat transfer.

    a. −1.33   **c.** −1.25

    b. 1.33   **d.** 1.25

7.6 For a closed system, as the exergy reference environment temperature _____, the exergy destruction rate decreases.

7.7 Air within a piston–cylinder assembly undergoes an expansion process from an initial volume of 0.5 m³ to 1 m³. The dead state pressure is 100 kPa. If the work associated with the process is 60 kJ, determine the exergy transfer accompanying work, in kJ.

    a. −110   **c.** 10

    b. 55   **d.** 59.5

7.8 In regard to the exergy of a system, when the system is at rest relative to its environment, the values of its kinetic and potential energies are _____.

7.9 Air flows through a turbine with an exergetic efficiency of 65%. If the air's specific flow exergy decreases by 300 kJ/kg as it travels through the turbine, determine the work in kJ per kg of air flowing.

    a. 462   **c.** 195

    b. −462   **d.** −195

Indicate whether the following statements are true or false. Explain.

7.10 The exergetic efficiency equations for turbines and compressors are the same.

7.11 The exergy of a closed system depends on several factors, including entropy production and enthalpy.

7.12 If a system is at rest above the reference environment and it is not permitted to change its height, there are no contributions to the total exergy of the system from potential or kinetic energy.

7.13 Exergetic efficiencies are also known as first law efficiencies.

7.14 Exergy is destroyed by irreversibilities.

7.15 Exergy transfer accompanying heat transfer is a function of the environment temperature and the temperature of the boundary where heat transfer occurs.

7.16 The *well-to-wheel* efficiency compares different options for generating electricity used in industry, business, and the home.

7.17 Like entropy, exergy is produced by action of irreversibilities.

7.18 To define exergy, we think of two systems: a system of interest and an *exergy reference environment*.

7.19 If unit costs are based on exergy, we expect the unit cost of the electricity developed by a turbine-generator to be greater than the unit cost of the high-pressure steam provided to the turbine.

7.20 The thermomechanical exergy at a state of a system can be thought of as the magnitude of the minimum theoretical work input required to bring the system from the dead state to the given state.

7.21 When products of combustion are at a temperature significantly greater than required by a specified task, we say the task is well matched to the fuel source.

7.22 The energy of an isolated system must remain constant, but its exergy can only increase.

7.23 Mass, volume, energy, entropy, and exergy are all intensive properties.

7.24 Exergy destruction is proportional to entropy production.

PROBLEMS: DEVELOPING ENGINEERING SKILLS

C Problem may require use of appropriate computer software in order to complete.

Exploring Exergy Concepts

7.1 By inspection of **Fig. P7.1** giving a *T*–υ diagram for water, indicate whether exergy would increase, decrease, or remain the same in (a) Process 1-2, (b) Process 3-4, (c) Process 5-6. Explain.

7.2 By inspection of **Fig. P7.2** giving a *T*–*s* diagram for R-134a, indicate whether exergy would increase, decrease, or remain the same in (a) Process 1–2, (b) Process 3–4, (c) Process 5–6. Explain.

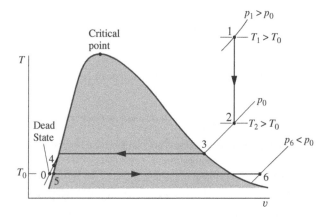

FIGURE P7.1

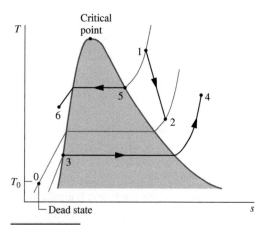

FIGURE P7.2

7.3 An ideal gas is stored in a closed vessel at pressure p and temperature T.

    a. If $T = T_0$, derive an expression for the specific exergy in terms of p, p_0, T_0, and the gas constant R.

    b. If $p = p_0$, derive an expression for the specific exergy in terms of T, T_0, and the specific heat c_p, which can be taken as constant.

Ignore the effects of motion and gravity.

7.4 Consider an evacuated tank of volume V. For the space inside the tank as the system, show that the exergy is given by $\mathsf{E} = p_0 V$. Discuss.

7.5 Equal molar amounts of carbon monoxide and neon are maintained at the same temperature and pressure. Which has the greater value for exergy relative to the same reference environment? Assume the ideal gas model with constant c_v for each gas. There are no significant effects of motion and gravity.

7.6 Two solid blocks, each having mass m and specific heat c, and initially at temperatures T_1 and T_2, respectively, are brought into contact, insulated on their outer surfaces, and allowed to come into thermal equilibrium.

    a. Derive an expression for the exergy destruction in terms of m, c, T_1, T_2, and the temperature of the environment, T_0.

    b. Demonstrate that the exergy destruction cannot be negative.

    c. What is the source of exergy destruction in this case?

7.7 A system undergoes a refrigeration cycle while receiving Q_C by heat transfer at temperature T_C and discharging energy Q_H by heat transfer at a higher temperature T_H. There are no other heat transfers.

    a. Using energy and exergy balances, show that the net work input to the cycle cannot be zero.

    b. Show that the coefficient of performance of the cycle can be expressed as

$$\beta = \left(\frac{T_C}{T_H - T_C} \right)\left(1 - \frac{T_H \mathsf{E}_d}{T_0(Q_H - Q_C)} \right)$$

    where E_d is the exergy destruction and T_0 is the temperature of the exergy reference environment.

    c. Using the result of part (b), obtain an expression for the maximum theoretical value for the coefficient of performance.

7.8 When matter flows across the boundary of a control volume, an energy transfer by work, called *flow work*, occurs. The rate is $\dot{m}(pv)$ where $\dot{m}$, p, and v denote the mass flow rate, pressure, and specific volume, respectively, of the matter crossing the boundary (see Sec. 4.4.2). Show that the *exergy transfer accompanying flow work* is given by $\dot{m}(pv - p_0 v)$, where p_0 is the pressure at the dead state.

7.9 An ideal gas with constant specific heat ratio k enters a turbine operating at steady state at T_1 and p_1 and expands adiabatically to T_2 and p_2. When would the value of the exergetic turbine efficiency exceed the value of the isentropic turbine efficiency? Discuss. Ignore the effects of motion and gravity.

7.10 **C** For an ideal gas with constant specific heat ratio k, show that in the absence of significant effects of motion and gravity the specific flow exergy can be expressed as

$$\frac{e_f}{c_p T_0} = \frac{T}{T_0} - 1 - \ln\frac{T}{T_0} + \ln\left(\frac{p}{p_0} \right)^{(k-1)/k}$$

    a. For $k = 1.2$ develop plots of $e_f/c_p T_0$ versus for T/T_0 for $p/p_0 = 0.25, 0.5, 1, 2, 4$. Repeat for $k = 1.3$ and 1.4.

    b. The specific flow exergy can take on negative values when $p/p_0 < 1$. What does a negative value mean physically?

Evaluating Exergy

7.11 A domestic water heater holds 189 L of water at 60°C, 1 atm. Determine the exergy of the hot water, in kJ. To what elevation, in m, would a 1000-kg mass have to be raised from zero elevation relative to the reference environment for its exergy to equal that of the hot water? Let $T_0 = 298$ K, $p_0 = 1$ atm, $g = 9.81$ m/s^2.

7.12 Determine the exergy, in kJ, at 0.7 bar, 90°C for 1 kg of (a) water, (b) Refrigerant 134a, (c) air as an ideal gas with c_p constant. In each case, the mass is at rest and zero evaluation relative to an exergy reference environment for which $T_0 = 20$°C, $p_0 = 1$ bar.

7.13 **C** A vessel contains carbon dioxide. Using the ideal gas model

    a. determine the specific exergy of the gas, in kJ/kg, at $p = 620$ kPa and $T = 90$°C.

    b. plot the specific exergy of the gas, in kJ/kg, versus pressure ranging from 100 to 620 kPa, for $T = 25$°C.

    c. plot the specific exergy of the gas, in kJ/kg, versus temperature ranging from 25°C to 90°C, for $p = 100$ kPa.

The gas is at rest and zero elevation relative to an exergy reference environment for which $T_0 = 50$°C, $p_0 = 100$ kPa.

7.14 **C** A vessel contains 1 kg of air at pressure p and 90°C. Using the ideal gas model, plot the specific exergy of the air, in kJ/kg, for p ranging from 0.5 to 2 bar. The air is at rest and negligible elevation relative to an exergy reference environment for which $T_0 = 15$°C, $p_0 = 1$ bar.

7.15 Determine the exergy, in kJ, of the contents of a 2-m^3 storage tank, if the tank is filled with

    a. air as an ideal gas at 400°C and 0.35 bar.

    b. water vapor at 400°C and 0.35 bar.

Ignore the effects of motion and gravity and let $T_0 = 17$°C, $p_0 = 1$ atm.

7.16 A concrete slab measuring 0.3 m × 4 m × 6 m, initially at 298 K, is exposed to the sun for several hours, after which its temperature is 301 K. The density of the concrete is 2300 kg/m^3 and its specific heat is $c = 0.88$ kJ/kg · K. (a) Determine the increase in exergy of the slab, in kJ. (b) To what elevation, in m, would a 1000-kg mass have to be raised from zero elevation relative to the reference environment for its exergy to equal the exergy increase of the slab? Let $T_0 = 298$ K, $p_0 = 1$ atm, $g = 9.81$ m/s^2.

7.17 Refrigerant 134a vapor initially at 1 bar and 20°C fills a rigid vessel. The vapor is cooled until the temperature becomes −32°C. There is no work during the process. For the refrigerant, determine the heat transfer per unit mass and the change in specific exergy, each in kJ/kg. Comment. Let $T_0 = 20$°C, $p_0 = 0.1$ MPa and ignore the effects of motion and gravity.

7.18 As shown in **Fig. P7.18**, one kilogram of water undergoes a process from an initial state where the water is saturated vapor at 100°C, the velocity is 25 m/s, and the elevation is 5 m to a final state where the water is saturated liquid at 5°C, the velocity is 22 m/s, and the elevation is 1 m. Determine in kJ, (a) the exergy at the initial state, (b) the exergy at the final state, and (c) the change in exergy. Let $T_0 = 25$°C, $p_0 = 1$ atm, and $g = 9.8$ m/s^2.

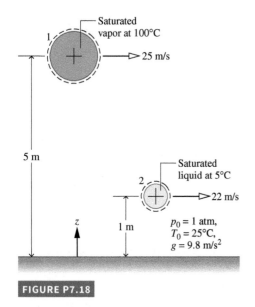

FIGURE P7.18

7.19 Two kilograms of air initially at 90°C and 350 kPa undergo two processes in series:

Process 1–2: Isothermal to $p_2 = 70$ kPa

Process 2–3: Constant pressure to $T_3 = -20°C$

Employing the ideal gas model

    a. represent each process on a p–v diagram and indicate the dead state.

    b. determine the change in exergy for each process, in kJ.

Let $T_0 = 25°C$, $p_0 = 100$ kPa and ignore the effects of motion and gravity.

7.20 Consider 100 kg of steam initially at 20 bar and 240°C as the system. Determine the change in exergy, in kJ, for each of the following processes:

    a. The system is heated at constant pressure until its volume doubles.

    b. The system expands isothermally until its volume doubles.

Let $T_0 = 20°C$, $p_0 = 1$ bar and ignore the effects of motion and gravity.

Applying the Exergy Balance: Closed Systems

7.21 As shown in **Fig. P7.21**, 1.11 kg of Refrigerant 134a is contained in a rigid, insulated vessel. The refrigerant is initially saturated vapor at −28°C. The vessel is fitted with a paddle wheel from which a mass is suspended. As the mass descends a certain distance, the refrigerant is stirred until it attains a final equilibrium state at a pressure of 1.4 bar. The only significant changes in state are experienced by the refrigerant and the suspended mass. Determine, in kJ,

    a. the change in exergy of the refrigerant.

    b. the change in exergy of the suspended mass.

    c. the change in exergy of an isolated system of the vessel and pulley-mass assembly.

    d. the destruction of exergy within the isolated system.

Let $T_0 = 293$ K(20°C), $p_0 = 1$ bar.

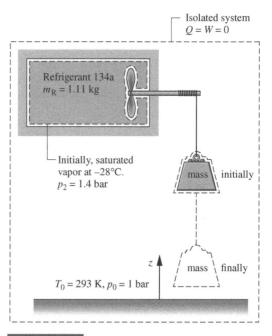

FIGURE P7.21

7.22 A rigid, insulated tank contains 0.6 kg of air, initially at 200 kPa, 20°C. The air is stirred by a paddle wheel until its pressure is 250 kPa. Using the ideal gas model with $c_v = 0.72$ kJ/kg · K, determine, in kJ, (a) the work, (b) the change in exergy of the air, and (c) the amount of exergy destroyed. Ignore the effects of motion and gravity, and let $T_0 = 20°C$, $p_0 = 100$ kPa.

7.23 As shown in **Fig. P7.23**, 1 kg of ammonia is contained in a well-insulated piston–cylinder assembly fitted with an electrical resistor of negligible mass. The ammonia is initially at 140 kPa and a quality of 80%. The resistor is activated until the volume of the ammonia increases by 25%, while its pressure varies negligibly. Determine, in kJ,

    a. the amount of energy transfer by electrical work and the accompanying exergy transfer.

    b. the amount of energy transfer by work to the piston and the accompanying exergy transfer.

    c. the change in exergy of the ammonia.

    d. the amount of exergy destruction.

Ignore the effects of motion and gravity and let $T_0 = 15°C$, $p_0 = 1$ bar.

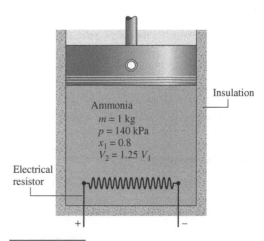

FIGURE P7.23

7.24 One kmol of carbon dioxide gas is contained in a 3 m³ rigid, insulated vessel initially at 4 bar. An electric resistor of negligible mass transfers energy to the gas at a constant rate of 12 kW for 1 min. Employing the ideal gas model and ignoring the effects of motion and gravity, determine (a) the change in exergy of the gas, (b) the electrical work, and (c) the exergy destruction, each in kJ. Let $T_0 = 20°C$, $p_0 = 1$ bar.

7.25 As shown in Fig. P7.25, a 0.3 kg metal bar initially at 1200 K is removed from an oven and quenched by immersing it in a closed tank containing 9 kg of water initially at 300 K. Each substance can be modeled as incompressible. An appropriate constant specific heat for the water is $c_w = 4.2$ kJ/kg · K, and an appropriate value for the metal is $c_m = 0.42$ kJ/kg · K. Heat transfer from the tank contents can be neglected. Determine the exergy destruction, in kJ. Let $T_0 = 25°C$.

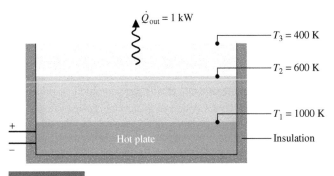

Metal bar:
$T_{mi} = 1200$ K
$c_m = 0.42$ kJ/kg·K
$m_m = 0.3$ kg

Water:
$T_{wi} = 300$ K
$c_w = 4.2$ kJ/kg·K
$m_w = 9$ kg

FIGURE P7.25

7.26 Figure P7.26 provides steady-state data for a composite of a hot plate and two solid layers. Perform a full exergy accounting, in kW, of the electrical power provided to the composite, including the exergy transfer accompanying heat transfer from the composite and the destruction of exergy in the hot plate and each of the two layers. Let $T_0 = 300$ K.

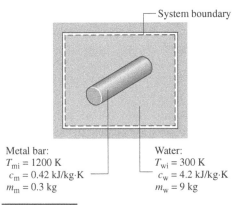

$\dot{Q}_{out} = 1$ kW
$T_3 = 400$ K
$T_2 = 600$ K
$T_1 = 1000$ K
Insulation
Hot plate

FIGURE P7.26

7.27 Figure P7.27 provides steady-state data for the outer wall of a dwelling on a day when the indoor temperature is maintained at 20°C and the outdoor temperature is 0°C. The heat transfer rate through the wall is 1100 W. Determine, in W, the rate of exergy destruction (a) within the wall, and (b) within the enlarged system shown on the figure by the dashed line. Comment. Let $T_0 = 0°C$.

7.28 A gearbox operating at steady state receives 2 kW along the input shaft and delivers 1.4 kW along the output shaft. The outer surface of the gearbox is at 40°C. For the gearbox, (a) determine, in kJ/s, the rate of heat transfer and (b) perform a full exergy accounting, in kJ/s, of the input power. Let $T_0 = 20°C$.

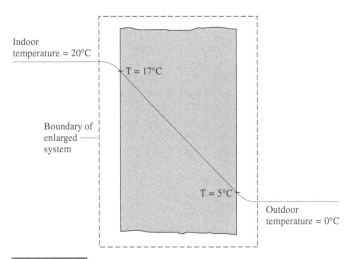

Indoor temperature = 20°C
Boundary of enlarged system
T = 17°C
T = 5°C
Outdoor temperature = 0°C

FIGURE P7.27

7.29 A gearbox operating at steady state receives 15 kW along its input shaft, delivers power along its output shaft, and is cooled on its outer surface according to hA($T_b - T_0$), where $T_b = 45°C$ is the temperature of the outer surface and $T_0 = 5°C$ is the temperature of the surroundings far from the gearbox. The product of the heat transfer coefficient h and outer surface area A is 0.018 kW/K. For the gearbox, determine, in hp, a full exergy accounting of the input power. Let $T_0 = 5°C$.

7.30 At steady state, an electric motor develops power along its output shaft of 0.4 kW while drawing 4 amps at 120 V. The outer surface of the motor is at 50°C. For the motor, (a) determine, in kW, the rate of heat transfer and (b) perform a full exergy accounting, in kW, of the electrical power input. Let $T_0 = 15°C$.

7.31 An electric water heater having a 200-L capacity heats water from 23 to 55°C. Heat transfer from the outside of the water heater is negligible, and the states of the electrical heating element and the tank holding the water do not change significantly. Perform a full exergy accounting, in kJ, of the electricity supplied to the water heater. Model the water as incompressible with a specific heat $c = 4.18$ kJ/kg · K. Let $T_0 = 23°C$.

7.32 A thermal reservoir at 1200 K is separated from another thermal reservoir at 300 K by a cylindrical rod insulated on its lateral surfaces. At steady state, energy transfer by conduction takes place through the rod. The rod diameter is 2 cm, the length is L, and the thermal conductivity is 0.4 kW/m · K. Plot the following quantities, each in kW, versus L ranging from 0.01 to 1 m: the rate of conduction through the rod, the rates of exergy transfer accompanying heat transfer into and out of the rod, and the rate of exergy destruction. Let $T_0 = 300$ K.

7.33 Two kilograms of a two-phase liquid–vapor mixture of water initially at 300°C and $x_1 = 0.5$ undergo the two different processes described below. In each case, the mixture is brought from the initial state to a saturated vapor state, while the volume remains constant. For each process, determine the change in exergy of the water, the net amounts of exergy transfer by work and heat, and the amount of exergy destruction, each in kJ. Let $T_0 = 300$ K, $p_0 = 1$ bar, and ignore the effects of motion and gravity. Comment on the difference between the exergy destruction values.

a. The process is brought about adiabatically by stirring the mixture with a paddle wheel.

b. The process is brought about by heat transfer from a thermal reservoir at 630 K. The temperature of the water at the location where the heat transfer occurs is 630 K.

7.34 As shown in Fig. P7.34, a silicon chip measuring 5 mm on a side and 1 mm in thickness is embedded in a ceramic substrate. At steady state, the chip has an electrical power input of 0.225 W. The top surface of the chip is exposed to a coolant whose temperature is 20°C. The heat transfer coefficient for convection between the chip and the coolant is 150 W/m² · K. Heat transfer by conduction between the chip and the substrate is negligible. Determine (a) the surface temperature of the chip, in °C, and (b) the rate of exergy destruction within the chip, in W. What causes the exergy destruction in this case? Let $T_0 = 293$ K.

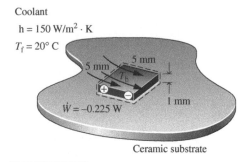

Coolant
h = 150 W/m² · K
$T_f = 20°$ C

5 mm
5 mm
T_b
$\dot{W} = -0.225$ W
1 mm

Ceramic substrate

FIGURE P7.34

Applying the Exergy Balance:
Control Volumes at Steady State

7.35 Determine the specific exergy and the specific flow exergy, each in kJ/kg, for steam at 4.1 MPa, 480°C, with V = 45 m/s and z = 15 m. The velocity and elevation are relative to an exergy reference environment for which $T_0 = 20°$C, $p_0 = 1$ bar, and g = 9.8 m/s².

7.36 At steady state, hot gaseous products of combustion cool from 1540°C to 130°C as they flow through a pipe. Owing to negligible fluid friction, the flow occurs at nearly constant pressure. Applying the ideal gas model with $c_p = 1.04$ kJ/kg · K, determine the exergy transfer accompanying heat transfer from the gas, in kJ/kg of gas flowing. Let $T_0 = 15°$C and ignore the effects of motion and gravity.

7.37 🅒 Steam at 6900 kPa, 316°C enters a valve operating at steady state and undergoes a throttling process.

   a. Determine the exit temperature, in °C, and the exergy destruction rate, in kJ/kg of steam flowing, for an exit pressure of 3450 kPa.

   b. Plot the exit temperature, in °C, and the exergy destruction rate, in kJ/kg of steam flowing, each versus exit pressure ranging from 3450 to 6900 kPa.

Let $T_0 = 20°$C, $p_0 = 100$ kPa.

7.38 Water vapor enters a valve with a mass flow rate of 2.7 kg/s at a temperature of 280°C and a pressure of 30 bar and undergoes a throttling process to 20 bar.

   a. Determine the flow exergy rates at the valve inlet and exit and the rate of exergy destruction, each in kW.

   b. Evaluating exergy at 8 cents per kW · h, determine the annual cost associated with the exergy destruction, assuming 8000 hours of operation annually.

Let $T_0 = 25°$C, $p_0 = 1$ atm.

7.39 Air enters a turbine operating at steady state with a pressure of 520 kPa, a temperature of 450 K, and a velocity of 120 m/s. At the turbine exit, the conditions are 100 kPa, 330 K, and 30 m/s. Heat transfer from the turbine to its surroundings takes place at an average surface temperature of 345 K. The rate of heat transfer is 4.65 kJ/kg of air passing through the turbine. For the turbine, determine the work developed and the exergy destruction, each in kJ/kg of air flowing. Let $T_0 = 5°$C, $p_0 = 100$ kPa.

7.40 Air enters a compressor operating at steady state at $T_1 = 300$ K, $p_1 = 1$ bar with a velocity of 70 m/s. At the exit, $T_2 = 540$ K, $p_2 = 5$ bar and the velocity is 150 m/s. The air can be modeled as an ideal gas with $c_p = 1.01$ kJ/kg · K. Stray heat transfer can be ignored. Determine, in kJ per kg of air flowing, (a) the power required by the compressor and (b) the rate of exergy destruction within the compressor. Let $T_0 = 300$ K, $p_0 = 1$ bar. Ignore the effects of motion and gravity.

7.41 Steam enters a turbine operating at steady state at 6 MPa, 500°C with a mass flow rate of 400 kg/s. Saturated vapor exits at 8 kPa. Heat transfer from the turbine to its surroundings takes place at a rate of 8 MW at an average surface temperature of 180°C. The effects of motion and gravity are negligible.

   a. For a control volume enclosing the turbine, determine the power developed and the rate of exergy destruction, each in MW.

   b. If the turbine is located in a facility where the ambient temperature is 27°C, determine the rate of exergy destruction for an enlarged control volume that includes the turbine and its immediate surroundings so the heat transfer takes place at the ambient temperature. Explain why the exergy destruction values of parts (a) and (b) differ.

Let $T_0 = 300$ K, $p_0 = 100$ kPa.

7.42 Refrigerant 134a at −10°C, 1.4 bar, and a mass flow rate of 280 kg/h enters an insulated compressor operating at steady state and exits at 9 bar. The isentropic compressor efficiency is 82%. Determine

   a. the temperature of the refrigerant exiting the compressor, in °C.

   b. the power input to the compressor, in kW.

   c. the rate of exergy destruction, in kW.

Ignore the effects of motion and gravity and let $T_0 = 20°$C, $p_0 = 1$ bar.

7.43 For the *vortex tube* of Example 6.7, determine the rate of exergy destruction, in kJ/kg of air entering. Referring to this value for exergy destruction, comment on the inventor's claim. Let $T_0 = 295$ K, $p_0 = 1$ bar.

7.44 🅒 Steam enters an insulated turbine operating at steady state at 700 kPa, 260°C, with a mass flow rate of 38 kg/s and expands to a pressure of 1 bar. The isentropic turbine efficiency is 80%. If exergy is valued at 8 cents per kW · h, determine

   a. the value of the power produced, in $/h.

   b. the cost of the exergy destroyed, in $/h.

   c. Plot the values of the power produced and the exergy destroyed, each in $/h, versus isentropic efficiency ranging from 80 to 100%.

Ignore the effects of motion and gravity. Let $T_0 = 20°$C, $p_0 = 1$ bar.

7.45 Figure P7.45 shows a device to develop power using a heat transfer from a high-temperature industrial process together with a steam input. The figure provides data for steady-state operation. All surfaces are well insulated, except for the one at 527°C, across which heat transfer occurs at a rate of 4.21 kW. The device develops power at a rate of 6 kW. Determine, in kW,

   a. the rate exergy enters accompanying heat transfer.

   b. the *net* rate exergy is carried in by the steam, $(\dot{E}_{f1} - \dot{E}_{f2})$.

   c. the rate of exergy destruction within the device.

Ignore the effects of motion and gravity and let $T_0 = 293$ K, $p_0 = 1$ bar.

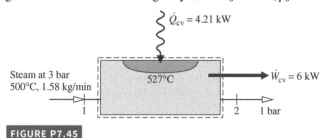

$\dot{Q}_{cv} = 4.21$ kW

Steam at 3 bar
500°C, 1.58 kg/min
527°C
$\dot{W}_{cv} = 6$ kW
1
2   1 bar

FIGURE P7.45

7.46 Water at $T_1 = 40°C$, $p_1 = 210$ kPa enters a counterflow heat exchanger operating at steady state with a mass flow rate of 45 kg/s and exits at $T_2 = 94°C$ with closely the same pressure. Air enters in a separate stream at $T_3 = 280°C$ and exits at $T_4 = 60°C$ with no significant change in pressure. Air can be modeled as an ideal gas and stray heat transfer can be ignored. Determine (a) the mass flow rate of the air, in kg/s, and (b) the rate of exergy destruction within the heat exchanger, in kJ/s. Ignore the effects of motion and gravity and let $T_0 = 15°C$, $p_0 = 1$ bar.

7.47 Liquid water enters a heat exchanger operating at steady state at $T_1 = 15°C$, $p_1 = 1$ bar and exits at $T_2 = 70°C$ with a negligible change in pressure. In a separate stream, steam enters at $T_3 = 140$ kPa, $x_3 = 92\%$ and exits at $T_4 = 60°C$, $p_4 = 124$ kPa. Stray heat transfer and the effects of motion and gravity are negligible. Let $T_0 = 15°C$, $p_0 = 1$ bar. Determine (a) the ratio of the mass flow rates of the two streams and (b) the rate of exergy destruction, in kJ/kg of steam entering the heat exchanger.

7.48 Argon enters a nozzle operating at steady state at 1400 K, 380 kPa with a velocity of 15 m/s and exits the nozzle at 950 K, 140 kPa. Stray heat transfer can be ignored. Modeling argon as an ideal gas with $k = 1.77$, determine (a) the velocity at the exit, in m/s, and (b) the rate of exergy destruction, in kJ per kg of argon flowing. Let $T_0 = 298$ K, $p_0 = 1$ bar.

7.49 Nitrogen (N_2) enters a well-insulated nozzle operating at steady state at 520 kPa, 670 K, 20 m/s. At the nozzle exit, the pressure is 140 kPa. The isentropic nozzle efficiency is 90%. For the nozzle, determine the exit velocity, in m/s, and the exergy destruction rate, in kJ/kg of nitrogen flowing. Let $T_0 = 20°C$, $p_0 = 100$ kPa.

7.50 Steady-state operating data are shown in **Fig. P7.50** for an open feedwater heater. Heat transfer from the feedwater heater to its surroundings occurs at an average outer surface temperature of 50°C at a rate of 100 kW. Ignore the effects of motion and gravity and let $T_0 = 25°C$, $p_0 = 1$ bar. Determine

a. the ratio of the incoming mass flow rates, $\dot{m}_1/\dot{m}_2$.

b. the rate of exergy destruction, in kW.

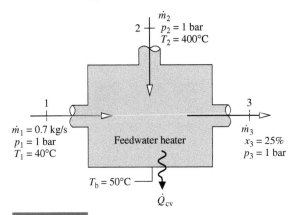

FIGURE P7.50

7.51 **Figure P7.51** provides steady-state operating data for a mixing chamber in which entering liquid and vapor streams of water mix to form an exiting saturated liquid stream. Heat transfer from the mixing chamber to its surroundings occurs at an average surface temperature of 38°C. The effects of motion and gravity are negligible. Let $T_0 = 20°C$, $p_0 = 1$ bar. For the mixing chamber, determine, each in kJ/s, (a) the rate of heat transfer and the accompanying rate of exergy transfer and (b) the rate of exergy destruction.

7.52 Liquid water at 140 kPa, 10°C enters a mixing chamber operating at steady state with a mass flow rate of 2.3 kg/s and mixes with a separate stream of steam entering at 140 kPa, 120°C with a mass flow rate of 0.17 kg/s. A single mixed stream exits at

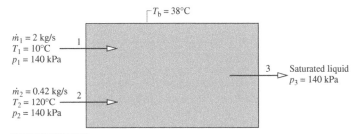

FIGURE P7.51

140 kPa, 55°C. Heat transfer from the mixing chamber occurs to its surroundings. Neglect the effects of motion and gravity and let $T_0 = 20°C$, $p_0 = 1$ bar. Determine the rate of exergy destruction, in kg/s, for a control volume including the mixing chamber and enough of its immediate surroundings that heat transfer occurs at 20°C.

7.53 Steam at 30 bar and 700°C is available at one location in an industrial plant. At another location, steam at 20 bar and 400°C is required for use in a certain process. An engineer suggests that steam at this condition can be provided by allowing the higher-pressure steam to expand through a valve to 20 bar and then cool to 400°C through a heat exchanger with heat transfer to the surroundings, which are at 20°C.

a. Evaluate this suggestion by determining the associated exergy destruction rate per mass flow rate of steam (kJ/kg) for the valve and heat exchanger. Discuss.

b. Evaluating exergy at 8 cents per kW · h and assuming continuous operation at steady state, determine the total annual cost, in $, of the exergy destruction for a mass flow rate of 1 kg/s.

c. Suggest an alternative method for obtaining steam at the desired condition that would be preferable thermodynamically, and determine the total annual cost, in $, of the exergy destruction for a mass flow rate of 1 kg/s. Let $T_0 = 20°C$, $p_0 = 1$ atm.

7.54 A gas turbine operating at steady state is shown in **Fig. P7.54**. Air enters the compressor with a mass flow rate of 5 kg/s at 0.95 bar and 22°C and exits at 5.7 bar. The air then passes through a heat exchanger before entering the turbine at 1100 K, 5.7 bar. Air exits the turbine at 0.95 bar. The compressor and turbine operate adiabatically and the effects of motion and gravity can be ignored. The compressor and turbine isentropic efficiencies are 82 and 85%, respectively. Using the ideal gas model for air, determine, each in kW,

a. the *net* power developed.

b. the rates of exergy destruction for the compressor and turbine.

c. the *net* rate exergy is carried out of the plant at the turbine exit, $(\dot{E}_{f4} - \dot{E}_{f1})$.

Let $T_0 = 22°C$, $p_0 = 0.95$ bar.

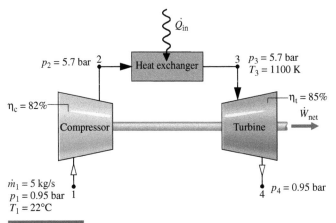

FIGURE P7.54

7.55 At steady state, steam with a mass flow rate of 4.5 kg/s enters a turbine at 430°C and 4140 kPa and expands to 414 kPa. The power developed by the turbine is 2126 kW. The steam then passes through a counterflow heat exchanger with a negligible change in pressure, exiting at 430°C. Air enters the heat exchanger in a separate stream at 1.1 bar, 550°C and exits at 1 bar, 330°C. The effects of motion and gravity can be ignored and there is no significant heat transfer between either component and its surroundings. Determine

  a. the mass flow rate of air, in kg/s.

  b. the rates of exergy destruction in the turbine and heat exchanger, each in kJ/s.

Evaluating exergy at 8 cents per kW · h, determine the hourly cost of each of the exergy destructions found in part (b). Let $T_0 = 4°C$, $p_0 = 1$ bar.

7.56 Figure P7.56 shows a gas turbine power plant operating at steady state consisting of a compressor, a heat exchanger, and a turbine. Air enters the compressor with a mass flow rate of 3.9 kg/s at 0.95 bar, 22°C and exits the turbine at 0.95 bar, 421°C. Heat transfer to the air as it flows through the heat exchanger occurs at an average temperature of 488°C. The compressor and turbine operate adiabatically. Using the ideal gas model for the air, and neglecting the effects of motion and gravity, determine, in MW,

  a. the rate of exergy transfer accompanying heat transfer to the air flowing through the heat exchanger.

  b. the *net* rate exergy is carried out of the plant at the turbine exit, $(\dot{E}_{f2} - \dot{E}_{f1})$.

  c. the rate of exergy destruction within the power plant.

  d. Using the results of parts (a)–(c), perform a full exergy accounting of the exergy supplied to the power plant accompanying heat transfer. Comment.

Let $T_0 = 295$ K(22°C), $p_0 = 0.95$ bar.

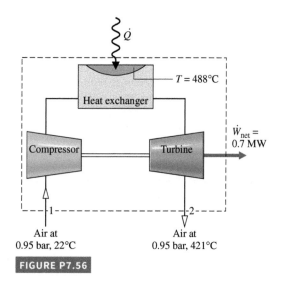

Air at
0.95 bar, 22°C

Air at
0.95 bar, 421°C

FIGURE P7.56

7.57 Figure P7.57 shows a power-generating system at steady state. Saturated liquid water enters at 80 bar with a mass flow rate of 94 kg/s. Saturated liquid exits at 0.08 bar with the same mass flow rate. As indicated by arrows, three heat transfers occur, each at a specified temperature in the direction of the arrow: The first adds 135 MW at 295°C, the second adds 55 MW at 375°C, and the third removes energy at 20°C. The system generates power at the rate of 80 MW. The effects of motion and gravity can be ignored. Let $T_0 = 20°C$, $p_0 = 1$ atm. Determine, in MW, (a) the rate of heat transfer $\dot{Q}_3$ and the accompanying rate of exergy transfer and (b) a full exergy accounting of the *total* exergy supplied to the system with the two heat additions and with the *net* exergy, $(\dot{E}_{f1} - \dot{E}_{f2})$, carried in by the water stream as it passes from inlet to exit.

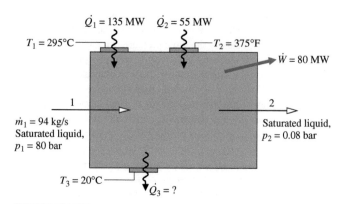

FIGURE P7.57

7.58 Figure P7.58 shows a gas turbine power plant using air as the working fluid. The accompanying table gives steady-state operating data. Air can be modeled as an ideal gas. Stray heat transfer and the effects of motion and gravity can be ignored. Let $T_0 = 290$ K, $p_0 = 100$ kPa. Determine, each in kJ per kg of air flowing, (a) the *net* power developed, (b) the *net* exergy increase of the air passing through the heat exchanger, $(e_{f3} - e_{f2})$, and (c) a full exergy accounting based on the exergy supplied to the plant found in part (b). Comment.

State	p(kPa)	T(K)	h(kJ/kg)	s°(kJ/kg · K)[a]
1	100	290	290.16	1.6680
2	500	505	508.17	2.2297
3	500	875	904.99	2.8170
4	100	635	643.93	2.4688

[a]$s°$ is the variable appearing in Eq. 6.20a and Table A-22.

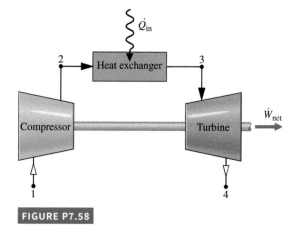

FIGURE P7.58

7.59 Carbon dioxide (CO_2) gas enters a turbine operating at steady state at 50 bar, 500 K with a velocity of 50 m/s. The inlet area is 0.02 m^2. At the exit, the pressure is 20 bar, the temperature is 440 K, and the velocity is 10 m/s. The power developed by the turbine is 3 MW, and heat transfer occurs across a portion of the surface where the average temperature is 462 K. Assume ideal gas behavior for the carbon dioxide and neglect the effect of gravity. Let $T_0 = 298$ K, $p_0 = 1$ bar.

  a. Determine the rate of heat transfer, in kW.

  b. Perform a full exergy accounting, in kW, based on the *net* rate exergy is carried into the turbine by the carbon dioxide.

7.60 Air is compressed in an axial-flow compressor operating at steady state from 27°C, 1 bar to a pressure of 2.1 bar. The work required is 94.6 kJ per kg of air flowing. Heat transfer from the compressor occurs at an average surface temperature of 40°C at the rate of 14 kJ per kg of air flowing. The effects of motion and gravity can be ignored. Let $T_0 = 20°C$, $p_0 = 1$ bar. Assuming ideal gas behavior, (a) determine the temperature of the air at the exit, in °C, (b) determine the rate of exergy destruction within the compressor, in kJ per kg of air flowing, and (c) perform a full exergy accounting, in kJ per kg of air flowing, based on work input.

7.61 **Figure P7.61** shows liquid water at 550 kPa, 150°C entering a flash chamber though a valve at the rate of 10 kg/s. At the valve exit, the pressure is 290 kPa. Saturated liquid at 275 kPa exits from the bottom of the flash chamber and saturated vapor at 275 kPa exits from near the top. The vapor stream is fed to a steam turbine having an isentropic efficiency of 90% and an exit pressure of 14 kPa. For steady-state operation, negligible heat transfer with the surroundings, and no significant effects of motion and gravity, perform a full exergy accounting, in kJ/s, of the *net* rate at which exergy is supplied: $(\dot{E}_{f1} - \dot{E}_{f3} - \dot{E}_{f5})$. Let $T_0 = 280$ K, $p_0 = 1$ bar.

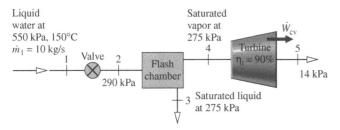

FIGURE P7.61

7.62 A compressor fitted with a water jacket and operating at steady state takes in air with a volumetric flow rate of 900 m³/h at 22°C, 0.95 bar and discharges air at 317°C, 8 bar. Cooling water enters the water jacket at 20°C, 100 kPa with a mass flow rate of 1400 kg/h and exits at 30°C and essentially the same pressure. There is no significant heat transfer from the outer surface of the water jacket to its surroundings, and the effects of motion and gravity can be ignored. For the water-jacketed compressor, perform a full exergy accounting of the power input. Let $T_0 = 20°C$, $p_0 = 1$ bar.

7.63 **Figure P7.63** provides steady-state operating data for a throttling valve in parallel with a steam turbine having an isentropic turbine efficiency of 90%. The streams exiting the valve and the turbine mix in a mixing chamber. Heat transfer with the surroundings and the effects of motion and gravity can be neglected. Determine

a. the power developed by the turbine, in kJ/s.

b. the mass flow rates through the turbine and valve, each in kg/s.

c. a full exergy accounting, in kJ/s, of the *net* rate at which exergy is supplied: $(\dot{E}_{f1} - \dot{E}_{f4})$.

Let $T_0 = 280$ K, $p_0 = 1$ bar.

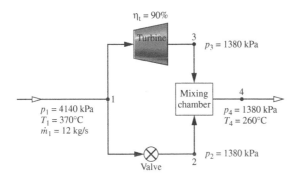

FIGURE P7.63

Using Exergetic Efficiencies

7.64 Plot the exergetic efficiency given by Eq. 7.21b versus T_u/T_0 for $T_s/T_0 = 8.0$ and $\eta = 0.4, 6, 0.8, 1.0$. What can be learned from the plot when T_u/T_0 is fixed? When ε is fixed? Discuss.

7.65 For the heat exchanger of Example 7.6, evaluate the exergetic efficiency given by Eq. 7.27 with states numbered for the case at hand.

7.66 Steam enters a turbine operating at steady state at $p_1 = 12$ MPa, $T_1 = 700°C$ and exits at $p_2 = 0.6$ MPa. The isentropic turbine efficiency is 88%. Property data are provided in the accompanying table. Stray heat transfer and the effects of motion and gravity are negligible. Let $T_0 = 300$ K, $p_0 = 100$ kPa. Determine (a) the power developed and the rate of exergy destruction, each in kJ per kg of steam flowing, and (b) the exergetic turbine efficiency.

State	p (MPa)	$T(°C)$	h (kJ/kg)	s (kJ/kg · K)
Turbine inlet	12	700	3858.4	7.0749
Turbine exit	0.6	($n_t = 88\%$)	3017.5	7.2938

7.67 Saturated liquid water at 0.01 MPa enters a power plant pump operating at a steady state. Liquid water exits the pump at 10 MPa. The isentropic pump efficiency is 90%. Property data are provided in the accompanying table. Stray heat transfer and the effects of motion and gravity are negligible. Let $T_0 = 300$ K, $p_0 = 100$ kPa. Determine (a) the power required by the pump and the rate of exergy destruction, each in kJ per kg of water flowing, and (b) the exergetic pump efficiency.

State	p (MPa)	h (kJ/kg)	s (kJ/kg · K)
Pump inlet	0.01	191.8	0.6493
Pump exit	10	204.5	0.6531

7.68 **Figure P7.68** provides two options for generating hot water at steady state. In (a), water heating is achieved by utilizing *industrial waste heat* supplied at a temperature of 500 K. In (b), water heating is achieved by an electrical resistor. For each case, devise and evaluate an exergetic efficiency. Compare the calculated efficiency values and comment. Stray heat transfer and the effects of motion and gravity are negligible. Let $T_0 = 20°C$, $p_0 = 1$ bar.

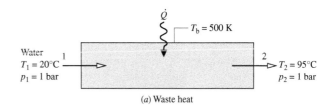

(a) Waste heat

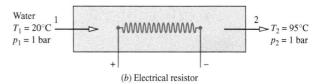

(b) Electrical resistor

FIGURE P7.68

7.69 Hydrogen at 25 bar, 450°C enters a turbine and expands to 2 bar, 160°C with a mass flow rate of 0.2 kg/s. The turbine operates at steady state with negligible heat transfer with its surroundings. Assuming the ideal gas model with $k = 1.37$ and ignoring the effects of motion and gravity, determine

a. the isentropic turbine efficiency.

b. the exergetic turbine efficiency.

Let $T_0 = 25°C$, $p_0 = 1$ atm.

7.70 Air enters an insulated turbine operating at steady state with a pressure of 4 bar, a temperature of 450 K, and a volumetric flow rate of 5 m^3/s. At the exit, the pressure is 1 bar. The isentropic turbine efficiency is 84%. Assuming the ideal gas model and ignoring the effects of motion and gravity, determine

a. the power developed and the exergy destruction rate, each in kW.

b. the exergetic turbine efficiency.

Let $T_0 = 20°C$, $p_0 = 1$ bar.

7.71 Steam at 1.4 MPa, 350°C enters a turbine operating at steady state with a mass flow rate of 0.12 kg/s and exits at 100 kPa, 115°C. Stray heat transfer and the effects of motion and gravity can be ignored. Let $T_0 = 20°C$, $p_0 = 100$ kPa. Determine for the turbine (a) the power developed and the rate of exergy destruction, each in kJ/s, and (b) the isentropic and exergetic turbine efficiencies.

7.72 Water vapor at 6 MPa, 600°C enters a turbine operating at steady state and expands adiabatically to 10 kPa. The mass flow rate is 2 kg/s and the isentropic turbine efficiency is 94.7%. Kinetic and potential energy effects are negligible. Determine

a. the power developed by the turbine, in kW.

b. the rate at which exergy is destroyed within the turbine, in kW.

c. the exergetic turbine efficiency.

Let $T_0 = 298$ K, $p_0 = 1$ atm.

7.73 Figure P7.73 shows a turbine operating at steady state with steam entering at $p_1 = 30$ bar, $T_1 = 350°C$ and a mass flow rate of 30 kg/s. Process steam is extracted at $p_2 = 5$ bar, $T_2 = 200°C$. The remaining steam exits at $p_3 = 0.15$ bar, $x_3 = 90\%$, and a mass flow rate of 25 kg/s. Stray heat transfer and the effects of motion and gravity are negligible. Let $T_0 = 25°C$, $p_0 = 1$ bar. The accompanying table provides property data at key states. For the turbine, determine the power developed and rate of exergy destruction, each in MW. Also devise and evaluate an exergetic efficiency for the turbine.

State	p (bar)	T (°C)	h (kJ/kg)	s (kJ/kg · K)
1	30	350	3115.3	6.7428
2	5	200	2855.4	7.0592
3	0.15	(x = 90%)	2361.7	7.2831

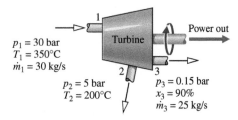

$p_1 = 30$ bar
$T_1 = 350°C$
$\dot{m}_1 = 30$ kg/s

Power out
Turbine

$p_2 = 5$ bar
$T_2 = 200°C$

$p_3 = 0.15$ bar
$x_3 = 90\%$
$\dot{m}_3 = 25$ kg/s

FIGURE P7.73

7.74 **C** Steam at 2760 kPa, 315°C enters a well-insulated turbine operating at steady state and exits as saturated vapor at a pressure p.

a. For $p = 345$ kPa, determine the exergy destruction rate, in kJ/kg of steam expanding through the turbine, and the turbine exergetic and isentropic efficiencies.

b. Plot the exergy destruction rate, in kJ/kg of steam flowing, and the exergetic efficiency and isentropic efficiency, each versus pressure p ranging from 7 to 350 kPa.

Ignore the effects of motion and gravity and let $T_0 = 15°C$, $p_0 = 1$ bar.

7.75 **C** Saturated water vapor at 2760 kPa enters an insulated turbine operating at steady state. A two-phase liquid–vapor mixture exits at

4 kPa. Plot each of the following versus the steam quality at the turbine exit ranging from 75 to 100%

a. the power developed and the rate of exergy destruction, each in kJ/kg of steam flowing.

b. the isentropic turbine efficiency.

c. the exergetic turbine efficiency.

Let $T_0 = 15°C$, $p_0 = 1$ bar. Ignore the effects of motion and gravity.

7.76 **C** Oxygen (O_2) enters an insulated turbine operating at steady state at 900°C and 3 MPa and exhausts at 400 kPa. The mass flow rate is 0.75 kg/s. Plot each of the following versus the turbine exit temperature, in °C:

a. the power developed, in kW.

b. the rate of exergy destruction in the turbine, in kW.

c. the exergetic turbine efficiency.

For the oxygen, use the ideal gas model with $k = 1.395$. Ignore the effects of motion and gravity. Let $T_0 = 30°C$, $p_0 = 1$ bar.

7.77 A pump operating at steady state takes in saturated liquid water at 34 kPa and discharges water at 10 MPa. The isentropic pump efficiency is 75%. Heat transfer with the surroundings and the effects of motion and gravity can be neglected. If $T_0 = 20°C$, determine for the pump

a. exergy destruction, in kJ/kg of water flowing.

b. the exergetic efficiency.

7.78 Refrigerant 134a as saturated vapor at −10°C enters a compressor operating at steady state with a mass flow rate of 0.3 kg/s. At the compressor exit the pressure of the refrigerant is 5 bar. Stray heat transfer and the effects of motion and gravity can be ignored. If the rate of exergy destruction within the compressor much be kept less than 2.4 kW, determine the allowed ranges for (a) the power required by the compressor, in kW, and (b) the isentropic and exergetic compressor efficiencies. Let $T_0 = 298$K, $p_0 = 1$ bar.

7.79 A compressor operating at steady state takes in 1 kg/s of air at 1 bar and 25°C and compresses it to 8 bar and 160°C. The power input to the compressor is 230 kW, and heat transfer occurs from the compressor to the surroundings at an average surface temperature of 50°C.

a. Perform a full exergy accounting of the power input to the compressor.

b. Devise and evaluate an exergetic efficiency for the compressor.

c. Evaluating exergy at 8 cents per kW · h, determine the hourly costs of the power input, exergy loss associated with heat transfer, and exergy destruction.

Neglect the effects of motion and gravity. Let $T_0 = 25°C$, $p_0 = 1$ bar.

7.80 A counterflow heat exchanger operates at steady state. Air flows on both sides with a mass flow rate of 1 kg/s. On one side, air enters at 470 K, 414 kPa and exits at 555 K, 345 kPa. On the other side, air enters at 720 K, 110 kPa and exits at 640 K and 100 kPa. Heat transfer between the heat exchanger and its surroundings can be ignored, as can the effects of motion and gravity. Evaluate for the heat exchanger

a. the rate of exergy destruction, in kJ/s.

b. the exergetic efficiency given by Eq. 7.27.

Let $T_0 = 15°C$, $p_0 = 1$ bar.

7.81 Saturated water vapor at 1 bar enters a direct-contact heat exchanger operating at steady state and mixes with a stream of liquid water entering at 25°C, 1 bar. A two-phase liquid–vapor mixture exits at 1 bar. The entering streams have equal mass flow rates. Neglecting heat transfer with the surroundings and the effects of motion and gravity, determine for the heat exchanger

   a. the rate of exergy destruction, in kJ per kg of mixture exiting.

   b. the exergetic efficiency given by Eq. 7.29.

Let $T_0 = 20°C$, $p_0 = 1$ bar.

7.82 Refrigerant 134a enters a counterflow heat exchanger operating at steady state at $-20°C$ and a quality of 35% and exits as saturated vapor at $-20°C$. Air enters as a separate stream with a mass flow rate of 4 kg/s and is cooled at a constant pressure of 1 bar from 300 to 260 K. Heat transfer between the heat exchanger and its surroundings can be ignored, as can the effects of motion and gravity.

   a. As in Fig. E7.6, sketch the variation with position of the temperature of each stream. Locate T_0 on the sketch.

   b. Determine the rate of exergy destruction within the heat exchanger, in kW.

   c. Devise and evaluate an exergetic efficiency for the heat exchanger.

Let $T_0 = 300$ K, $p_0 = 1$ bar.

7.83 Liquid water at 95°C, 1 bar enters a direct-contact heat exchanger operating at steady state and mixes with a stream of liquid water entering at 15°C, 1 bar. A single liquid stream exits at 1 bar. The entering streams have equal mass flow rates. Neglecting heat transfer with the surroundings and the effects of motion and gravity, determine for the heat exchanger

   a. the rate of exergy destruction, in kJ per kg of liquid exiting.

   b. the exergetic efficiency given by Eq. 7.29.

Let $T_0 = 15°C$, $p_0 = 1$ bar.

7.84 Figure P7.84 shows a *cogeneration* system producing two useful products: net power and process steam. The accompanying table provides steady-state mass flow rate, temperature, pressure, and flow exergy data at the ten numbered states on the figure. Stray heat transfer and the effects of motion and gravity can be ignored. Let $T_0 = 298.15$ K, $p_0 = 1.013$ bar. Determine, in MW

   a. the *net* rate exergy is carried out with the process steam, $(\dot{E}_{f9} - \dot{E}_{f8})$.

   b. the *net* rate exergy is carried out with the combustion products, $(\dot{E}_{f7} - \dot{E}_{f1})$.

   c. the rates of exergy destruction in the air preheater, heat-recovery steam generator, and the combustion chamber.

Devise and evaluate an exergetic efficiency for the overall cogeneration system.

State	Substance	Mass Flow Rate (kg/s)	Temperature (K)	Pressure (bar)	Flow Exergy Rate, $\dot{E}_f$ (MW)
1	Air	91.28	298.15	1.013	0.00
2	Air	91.28	603.74	10.130	27.54
3	Air	91.28	850.00	9.623	41.94
4	Combustion products	92.92	1520.00	9.142	101.45
5	Combustion products	92.92	1006.16	1.099	38.78
6	Combustion products	92.92	779.78	1.066	21.75
7	Combustion products	92.92	426.90	1.013	2.77
8	Water	14.00	298.15	20.000	0.06
9	Water	14.00	485.57	20.000	12.81
10	Methane	1.64	298.15	12.000	84.99

7.85 Figure P7.85 shows a *combined* gas turbine–vapor power plant operating at steady state. The gas turbine is numbered 1–5. The vapor power plant is numbered 6–9. The accompanying table gives data at these numbered states. The total *net* power output is 45 MW and the mass flow rate of the water flowing through the vapor power plant is 15.6 kg/s. Air flows through the gas turbine power plant, and the ideal gas model applies to the air. Stray heat transfer and the effects of motion and gravity can be ignored. Let $T_0 = 300$ K, $p_0 = 100$ kPa. Determine

   a. the mass flow rate of the air flowing through the gas turbine, in kg/s.

   b. the *net* rate exergy is carried out with the exhaust air stream, $(\dot{E}_{f5} - \dot{E}_{f1})$ in MW.

   c. the rate of exergy destruction in the compressor and pump, each in MW.

   d. the *net* rate of exergy increase of the air flowing through the combustor, $(\dot{E}_{f5} - \dot{E}_{f1})$, in MW.

Devise and evaluate an exergetic efficiency for the overall combined power plant.

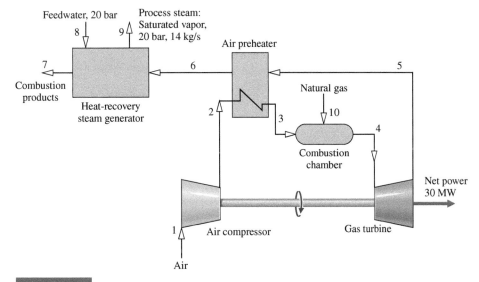

Feedwater, 20 bar

Process steam: Saturated vapor, 20 bar, 14 kg/s

Air preheater

Combustion products

Heat-recovery steam generator

Natural gas

Combustion chamber

Net power 30 MW

Air compressor

Gas turbine

Air

FIGURE P7.84

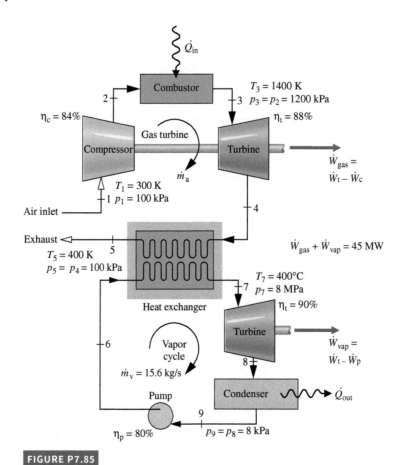

Gas Turbine			Vapor Cycle		
State	h(kJ/kg)	$s°$(kJ/kg · K)[a]	State	h(kJ/kg)	s(kJ/kg · K)
1	300.19	1.7020	6	183.96	0.5975
2	669.79	2.5088	7	3138.30	6.3634
3	1515.42	3.3620	8	2104.74	6.7282
4	858.02	2.7620	9	173.88	0.5926
5	400.98	1.9919			

[a]$s°$ is the variable appearing in Eq. 6.20a and Table A-22.

Considering Thermoeconomics

7.86 The rate of heat transfer from the outer surface of an electric water heater to the surroundings is given by $hA(T_b - T_f)$ where $hA = 8.9 \times 10^{-3}$ kW/K, T_b is the surface temperature, in K, and $T_f = 293$ K is the temperature of the surroundings at a distance. Evaluating electricity at 8 cents per kW · h

    a. determine the cost of the heat loss, in $ per year, when $T_b = 297$ K.

    b. plot the cost of the heat loss, in $ per year, versus T_b ranging from 297 K to 317 K.

Let $T_0 = 293$ K.

7.87 Reconsider Example 7.10 for a turbine exit state fixed by $p_2 = 2$ bar, $h_2 = 2723.7$ kJ/kg, $s_2 = 7.1699$ kJ/kg · K. The cost of owning and operating the turbine is $\dot{Z}_t = 7.2\,\dot{W}_e$, in $/h, where $\dot{W}_e$ is in MW. All other data remain unchanged. Determine

    a. the power developed by the turbine, in MW.

    b. the exergy destroyed within the turbine, in MW.

    c. the exergetic turbine efficiency.

    d. the unit cost of the turbine power, in cents per kW · h of exergy.

7.88 A cogeneration system operating at steady state is shown schematically in **Fig. P7.88**. The exergy transfer rates of the entering and exiting streams are shown on the figure, in MW. The fuel, produced by reacting coal with steam, has a unit cost of 5.85 cents per kW · h of exergy. The cost of owning and operating the system is $1800/h. The feedwater and combustion air enter with negligible exergy and cost. Expenses related to proper disposal of the combustion products are included with the cost of owning and operating the system.

    a. Determine the rate of exergy destruction within the cogeneration system, in MW.

    b. Devise and evaluate an exergetic efficiency for the system.

    c. Assuming the power and steam each have the same unit cost based on exergy, evaluate the unit cost, in cents per kW · h. Also evaluate the cost rates of the power and steam, each in $/h.

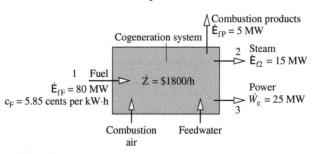

7.89 Figure P7.89 shows a boiler at steady state. Steam having a specific flow exergy of 1300 kJ/kg exits the boiler at a mass flow rate of 5.69×10^4 kg/h. The cost of owning and operating the boiler is \$91/h. The ratio of the exiting steam exergy to the entering fuel exergy is 0.45. The unit cost of the fuel based on exergy is \$1.50 per 10^6 kJ. If the cost rates of the combustion air, feedwater, heat transfer with the surroundings, and exiting combustion products are ignored, develop

 a. an expression for the unit cost based on exergy of the steam exiting the boiler.

 b. Using the result of part (a), determine the unit cost of the steam, in cents per kg of steam flowing.

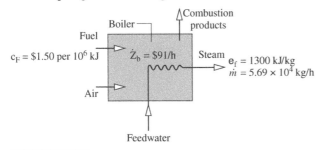

Fuel — $c_F = \$1.50$ per 10^6 kJ

Boiler — $\dot{Z}_b = \$91/h$

Combustion products

Steam — $e_f = 1300$ kJ/kg, $\dot{m} = 5.69 \times 10^4$ kg/h

Air

Feedwater

FIGURE P7.89

7.90 [C] The table below gives alternative specifications for the state of the process steam exiting the turbine of Example 7.10. The cost of owning and operating the turbine, in \$/h, varies with the power $\dot{W}_e$, in MW, according to $\dot{Z}_t = 7.2\dot{W}_e$. All other data remain unchanged.

p_2 (bar)	40	30	20	9	5	2	1
T_2(°C)	436	398	349	262	205	128	sat

Plot versus p_2, in bar

 a. the power $\dot{W}_e$, in MW.

 b. the unit costs of the power and process steam, each in cents per kW · h of exergy.

 c. the unit cost of the process steam, in cents per kg of steam flowing.

7.91 At steady state, a turbine with an exergetic efficiency of 85% develops 18×10^7 kW · h of work annually (8000 operating hours). The annual cost of owning and operating the turbine is $\$5.0 \times 10^5$. The steam entering the turbine has a specific flow exergy of 1500 kJ/kg, a mass flow rate of 40 kg/s, and is valued at \$0.0182 per kW · h of exergy.

 a. Evaluate the unit cost of the power developed, in \$ per kW · h.

 b. Evaluate the unit cost based on exergy of the steam entering and exiting the turbine, each in cents per kg of steam flowing through the turbine.

DESIGN & OPEN-ENDED PROBLEMS: EXPLORING ENGINEERING PRACTICE

7.1D A utility charges households the same per kW · h for space heating via steam radiators as it does for electricity. Critically evaluate this costing practice and prepare a memorandum summarizing your principal conclusions.

7.2D Many appliances, including ovens, stoves, clothes dryers, and hot-water heaters, offer a choice between electric and gas operation. Select an appliance that offers this choice and perform a detailed comparison between the two options, including but not necessarily limited to a *life-cycle* exergy analysis and an economic analysis accounting for purchase, installation, operating, maintenance, and disposal costs. Present your finding in a poster presentation.

7.3D Proposals for water-fueled cars have appeared on the Internet. One inventor claims that the car's battery starts the engine, but once the engine is running the car's alternator powers an on-board unit that extracts hydrogen from water. It is further claimed that the hydrogen fuels the engine that powers the alternator. As long as there is water available on-board, the inventor says the entire system is self-sufficient. Investigate the feasibility of water-fueled cars. Write a report, including at least three references.

7.4D You have been asked to testify before an agricultural committee of your state legislature that is crafting regulations pertaining to the production of methane using poultry waste. Develop a policy briefing providing a balanced assessment of issues relating to methane production from this source, including engineering, public health, environmental, and economic considerations.

7.5D Heat pump water heating systems have been introduced that often operate more efficiently than conventional electrical and gas-fueled water heaters depending on factors including ambient temperature and relative humidity conditions, water usage, and setpoint temperature. For a family of four in your locale, investigate the feasibility of using a heat pump water heating system. Include a detailed economic evaluation accounting for equipment, installation, and operating costs compared to those for a conventional electric resistance water heater. Place your findings on a poster suitable for presentation at a technical conference.

7.6D *Anaerobic digestion* is a proven means of producing methane from livestock waste. To provide for the space heating, water heating, and cooking needs of a typical farm dwelling in your locale, determine the size of the anaerobic digester and the number of waste-producing animals required. Select animals from poultry, swine, and cattle, as appropriate. Place your findings in a report, including an economic evaluation and at least three references.

7.7D Tankless microwave water heating systems have been introduced that not only quickly provide hot water but also significantly reduce the exergy destruction inherent in domestic water heating with conventional electrical and gas-fueled water heaters. For a 230 m² dwelling in your locale, investigate the feasibility of using a microwave water heating system. Include a detailed economic evaluation accounting for equipment, installation, and operating costs. Place your findings in a memorandum.

7.8D In the 1840s, British engineers developed *atmospheric railways* that featured a large-diameter tube located between the tracks and stretching the entire length of the railroad. Pistons attached by struts to the rail cars moved inside the tube. As shown in **Fig. P7.8D**, piston motion was achieved by maintaining a vacuum ahead of the piston while the atmosphere was allowed to act behind it. Although several such railways came into use, limitations of the technology then available eventually ended this mode of transportation. Investigate the feasibility of combining the atmospheric railway concept with today's technology to develop rail service for commuting within urban areas. Write a report including at least three references.

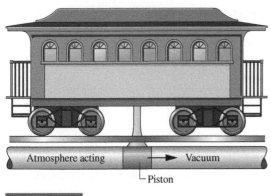

Atmosphere acting

Vacuum

Piston

7.9D Energy and exergy flow diagrams, called Sankey diagrams, use a primarily graphical approach to implement first and second law reasoning regarding a system's performance, including a concise display of energy and exergy efficiencies. Investigate this approach to system analysis and prepare a poster presentation, including at least three references, in which the roles of the energy and exergy flow diagrams are discussed and compared.

7.10D The objective of this project is to design a low-cost, electric-powered, portable or wearable consumer product that meets a need you have identified. In performing every function, the electricity required must come fully from *human motion*. No electricity from batteries and/or wall sockets is allowed. Additionally, the product must not be intrusive or interfere with the normal activities of the user, alter his/her gait or range of motion, lead to possible physical disability, or induce accidents leading to injury. The product cannot resemble any existing product unless it has a valuable new feature, significantly reduces cost, or provides some other meaningful advantage. The final report will include schematics, circuit diagrams, a parts list, and a suggested retail cost based on comprehensive costing.

CHAPTER 8

Vapor Power Systems

Engineering Context

In the twenty-first century we will be challenged to responsibly provide for our growing power needs. The scope of the challenge and how we will address it are discussed in "Introducing Power Generation" at the beginning of this chapter. You are encouraged to study this introduction before considering the several types of power-generating systems discussed in the present chapter and the next. In these chapters, we describe some of the practical arrangements employed for power production and illustrate how such power plants can be modeled thermodynamically. The discussion is organized into three main areas of application: vapor power plants, gas turbine power plants, and internal combustion engines. These power systems produce much of the electrical and mechanical power used worldwide. The **objective** of this chapter is to study *vapor* power plants in which the *working fluid* is alternately vaporized and condensed. Chapter 9 is concerned with gas turbines and internal combustion engines in which the working fluid remains a gas.

LEARNING OUTCOMES

When you complete your study of this chapter, you will be able to...

- Explain the basic principles of vapor power plants.
- Develop and analyze thermodynamic models of vapor power plants based on the Rankine cycle and its modifications, including
  - Sketching schematic and accompanying *T–s* diagrams.
  - Evaluating property data at principal states in the cycle.
  - Applying mass, energy, and entropy balances for the basic processes.
  - Determining power cycle performance, thermal efficiency, net power output, and mass flow rates.
- Describe the effects of varying key parameters on Rankine cycle performance.
- Discuss the principal sources of exergy destruction and loss in vapor power plants.

Introducing Power Generation

An exciting and urgent engineering challenge in the decades immediately ahead is to responsibly meet our national power needs. The challenge has its roots in the declining economically recoverable supplies of nonrenewable energy resources, effects of global climate change, and burgeoning population. In this introduction, we consider both conventional and emerging means for generating power. The present discussion also serves to introduce Chaps. 8 and 9, which detail vapor and gas power systems, respectively.

TABLE 8.1

Current U.S. Electricity Generation by Source

Natural gas	33.8%
Coal	30.4%
Nuclear	19.7%
Hydroelectric (conventional)	6.5%
Other renewables[a]	8.4%
Petroleum (total)	0.6%
Others	0.5%

[a]Renewable sources including solar, excluding hydroelectric

Source: United States Energy Information Administration, 2016, Net Generation by Energy Source: Total (All Sectors)

Table 1.1 used to calcuate these percentages is located on page 15 of the report posted at the above website, report is titled "Electric Power Monthly with Data for January 2017" and it was published in March 2017

Table 1.1. Net Generation by Energy Source: Total (All Sectors), 2007-January 2017 (Thousand Megawatthours)

Today   An important feature of a prudent national energy posture is a wide range of sources for the power generation mix, thus avoiding vulnerabilities that can accompany overreliance on too few energy sources. This feature is seen in Table 8.1, which gives a snapshot of the sources for nearly all U.S. electricity today. The table shows major (but declining) dependance on coal for electricity generation. Natural gas and nuclear are also significant sources. All three are nonrenewable.

The United States has abundant coal reserves and a rail system allowing for smooth distribution of coal to electricity producers. This good news is tempered by significant human-health and environmental-impact issues associated with coal. Coal use in power generation is discussed further in Secs. 8.3 and 8.5.3.

Use of natural gas has been growing in the United States because it is competitive cost-wise with coal and has fewer adverse environmental effects related to combustion. Natural gas not only provides for home heating needs but also supports a broad deployment of natural gas–fueled power plants. Natural gas proponents stress its value as a transitional fuel as we move away from coal and toward greater reliance on renewables. Some advocate greater use of natural gas in transportation. North American natural gas supplies seem ample for years to come. This includes natural gas from deep-water ocean sites and shale deposits, each of which has environmental-impact issues associated with gas extraction. For instance, the hydraulic drilling technique known as *fracking* used to obtain gas from shale deposits produces huge amounts of briny, chemically laden wastewater that can affect human health and the environment if not properly managed.

The share of nuclear power in U.S. electricity generation is holding steady at about 20%. In the 1950s, nuclear power was widely expected to be a dominant source of electricity by the year 2000. However, persistent concerns over reactor safety, an unresolved radioactive waste–disposal issue, and construction costs in the billions of dollars have resulted in a much smaller deployment of nuclear power than many had anticipated.

In some regions of the United States, hydroelectric power plants contribute significantly to meeting electricity needs. Although hydropower is a renewable source, it is not free of environmental impacts—for example, adverse effects on aquatic life in dammed rivers. The current share of wind, solar, geothermal, and other renewable sources in electricity generation is small but growing. Oil currently contributes only modestly.

Oil, natural gas, coal, and fissionable material are all in danger of reaching global production peaks in the foreseeable future and then entering periods of decline. Diminishing supplies will make these nonrenewable energy resources ever more costly. Increasing global demands for oil and fissionable material also pose national security concerns owing to the need for their importation to the United States.

Table 8.1 shows the United States currently has a range of sources for electricity generation and does not err by relying on too few. But in years ahead a gradual shift to a mix more reliant on renewable resources will be necessary.

Tomorrow   Looming scarcities of nonrenewable energy resources and their adverse effects on human health and the environment have sparked interest in broadening the ways in which we provide for our electricity needs—especially increasing use of renewable resources. Yet power production in the first half of the twenty-first century will rely primarily on means already available. Analysts say there are no technologies just over the horizon that will make much impact. Moreover, new technologies typically require decades and vast expenditures to establish.

Table 8.2 summarizes the types of power plants that will provide the electricity needed by our population up to mid-century, when electric power is expected to play an even larger role than now and new patterns of behavior affecting energy are likely (see Table 1.2).

There are several noteworthy features of Table 8.2. Seven of the twelve power plant types listed use renewable sources of energy. The five using nonrenewable sources include the three contributing most significantly to our current power mix (coal, natural gas, and nuclear). Four power plant types involve combustion—coal, natural gas, oil, and biomass—and thus require effective means for controlling gaseous emissions and power plant waste.

The twelve power plant types of Table 8.2 are unlikely to share equally in meeting national needs. In the immediate future coal, natural gas, and nuclear will continue to be major contributors while renewables will continue to lag. Gradually, the respective contributions are expected to shift to greater deployment of plants using renewables. This will be driven by state and national mandates requiring as much as 50% of electricity from renewables by 2030.

Of the emerging large-scale power plant types using a renewable source, wind is currently the most promising. Excellent wind resources exist in several places in the United States, both on land and offshore. The cost of wind-generated electricity is becoming competitive with coal-generated power. Other nations with active wind-power programs have supplied over 30% of their total electricity needs by wind, providing models for what might be possible in the United States. Still, wind turbines are not without environmental concerns. They are considered noisy by some and unsightly by others. Another concern is fatalities of birds and bats at wind-turbine sites.

Owing to higher costs, solar power is currently lagging behind wind power. Yet promising sites for solar power plants exist in many locations, especially in the Southwest. Active research and development efforts are focused on ways to reduce costs.

Geothermal plants use steam and hot water from deep hydrothermal reservoirs to generate electricity. Geothermal power plants exist in several states, including California, Nevada, Utah, and Hawaii. While geothermal power has considerable potential, its deployment has been inhibited by exploration, drilling, and extraction costs. The relatively low temperature of geothermal water also limits the extent to which electricity can be generated economically.

Although fuel cells are the subject of active research and development programs worldwide for stationary power generation and transportation, they are not yet widely deployed owing to costs. For more on fuel cells, see Sec. 13.4.

TABLE 8.2

Large-Scale Electric Power Generation through 2050 from Renewable and Nonrenewable Sources[a]

Power Plant Type	Nonrenewable Source	Renewable Source	Thermodynamic Cycle
Coal-fueled	Yes		Rankine
Natural gas–fueled	Yes		Brayton[b]
Nuclear-fueled	Yes		Rankine
Oil-fueled	Yes		Rankine[c]
Biomass-fueled		Yes	Rankine
Geothermal		Yes	Rankine
Solar-concentrating		Yes	Rankine
Hydroelectric		Yes	None
Wind		Yes	None
Solar-photovoltaic		Yes	None
Fuel cells	Yes		None
Currents, tides, and waves		Yes	None

[a]For current information about these power plant types, visit www.energy.gov/energysources. The Rankine cycle is the subject of the current chapter.

[b]Brayton cycle applications are considered in Chap. 9. For electricity generation, natural gas is primarily used with gas turbine power plants based on the Brayton cycle.

[c]Petroleum-fueled *reciprocating internal combustion* engines, discussed in Chap. 9, also generate electricity.

Power plants using current, tides, and waves are included in Table 8.2 because their potential for power generation is so vast. But thus far engineering embodiments are few, and this technology is not expected to mature in time to help much in the decades immediately ahead.

The discussion of Table 8.2 concludes with a guide for navigating the parts of this book devoted to power generation. In Table 8.2, seven of the power plant types are identified with thermodynamic cycles. Those based on the Rankine cycle are considered in this chapter. Natural gas–fueled gas turbines based on the Brayton cycle are considered in Chap. 9, together with power generation by *reciprocating internal combustion* engines. Fuel cells are discussed in Sec. 13.4.

Power Plant Policy Making

Power plants not only require huge investments but also have useful lives measured in decades. Accordingly, decisions about constructing power plants must consider the present *and* look to the future.

Thinking about power plants is best done on a *life-cycle* basis, not with a narrow focus on the plant operation phase alone. The life cycle begins with extracting resources required by the plant from the earth and ends with the plant's eventual retirement from service. See Table 8.3.

To account accurately for total power plant cost, costs incurred in *all* phases should be considered, including costs of acquiring natural resources, plant construction, power plant furnishing, remediation of effects on the environment and human health, and eventual retirement. The extent of governmental subsidies should be carefully weighed in making an equitable assessment of cost.

Capture, treatment, and proper disposal of effluents and waste products, including long-term storage where needed, must be a focus in power plant planning. None of the power plants listed in Table 8.2 is exempt from such scrutiny. While carbon dioxide production is particularly significant for power plants involving combustion, every plant type listed has carbon dioxide production in at least some of its life-cycle phases. The same can be said for other environmental and human-health impacts ranging from adverse land use to contamination of drinking water.

Public policy makers today have to consider not only the best ways to provide a reliable power *supply* but also how to do so judiciously. They should revisit entrenched regulations and practices suited to power generation and use in twentieth-century power generation but that now may stifle innovation. They also should be prepared to innovate when opportunities arise.

Policy makers must think critically about how to promote increased efficiency. Yet they must be watchful of a *rebound* effect sometimes observed when a resource, coal for instance, is used more efficiently to develop a product, electricity for instance. Efficiency-induced cost reductions can spur such demand for the product that little or no reduction in consumption of the resource occurs. With exceptional product demand, resource consumption can even rebound to a greater level than before.

Decision making in such a constrained social and technical environment is clearly a balancing act. Still, wise planning, including rationally decreasing waste and increasing efficiency, will allow us to stretch diminishing stores of nonrenewable energy resources, gain time to deploy renewable energy technologies, avoid construction of many new power plants, and reduce our contribution to global climate change, all while maintaining the lifestyle we enjoy.

TABLE 8.3

Power Plant Life-Cycle Snapshot

1. Mining, pumping, processing, and transporting
   (a) energy resources: coal, natural gas, oil, fissionable material, as appropriate.
   (b) commodities required for fabricating plant components and plant construction.

2. Remediation of environmental impacts related to the above.

3. Fabrication of plant components: boilers, pumps, reactors, solar arrays, steam and wind turbines, interconnections between components, and others.

4. Plant construction and connection to the power grid.

5. Plant operation: power production over several decades.

6. Capture, treatment, and disposal of effluents and waste products, including long-term storage when needed.

7. Retirement from service and site restoration when the useful life is over.

Horizons

Reducing Carbon Dioxide Through Emissions Trading

Policy makers in 9 northeastern states (Connecticut, Delaware, Maine, Maryland, Massachusetts, New Hampshire, New York, Rhode Island, and Vermont) with a total population over 41 million boldly established the nation's first *cap-and-trade* program, called the Regional Greenhouse Gas Initiative (RGGI), to harness the economic forces of the marketplace to reduce carbon dioxide emitted from power plants. The aim of these states is to spur a shift in the region's electricity supply toward more efficient generation and greater use of renewable energy technology.

The group of 9 states agreed to cap the total level of CO_2 emitted annually from power plants in the region starting in 2009 and continuing through 2014. Due to better than anticipated CO_2 reductions in 2012, the RGGI lowered the 2014 CO_2 target cap to 91 millions tons. To encourage innovation, the total CO_2 level will be reduced 2.5% annually until 2020. Power plant operators purchase *allowances* (or credits), which represent a permit to emit a specific amount of CO_2, to cover their expected CO_2 emissions. Proceeds of the sale of allowances are intended to support efforts in the region to foster energy efficiency and renewable energy technology. A utility that emits less than its projected allotment can sell unneeded allowances to utilities unable to meet their obligations. This is called a trade. In effect, the buyer pays a charge for polluting while the seller is rewarded for polluting less.

Costing carbon dioxide creates an economic incentive for decreasing such emissions. Accordingly, cap-and-trade provides a pathway for utilities to reduce carbon dioxide emitted from their power plants cost-effectively. Other state and national cap-and-trade programs have been proposed but not yet accepted. If it continues its success, the RGGI cap-and-trade program could spur other regions and the nation as a whole to adopt such programs.

Power Transmission and Distribution Our society must not only generate the electricity required for myriad uses but also provide it to consumers. The interface between these closely linked activities has not always been smooth. The U.S. power grid transmitting and distributing electricity to consumers has changed little for several decades, while the number of consumers and their power needs have changed greatly. This has induced significant systemic issues. The current grid is increasingly a twentieth-century relic, susceptible to power outages threatening safety and security and costing the economy billions annually.

The principal difference between the current grid and the grid of the future is a change from an electricity transmission and distribution focus to an electricity management focus that accommodates multiple power generation technologies and fosters more efficient electricity use. A twenty-first-century grid will be equipped for real-time information, conducive to lightning decision making and response, and capable of providing consumers with quality, reliable, and affordable electricity anywhere and anytime.

This is a tall order, yet it has driven utilities and government to think deeply about how to bring electricity generation, transmission, and distribution into the twenty-first century and the digital age. The result is an electricity *superhighway* called the *smart grid*. See the *Horizons* feature that follows.

Horizons

Our Electricity Superhighway

The smart grid is envisioned as an intelligent system that accepts electricity from any source—renewable and nonrenewable, centralized and distributed (decentralized)—and delivers it locally, regionally, or across the nation. A robust and dynamic communications network will be at its core, enabling high-speed two-way data flow between power provider and end user. The grid will provide consumers at every level—industry, business, and the home—with information needed for decisions on when, where, and how to use electricity. Using *smart* meters and programmable controls, consumers will manage power use in keeping with individual requirements and lifestyle choices, yet in harmony with community, regional, and national priorities.

Other smart grid features include the ability to

- Respond to and manage *peak loads* responsibly
- Identify outages and their causes promptly
- Reroute power to meet changing demand automatically
- Use a mix of available power sources, including *distributed* generation, amicably and cost-effectively

And all the while it will foster exceptional performance in electricity generation, transmission, distribution, and end use.

The smart grid will accommodate emerging power technologies such as wind and solar, emerging large-scale power consumers such as plug-in and all-electric vehicles, and technologies yet to be invented. A more efficient and better-managed grid will meet the increase in electricity demand anticipated by 2050 without the need to build as many new fossil-fueled or nuclear power plants along the way. Fewer plants mean less carbon dioxide, other emissions, and solid waste.

Considering Vapor Power Systems

8.1 Introducing Vapor Power Plants

Referring again to Table 8.2, seven of the power plant types listed require a thermodynamic cycle, and six of these are identified with the *Rankine cycle*. The Rankine cycle is the basic building block of vapor power plants, which are the focus of this chapter.

The components of four alternative vapor power plant configurations are shown schematically in **Fig. 8.1**. In order, these plants are (a) fossil-fueled, (b) nuclear-fueled, (c) solar thermal, and (d) geothermal. In Fig. 8.1*a*, the overall plant is broken into four major subsystems identified

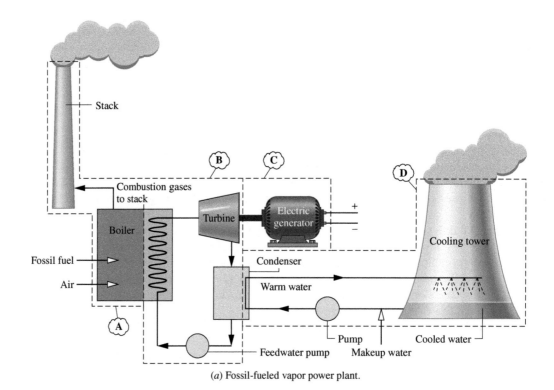

(a) Fossil-fueled vapor power plant.

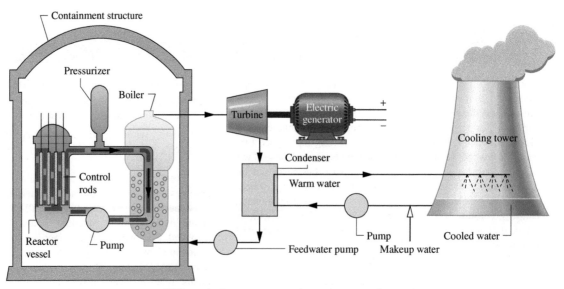

(b) Pressurized-water reactor nuclear vapor power plant.

Fig. 8.1 Components of alternative vapor power plants (not to scale).

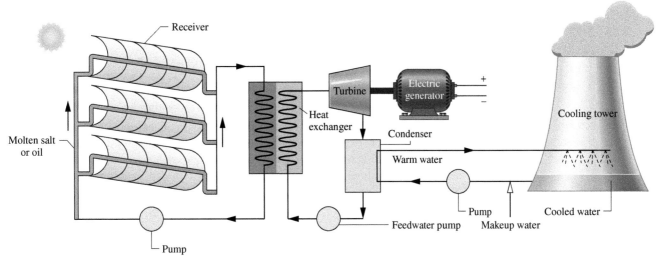

(*c*) Concentrating solar thermal vapor power plant.

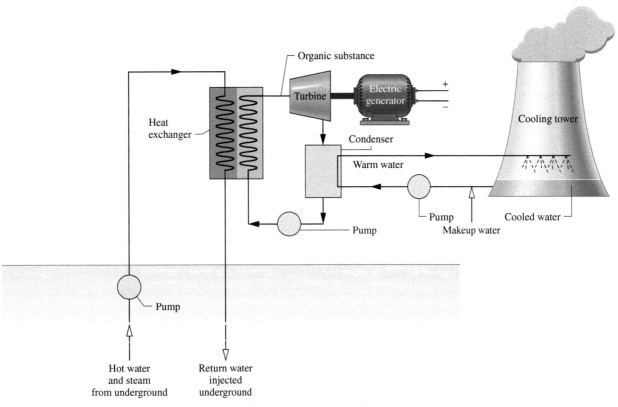

(*d*) Geothermal vapor power plant.

Fig. 8.1 (*Continued*)

by the letters **A** through **D**. These letters have been omitted in the other three configurations for simplicity. The discussions of this chapter focus on subsystem **B**, where the energy conversion from *heat* to *work* occurs. The function of subsystem **A** is to supply the energy needed to vaporize the power plant *working fluid* into the vapor required by the turbine of subsystem **B**. The principal difference in the four power plant configurations shown in Fig. 8.1 is the way working fluid vaporization is accomplished by action of subsystem **A**:

- Vaporization is accomplished in fossil-fueled plants by heat transfer *to* water passing through the boiler tubes *from* hot gases produced in the combustion of the fuel, as shown in Fig. 8.1*a*. This is also seen in plants fueled by biomass, municipal waste (trash), and mixtures of coal and biomass.

- In nuclear plants, energy required for vaporizing the cycle working fluid originates in a controlled nuclear reaction occurring in a reactor-containment structure. The *pressurized-water* reactor shown in Fig. 8.1b has two water loops: One loop circulates water through the reactor core and a boiler within the containment structure; this water is kept under pressure so it heats but does not boil. A separate loop carries steam from the boiler to the turbine. *Boiling-water* reactors (not shown in Fig. 8.1) have a single loop that boils water flowing through the core and carries steam directly to the turbine.

- Solar power plants have receivers for collecting and concentrating solar radiation. As shown in Fig. 8.1c, a suitable substance, molten salt or oil, flows through the receiver, where it is heated, directed to an interconnecting heat exchanger that replaces the boiler of the fossil- and nuclear-fueled plants, and finally returned to the receiver. The heated molten salt or oil provides energy required to vaporize water flowing in the other stream of the heat exchanger. This steam is provided to the turbine.

- The geothermal power plant shown in Fig. 8.1d also uses an interconnecting heat exchanger. In this case hot water and steam from deep below Earth's surface flows on one side of the heat exchanger. A *secondary* working fluid having a lower boiling point than the water, such as isobutane or another organic substance, vaporizes on the other side of the heat exchanger. The secondary working fluid vapor is provided to the turbine.

Referring again to Fig. 8.1a as representative, let's consider the other subsystems, beginning with system **B**. Regardless of the source of the energy required to vaporize the working fluid and the type of working fluid, the vapor produced passes through the turbine, where it expands to lower pressure, developing power. The turbine power shaft is connected to an electric generator (subsystem **C**). The vapor exiting the turbine passes through the condenser, where it condenses on the outside of tubes carrying cooling water.

The cooling water circuit comprises subsystem **D**. For the plant shown, cooling water is sent to a cooling tower, where energy received from steam condensing in the condenser is rejected into the atmosphere. Cooling water then returns to the condenser.

Concern for the environment governs what is allowable in the interactions between subsystem **D** and its surroundings. One of the major difficulties in finding a site for a vapor power plant is access to sufficient quantities of condenser cooling water. To reduce cooling-water needs, harm to aquatic life in the vicinity of the plant, and other *thermal pollution* effects, large-scale power plants typically employ cooling towers.

Fuel processing and handling are significant issues for both fossil-fueled and nuclear-fueled plants because of human-health and environmental-impact considerations. Fossil-fueled plants must observe increasingly stringent limits on smokestack emissions and disposal of toxic solid waste. Nuclear-fueled plants are saddled with a significant radioactive waste–disposal problem. Still, all four of the power plant configurations considered in Fig. 8.1 have environmental, health, and land-use issues related to various stages of their life cycles, including how they are manufactured, installed, operated, and ultimately disposed.

8.2 The Rankine Cycle

Rankine cycle

Referring to subsystem **B** of Fig. 8.1a again, observe that each unit of mass of working fluid periodically undergoes a thermodynamic cycle as it circulates through the series of interconnected components. This cycle is the **Rankine cycle**.

Important concepts introduced in previous chapters for thermodynamic *power* cycles generally also apply to the Rankine cycle:

- The first law of thermodynamics requires that the *net* work developed by a system undergoing a power cycle must equal the *net* energy added by heat transfer to the system (Sec. 2.6.2).

- The second law of thermodynamics requires that the *thermal efficiency* of a power cycle must be less than 100% (Sec. 5.6.1).

It is recommended that you review this material as needed.

Discussions in previous chapters also have shown that improved thermodynamic performance accompanies the reduction of irreversibilities and losses. The extent to which irreversibilities and losses can be reduced in vapor power plants depends on several factors, however, including limits imposed by thermodynamics *and* by economics.

Power Cycles

8.2.1 Modeling the Rankine Cycle

The processes taking place in a vapor power plant are sufficiently complicated that idealizations are required to develop thermodynamic models of plant components and the overall plant. Depending on the objective, models can range from highly detailed computer models to very simple models requiring a hand calculator at most.

Study of such models, even simplified ones, can lead to valuable conclusions about the performance of the corresponding actual plants. Thermodynamic models allow at least *qualitative* deductions about how changes in major operating parameters affect actual performance. They also provide uncomplicated settings in which to investigate the functions and benefits of features intended to improve overall performance.

Whether the aim is a detailed or simplified model of a vapor power plant adhering to the Rankine cycle, all of the fundamentals required for thermodynamic analysis have been introduced in previous chapters. They include the conservation of mass and conservation of energy principles, the second law of thermodynamics, and use of thermodynamic data. These principles apply to individual plant components such as turbines, pumps, and heat exchangers as well as to the overall cycle.

Let us now turn to the thermodynamic modeling of subsystem **B** of Fig. 8.1*a*. The development begins by considering, in turn, the four principal components: turbine, condenser, pump, and boiler. Then we consider important performance parameters. Since most large-scale vapor power plants use water as the working fluid, water is featured in the following discussions. For ease of presentation, we also focus on fossil-fuel plants, recognizing that major findings apply to the other types of power plants shown in Fig. 8.1.

The principal work and heat transfers of subsystem **B** are illustrated in Fig. 8.2. In subsequent discussions, these energy transfers are taken to be *positive in the directions of the arrows*. The unavoidable stray heat transfer that takes place between the plant components and their surroundings is neglected here for simplicity. Kinetic and potential energy changes are also ignored. Each component is regarded as operating at steady state. Using the conservation of mass and conservation of energy principles together with these idealizations, we develop expressions for the energy transfers shown on Fig. 8.2 beginning at state 1 and proceeding through each component in turn.

TAKE NOTE...

When analyzing vapor power cycles, we take energy transfers as positive in the directions of arrows on system schematics and write energy balances accordingly.

Turbines Tabs a, b, and c

Turbine Vapor from the boiler at state 1, having an elevated temperature and pressure, expands through the turbine to produce work and then is discharged to the condenser at state 2 with relatively low pressure. Neglecting heat transfer with the surroundings, the mass and energy rate balances for a control volume around the turbine reduce at steady state to give

$$0 = \dot{Q}_{cv}^{\,0} - \dot{W}_t + \dot{m}\left[h_1 - h_2 + \frac{V_1^2 - V_2^2}{2}^{\,0} + g(z_1 - z_2)^{\,0} \right]$$

or

$$\frac{\dot{W}_t}{\dot{m}} = h_1 - h_2 \qquad (8.1)$$

where $\dot{m}$ denotes the mass flow rate of the cycle working fluid, and $\dot{W}_t/\dot{m}$ is the work developed per unit of mass of vapor passing through the turbine. As noted above, kinetic and potential energy changes are ignored.

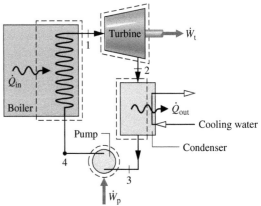

Fig. 8.2 Principal work and heat transfers of subsystem **B**.

Condenser   In the condenser there is heat transfer from the working fluid to cooling water flowing in a separate stream. The working fluid condenses and the temperature of the cooling water increases. At steady state, mass and energy rate balances for a control volume enclosing the condensing side of the heat exchanger give

$$\frac{\dot{Q}_{out}}{\dot{m}} = h_2 - h_3 \qquad (8.2)$$

where $\dot{Q}_{out}/\dot{m}$ is the energy transferred by heat *from* the working fluid to the cooling water per unit mass of working fluid passing through the condenser. This energy transfer is positive in the direction of the arrow on Fig. 8.2.

Pumps Tabs a, b, and c

Pump   The liquid condensate leaving the condenser at 3 is pumped from the condenser into the higher-pressure boiler. Taking a control volume around the pump and assuming no heat transfer with the surroundings, mass and energy rate balances give

$$\frac{\dot{W}_p}{\dot{m}} = h_4 - h_3 \qquad (8.3)$$

where $\dot{W}_p/\dot{m}$ is the work *input* per unit of mass passing through the pump. This energy transfer is positive in the direction of the arrow on Fig. 8.2.

feedwater

Boiler   The working fluid completes a cycle as the liquid leaving the pump at 4, called the boiler **feedwater**, is heated to saturation and evaporated in the boiler. Taking a control volume enclosing the boiler tubes and drums carrying the feedwater from state 4 to state 1, mass and energy rate balances give

$$\frac{\dot{Q}_{in}}{\dot{m}} = h_1 - h_4 \qquad (8.4)$$

where $\dot{Q}_{in}/\dot{m}$ is the energy transferred by heat from the energy source *into* the working fluid per unit mass passing through the boiler.

Performance Parameters   The thermal efficiency gauges the extent to which the energy input to the working fluid passing through the boiler is converted to the *net* work output. Using the quantities and expressions just introduced, the **thermal efficiency** of the power cycle of Fig. 8.2 is

thermal efficiency

$$\eta = \frac{\dot{W}_t/\dot{m} - \dot{W}_p/\dot{m}}{\dot{Q}_{in}/\dot{m}} = \frac{(h_1 - h_2) - (h_4 - h_3)}{h_1 - h_4} \qquad (8.5a)$$

The net work output equals the net heat input. Thus, the thermal efficiency can be expressed alternatively as

$$\eta = \frac{\dot{Q}_{in}/\dot{m} - \dot{Q}_{out}/\dot{m}}{\dot{Q}_{in}/\dot{m}} = 1 - \frac{\dot{Q}_{out}/\dot{m}}{\dot{Q}_{in}/\dot{m}}$$

$$= 1 - \frac{(h_2 - h_3)}{(h_1 - h_4)} \qquad (8.5b)$$

heat rate

The **heat rate** is the amount of energy added by heat transfer to the cycle, usually in kJ, to produce a unit of net work output, usually in kW · h. Accordingly, the heat rate, which is inversely proportional to the thermal efficiency, has units of $\dfrac{\text{kJ}}{\text{kW} \cdot \text{h}}$.

Another parameter used to describe power plant performance is the **back work ratio**, or bwr, defined as the ratio of the pump work input to the work developed by the turbine. With Eqs. 8.1 and 8.3, the back work ratio for the power cycle of Fig. 8.2 is

$$\boxed{\text{bwr} = \frac{\dot{W}_p/\dot{m}}{\dot{W}_t/\dot{m}} = \frac{(h_4 - h_3)}{(h_1 - h_2)}}$$ (8.6) **back work ratio**

Examples to follow illustrate that the change in specific enthalpy for the expansion of vapor through the turbine is normally many times greater than the increase in enthalpy for the liquid passing through the pump. Hence, the back work ratio is characteristically quite low for vapor power plants.

Provided states 1 through 4 are fixed, Eqs. 8.1 through 8.6 can be applied to determine the thermodynamic performance of a simple vapor power plant. Since these equations have been developed from mass and energy rate balances, they apply equally for actual performance when irreversibilities are present and for idealized performance in the absence of such effects. It might be surmised that the irreversibilities of the various power plant components can affect overall performance, and this is the case. Accordingly, it is instructive to consider an idealized cycle in which irreversibilities are assumed absent, for such a cycle establishes an *upper limit* on the performance of the Rankine cycle. The ideal cycle also provides a simple setting in which to study various aspects of vapor power plant performance. The ideal Rankine cycle is the subject of Sec. 8.2.2.

Energy & Environment

The United States currently relies on relatively abundant coal supplies to generate about 30% of its electricity (Table 8.1). A large fleet of coal-burning vapor power plants reliably provides comparatively inexpensive electricity to homes, businesses, and industry. Yet this good news is eroded by human-health and environmental-impact problems linked to coal combustion. Impacts accompany coal extraction, power generation, and waste disposal. Analysts say the cost of coal-derived electricity would be much higher if the full costs related to these adverse aspects of coal use were included.

Coal extraction practices such as mountaintop mining, where tops of mountains are sheared off to get at underlying coal, are a particular concern when removed rock, soil, and mining debris are discarded into streams and valleys below, marring natural beauty, affecting water quality, and devastating CO_2-trapping forests. Moreover, fatalities and critical injuries of coal miners while working to extract coal are widely seen as deplorable.

Gases formed in coal combustion include sulfur dioxide and oxides of nitrogen, which contribute to acid rain and smog. Fine particles and mercury, which more directly affect human health, are other unwanted outcomes of coal use. Coal combustion is also a major contributor to global climate change, primarily through carbon dioxide emissions. At the national level, controls are required for sulfur dioxide, nitric oxides, and fine particles but not currently required for mercury and carbon dioxide.

Solid waste is another major problem area. Solid waste from coal combustion is one of the largest waste streams produced in the United States. Solid waste includes sludge from smokestack

scrubbers and fly ash, a by-product of pulverized coal combustion. While some of this waste is diverted to make commercial products, including cement, road de-icer, and synthetic gypsum used for drywall and as a fertilizer, vast amounts of waste are stored in landfills and pools containing watery slurries. Leakage from these impoundments can contaminate drinking water supplies. Watery waste accidentally released from holding pools causes widespread devastation and elevated levels of dangerous substances in surrounding areas. Some observers contend much more should be done to regulate health- and environment-endangering gas emissions and solid waste from coal-fired power plants and other industrial sites.

The more efficiently each ton of coal is utilized to generate power, the less CO_2, other combustion gases, and solid waste will be produced. Accordingly, improving efficiency is a well-timed pathway for continued coal use in the twenty-first century. Gradual replacement of existing power plants, beginning with those several decades old, by more efficient plants will reduce to some extent gas emissions and solid waste related to coal use.

Various advanced technologies also aim to foster coal use—but used more responsibly. They include *supercritical* vapor power plants (Sec. 8.3), *carbon capture and storage* (Sec. 8.5.3), and *integrated gasification combined cycle* (IGCC) power plants (Sec. 9.10). Owing to our large coal reserves and the critical importance of electricity to our society, major governmental and private-sector initiatives are in progress to develop additional technologies that promote responsible coal use.

8.2.2 Ideal Rankine Cycle

If the working fluid passes through the various components of the simple vapor power cycle without irreversibilities, frictional pressure drops would be absent from the boiler and condenser, and the working fluid would flow through these components at constant pressure. Also, in the absence of irreversibilities and heat transfer with the surroundings, the processes through the turbine and pump would be isentropic. A cycle adhering to these idealizations is the **ideal Rankine cycle** shown in **Fig. 8.3**.

ideal Rankine cycle

Referring to Fig. 8.3, we see that the working fluid undergoes the following series of internally reversible processes:

Process 1–2 Isentropic expansion of the working fluid through the turbine from saturated vapor at state 1 to the condenser pressure.

Process 2–3 Heat transfer *from* the working fluid as it flows at constant pressure through the condenser exiting as saturated liquid at state 3.

Process 3–4 Isentropic compression in the pump to state 4 in the compressed liquid region.

Process 4–1 Heat transfer *to* the working fluid as it flows at constant pressure through the boiler to complete the cycle.

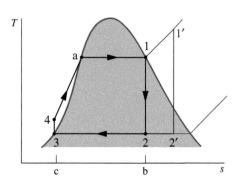

Fig. 8.3 Temperature–entropy diagram of the ideal Rankine cycle.

The ideal Rankine cycle also includes the possibility of superheating the vapor, as in cycle 1′–2′–3–4–1′. The importance of superheating is discussed in Sec. 8.3.

Since the ideal Rankine cycle consists of internally reversible processes, areas under the process lines of Fig. 8.3 can be interpreted as heat transfers per unit of mass flowing. Applying Eq. 6.49, area 1–b–c–4–a–1 represents the heat transfer to the working fluid passing through the boiler and area 2–b–c–3–2 is the heat transfer from the working fluid passing through the condenser, each per unit of mass flowing. The enclosed area 1–2–3–4–a–1 can be interpreted as the net heat input or, equivalently, the net work output, each per unit of mass flowing.

Because the pump is idealized as operating without irreversibilities, Eq. 6.51b can be invoked as an alternative to Eq. 8.3 for evaluating the pump work. That is,

Animation

Rankine Cycle Tabs a and b

$$\left(\frac{\dot{W}_\mathrm{p}}{\dot{m}}\right)_{\substack{\text{int}\\\text{rev}}} = \int_3^4 v\, dp \tag{8.7a}$$

where the minus sign has been dropped for consistency with the positive value for pump work in Eq. 8.3. The subscript "int rev" signals that this expression is restricted to an internally reversible process through the pump. An "int rev" designation is not required by Eq. 8.3, however, because it is obtained with the conservation of mass and energy principles and thus is generally applicable.

Evaluation of the integral of Eq. 8.7a requires a relationship between the specific volume and pressure for Process 3–4. Because the specific volume of the liquid normally varies only slightly as the liquid flows from the inlet to the exit of the pump, a plausible approximation to the value of the integral can be had by taking the specific volume at the pump inlet, v_3, as constant for the process. Then

TAKE NOTE...

For cycles, we modify the problem-solving methodology: The Analysis begins with a systematic evaluation of required property data at each numbered state. This reinforces what we know about the components, since given information and assumptions are required to fix the states.

$$\left(\frac{\dot{W}_\mathrm{p}}{\dot{m}}\right)_\mathrm{s} \approx v_3(p_4 - p_3) \tag{8.7b}$$

where the subscript s signals the *isentropic*—internally reversible *and* adiabatic—process of the liquid flowing through the pump.

The next example illustrates the analysis of an ideal Rankine cycle.

▶▶▶ EXAMPLE 8.1 ▶ ·

Analyzing an Ideal Rankine Cycle

Steam is the working fluid in an ideal Rankine cycle. Saturated vapor enters the turbine at 8.0 MPa and saturated liquid exits the condenser at a pressure of 0.008 MPa. The *net* power output of the cycle is 100 MW. Determine for the cycle **(a)** the thermal efficiency, **(b)** the back

work ratio, **(c)** the mass flow rate of the steam, in kg/h, **(d)** the rate of heat transfer, $\dot{Q}_\mathrm{in}$, into the working fluid as it passes through the boiler, in MW, **(e)** the rate of heat transfer, $\dot{Q}_\mathrm{out}$, from the condensing steam as it passes through the condenser, in MW, **(f)** the mass flow rate of the condenser cooling water, in kg/h, if cooling water enters the condenser at 15°C and exits at 35°C.

SOLUTION

Known An ideal Rankine cycle operates with steam as the working fluid. The boiler and condenser pressures are specified, and the net power output is given.

Find Determine the thermal efficiency, the back work ratio, the mass flow rate of the steam, in kg/h, the rate of heat transfer to the working fluid as it passes through the boiler, in MW, the rate of heat transfer from the condensing steam as it passes through the condenser, in MW, the mass flow rate of the condenser cooling water, which enters at 15°C and exits at 35°C.

Schematic and Given Data:

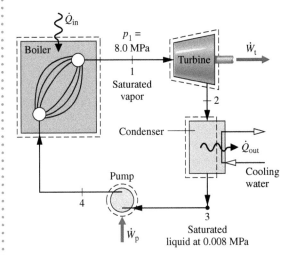

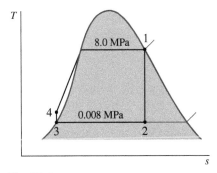

Fig. E8.1

Engineering Model

1. Each component of the cycle is analyzed as a control volume at steady state. The control volumes are shown on the accompanying sketch by dashed lines.

2. All processes of the working fluid are internally reversible.

3. The turbine and pump operate adiabatically.

4. Kinetic and potential energy effects are negligible.

5. Saturated vapor enters the turbine. Condensate exits the condenser as saturated liquid.

❶ **Analysis** To begin the analysis, we fix each of the principal states located on the accompanying schematic and T–s diagrams. Starting at the inlet to the turbine, the pressure is 8.0 MPa and

the steam is a saturated vapor, so from Table A-3, $h_1 = 2758.0$ kJ/kg and $s_1 = 5.7432$ kJ/kg · K.

State 2 is fixed by $p_2 = 0.008$ MPa and the fact that the specific entropy is constant for the adiabatic, internally reversible expansion through the turbine. Using saturated liquid and saturated vapor data from Table A-3, we find that the quality at state 2 is

$$x_2 = \frac{s_2 - s_\mathrm{f}}{s_\mathrm{g} - s_\mathrm{f}} = \frac{5.7432 - 0.5926}{7.6361} = 0.6745$$

The enthalpy is then

$$h_2 = h_\mathrm{f} + x_2 h_\mathrm{fg} = 173.88 + (0.6745)2403.1$$
$$= 1794.8 \text{ kJ/kg}$$

State 3 is saturated liquid at 0.008 MPa, so $h_3 = 173.88$ kJ/kg.

State 4 is fixed by the boiler pressure p_4 and the specific entropy $s_4 = s_3$. The specific enthalpy h_4 can be found by interpolation in the compressed liquid tables. However, because compressed liquid data are relatively sparse, it is more convenient to solve Eq. 8.3 for h_4, using Eq. 8.7b to approximate the pump work. With this approach

$$h_4 = h_3 + \dot{W}_\mathrm{p}/\dot{m} = h_3 + v_3(p_4 - p_3)$$

By inserting property values from Table A-3

$$h_4 = 173.88 \text{ kJ/kg} + (1.0084 \times 10^{-3} \text{ m}^3\text{/kg})$$
$$\times (8.0 - 0.008)\text{MPa} \left| \frac{10^6 \text{ N/m}^2}{1 \text{ MPa}} \right| \left| \frac{1 \text{ kJ}}{10^3 \text{ N} \cdot \text{m}} \right|$$
$$= 173.88 + 8.06 = 181.94 \text{ kJ/kg}$$

a. The *net* power developed by the cycle is

$$\dot{W}_\mathrm{cycle} = \dot{W}_\mathrm{t} - \dot{W}_\mathrm{p}$$

Mass and energy rate balances for control volumes around the turbine and pump give, respectively,

$$\frac{\dot{W}_\mathrm{t}}{\dot{m}} = h_1 - h_2 \qquad \text{and} \qquad \frac{\dot{W}_\mathrm{p}}{\dot{m}} = h_4 - h_3$$

where $\dot{m}$ is the mass flow rate of the steam. The rate of heat transfer to the working fluid as it passes through the boiler is determined using mass and energy rate balances as

$$\frac{\dot{Q}_\mathrm{in}}{\dot{m}} = h_1 - h_4$$

The thermal efficiency is then

$$\eta = \frac{\dot{W}_\mathrm{t} - \dot{W}_\mathrm{p}}{\dot{Q}_\mathrm{in}} = \frac{(h_1 - h_2) - (h_4 - h_3)}{h_1 - h_4}$$
$$= \frac{[(2758.0 - 1794.8) - (181.94 - 173.88)] \text{ kJ/kg}}{(2758.0 - 181.94) \text{ kJ/kg}}$$
$$= 0.371 \,(37.1\%)$$

b. The back work ratio is

❷ $$\text{bwr} = \frac{\dot{W}_\mathrm{p}}{\dot{W}_\mathrm{t}} = \frac{h_4 - h_3}{h_1 - h_2} = \frac{(181.94 - 173.88) \text{ kJ/kg}}{(2758.0 - 1794.8) \text{ kJ/kg}}$$
$$= \frac{8.06}{963.2} = 8.37 \times 10^{-3} \,(0.84\%)$$

c. The mass flow rate of the steam can be obtained from the expression for the net power given in part (a). Thus,

$$\dot{m} = \frac{\dot{W}_{cycle}}{(h_1 - h_2) - (h_4 - h_3)}$$

$$= \frac{(100 \text{ MW}) \, |10^3 \text{ kW/MW}||3600 \text{ s/h}|}{(963.2 - 8.06) \text{ kJ/kg}}$$

$$= 3.77 \times 10^5 \text{ kg/h}$$

d. With the expression for $\dot{Q}_{in}$ from part (a) and previously determined specific enthalpy values

$$\dot{Q}_{in} = \dot{m}(h_1 - h_4)$$

$$= \frac{(3.77 \times 10^5 \text{ kg/h})(2758.0 - 181.94) \text{ kJ/kg}}{|3600 \text{ s/h}||10^3 \text{ kW/MW}|}$$

$$= 269.77 \text{ MW}$$

e. Mass and energy rate balances applied to a control volume enclosing the steam side of the condenser give

$$\dot{Q}_{out} = \dot{m}(h_2 - h_3)$$

$$= \frac{(3.77 \times 10^5 \text{ kg/h})(1794.8 - 173.88) \text{ kJ/kg}}{|3600 \text{ s/h}||10^3 \text{ kW/MW}|}$$

$$= 169.75 \text{ MW}$$

➌ Note that the ratio of $\dot{Q}_{out}$ to $\dot{Q}_{in}$ is 0.629 (62.9%). Alternatively, $\dot{Q}_{out}$ can be determined from an energy rate balance on the *overall* vapor power plant. At steady state, the net power developed equals the net rate of heat transfer to the plant

$$\dot{W}_{cycle} = \dot{Q}_{in} - \dot{Q}_{out}$$

Rearranging this expression and inserting values

$$\dot{Q}_{out} = \dot{Q}_{in} - \dot{W}_{cycle} = 269.77 \text{ MW} - 100 \text{ MW} = 169.77 \text{ MW}$$

The slight difference from the above value is due to round-off.

f. Taking a control volume around the condenser, the mass and energy rate balances give at steady state

$$0 = \dot{Q}_{cv}^{\,0} - \dot{W}_{cv}^{\,0} + \dot{m}_{cw}(h_{cw,\,in} - h_{cw,\,out}) + \dot{m}(h_2 - h_3)$$

where $\dot{m}_{cw}$ is the mass flow rate of the cooling water. Solving for $\dot{m}_{cw}$

$$\dot{m}_{cw} = \frac{\dot{m}(h_2 - h_3)}{(h_{cw,\,out} - h_{cw,\,in})}$$

The numerator in this expression is evaluated in part (e). For the cooling water, $h \approx h_f(T)$, so with saturated liquid enthalpy values from Table A-2 at the entering and exiting temperatures of the cooling water

$$\dot{m}_{cw} = \frac{(169.75 \text{ MW}) \, |10^3 \text{ kW/MW}||3600 \text{ s/h}|}{(146.68 - 62.99) \text{ kJ/kg}} = 7.3 \times 10^6 \text{ kg/h}$$

➊ Note that a slightly revised problem-solving methodology is used in this example problem: We begin with a systematic evaluation of the specific enthalpy at each numbered state.

➋ Note that the back work ratio is relatively low for the Rankine cycle. In the present case, the work required to operate the pump is less than 1% of the turbine output.

➌ In this example, 62.9% of the energy added to the working fluid by heat transfer is subsequently discharged to the cooling water. Although considerable energy is carried away by the cooling water, its exergy is small because the water exits at a temperature only a few degrees greater than that of the surroundings. See Sec. 8.6 for further discussion.

SKILLS DEVELOPED

Ability to…

- sketch the T–s diagram of the basic Rankine cycle.
- fix each of the principal states and retrieve necessary property data.
- apply mass and energy balances.
- calculate performance parameters for the cycle.

Quick Quiz

If the mass flow rate of steam were 150 kg/s, what would be the net power, in MW, and the thermal efficiency? Ans. 143.2 MW, 37.1%.

8.2.3 Effects of Boiler and Condenser Pressures on the Rankine Cycle

In discussing Fig. 5.12 (Sec. 5.9.1), we observed that the thermal efficiency of power cycles tends to increase as the average temperature at which energy is added by heat transfer increases and/or the average temperature at which energy is rejected by heat transfer decreases. (For elaboration, see box.) Let us apply this idea to study the effects on performance of the ideal Rankine cycle of changes in the boiler and condenser pressures. Although these findings are obtained with reference to the ideal Rankine cycle, they also hold qualitatively for actual vapor power plants.

Figure 8.4*a* shows two ideal cycles having the same condenser pressure but different boiler pressures. By inspection, the average temperature of heat addition is seen to be greater for the higher-pressure cycle 1′–2′–3–4′–1′ than for cycle 1–2–3–4–1. It follows that increasing the boiler pressure of the ideal Rankine cycle tends to increase the thermal efficiency.

Considering the Effect of Temperature on Thermal Efficiency

Since the ideal Rankine cycle consists entirely of internally reversible processes, an expression for thermal efficiency can be obtained in terms of *average* temperatures during the heat interaction processes. Let us begin the development of this expression by recalling that areas under the process lines of Fig. 8.3 can be interpreted as the heat transfer per unit of mass flowing through the respective components. For example, the total area 1–b–c–4–a–1 represents the heat transfer into the working fluid per unit of mass passing through the boiler. In symbols,

$$\left(\frac{\dot{Q}_{in}}{\dot{m}}\right)_{\substack{int\\rev}} = \int_4^1 T\,ds = \text{area } 1\text{–b–c–4–a–1}$$

The integral can be written in terms of an average temperature of heat addition, $\bar{T}_{in}$, as follows:

$$\left(\frac{\dot{Q}_{in}}{\dot{m}}\right)_{\substack{int\\rev}} = \bar{T}_{in}(s_1 - s_4)$$

where the overbar denotes *average*. Similarly, area 2–b–c–3–2 represents the heat transfer from the condensing steam per unit of mass passing through the condenser

$$\left(\frac{\dot{Q}_{out}}{\dot{m}}\right)_{\substack{int\\rev}} = T_{out}(s_2 - s_3) = \text{area } 2\text{–b–c–3–2}$$
$$= T_{out}(s_1 - s_4)$$

where T_{out} denotes the temperature on the steam side of the condenser of the ideal Rankine cycle pictured in Fig. 8.3. The thermal efficiency of the ideal Rankine cycle can be expressed in terms of these heat transfers as

$$\eta_{ideal} = 1 - \frac{(\dot{Q}_{out}/\dot{m})_{\substack{int\\rev}}}{(\dot{Q}_{in}/\dot{m})_{\substack{int\\rev}}} = 1 - \frac{T_{out}}{\bar{T}_{in}} \qquad (8.8)$$

By the study of Eq. 8.8, we conclude that the thermal efficiency of the ideal cycle tends to increase as the average temperature at which energy is added by heat transfer increases and/or the temperature at which energy is rejected decreases. With similar reasoning, these conclusions can be shown to apply to the other ideal cycles considered in this chapter and the next.

Figure 8.4*b* shows two cycles with the same boiler pressure but two different condenser pressures. One condenser operates at atmospheric pressure and the other at *less than* atmospheric pressure. The temperature of heat rejection for cycle 1–2–3–4–1 condensing at atmospheric pressure is 100°C. The temperature of heat rejection for the lower-pressure cycle 1–2″–3″–4″–1 is correspondingly lower, so this cycle has the greater thermal efficiency. It follows that decreasing the condenser pressure tends to increase the thermal efficiency.

The lowest feasible condenser pressure is the saturation pressure corresponding to the ambient temperature, for this is the lowest possible temperature for heat rejection to the surroundings. The goal of maintaining the lowest practical turbine exhaust (condenser) pressure is a primary reason for including the condenser in a power plant. Liquid water at atmospheric

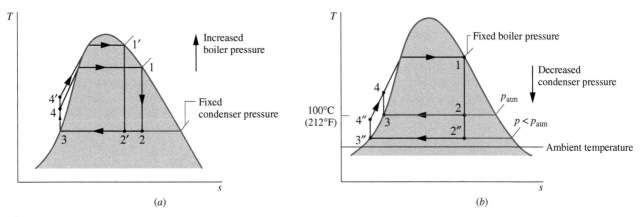

Fig. 8.4 Effects of varying operating pressures on the ideal Rankine cycle. (*a*) Effect of boiler pressure. (*b*) Effect of condenser pressure.

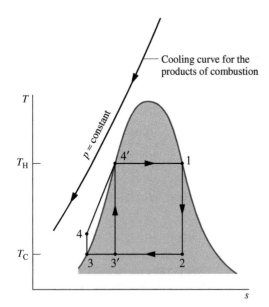

T

T_H

T_C

— Cooling curve for the products of combustion

$p \approx$ constant

4'   1

4

3   3'   2

s

Fig. 8.5 Illustration used to compare the ideal Rankine cycle with the Carnot cycle.

pressure could be drawn into the boiler by a pump, and steam could be discharged directly to the atmosphere at the turbine exit. However, by including a condenser in which the steam side is operated at a pressure *below atmospheric*, the turbine has a lower-pressure region in which to discharge, resulting in a significant increase in net work and thermal efficiency. The addition of a condenser also allows the working fluid to flow in a closed loop. This arrangement permits continual circulation of the working fluid, so purified water that is less corrosive than tap water can be used economically.

Comparison with Carnot Cycle   Referring to Fig. 8.5, the ideal Rankine cycle 1–2–3–4–4′–1 has a lower thermal efficiency than the Carnot cycle 1–2–3′–4′–1 having the same maximum temperature T_H and minimum temperature T_C because the average temperature between 4 and 4′ is less than T_H. Despite the greater thermal efficiency of the Carnot cycle, it has shortcomings as a model for the simple fossil-fueled vapor power cycle. First, heat transfer to the working fluid of such vapor power plants is obtained from hot products of combustion cooling at approximately constant pressure. To exploit fully the energy released on combustion, the hot products should be cooled as much as possible. The first portion of the heating process of the Rankine cycle shown in Fig. 8.5, Process 4–4′, is achieved by cooling the combustion products *below* the maximum temperature T_H. With the Carnot cycle, however, the combustion products would be cooled *at the most* to T_H. Thus, a smaller portion of the energy released on combustion would be used. Another shortcoming of the Carnot vapor power cycle involves the pumping process. Note that state 3′ of Fig. 8.5 is a two-phase liquid–vapor mixture. Significant practical problems are encountered in developing pumps that handle two-phase mixtures, as would be required by Carnot cycle 1–2–3′–4′–1. It is better to condense the vapor completely and handle only liquid in the pump, as is done in the Rankine cycle. Pumping from 3 to 4 and heating without work from 4 to 4′ are processes that can be readily achieved in practice.

8.2.4 Principal Irreversibilities and Losses

Irreversibilities and losses are associated with each of the four subsystems designated in Fig. 8.1*a* by **A**, **B**, **C**, and **D**. Some of these effects have much greater influence on overall power plant performance than others. In this section, we consider irreversibilities and losses associated with the working fluid as it flows around the closed loop of subsystem **B**: the Rankine cycle. These effects are broadly classified as *internal* or *external* depending on whether they occur within subsystem **B** or its surroundings.

Internal Effects

Turbine   The principal internal irreversibility experienced by the working fluid is associated with expansion through the turbine. Heat transfer from the turbine to its surroundings is a loss; but since it is of secondary importance, this loss is ignored in subsequent discussions. As illustrated by Process 1–2 of Fig. 8.6, actual adiabatic expansion through the turbine is accompanied by an increase in entropy. The work developed in this process per unit of mass flowing is *less* than that for the corresponding isentropic expansion 1–2s. Isentropic turbine efficiency, η_t, introduced in Sec. 6.12.1, accounts for the effect of irreversibilities within the turbine in terms of actual and isentropic work amounts. Designating states as in Fig. 8.6, the isentropic turbine efficiency is

Animation

Turbines Tab e

$$\eta_t = \frac{(\dot{W}_t/\dot{m})}{(\dot{W}_t/\dot{m})_s} = \frac{h_1 - h_2}{h_1 - h_{2s}} \qquad (8.9)$$

where the numerator is the actual work developed per unit of mass flowing through the turbine and the denominator is the work per unit of mass flowing for an isentropic expansion from the turbine inlet state to the turbine exhaust pressure. Irreversibilities within the turbine reduce the net power output of the plant and thus thermal efficiency.

Pump   The work input to the pump required to overcome irreversibilities also reduces the net power output of the plant. As illustrated by Process 3–4 of Fig. 8.6, the actual pumping process is accompanied by an increase in entropy. For this process; the work *input* per unit of mass flowing is *greater* than that for the corresponding isentropic process 3–4s. As for the turbine, heat transfer is ignored as secondary. Isentropic pump efficiency, η_p, introduced in Sec. 6.12.3, accounts for the effect of irreversibilities within the pump in terms of actual and isentropic work amounts. Designating states as in Sec. 8.6, the isentropic pump efficiency is

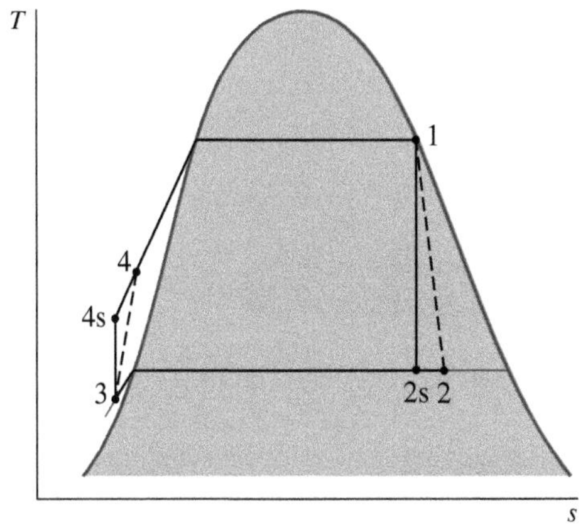

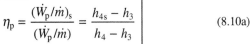

$$\eta_p = \frac{(\dot{W}_p/\dot{m})_s}{(\dot{W}_p/\dot{m})} = \frac{h_{4s} - h_3}{h_4 - h_3} \qquad (8.10a)$$

Fig. 8.6 Temperature–entropy diagram showing the effects of turbine and pump irreversibilities.

In Eq. 8.10a the pump work for the isentropic process appears in the numerator. The actual pump work, being the larger magnitude, is the denominator.

The pump work for the isentropic process can be evaluated using Eq. 8.7b to give an alternative expression for the isentropic pump efficiency:

$$\eta_p = \frac{(\dot{W}_p/\dot{m})_s}{(\dot{W}_p/\dot{m})} = \frac{v_3(p_4 - p_3)}{h_4 - h_3} \qquad (8.10b)$$

Pumps Tab e

Because pump work is much less than turbine work, irreversibilities in the pump have a much smaller impact on thermal efficiency than irreversibilities in the turbine.

Other Effects   Frictional effects resulting in pressure reductions are additional sources of internal irreversibility as the working fluid flows through the boiler, condenser, and piping connecting the several components. Detailed thermodynamic analyses account for these effects. For simplicity, they are ignored in subsequent discussions. In keeping with this, Fig. 8.6 shows no pressure drops for flow through the boiler and condenser or between plant components.

Another detrimental effect on plant performance can be noted by comparing the ideal cycle of Fig. 8.6 with the counterpart ideal cycle of Fig. 8.3. In Fig. 8.6, pump inlet state 3 falls in the liquid region and is not saturated liquid as in Fig. 8.3, giving lower average temperatures of heat addition and rejection. The overall effect typically is a *lower* thermal efficiency in the case of Fig. 8.6 compared to that of Fig. 8.3.

External Effects   The turbine and pump irreversibilities considered above are *internal* irreversibilities experienced by the working fluid flowing around the closed loop of the Rankine cycle. They have detrimental effects on power plant performance. Yet the most significant source of irreversibility *by far* for a fossil-fueled vapor power plant is associated with combustion of the fuel and subsequent heat transfer from hot combustion gases to the cycle working fluid. As combustion and subsequent heat transfer occur in the surroundings of subsystem **B** of Fig. 8.1*a*, they are classified here as *external*. These effects are considered quantitatively in Sec. 8.6 and Chap. 13 using the exergy concept.

Another effect occurring in the surroundings of subsystem **B** is energy discharged by heat transfer to cooling water as the working fluid condenses. The significance of this loss is *far less* than suggested by the magnitude of the energy discharged. Although cooling water carries away considerable energy, the *utility* of this energy is extremely limited when condensation occurs near ambient temperature and the temperature of the cooling water increases only by a few degrees above the ambient during flow through the condenser. Such cooling water has little thermodynamic or economic value. Instead, the slightly warmed cooling water is normally *disadvantageous* for plant operators in terms of cost because operators must provide responsible means for disposing of the energy gained by cooling water in flow through the condenser—they provide a cooling tower, for instance. The limited utility of condenser cooling water is demonstrated quantitatively in Sec. 8.6 using the exergy concept.

Rankine Cycle Tab c

Finally, stray heat transfers from the outer surfaces of plant components have detrimental effects on performance since they reduce the extent of conversion from heat to work. Such heat transfers are secondary effects ignored in subsequent discussions.

In the next example, the ideal Rankine cycle of Example 8.1 is modified to show the effects of turbine and pump isentropic efficiencies on performance.

▶ EXAMPLE 8.2 ▶

Analyzing a Rankine Cycle with Irreversibilities

Reconsider the vapor power cycle of Example 8.1, but include in the analysis that the turbine and the pump each have an isentropic efficiency of 85%. Determine for the modified cycle **(a)** the thermal efficiency, **(b)** the mass flow rate of steam, in kg/h, for a net power output of 100 MW, **(c)** the rate of heat transfer $\dot{Q}_{in}$ into the working fluid as it passes through the boiler, in MW, **(d)** the rate of heat transfer $\dot{Q}_{out}$ from the condensing steam as it passes through the condenser, in MW, **(e)** the mass flow rate of the condenser cooling water, in kg/h, if cooling water enters the condenser at 15°C and exits as 35°C.

SOLUTION

Known A vapor power cycle operates with steam as the working fluid. The turbine and pump both have efficiencies of 85%.

Find Determine the thermal efficiency, the mass flow rate, in kg/h, the rate of heat transfer to the working fluid as it passes through the boiler, in MW, the heat transfer rate from the condensing steam as it passes through the condenser, in MW, and the mass flow rate of the condenser cooling water, in kg/h.

Schematic and Given Data:

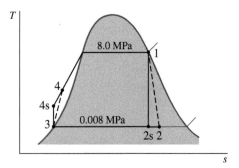

Fig. E8.2

Engineering Model

1. Each component of the cycle is analyzed as a control volume at steady state.

2. The working fluid passes through the boiler and condenser at constant pressure. Saturated vapor enters the turbine. The condensate is saturated at the condenser exit.

3. The turbine and pump each operate adiabatically with an efficiency of 85%.

4. Kinetic and potential energy effects are negligible.

Analysis Owing to the presence of irreversibilities during the expansion of the steam through the turbine, there is an increase in specific entropy from turbine inlet to exit, as shown on the accompanying T–s diagram. Similarly, there is an increase in specific entropy from pump inlet to exit. Let us begin the analysis by fixing each of the principal states. State 1 is the same as in Example 8.1, so $h_1 = 2758.0$ kJ/kg and $s_1 = 5.7432$ kJ/kg · K.

The specific enthalpy at the turbine exit, state 2, can be determined using the isentropic turbine efficiency, Eq. 8.9,

$$\eta_t = \frac{\dot{W}_t/\dot{m}}{(\dot{W}_t/\dot{m})_s} = \frac{h_1 - h_2}{h_1 - h_{2s}}$$

where h_{2s} is the specific enthalpy at state 2s on the accompanying T–s diagram. From the solution to Example 8.1, $h_{2s} = 1794.8$ kJ/kg. Solving for h_2 and inserting known values

$$
\begin{aligned}
h_2 &= h_1 - \eta_t(h_1 - h_{2s}) \\
&= 2758 - 0.85(2758 - 1794.8) = 1939.3 \text{ kJ/kg}
\end{aligned}
$$

State 3 is the same as in Example 8.1, so $h_3 = 173.88$ kJ/kg.

To determine the specific enthalpy at the pump exit, state 4, reduce mass and energy rate balances for a control volume around the pump to obtain $\dot{W}_p/\dot{m} = h_4 - h_3$. On rearrangement, the specific enthalpy at state 4 is

$$h_4 = h_3 + \dot{W}_p/\dot{m}$$

To determine h_4 from this expression requires the pump work. Pump work can be evaluated using the isentropic pump efficiency in the form of Eq. 8.10b: Solving for $\dot{W}_p/\dot{m}$ results in

$$\frac{\dot{W}_p}{\dot{m}} = \frac{v_3(p_4 - p_3)}{\eta_p}$$

The numerator of this expression was determined in the solution to Example 8.1. Accordingly,

$$\frac{\dot{W}_p}{\dot{m}} = \frac{8.06 \text{ kJ/kg}}{0.85} = 9.48 \text{ kJ/kg}$$

The specific enthalpy at the pump exit is then

$$h_4 = h_3 + \dot{W}_p/\dot{m} = 173.88 + 9.48 = 183.36 \text{ kJ/kg}$$

a. The net power developed by the cycle is

$$\dot{W}_{cycle} = \dot{W}_t - \dot{W}_p = \dot{m}[(h_1 - h_2) - (h_4 - h_3)]$$

The rate of heat transfer to the working fluid as it passes through the boiler is

$$\dot{Q}_{in} = \dot{m}(h_1 - h_4)$$

Thus, the thermal efficiency is

$$\eta = \frac{(h_1 - h_2) - (h_4 - h_3)}{h_1 - h_4}$$

Inserting values

$$\eta = \frac{(2758 - 1939.3) - 9.48}{2758 - 183.36} = 0.314 \text{ (31.4\%)}$$

b. With the net power expression of part (a), the mass flow rate of the steam is

$$\dot{m} = \frac{\dot{W}_{cycle}}{(h_1 - h_2) - (h_4 - h_3)}$$

$$= \frac{(100\ MW)|3600\ s/h||10^3\ kW/MW|}{(818.7 - 9.48)\ kJ/kg} = 4.449 \times 10^5\ kg/h$$

c. With the expression for $\dot{Q}_{in}$ from part (a) and previously determined specific enthalpy values

$$\dot{Q}_{in} = \dot{m}(h_1 - h_4)$$

$$= \frac{(4.449 \times 10^5\ kg/h)(2758 - 183.36)\ kJ/kg}{|3600\ s/h||10^3\ kW/MW|} = 318.2\ MW$$

d. The rate of heat transfer from the condensing steam to the cooling water is

$$\dot{Q}_{out} = \dot{m}(h_2 - h_3)$$

$$= \frac{(4.449 \times 10^5\ kg/h)(1939.3 - 173.88)\ kJ/kg}{|3600\ s/h||10^3\ kW/MW|} = 218.2\ MW$$

e. The mass flow rate of the cooling water can be determined from

$$\dot{m}_{cw} = \frac{\dot{m}(h_2 - h_3)}{(h_{cw,\ out} - h_{cw,\ in})}$$

$$= \frac{(218.2\ MW)|10^3\ kW/MW||3600\ s/h|}{(146.68 - 62.99)\ kJ/kg} = 9.39 \times 10^6\ kg/h$$

SKILLS DEVELOPED

Ability to...

- sketch the *T–s* diagram of the Rankine cycle with turbine and pump irreversibilities.
- fix each of the principal states and retrieve necessary property data.
- apply mass, energy, and entropy principles.
- calculate performance parameters for the cycle.

Quick Quiz

If the mass flow rate of steam were 150 kg/s, what would be the pump power required, in kW, and the back work ratio? Ans. 1422 kW, 0.0116.

Discussion of Examples 8.1 and 8.2   The effect of irreversibilities within the turbine and pump can be gauged by comparing values from Example 8.2 with their counterparts in Example 8.1. In Example 8.2, the turbine work per unit of mass is *less* and the pump work per unit of mass is *greater* than in Example 8.1, as can be confirmed using data from these examples. The thermal efficiency in Example 8.2 is *less* than in the ideal case of Example 8.1. For a fixed net power output (100 MW), the smaller net work output per unit mass in Example 8.2 dictates a greater mass flow rate of steam than in Example 8.1. The magnitude of the heat transfer to cooling water is also greater in Example 8.2 than in Example 8.1; consequently, a greater mass flow rate of cooling water is required.

8.3 Improving Performance—Superheat, Reheat, and Supercritical

The representations of the vapor power cycle considered thus far do not depict actual vapor power plants faithfully, for various modifications are usually incorporated to improve overall performance. In this section we consider cycle modifications known as *superheat* and *reheat*. Both features are normally incorporated into vapor power plants. We also consider supercritical steam generation.

Let us begin the discussion by noting that an increase in the boiler pressure or a decrease in the condenser pressure may result in a reduction of the steam quality at the exit of the turbine. This can be seen by comparing states 2′ and 2″ of Figs. 8.4a and 8.4b to the corresponding state 2 of each diagram. If the quality of the mixture passing through the turbine becomes too low, the impact of liquid droplets in the flowing liquid–vapor mixture can erode the turbine blades, causing a decrease in the turbine efficiency and an increased need for maintenance. Accordingly, common practice is to maintain at least 90% quality ($x \geq 0.9$) at the turbine exit. The cycle modifications known as *superheat* and *reheat* permit advantageous operating pressures in the boiler and condenser and yet avoid the problem of low quality of the turbine exhaust.

Animation

Rankine Cycle Tab b

superheat

Superheat

First, let us consider **superheat**. As we are not limited to having saturated vapor at the turbine inlet, further energy can be added by heat transfer to the steam, bringing it to a superheated vapor condition at the turbine inlet. This is accomplished in a separate heat exchanger called a superheater. The combination of boiler and superheater is referred to as a *steam generator*. Figure 8.3 shows an ideal Rankine cycle with superheated vapor at the turbine inlet: cycle $1'–2'–3–4–1'$. The cycle with superheat has a higher average temperature of heat addition than the cycle without superheating (cycle $1–2–3–4–1$), so the thermal efficiency is higher. Moreover, the quality at turbine exhaust state $2'$ is greater than at state 2, which would be the turbine exhaust state without superheating. Accordingly, superheating also tends to alleviate the problem of low steam quality at the turbine exhaust. With sufficient superheating, the turbine exhaust state may even fall in the superheated vapor region.

reheat

Reheat

A further modification normally employed in vapor power plants is **reheat**. With reheat, a power plant can take advantage of the increased efficiency that results with higher boiler pressures and yet avoid low-quality steam at the turbine exhaust. In the ideal reheat cycle shown in **Fig. 8.7**, steam does not expand to the condenser pressure in a single stage. Instead, steam expands through a first-stage turbine (Process 1–2) to some pressure between the steam generator and condenser pressures. Steam is then reheated in the steam generator (Process 2–3). Ideally, there would be no pressure drop as the steam is reheated. After reheating, the steam expands in a second-stage turbine to the condenser pressure (Process 3–4). Observe that with reheat the quality of the steam at the turbine exhaust is increased. This can be seen from the $T–s$ diagram of Fig. 8.7 by comparing state 4 with state $4'$, the turbine exhaust state without reheating.

TAKE NOTE...

When computing the thermal efficiency of a reheat cycle, it is necessary to account for the work output of both turbine stages as well as the total heat addition occurring in the vaporization/superheating and reheating processes. This calculation is illustrated in Example 8.3.

Supercritical

The temperature of the steam entering the turbine is restricted by metallurgical limitations imposed by materials used to fabricate the superheater, reheater, and turbine. High pressure in the steam generator also requires piping that can withstand great stresses at elevated temperatures. Still, improved materials and fabrication methods have gradually permitted significant increases in maximum allowed cycle temperature and steam generator pressure with corresponding increases in thermal efficiency that save fuel and reduce environmental impact. This progress now allows vapor power plants to operate with steam generator pressures exceeding the critical pressure of water (22.1 MPa). These plants are known as **supercritical** vapor power plants.

supercritical

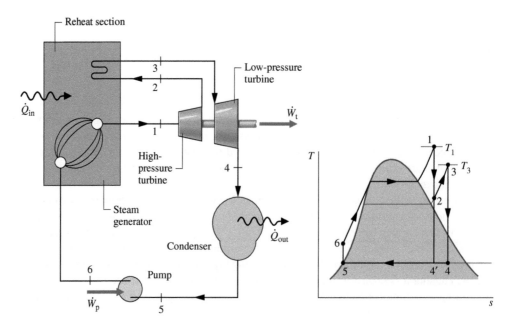

Fig. 8.7 Ideal reheat cycle.

Figure 8.8 shows a supercritical ideal reheat cycle. As indicated by Process 6–1, steam generation occurs at a pressure above the critical pressure. No pronounced phase change occurs during this process, and a conventional boiler is not used. Instead, water flowing through tubes is gradually heated from liquid to vapor without the bubbling associated with boiling. In such cycles, heating is provided by combustion of pulverized coal with air.

Today's supercritical vapor power plants produce steam at pressures and temperatures near 30 MPa and 600°C, respectively, permitting thermal efficiencies up to 47%. As superalloys with improved high-temperature limit and corrosion resistance become commercially available, *ultra*-supercritical plants may produce steam at 35 MPa and 700°C with thermal efficiencies exceeding 50%. Subcritical plants have efficiencies only up to about 40%.

While installation costs of supercritical plants are somewhat higher per unit of power generated than subcritical plants, fuel costs of supercritical plants are considerably lower owing to increased thermal efficiency. Since less fuel is used for a given power output, supercritical plants produce less carbon dioxide, other combustion gases, and solid waste than subcritical plants. The evolution of supercritical power plants from subcritical counterparts provides a case study on how advances in technology enable increases in thermodynamic efficiency with accompanying fuel savings and reduced environmental impact cost-effectively.

In the next example, the ideal Rankine cycle of Example 8.1 is modified to include superheat and reheat.

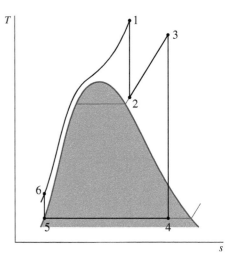

Fig. 8.8 Supercritical ideal reheat cycle.

> ▶ ▶ **EXAMPLE 8.3** ▶

Evaluating Performance of an Ideal Reheat Cycle

Steam is the working fluid in an ideal Rankine cycle with superheat and reheat. Steam enters the first-stage turbine at 8.0 MPa, 480°C, and expands to 0.7 MPa. It is then reheated to 440°C before entering the second-stage turbine, where it expands to the condenser pressure of 0.008 MPa. The *net* power output is 100 MW. Determine **(a)** the thermal efficiency of the cycle, **(b)** the mass flow rate of steam, in kg/h, **(c)** the rate of heat transfer $\dot{Q}_{out}$ from the condensing steam as it passes through the condenser, in MW. Discuss the effects of reheat on the vapor power cycle.

SOLUTION

Known An ideal reheat cycle operates with steam as the working fluid. Operating pressures and temperatures are specified, and the net power output is given.

Find Determine the thermal efficiency, the mass flow rate of the steam, in kg/h, and the heat transfer rate from the condensing steam as it passes through the condenser, in MW. Discuss.

Schematic and Given Data:

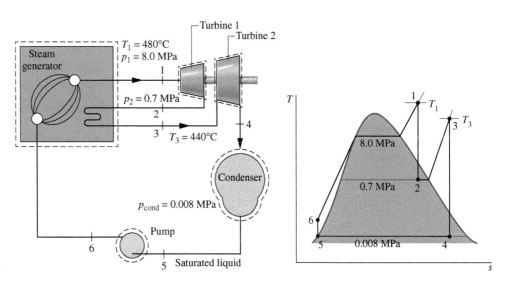

Fig. E8.3

Engineering Model

1. Each component in the cycle is analyzed as a control volume at steady state. The control volumes are shown on the accompanying sketch by dashed lines.

2. All processes of the working fluid are internally reversible.

3. The turbine and pump operate adiabatically.

4. Condensate exits the condenser as saturated liquid.

5. Kinetic and potential energy effects are negligible.

Analysis   To begin, we fix each of the principal states. Starting at the inlet to the first turbine stage, the pressure is 8.0 MPa and the temperature is 480°C, so the steam is a superheated vapor. From Table A-4, $h_1 = 3348.4$ kJ/kg and $s_1 = 6.6586$ kJ/kg · K.

State 2 is fixed by $p_2 = 0.7$ MPa and $s_2 = s_1$ for the isentropic expansion through the first-stage turbine. Using saturated liquid and saturated vapor data from Table A-3, the quality at state 2 is

$$x_2 = \frac{s_2 - s_f}{s_g - s_f} = \frac{6.6586 - 1.9922}{6.708 - 1.9922} = 0.9895$$

The specific enthalpy is then

$$h_2 = h_f + x_2 h_{fg}$$
$$= 697.22 + (0.9895)2066.3 = 2741.8 \text{ kJ/kg}$$

State 3 is superheated vapor with $p_3 = 0.7$ MPa and $T_3 = 440$°C, so from Table A-4, $h_3 = 3353.3$ kJ/kg and $s_3 = 7.7571$ kJ/kg · K.

To fix state 4, use $p_4 = 0.008$ MPa and $s_4 = s_3$ for the isentropic expansion through the second-stage turbine. With data from Table A-3, the quality at state 4 is

$$x_4 = \frac{s_4 - s_f}{s_g - s_f} = \frac{7.7571 - 0.5926}{8.2287 - 0.5926} = 0.9382$$

The specific enthalpy is

$$h_4 = 173.88 + (0.9382)2403.1 = 2428.5 \text{ kJ/kg}$$

State 5 is saturated liquid at 0.008 MPa, so $h_5 = 173.88$ kJ/kg. Finally, the state at the pump exit is the same as in Example 8.1, so $h_6 = 181.94$ kJ/kg.

a. The *net* power developed by the cycle is

$$\dot{W}_{cycle} = \dot{W}_{t1} + \dot{W}_{t2} - \dot{W}_p$$

Mass and energy rate balances for the two turbine stages and the pump reduce to give, respectively,

Turbine 1:	$\dot{W}_{t1}/\dot{m} = h_1 - h_2$
Turbine 2:	$\dot{W}_{t2}/\dot{m} = h_3 - h_4$
Pump:	$\dot{W}_p/\dot{m} = h_6 - h_5$

where $\dot{m}$ is the mass flow rate of the steam.

The total rate of heat transfer to the working fluid as it passes through the boiler–superheater and reheater is

$$\frac{\dot{Q}_{in}}{\dot{m}} = (h_1 - h_6) + (h_3 - h_2)$$

Using these expressions, the thermal efficiency is

$$\eta = \frac{(h_1 - h_2) + (h_3 - h_4) - (h_6 - h_5)}{(h_1 - h_6) + (h_3 - h_2)}$$
$$= \frac{(3348.4 - 2741.8) + (3353.3 - 2428.5) - (181.94 - 173.88)}{(3348.4 - 181.94) + (3353.3 - 2741.8)}$$
$$= \frac{606.6 + 924.8 - 8.06}{3166.5 + 611.5} = \frac{1523.3 \text{ kJ/kg}}{3778 \text{ kJ/kg}} = 0.403(40.3\%)$$

b. The mass flow rate of the steam can be obtained with the expression for net power given in part (a).

$$\dot{m} = \frac{\dot{W}_{cycle}}{(h_1 - h_2) + (h_3 - h_4) - (h_6 - h_5)}$$
$$= \frac{(100 \text{ MW})|3600 \text{ s/h}||10^3 \text{ kW/MW}|}{(606.6 + 924.8 - 8.06) \text{ kJ/kg}} = 2.363 \times 10^5 \text{ kg/h}$$

c. The rate of heat transfer from the condensing steam to the cooling water is

$$\dot{Q}_{out} = \dot{m}(h_4 - h_5)$$
$$= \frac{2.363 \times 10^5 \text{ kg/h}(2428.5 - 173.88) \text{ kJ/kg}}{|3600 \text{ s/h}||10^3 \text{ kW/MW}|} = 148 \text{ MW}$$

To see the effects of reheat, we compare the present values with their counterparts in Example 8.1. With superheat and reheat, the thermal efficiency is increased over that of the cycle of Example 8.1. For a specified net power output (100 MW), a larger thermal efficiency means that a smaller mass flow rate of steam is required. Moreover, with a greater thermal efficiency the rate of heat transfer to the cooling water is also less, resulting in a reduced demand for cooling water. With reheating, the steam quality at the turbine exhaust is substantially increased over the value for the cycle of Example 8.1.

SKILLS DEVELOPED

Ability to...

- sketch the *T–s* diagram of the ideal Rankine cycle with reheat.
- fix each of the principal states and retrieve necessary property data.
- apply mass and energy balances.
- calculate performance parameters for the cycle.

Quick Quiz

What is the rate of heat addition for the reheat process, in MW, and what percent is that value of the total heat addition to the cycle? Ans. 40.1 MW, 16.2%.

The following example illustrates the effect of turbine irreversibilities on the ideal reheat cycle of Example 8.3.

Evaluating Performance of a Reheat Cycle with Turbine Irreversibility

Reconsider the reheat cycle of Example 8.3 but include in the analysis that each turbine stage has the same isentropic efficiency. **(a)** If η_t = 85%, determine the thermal efficiency. **(b)** Plot the thermal efficiency versus turbine stage isentropic efficiency ranging from 85 to 100%.

SOLUTION

Known A reheat cycle operates with steam as the working fluid. Operating pressures and temperatures are specified. Each turbine stage has the same isentropic efficiency.

Find If η_t = 85%, determine the thermal efficiency. Also plot the thermal efficiency versus turbine stage isentropic efficiency ranging from 85 to 100%.

Schematic and Given Data:

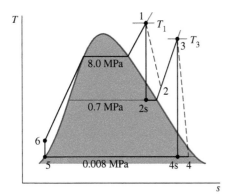

Fig. E8.4a

Engineering Model

1. As in Example 8.3, each component is analyzed as a control volume at steady state.

2. Except for the two turbine stages, all processes are internally reversible.

3. The turbine and pump operate adiabatically.

4. The condensate exits the condenser as saturated liquid.

5. Kinetic and potential energy effects are negligible.

Analysis

a. From the solution to Example 8.3, the following specific enthalpy values are known, in kJ/kg: h_1 = 3348.4, h_{2s} = 2741.8, h_3 = 3353.3, h_{4s} = 2428.5, h_5 = 173.88, h_6 = 181.94.

The specific enthalpy at the exit of the first-stage turbine, h_2, can be determined by solving the expression for the turbine isentropic efficiency, Eq. 8.9, to obtain

$$h_2 = h_1 - \eta_t(h_1 - h_{2s})$$

$$= 3348.4 - 0.85(3348.4 - 2741.8) = 2832.8 \text{ kJ/kg}$$

The specific enthalpy at the exit of the second-stage turbine can be found similarly:

$$h_4 = h_3 - \eta_t(h_3 - h_{4s})$$

$$= 3353.3 - 0.85(3353.3 - 2428.5) = 2567.2 \text{ kJ/kg}$$

The thermal efficiency is then

$$\eta = \frac{(h_1 - h_2) + (h_3 - h_4) - (h_6 - h_5)}{(h_1 - h_6) + (h_3 - h_2)} \qquad ❶$$

$$= \frac{(3348.4 - 2832.8) + (3353.3 - 2567.2) - (181.94 - 173.88)}{(3348.4 - 181.94) + (3353.3 - 2832.8)}$$

$$= \frac{1293.6 \text{ kJ/kg}}{3687.0 \text{ kJ/kg}} = 0.351 \ (35.1\%)$$

b. The *IT* code for the solution follows, where etat1 is η_{t1}, etat2 is η_{t2}, eta is η, Wnet = $\dot{W}_{net}/\dot{m}$, and Qin = $\dot{Q}_{in}/\dot{m}$.

```
// Fix the states
T1 = 480// °C
p1 = 80 // bar
h1 = h_PT ("Water/Steam", p1, T1)
s1 = s_PT ("Water/Steam", p1, T1)
p2 = 7 // bar
h2s = h_Ps ("Water/Steam", p2, s1)
etat1 = 0.85
h2 = h1 - etat1 * (h1 - h2s)
T3 = 440 // °C
p3 = p2
h3 = h_PT ("Water/Steam", p3, T3)
s3 = s_PT ("Water/Steam", p3, T3)
p4 = 0.08//bar
h4s = h_Ps ("Water/Steam", p4, s3)
etat2 = etat1
h4 = h3 - etat2 * (h3 - h4s)
p5 = p4
h5 = hsat_Px ("Water/Steam", p5, 0) // kJ/kg
v5 = vsat_Px ("Water/Steam", p5, 0) // m³/kg
p6 = p1
h6 = h5 + v5 * (p6 - p5) * 100// The 100 in this
expression is a unit conversion factor.
// Calculate thermal efficiency
Wnet = (h1 - h2) + (h3 - h4) - (h6 - h5)
Qin = (h1 - h6) + (h3 - h2)
eta = Wnet/Qin
```

Using the **Explore** button, sweep eta from 0.85 to 1.0 in steps of 0.01. Then, using the **Graph** button, obtain the following plot:

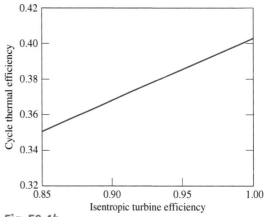

Fig. E8.4b

From Fig. E8.4*b*, we see that the cycle thermal efficiency increases from 0.351 to 0.403 as turbine stage isentropic efficiency increases from 0.85 to 1.00, as expected based on the results of Example 8.3 and part (a) of the present example. Turbine isentropic efficiency is seen to have a significant effect on cycle thermal efficiency.

---

❶ Owing to the irreversibilities present in the turbine stages, the net work per unit of mass developed in the present case is significantly less than in the case of Example 8.3. The thermal efficiency is also considerably less.

SKILLS DEVELOPED

Ability to...

• sketch the *T*–*s* diagram of the Rankine cycle with reheat, including turbine and pump irreversibilities.

• fix each of the principal states and retrieve necessary property data.

• apply mass, energy, and entropy principles.

• calculate performance parameters for the cycle.

Quick Quiz

If the temperature T_3 were increased to 480°C, would you expect the thermal efficiency to increase, decrease, or stay the same? Ans. Increase.

8.4 Improving Performance—Regenerative Vapor Power Cycle

Another commonly used method for increasing the thermal efficiency of vapor power plants is *regenerative feedwater heating*, or simply **regeneration**. This is the subject of the present section.

regeneration

To introduce the principle underlying regenerative feedwater heating, consider Fig. 8.3 once again. In cycle 1–2–3–4–a–1, the working fluid enters the boiler as a compressed liquid at state 4 and is heated while in the liquid phase to state a. With regenerative feedwater heating, the working fluid enters the boiler at a state *between* 4 and a. As a result, the average temperature of heat addition is increased, thereby tending to increase the thermal efficiency.

8.4.1 Open Feedwater Heaters

open feedwater heater

Let us consider how regeneration can be accomplished using an **open feedwater heater**, a type of direct-contact heat exchanger in which streams at different temperatures mix to form a stream at an intermediate temperature. Shown in **Fig. 8.9** are the schematic diagram and

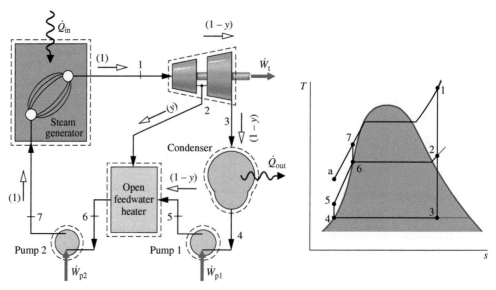

Fig. 8.9 Regenerative vapor power cycle with one open feedwater heater.

the associated *T–s* diagram for a regenerative vapor power cycle having one open feedwater heater. For this cycle, the working fluid passes isentropically through the turbine stages and pumps, and flow through the steam generator, condenser, and feedwater heater takes place with no pressure drop in any of these components. Still, there is a source of irreversibility owing to mixing within the feedwater heater.

Steam enters the first-stage turbine at state 1 and expands to state 2, where a fraction of the total flow is *extracted*, or *bled*, into an open feedwater heater operating at the extraction pressure, p_2. The rest of the steam expands through the second-stage turbine to state 3. This portion of the total flow is condensed to saturated liquid, state 4, and then pumped to the extraction pressure and introduced into the feedwater heater at state 5. A single mixed stream exits the feedwater heater at state 6. For the case shown in Fig. 8.9, the mass flow rates of the streams entering the feedwater heater are such that state 6 is saturated liquid at the extraction pressure. The liquid at state 6 is then pumped to the steam generator pressure and enters the steam generator at state 7. Finally, the working fluid is heated from state 7 to state 1 in the steam generator.

Referring to the *T–s* diagram of the cycle, note that the heat addition would take place from state 7 to state 1, rather than from state a to state 1, as would be the case without regeneration. Accordingly, the amount of energy that must be supplied from the combustion of a fossil fuel, or another source, to vaporize and superheat the steam would be reduced. This is the desired outcome. Only a portion of the total flow expands through the second-stage turbine (Process 2–3), however, so less work would be developed as well. In practice, operating conditions are such that the reduction in heat added more than offsets the decrease in net work developed, resulting in an increased thermal efficiency in regenerative power plants.

Cycle Analysis   Consider next the thermodynamic analysis of the regenerative cycle illustrated in Fig. 8.9. An important initial step in analyzing any regenerative vapor cycle is the evaluation of the mass flow rates through each of the components. Taking a single control volume enclosing both turbine stages, the mass rate balance reduces at steady state to

$$\dot{m}_2 + \dot{m}_3 = \dot{m}_1$$

where $\dot{m}_1$ is the rate at which mass enters the first-stage turbine at state 1, $\dot{m}_2$ is the rate at which mass is extracted and exits at state 2, and $\dot{m}_3$ is the rate at which mass exits the second-stage turbine at state 3. Dividing by $\dot{m}_1$ places this on the basis of a *unit of mass* passing through the first-stage turbine

$$\frac{\dot{m}_2}{\dot{m}_1} + \frac{\dot{m}_3}{\dot{m}_1} = 1$$

Denoting the fraction of the total flow extracted at state 2 by y ($y = \dot{m}_2/\dot{m}_1$), the fraction of the total flow passing through the second-stage turbine is

$$\frac{\dot{m}_3}{\dot{m}_1} = 1 - y \tag{8.11}$$

The fractions of the total flow at various locations are indicated in parentheses on Fig. 8.9.

The fraction y can be determined by applying the conservation of mass and conservation of energy principles to a control volume around the feedwater heater. Assuming no heat transfer between the feedwater heater and its surroundings and ignoring kinetic and potential energy effects, the mass and energy rate balances reduce at steady state to give

$$0 = y h_2 + (1 - y)h_5 - h_6$$

Solving for y

$$\boxed{y = \frac{h_6 - h_5}{h_2 - h_5}} \tag{8.12}$$

Equation 8.12 allows the fraction y to be determined when states 2, 5, and 6 are fixed.

Expressions for the principal work and heat transfers of the regenerative cycle can be determined by applying mass rate balances to control volumes around the individual components.

Beginning with the turbine, the total work is the sum of the work developed by each turbine stage. Neglecting kinetic and potential energy effects and assuming no heat transfer with the surroundings, we can express the total turbine work on the basis of a unit of mass passing through the first-stage turbine as

$$\frac{\dot{W}_t}{\dot{m}_1} = (h_1 - h_2) + (1 - y)(h_2 - h_3) \tag{8.13}$$

The total pump work is the sum of the work required to operate each pump individually. On the basis of a unit of mass passing through the first-stage turbine, the total pump work is

$$\frac{\dot{W}_p}{\dot{m}_1} = (h_7 - h_6) + (1 - y)(h_5 - h_4) \tag{8.14}$$

The energy added by heat transfer to the working fluid passing through the steam generator, per unit of mass expanding through the first-stage turbine, is

$$\frac{\dot{Q}_{in}}{\dot{m}_1} = h_1 - h_7 \tag{8.15}$$

and the energy rejected by heat transfer to the cooling water is

$$\frac{\dot{Q}_{out}}{\dot{m}_1} = (1 - y)(h_3 - h_4) \tag{8.16}$$

The following example illustrates the analysis of a regenerative cycle with one open feedwater heater, including the evaluation of properties at state points around the cycle and the determination of the fractions of the total flow at various locations.

▶▶▶ EXAMPLE 8.5 ▶

Considering a Regenerative Cycle with Open Feedwater Heater

Consider a regenerative vapor power cycle with one open feedwater heater. Steam enters the turbine at 8.0 MPa, 480°C and expands to 0.7 MPa, where some of the steam is extracted and diverted to the open feedwater heater operating at 0.7 MPa. The remaining steam expands through the second-stage turbine to the condenser pressure of 0.008 MPa. Saturated liquid exits the open feedwater heater at 0.7 MPa. The isentropic efficiency of each turbine stage is 85% and each pump operates isentropically. If the net power output of the cycle is 100 MW, determine (a) the thermal efficiency and (b) the mass flow rate of steam entering the first turbine stage, in kg/h.

SOLUTION

Known A regenerative vapor power cycle operates with steam as the working fluid. Operating pressures and temperatures are specified; the isentropic efficiency of each turbine stage and the net power output are also given.

Find Determine the thermal efficiency and the mass flow rate into the turbine, in kg/h.

Schematic and Given Data:

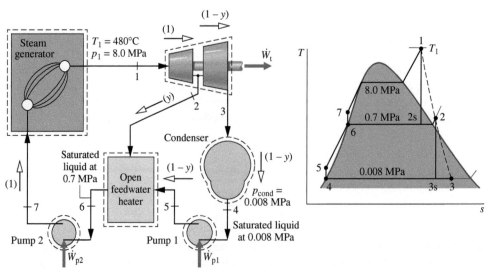

Fig. E8.5

Engineering Model

1. Each component in the cycle is analyzed as a steady-state control volume. The control volumes are shown in the accompanying sketch by dashed lines.

2. All processes of the working fluid are internally reversible, except for the expansions through the two turbine stages and mixing in the open feedwater heater.

3. The turbines, pumps, and feedwater heater operate adiabatically.

4. Kinetic and potential energy effects are negligible.

5. Saturated liquid exits the open feedwater heater, and saturated liquid exits the condenser.

Analysis   The specific enthalpy at states 1 and 4 can be read from the steam tables. The specific enthalpy at state 2 is evaluated in the solution to Example 8.4. The specific entropy at state 2 can be obtained from the steam tables using the known values of enthalpy and pressure at this state. In summary, $h_1 = 3348.4$ kJ/kg, $h_2 = 2832.8$ kJ/kg, $s_2 = 6.8606$ kJ/kg $\cdot$ K, $h_4 = 173.88$ kJ/kg.

The specific enthalpy at state 3 can be determined using the isentropic efficiency of the second-stage turbine

$$h_3 = h_2 - \eta_t(h_2 - h_{3s})$$

With $s_{3s} = s_2$, the quality at state 3s is $x_{3s} = 0.8208$; using this, we get $h_{3s} = 2146.3$ kJ/kg. Hence,

$$h_3 = 2832.8 - 0.85(2832.8 - 2146.3) = 2249.3 \text{ kJ/kg}$$

State 6 is saturated liquid at 0.7 MPa. Thus, $h_6 = 697.22$ kJ/kg.

Since the pumps operate isentropically, the specific enthalpy values at states 5 and 7 can be determined as

$$h_5 = h_4 + v_4(p_5 - p_4)$$
$$= 173.88 + (1.0084 \times 10^{-3})(\text{m}^3/\text{kg})(0.7 - 0.008) \text{ MPa}$$
$$\times \left|\frac{10^6 \text{ N/m}^2}{1 \text{ MPa}}\right|\left|\frac{1 \text{ kJ}}{10^3 \text{ N} \cdot \text{m}}\right| = 174.6 \text{ kJ/kg}$$
$$h_7 = h_6 + v_6(p_7 - p_6)$$
$$= 697.22 + (1.1080 \times 10^{-3})(8.0 - 0.7)|10^3| = 705.3 \text{ kJ/kg}$$

Applying mass and energy rate balances to a control volume enclosing the open feedwater heater, we find the fraction y of the flow extracted at state 2 from

$$y = \frac{h_6 - h_5}{h_2 - h_5} = \frac{697.22 - 174.6}{2832.8 - 174.6} = 0.1966$$

a. On the basis of a unit of mass passing through the first-stage turbine, the total turbine work output is

$$\frac{\dot{W}_t}{\dot{m}_1} = (h_1 - h_2) + (1 - y)(h_2 - h_3)$$
$$= (3348.4 - 2832.8) + (0.8034)(2832.8 - 2249.3)$$
$$= 984.4 \text{ kJ/kg}$$

The total pump work per unit of mass passing through the first-stage turbine is

$$\frac{\dot{W}_p}{\dot{m}_1} = (h_7 - h_6) + (1 - y)(h_5 - h_4)$$
$$= (705.3 - 697.22) + (0.8034)(174.6 - 173.88) = 8.7 \text{ kJ/kg}$$

The heat added in the steam generator per unit of mass passing through the first-stage turbine is

$$\frac{\dot{Q}_{in}}{\dot{m}_1} = h_1 - h_7 = 3348.4 - 705.3 = 2643.1 \text{ kJ/kg}$$

The thermal efficiency is then

$$\eta = \frac{\dot{W}_t/\dot{m}_1 - \dot{W}_p/\dot{m}_1}{\dot{Q}_{in}/\dot{m}_1} = \frac{984.4 - 8.7}{2643.1} = 0.369 \ (36.9\%)$$

b. The mass flow rate of the steam entering the turbine, $\dot{m}_1$, can be determined using the given value for the net power output, 100 MW. Since

$$\dot{W}_{cycle} = \dot{W}_t - \dot{W}_p$$

and

$$\frac{\dot{W}_t}{\dot{m}_1} = 984.4 \text{ kJ/kg} \qquad \text{and} \qquad \frac{\dot{W}_p}{\dot{m}_1} = 8.7 \text{ kJ/kg}$$

it follows that

$$\dot{m}_1 = \frac{(100 \text{ MW})|3600 \text{ s/h}|}{(984.4 - 8.7)\text{ kJ/kg}}\left|\frac{10^3 \text{ kJ/s}}{1 \text{ MW}}\right| = 3.69 \times 10^5 \text{ kg/h}$$

---

❶ Note that the fractions of the total flow at various locations are labeled on the figure.

SKILLS DEVELOPED

Ability to...

- sketch the T–s diagram of the regenerative vapor power cycle with one open feedwater heater.

- fix each of the principal states and retrieve necessary property data.

- apply mass, energy, and entropy principles.

- calculate performance parameters for the cycle.

Quick Quiz

If the mass flow rate of steam entering the first-stage turbine were 150 kg/s, what would be the net power, in MW, and the fraction of steam extracted, y? Ans. 146.4 MW, 0.1966.

8.4.2 Closed Feedwater Heaters

Regenerative feedwater heating also can be accomplished with **closed feedwater heaters**.   **closed feedwater heaters**
Closed heaters are shell-and-tube-type recuperators in which the feedwater temperature increases as the extracted steam condenses on the outside of the tubes carrying the feedwater.
Since the two streams do not mix, they can be at different pressures.

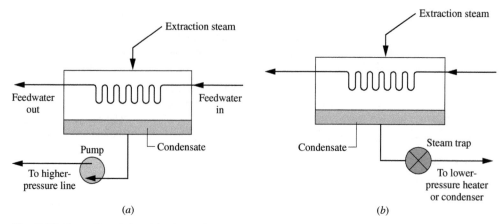

Fig. 8.10 Examples of closed feedwater heaters.

The diagrams of **Fig. 8.10** show two different schemes for removing the condensate from closed feedwater heaters. In Fig. 8.10*a*, this is accomplished by means of a pump whose function is to pump the condensate forward to a higher-pressure point in the cycle. In Fig. 8.10*b*, the condensate is allowed to expand through a *trap* into a feedwater heater operating at a lower pressure or into the condenser. A trap is a type of valve that permits only liquid to pass through to a region of lower pressure.

A regenerative vapor power cycle having one closed feedwater heater with the condensate trapped into the condenser is shown schematically in **Fig. 8.11**. For this cycle, the working fluid passes isentropically through the turbine stages and pumps. Except for expansion through the trap, there are no pressure drops accompanying flow through other components. The *T–s* diagram shows the principal states of the cycle.

The total steam flow expands through the first-stage turbine from state 1 to state 2. At this location, a fraction of the flow is bled into the closed feedwater heater, where it condenses. Saturated liquid at the extraction pressure exits the feedwater heater at state 7. The condensate is then trapped into the condenser, where it is reunited with the portion of the total flow passing through the second-stage turbine. The expansion from state 7 to state 8 through the trap is irreversible, so it is shown by a dashed line on the *T–s* diagram. The total flow exiting the condenser as saturated liquid at state 4 is pumped to the steam generator pressure and enters the feedwater heater at state 5. The temperature of the feedwater is increased in passing through the feedwater heater. The feedwater then exits at state 6. The cycle is completed as the working fluid is heated in the steam generator at constant pressure from state 6 to state 1. Although the

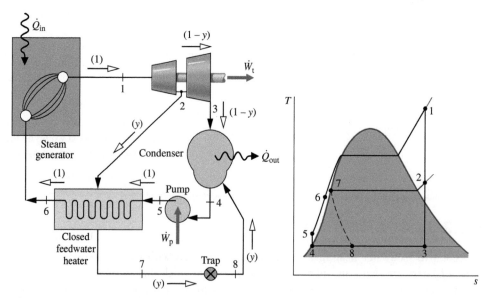

Fig. 8.11 Regenerative vapor power cycle with one closed feedwater heater.

closed heater shown on the figure operates with no pressure drop in either stream, there is a source of irreversibility due to the stream-to-stream temperature difference.

Cycle Analysis   The schematic diagram of the cycle shown in Fig. 8.11 is labeled with the fractions of the total flow at various locations. This is usually helpful in analyzing such cycles. The fraction of the total flow extracted, y, can be determined by applying the conservation of mass and conservation of energy principles to a control volume around the closed heater. Assuming no heat transfer between the feedwater heater and its surroundings and neglecting kinetic and potential energy effects, the mass and energy rate balances reduce at steady state to give

$$0 = y(h_2 - h_7) + (h_5 - h_6)$$

Solving for y

$$y = \frac{h_6 - h_5}{h_2 - h_7} \tag{8.17}$$

Assuming a throttling process for expansion across the trap, state 8 is fixed using $h_8 = h_7$. The principal work and heat transfers are evaluated as discussed previously.

8.4.3 Multiple Feedwater Heaters

The thermal efficiency of the regenerative cycle can be increased by incorporating several feedwater heaters at suitably chosen pressures. The number of feedwater heaters used is based on economic considerations, since incremental increases in thermal efficiency achieved with each additional heater must justify the added capital costs (heater, piping, pumps, etc.). Power plant designers use computer programs to simulate the thermodynamic and economic performance of different designs to help them decide on the number of heaters to use, the types of heaters, and the pressures at which they should operate.

Figure 8.12 shows the layout of a power plant with three closed feedwater heaters and one open heater. Power plants with multiple feedwater heaters ordinarily have at least one open feedwater heater operating at a pressure greater than atmospheric pressure so that oxygen and other dissolved gases can be vented from the cycle. This procedure, known as **deaeration**,   *deaeration* is needed to maintain the purity of the working fluid in order to minimize corrosion. Actual power plants have many of the same basic features as the one shown in the figure.

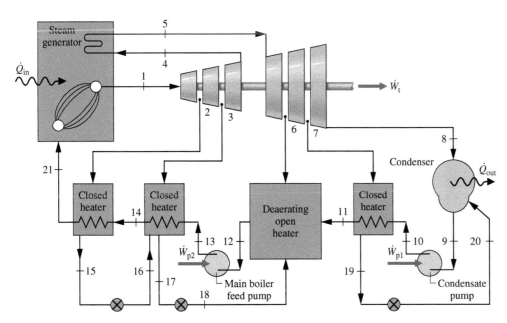

Fig. 8.12 Example of a power plant layout.

In analyzing regenerative vapor power cycles with multiple feedwater heaters, it is good practice to base the analysis on a unit of mass entering the first-stage turbine. To clarify the quantities of matter flowing through the various plant components, the fractions of the total flow removed at each extraction point and the fraction of the total flow remaining at each state point in the cycle should be labeled on a schematic diagram of the cycle. The fractions extracted are determined from mass and energy rate balances for control volumes around each of the feedwater heaters, starting with the highest-pressure heater and proceeding to each lower-pressure heater in turn. This procedure is used in the next example that involves a reheat–regenerative vapor power cycle with two feedwater heaters, one open feedwater heater and one closed feedwater heater.

▶ EXAMPLE 8.6 ▶

Considering a Reheat–Regenerative Cycle with Two Feedwater Heaters

Consider a reheat–regenerative vapor power cycle with two feedwater heaters, a closed feedwater heater and an open feedwater heater. Steam enters the first turbine at 8.0 MPa, 480°C and expands to 0.7 MPa. The steam is reheated to 440°C before entering the second turbine, where it expands to the condenser pressure of 0.008 MPa. Steam is extracted from the first turbine at 2 MPa and fed to the closed feedwater heater. Feedwater leaves the closed heater at 205°C and 8.0 MPa, and condensate exits as saturated liquid at 2 MPa. The condensate is trapped into the open feedwater heater. Steam extracted from the second turbine at 0.3 MPa is also fed into the open feedwater heater, which operates at 0.3 MPa. The stream exiting the open feedwater heater is saturated liquid at 0.3 MPa. The *net* power output of the cycle is 100 MW. There is no stray heat transfer from any component to its surroundings. If the working fluid experiences no irreversibilities as it passes through

the turbines, pumps, steam generator, reheater, and condenser, determine **(a)** the thermal efficiency, **(b)** the mass flow rate of the steam entering the first turbine, in kg/h.

SOLUTION

Known A reheat–regenerative vapor power cycle operates with steam as the working fluid. Operating pressures and temperatures are specified, and the net power output is given.

Find Determine the thermal efficiency and the mass flow rate entering the first turbine, in kg/h.

Schematic and Given Data:

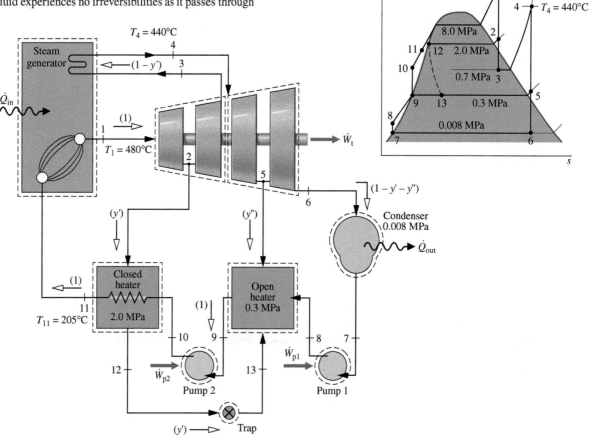

Fig. E8.6

Engineering Model

1. Each component in the cycle is analyzed as a control volume at steady state. The control volumes are shown on the accompanying sketch by dashed lines.

2. There is no stray heat transfer from any component to its surroundings.

3. The working fluid undergoes internally reversible processes as it passes through the turbines, pumps, steam generator, reheater, and condenser.

4. The expansion through the trap is a *throttling* process.

5. Kinetic and potential energy effects are negligible.

6. Condensate exits the closed feedwater heater as a saturated liquid at 2 MPa. Feedwater exits the open feedwater heater as a saturated liquid at 0.3 MPa. Condensate exits the condenser as a saturated liquid.

Analysis   Let us determine the specific enthalpies at the principal states of the cycle. State 1 is the same as in Example 8.3, so $h_1 = 3348.4$ kJ/kg and $s_1 = 6.6586$ kJ/kg · K.

State 2 is fixed by $p_2 = 2.0$ MPa and the specific entropy s_2, which is the same as that of state 1. Interpolating in Table A-4, we get $h_2 = 2963.5$ kJ/kg. The state at the exit of the first turbine is the same as at the exit of the first turbine of Example 8.3, so $h_3 = 2741.8$ kJ/kg.

State 4 is superheated vapor at 0.7 MPa, 440°C. From Table A-4, $h_4 = 3353.3$ kJ/kg and $s_4 = 7.7571$ kJ/kg · K. Interpolating in Table A-4 at $p_5 = 0.3$ MPa and $s_5 = s_4 = 7.7571$ kJ/kg · K, the enthalpy at state 5 is $h_5 = 3101.5$ kJ/kg.

Using $s_6 = s_4$, the quality at state 6 is found to be $x_6 = 0.9382$. So

$$h_6 = h_f + x_6 h_{fg}$$
$$= 173.88 + (0.9382)2403.1 = 2428.5 \text{ kJ/kg}$$

At the condenser exit, $h_7 = 173.88$ kJ/kg. The specific enthalpy at the exit of the first pump is

$$h_8 = h_7 + v_7(p_8 - p_7)$$
$$= 173.88 + (1.0084)(0.3 - 0.008) = 174.17 \text{ kJ/kg}$$

The required unit conversions were considered in previous examples.

The liquid leaving the open feedwater heater at state 9 is saturated liquid at 0.3 MPa. The specific enthalpy is $h_9 = 561.47$ kJ/kg. The specific enthalpy at the exit of the second pump is

$$h_{10} = h_9 + v_9(p_{10} - p_9)$$
$$= 561.47 + (1.0732)(8.0 - 0.3) = 569.73 \text{ kJ/kg}$$

The condensate leaving the closed heater is saturated liquid at 2 MPa. From Table A-3, $h_{12} = 908.79$ kJ/kg. The fluid passing through the trap undergoes a throttling process, so $h_{13} = 908.79$ kJ/kg.

The specific enthalpy of the feedwater exiting the closed heater at 8.0 MPa and 205°C is found using Eq. 3.13 as

$$h_{11} = h_f + v_f(p_{11} - p_{sat})$$
$$= 875.1 + (1.1646)(8.0 - 1.73) = 882.4 \text{ kJ/kg}$$

where h_f and v_f are the saturated liquid specific enthalpy and specific volume at 205°C, respectively, and p_{sat} is the saturation pressure in MPa at this temperature. Alternatively, h_{11} can be found from Table A-5.

The schematic diagram of the cycle is labeled with the fractions of the total flow into the turbine that remain at various locations. The fractions of the total flow diverted to the closed heater and open heater,

respectively, are $y' = \dot{m}_2/\dot{m}_1$ and $y'' = \dot{m}_5/\dot{m}_1$, where $\dot{m}_1$ denotes the mass flow rate entering the first turbine.

The fraction y' can be determined by application of mass and energy rate balances to a control volume enclosing the closed heater. The result is

$$y' = \frac{h_{11} - h_{10}}{h_2 - h_{12}} = \frac{882.4 - 569.73}{2963.5 - 908.79} = 0.1522$$

The fraction y'' can be determined by application of mass and energy rate balances to a control volume enclosing the open heater, resulting in

$$0 = y''h_5 + (1 - y' - y'')h_8 + y'h_{13} - h_9$$

Solving for y''

$$y'' = \frac{(1 - y')h_8 + y'h_{13} - h_9}{h_8 - h_5}$$
$$= \frac{(0.8478)174.17 + (0.1522)908.79 - 561.47}{174.17 - 3101.5}$$
$$= 0.0941$$

a. The following work and heat transfer values are expressed on the basis of a unit mass entering the first turbine. The work developed by the first turbine per unit of mass entering is the sum

$$\frac{\dot{W}_{t1}}{\dot{m}_1} = (h_1 - h_2) + (1 - y')(h_2 - h_3)$$
$$= (3348.4 - 2963.5) + (0.8478)(2963.5 - 2741.8)$$
$$= 572.9 \text{ kJ/kg}$$

Similarly, for the second turbine

$$\frac{\dot{W}_{t2}}{\dot{m}_1} = (1 - y')(h_4 - h_5) + (1 - y' - y'')(h_5 - h_6)$$
$$= (0.8478)(3353.3 - 3101.5) + (0.7537)(3101.5 - 2428.5)$$
$$= 720.7 \text{ kJ/kg}$$

For the first pump

$$\frac{\dot{W}_{p1}}{\dot{m}_1} = (1 - y' - y'')(h_8 - h_7)$$
$$= (0.7537)(174.17 - 173.88) = 0.22 \text{ kJ/kg}$$

and for the second pump

$$\frac{\dot{W}_{p2}}{\dot{m}_1} = (h_{10} - h_9)$$
$$= 569.73 - 561.47 = 8.26 \text{ kJ/kg}$$

The total heat added is the sum of the energy added by heat transfer during boiling/superheating and reheating. When expressed on the basis of a unit of mass entering the first turbine, this is

$$\frac{\dot{Q}_{in}}{\dot{m}_1} = (h_1 - h_{11}) + (1 - y')(h_4 - h_3)$$
$$= (3348.4 - 882.4) + (0.8478)(3353.3 - 2741.8)$$
$$= 2984.4 \text{ kJ/kg}$$

With the foregoing values, the thermal efficiency is

$$\eta = \frac{\dot{W}_{t1}/\dot{m}_1 + \dot{W}_{t2}/\dot{m}_1 - \dot{W}_{p1}/\dot{m}_1 - \dot{W}_{p2}/\dot{m}_1}{\dot{Q}_{in}/\dot{m}_1}$$
$$= \frac{572.9 + 720.7 - 0.22 - 8.26}{2984.4} = 0.431(43.1\%)$$

b. The mass flow rate entering the first turbine can be determined using the given value of the net power output. Thus,

$$\dot{m}_1 = \frac{\dot{W}_{cycle}}{\dot{W}_{t1}/\dot{m}_1 + \dot{W}_{t2}/\dot{m}_1 - \dot{W}_{p1}/\dot{m}_1 - \dot{W}_{p2}/\dot{m}_1}$$

❶ $$= \frac{(100 \text{ MW})|3600 \text{ s/h}||10^3 \text{ kW/MW}|}{1285.1 \text{ kJ/kg}} = 2.8 \times 10^5 \text{ kg/h}$$

❶ Compared to the corresponding values determined for the simple Rankine cycle of Example 8.1, the thermal efficiency of the present regenerative cycle is substantially greater and the mass flow rate is considerably less.

SKILLS DEVELOPED

Ability to...

- sketch the *T–s* diagram of the reheat–regenerative vapor power cycle with one closed and one open feedwater heater.
- fix each of the principal states and retrieve necessary property data.
- apply mass, energy, and entropy principles.
- calculate performance parameters for the cycle.

Quick Quiz

If each turbine stage had an isentropic efficiency of 85%, at which numbered states would the specific enthalpy values change? Ans. 2, 3, 5, and 6.

8.5 Other Vapor Power Cycle Aspects

In this section we consider vapor power cycle aspects related to working fluids, cogeneration systems, and carbon capture and storage.

8.5.1 Working Fluids

Demineralized water is used as the working fluid in the vast majority of vapor power systems because it is plentiful, low cost, nontoxic, chemically stable, and relatively noncorrosive. Water also has a large change in specific enthalpy as it vaporizes at typical steam generator pressures, which tends to limit the mass flow rate for a desired power output. With water, the pumping power is characteristically low, and the techniques of superheat, reheat, and regeneration are effective for improving power plant performance.

The high critical pressure of water (22.1 MPa) has posed a challenge to engineers seeking to improve thermal efficiency by increasing steam generator pressure and thus average temperature of heat addition. See the discussion of supercritical cycles in Sec. 8.3.

Although water has some shortcomings as a working fluid, no other single substance is more satisfactory for large electrical generating plants. Still, vapor cycles intended for special applications employ working fluids more in tune with the application at hand than water.

organic Rankine cycles   **Organic Rankine cycles** employ organic substances as working fluids, including pentane, mixtures of hydrocarbons, commonly used refrigerants, ammonia, and silicon oil. The organic working fluid is typically selected to meet the requirements of the particular application. For instance, the relatively low boiling point of these substances allows the Rankine cycle to produce power from low-temperature sources, including industrial *waste heat*, geothermal hot water, and fluids heated by concentrating-solar collectors.

binary vapor cycle   A **binary vapor cycle** couples two vapor cycles so the energy discharged by heat transfer from one cycle is the input for the other. Different working fluids are used in these cycles, one having advantageous high-temperature characteristics and another with complementary characteristics at the low-temperature end of the overall operating range. Depending on the application, these working fluids might include water and organic substances. The result is a combined cycle having a high average temperature of heat addition and a low average temperature of heat rejection and, thus, a thermal efficiency greater than either cycle has individually.

Figure 8.13 shows the schematic and accompanying *T–s* diagram of a binary vapor cycle. In this arrangement, two ideal Rankine cycles are combined using an interconnecting heat exchanger that serves as the condenser for the higher-temperature cycle (topping cycle) and boiler for the lower-temperature cycle (bottoming cycle). Heat rejected from the topping cycle provides the heat input for the bottoming cycle.

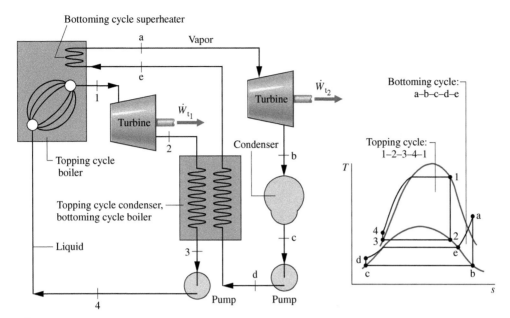

Fig. 8.13 Binary vapor cycle.

8.5.2 Cogeneration

Our society can use fuel more effectively through greater use of **cogeneration** systems, also known as combined heat and power systems. Cogeneration systems are integrated systems that simultaneously yield two valuable products, electricity and steam (or hot water), from a single fuel input. Cogeneration systems typically provide cost savings relative to producing power and steam (or hot water) in separate systems. Costing of cogeneration systems is introduced in Sec. 7.7.3.

cogeneration

Cogeneration systems are widely deployed in industrial plants, refineries, paper mills, food processing plants, and other facilities requiring process steam, hot water, and electricity for machines, lighting, and other purposes. **District heating** is another important cogeneration application. District heating plants are located within communities to provide steam or hot water for space heating and other thermal needs together with electricity for domestic, commercial, and industrial use. For instance, in New York City, district heating plants provide heating to Manhattan buildings while also generating electricity for various uses.

district heating

Cogeneration systems can be based on vapor power plants, gas turbine power plants, reciprocating internal combustion engines, and fuel cells. In this section, we consider vapor power–based cogeneration and, for simplicity, only district heating plants. The particular district heating systems considered have been selected because they are well suited for introducing the subject. Gas turbine–based cogeneration is considered in Sec. 9.9.2. The possibility of fuel cell–based cogeneration is considered in Sec. 13.4.

Back-Pressure Plants   A *back-pressure* district heating plant is shown in Fig. 8.14a. The plant resembles the simple Rankine cycle plant considered in Sec. 8.2 but with an important difference: In this case, energy released when the cycle working fluid condenses during flow through the condenser is harnessed to produce steam for export to the nearby community for various uses. The steam comes at the expense of the potential for power, however.

The power generated by the plant is linked to the district heating need for steam and is determined by the pressure at which the cycle working fluid condenses, called the *back pressure*. For instance, if steam as saturated vapor at 100°C is needed by the community, the cycle working fluid, assumed here to be demineralized water, must condense at a temperature greater than 100°C and thus at a back pressure greater than 1 atm. Accordingly, for fixed turbine inlet conditions and mass flow rate, the power produced in district heating is necessarily less than when condensation occurs well below 1 atm as it does in a plant fully dedicated to power generation.

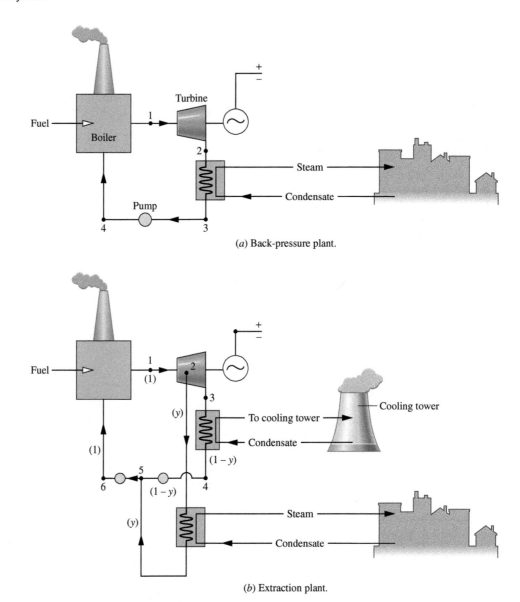

(a) Back-pressure plant.

(b) Extraction plant.

Fig. 8.14 Vapor cycle district heating plants.

Extraction Plants   An extraction district heating plant is shown in Fig. 8.14b. The figure is labeled (in parentheses) with fractions of the total flow entering the turbine remaining at various locations; in this respect the plant resembles the regenerative vapor power cycles considered in Sec. 8.4. Steam extracted from the turbine is used to service the district heating need. Differing heating needs can be flexibly met by varying the fraction of the steam extracted, denoted by y. For fixed turbine inlet conditions and mass flow rate, an increase in the fraction y to meet a greater district heating need is met by a reduction in power generated. When there is no demand for district heating, the full amount of steam generated in the boiler expands through the turbine, producing greatest power under the specified conditions. The plant then resembles the simple Rankine cycle of Sec. 8.2.

8.5.3 Carbon Capture and Storage

The concentration of carbon dioxide in the atmosphere has increased significantly since pre-industrial times. Some of the increase is traceable to burning fossil fuels. Coal-fired vapor power plants are major sources. Evidence is mounting that excessive CO_2 in the atmosphere contributes to global climate change, and there is growing agreement that measures must be taken to reduce such emissions.

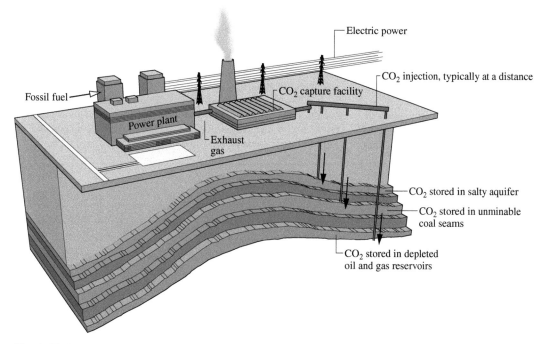

Fig. 8.15 Carbon capture and storage: power plant application.

Carbon dioxide emissions can be reduced by using fossil fuels more efficiently and avoiding wasteful practices. Moreover, if utilities use fewer fossil-fueled plants and more wind, hydro-power, and solar plants, less carbon dioxide will come from this sector. Practicing greater effi-ciency, eliminating wasteful practices, and using more renewable energy are important pathways for controlling CO_2. Yet these strategies are insufficient.

Since they will be plentiful for several decades, fossil fuels will continue to be used for generating electricity and meeting industrial needs. Accordingly, reducing CO_2 emissions at the *plant level* is imperative. One option is greater use of low-carbon fuels—more natural gas and less coal, for example. Another option involves removal of carbon dioxide from the exhaust gas of power plants, oil and gas refineries, and other industrial sources followed by storage (sequestration) of captured CO_2.

Figure 8.15 illustrates a type of carbon dioxide storage method actively under consid-eration today. Captured CO_2 is injected into depleted oil and gas reservoirs, unminable coal seams, deep salty aquifers, and other geological structures. Storage in oceans by injecting CO_2 to great depths from offshore pumping stations is another method under consideration.

Deployment of CO_2 capture and storage technology faces major hurdles, including uncer-tainty over how long injected gas will remain stored and possible collateral environmental im-pact when so much gas is stored in nature. Another technical challenge is the development of effective means for separating CO_2 from voluminous power plant and industrial gas streams.

Expenditures of energy resources and money required to capture CO_2, transport it to stor-age sites, and place it into storage will be significant. Yet with our current knowledge, carbon capture and storage is the principal strategy available today for reducing carbon dioxide emis-sions at the plant level. This area clearly is ripe for innovation.

Horizons

What to Do About That CO_2?

The race is on to find alternatives to storage of carbon dioxide captured from power plant exhaust gas streams and other sources. Analysts say there may be no better alternative, but storage does not have to be the fate of *all* of the captured CO_2 if there are com-mercial applications for *some* of it.

One use of carbon dioxide is for *enhanced oil recovery*—namely, to increase the amount of oil extractable from oil fields. By injecting CO_2 at high pressure into an oil-bearing underground layer, difficult-to-extract oil is forced to the surface. Proponents contend that widespread application of captured carbon dioxide for oil recov-ery will provide a source of income instead of cost, which is the case when the CO_2 is simply stored underground. Some envision a lively commerce involving export of liquefied carbon dioxide by ship *from* industrialized, oil-importing nations *to* oil-producing nations.

Another possible commercial use proposed for captured carbon dioxide is to produce algae, a tiny single-cell plant. When supplied with carbon dioxide, algae held in bioreactors absorb the carbon dioxide via photosynthesis, spurring algae growth. Carbon-enriched algae can be processed into transportation fuels, providing substitutes for gasoline and a source of income.

Researchers are also working on other ways to turn captured carbon dioxide into fuel. One approach attempts to mimic processes occurring in living things, whereby carbon atoms, extracted from carbon dioxide, and hydrogen atoms, extracted from water, are combined to create hydrocarbon molecules. Another approach uses solar radiation to split carbon dioxide into carbon monoxide and oxygen, and split water into hydrogen and oxygen. These building blocks can be combined into liquid fuels, researchers say.

Algae growth and fuel production using carbon dioxide are each in early stages of development. Still, such concepts suggest potential commercial uses for captured carbon dioxide and inspire hope of others yet to be imagined.

8.6 Case Study: Exergy Accounting of a Vapor Power Plant

The discussions to this point show that a useful picture of power plant performance can be obtained with the conservation of mass and conservation of energy principles. However, these principles provide only the *quantities* of energy transferred to and from the plant and do not consider the *utility* of the different types of energy transfer. For example, with the conservation principles alone, a unit of energy exiting as generated electricity is regarded as equivalent to a unit of energy exiting in relatively low-temperature cooling water, even though the electricity has greater utility and economic value. Also, nothing can be learned with the conservation principles alone about the relative significance of the irreversibilities present in the various plant components and the losses associated with those components. The method of exergy analysis introduced in Chap. 7 allows issues such as these to be considered quantitatively.

TAKE NOTE...

Chapter 7 is a prerequisite for the study of this section.

Exergy Accounting   In this section we account for the exergy entering a power plant with the fuel. (Means for evaluating the fuel exergy are introduced in Sec. 13.6.) A portion of the fuel exergy is ultimately returned to the plant surroundings as the net work developed. However, the largest part is either destroyed by irreversibilities within the various plant components or carried from the plant by cooling water, stack gases, and unavoidable heat transfers with the surroundings. These considerations are illustrated in the present section by three solved examples, treating respectively the boiler, turbine and pump, and condenser of a simple vapor power plant.

The irreversibilities present in each power plant component exact a tariff on the exergy supplied to the plant, as measured by the exergy destroyed in that component. The component levying the greatest tariff is the boiler, for a significant portion of the exergy entering the plant with the fuel is destroyed by irreversibilities within it. There are two main sources of irreversibility in the boiler: (1) the irreversible heat transfer occurring between the hot combustion gases and the working fluid of the vapor power cycle flowing through the boiler tubes, and (2) the combustion process itself. To simplify the present discussion, the boiler is considered to consist of a combustor unit in which fuel and air are burned to produce hot combustion gases, followed by a heat exchanger unit where the cycle working fluid is vaporized as the hot gases cool. This idealization is illustrated in **Fig. 8.16**.

For purposes of illustration, let us assume that 30% of the exergy entering the combustion unit with the fuel is destroyed by the combustion irreversibility and 1% of the fuel exergy exits the heat exchanger unit with the stack gases. The corresponding values for an actual power plant might differ from these nominal values. However, they provide characteristic values for discussion. (Means for evaluating the combustion exergy destruction and the exergy accompanying the exiting stack gases are introduced in Chap. 13.)

Using the foregoing values for the combustion exergy destruction and stack gas loss, it follows that a *maximum* of 69% of the fuel exergy remains for transfer from the hot combustion gases to the cycle working fluid. It is from this portion of the fuel exergy that the net work developed

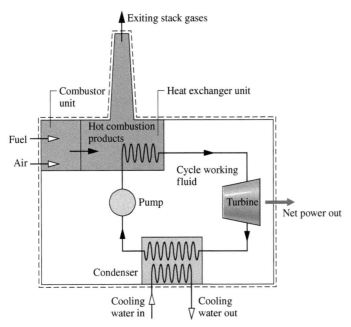

Fig. 8.16 Power plant schematic for the exergy analysis case study.

by the plant is obtained. In Examples 8.7 through 8.9, we account for the exergy supplied by the hot combustion gases passing through the heat exchanger unit. The principal results of this series of examples are reported in Table 8.4. Carefully note that the values of Table 8.4 are keyed to the vapor power plant of Example 8.2 and thus have only qualitative significance for vapor power plants in general.

Case Study Conclusions   The entries of Table 8.4 suggest some general observations about vapor power plant performance. First, the table shows that the exergy destructions are more significant than the plant losses. The largest portion of the exergy entering the plant with the fuel is destroyed, with exergy destruction in the boiler overshadowing all others. By contrast, the loss associated with heat transfer to the cooling water is relatively unimportant.

TABLE 8.4

Vapor Power Plant Exergy Accounting[a]

Outputs	
Net power out[b]	30%
Losses	
Condenser cooling water[c]	1%
Stack gases (assumed)	1%
Exergy destruction	
Boiler	
Combustion unit (assumed)	30%
Heat exchanger unit[d]	30%
Turbine[e]	5%
Pump[f]	—
Condenser[g]	3%
Total	100%

[a]All values are expressed as a percentage of the exergy carried into the plant with the fuel. Values are rounded to the nearest full percent. Exergy losses associated with stray heat transfer from plant components are ignored.

[b]Example 8.8.
[c]Example 8.9.
[d]Example 8.7.
[e]Example 8.8.
[f]Example 8.8.
[g]Example 8.9.

The cycle thermal efficiency (calculated in the solution to Example 8.2) is 31.4%, so over two-thirds (68.6%) of the *energy* supplied to the cycle working fluid is subsequently carried out by the condenser cooling water. By comparison, the amount of *exergy* carried out is virtually negligible because the temperature of the cooling water is raised only a few degrees over that of the surroundings and thus has limited utility. The loss amounts to only 1% of the exergy entering the plant with the fuel. Similarly, losses accompanying unavoidable heat transfer with the surroundings and the exiting stack gases typically amount only to a few percent of the exergy entering the plant with the fuel and are generally overstated when considered from the perspective of energy alone.

An exergy analysis allows the sites where destructions or losses occur to be identified and rank ordered for significance. This knowledge is useful in directing attention to aspects of plant performance that offer the greatest opportunities for improvement through the application of practical engineering measures. However, the decision to adopt any particular modification is governed by economic considerations that take into account both economies in fuel use and the costs incurred to achieve those economies.

The calculations presented in the following examples illustrate the application of exergy principles through the analysis of a simple vapor power plant. There are no fundamental difficulties, however, in applying the methodology to actual power plants, including consideration of the combustion process. The same procedures also can be used for exergy accounting of the gas turbine power plants considered in Chap. 9 and other types of thermal systems.

The following example illustrates the exergy analysis of the heat exchanger unit of the boiler of the case study vapor power plant.

▶ ▶ EXAMPLE 8.7 ▶ . . . .

Vapor Cycle Exergy Analysis—Heat Exchanger Unit

The heat exchanger unit of the boiler of Example 8.2 has a stream of water entering as a liquid at 8.0 MPa and exiting as a saturated vapor at 8.0 MPa. In a separate stream, gaseous products of combustion cool at a constant pressure of 1 atm from 1107 to 547°C. The gaseous stream can be modeled as air as an ideal gas. Let $T_0 = 22°C$, $p_0 = 1$ atm. Determine **(a)** the net rate at which exergy is carried into the heat exchanger unit by the gas stream, in MW, **(b)** the net rate at which exergy is carried from the heat exchanger by the water stream, in MW, **(c)** the rate of exergy destruction, in MW, **(d)** the exergetic efficiency given by Eq. 7.27.

SOLUTION

Known A heat exchanger at steady state has a water stream entering and exiting at known states and a separate gas stream entering and exiting at known states.

Find Determine the net rate at which exergy is carried into the heat exchanger by the gas stream, in MW, the net rate at which exergy is carried from the heat exchanger by the water stream, in MW, the rate of exergy destruction, in MW, and the exergetic efficiency.

Schematic and Given Data:

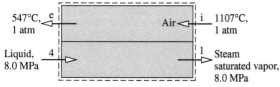

Fig. E8.7

Engineering Model

1. The control volume shown in the accompanying figure operates at steady state with $\dot{Q}_{cv} = \dot{W}_{cv} = 0$.

2. Kinetic and potential energy effects can be ignored.

3. The gaseous combustion products are modeled as air as an ideal gas.

4. The air and the water each pass through the steam generator at constant pressure.

5. Only 69% of the exergy entering the plant with the fuel remains after accounting for the stack loss and combustion exergy destruction.

6. $T_0 = 22°C$, $p_0 = 1$ atm.

Analysis The analysis begins by evaluating the mass flow rate of the air in terms of the mass flow rate of the water. The air and water pass through the boiler in separate streams. Hence, at steady state the conservation of mass principle requires

$$\dot{m}_i = \dot{m}_e \quad \text{(air)}$$
$$\dot{m}_4 = \dot{m}_1 \quad \text{(water)}$$

Using these relations, an energy rate balance for the overall control volume reduces at steady state to

$$0 = \dot{\cancel{Q}}_{cv}^0 - \dot{\cancel{W}}_{cv}^0 + \dot{m}_a(h_i - h_e) + \dot{m}(h_4 - h_1)$$

where $\dot{Q}_{cv} = \dot{W}_{cv} = 0$ by assumption 1, and the kinetic and potential energy terms are dropped by assumption 2. In this equation $\dot{m}_a$ and $\dot{m}$ denote, respectively, the mass flow rates of the air and water.

On solving

$$\frac{\dot{m}_a}{\dot{m}} = \frac{h_1 - h_4}{h_i - h_e}$$

The solution to Example 8.2 gives $h_1 = 2758$ kJ/kg and $h_4 = 183.36$ kJ/kg. From Table A-22, $h_i = 1491.44$ kJ/kg and $h_e = 843.98$ kJ/kg. Hence,

$$\frac{\dot{m}_a}{\dot{m}} = \frac{2758 - 183.36}{1491.44 - 843.98} = 3.977 \frac{\text{kg (air)}}{\text{kg (steam)}}$$

From Example 8.2, $\dot{m} = 4.449 \times 10^5$ kg/h. Thus, $\dot{m}_a = 17.694 \times 10^5$ kg/h.

a. The net rate at which exergy is carried into the heat exchanger unit by the gaseous stream can be evaluated using Eq. 7.18:

$$\begin{bmatrix} \text{net rate at which exergy} \\ \text{is carried in by the} \\ \text{gaseous stream} \end{bmatrix} = \dot{m}_a(\mathsf{e}_{fi} - \mathsf{e}_{fe})$$

$$= \dot{m}_a[h_i - h_e - T_0(s_i - s_e)]$$

Since the gas pressure remains constant, Eq. 6.20a giving the change in specific entropy of an ideal gas reduces to $s_i - s_e = s_i^\circ - s_e^\circ$. Thus, with h and s° values from Table A-22,

$$\dot{m}_a(\mathsf{e}_{fi} - \mathsf{e}_{fe}) = (17.694 \times 10^5 \text{ kg/h})[(1491.44 - 843.98) \text{ kJ/kg}$$
$$- (295 \text{ K})(3.34474 - 2.74504) \text{ kJ/kg} \cdot \text{K}]$$

$$= \frac{8.326 \times 10^8 \text{ kJ/h}}{|3600 \text{ s/h}|} \left| \frac{1 \text{ MW}}{10^3 \text{ kJ/s}} \right| = 231.28 \text{ MW}$$

b. The net rate at which exergy is carried out of the boiler by the water stream is determined similarly:

$$\begin{bmatrix} \text{net rate at which exergy} \\ \text{is carried out by the} \\ \text{water stream} \end{bmatrix} = \dot{m}(\mathsf{e}_{f1} - \mathsf{e}_{f4})$$

$$= \dot{m}[h_1 - h_4 - T_0(S_1 - S_4)]$$

From Table A-3, $s_1 = 5.7432$ kJ/kg $\cdot$ K. Double interpolation in Table A-5 at 8.0 MPa and $h_4 = 183.36$ kJ/kg gives $s_4 = 0.5957$ kJ/kg $\cdot$ K. Substituting known values

$$\dot{m}(\mathsf{e}_{f1} - \mathsf{e}_{f4}) = (4.449 \times 10^5)[(2758 - 183.36)$$
$$- 295(5.7432 - 0.5957)]$$

①

$$= \frac{4.699 \times 10^8 \text{ kJ/h}}{|3600 \text{ s/h}|} \left| \frac{1 \text{ MW}}{10^3 \text{ kJ/s}} \right| = 130.53 \text{ MW}$$

c. The rate of exergy destruction can be evaluated by reducing the exergy rate balance to obtain

②

$$\dot{E}_d = \dot{m}_a(\mathsf{e}_{fi} - \mathsf{e}_{fe}) + \dot{m}(\mathsf{e}_{f4} - \mathsf{e}_{f1})$$

With the results of parts (a) and (b)

③

$$\dot{E}_d = 231.28 \text{ MW} - 130.53 \text{ MW} = 100.75 \text{ MW}$$

d. The exergetic efficiency given by Eq. 7.27 is

$$\varepsilon = \frac{\dot{m}(\mathsf{e}_{f1} - \mathsf{e}_{f4})}{\dot{m}_a(\mathsf{e}_{fi} - \mathsf{e}_{fe})} = \frac{130.53 \text{ MW}}{231.28 \text{ MW}} = 0.564 \ (56.4\%)$$

This calculation indicates that 43.6% of the exergy supplied to the heat exchanger unit by the cooling combustion products is destroyed. However, since only 69% of the exergy entering the plant with the fuel is assumed to remain after the stack loss and combustion exergy destruction are accounted for (assumption 5), it can be concluded that $0.69 \times 43.6\% = 30\%$ of the exergy entering the plant with the fuel is destroyed within the heat exchanger. This is the value listed in Table 8.4.

---

① Since energy is conserved, the rate at which energy is transferred *to* the water as it flows through the heat exchanger *equals* the rate at which energy is transferred *from* the gas passing through the heat exchanger. By contrast, the rate at which exergy is transferred *to* the water is *less* than the rate at which exergy is transferred *from* the gas by the rate at which exergy is *destroyed* within the heat exchanger.

② The rate of exergy destruction can be determined alternatively by evaluating the rate of entropy production, $\dot{\sigma}_{cv}$, from an entropy rate balance and multiplying by T_0 to obtain $\dot{E}_d = T_0\dot{\sigma}_{cv}$.

③ Underlying the assumption that each stream passes through the heat exchanger at constant pressure is the neglect of friction as an irreversibility. Thus, the only contributor to exergy destruction in this case is heat transfer from the higher-temperature combustion products to the vaporizing water.

SKILLS DEVELOPED

Ability to...

- perform exergy analysis of a power plant steam generator.

Quick Quiz

If the gaseous products of combustion are cooled to 517°C ($h_e = 810.99$ kJ/kg), what is the mass flow rate of the gaseous products, in kg/h? Ans. 16.83×10^5 kg/h.

In the next example, we determine the exergy destruction rates in the turbine and pump of the case study vapor power plant.

▶▶ **EXAMPLE 8.8** ▶

Vapor Cycle Exergy Analysis—Turbine and Pump

Reconsider the turbine and pump of Example 8.2. Determine for each of these components the rate at which exergy is destroyed, in MW. Express each result, and the net power output of the plant, as a percentage of the exergy entering the plant with the fuel. Let $T_0 = 22$°C, $p_0 = 1$ atm.

SOLUTION

Known A vapor power cycle operates with steam as the working fluid. The turbine and pump each have an isentropic efficiency of 85%.

Find For the turbine and the pump individually, determine the rate at which exergy is destroyed, in MW. Express each result, and the net power output, as a percentage of the exergy entering the plant with the fuel.

Schematic and Given Data:

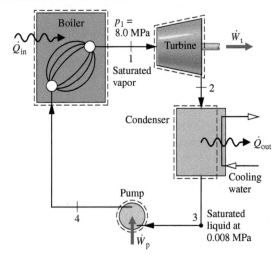

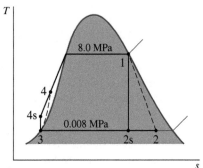

Fig. E8.8

Engineering Model

1. The turbine and the pump can each be analyzed as a control volume at steady state.

2. The turbine and pump operate adiabatically and each has an isentropic efficiency of 85%.

3. Kinetic and potential energy effects are negligible.

4. Only 69% of the exergy entering the plant with the fuel remains after accounting for the stack loss and combustion exergy destruction.

5. $T_0 = 22°C$, $p_0 = 1$ atm.

Analysis The rate of exergy destruction can be found by reducing the exergy rate balance or by use of the relationship $\dot{E}_d = T_0 \dot{\sigma}_{cv}$, where $\dot{\sigma}_{cv}$ is the rate of entropy production from an entropy rate balance. With either approach, the rate of exergy destruction for the turbine can be expressed as

$$\dot{E}_d = \dot{m} T_0 (s_2 - s_1)$$

From Table A-3, $s_1 = 5.7432$ kJ/kg · K. Using $h_2 = 1939.3$ kJ/kg from the solution to Example 8.2, the value of s_2 can be determined from Table A-3 as $s_2 = 6.2021$ kJ/kg · K. Substituting values

$$\dot{E}_d = (4.449 \times 10^5 \text{ kg/h})(295 \text{ K})(6.2021 - 5.7432)(\text{kJ/kg} \cdot \text{K})$$

$$= \left(0.602 \times 10^8 \frac{\text{kJ}}{\text{h}} \right) \left| \frac{1 \text{ h}}{3600 \text{ s}} \right| \left| \frac{1 \text{ MW}}{10^3 \text{ kJ/s}} \right| = 16.72 \text{ MW}$$

From the solution to Example 8.7, the net rate at which exergy is supplied by the cooling combustion gases is 231.28 MW. The turbine rate of exergy destruction expressed as a percentage of this is $(16.72/231.28)(100\%) = 7.23\%$. However, since only 69% of the entering fuel exergy remains after the stack loss and combustion exergy destruction are accounted for, it can be concluded that $0.69 \times 7.23\% = 5\%$ of the exergy entering the plant with the fuel is destroyed within the turbine. This is the value listed in Table 8.4.

Similarly, the exergy destruction rate for the pump is

$$\dot{E}_d = \dot{m} T_0 (s_4 - s_3)$$

With s_3 from Table A-3 and s_4 from the solution to Example 8.7

$$\dot{E}_d = (4.449 \times 10^5 \text{ kg/h})(295 \text{ K})(0.5957 - 0.5926)(\text{kJ/kg} \cdot \text{K})$$

$$= \left(4.07 \times 10^5 \frac{\text{kJ}}{\text{h}} \right) \left| \frac{1 \text{ h}}{3600 \text{ s}} \right| \left| \frac{1 \text{ MW}}{10^3 \text{ kJ/s}} \right| = 0.11 \text{ MW}$$

Expressing this as a percentage of the exergy entering the plant as calculated above, we have $(0.11/231.28)(69\%) = 0.03\%$. This value is rounded to zero in Table 8.4.

The net power output of the vapor power plant of Example 8.2 is 100 MW. Expressing this as a percentage of the rate at which exergy is carried into the plant with the fuel, $(100/231.28)(69\%) = 30\%$, as shown in Table 8.4.

SKILLS DEVELOPED

Ability to...

• perform exergy analysis of a power plant turbine and pump.

Quick Quiz

What is the exergetic efficiency of the power plant? Ans. 30%.

The following example illustrates the exergy analysis of the condenser of the case study vapor power plant.

▶ ▶ ▶ **EXAMPLE 8.9** ▶

Vapor Cycle Exergy Analysis—Condenser

The condenser of Example 8.2 involves two separate water streams. In one stream a two-phase liquid–vapor mixture enters at 0.008 MPa and exits as a saturated liquid at 0.008 MPa. In the other stream, cooling water enters at 15°C and exits at 35°C. **(a)** Determine the net rate at which exergy is carried from the condenser by the cooling water, in MW. Express this result as a percentage of the exergy entering the plant with the fuel. **(b)** Determine for the condenser the rate of exergy destruction, in MW. Express this result as a percentage of the exergy entering the plant with the fuel. Let $T_0 = 22°C$ and $p_0 = 1$ atm.

SOLUTION

Known   A condenser at steady state has two streams: (1) a two-phase liquid–vapor mixture entering and condensate exiting at known states and (2) a separate cooling water stream entering and exiting at known temperatures.

Find   Determine the net rate at which exergy is carried from the condenser by the cooling water stream and the rate of exergy destruction for the condenser. Express both quantities in MW and as percentages of the exergy entering the plant with the fuel.

Schematic and Given Data:

Saturated liquid, 0.008 MPa   3

Two-phase liquid–vapor mixture, 0.008 MPa   2

Liquid, 15°C   i

Liquid, 35°C   e

Fig. E8.9

Engineering Model

1. The control volume shown on the accompanying figure operates at steady state with $\dot{Q}_{cv} = \dot{W}_{cv} = 0$.

2. Kinetic and potential energy effects can be ignored.

3. Only 69% of the fuel exergy remains after accounting for the stack loss and combustion exergy destruction.

4. $T_0 = 22°C$, $p_0 = 1$ atm.

Analysis

a. The net rate at which exergy is carried out of the condenser can be evaluated by using Eq. 7.18:

$$\begin{bmatrix} \text{net rate at which exergy} \\ \text{is carried out by the} \\ \text{cooling water} \end{bmatrix} = \dot{m}_{cw}(\mathsf{e}_{fe} - \mathsf{e}_{fi})$$

$$= \dot{m}_{cw}[h_e - h_i - T_0(s_e - s_i)]$$

where $\dot{m}_{cw}$ is the mass flow rate of the cooling water from the solution to Example 8.2. With saturated liquid values for specific enthalpy and entropy from Table A-2 at the specified inlet and exit temperatures of the cooling water

$$\dot{m}_{cw}(\mathsf{e}_{fe} - \mathsf{e}_{fi}) = (9.39 \times 10^6 \text{ kg/h})[(146.68 - 62.99) \text{ kJ/kg}$$
$$- (295 \text{ K})(0.5053 - 0.2245) \text{ kJ/kg} \cdot \text{K}]$$

$$= \frac{8.019 \times 10^6 \text{ kJ/h}}{|3600 \text{ s/h}|} \left| \frac{1 \text{ MW}}{10^3 \text{ kJ/s}} \right| = 2.23 \text{ MW}$$

Expressing this as a percentage of the exergy entering the plant with the fuel, we get $(2.23/231.28)(69\%) = 1\%$. This is the value listed in Table 8.4.

b. The rate of exergy destruction for the condenser can be evaluated by reducing the exergy rate balance. Alternatively, the relationship $\dot{E}_d = T_0\dot{\sigma}_{cv}$ can be employed, where $\dot{\sigma}_{cv}$ is the time rate of entropy production for the condenser determined from an entropy rate balance. With either approach, the rate of exergy destruction can be expressed as

$$\dot{E}_d = T_0[\dot{m}(s_3 - s_2) + \dot{m}_{cw}(s_e - s_i)]$$

Substituting values

$$\dot{E}_d = 295[(4.449 \times 10^5)(0.5926 - 6.2021)$$
$$+ (9.39 \times 10^6)(0.5053 - 0.2245)]$$
$$= \frac{416.1 \times 10^5 \text{ kJ/h}}{|3600 \text{ s/h}|} \left| \frac{1 \text{ MW}}{10^3 \text{ kJ/s}} \right| = 11.56 \text{ MW}$$

Expressing this as a percentage of the exergy entering the plant with the fuel, we get $(11.56/231.28)(69\%) = 3\%$. This is the value listed in Table 8.4.

SKILLS DEVELOPED

Ability to...

• perform exergy analysis of a power plant condenser.

Quick Quiz

Referring to data from Example 8.2, what percent of the *energy* supplied to the steam passing through the steam generator is carried out by the cooling water? Ans. 68.6%.

CHAPTER SUMMARY AND STUDY GUIDE

This chapter begins with an introduction to power generation that surveys current U.S. power generation by source and looks ahead to power generation needs in decades to come. These discussions provide context for the study of vapor power plants in this chapter and gas power plants in Chap. 9.

In Chap. 8 we have considered practical arrangements for vapor power plants, illustrated how vapor power plants are modeled thermodynamically, and considered the principal irreversibilities and losses associated with such plants. The main components of *simple* vapor power plants are modeled by the Rankine cycle.

In this chapter, we also have introduced modifications to the simple vapor power cycle aimed at improving overall performance. These include superheat, reheat, regeneration, supercritical operation, cogeneration, and binary cycles. We have also included a case study to illustrate the application of exergy analysis to vapor power plants.

The following checklist provides a study guide for this chapter. When your study of the text and end-of-chapter exercises has been completed you should be able to

• write out the meanings of the terms listed in the margin throughout the chapter and explain each of the related concepts. The subset of key concepts listed below is particularly important.

- sketch schematic diagrams and accompanying *T–s* diagrams of Rankine, reheat, and regenerative vapor power cycles.
- apply conservation of mass and energy, the second law, and property data to determine power cycle performance, including thermal efficiency, net power output, and mass flow rates.

- discuss the effects on Rankine cycle performance of varying steam generator pressure, condenser pressure, and turbine inlet temperature.
- discuss the principal sources of exergy destruction and loss in vapor power plants.

KEY ENGINEERING CONCEPTS

Rankine cycle	superheat	closed feedwater heater
feedwater	reheat	organic Rankine cycle
thermal efficiency	supercritical	binary vapor cycle
back work ratio	regeneration	cogeneration
ideal Rankine cycle	open feedwater heater	district heating

KEY EQUATIONS

$\eta = \dfrac{\dot{W}_t/\dot{m} - \dot{W}_p/\dot{m}}{\dot{Q}_{in}/\dot{m}} = \dfrac{(h_1 - h_2) - (h_4 - h_3)}{h_1 - h_4}$	(8.5a)	Thermal efficiency for the Rankine cycle of Fig. 8.2
$\text{bwr} = \dfrac{\dot{W}_p/\dot{m}}{\dot{W}_t/\dot{m}} = \dfrac{(h_4 - h_3)}{h_1 - h_2}$	(8.6)	Back work ratio for the Rankine cycle of Fig. 8.2
$\left(\dfrac{\dot{W}_p}{\dot{m}}\right)_s \approx v_3(p_4 - p_3)$	(8.7b)	Approximation for the pump work of the ideal Rankine cycle of Fig. 8.3

EXERCISES: THINGS ENGINEERS THINK ABOUT

8.1 What are electrical *brownouts* and *blackouts* and what causes them?

8.2 Referring to your monthly electric bills, make a plot of your monthly kW · h usage for the last year. Comment.

8.3 How much does the temperature of condenser cooling water vary in your area throughout the year? How would that affect the thermal efficiency of local power plants?

8.4 When does *peak demand* for electricity usually occur in your service area?

8.5 What is a *baseload* power plant?

8.6 Power plants located in arid areas may accomplish condensation by air cooling instead of cooling with water flowing through the condenser. How might air cooling affect the thermal efficiency?

8.7 What type of power plant produces the electricity used in your residence?

8.8 Brainstorm some ways to use the cooling water exiting the condenser of a large power plant.

8.9 Why is it important for power plant operators to keep pipes circulating water through plant components free from *fouling*?

8.10 There are many locations where solar energy is plentiful, yet to date there is no widespread use of solar energy for electric power generation. What issues have kept solar energy from being widely used?

8.11 Decades of coal mining have left piles of waste coal, or *culm*, in many locations through the United States. What effects does culm have on human health and the environment?

8.12 How do operators of electricity-generating plants detect and respond to changes in consumer demand throughout the day?

8.13 What is an energy *orb*?

CHECKING UNDERSTANDING

8.1 The component of the Rankine cycle in which the working fluid vaporizes is the

    a. boiler   c. pump

    b. condenser   d. turbine

8.2 The ratio of the pump work input to the work developed by the turbine is

    a. back work ratio   c. net work

    b. isentropic efficiency   d. thermal efficiency

8.3 A shell-and-tube-type recuperator in which the feedwater temperature increases as the extracted steam condenses on the outside of the tubes carrying the feedwater is a _____.

8.4 The processes associated with the ideal Rankine cycle are

    a. two adiabatic processes, two isentropic processes.

    b. two constant-volume processes, two isentropic processes.

    c. two constant-temperature processes, two isentropic processes.

    d. two constant-pressure processes, two isentropic processes.

8.5 A district heating plant whose net power output is linked to the district heating need for steam and is determined by the pressure at which the cycle working fluid condenses is _____.

8.6 The performance parameter that compares work associated with an actual adiabatic expansion through a turbine with that of the corresponding isentropic expansion is _____.

8.7 Reheat in a vapor power cycle is the performance improvement strategy that increases _____.

8.8 A direct-contact–type heat exchanger found in regenerative vapor power cycles in which streams at different temperatures mix to form a stream at an intermediate temperature is an _____.

8.9 The component of a regenerative vapor power cycle that permits only liquid to pass through to a region of lower pressure is a _____.

8.10 Identifying the disposition of the exergy entering a power plant with the fuel is called _____.

8.11 The component of a Rankine cycle in which the working fluid rejects energy by heat transfer is the _____.

8.12 Vapor power plants that operate with steam generator pressures exceeding the critical pressure of water are

  a. ideal vapor power plants.

  b. regenerative vapor power plants.

  c. reheat vapor power plants.

  d. supercritical vapor power plants.

8.13 With _____, steam expands through a first-stage turbine to an intermediate pressure, returns through a steam generator, and then expands through a second-stage turbine to the condenser pressure.

8.14 The component of the Rankine cycle that requires a power input is the _____.

8.15 The two main sources of irreversibilities in the boiler are _____ and _____.

Indicate whether the following statements are true or false. Explain.

8.16 Lowering condenser pressure lowers the average temperature of heat rejection from the Rankine cycle.

8.17 The total cost associated with a power plant considers only construction, operation, maintenance, and retirement.

8.18 Power plant types that generate electric power include coal-fueled, nuclear-fueled, solar-concentrating, and wind power plants.

8.19 Extraction district heating plants adapt to differing heating needs by varying the fraction of steam extracted from the turbine.

8.20 The current power grid focuses on electricity transmission and distribution.

8.21 For a vapor power cycle with a turbine that produces 5 MW and a pump that requires 100 kW, the net power produced by the cycle is 5100 kW.

8.22 A steam generator is the combination of a boiler and a superheater.

8.23 For a vapor power cycle with $\dot{Q}_{in}$ = 10 MW and $\dot{W}_{cycle}$ = 2 MW, $\dot{Q}_{out}$ = 12 MW.

8.24 Entropy must increase as steam expands through an actual adiabatic turbine.

8.25 A vapor power plant that operates with a steam generator at a pressure of 19 MPa is a supercritical vapor power plant.

PROBLEMS: DEVELOPING ENGINEERING SKILLS

C Problem may require use of appropriate computer software in order to complete.

Analyzing Rankine Cycles

8.1 Water is the working fluid in an ideal Rankine cycle. The condenser pressure is 6 kPa, and saturated vapor enters the turbine at 10 MPa. Determine the heat transfer rates, in kJ per kg of steam flowing, for the working fluid passing through the boiler and condenser and calculate the thermal efficiency.

8.2 Steam is the working fluid in the ideal Rankine cycle 1–2–3–4–1 and in the Carnot cycle 1–2–3′–4′–1 that both operate between pressures of 1.5 bar and 60 bar as shown in the *T–s* diagram in **Fig. P8.2**. Both cycles incorporate the steady flow devices shown in Fig. 8.2. For each cycle determine (a) the net power developed per unit mass of steam flowing, in kJ/kg, and (b) the thermal efficiency. Compare results and comment.

8.3 Water is the working fluid in an ideal Rankine cycle. Superheated vapor enters the turbine at 10 MPa, 480°C, and the condenser pressure is 6 kPa. Determine for the cycle

  a. the rate of heat transfer to the working fluid passing through the steam generator, in kJ per kg of steam flowing.

  b. the thermal efficiency.

  c. the rate of heat transfer from the working fluid passing through the condenser to the cooling water, in kJ per kg of steam flowing.

8.4 Water is the working fluid in an ideal Rankine cycle. Saturated vapor enters the turbine at 18 MPa, and the condenser pressure is

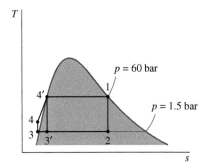

State	p (bar)	h (kJ/kg)	x	v (m³/kg)
1	60	2784.3	1	
2	1.5	2180.6	0.7696	
3′	1.5	1079.8	0.2752	
3	1.5	467.11	0	0.0010528
4	60	473.27	—	
4′	60	1213.4	0	

FIGURE P8.2

6 kPa. The mass flow rate of steam entering the turbine is 150 kg/s. Determine

    a. the net power developed, in kW.

    b. the rate of heat transfer to the steam passing through the boiler, in kW.

    c. the thermal efficiency.

    d. the mass flow rate of condenser cooling water, in kg/s, if the cooling water undergoes a temperature increase of 18°C with negligible pressure change in passing through the condenser.

8.5 Water is the working fluid in a Carnot vapor power cycle. Saturated liquid enters the boiler at 16 MPa, and saturated vapor enters the turbine. The condenser pressure is 8 kPa. The mass flow rate of steam entering the turbine is 120 kg/s. Determine

    a. the thermal efficiency.

    b. the back work ratio.

    c. the net power developed, in kW.

    d. the rate of heat transfer from the working fluid passing through the condenser, in kW.

8.6 Water is the working fluid in an ideal Rankine cycle. The pressure and temperature at the turbine inlet are 11 MPa and 590°C, respectively, and the condenser pressure is 6.9 kPa The mass flow rate of steam entering the turbine is 635×10^3 kg/h. The cooling water experiences a temperature increase from 16 to 27°C, with negligible pressure drop, as it passes through the condenser. Determine for the cycle

    a. the net power developed, in kJ/h.

    b. the thermal efficiency.

    c. the mass flow rate of cooling water, in kg/h.

8.7 In an ideal Rankine cycle, saturated vapor enters the steam turbine at 20 bar and leaves it at 0.08 bar. It then enters the condenser, where it is condensed to saturated liquid water at 0.08 bar. Determine

    a. net work per kg of steam flowing.

    b. the thermal efficiency of the cycle.

8.8 A nuclear power plant based on the Rankine cycle operates with a boiling-water reactor to develop net cycle power of 3 MW. Steam exits the reactor core at 100 bar, 520°C and expands through the turbine to the condenser pressure of 1 bar. Saturated liquid exits the condenser and is pumped to the reactor pressure of 100 bar. Isentropic efficiencies of the turbine and pump are 81% and 78%, respectively. Cooling water enters the condenser at 15°C with a mass flow rate of 114.79 kg/s. Determine

    a. the thermal efficiency.

    b. the temperature of the cooling water exiting the condenser, in °C.

8.9 Water is used as the working fluid in an ideal Rankine cycle. The steam is supplied as superheated steam at 30 bar, 440°C. The condenser pressure is 0.5 bar. If isentropic efficiency of turbine and pump are 90% and 85%, respectively, determine the thermal efficiency of the cycle.

8.10 The ideal Rankine cycle 1–2–3–4–1 of Problem 8.2 is modified to include the effects of irreversibilities in the adiabatic expansion and compression processes as shown in the *T–s* diagram in **Fig. P8.10**. Let $T_0 = 300$ K, $p_0 = 1$ bar. Determine

    a. the isentropic turbine efficiency.

    b. the rate of exergy destruction per unit mass of steam flowing in the turbine, in kJ/kg.

    c. the isentropic pump efficiency.

    d. the thermal efficiency.

8.11 **C** Steam enters the turbine of a simple vapor power plant with a pressure of 10 MPa and temperature T, and expands adiabatically to 6 kPa. The isentropic turbine efficiency is 85%.

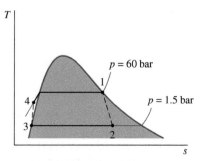

State	p (bar)	h (kJ/kg)	x	v (m³/kg)	s (kJ/kg·K)
1	60	2784.3	1		5.8892
2	1.5	2262.8	0.8065		6.1030
3	1.5	467.11	0	0.0010528	
4	60	474.14	—		

FIGURE P8.10

Saturated liquid exits the condenser at 6 kPa and the isentropic pump efficiency is 82%.

    a. For $T = 580$°C, determine the turbine exit quality and the cycle thermal efficiency.

    b. Plot the quantities of part (a) versus T ranging from 580 to 700°C.

8.12 Reconsider the cycle of Problem 8.6, but include in the analysis that the turbine and pump have isentropic efficiencies of 88%. The mass flow rate is unchanged. Determine for the modified cycle

    a. the net power developed, in kJ/h.

    b. the rate of heat transfer to the working fluid passing through the steam generator, in kJ/h.

    c. the thermal efficiency.

    d. the volumetric flow rate of cooling water entering the condenser, in m³/min.

8.13 Water is the working fluid in a Rankine cycle. Superheated vapor enters the turbine at 10 MPa, 480°C, and the condenser pressure is 6 kPa. The turbine and pump have isentropic efficiencies of 80 and 70%, respectively. Determine for the cycle

    a. the rate of heat transfer to the working fluid passing through the steam generator, in kJ per kg of steam flowing.

    b. the thermal efficiency.

    c. the rate of heat transfer from the working fluid passing through the condenser to the cooling water, in kJ per kg of steam flowing.

8.14 Water is the working fluid in a Rankine cycle. Superheated vapor enters the turbine at 8 MPa, 560°C with a mass flow rate of 7.8 kg/s and exits at 8 kPa. Saturated liquid enters the pump at 8 kPa. The isentropic turbine efficiency is 88%, and the isentropic pump efficiency is 82%. Cooling water enters the condenser at 18°C and exits at 36°C with no significant change in pressure. Determine

    a. the net power developed, in kW.

    b. the thermal efficiency.

    c. the mass flow rate of cooling water, in kg/s.

8.15 **Figure P8.15** provides steady-state operating data for a vapor power plant using water as the working fluid. The mass flow rate of water is 15 kg/s. The turbine and pump operate adiabatically but not reversibly. Determine

    a. the thermal efficiency.

    b. the rates of heat transfer $\dot{Q}_{in}$ and $\dot{Q}_{out}$, each in kW.

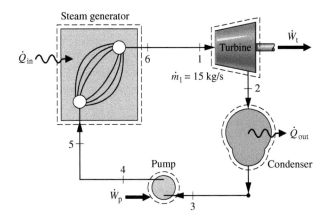

Steam generator

$\dot{Q}_{in}$

$\dot{W}_t$

Turbine

$\dot{m}_1 = 15$ kg/s

$\dot{Q}_{out}$

Condenser

Pump

$\dot{W}_p$

State	p	T (°C)	h (kJ/kg)
1	4 MPa	440	3307.1
2	10 kPa	- - -	1633.3
3	10 kPa	Sat.	191.83
4	7.5 MPa	- - -	199.4
5	6 MPa	36	150.86
6	4 MPa	540	3536.9

FIGURE P8.15

8.16 Superheated steam at 8 MPa and 480°C leaves the steam generator of a vapor power plant. Heat transfer and frictional effects in the line connecting the steam generator and the turbine reduce the pressure and temperature at the turbine inlet to 7.6 MPa and 440°C, respectively. The pressure at the exit of the turbine is 10 kPa, and the turbine operates adiabatically. Liquid leaves the condenser at 8 kPa, 36°C. The pressure is increased to 8.6 MPa across the pump. The turbine and pump isentropic efficiencies are 88%. The mass flow rate of steam is 79.53 kg/s. Determine

   a. the net power output, in kW.

   b. the thermal efficiency.

   c. the rate of heat transfer from the line connecting the steam generator and the turbine, in kW.

   d. the mass flow rate of condenser cooling water, in kg/s, if the cooling water enters at 15°C and exits at 35°C with negligible pressure change.

8.17 Steam enters the turbine of a vapor power plant at 4.1 MPa, 538°C and exits as a two-phase liquid–vapor mixture at temperature T. Condensate exits the condenser at a temperature 2.78°C lower than T and is pumped to 4.1 MPa. The turbine and pump isentropic efficiencies are 90 and 80%, respectively. The net power developed is 1 MW.

   a. For $T = 27$°C, determine the steam quality at the turbine exit, the steam mass flow rate, in kg/h, and the thermal efficiency.

   b. Plot the quantities of part (a) versus T ranging from 27°C to 41°C.

8.18 Superheated steam at 18 MPa, 560°C, enters the turbine of a vapor power plant. The pressure at the exit of the turbine is 0.06 bar, and liquid leaves the condenser at 0.045 bar, 26°C. The pressure is increased to 18.2 MPa across the pump. The turbine and pump have isentropic efficiencies of 82 and 77%, respectively. For the cycle, determine

   a. the net work per unit mass of steam flow, in kJ/kg.

   b. the heat transfer to steam passing through the boiler, in kJ per kg of steam flowing.

   c. the thermal efficiency.

   d. the heat transfer to cooling water passing through the condenser, in kJ per kg of steam condensed.

8.19 In the preliminary design of a power plant, water is chosen as the working fluid and it is determined that the turbine inlet temperature may not exceed 520°C. Based on expected cooling water temperatures, the condenser is to operate at a pressure of 0.06 bar. Determine the steam generator pressure required if the isentropic turbine efficiency is 80% and the quality of steam at the turbine exit must be at least 90%.

Considering Reheat and Supercritical Cycles

8.20 Steam at 10 MPa, 600°C enters the first-stage turbine of an ideal Rankine cycle with reheat. The steam leaving the reheat section of the steam generator is at 500°C, and the condenser pressure is 6 kPa. If the quality at the exit of the second-stage turbine is 90%, determine the cycle thermal efficiency.

8.21 Water is the working fluid in an ideal Rankine cycle with reheat. Superheated vapor enters the turbine at 10 MPa, 480°C, and the condenser pressure is 6 kPa. Steam expands through the first-stage turbine to 0.7 MPa and then is reheated to 480°C. Determine for the cycle

   a. the rate of heat addition, in kJ per kg of steam entering the first-stage turbine.

   b. the thermal efficiency.

   c. the rate of heat transfer from the working fluid passing through the condenser to the cooling water, in kJ per kg of steam entering the first-stage turbine.

8.22 For the cycle of Problem 8.21, reconsider the analysis assuming the pump and each turbine stage has an isentropic efficiency of 80%. Answer the same questions as in Problem 8.21 for the modified cycle.

8.23 Steam heated at constant pressure in a steam generator enters the first stage of a supercritical reheat cycle at 28 MPa, 520°C. Steam exiting the first-stage turbine at 6 MPa is reheated at constant pressure to 500°C. Each turbine stage has an isentropic efficiency of 78% while the pump has an isentropic efficiency of 82%. Saturated liquid exits the condenser that operates at constant pressure, p.

   a. For $p = 6$ kPa, determine the quality of the steam exiting the second stage of the turbine and the thermal efficiency.

   b. Plot the quantities of part (a) versus p ranging from 4 kPa to 70 kPa.

8.24 Steam at 33 MPa, 538°C enters the first stage of a supercritical reheat cycle including two turbine stages. The steam exiting the first-stage turbine at 4.1 MPa is reheated at constant pressure to 538°C. Each turbine stage and the pump has an isentropic efficiency of 85%. The condenser pressure is 6.9 kPa. If the net power output of the cycle is 100 MW, determine

   a. the rate of heat transfer to the working fluid passing through the steam generator, in MW.

   b. the rate of heat transfer from the working fluid passing through the condenser, in MW.

   c. the cycle thermal efficiency.

8.25 An ideal Rankine cycle with reheat uses water as the working fluid. The conditions at the inlet to the first turbine stage are 11 MPa, 650°C and the steam is reheated between the turbine stages to 650°C. For a condenser pressure of 6.9 kPa, plot the cycle thermal efficiency versus reheat pressure for pressures ranging from 0.4 to 8.3 MPa.

Analyzing Regenerative Cycles

8.26 Water is the working fluid in an ideal regenerative Rankine cycle. Superheated vapor enters the turbine at 10 MPa, 480°C, and the condenser pressure is 6 kPa. Steam expands through the first-stage turbine to 0.7 MPa, where some of the steam is extracted and diverted to an open feedwater heater operating at 0.7 MPa. The remaining steam expands through the second-stage turbine to the condenser pressure of 6 kPa. Saturated liquid exits the feedwater heater at 0.7 MPa. Determine for the cycle

   a. the rate of heat addition, in kJ per kg of steam entering the first-stage turbine.

   b. the thermal efficiency.

   c. the rate of heat transfer from the working fluid passing through the condenser to the cooling water, in kJ per kg of steam entering the first-stage turbine.

8.27 For the cycle of Problem 8.26, reconsider the analysis assuming the pump and each turbine stage has an isentropic efficiency of 76%. Answer the same questions as in Problem 8.26 for the modified cycle.

8.28 A power plant operates on a regenerative vapor power cycle with one open feedwater heater. Steam enters the first turbine stage at 12 MPa, 520°C and expands to 1 MPa, where some of the steam is extracted and diverted to the open feedwater heater operating at 1 MPa. The remaining steam expands through the second turbine stage to the condenser pressure of 6 kPa. Saturated liquid exits the open feedwater heater at 1 MPa. For isentropic processes in the turbines and pumps, determine for the cycle (a) the thermal efficiency and (b) the mass flow rate into the first turbine stage, in kg/h, for a net power output of 330 MW.

8.29 **C** Reconsider the cycle of Problem 8.28 as the feedwater heater pressure takes on other values. Plot the thermal efficiency and the rate of exergy destruction within the feedwater heater, in kW, versus the feedwater heater pressure ranging from 0.5 to 10 MPa. Let $T_0 = 293$ K.

8.30 Compare the results of Problem 8.28 with those for an ideal Rankine cycle having the same turbine inlet conditions and condenser pressure, but no regenerator.

8.31 **C** For the cycle of Problem 8.28, investigate the effects on cycle performance as the feedwater heater pressure takes on other values. Construct suitable plots and discuss. Assume that each turbine stage and each pump has an isentropic efficiency of 80%.

8.32 Water is the working fluid in an ideal regenerative Rankine cycle with one open feedwater heater. Superheated vapor enters the first-stage turbine at 16 MPa, 560°C, and the condenser pressure is 8 kPa.

The mass flow rate of steam entering the first-stage turbine is 120 kg/s. Steam expands through the first-stage turbine to 1 MPa where some of the steam is extracted and diverted to an open feedwater heater operating at 1 MPa. The remainder expands through the second-stage turbine to the condenser pressure of 8 kPa. Saturated liquid exits the feedwater heater at 1 MPa. Determine

   a. the net power developed, in kW.

   b. the rate of heat transfer to the steam passing through the boiler, in kW.

   c. the thermal efficiency.

   d. the mass flow rate of condenser cooling water, in kg/s, if the cooling water undergoes a temperature increase of 18°C with negligible pressure change in passing through the condenser.

8.33 Reconsider the cycle of Problem 8.32, but include in the analysis that each turbine stage and pump has an isentropic efficiency of 85%.

8.34 Water is the working fluid in an ideal regenerative Rankine cycle with one closed feedwater heater. Superheated vapor enters the turbine at 10 MPa, 480°C, and the condenser pressure is 6 kPa. Steam expands through the first-stage turbine where some is extracted and diverted to a closed feedwater heater at 0.7 MPa. Condensate drains from the feedwater heater as saturated liquid at 0.7 MPa and is trapped into the condenser. The feedwater leaves the heater at 10 MPa and a temperature equal to the saturation temperature at 0.7 MPa. Determine for the cycle

   a. the rate of heat transfer to the working fluid passing through the steam generator, in kJ per kg of steam entering the first-stage turbine.

   b. the thermal efficiency.

   c. the rate of heat transfer from the working fluid passing through the condenser to the cooling water, in kJ per kg of steam entering the first-stage turbine.

8.35 For the cycle of Problem 8.34, reconsider the analysis assuming the pump and each turbine stage have isentropic efficiencies of 80%. Answer the same questions as in Problem 8.34 for the modified cycle.

8.36 Consider a regenerative vapor power cycle with two feedwater heaters, a closed one and an open one, as shown in **Fig. P8.36**. Steam

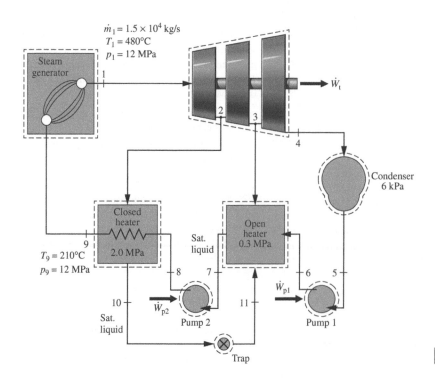

FIGURE P8.36

enters the first turbine stage at 12 MPa, 480°C, and expands to 2 MPa. Some steam is extracted at 2 MPa and fed to the closed feedwater heater. The remainder expands through the second-stage turbine to 0.3 MPa, where an additional amount is extracted and fed into the open feedwater heater operating at 0.3 MPa. The steam expanding through the third-stage turbine exits at the condenser pressure of 6 kPa.

Feedwater leaves the closed heater at 210°C, 12 MPa, and condensate exiting as saturated liquid at 2 MPa is trapped into the open feedwater heater. Saturated liquid at 0.3 MPa leaves the open feedwater heater. Assume all pumps and turbine stages operate isentropically. Determine for the cycle

a. the rate of heat transfer to the working fluid passing through the steam generator, in kJ per kg of steam entering the first-stage turbine.

b. the thermal efficiency.

c. the rate of heat transfer from the working fluid passing through the condenser to the cooling water, in kJ per kg of steam entering the first-stage turbine.

8.37 For the cycle of Problem 8.36, reconsider the analysis assuming the pump and each turbine stage has an isentropic efficiency of 80%. Answer the same questions as in Problem 8.36 for the modified cycle.

8.38 As indicated in **Fig. P8.38**, a power plant similar to that in Fig. 8.11 operates on a regenerative vapor power cycle with one closed feedwater heater. Steam enters the first turbine stage at state 1 where pressure is 12 MPa and temperature is 560°C. Steam expands to state 2 where pressure is 1 MPa and some of the steam is extracted and diverted to the closed feedwater heater. Condensate exits the feedwater heater at state 7 as saturated liquid at a pressure of 1 MPa, undergoes a throttling process through a trap to a pressure of 6 kPa at state 8, and then enters the condenser. The remaining steam expands through the second turbine stage to a pressure of 6 kPa at state 3 and then enters the condenser. Saturated liquid feedwater exiting the condenser at state 4 at a pressure of 6 kPa enters a pump and exits the pump at a pressure of 12 MPa. The feedwater then flows through the closed

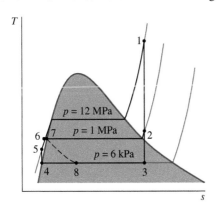

State	p (kPa)	T (°C)	h (kJ/kg)	s (kJ/kg·K)	x
1	12,000	560	3506.2	6.6840	
2	1,000		2823.3	6.6840	
3	6		2058.2	6.6840	0.7892
4	6		151.53	0.5210	0
5	12,000		163.60	0.5210	
6	12,000		606.61	1.7808	
7	1,000		762.81	2.1387	0
8	6		762.81	2.4968	0.2530

FIGURE P8.38

feedwater heater, exiting at state 6 with a pressure of 12 MPa. The net power output for the cycle is 330 MW. For isentropic processes in each turbine stage and the pump, determine

a. the cycle thermal efficiency.

b. the mass flow rate into the first turbine stage, in kg/s.

c. the rate of entropy production in the closed feedwater heater, in kW/K.

d. the rate of entropy production in the steam trap, in kW/K.

8.39 **C** For the cycle of Problem 8.36, investigate the effects on cycle performance as the higher extraction pressure takes on other values. The operating conditions for the open feedwater heater are unchanged from those in Problem 8.36. Assume that condensate drains from the closed feedwater heater as saturated liquid at the higher extraction pressure. Also, feedwater leaves the heater at 12 MPa and a temperature equal to the saturation temperature at the extraction pressure. Construct suitable plots and discuss.

8.40 Water is the working fluid in an ideal regenerative Rankine cycle with one closed feedwater heater. Superheated vapor enters the turbine at 16 MPa, 560°C, and the condenser pressure is 8 kPa. The cycle has a closed feedwater heater using extracted steam at 1 MPa. Condensate drains from the feedwater heater as saturated liquid at 1 MPa and is trapped into the condenser. The feedwater leaves the heater at 16 MPa and a temperature equal to the saturation temperature at 1 MPa. The mass flow rate of steam entering the first-stage turbine is 120 kg/s. Determine

a. the net power developed, in kW.

b. the rate of heat transfer to the steam passing through the boiler, in kW.

c. the thermal efficiency.

d. the mass flow rate of condenser cooling water, in kg/s, if the cooling water undergoes a temperature increase of 18°C with negligible pressure change in passing through the condenser.

8.41 Reconsider the cycle of Problem 8.40, but include in the analysis that each turbine stage and the pump have an isentropic efficiency of 78%. Answer the same questions as in Problem 8.40 for the modified cycle.

8.42 A power plant operates on a regenerative vapor power cycle with two feedwater heaters. Steam enters the first turbine stage as 12 MPa, 520°C and expands in three stages to the condenser pressure of 6 kPa. Between the first and second stages, some steam is diverted to a closed feedwater heater at 1 MPa, with saturated liquid condensate being pumped ahead into the boiler feedwater line. The feedwater leaves the closed heater at 12 MPa, 170°C. Steam is extracted between the second and third turbine stages at 0.15 MPa and fed into an open feedwater heater operating at that pressure. Saturated liquid at 0.15 MPa leaves the open feedwater heater. For isentropic processes in the pumps and turbines, determine for the cycle (a) the thermal efficiency and (b) the mass flow rate into the first-stage turbine, in kg/h, if the net power developed is 320 MW.

8.43 Reconsider the cycle of Problem 8.42, but include in the analysis that each turbine stage has an isentropic efficiency of 82% and each pump an efficiency of 100%.

8.44 Water is the working fluid in a Rankine cycle modified to include one closed feedwater heater and one open feedwater heater. Superheated vapor enters the turbine at 16 MPa, 560°C, and the condenser pressure is 8 kPa. The mass flow rate of steam entering the first-stage turbine is 120 kg/s. The closed feedwater heater uses extracted steam at 4 MPa, and the open feedwater heater uses extracted steam at 0.3 MPa. Saturated liquid condensate drains from the closed feedwater heater at 4 MPa and is trapped into the open feedwater heater. The feedwater leaves the closed heater at 16 MPa and a temperature equal to the saturation temperature at 4 MPa. Saturated liquid leaves the open heater at 0.3 MPa. Assume all turbine stages and pumps operate isentropically. Determine

a. the net power developed, in kW.

b. the rate of heat transfer to the steam passing through the steam generator, in kW.

c. the thermal efficiency.

d. the mass flow rate of condenser cooling water, in kg/s, if the cooling water undergoes a temperature increase of 18°C with negligible pressure change in passing through the condenser.

8.45 Reconsider the cycle of Problem 8.44, but include in the analysis that the isentropic efficiencies of the turbine stages and pumps are 85%.

8.46 Consider a regenerative vapor power cycle with two feedwater heaters, a closed one and an open one, and reheat. Steam enters the first turbine stage at 12 MPa, 480°C, and expands to 2 MPa. Some steam is extracted at 2 MPa and fed to the closed feedwater heater. The remainder is reheated at 2 MPa to 440°C and then expands through the second-stage turbine to 0.3 MPa, where an additional amount is extracted and fed into the open feedwater heater operating at 0.3 MPa. The steam expanding through the third-stage turbine exits at the condenser pressure of 6 kPa. Feedwater leaves the closed heater at 210°C, 12 MPa, and condensate exiting as saturated liquid at 2 MPa is trapped into the open feedwater heater. Saturated liquid at 0.3 MPa leaves the open feedwater heater. Assume all pumps and turbine stages operate isentropically. Determine for the cycle

a. the heat transfer to the working fluid passing through the steam generator, in kJ per kg of steam entering the first-stage turbine.

b. the thermal efficiency.

c. the heat transfer from the working fluid passing through the condenser to the cooling water, in kJ per kg of steam entering the first-stage turbine.

8.47 Reconsider the cycle of Problem 8.46, but include in the analysis that the turbine stage and pumps all have isentropic efficiencies of 80%. Answer the same questions about the modified cycle as in Problem 8.46.

8.48 **C** For the cycle of Problem 8.47, plot thermal efficiency versus turbine stage and pump isentropic efficiencies for values ranging from 80 to 100%. Discuss.

8.49 Steam enters the first turbine stage of a vapor power cycle with reheat and regeneration at 32 MPa, 600°C, and expands to 8 MPa. A portion of the flow is diverted to a closed feedwater heater at 8 MPa, and the remainder is reheated to 560°C before entering the second turbine stage. Expansion through the second turbine stage occurs to 1 MPa, where another portion of the flow is diverted to a second closed feedwater heater at 1 MPa. The remainder of the flow expands through the third turbine stage to 0.15 MPa, where a portion of the flow is diverted to an open feedwater heater operating at 0.15 MPa, and the rest expands through the fourth turbine stage to the condenser pressure of 6 kPa. Condensate leaves each closed feedwater heater as saturated liquid at the respective extraction pressure. The feedwater streams leave each closed feedwater heater at a temperature equal to the saturation temperature at the respective extraction pressure. The condensate streams from the closed heaters each pass through traps into the next lower-pressure feedwater heater. Saturated liquid exiting the open heater is pumped to the steam generator pressure. If each turbine stage has an isentropic efficiency of 85% and the pumps operate isentropically

a. sketch the layout of the cycle and number the principal state points.

b. determine the thermal efficiency of the cycle.

c. calculate the mass flow rate into the first turbine stage, in kg/h, for a net power output of 500 MW.

8.50 Steam enters the first turbine stage of a vapor power plant with reheat and regeneration at 12.4 MPa, 590°C and expands in five stages to a condenser pressure of 6.9 kPa. Reheat is at 690 kPa to 538°C. The cycle includes three feedwater heaters. Closed heaters operate at 4.1 MPa and 1.1 MPa, with the drains from each trapped into the next lower-pressure feedwater heater. The feedwater leaving each closed heater is at the saturation temperature corresponding to the extraction pressure. An open feedwater heater operates at 0.14 MPa. The pumps operate isentropically, and each turbine stage has an isentropic efficiency of 88%.

a. Sketch the layout of the cycle and number the principal state points.

b. Determine the thermal efficiency of the cycle.

c. Determine the heat rate, in kJ/kW · h.

d. Calculate the mass flow rate into the first turbine stage, in kg/h, for a net power output of 3.2×10^9 kJ/h.

Considering Other Vapor Cycle Aspects

8.51 A binary vapor power cycle consists of two ideal Rankine cycles with steam and ammonia as the working fluids. In the steam cycle, superheated vapor enters the turbine at 6 MPa, 640°C, and saturated liquid exits the condenser at 60°C, The heat rejected from the steam cycle is provided to the ammonia cycle, producing saturated vapor at 50°C, which enters the ammonia turbine. Saturated liquid leaves the ammonia condenser at 1 MPa. For a net power output of 20 MW from the binary cycle, determine

a. the power output of the steam and ammonia turbines, respectively, in MW.

b. the rate of heat addition to the binary cycle, in MW.

c. the thermal efficiency.

8.52 Figure P8.52 shows a vapor power cycle with reheat and regeneration. The steam generator produces vapor at 6.9 MPa, 430°C. Some of this steam expands through the first turbine stage to 690 kPa and the remainder is directed to the heat exchanger. The steam exiting the first turbine stage enters the flash chamber.

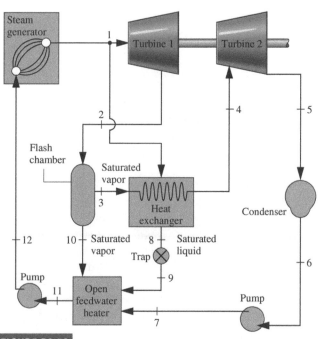

FIGURE P8.52

Saturated vapor and saturated liquid at 690 kPa exit the flash chamber as separate streams. The vapor is reheated in the heat exchanger to 277°C before entering the second turbine stage. The open feedwater heater operates at 690 kPa, and the condenser pressure is 6.9 kPa. Each turbine stage has an isentropic efficiency of 88% and the pumps operate isentropically. For a net power output of 5.27×10^9 kJ/h, determine

a. the mass flow rate through the steam generator, in kg/h.

b. the thermal efficiency of the cycle.

c. the rate of heat transfer to the cooling water passing through the condenser, in kJ/h.

8.53 Figure P8.53 provides steady-state operating data for a cogeneration cycle that generates electricity and provides heat for campus buildings. Steam at 1.5 MPa, 280°C, enters a two-stage turbine with a mass flow rate of 1 kg/s. A fraction of the total flow, 0.15, is extracted between the two stages at 0.2 MPa to provide for building heating, and the remainder expands through the second stage to the condenser pressure of 0.1 bar. Condensate returns from the campus buildings at 0.1 MPa, 60°C and passes through a trap into the condenser, where it is reunited with the main feedwater flow. Saturated liquid leaves the condenser at 0.1 bar. Determine

a. the rate of heat transfer to the working fluid passing through the boiler, in kW.

b. the net power developed, in kW.

c. the rate of heat transfer for building heating, in kW.

d. the rate of heat transfer to the cooling water passing through the condenser, in kW.

8.54 Consider a cogeneration system operating as shown in Fig. P8.54. Steam enters the first turbine stage at 6 MPa, 540°C. Between the first and second stages, 45% of the steam is extracted at 500 kPa and diverted to a process heating load of 5×10^8 kJ/h. Condensate exits the process heat exchanger at 450 kPa with specific enthalpy of 589.13 kJ/kg and is mixed with liquid exiting the lower-pressure pump at 450 kPa. The entire flow is then pumped to the steam generator

pressure. At the inlet to the steam generator the specific enthalpy is 469.91 kJ/kg. Saturated liquid at 60 kPa leaves the condenser. The turbine stages and the pumps operate with isentropic efficiencies of 82% and 88%, respectively. Determine

a. the mass flow rate of steam entering the first turbine stage, in kg/s.

b. the net power developed by the cycle, in MW.

c. the rate of entropy production in the turbine, in kW/K.

8.55 Figure P8.55 shows the schematic diagram of a cogeneration cycle. In the steam cycle, superheated vapor enters the turbine with a mass flow rate of 5 kg/s at 40 bar, 440°C and expands isentropically to 1.5 bar. Half of the flow is extracted at 1.5 bar and used for industrial process heating. The rest of the steam passes through a heat exchanger, which serves as the boiler of the Refrigerant 134a cycle and the condenser of the steam cycle. The condensate leaves the heat exchanger as saturated liquid at 1 bar, where it is combined with the return flow from the process, at 60°C and 1 bar, before being pumped isentropically to the steam generator pressure. The Refrigerant 134a cycle is an ideal Rankine cycle with refrigerant entering the turbine at 16 bar, 100°C and saturated liquid leaving the condenser at 9 bar. Determine, in kW

a. the rate of heat transfer to the working fluid passing through the steam generator of the steam cycle.

b. the net power output of the binary cycle.

c. the rate of heat transfer to the industrial process.

Vapor Cycle Exergy Analysis

8.56 In a *cogeneration* system, a Rankine cycle operates with steam entering the turbine at 5.5 MPa, 370°C, and a condenser pressure of 1.24 MPa. The isentropic turbine efficiency is 80%. Energy rejected by the condensing steam is transferred to a separate process stream of water entering at 120°C, 1 MPa and exiting as saturated vapor at 1 MPa. Determine the mass flow rate, in kg/h, for the working fluid of the Rankine cycle if the mass flow rate of the process stream is 22700 kg/h. Devise and evaluate an exergetic efficiency for the overall cogeneration system. Let $T_0 = 21$°C, $p_0 = 101.3$ kPa.

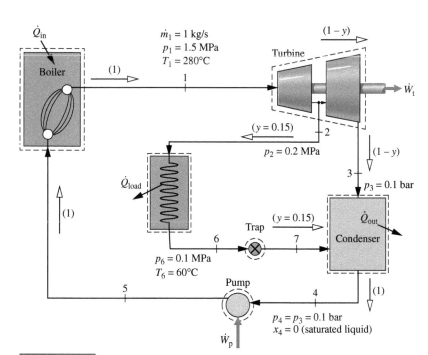

State	p	T (°C)	h (kJ/kg)
1	1.5 MPa	280	2992.7
2	0.2 MPa	sat	2652.9
3	0.1 bar	sat	2280.4
4	0.1 bar	sat	191.83
5	1.5 MPa	- - -	193.34
6	0.1 MPa	60	251.13
7	0.1 bar	- - -	251.13

FIGURE P8.53

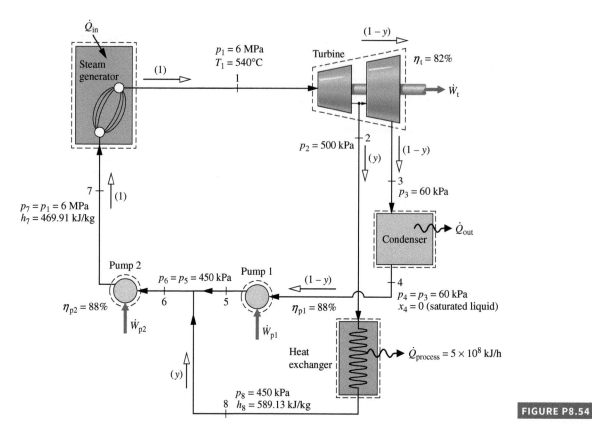

FIGURE P8.54

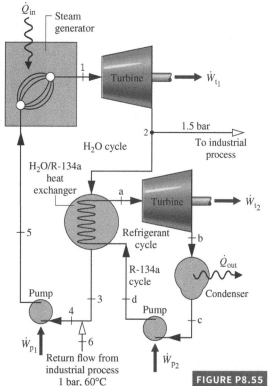

FIGURE P8.55

8.57 The steam generator of a vapor power plant can be considered for simplicity to consist of a combustor unit in which fuel and air are burned to produce hot combustion gases, followed by a heat exchanger unit where the cycle working fluid is vaporized and superheated as the hot gases cool. Consider water as the working fluid undergoing the cycle of Problem 8.13. Hot combustion gases, which are assumed to have the properties of air, enter the heat exchanger portion of the steam generator

at 1200 K and exit at 600 K with a negligible change in pressure. Determine for the heat exchanger unit

    a. the net rate at which exergy is carried in by the gas stream, in kJ per kg of steam flowing.

    b. the net rate at which exergy is carried out by the water stream, in kJ per kg of steam flowing.

    c. the rate of exergy destruction, in kW.

    d. the exergetic efficiency given by Eq. 7.27.

Let $T_0 = 15°C$, $p_0 = 0.1$ MPa.

8.58 In the steam generator of the cycle of Problem 8.12, the energy input to the working fluid is provided by heat transfer from hot gaseous products of combustion, which cool as a separate stream from 190°C to 810°C with a negligible pressure drop. The gas stream can be modeled as air as an ideal gas. Determine, in kJ/h, the rate of exergy destruction in the

    a. heat exchanger unit of the steam generator.

    b. turbine and pump.

    c. condenser.

Also, calculate the net rate at which exergy is carried away by the cooling water passing through the condenser, in kJ/h. Let $T_0 = 16°C$, $p_0 = 101.3$ kPa.

8.59 Determine the rate of exergy input, in kJ/h, to the working fluid passing through the steam generator in Problem 8.50. Perform calculations to account for all outputs, losses, and destructions of this exergy. Let $T_0 = 15°C$, $p_0 = 1$ bar.

8.60 Determine the rate of exergy transfer, in kJ per kg of steam entering the first-stage turbine, to the working fluid passing through the steam generator in Problem 8.32. Perform calculations to account for all outputs, losses, and destructions of this exergy. Let $T_0 = 15°C$, $p_0 = 0.1$ MPa.

8.61 Steam enters the turbine of a vapor power plant at 482°C, 3.5 MPa and expands adiabatically, exiting at 6.9 kPa with a quality of 97%. Condensate leaves the condenser as saturated liquid at 6.9 kPa. The isentropic pump efficiency is 80%. The specific exergy of the fuel entering the combustor unit of the steam generator is estimated to be 51920 kJ/kg, and no exergy is carried in by the combustion air. The exergy of the stack gases exiting the steam generator is estimated to be 779 kJ per kg of fuel entering. The mass flow rate of the steam is 15.1 kg per kg of fuel entering the steam generator. Cooling water enters the condenser at $T_0 = 25°C$, $p_0 = 101.3$ kPa and exits at 32°C, 101.3 kPa. Determine, as percentages of the exergy entering with the fuel, the

a. exergy exiting with the stack gases.

b. exergy destroyed in the steam generator.

c. net power developed by the cycle.

d. exergy destroyed in the turbine and the pump.

e. exergy exiting with the cooling water.

f. exergy destroyed in the condenser.

8.62 Steam enters the turbine of a vapor power plant at 100 bar, 520°C and expands adiabatically, exiting at 0.08 bar with a quality of 90%. Condensate leaves the condenser as saturated liquid at

0.08 bar. Liquid exits the pump at 100 bar, 43°C. The specific exergy of the fuel entering the combustor unit of the steam generator is estimated to be 14,700 kJ/kg. No exergy is carried in by the combustion air. The exergy of the stack gases leaving the steam generator is estimated to be 150 kJ per kg of fuel. The mass flow rate of the steam is 3.92 kg per kg of fuel. Cooling water enters the condenser at $T_0 = 20°C$, $p_0 = 1$ bar and exits at 35°C, 1 bar. Develop a full accounting of the exergy entering the plant with the fuel.

8.63 Figure P8.63 provides steady-state operating data for a cogeneration cycle that generates electricity and provides heat for campus buildings. Steam at 1.5 MPa, 280°C, enters a two-stage turbine with a mass flow rate of 1 kg/s. Steam is extracted between the two stages at 0.2 MPa with a mass flow rate of 0.15 kg/s to provide for building heating, while the remainder expands through the second turbine stage to the condenser pressure of 0.1 bar with mass flow rate of 0.85 kg/s. The campus load heat exchanger in the schematic represents all of the heat transfer to the campus buildings. For the purposes of this analysis, assume that the heat transfer in the campus load heat exchanger occurs at an average boundary temperature of 110°C. Condensate returns from the campus buildings at 0.1 MPa, 60°C and passes

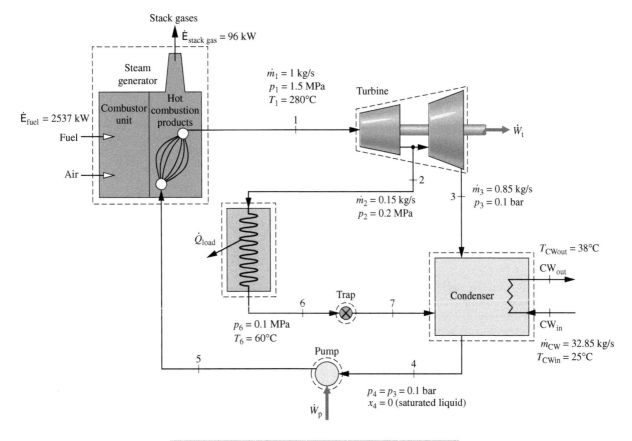

State	p	T (°C)	h (kJ/kg)	s (kJ/kg·K)
1	1.5 MPa	280	2992.7	6.8381
2	0.2 MPa	sat	2652.9	6.9906
3	0.1 bar	sat	2280.4	7.1965
4	0.1 bar	sat	191.83	0.6493
5	1.5 MPa	- - -	193.34	0.6539
6	0.1 MPa	60	251.13	0.8312
7	0.1 bar	- - -	251.13	0.8352
Cooling water$_{in}$	- - -	25	104.89	0.3674
Cooling water$_{out}$	- - -	38	159.21	0.5458

FIGURE P8.63

through a trap into the condenser, where it is reunited with the main feedwater flow. The cooling water has a mass flow rate of 32.85 kg/s entering the condenser at 25°C and exiting the condenser at 38°C. The working fluid leaves the condenser as saturated liquid at 0.1 bar. The rate of exergy input with fuel entering the combustor unit of the steam generator is 2537 kW, and no exergy is carried in by the combustion air. The rate of exergy loss with the stack gases exiting the steam generator is 96 kW. Let $T_0 = 25°C$, $p_0 = 0.1$ MPa. Determine, as percentages of the rate of exergy input with fuel entering the combustor unit, all outputs, losses, and destructions of this exergy for the cogeneration cycle.

DESIGN & OPEN-ENDED PROBLEMS: EXPLORING ENGINEERING PRACTICE

8.1D Develop the preliminary design for a power plant to produce 1000 MW of electricity using water as the working fluid and a supercritical pressure of 24 MPa and with a maximum cycle temperature of 600°C. Include in your specifications whether reheat and regeneration should be included, and if so, at what pressures. Show a schematic of your cycle layout and include in your report appropriate analysis, including computer plots, to justify the configuration you recommend.

8.2D Identify a major electrical power failure that occurred recently in your locale. Research the circumstances associated with the failure and the steps taken to avoid such an event in the future. Summarize your findings and lessons learned from this failure in a one-page executive summary.

8.3D Early commercial vapor power plants operated with turbine inlet conditions of about 12 bar and 200°C. Plants are under development today that can operate at over 34 MPa, with turbine inlet temperatures of 650°C or higher. How have steam generator and turbine designs changed over the years to allow for such increases in pressure and temperature? Discuss.

8.4D Water management is an important aspect of electric power production. Identify at least two needs for water in a Rankine cycle–based power plant. Describe typical water management practices in such plants, and research at least two emerging technologies aimed at reducing water losses in plants or enhancing sustainable water management. Summarize your findings in a report with at least three references.

8.5D Consider the feasibility of using *biomass* to fuel a 200-MW electric power plant in a rural area in your locale. Analyze the advantages and disadvantages of biomass in comparison with coal and natural gas. Include in your analysis material handling issues, plant operations, environmental considerations, and costs. Prepare a slide presentation of your recommendations.

8.6D Identify three types of renewable energy electricity-generating plants of interest to you. For an actual plant of each type, determine its location and electricity production capacity. For each plant create a schematic accompanied by a brief narrative explaining how renewable energy is converted to electricity. Also for each plant determine if it will require energy storage (See. 2.6) and the extent of any economic incentives mandated by legislative action in the last 10 years. Summarize your findings in a PowerPoint presentation.

8.7D A 450 m^2 research outpost is being designed for studying global change in Antarctica. The outpost will house five scientists along with their communication and research equipment. Develop the preliminary design of an array of *photovoltaic cells* to provide all necessary power from solar energy during months of continuous sunlight. Specify the number and type of cells needed and present schematics of the system you propose.

8.8D *Concurrent engineering design* considers all phases of a product's *life cycle* holistically with the aim of arriving at an acceptable final design more quickly and with less cost than achievable in a sequential approach. A tenet of concurrent design is the use of a multi-talented design team having technical *and* nontechnical expertise. For power plant design, technical expertise necessarily includes the skills of several engineering disciplines. Determine the makeup of the design team and the skill set each member provides for the concurrent design of a power plant selected from those listed in Table 8.2. Summarize your findings in a poster presentation suitable for presentation at a technical conference.

8.9D Some observers contend that *enhanced oil recovery* is a viable commercial use for carbon dioxide captured from the exhaust gas of coal-fired power plants and other industrial sources. Proponents envision that this will foster transport of carbon dioxide by ship from industrialized, oil-importing nations to less industrialized, oil-producing nations. They say such commerce will require innovations in ship design. Develop a conceptual design of a carbon dioxide transport ship. Consider only major issues, including but not limited to power plant type, cargo volume, means for loading and unloading carbon dioxide, minimization of carbon dioxide loss to the atmosphere, and costs. Include figures and sample calculations as appropriate. Explain how your carbon dioxide transport ship differs from ships transporting natural gas.

8.10D Figure P8.10D shows a *thermosyphon* Rankine power plant having one end immersed in a *solar pond* and the other exposed to the ambient air. At the lower end, a working fluid is vaporized by heat transfer from the hot water of the solar pond. The vapor flows upward through an axialflow turbine to the upper end, where condensation occurs by heat transfer to the ambient air. The condensate drains to the lower end by gravity. Evaluate the feasibility of producing power in the 0.1–0.25 kW range using such a device. What working fluids would be suitable? Write a report of your findings.

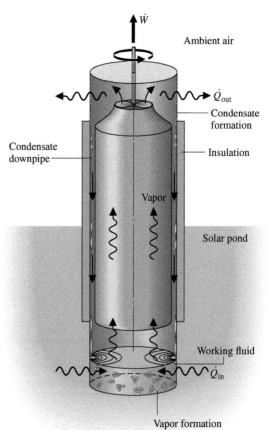

FIGURE P8.10D

CHAPTER 9

Gas Power Systems

Engineering Context

An introduction to power generation systems is provided in Chap. 8. The introduction surveys current U.S. power generation by source and looks ahead to power generation needs in the next few decades. Because the introduction provides the context for a study of power systems generally, we recommend that you review it before continuing with the present chapter dealing with gas power systems.

While the focus of Chap. 8 is on vapor power systems in which the working fluids are alternatively vaporized and condensed, the **objective** of this chapter is to study power systems utilizing working fluids that are always a gas. Included in this group are gas turbines and internal combustion engines of the spark-ignition and compression-ignition types. In the first part of the chapter, internal combustion engines are considered. Gas turbine power plants are discussed in the second part of the chapter. The chapter concludes with a brief study of compressible flow in nozzles and diffusers, which are components in gas turbines for aircraft propulsion and other devices of practical importance.

LEARNING OUTCOMES

When you complete your study of this chapter, you will be able to...

- Conduct air-standard analyses of internal combustion engines based on the Otto, Diesel, and dual cycles, including the ability to
  - Sketch p–v and T–s diagrams and evaluate property data at principal states.
  - Apply energy, entropy, and exergy balances.
  - Determine net power output, thermal efficiency, and mean effective pressure.
- Conduct air-standard analyses of gas turbine power plants based on the Brayton cycle and its modifications, including the ability to
  - Sketch T–s diagrams and evaluate property data at principal states.
  - Apply mass, energy, entropy, and exergy balances.
  - Determine net power output, thermal efficiency, back work ratio, and the effects of compressor pressure ratio on performance.
- Analyze subsonic and supersonic flows through nozzles and diffusers, including the ability to
  - Describe the effects of area change on flow properties and the effects of back pressure on mass flow rate.
  - Explain the occurrence of choking and normal shocks.
  - Analyze the flow of ideal gases with constant specific heats.

Considering Internal Combustion Engines

This part of the chapter deals with *internal* combustion engines. Although most gas turbines are also internal combustion engines, the name is usually applied to *reciprocating* internal combustion engines of the type commonly used in automobiles, trucks, and buses. These engines differ from the power plants considered in Chap. 8 because the processes occur within reciprocating piston–cylinder arrangements and not in interconnected series of different components.

spark-ignition

compression-ignition

Two principal types of reciprocating internal combustion engines are the **spark-ignition** engine and the **compression-ignition** engine. In a spark-ignition engine, a mixture of fuel and air is ignited by a spark plug. In a compression-ignition engine, air is compressed to a high enough pressure and temperature that combustion occurs spontaneously when fuel is injected. Spark-ignition engines have advantages for applications requiring power up to about 225 kW. Because they are relatively light and lower in cost, spark-ignition engines are particularly suited for use in automobiles. Compression-ignition engines are normally preferred for applications when fuel economy and relatively large amounts of power are required (heavy trucks and buses, locomotives and ships, auxiliary power units). In the middle range, spark-ignition and compression-ignition engines are used.

9.1 Introducing Engine Terminology

Figure 9.1 is a sketch of a reciprocating internal combustion engine consisting of a piston that moves within a cylinder fitted with two valves. The sketch is labeled with some special terms. The *bore* of the cylinder is its diameter. The *stroke* is the distance the piston moves in one direction. The piston is said to be at *top dead center* when it has moved to a position where the cylinder volume is a minimum. This minimum volume is known as the *clearance* volume. When the piston has moved to the position of maximum cylinder volume, the piston

compression ratio

is at *bottom dead center*. The volume swept out by the piston as it moves from the top dead center to the bottom dead center position is called the *displacement volume*. The **compression ratio** r is defined as the volume at bottom dead center divided by the volume at top dead center. The reciprocating motion of the piston is converted to rotary motion by a crank mechanism.

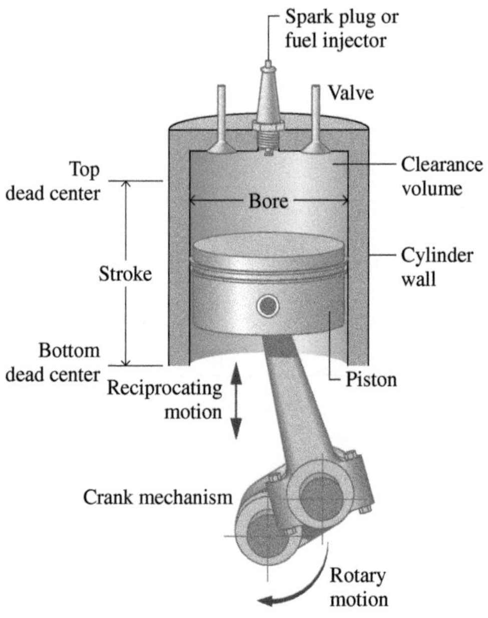

In a *four-stroke* internal combustion engine, the piston executes four distinct strokes within the cylinder for every two revolutions of the crankshaft. **Figure 9.2** gives a pressure–volume diagram such as might be displayed electronically.

1. With the intake valve open, the piston makes an *intake stroke* to draw a fresh charge into the cylinder. For spark-ignition engines, the charge is a combustible mixture of fuel and air. Air alone is the charge in compression-ignition engines.

2. With both valves closed, the piston undergoes a *compression stroke*, raising the temperature and pressure of the charge. This requires work input from the piston to the cylinder contents. A combustion process is then initiated, resulting in a high-pressure, high-temperature gas mixture. Combustion is induced near the end of the compression stroke in spark-ignition engines by the spark plug. In compression-ignition engines, combustion is initiated by injecting fuel into the hot compressed air, beginning near the end of the compression stroke and continuing through the first part of the expansion.

Fig. 9.1 Nomenclature for reciprocating piston–cylinder engines.

3. A *power* stroke follows the compression stroke, during which the gas mixture expands and work is done on the piston as it returns to bottom dead center.

4. The piston then executes an *exhaust stroke* in which the burned gases are purged from the cylinder through the open exhaust valve.

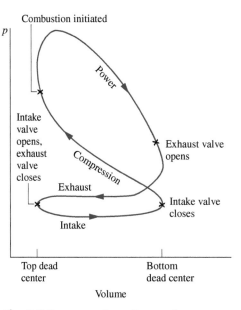

Fig. 9.2 Pressure–volume diagram for a reciprocating internal combustion engine.

Smaller engines operate on *two-stroke* cycles. In two-stroke engines, the intake, compression, expansion, and exhaust operations are accomplished in one revolution of the crankshaft. Although internal combustion engines undergo *mechanical* cycles, the cylinder contents do not execute a *thermodynamic* cycle, for matter is introduced with one composition and is later discharged at a different composition.

A parameter used to describe the performance of reciprocating piston engines is the *mean effective pressure,* or mep. The **mean effective pressure** is the theoretical constant pressure that, if it acted on the piston during the power stroke, would produce the same *net* work as actually developed in one cycle. That is,

$$\boxed{\text{mep} = \frac{\text{net work for one cycle}}{\text{displacement volume}}} \qquad (9.1)$$

For two engines of equal displacement volume, the one with a higher mean effective pressure would produce the greater net work and, if the engines run at the same speed, greater power.

    mean effective pressure

Air-Standard Analysis   A detailed study of the performance of a reciprocating internal combustion engine would take into account many features. These would include the combustion process occurring within the cylinder and the effects of irreversibilities associated with friction and with pressure and temperature gradients. Heat transfer between the gases in the cylinder and the cylinder walls and the work required to charge the cylinder and exhaust the products of combustion also would be considered. Owing to these complexities, accurate modeling of reciprocating internal combustion engines normally involves computer simulation. To conduct *elementary* thermodynamic analyses of internal combustion engines, considerable simplification is required. One procedure is to employ an **air-standard analysis** having the following elements:

    air-standard analysis: internal combustion engines

- A fixed amount of air modeled as an ideal gas is the working fluid. See Table 9.1 for a review of ideal gas relations.

- The combustion process is replaced by a heat transfer from an external source.

- There are no exhaust and intake processes as in an actual engine. The cycle is completed by a constant-volume heat transfer process taking place while the piston is at the bottom dead center position.

- All processes are internally reversible.

In addition, in a **cold air-standard analysis**, the specific heats are assumed constant at their ambient temperature values. With an air-standard analysis, we avoid dealing with the complexities of the combustion process and the change of composition during combustion. A comprehensive analysis requires that such complexities be considered, however. For a discussion of combustion, see Chap. 13.

    cold air-standard analysis

Although an air-standard analysis simplifies the study of internal combustion engines considerably, values for the mean effective pressure and operating temperatures and pressures calculated on this basis may depart significantly from those of actual engines. Accordingly, air-standard analysis allows internal combustion engines to be examined only qualitatively. Still, insights concerning actual performance can result with such an approach.

TABLE 9.1

Ideal Gas Model Review

Equations of state:

$$pv = RT \qquad (3.32)$$

$$pV = mRT \qquad (3.33)$$

Changes in u and h:

$$u(T_2) - u(T_1) = \int_{T_1}^{T_2} c_v(T)\ dT \qquad (3.40)$$

$$h(T_2) - h(T_1) = \int_{T_1}^{T_2} c_p(T)\ dT \qquad (3.43)$$

Constant Specific Heats	Variable Specific Heats
$u(T_2) - u(T_1) = c_v(T_2 - T_1)$ (3.50) $h(T_2) - h(T_1) = c_p(T_2 - T_1)$ (3.51) See Tables A-20, A-21 for data.	$u(T)$ and $h(T)$ are evaluated from appropriate tables: Tables A-22 for air (mass basis) and A-23 for other gases (molar basis).

Changes in s:

$$s(T_2, v_2) - s(T_1, v_1) =$$
$$\int_{T_1}^{T_2} c_v(T)\frac{dT}{T} + R\ \ln\frac{v_2}{v_1} \qquad (6.17)$$

$$s(T_2, p_2) - s(T_1, p_1) =$$
$$\int_{T_1}^{T_2} c_p(T)\frac{dT}{T} - R\ \ln\frac{p_2}{p_1} \qquad (6.18)$$

Constant Specific Heats	Variable Specific Heats
$s(T_2, v_2) - s(T_1, v_1) =$ $\qquad c_v\ln\dfrac{T_2}{T_1} + R\ \ln\dfrac{v_2}{v_1}$ (6.21) $s(T_2, p_2) - s(T_1, p_1) =$ $\qquad c_p\ln\dfrac{T_2}{T_1} - R\ \ln\dfrac{p_2}{p_1}$ (6.22) See Tables A-20, A-21 for data.	$s(T_2, p_2) - s(T_1, p_1) =$ $\qquad s^\circ(T_2) - s^\circ(T_1) - R\ \ln\dfrac{p_2}{p_1}$ (6.20a) where $s^\circ(T)$ is evaluated from appropriate tables: Tables A-22 for air (mass basis) and A-23 for other gases (molar basis).

Relating states of equal specific entropy: $\Delta s = 0$:

Constant Specific Heats	Variable Specific Heats — Air Only
$\dfrac{T_2}{T_1} = \left(\dfrac{p_2}{p_1}\right)^{(k-1)/k}$ (6.43) $\dfrac{T_2}{T_1} = \left(\dfrac{v_1}{v_2}\right)^{k-1}$ (6.44) $\dfrac{p_2}{p_1} = \left(\dfrac{v_1}{v_2}\right)^{k}$ (6.45) where $k = c_p/c_v$ is given in Table A-20 for several gases.	$\dfrac{p_2}{p_1} = \dfrac{p_{r2}}{p_{r1}}$ (air only) (6.41) $\dfrac{v_2}{v_1} = \dfrac{v_{r2}}{v_{r1}}$ (air only) (6.42) where p_r and v_r are provided for air in Table A-22.

Energy & Environment

Some observers caution that much of what remains of the world's stock of *readily accessible* oil will be consumed within the next 50 years. We are not actually running out of oil, they say, but running out of oil that can be produced *inexpensively*. Accordingly, unless demand is reduced, oil prices may increase greatly, ending the era of cheap oil enjoyed for decades and posing a stiff challenge for societies over the globe.

In the United States transportation accounts for about 70% of annual oil consumption, and so transportation is a sector ripe for reducing our need for oil. One way forward is greater use of hybrid-electric and all-electric vehicles. Still, we can continue to drive cars with internal combustion engines by using improved engine and transmission technologies, tires with less rolling resistance, and enhanced aerodynamics. These cars will be manufactured from lightweight, high-strength materials and also may have a *start–stop* feature that automatically shuts down and restarts the engine to minimize engine idling when the cars are motionless, thereby reducing both fuel use and emissions.

9.2 Air-Standard Otto Cycle

In the remainder of this part of the chapter, we consider three cycles that adhere to air-standard cycle idealizations: the Otto, Diesel, and dual cycles. These cycles differ from each other only in the way the heat addition process that replaces combustion in the actual cycle is modeled.

The air-standard Otto cycle is an ideal cycle that assumes heat addition occurs instantaneously while the piston is at top dead center. The **Otto cycle** is shown on the *p–v* and *T–s* diagrams of **Fig. 9.3.** The cycle consists of four internally reversible processes in series:

Otto cycle

- *Process 1–2:* An isentropic compression of the air as the piston moves from bottom dead center to top dead center.
- *Process 2–3:* A constant-volume heat transfer to the air from an external source while the piston is at top dead center. This process is intended to represent the ignition of the fuel–air mixture and the subsequent rapid burning.
- *Process 3–4:* An isentropic expansion (power stroke).
- *Process 4–1:* Completes the cycle by a constant-volume process in which heat is rejected from the air while the piston is at bottom dead center.

TAKE NOTE...

For internally reversible processes of closed systems, see Secs. 2.2.5 and 6.6.1 for discussions of area interpretations of work and heat transfer on *p–v* and *T–s* diagrams, respectively.

Since the air-standard Otto cycle is composed of internally reversible processes, areas on the *T–s* and *p–v* diagrams of Fig. 9.3 can be interpreted as heat and work, respectively. On the *T–s* diagram, area 2–3–a–b–2 represents heat added per unit of mass and area 1–4–a–b–1 the heat rejected per unit of mass. On the *p–v* diagram, area 1–2–a–b–1 represents work input per unit of mass during the compression process and area 3–4–b–a–3 is work done per unit of mass in the expansion process. The enclosed area of each figure can be interpreted as the net work output or, equivalently, the net heat added.

Animation

Otto Cycle Tabs a and b

Cycle Analysis   The air-standard Otto cycle consists of two processes in which there is work but no heat transfer, Processes 1–2 and 3–4, and two processes in which there is heat transfer but no work, Processes 2–3 and 4–1. Expressions for these energy transfers are obtained by reducing the closed system energy balance assuming that changes in kinetic and potential energy can be ignored. The results are

TAKE NOTE...

When analyzing air-standard cycles, it is frequently convenient to regard all work and heat transfers as positive quantities and write the energy balance accordingly.

$$
\frac{W_{12}}{m} = u_2 - u_1, \qquad \frac{W_{34}}{m} = u_3 - u_4
$$
$$
\frac{Q_{23}}{m} = u_3 - u_2, \qquad \frac{Q_{41}}{m} = u_4 - u_1
$$

(9.2)

Carefully note that in writing Eqs. 9.2, we have departed from our usual sign convention for heat and work. Thus, W_{12}/m is a positive number representing the work *input* during

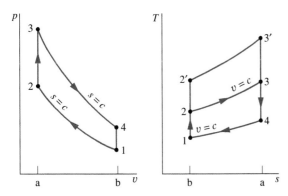

Fig. 9.3 *p–v* and *T–s* diagrams of the air-standard Otto cycle.

compression and Q_{41}/m is a positive number representing the heat *rejected* in Process 4–1. The *net work* of the cycle is expressed as

$$\frac{W_{cycle}}{m} = \frac{W_{34}}{m} - \frac{W_{12}}{m} = (u_3 - u_4) - (u_2 - u_1)$$

Alternatively, the net work can be evaluated as the *net heat added*

$$\frac{W_{cycle}}{m} = \frac{Q_{23}}{m} - \frac{Q_{41}}{m} = (u_3 - u_2) - (u_4 - u_1)$$

which, on rearrangement, can be placed in the same form as the previous expression for net work.

The thermal efficiency is the ratio of the net work of the cycle to the heat added.

$$\eta = \frac{(u_3 - u_2) - (u_4 - u_1)}{u_3 - u_2} = 1 - \frac{u_4 - u_1}{u_3 - u_2} \qquad (9.3)$$

When air table data are used to conduct an analysis involving an air-standard Otto cycle, the specific internal energy values required by Eq. 9.3 can be obtained from Table A-22. The following relationships based on Eq. 6.42 apply for the isentropic processes 1–2 and 3–4

$$v_{r2} = v_{r1}\left(\frac{V_2}{V_1}\right) = \frac{v_{r1}}{r} \qquad (9.4)$$

$$v_{r4} = v_{r3}\left(\frac{V_4}{V_3}\right) = rv_{r3} \qquad (9.5)$$

compression ratio where r denotes the **compression ratio**. Note that since $V_3 = V_2$ and $V_4 = V_1$, $r = V_1/V_2 = V_4/V_3$. The parameter v_r is tabulated versus temperature for air in Table A-22.

When the Otto cycle is analyzed on a cold air-standard basis, the following expressions based on Eq. 6.44 would be used for the isentropic processes in place of Eqs. 9.4 and 9.5, respectively,

$$\frac{T_2}{T_1} = \left(\frac{V_1}{V_2}\right)^{k-1} = r^{k-1} \qquad \text{(constant } k) \qquad (9.6)$$

$$\frac{T_4}{T_3} = \left(\frac{V_3}{V_4}\right)^{k-1} = \frac{1}{r^{k-1}} \qquad \text{(constant } k) \qquad (9.7)$$

where k is the specific heat ratio, $k = c_p/c_v$.

Effect of Compression Ratio on Performance By referring to the T–s diagram of Fig. 9.3, we can conclude that the Otto cycle thermal efficiency increases as the compression ratio increases. An increase in the compression ratio changes the cycle from 1–2–3–4–1 to 1–2′–3′–4–1. Since the average temperature of heat addition is greater in the latter cycle and both cycles have the same heat rejection process, cycle 1–2′–3′–4–1 would have the greater thermal efficiency. The increase in thermal efficiency with compression ratio is also brought out simply by the following development on a cold air-standard basis. For constant c_v, Eq. 9.3 becomes

$$\eta = 1 - \frac{c_v(T_4 - T_1)}{c_v(T_3 - T_2)}$$

On rearrangement

$$\eta = 1 - \frac{T_1}{T_2}\left(\frac{T_4/T_1 - 1}{T_3/T_2 - 1}\right)$$

From Eqs. 9.6 and 9.7 above, $T_4/T_1 = T_3/T_2$, so

$$\eta = 1 - \frac{T_1}{T_2}$$

Finally, introducing Eq. 9.6

$$\boxed{\eta = 1 - \frac{1}{r^{k-1}} \qquad \text{(cold air-standard basis)}} \qquad (9.8)$$

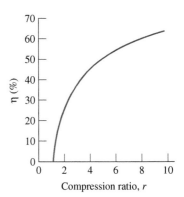

Fig. 9.4 Thermal efficiency of the cold air-standard Otto cycle, $k = 1.4$.

Equation 9.8 indicates that the cold air-standard Otto cycle thermal efficiency is a function of compression ratio and k. This relationship is shown in **Fig. 9.4** for $k = 1.4$, representing ambient air.

The foregoing discussion suggests that it is advantageous for internal combustion engines to have high compression ratios, and this is the case. The possibility of autoignition, or "knock," places an upper limit on the compression ratio of spark-ignition engines, however. After the spark has ignited a portion of the fuel–air mixture, the rise in pressure accompanying combustion compresses the remaining charge. Autoignition can occur if the temperature of the unburned mixture becomes too high before the mixture is consumed by the flame front. Since the temperature attained by the air–fuel mixture during the compression stroke increases as the compression ratio increases, the likelihood of autoignition occurring increases with the compression ratio. Autoignition may result in high-pressure waves in the cylinder (manifested by a knocking or pinging sound) that can lead to loss of power as well as engine damage.

Owing to performance limitations, such as autoignition, the compression ratios of spark-ignition engines using the *unleaded* fuel required today because of air pollution concerns are in the range 9.5 to 11.5, approximately. Higher compression ratios can be achieved in compression-ignition engines because only air is compressed. Compression ratios in the range 12 to 20 are typical. Compression-ignition engines also can use less refined fuels having higher ignition temperatures than the volatile fuels required by spark-ignition engines.

In the next example, we illustrate the analysis of the air-standard Otto cycle. Results are compared with those obtained on a cold air-standard basis.

▶▶▶ **EXAMPLE 9.1** ▶

Analyzing the Otto Cycle

The temperature at the beginning of the compression process of an air-standard Otto cycle with a compression ratio of 8 is 300°K, the pressure is 1 bar, and the cylinder volume is 560 cm³. The maximum temperature during the cycle is 2000°K. Determine (**a**) the temperature and pressure at the end of each process of the cycle, (**b**) the thermal efficiency, and (**c**) the mean effective pressure, in bars.

SOLUTION

Known An air-standard Otto cycle with a given value of compression ratio is executed with specified conditions at the beginning of the compression stroke and a specified maximum temperature during the cycle.

Find Determine the temperature and pressure at the end of each process, the thermal efficiency, and mean effective pressure, in bars.

Schematic and Given Data:

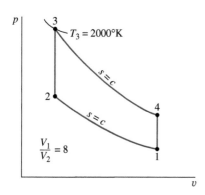

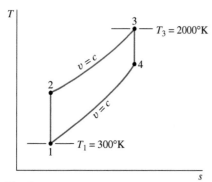

Fig. E9.1

Engineering Model

1. The air in the piston–cylinder assembly is the closed system.
2. The compression and expansion processes are adiabatic.
3. All processes are internally reversible.
4. The air is modeled as an ideal gas.
5. Kinetic and potential energy effects are negligible.

Analysis

a. The analysis begins by determining the temperature, pressure, and specific internal energy at each principal state of the cycle. At $T_1 = 300°K$, Table A-22 gives $u_1 = 214.07$ kJ/kg and $v_{r1} = 621.2$.

For the isentropic compression Process 1–2

$$v_{r2} = \frac{V_2}{V_1}v_{r1} = \frac{v_{r1}}{r} = \frac{621.2}{8} = 77.65$$

Interpolating with v_{r2} in Table A-22, we get $T_2 = 673$ K and $u_2 = 491.2$ kJ/kg. With the ideal gas equation of state

$$p_2 = p_1\frac{T_2}{T_1}\frac{V_1}{V_2} = (1\text{ bar})\left(\frac{673°K}{300°K}\right)8 = 17.95\text{ bars}$$

The pressure at state 2 can be evaluated alternatively by using the isentropic relationship, $p_2 = p_1(p_{r2}/p_{r1})$.

Since Process 2–3 occurs at constant volume, the ideal gas equation of state gives

$$p_3 = p_2\frac{T_3}{T_2} = (17.96\text{ bars})\left(\frac{2000\text{ K}}{673\text{ K}}\right) = 53.3\text{ bars}$$

At $T_3 = 2000$ K, Table A-22 gives $u_3 = 1678.7$ kJ/kg and $v_{r3} = 2.776$.

For the isentropic expansion process 3–4

$$v_{r4} = v_{r3}\frac{V_4}{V_3} = v_{r3}\frac{V_1}{V_2} = 2.766(8) = 22.21$$

Interpolating in Table A-22 with v_{r4} gives $T_4 = 1043$ K, $u_4 = 795.8$ kJ/kg. The pressure at state 4 can be found using the isentropic relationship $p_4 = p_3(p_{r4}/p_{r3})$ or the ideal gas equation of state applied at states 1 and 4. With $V_4 = V_1$, the ideal gas equation of state gives

$$p_4 = p_1\frac{T_4}{T_1} = (1\text{ bar})\left(\frac{1043\text{ K}}{300\text{ K}}\right) = 3.48\text{ bars}$$

b. The thermal efficiency is

$$\eta = 1 - \frac{Q_{41}/m}{Q_{23}/m} = 1 - \frac{u_4 - u_1}{u_3 - u_2}$$

$$= 1 - \frac{795.8 - 214.07}{1678.7 - 491.2} = 0.51\ (51\%)$$

c. To evaluate the mean effective pressure requires the net work per cycle. That is

$$W_{cycle} = m[(u_3 - u_4) - (u_2 - u_1)]$$

where m is the mass of the air, evaluated from the ideal gas equation of state as follows:

$$m = \frac{p_1V_1}{(\bar{R}/M)T_1}$$

$$= \frac{(1\text{ bar})(560\text{ cm}^3)}{\left(\dfrac{8.314}{28.97}\dfrac{\text{kJ}}{\text{kg}\cdot\text{K}}\right)(300\text{K})}\left(\frac{1\text{ m}^3}{10^6\text{ cm}^3}\right)\left(\frac{10^5\text{ N/m}^2}{1\text{ bar}}\right)\left(\frac{1\text{ kJ}}{10^3\text{N}\cdot\text{m}}\right)$$

$$= 6.5 \times 10^{-4}\text{ kg}$$

Inserting values into the expression for W_{cycle}

$$W_{cycle} = (6.5 \times 10^{-4}\text{ kg})\,[(1678.7 - 795.8) - (491.2 - 214.07)]\text{ kJ/kg}$$
$$= 0.394\text{ kJ}$$

The displacement volume is $V_1 - V_2$, so the mean effective pressure is given by

$$\text{mep} = \frac{W_{cycle}}{V_1 - V_2} = \frac{W_{cycle}}{V_1(1 - V_2/V_1)}$$

❶
$$= \frac{0.394\text{ kJ}}{(5601.3)/1 - \text{kg}}\left(\frac{10^6\text{ cm}^3}{1\text{ m}^3}\right)\left(\frac{10^3\text{N}\cdot\text{M}}{1\text{ kJ}}\right)\left(\frac{1\text{ bar}}{10^5\text{N}\cdot\text{M}}\right)$$

$$= 8.04\text{ bar}$$

---

❶ This solution utilizes Table A-22 for air, which accounts explicitly for the variation of the specific heats with temperature. A solution also can be developed on a cold air-standard basis in which constant specific heats are assumed. This solution is left as an exercise, but for comparison the results are presented for the case $k = 1.4$ in the following table:

Parameter	Air-Standard Analysis	Cold Air-Standard Analysis, $k = 1.4$
T_2	673 K	689 K
T_3	2000 K	2000 K
T_4	1043 K	871 K
η	0.51 (51%)	0.565 (56.5%)
mep	8.03 bar	7.05 bar

SKILLS DEVELOPED

Ability to...

- sketch the Otto cycle p–v and T–s diagrams.
- evaluate temperatures and pressures at each principal state and retrieve necessary property data.
- calculate thermal efficiency and mean effective pressure.

Quick Quiz

Determine the heat addition and the heat rejection for the cycle, each in kJ. Ans. $Q_{23} = 0.791$ kJ, $Q_{41} = 0.388$ kJ.

9.3 Air-Standard Diesel Cycle

The air-standard Diesel cycle is an ideal cycle that assumes heat addition occurs during a constant-pressure process that starts with the piston at top dead center. The **Diesel cycle** is shown on p–v and T–s diagrams in **Fig. 9.5**. The cycle consists of four internally reversible processes in series. The first process from state 1 to state 2 is the same as in the Otto cycle: an isentropic compression. Heat is not transferred to the working fluid at constant volume as in the Otto cycle, however. In the Diesel cycle, heat is transferred to the working fluid at *constant pressure*. Process 2–3 also makes up the first part of the power stroke. The isentropic expansion from state 3 to state 4 is the remainder of the power stroke. As in the Otto cycle, the cycle is completed by constant-volume Process 4–1 in which heat is rejected from the air while the piston is at bottom dead center. This process replaces the exhaust and intake processes of the actual engine.

Diesel cycle

Since the air-standard Diesel cycle is composed of internally reversible processes, areas on the T–s and p–v diagrams of Fig. 9.5 can be interpreted as heat and work, respectively. On the T–s diagram, area 2–3–a–b–2 represents heat added per unit of mass and area 1–4–a–b–1 is heat rejected per unit of mass. On the p–v diagram, area 1–2–a–b–1 is work input per unit of mass during the compression process. Area 2–3–4–b–a–2 is the work done per unit of mass as the piston moves from top dead center to bottom dead center. The enclosed area of each figure is the net work output, which equals the net heat added.

Animation

Diesel Cycle Tabs a and b

Cycle Analysis   In the Diesel cycle the heat addition takes place at constant pressure. Accordingly, Process 2–3 involves both work and heat. The work is given by

$$\frac{W_{23}}{m} = \int_2^3 p \; dv = p_2(v_3 - v_2) \tag{9.9}$$

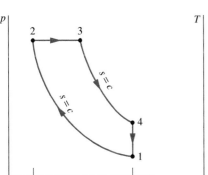

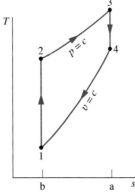

Fig. 9.5 p–v and T–s diagrams of the air-standard Diesel cycle.

The heat added in Process 2–3 can be found by applying the closed system energy balance

$$m(u_3 - u_2) = Q_{23} - W_{23}$$

Introducing Eq. 9.9 and solving for the heat transfer

$$\frac{Q_{23}}{m} = (u_3 - u_2) + p(v_3 - v_2) = (u_3 + pv_3) - (u_2 + pv_2)$$
$$= h_3 - h_2 \qquad (9.10)$$

where the specific enthalpy is introduced to simplify the expression. As in the Otto cycle, the heat rejected in Process 4–1 is given by

$$\frac{Q_{41}}{m} = u_4 - u_1$$

The thermal efficiency is the ratio of the net work of the cycle to the heat added:

$$\boxed{\eta = \frac{W_{\text{cycle}}/m}{Q_{23}/m} = 1 - \frac{Q_{41}/m}{Q_{23}/m} = 1 - \frac{u_4 - u_1}{h_3 - h_2}} \qquad (9.11)$$

As with the Otto cycle, the thermal efficiency of the Diesel cycle increases with the compression ratio.

To evaluate the thermal efficiency from Eq. 9.11 requires values for u_1, u_4, h_2, and h_3 or equivalently the temperatures at the principal states of the cycle. Let us consider next how these temperatures are evaluated. For a given initial temperature T_1 and compression ratio r, the temperature at state 2 can be found using the following isentropic relationship and v_r data:

$$v_{r2} = \frac{V_2}{V_1} v_{r1} = \frac{1}{r} v_{r1}$$

To find T_3, note that the ideal gas equation of state reduces with $p_3 = p_2$ to give

$$T_3 = \frac{V_3}{V_2} T_2 = r_c T_2$$

cutoff ratio   where $r_c = V_3/V_2$, called the **cutoff ratio**, has been introduced.

Since $V_4 = V_1$, the volume ratio for the isentropic process 3–4 can be expressed as

$$\frac{V_4}{V_3} = \frac{V_4}{V_2} \frac{V_2}{V_3} = \frac{V_1}{V_2} \frac{V_2}{V_3} = \frac{r}{r_c} \qquad (9.12)$$

where the compression ratio r and cutoff ratio r_c have been introduced for conciseness.

Using Eq. 9.12 together with v_{r3} at T_3, the temperature T_4 can be determined by interpolation once v_{r4} is found from the isentropic relationship

$$v_{r4} = \frac{V_4}{V_3} v_{r3} = \frac{r}{r_c} v_{r3}$$

In a *cold air-standard analysis*, the appropriate expression for evaluating T_2 is provided by

$$\frac{T_2}{T_1} = \left(\frac{V_1}{V_2}\right)^{k-1} = r^{k-1} \qquad (\text{constant } k)$$

The temperature T_4 is found similarly from

$$\frac{T_4}{T_3} = \left(\frac{V_3}{V_4}\right)^{k-1} = \left(\frac{r_c}{r}\right)^{k-1} \qquad (\text{constant } k)$$

where Eq. 9.12 has been used to replace the volume ratio.

Effect of Compression Ratio on Performance   As with the Otto cycle, the thermal efficiency of the Diesel cycle increases with increasing compression ratio. This can be brought out simply using a *cold* air-standard analysis. On a cold air-standard basis, the thermal efficiency of the Diesel cycle can be expressed as

$$\eta = 1 - \frac{1}{r^{k-1}}\left[\frac{r_c^k - 1}{k(r_c - 1)}\right] \quad \text{(cold air-standard basis)} \quad (9.13)$$

where r is the compression ratio and r_c is the cutoff ratio. The derivation is left as an exercise. This relationship is shown in Fig. 9.6 for $k = 1.4$. Equation 9.13 for the Diesel cycle differs from Eq. 9.8 for the Otto cycle only by the term in brackets, which for $r_c > 1$ is greater than unity. Thus, when the compression ratio is the same, the thermal efficiency of the cold air-standard Diesel cycle is less than that of the cold air-standard Otto cycle.

In the next example, we illustrate the analysis of the air-standard Diesel cycle.

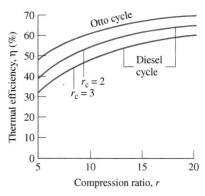

Fig. 9.6 Thermal efficiency of the cold air-standard Diesel cycle, $k = 1.4$.

> ▶ ▶ ▶ **EXAMPLE 9.2** ▶ ·

Analyzing the Diesel Cycle

At the beginning of the compression process of an air-standard Diesel cycle operating with a compression ratio of 18, the temperature is 300 K and the pressure is 0.1 MPa. The cutoff ratio for the cycle is 2. Determine **(a)** the temperature and pressure at the end of each process of the cycle, **(b)** the thermal efficiency, **(c)** the mean effective pressure, in MPa.

SOLUTION

Known   An air-standard Diesel cycle is executed with specified conditions at the beginning of the compression stroke. The compression and cutoff ratios are given.

Find   Determine the temperature and pressure at the end of each process, the thermal efficiency, and mean effective pressure.

Schematic and Given Data:

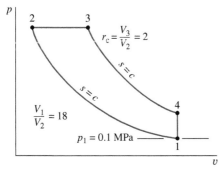

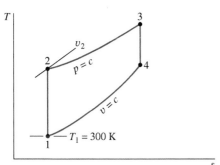

Fig. E9.2

Engineering Model

1. The air in the piston–cylinder assembly is the closed system.
2. The compression and expansion 3–4 are adiabatic.
3. All processes are internally reversible.
4. The air is modeled as an ideal gas.
5. Kinetic and potential energy effects are negligible.

Analysis

a. The analysis begins by determining properties at each principal state of the cycle. With $T_1 = 300$ K, Table A-22 gives $u_1 = 214.07$ kJ/kg and $v_{r1} = 621.2$. For the isentropic compression process 1–2

$$v_{r2} = \frac{V_2}{V_1}v_{r1} = \frac{v_{r1}}{r} = \frac{621.2}{18} = 34.51$$

Interpolating in Table A-22, we get $T_2 = 898.3$ K and $h_2 = 930.98$ kJ/kg. With the ideal gas equation of state

$$p_2 = p_1\frac{T_2}{T_1}\frac{V_1}{V_2} = (0.1)\left(\frac{898.3}{300}\right)(18) = 5.39 \text{ MPa}$$

The pressure at state 2 can be evaluated alternatively using the isentropic relationship, $p_2 = p_1(p_{r2}/p_{r1})$.

Since Process 2–3 occurs at constant pressure, the ideal gas equation of state gives

$$T_3 = \frac{V_3}{V_2}T_2$$

Introducing the cutoff ratio, $r_c = V_3/V_2$

$$T_3 = r_c T_2 = 2(898.3) = 1796.6 \text{ K}$$

From Table A-22, $h_3 = 1999.1$ kJ/kg and $v_{r3} = 3.97$. For the isentropic expansion process 3–4

$$v_{r4} = \frac{V_4}{V_3}v_{r3} = \frac{V_4}{V_2}\frac{V_2}{V_3}v_{r3}$$

Introducing $V_4 = V_1$, the compression ratio r, and the cutoff ratio r_c, we have

$$v_{r4} = \frac{r}{r_c} v_{r3} = \frac{18}{2}(3.97) = 35.73$$

Interpolating in Table A-22 with v_{r4}, we get $u_4 = 664.3$ kJ/kg and $T_4 = 887.7$ K. The pressure at state 4 can be found using the isentropic relationship $p_4 = p_3(p_{r4}/p_{r3})$ or the ideal gas equation of state applied at states 1 and 4. With $V_4 = V_1$, the ideal gas equation of state gives

$$p_4 = p_1\frac{T_4}{T_1} = (0.1 \text{ MPa})\left(\frac{887.7 \text{ K}}{300 \text{ K}}\right) = 0.3 \text{ MPa}$$

b. The thermal efficiency is found using

$$\eta = 1 - \frac{Q_{41}/m}{Q_{23}/m} = 1 - \frac{u_4 - u_1}{h_3 - h_2}$$

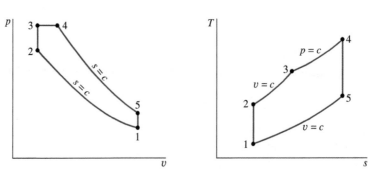

$$= 1 - \frac{664.3 - 214.07}{1999.1 - 930.98} = 0.578(57.8\%)$$

c. The mean effective pressure written in terms of specific volumes is

$$\text{mep} = \frac{W_{\text{cycle}}/m}{v_1 - v_2} = \frac{W_{\text{cycle}}/m}{v_1(1 - 1/r)}$$

The net work of the cycle equals the net heat added

$$\frac{W_{\text{cycle}}}{m} = \frac{Q_{23}}{m} - \frac{Q_{41}}{m} = (h_3 - h_2) - (u_4 - u_1)$$

$$= (1999.1 - 930.98) - (664.3 - 214.07)$$

$$= 617.9 \text{ kJ/kg}$$

The specific volume at state 1 is

$$v_1 = \frac{(\bar{R}/M)T_1}{p_1} = \frac{\left(\dfrac{8314 \text{ N} \cdot \text{m}}{28.97 \text{ kg} \cdot \text{K}}\right)(300 \text{ K})}{10^5 \text{ N/m}^2} = 0.861 \text{ m}^3/\text{kg}$$

Inserting values

$$\text{mep} = \frac{617.9 \text{ kJ/kg}}{0.861(1 - 1/18) \text{ m}^3/\text{kg}}\left|\frac{10^3 \text{ N} \cdot \text{m}}{1 \text{ kJ}}\right|\left|\frac{1 \text{ MPa}}{10^6 \text{ N/m}^2}\right|$$

$$= 0.76 \text{ MPa}$$

❶ This solution uses the air tables, which account explicitly for the variation of the specific heats with temperature. Note that Eq. 9.13 based on the assumption of *constant* specific heats has not been used to determine the thermal efficiency. The cold air-standard solution of this example is left as an exercise.

SKILLS DEVELOPED

Ability to...

- sketch the Diesel cycle p–v and T–s diagrams.
- evaluate temperatures and pressures at each principal state and retrieve necessary property data.
- calculate the thermal efficiency and mean effective pressure.

Quick Quiz

If the mass of air is 0.0123 kg, what is the *displacement volume*, in L? Ans. 10 L.

9.4 Air-Standard Dual Cycle

dual cycle

The pressure–volume diagrams of actual internal combustion engines are not described well by the Otto and Diesel cycles. An air-standard cycle that can be made to approximate the pressure variations more closely is the *air-standard dual cycle*. The **dual cycle** is shown in **Fig. 9.7**. As in the Otto and Diesel cycles, Process 1–2 is an isentropic compression. The heat addition occurs in two steps, however: Process 2–3 is a constant-volume heat addition; Process 3–4 is

Fig. 9.7 p–v and T–s diagrams of the air-standard dual cycle.

a constant-pressure heat addition. Process 3–4 also makes up the first part of the power stroke. The isentropic expansion from state 4 to state 5 is the remainder of the power stroke. As in the Otto and Diesel cycles, the cycle is completed by a constant-volume heat rejection process, Process 5–1. Areas on the T–s and p–v diagrams can be interpreted as heat and work, respectively, as in the cases of the Otto and Diesel cycles.

Cycle Analysis

Since the dual cycle is composed of the same types of processes as the Otto and Diesel cycles, we can simply write down the appropriate work and heat transfer expressions by reference to the corresponding earlier developments. Thus, during the isentropic compression process 1–2 there is no heat transfer, and the work is

$$\frac{W_{12}}{m} = u_2 - u_1$$

As for the corresponding process of the Otto cycle, in the constant-volume portion of the heat addition process, Process 2–3, there is no work, and the heat transfer is

$$\frac{Q_{23}}{m} = u_3 - u_2$$

In the constant-pressure portion of the heat addition process, Process 3–4, there is both work and heat transfer, as with the corresponding process of the Diesel cycle

$$\frac{W_{34}}{m} = p(v_4 - v_3) \qquad \text{and} \qquad \frac{Q_{34}}{m} = h_4 - h_3$$

During the isentropic expansion process 4–5 there is no heat transfer, and the work is

$$\frac{W_{45}}{m} = u_4 - u_5$$

Finally, the constant-volume heat rejection process 5–1 that completes the cycle involves heat transfer but no work:

$$\frac{Q_{51}}{m} = u_5 - u_1$$

The thermal efficiency is the ratio of the net work of the cycle to the *total* heat added:

$$\eta = \frac{W_{cycle}/m}{(Q_{23}/m + Q_{34}/m)} = 1 - \frac{Q_{51}/m}{(Q_{23}/m + Q_{34}/m)}$$

$$= 1 - \frac{(u_5 - u_1)}{(u_3 - u_2) + (h_4 - h_3)} \tag{9.14}$$

The example to follow provides an illustration of the analysis of an air-standard dual cycle. The analysis exhibits many of the features found in the Otto and Diesel cycle examples considered previously.

> ▶ ▶ **EXAMPLE 9.3** ▶

Analyzing the Dual Cycle

At the beginning of the compression process of an air-standard dual cycle with a compression ratio of 18, the temperature is 300 K and the pressure is 0.1 MPa. The pressure ratio for the constant volume part of the heating process is 1.5:1. The volume ratio for the constant pressure part of the heating process is 1.2:1. Determine **(a)** the thermal efficiency and **(b)** the mean effective pressure, in MPa.

SOLUTION

Known An air-standard dual cycle is executed in a piston–cylinder assembly. Conditions are known at the beginning of the compression process, and necessary volume and pressure ratios are specified.

Find Determine the thermal efficiency and the mep, in MPa.

Schematic and Given Data:

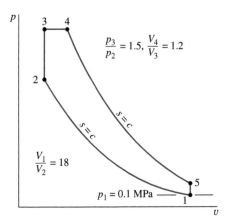

$$\frac{p_3}{p_2} = 1.5, \frac{V_4}{V_3} = 1.2$$

$$\frac{V_1}{V_2} = 18$$

$$p_1 = 0.1 \text{ MPa}$$

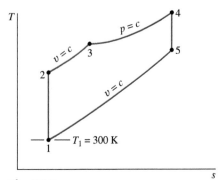

$$T_1 = 300 \text{ K}$$

Fig. E9.3

Engineering Model

1. The air in the piston–cylinder assembly is the closed system.
2. The compression and expansion 4–5 are adiabatic.
3. All processes are internally reversible.
4. The air is modeled as an ideal gas.
5. Kinetic and potential energy effects are negligible.

Analysis The analysis begins by determining properties at each principal state of the cycle. States 1 and 2 are the same as in Example 9.2, so $u_1 = 214.07$ kJ/kg, $T_2 = 898.3$ K, $u_2 = 673.2$ kJ/kg. Since Process 2–3 occurs at constant volume, the ideal gas equation of state reduces to give

$$T_3 = \frac{p_3}{p_2} T_2 = (1.5)(898.3) = 1347.5 \text{ K}$$

Interpolating in Table A-22, we get $h_3 = 1452.6$ kJ/kg and $u_3 = 1065.8$ kJ/kg.

Since Process 3–4 occurs at constant pressure, the ideal gas equation of state reduces to give

$$T_4 = \frac{V_4}{V_3} T_3 = (1.2)(1347.5) = 1617 \text{ K}$$

From Table A-22, $h_4 = 1778.3$ kJ/kg and $v_{r4} = 5.609$.
Process 4–5 is an isentropic expansion, so

$$v_{r5} = v_{r4} \frac{V_5}{V_4}$$

The volume ratio V_5/V_4 required by this equation can be expressed as

$$\frac{V_5}{V_4} = \frac{V_5}{V_3} \frac{V_3}{V_4}$$

With $V_5 = V_1$, $V_2 = V_3$, and given volume ratios

$$\frac{V_5}{V_4} = \frac{V_1}{V_2} \frac{V_3}{V_4} = 18 \left(\frac{1}{1.2} \right) = 15$$

Inserting this in the above expression for v_{r5}

$$v_{r5} = (5.609)(15) = 84.135$$

Interpolating in Table A-22, we get $u_5 = 475.96$ kJ/kg.

a. The thermal efficiency is

$$\eta = 1 - \frac{Q_{51}/m}{(Q_{23}/m + Q_{34}/m)} = 1 - \frac{(u_5 - u_1)}{(u_3 - u_2) + (h_4 - h_3)}$$

$$= 1 - \frac{(475.96 - 214.07)}{(1065.8 - 673.2) + (1778.3 - 1452.6)}$$

$$= 0.635 (63.5\%)$$

b. The mean effective pressure is

$$\text{mep} = \frac{W_{\text{cycle}}/m}{v_1 - v_2} = \frac{W_{\text{cycle}}/m}{v_1(1 - 1/r)}$$

The net work of the cycle equals the net heat added, so

$$\text{mep} = \frac{(u_3 - u_2) + (h_4 - h_3) - (u_5 - u_1)}{v_1(1 - 1/r)}$$

The specific volume at state 1 is evaluated in Example 9.2 as $v_1 = 0.861$ m³/kg. Inserting values into the above expression for mep

$$\text{mep} = \frac{[(1065.8 - 673.2) + (1778.3 - 1452.6) - (475.96 - 214.07)]}{0.861(1 - 1/18) \text{ m}^3/\text{kg}} \times \left(\frac{\text{kJ}}{\text{kg}} \right) \left| \frac{10^3 \text{ N} \cdot \text{m}}{1 \text{ kJ}} \right| \left| \frac{1 \text{ MPa}}{10^6 \text{ N/m}^2} \right|$$

$$= 0.56 \text{ MPa}$$

SKILLS DEVELOPED

Ability to...

• sketch the dual cycle p–v and T–s diagrams.

• evaluate temperatures and pressures at each principal state and retrieve necessary property data.

• calculate the thermal efficiency and mean effective pressure.

Quick Quiz

Evaluate the total heat addition and the net work of the cycle, each in kJ per kg of air. Ans. $Q_{\text{in}}/m = 718$ **kJ/kg,** $W_{\text{cycle}}/m = 456$ **kJ/kg.**

Are Biodiesel-Fueled Vehicles in Our Future?

Diesel-powered vehicles may be more prevalent in the United States in coming years. Diesel engines are more powerful and about one-third more fuel-efficient than similar-sized gasoline engines currently dominating the U.S. market. Ultra-low sulfur diesel fuel, commonly available today at U.S. fuel pumps, and improved exhaust treatment allow diesel-powered vehicles to meet the same emissions standards as gasoline vehicles.

Biodiesel also can be used as fuel. Biodiesel is domestically produced from nonpetroleum, renewable sources, including vegetable oils (from soybeans, rapeseeds, sunflower seeds, and jatropha seeds), animal fats, and algae. Waste vegetable oil from industrial deep fryers, snack food factories, and restaurants can be converted to biodiesel.

Petroleum-based diesel fuel blended with 5% biodiesel, called B5, is said to be safe for use in diesel-powered cars. Other blends may damage the engine or void the warranty, automakers caution. And although biofuels provide benefits, producing them from corn and other crops used by millions for food and/or with methods leading to environmental damage, such as deforestation, places a cloud over their widespread use for transportation.

Finally, despite having many pluses, the extent to which diesel-powered cars will be seen on our roadways is uncertain, for they face stiff competition from cars with advanced engines fueled by gasoline and from hybrid-electric and all-electric vehicles.

Considering Gas Turbine Power Plants

This part of the chapter deals with gas turbine power plants. Gas turbines tend to be lighter and more compact than the vapor power plants studied in Chap. 8. The favorable power-output-to-weight ratio of gas turbines makes them well suited for transportation applications (aircraft propulsion, marine power plants, and so on). In recent years, gas turbines also have contributed an increasing share of U.S. electric power needs.

Today's electric power-producing gas turbines are almost exclusively fueled by natural gas. However, depending on the application, other fuels can be used by gas turbines, including distillate fuel oil; propane; gases produced from landfills, sewage treatment plants, and animal waste; and *syngas* (synthesis gas) obtained by gasification of coal (see Sec. 9.10).

9.5 Modeling Gas Turbine Power Plants

Gas turbine power plants may operate on either an open or closed basis. The *open* mode pictured in **Fig. 9.8a** is more common. This is an engine in which atmospheric air is continuously drawn into the compressor, where it is compressed to a high pressure. The air then

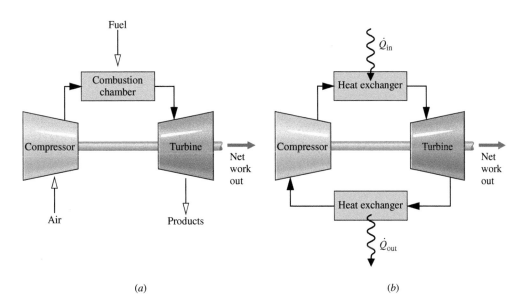

(a)   *(b)*

Fig. 9.8 Simple gas turbine. (*a*) Open to the atmosphere. (*b*) Closed.

enters a combustion chamber, or combustor, where it is mixed with fuel and combustion occurs, resulting in combustion products at an elevated temperature. The combustion products expand through the turbine and are subsequently discharged to the surroundings. Part of the turbine work developed is used to drive the compressor; the remainder is available to generate electricity, to propel a vehicle, or for other purposes.

In the *closed* mode pictured in Fig. 9.8*b*, the working fluid receives an energy input by heat transfer from an external source, for example, a gas-cooled nuclear reactor. The gas exiting the turbine is passed through a heat exchanger, where it is cooled prior to reentering the compressor.

An idealization often used in the study of open gas turbine power plants is that of an **air-standard analysis**. In an air-standard analysis two assumptions are always made:

air-standard analysis: gas turbines

- The working fluid is air, which behaves as an ideal gas.
- The temperature rise that would be brought about by combustion is accomplished by a heat transfer from an external source.

With an air-standard analysis, we avoid dealing with the complexities of the combustion process and the change of composition during combustion. Accordingly, an air-standard analysis simplifies study of gas turbine power plants considerably, but numerical values calculated on this basis may provide only qualitative indications of power plant performance. Still, we can learn important aspects of gas turbine operation using an air-standard analysis; see Sec. 9.6 for further discussion supported by solved examples.

Energy & Environment

Natural gas is widely used for power generation by gas turbines, industrial and home heating, and chemical processing. The versatility of natural gas is matched by its relative abundance in North America, including natural gas extracted from deep-water ocean sites and shale deposits. Pipeline delivery of natural gas across the nation and from Canada has occurred for decades. Natural gas is also delivered overseas by ship.

Turning natural gas into a liquid, and thereby reducing specific volume significantly, is the only practical way to transfer it overseas using ships. Liquefied natural gas (LNG) is stored in tanks onboard ships at about −163°C (−260°F). To reduce heat transfer to the cargo LNG from outside sources, the tanks are insulated and the ships have double hulls with ample space between them. Still, a fraction of the cargo evaporates during long voyages. Such

boil-off gas is commonly used to fuel the ship's propulsion system and meet other onboard energy needs. When tankers arrive at their destinations, LNG is converted to gas by heating it. The gas is then sent via pipeline to storage tanks onshore for distribution to customers.

Shipboard delivery of LNG has some minuses. Owing to cumulative effects during the LNG delivery chain, considerable exergy is destroyed and lost in liquefying gas at the beginning of the chain, transporting LNG by ship, and regasifying it when port is reached. If comparatively warm seawater is used to regasify LNG, environmentalists worry about the effect on aquatic life nearby. Some observers are also concerned about safety, especially when huge quantities of gas are stored at ports in major urban areas.

9.6 Air-Standard Brayton Cycle

A schematic diagram of an air-standard gas turbine is shown in Fig. 9.9. The directions of the principal energy transfers are indicated on this figure by arrows. In accordance with the assumptions of an air-standard analysis, the temperature rise that would be achieved in the combustion process is brought about by a heat transfer to the working fluid from an external source and the working fluid is considered to be air as an ideal gas. With the air-standard idealizations, air would be drawn into the compressor at state 1 from the surroundings and later returned to the surroundings at state 4 with a temperature greater than the ambient temperature. After interacting with the surroundings, each unit mass of discharged air would eventually return to the same state as the air entering the compressor, so we may think of the air passing through the components of the gas turbine as undergoing a thermodynamic cycle. A simplified representation of the states visited by the air in such a cycle can be devised by regarding the turbine exhaust air as restored to the compressor inlet state by passing through a heat exchanger where heat rejection to the surroundings occurs. The cycle that results with this further idealization is called the air-standard **Brayton cycle**.

Brayton cycle

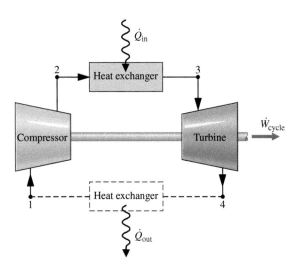

Fig. 9.9 Air-standard gas turbine cycle.

9.6.1 Evaluating Principal Work and Heat Transfers

The following expressions for the work and heat transfers of energy that occur at steady state are readily derived by reduction of the control volume mass and energy rate balances. These energy transfers are positive in the directions of the arrows in Fig. 9.9. Assuming the turbine operates adiabatically and with negligible effects of kinetic and potential energy, the work developed per unit of mass flowing is

$$\frac{\dot{W}_t}{\dot{m}} = h_3 - h_4 \tag{9.15}$$

where $\dot{m}$ denotes the mass flow rate. With the same assumptions, the compressor work per unit of mass flowing is

$$\frac{\dot{W}_c}{\dot{m}} = h_2 - h_1 \tag{9.16}$$

The symbol $\dot{W}_c$ denotes work *input* and takes on a positive value. The heat added to the cycle per unit of mass is

$$\frac{\dot{Q}_{in}}{\dot{m}} = h_3 - h_2 \tag{9.17}$$

The heat rejected per unit of mass is

$$\frac{\dot{Q}_{out}}{\dot{m}} = h_4 - h_1 \tag{9.18}$$

where $\dot{Q}_{out}$ is positive in value.

The thermal efficiency of the cycle in Fig. 9.9 is

$$\eta = \frac{\dot{W}_t/\dot{m} - \dot{W}_c/\dot{m}}{\dot{Q}_{in}/\dot{m}} = \frac{(h_3 - h_4) - (h_2 - h_1)}{h_3 - h_2} \tag{9.19}$$

The **back work ratio** for the cycle is

back work ratio

$$\text{bwr} = \frac{\dot{W}_c/\dot{m}}{\dot{W}_t/\dot{m}} = \frac{h_2 - h_1}{h_3 - h_4} \tag{9.20}$$

For the same pressure rise, a gas turbine compressor would require a much greater work input per unit of mass flow than the pump of a vapor power plant because the average specific volume of the gas flowing through the compressor would be many times greater than that of the liquid passing through the pump (see discussion of Eq. 6.51b in Sec. 6.13). Hence, a relatively large portion of the work developed by the turbine is required to drive the compressor. Typical back work ratios of gas turbines range from 40 to 80%. In comparison, the back work ratios of vapor power plants are normally only 1 or 2%.

If the temperatures at the numbered states of the cycle are known, the specific enthalpies required by the foregoing equations are readily obtained from the ideal gas table for air, Table A-22. Alternatively, with the sacrifice of some accuracy, the variation of the specific heats with temperature can be ignored and the specific heats taken as constant. The air-standard analysis is then referred to as a *cold air-standard analysis*. As illustrated by the discussion of internal combustion engines given previously, the chief advantage of the assumption of constant specific heats is that simple expressions for quantities such as thermal efficiency can be derived, and these can be used to deduce qualitative indications of cycle performance without involving tabular data.

Since Eqs. 9.15 through 9.20 have been developed from mass and energy rate balances, they apply equally when irreversibilities are present and in the absence of irreversibilities. Although irreversibilities and losses associated with the various power plant components have a pronounced effect on overall performance, it is instructive to consider an idealized cycle in which they are assumed absent. Such a cycle establishes an upper limit on the performance of the air-standard Brayton cycle. This is considered next.

9.6.2 Ideal Air-Standard Brayton Cycle

Ignoring irreversibilities as the air circulates through the various components of the Brayton cycle, there are no frictional pressure drops, and the air flows at constant pressure through the heat exchangers. If stray heat transfers to the surroundings are also ignored, the processes through the turbine and compressor are isentropic. The ideal cycle shown on the p–v and T–s diagrams in **Fig. 9.10** adheres to these idealizations.

Areas on the T–s and p–v diagrams of Fig. 9.10 can be interpreted as heat and work, respectively, per unit of mass flowing. On the T–s diagram, area 2–3–a–b–2 represents the heat added per unit of mass and area 1–4–a–b–1 is the heat rejected per unit of mass. On the p–v diagram, area 1–2–a–b–1 represents the compressor work input per unit of mass and area 3–4–b–a–3 is the turbine work output per unit of mass. The enclosed area on each figure can be interpreted as the net work output or, equivalently, the net heat added.

When air table data are used to conduct an analysis involving the ideal Brayton cycle, the following relationships, based on Eq. 6.41, apply for the isentropic processes 1–2 and 3–4:

$$p_{r2} = p_{r1} \frac{p_2}{p_1} \tag{9.21}$$

$$p_{r4} = p_{r3} \frac{p_4}{p_3} = p_{r3} \frac{p_1}{p_2} \tag{9.22}$$

where p_2/p_1 is the *compressor pressure ratio*. Recall that p_r is tabulated versus temperature in Table A-22. Since air flows through the heat exchangers of the ideal cycle at constant pressure, it follows that $p_4/p_3 = p_1/p_2$. This relationship has been used in writing Eq. 9.22.

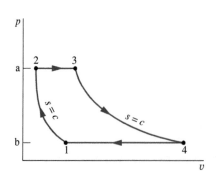
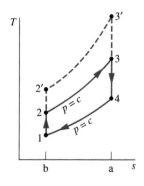

Fig. 9.10 Air-standard ideal Brayton cycle.

When an ideal Brayton cycle is analyzed on a cold air-standard basis, the specific heats are taken as constant. Equations 9.21 and 9.22 are then replaced, respectively, by the following expressions, based on Eq. 6.43,

$$T_2 = T_1 \left(\frac{p_2}{p_1} \right)^{(k-1)/k} \tag{9.23}$$

$$T_4 = T_3 \left(\frac{p_4}{p_3} \right)^{(k-1)/k} = T_3 \left(\frac{p_1}{p_2} \right)^{(k-1)/k} \tag{9.24}$$

> ▶ **Animation**
>
> **Brayton Cycle Tab a**

where k is the specific heat ratio, $k = c_p/c_v$.

In the next example, we illustrate the analysis of an ideal air-standard Brayton cycle and compare results with those obtained on a cold air-standard basis.

▶ ▶ ▶ EXAMPLE 9.4 ▶

Analyzing the Ideal Brayton Cycle

Air enters the compressor of an ideal air-standard Brayton cycle at 100 kPa, 300 K, with a volumetric flow rate of 5 m³/s. The compressor pressure ratio is 10. The turbine inlet temperature is 1400 K. Determine **(a)** the thermal efficiency of the cycle, **(b)** the back work ratio, **(c)** the *net* power developed, in kW.

SOLUTION

Known   An ideal air-standard Brayton cycle operates with given compressor inlet conditions, given turbine inlet temperature, and a known compressor pressure ratio.

Find   Determine the thermal efficiency, the back work ratio, and the *net* power developed, in kW.

Schematic and Given Data:

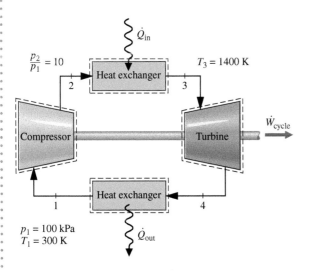

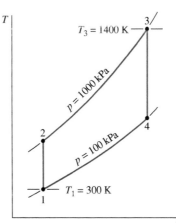

Fig. E9.4

Engineering Model

1. Each component is analyzed as a control volume at steady state. The control volumes are shown on the accompanying sketch by dashed lines.

2. The turbine and compressor processes are isentropic.

3. There are no pressure drops for flow through the heat exchangers.

4. Kinetic and potential energy effects are negligible.

5. The working fluid is air modeled as an ideal gas.

❶ Analysis   The analysis begins by determining the specific enthalpy at each numbered state of the cycle. At state 1, the temperature is 300 K. From Table A-22, $h_1 = 300.19$ kJ/kg and $p_{r1} = 1.386$.

Since the compressor process is isentropic, the following relationship can be used to determine h_2:

$$p_{r2} = \frac{p_2}{p_1} p_{r1} = (10)(1.386) = 13.86$$

Then, interpolating in Table A-22, we obtain $h_2 = 579.9$ kJ/kg.

The temperature at state 3 is given as $T_3 = 1400$ K. With this temperature, the specific enthalpy at state 3 from Table A-22 is $h_3 = 1515.4$ kJ/kg. Also, $p_{r3} = 450.5$.

The specific enthalpy at state 4 is found by using the isentropic relationship

$$p_{r4} = p_{r3}\frac{p_4}{p_3} = (450.5)(1/10) = 45.05$$

Interpolating in Table A-22, we get $h_4 = 808.5$ kJ/kg.

a. The thermal efficiency is

$$
\eta = \frac{(\dot{W}_t/\dot{m}) - (\dot{W}_c/\dot{m})}{\dot{Q}_{in}/\dot{m}}
$$

$$
= \frac{(h_3 - h_4) - (h_2 - h_1)}{h_3 - h_2} = \frac{(1515.4 - 808.5) - (579.9 - 300.19)}{1515.4 - 579.9}
$$

$$
= \frac{706.9 - 279.7}{935.5} = 0.457 \ (45.7\%)
$$

b. The back work ratio is

2

$$
\text{bwr} = \frac{\dot{W}_c/\dot{m}}{\dot{W}_t/\dot{m}} = \frac{h_2 - h_1}{h_3 - h_4} = \frac{279.7}{706.9} = 0.396 \ (39.6\%)
$$

c. The net power developed is

$$
\dot{W}_{cycle} = \dot{m}[(h_3 - h_4) - (h_2 - h_1)]
$$

To evaluate the net power requires the mass flow rate $\dot{m}$, which can be determined from the volumetric flow rate and specific volume at the compressor inlet as follows:

$$
\dot{m} = \frac{(AV)_1}{v_1}
$$

Since $v_1 = (\bar{R}/M)T_1/p_1$, this becomes

$$
\dot{m} = \frac{(AV)_1 \, p_1}{(\bar{R}/M)T_1} = \frac{(5 \text{ m}^3/\text{s})(100 \times 10^3 \text{ N/m}^2)}{\left(\dfrac{8314 \text{ N} \cdot \text{m}}{28.97 \text{ kg} \cdot \text{k}}\right)(300 \text{ K})}
$$

$$
= 5.807 \text{ kg/s}
$$

Finally,

$$
\dot{W}_{cycle} = (5.807 \text{ kg/s})(706.9 - 279.7)\left(\frac{\text{kJ}}{\text{kg}}\right)\left|\frac{1 \text{ kW}}{1 \text{ kJ/s}}\right| = 2481 \text{ kW}
$$

---

1 The use of the ideal gas table for air is featured in this solution. A solution also can be developed on a cold air-standard basis in which constant specific heats are assumed. The details are left as an exercise, but for comparison the results are presented for the case $k = 1.4$ in the following table:

Parameter	Air-Standard Analysis	Cold Air-Standard Analysis, $k = 1.4$
T_2	574.1 K	579.2 K
T_4	787.7 K	725.1 K
η	0.457	0.482
bwr	0.396	0.414
$\dot{W}_{cycle}$	2481 kW	2308 kW

2 The value of the back work ratio in the present gas turbine case is significantly greater than the back work ratio of the simple vapor power cycle of Example 8.1.

SKILLS DEVELOPED

Ability to...

- sketch the schematic of the basic air-standard gas turbine and the T–s diagram for the corresponding ideal Brayton cycle.

- evaluate temperatures and pressures at each principal state and retrieve necessary property data.

- calculate the thermal efficiency and back work ratio.

Quick Quiz

Determine the rate of heat transfer to the air passing through the combustor, in kW. Ans. 5432 kW.

Effect of Compressor Pressure Ratio on Performance   Conclusions that are qualitatively correct for actual gas turbines can be drawn from a study of the ideal Brayton cycle. The first of these conclusions is that the thermal efficiency increases with increasing pressure ratio across the compressor.

> **FOR EXAMPLE**
>
> Referring again to the T–s diagram of Fig. 9.10, we see that an increase in the compressor pressure ratio changes the cycle from 1–2–3–4–1 to 1–2′–3′–4–1. Since the average temperature of heat addition is greater in the latter cycle and both cycles have the same heat rejection process, cycle 1–2′–3′–4–1 would have the greater thermal efficiency.

The increase in thermal efficiency with the pressure ratio across the compressor is also brought out simply by the following development, in which the specific heat c_p and, thus, the specific heat ratio k are assumed constant. For constant c_p, Eq. 9.19 becomes

$$
\eta = \frac{c_p(T_3 - T_4) - c_p(T_2 - T_1)}{c_p(T_3 - T_2)} = 1 - \frac{(T_4 - T_1)}{(T_3 - T_2)}
$$

Or, on further rearrangement,

$$\eta = 1 - \frac{T_1}{T_2}\left(\frac{T_4/T_1 - 1}{T_3/T_2 - 1}\right)$$

From Eqs. 9.23 and 9.24, $T_4/T_1 = T_3/T_2$, so

$$\eta = 1 - \frac{T_1}{T_2}$$

Finally, introducing Eq. 9.23,

$$\boxed{\eta = 1 - \frac{1}{(p_2/p_1)^{(k-1)/k}} \qquad \text{(cold air-standard basis)}} \qquad (9.25)$$

By inspection of Eq. 9.25, it can be seen that the cold air-standard ideal Brayton cycle thermal efficiency increases with increasing pressure ratio across the compressor.

As there is a limit imposed by metallurgical considerations on the maximum allowed temperature at the turbine inlet, it is instructive to consider the effect of increasing compressor pressure ratio on thermal efficiency when the turbine inlet temperature is restricted to the maximum allowable temperature. We do this using **Figs. 9.11** and **9.12**.

The *T–s* diagrams of two ideal Brayton cycles having the same turbine inlet temperature but different compressor pressure ratios are shown in Fig. 9.11. Cycle A has a greater compressor pressure ratio than cycle B and thus the greater thermal efficiency. However, cycle B has a larger enclosed area and thus the greater net work developed per unit of mass flow. Accordingly, for cycle A to develop the same net *power* output as cycle B, a larger mass flow rate would be required, which might dictate a larger system.

These considerations are important for gas turbines intended for use in vehicles where engine weight must be kept small. For such applications, it is desirable to operate near the compressor pressure ratio that yields the most work per unit of mass flow and not the pressure ratio for the greatest thermal efficiency. To illustrate this, see Fig. 9.12 showing the variations with increasing compressor pressure ratio of thermal efficiency and net work per unit of mass flow. While thermal efficiency increases with pressure ratio, the net work per unit of mass curve has a maximum value at a pressure ratio of about 21. Also observe that the curve is relatively flat in the vicinity of the maximum. Thus, for vehicle design purposes a wide range of compressor pressure ratio values may be considered as *nearly optimal* from the standpoint of maximum work per unit of mass.

Example 9.5 provides an illustration of the determination of the compressor pressure ratio for maximum net work per unit of mass flow for the cold air-standard Brayton cycle.

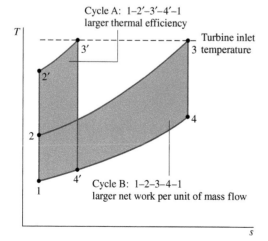

Fig. 9.11 Ideal Brayton cycles with different compressor pressure ratios and the same turbine inlet temperature.

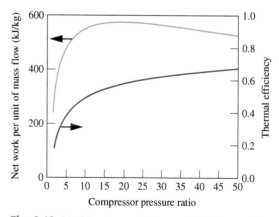

Fig. 9.12 Ideal Brayton cycle thermal efficiency and net work per unit of mass flow versus compressor pressure ratio for $k = 1.4$, a turbine inlet temperature of 1700 K, and a compressor inlet temperature of 300 K.

> ►►► **EXAMPLE 9.5** ► •

Determining Compressor Pressure Ratio for Maximum Net Work

Determine the pressure ratio across the compressor of an ideal Brayton cycle for the maximum net work output per unit of mass flow if the state at the compressor inlet and the temperature at the turbine inlet are fixed. Use a cold air-standard analysis and ignore kinetic and potential energy effects. Discuss.

SOLUTION

Known   An ideal Brayton cycle operates with a specified state at the inlet to the compressor and a specified turbine inlet temperature.

Find   Determine the pressure ratio across the compressor for the maximum net work output per unit of mass flow, and discuss the result.

Schematic and Given Data:

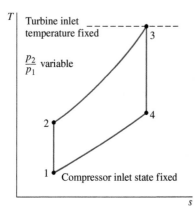

Fig. E9.5

Engineering Model

1. Each component is analyzed as a control volume at steady state.

2. The turbine and compressor processes are isentropic.

3. There are no pressure drops for flow through the heat exchangers.

4. Kinetic and potential energy effects are negligible.

5. The working fluid is air modeled as an ideal gas.

6. The specific heat c_p and thus the specific heat ratio k are constant.

Analysis   The net work of the cycle per unit of mass flow is

$$\frac{\dot{W}_{cycle}}{\dot{m}} = (h_3 - h_4) - (h_2 - h_1)$$

Since c_p is constant (assumption 6),

$$\frac{\dot{W}_{cycle}}{\dot{m}} = c_p[(T_3 - T_4) - (T_2 - T_1)]$$

Or on rearrangement

$$\frac{\dot{W}_{cycle}}{\dot{m}} = c_p T_1\left(\frac{T_3}{T_1} - \frac{T_4}{T_3}\frac{T_3}{T_1} - \frac{T_2}{T_1} + 1\right)$$

Replacing the temperature ratios T_2/T_1 and T_4/T_3 by using Eqs. 9.23 and 9.24, respectively, gives

$$\frac{\dot{W}_{cycle}}{\dot{m}} = c_p T_1\left[\frac{T_3}{T_1} - \frac{T_3}{T_1}\left(\frac{p_1}{p_2}\right)^{(k-1)/k} - \left(\frac{p_2}{p_1}\right)^{(k-1)/k} + 1\right]$$

From this expression it can be concluded that for specified values of T_1, T_3, and c_p, the value of the net work output per unit of mass flow varies with the pressure ratio p_2/p_1 only.

To determine the pressure ratio that maximizes the net work output per unit of mass flow, first form the derivative

$$\frac{\partial(\dot{W}_{cycle}/\dot{m})}{\partial(p_2/p_1)} = \frac{\partial}{\partial(p_2/p_1)}\left\{c_p T_1\left[\frac{T_3}{T_1} - \frac{T_3}{T_1}\left(\frac{p_1}{p_2}\right)^{(k-1)/k}\right.\right.$$

$$\left.\left. - \left(\frac{p_2}{p_1}\right)^{(k-1)/k} + 1\right]\right\}$$

$$= c_p T_1\left(\frac{k-1}{k}\right)\left[\left(\frac{T_3}{T_1}\right)\left(\frac{p_1}{p_2}\right)^{-1/k}\left(\frac{p_1}{p_2}\right)^2 - \left(\frac{p_2}{p_1}\right)^{-1/k}\right]$$

$$= c_p T_1\left(\frac{k-1}{k}\right)\left[\left(\frac{T_3}{T_1}\right)\left(\frac{p_1}{p_2}\right)^{(2k-1)/k} - \left(\frac{p_2}{p_1}\right)^{-1/k}\right]$$

When the partial derivative is set to zero, the following relationship is obtained:

$$\frac{p_2}{p_1} = \left(\frac{T_3}{T_1}\right)^{k/[2(k-1)]} \tag{a}$$

By checking the sign of the second derivative, we can verify that the net work per unit of mass flow is a maximum when this relationship is satisfied.

For gas turbines intended for transportation, it is desirable to operate near the compressor pressure ratio that yields the most work per unit of mass flow. The present example shows how the maximum net work per unit of mass flow is determined on a cold air-standard basis when the state at the compressor inlet and turbine inlet temperature are fixed.

SKILLS DEVELOPED

Ability to…

• complete the detailed derivation of a thermodynamic expression.

• use calculus to maximize a function.

Quick Quiz

For an ideal cold air-standard Brayton cycle with a compressor inlet temperature of 300 K and a maximum cycle temperature of 1700 K, use Eq. (a) above to find the compressor pressure ratio that maximizes the net power output per unit mass flow. Assume $k = 1.4$. **Ans. 21.** (Value agrees with Fig. 9.12.)

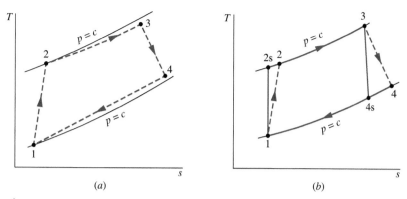

Fig. 9.13 Effects of irreversibilities on the air-standard gas turbine.

9.6.3 Considering Gas Turbine Irreversibilities and Losses

The principal state points of an air-standard gas turbine might be shown more realistically as in **Fig. 9.13a**. Because of frictional effects within the compressor and turbine, the working fluid would experience increases in specific entropy across these components. Owing to friction, there also would be pressure drops as the working fluid passes through the heat exchangers. However, because frictional pressure drops in the heat exchangers are less significant sources of irreversibility, we ignore them in subsequent discussions and for simplicity show the flow through the heat exchangers as occurring at constant pressure. This is illustrated by **Fig. 9.13b**. Stray heat transfers from the power plant components to the surroundings represent losses, but these effects are usually of secondary importance and are also ignored in subsequent discussions.

As the effect of irreversibilities in the turbine and compressor becomes more pronounced, the work developed by the turbine decreases and the work input to the compressor increases, resulting in a marked decrease in the net work of the power plant. Accordingly, if appreciable net work is to be developed by the plant, relatively high isentropic turbine and compressor efficiencies are required.

After decades of developmental effort, efficiencies of 80 to 90% can now be achieved for the turbines and compressors in gas turbine power plants. Designating the states as in Fig. 9.13b, the isentropic turbine and compressor efficiencies are given by

$$\eta_t = \frac{(\dot{W}_t/\dot{m})}{(\dot{W}_t/\dot{m})_s} = \frac{h_3 - h_4}{h_3 - h_{4s}}$$

$$\eta_c = \frac{(\dot{W}_c/\dot{m})_s}{(\dot{W}_c/\dot{m})} = \frac{h_{2s} - h_1}{h_2 - h_1}$$

Among the irreversibilities of actual gas turbine power plants, irreversibilities within the turbine and compressor *are* important, but the most significant *by far* is combustion irreversibility. An air-standard analysis does not allow combustion irreversibility to be evaluated, however, and means introduced in Chap. 13 must be applied.

Example 9.6 brings out the effect of turbine and compressor irreversibilities on plant performance.

Brayton Cycle Tab b

TAKE NOTE...

Isentropic turbine and compressor efficiencies are introduced in Sec. 6.12. See discussions of Eqs. 6.46 and 6.48, respectively.

► EXAMPLE 9.6 ►

Evaluating Performance of a Brayton Cycle with Irreversibilities

Reconsider Example 9.4, but include in the analysis that the turbine and compressor each have an isentropic efficiency of 80%. Determine for the modified cycle **(a)** the thermal efficiency of the cycle, **(b)** the back work ratio, **(c)** the *net* power developed, in kW.

SOLUTION

Known   An air-standard Brayton cycle operates with given compressor inlet conditions, given turbine inlet temperature, and known compressor pressure ratio. The compressor and turbine each have an isentropic efficiency of 80%.

Find Determine the thermal efficiency, the back work ratio, and the net power developed, in kW.

Schematic and Given Data:

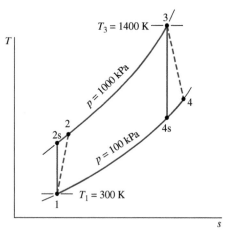

Fig. E9.6

Engineering Model

1. Each component is analyzed as a control volume at steady state.
2. The compressor and turbine are adiabatic.
3. There are no pressure drops for flow through the heat exchangers.
4. Kinetic and potential energy effects are negligible.
5. The working fluid is air modeled as an ideal gas.

Analysis

a. The thermal efficiency is given by

$$\eta = \frac{(\dot{W}_t/\dot{m}) - (\dot{W}_c/\dot{m})}{\dot{Q}_{in}/\dot{m}}$$

The work terms in the numerator of this expression are evaluated using the given values of the compressor and turbine isentropic efficiencies as follows:

The turbine work per unit of mass is

$$\frac{\dot{W}_t}{\dot{m}} = \eta_t \left(\frac{\dot{W}_t}{\dot{m}}\right)_s$$

where η_t is the turbine efficiency. The value of $(\dot{W}_t/\dot{m})_s$ is determined in the solution to Example 9.4 as 706.9 kJ/kg. Thus,

❶
$$\frac{\dot{W}_t}{\dot{m}} = 0.8(706.9) = 565.5 \text{ kJ/kg}$$

For the compressor, the work per unit of mass is

$$\frac{\dot{W}_c}{\dot{m}} = \frac{(\dot{W}_c/\dot{m})_s}{\eta_c}$$

where η_c is the compressor efficiency. The value of $(\dot{W}_c/\dot{m})_s$ is determined in the solution to Example 9.4 as 279.7 kJ/kg, so

$$\frac{\dot{W}_c}{\dot{m}} = \frac{279.7}{0.8} = 349.6 \text{ kJ/kg}$$

The specific enthalpy at the compressor exit, h_2, is required to evaluate the denominator of the thermal efficiency expression. This enthalpy can be determined by solving

$$\frac{\dot{W}_c}{\dot{m}} = h_2 - h_1$$

to obtain

$$h_2 = h_1 + \dot{W}_c/\dot{m}$$

Inserting known values

$$h_2 = 300.19 + 349.6 = 649.8 \text{ kJ/kg}$$

The heat transfer to the working fluid per unit of mass flow is then

$$\frac{\dot{Q}_{in}}{\dot{m}} = h_3 - h_2 = 1515.4 - 649.8 = 865.6 \text{ kJ/kg}$$

where h_3 is from the solution to Example 9.4.
Finally, the thermal efficiency is

$$\eta = \frac{565.5 - 349.6}{865.6} = 0.249 \, (24.9\%)$$

b. The back work ratio is

$$\text{bwr} = \frac{\dot{W}_c/\dot{m}}{\dot{W}_t/\dot{m}} = \frac{349.6}{565.5} = 0.618 \, (61.8\%)$$

c. The mass flow rate is the same as in Example 9.4. The net power developed by the cycle is then

❷
$$\dot{W}_{cycle} = \left(5.807 \frac{\text{kg}}{\text{s}}\right)(565.5 - 349.6)\frac{\text{kJ}}{\text{kg}}\left|\frac{1 \text{ kW}}{1 \text{ kJ/s}}\right| = 1254 \text{ kW}$$

---

❶ The solution to this example on a cold air-standard basis is left as an exercise.

❷ Irreversibilities within the turbine and compressor have a significant impact on the performance of gas turbines. This is brought out by comparing the results of the present example with those of Example 9.4. Irreversibilities result in an increase in the work of compression and a reduction in work output of the turbine. The back work ratio is greatly increased and the thermal efficiency significantly decreased. Still, we should recognize that the most significant irreversibility of gas turbines *by far* is combustion irreversibility.

SKILLS DEVELOPED

Ability to…

- sketch the schematic of the basic air-standard gas turbine and the *T–s* diagram for the corresponding Brayton cycle with compressor and turbine irreversibilities.
- evaluate temperatures and pressures at each principal state and retrieve necessary property data.
- calculate the thermal efficiency and back work ratio.

Quick Quiz

What would be the thermal efficiency and back work ratio if the isentropic turbine efficiency were 70% keeping isentropic compressor efficiency and other given data the same? Ans. $\eta = 16.8\%$, bwr = 70.65%.

9.7 Regenerative Gas Turbines

The turbine exhaust temperature of a simple gas turbine is normally well above the ambient temperature. Accordingly, the hot turbine exhaust gas has significant thermodynamic utility (exergy) that would be irrevocably lost were the gas discarded directly to the surroundings. One way of utilizing this potential is by means of a heat exchanger called a **regenerator**, which allows the air exiting the compressor to be *preheated* before entering the combustor, thereby reducing the amount of fuel that must be burned in the combustor. The combined cycle arrangement considered in Sec. 9.9 is another way to utilize the hot turbine exhaust gas.

    An air-standard Brayton cycle modified to include a regenerator is illustrated in **Fig. 9.14**. The regenerator shown is a counterflow heat exchanger through which the hot turbine exhaust gas and the cooler air leaving the compressor pass in opposite directions. Ideally, no frictional pressure drop occurs in either stream. The turbine exhaust gas is cooled from state 4 to state y, while the air exiting the compressor is heated from state 2 to state x. Hence, a heat transfer from a source external to the cycle is required only to increase the air temperature from state x to state 3, rather than from state 2 to state 3, as would be the case without regeneration. The heat added per unit of mass is then given by

$$\frac{\dot{Q}_{in}}{\dot{m}} = h_3 - h_x \qquad (9.26)$$

The net work developed per unit of mass flow is not altered by the addition of a regenerator. Thus, since the heat added is reduced, the thermal efficiency increases.

Regenerator Effectiveness   From Eq. 9.26 it can be concluded that the external heat transfer required by a gas turbine power plant decreases as the specific enthalpy h_x increases and thus as the temperature T_x increases. Evidently, there is an incentive in terms of fuel saved for selecting a regenerator that provides the greatest practical value for this temperature. To consider the *maximum* theoretical value for T_x, refer to **Fig. 9.15**, which shows temperature variations of the hot and cold streams of a counterflow heat exchanger.

- First, refer to Fig. 9.15*a*. Since a finite temperature difference between the streams is required for heat transfer to occur, the temperature of the cold stream at each location, denoted by the coordinate z, is less than that of the hot stream. In particular, the

regenerator

TAKE NOTE...

The gas turbine of Fig. 9.14 is called *ideal* because flow through the turbine and compressor occurs isentropically and there are no frictional pressure drops. Still, heat transfer between the counterflow streams of the regenerator *is* a source of irreversibility.

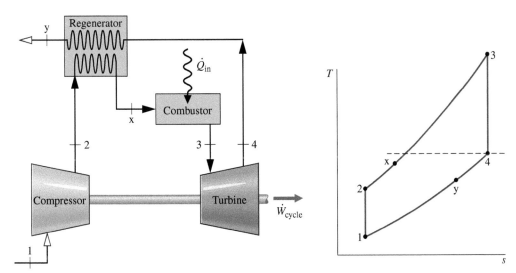

Fig. 9.14 Ideal regenerative air-standard gas turbine cycle.

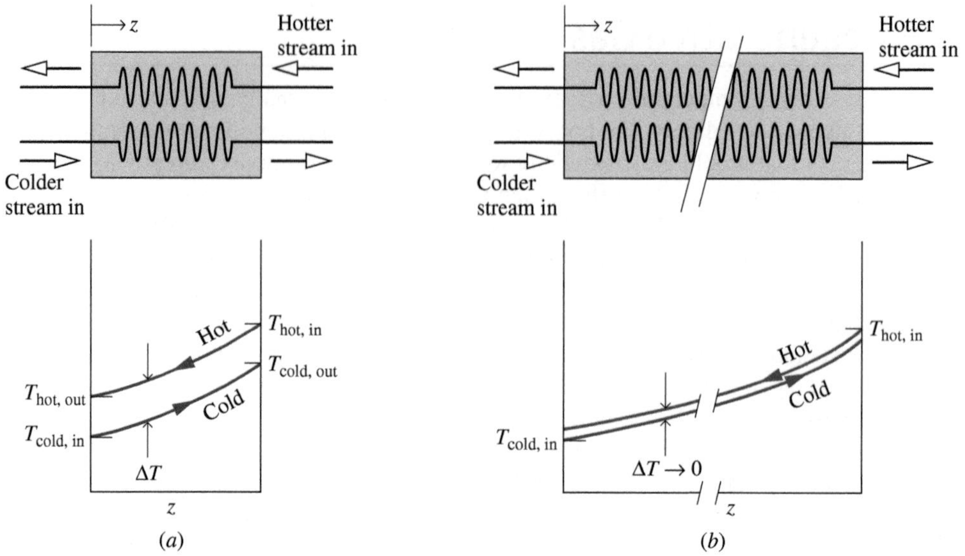

Fig. 9.15 Temperature distributions in counterflow heat exchangers. (*a*) Actual. (*b*) Reversible.

temperature of the colder stream as it exits the heat exchanger is less than the temperature of the incoming hot stream. If the heat transfer area were increased, providing more opportunity for heat transfer between the two streams, there would be a smaller temperature difference at each location.

• In the limiting case of infinite heat transfer area, the temperature difference would approach zero at all locations, as illustrated in Fig. 9.15*b*, and the heat transfer would approach reversibility. In this limit, the exit temperature of the colder stream would approach the temperature of the incoming hot stream. Thus, the highest possible temperature that could be achieved by the colder stream is the temperature of the incoming hot gas.

Referring again to the regenerator of Fig. 9.14, we can conclude from the discussion of Fig. 9.15 that the maximum theoretical value for the temperature T_x is the turbine exhaust temperature T_4, obtained if the regenerator were operating reversibly. The *regenerator effectiveness*, η_{reg}, is a parameter that gauges the departure of an actual regenerator from such an ideal regenerator. The **regenerator effectiveness** is defined as the ratio of the actual enthalpy increase of the air flowing through the compressor side of the regenerator to the maximum theoretical enthalpy increase. That is,

regenerator effectiveness

$$\eta_{reg} = \frac{h_x - h_2}{h_4 - h_2} \tag{9.27}$$

As heat transfer approaches reversibility, h_x approaches h_4 and η_{reg} tends to unity (100%).

In practice, regenerator effectiveness values typically range from 60 to 80%, and thus the temperature T_x of the air exiting on the compressor side of the regenerator is normally well below the turbine exhaust temperature. To increase the effectiveness above this range would require greater heat transfer area, resulting in equipment costs that might cancel any advantage due to fuel savings. Moreover, the greater heat transfer area that would be required for a larger effectiveness can result in a significant frictional pressure drop for flow through the regenerator, thereby affecting overall performance. The decision to add a regenerator is influenced by considerations such as these, and the final decision is primarily an economic one.

In Example 9.7, we analyze an air-standard Brayton cycle with regeneration and explore the effect on thermal efficiency as the regenerator effectiveness varies.

EXAMPLE 9.7 ► ●

Evaluating Thermal Efficiency of a Brayton Cycle with Regeneration

A regenerator is incorporated in the cycle of Example 9.4. **(a)** Determine the thermal efficiency for a regenerator effectiveness of 80%. **(b)** Plot the thermal efficiency versus regenerator effectiveness ranging from 0 to 80%.

SOLUTION

Known   A regenerative gas turbine operates with air as the working fluid. The compressor inlet state, turbine inlet temperature, and compressor pressure ratio are known.

Find   For a regenerator effectiveness of 80%, determine the thermal efficiency. Also plot the thermal efficiency versus the regenerator effectiveness ranging from 0 to 80%.

Schematic and Given Data:

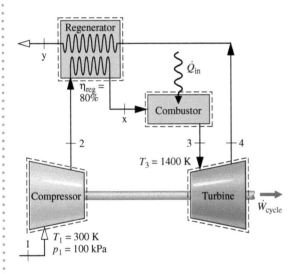

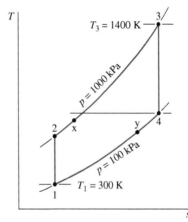

Fig. E9.7a

Solving for h_x

$$h_x = \eta_{reg}(h_4 - h_2) + h_2$$
$$= (0.8)(808.5 - 579.9) + 579.9 = 762.8 \text{ kJ/kg}$$

With the specific enthalpy values determined above, the thermal efficiency is

① $$\eta = \frac{(\dot{W}_t/\dot{m}) - (\dot{W}_c/\dot{m})}{(\dot{Q}_{in}/\dot{m})} = \frac{(h_3 - h_4) - (h_2 - h_1)}{(h_3 - h_x)}$$

$$= \frac{(1515.4 - 808.5) - (579.9 - 300.19)}{(1515.4 - 762.8)}$$

② $$= 0.568 \ (56.8\%)$$

Engineering Model

1. Each component is analyzed as a control volume at steady state. The control volumes are shown on the accompanying sketch by dashed lines.

2. The compressor and turbine processes are isentropic.

3. There are no pressure drops for flow through the heat exchangers.

4. The regenerator effectiveness is 80% in part (a).

5. Kinetic and potential energy effects are negligible.

6. The working fluid is air modeled as an ideal gas.

Analysis

a. The specific enthalpy values at the numbered states on the *T–s* diagram are the same as those in Example 9.4: $h_1 = 300.19$ kJ/kg, $h_2 = 579.9$ kJ/kg, $h_3 = 1515.4$ kJ/kg, $h_4 = 808.5$ kJ/kg.

To find the specific enthalpy h_x, the regenerator effectiveness is used as follows: By definition

$$\eta_{reg} = \frac{h_x - h_2}{h_4 - h_2}$$

b. The *IT* code for the solution follows, where η_{reg} is denoted as etareg, η is eta, $\dot{W}_{comp}/\dot{m}$ is Wcomp, and so on.

```
// Fix the states
T1 = 300//K
p1 = 100//kPa
h1 = h_T("Air", T1)
s1 = s_TP("Air", T1, p1)

p2 = 1000//kPa
s2 = s_TP("Air", T2, p2)
s2 = s1
h2 = h_T("Air", T2)

T3 = 1400//K
p3 = p2
h3 = h_T("Air", T3)
s3 = s_TP("Air", T3, p3)

p4 = p1
s4 = s_TP("Air", T4, p4)
```

```
s4 = s3
h4 = h_T("Air", T4)
etareg = 0.8
hx = etareg*(h4 - h2) + h2

// Thermal efficiency
Wcomp = h2 - h1
Wturb = h3 - h4
Qin = h3 - hx
eta = (Wturb - Wcomp) / Qin
```

Using the **Explore** button, sweep etareg from 0 to 0.8 in steps of 0.01. Then, using the **Graph** button, obtain the following plot:

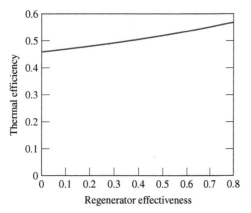

Fig. E9.7b

③ From this plot, we see that the cycle thermal efficiency increases from 0.456, which agrees closely with the result of Example 9.4

(no regenerator), to 0.567 for a regenerator effectiveness of 80%, which agrees closely with the result of part (a). Regenerator effectiveness is seen to have a significant effect on cycle thermal efficiency.

① The values for work per unit of mass flow of the compressor and turbine are unchanged by the addition of the regenerator. Thus, the back work ratio and net work output are not affected by this modification.

② Comparing the present thermal efficiency value with the one determined in Example 9.4, it should be evident that the thermal efficiency can be increased significantly by means of regeneration.

③ The regenerator allows improved fuel utilization to be achieved by transferring a portion of the exergy in the hot turbine exhaust gas to the cooler air flowing on the other side of the regenerator.

SKILLS DEVELOPED

Ability to...

- sketch the schematic of the regenerative gas turbine and the *T–s* diagram for the corresponding air-standard cycle.
- evaluate temperatures and pressures at each principal state and retrieve necessary property data.
- calculate the thermal efficiency.

Quick Quiz

What would be the thermal efficiency if the regenerator effectiveness were 100%? Ans. 60.4%.

9.8 # Regenerative Gas Turbines with Reheat and Intercooling

Two modifications of the basic gas turbine that increase the net work developed are multistage expansion with *reheat* and multistage compression with *intercooling*. When used in conjunction with regeneration, these modifications can result in substantial increases in thermal efficiency. The concepts of reheat and intercooling are introduced in this section.

9.8.1 Gas Turbines with Reheat

For metallurgical reasons, the temperature of the gaseous combustion products entering the turbine must be limited. This temperature can be controlled by providing air in excess of the amount required to burn the fuel in the combustor (see Chap. 13). As a consequence, the gases exiting the combustor contain sufficient air to support the combustion of additional fuel. Some gas turbine power plants take advantage of the excess air by means of a multistage turbine with **reheat** a **reheat combustor** between the stages. With this arrangement the net work per unit of mass flow can be increased. Let us consider reheat from the vantage point of an air-standard analysis.

The basic features of a two-stage gas turbine with reheat are brought out by considering an ideal air-standard Brayton cycle modified as shown in **Fig. 9.16**. After expansion from state 3 to state a in the first turbine, the gas is reheated at constant pressure from state a to state b. The expansion is then completed in the second turbine from state b to state 4. The ideal Brayton cycle

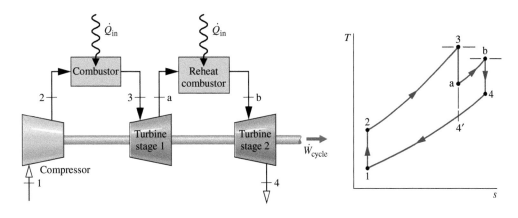

Fig. 9.16 Ideal gas turbine with reheat.

without reheat, 1–2–3–4′–1, is shown on the same T–s diagram for comparison. Because lines of constant pressure on a T–s diagram diverge slightly with increasing entropy, the total work of the two-stage turbine is greater than that of a single expansion from state 3 to state 4′. Thus, the *net* work for the reheat cycle is greater than that of the cycle without reheat. Despite the increase in net work with reheat, the cycle thermal efficiency would not necessarily increase because a greater total heat addition would be required. However, the temperature at the exit of the turbine is higher with reheat than without reheat, so the potential for regeneration is enhanced.

When reheat and regeneration are used together, the thermal efficiency can increase significantly. The following example provides an illustration.

▶ ▶ ▶ **EXAMPLE 9.8** ▶ ·

Determining Thermal Efficiency of a Brayton Cycle with Reheat and Regeneration

Consider a modification of the cycle of Example 9.4 involving reheat and regeneration. Air enters the compressor at 100 kPa, 300 K and is compressed to 1000 kPa. The temperature at the inlet to the first turbine stage is 1400 K. The expansion takes place isentropically in two stages, with reheat to 1400 K between the stages at a constant pressure of 300 kPa. A regenerator having an effectiveness of 100% is also incorporated in the cycle. Determine the thermal efficiency.

SOLUTION

Known An ideal air-standard gas turbine cycle operates with reheat and regeneration. Temperatures and pressures at principal states are specified.

Find Determine the thermal efficiency.

Schematic and Given Data:

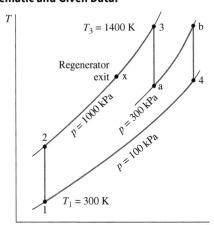

Fig. E9.8

Engineering Model

1. Each component of the power plant is analyzed as a control volume at steady state.

2. The compressor and turbine processes are isentropic.

3. There are no pressure drops for flow through the heat exchangers.

4. The regenerator effectiveness is 100%.

5. Kinetic and potential energy effects are negligible.

6. The working fluid is air modeled as an ideal gas.

Analysis We begin by determining the specific enthalpies at each principal state of the cycle. States 1, 2, and 3 are the same as in Example 9.4: $h_1 = 300.19$ kJ/kg, $h_2 = 579.9$ kJ/kg, $h_3 = 1515.4$ kJ/kg. The temperature at state b is the same as at state 3, so $h_b = h_3$.

Since the first turbine process is isentropic, the enthalpy at state a can be determined using p_r data from Table A-22 and the relationship

$$p_{ra} = p_{r3}\frac{p_a}{p_3} = (450.5)\frac{300}{1000} = 135.15$$

Interpolating in Table A-22, we get $h_a = 1095.9$ kJ/kg.

The second turbine process is also isentropic, so the enthalpy at state 4 can be determined similarly. Thus,

$$p_{r4} = p_{rb}\frac{p_4}{p_b} = (450.5)\frac{100}{300} = 150.17$$

Interpolating in Table A-22, we obtain $h_4 = 1127.6$ kJ/kg. Since the regenerator effectiveness is 100%, $h_x = h_4 = 1127.6$ kJ/kg.

The thermal efficiency calculation must take into account the compressor work, the work of *each* turbine, and the *total* heat added. Thus, on a unit mass basis

$$\eta = \frac{(h_3 - h_a) + (h_b - h_4) - (h_2 - h_1)}{(h_3 - h_x) + (h_b - h_a)}$$

$$= \frac{(1515.4 - 1095.9) + (1515.4 - 1127.6) - (579.9 - 300.19)}{(1515.4 - 1127.6) + (1515.4 - 1095.9)}$$

1 $= 0.654 \ (65.4\%)$

1 Comparing the present value with the thermal efficiency determined in part (a) of Example 9.4, we can conclude that the use of reheat coupled with regeneration can result in a substantial increase in thermal efficiency.

9.8.2 Compression with Intercooling

The net work output of a gas turbine also can be increased by reducing the compressor work input. This can be accomplished by means of multistage compression with intercooling. The present discussion provides an introduction to this subject.

Let us first consider the work input to compressors at steady state, assuming that irreversibilities are absent and changes in kinetic and potential energy from inlet to exit are negligible. The *p–v* diagram of **Fig. 9.17** shows two alternative compression paths from a specified state 1 to a specified final pressure p_2. Path 1–2′ is for an adiabatic compression. Path 1–2 corresponds to a compression with heat transfer *from* the working fluid to the surroundings. The area to the left of each curve equals the *magnitude* of the work per unit mass of the respective process (see Sec. 6.13.2). The smaller area to the left of Process 1–2 indicates that the work of this process is less than for the adiabatic compression from 1 to 2′. This suggests that cooling a gas *during* compression is advantageous in terms of the work-input requirement.

Although cooling a gas *as it is compressed* would reduce the work, a heat transfer rate high enough to effect a significant reduction in work is difficult to achieve in practice. A practical alternative is to separate the work and heat interactions into separate processes by letting compression take place in stages with heat exchangers, called **intercoolers**, cooling the gas between stages. **Figure 9.18** illustrates a two-stage compressor with an intercooler. The accompanying *p–v* and *T–s* diagrams show the states for internally reversible processes:

- Process 1–c is an isentropic compression from state 1 to state c where the pressure is p_i.
- Process c–d is constant-pressure cooling from temperature T_c to T_d.
- Process d–2 is an isentropic compression to state 2.

The work input per unit of mass flow is represented on the *p–v* diagram by shaded area 1–c–d–2–a–b–1. Without intercooling the gas would be compressed isentropically in a single stage from state 1 to state 2′ and the work would be represented by enclosed area 1–2′–a–b–1. The crosshatched area on the *p–v* diagram represents the reduction in work that would be achieved with intercooling.

Some large compressors have several stages of compression with intercooling between stages. The determination of the number of stages and the conditions at which to operate the various intercoolers is a problem in optimization. The use of multistage compression with intercooling in a gas turbine power plant increases the net work developed by reducing the compression work. By itself, though, compression with intercooling would not necessarily increase the thermal efficiency of a gas turbine because the temperature of the air entering the combustor would be reduced (compare temperatures at states 2′ and 2 on the *T–s* diagram of Fig. 9.18). A lower temperature at the combustor inlet would require additional heat transfer to achieve the desired turbine inlet temperature. The lower temperature at the compressor exit

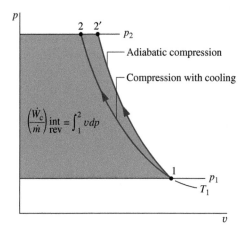

Fig. 9.17 Internally reversible compression processes between two fixed pressures.

intercooler

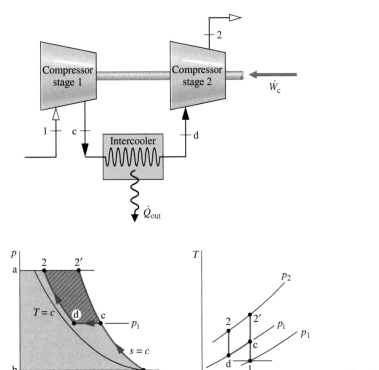

Fig. 9.18 Two-stage compression with intercooling.

enhances the potential for regeneration, however, so when intercooling is used in conjunction with regeneration, an appreciable increase in thermal efficiency can result.

In the next example, we analyze a two-stage compressor with intercooling between the stages. Results are compared with those for a single stage of compression.

► ► ► EXAMPLE 9.9 ►

Evaluating a Two-Stage Compressor with Intercooling

Air is compressed from 100 kPa, 300 K to 1000 kPa in a two-stage compressor with intercooling between stages. The intercooler pressure is 300 kPa. The air is cooled back to 300 K in the intercooler before entering the second compressor stage. Each compressor stage is isentropic. For steady-state operation and negligible changes in kinetic and potential energy from inlet to exit, determine (a) the temperature at the exit of the second compressor stage and (b) the total compressor work input per unit of mass flow. (c) Repeat for a single stage of compression from the given inlet state to the final pressure.

SOLUTION

Known Air is compressed at steady state in a two-stage compressor with intercooling between stages. Operating pressures and temperatures are given.

Find Determine the temperature at the exit of the second compressor stage and the total work input per unit of mass flow. Repeat for a single stage of compression.

Schematic and Given Data:

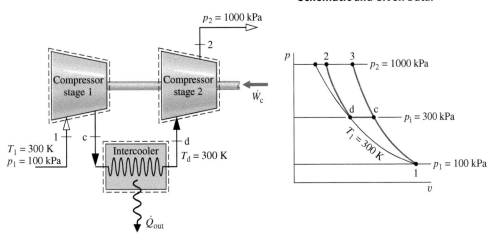

Fig. E9.9

Engineering Model

1. The compressor stages and intercooler are analyzed as control volumes at steady state. The control volumes are shown on the accompanying sketch by dashed lines.

2. The compression processes are isentropic.

3. There is no pressure drop for flow through the intercooler.

4. Kinetic and potential energy effects are negligible.

5. The air is modeled as an ideal gas.

Analysis

a. The temperature at the exit of the second compressor stage, T_2, can be found using the following relationship for the isentropic process d–2:

$$p_{r2} = p_{rd} \frac{p_2}{p_d}$$

With p_{rd} at $T_d = 300$ K from Table A-22, $p_2 = 1000$ kPa, and $p_d = 300$ kPa,

$$p_{r2} = (1.386) \frac{1000}{300} = 4.62$$

Interpolating in Table A-22, we get $T_2 = 422$ K and $h_2 = 423.8$ kJ/kg.

b. The total compressor work input per unit of mass is the sum of the work inputs for the two stages. That is,

$$\frac{\dot{W}_c}{\dot{m}} = (h_c - h_1) + (h_2 - h_d)$$

From Table A-22 at $T_1 = 300$ K, $h_1 = 300.19$ kJ/kg. Since $T_d = T_1$, $h_d = 300.19$ kJ/kg. To find h_c, use p_r data from Table A-22 together with $p_1 = 100$ kPa and $p_c = 300$ kPa to write

$$p_{rc} = p_{r1} \frac{p_c}{p_1} = (1.386) \frac{300}{100} = 4.158$$

Interpolating in Table A-22, we obtain $h_c = 411.3$ kJ/kg. Hence, the total compressor work per unit of mass is

$$\frac{\dot{W}_c}{\dot{m}} = (411.3 - 300.19) + (423.8 - 300.19) = 234.7 \text{ kJ/kg}$$

c. For a single isentropic stage of compression, the exit state would be state 3 located on the accompanying p–v diagram. The temperature at this state can be determined using

$$p_{r3} = p_{r1} \frac{p_3}{p_1} = (1.386) \frac{1000}{100} = 13.86$$

Interpolating in Table A-22, we get $T_3 = 574$ K and $h_3 = 579.9$ kJ/kg. The work input for a single stage of compression is then

$$\frac{\dot{W}_c}{\dot{m}} = h_3 - h_1 = 579.9 - 300.19 = 279.7 \text{ kJ/kg}$$

This calculation confirms that a smaller work input is required with two-stage compression and intercooling than with a single stage of compression. With intercooling, however, a much lower gas temperature is achieved at the compressor exit.

SKILLS DEVELOPED

Ability to…

- sketch the schematic of a two-stage compressor with intercooling between the stages and the corresponding T–s diagram.

- evaluate temperatures and pressures at each principal state and retrieve necessary property data.

- apply energy and entropy balances.

Quick Quiz

In this case, what is the percentage reduction in compressor work with two-stage compression and intercooling compared to a single stage of compression? Ans. 16.1%.

Referring again to Fig. 9.18, the size of the crosshatched area on the p–v diagram representing the reduction in work with intercooling depends on both the temperature T_d at the exit of the intercooler and the intercooler pressure p_i. By properly selecting T_d and p_i, the total work input to the compressor can be minimized. For example, if the pressure p_i is specified, the work input would decrease (crosshatched area would increase) as the temperature T_d approaches T_1, the temperature at the inlet to the compressor. For air entering the compressor from the surroundings, T_1 would be the limiting temperature that could be achieved at state d through heat transfer with the surroundings only. Also, for a specified value of the temperature T_d, the pressure p_i can be selected so that the total work input is a minimum (crosshatched area is a maximum).

Example 9.10 provides an illustration of the determination of the intercooler pressure for minimum total work using a cold air-standard analysis.

► ► ► **EXAMPLE 9.10** ► · · · · · · · · · · · ·

Determining Intercooler Pressure for Minimum Total Compressor Work

For a two-stage compressor with fixed inlet state and exit pressure, conduct a cold air-standard analysis to express in terms of known property values the intercooler pressure required for the minimum total compressor work per unit of mass flowing. Assume steady-state operation and the following idealizations:

Each compression process is isentropic. There is no pressure drop through the intercooler. The temperature at the inlet to the second compressor is greater than, or equal to, that at the inlet to the first compressor. Kinetic and potential energy effects are negligible.

SOLUTION

Known A two-stage compressor with intercooling operates at steady state under specified conditions.

Find Determine the intercooler pressure for minimum total compressor work input, per unit of mass flowing.

Schematic and Given Data:

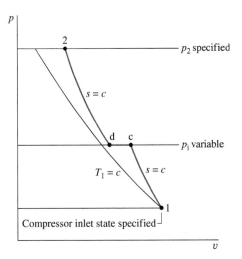

Fig. E9.10

Engineering Model

1. The compressor stages and intercooler are analyzed as control volumes at steady state.

2. The compression processes are isentropic.

3. There is no pressure drop for flow through the intercooler.

4. The temperature at the inlet to the second compressor stage obeys $T_d \geq T_1$.

5. Kinetic and potential energy effects are negligible.

6. The working fluid is air modeled as an ideal gas.

7. The specific heat c_p and thus the specific heat ratio k are constant.

Analysis The total compressor work input per unit of mass flow is

$$\frac{\dot{W}_c}{\dot{m}} = (h_c - h_1) + (h_2 - h_d)$$

Since c_p is constant,

$$\frac{\dot{W}_c}{\dot{m}} = c_p(T_c - T_1) + c_p(T_2 - T_d)$$

$$= c_p T_1\left(\frac{T_c}{T_1} - 1\right) + c_p T_d\left(\frac{T_2}{T_d} - 1\right)$$

Since the compression processes are isentropic and the specific heat ratio k is constant, the pressure and temperature ratios across the compressor stages are related, respectively, by

$$\frac{T_c}{T_1} = \left(\frac{p_i}{p_1}\right)^{(k-1)/k} \quad \text{and} \quad \frac{T_2}{T_d} = \left(\frac{p_2}{p_i}\right)^{(k-1)/k}$$

Collecting results

$$\frac{\dot{W}_c}{\dot{m}} = c_p T_1\left[\left(\frac{p_i}{p_1}\right)^{(k-1)/k} - 1\right] + c_p T_d\left[\left(\frac{p_2}{p_i}\right)^{(k-1)/k} - 1\right]$$

Hence, for specified values of T_1, T_d, p_1, p_2, and c_p, the value of the total compressor work input varies with the intercooler pressure only. To determine the pressure p_i that minimizes the total work, form the derivative

$$\frac{\partial(\dot{W}_c/\dot{m})}{\partial p_i} = c_p T_1\frac{\partial}{\partial p_i}\left(\frac{p_i}{p_1}\right)^{(k-1)/k} + c_p T_d\frac{\partial}{\partial p_i}\left(\frac{p_2}{p_i}\right)^{(k-1)/k}$$

$$= c_p T_1\left(\frac{k-1}{k}\right)\left[\left(\frac{p_i}{p_1}\right)^{-1/k}\left(\frac{1}{p_1}\right) - \frac{T_d}{T_1}\left(\frac{p_2}{p_i}\right)^{-1/k}\left(\frac{p_2}{p_i^2}\right)\right]$$

$$= c_p T_1\left(\frac{k-1}{k}\right)\frac{1}{p_i}\left[\left(\frac{p_i}{p_1}\right)^{(k-1)/k} - \frac{T_d}{T_1}\left(\frac{p_2}{p_i}\right)^{(k-1)/k}\right]$$

Setting the partial derivative to zero, we get

$$\frac{p_i}{p_1} = \left(\frac{p_2}{p_i}\right)\left(\frac{T_d}{T_1}\right)^{k/(k-1)} \tag{a}$$

Alternatively,

❶ $$p_i = \sqrt{p_1 p_2\left(\frac{T_d}{T_1}\right)^{k/(k-1)}} \tag{b}$$

By checking the sign of the second derivative, it can be verified that the total compressor work is a minimum.

---

❶ Observe that when $T_d = T_1$, $p_i = \sqrt{p_1 p_2}$.

SKILLS DEVELOPED

Ability to…

• complete the detailed derivation of a thermodynamic expression.

• use calculus to minimize a function.

Quick Quiz

If $p_1 = 1$ bar, $p_2 = 12$ bar, $T_d = T_1 = 300$ K, and $k = 1.4$, determine the intercooler pressure for minimum total compressor work, in bar, and the accompanying temperature at the exit of each compressor stage, in K. Ans. 3.46 bar, 428 K.

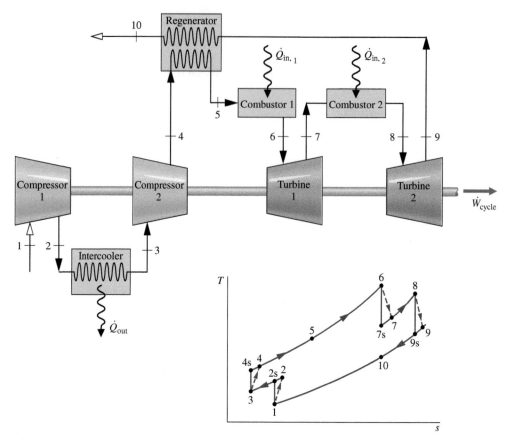

Fig. 9.19 Regenerative gas turbine with intercooling and reheat.

9.8.3 Reheat and Intercooling

Reheat between turbine stages and intercooling between compressor stages provide two important advantages: The net work output is increased, and the potential for regeneration is enhanced. Accordingly, when reheat and intercooling are used together with regeneration, a substantial improvement in performance can be realized. One arrangement incorporating reheat, intercooling, and regeneration is shown in Fig. 9.19. This gas turbine has two stages of compression and two turbine stages. The accompanying T–s diagram is drawn to indicate irreversibilities in the compressor and turbine stages. The pressure drops that would occur as the working fluid passes through the intercooler, regenerator, and combustors are not shown.

Example 9.11 illustrates the analysis of a regenerative gas turbine with intercooling and reheat.

▶▶▶ EXAMPLE 9.11 ▶ •

Analyzing a Regenerative Gas Turbine with Intercooling and Reheat

A regenerative gas turbine with intercooling and reheat operates at steady state. Air enters the compressor at 100 kPa, 300 K with a mass flow rate of 5.807 kg/s. The pressure ratio across the two-stage compressor is 10. The pressure ratio across the two-stage turbine is also 10. The intercooler and reheater each operate at 300 kPa. At the inlets to the turbine stages, the temperature is 1400 K. The temperature at the inlet to the second compressor stage is 300 K. The isentropic efficiency of each compressor and turbine stage is 80%. The regenerator effectiveness is 80%. Determine **(a)** the thermal efficiency, **(b)** the back work ratio, **(c)** the net

power developed, in kW, **(d)** the total rate energy is added by heat transfer, in kW.

SOLUTION

Known An air-standard regenerative gas turbine with intercooling and reheat operates at steady state. Operating pressures and temperatures are specified. Turbine and compressor isentropic efficiencies are given and the regenerator effectiveness is known.

Find   Determine the thermal efficiency, back work ratio, net power developed, in kW, and total rate energy is added by heat transfer, in kW.

Schematic and Given Data:

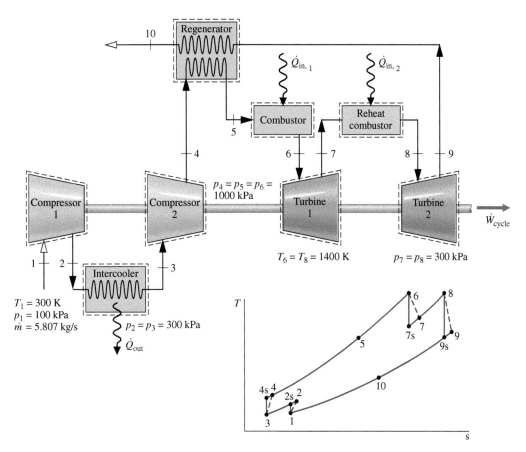

$T_1 = 300$ K
$p_1 = 100$ kPa
$\dot{m} = 5.807$ kg/s

$p_2 = p_3 = 300$ kPa

$p_4 = p_5 = p_6 = 1000$ kPa

$T_6 = T_8 = 1400$ K

$p_7 = p_8 = 300$ kPa

Fig. E9.11

Engineering Model

1. Each component is analyzed as a control volume at steady state. The control volumes are shown on the accompanying sketch by dashed lines.

2. There are no pressure drops for flow through the heat exchangers.

3. The compressor and turbine are adiabatic.

4. Kinetic and potential energy effects are negligible.

5. The working fluid is air modeled as an ideal gas.

Analysis   Specific enthalpy values at states shown in Fig. E9.11 are provided in the following table. See note ❶ for discussion.

State	h (kJ/kg)	State	h (kJ/kg)
1	300.19	6	1515.4
2s	411.3	7s	1095.9
2	439.1	7	1179.8
3	300.19	8	1515.4
4s	423.8	9s	1127.6
4	454.7	9	1205.2
5	1055.1		

a. The thermal efficiency must take into account the work of both turbine stages, the work of both compressor stages, and the total heat added. The total turbine work per unit of mass flow is

$$\frac{\dot{W}_t}{\dot{m}} = (h_6 - h_7) + (h_8 - h_9)$$

$$= (1515.4 - 1179.8) + (1515.4 - 1205.2) = 645.8 \text{ kJ/kg}$$

The total compressor work input per unit of mass flow is

$$\frac{\dot{W}_c}{\dot{m}} = (h_2 - h_1) + (h_4 - h_3)$$

$$= (439.1 - 300.19) + (454.7 - 300.19) = 293.4 \text{ kJ/kg}$$

The total heat added per unit of mass flow is

$$\frac{\dot{Q}_{in}}{\dot{m}} = (h_6 - h_5) + (h_8 - h_7)$$

$$= (1515.4 - 1055.1) + (1515.4 - 1179.8) = 795.9 \text{ kJ/kg}$$

Calculating the thermal efficiency

$$\eta = \frac{645.8 - 293.4}{795.9} = 0.443 \ (44.3\%)$$

b. The back work ratio is

$$\text{bwr} = \frac{\dot{W}_c/\dot{m}}{\dot{W}_t/\dot{m}} = \frac{293.4}{645.8} = 0.454 \ (45.4\%)$$

c. The net power developed is

$$\dot{W}_{\text{cycle}} = \dot{m}(\dot{W}_t/\dot{m} - \dot{W}_c/\dot{m})$$

②
$$= \left(5.807\frac{\text{kg}}{\text{s}}\right)(645.8 - 293.4)\frac{\text{kJ}}{\text{kg}}\left|\frac{1\,\text{kW}}{1\,\text{kJ/s}}\right| = 2046\,\text{kW}$$

d. The total rate energy is added by heat transfer is obtained using the specified mass flow rate and data from part (a)

$$\dot{Q}_{\text{in}} = \dot{m}(\dot{Q}_{\text{in}}/\dot{m})$$

③
$$= \left(5.807\frac{\text{kg}}{\text{s}}\right)\left(795.9\frac{\text{kJ}}{\text{kg}}\right)\left|\frac{1\,\text{kW}}{1\,\text{kJ/s}}\right| = 4622\,\text{kW}$$

---

① The enthalpies at states 1, 2s, 3, and 4s are obtained from the solution to Example 9.9 where these states are designated as 1, c, d, and 2, respectively. Thus, $h_1 = h_3 = 300.19$ kJ/kg, $h_{2s} = 411.3$ kJ/kg, $h_{4s} = 423.8$ kJ/kg.

The specific enthalpies at states 6, 7s, 8, and 9s are obtained from the solution to Example 9.8, where these states are designated as 3, a, b, and 4, respectively. Thus, $h_6 = h_8 = 1515.4$ kJ/kg, $h_{7s} = 1095.9$ kJ/kg, $h_{9s} = 1127.6$ kJ/kg.

Specific enthalpies at states 2 and 4 are obtained using the isentropic efficiency of the first and second compressor stages, respectively. Specific enthalpy at state 5 is obtained using the regenerator effectiveness. Finally, specific enthalpies at states 7

and 9 are obtained using the isentropic efficiency of the first and second turbine stages, respectively.

② Comparing the thermal efficiency, back work ratio, and net power values of the current example with the corresponding values of Example 9.6, it should be evident that gas turbine power plant performance can be increased significantly by coupling reheat and intercooling with regeneration.

③ With the results of parts (c) and (d), we get $\eta = 0.443$, which agrees with the value obtained in part (a), as expected. Since the mass flow rate is constant throughout the system, the thermal efficiency can be calculated alternatively using energy transfers on a per unit mass of air flowing basis, in kJ/kg, or on a time rate basis, in kW.

SKILLS DEVELOPED

Ability to...

- sketch the schematic of the regenerative gas turbine with intercooling and reheat and the *T–s* diagram for the corresponding air-standard cycle.

- evaluate temperatures and pressures at each principal state and retrieve necessary property data.

- calculate the thermal efficiency, back work ratio, net power developed, and total rate energy is added by heat transfer.

Quick Quiz

Verify the specific enthalpy values specified at states 4, 5, and 9 in the table of data provided.

9.8.4 Ericsson and Stirling Cycles

As illustrated by Example 9.11, significant increases in the thermal efficiency of gas turbine power plants can be achieved through intercooling, reheat, and regeneration. There is an economic limit to the number of stages that can be employed, and normally there would be no more than two or three. Nonetheless, it is instructive to consider the situation where the number of stages of both intercooling and reheat becomes indefinitely large.

Ericsson Cycle Figure 9.20*a* shows an *ideal* regenerative gas turbine cycle with several stages of compression and expansion and a regenerator whose effectiveness is 100%. As in Fig. 9.8*b*, this is a *closed* gas turbine cycle. Each intercooler is assumed to return the working fluid to the temperature T_C at the inlet to the first compression stage, state 1, and each reheater restores the working fluid to the temperature T_H at the inlet to the first turbine stage, state 3. The regenerator allows the heat input for Process 2–3 to be obtained from the heat rejected in Process 4–1. Accordingly, all the heat added *externally* occurs in the reheaters, and all the heat rejected to the surroundings takes place in the intercoolers.

In the limit, as an infinite number of reheat and intercooler stages is employed, all heat added occurs while the working fluid expands at its highest temperature, T_H, and all heat rejected takes place while the working fluid is compressed at its lowest temperature, T_C. The limiting cycle, shown in **Fig. 9.20*b***, is called the **Ericsson cycle**.

Ericsson cycle

Since irreversibilities are presumed absent and all heat is supplied and rejected isothermally, the thermal efficiency of the Ericsson cycle equals that of *any* reversible power cycle operating with heat addition at the temperature T_H and heat rejection at the temperature T_C: $\eta_{\text{max}} = 1 - T_C/T_H$. This expression is applied in Secs. 5.10 and 6.6 to evaluate the thermal efficiency of Carnot power cycles. Although the details of the Ericsson cycle differ from those

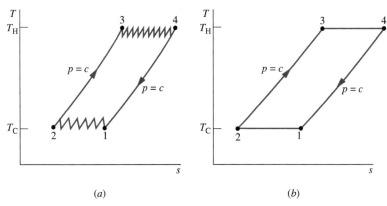

Fig. 9.20 Ericsson cycle as a limit of ideal gas turbine operation using multistage compression with intercooling, multistage expansion with reheating, and regeneration.

of the Carnot cycle, both cycles have the same value of thermal efficiency when operating between the temperatures T_H and T_C.

Stirling Cycle   Another cycle that employs a regenerator is the *Stirling* cycle, shown on the p–v and T–s diagrams of Fig. 9.21. The cycle consists of four internally reversible processes in series: isothermal compression from state 1 to state 2 at temperature T_C, constant-volume heating from state 2 to state 3, isothermal expansion from state 3 to state 4 at temperature T_H, and constant-volume cooling from state 4 to state 1 to complete the cycle.

A regenerator whose effectiveness is 100% allows the heat rejected during Process 4–1 to provide the heat input in Process 2–3. Accordingly, all the heat added to the working fluid externally takes place in the isothermal process 3–4 and all the heat rejected to the surroundings occurs in the isothermal process 1–2.

It can be concluded, therefore, that the thermal efficiency of the Stirling cycle is given by the same expression as for the Carnot and Ericsson cycles. Since all three cycles are *reversible*, we can imagine them as being executed in various ways, including use of gas turbines and piston–cylinder engines. In each embodiment, however, practical issues prevent it from actually being realized.

Stirling Engine   The Ericsson and Stirling cycles are principally of theoretical interest as examples of cycles that exhibit the same thermal efficiency as the Carnot cycle. However, a practical engine of the piston–cylinder type that operates on a *closed* regenerative cycle having features in common with the Stirling cycle has been under study for years. This engine is known as a **Stirling engine**. The Stirling engine offers the opportunity for high efficiency together with reduced emissions from combustion products because combustion takes place externally and not within the cylinder as for internal combustion engines. In the Stirling engine, energy is transferred to the working fluid from products of combustion, which are kept separate. It is an *external* combustion engine.

Stirling engine

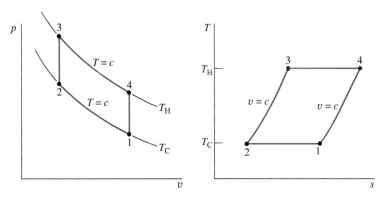

Fig. 9.21 p–v and T–s diagrams of the Stirling cycle.

9.9 Gas Turbine–Based Combined Cycles

In this section, gas turbine–based combined cycles are considered for power generation. Cogeneration, including district heating, is also considered. These discussions complement those of Sec. 8.5, where vapor power systems performing similar functions are introduced.

The present applications build on recognizing that the exhaust gas temperature of a simple gas turbine is typically well above ambient temperature and thus hot gas exiting the turbine has significant thermodynamic utility that might be harnessed economically. This observation provides the basis for the regenerative gas turbine cycle introduced in Sec. 9.7 and for the current applications.

9.9.1 Combined Gas Turbine–Vapor Power Cycle

A combined cycle couples two power cycles such that the energy discharged by heat transfer from one cycle is used partly or wholly as the heat input for the other cycle. This is illustrated by the combined cycle involving gas and vapor power turbines shown in **Fig. 9.22**. The gas and vapor power cycles are combined using an interconnecting heat-recovery steam generator that serves as the boiler for the vapor power cycle.

The combined cycle has the gas turbine's high average temperature of heat addition and the vapor cycle's low average temperature of heat rejection and, thus, a thermal efficiency greater than either cycle would have individually. For many applications combined cycles are a good choice, and they are increasingly being used worldwide for electric power generation.

With reference to Fig. 9.22, the thermal efficiency of the combined cycle is

$$\eta = \frac{\dot{W}_{gas} + \dot{W}_{vap}}{\dot{Q}_{in}} \tag{9.28}$$

where $\dot{W}_{gas}$ is the *net* power developed by the gas turbine and $\dot{W}_{vap}$ is the *net* power developed by the vapor cycle. $\dot{Q}_{in}$ denotes the *total* rate of heat transfer to the combined cycle, including additional heat transfer, if any, to superheat the vapor entering the vapor turbine. The

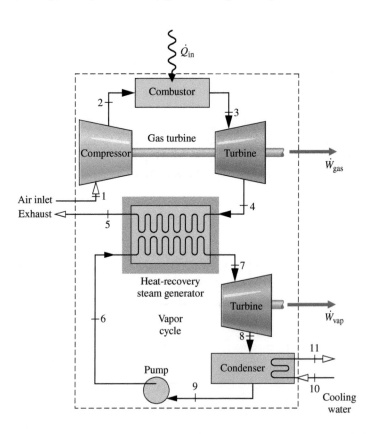

Fig. 9.22 Combined gas turbine–vapor power plant.

evaluation of the quantities appearing in Eq. 9.28 follows the procedures described in the sections on vapor cycles and gas turbines.

The relation for the energy transferred from the gas cycle to the vapor cycle for the system of Fig. 9.22 is obtained by applying the mass and energy rate balances to a control volume enclosing the heat-recovery steam generator. For steady-state operation, negligible heat transfer with the surroundings, and no significant changes in kinetic and potential energy, the result is

$$\dot{m}_v(h_7 - h_6) = \dot{m}_g(h_4 - h_5) \tag{9.29}$$

where $\dot{m}_g$ and $\dot{m}_v$ are the mass flow rates of the gas and vapor, respectively.

As witnessed by relations such as Eqs. 9.28 and 9.29, combined cycle performance can be analyzed using mass and energy balances. To complete the analysis, however, the second law is required to assess the impact of irreversibilities and the true magnitudes of losses. Among the irreversibilities, the most significant is the exergy destroyed by combustion. About 30% of the exergy entering the combustor with the fuel is destroyed by combustion irreversibility. An analysis of the gas turbine on an air-standard basis does not allow this exergy destruction to be evaluated, however, and means introduced in Chap. 13 must be applied for this purpose.

Energy & Environment

Advanced combined-cycle H-class power plants capable of achieving the long-elusive 60% combined-cycle thermal efficiency level are a reality. H-class power plants integrate a gas turbine, steam turbine, steam generator, and heat-recovery steam generator. They are capable of net power outputs up to 600 MW, while significantly saving fuel, reducing carbon dioxide emissions, and meeting low nitric oxide standards.

Before the H-class breakthrough, gas turbine manufacturers had struggled against a temperature-imposed barrier that limited thermal efficiency for gas turbine–based power systems. For years, the barrier was a gas turbine inlet temperature of about 1260°C (2300°F). Above that level, available cooling technologies were unable to protect turbine blades and other key components from thermal degradation. Since higher temperatures go hand-in-hand with higher thermal efficiencies, the perceived temperature barrier limited the efficiency achievable.

Two developments were instrumental in allowing combined-cycle thermal efficiencies of 60% or more: steam cooling of both stationary and rotating blades and blades made from single crystals.

- In steam cooling, relatively low-temperature steam generated in the companion vapor power plant is fed to channels in the blades of the high-temperature stages of the gas turbine, thereby cooling the blades while producing superheated steam for use in the vapor plant, adding to overall cycle efficiency. Innovative coatings, typically ceramic composites, also help components withstand very high gas temperatures.

- H-class gas turbines also have *single-crystal* blades. Conventionally cast blades are *poly*crystalline. They consist of a multitude of small *grains* (crystals) with interfaces between the grains called grain boundaries. Adverse physical events such as corrosion and *creep* originating at grain boundaries greatly shorten blade life and impose limits on allowed turbine temperatures. Having no grain boundaries, single-crystal blades are far more durable and less prone to thermal degradation.

Not content with their achievements, major gas turbine manufacturers are striving to reach even higher turbine inlet temperatures.

The next example illustrates the use of mass and energy balances, the second law, and property data to analyze combined cycle performance.

▶ **EXAMPLE 9.12** ▶

Energy and Exergy Analyses of a Combined Gas Turbine-Vapor Power Plant

A combined gas turbine–vapor power plant has a net power output of 45 MW. Air enters the compressor of the gas turbine at 100 kPa, 300 K, and is compressed to 1200 kPa. The isentropic efficiency of the compressor is 84%. The condition at the inlet to the turbine is 1200 kPa, 1400 K. Air expands through the turbine, which has an isentropic efficiency of 88%, to a pressure of 100 kPa. The air then passes through the interconnecting heat-recovery steam generator and is finally discharged at 400 K. Steam enters the turbine of the vapor power cycle at 8 MPa, 400°C, and expands to the condenser pressure of 8 kPa. Water enters the pump as saturated liquid at 8 kPa. The turbine and pump of the vapor cycle have isentropic efficiencies of 90 and 80%, respectively.

a. Determine the mass flow rates of the air and the steam, each in kg/s; the net power developed by the gas turbine and vapor power cycle, each in MW; and the thermal efficiency.

b. Develop a full accounting of the *net* rate of exergy increase as the air passes through the gas turbine combustor. Discuss.

Let $T_0 = 300$ K, $p_0 = 100$ kPa.

SOLUTION

Known   A combined gas turbine–vapor power plant operates at steady state with a known net power output. Operating pressures and temperatures are specified. Turbine, compressor, and pump efficiencies are also given.

Find   Determine the mass flow rate of each working fluid, in kg/s; the net power developed by each cycle, in MW; and the thermal efficiency. Develop a full accounting of the exergy increase of the air passing through the gas turbine combustor and discuss the results.

Schematic and Given Data:

4. There are no pressure drops for flow through the combustor, heat-recovery steam generator, and condenser.

5. An air-standard analysis is used for the gas turbine.

6. $T_0 = 300$ K, $p_0 = 100$ kPa.

Analysis   The property data given in the table below are determined using procedures illustrated in previously solved examples of Chaps. 8 and 9. The details are left as an exercise.

Gas Turbine		
State	h (kJ/kg)	s° (kJ/kg · K)
1	300.19	1.7020
2	669.79	2.5088
3	1515.42	3.3620
4	858.02	2.7620
5	400.98	1.9919

Vapor Cycle		
State	h (kJ/kg)	s (kJ/kg · K)
6	183.96	0.5975
7	3138.30	6.3634
8	2104.74	6.7282
9	173.88	0.5926

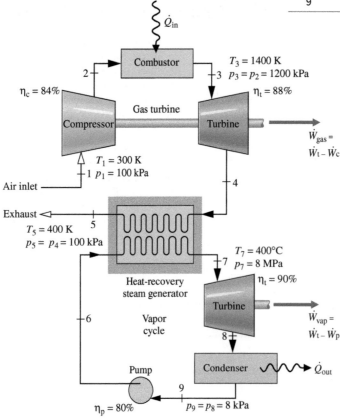

Fig. E9.12

Engineering Model

1. Each component on the accompanying sketch is analyzed as a control volume at steady state.

2. The turbines, compressor, pump, and interconnecting heat-recovery steam generator operate adiabatically.

3. Kinetic and potential energy effects are negligible.

Energy Analysis

a. To determine the mass flow rates of the vapor, $\dot{m}_v$, and the air, $\dot{m}_g$, begin by applying mass and energy rate balances to the interconnecting heat-recovery steam generator to obtain

$$0 = \dot{m}_g(h_4 - h_5) + \dot{m}_v(h_6 - h_7)$$

or

$$\frac{\dot{m}_{\mathrm{v}}}{\dot{m}_{\mathrm{g}}} = \frac{h_4 - h_5}{h_7 - h_6} = \frac{858.02 - 400.98}{3138.3 - 183.96} = 0.1547$$

Mass and energy rate balances applied to the gas turbine and vapor power cycles give the net power developed by each, respectively,

$$\dot{W}_{\mathrm{gas}} = \dot{m}_{\mathrm{g}}[(h_3 - h_4) - (h_2 - h_1)]$$
$$\dot{W}_{\mathrm{vap}} = \dot{m}_{\mathrm{v}}[(h_7 - h_8) - (h_6 - h_9)]$$

With $\dot{W}_{\mathrm{net}} = \dot{W}_{\mathrm{gas}} + \dot{W}_{\mathrm{vap}}$

$$\dot{W}_{\mathrm{net}} = \dot{m}_{\mathrm{g}}\left\{[(h_3 - h_4) - (h_2 - h_1)] + \frac{\dot{m}_{\mathrm{v}}}{\dot{m}_{\mathrm{g}}}[(h_7 - h_8) - (h_6 - h_9)]\right\}$$

Solving for $\dot{m}_{\mathrm{g}}$, and inserting $\dot{W}_{\mathrm{net}} = 45 \text{ MW} = 45{,}000 \text{ kJ/s}$ and $\dot{m}_{\mathrm{v}}/\dot{m}_{\mathrm{a}} = 0.1547$, we get

$$\dot{m}_{\mathrm{g}} = \frac{45{,}000 \text{ kJ/s}}{\{[(1515.42 - 858.02) - (669.79 - 300.19)] + 0.1547[(3138.3 - 2104.74) - (183.96 - 173.88)]\} \text{ kJ/kg}}$$

$$= 100.87 \text{ kg/s}$$

and

$$\dot{m}_{\mathrm{v}} = (0.1547)\dot{m}_{\mathrm{g}} = 15.6 \text{ kg/s}$$

Using these mass flow rate values and specific enthalpies from the table above, the net power developed by the gas turbine and vapor power cycles, respectively, is

$$\dot{W}_{\mathrm{gas}} = \left(100.87 \frac{\text{kg}}{\text{s}}\right)\left(287.8 \frac{\text{kJ}}{\text{kg}}\right)\left|\frac{1 \text{ MW}}{10^3 \text{ kJ/s}}\right| = 29.03 \text{ MW}$$

$$\dot{W}_{\mathrm{vap}} = \left(15.6 \frac{\text{kg}}{\text{s}}\right)\left(1023.5 \frac{\text{kJ}}{\text{kg}}\right)\left|\frac{1 \text{ MW}}{10^3 \text{ kJ/s}}\right| = 15.97 \text{ MW}$$

The thermal efficiency is given by Eq. 9.28. The net power output is specified in the problem statement as 45 MW. Thus, only $\dot{Q}_{\mathrm{in}}$ must be determined. Applying mass and energy rate balances to the combustor, we get

$$\dot{Q}_{\mathrm{in}} = \dot{m}_{\mathrm{g}}(h_3 - h_2)$$

$$= \left(100.87 \frac{\text{kg}}{\text{s}}\right)(1515.42 - 669.79)\frac{\text{kJ}}{\text{kg}}\left|\frac{1 \text{ MW}}{10^3 \text{ kJ/s}}\right|$$

$$= 85.3 \text{ MW}$$

① Finally, thermal efficiency is

$$\eta = \frac{45 \text{ MW}}{85.3 \text{ MW}} = 0.528 \ (52.8\%)$$

Exergy Analysis

b. The *net* rate of exergy increase of the air passing through the combustor is (Eq. 7.18)

$$\dot{E}_{\mathrm{f3}} - \dot{E}_{\mathrm{f2}} = \dot{m}_{\mathrm{g}}[h_3 - h_2 - T_0(s_3 - s_2)]$$
$$= \dot{m}_{\mathrm{g}}[h_3 - h_2 - T_0(s_3^\circ - s_2^\circ - R \ln p_3/p_2)]$$

With assumption 4, we have

$$\dot{E}_{\mathrm{f3}} - \dot{E}_{\mathrm{f2}} = \dot{m}_{\mathrm{g}}\left[h_3 - h_2 - T_0\left(s_3^\circ - s_2^\circ - R \ln \overset{0}{\frac{p_3}{p_1}}\right)\right]$$

$$= \left(100.87 \frac{\text{kJ}}{\text{s}}\right)\left[(1515.42 - 669.79)\frac{\text{kJ}}{\text{kg}}\right.$$

$$\left. - 300 \text{ K} (3.3620 - 2.5088)\frac{\text{kJ}}{\text{kg} \cdot \text{K}}\right]$$

$$= 59{,}480 \frac{\text{kJ}}{\text{s}}\left|\frac{1 \text{ MW}}{10^3 \text{ kJ/s}}\right| = 59.48 \text{ MW}$$

The *net* rate exergy is carried out of the plant by the exhaust air stream at 5 is

$$\dot{E}_{\mathrm{f5}} - \dot{E}_{\mathrm{f1}} = \dot{m}_{\mathrm{g}}\left[h_5 - h_1 - T_0\left(s_5^\circ - s_1^\circ - R \ln \overset{0}{\frac{p_5}{p_1}}\right)\right]$$

$$= \left(100.87 \frac{\text{kg}}{\text{s}}\right)[(400.98 - 300.19)$$

$$- 300(1.9919 - 1.7020)]\left(\frac{\text{kJ}}{\text{kg}}\right)\left|\frac{1 \text{ MW}}{10^3 \text{ kJ/s}}\right|$$

$$= 1.39 \text{ MW}$$

The *net* rate exergy is carried out of the plant as water passes through the condenser is

$$\dot{E}_{\mathrm{f8}} - \dot{E}_{\mathrm{f9}} = \dot{m}_{\mathrm{v}}[h_8 - h_9 - T_0(s_8 - s_9)]$$

$$= \left(15.6 \frac{\text{kg}}{\text{s}}\right)\left[(2104.74 - 173.88)\frac{\text{kJ}}{\text{kg}}\right.$$

$$\left. - 300 \text{ K}(6.7282 - 0.5926)\frac{\text{kJ}}{\text{kg} \cdot \text{K}}\right]\left|\frac{1 \text{ MW}}{10^3 \text{ kJ/s}}\right|$$

$$= 1.41 \text{ MW}$$

The rates of exergy destruction for the air turbine, compressor, steam turbine, pump, and heat-recovery steam generator are evaluated using $\dot{E}_{\mathrm{d}} = T_0 \dot{\sigma}_{\mathrm{cv}}$, respectively, as follows:

Air turbine:

② $\dot{E}_{\mathrm{d}} = \dot{m}_{\mathrm{g}} T_0(s_4 - s_3)$

$$= \dot{m}_{\mathrm{g}} T_0(s_4^\circ - s_3^\circ - R \ln p_4/p_3)$$

$$= \left(100.87 \frac{\text{kg}}{\text{s}}\right)(300 \text{ K})\left[(2.7620 - 3.3620)\frac{\text{kJ}}{\text{kg} \cdot \text{K}}\right.$$

$$\left. - \left(\frac{8.314}{28.97} \frac{\text{kJ}}{\text{kg} \cdot \text{K}}\right)\ln\left(\frac{100}{1200}\right)\right]\left|\frac{1 \text{ MW}}{10^3 \text{ kJ/s}}\right|$$

$$= 3.42 \text{ MW}$$

Compressor:

$$\dot{E}_d = \dot{m}_g T_0(s_2 - s_1)$$

$$= \dot{m}_g T_0(s_2^\circ - s_1^\circ - R \ln p_2/p_1)$$

$$= (100.87)(300)\left[(2.5088 - 1.7020) - \frac{8.314}{28.97}\ln\left(\frac{1200}{100}\right)\right]\left|\frac{1}{10^3}\right|$$

$$= 2.83 \text{ MW}$$

Steam turbine:

$$\dot{E}_d = \dot{m}_v T_0(s_8 - s_7)$$

$$= (15.6)(300)(6.7282 - 6.3634)\left|\frac{1}{10^3}\right|$$

$$= 1.71 \text{ MW}$$

Pump:

$$\dot{E}_d = \dot{m}_v T_0(s_6 - s_9)$$

$$= (15.6)(300)(0.5975 - 0.5926)\left|\frac{1}{10^3}\right|$$

$$= 0.02 \text{ MW}$$

Heat-recovery steam generator:

$$\dot{E}_d = T_0[\dot{m}_g(s_5 - s_4) + \dot{m}_v(s_7 - s_6)]$$

$$= (300 \text{ K})\left[\left(100.87\frac{\text{kg}}{\text{s}}\right)(1.9919 - 2.7620)\frac{\text{kJ}}{\text{kg}\cdot\text{K}}\right.$$

$$\left. + \left(15.6\frac{\text{kg}}{\text{s}}\right)(6.3634 - 0.5975)\frac{\text{kJ}}{\text{kg}\cdot\text{K}}\right]\left|\frac{1 \text{ MW}}{10^3 \text{ kJ/s}}\right|$$

$$= 3.68 \text{ MW}$$

❸ The results are summarized by the following exergy rate *balance sheet* in terms of exergy magnitudes on a rate basis:

Net exergy increase of the gas passing through the combustor:	59.48 MW	100%	(70%)*
Disposition of the exergy:			
• Net power developed			
gas turbine cycle	29.03 MW	48.8%	(34.2%)
vapor cycle	15.97 MW	26.8%	(18.8%)
Subtotal	45.00 MW	75.6%	(53.0%)
• Net exergy lost			
with exhaust gas at state 5	1.39 MW	2.3%	(1.6%)
from water passing through condenser	1.41 MW	2.4%	(1.7%)
• Exergy destruction			
air turbine	3.42 MW	5.7%	(4.0%)
compressor	2.83 MW	4.8%	(3.4%)
steam turbine	1.71 MW	2.9%	(2.0%)
pump	0.02 MW	—	—
heat-recovery steam generator	3.68 MW	6.2%	(4.3%)

*Estimation based on fuel exergy. For discussion, see note 3.

The subtotals given in the table under the *net power developed* heading indicate that the combined cycle is effective in generating power from the exergy supplied. The table also indicates the relative significance of the exergy destructions in the turbines, compressor, pump, and heat-recovery steam generator, as well as the relative significance of the exergy losses. Finally, the table indicates that the total of the exergy destructions overshadows the losses. While the energy analysis of part (a) yields valuable results about combined-cycle performance, the exergy analysis of part (b) provides insights about the effects of irreversibilities and true magnitudes of the losses that cannot be obtained using just energy.

---

❶ For comparison, note that the combined-cycle thermal efficiency in this case is much greater than those of the stand-alone regenerative vapor and gas cycles considered in Examples 8.5 and 9.11, respectively.

❷ The development of the appropriate expressions for the rates of entropy production in the turbines, compressor, pump, and heat-recovery steam generator is left as an exercise.

❸ In this exergy balance sheet, the percentages shown in parentheses are estimates based on the fuel exergy. Although combustion is the most significant source of irreversibility, the exergy destruction due to combustion cannot be evaluated using an air-standard analysis. Calculations of exergy destruction due to combustion (Chap. 13) reveal that approximately 30% of the exergy entering the combustor with the fuel would be destroyed, leaving about 70% of the fuel exergy for subsequent use. Accordingly, the value 59.48 MW for the net exergy increase of the air passing through the combustor is assumed to be 70% of the fuel exergy supplied. All other percentages in parentheses are obtained by multiplying the corresponding percentages, based on the exergy increase of the air passing through the combustor, by the factor 0.7. Since they account for combustion irreversibility, the table values in parentheses give the more accurate picture of combined cycle performance.

SKILLS DEVELOPED

Ability to...

- apply mass and energy balances.
- determine thermal efficiency.
- evaluate exergy quantities.
- develop an exergy accounting.

Quick Quiz

Determine the *net* rate energy is carried out of the plant by heat transfer as water passes through the condenser, in MW, and comment. Ans. 30.12 MW. The significance of this energy loss is *far less* than indicated by the answer. In terms of exergy, the loss at the condenser is 1.41 MW [see part (b)], which better measures the limited utility of the relatively low-temperature water flowing through the condenser.

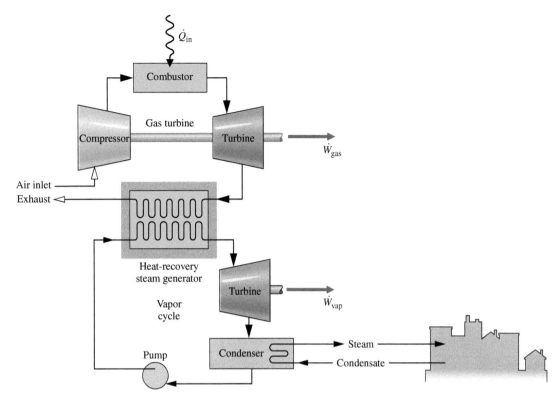

Fig. 9.23 Combined-cycle district heating plant.

9.9.2 Cogeneration

Cogeneration systems are integrated systems that yield two valuable products simultaneously from a single fuel input, electricity and steam (or hot water), achieving cost savings. Cogeneration systems have numerous industrial and commercial applications. District heating is one of these.

District heating plants are located within communities to provide steam or hot water for space heating and other needs together with electricity for domestic, commercial, and industrial use. Vapor cycle–based district heating plants are considered in Sec. 8.5.

Building on the combined gas turbine–vapor power cycle introduced in Sec. 9.9.1, Fig. 9.23 illustrates a district heating system consisting of a gas turbine cycle partnered with a vapor power cycle operating in the *back-pressure* mode discussed in Sec. 8.5.3. In this embodiment, steam (or hot water) is provided from the condenser to service the district heating load.

Referring again to Fig. 9.23, if the condenser is omitted, steam is supplied directly from the steam turbine to service the district heating load; condensate is returned to the heat-recovery steam generator. If the steam turbine is also omitted, steam passes directly from the heat-recovery unit to the community and back again; power is generated by the gas turbine alone.

9.10 Integrated Gasification Combined-Cycle Power Plants

For decades vapor power plants fueled by coal have been the workhorses of U.S. electricity generation (see Chap. 8). However, human health and environmental impact issues linked to coal combustion have placed this type of power generation under a cloud. In light of our large coal reserves and the critical importance of electricity to our society, major governmental and private-sector efforts are aimed at developing alternative power generation technologies

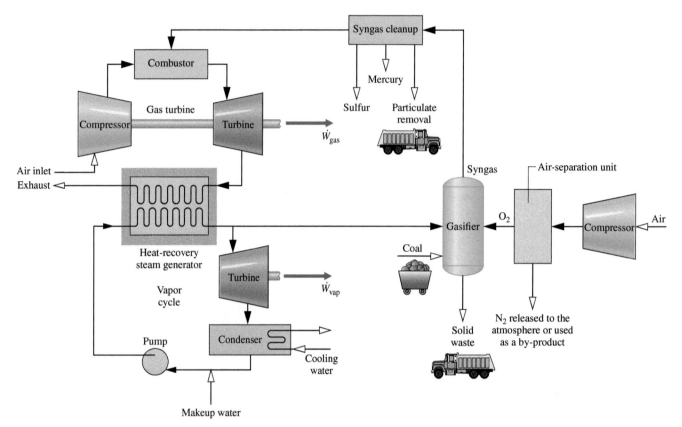

Fig. 9.24 Integrated gasification combined-cycle power plant.

using coal but with fewer adverse effects. In this section, we consider one such technology: integrated gasification combined-cycle (IGCC) power plants.

An IGCC power plant integrates a coal gasifier with a combined gas turbine–vapor power plant like that considered in Sec. 9.9. Key elements of an IGCC plant are shown in **Fig. 9.24**. Gasification is achieved through controlled combustion of coal with oxygen in the presence of steam to produce *syngas* (synthesis gas) and solid waste. Oxygen is provided to the gasifier by the companion air-separation unit. Syngas exiting the gasifier is mainly composed of carbon monoxide and hydrogen. The syngas is cleaned of pollutants and then fired in the gas turbine combustor. The performance of the combined cycle follows the discussion provided in Sec. 9.9.

In IGCC plants, pollutants (sulfur compounds, mercury, and particulates) are removed *before* combustion when it is more effective to do so, rather than after combustion as in conventional coal-fueled power plants. While IGCC plants emit fewer sulfur dioxide, nitric oxide, mercury, and particulate emissions than comparable conventional coal plants, abundant solid waste is still produced that must be responsibly managed.

Taking a closer look at Fig. 9.24, better IGCC plant performance can be realized through tighter integration between the air-separation unit and combined cycle. For instance, by providing compressed air from the gas turbine compressor to the air-separation unit, the compressor feeding ambient air to the air-separation unit can be eliminated or reduced in size. Also, by injecting nitrogen produced by the separation unit into the air stream entering the combustor, mass flow rate through the turbine increases and therefore greater power is developed.

Only a few IGCC plants have been constructed worldwide thus far. Accordingly, only time will tell if this technology will make significant inroads against coal-fired vapor power plants, including the newest generation of supercritical plants. Proponents point to increased combined cycle thermal efficiency as a way to extend the viability of U.S. coal reserves. Others say investment might be better directed to technologies fostering use of renewable sources of energy for power generation than to technologies fostering use of coal, which has so many adverse effects related to its utilization.

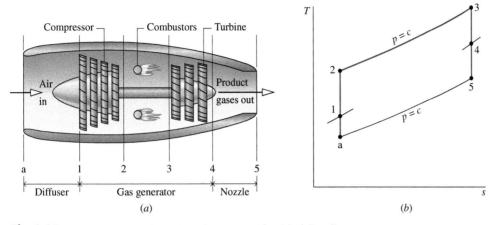

Fig. 9.25 Turbojet engine schematic and accompanying ideal T–s diagram.

9.11 Gas Turbines for Aircraft Propulsion

Gas turbines are particularly suited for aircraft propulsion because of their favorable power-to-weight ratios. The **turbojet engine** is commonly used for this purpose. As illustrated in Fig. 9.25a, this type of engine consists of three main sections: the diffuser, the gas generator, and the nozzle.

    The diffuser placed before the compressor decelerates the incoming air relative to the engine. A pressure rise known as the **ram effect** is associated with this deceleration. The gas generator section consists of a compressor, combustor, and turbine, with the same functions as the corresponding components of a stationary gas turbine power plant. In a turbojet engine, the turbine power output need only be sufficient to drive the compressor and auxiliary equipment, however.

    Combustion gases leave the turbine at a pressure significantly greater than atmospheric and expand through the nozzle to a high velocity before being discharged to the surroundings. The overall change in the velocity of the gases relative to the engine gives rise to the propulsive force, or **thrust**.

    Some turbojets are equipped with an **afterburner**, as shown in Fig. 9.26. This is essentially a reheat device in which additional fuel is injected into the gas exiting the turbine and burned, producing a higher temperature at the nozzle inlet than would be achieved otherwise. As a consequence, a greater nozzle exit velocity is attained, resulting in increased thrust.

turbojet engine

ram effect

thrust

afterburner

Turbojet Analysis   The T–s diagram of the processes in an ideal turbojet engine is shown in Fig. 9.25b. In accordance with the assumptions of an air-standard analysis, the working fluid is air

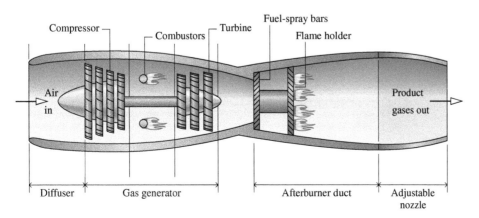

Fig. 9.26 Schematic of a turbojet engine with afterburner.

modeled as an ideal gas. The diffuser, compressor, turbine, and nozzle processes are isentropic, and the combustor operates at constant pressure.

TAKE NOTE...

Thrust is the forward-directed force developed due to the change in momentum of the gases flowing through the turbojet engine. See Sec. 9.12.1 for the *momentum equation*.

- Process a–1 shows the pressure rise that occurs in the diffuser as the air decelerates isentropically through this component.
- Process 1–2 is an isentropic compression.
- Process 2–3 is a constant-pressure heat addition.
- Process 3–4 is an isentropic expansion through the turbine during which work is developed.
- Process 4–5 is an isentropic expansion through the nozzle in which the air accelerates and the pressure decreases.

Owing to irreversibilities in an actual engine, there would be increases in specific entropy across the diffuser, compressor, turbine, and nozzle. In addition, there would be a combustion irreversibility and a pressure drop through the combustor of the actual engine. Further details regarding flow through nozzles and diffusers are provided in Secs. 9.13 and 9.14. The subject of combustion is discussed in Chap. 13.

In a typical thermodynamic analysis of a turbojet on an air-standard basis, the following quantities might be known: the velocity at the diffuser inlet, the compressor pressure ratio, and the turbine inlet temperature. The objective of the analysis might then be to determine the velocity at the nozzle exit. Once the nozzle exit velocity is known, the *thrust* can be determined.

All principles required for the thermodynamic analysis of turbojet engines on an air-standard basis have been introduced. Example 9.13 provides an illustration.

> ▶ ▶ **EXAMPLE 9.13** ▶ •

Analyzing a Turbojet Engine

Air enters a turbojet engine at 0.8 bar, 240 K, and an inlet velocity of 1000 km/h (278 m/s). The pressure ratio across the compressor is 8. The turbine inlet temperature is 1200 K and the pressure at the nozzle exit is 0.8 bar. The work developed by the turbine equals the compressor work input. The diffuser, compressor, turbine, and nozzle processes are isentropic, and there is no pressure drop for flow through the combustor. For operation at steady state, determine the velocity at the nozzle exit and the pressure at each principal state. Neglect kinetic energy at the exit of all components except the nozzle and neglect potential energy throughout.

SOLUTION

Known An ideal turbojet engine operates at steady state. Key operating conditions are specified.

Find Determine the velocity at the nozzle exit, in m/s, and the pressure, in bar, at each principal state.

Schematic and Given Data:

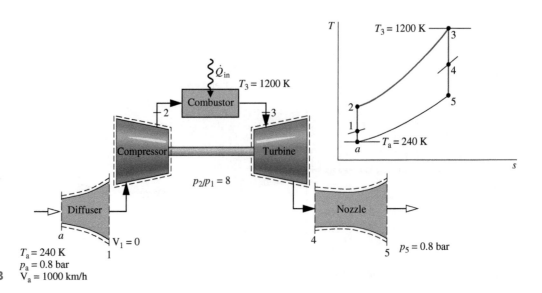

Fig. E9.13

Engineering Model

1. Each component is analyzed as a control volume at steady state. The control volumes are shown on the accompanying sketch by dashed lines.

2. The diffuser, compressor, turbine, and nozzle processes are isentropic.

3. There is no pressure drop for flow through the combustor.

4. The turbine work output equals the work required to drive the compressor.

5. Except at the inlet and exit of the engine, kinetic energy effects can be ignored. Potential energy effects are negligible throughout.

6. The working fluid is air modeled as an ideal gas.

Analysis To determine the velocity at the exit to the nozzle, the mass and energy rate balances for a control volume enclosing this component reduce at steady state to give

$$0 = \cancel{\dot{Q}_{cv}}^0 - \cancel{\dot{W}_{cv}}^0 + \dot{m}\left[(h_4 - h_5) + \left(\frac{\cancel{V_4^2}^0 - V_5^2}{2}\right) + g(\cancel{z_4}^0 \cancel{z_5}^0)\right]$$

where $\dot{m}$ is the mass flow rate. The inlet kinetic energy is dropped by assumption 5. Solving for V_5

$$V_5 = \sqrt{2(h_4 - h_5)}$$

This expression requires values for the specific enthalpies h_4 and h_5 at the nozzle inlet and exit, respectively. With the operating parameters specified, the determination of these enthalpy values is accomplished by analyzing each component in turn, beginning with the diffuser. The pressure at each principal state can be evaluated as a part of the analyses required to find the enthalpies h_4 and h_5.

Mass and energy rate balances for a control volume enclosing the diffuser reduce to give

$$h_1 = h_a + \frac{V_a^2}{2}$$

With h_a from Table A-22 and the given value of V_a

① $h_1 = 240.02 \text{ kJ/kg} + \left[\frac{(278)^2}{2}\right]\left(\frac{m^2}{s^2}\right)\left|\frac{1 \text{ N}}{1 \text{ kg} \cdot \text{m/s}^2}\right|\left|\frac{1 \text{ kJ}}{10^3 \text{ N} \cdot \text{m}}\right|$

$$= 278.7 \text{ kJ/kg}$$

Interpolating in Table A-22 gives $p_{r1} = 1.070$. The flow through the diffuser is isentropic, so pressure p_1 is

$$p_1 = \frac{p_{r1}}{p_{ra}} p_a$$

With p_r data from Table A-22 and the known value of p_a

$$p_1 = \frac{1.070}{0.6355}(0.8 \text{ bar}) = 1.347 \text{ bar}$$

Using the given compressor pressure ratio, the pressure at state 2 is $p_2 = 8(1.347 \text{ bar}) = 10.78 \text{ bar}$.

The flow through the compressor is also isentropic. Thus

$$p_{r2} = p_{r1}\frac{p_2}{p_1} = 1.070(8) = 8.56$$

Interpolating in Table A-22, we get $h_2 = 505.5 \text{ kJ/kg}$.

At state 3 the temperature is given as $T_3 = 1200 \text{ K}$. From Table A-22, $h_3 = 1277.79 \text{ kJ/kg}$. By assumption 3, $p_3 = p_2$. The work

developed by the turbine is just sufficient to drive the compressor (assumption 4). That is

$$\frac{\dot{W}_t}{\dot{m}} = \frac{\dot{W}_c}{\dot{m}}$$

or

$$h_3 - h_4 = h_2 - h_1$$

Solving for h_4

$$h_4 = h_3 + h_1 - h_2 = 1277.79 + 278.7 - 505.5$$

$$= 1051 \text{ kJ/kg}$$

Interpolating in Table A-22 with h_4 gives $p_{r4} = 116.0$.

The expansion through the turbine is isentropic, so

$$p_4 = p_3 \frac{p_{r4}}{p_{r3}}$$

With $p_3 = p_2$ and p_r data from Table A-22

$$p_4 = (10.78 \text{ bar})\frac{116.0}{238.0} = 5.25 \text{ bar}$$

The expansion through the nozzle is isentropic to $p_5 = 0.8$ bar. Thus

$$p_{r5} = p_{r4}\frac{p_5}{p_4} = (116.0)\frac{0.8}{5.25} = 17.68$$

From Table A-22, $h_5 = 621.3 \text{ kJ/kg}$, which is the remaining specific enthalpy value required to determine the velocity at the nozzle exit.

Using the values for h_4 and h_5 determined above, the velocity at the nozzle exit is

$$V_5 = \sqrt{2(h_4 - h_5)}$$

$$= \sqrt{2(1051 - 621.3)\frac{\text{kJ}}{\text{kg}}\left(\frac{1 \text{ kg m/s}^2}{1 \text{ N}}\right)\left(\frac{10^3 \text{ N} \cdot \text{m}}{1 \text{ kJ}}\right)}$$

② $= 927 \text{ m/s}$

---

① Note the unit conversions required here and in the calculation of V_5.

② The increase in the velocity of the air as it passes through the engine gives rise to the thrust produced by the engine. A detailed analysis of the forces acting on the engine requires Newton's second law of motion in a form suitable for control volumes (see Sec. 9.12.1).

SKILLS DEVELOPED

Ability to...

- sketch the schematic of the turbojet engine and the T–s diagram for the corresponding air-standard cycle.

- evaluate temperatures and pressures at each principal state and retrieve necessary property data.

- apply mass, energy, and entropy balances.

- calculate the nozzle exit velocity.

Quick Quiz

Referring to Eq. 6.47, what would be the nozzle exit velocity, in m/s, if the nozzle isentropic efficiency were 90%? Ans. 880 m/s.

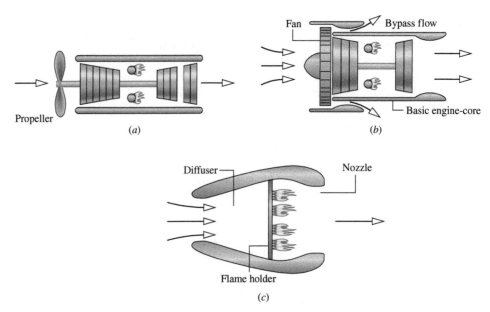

Fig. 9.27 Other examples of aircraft engines. (*a*) Turboprop. (*b*) Turbofan. (*c*) Ramjet.

Other Applications Other related applications of the gas turbine include *turboprop* and *turbofan* engines. The turboprop engine shown in **Fig. 9.27a** consists of a gas turbine in which the gases are allowed to expand through the turbine to atmospheric pressure. The net power developed is directed to a propeller, which provides thrust to the aircraft. Turboprops are able to achieve speeds up to about 925 km/h (575 miles/h). In the turbofan shown in **Fig. 9.27b**, the core of the engine is much like a turbojet, and some thrust is obtained from expansion through the nozzle. However, a set of large-diameter blades attached to the front of the engine accelerates air around the core. This *bypass flow* provides additional thrust for takeoff, whereas the core of the engine provides the primary thrust for cruising. Turbofan engines are commonly used for commercial aircraft with flight speeds of up to about 1000 km/h (620 miles/h). A particularly simple type of engine known as a ramjet is shown in **Fig. 9.27c**. This engine requires neither a compressor nor a turbine. A sufficient pressure rise is obtained by decelerating the high-speed incoming air in the diffuser (ram effect). For the ramjet to operate, therefore, the aircraft must already be in flight at high speed. The combustion products exiting the combustor are expanded through the nozzle to produce the thrust.

In each of the engines mentioned thus far, combustion of the fuel is supported by air brought into the engines from the atmosphere. For very high-altitude flight and space travel, where this is no longer possible, *rockets* may be employed. In these applications, both fuel and an oxidizer (such as liquid oxygen) are carried onboard the craft. Thrust is developed when the high-pressure gases obtained on combustion are expanded through a nozzle and discharged from the rocket.

Considering Compressible Flow Through Nozzles and Diffusers

In many applications of engineering interest, gases move at relatively high velocities and exhibit appreciable changes in specific volume (density). The flows through the nozzles and diffusers of jet engines discussed in Sec. 9.11 are important examples. Other examples are the flows through wind tunnels, shock tubes, and steam ejectors. These flows are known as **compressible flows**. In this part of the chapter, we introduce some of the principles involved in analyzing compressible flows.

compressible flow

9.12 Compressible Flow Preliminaries

Concepts introduced in this section play important roles in the study of compressible flows. The momentum equation is introduced in a form applicable to the analysis of control volumes at steady state. The velocity of sound is also defined, and the concepts of Mach number and stagnation state are discussed.

9.12.1 Momentum Equation for Steady One-Dimensional Flow

The analysis of compressible flows requires the principles of conservation of mass and energy, the second law of thermodynamics, and relations among the thermodynamic properties of the flowing gas. In addition, Newton's second law of motion is required. Application of Newton's second law of motion to systems of fixed mass (closed systems) involves the familiar form

$$\mathbf{F} = m\mathbf{a}$$

where $\mathbf{F}$ is the resultant force acting *on* a system of mass m and $\mathbf{a}$ is the acceleration. The object of the present discussion is to introduce Newton's second law of motion in a form appropriate for the study of the control volumes considered in subsequent discussions.

Consider the control volume shown in Fig. 9.28, which has a single inlet, designated by 1, and a single exit, designated by 2. The flow is assumed to be one-dimensional at these locations. The energy and entropy rate equations for such a control volume have terms that account for energy and entropy transfers, respectively, at the inlets and exits. Momentum also can be carried into or out of the control volume at the inlets and exits, and such transfers can be accounted for as

$$\begin{bmatrix} \text{time rate of momentum} \\ \text{transfer into or} \\ \text{out of a control volume} \\ \text{accompanying mass flow} \end{bmatrix} = \dot{m}\mathbf{V} \qquad (9.30)$$

In this expression, the momentum per unit of mass flowing across the boundary of the control volume is given by the velocity vector $\mathbf{V}$. In accordance with the one-dimensional flow model, the vector is normal to the inlet or exit and oriented in the direction of flow.

In words, Newton's second law of motion for control volumes is

$$\begin{bmatrix} \text{time rate of change} \\ \text{or momentum contained} \\ \text{within the control volume} \end{bmatrix} = \begin{bmatrix} \text{resultant force} \\ \text{acting } on \text{ the} \\ \text{control volume} \end{bmatrix} + \begin{bmatrix} \text{net rate at which momentum is} \\ \text{transferred into the control} \\ \text{volume accompanying mass flow} \end{bmatrix}$$

At steady state, the total amount of momentum contained in the control volume is constant with time. Accordingly, when applying Newton's second law of motion to control volumes at steady state, it is necessary to consider only the momentum accompanying

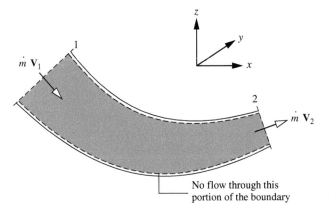

Fig. 9.28 One-inlet, one-exit control volume at steady state labeled with momentum transfers accompanying mass flow.

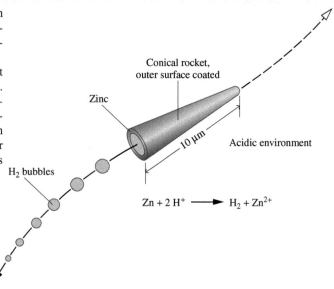

Hydrogen Bubble-Propelled Microrockets Have Promise for Biomedical and Industrial Applications

Tiny cone-shaped devices displaying autonomous motion when in extreme acidic environments may have diverse applications ranging from targeted drug delivery in the human body to the monitoring of industrial processes, researchers say.

The hollow cones are fabricated from zinc and are only about 10 micrometers long, much too small to be seen by the human eye. When a cone is placed in an acidic environment, the zinc loses electrons. As hydrogen ions in the acid take up these electrons, hydrogen bubbles form on the cone's inner zinc surface. Finally, as shown in the accompanying figure, the bubbles are emitted from the larger end of the conical rocket, and the rocket is *thrust forward* at speeds as high as 100 cone-lengths per second.

---

the incoming and outgoing streams of matter and the forces acting on the control volume. Newton's law then states that the resultant force **F** acting *on* the control volume equals the difference between the rates of momentum exiting and entering the control volume accompanying mass flow. This is expressed by the following **steady-state form of the momentum equation**:

steady-state momentum equation

$$\mathbf{F} = \dot{m}_2 \mathbf{V}_2 - \dot{m}_1 \mathbf{V}_1 = \dot{m}(\mathbf{V}_2 - \mathbf{V}_1) \tag{9.31}$$

TAKE NOTE...

The resultant force **F** includes the forces due to pressure acting at the inlet and exit, forces acting on the portion of the boundary through which there is no mass flow, and the force of gravity.

Since $\dot{m}_1 = \dot{m}_2$ at steady state, the common mass flow is designated in this expression simply as $\dot{m}$. The expression of Newton's second law of motion given by Eq. 9.31 suffices for subsequent discussions. More general control volume formulations are normally provided in fluid mechanics texts.

9.12.2 Velocity of Sound and Mach Number

A sound wave is a small pressure disturbance that propagates through a gas, liquid, or solid at a velocity c that depends on the properties of the medium. In this section we obtain an expression that relates the *velocity of sound*, or sonic velocity, to other properties. The velocity of sound is an important property in the study of compressible flows.

Modeling Pressure Waves Let us begin by referring to Fig. 9.29a, which shows a pressure wave moving to the right with a velocity of magnitude c. The wave is generated by a small displacement of the piston. As shown on the figure, the pressure, density, and temperature in the region to the left of the wave depart from the respective values of the undisturbed fluid to the right of the wave, which are designated simply p, ρ, and T. After the wave has passed, the fluid to its left is in steady motion with a velocity of magnitude ΔV.

Figure 9.29a shows the wave from the point of view of a stationary observer. It is easier to analyze this situation from the point of view of an observer at rest relative to the wave, as shown in Fig. 9.29b. By adopting this viewpoint, a steady-state analysis can be applied to the control volume identified on the figure. To an observer at rest relative to the wave, it appears as though the fluid is moving toward the stationary wave from the right with velocity c, pressure p, density ρ, and temperature T and moving away on the left with velocity $c - \Delta V$, pressure $p + \Delta p$, density $\rho + \Delta \rho$, and temperature $T + \Delta T$.

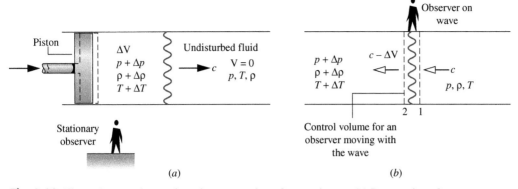

Fig. 9.29 Illustrations used to analyze the propagation of a sound wave. (*a*) Propagation of a pressure wave through a quiescent fluid relative to a stationary observer. (*b*) Observer at rest relative to the wave.

At steady state, the conservation of mass principle for the control volume reduces to $\dot{m}_1 = \dot{m}_2$, or

$$\rho A c = (\rho + \Delta\rho)A(c - \Delta V)$$

On rearrangement

$$0 = c\Delta\rho - \rho\Delta V - \Delta\rho\Delta V^{\,0} \qquad (9.32)$$

If the disturbance is *weak*, the third term on the right of Eq. 9.32 can be neglected, leaving

$$\Delta V = (c/\rho)\Delta\rho \qquad (9.33)$$

Next, the momentum equation, Eq. 9.31, is applied to the control volume under consideration. Since the thickness of the wave is small, shear forces at the wall are negligible. The effect of gravity is also ignored. Hence, the only significant forces acting on the control volume in the direction of flow are the forces due to pressure at the inlet and exit. With these idealizations, the component of the momentum equation in the direction of flow reduces to

$$\begin{aligned} pA - (p + \Delta p)A &= \dot{m}(c - \Delta V) - \dot{m}c \\ &= \dot{m}(c - \Delta V - c) \\ &= (\rho A c)(-\Delta V) \end{aligned}$$

or

$$\Delta p = \rho c\, \Delta V \qquad (9.34)$$

Combining Eqs. 9.33 and 9.34 and solving for c

$$c = \sqrt{\frac{\Delta p}{\Delta\rho}} \qquad (9.35)$$

Sound Waves   For sound waves the differences in pressure, density, and temperature across the wave are quite small. In particular, $\Delta\rho \ll \rho$, justifying the neglect of the third term of Eq. 9.32. Furthermore, the ratio $\Delta p/\Delta\rho$ in Eq. 9.35 can be interpreted as the derivative of pressure with respect to density across the wave. Experiments also indicate that the relation between pressure and density across a sound wave is nearly *isentropic*. The expression for the **velocity of sound** then becomes

velocity of sound

$$\boxed{c = \sqrt{\left(\frac{\partial p}{\partial \rho}\right)_s}} \qquad (9.36a)$$

or in terms of specific volume

$$c = \sqrt{-v^2 \left(\frac{\partial p}{\partial v}\right)_s}$$

(9.36b)

The velocity of sound is an intensive property whose value depends on the state of the medium through which sound propagates. Although we have assumed that sound propagates isentropically, the medium itself may be undergoing any process.

Means for evaluating the velocity of sound c for gases, liquids, and solids are introduced in Sec. 11.5. The special case of an ideal gas will be considered here because it is used extensively later in the chapter. For this case, the relationship between pressure and specific volume of an ideal gas at fixed entropy is $pv^k =$ constant, where k is the specific heat ratio (Sec. 6.11.2). Thus, $(\partial p / \partial v)_s = -kp/v$, and Eq. 9.36b gives $c = \sqrt{kpv}$. Or, with the ideal gas equation of state

$$c = \sqrt{kRT} \qquad \text{(ideal gas)}$$

(9.37)

FOR EXAMPLE

To illustrate the use of Eq. 9.37, let us calculate the velocity of sound in air at 300 K and 650 K. From Table A-20 at 300 K, $k = 1.4$. Thus,

$$c = \sqrt{1.4 \left(\frac{8314 \text{ N} \cdot \text{m}}{28.97 \text{ kg} \cdot \text{K}}\right)(300 \text{ K}) \left|\frac{1 \text{ kg} \cdot \text{m/s}^2}{1 \text{ N}}\right|} = 347 \frac{\text{m}}{\text{s}}$$

At 650 K, $k = 1.37$, and $c = 506$ m/s (1660 ft/s), as can be verified.

Mach Number

Mach number

In subsequent discussions, the ratio of the velocity V at a state in a flowing fluid to the value of the sonic velocity c at the same state plays an important role. This ratio is called the **Mach number** M

$$M = \frac{\text{V}}{c}$$

(9.38)

supersonic

subsonic

When $M > 1$, the flow is said to be **supersonic**; when $M < 1$, the flow is **subsonic**; and when $M = 1$, the flow is *sonic*. The term *hypersonic* is used for flows with Mach numbers much greater than one, and the term *transonic* refers to flows where the Mach number is close to unity.

BioConnections

For centuries, physicians have used sounds emanating from the human body to aid diagnosis. Most of us are familiar with the stethoscope, which has been used by medical practitioners for more effective sound detection since the early nineteenth century.

Another commonly encountered sound-related diagnostic method in medical practice is the use of sound at frequencies above that audible by the human ear, known as *ultrasound. Ultrasound imaging* allows physicians to peer inside the body and evaluate solid structures in the abdominal cavity. Ultrasound devices beam sound waves into the body and collect return echoes as the beam encounters regions of differing density. The reflected sound waves produce shadow pictures of structures below the skin on a monitor screen. Pictures show the shape, size, and movement of target objects in the path of the beam.

Obstetricians commonly use ultrasound to assess the fetus during pregnancy. Emergency physicians use ultrasound to assess abdominal pain or other concerns. Ultrasound is also used to break up small kidney stones.

Cardiologists use an ultrasound application known as *echocardiography* to evaluate the heart and its valve function, measure the amount of blood pumped with each stroke, and detect blood clots in veins and artery blockage. Among several uses of echocardiography are the *stress* test, where the echocardiogram is done both before and after exercise, and the *transesophageal* echo test, where the probe is passed down the esophagus to locate it closer to the heart, thereby enabling clearer pictures of the heart without interference from the ribs or lungs.

9.12.3 Determining Stagnation State Properties

When dealing with compressible flows, it is often convenient to work with properties evaluated at a reference state known as the **stagnation state**. The stagnation state is the state a flowing fluid would attain if it were decelerated to zero velocity isentropically. We might imagine this as taking place in a diffuser operating at steady state. By reducing an energy balance for such a diffuser, it can be concluded that the enthalpy at the stagnation state associated with an actual state in the flow where the specific enthalpy is h and the velocity is V is given by

stagnation state

$$\boxed{h_{\mathrm{o}} = h + \frac{\mathrm{V}^2}{2}}$$

(9.39)   **stagnation enthalpy**

The enthalpy designated here as h_{o} is called the **stagnation enthalpy**. The pressure p_{o} and temperature T_{o} at a stagnation state are called the **stagnation pressure** and **stagnation temperature**, respectively.

stagnation pressure and temperature

9.13 Analyzing One-Dimensional Steady Flow in Nozzles and Diffusers

Although the subject of compressible flow arises in a great many important areas of engineering application, the remainder of this presentation is concerned only with flow through nozzles and diffusers. Texts dealing with compressible flow should be consulted for discussion of other areas of application.

In the present section we determine the shapes required by nozzles and diffusers for subsonic and supersonic flow. This is accomplished using mass, energy, entropy, and momentum principles, together with property relationships. In addition, we study how the flow through nozzles is affected as conditions at the nozzle exit are changed. The presentation concludes with an analysis of normal shocks, which can exist in supersonic flows.

9.13.1 Exploring the Effects of Area Change in Subsonic and Supersonic Flows

The objective of the present discussion is to establish criteria for determining whether a nozzle or diffuser should have a converging, diverging, or converging–diverging shape. This is accomplished using differential equations relating the principal variables that are obtained using mass and energy balances together with property relations, as considered next.

Governing Differential Equations   Let us begin by considering a control volume enclosing a nozzle or diffuser. At steady state, the mass flow rate is constant, so

$$\rho \mathrm{AV} = \text{constant}$$

In differential form

$$d(\rho \mathrm{AV}) = 0$$
$$\mathrm{AV}\, d\rho + \rho \mathrm{A}\, d\mathrm{V} + \rho \mathrm{V}\, d\mathrm{A} = 0$$

or on dividing each term by $\rho \mathrm{AV}$

$$\frac{d\rho}{\rho} + \frac{d\mathrm{V}}{\mathrm{V}} + \frac{d\mathrm{A}}{\mathrm{A}} = 0$$

(9.40)

Assuming $\dot{Q}_{cv} = \dot{W}_{cv} = 0$ and negligible potential energy effects, an energy rate balance reduces to give

$$h_2 + \frac{V_2^2}{2} = h_1 + \frac{V_1^2}{2}$$

Introducing Eq. 9.39, it follows that the stagnation enthalpies at states 1 and 2 are equal: $h_{o2} = h_{o1}$. Since any state downstream of the inlet can be regarded as state 2, the following relationship between the specific enthalpy and kinetic energy must be satisfied at each state:

$$h + \frac{V^2}{2} = h_{o1} \quad \text{(constant)}$$

In differential form this becomes

$$dh = -V\,dV \tag{9.41}$$

This equation shows that if the velocity increases (decreases) in the direction of flow, the specific enthalpy must decrease (increase) in the direction of flow.

In addition to Eqs. 9.40 and 9.41 expressing conservation of mass and energy, relationships among properties must be taken into consideration. Assuming the flow occurs isentropically, the property relation (Eq. 6.10b)

$$T\,ds = dh - \frac{dp}{\rho}$$

reduces to give

$$dh = \frac{1}{\rho}dp \tag{9.42}$$

This equation shows that when pressure increases or decreases in the direction of flow, the specific enthalpy changes in the same way.

Forming the differential of the property relation $p = p(\rho, s)$

$$dp = \left(\frac{\partial p}{\partial \rho}\right)_s d\rho + \left(\frac{\partial p}{\partial s}\right)_\rho ds$$

The second term vanishes in isentropic flow. Introducing Eq. 9.36a, we have

$$dp = c^2 d\rho \tag{9.43}$$

which shows that when pressure increases or decreases in the direction of flow, density changes in the same way.

Additional conclusions can be drawn by combining the above differential equations. Combining Eqs. 9.41 and 9.42 results in

$$\frac{1}{\rho}dp = -V\,dV \tag{9.44}$$

which shows that if the velocity increases (decreases) in the direction of flow, the pressure must decrease (increase) in the direction of flow.

Eliminating dp between Eqs. 9.43 and 9.44 and combining the result with Eq. 9.40 gives

$$\frac{dA}{A} = -\frac{dV}{V}\left[1 - \left(\frac{V}{c}\right)^2\right]$$

or with the *Mach number M*

$$\boxed{\frac{dA}{A} = -\frac{dV}{V}(1 - M^2)} \tag{9.45}$$

TAKE NOTE...

Engineering Model:

- Control volume at steady state.
- $\dot{Q}_{cv} = \dot{W}_{cv} = 0$.
- Negligible potential energy.
- Isentropic flow.

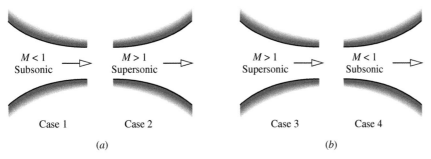

Fig. 9.30 Effects of area change in subsonic and supersonic flows. (*a*) Nozzles: V increases; h, p, and ρ decrease. (*b*) Diffusers: V decreases; h, p, and ρ increase.

Variation of Area with Velocity   Equation 9.45 shows how area must vary with velocity. The following four cases can be identified:

Case 1   Subsonic nozzle. $d\mathrm{V} > 0$, $M < 1 \Rightarrow dA < 0$: The duct *converges* in the direction of flow.

Case 2   Supersonic nozzle. $d\mathrm{V} > 0$, $M > 1 \Rightarrow dA > 0$: The duct *diverges* in the direction of flow.

Case 3   Supersonic diffuser. $d\mathrm{V} < 0$, $M > 1 \Rightarrow dA < 0$: The duct *converges* in the direction of flow.

Case 4   Subsonic diffuser. $d\mathrm{V} < 0$, $M < 1 \Rightarrow dA > 0$: The duct *diverges* in the direction of flow.

The forgoing conclusions concerning the nature of the flow in subsonic and supersonic nozzles and diffusers are summarized in **Fig. 9.30**. From Fig. 9.30*a*, we see that to accelerate a fluid flowing subsonically, a converging nozzle must be used, but once $M = 1$ is achieved, further acceleration can occur only in a diverging nozzle. From Fig. 9.30*b*, we see that a converging diffuser is required to decelerate a fluid flowing supersonically, but once $M = 1$ is achieved, further deceleration can occur only in a diverging diffuser. These findings suggest that a Mach number of unity can occur only at the location in a nozzle or diffuser where the cross-sectional area is a minimum. This location of minimum area is called the **throat**.   throat

The developments of this section have not required the specification of an equation of state; thus, the conclusions hold for all gases. Moreover, while the conclusions have been drawn for isentropic flow through nozzles and diffusers, they are at least qualitatively valid for many actual such applications and, in particular, for nozzles and diffusers having high *isentropic efficiencies* (Sec. 6.12).

9.13.2 Effects of Back Pressure on Mass Flow Rate

In the present discussion we consider the effect of varying the *back pressure* on the rate of mass flow through nozzles. The **back pressure** is the pressure in the exhaust region outside   back pressure
the nozzle. The case of converging nozzles is taken up first and then converging–diverging nozzles are considered.

Converging Nozzles   Figure 9.31 shows a converging duct with stagnation conditions at the inlet, discharging into a region in which the back pressure p_B can be varied. For the series of cases labeled a through e, let us consider how the mass flow rate $\dot{m}$ and nozzle exit pressure p_E vary as the back pressure is decreased while keeping the inlet conditions fixed.

When $p_B = p_E = p_o$, there is no flow, so $\dot{m} = 0$. This corresponds to case a of Fig. 9.31. If the back pressure p_B is decreased, as in cases b and c, there will be flow through the nozzle. As long as the flow is subsonic at the exit, information about changing conditions in the exhaust region can be transmitted upstream. Decreases in back pressure thus result in greater mass flow rates and new pressure variations within the nozzle. In each instance, the velocity is subsonic throughout the nozzle and the exit pressure equals the back pressure. The exit Mach number increases as p_B decreases, however, and eventually a Mach number of unity will be attained at the nozzle exit. The corresponding pressure is denoted by p^*, called the *critical pressure*. This case is represented by d on Fig. 9.31.

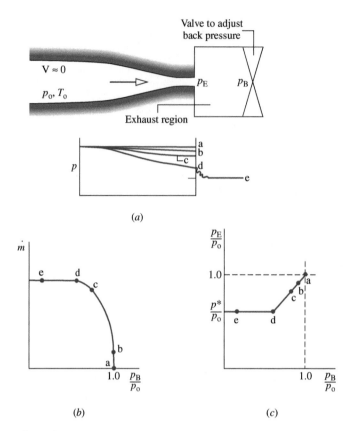

Fig. 9.31 Effect of back pressure on the operation of a converging nozzle.

Recalling that the Mach number cannot increase beyond unity in a converging section, let us consider next what happens when the back pressure is reduced further to a value less than p^*, such as represented by case e. Since the velocity at the exit equals the velocity of sound, information about changing conditions in the exhaust region no longer can be transmitted upstream past the exit plane. Accordingly, reductions in p_B below p^* have no effect on flow conditions in the nozzle. Neither the pressure variation within the nozzle nor the mass flow rate is affected. Under these conditions, the nozzle is said to be **choked**. When a nozzle is choked, the mass flow rate is the *maximum possible for the given stagnation conditions*. For p_B less than p^*, the flow expands outside the nozzle to match the lower back pressure, as shown by case e of Fig. 9.31. The pressure variation outside the nozzle cannot be predicted using the one-dimensional flow model.

choked flow: converging nozzle

Converging–Diverging Nozzles Figure 9.32 illustrates the effects of varying back pressure on a *converging–diverging* nozzle. The series of cases labeled a through j is considered next.

- Let us first discuss the cases designated a, b, c, and d. Case a corresponds to $p_B = p_E = p_o$ for which there is no flow. When the back pressure is slightly less than p_o (case b), there is some flow, and the flow is subsonic throughout the nozzle. In accordance with the discussion of Fig. 9.30, the greatest velocity and lowest pressure occur at the throat, and the diverging portion acts as a diffuser in which pressure increases and velocity decreases in the direction of flow. If the back pressure is reduced further, corresponding to case c, the mass flow rate and velocity at the throat are greater than before. Still, the flow remains subsonic throughout and qualitatively the same as case b. As the back pressure is reduced, the Mach number at the throat increases, and eventually a Mach number of unity is attained there (case d). As before, the greatest velocity and lowest pressure occur at the throat, and the diverging portion remains a subsonic diffuser. However, because the throat velocity is sonic, the nozzle is now **choked**: The *maximum mass flow rate has been attained for the given stagnation conditions*. Further reductions in back pressure cannot result in an increase in the mass flow rate.

choked flow: converging–diverging nozzle

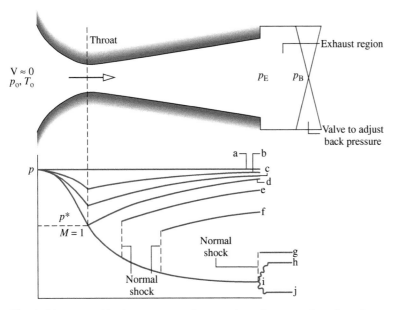

Fig. 9.32 Effect of back pressure on the operation of a converging–diverging nozzle.

- When the back pressure is reduced below that corresponding to case d, the flow through the converging portion and at the throat remains unchanged. Conditions within the diverging portion can be altered, however, as illustrated by cases e, f, and g. In case e, the fluid passing the throat continues to expand and becomes supersonic in the diverging portion just downstream of the throat; but at a certain location an abrupt change in properties occurs. This is called a **normal shock**. Across the shock, there is a rapid and irreversible increase in pressure, accompanied by a rapid decrease from supersonic to subsonic flow. Downstream of the shock, the diverging duct acts as a subsonic diffuser in which the fluid continues to decelerate and the pressure increases to match the back pressure imposed at the exit. If the back pressure is reduced further (case f), the location of the shock moves downstream, but the flow remains qualitatively the same as in case e. With further reductions in back pressure, the shock location moves farther downstream of the throat until it stands at the exit (case g). In this case, the flow throughout the nozzle is isentropic, with subsonic flow in the converging portion, $M = 1$ at the throat, and supersonic flow in the diverging portion. Since the fluid leaving the nozzle passes through a shock, it is subsonic just downstream of the exit plane.

  normal shock

- Finally, let us consider cases h, i, and j where the back pressure is less than that corresponding to case g. In each of these cases, the flow through the nozzle is not affected. The adjustment to changing back pressure occurs outside the nozzle. In case h, the pressure decreases continuously as the fluid expands isentropically through the nozzle and then increases to the back pressure outside the nozzle. The compression that occurs outside the nozzle involves *oblique shock waves*. In case i, the fluid expands isentropically to the back pressure and no shocks occur within or outside the nozzle. In case j, the fluid expands isentropically through the nozzle and then expands outside the nozzle to the back pressure through *oblique expansion* waves. Once $M = 1$ is achieved at the throat, the mass flow rate is fixed at the maximum value for the given stagnation conditions, so the mass flow rate is the same for back pressures corresponding to cases d through j. The pressure variations outside the nozzle involving oblique waves cannot be predicted using the one-dimensional flow model.

9.13.3 Flow Across a Normal Shock

We have seen that under certain conditions a rapid and abrupt change of state called a shock takes place in the diverging portion of a supersonic nozzle. In a *normal* shock, this change of state occurs across a plane normal to the direction of flow. The object of the present discussion is to develop means for determining the change of state across a normal shock.

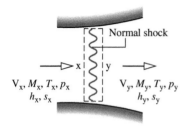

Fig. 9.33 Control volume enclosing a normal shock.

Modeling Normal Shocks A control volume enclosing a normal shock is shown in **Fig. 9.33**. The control volume is assumed to be at steady state with $\dot{W}_{cv} = 0$, $\dot{Q}_{cv} = 0$ and negligible effects of potential energy. The thickness of the shock is very small (on the order of 10^{-5} cm). Thus, there is no significant change in flow area across the shock, even though it may occur in a diverging passage, and the forces acting at the wall can be neglected relative to the pressure forces acting at the upstream and downstream locations denoted by x and y, respectively.

The upstream and downstream states are related by the following equations:

Mass:

$$\rho_x V_x = \rho_y V_y \tag{9.46}$$

Energy:

$$h_x + \frac{V_x^2}{2} = h_y + \frac{V_y^2}{2} \tag{9.47a}$$

or

$$h_{ox} = h_{oy} \tag{9.47b}$$

Momentum:

$$p_x - p_y = \rho_y V_y^2 - \rho_x V_x^2 \tag{9.48}$$

Entropy:

$$s_y - s_x = \dot{\sigma}_{cv}/\dot{m} \tag{9.49}$$

When combined with property relations for the particular fluid under consideration, Eqs. 9.46, 9.47, and 9.48 allow the downstream conditions to be determined for specified upstream conditions. Equation 9.49, which corresponds to Eq. 6.39, leads to the important conclusion that the downstream state *must* have greater specific entropy than the upstream state, or $s_y > s_x$.

Fanno and Rayleigh Lines The mass and energy equations, Eqs. 9.46 and 9.47, can be combined with property relations for the particular fluid to give an equation that when plotted on an *h–s* diagram is called a **Fanno line**. Similarly, the mass and momentum equations, Eqs. 9.46 and 9.48, can be combined to give an equation that when plotted on an *h–s* diagram is called a **Rayleigh line**. Fanno and Rayleigh lines are sketched on *h–s* coordinates in **Fig. 9.34**. It can be shown that the point of maximum entropy on each line, points a and b, corresponds to $M = 1$. It also can be shown that the upper and lower branches of each line correspond, respectively, to subsonic and supersonic velocities.

The downstream state y must satisfy the mass, energy, and momentum equations simultaneously, so state y is fixed by the intersection of the Fanno and Rayleigh lines passing through

Fanno line

Rayleigh line

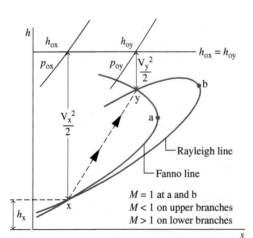

$M = 1$ at a and b
$M < 1$ on upper branches
$M > 1$ on lower branches

Fig. 9.34 Intersection of Fanno and Rayleigh lines as a solution to the normal shock equations.

state x. Since $s_y > s_x$, it can be concluded that the flow across the shock can only pass *from* x *to* y. Accordingly, the velocity changes from supersonic before the shock ($M_x > 1$) to subsonic after the shock ($M_y < 1$). This conclusion is consistent with the discussion of cases e, f, and g in Fig. 9.32. A significant increase in pressure across the shock accompanies the decrease in velocity. Figure 9.34 also locates the stagnation states corresponding to the states upstream and downstream of the shock. The stagnation enthalpy does not change across the shock, but there is a marked decrease in stagnation pressure associated with the irreversible process occurring in the normal shock region.

9.14 Flow in Nozzles and Diffusers of Ideal Gases with Constant Specific Heats

The discussion of flow in nozzles and diffusers presented in Sec. 9.13 requires no assumption regarding the equation of state, and therefore the results obtained hold generally. Attention is now restricted to ideal gases with constant specific heats. This case is appropriate for many practical problems involving flow through nozzles and diffusers. The assumption of constant specific heats also allows the derivation of relatively simple closed-form equations.

9.14.1 Isentropic Flow Functions

Let us begin by developing equations relating a state in a compressible flow to the corresponding stagnation state. For the case of an ideal gas with constant c_p, Eq. 9.39 becomes

$$T_o = T + \frac{V^2}{2c_p}$$

where T_o is the stagnation temperature. Using Eq. 3.47a, $c_p = kR/(k-1)$, together with Eqs. 9.37 and 9.38, the relation between the temperature T and the Mach number M of the flowing gas and the corresponding stagnation temperature T_o is

$$\boxed{\frac{T_o}{T} = 1 + \frac{k-1}{2}M^2} \tag{9.50}$$

With Eq. 6.43, a relationship between the temperature T and pressure p of the flowing gas and the corresponding stagnation temperature T_o and the stagnation pressure p_o is

$$\frac{p_o}{p} = \left(\frac{T_o}{T}\right)^{k/(k-1)}$$

Introducing Eq. 9.50 into this expression gives

$$\boxed{\frac{p_o}{p} = \left(1 + \frac{k-1}{2}M^2\right)^{k/(k-1)}} \tag{9.51}$$

Although sonic conditions may not actually be attained in a particular flow, it is convenient to have an expression relating the area A at a given section to the area A* that *would be* required for sonic flow ($M = 1$) at the same mass flow rate and stagnation state. These areas are related through

$$\rho A V = \rho^* A^* V^*$$

where ρ^* and V^* are the density and velocity, respectively, when $M = 1$. Introducing the ideal gas equation of state, together with Eqs. 9.37 and 9.38, and solving for A/A^*

$$\frac{A}{A^*} = \frac{1}{M}\frac{p^*}{p}\left(\frac{T}{T^*}\right)^{1/2} = \frac{1}{M}\frac{p^*/p_o}{p/p_o}\left(\frac{T/T_o}{T^*/T_o}\right)^{1/2}$$

where T^* and p^* are the temperature and pressure, respectively, when $M = 1$. Then with Eqs. 9.50 and 9.51

$$\frac{A}{A^*} = \frac{1}{M}\left[\left(\frac{2}{k+1}\right)\left(1 + \frac{k-1}{2}M^2\right)\right]^{(k+1)/2(k-1)} \tag{9.52}$$

The variation of A/A^* with M is given in **Fig. 9.35** for $k = 1.4$. The figure shows that a unique value of A/A^* corresponds to any choice of M. However, for a given value of A/A^* other than unity, there are two possible values for the Mach number, one subsonic and one supersonic. This is consistent with the discussion of Fig. 9.30, where it was found that a converging–diverging passage with a section of minimum area is required to accelerate a flow from subsonic to supersonic velocity.

Equations 9.50, 9.51, and 9.52 allow the ratios T/T_o, p/p_o, and A/A^* to be computed and tabulated with the Mach number as the single independent variable for a specified value of k. **Table 9.2** provides a tabulation of this kind for $k = 1.4$. Such a table facilitates the analysis of flow through nozzles and diffusers. Equations 9.50, 9.51, and 9.52 also can be readily evaluated using calculators and computer software such as *Interactive Thermodynamics: IT*.

In Example 9.14, we consider the effect of back pressure on flow in a converging nozzle. The *first step* of the analysis is to check whether the flow is choked.

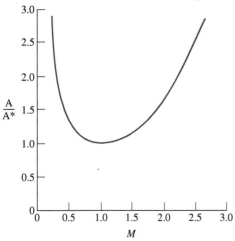

Fig. 9.35 Variation of A/A^* with Mach number in isentropic flow for $k = 1.4$.

TABLE 9.2

Isentropic Flow Functions for an Ideal Gas with $k = 1.4$

M	T/T_o	p/p_o	A/A^*
0	1.000 00	1.000 00	∞
0.10	0.998 00	0.993 03	5.8218
0.20	0.992 06	0.972 50	2.9635
0.30	0.982 32	0.939 47	2.0351
0.40	0.968 99	0.895 62	1.5901
0.50	0.952 38	0.843 02	1.3398
0.60	0.932 84	0.784 00	1.1882
0.70	0.910 75	0.720 92	1.094 37
0.80	0.886 52	0.656 02	1.038 23
0.90	0.860 58	0.591 26	1.008 86
1.00	0.833 33	0.528 28	1.000 00
1.10	0.805 15	0.468 35	1.007 93
1.20	0.776 40	0.412 38	1.030 44
1.30	0.747 38	0.360 92	1.066 31
1.40	0.718 39	0.314 24	1.1149
1.50	0.689 65	0.272 40	1.1762
1.60	0.661 38	0.235 27	1.2502
1.70	0.633 72	0.202 59	1.3376
1.80	0.606 80	0.174 04	1.4390
1.90	0.580 72	0.149 24	1.5552
2.00	0.555 56	0.127 80	1.6875
2.10	0.531 35	0.109 35	1.8369
2.20	0.508 13	0.093 52	2.0050
2.30	0.485 91	0.079 97	2.1931
2.40	0.464 68	0.068 40	2.4031

▶▶▶ **EXAMPLE 9.14** ▶ ·

Determining the Effect of Back Pressure: Converging Nozzle

A converging nozzle has an exit area of 0.001 m^2. Air enters the nozzle with negligible velocity at a pressure of 1.0 MPa and a temperature of 360 K. For isentropic flow of an ideal gas with $k = 1.4$, determine the mass flow rate, in kg/s, and the exit Mach number for back pressures of **(a)** 500 kPa and **(b)** 784 kPa.

SOLUTION

Known Air flows isentropically from specified stagnation conditions through a converging nozzle with a known exit area.

Find For back pressures of 500 and 784 kPa, determine the mass flow rate, in kg/s, and the exit Mach number.

Schematic and Given Data:

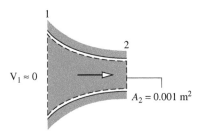

$T_1 = T_0 = 360$ K
$p_1 = p_0 = 1.0$ MPa

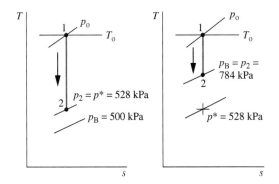

Fig. E9.14

Engineering Model

1. The control volume shown in the accompanying sketch operates at steady state.

2. The air is modeled as an ideal gas with $k = 1.4$.

3. Flow through the nozzle is isentropic.

Analysis The *first step* is to check whether the flow is choked. With $k = 1.4$ and $M = 1.0$, Eq. 9.51 gives $p^*/p_0 = 0.528$. Since ❶ $p_0 = 1.0$ MPa, the critical pressure is $p^* = 528$ kPa. Thus, for back pressures of 528 kPa or less, the Mach number is unity at the exit and the nozzle is choked.

a. From the above discussion, it follows that for a back pressure of 500 kPa, the nozzle is choked. At the exit, $M_2 = 1.0$ and the exit pressure equals the critical pressure, $p_2 = 528$ kPa. The mass flow rate is the maximum value that can be attained for the given

stagnation properties. With the ideal gas equation of state, the mass flow rate is

$$\dot{m} = \rho_2 A_2 V_2 = \frac{p_2}{RT_2} A_2 V_2$$

The exit area A_2 required by this expression is specified as 10^{-3} m^2. Since $M = 1$ at the exit, the exit temperature T_2 can be found from Eq. 9.50, which on rearrangement gives

$$T_2 = \frac{T_0}{1 + \frac{k-1}{2} M^2} = \frac{360 \text{ K}}{1 + \left(\frac{1.4-1}{2}\right)(1)^2} = 300 \text{ K}$$

Then, with Eq. 9.37, the exit velocity V_2 is

$$V_2 = \sqrt{kRT_2}$$

$$= \sqrt{1.4\left(\frac{8314 \text{ N} \cdot \text{m}}{28.97 \text{ kg} \cdot \text{K}}\right)(300 \text{ K})\left|\frac{1 \text{ kg} \cdot \text{m/s}^2}{1 \text{ N}}\right|} = 347.2 \text{ m/s}$$

Finally

$$\dot{m} = \frac{(528 \times 10^3 \text{ N/m}^2)(10^{-3} \text{ m}^2)(347.2 \text{ m/s})}{\left(\frac{8314 \text{ N} \cdot \text{m}}{28.97 \text{ kg} \cdot \text{K}}\right)(300 \text{ K})} = 2.13 \text{ kg/s}$$

b. Since the back pressure of 784 kPa is greater than the critical pressure determined above, the flow throughout the nozzle is subsonic and the exit pressure equals the back pressure, $p_2 = 784$ kPa. The exit Mach number can be found by solving Eq. 9.51 to obtain

$$M_2 = \left\{\frac{2}{k-1}\left[\left(\frac{p_0}{p_2}\right)^{(k-1)/k} - 1\right]\right\}^{1/2}$$

Inserting values

$$M_2 = \left\{\frac{2}{1.4-1}\left[\left(\frac{1 \times 10^6}{7.84 \times 10^5}\right)^{0.286} - 1\right]\right\}^{1/2} = 0.6$$

With the exit Mach number known, the exit temperature T_2 can be found from Eq. 9.50 as 336 K. The exit velocity is then

$$V_2 = M_2 c_2 = M_2 \sqrt{kRT_2} = 0.6\sqrt{1.4\left(\frac{8314}{28.97}\right)(336)}$$

$$= 220.5 \text{ m/s}$$

The mass flow rate is

$$\dot{m} = \rho_2 A_2 V_2 = \frac{p_2}{RT_2} A_2 V_2 = \frac{(784 \times 10^3)(10^{-3})(220.5)}{(8314/28.97)(336)}$$

$$= 1.79 \text{ kg/s}$$

① The use of Table 9.2 reduces some of the computation required in the solution. It is left as an exercise to develop a solution using this table. Also, observe that the first step of the analysis is to check whether the flow is choked.

SKILLS DEVELOPED

Ability to…

• apply the ideal gas model with constant k in the analysis of isentropic flow through a converging nozzle.

• understand when choked flow occurs in a converging nozzle for different back pressures.

• determine conditions at the throat and the mass flow rate for different back pressures and a fixed stagnation state.

Quick Quiz

Using the isentropic flow functions in Table 9.2, determine the exit temperature and Mach number for a back pressure of 843 kPa. Ans. 342.9 K, 0.5.

9.14.2 Normal Shock Functions

Next, let us develop closed-form equations for normal shocks for the case of an ideal gas with constant specific heats. For this case, it follows from the energy equation, Eq. 9.47b, that there is no change in stagnation temperature across the shock, $T_{ox} = T_{oy}$. Then, with Eq. 9.50, the following expression for the ratio of temperatures across the shock is obtained:

$$\frac{T_y}{T_x} = \frac{1 + \frac{k-1}{2}M_x^2}{1 + \frac{k-1}{2}M_y^2} \tag{9.53}$$

Rearranging Eq. 9.48

$$p_x + \rho_x V_x^2 = p_y + \rho_y V_y^2$$

Introducing the ideal gas equation of state, together with Eqs. 9.37 and 9.38, the ratio of the pressure downstream of the shock to the pressure upstream is

$$\frac{p_y}{p_x} = \frac{1 + kM_x^2}{1 + kM_y^2} \tag{9.54}$$

Similarly, Eq. 9.46 becomes

$$\frac{p_y}{p_x} = \sqrt{\frac{T_y}{T_x}}\frac{M_x}{M_y}$$

The following equation relating the Mach numbers M_x and M_y across the shock can be obtained when Eqs. 9.53 and 9.54 are introduced in this expression:

$$M_y^2 = \frac{M_x^2 + \frac{2}{k-1}}{\frac{2k}{k-1}M_x^2 - 1} \tag{9.55}$$

The ratio of stagnation pressures across a shock p_{oy}/p_{ox} is often useful. It is left as an exercise to show that

$$\frac{p_{oy}}{p_{ox}} = \frac{M_x}{M_y}\left(\frac{1 + \frac{k-1}{2}M_y^2}{1 + \frac{k-1}{2}M_x^2}\right)^{(k+1)/2(k-1)} \tag{9.56}$$

TABLE 9.3

Normal Shock Functions for an Ideal Gas with $k = 1.4$

M_x	M_y	p_y/p_x	T_y/T_x	p_{oy}/p_{ox}
1.00	1.000 00	1.0000	1.0000	1.000 00
1.10	0.911 77	1.2450	1.0649	0.998 92
1.20	0.842 17	1.5133	1.1280	0.992 80
1.30	0.785 96	1.8050	1.1909	0.979 35
1.40	0.739 71	2.1200	1.2547	0.958 19
1.50	0.701 09	2.4583	1.3202	0.929 78
1.60	0.668 44	2.8201	1.3880	0.895 20
1.70	0.640 55	3.2050	1.4583	0.855 73
1.80	0.616 50	3.6133	1.5316	0.812 68
1.90	0.595 62	4.0450	1.6079	0.767 35
2.00	0.577 35	4.5000	1.6875	0.720 88
2.10	0.561 28	4.9784	1.7704	0.674 22
2.20	0.547 06	5.4800	1.8569	0.628 12
2.30	0.534 41	6.0050	1.9468	0.583 31
2.40	0.523 12	6.5533	2.0403	0.540 15
2.50	0.512 99	7.1250	2.1375	0.499 02
2.60	0.503 87	7.7200	2.2383	0.460 12
2.70	0.495 63	8.3383	2.3429	0.423 59
2.80	0.488 17	8.9800	2.4512	0.389 46
2.90	0.481 38	9.6450	2.5632	0.357 73
3.00	0.475 19	10.333	2.6790	0.328 34
4.00	0.434 96	18.500	4.0469	0.138 76
5.00	0.415 23	29.000	5.8000	0.061 72
10.00	0.387 57	116.50	20.388	0.003 04
∞	0.377 96	∞	∞	0.0

Since there is no area change across a shock, Eqs. 9.52 and 9.56 combine to give

$$\frac{A_x^*}{A_y^*} = \frac{p_{oy}}{p_{ox}} \tag{9.57}$$

For specified values of M_x and specific heat ratio k, the Mach number downstream of a shock can be found from Eq. 9.55. Then, with M_x, M_y, and k known, the ratios T_y/T_x, p_y/p_x, and p_{oy}/p_{ox} can be determined from Eqs. 9.53, 9.54, and 9.56. Accordingly, tables can be set up giving M_y, T_y/T_x, p_y/p_x, and p_{oy}/p_{ox} versus the Mach number M_x as the single independent variable for a specified value of k. Table 9.3 is a tabulation of this kind for $k = 1.4$.

In the next example, we consider the effect of back pressure on flow in a converging–diverging nozzle. Key elements of the analysis include determining whether the flow is choked and if a normal shock exists.

EXAMPLE 9.15 ▶

Effect of Back Pressure: Converging–Diverging Nozzle

A converging–diverging nozzle operating at steady state has a throat area of 6.25 cm² and an exit area of 15 cm². Air enters the nozzle with a negligible velocity at a pressure of 6.8 bar and a temperature of 280 K. For air as an ideal gas with $k = 1.4$, determine the mass flow rate, in kg/s, the exit pressure, in bar, and exit Mach number for each of the five following cases. (**a**) Isentropic flow with $M = 0.7$ at the throat. (**b**) Isentropic flow with $M = 1$ at the throat and the diverging portion acting as a diffuser. (**c**) Isentropic flow with $M = 1$ at the throat and the diverging portion acting as a nozzle. (**d**) Isentropic flow through the nozzle with

a normal shock standing at the exit. (**e**) A normal shock stands in the diverging section at a location where the area is 12.5 cm². Elsewhere in the nozzle, the flow is isentropic.

SOLUTION

Known Air flows from specified stagnation conditions through a converging–diverging nozzle having a known throat and exit area.

Find The mass flow rate, exit pressure, and exit Mach number are to be determined for each of five cases.

Schematic and Given Data:

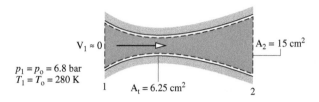

$V_1 \approx 0$
$A_2 = 15 \text{ cm}^2$
$p_1 = p_0 = 6.8 \text{ bar}$
$T_1 = T_0 = 280 \text{ K}$
1   $A_t = 6.25 \text{ cm}^2$ 2

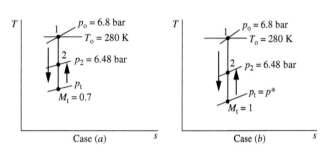

Case (*a*): $p_0 = 6.8$ bar, $T_0 = 280$ K, $p_2 = 6.48$ bar, p_t, $M_t = 0.7$

Case (*b*): $p_0 = 6.8$ bar, $T_0 = 280$ K, $p_2 = 6.48$ bar, $p_t = p^*$, $M_t = 1$

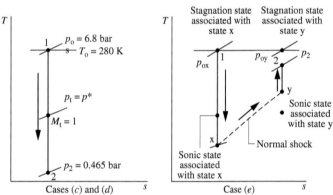

Cases (*c*) and (*d*): $p_0 = 6.8$ bar, $T_0 = 280$ K, $p_t = p^*$, $M_t = 1$, $p_2 = 0.465$ bar

Case (*e*): Stagnation state associated with state x; Stagnation state associated with state y; p_{ox}; 1; p_{oy}; 2; p_2; y; Sonic state associated with state y; Normal shock; x; Sonic state associated with state x

Fig. E9.15

Engineering Model

1. The control volume shown in the accompanying sketch operates at steady state. The T–s diagrams provided locate states within the nozzle.

2. The air is modeled as an ideal gas with $k = 1.4$.

3. Flow through the nozzle is isentropic throughout, except for case e, where a shock stands in the diverging section.

Analysis

a. The accompanying T–s diagram shows the states visited by the gas in this case. The following are known: the Mach number at the throat, $M_t = 0.7$, the throat area, $A_t = 6.25 \text{ cm}^2$, and the exit area, $A_2 = 15 \text{ cm}^2$. The exit Mach number M_2, exit temperature T_2, and exit pressure p_2 can be determined using the identity

$$\frac{A_2}{A^*} = \frac{A_2}{A_t}\frac{A_t}{A^*}$$

With $M_t = 0.7$, Table 9.2 gives $A_t/A^* = 1.09437$. Thus

$$\frac{A_2}{A^*} = \left(\frac{15 \text{ cm}^2}{6.25 \text{ cm}^2}\right)(1.09437) = 2.6265$$

The flow throughout the nozzle, including the exit, is subsonic. Accordingly, with this value for A_2/A^*, Table 9.2 gives $M_2 \approx 0.24$. For $M_2 = 0.24$, $T_2/T_0 = 0.988$, and $p_2/p_0 = 0.959$. Since the stagnation temperature and pressure are 280 K and 6.8 bar, respectively, it follows that $T_2 = 277$ K and $p_2 = 6.52$ bar.

The velocity at the exit is

$$V_2 = M_2 c_2 = M_2\sqrt{kRT_2}$$

$$= 0.24\sqrt{1.4\left(\frac{8314}{28.97}\frac{\text{kJ}}{\text{kg}\cdot\text{R}}\right)(277 \text{ K})\left|\frac{1 \text{ kg}\cdot\text{m/s}^2}{1 \text{ N}}\right|\left|\frac{10^3 \text{ N}\cdot\text{m}}{1 \text{ kJ}}\right|}$$

$$= 80.1 \text{ m/s}$$

The mass flow rate is

$$\dot{m} = \rho_2 A_2 V_2 = \frac{p_2}{RT_2}A_2 V_2$$

$$= \frac{(6.52 \text{ bar})(15 \text{ cm}^2)(80.1 \text{ m/s})}{\left(\frac{8314 \text{ kJ}}{28.97 \text{ kg}\cdot\text{K}}\right)(277°\text{K})} = 0.985 \text{ kg/s}$$

b. The accompanying T–s diagram shows the states visited by the gas in this case. Since $M = 1$ at the throat, we have $A_t = A^*$, and thus, $A_2/A^* = 2.4$. Table 9.2 gives two Mach numbers for this ratio: $M \approx 0.26$ and $M \approx 2.4$. The diverging portion acts as a diffuser in the present part of the example; accordingly, the subsonic value is appropriate. The supersonic value is appropriate in part (c).

Thus, from Table 9.2 we have at $M_2 = 0.26$, $T_2/T_0 = 0.986$, and $p_2/p_0 = 0.953$. Since $T_0 = 280$ K and $p_0 = 6.8$ bar, it follows that $T_2 = 276$ K and $p_2 = 6.48$ bar.

The velocity at the exit is

$$V_2 = M_2 c_2 = M_2\sqrt{kRT_2}$$

$$= 0.26\sqrt{(1.4)\left(\frac{83.14}{28.97}\right)(276)(10^3)} = 86.58 \text{ m/s}$$

The mass flow rate is

$$\dot{m} = \frac{p_2}{RT_2}A_2 V_2 = \frac{(6.48)(15)(86.58)}{\left(\frac{8.314}{28.97}\right)(276)} = 1.062 \text{ kg/s}$$

This is the maximum mass flow rate for the specified geometry and stagnation conditions: the flow is choked.

c. The accompanying T–s diagram shows the states visited by the gas in this case. As discussed in part (b), the exit Mach number in the present part of the example is $M_2 = 2.4$. Using this, Table 9.2 gives $p_2/p_0 = 0.0684$. With $p_0 = 6.8$ bar, the pressure at the exit is $p_2 = 0.465$ bar. Since the nozzle is choked, the mass flow rate is the same as found in part (b).

d. Since a normal shock stands at the exit and the flow upstream of the shock is isentropic, the Mach number M_x and the pressure p_x correspond to the values found in part (c), $M_x = 2.4$, $p_x = 0.465$ bar. Then, from Table 9.3, $M_y \approx 0.52$ and $p_y/p_x = 6.5533$. The pressure downstream of the shock is thus 3.047 bar. This is the exit pressure. The mass flow is the same as found in part (b).

e. The accompanying T–s diagram shows the states visited by the gas. It is known that a shock stands in the diverging portion where the area is $A_x = 12.5 \text{ cm}^2$. Since a shock occurs, the flow is sonic at the throat, so $A_x^* = A_t = 6.25 \text{ cm}^2$. The Mach number M_x can then be found from Table 9.2, by using $A_x/A_x^* = 2$, as $M_x = 2.2$.

The Mach number at the exit can be determined using the identity

$$\frac{A_2}{A_y^*} = \left(\frac{A_2}{A_x^*}\right)\left(\frac{A_x^*}{A_y^*}\right)$$

Introducing Eq. 9.57 to replace A_x^*/A_y^*, this becomes

$$\frac{A_2}{A_y^*} = \left(\frac{A_2}{A_x^*}\right)\left(\frac{p_{oy}}{p_{ox}}\right)$$

where p_{ox} and p_{oy} are the stagnation pressures before and after the shock, respectively. With $M_x = 2.2$, the ratio of stagnation pressures is obtained from Table 9.3 as $p_{oy}/p_{ox} = 0.62812$. Thus

$$\frac{A_2}{A_y^*} = \left(\frac{15 \text{ cm}^2}{6.25 \text{ cm}^2}\right)(0.62812) = 1.51$$

Using this ratio and noting that the flow is subsonic after the shock, Table 9.2 gives $M_2 \approx 0.43$, for which $p_2/p_{oy} = 0.88$.

The pressure at the exit can be determined using the identity

$$p_2 = \left(\frac{p_2}{p_{oy}}\right)\left(\frac{p_{oy}}{p_{ox}}\right)p_{ox} = (0.88)(0.628)(6.8 \text{ bar}) = 3.76 \text{ bar}$$

Since the flow is choked, the mass flow rate is the same as that found in part (b).

 Part (a) of the present example corresponds to the cases labeled b and c on Fig. 9.30. Part (c) corresponds to case d of Fig. 9.30. Part (d) corresponds to case g of Fig. 9.30 and part (e) corresponds to cases e and f.

SKILLS DEVELOPED

Ability to…

- analyze isentropic flow through a converging–diverging nozzle for an ideal gas with constant k.
- understand the occurrence of choked flow and normal shocks in a converging–diverging nozzle for different back pressures.
- analyze the flow through a converging–diverging nozzle when normal shocks are present for an ideal gas with constant k.

Quick Quiz

What is the stagnation temperature, in °R, corresponding to the exit state for case (e)? Ans. 280°K.

CHAPTER SUMMARY AND STUDY GUIDE

In this chapter, we have studied the thermodynamic modeling of internal combustion engines, gas turbine power plants, and compressible flow in nozzles and diffusers. The modeling of cycles is based on the use of air-standard analysis, where the working fluid is considered to be air modeled as an ideal gas.

The processes in internal combustion engines are described in terms of three air-standard cycles: the Otto, Diesel, and dual cycles, which differ from each other only in the way the heat addition process is modeled. For these cycles, we have evaluated the principal work and heat transfers along with two important performance parameters: the mean effective pressure and the thermal efficiency. The effect of varying compression ratio on cycle performance is also investigated.

The performance of simple gas turbine power plants is described in terms of the air-standard Brayton cycle. For this cycle, we evaluate the principal work and heat transfers along with two important performance parameters: the back work ratio and the thermal efficiency. We also consider the effects on performance of irreversibilities and of varying compressor pressure ratio. Three modifications of the simple cycle to improve performance are introduced: regeneration, reheat, and compression with intercooling. Applications related to gas turbines are also considered, including combined gas turbine–vapor power cycles, integrated gasification combined-cycle (IGCC) power plants, and gas turbines for aircraft propulsion. In addition, the Ericsson and Stirling cycles are introduced.

The chapter concludes with the study of compressible flow through nozzles and diffusers. We begin by introducing the momentum equation for steady, one-dimensional flow, the velocity of sound, and the stagnation state. We then consider the effects of area change and back pressure on performance in both subsonic and supersonic flows. Choked flow and the presence of normal shocks in such flows are investigated. Tables are introduced to facilitate analysis for the case of ideal gases with constant specific heat ratio, $k = 1.4$.

The following list provides a study guide for this chapter. When your study of the text and end-of-chapter exercises has been completed, you should be able to

- write out the meanings of the terms listed in the margin throughout the chapter and understand each of the related concepts. The subset of key concepts listed under "Key Engineering Concepts" is particularly important.
- sketch p–v and T–s diagrams of the Otto, Diesel, and dual cycles. Apply the closed system energy balance and the second law along with property data to determine the performance of these cycles, including mean effective pressure, thermal efficiency, and the effects of varying compression ratio.
- sketch schematic diagrams and accompanying T–s diagrams of the Brayton cycle and modifications involving regeneration, reheat, and compression with intercooling. In each case, be able to apply mass and energy balances, the second law, and property data to determine gas turbine power cycle performance, including thermal efficiency, back work ratio, net power output, and the effects of varying compressor pressure ratio.
- analyze the performance of gas turbine–related applications involving combined gas turbine–vapor power plants, IGCC power plants, and aircraft propulsion. You also should be able to apply the principles of this chapter to Ericsson and Stirling cycles.
- discuss for nozzles and diffusers the effects of area change in subsonic and supersonic flows, the effects of back pressure on mass flow rate, and the appearance and consequences of choking and normal shocks.
- analyze the flow in nozzles and diffusers of ideal gases with constant specific heats, as in Examples 9.14 and 9.15.

KEY ENGINEERING CONCEPTS

mean effective pressure	regenerator effectiveness	velocity of sound
air-standard analysis	reheat	Mach number
Otto cycle	intercooler	subsonic and supersonic flow
Diesel cycle	combined cycle	stagnation state
dual cycle	turbojet engine	choked flow
Brayton cycle	compressible flow	normal shock
regenerator	momentum equation	

KEY EQUATIONS

$$\text{mep} = \frac{\text{net work for one cycle}}{\text{displacement volume}}$$	(9.1)	Mean effective pressure for reciprocating piston engines

Otto Cycle

$$\eta = \frac{(u_3 - u_2) - (u_4 - u_1)}{u_3 - u_2} = 1 - \frac{u_4 - u_1}{u_3 - u_2}$$	(9.3)	Thermal efficiency (Fig. 9.3)
$$\eta = 1 - \frac{1}{r^{k-1}}$$	(9.8)	Thermal efficiency (cold air-standard basis)

Diesel Cycle

$$\eta = \frac{W_{\text{cycle}}/m}{Q_{23}/m} = 1 - \frac{Q_{41}/m}{Q_{23}/m} = 1 - \frac{u_4 - u_1}{h_3 - h_2}$$	(9.11)	Thermal efficiency (Fig. 9.5)
$$\eta = 1 - \frac{1}{r^{k-1}}\left[\frac{r_c^k - 1}{k(r_c - 1)}\right]$$	(9.13)	Thermal efficiency (cold air-standard basis)

Brayton Cycle

$$\eta = \frac{\dot{W}_t/\dot{m} - \dot{W}_c/\dot{m}}{\dot{Q}_{\text{in}}/\dot{m}} = \frac{(h_3 - h_4) - (h_2 - h_1)}{h_3 - h_2}$$	(9.19)	Thermal efficiency (Fig. 9.9)
$$\text{bwr} = \frac{\dot{W}_c/\dot{m}}{\dot{W}_t/\dot{m}} = \frac{h_2 - h_1}{h_3 - h_4}$$	(9.20)	Back work ratio (Fig. 9.9)
$$\eta = 1 - \frac{1}{(p_2/p_1)^{(k-1)/k}}$$	(9.25)	Thermal efficiency (cold air-standard basis)
$$\eta_{\text{reg}} = \frac{h_x - h_2}{h_4 - h_2}$$	(9.27)	Regenerator effectiveness for the regenerative gas turbine cycle (Fig. 9.14)

Compressible Flow in Nozzles and Diffusers

$$\mathbf{F} = \dot{m}(\mathbf{V}_2 - \mathbf{V}_1)$$	(9.31)	Momentum equation for steady-state one-dimensional flow
$$c = \sqrt{kRT}$$	(9.37)	Ideal gas velocity of sound
$$M = \mathrm{V}/c$$	(9.38)	Mach number
$$h_o = h + \mathrm{V}^2/2$$	(9.39)	Stagnation enthalpy
$$\frac{T_0}{T} = 1 + \frac{k-1}{2}M^2$$	(9.50)	Isentropic flow function relating temperature and stagnation temperature (constant k)
$$\frac{p_0}{p} = \left(\frac{T_0}{T}\right)^{k/(k-1)} = \left(1 + \frac{k-1}{2}M^2\right)^{k/(k-1)}$$	(9.51)	Isentropic flow function relating pressure and stagnation pressure (constant k)

EXERCISES: THINGS ENGINEERS THINK ABOUT

9.1 Diesel engines are said to produce higher *torque* than gasoline engines. What does that mean?

9.2 Formula One race cars have 2.4 liter engines. What does that signify? How is your car's engine sized in liters?

9.3 What is *metal dusting*, which has been observed in the production of syngas and in other chemical processes?

9.4 A car magazine says that your car's engine has more power when the ambient temperature is low. Do you agree?

9.5 You jump off a raft into the water in the middle of a lake. What direction does the raft move? Explain.

9.6 Why are the external surfaces of a lawn mower engine covered with fins?

9.7 The term *regeneration* is used to describe the use of regenerative feedwater heaters in vapor power plants and regenerative heat exchangers in gas turbines. In what ways are the purposes of these devices similar? How do they differ?

9.8 While advanced combined-cycle power plants are able to achieve a thermal efficiency of 60%, what other performance features do such plants enjoy that owner-operators may value even more highly?

9.9 What is the purpose of the gas turbine–powered *auxiliary power units* commonly seen at airports near commercial aircraft?

9.10 Automakers have developed prototype gas turbine–powered vehicles, but the vehicles have not been generally marketed to consumers. Why?

9.11 In making a quick stop at a friend's home, is it better to let your car's engine idle or turn it off and restart when you leave?

9.12 What is the difference between the *diesel* and *gasoline* fuel used in internal combustion engines?

CHECKING UNDERSTANDING

9.1 The thermal efficiency expression given by Eq. 5.9 applies for the (a) Carnot cycle only, (b) Carnot, Otto, and Diesel cycles, (c) Carnot, Ericsson, and Stirling cycles, (d) Carnot, dual, and ideal Brayton cycles.

9.2 In Fig. 9.19, an intercooler separates the two compressor stages. How does the intercooler contribute to improving overall system performance?

9.3 For a specified compression ratio, and assuming a cold air-standard analysis for simplicity, which cycle has the greater thermal efficiency: an Otto cycle or a Diesel cycle?

9.4 The compression processes of the Otto and ideal Brayton cycles are each represented by an isentropic process; yet the way compression occurs in these cycles differ. Explain.

9.5 The value of the back work ratio of a Brayton cycle is typically (a) much less than for a Rankine cycle, (b) much greater than for a Rankine cycle, (c) about the same as for a Rankine cycle, (d) cannot be determined without more information.

9.6 How is combustion initiated in a conventional gasoline-fueled internal combustion engine?

9.7 In a *cold air-standard analysis*, what is assumed about the specific heats and specific heat ratio?

9.8 Referring to cycle 1–2–3–4–1 of Fig. 9.10, the net work per unit of mass flowing is represented on the p–v diagram by area _____. The heat rejected per unit of mass flowing is represented on the T–s diagram by area _____.

9.9 Referring to Fig. 9.18, if the temperature at the exit of the intercooler, state d, is the same as at state 1, locate the new states d and 2 on the p–v and T–s diagrams for the two-stage compressor, keeping states 1, c, and 2′ the same.

9.10 Sketch a Carnot gas power cycle on the p–v and T–s diagrams of Fig. 9.21 for the Stirling gas power cycle, assuming each cycle has the same isothermal heat addition process at temperature T_H: Process 3–4. How do the thermal efficiencies of these cycles compare?

9.11 When a regenerator is introduced in a simple Brayton cycle, the thermal efficiency (a) increases, (b) decreases, (c) increases or decreases depending on the regenerator effectiveness, (d) remains the same.

9.12 Sketch the T–s diagram for the turbojet engine shown in Fig. 9.25a using the following assumptions: The diffuser and nozzle processes are isentropic, the compressor and turbine have isentropic

efficiencies of 85% and 90%, respectively, and there is a 5% pressure drop for flow through the combustor.

9.13 Referring to Fig. 9.17, sketch an internally reversible process from state 1 to pressure p_2 representing compression *with heating*. How does the magnitude of the compressor work input per unit of mass flowing for this case compare to that for compression with cooling?

9.14 Referring to the heat exchanger shown by the dashed line in Fig. 9.9, is this an actual or virtual component of the air-standard Brayton cycle? What is its purpose?

9.15 In the diffuser of a turbojet engine the incoming air is decelerated and its pressure (a) increases, (b) decreases, (c) remains the same.

9.16 **Figure P9.16C** shows an isentropic expansion through a turbine at steady state. The area on the diagram that represents the work developed by the turbine per unit of mass flowing is _____.

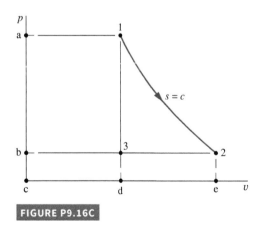

FIGURE P9.16C

9.17 From inspection of Eq. 9.36b, in which phase of a substance do you expect the velocity of sound to be greatest: gas, liquid, or solid? Explain.

9.18 Referring to Fig. 9.32, identify cases involving entropy production (a) within the nozzle, (b) within the exhaust region.

9.19 Can a normal shock occur in a converging channel? Explain.

9.20 Consider a jet engine operating at steady state while on a test stand. The test stand imposes a force on the engine (a) in the direction of flow, (b) opposite to the direction of flow. Explain.

Indicate whether the following statements are true or false. Explain.

9.21 Although the exhaust gas temperature of a simple gas turbine is typically well above the ambient temperature, the exhaust gas is normally discarded to the surroundings for operational simplicity.

9.22 In a gas turbine operating on a *closed* basis the working fluid receives an energy input by heat transfer from an external source.

9.23 In a *two-stroke* internal combustion engine, the intake, compression, expansion, and exhaust operations are accomplished in two revolutions of crankshaft.

9.24 The thermal efficiency of a power cycle formed by combining a gas turbine power cycle and a vapor power cycle is the sum of the individual thermal efficiencies.

9.25 An ideal gas ($k = 1.4$) has a velocity of 200 m/s, a temperature of 335 K, and a pressure of 8 bar. The corresponding stagnation temperature is less than 335 K.

PROBLEMS: DEVELOPING ENGINEERING SKILLS

C Problem may require use of appropriate computer software in order to complete.

Otto, Diesel, and Dual Cycles

9.1 At the beginning of the compression process of an air-standard Otto cycle, $p_1 = 1$ bar and $T_1 = 300$ K. The compression ratio is 8.5 and the heat addition per unit mass of air is 1400 kJ/kg. Determine (a) the net work, in kJ/kg, (b) the thermal efficiency of the cycle, (c) the mean effective pressure, in bar, (d) the maximum temperature of the cycle, in K.

9.2 **C** At the beginning of the compression process of an air-standard Otto cycle, $p_1 = 100$ kPa and $T_1 = 300$ K. The heat addition per unit mass of air is 1350 kJ/kg. Plot each of the following versus compression ratio ranging from 1 to 12: (a) the net work, in kJ/kg, (b) the thermal efficiency of the cycle, (c) the mean effective pressure, in kPa, (d) the maximum temperature of the cycle, in K.

9.3 At the beginning of the compression process of an air-standard Otto cycle, $p_1 = 1$ bar, $T_1 = 290$ K, $V_1 = 400$ cm³. The maximum temperature in the cycle is 2200 K and the compression ratio is 8. Determine

    a. the heat addition, in kJ.

    b. the net work, in kJ.

    c. the thermal efficiency.

    d. the mean effective pressure, in bar.

9.4 **C** Plot each of the quantities specified in parts (a) through (d) of Problem 9.3 versus the compression ratio ranging from 2 to 12.

9.5 **C** An air-standard Otto cycle has a compression ratio of 8 and the temperature and pressure at the beginning of the compression process are 300 K and 100 kPa, respectively. The mass of air is 6.8×10^{-4} kg. The heat addition is 0.9 kJ. Determine

    a. the maximum temperature, in K.

    b. the maximum pressure, in kPa.

    c. the thermal efficiency.

    d. To investigate the effects of varying compression ratio, plot each of the quantities calculated in parts (a) through (c) for compression ratios ranging from 2 to 12.

9.6 Solve Problem 9.5 on a cold air-standard basis with specific heats evaluated at 300 K.

9.7 **C** At the beginning of the compression process in an air-standard Otto cycle, $p_1 = 100$ kPa and $T_1 = 294$ K. Plot the thermal efficiency and mean effective pressure, in kPa, for maximum cycle temperatures ranging from 1100 to 2800 K and compression ratios of 6, 8, and 10.

9.8 Solve Problem 9.7 on a cold air-standard basis using $k = 1.4$.

9.9 An air-standard Otto cycle has a compression ratio of 7.5. At the beginning of compression, $p_1 = 85$ kPa and $T_1 = 32°C$. The mass of air is 2 g, and the maximum temperature in the cycle is 960 K. Determine

    a. the heat rejection, in kJ.

    b. the net work, in kJ.

    c. the thermal efficiency.

    d. the mean effective pressure, in kPa.

9.10 A four-cylinder, four-stroke internal combustion engine operates at 2700 RPM. The processes within each cylinder are modeled as an air-standard Otto cycle with a pressure of 100 kPa, a temperature of 25°C, and a volume of 5.4×10^{-4} m³ at the beginning of compression. The compression ratio is 10, and maximum pressure in the cycle is 7500 kPa. Determine, using a cold air-standard analysis with $k = 1.4$, the power developed by the engine, in horsepower, and the mean effective pressure, in kPa.

9.11 Consider a modification of the air-standard Otto cycle in which the isentropic compression and expansion processes are each replaced with polytropic processes having $n = 1.3$. The compression ratio is 9 for the modified cycle. At the beginning of compression, $p_1 = 1$ bar and $T_1 = 300$ K and $V_1 = 2270$ cm³. The maximum temperature during the cycle is 2000 K. Determine

    a. the heat transfer and work in kJ, for each process in the modified cycle.

    b. the thermal efficiency.

    c. the mean effective pressure, in bar.

9.12 A four-cylinder, four-stroke internal combustion engine has a bore of 65 mm and a stroke of 53 mm. The clearance volume is 12% of the cylinder volume at bottom dead center and the crankshaft rotates at 3600 RPM. The processes within each cylinder are modeled as an air-standard Otto cycle with a pressure of 100 kPa and a temperature of 38°C at the beginning of compression. The maximum temperature in the cycle is 2890 K. Based on this model, calculate the net work per cycle, in kJ, and the power developed by the engine, in kW.

9.13 The pressure-specific volume diagram of the air-standard *Lenoir* cycle is shown in Fig. P9.13. The cycle consists of constant volume heat addition, isentropic expansion, and constant pressure compression. For the cycle, $p_1 = 100$ kPa and $T_1 = 300$ K. The mass of air is 1.92×10^{-3} kg, and the maximum cycle temperature is 890 K. Assuming $c_v = 0.1716$ kJ/kg · K, determine for the cycle

    a. the net work, in kJ.

    b. the thermal efficiency.

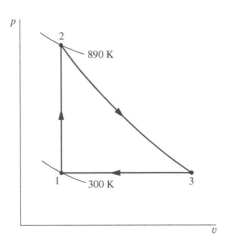

FIGURE P9.13

9.14 The pressure and temperature at the beginning of compression of an air-standard Diesel cycle are 95 kPa and 300 K, respectively. At the end of the heat addition, the pressure is 7.2 MPa and the temperature is 2150 K. Determine

  a. the compression ratio.

  b. the cutoff ratio.

  c. the thermal efficiency of the cycle.

  d. the mean effective pressure, in kPa.

9.15 Solve Problem 9.14 on a cold air-standard basis with specific heats evaluated at 300 K.

9.16 **C** The conditions at the beginning of compression in an air-standard Diesel cycle are fixed by $p_1 = 200$ kPa, $T_1 = 380$ K. The compression ratio is 20 and the cutoff ratio is 1.8. For $k = 1.4$, determine

  a. the maximum temperature, in K.

  b. the heat addition per unit mass, in kJ/kg.

  c. the net work per unit mass, in kJ/kg.

  d. the thermal efficiency.

  e. the mean effective pressure, in kPa.

  f. To investigate the effects of varying compression ratio, plot each of the quantities calculated in parts (a) through (e) for compression ratios ranging from 5 to 25.

9.17 Consider an air-standard Diesel cycle. Operating data at principal states in the cycle are given in the table below. The states are numbered as in Fig. 9.5. Determine

  a. the cutoff ratio.

  b. the heat addition per unit mass, in kJ/kg.

  c. the net work per unit mass, in kJ/kg.

  d. the thermal efficiency.

State	T (K)	p (kPa)	u (kJ/kg)	h (kJ/kg)
1	380	100	271.69	380.77
2	1096.6	5197.6	842.40	1157.18
3	1864.2	5197.6	1548.47	2082.96
4	875.2	230.1	654.02	905.26

9.18 **C** An air-standard Diesel cycle has a compression ratio of 16 and a cutoff ratio of 2. At the beginning of compression, $p_1 = 100$ kPa, $V_1 = 0.01$ m^3, and $T_1 = 300$ K. Calculate

  a. the heat added, in kJ.

  b. the maximum temperature in the cycle, in K.

  c. the thermal efficiency.

  d. the mean effective pressure, in kPa

  e. To investigate the effects of varying compression ratio, plot each of the quantities calculated in parts (a) through (d) for compression ratios ranging from 5 to 18 and for cutoff ratios of 1.5, 2, and 2.5.

9.19 The displacement volume of an internal combustion engine is 5.6 liters. The processes within each cylinder of the engine are modeled as an air-standard Diesel cycle with a cutoff ratio of 2.4. The state of the air at the beginning of compression is fixed by $p_1 = 95$ kPa, $T_1 = 27°$C, and $V_1 = 6.0$ liters. Determine the net work per cycle, in kJ, the power developed by the engine, in kW, and the thermal efficiency, if the cycle is executed 1500 times per min.

9.20 At the beginning of the compression process of an air-standard Diesel cycle, $p_1 = 95$ kPa and $T_1 = 300$ K. The maximum temperature is 1800 K and the mass of air is 12 g. For compression ratios of 15, 18, and 21, determine the net work developed, in kJ, the thermal efficiency, and the mean effective pressure, in kPa.

9.21 At the beginning of compression in an air-standard Diesel cycle, $p_1 = 170$ kPa, $V_1 = 0.016$ m^3, and $T_1 = 315$ K. The compression ratio is 15 and the maximum cycle temperature is 1400 K. Determine

  a. the mass of air, in kg.

  b. the heat addition and heat rejection per cycle, each in kJ.

  c. the net work, in kJ, and the thermal efficiency.

9.22 **C** At the beginning of the compression process in an air-standard Diesel cycle, $p_1 = 1$ bar and $T_1 = 300$ K. For maximum cycle temperatures of 1200, 1500, 1800, and 2100 K, plot the heat addition per unit of mass, in kJ/kg, the net work per unit of mass, in kJ/kg, the mean effective pressure, in bar, and the thermal efficiency, each versus compression ratio ranging from 5 to 20.

9.23 **C** An air-standard Diesel cycle has a maximum temperature of 1800 K. At the beginning of compression, $p_1 = 95$ kPa and $T_1 = 300$ K. The mass of air is 12 g. For compression ratios ranging from 15 to 25, plot

  a. the net work of the cycle, in kJ.

  b. the thermal efficiency.

  c. the mean effective pressure, in kPa.

9.24 **C** The state at the beginning of compression of an air-standard Diesel cycle is fixed by $p_1 = 100$ kPa and $T_1 = 310$ K. The compression ratio is 15. For cutoff ratios ranging from 1.5 to 2.5, plot

  a. the maximum temperature, in K.

  b. the pressure at the end of the expansion, in kPa.

  c. the net work per unit mass of air, in kJ/kg.

  d. the thermal efficiency.

9.25 An air-standard dual cycle has a compression ratio of 10. At the beginning of compression, $p_1 = 100$ kPa, $T_1 = 298$ K, and $V_1 = 12$ L. The heat addition is 20 kJ, with one half added at constant volume and one-half added at constant pressure. Determine

  a. the temperatures at the end of each heat addition process, in K.

  b. the net work of the cycle per unit mass of air, in kJ/kg.

  c. the thermal efficiency.

  d. the mean effective pressure, in kPa.

9.26 Consider an air-standard dual cycle. Operating data at principal states in the cycle are given in the table below. The states are numbered as in Fig. 9.7. If the mass of air is 0.05 kg, determine

  a. the cutoff ratio.

  b. the heat addition to the cycle, in kJ.

c. the heat rejection from the cycle, in kJ.

d. the net work, in kJ.

e. the thermal efficiency.

State	T (K)	p (kPa)	u (kJ/kg)	h (kJ/kg)
1	300	95	214.07	300.19
2	862.4	4372.8	643.35	890.89
3	1800	9126.9	1487.2	2003.3
4	1980	9126.9	1659.5	2227.1
5	840.3	265.7	625.19	866.41

9.27 [C] The pressure and temperature at the beginning of compression in an air-standard dual cycle are 95 kPa and 300 K, respectively. The compression ratio is 15 and the heat addition per unit mass of air is 1860 kJ/kg. At the end of the constant volume heat addition process, the pressure is 8.2 MPa. Determine

a. the net work of the cycle per unit mass of air, in kJ/kg.

b. the heat rejection for the cycle per unit mass of air, in kJ/kg.

c. the thermal efficiency.

d. the cutoff ratio.

e. To investigate the effects of varying compression ratio, plot each of the quantities calculated in parts (a) through (d) for compression ratios ranging from 10 to 28.

9.28 [C] An air-standard dual cycle has a compression, ratio of 16. At the beginning of compression, $p_1 = 100$ kPa, $V_1 = 0.01$ m^3, and $T_1 = 10°C$. The pressure doubles during the constant volume heat addition process. For a maximum cycle temperature of 1670 K, determine

a. the heat addition to the cycle, in kJ.

b. the net work of the cycle, in kJ.

c. the thermal efficiency.

d. the mean effective pressure, in kPa.

e. To investigate the effects of varying maximum cycle temperature, plot each of the quantities calculated in parts (a) through (d) for maximum cycle temperatures ranging from 1670 to 2220 K.

9.29 [C] At the beginning of the compression process in an air-standard dual cycle, $p_1 = 1$ bar and $T_1 = 300$ K. The total heat addition is 1000 kJ/kg. Plot the net work per unit of mass, in kJ/kg, the mean effective pressure, in bar, and the thermal efficiency versus compression ratio for different fractions of constant-volume and constant-pressure heat addition. Consider compression ratio ranging from 10 to 20.

Brayton Cycle

9.30 An ideal air-standard Brayton cycle operating at steady state produces 10 MW of power. Operating data at principal states in the cycle are given in the table below. The states are numbered as in Fig. 9.9. Sketch the *T–s* diagram for the cycle and determine

a. the mass flow rate of air, in kg/s.

b. the rate of heat transfer, in kW, to the working fluid passing through the heat exchanger.

c. the thermal efficiency.

State	p (kPa)	T (K)	h (kJ/kg)
1	100	300	300.19
2	1200	603.5	610.65
3	1200	1450	1575.57
4	100	780.7	800.78

9.31 For an ideal Brayton cycle on a cold air-standard basis show that

a. the back work ratio is given by

$$bwr = T_1/T_4$$

where T_1 is the temperature at the compressor inlet and T_4 is the temperature at the turbine exit.

b. the temperature at the compressor exit that maximizes the net work developed per unit of mass flowing is given by

$$T_2 = (T_1 \, T_3)^{1/2}$$

where T_1 is the temperature at the compressor inlet and T_3 is the temperature at the turbine inlet.

9.32 Air enters the compressor of an ideal cold air-standard Brayton cycle at 120 kPa, 350 K, with a mass flow rate of 8 kg/s. The compressor pressure ratio is 12, and the turbine inlet temperature is 1480 K. For $k = 1.4$, calculate

a. the thermal efficiency of the cycle.

b. the back work ratio.

c. the net power developed, in kW.

9.33 The rate of heat addition to an air-standard Brayton cycle is 9.9×10^5 kW. The pressure ratio for the cycle is 14 and the minimum and maximum temperatures are 300 K and 1670 K, respectively. Determine

a. the thermal efficiency of the cycle.

b. the mass flow rate of air, in kg/s.

c. the net power developed by the cycle, in kJ/s.

9.34 Solve Problem 9.33 on a cold air-standard basis with specific heats evaluated at 300 K.

9.35 [C] Consider an ideal air-standard Brayton cycle with minimum and maximum temperatures of 300 K and 1500 K, respectively. The pressure ratio is that which maximizes the net work developed by the cycle per unit of mass of air flow. On a cold air-standard basis, calculate

a. the compressor and turbine work per unit of mass of air flow, each in kJ/kg.

b. the thermal efficiency of the cycle.

c. Plot the thermal efficiency versus the maximum cycle temperature ranging from 1200 to 1800 K.

9.36 [C] The compressor inlet temperature of an ideal air-standard Brayton cycle is 300 K and the maximum allowable turbine inlet temperature is 1440 K. Plot the net work developed per unit mass of air flow, in kJ/kg, and the thermal efficiency versus compressor pressure ratio for pressure ratios ranging from 12 to 24. Using your plots, estimate the pressure ratio for maximum net work and the corresponding value of thermal efficiency. Compare the results to those obtained in analyzing the cycle on a cold air-standard basis.

9.37 Air enters the compressor of an air-standard Brayton cycle with a volumetric flow rate of 50 m^3/s at 1 bar, 285 K. The compressor pressure ratio is 18, and the maximum cycle temperature is 2050 K. For the compressor, the isentropic efficiency is 91% and for the turbine the isentropic efficiency is 94%. Determine

a. the net power developed, in MW.

b. the rate of heat addition in the combustor, in MW.

c. the thermal efficiency of the cycle.

9.38 Air enters the compressor of a simple gas turbine at $p_1 = 95$ kPa, $T_1 = 300$ K. The isentropic efficiencies of the compressor and turbine are 83 and 87%, respectively. The compressor pressure ratio is 14 and the temperature at the turbine inlet is 1390 K. The net power developed is 1465 kW. On the basis of an air-standard analysis, calculate

a. the volumetric flow rate of the air entering the compressor, in m^3/s.

b. the temperatures at the compressor and turbine exits, each in K.

c. the thermal efficiency of the cycle.

9.39 Solve Problem 9.38 on a cold air-standard basis with specific heats evaluated at 300 K.

9.40 Air enters the compressor of a simple gas turbine at 100 kPa, 300 K, with a volumetric flow rate of 5 m^3/s. The compressor pressure ratio is 10 and its isentropic efficiency is 85%. At the inlet to the turbine, the pressure is 950 kPa, and the temperature is 1400 K. The turbine has an isentropic efficiency of 88% and the exit pressure is 100 kPa. On the basis of an air-standard analysis,

a. develop a full accounting of the *net* exergy increase of the air passing through the gas turbine combustor, in kW.

b. devise and evaluate an exergetic efficiency for the gas turbine cycle.

Let T_0 = 300 K, p_0 = 100 kPa.

9.41 Air enters the compressor of a simple gas turbine at 100 kPa, 27°C, and exits at 600 kPa, 268°C. The air enters the turbine at 838°C, 600 kPa and expands to 492°C, 100 kPa. The compressor and turbine operate adiabatically, and kinetic and potential energy effects are negligible. On the basis of an air-standard analysis,

a. develop a full accounting of the *net* exergy increase of the air passing through the gas turbine combustor, in kJ/kg.

b. devise and evaluate an exergetic efficiency for the gas turbine cycle.

Let T_0 = 27°C, p_0 = 100 kPa.

Regeneration, Reheat, and Compression with Intercooling

9.42 An ideal air-standard regenerative Brayton cycle produces 10 MW of power. Operating data at principal states in the cycle are given in the table below. The states are numbered as in Fig. 9.14. Sketch the *T–s* diagram and determine

a. the mass flow rate of air, in kg/s.

b. the rate of heat transfer, in kW, to the working fluid passing through the combustor.

c. the thermal efficiency.

State	p (kPa)	T (K)	h (kJ/kg)
1	100	300	300.19
2	1200	603.5	610.65
x	1200	780.7	800.78
3	1200	1450	1575.57
4	100	780.7	800.78
y	100	603.5	610.65

9.43 The cycle of Problem 9.42 is modified to include the effects of irreversibilities in the adiabatic expansion and compression processes. The regenerator effectiveness is 100%. If the states at the compressor and turbine inlets remain unchanged, the cycle produces 10 MW of power, and the compressor and turbine isentropic efficiencies are both 80%, determine

a. the pressure, in kPa, temperature, in K, and enthalpy, in kJ/kg, at each principal state of the cycle and sketch the *T–s* diagram.

b. the mass flow rate of air, in kg/s.

c. the rate of heat transfer, in kW, to the working fluid passing through the combustor.

d. the thermal efficiency.

9.44 Air enters the compressor of a cold air-standard Brayton cycle with regeneration at 100 kPa, 300 K, with a mass flow rate of 6 kg/s. The compressor pressure ratio is 10, and the turbine inlet temperature is 1400 K. The turbine and compressor each have isentropic efficiencies of 80% and the regenerator effectiveness is 80%. For k = 14 calculate

a. the thermal efficiency of the cycle.

b. the back work ratio.

c. the net power developed, in kW.

d. the rate of exergy destruction in the regenerator, in kW, for T_0 = 300 K.

9.45 **C** Air enters the compressor of a regenerative air-standard Brayton cycle with a volumetric flow rate of 60 m^3/s at 0.8 bar, 280 K. The compressor pressure ratio is 20, and the maximum cycle temperature is 2100 K. For the compressor, the isentropic efficiency is 92% and for the turbine the isentropic efficiency is 95%. For a regenerator effectiveness of 85%, determine

a. the net power developed, in MW.

b. the rate of heat addition in the combustor, in MW.

c. the thermal efficiency of the cycle.

d. Plot the quantities calculated in parts (a) through (c) for regenerator effectiveness values ranging from 0 to 100%. Discuss.

9.46 **C** Reconsider Problem 9.38, but include a regenerator in the cycle. For regenerator effectiveness values ranging from 0 to 100%, plot

a. the thermal efficiency.

b. the percent decrease in heat addition to the air.

9.47 An air-standard Brayton cycle has a compressor pressure ratio of 10. Air enters the compressor at p_1 = 100 kPa, T_1 = 20°C with a mass flow rate of 11 kg/s. The turbine inlet temperature is 1222 K. Calculate the thermal efficiency and the net power developed, in kW, if

a. the turbine and compressor isentropic efficiencies are each 100%.

b. the turbine and compressor isentropic efficiencies are 88 and 84%, respectively.

c. the turbine and compressor isentropic efficiencies are 88 and 84%, respectively, and a regenerator with an effectiveness of 80% is incorporated.

9.48 Air enters the compressor of a regenerative gas turbine with a volumetric flow rate of 150 m^3/s at 100 kPa, 25°C, and is compressed to 410 kPa. The air then passes through the regenerator and exits at 620 K. The temperature at the turbine inlet is 950 K. The compressor and turbine each has an isentropic efficiency of 84%. Using an air-standard analysis, calculate

a. the thermal efficiency of the cycle.

b. the regenerator effectiveness.

c. the net power output, in kJ/s.

9.49 Air enters the turbine of a gas turbine at 1100 kPa, 1100 K, and expands to 100 kPa in two stages. Between the stages, the air is reheated at a constant pressure of 325 kPa to 1100 K. The expansion through each turbine stage is isentropic. Determine, in kJ per kg of air flowing

a. the work developed by each stage.

b. the heat transfer for the reheat process.

c. the increase in net work as compared to a single stage of expansion with no reheat.

9.50 Air enters the turbine of a gas turbine at 1100 kPa, 1100 K, and expands to 100 kPa in two stages. Between the stages, the air is reheated at a constant pressure of 325 kPa to 1100 K. Each turbine stage has isentropic efficiencies of 80%. Determine

a. the work developed by each stage, in kJ per kg of air flowing.

b. the heat transfer for the reheat process, in kJ per kg of air flowing.

c. the percentage increase in net work as compared to a single stage of expansion with no reheat, if the isentropic efficiency of the single stage turbine is 80%.

9.51 Air enters the compressor of a cold air-standard Brayton cycle with regeneration and reheat at 100 kPa, 300 K, with a mass flow rate of 6 kg/s. The compressor pressure ratio is 10, and the inlet temperature for each turbine stage is 1400 K. The turbine stages and compressor each have isentropic efficiencies of 80% and the regenerator effectiveness is 80%. For $k = 1.4$, calculate

a. the thermal efficiency of the cycle.

b. the back work ratio.

c. the net power developed, in kW.

d. the rates of exergy destruction in the compressor and each turbine stage as well as the regenerator, in kW, for $T_0 = 300$ K.

9.52 If the inlet state and the exit pressure are specified for a two-stage turbine with reheat between the stages and operating at steady state, show that the maximum total work output is obtained when the pressure ratio is the same across each stage. Use a cold air-standard analysis assuming that each compression process is isentropic, there is no pressure drop through the reheater, and the temperature at the inlet to each turbine stage is the same. Kinetic and potential energy effects can be ignored.

9.53 A two-stage air compressor operates at steady state, compressing 0.15 m³/min of air from 100 kPa, 300 K, to 1100 kPa. An intercooler between the two stages cools the air to 300 K at a constant pressure of 325 kPa. The compression processes are isentropic. Calculate the power required to run the compressor, in kW, and compare the result to the power required for isentropic compression from the same inlet state to the same final pressure.

9.54 Air enters a two-stage compressor operating at steady state at 1 bar, 290 K. The overall pressure ratio across the stages is 16 and each stage operates isentropically. Intercooling occurs at the pressure that minimizes total compressor work, as determined in Example 9.10. Air exits the intercooler at 290 K. Assuming ideal gas behavior with $k = 1.4$, determine

a. the intercooler pressure, in bar, and the heat transfer, in kJ per kg of air flowing.

b. the work required for each compressor stage, in kJ per kg of air flowing.

9.55 A two-stage air compressor operates at steady state, compressing 0.15 m³/s of air from 100 kPa, 300 K, to 1100 kPa. An intercooler between the two stages cools the air to 300 K at a constant pressure of 325 kPa. Each compressor stage has an isentropic efficiency of 80%. Calculate the power required to run the compressor, in kW, and compare the result to the power required for a single stage compression from the same inlet state to the same final pressure, if the isentropic efficiency of the single stage compressor is 80%.

9.56 Air enters a compressor operating at steady state at 96 kPa, 16°C, with a volumetric flow rate of 162 m³/min. The compression occurs in two stages, with each stage being a polytropic process with $n = 1.27$. The air is cooled to 27°C between the stages by an intercooler operating at 310 kPa Air exits the compressor at 1 MPa. Determine, in kJ per min,

a. the power and heat transfer rate for each compressor stage.

b. the heat transfer rate for the intercooler.

9.57 An air-standard regenerative Brayton cycle operating at steady state with intercooling and reheat produces 10 MW of power. Operating data at principal states in the cycle are given in the table below. The states are numbered as in Fig. 9.19. Sketch the T–s diagram for the cycle and determine

a. the mass flow rate of air, in kg/s.

b. the rate of heat transfer, in kW, to the working fluid passing through each combustor.

c. the thermal efficiency.

State	p (kPa)	T (K)	h (kJ/kg)
1	100	300	300.19
2	300	410.1	411.22
3	300	300	300.19
4	1200	444.8	446.50
5	1200	1111.0	1173.84
6	1200	1450	1575.57
7	300	1034.3	1085.31
8	300	1450	1575.57
9	100	1111.0	1173.84
10	100	444.8	446.50

Other Gas Power System Applications

9.58 Air at 30 kPa, 240 K, and 200 m/s enters a turbojet engine in flight. The air mass flow rate is 26 kg/s. The compressor pressure ratio is 11, the turbine inlet temperature is 1360 K, and air exits the nozzle at 30 kPa. The diffuser and nozzle processes are isentropic, the compressor and turbine have isentropic efficiencies of 85% and 88%, respectively, and there is no pressure drop for flow through the combustor. Kinetic energy is negligible everywhere except at the diffuser inlet and the nozzle exit. On the basis of air-standard analysis, determine

a. the pressures, in kPa, and temperatures, in K, at each principal state.

b. the rate of heat addition to the air passing through the combustor, in kJ/s.

c. the velocity at the nozzle exit, in m/s.

9.59 Air enters the diffuser of a turbojet engine with a mass flow rate of 39 kg/s at 62 kPa, 233 K, and a velocity of 225 m/s. The pressure ratio for the compressor is 12, and its isentropic efficiency is 88%. Air enters the turbine at 1333 K with the same pressure as at the exit of the compressor. Air exits the nozzle at 62 kPa. The diffuser operates isentropically and the nozzle and turbine have isentropic efficiencies of 92% and 90%, respectively. On the basis of an air-standard analysis, calculate

a. the rate of heat addition, in kJ/h.

b. the pressure at the turbine exit, in kPa.

c. the compressor power input, in kJ/h.

d. the velocity at the nozzle exit, in m/s.

Neglect kinetic energy except at the diffuser inlet and the nozzle exit.

9.60 Consider the addition of an afterburner to the turbojet in Problem 9.58 that raises the temperature at the inlet of the nozzle to 1260 K. Determine the velocity at the nozzle exit, in m/s.

9.61 Consider the addition of an afterburner to the turbojet in Problem 9.59 that raises the temperature at the inlet of the nozzle to 1222 K. Determine the velocity at the nozzle exit, in m/s.

9.62 Air enters the diffuser of a ramjet engine at 40 kPa, 233 K, with a velocity of 480 m/s, and decelerates essentially to zero velocity. After combustion, the gases reach a temperature of 1111 K before being discharged through the nozzle at 40 kPa. On the basis of an air-standard analysis, determine

a. the pressure at the diffuser exit, in kPa.

b. the velocity at the nozzle exit, in m/s.

Neglect kinetic energy except at the diffuser inlet and the nozzle exit.

9.63 Air enters the diffuser of a ramjet engine at 40 kPa, 240 K, with a velocity of 2500 km/h and decelerates to negligible velocity. On the basis of an air-standard analysis, the heat addition is 1080 kJ per kg of air passing through the engine. Air exits the nozzle at 40 kPa. Determine

a. the pressure at the diffuser exit, in kPa.

b. the velocity at the nozzle exit, in m/s.

Neglect kinetic energy except at the diffuser inlet and the nozzle exit.

9.64 A turboprop engine (Fig. 9.27a) consists of a diffuser, compressor, combustor, turbine, and nozzle. The turbine drives a propeller as well as the compressor. Air enters the diffuser with a volumetric flow rate of 83.7 m³/s at 40 kPa, 240 K, and a velocity of 180 m/s, and decelerates essentially to zero velocity. The compressor pressure ratio is 10 and the compressor has an isentropic efficiency of 85%. The turbine inlet temperature is 1140 K, and its isentropic efficiency is 85%. The turbine exit pressure is 50 kPa. Combustion occurs at constant pressure. Flow through the diffuser and nozzle is isentropic. Using an air-standard analysis, determine

a. the power delivered to the propeller, in MW.

b. the velocity at the nozzle exit, in m/s.

Neglect kinetic energy except at the diffuser inlet and the nozzle exit.

9.65 A turboprop engine consists of a diffuser, compressor, combustor, turbine, and nozzle. The turbine drives a propeller as well as the compressor. Air enters the diffuser at 83 kPa, 256 K, with a volumetric flow rate of 850 m³/min and a velocity of 156 m/s. In the diffuser, the air decelerates isentropically to negligible velocity. The compressor pressure ratio is 9, and the turbine inlet temperature is 1170 K. The turbine exit pressure is 172.5 kPa, and the air expands to 83 kPa through a nozzle. The compressor and turbine each has an isentropic efficiency of 87%, and the nozzle has an isentropic efficiency of 95%. Using an air-standard analysis, determine

a. the power delivered to the propeller, in kW.

b. the velocity at the nozzle exit, in m/s.

Neglect kinetic energy except at the diffuser inlet and the nozzle exit.

9.66 A combined gas turbine–vapor power plant operates as shown in **Fig. P9.66**. Pressure and temperature data are given at principal states, and the net power developed by the gas turbine is 147 MW. Using air-standard analysis for the gas turbine, determine

a. the net power, in MW, developed by the power plant.

b. the overall thermal efficiency of the plant.

Stray heat transfer and kinetic and potential energy effects can be ignored.

9.67 A combined gas turbine–vapor power plant operates as in Fig. 9.22. Steady-state data at principal states of the combined cycle are given in the table below. An air-standard anaylsis is assumed for the gas turbine in which the air passing through the combustor receives energy by heat transfer at a rate of 50 MW. Except for the combustor,

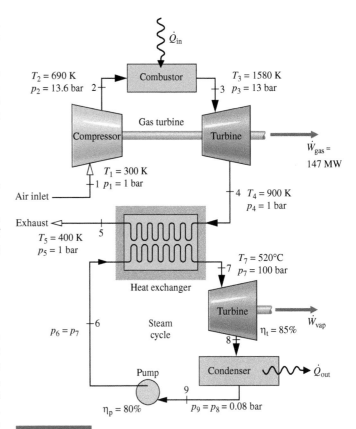

FIGURE P9.66

all components operate adiabatically. Kinetic and potential energy effects are negligible. Determine.

a. the mass flow rates of the air, steam, and cooling water, each in kg/s.

b. the net power developed by the gas turbine cycle and the vapor cycle, respectively, each in MW.

c. the thermal efficiency of the combined cycle.

State	p (bar)	T (°C)	h (kJ/kg)
1	1	25	298.2
2	14	—	691.4
3	14	1250	1663.9
4	1	—	923.2
5	1	200	475.3
6	125	—	204.5
7	125	500	3341.8
8	0.1	—	2175.6
9	0.1	—	191.8
10	—	20	84.0
11	—	35	146.7

9.68 A combined gas turbine–vapor power plant (Fig. 9.22) has a net power output of 100 MW. Air enters the compressor of the gas turbine at 100 kPa, 300 K, and is compressed to 1200 kPa. The isentropic efficiency of the compressor is 84%. The conditions at the inlet to the turbine are 1200 kPa and 1400 K. Air expands through the turbine, which has an isentropic efficiency of 88%, to a pressure of 100 kPa. The air then passes through the interconnecting heat exchanger, and is finally discharged at 480 K. Steam enters the turbine of the vapor power cycle at 8 MPa, 400°C and expands to the condenser pressure of 8 kPa. Water enters the pump as saturated liquid at 8 kPa. The turbine and pump have isentropic efficiencies of 90 and 80%, respectively. Determine

   a. the mass flow rates of air and steam, each in kg/s.

   b. the thermal efficiency of the combined cycle.

   c. a full accounting of the *net* exergy increase of the air passing through the combustor of the gas turbine, $\dot{m}_{air}[e_{f3} - e_{f2}]$, in MW. Discuss.

Let $T_0 = 300$ K, $p_0 = 100$ kPa.

9.69 Hydrogen enters the turbine of an Ericsson cycle at 920 K, 15 bar with a mass flow rate of 1 kg/s. At the inlet to the compressor the condition is 300 K and 1.5 bar. Assuming the ideal gas model and ignoring kinetic and potential energy effects, determine

   a. the net power developed, in kW.

   b. the thermal efficiency.

   c. the back work ratio.

9.70 Air is the working fluid in an Ericsson cycle. Expansion through the turbine takes place at a constant temperature of 1250 K. Heat transfer from the compressor occurs at 310 K. The compressor pressure ratio is 12. Determine

   a. the net work, in kJ/kg of air flowing.

   b. the thermal efficiency.

9.71 Nitrogen (N_2) is the working fluid of a Stirling cycle with a compression ratio of nine. At the beginning of the isothermal compression, the temperature, pressure, and volume are 310 K, 1 bar, 0.008 m³, and respectively. The temperature during the isothermal expansion is 1000 K. Determine

   a. the net work, in kJ.

   b. the thermal efficiency.

   c. the mean effective pressure, in bar.

Compressible Flow

9.72 Calculate the thrust developed by the turbojet engine in Problem 9.58, in kN.

9.73 Calculate the thrust developed by the turbojet engine in Problem 9.60, in kN.

9.74 Air enters the diffuser of a turbojet engine at 18 kPa, 216 K, with a volumetric flow rate of 230 m³/s and a velocity of 265 m/s. The compressor pressure ratio is 15, and its isentropic efficiency is 87%. Air enters the turbine at 1360 K and the same pressure as at the exit of the compressor. The turbine isentropic efficiency is 89%, and the nozzle isentropic efficiency is 97%. The pressure at the nozzle exit is 18 kPa. On the basis of an air-standard analysis, calculate the thrust, in kN.

9.75 Calculate the ratio of the thrust developed to the mass flow rate of air, in N per kg/s, for the ramjet engine in Problem 9.63.

9.76 Air flows at steady state through a horizontal well-insulated, constant-area duct of diameter 0.25 m. At the inlet, $p_1 = 2.4$ bar, $T_1 = 430$ K. The temperature of the air leaving the duct is 370 K. The mass flow rate is 600 kg/min. Determine the magnitude, in N, of the net horizontal force exerted by the duct wall on the air. In which direction does the force act?

9.77 Liquid water at 21°C flows at steady state through a 5 cm-diameter horizontal pipe. The mass flow rate is 12 kg/s. The pressure decreases by 14 kPa from inlet to exit of the pipe. Determine the magnitude, in N, and direction of the horizontal force required to hold the pipe in place.

9.78 Air enters a horizontal, well-insulated nozzle operating at steady state at 10 bar, 460 K, with a velocity of 45 m/s and exits at 5 bar, 400 K. The mass flow rate is 0.8 kg/s. Determine the net force, in N, exerted by the air on the duct in the direction of flow.

9.79 Using the ideal gas model, determine the sonic velocity of

   a. air at 16°C.

   b. oxygen (O_2) at 500 K.

   c. argon at 300 K.

9.80 A flash of lightning is sighted and 3 seconds later thunder is heard. Approximately how far away was the lightning strike?

9.81 An ideal gas flows through a duct. At a particular location, the temperature, pressure, and velocity are known. Determine the Mach number, stagnation temperature, in °K and the stagnation pressure, in kPa for

   a. air at 154°C, 690 kPa, and a velocity of 420 m/s.

   b. helium at 290 K, 140 kPa, and a velocity of 270 m/s.

   c. nitrogen at 333 K, 345 kPa, and a velocity of 150 m/s.

9.82 Steam flows through a passageway, and at a particular location the pressure is 3 bar, the temperature is 280°C, and the velocity is 690 m/s. Determine the corresponding specific stagnation enthalpy, in kJ/kg, and stagnation temperature, in °C, if the stagnation pressure is 7 bar.

9.83 Consider isentropic flow of an ideal gas with constant k.

   a. Show that $\dfrac{T^*}{T_0} = \dfrac{2}{k+1}$ and $\dfrac{p^*}{p_0} = \left(\dfrac{2}{k+1}\right)^{k/(k-1)}$ where T^* and p^* are the temperature and pressure, respectively, at the state where Mach number is unity, and T_0 and p_0 are the temperature and pressure, respectively, at the stagnation state.

   b. Using the results of part (a), evaluate T^* and p^* for Example 9.14, in K and kPa, respectively.

9.84 Consider isentropic flow of an ideal gas with constant k through a converging nozzle from a large tank at 500 K, 8 bar. Using the results of Problem 9.83(a) with k at 500 K, evaluate the temperature, in K, and pressure, in bar, at the state where Mach number is unity for

   a. air.

   b. oxygen, O_2.

   c. carbon dioxide, CO_2.

9.85 An ideal gas mixture with $k = 1.32$ and a molecular weight of 23 is supplied to a converging nozzle at $p_0 = 4$ bar, $T_0 = 680$ K, which discharges into a region where the pressure is 1 bar. The exit area is 35 cm². For steady isentropic flow through the nozzle, determine

   a. the exit temperature of the gas, in K.

   b. the exit velocity of the gas, in m/s.

   c. the mass flow rate, in kg/s.

9.86 Air as an ideal gas with $k = 1.4$ enters a converging–diverging nozzle operating at steady state and expands isentropically as shown in **Fig. P9.86.** Using data from the figure and from Table 9.2 as needed, determine

   a. the stagnation pressure, in kPa, and the stagnation temperature, in K.

   b. the throat area, in mm².

   c. the exit area, in mm².

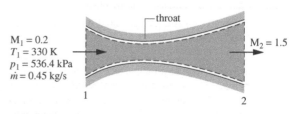

$M_1 = 0.2$
$T_1 = 330$ K
$p_1 = 536.4$ kPa
$\dot{m} = 0.45$ kg/s

$M_2 = 1.5$

FIGURE P9.86

9.87 A converging–diverging nozzle operating at steady state has a throat area of 3 cm^2 and an exit area of 6 cm^2. Air as an ideal gas with $k = 1.4$ enters the nozzle at 8 bar, 400 K, and a Mach number of 0.2, and flows isentropically throughout. If the nozzle is choked, and the diverging portion acts as a supersonic nozzle, determine the mass flow rate, in kg/s, and the Mach number, pressure, in bar, and temperature, in K, at the exit. Repeat if the diverging portion acts as a subsonic diffuser. In each analysis use data from Table 9.2 as needed.

9.88 Air enters a nozzle operating at steady state at 310 kPa, 445 K, with a velocity of 146 m/s, and expands isentropically to an exit velocity of 457 m/s. Determine

　　a. the exit pressure, in kPa.

　　b. the ratio of the exit area to the inlet area.

　　c. whether the nozzle is diverging only, converging only, or converging–diverging in cross section.

9.89 Steam expands isentropically through a converging nozzle operating at steady state from a large tank at 1.83 bar, 280°C. The mass flow rate is 2 kg/s, the flow is choked, and the exit plane pressure is 1 bar. Using steam table data as needed, determine the diameter of the nozzle, in cm, at locations where the pressure is 1.5 bar, and 1 bar, respectively.

9.90 Air enters a converging nozzle operating at steady state with negligible velocity at 10 bar, 360 K and exits at 5.28 bar. The exit area is 0.001 m^2 and the isentropic nozzle efficiency is 98%. The air is modeled as an ideal gas with $k = 1.4$. Potential energy effects are negligible. Determine at the nozzle exit

　　a. the velocity, in m/s.

　　b. the temperature, in K.

　　c. the Mach number.

　　d. the stagnation pressure, in bar.

　　e. Also evaluate the mass flow rate, in kg/s.

9.91 Air as an ideal gas with $k = 1.4$ enters a converging diverging duct with a Mach number of 2. At the inlet, the pressure is 180 kPa and the temperature is 250 K. A normal shock stands at a location in the converging section of the duct, with $M_x = 1.5$. At the exit of the duct, the pressure is 1030 kPa. The flow is isentropic everywhere except in the immediate vicinity of the shock. Determine temperature, in K, and the Mach number at the exit.

9.92 A converging–diverging nozzle operates at steady state. Air as an ideal gas with $k = 1.4$ enters the nozzle at 500 K, 6 bar, and a Mach number of 0.3. A normal shock stands in the diverging section at a location where the Mach number is 1.40. The cross-sectional areas of the throat and the exit plane are 4 cm^2 and 6 cm^2, respectively. The flow is

isentropic, except where the shock stands. Determine the exit pressure, in bar, and the mass flow rate, in kg/s.

9.93 Air as an ideal gas with $k = 1.4$ enters a converging–diverging channel at a Mach number of 1.6. A normal shock stands at the inlet to the channel. Downstream of the shock the flow is isentropic; the Mach number is unity at the throat; and the air exits at 138 kPa, 390 K, with negligible velocity. If the mass flow rate is 20 kg/s, determine the inlet and throat areas, in m^2.

9.94 A converging–diverging nozzle operates at steady state with a mass flow rate of 0.3 kg/s. Air as an ideal gas with $k = 1.4$ flows through the nozzle, discharging to the atmosphere at 100 kPa and 300 K. A normal shock stands at the exit plane with $M_x = 2$. Up to the shock, the flow is isentropic. Determine

　　a. the stagnation pressure p_{ox}, in kPa.

　　b. the stagnation temperature T_{ox}, in K.

　　c. the nozzle exit area, in mm^2.

9.95 An ideal gas expands isentropically through a converging nozzle from a large tank at 0.8 MPa, 333 K, and discharges into a region at 0.4 MPa. Determine the mass flow rate, in kg/s, for an exit flow area of 6.25 cm^2 if the gas is

　　a. air, with $k = 1.4$.

　　b. carbon dioxide, with $k = 1.26$.

　　c. argon, with $k = 1.667$.

9.96 Air at 3.5 bar, 520 K, and a Mach number of 0.3 enters a converging–diverging nozzle operating at steady state. A normal shock stands in the diverging section at a location where the Mach number is $M_x = 1.7$. The flow is isentropic, except where the shock stands. If the air behaves as an ideal gas with $k = 1.4$, determine

　　a. the stagnation temperature T_{ox}, in K.

　　b. the stagnation pressure p_{ox}, in bar.

　　c. the pressure p_x, in bar.

　　d. the pressure p_y, in bar.

　　e. the stagnation pressure p_{oy}, in bar.

　　f. the stagnation temperature T_{oy}, in K.

If the throat area is 7.5×10^{-4} m^2, and the exit plane pressure is 2.5 bar, determine the mass flow rate, in kg/s, and the exit area, in m^2.

9.97 Air as an ideal gas with $k = 1.4$ enters a diffuser operating at steady state at 4 bar, 290 K, with a velocity of 512 m/s. Assuming isentropic flow, plot the velocity, in m/s, the Mach number, and the area ratio A/A* for locations in the flow corresponding to pressures ranging from 4 to 14 bar.

DESIGN & OPEN-ENDED PROBLEMS: EXPLORING ENGINEERING PRACTICE

9.1D With the development of high-strength metal–ceramic materials, internal combustion engines can now be built with no cylinder wall cooling. These *adiabatic engines* operate with cylinder wall temperatures as high as 927°C. What are the important considerations for adiabatic engine design? Are adiabatic engines likely to find widespread application? Discuss.

9.2D Automotive gas turbines have been under development for decades but have not been commonly used in automobiles. Yet helicopters routinely use gas turbines. Explore why different types of engines are used in these respective applications. Compare selection factors such as performance, power-to-weight ratio, space requirements, fuel availability, and environmental impact. Summarize your findings in a report with at least three references.

9.3D Investigate the following technologies: plug-in hybrid vehicles, all-electric vehicles, hydrogen fuel cell vehicles, diesel-powered vehicles, natural gas–fueled vehicles, and ethanol-fueled vehicles, and make recommendations on which of these technologies should receive federal research, development, and deployment support over the next decade. Base your recommendation on the result of a decision matrix method such as the *Pugh method* to compare the various technologies. Clearly identify and justify the criteria used for the comparison and the logic behind the scoring process. Prepare a 15-minute briefing and an executive summary suitable for a conference with your local congressperson.

9.4D *Steam-injected* gas turbines use hot turbine exhaust gases to produce steam that is injected directly into the gas turbine system.

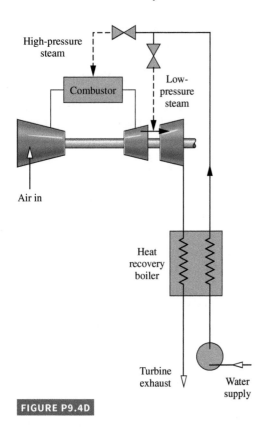

High-pressure
steam

Combustor

Low-
pressure
steam

Air in

Heat
recovery
boiler

Turbine
exhaust

Water
supply

FIGURE P9.4D

Figure P9.4D illustrates two possible approaches to steam injection. In one approach, steam is injected directly into the combustor. In the other approach, steam is injected into the low-pressure turbine stages.

a. What are the relative advantages and disadvantages of these steam injection approaches? How does steam injection lead to better power cycle performance?

b. Each of the steam injection approaches is inherently simpler than combined cycles like the one illustrated in Fig. 9.23. What are the other advantages and disadvantages of direct steam injection compared to combined cycles?

9.5D Owing to its very low temperature, liquid natural gas (LNG) transported by ship has considerable thermo-mechanical exergy. Yet when LNG is regasified in heat exchangers where seawater is the other stream, that exergy is largely destroyed. Conduct a search of the patent literature for methods to recover a substantial portion of LNG exergy during regasification. Consider patents both granted and pending. Critically evaluate the technical merit and economic feasibility of two different methods found in your search. Report your conclusions in an executive summary and PowerPoint presentation.

9.6D Factories requiring compressed air and process heat commonly run electrically driven air compressors and natural gas-fired boilers to meet these respective needs. As an alternative, commercially available natural gas-fueled engine-driven compressor systems can simultaneously provide for both needs. Determine if utility rates in your locale are favorable for the adoption of such systems. Prepare a memorandum explaining your conclusions.

9.7D A manufacturing company currently purchases 2.4×10^5 MW · h of electricity annually from the local utility company. An aging boiler on the premises annually provides 4×10^8 kg of process steam at 20 bar. Consider the feasibility of acquiring a *cogeneration* system to meet these needs. The system would employ a natural gas–fueled gas turbine to produce the electricity and a heat-recovery steam generator to produce the steam. Using *thermoeconomic* principles, investigate the economic

issues that should be considered in making a recommendation about the proposed cogeneration system. Write a report of your findings.

9.8D **Figure P9.8D** shows two cold air-standard cycles: 1–2–3–4′–1 is an Otto cycle and 1–2–3–4–5–1 is an *over-expanded* variation of the Otto cycle. The over-expanded cycle is of interest today because it provides a model for the engine used in various production hybrid-electric vehicles.

a. Develop the following expression for the ratio of the thermal efficiency of the over-expanded cycle to the thermal efficiency of the Otto cycle, η_{otto}, given by Eq. 9.8.

$$\eta/\eta_{otto} =$$
$$\left[1 - \frac{1}{(r^*r)^{k-1}} \left\{ 1 + \left(\frac{c_v T_1}{q} \right) r^{k-1} [1 - k(r^*)^{k-1} + (k-1)(r^*)^k] \right\} \right] \cdot$$
$$/\eta_{otto}$$

b. Plot the ratio obtained in part (a) versus r^* ranging from 1 to 3 for the case $r = 8$, $k = 1.3$, and $q/c_v T_1 = 8.1$. Also plot the ratio of the mean effective pressure (mep) of the over-expanded cycle to the mep of the Otto cycle versus r^*.

c. Referring to your plots, together with information obtained from an Internet search, draw conclusions about the performance of the engines used in hybrid-electric vehicles modeled here by the over-expanded cycle.

Prepare a memorandum including the derivation of part (a), the plots of part (b), and the conclusions of part (c) relating to actual engine performance.

9.9D Develop the preliminary specifications for a 160-MW closed-cycle gas turbine power plant. Consider carbon dioxide and helium as possible gas turbine working fluids that circulate through a nuclear power unit where they absorb energy from the nuclear reaction. Sketch the schematic of your proposed cycle. For the two working fluids, compare the operating pressures and temperatures and estimate the expected performance of the cycle. Write a report that includes your analysis and design, and recommend a working fluid. Include at least three references.

9.10D **Figure P9.10D** provides the schematic of an internal combustion automobile engine fitted with two Rankine vapor power bottoming cycles: a high-temperature cycle 1–2–3–4–1 and a low-temperature cycle 5–6–7–8–5. These cycles develop additional power using *waste heat* derived from exhaust gas and engine coolant. Using

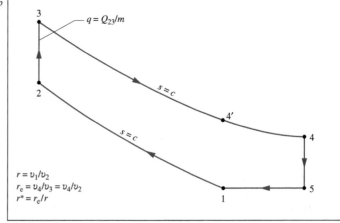

p

3

$q = Q_{23}/m$

2

$s = c$

4′

4

$s = c$

5

$r = v_1/v_2$
$r_e = v_4/v_3 = v_4/v_2$
$r^* = r_e/r$

1

v

FIGURE P9.8D

operating data for a commercially available car having a conventional four-cylinder internal combustion engine with a size of 2.5 liters or less, specify cycle working fluids and state data at key points sufficient to produce at least 15 hp more power. Apply engineering modeling compatible with that used in the text for Rankine cycles and air-standard analysis of internal combustion engines. Write a final report justifying your specifications together with supporting calculations. Provide a critique of the use of such bottoming cycles on car engines and a recommendation about whether such technology should be actively pursued by automakers.

9.11D An ideal gas whose specific heat ratio is k flows adiabatically *with friction* through a nozzle, entering with negligible velocity at temperature T_0 and pressure p_0. Operation is at steady state and potential energy effects can be ignored.

a. Showing all details, develop the following expression giving the mass flow rate per unit of nozzle flow area in dimensionless form as a function of local Mach number and k.

$$\frac{\dot{m}}{A}\frac{\sqrt{T_0}}{p_0}\sqrt{\frac{R}{k}} = \frac{M\left[1 + \left(\frac{k-1}{2}\right)M^2\left(1 - \frac{1}{\eta}\right)\right]^{k/(k-1)}}{\left[1 + \left(\frac{k-1}{2}\right)M^2\right]^{(k+1)/2(k-1)}}$$

where η denotes the isentropic nozzle efficiency, which is assumed constant.

b. Using the result of part (a), investigate nozzle performance for three isentropic nozzle efficiency values in the range 0.98 to 1.0. For each value, determine the Mach number at the throat and where Mach number unity occurs within the nozzle. Interpret the effect of friction in such a nozzle flow.

Report your results in a 15-minute PowerPoint presentation suitable for your class in thermodynamics.

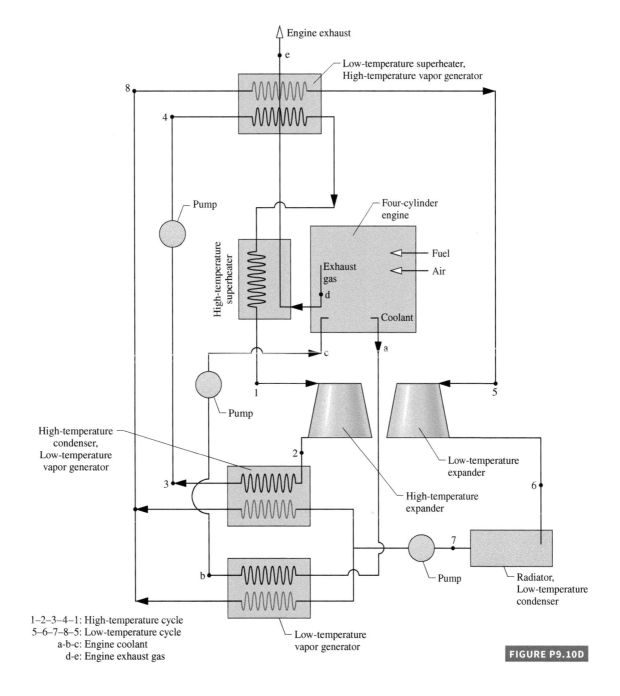

1–2–3–4–1: High-temperature cycle
5–6–7–8–5: Low-temperature cycle
a-b-c: Engine coolant
d-e: Engine exhaust gas

FIGURE P9.10D

CHAPTER 10

Refrigeration and Heat Pump Systems

Engineering Context

Refrigeration systems for food preservation and air conditioning play prominent roles in our everyday lives. Heat pumps also are used for heating buildings and for producing industrial process heat. There are many other examples of commercial and industrial uses of refrigeration, including air separation to obtain liquid oxygen and liquid nitrogen, liquefaction of natural gas, and production of ice.

To achieve refrigeration by most conventional means requires an electric power input. Heat pumps also require power to operate. Referring again to Table 8.1, we see that in the United States electricity is obtained today primarily from coal, natural gas, and nuclear, all of which are nonrenewable. These nonrenewables have significant adverse effects on human health and the environment associated with their use. Depending on the type of resource, such effects are related to extraction from the earth, processing and distribution, emissions during power production, and waste products.

Ineffective refrigeration and heat pump systems, excessive building cooling and heating, and other wasteful practices and lifestyle choices not only misuse increasingly scarce nonrenewable resources but also endanger our health and burden the environment. Accordingly, refrigeration and heat pump systems are areas of application where more effective systems and practices can significantly improve our national energy posture.

The **objective** of this chapter is to describe some of the common types of refrigeration and heat pump systems presently in use and to illustrate how such systems can be modeled thermodynamically. The three principal types described are the vapor-compression, absorption, and reversed Brayton cycles. As for the power systems studied in Chaps. 8 and 9, both vapor and gas systems are considered. In vapor systems, the refrigerant is alternately vaporized and condensed. In gas refrigeration systems, the refrigerant remains a gas.

LEARNING OUTCOMES

When you complete your study of this chapter, you will be able to...

- Demonstrate understanding of basic vapor-compression refrigeration and heat pump systems.
- Develop and analyze thermodynamic models of vapor-compression systems and their modifications, including
  - Sketching schematic and accompanying *T–s* diagrams.
  - Evaluating property data at principal states of the systems.
  - Applying mass, energy, entropy, and exergy balances for the basic processes.
  - Determining refrigeration and heat pump system performance, coefficient of performance, and capacity.
- Explain the effects on vapor-compression system performance of varying key parameters.
- Demonstrate understanding of the operating principles of absorption and gas refrigeration systems and perform thermodynamic analysis of gas systems.

10.1 Vapor Refrigeration Systems

The purpose of a refrigeration system is to maintain a *cold* region at a temperature below the temperature of its surroundings. This is commonly achieved using the vapor refrigeration systems that are the subject of the present section.

10.1.1 Carnot Refrigeration Cycle

To introduce some important aspects of vapor refrigeration, let us begin by considering a Carnot vapor refrigeration cycle. This cycle is obtained by reversing the Carnot vapor power cycle introduced in Sec. 5.10. **Figure 10.1** shows the schematic and accompanying T–s diagram of a Carnot refrigeration cycle operating between a region at temperature T_C and another region at a higher temperature T_H. The cycle is executed by a refrigerant circulating steadily through a series of components. All processes are internally reversible. Also, since heat transfers between the refrigerant and each region occur with no temperature differences, there are no external irreversibilities. The energy transfers shown on the diagram are positive in the directions indicated by the arrows.

Let us follow the refrigerant as it passes steadily through each of the components in the cycle, beginning at the inlet to the evaporator. The refrigerant enters the evaporator as a two-phase liquid–vapor mixture at state 4. In the evaporator some of the refrigerant changes phase from liquid to vapor as a result of heat transfer from the region at temperature T_C to the refrigerant. The temperature and pressure of the refrigerant remain constant during the process from state 4 to state 1. The refrigerant is then compressed adiabatically from state 1, where it is a two-phase liquid–vapor mixture, to state 2, where it is a saturated vapor. During this process, the temperature of the refrigerant increases from T_C to T_H, and the pressure also increases. The refrigerant passes from the compressor into the condenser, where it changes phase from saturated vapor to saturated liquid as a result of heat transfer to the region at temperature T_H. The temperature and pressure remain constant in the process from state 2 to state 3. The refrigerant returns to the state at the inlet of the evaporator by expanding adiabatically through a turbine. In this process, from state 3 to state 4, the temperature decreases from T_H to T_C, and there is a decrease in pressure.

Since the Carnot vapor refrigeration cycle is made up of internally reversible processes, areas on the T–s diagram can be interpreted as heat transfers. Referring to Fig. 10.1, area 1–a–b–4–1 is the heat added to the refrigerant from the cold region per unit mass of refrigerant

TAKE NOTE...

See Sec. 6.13.1 for the area interpretation of heat transfer on a T–s diagram for the case of internally reversible flow though a control volume at steady state.

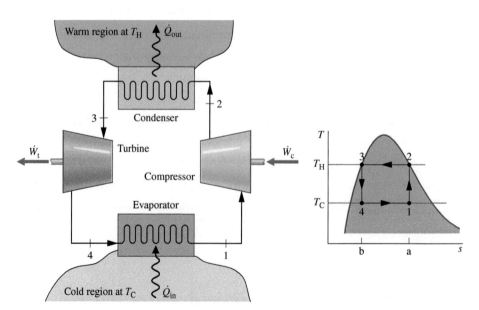

Fig. 10.1 Carnot vapor refrigeration cycle.

flowing. Area 2–a–b–3–2 is the heat rejected from the refrigerant to the warm region per unit mass of refrigerant flowing. The enclosed area 1–2–3–4–1 is the *net* heat transfer *from* the refrigerant. The net heat transfer *from* the refrigerant equals the net work done *on* the refrigerant. The net work is the difference between the compressor work input and the turbine work output.

The *coefficient of performance β* of *any* refrigeration cycle is the ratio of the refrigeration effect to the net work input required to achieve that effect. For the Carnot vapor refrigeration cycle shown in Fig. 10.1, the coefficient of performance is

$$\beta_{max} = \frac{\dot{Q}_{in}/\dot{m}}{\dot{W}_c/\dot{m} - \dot{W}_t/\dot{m}} = \frac{\text{area } 1-a-b-4-1}{\text{area } 1-2-3-4-1} = \frac{T_C(s_a - s_b)}{(T_H - T_C)(s_a - s_b)}$$

which reduces to

$$\beta_{max} = \frac{T_C}{T_H - T_C} \qquad (10.1)$$

This equation, which corresponds to Eq. 5.10, represents the *maximum* theoretical coefficient of performance of any refrigeration cycle operating between regions at T_C and T_H.

10.1.2 Departures from the Carnot Cycle

Actual vapor refrigeration systems depart significantly from the Carnot cycle considered above and have coefficients of performance lower than would be calculated from Eq. 10.1. Three ways actual systems depart from the Carnot cycle are considered next.

- One of the most significant departures is related to the heat transfers between the refrigerant and the two regions. In actual systems, these heat transfers are not accomplished reversibly as presumed above. In particular, to achieve a rate of heat transfer sufficient to maintain the temperature of the cold region at T_C with a practical-sized evaporator requires the temperature of the refrigerant in the evaporator, T_C', to be several degrees *below* T_C. This is illustrated by the placement of the temperature T_C' on the *T–s* diagram of **Fig. 10.2**. Similarly, to obtain a sufficient heat transfer rate from the refrigerant to the warm region requires that the refrigerant temperature in the condenser, T_H', be several degrees *above* T_H. This is illustrated by the placement of the temperature T_H' on the *T–s* diagram of Fig. 10.2.

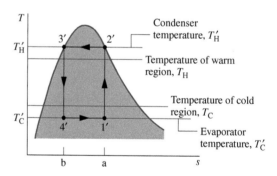

Fig. 10.2 Comparison of the condenser and evaporator temperatures with those of the warm and cold regions.

  Maintaining the refrigerant temperatures in the heat exchangers at T_C' and T_H' rather than at T_C and T_H, respectively, has the effect of reducing the coefficient of performance. This can be seen by expressing the coefficient of performance of the refrigeration cycle designated by 1'–2'–3'–4'–1' on Fig. 10.2 as

$$\beta' = \frac{\text{area } 1'-a-b-4'-1}{\text{area } 1'-2'-3'-4'-1'} = \frac{T_C'}{T_H' - T_C'} \qquad (10.2)$$

Comparing the areas underlying the expressions for β_{max} and β' given above, we conclude that the value of β' is less than β_{max}. This conclusion about the effect of refrigerant temperature on the coefficient of performance also applies to other refrigeration cycles considered in the chapter.

- Even when the temperature differences between the refrigerant and warm and cold regions are taken into consideration, there are other features that make the vapor refrigeration cycle of Fig. 10.2 impractical as a prototype. Referring again to the figure, note that the compression process from state 1' to state 2' occurs with the refrigerant as a

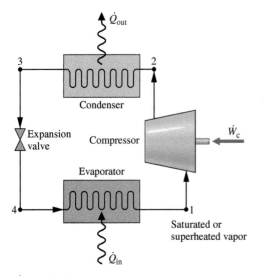

Fig. 10.3 Components of a vapor-compression refrigeration system.

two-phase liquid–vapor mixture. This is commonly referred to as *wet compression*. Wet compression is normally avoided because the presence of liquid droplets in the flowing liquid–vapor mixture can damage the compressor. In actual systems, the compressor handles vapor only. This is known as *dry compression*.

● Another feature that makes the cycle of Fig. 10.2 impractical is the expansion process from the saturated liquid state 3′ to the low-quality, two-phase liquid–vapor mixture state 4′. This expansion typically produces a relatively small amount of work compared to the work input in the compression process. The work developed by an actual turbine would be smaller yet because turbines operating under these conditions have low isentropic efficiencies. Accordingly, the work output of the turbine is normally sacrificed by substituting a simple throttling valve for the expansion turbine, with consequent savings in initial and maintenance costs. The components of the resulting cycle are illustrated in **Fig. 10.3**, where dry compression is presumed. This cycle, known as the *vapor-compression refrigeration cycle*, is the subject of the section to follow.

10.2 Analyzing Vapor-Compression Refrigeration Systems

vapor-compression refrigeration

Vapor-compression refrigeration systems are the most common refrigeration systems in use today. The objective of this section is to introduce some important features of systems of this type and to illustrate how they are modeled thermodynamically.

10.2.1 Evaluating Principal Work and Heat Transfers

Let us consider the steady-state operation of the vapor-compression system illustrated in Fig. 10.3. Shown on the figure are the principal work and heat transfers, which are positive in the directions of the arrows. Kinetic and potential energy changes are neglected in the following analyses of the components. We begin with the evaporator, where the desired refrigeration effect is achieved.

● As the refrigerant passes through the evaporator, heat transfer from the refrigerated space results in the vaporization of the refrigerant. For a control volume enclosing the refrigerant side of the evaporator, the mass and energy rate balances reduce to give the rate of heat transfer per unit mass of refrigerant flowing as

$$\frac{\dot{Q}_{in}}{\dot{m}} = h_1 - h_4 \tag{10.3}$$

refrigeration capacity
ton of refrigeration

where $\dot{m}$ is the mass flow rate of the refrigerant. The heat transfer rate $\dot{Q}_{in}$ is referred to as the **refrigeration capacity**. In the SI unit system, the capacity is normally expressed in kW. Another commonly used unit for the refrigeration capacity is the **ton of refrigeration**, which is about 211 kJ/min.

● The refrigerant leaving the evaporator is compressed to a relatively high pressure and temperature by the compressor. Assuming no heat transfer to or from the compressor, the mass and energy rate balances for a control volume enclosing the compressor give

$$\frac{\dot{W}_c}{\dot{m}} = h_2 - h_1 \tag{10.4}$$

where $\dot{W}_c/\dot{m}$ is the rate of power *input* per unit mass of refrigerant flowing.

- Next, the refrigerant passes through the condenser, where the refrigerant condenses and there is heat transfer from the refrigerant to the cooler surroundings. For a control volume enclosing the refrigerant side of the condenser, the rate of heat transfer *from* the refrigerant per unit mass of refrigerant flowing is

$$\frac{\dot{Q}_{\text{out}}}{\dot{m}} = h_2 - h_3 \qquad (10.5)$$

- Finally, the refrigerant at state 3 enters the expansion valve and expands to the evaporator pressure. This process is usually modeled as a *throttling* process for which

$$h_4 = h_3 \qquad (10.6)$$

The refrigerant pressure decreases in the irreversible adiabatic expansion, and there is an accompanying increase in specific entropy. The refrigerant exits the valve at state 4 as a two-phase liquid–vapor mixture.

In the vapor-compression system, the net power input is equal to the compressor power, since the expansion valve involves no power input or output. Using the quantities and expressions introduced above, the coefficient of performance of the vapor-compression refrigeration system of Fig. 10.3 is

$$\boxed{\beta = \frac{\dot{Q}_{\text{in}}/\dot{m}}{\dot{W}_{\text{c}}/\dot{m}} = \frac{h_1 - h_4}{h_2 - h_1}} \qquad (10.7)$$

Provided states 1 through 4 are fixed, Eqs. 10.3 through 10.7 can be used to evaluate the principal work and heat transfers and the coefficient of performance of the vapor-compression system shown in Fig. 10.3. Since these equations have been developed by reducing mass and energy rate balances, they apply equally for actual performance when irreversibilities are present in the evaporator, compressor, and condenser and for idealized performance in the absence of such effects. Although irreversibilities in the evaporator, compressor, and condenser can have a pronounced effect on overall performance, it is instructive to consider an idealized cycle in which they are assumed absent. Such a cycle establishes an upper limit on the performance of the vapor-compression refrigeration cycle. It is considered next.

10.2.2 Performance of Ideal Vapor-Compression Systems

If irreversibilities within the evaporator and condenser are ignored, there are no frictional pressure drops, and the refrigerant flows at constant pressure through the two heat exchangers. If compression occurs without irreversibilities, and stray heat transfer to the surroundings is also ignored, the compression process is isentropic. With these considerations, the vapor-compression refrigeration cycle labeled 1–2s–3–4–1 on the *T–s* diagram of Fig. 10.4 results. The cycle consists of the following series of processes:

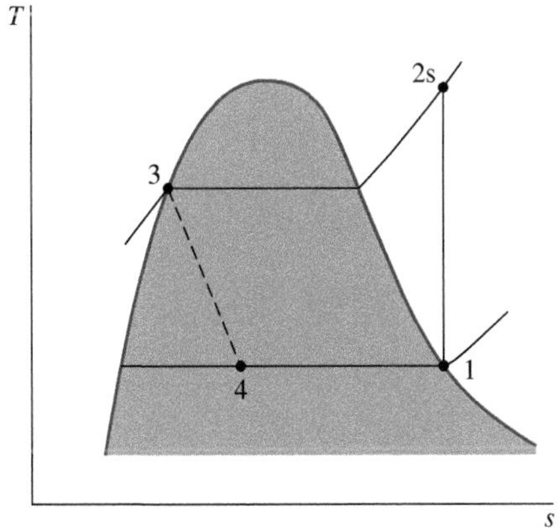

Process 1–2s *Isentropic* compression of the refrigerant from state 1 to the condenser pressure at state 2s.

Process 2s–3 Heat transfer *from* the refrigerant as it flows at constant pressure through the condenser. The refrigerant exits as a liquid at state 3.

Process 3–4 *Throttling* process from state 3 to a two-phase liquid–vapor mixture at 4.

Process 4–1 Heat transfer *to* the refrigerant as it flows at constant pressure through the evaporator to complete the cycle.

Fig. 10.4 *T–s* diagram of an ideal vapor-compression cycle.

All processes of the cycle shown in Fig. 10.4 are internally reversible except for the throttling process. Despite the inclusion of this irreversible process, the cycle is commonly referred to as the **ideal vapor-compression cycle**.

ideal vapor-compression cycle

The following example illustrates the application of the first and second laws of thermodynamics along with property data to analyze an ideal vapor-compression cycle.

▶▶▶ **EXAMPLE 10.1** ▶ •

Analyzing an Ideal Vapor-Compression Refrigeration Cycle

Refrigerant 134a is the working fluid in an ideal vapor-compression refrigeration cycle that communicates thermally with a cold region at 0°C and a warm region at 26°C. Saturated vapor enters the compressor at 0°C and saturated liquid leaves the condenser at 26°C. The mass flow rate of the refrigerant is 0.08 kg/s. Determine **(a)** the compressor power, in kW, **(b)** the refrigeration capacity, in tons, **(c)** the coefficient of performance, and **(d)** the coefficient of performance of a Carnot refrigeration cycle operating between warm and cold regions at 26 and 0°C, respectively.

SOLUTION

Known An ideal vapor-compression refrigeration cycle operates with Refrigerant 134a. The states of the refrigerant entering the compressor and leaving the condenser are specified, and the mass flow rate is given.

Find Determine the compressor power, in kW, the refrigeration capacity, in tons, coefficient of performance, and the coefficient of performance of a Carnot vapor refrigeration cycle operating between warm and cold regions at the specified temperatures.

Schematic and Given Data:

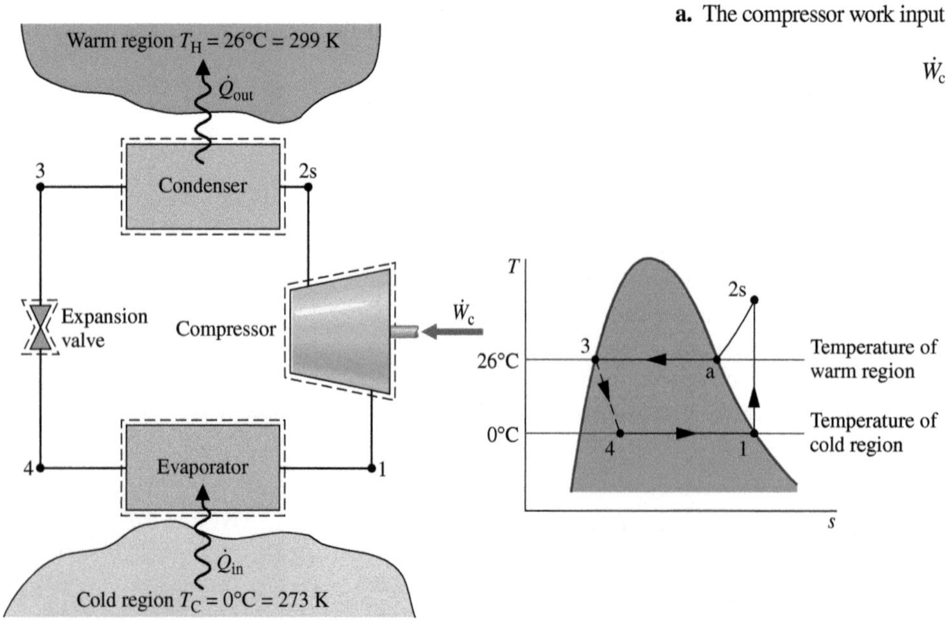

Warm region $T_H = 26°C = 299$ K

$\dot{Q}_{out}$

3 — Condenser — 2s

Expansion valve — Compressor — $\dot{W}_c$

26°C

0°C

4 — Evaporator — 1

$\dot{Q}_{in}$

Cold region $T_C = 0°C = 273$ K

Fig. E10.1

T

2s

3 ← a

Temperature of warm region

4 → 1

Temperature of cold region

s

2. Except for the expansion through the valve, which is a throttling process, all processes of the refrigerant are internally reversible.

3. The compressor and expansion valve operate adiabatically.

4. Kinetic and potential energy effects are negligible.

5. Saturated vapor enters the compressor, and saturated liquid leaves the condenser.

Analysis Let us begin by fixing each of the principal states located on the accompanying schematic and $T–s$ diagrams. At the inlet to the compressor, the refrigerant is a saturated vapor at 0°C, so from Table A-10, $h_1 = 247.23$ kJ/kg and $s_1 = 0.9190$ kJ/kg · K.

❶ The pressure at state 2s is the saturation pressure corresponding to 26°C, or $p_2 = 6.853$ bar. State 2s is fixed by p_2 and the fact that the specific entropy is constant for the adiabatic, internally reversible compression process. The refrigerant at state 2s is a superheated vapor with $h_{2s} = 264.7$ kJ/kg. State 3 is saturated liquid at 26°C, so $h_3 = 85.75$ kJ/kg. The expansion through the valve is a throttling process (assumption 2), so $h_4 = h_3$.

a. The compressor work input is

$$\dot{W}_c = \dot{m}(h_{2s} - h_1)$$

where $\dot{m}$ is the mass flow rate of refrigerant. Inserting values

$$\dot{W}_c = (0.08 \text{ kg/s})(264.7 - 247.23) \text{ kJ/kg}\left|\frac{1 \text{ kW}}{1 \text{ kJ/s}}\right|$$

$$= 1.4 \text{ kW}$$

Engineering Model

1. Each component of the cycle is analyzed as a control volume at steady state. The control volumes are indicated by dashed lines on the accompanying sketch.

b. The refrigeration capacity is the heat transfer rate to the refrigerant passing through the evaporator. This is given by

$$\dot{Q}_{in} = \dot{m}(h_1 - h_4)$$

$$= (0.08 \text{ kg/s})|60 \text{ s/min}|(247.23 - 85.75) \text{ kJ/kg}\left|\frac{1 \text{ ton}}{211 \text{ kJ/min}}\right|$$

$$= 3.67 \text{ ton}$$

c. The coefficient of performance β is

$$\beta = \frac{\dot{Q}_{in}}{\dot{W}_c} = \frac{h_1 - h_4}{h_{2s} - h_1} = \frac{247.23 - 85.75}{264.7 - 247.23} = 9.24$$

d. For a Carnot vapor refrigeration cycle operating at $T_H = 299$ K and $T_C = 273$ K, the coefficient of performance determined from Eq. 10.1 is

❷
$$\beta_{max} = \frac{T_C}{T_H - T_C} = 10.5$$

---

❶ The value for h_{2s} can be obtained by double interpolation in Table A-12 or by using *Interactive Thermodynamics: IT*.

❷ As expected, the ideal vapor-compression cycle has a lower coefficient of performance than a Carnot cycle operating between the temperatures of the warm and cold regions. The smaller value can be attributed to the effects of the external irreversibility associated with desuperheating the refrigerant in the condenser (Process 2s–a on the *T–s* diagram) and the internal irreversibility of the throttling process.

Quick Quiz

Keeping all other given data the same, determine the mass flow rate of refrigerant, in kg/s, for a 10-ton refrigeration capacity. Ans. 0.218 kg/s.

10.2.3 Performance of Actual Vapor-Compression Systems

Figure 10.5 illustrates several features exhibited by *actual* vapor-compression systems. As shown in the figure, the heat transfers between the refrigerant and the warm and cold regions are not accomplished reversibly: The refrigerant temperature in the evaporator is less than the cold region temperature, T_C, and the refrigerant temperature in the condenser is greater than the warm region temperature, T_H. Such irreversible heat transfers have a significant effect on performance. In particular, the coefficient of performance decreases as the average temperature of the refrigerant in the evaporator decreases and as the average temperature of the refrigerant in the condenser increases. Example 10.2 provides an illustration.

Vapor Compression Refrigeration Cycle (VCRC) Tab b

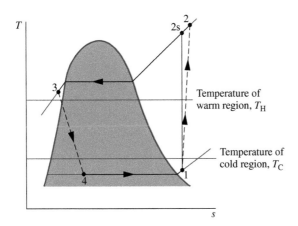

Fig. 10.5 *T–s* diagram of an actual vapor-compression cycle.

Considering the Effect of Irreversible Heat Transfer on Performance

Modify Example 10.1 to allow for temperature differences between the refrigerant and the warm and cold regions as follows. Saturated vapor enters the compressor at −10°C. Saturated liquid leaves the condenser at a pressure of 9 bar. Determine for the modified vapor-compression refrigeration cycle **(a)** the compressor power, in kW, **(b)** the refrigeration capacity, in tons, **(c)** the coefficient of performance. Compare results with those of Example 10.1.

SOLUTION

Known A modified vapor-compression refrigeration cycle operates with Refrigerant 134a as the working fluid. The evaporator temperature and condenser pressure are specified, and the mass flow rate is given.

Find Determine the compressor power, in kW, the refrigeration capacity, in tons, and the coefficient of performance. Compare results with those of Example 10.1.

Schematic and Given Data:

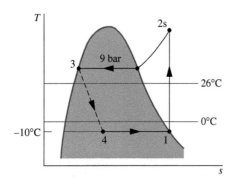

Fig. E10.2

Engineering Model

1. Each component of the cycle is analyzed as a control volume at steady state. The control volumes are indicated by dashed lines on the sketch accompanying Example 10.1.

2. Except for the process through the expansion valve, which is a throttling process, all processes of the refrigerant are internally reversible.

3. The compressor and expansion valve operate adiabatically.

4. Kinetic and potential energy effects are negligible.

5. Saturated vapor enters the compressor, and saturated liquid exits the condenser.

Analysis Let us begin by fixing each of the principal states located on the accompanying T–s diagram. Starting at the inlet to the compressor, the refrigerant is a saturated vapor at −10°C, so from Table A-10, $h_1 = 241.35$ kJ/kg and $s_1 = 0.9253$ kJ/kg · K.

The superheated vapor at state 2s is fixed by $p_2 = 9$ bar and the fact that the specific entropy is constant for the adiabatic, internally reversible compression process. Interpolating in Table A-12 gives $h_{2s} = 272.39$ kJ/kg.

State 3 is a saturated liquid at 9 bar, so $h_3 = 99.56$ kJ/kg. The expansion through the valve is a throttling process; thus, $h_4 = h_3$.

a. The compressor power input is

$$\dot{W}_c = \dot{m}(h_{2s} - h_1)$$

where $\dot{m}$ is the mass flow rate of refrigerant. Inserting values

$$\dot{W}_c = (0.08 \text{ kg/s})(272.39 - 241.35) \text{ kJ/kg} \left| \frac{1 \text{ kW}}{1 \text{ kJ/s}} \right|$$

$$= 2.48 \text{ kW}$$

b. The refrigeration capacity is

$$\dot{Q}_{in} = \dot{m}(h_1 - h_4)$$

$$= (0.08 \text{ kg/s}) |60 \text{ s/min}| (241.35 - 99.56) \text{ kJ/kg} \left| \frac{1 \text{ ton}}{211 \text{ kJ/min}} \right|$$

$$= 3.23 \text{ ton}$$

c. The coefficient of performance β is

$$\beta = \frac{\dot{Q}_{in}}{\dot{W}_c} = \frac{h_1 - h_4}{h_{2s} - h_1} = \frac{241.35 - 99.56}{272.39 - 241.35} = 4.57$$

Comparing the results of the present example with those of Example 10.1, we see that the power input required by the compressor is greater in the present case. Furthermore, the refrigeration capacity and coefficient of performance are smaller in this example than in Example 10.1. This illustrates the considerable influence on performance of irreversible heat transfer between the refrigerant and the cold and warm regions.

SKILLS DEVELOPED

Ability to...

- sketch the T–s diagram of the ideal vapor-compression refrigeration cycle.

- fix each of the principal states and retrieve necessary property data.

- calculate compressor power, refrigeration capacity, and coefficient of performance.

Quick Quiz

Determine the rate of heat transfer from the refrigerant passing through the condenser to the surroundings, in kW. Ans. 13.83 kW.

Referring again to Fig. 10.5, we can identify another key feature of actual vapor-compression system performance. This is the effect of irreversibilities during compression, suggested by the use of a dashed line for the compression process from state 1 to state 2. The dashed line is drawn to show the increase in specific entropy that accompanies an *adiabatic irreversible* compression. Comparing cycle 1–2–3–4–1 with cycle 1–2s–3–4–1, the refrigeration capacity would be the same for each, but the work input would be greater in the case of irreversible compression than in the ideal cycle. Accordingly, the coefficient of performance of cycle 1–2–3–4–1 is less than that of cycle 1–2s–3–4–1. The effect of irreversible compression can be accounted for by using the isentropic compressor efficiency, which for states designated as in Fig. 10.5 is given by

$$\eta_c = \frac{(\dot{W}_c/\dot{m})_s}{(\dot{W}_c/\dot{m})} = \frac{h_{2s} - h_1}{h_2 - h_1}$$

Additional departures from ideality stem from frictional effects that result in pressure drops as the refrigerant flows through the evaporator, condenser, and piping connecting the various components. These pressure drops are not shown on the T–s diagram of Fig. 10.5 and are ignored in subsequent discussions for simplicity.

Finally, two additional features exhibited by actual vapor-compression systems are shown in Fig. 10.5. One is the superheated vapor condition at the evaporator exit (state 1), which differs from the saturated vapor condition shown in Fig. 10.4. Another is the subcooling of the condenser exit state (state 3), which differs from the saturated liquid condition shown in Fig. 10.4.

Example 10.3 illustrates the effects of irreversible compression and condenser exit subcooling on the performance of the vapor-compression refrigeration system.

Vapor Compression Refrigeration Cycle (VCRC) Tabs c and d

▶ EXAMPLE 10.3 ▶ ·

Analyzing an Actual Vapor-Compression Refrigeration Cycle

Reconsider the vapor-compression refrigeration cycle of Example 10.2, but include in the analysis that the compressor has an isentropic efficiency of 80%. Also, let the temperature of the liquid leaving the condenser be 30°C. Determine for the modified cycle **(a)** the compressor power, in kW, **(b)** the refrigeration capacity, in tons, **(c)** the coefficient of performance, and **(d)** the rates of exergy destruction within the compressor and expansion valve, in kW, for $T_0 = 299$ K (26°C).

SOLUTION

Known A vapor-compression refrigeration cycle has an isentropic compressor efficiency of 80%.

Find Determine the compressor power, in kW, the refrigeration capacity, in tons, the coefficient of performance, and the rates of exergy destruction within the compressor and expansion valve, in kW.

Schematic and Given Data:

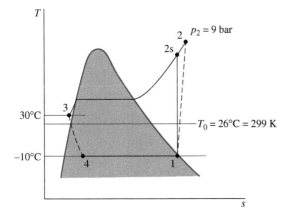

Fig. E10.3

Engineering Model

1. Each component of the cycle is analyzed as a control volume at steady state.

2. There are no pressure drops through the evaporator and condenser.

3. The compressor operates adiabatically with an isentropic efficiency of 80%. The expansion through the valve is a throttling process.

4. Kinetic and potential energy effects are negligible.

5. Saturated vapor at −10°C enters the compressor, and liquid at 30°C leaves the condenser.

6. The environment temperature for calculating exergy is $T_0 = 299$ K (26°C).

Analysis Let us begin by fixing the principal states. State 1 is the same as in Example 10.2, so $h_1 = 241.35$ kJ/kg and $s_1 = 0.9253$ kJ/kg · K.

Owing to the presence of irreversibilities during the adiabatic compression process, there is an increase in specific entropy from compressor inlet to exit. The state at the compressor exit, state 2, can be fixed using the isentropic compressor efficiency

$$\eta_c = \frac{(\dot{W}_c/\dot{m})_s}{\dot{W}_c/\dot{m}} = \frac{(h_{2s} - h_1)}{(h_2 - h_1)}$$

where h_{2s} is the specific enthalpy at state 2s, as indicated on the accompanying T–s diagram. From the solution to Example 10.2, $h_{2s} = 272.39$ kJ/kg. Solving for h_2 and inserting known values

$$h_2 = \frac{h_{2s} - h_1}{\eta_c} + h_1 = \frac{(272.39 - 241.35)}{(0.80)} + 241.35 = 280.15 \text{ kJ/kg}$$

State 2 is fixed by the value of specific enthalpy h_2 and the pressure, $p_2 = 9$ bar. Interpolating in Table A-12, the specific entropy is $s_2 = 0.9497$ kJ/kg · K.

The state at the condenser exit, state 3, is in the liquid region. The specific enthalpy is approximated using Eq. 3.14, together with saturated liquid data at 30°C, as follows: $h_3 \approx h_f = 91.49$ kJ/kg. Similarly, with Eq. 6.5, $s_3 \approx s_f = 0.3396$ kJ/kg · K.

The expansion through the valve is a throttling process; thus, $h_4 = h_3$. The quality and specific entropy at state 4 are, respectively,

$$x_4 = \frac{h_4 - h_{f4}}{h_{g4} - h_{f4}} = \frac{91.49 - 36.97}{204.39} = 0.2667$$

and

$$s_4 = s_{f4} + x_4(s_{g4} - s_{f4})$$
$$= 0.1486 + (0.2667)(0.9253 - 0.1486) = 0.3557 \text{ kJ/kg} \cdot \text{K}$$

a. The compressor power is

$$\dot{W}_c = \dot{m}(h_2 - h_1)$$

$$= (0.08 \text{ kg/s})(280.15 - 241.35) \text{kJ/kg} \left| \frac{1 \text{ kW}}{1 \text{ kJ/s}} \right| = 3.1 \text{ kW}$$

b. The refrigeration capacity is

$$\dot{Q}_{in} = \dot{m}(h_1 - h_4)$$

$$= (0.08 \text{ kg/s}) |60 \text{ s/min}| (241.35 - 91.49) \text{ kJ/kg} \left| \frac{1 \text{ ton}}{211 \text{ kJ/min}} \right|$$

$$= 3.41 \text{ ton}$$

c. The coefficient of performance is

❶ $$\beta = \frac{(h_1 - h_4)}{(h_2 - h_1)} = \frac{(241.35 - 91.49)}{(280.15 - 241.35)} = 3.86$$

d. The rates of exergy destruction in the compressor and expansion valve can be found by reducing the exergy rate balance or using the relationship $\dot{E}_d = T_0\dot{\sigma}_{cv}$, where $\dot{\sigma}_{cv}$ is the rate of entropy production from an entropy rate balance. With either approach, the rates of exergy destruction for the compressor and valve are, respectively,

$$(\dot{E}_d)_c = \dot{m}T_0(s_2 - s_1) \quad \text{and} \quad (\dot{E}_d)_{valve} = \dot{m}T_0(s_4 - s_3)$$

Substituting values

❷ $$(\dot{E}_d)_c = \left(0.08 \frac{\text{kg}}{\text{s}}\right)(299 \text{ K})(0.9497 - 0.9253)\frac{\text{kJ}}{\text{kg} \cdot \text{K}} \left| \frac{1 \text{ kW}}{1 \text{ kJ/s}} \right|$$

$$= 0.58 \text{ kW}$$

and

$$(\dot{E}_d)_{valve} = (0.08)(299)(0.3557 - 0.3396) = 0.39 \text{ kW}$$

---

❶ While the refrigeration capacity is greater than in Example 10.2, irreversibilities in the compressor result in an increase in compressor power compared to isentropic compression. The overall effect is a lower coefficient of performance than in Example 10.2.

❷ The exergy destruction rates calculated in part (d) measure the effect of irreversibilities as the refrigerant flows through the compressor and valve. The percentages of the power input (exergy input) to the compressor destroyed in the compressor and valve are 18.7 and 12.6%, respectively.

SKILLS DEVELOPED

Ability to…

• sketch the *T–s* diagram of the vapor-compression refrigeration cycle with irreversibilities in the compressor and subcooled liquid exiting the condenser.

• fix each of the principal states and retrieve necessary property data.

• calculate compressor power, refrigeration capacity, and coefficient of performance.

• calculate exergy destruction in the compressor and expansion valve.

Quick Quiz

What would be the coefficient of performance if the isentropic compressor efficiency were 100%? Ans. 4.83.

10.2.4 The *p–h* Diagram

p–h **diagram**

A thermodynamic property diagram widely used in the refrigeration field is the pressure–enthalpy or *p–h* diagram. **Figure 10.6** shows the main features of such a property diagram. The principal states of the vapor-compression cycles of Fig. 10.5 are located on this *p–h* diagram. It is

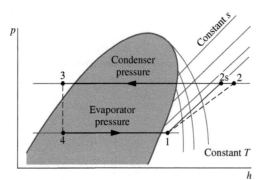

Fig. 10.6 Principal features of the pressure–enthalpy diagram for a typical refrigerant, with vapor-compression cycles superimposed.

left as an exercise to sketch the cycles of Examples 10.1, 10.2, and 10.3 on *p–h* diagrams. Property tables and *p–h* diagrams for many refrigerants are given in handbooks dealing with refrigeration.

The *p–h* diagrams for two refrigerants, CO_2 (R-744) and R-410A, are included as Figs. A-10 and A-11, respectively, in the Appendix. The ability to locate states on property diagrams is an important skill that is used selectively in end-of-chapter problems.

10.3 Selecting Refrigerants

Refrigerant selection for a wide range of refrigeration and air-conditioning applications is generally based on three factors: performance, safety, and environmental impact. The term *performance* refers to providing the required cooling or heating capacity reliably and cost-effectively. Safety refers to avoiding hazards such as toxicity and flammability. Finally, environmental impact primarily refers to using refrigerants that do not harm the stratospheric ozone layer or contribute significantly to global climate change. We begin by considering some performance aspects.

The temperatures of the refrigerant in the evaporator and condenser of vapor-compression cycles are governed by the temperatures of the cold and warm regions, respectively, with which the system interacts thermally. This, in turn, determines the operating pressures in the evaporator and condenser. Consequently, the selection of a refrigerant is based partly on the suitability of its pressure–temperature relationship in the range of the particular application. It is generally desirable to avoid excessively low pressures in the evaporator and excessively high pressures in the condenser. Other considerations in refrigerant selection include chemical stability, corrosiveness, and cost. The type of compressor also affects the choice of refrigerant. Centrifugal compressors are best suited for low evaporator pressures and refrigerants with large specific volumes at low pressure. Reciprocating compressors perform better over large pressure ranges and are better able to handle low specific volume refrigerants.

Refrigerant Types and Characteristics   Prior to the 1930s, accidents were prevalent among those who worked closely with refrigerants due to the toxicity and flammability of most refrigerants at the time. Because of such hazards, two classes of synthetic refrigerants were developed, each containing chlorine and possessing highly stable molecular structures: CFCs (chlorofluorocarbons) and HCFCs (hydrochlorofluorocarbons). These refrigerants were widely known as "freons," the common trade name.

In the early 1930s, CFC production began with R-11, R-12, R-113, and R-114. In 1936, the first HCFC refrigerant, R-22, was introduced. Over the next several decades, nearly all of the synthetic refrigerants used in the United States were either CFCs or HCFCs, with R-12 being most commonly used.

To keep order with so many new refrigerants having complicated names, the "R" numbering system was established in 1956 by DuPont and persists today as the industry standard. Table 10.1 lists information including refrigerant number, chemical composition, and global warming potential for selected refrigerants.

Environmental Considerations   After decades of use, compelling scientific data indicating that release of chlorine-containing refrigerants into the atmosphere is harmful became widely recognized. Concerns focused on released refrigerants depleting the stratospheric ozone layer and contributing to global climate change. Because of the molecular stability of the CFC and HCFC molecules, their adverse effects are long-lasting.

In 1987, an international agreement, the Montreal Protocol, was agreed upon to ban production of certain chlorine-containing refrigerants. In response, a new class of chlorine-free refrigerants was developed: the HFCs (hydrofluorocarbons). One of these, R-134a, has been used for over 25 years as the primary replacement for R-12. Although R-134a and other HFC refrigerants do not contribute to atmospheric ozone depletion, they do contribute to global climate change. Owing to a relatively high Global Warming Potential of about 1300 for R-134a, we will soon see reductions in its use in the United States which follows complete phase-out within European automotive applications. Carbon dioxide (R-744) and HFO-1234yf, from a

TABLE 10.1

Refrigerant Data Including Global Warming Potential (GWP)

Refrigerant Number	Type	Chemical Formula	Approx. GWP[a]
R-12	CFC	CCl_2F_2	10200
R-11	CFC	CCl_3F	4660
R-114	CFC	$CClF_2CClF_2$	8590
R-113	CFC	CCl_2FCClF_2	5820
R-22	HCFC	$CHClF_2$	1760
R-134a	HFC	CH_2FCF_3	1300
R-1234yf	HFO	$CF_3CF=CH_2$	<1
R-410A	HFC blend	R-32, R-125 (50/50 Weight %)	1924
R-407C	HFC blend	R-32, R-125, R-134a (23/25/52 Weight %)	1624
R-744 (carbon dioxide)	Natural	CO_2	1
R-717 (ammonia)	Natural	NH_3	0
R-290 (propane)	Natural	C_3H_8	3
R-50 (methane)	Natural	CH_4	28
R-600 (butane)	Natural	C_4H_{10}	3

[a]The Global Warming Potential (GWP) depends on the time period over which the potential influence on global warming is estimated. The values listed are based on a 100-year time period, which is an interval favored by some regulators.

Source: Intergovernmental Panel on Climate Change (IPCC) *Fifth Amendment Report: Climate Change 2014.*

new refrigerant class called hydrofluoroolefin, are possible replacements for R-134a in automotive systems. See Sec. 10.7.3 for discussion of CO_2-charged automotive air-conditioning systems.

Another refrigerant that has been used extensively in air-conditioning and refrigeration systems for decades, R-22, is being phased out under a 1995 amendment to the international agreement on refrigerants because of its chlorine content. Effective in 2010, R-22 cannot be installed in new systems. However, recovered and recycled R-22 can be used to service existing systems until supplies are no longer available. As R-22 is phased out, replacement refrigerants are being used, including R-410A and R-407C, both HFC blends.

Natural Refrigerants   Nonsynthetic, naturally occurring substances also can be used as refrigerants. Called *natural* refrigerants, they include carbon dioxide, ammonia, and hydrocarbons. As indicated by Table 10.1, natural refrigerants typically have low Global Warming Potentials.

Ammonia (R-717), which was widely used in the early development of vapor-compression refrigeration, continues to serve today as a refrigerant for large systems used by the food

Energy & Environment

The European Union announced in 2011 that the use of HFC-134a in automobile air conditioners would be banned by 2017 due to its damaging effect on the ozone layer. A replacement refrigerant, R-1234yf, has been developed with a much lower Global Warming Potential (GWP): 350 times lower that that of R-134a. However, the findings of research conducted by one European automaker challenge the safety of this replacement. Crash testing has shown that R-1234yf is highly flammable and can release toxic chemicals upon combustion. This research has resulted in some automobile manufacturers withdrawing support from R-1234yf, although the European Union states that the information is not likely to change the implementation plan for the new refrigerant. In response to the research findings, hundreds of additional safety tests to confirm the safety of R-1234yf have been conducted in the United States. Some of the research findings report that refrigerant ignition only occurs in cases where significant modifications to vehicle hardware and software have occurred. In the United Sates, the risk of passenger exposure to a vehicle fire based on the use of R-1234yf has been determined to be very low and over 100,000 vehicles in the United States and Europe are on the roads with refrigerant R-1234yf.

industry and in other industrial applications. In the past two decades, ammonia has been increasingly used because of the R-12 phaseout and is receiving even greater interest today due to the R-22 phaseout. Ammonia is also used in the absorption systems discussed in Sec. 10.5.

Hydrocarbons, such as propane (R-290), are used worldwide in various refrigeration and air-conditioning applications including commercial and household appliances. In the United States, safety concerns limit propane use to niche markets like industrial process refrigeration. Other hydrocarbons—methane (R-50) and butane (R-600)—are also under consideration for use as refrigerants.

Refrigeration with No Refrigerant Needed Alternative cooling technologies aim to achieve a refrigerating effect without use of refrigerants, thereby avoiding adverse effects associated with release of refrigerants to the atmosphere. One such technology is thermoelectric cooling. See the box.

New Materials May Improve Thermoelectric Cooling

You can buy a thermoelectric cooler powered from the cigarette lighter outlet of your car. The same technology is used in space flight applications and in power amplifiers and microprocessors.

Figure 10.7 shows a thermoelectric cooler separating a cold region at temperature T_C and a warm region at temperature T_H. The cooler is formed from two *n-type* and two *p-type* semiconductors with low thermal conductivity, five metallic interconnects with high electrical conductivity and high thermal conductivity, two electrically insulating ceramic substrates, and a power source. When power is provided by the source, current flows through the resulting electric circuit, giving a refrigeration effect: a heat transfer of energy *from* the cold region. This is known as the *Peltier* effect.

The p-type semiconductor material in the right leg of the cooler shown in Fig. 10.7 has electron vacancies called *holes*. Electrons move through this material by filling individual holes, slowing electron motion. In the adjacent n-type semiconductor, no holes exist in its material structure, so electrons move freely and more rapidly through that material. When power is provided by the power source, positively charged holes move in the direction of current while negatively charged electrons move opposite to the

current; each transfers energy from the cold region to the warm region.

The process of Peltier refrigeration may be understood by following the journey of an electron as it travels from the negative terminal of the power source to the positive terminal. On flowing through the metallic interconnect and into the p-type material, the electron slows and loses energy, causing the surrounding material to warm. At the other end of the p-type material, the electron accelerates as it enters the metallic interconnect and then the n-type material. The accelerating electron acquires energy from the surrounding material and causes the end of the p-type leg to cool. While the electron traverses the p-type material from the hot end to the cold end, holes are moving from the cold end to the hot end, transferring energy away from the cold end. While traversing the n-type material from the cold end to the hot end, the electron also transfers energy away from the cold end to the hot end. When it reaches the warm end of the n-type leg, the electron flows through the metallic interconnect and enters the next p-type material, where it slows and again loses energy. This scenario repeats itself at each pair of p-type and n-type legs, resulting in more removal of energy from the cold end and its deposit at the hot end. Thus, the overall effect of the thermoelectric cooler is heat transfer from the cold region to the warm region.

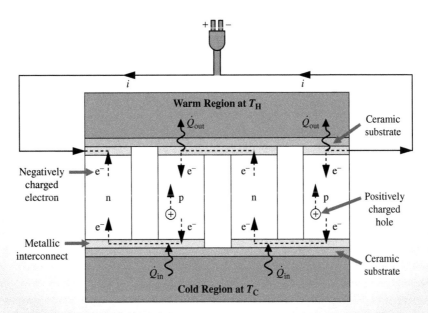

Fig. 10.7 Schematic of a thermoelectric cooler.

These simple coolers have no moving parts at a macroscopic level and are compact. They are reliable and quiet. They also use no refrigerants that harm the ozone layer or contribute to global climate change. Despite such advantages, thermoelectric coolers have found only specialized application because of low coefficients of performance compared to vapor-compression systems. However, new materials and production methods may make this type of cooler more effective, material scientists report.

As shown in Fig. 10.7, at the core of a thermoelectric cooler are two dissimilar materials, in this case n-type and p-type semiconductors. To be effective for thermoelectric cooling, the materials must have low thermal conductivity and high electrical conductivity, a rare combination in nature. However, new materials with novel microscopic structures at the *nanometer* level may lead to improved cooler performance. With nanotechnology and other advanced techniques, material scientists are striving to find materials with the favorable characteristics needed to improve the performance of thermoelectric cooling devices.

10.4 Other Vapor-Compression Applications

The basic vapor-compression refrigeration cycle can be adapted for special applications. Three are presented in this section. The first is *cold storage*, which is a thermal energy storage approach that involves chilling water or making ice. The second is a *combined-cycle* arrangement where refrigeration at relatively low temperature is achieved through a series of vapor-compression systems, with each normally employing a different refrigerant. In the third, compression work is reduced through *multistage compression* with *intercooling* between stages. The second and third applications considered are analogous to power cycle applications considered in Chaps. 8 and 9.

10.4.1 Cold Storage

Chilling water or making ice during *off-peak* periods, usually overnight or over weekends, and storing chilled water/ice in tanks until needed for cooling is known as *cold storage*. Cold storage is an aspect of thermal energy storage considered in the box in Section 3.8. Applications of cold storage include cooling of office and commercial buildings, medical centers, college campus buildings, and shopping malls.

Figure 10.8 illustrates a cold storage system intended for the comfort cooling of an occupied space. It consists of a vapor-compression refrigeration unit, ice maker and ice storage tank, and coolant loop. Running at night, when less power is required for its operation due to cooler ambient temperatures and when electricity rates are lowest, the refrigeration unit freezes water. The ice produced is stored in the accompanying tank. When cooling is required by building occupants during the day, the temperature of circulating building air is reduced as it passes over coils carrying chilled coolant flowing from the ice storage tank. Depending on local climate, some moisture also may be removed or added (see Secs. 12.8.3 and 12.8.4). Cool storage can provide all cooling required by the occupants or work in tandem with vapor-compression or other comfort cooling systems to meet needs.

10.4.2 Cascade Cycles

Combined-cycle arrangements for refrigeration are called *cascade* cycles. In Fig. 10.9 a cascade cycle is shown in which *two* vapor-compression refrigeration cycles, labeled A and B, are arranged in series with a counterflow heat exchanger linking them. In the intermediate heat exchanger, the energy rejected during condensation of the refrigerant in the lower-temperature cycle A is used to evaporate the refrigerant in the higher-temperature cycle B. The desired refrigeration effect occurs in the low-temperature evaporator, and heat rejection from the overall cycle occurs in the high-temperature condenser. The coefficient of performance is the ratio of the refrigeration effect to the *total* work input

$$\beta = \frac{\dot{Q}_{in}}{\dot{W}_{cA} + \dot{W}_{cB}}$$

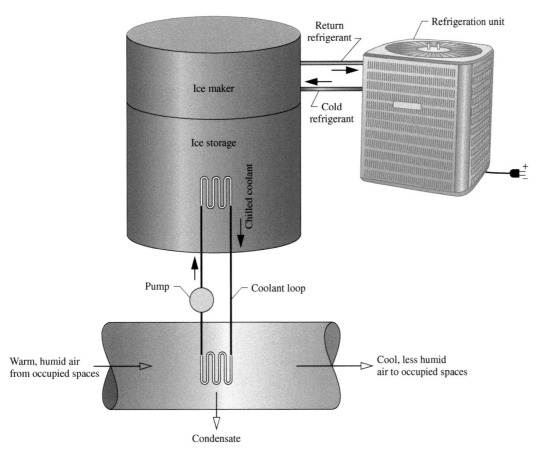

Fig. 10.8 Cold storage applied to comfort cooling.

The mass flow rates in cycles A and B are normally different. However, the mass flow rates are related by mass and energy rate balances on the interconnecting counterflow heat exchanger serving as the condenser for cycle A and the evaporator for cycle B. Although only two cycles are shown in Fig. 10.9, cascade cycles may employ three or more individual cycles.

A significant feature of the cascade system illustrated in Fig. 10.9 is that the refrigerants in the two or more stages can be selected to have advantageous evaporator and condenser pressures in the two or more temperature ranges. In a double cascade system, a refrigerant would be selected for cycle A that has a saturation pressure–temperature relationship that allows refrigeration at a relatively low temperature without excessively low evaporator pressures. The refrigerant for cycle B would have saturation characteristics that permit condensation at the required temperature without excessively high condenser pressures.

10.4.3 Multistage Compression with Intercooling

The advantages of multistage compression with intercooling between stages have been cited in Sec. 9.8, dealing with gas power systems. Intercooling is achieved in gas power systems by heat transfer to the lower-temperature surroundings. In refrigeration systems, the refrigerant temperature is below that of the surroundings for much of the cycle, so other means must be employed to accomplish intercooling and achieve the attendant savings in the required compressor work input. An arrangement for two-stage compression using the refrigerant itself for intercooling is shown in **Fig. 10.10**. The principal states of the refrigerant for an ideal cycle are shown on the accompanying *T–s* diagram.

Intercooling is accomplished in this cycle by means of a direct-contact heat exchanger. Relatively low-temperature saturated vapor enters the heat

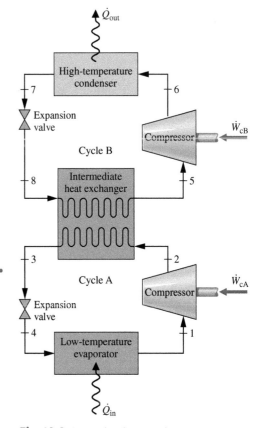

Fig. 10.9 Example of a cascade vapor-compression refrigeration cycle.

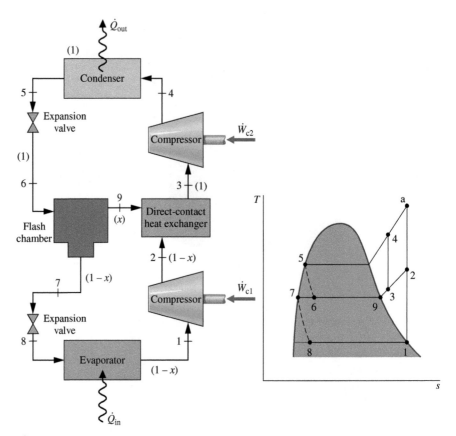

Fig. 10.10 Refrigeration cycle with two stages of compression and flash intercooling.

exchanger at state 9, where it mixes with higher-temperature refrigerant leaving the first compression stage at state 2. A single mixed stream exits the heat exchanger at an intermediate temperature at state 3 and is compressed in the second compressor stage to the condenser pressure at state 4. Less work is required per unit of mass flow for compression from 1 to 2 followed by compression from 3 to 4 than for a single stage of compression 1–2–a. Since the refrigerant temperature entering the condenser at state 4 is lower than for a single stage of compression in which the refrigerant would enter the condenser at state a, the external irreversibility associated with heat transfer in the condenser is also reduced.

A central role is played in the cycle of Fig. 10.10 by a liquid–vapor separator, called a **flash chamber**. Refrigerant exiting the condenser at state 5 expands through a valve and enters the flash chamber at state 6 as a two-phase liquid–vapor mixture with quality x. In the flash chamber, the liquid and vapor components separate into two streams. Saturated vapor exiting the flash chamber enters the heat exchanger at state 9, where intercooling is achieved as discussed above. Saturated liquid exiting the flash chamber at state 7 expands through a second valve into the evaporator. On the basis of a unit of mass flowing through the condenser, the fraction of the vapor formed in the flash chamber equals the quality x of the refrigerant at state 6. The fraction of the liquid formed is then $(1 - x)$. The fractions of the total flow at various locations are shown in parentheses on Fig. 10.10.

flash chamber *(margin note)*

10.5 Absorption Refrigeration

absorption refrigeration *(margin note)*

Absorption refrigeration cycles are the subject of this section. These cycles have some features in common with the vapor-compression cycles considered previously but differ in two important respects:

- One is the nature of the compression process. Instead of compressing a vapor between the evaporator and the condenser, the refrigerant of an absorption system is absorbed by a secondary substance, called an absorbent, to form a *liquid solution*. The liquid solution is then *pumped* to the higher pressure. Because the average specific volume of the liquid solution is much less than that of the refrigerant vapor, significantly less work is required (see the discussion of Eq. 6.51b in Sec. 6.13.2). Accordingly, absorption refrigeration systems have the advantage of relatively small work input compared to vapor-compression systems.

- The other main difference between absorption and vapor-compression systems is that some means must be introduced in absorption systems to retrieve the refrigerant vapor from the liquid solution before the refrigerant enters the condenser. This involves heat transfer from a relatively high-temperature source. Steam or waste heat that otherwise would be discharged to the surroundings without use is particularly economical for this purpose. Natural gas or some other fuel can be burned to provide the heat source, and there have been practical applications of absorption refrigeration using alternative energy sources such as solar and geothermal energy.

The principal components of an absorption refrigeration system are shown schematically in Fig. 10.11. In this case, ammonia is the refrigerant and water is the absorbent. Ammonia circulates through the condenser, expansion valve, and evaporator as in a vapor-compression system. However, the compressor is replaced by the absorber, pump, generator, and valve shown on the right side of the diagram.

- In the *absorber*, ammonia vapor coming from the evaporator at state 1 is absorbed by liquid water. The formation of this liquid solution is exothermic. Since the amount of ammonia that can be dissolved in water increases as the solution temperature decreases, cooling water is circulated around the absorber to remove the energy released as ammonia goes into solution and to maintain the temperature in the absorber as low as possible. The strong ammonia–water solution leaves the absorber at point a and enters the *pump*, where its pressure is increased to that of the generator.

- In the *generator*, heat transfer from a high-temperature source drives ammonia vapor out of the solution (an endothermic process), leaving a weak ammonia–water solution in the generator. The vapor liberated passes to the condenser at state 2, and the remaining weak solution at c flows back to the absorber through a *valve*. The only work input is the power required to operate the pump, and this is small in comparison to the work that would be required to compress refrigerant vapor between the same pressure levels. However, costs associated with the heat source and extra equipment not required by vapor-compressor systems can cancel the advantage of a smaller work input.

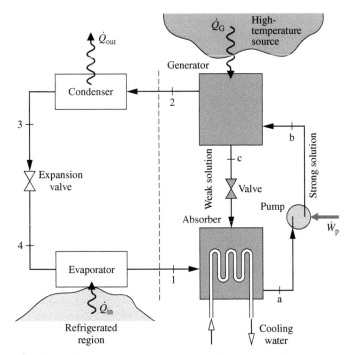

Fig. 10.11 Simple ammonia–water absorption refrigeration system.

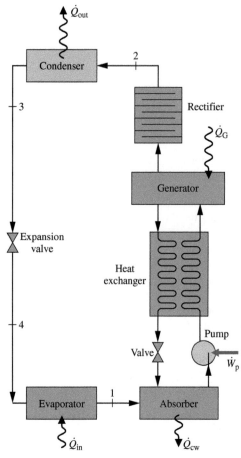

Fig. 10.12 Modified ammonia–water absorption system.

Animation

Heat Pump Cycle

Ammonia–water systems normally employ several modifications of the simple absorption cycle considered above. Two common modifications are illustrated in **Fig. 10.12**. In this cycle, a heat exchanger is included between the generator and the absorber that allows the strong water–ammonia solution entering the generator to be preheated by the weak solution returning from the generator to the absorber, thereby reducing the heat transfer to the generator, $\dot{Q}_G$. The other modification shown on the figure is the *rectifier* placed between the generator and the condenser. The function of the rectifier is to remove any traces of water from the refrigerant before it enters the condenser. This eliminates the possibility of ice formation in the expansion valve and the evaporator.

Another type of absorption system uses *lithium bromide* as the absorbent and *water* as the refrigerant. The basic principle of operation is the same as for ammonia–water systems. To achieve refrigeration at lower temperatures than are possible with water as the refrigerant, a lithium bromide–water absorption system may be combined with another cycle using a refrigerant with good low-temperature characteristics, such as ammonia, to form a cascade refrigeration system.

10.6 Heat Pump Systems

The objective of a heat pump is to maintain the temperature within a dwelling or other building above the temperature of the surroundings or to provide a heat transfer for certain industrial processes that occur at elevated temperatures. Heat pump systems have many features in common with the refrigeration systems considered thus far and may be of the vapor-compression or absorption type. Vapor-compression heat pumps are well suited for space heating applications and are commonly used for this purpose. Absorption heat pumps have been developed for industrial applications and are also increasingly being used for space heating. To introduce some aspects of heat pump operation, let us begin by considering the Carnot heat pump cycle.

10.6.1 Carnot Heat Pump Cycle

By simply changing our viewpoint, we can regard the cycle shown in Fig. 10.1 as a *heat pump*. The objective of the cycle now, however, is to deliver the heat transfer $\dot{Q}_{out}$ to the warm region, which is the space to be heated. At steady state, the rate at which energy is supplied to the warm region by heat transfer is the sum of the energy supplied to the working fluid from the cold region, $\dot{Q}_{in}$, and the net rate of work input to the cycle, $\dot{W}_{net}$. That is,

$$\dot{Q}_{out} = \dot{Q}_{in} + \dot{W}_{net} \tag{10.8}$$

The *coefficient of performance* of *any* heat pump cycle is defined as the ratio of the heating effect to the net work required to achieve that effect. For the Carnot heat pump cycle of Fig. 10.1

$$\gamma_{max} = \frac{\dot{Q}_{out}/\dot{m}}{\dot{W}_c/\dot{m} - \dot{W}_t/\dot{m}} = \frac{\text{area } 2\text{-a-b-3-2}}{\text{area } 1\text{-2-3-4-1}}$$

which reduces to

$$\gamma_{max} = \frac{T_H(s_a - s_b)}{(T_H - T_C)(s_a - s_b)} = \frac{T_H}{T_H - T_C} \tag{10.9}$$

This equation, which corresponds to Eq. 5.11, represents the *maximum* theoretical coefficient of performance for any heat pump cycle operating between two regions at temperatures T_C and T_H. Actual heat pump systems have coefficients of performance that are lower than calculated from Eq. 10.9.

A study of Eq. 10.9 shows that as the temperature T_C of the cold region decreases, the coefficient of performance of the Carnot heat pump decreases. This trait is also exhibited by actual heat pump systems and suggests why heat pumps in which the role of the cold region is played by the local atmosphere (air-source heat pumps) normally require backup systems to provide heating on days when the ambient temperature becomes very low. If sources such as well water or the ground itself are used, relatively high coefficients of performance can be achieved despite low ambient air temperatures, and backup systems may not be required.

10.6.2 Vapor-Compression Heat Pumps

Actual heat pump systems depart significantly from the Carnot cycle model. Most systems in common use today are of the vapor-compression type. The method of analysis of *vapor-compression heat pumps* is the same as that of vapor-compression refrigeration cycles considered previously. Also, the previous discussions concerning the departure of actual systems from ideality apply for vapor-compression heat pump systems as for vapor-compression refrigeration cycles.

As illustrated by **Fig. 10.13**, a typical **vapor-compression heat pump** for space heating has the same basic components as the vapor-compression refrigeration system: compressor, condenser, expansion valve, and evaporator. The objective of the system is different, however. In a heat pump system, $\dot{Q}_{in}$ comes from the surroundings, and $\dot{Q}_{out}$ is directed to the dwelling as the desired effect. A net work input is required to accomplish this effect.

vapor-compression heat pump

The coefficient of performance of a simple vapor-compression heat pump with states as designated on Fig. 10.13 is

$$\gamma = \frac{\dot{Q}_{out}/\dot{m}}{\dot{W}_c/\dot{m}} = \frac{h_2 - h_3}{h_2 - h_1} \qquad (10.10)$$

The value of γ can never be less than unity.

Many possible sources are available for heat transfer to the refrigerant passing through the evaporator, including outside air; the ground; and lake, river, or well water. Liquid

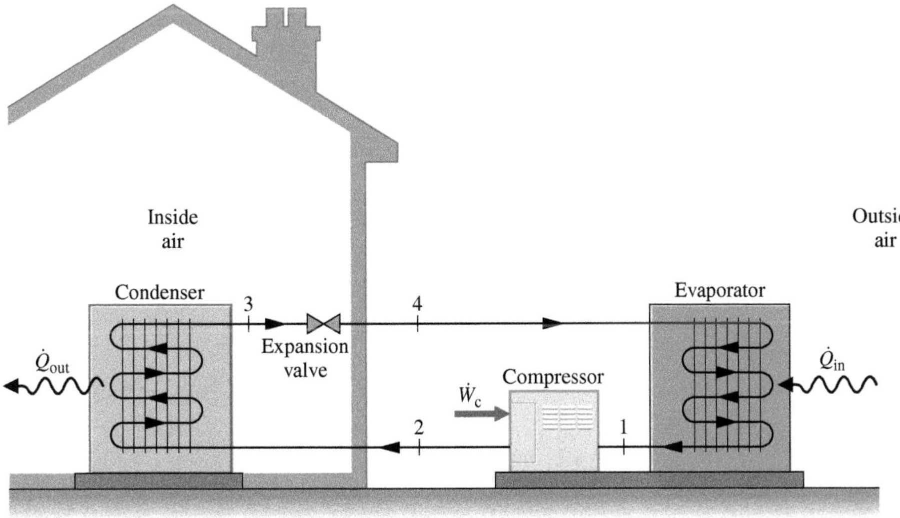

Fig. 10.13 Air-source vapor-compression heat pump system.

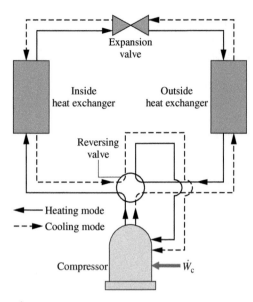

Fig. 10.14 Example of an air-to-air reversing heat pump.

air-source heat pump

circulated through a solar collector and stored in an insulated tank also can be used as a source for a heat pump. Industrial heat pumps employ waste heat or warm liquid or gas streams as the low-temperature source and are capable of achieving relatively high condenser temperatures.

In the most common type of vapor-compression heat pump for space heating, the evaporator communicates thermally with the outside air. Such **air-source heat pumps** also can be used to provide cooling in the summer with the use of a reversing valve, as illustrated in **Fig. 10.14**. The solid lines show the flow path of the refrigerant in the heating mode, as described previously. To use the same components as an air conditioner, the valve is actuated, and the refrigerant follows the path indicated by the dashed line. In the cooling mode, the outside heat exchanger becomes the condenser, and the inside heat exchanger becomes the evaporator. Although heat pumps can be more costly to install and operate than other direct heating systems, they can be competitive when the potential for dual use is considered.

Example 10.4 illustrates use of the first and second laws of thermodynamics together with property data to analyze the performance of an actual heat pump cycle, including cost of operation.

▶ ▶ ▶ EXAMPLE 10.4 ▶ ·

Analyzing an Actual Vapor-Compression Heat Pump Cycle

Refrigerant 134a is the working fluid in an electric-powered, air-source heat pump that maintains the inside temperature of a building at 22°C for a week when the average outside temperature is 5°C. Saturated vapor enters the compressor at −8°C and exits at 50°C, 10 bar. Saturated liquid exits the condenser at 10 bar. The refrigerant mass flow rate is 0.2 kg/s for steady-state operation. Determine **(a)** the compressor power, in kW, **(b)** the isentropic compressor efficiency, **(c)** the heat transfer rate provided to the building, in kW, **(d)** the coefficient of performance, and **(e)** the total cost of electricity, in $, for 80 hours of operation during that week, evaluating electricity at 15 cents per kW · h.

SOLUTION

Known A heat pump cycle operates with Refrigerant 134a. The states of the refrigerant entering and exiting the compressor and leaving the condenser are specified. The refrigerant mass flow rate and interior and exterior temperatures are given.

Find Determine the compressor power, the isentropic compressor efficiency, the heat transfer rate to the building, the coefficient of performance, and the cost to operate the electric heat pump for 80 hours of operation.

Schematic and Given Data:

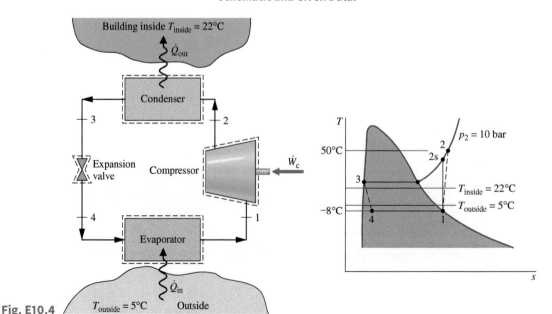

Fig. E10.4

Engineering Model

1. Each component of the cycle is analyzed as a control volume at steady state.

2. There are no pressure drops through the evaporator and condenser.

3. The compressor operates adiabatically. The expansion through the valve is a throttling process.

4. Kinetic and potential energy effects are negligible.

5. Saturated vapor enters the compressor and saturated liquid exits the condenser.

6. For costing purposes, conditions provided are representative of the entire week of operation and the value of electricity is 15 cents per kW · h.

Analysis   Let us begin by fixing the principal states located on the accompanying schematic and T–s diagram. State 1 is saturated vapor at $-8°C$; thus h_1 and s_1 are obtained directly from Table A-10. State 2 is superheated vapor; knowing T_2 and p_2, h_2 is obtained from Table A-12. State 3 is saturated liquid at 10 bar and h_3 is obtained from Table A-11. Finally, expansion through the valve is a throttling process; therefore, $h_4 = h_3$. A summary of property values at these states is provided in the following table:

State	$T(°C)$	p(bar)	h(kJ/kg)	s(kJ/kg · K)
1	−8	2.1704	242.54	0.9239
2	50	10	280.19	—
3	—	10	105.29	—
4	—	2.1704	105.29	—

a. The compressor power is

$$\dot{W}_c = \dot{m}(h_2 - h_1) = 0.2\frac{kg}{s}(280.19 - 242.54)\frac{kJ}{kg}\left|\frac{1\ kW}{1\ kJ/s}\right| = 7.53\ kW$$

b. The isentropic compressor efficiency is

$$\eta_c = \frac{(\dot{W}_c/\dot{m})_s}{(\dot{W}_c/\dot{m})} = \frac{(h_{2s} - h_1)}{(h_2 - h_1)}$$

where h_{2s} is the specific entropy at state 2s, as indicated on the accompanying T–s diagram. State 2s is fixed using p_2 and $s_{2s} = s_1$.

Interpolating in Table A-12, $h_{2s} = 274.18$ kJ/kg. Solving for compressor efficiency

$$\eta_c = \frac{(h_{2s} - h_1)}{(h_2 - h_1)} = \frac{(274.18 - 242.54)}{(280.19 - 242.54)} = 0.84\ (84\%)$$

c. The heat transfer rate provided to the building is

$$\dot{Q}_{out} = \dot{m}(h_2 - h_3) = \left(0.2\frac{kg}{s}\right)(280.19 - 105.29)\frac{kJ}{kg}\left|\frac{1\ kW}{1\ kJ/s}\right|$$

$$= 34.98\ kW$$

d. The heat pump coefficient of performance is

$$\gamma = \frac{\dot{Q}_{out}}{\dot{W}_c} = \frac{34.98\ kW}{7.53\ kW} = 4.65$$

e. Using the result from part (a) together with the given cost and use data

[electricity cost for 80 hours of operation]

$$= (7.53\ kW)(80\ h)\left(0.15\frac{\$}{kW \cdot h}\right) = \$90.36$$

SKILLS DEVELOPED

Ability to...

- sketch the T–s diagram of the vapor-compression heat pump cycle with irreversibilities in the compressor.
- fix each of the principal states and retrieve necessary property data.
- calculate the compressor power, heat transfer rate delivered, and coefficient of performance.
- calculate isentropic compressor efficiency.
- conduct an elementary economic evaluation.

Quick Quiz

If the cost of electricity is 10 cents per kW · h, which is the U.S. average for the period under consideration, evaluate the cost to operate the heat pump, in $, keeping all other data the same. **Ans. $60.24.**

10.7 Gas Refrigeration Systems

All refrigeration systems considered thus far involve changes in phase. Let us now turn to **gas refrigeration systems** in which the working fluid remains a gas throughout. Gas refrigeration systems have a number of important applications. They are used to achieve very low temperatures for the liquefaction of air and other gases and for other specialized applications such as aircraft cabin cooling. The Brayton refrigeration cycle illustrates an important type of gas refrigeration system.

gas refrigeration systems

10.7.1 Brayton Refrigeration Cycle

The **Brayton refrigeration cycle** is the reverse of the closed Brayton power cycle introduced in Sec. 9.6. A schematic of the reversed Brayton cycle is provided in **Fig. 10.15a**. The refrigerant gas, which may be air, enters the compressor at state 1, where the temperature is

Brayton refrigeration cycle

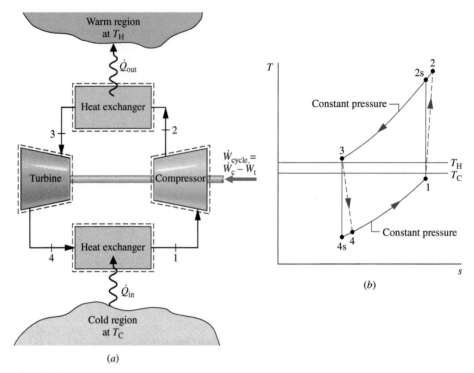

Fig. 10.15 Brayton refrigeration cycle.

somewhat below the temperature of the cold region, T_C, and is compressed to state 2. The gas is then cooled to state 3, where the gas temperature approaches the temperature of the warm region, T_H. Next, the gas is expanded to state 4, where the temperature, T_4, is well below that of the cold region. Refrigeration is achieved through heat transfer from the cold region to the gas as it passes from state 4 to state 1, completing the cycle. The T–s diagram in Fig. 10.15b shows an *ideal* Brayton refrigeration cycle, denoted by 1–2s–3–4s–1, in which all processes are assumed to be internally reversible and the processes in the turbine and compressor are adiabatic. Also shown is the cycle 1–2–3–4–1, which suggests the effects of irreversibilities during adiabatic compression and expansion. Frictional pressure drops have been ignored.

Cycle Analysis The method of analysis of the Brayton refrigeration cycle is similar to that of the Brayton power cycle. Thus, at steady state the work of the compressor and the turbine per unit of mass flow are, respectively,

$$\frac{\dot{W}_c}{\dot{m}} = h_2 - h_1 \quad \text{and} \quad \frac{\dot{W}_t}{\dot{m}} = h_3 - h_4$$

In obtaining these expressions, heat transfer with the surroundings and changes in kinetic and potential energy have been ignored. The magnitude of the work developed by the turbine of a Brayton refrigeration cycle is typically significant relative to the compressor work input.

Heat transfer from the cold region to the refrigerant gas circulating through the low-pressure heat exchanger, the refrigeration effect, is

$$\frac{\dot{Q}_{in}}{\dot{m}} = h_1 - h_4$$

The coefficient of performance is the ratio of the refrigeration effect to the net work input:

$$\beta = \frac{\dot{Q}_{in}/\dot{m}}{\dot{W}_c/\dot{m} - \dot{W}_t/\dot{m}} = \frac{(h_1 - h_4)}{(h_2 - h_1) - (h_3 - h_4)} \tag{10.11}$$

In the next example, we illustrate the analysis of an ideal Brayton refrigeration cycle.

▶ ▶ **EXAMPLE 10.5** ▶ ·

Ideal Brayton Refrigeration Cycle

Air enters the compressor of an ideal Brayton refrigeration cycle at 1 bar, 270°K, with a volumetric flow rate of 1.4 m³/s. If the compressor pressure ratio is 3 and the turbine inlet temperature is 300°K, determine **(a)** the *net* power input, in kW, **(b)** the refrigeration capacity, in kW, **(c)** the coefficient of performance.

SOLUTION

Known   An ideal Brayton refrigeration cycle operates with air. Compressor inlet conditions, the turbine inlet temperature, and the compressor pressure ratio are given.

Find   Determine the *net* power input, in kW, the refrigeration capacity, in kW, and the coefficient of performance.

Schematic and Given Data:

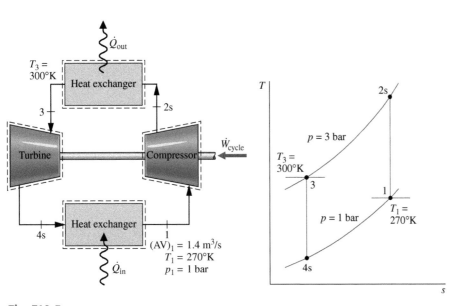

Fig. E10.5

Engineering Model

1. Each component of the cycle is analyzed as a control volume at steady state. The control volumes are indicated by dashed lines on the accompanying sketch.

2. The turbine and compressor processes are isentropic.

3. There are no pressure drops through the heat exchangers.

4. Kinetic and potential energy effects are negligible.

5. The working fluid is air modeled as an ideal gas.

Analysis   The analysis begins by determining the specific enthalpy at each numbered state of the cycle. At state 1, the temperature is 270 K From Table A-22, $h_1 = 270.11$ kJ/kg, $p_{r1} = 0.9590$. Since the compressor process is isentropic, h_{2s} can be determined by first evaluating p_r at state 2s. That is

$$p_{r2} = \frac{p_2}{p_1} p_{r1} = (3)(0.9590) = 2.877$$

Then, interpolating in Table A-22, we get $h_{2s} = 370.1$ kJ/kg.

The temperature at state 3 is given as $T_3 = 300°$K. From Table A-22, $h_3 = 300.19$ kJ/kg, $p_{r3} = 1.3860$. The specific enthalpy at state 4s is found by using the isentropic relation

$$p_{r4} = p_{r3} \frac{p_4}{p_3} = (1.3860)(1/3) = 0.462$$

Interpolating in Table A-22, we obtain $h_{4s} = 219.0$ kJ/kg.

a. The net power input is

$$\dot{W}_{cycle} = \dot{m}[(h_{2s} - h_1) - (h_3 - h_{4s})]$$

This requires the mass flow rate $\dot{m}$, which can be determined from the volumetric flow rate and the specific volume at the compressor inlet:

$$\dot{m} = \frac{(AV)_1}{v_1}$$

Since $v_1 = (\bar{R}/M)T_1/p_1$

$$\dot{m} = \frac{(AV)_1 p_1}{(\bar{R}/M)T_1}$$

$$= \frac{(1.4 \text{ m}^3/\text{s})(1 \text{ bar})}{\left(\dfrac{8.314 \text{ kJ}}{28.97}\right)(270 \text{ K})} \left(\frac{10^5 \text{ N/m}^2}{1 \text{ bar}}\right)\left(\frac{1 \text{ kJ}}{10^3 \text{ N} \cdot \text{m}}\right)$$

$$= 1.807 \text{ kg/s}$$

Finally

$$\dot{W}_{cycle} = (1.807 \text{ kg/s})[(370.1 - 270.11) - (300.19 - 219.0)] \text{ kJ/kg} \left(\frac{1 \text{ kW}}{1 \text{ kJ/s}}\right)$$

$$= 33.97 \text{ kW}$$

b. The refrigeration capacity is

$$\dot{Q}_{in} = \dot{m}(h_1 - h_{4s})$$

$$= (1.807 \text{ kg/s})[(270.11 - 219)] \text{ kJ/kg} \left(\frac{1 \text{ kW}}{1 \text{ kJ/s}}\right)$$

$$= 92.36 \text{ kW}$$

c. The coefficient of performance is

$$\beta = \frac{\dot{Q}_{in}}{\dot{W}_{cycle}} = \frac{92.36}{33.97} = 2.72$$

Irreversibilities within the compressor and turbine serve to decrease the coefficient of performance significantly from that of the corresponding ideal cycle because the compressor

work requirement is increased and the turbine work output is decreased. This is illustrated in the example to follow.

SKILLS DEVELOPED

Ability to...

- sketch the *T–s* diagram of the ideal Brayton refrigeration cycle.
- fix each of the principal states and retrieve necessary property data.
- calculate refrigeration capacity and coefficient of performance.

Quick Quiz

Determine the refrigeration capacity in tons of refrigeration. Ans. 25.53 ton.

Example 10.6 illustrates the effects of irreversible compression and turbine expansion building on Example 10.5, on the performance of Brayton cycle refrigeration. For this, we apply the isentropic compressor and turbine efficiencies introduced in Sec. 6.12.

> ▶▶▶ **EXAMPLE 10.6** ▶ •

Brayton Refrigeration Cycle with Irreversibilities

Reconsider Example 10.4, but include in the analysis that the compressor and turbine each have an isentropic efficiency of 80%. Determine for the modified cycle **(a)** the net power input, in kW, **(b)** the refrigeration capacity, in kW, **(c)** the coefficient of performance, and interpret its value.

SOLUTION

Known A Brayton refrigeration cycle operates with air. Compressor inlet conditions, the turbine inlet temperature, and the compressor pressure ratio are given. The compressor and turbine each have an efficiency of 80%.

Find Determine the net power input and the refrigeration capacity, each in kW. Also, determine the coefficient of performance and interpret its value.

Schematic and Given Data:

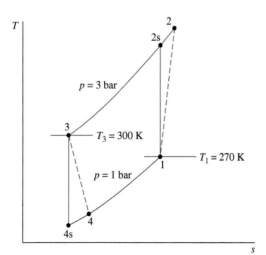

Fig. E10.6

Engineering Model

1. Each component of the cycle is analyzed as a control volume at steady state.

2. The compressor and turbine are adiabatic.

3. There are no pressure drops through the heat exchangers.

4. Kinetic and potential energy effects are negligible.

5. The working fluid is air modeled as an ideal gas.

Analysis

a. The power input to the compressor is evaluated using the isentropic compressor efficiency, η_c. That is

$$\frac{\dot{W}_c}{\dot{m}} = \frac{(\dot{W}_c/\dot{m})_s}{\eta_c}$$

The value of the work per unit mass for the isentropic compression, $(\dot{W}_c/\dot{m})_s$, is determined with data from the solution in Example 10.4 as 99.99 kJ/kg. The actual power required is then

$$\dot{W}_c = \frac{\dot{m}(\dot{W}_c/\dot{m})_s}{\eta_c} = \frac{(1.807 \text{ kg/s})(99.99 \text{ kJ/kg})}{(0.8)}$$

$$= 225.9 \text{ kW}$$

The turbine power output is determined in a similar manner, using the turbine isentropic efficiency η_t. Thus, $\dot{W}_t/\dot{m} = \eta_t(\dot{W}_t/\dot{m})_s$. Using data from the solution to Example 10.4 gives $(\dot{W}_t/\dot{m})_s = 81.19 \text{ kJ/kg}$. The actual turbine work is then

$$\dot{W}_t = \dot{m}\eta_t(\dot{W}_t/\dot{m})_s = (1.807 \text{ kg/s})(0.8)(81.19 \text{ kJ/kg})$$

$$= 117.4 \text{ kW}$$

The *net* power input to the cycle is

$$\dot{W}_{cycle} = 225.9 - 117.4 = 108.5 \text{ kW}$$

b. The specific enthalpy at the turbine exit, h_4, is required to evaluate the refrigeration capacity. This enthalpy can be determined by solving $\dot{W}_t = \dot{m}(h_3 - h_4)$ to obtain $h_4 = h_3 - \dot{W}_t/\dot{m}$. Inserting known values

$$h_4 = 300.19 - \left(\frac{117.4}{1.807}\right) = 235.2 \text{ kJ/kg}$$

The refrigeration capacity is then

$$\dot{Q}_{in} = \dot{m}(h_1 - h_4) = (1.807)(270.11 - 235.2) = 63.08 \text{ kW}$$

c. The coefficient of performance is

$$\beta = \frac{\dot{Q}_{in}}{\dot{W}_{cycle}} = \frac{63.08}{108.5} = 0.581$$

The value of the coefficient of performance in this case is less than unity. This means that the refrigeration effect is smaller than the net work required to achieve it. Additionally, note that irreversibilities in the compressor and turbine have a significant effect on the performance of gas refrigeration systems. This is brought out by comparing the results of the present example with those of

Example 10.4. Irreversibilities result in an increase in the work of compression and a reduction in the work output of the turbine. The refrigeration capacity is also reduced. The overall effect is that the coefficient of performance is decreased significantly.

SKILLS DEVELOPED

Ability to...

- sketch the *T–s* diagram of the Brayton refrigeration cycle with irreversibilities in the turbine and compressor.

- fix each of the principal states and retrieve necessary property data.

- calculate refrigeration capacity and coefficient of performance.

Quick Quiz

Determine the coefficient of performance for a Carnot refrigeration cycle operating between reservoirs at 266.7 K and 300 K. Ans. 8.

10.7.2 Additional Gas Refrigeration Applications

To obtain even moderate refrigeration capacities with the Brayton refrigeration cycle, equipment capable of achieving relatively high pressures and volumetric flow rates is needed. For most applications involving air conditioning and for ordinary refrigeration processes, vapor-compression systems can be built more cheaply and can operate with higher coefficients of performance than gas refrigeration systems. With suitable modifications, however, gas refrigeration systems can be used to achieve temperatures of about −150°C, which are well below the temperatures normally obtained with vapor systems.

Figure 10.16 shows the schematic and *T–s* diagram of an ideal Brayton cycle modified by introduction of a heat exchanger. The heat exchanger allows the air exiting the compressor at state 2 to cool *below* the warm region temperature T_H, giving a low turbine inlet temperature, T_3. Without the heat exchanger, air could be cooled only close to T_H, as represented on the figure by state a. In the subsequent expansion through the turbine, the air achieves a much lower temperature at state 4 than would have been possible without the heat exchanger. Accordingly, the refrigeration effect, achieved from state 4 to state b, occurs at a correspondingly lower average temperature.

An example of the application of gas refrigeration to cabin cooling in an aircraft is illustrated in **Fig. 10.17**. As shown in the figure, a small amount of high-pressure air is extracted from the

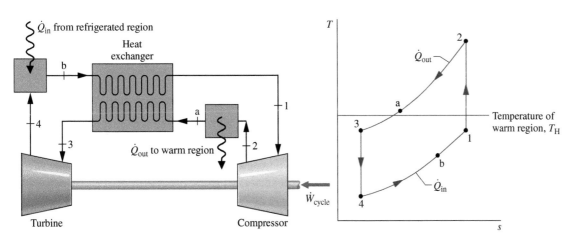

Fig. 10.16 Brayton refrigeration cycle modified with a heat exchanger.

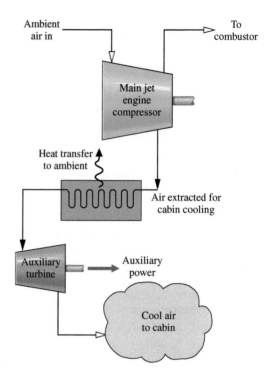

Fig. 10.17 An application of gas refrigeration to aircraft cabin cooling.

main jet engine compressor and cooled by heat transfer to the ambient. The high-pressure air is then expanded through an auxiliary turbine to the pressure maintained in the cabin. The air temperature is reduced in the expansion and thus is able to fulfill its cabin cooling function. As an additional benefit, the turbine expansion can provide some of the auxiliary power needs of the aircraft.

Size and weight are important considerations in the selection of equipment for use in aircraft. Open-cycle systems, like the example given here, utilize *compact* high-speed rotary turbines and compressors. Furthermore, since the air for cooling comes directly from the surroundings, there are fewer heat exchangers than would be needed if a separate refrigerant were circulated in a closed vapor-compression cycle.

10.7.3 Automotive Air Conditioning Using Carbon Dioxide

Owing primarily to environmental concerns, automotive air-conditioning systems using CO_2 are currently under active consideration. Carbon dioxide causes no harm to the ozone layer, and its Global Warming Potential of 1 is small compared to that of R-134a, commonly used in automotive air-conditioning systems. Carbon dioxide is nontoxic and nonflammable. As it is abundant in the atmosphere and the exhaust gas of coal-burning power and industrial plants, CO_2 is a relatively inexpensive choice as a refrigerant. Still, automakers considering a shift to CO_2 away from R-134a must weigh system performance, equipment costs, and other key issues before embracing such a change in longstanding practice.

Figure 10.18 shows the schematic of a CO_2-charged automotive air-conditioning system with an accompanying T–s diagram labeled with the critical temperature T_c and critical pressure p_c of CO_2: 31°C and 72.9 atm, respectively. The system combines aspects of gas refrigeration with aspects of vapor-compression refrigeration. Let us follow the CO_2 as it passes steadily through each of the components, beginning with the inlet to the compressor.

Carbon dioxide enters the compressor as superheated vapor at state 1 and is compressed to a much higher temperature and pressure at state 2. The CO_2 passes from the compressor into the gas cooler, where it cools at constant pressure to state 3 as a result of heat transfer to the ambient. The temperature at state 3 approaches that of the ambient, denoted on the figure by T_H. CO_2 is further cooled in the interconnecting heat exchanger at constant pressure to state 4, where the temperature is below that of ambient. Cooling is provided by low-temperature CO_2 in the other stream of the heat exchanger. During this portion of the refrigeration cycle, the processes are like those seen in gas refrigeration.

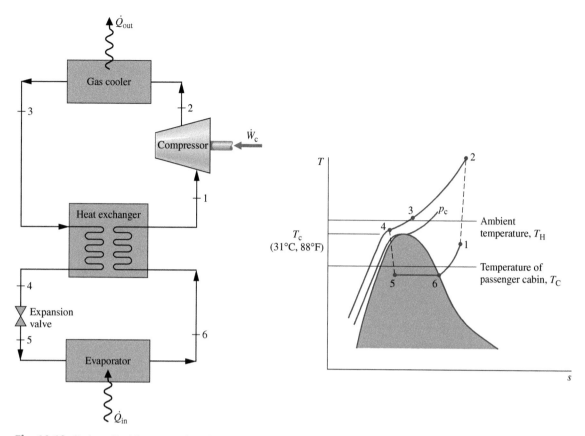

Fig. 10.18 Carbon dioxide automotive air-conditioning system.

This similarity ends abruptly as the CO_2 next expands through the valve to state 5 in the liquid–vapor region and then enters the evaporator, where it is vaporized to state 6 by heat transfer from the passenger cabin at temperature T_C, thereby cooling the passenger cabin. These processes are like those seen in vapor-compression refrigeration systems. Finally, at state 6 the CO_2 enters the heat exchanger, exiting at state 1. The heat exchanger increases the cycle's performance in two ways: by delivering lower-quality two-phase mixture at state 5, increasing the refrigerating effect through the evaporator, and by producing higher-temperature superheated vapor at state 1, reducing the compressor power required.

CHAPTER SUMMARY AND STUDY GUIDE

In this chapter we have considered refrigeration and heat pump systems, including vapor systems where the refrigerant is alternately vaporized and condensed, and gas systems where the refrigerant remains a gas. The three principal types of systems discussed are the vapor-compression, absorption, and reversed Brayton cycles.

The performance of simple vapor refrigeration systems is described in terms of the vapor-compression cycle. For this cycle, we have evaluated the principal work and heat transfers along with two important performance parameters: the coefficient of performance and the refrigeration capacity. We have considered the effect on performance of irreversibilities during the compression process and in the expansion across the valve, as well as the effect of irreversible heat transfer between the refrigerant and the warm and cold regions. Variations of the basic vapor-compression refrigeration cycle also have been considered, including cold storage, cascade cycles, and multistage compression with intercooling. A discussion of vapor-compression heat pump systems is also provided.

Qualitative discussions are presented of refrigerant properties and of considerations in selecting refrigerants. Absorption refrigeration and heat pump systems are also discussed qualitatively. The chapter concludes with a study of gas refrigeration systems.

The following list provides a study guide for this chapter. When your study of the text and end-of-chapter exercises has been completed, you should be able to

- write out the meanings of the terms listed in the margin throughout the chapter and understand each of the related concepts. The subset of key concepts listed below is particularly important.

- sketch the T–s diagrams of vapor-compression refrigeration and heat pump cycles and of Brayton refrigeration cycles correctly showing the relationship of the refrigerant temperature to the temperatures of the warm and cold regions.

- apply the first and second laws along with property data to determine the performance of vapor-compression refrigeration and heat pump cycles and of Brayton refrigeration cycles, including evaluation of the power required, the coefficient of performance, and the capacity.

- sketch schematic diagrams of vapor-compression cycle modifications, including cascade cycles and multistage compression with intercooling between the stages. In each case be able to retrieve property data and apply mass and energy balances and the second law to determine performance.

- explain the operation of absorption refrigeration systems.

KEY ENGINEERING CONCEPTS

vapor-compression refrigeration
refrigeration capacity

ton of refrigeration
absorption refrigeration

vapor-compression heat pump
Brayton refrigeration cycle

KEY EQUATIONS

$\beta_{max} = \dfrac{T_C}{T_H - T_C}$	(10.1)	Coefficient of performance of the Carnot refrigeration cycle (Fig. 10.1)
$\beta = \dfrac{\dot{Q}_{in}/\dot{m}}{\dot{W}_c/\dot{m}} = \dfrac{h_1 - h_4}{h_2 - h_1}$	(10.7)	Coefficient of performance of the vapor-compression refrigeration cycle (Fig. 10.3)
$\gamma_{max} = \dfrac{T_H}{T_H - T_C}$	(10.9)	Coefficient of performance of the Carnot heat pump cycle (Fig. 10.1)
$\gamma = \dfrac{\dot{Q}_{out}/\dot{m}}{\dot{W}_c/\dot{m}} = \dfrac{h_2 - h_3}{h_2 - h_1}$	(10.10)	Coefficient of performance of the vapor-compression heat pump cycle (Fig. 10.13)
$\beta = \dfrac{\dot{Q}_{in}/\dot{m}}{\dot{W}_c/\dot{m} - \dot{W}_t/\dot{m}} = \dfrac{(h_1 - h_4)}{(h_2 - h_1) - (h_3 - h_4)}$	(10.11)	Coefficient of performance of the Brayton refrigeration cycle (Fig. 10.15)

EXERCISES: THINGS ENGINEERS THINK ABOUT

10.1 What are the temperatures inside the fresh food and freezer compartments of your refrigerator? Do you know what values are *recommended* for these temperatures?

10.2 You have a refrigerator located in your garage. Does it perform differently in the summer than in the winter? Explain.

10.3 Abbe installs a dehumidifier to dry the walls of a small, closed basement room. When she enters the room later, it feels warm. Why?

10.4 Why does the indoor unit of a central air conditioning system have a drain hose?

10.5 If it takes about 335 kJ to freeze 1 kg of water, how much ice could an ice maker having a 1-ton refrigeration capacity produce in 24 hours?

10.6 Why do refrigerator manufacturers recommend that owners clean the back of some domestic refrigerators?

10.7 Would water be a suitable working fluid for use in a refrigerator?

10.8 What are *n-type* and *p-type* semiconductors?

10.9 Your new air conditioner has a label that lists an *EER* of 10 kJ/h per watt. What does that mean?

10.10 If your car's air conditioner discharges only warm air while operating, what might be wrong with it?

10.11 Can the coefficient of performance of a heat pump have a value less than one?

10.12 If the heat exchanger is omitted from the system of Fig. 10.16, what is the effect on the coefficient of performance?

CHECKING UNDERSTANDING

Match the appropriate definition in the right column with each term in the left column.

10.1 Refrigeration Capacity

10.2 Vapor-Compression Heat Pump

10.3 Flash Chamber

10.4 Thermoelectric Cooling

10.5 Global Warming Potential

A. Simplified index used to estimate the potential influence on global warming caused by the release of a specified gas.

B. A liquid–vapor separator.

C. The heat transfer from the refrigerated space resulting in the evaporation vaporization of the refrigerant.

D. A cooling technology that does not use a refrigerant.

E. System used for heating that consists of the same basic components as a vapor-compression refrigeration system.

10.6 The *Carnot* refrigeration cycle consists of internally reversible processes and areas on a _____ can be interpreted as heat transfers.

10.7 Why is *wet compression* avoided within refrigeration cycle compression?

10.8 The refrigerant temperature in the condenser is several degrees _____ the temperature of the warm region, T_H.

10.9 Hydrocarbons such as butane and methane are examples of _____ refrigerants.

10.10 Which component of an air-source heat pump allows it to be used for both heating in the winter and cooling in the summer?

10.11 In the condensing process of an actual vapor-compression system, what effect can subcooling the refrigerant have on external irreversibilities on the system?

10.12 Comparing centrifugal and reciprocating compressors, centrifugal compressors are better suited for _____ evaporating pressures and refrigerants with _____ specific volumes at low pressure.

10.13 Cold storage involves creating and storing chilled water or ice during _____ periods.

10.14 The essential elements within a thermoelectric cooler are two _____ metals.

10.15 Which component within an absorption refrigeration system requires heat transfer from a high temperature source?

Indicate whether the following statements are true or false. Explain.

10.16 The choice of refrigerant affects the type of compressor used.

10.17 The desuperheating section of the refrigerant condenser generally introduces external irreversibilities for the refrigeration system.

10.18 In a vapor-compression refrigeration system, the net power input is equal to the sum of the compressor power input and turbine power output.

10.19 The refrigerant flowing through the compressor in a vapor-compression refrigeration system is generally in the super-heated vapor phase.

10.20 Refrigerant exits the expansion valve of a vapor-compression refrigeration system as a two-phase liquid–vapor mixture.

10.21 The coefficient of performance of a vapor-compression system tends to increase as the evaporating temperature decreases and the condensing temperature increases.

10.22 Refrigerant-134a is an example of a natural refrigerant.

10.23 The core materials used in thermoelectric coolers must have low thermal conductivity and high electrical conductivity.

10.24 In a cascade vapor-compression refrigeration cycle, each stage can use a unique refrigerant based on suitable evaporator and condenser pressures.

10.25 A vapor-compression refrigerator located in a garage performs differently depending on the variation of garage indoor air temperature.

PROBLEMS: DEVELOPING ENGINEERING SKILLS

C Problem may require use of appropriate computer software in order to complete.

Vapor Refrigeration Systems

10.1 Refrigerant 22 is the working fluid in a Carnot refrigeration cycle operating at steady state. The refrigerant enters the condenser as saturated vapor at 32°C and exits as saturated liquid. The evaporator operates at 0°C. What is the coefficient of performance of the cycle? Determine, in kJ per gram of refrigerant flowing

    a. the work input to the compressor.

    b. the work developed by the turbine.

    c. the heat transfer to the refrigerant passing through the evaporator.

10.2 A Carnot vapor refrigeration cycle operates between thermal reservoirs at 4°C and 32°C. For (a) Refrigerant 134a, (b) propane, (c) water, (d) Refrigerant 22, and (e) ammonia as the working fluid, determine the operating pressures in the condenser and evaporator, in kPa, and the coefficient of performance.

10.3 A Carnot vapor refrigeration cycle is used to maintain a cold region at 4°C when the ambient temperature is 32°C. Refrigerant 134a enters the condenser as saturated vapor at 965 kPa and leaves as saturated liquid at the same pressure. The evaporator pressure is 276 kPa. The mass flow rate of refrigerant is 0.05 kg/s. Calculate

    a. the compressor and turbine power, each in kW.

    b. the coefficient of performance.

10.4 An ideal vapor-compression refrigeration cycle operates at steady state with Refrigerant 134a as the working fluid. Saturated vapor enters the compressor at 2 bar, and saturated liquid exits the condenser at 8 bar. The mass flow rate of refrigerant is 7 kg/min. Determine

    a. the compressor power, in kW.

    b. the refrigerating capacity, in tons.

    c. the coefficient of performance.

10.5 **C** Plot each of the quantities in Problem 10.4 versus evaporator temperature for evaporator pressures ranging from 0.6 to 4 bar, while the condenser pressure remains fixed at 8 bar.

10.6 Refrigerant 134a is the working fluid in an ideal vapor-compression refrigeration cycle operating at steady state. Refrigerant enters the compressor at 1 bar, −15°C, and the condenser pressure is 8 bar. Liquid exits the condenser at 30°C. The mass flow rate of refrigerant is 8 kg/min. Determine

    a. the compressor power, in kW.

    b. the refrigeration capacity, in tons.

    c. the coefficient of performance.

10.7 An ideal vapor-compression refrigeration system operates at steady state with Refrigerant 134a as the working fluid. Superheated vapor enters the compressor at 70 kPa, −18°C, and saturated liquid leaves the condenser at 1240 kPa. The refrigeration capacity is 8 tons. Determine

    a. the compressor power, in kW.

    b. the rate of heat transfer from the working fluid passing through the condenser, in kW.

    c. the coefficient of performance.

10.8 Refrigerant 22 enters the compressor of an ideal vapor-compression refrigeration system as saturated vapor at −40°C with a volumetric flow rate of 15 m³/min. The refrigerant leaves the condenser at 32°C, 9 bar. Determine

    a. the compressor power, in kW.

    b. the refrigerating capacity, in tons.

    c. the coefficient of performance.

10.9 Ammonia with a mass flow rate of 5 kg/min is the working fluid within an ideal vapor-compression refrigeration cycle. Saturated vapor enters the compressor and saturated liquid exits the condenser. The evaporator temperature is −10°C and the condenser pressure is 10 bar. Determine

    a. the coefficient of performance.

    b. the refrigerating capacity, in tons.

10.10 An ideal vapor-compression refrigeration cycle, with ammonia as the working fluid, has an evaporator temperature of −22°C and a condenser pressure of 14 bar. Saturated vapor enters the compressor, and saturated liquid exits the condenser. The mass flow rate of the refrigerant is 5 kg/min. Determine

    a. the coefficient of performance.

    b. the refrigerating capacity, in tons.

10.11 Refrigerant 22 enters the compressor of an ideal vapor-compression refrigeration system as saturated vapor at −40°C with a volumetric flow rate of 15 m^3/min. The refrigerant leaves the condenser at 19°C, 9 bar. Determine

    a. the compressor power, in kW.

    b. the refrigerating capacity, in tons.

    c. the coefficient of performance.

    d. the rate of entropy production for the cycle, in kW/K.

10.12 A vapor-compression refrigeration cycle operates at steady state with Refrigerant 134a as the working fluid. Saturated vapor enters the compressor at 1 bar, and saturated liquid exits the condenser at 9 bar. The isentropic compressor efficiency is 78%. The mass flow rate of refrigerant is 6 kg/min. Determine

    a. the compressor power, in kW.

    b. the refrigeration capacity, in tons.

    c. the coefficient of performance.

10.13 Modify the cycle in Problem 10.7 to have an isentropic compressor efficiency of 83% and let the temperature of the liquid leaving the condenser be 38°C. Determine, for the modified cycle,

    a. the compressor power, in kW.

    b. the rate of heat transfer from the working fluid passing through the condenser, in kW.

    c. the coefficient of performance.

    d. the rates of exergy destruction in the compressor and expansion valve, each in kW, for $T_0 = 38$°C.

10.14 Data for steady-state operation of a vapor-compression refrigeration cycle with Refrigerant 134a as the working fluid are given in the table below. The states are numbered as in Fig. 10.3. The refrigeration capacity is 4.6 tons. Ignoring heat transfer between the compressor and its surroundings, sketch the *T–s* diagram of the cycle and determine

    a. the mass flow rate of the refrigerant, in kg/min.

    b. the isentropic compressor efficiency.

    c. the coefficient of performance.

    d. the rates of exergy destruction in the compressor and expansion valve, each in kW.

    e. the net changes in flow exergy rate of the refrigerant passing through the evaporator and condenser, respectively, each in kW.

Let $T_0 = 21$°C, $p_0 = 1$ bar.

State	p (bar)	T (°C)	h (kJ/kg)	s (kJ/kg · K)
1	1.4	−10	243.40	0.9606
2	7	58.5	295.13	1.0135
3	7	24	82.90	0.3113
4	1.4	−18.8	82.90	0.33011

10.15 A vapor-compression refrigeration system, using ammonia as the working fluid, has evaporator and condenser pressures of 1380 kPa and 207 kPa, respectively. The refrigerant passes through

each heat exchanger with a negligible pressure drop. At the inlet and exit of the compressor, the temperatures are −12°C and 150°C, respectively. The heat transfer rate from the working fluid passing through the condenser is 15 kW, and liquid exits at 1380 kPa, 32°C. If the compressor operates adiabatically, determine

    a. the compressor power input, in kW.

    b. the coefficient of performance.

10.16 **C** Consider the following vapor-compression refrigeration cycle used to maintain a cold region at temperature T_C when the ambient temperature is 27°C: Saturated vapor enters the compressor at −9°C below T_C, and the compressor operates adiabatically with an isentropic efficiency of 80%. Saturated liquid exits the condenser at 35°C. There are no pressure drops through the evaporator or condenser, and the refrigerating capacity is 1 ton. Plot refrigerant mass flow rate, in kg/s, coefficient of performance, and *refrigerating efficiency*, versus T_C ranging from 4°C to −32°C if the refrigerant is

    a. Refrigerant 134a.   **c.** Refrigerant 22.

    b. propane.   **d.** ammonia.

The refrigerating efficiency is defined as the ratio of the cycle coefficient of performance to the coefficient of performance of a Carnot refrigeration cycle operating between thermal reservoirs at the ambient temperature and the temperature of the cold region.

10.17 In a vapor-compression refrigeration cycle, ammonia exits the evaporator as saturated vapor at −20°C. The refrigerant enters the condenser at 18 bar and 180°C, and saturated liquid exits at 18 bar. There is no significant heat transfer between the compressor and its surroundings, and the refrigerant passes through the evaporator with a negligible change in pressure. If the refrigerating capacity is 180 kW, determine

    a. the mass flow rate of refrigerant, in kg/s.

    b. the power input to the compressor, in kW.

    c. the coefficient of performance.

    d. the isentropic compressor efficiency.

10.18 A vapor-compression refrigeration system with a capacity of 10 tons has superheated Refrigerant 134a vapor entering the compressor at 15°C, 4 bar, and exiting at 12 bar. The compression process can be modeled by $pv^{1.01} = constant$. At the condenser exit, the pressure is 11.6 bar, and the temperature is 44°C. The condenser is water-cooled, with water entering at 20°C and leaving at 30°C with a negligible change in pressure. Heat transfer from the outside of the condenser can be neglected. Determine

    a. the mass flow rate of the refrigerant, in kg/s.

    b. the power input and the heat transfer rate for the compressor, each in kW.

    c. the coefficient of performance.

    d. the mass flow rate of the cooling water, in kg/s.

    e. the rates of exergy destruction in the condenser and expansion valve, each expressed as a percentage of the power input. Let $T_0 = 20$°C.

10.19 The capacity of a propane vapor-compression refrigeration system is 5 tons. Saturated vapor at −18°C enters the compressor, and superheated vapor leaves at 50°C, 1240 kPa. Heat transfer from the compressor to its surroundings occurs at a rate of 8 kJ/kg of refrigerant passing through the compressor. Liquid refrigerant enters the expansion valve at 30°C, 1240 kPa. The condenser is water-cooled, with water entering at 18°C and leaving at 27°C with a negligible change in pressure. Determine

    a. the compressor power input, in kW.

    b. the mass flow rate of cooling water through the condenser, in kg/s.

    c. the coefficient of performance.

10.20 A window-mounted air conditioner supplies 20 m³/min of air at 16°C, 1 bar to a room. Air returns from the room to the evaporator of the unit at 25°C. The air conditioner operates at steady state on a vapor-compression refrigeration cycle with Refrigerant-22 entering the compressor at 3 bar, 10°C. Saturated liquid refrigerant at 8 bar leaves the condenser. The compressor has an isentropic efficiency of 75%, and refrigerant exits the compressor at 8 bar. Determine the compressor power, in kW, the refrigeration capacity, in tons, and the coefficient of performance.

10.21 A vapor-compression refrigeration system for a household refrigerator has a refrigerating capacity of 0.3 kW. Refrigerant enters the evaporator at −23°C and exits at −18°C. The isentropic compressor efficiency is 80%. The refrigerant condenses at 35°C and exits the condenser subcooled at 32°C. There are no significant pressure drops in the flows through the evaporator and condenser. Determine the evaporator and condenser pressures, each in kPa, the mass flow rate of refrigerant, in kg/s, the compressor power input, in kW, and the coefficient of performance for (a) Refrigerant 134a and (b) propane as the working fluid.

10.22 A vapor-compression air-conditioning system operates at steady state as shown in Fig. P10.22. The system maintains a cool region at 15°C and discharges energy by heat transfer to the surroundings at 32°C. Refrigerant 134a enters the compressor as a saturated vapor at 4°C and is compressed adiabatically to 1103 kPa. The isentropic compressor efficiency is 80%. Refrigerant exits the condenser as a saturated liquid at 1103 kPa. The mass flow rate of the refrigerant is 0.07 kg/s. Kinetic and potential energy changes are negligible as are changes in pressure for flow through the evaporator and condenser. Determine

a. the power required by the compressor, in kJ/s.

b. the coefficient of performance.

c. the rates of exergy destruction in the compressor and expansion valve, each in kJ/s.

d. the rates of exergy destruction and exergy transfer accompanying heat transfer, in kJ/s, for a control volume comprising the evaporator and a portion of the cool region such that heat transfer takes place at $T_C = 288$ K (15°C).

e. the rates of exergy destruction and exergy transfer accompanying heat transfer, each in kJ/s, for a control volume enclosing the condenser and a portion of the surroundings such that heat transfer takes place at $T_H = 305$ K (32°C).

Let $T_0 = 305$ K.

10.23 [C] A vapor-compression refrigeration cycle with Refrigerant 134a as the working fluid operates with an evaporator temperature of 10°C and a condenser pressure of 1240 kPa. Refrigerant enters the condenser at 60°C and exits as saturated liquid. The cycle has a refrigeration capacity of 5 tons. Determine

a. the refrigerant mass flow rate, in kg/s.

b. the compressor isentropic efficiency.

c. the compressor power, in kW.

d. the coefficient of performance.

Plot each of the quantities calculated in parts (b) through (d) for compressor exit temperatures varying from 54°C to 60°C.

Cascade and Multistage Systems

10.24 A vapor-compression refrigeration system uses the arrangement shown in Fig. 10.8 for two-stage compression with intercooling between the stages. Refrigerant 134a is the working fluid. Saturated vapor at −32°C enters the first compressor stage. The flash chamber and direct contact heat exchanger operate at 5 bar, and the condenser pressure is 14 bar. Saturated liquid streams at 14 and 5 bar enter the high- and low-pressure expansion valves, respectively. If each compressor operates isentropically and the refrigerating capacity of the system is 12 tons, determine

a. the power input to each compressor, in kW.

b. the coefficient of performance.

10.25 [C] Figure P10.25 shows a two-stage vapor-compression refrigeration system with ammonia as the working fluid. The system uses a direct contact heat exchanger to achieve intercooling. The evaporator has a refrigerating capacity of 30 tons and produces −30°C saturated vapor at its exit. In the first compressor stage, the refrigerant is compressed adiabatically to 0.55 MPa, which is the pressure in the direct contact heat exchanger. Saturated vapor at 0.55 MPa enters the second compressor stage and is compressed adiabatically to 1.7 MPa. Each compressor stage has an isentropic efficiency of 85%. There are no significant pressure drops as the refrigerant passes through the heat exchangers. Saturated liquid enters each expansion valve. Determine

a. the ratio of mass flow rates, $\dot{m}_3/\dot{m}_1$.

b. the power input to each compressor stage, in kW.

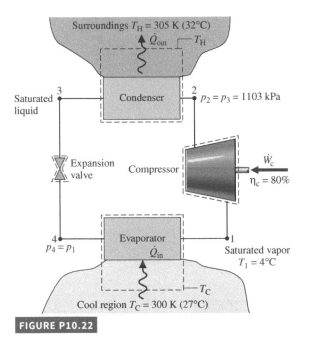

FIGURE P10.22

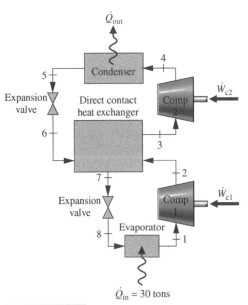

FIGURE P10.25

c. the coefficient of performance.

d. Plot each of the quantities calculated in parts (a)–(c) versus the direct-contact heat exchanger pressure ranging from 0.14 to 1.4 MPa. Discuss.

10.26 Figure P10.26 shows a two-stage, vapor-compression refrigeration system with two evaporators and a direct contact heat exchanger. Saturated vapor from evaporator 1 enters compressor 1 at 0.12 MPa and exits at 0.5 MPa. Evaporator 2 operates at 0.5 MPa, with saturated vapor exiting at state 8. The condenser pressure is 1.4 MPa, and saturated liquid refrigerant exits the condenser. Each compressor stage has an isentropic efficiency of 80%. The refrigeration capacity of each evaporator is shown on the figure. Sketch the T–s diagram of the cycle and determine

a. the temperatures, in °C, of the refrigerant in each evaporator.

b. the power input to each compressor stage, in kW.

c. the overall coefficient of performance.

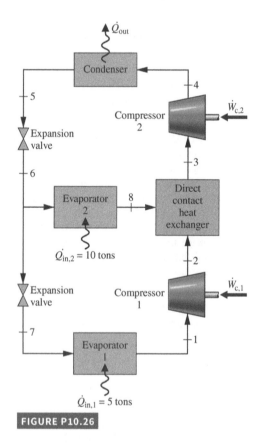

FIGURE P10.26

10.27 Figure P10.27 shows the schematic diagram of a vapor-compression refrigeration system with two evaporators using Refrigerant 134a as the working fluid. This arrangement is used to achieve refrigeration at two different temperatures with a single compressor and a single condenser. The low-temperature evaporator operates at −18°C with saturated vapor at its exit and has a refrigerating capacity of 3 tons. The higher-temperature evaporator produces saturated vapor at 3.2 bar at its exit and has a refrigerating capacity of 2 tons. Compression is isentropic to the condenser pressure of 10 bar. There are no significant pressure drops in the flows through the condenser and the two evaporators, and the refrigerant leaves the condenser as saturated liquid at 10 bar. Calculate

a. the mass flow rate of refrigerant through each evaporator, in kg/min.

b. the compressor power input, in kW.

c. the rate of heat transfer from the refrigerant passing through the condenser, in kW.

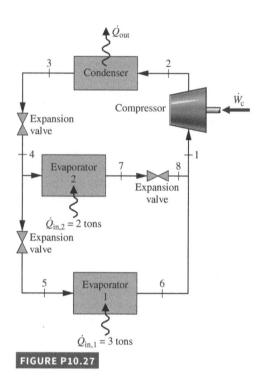

FIGURE P10.27

10.28 An ideal vapor-compression refrigeration cycle is modified to include a counterflow heat exchanger, as shown in **Fig. P10.28**. Ammonia leaves the evaporator as saturated vapor at 2.0 bar and is heated at constant pressure to 10°C before entering the compressor. Following isentropic compression to 16 bar, the refrigerant passes through

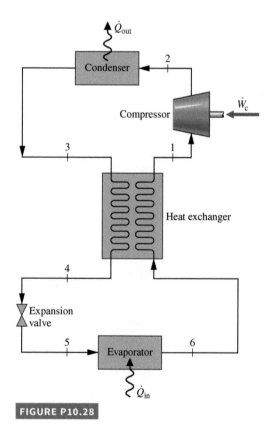

FIGURE P10.28

the condenser, exiting at 36°C, 16 bar. The liquid then passes through the heat exchanger, entering the expansion valve at 16 bar. If the mass flow rate of refrigerant is 14 kg/min, determine

a. the refrigeration capacity, in tons of refrigeration.

b. the compressor power input, in kW.

c. the coefficient of performance.

Discuss possible advantages and disadvantages of this arrangement.

Vapor-Compression Heat Pump Systems

10.29 **Figure P10.29** gives data for an ideal vapor-compression heat pump cycle operating at steady state with Refrigerant 134a as the working fluid. The heat pump provides heating at a rate of 15 kW to maintain the interior of a building at 20°C when the outside temperature is 5°C. Sketch the *T–s* diagram for the cycle and determine the

a. temperatures at the principal states of the cycle, each in °C.

b. power input to the compressor, in kW.

c. coefficient of performance.

d. coefficient of performance for a Carnot heat pump cycle operating between reservoirs at the building interior and outside temperatures, respectively.

Compare the coefficients of performance determined in (c) and (d). Discuss.

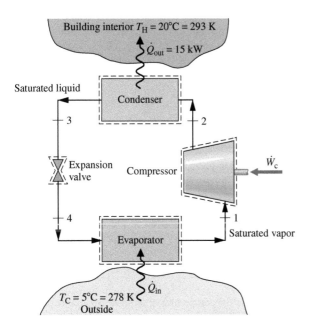

State	*p* (bar)	*h* (kJ/kg)
1	2.4	244.09
2	8	268.97
3	8	93.42
4	2.4	93.42

FIGURE P10.29

10.30 Refrigerant 134a is the working fluid in a vapor-compression heat pump system with a heating capacity of 63,300 kJ/h. The condenser operates at 1.4 MPa, and the evaporator temperature is −18°C. The refrigerant is a saturated vapor at the evaporator exit and a liquid at 43°C at the condenser exit. Pressure drops in the flows

through the evaporator and condenser are negligible. The compression process is adiabatic, and the temperature at the compressor exit is 82°C. Determine

a. the mass flow rate of refrigerant, in kg/min.

b. the compressor power input, in kW.

c. the isentropic compressor efficiency.

d. the coefficient of performance.

10.31 Refrigerant 134a is the working fluid in a vapor-compression heat pump that provides 50 kW to heat a dwelling on a day when the outside temperature is below freezing. Saturated vapor enters the compressor at 1.8 bar, and saturated liquid exits the condenser, which operates at 10 bar. Determine, for isentropic compression,

a. the refrigerant mass flow rate, in kg/s.

b. the compressor power, in kW.

c. the coefficient of performance.

Recalculate the quantities in parts (b) and (c) for an isentropic compressor efficiency of 75%.

10.32 An office building requires a heat transfer rate of 20 kW to maintain the inside temperature at 21°C when the outside temperature is 0°C. A vapor-compression heat pump with Refrigerant 134a as the working fluid is to be used to provide the necessary heating. The compressor operates adiabatically with an isentropic efficiency of 82%. Specify appropriate evaporator and condenser pressures of a cycle for this purpose assuming $\Delta T_{cond} = \Delta T_{evap} = 10$°C, as shown in **Fig. P10.32**. The states are numbered as in Fig. 10.13. The refrigerant exits the evaporator as saturated vapor and exits the condenser as saturated liquid at the respective pressures. Determine the

a. mass flow rate of refrigerant, in kg/s.

b. compressor power, in kW.

c. coefficient of performance and compare with the coefficient of performance for a Carnot heat pump cycle operating between reservoirs at the inside and outside temperatures, respectively.

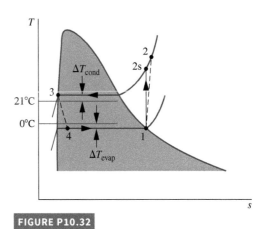

FIGURE P10.32

10.33 A process requires a heat transfer rate of 5.3×10^4 kJ/min at 77°C. It is proposed that a Refrigerant 134a vapor-compression heat pump be used to develop the process heat using a waste water stream at 52°C as the lower-temperature source. Specify appropriate evaporator and condenser pressures of a cycle for this purpose. Let the refrigerant be saturated vapor at the evaporator exit and saturated liquid at the condenser exit. Calculate

a. the mass flow rate of refrigerant, in kg/h.

b. the compressor power, in kW.

c. the coefficient of performance.

10.34 A vapor-compression heat pump with a heating capacity of 500 kJ/min is driven by a power cycle with a thermal efficiency of 25%. For the heat pump, Refrigerant 134a is compressed from saturated vapor at −10°C to the condenser pressure of 10 bar. The isentropic compressor efficiency is 80%. Liquid enters the expansion valve at 9.6 bar, 34°C. For the power cycle, 80% of the heat rejected is transferred to the heated space.

    a. Determine the power input to the heat pump compressor, in kW.

    b. Evaluate the ratio of the total rate that heat is delivered to the heated space to the rate of heat input to the power cycle. Discuss.

10.35 Refrigerant 134a enters the compressor of a vapor-compression heat pump at 0.1 MPa, −18°C and is compressed adiabatically to 1.1 MPa, 70°C. Liquid enters the expansion valve at 1.1 MPa, 35°C. At the valve exit, the pressure is 0.1 MPa.

    a. Determine the isentropic compressor efficiency.

    b. Determine the coefficient of performance.

    c. Perform a full exergy accounting of the compressor power input, in kJ per kg of refrigerant flowing. Discuss.

Let $T_0 = 267$ K.

10.36 A *geothermal* heat pump operating at steady state with Refrigerant-22 as the working fluid is shown schematically in **Fig. P10.36**. The heat pump uses 13°C water from wells as the thermal source. Operating data are shown on the figure for a day in which the outside air temperature is −6.7°C. Assume adiabatic operation of the compressor. For the heat pump, determine

    a. the volumetric flow rate of heated air to the house, in m³/min.

    b. the isentropic compressor efficiency.

    c. the compressor power, in kW.

    d. the coefficient of performance.

    e. the volumetric flow rate of water from the geothermal wells, in L/min.

For $T_0 = -6.7$°C, perform a full exergy accounting of the compressor power input, and devise and evaluate at second law efficiency for the heat pump system.

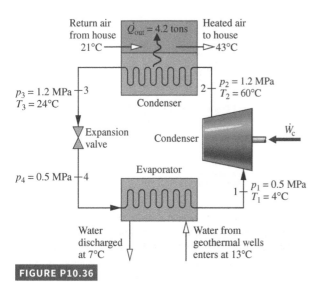

FIGURE P10.36

Gas Refrigeration Systems

10.37 Air enters the compressor of an ideal Brayton refrigeration cycle at 100 kPa, 270 K. The compressor pressure ratio is 3, and the temperature at the turbine inlet is 310 K. Determine

    a. the net work input, per unit mass of air flow, in kJ/kg.

    b. the refrigeration capacity, per unit mass of air flow, in kJ/kg.

    c. the coefficient of performance.

    d. the coefficient of performance of a Carnot refrigeration cycle operating between thermal reservoirs at $T_C = 270$ K and $T_H = 310$ K, respectively.

10.38 Air enters the compressor of a Brayton refrigeration cycle at 100 kPa, 270 K. The compressor pressure ratio is 3, and the temperature at the turbine inlet is 315 K. The compressor and turbine have isentropic efficiencies of 82% and 85%, respectively. Determine the

    a. net work input, per unit mass of air flow, in kJ/kg.

    b. exergy accounting of the net power input, in kJ per kg of air flowing. Discuss.

Let $T_0 = 315$ K.

10.39 An ideal Brayton refrigeration cycle has a compressor pressure ratio of 6. At the compressor inlet, the pressure and temperature of the entering air are 0.14 MPa and 256 K. The temperature at the inlet of the turbine is 340 K. For a refrigerating capacity of 15 tons, determine

    a. the mass flow rate, in kg/min.

    b. the net power input, in kJ/min.

    c. the coefficient of performance.

10.40 Reconsider Problem 10.39, but include in the analysis that the compressor and turbine have isentropic efficiencies of 78% and 92%, respectively.

10.41 The table below provides steady-state operating data for an ideal Brayton refrigeration cycle with air as the working fluid. The principal states are numbered as in Fig. 10.15. The volumetric flow rate at the turbine inlet is 0.4 m³/s. Sketch the T–s diagram for the cycle and determine the

    a. specific enthalpy, in kJ/kg, at the turbine exit.

    b. mass flow rate, in kg/s.

    c. net power input, in kW.

    d. refrigeration capacity, in kW.

    e. coefficient of performance.

State	p (kPa)	T (K)	h (kJ/kg)	p_r
1	140	270	270.11	0.9590
2	420	—	370.10	2.877
3	420	320	320.29	1.7375
4	140	—	?	—

10.42 Air enters the compressor of an ideal Brayton refrigeration cycle at 140 kPa, 270 K, and is compressed to 420 kPa. At the turbine inlet, the temperature is 320 K and the volumetric flow rate is 0.4 m³/s. Determine

    a. the mass flow rate, in kg/s.

    b. the net power input, in kW.

    c. the refrigerating capacity, in kW.

    d. the coefficient of performance.

10.43 **C** Air enters the compressor of a Brayton refrigeration cycle at 100 kPa, 260 K, and is compressed adiabatically to 300 kPa. Air enters the turbine at 300 kPa, 300 K, and expands adiabatically to 100 kPa. For the cycle

    a. determine the net work per unit mass of air flow, in kJ/kg, and the coefficient of performance if the compressor and turbine isentropic efficiencies are both 100%.

b. plot the net work per unit mass of air flow, in kJ/kg, and the coefficient of performance for equal compressor and turbine isentropic efficiencies ranging from 80 to 100%.

10.44 Consider a Brayton refrigeration cycle with a regenerative heat exchanger. Air enters the compressor at 267 K, 0.1 MPa and is compressed isentropically to 0.3 MPa. Compressed air enters the regenerative heat exchanger at 300 K and is cooled to 267 K before entering the turbine. The expansion through the turbine is isentropic. If the refrigeration capacity is 15 tons, calculate

a. the volumetric flow rate at the compressor inlet, in m^3/min.

b. the coefficient of performance.

10.45 Reconsider Problem 10.44, but include in the analysis that the compressor and turbine each have isentropic efficiencies of 88%. Answer the same questions for the modified cycle as in Problem 10.44.

10.46 Air at 2 bar, 380 K is extracted from a main jet engine compressor for cabin cooling. The extracted air enters a heat exchanger where it is cooled at constant pressure to 320 K through heat transfer with the ambient. It then expands adiabatically to 0.95 bar through a turbine and is discharged into the cabin. The turbine has an isentropic efficiency of 75%. If the mass flow rate of the air is 1.0 kg/s, determine

a. the power developed by the turbine, in kW.

b. the rate of heat transfer from the air to the ambient, in kW.

10.47 Air at 0.22 MPa, 378 K is extracted from a main jet engine compressor for cabin cooling. The extracted air enters a heat exchanger where it is cooled at constant pressure to 333 K through heat transfer with the ambient. It then expands adiabatically to 100 kPa through a

turbine and is discharged into the cabin at 278 K with a mass flow rate of 100 kg/min. Determine

a. the power developed by the turbine, in kW.

b. the isentropic turbine efficiency.

c. the rate of heat transfer from the air to the ambient, in kJ/min.

10.48 Air within a piston–cylinder assembly undergoes a *Stirling refrigeration cycle*, which is the reverse of the Stirling power cycle introduced in Sec. 9.8.4. At the beginning of the isothermal compression, the pressure and temperature are 100 kPa and 350 K, respectively. The compression ratio is 7, and the temperature during the isothermal expansion is 150 K. Determine the

a. heat transfer for the isothermal compression, in kJ per kg of air.

b. net work for the cycle, in kJ per kg of air.

c. coefficient of performance.

10.49 Air undergoes an *Ericsson refrigeration cycle*, which is the reverse of the Ericsson power cycle introduced in Sec. 9.11. At the beginning of the isothermal compression, the pressure and temperature are 100 kPa and 310 K, respectively. The pressure ratio during the isothermal compression is 3. During the isothermal expansion the temperature is 270 K. Determine

a. the heat transfer for the isothermal expansion, per unit mass of air flow, in kJ/kg.

b. the net work, per unit mass of air flow, in kJ/kg.

c. the coefficient of performance.

DESIGN & OPEN-ENDED PROBLEMS: EXPLORING ENGINEERING PRACTICE

10.1D Refrigerant 22 is widely used as the working fluid in air conditioners and industrial chillers. However, its use is likely to be phased out in the future due to concerns about ozone depletion. Investigate which environmentally acceptable working fluids are under consideration to replace Refrigerant 22 for these uses. Determine the design issues for air conditioners and chillers that would result from changing refrigerants. Write a report of your findings.

10.2D A vapor-compression refrigeration system using propane is being designed for a household food freezer. The refrigeration system must maintain a temperature of −18°C within the freezer compartment when the temperature of the room is 32°C. Under these conditions, the steady-state heat transfer rate from the room into the freezer compartment is 1580 kJ/h. Specify operating pressures and temperatures at key points within the refrigeration system and estimate the refrigerant mass flow rate and compressor power required. Investigate and report on safety issues related to the use of propane as a refrigerant.

10.3D In cases involving cardiac arrest, stroke, heart attack, and hyperthermia, hospital medical staff must move quickly to reduce the patient's body temperature by several degrees. A system for this purpose featuring a disposable plastic *body suit* is described in BIO-CONNECTIONS in Sec. 4.9. Conduct a search of the patent literature for alternative ways to achieve cooling of medically distressed individuals. Consider patents both granted and pending. Critically evaluate two different methods found in your search relative to each other and the body suit approach. Write a report including at least three references.

10.4D An air-conditioning system is under consideration that will use a vapor-compression ice maker during the nighttime, when electric rates are lowest, to store ice for meeting the daytime air-conditioning load. The maximum loads are 100 tons during the day and 50 tons at

night. Is it best to size the system to make enough ice at night to carry the entire daytime load or to use a smaller chiller that runs both day and night? Base your strategy on the day–night electric rate structure of your local electric utility company.

10.5D Identify and visit a local facility that uses cold thermal storage. Conduct a forensic study to determine if the cold storage system is well suited for the given application today. Consider costs, effectiveness in providing the desired cooling, contribution to global climate change, and other pertinent issues. Document the cold storage systems suitability for the application. If the cold storage system is not well suited, recommend system upgrades or an alternative approach for obtaining the desired cooling. Prepare a PowerPoint presentation of your findings.

10.6D A vertical, closed-loop *geothermal* heat pump is under consideration for a new 4500 m^2 school building. The design capacity is 100 tons for both heating and cooling. The local water table is 45 m, and the ground water temperature is 13°C. Specify a ground source heat pump as well as the number and depth of wells for this application and develop a layout of vertical wells and piping required by the system.

10.7D Food poisoning is on the rise and can be fatal. Many of those affected have eaten recently at a restaurant, café, or fast-food outlet serving food that has not been cooled properly by the food supplier or restaurant food-handlers. To be safe, foods should not be allowed to remain in the temperature range where bacteria most quickly multiply. Standard refrigerators typically do not have the ability to provide the rapid cooling needed to ensure dangerous levels of bacteria are not attained. A food processing company supplying a wide range of fish products to restaurants has requested your project group to provide advice on how to achieve best cooling practices in its factory.

In particular, you are asked to consider applicable health regulations, suitable equipment, typical operating costs, and other pertinent issues. A written report providing your recommendations is required, including an annotated list of food-cooling *Dos and Don'ts* for restaurants supplied by the company with fish.

10.8D According to researchers, advances in *nanomaterial* fabrication are leading to development of tiny thermoelectric modules that could be used in various applications, including integrating nanoscale cooling devices within the uniforms of firefighters, emergency workers, and military personnel; embedding thermoelectric modules in facades of a building; and using thermoelectric modules to recover waste heat in automobiles. Research two applications for this technology proposed within the past five years. Investigate the technical readiness and economic feasibility for each concept. Report your findings in an executive summary and a PowerPoint presentation with at least three references.

10.9D List the major design issues involved in using ammonia as the refrigerant in a system to provide chilled water at 4°C for air conditioning a college campus in your locale. Develop a layout of the equipment room and a schematic of the chilled water distribution system. Label the diagrams with key temperatures and make a list of capacities of each of the major pieces of equipment.

10.10D A vapor-compression refrigeration system operating continuously is being considered to provide a minimum of 80 tons of refrigeration for an industrial refrigerator maintaining a space at 2°C. The surroundings to which the system rejects energy by heat transfer reach a maximum temperature of 40°C. For effective heat transfer, the system requires a temperature difference of at least 20°C between the condensing refrigerant and surroundings and between the vaporizing refrigerant and refrigerated space. The project manager wishes to install a system that minimizes the annual cost for electricity (monthly electricity cost is fixed at 5.692 cents for the first 250 kW · h and 6.006 cents for any usage above 250 kW · h). You are asked to evaluate two alternative designs: a standard vapor-compression refrigeration cycle and a vapor-compression refrigeration cycle that employs a power-recovery turbine in lieu of an expansion valve. For each alternative, consider three refrigerants: ammonia, Refrigerant 22, and Refrigerant 134a. Based on electricity cost, recommend the better choice between the two alternatives and a suitable refrigerant. Other than electricity cost, what additional factors should the manager consider in making a final selection? Prepare a written report including results, conclusions, and recommendations.

CHAPTER 11

Thermodynamic Relations

Engineering Context

As seen in previous chapters, application of thermodynamic principles to engineering systems requires data for specific internal energy, enthalpy, entropy, and other properties. The **objective** of this chapter is to introduce thermodynamic relations that allow u, h, s, and other thermodynamic properties of simple compressible systems to be evaluated from data that are more readily measured. Primary emphasis is on systems involving a single chemical species such as water or a mixture such as air. An introduction to general property relations for mixtures and solutions is also included.

Means are available for determining pressure, temperature, volume, and mass experimentally. In addition, the relationships between the specific heats c_v and c_p and temperature at relatively low pressure are accessible experimentally. Values for certain other thermodynamic properties also can be measured without great difficulty. However, specific internal energy, enthalpy, and entropy are among those properties that are not easily obtained experimentally, so we resort to computational procedures to determine values for these.

LEARNING OUTCOMES

When you complete your study of this chapter, you will be able to...

- Calculate p–v–T data using equations of state involving two or more constants.
- Demonstrate understanding of exact differentials involving properties and utilize the property relations developed from the exact differentials summarized in Table 11.1.
- Evaluate Δu, Δh, and Δs, using the Clapeyron equation when considering phase change, and using equations of state and specific heat relations when considering single phases.
- Demonstrate understanding of how tables of thermodynamic properties are constructed.
- Evaluate Δh and Δs using generalized enthalpy and entropy departure charts.
- Apply mixture rules, such as Kay's rule, to relate pressure, volume, and temperature of mixtures.
- Apply thermodynamic relations to multicomponent systems.

11.1 Using Equations of State

An essential ingredient for the calculation of properties such as the specific internal energy, enthalpy, and entropy of a substance is an accurate representation of the relationship among pressure, specific volume, and temperature. The p–v–T relationship can be expressed alternatively: There are *tabular* representations, as exemplified by the steam tables. The relationship also can be expressed *graphically*, as in the p–v–T surface and compressibility factor charts. *Analytical* formulations, called **equations of state**, constitute a third general way of expressing the p–v–T relationship. Computer software such as *Interactive Thermodynamics: IT* also can be used to retrieve p–v–T data.

The virial equation and the ideal gas equation are examples of analytical equations of state introduced in previous sections of the book. Analytical formulations of the p–v–T relationship are particularly convenient for performing the mathematical operations required to calculate u, h, s, and other thermodynamic properties. The object of the present section is to expand on the discussion of p–v–T relations for simple compressible substances presented in Chap. 3 by introducing some commonly used equations of state.

equations of state

TAKE NOTE...

Using the generalized compressibility chart, virial equations of state and the ideal gas model are introduced in Chap. 3. See Secs. 3.11 and 3.12.

11.1.1 Getting Started

Recall from Sec. 3.11 that the **virial equation of state** can be derived from the principles of statistical mechanics to relate the p–v–T behavior of a gas to the forces between molecules. In one form, the compressibility factor Z is expanded in inverse powers of specific volume as

virial equation

$$Z = 1 + \frac{B(T)}{\overline{v}} + \frac{C(T)}{\overline{v}^2} + \frac{D(T)}{\overline{v}^3} + \cdots \qquad (11.1)$$

The coefficients B, C, D, etc. are called, respectively, the second, third, fourth, etc. virial coefficients. Each virial coefficient is a function of temperature alone. In principle, the virial coefficients are calculable if a suitable model for describing the forces of interaction between the molecules of the gas under consideration is known. Future advances in refining the theory of molecular interactions may allow the virial coefficients to be predicted with considerable accuracy from the fundamental properties of the molecules involved. However, at present, just the first few coefficients can be calculated and only for gases consisting of relatively simple molecules. Equation 11.1 also can be used in an empirical fashion in which the coefficients become parameters whose magnitudes are determined by fitting p–v–T data in particular realms of interest. Only a few coefficients can be found this way, and the result is a *truncated* equation valid only for certain states.

In the limiting case where the gas molecules are assumed not to interact in any way, the second, third, and higher terms of Eq. 11.1 vanish and the equation reduces to $Z = 1$. Since $Z = p\overline{v}/\overline{R}T$, this gives the ideal gas equation of state $p\overline{v} = \overline{R}/T$. The ideal gas equation of state provides an acceptable approximation at many states, including but not limited to states where the pressure is low relative to the critical pressure and/or the temperature is high relative to the critical temperature of the substance under consideration. At many other states, however, the ideal gas equation of state provides a poor approximation.

Over 100 equations of state have been developed in an attempt to improve on the ideal gas equation of state and yet avoid the complexities inherent in a full virial series. In general, these equations exhibit little in the way of fundamental physical significance and are mainly empirical in character. Most are developed for gases, but some describe the p–v–T behavior of the liquid phase, at least qualitatively. Every equation of state is restricted to particular states. This realm of applicability is often indicated by giving an interval of pressure, or density, where the equation can be expected to represent the p–v–T behavior faithfully. When it is not stated, the realm of applicability of a given equation can be approximated by expressing the equation in terms of the compressibility factor Z and the reduced properties p_R, T_R, v_R' and superimposing the result on a generalized compressibility chart or comparing with tabulated compressibility data obtained from the literature.

11.1.2 Two-Constant Equations of State

Equations of state can be classified by the number of adjustable constants they include. Let us consider some of the more commonly used equations of state in order of increasing complexity, beginning with two-constant equations of state.

van der Waals Equation An improvement over the ideal gas equation of state based on elementary molecular arguments was suggested in 1873 by van der Waals, who noted that gas molecules actually occupy more than the negligibly small volume presumed by the ideal gas model and also exert long-range attractive forces on one another. Thus, not all of the volume of a container would be available to the gas molecules, and the force they exert on the container wall would be reduced because of the attractive forces that exist between molecules. Based on these elementary molecular arguments, the **van der Waals equation of state** is

$$p = \frac{\bar{R}T}{\bar{v} - b} - \frac{a}{\bar{v}^2} \qquad (11.2)$$

van der Waals equation

The constant b is intended to account for the finite volume occupied by the molecules, the term $a/\bar{v}^2$ accounts for the forces of attraction between molecules, and $\bar{R}$ is the universal gas constant. Note than when a and b are set to zero, the ideal gas equation of state results.

The van der Waals equation gives pressure as a function of temperature and specific volume and thus is *explicit* in pressure. Since the equation can be solved for temperature as a function of pressure and specific volume, it is also explicit in temperature. However, the equation is cubic in specific volume, so it cannot generally be solved for specific volume in terms of temperature and pressure. The van der Waals equation is *not* explicit in specific volume.

Evaluating *a* and *b* The van der Waals equation is a *two-constant* equation of state. For a specified substance, values for the constants a and b can be found by fitting the equation to p–v–T data. With this approach several sets of constants might be required to cover all states of interest. Alternatively, a single set of constants for the van der Waals equation can be determined by noting that the critical isotherm passes through a point of inflection at the critical point, and the slope is zero there. Expressed mathematically, these conditions are, respectively

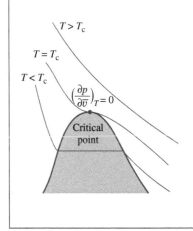

$$\left(\frac{\partial^2 p}{\partial \bar{v}^2}\right)_T = 0, \qquad \left(\frac{\partial p}{\partial \bar{v}}\right)_T = 0 \qquad \text{(critical point)} \qquad (11.3)$$

Although less overall accuracy normally results when the constants a and b are determined using critical point behavior than when they are determined by fitting p–v–T data in a particular region of interest, the advantage of this approach is that the van der Waals constants can be expressed in terms of the critical pressure p_c and critical temperature T_c, as demonstrated next.

For the van der Waals equation at the critical point

$$p_c = \frac{\bar{R}T_c}{\bar{v}_c - b} - \frac{a}{\bar{v}_c^2}$$

Applying Eqs. 11.3 with the van der Waals equation gives

$$\left(\frac{\partial^2 p}{\partial \bar{v}^2}\right)_T = \frac{2\bar{R}T_c}{(\bar{v}_c - b)^3} - \frac{6a}{\bar{v}_c^4} = 0$$

$$\left(\frac{\partial p}{\partial \bar{v}}\right)_T = -\frac{\bar{R}T_c}{(\bar{v}_c - b)^2} + \frac{2a}{\bar{v}_c^3} = 0$$

Solving the foregoing three equations for a, b, and $\bar{v}_c$ in terms of the critical pressure and critical temperature

$$a = \frac{27}{64} \frac{\bar{R}^2 T_c^2}{p_c} \tag{11.4a}$$

$$b = \frac{\bar{R} T_c}{8 p_c} \tag{11.4b}$$

$$\bar{v}_c = \frac{3}{8} \frac{\bar{R} T_c}{p_c} \tag{11.4c}$$

Values of the van der Waals constants a and b determined from Eqs. 11.4a and 11.4b for several common substances are given in Table A-24 for pressure in bar, specific volume in m³/kmol, and temperature in K.

Generalized Form Introducing the compressibility factor $Z = p\bar{v}/\bar{R}T$, the reduced temperature $T_R = T/T_c$, the pseudoreduced specific volume $v'_R = p_c \bar{v}/\bar{R}T_c$, and the foregoing expressions for a and b, the van der Waals equation can be written in terms of Z, v'_R, and T_R as

$$Z = \frac{v'_R}{v'_R - 1/8} - \frac{27/64}{T_R v'_R} \tag{11.5}$$

or alternatively in terms of Z, T_R, and p_R as

$$Z^3 - \left(\frac{p_R}{8T_R} + 1\right)Z^2 + \left(\frac{27 p_R}{64 T_R^2}\right)Z - \frac{27 p_R^2}{512 T_R^3} = 0 \tag{11.6}$$

The details of these developments are left as exercises. Equation 11.5 can be evaluated for specified values of v'_R and T_R and the resultant Z values located on a generalized compressibility chart to show approximately where the equation performs satisfactorily. A similar approach can be taken with Eq. 11.6.

The compressibility factor at the critical point yielded by the van der Waals equation is determined from Eq. 11.4c as

$$Z_c = \frac{p_c \bar{v}_c}{\bar{R} T_c} = 0.375$$

Actually, Z_c varies from about 0.23 to 0.33 for most substances (see Table A-1). Accordingly, with the set of constants given by Eqs. 11.4, the van der Waals equation is inaccurate in the vicinity of the critical point. Further study would show inaccuracy in other regions as well, so this equation is not suitable for many thermodynamic evaluations. The van der Waals equation is of interest to us primarily because it is the simplest model that accounts for the departure of actual gas behavior from the ideal gas equation of state.

Redlich–Kwong Equation Three other two-constant equations of state that have been widely used are the Berthelot, Dieterici, and Redlich–Kwong equations. The **Redlich–Kwong equation**, considered by many to be the best of the two-constant equations of state, is

Redlich–Kwong equation

$$\boxed{p = \frac{\bar{R}T}{\bar{v} - b} - \frac{a}{\bar{v}(\bar{v} + b)T^{1/2}}} \tag{11.7}$$

This equation, proposed in 1949, is mainly empirical in nature, with no rigorous justification in terms of molecular arguments. The Redlich–Kwong equation is explicit in pressure but not in specific volume or temperature. Like the van der Waals equation, the Redlich–Kwong equation is cubic in specific volume.

Although the Redlich–Kwong equation is somewhat more difficult to manipulate mathematically than the van der Waals equation, it is more accurate, particularly at higher pressures. The two-constant Redlich–Kwong equation performs better than some equations of state having several adjustable constants; still, two-constant equations of state tend to be limited in accuracy as pressure (or density) increases. Increased accuracy at such states normally requires equations with a greater number of adjustable constants. Modified forms of the Redlich–Kwong equation have been proposed to achieve improved accuracy.

Evaluating a and b   As for the van der Waals equation, the constants a and b in Eq. 11.7 can be determined for a specified substance by fitting the equation to p–v–T data, with several sets of constants required to represent accurately all states of interest. Alternatively, a single set of constants in terms of the critical pressure and critical temperature can be evaluated using Eqs. 11.3, as for the van der Waals equation. The result is

$$a = a' \frac{\overline{R}^2 T_c^{5/2}}{p_c} \quad \text{and} \quad b = b' \frac{\overline{R} T_c}{p_c} \tag{11.8}$$

where $a' = 0.42748$ and $b' = 0.08664$. Evaluation of these constants is left as an exercise. Values of the Redlich–Kwong constants a and b determined from Eqs. 11.8 for several common substances are given in Table A-24 for pressure in bar, specific volume in m³/kmol, and temperature in K.

Generalized Form   Introducing the compressibility factor Z, the reduced temperature T_R, the pseudoreduced specific volume v'_R, and the foregoing expressions for a and b, the Redlich–Kwong equation can be written as

$$Z = \frac{v'_R}{v'_R - b'} - \frac{a'}{(v'_R + b')T_R^{3/2}} \tag{11.9}$$

Equation 11.9 can be evaluated at specified values of v'_R and T_R and the resultant Z values located on a generalized compressibility chart to show the regions where the equation performs satisfactorily. With the constants given by Eqs. 11.8, the compressibility factor at the critical point yielded by the Redlich–Kwong equation is $Z_c = 0.333$, which is at the high end of the range of values for most substances, indicating that inaccuracy in the vicinity of the critical point should be expected.

In Example 11.1, the pressure of a gas is determined using three equations of state and the generalized compressibility chart. The results are compared.

▶▶ **EXAMPLE 11.1** ▶ ·

Comparing Equations of State

A cylindrical tank containing 4.0 kg of carbon monoxide gas at −50°C has an inner diameter of 0.2 m and a length of 1 m. Determine the pressure, in bar, exerted by the gas using **(a)** the generalized compressibility chart, **(b)** the ideal gas equation of state, **(c)** the van der Waals equation of state, **(d)** the Redlich–Kwong equation of state. Compare the results obtained.

Schematic and Given Data:

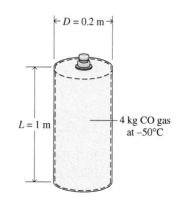

Fig. E11.1

SOLUTION

Known   A cylindrical tank of known dimensions contains 4.0 kg of CO gas at −50°C.

Find   Determine the pressure exerted by the gas using four alternative methods.

Engineering Model

1. As shown in the accompanying figure, the closed system is taken as the gas.

2. The system is at equilibrium.

Analysis The molar specific volume of the gas is required in each part of the solution. Let us begin by evaluating it. The volume occupied by the gas is

$$V = \left(\frac{\pi D^2}{4}\right)L = \frac{\pi(0.2 \text{ m})^2(1.0 \text{ m})}{4} = 0.0314 \text{ m}^3$$

The molar specific volume is then

$$\bar{v} = Mv = M\left(\frac{V}{m}\right) = \left(28 \frac{\text{kg}}{\text{kmol}}\right)\left(\frac{0.0314 \text{ m}^3}{4.0 \text{ kg}}\right) = 0.2198 \frac{\text{m}^3}{\text{kmol}}$$

a. From Table A-1 for CO, $T_c = 133$ K, $p_c = 35$ bar. Thus, the reduced temperature T_R and pseudoreduced specific volume v'_R are, respectively,

$$T_R = \frac{223 \text{ K}}{133 \text{ K}} = 1.68$$

$$v'_R = \frac{\bar{v}p_c}{\bar{R}T_c} = \frac{(0.2198 \text{ m}^3/\text{kmol})(35 \times 10^5 \text{ N/m}^2)}{(8314 \text{ N} \cdot \text{m/kmol} \cdot \text{K})(133 \text{ K})} = 0.696$$

Turning to Fig. A-2, $Z \approx 0.9$. Solving $Z = p\bar{v}/\bar{R}T$ for pressure and inserting known values

$$p = \frac{Z\bar{R}T}{\bar{v}} = \frac{0.9(8314 \text{ N} \cdot \text{m/kmol} \cdot \text{K})(223 \text{ K})}{(0.2198 \text{ m}^3/\text{kmol})}\left|\frac{1 \text{ bar}}{10^5 \text{ N/m}^2}\right|$$

$$= 75.9 \text{ bar}$$

b. The ideal gas equation of state gives

$$p = \frac{\bar{R}T}{\bar{v}} = \frac{(8314 \text{ N} \cdot \text{m/kmol} \cdot \text{K})(223 \text{ K})}{(0.2198 \text{ m}^3/\text{kmol})}\left|\frac{1 \text{ bar}}{10^5 \text{ N/m}^2}\right| = 84.4 \text{ bar}$$

c. For carbon monoxide, the van der Waals constants a and b given by Eqs. 11.4 can be read directly from Table A-24. Thus,

$$a = 1.474 \text{ bar}\left(\frac{\text{m}^3}{\text{kmol}}\right)^2 \quad \text{and} \quad b = 0.0395 \frac{\text{m}^3}{\text{kmol}}$$

Substituting into Eq. 11.2

$$p = \frac{\bar{R}T}{\bar{v} - b} - \frac{a}{\bar{v}^2}$$

$$= \frac{(8314 \text{ N} \cdot \text{m/kmol} \cdot \text{K})(223 \text{ K})}{(0.2198 - 0.0395)(\text{m}^3/\text{kmol})}\left|\frac{1 \text{ bar}}{10^5 \text{ N/m}^2}\right|$$

$$- \frac{1.474 \text{ bar}(\text{m}^3/\text{kmol})^2}{(0.2198 \text{ m}^3/\text{kmol})^2}$$

$$= 72.3 \text{ bar}$$

Alternatively, the values for v'_R and T_R obtained in the solution of part (a) can be substituted into Eq. 11.5, giving $Z = 0.86$. Then, with $p = Z\bar{R}T/\bar{v}$, $p = 72.5$ bar. The slight difference is attributed to roundoff.

d. For carbon monoxide, the Redlich–Kwong constants given by Eqs. 11.8 can be read directly from Table A-24. Thus

$$a = \frac{17.22 \text{ bar}(\text{m}^6)(\text{K})^{1/2}}{(\text{kmol})^2} \quad \text{and} \quad b = 0.02737 \text{ m}^3/\text{kmol}$$

Substituting into Eq. 11.7

$$p = \frac{\bar{R}T}{\bar{v} - b} - \frac{a}{\bar{v}(\bar{v} + b)T^{1/2}}$$

$$= \frac{(8314 \text{ N} \cdot \text{m/kmol} \cdot \text{K})(223 \text{ K})}{(0.2198 - 0.02737) \text{ m}^3/\text{kmol}}\left|\frac{1 \text{ bar}}{10^5 \text{ N/m}^2}\right|$$

$$- \frac{17.22 \text{ bar}}{(0.2198)(0.24717)(223)^{1/2}}$$

$$= 75.1 \text{ bar}$$

Alternatively, the values for v'_R and T_R obtained in the solution of part (a) can be substituted into Eq. 11.9, giving $Z = 0.89$. Then, with $p = Z\bar{R}T/\bar{v}$, $p = 75.1$ bar. In comparison to the value of part (a), the ideal gas equation of state predicts a pressure that is 11% higher and the van der Waals equation gives a value that is 5% lower. The Redlich–Kwong value is about 1% less than the value obtained using the compressibility chart.

SKILLS DEVELOPED

Ability to...

• determine pressure using the compressibility chart, ideal gas model, and the van der Waals and Redlich–Kwong equations of state.

• perform unit conversions correctly.

Quick Quiz

Using the given temperature and the pressure value determined in part (a), check the value of Z using Fig. A-2. Ans. $Z \approx 0.9$.

11.1.3 Multiconstant Equations of State

To fit the p–v–T data of gases over a wide range of states, Beattie and Bridgeman proposed in 1928 a pressure-explicit equation involving five constants in addition to the gas constant. The **Beattie–Bridgeman equation** can be expressed in a truncated virial form as

$$p = \frac{\bar{R}T}{\bar{v}} + \frac{\beta}{\bar{v}^2} + \frac{\gamma}{\bar{v}^3} + \frac{\delta}{\bar{v}^4}$$

(11.10)   **Beattie–Bridgeman equation**

where

$$\beta = B\bar{R}T - A - c\bar{R}/T^2$$
$$\gamma = -Bb\bar{R}T + Aa - Bc\bar{R}/T^2$$
$$\delta = Bbc\bar{R}/T^2$$

(11.11)

The five constants a, b, c, A, and B appearing in these equations are determined by curve fitting to experimental data.

Benedict, Webb, and Rubin extended the Beattie–Bridgeman equation of state to cover a broader range of states. The resulting equation, involving eight constants in addition to the gas constant, has been particularly successful in predicting the p–v–T behavior of *light hydrocarbons*. The **Benedict–Webb–Rubin equation** is

$$p = \frac{\bar{R}T}{\bar{v}} + \left(B\bar{R}T - A - \frac{C}{T^2} \right)\frac{1}{\bar{v}^2} + \frac{(b\bar{R}T - a)}{\bar{v}^3} + \frac{a\alpha}{\bar{v}^6} + \frac{c}{\bar{v}^3 T^2}\left(1 + \frac{\gamma}{\bar{v}^2} \right)\exp\left(-\frac{\gamma}{\bar{v}^2} \right)$$

(11.12)   **Benedict–Webb–Rubin equation**

Values of the constants appearing in Eq. 11.12 for five common substances are given in Table A-24 for pressure in bar, specific volume in m³/kmol, and temperature in K. Because Eq. 11.12 has been so successful, its realm of applicability has been extended by introducing additional constants.

Equations 11.10 and 11.12 are merely representative of multiconstant equations of state. Many other multiconstant equations have been proposed. With high-speed computers, equations having 50 or more constants have been developed for representing the p–v–T behavior of different substances.

11.2 Important Mathematical Relations

Values of two independent intensive properties are sufficient to fix the intensive state of a simple compressible system of specified mass and composition—for instance, temperature and specific volume (see Sec. 3.1). All other intensive properties can be determined as functions of the two independent properties: $p = p(T, v)$, $u = u(T, v)$, $h = h(T, v)$, and so on. These are all functions of two independent variables of the form $z = z(x, y)$, with x and y being the independent variables. It might also be recalled that the differential of every property is *exact* (Sec. 2.2.1). The differentials of nonproperties such as work and heat are inexact. Let us review briefly some concepts from calculus about functions of two independent variables and their differentials.

The **exact differential** of a function z, continuous in the variables x and y, is

TAKE NOTE...

The state principle for simple systems is introduced in Sec. 3.1.

exact differential

$$dz = \left(\frac{\partial z}{\partial x} \right)_y dx + \left(\frac{\partial z}{\partial y} \right)_x dy$$

(11.13a)

This can be expressed alternatively as

$$dz = M\,dx + N\,dy$$

(11.13b)

where $M = (\partial z/\partial x)_y$ and $N = (\partial z/\partial y)_x$. The coefficient M is the partial derivative of z with respect to x (the variable y being held constant). Similarly, N is the partial derivative of z with respect to y (the variable x being held constant).

If the coefficients M and N have continuous first partial derivatives, the order in which a second partial derivative of the function z is taken is immaterial. That is,

$$\frac{\partial}{\partial y}\left[\left(\frac{\partial z}{\partial x} \right)_y \right]_x = \frac{\partial}{\partial x}\left[\left(\frac{\partial z}{\partial y} \right)_x \right]_y$$

(11.14a)

or

$$\boxed{\left(\frac{\partial M}{\partial y}\right)_x = \left(\frac{\partial N}{\partial x}\right)_y}$$

(11.14b)

test for exactness   which can be called the **test for exactness**, as discussed next.

In words, Eqs. 11.14 indicate that the mixed second partial derivatives of the function z are equal. The relationship in Eqs. 11.14 is both a necessary and sufficient condition for the *exactness* of a differential expression, and it may therefore be used as a test for exactness. When an expression such as $M\,dx + N\,dy$ does not meet this test, no function z exists whose differential is equal to this expression. In thermodynamics, Eq. 11.14 is not generally used to test exactness but rather to develop additional property relations. This is illustrated in Sec. 11.3 to follow.

Two other relations among partial derivatives are listed next for which applications are found in subsequent sections of this chapter. These are

$$\boxed{\left(\frac{\partial x}{\partial y}\right)_z \left(\frac{\partial y}{\partial x}\right)_z = 1}$$

(11.15)

and

$$\boxed{\left(\frac{\partial y}{\partial z}\right)_x \left(\frac{\partial z}{\partial x}\right)_y \left(\frac{\partial x}{\partial y}\right)_z = -1}$$

(11.16)

> **FOR EXAMPLE**
>
> Consider the three quantities x, y, and z, any two of which may be selected as the independent variables. Thus, we can write $x = x(y, z)$ and $y = y(x, z)$. The differentials of these functions are, respectively,
>
> $$dx = \left(\frac{\partial x}{\partial y}\right)_z dy + \left(\frac{\partial x}{\partial z}\right)_y dz \qquad \text{and} \qquad dy = \left(\frac{\partial y}{\partial x}\right)_z dx + \left(\frac{\partial y}{\partial z}\right)_x dz$$
>
> Eliminating dy between these two equations results in
>
> $$\left[1 - \left(\frac{\partial x}{\partial y}\right)_z \left(\frac{\partial y}{\partial x}\right)_z\right] dx = \left[\left(\frac{\partial x}{\partial y}\right)_z \left(\frac{\partial y}{\partial z}\right)_x + \left(\frac{\partial x}{\partial z}\right)_y\right] dz$$
>
> (11.17)
>
> Since x and z can be varied independently, let us hold z constant and vary x. That is, let $dz = 0$ and $dx \neq 0$. It then follows from Eq. 11.17 that the coefficient of dx must vanish, so Eq. 11.15 must be satisfied. Similarly, when $dx = 0$ and $dz \neq 0$, the coefficient of dz in Eq. 11.17 must vanish. Introducing Eq. 11.15 into the resulting expression and rearranging gives Eq. 11.16. The details are left as an exercise.

Application   An equation of state $p = p(T, v)$ provides a specific example of a function of two independent variables. The partial derivatives $(\partial p/\partial T)_v$ and $(\partial p/\partial v)_T$ of $p(T, v)$ are important for subsequent discussions. The quantity $(\partial p/\partial T)_v$ is the partial derivative of p with respect to T (the variable v being held constant). This partial derivative represents the slope at a point on a line of constant specific volume (isometric) projected onto the p–T plane. Similarly, the partial derivative $(\partial p/\partial v)_T$ is the partial derivative of p with respect to v (the variable T being held constant). This partial derivative represents the slope at a point on a line of constant temperature (isotherm) projected on the p–v plane. The partial derivatives $(\partial p/\partial T)_v$ and $(\partial p/\partial v)_T$ are themselves intensive properties because they have unique values at each state.

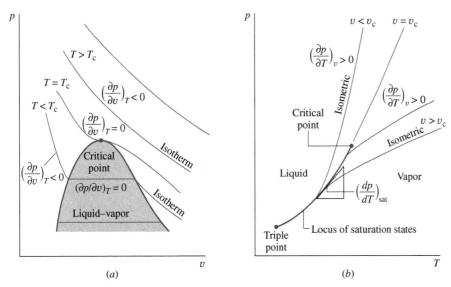

Fig. 11.1 Diagrams used to discuss $(\partial p/\partial v)_T$ and $(\partial p/\partial T)_v$. (a) p–v diagram. (b) Phase diagram.

The p–v–T surfaces given in Figs. 3.1 and 3.2 are graphical representations of functions of the form $p = p(v, T)$. **Figure 11.1** shows the liquid, vapor, and two-phase regions of a p–v–T surface projected onto the p–v and p–T planes. Referring first to Fig. 11.1a, note that several isotherms are sketched. In the single-phase regions, the partial derivative $(\partial p/\partial v)_T$ giving the slope is negative at each state along an isotherm except at the critical point, where the partial derivative vanishes. Since the isotherms are horizontal in the two-phase liquid–vapor region, the partial derivative $(\partial p/\partial v)_T$ vanishes there as well. For these states, pressure is independent of specific volume and is a function of temperature only: $p = p_{sat}(T)$.

Figure 11.1b shows the liquid and vapor regions with several isometrics (constant specific volume lines) superimposed. In the single-phase regions, the isometrics are nearly straight or are slightly curved and the partial derivative $(\partial p/\partial T)_v$ is positive at each state along the curves. For the two-phase liquid–vapor states corresponding to a specified value of temperature, the pressure is independent of specific volume and is determined by the temperature only. Hence, the slopes of isometrics passing through the two-phase states corresponding to a specified temperature are all equal, being given by the slope of the saturation curve at that temperature, denoted simply as $(dp/dT)_{sat}$. For these two-phase states, $(\partial p/\partial T)_v = (dp/dT)_{sat}$.

In this section, important aspects of functions of two variables have been introduced. The following example illustrates some of these ideas using the van der Waals equation of state.

▶ **EXAMPLE 11.2** ▶ . . . .

Applying Mathematical Relations

For the van der Waals equation of state, **(a)** determine an expression for the exact differential dp, **(b)** show that the mixed second partial derivatives of the result obtained in part (a) are equal, and **(c)** develop an expression for the partial derivative $(\partial v/\partial T)_p$.

SOLUTION

Known The equation of state is the van der Waals equation.

Find Determine the differential dp, show that the mixed second partial derivatives of dp are equal, and develop an expression for $(\partial v/\partial T)_p$.

Analysis

a. By definition, the differential of a function $p = p(T, v)$ is

$$dp = \left(\frac{\partial p}{\partial T}\right)_v dT + \left(\frac{\partial p}{\partial v}\right)_T dv$$

The partial derivatives appearing in this expression obtained from the van der Waals equation expressed as $p = RT/(v - b) - a/v^2$ are

$$M = \left(\frac{\partial p}{\partial T}\right)_v = \frac{R}{v - b}, \qquad N = \left(\frac{\partial p}{\partial v}\right)_T = -\frac{RT}{(v - b)^2} + \frac{2a}{v^3}$$

Accordingly, the differential takes the form

$$dp = \left(\frac{R}{v-b}\right)dT + \left[\frac{-RT}{(v-b)^2} + \frac{2a}{v^3}\right]dv$$

b. Calculating the mixed second partial derivatives

$$\left(\frac{\partial M}{\partial v}\right)_T = \frac{\partial}{\partial v}\left[\left(\frac{\partial p}{\partial T}\right)_v\right]_T = -\frac{R}{(v-b)^2}$$

$$\left(\frac{\partial N}{\partial T}\right)_v = \frac{\partial}{\partial T}\left[\left(\frac{\partial p}{\partial v}\right)_T\right]_v = -\frac{R}{(v-b)^2}$$

Thus, the mixed second partial derivatives are equal, as expected.

c. An expression for $(\partial v/\partial T)_p$ can be derived using Eqs. 11.15 and 11.16. Thus, with $x = p$, $y = v$, and $z = T$, Eq. 11.16 gives

$$\left(\frac{\partial v}{\partial T}\right)_p \left(\frac{\partial p}{\partial v}\right)_T \left(\frac{\partial T}{\partial p}\right)_v = -1$$

or

$$\left(\frac{\partial v}{\partial T}\right)_p = -\frac{1}{(\partial p/\partial v)_T (\partial T/\partial p)_v}$$

Then, with $x = T$, $y = p$, and $z = v$, Eq. 11.15 gives

$$\left(\frac{\partial T}{\partial p}\right)_v = \frac{1}{(\partial p/\partial T)_v}$$

Combining these results

$$\left(\frac{\partial v}{\partial T}\right)_p = \frac{(\partial p/\partial T)_v}{(\partial p/\partial v)_T}$$

The numerator and denominator of this expression have been evaluated in part (a), so

①
$$\left(\frac{\partial v}{\partial T}\right)_p = -\frac{R/(v-b)}{[-RT/(v-b)^2 + 2a/v^3]}$$

which is the desired result.

---

① Since the van der Waals equation is cubic in specific volume, it can be solved for v (T, p) at only certain states. Part (c) shows how the partial derivative $(\partial v/\partial T)_p$ can be evaluated at states where it exists.

SKILLS DEVELOPED

Ability to...

- use Eqs. 11.15 and 11.16 together with the van der Waals equation of state to develop a thermodynamic property relation.

Quick Quiz

Using the results obtained, develop an expression for $\left(\frac{\partial v}{\partial T}\right)_v$ of an ideal gas. Ans. v/T.

11.3 Developing Property Relations

In this section, several important property relations are developed, including the expressions known as the *Maxwell* relations. The concept of a *fundamental thermodynamic function* is also introduced. These results, which are important for subsequent discussions, are obtained for simple compressible systems of fixed chemical composition using the concept of an exact differential.

11.3.1 Principal Exact Differentials

The principal results of this section are obtained using Eqs. 11.18, 11.19, 11.22, and 11.23. The first two of these equations are derived in Sec. 6.3, where they are referred to as the *T ds equations*. For present purposes, it is convenient to express them as

$$du = T\,ds - p\,dv \tag{11.18}$$

$$dh = T\,ds + v\,dp \tag{11.19}$$

The other two equations used to obtain the results of this section involve, respectively, the specific **Helmholtz function** ψ defined by

Helmholtz function

$$\psi = u - Ts \tag{11.20}$$

and the specific **Gibbs function** g defined by

Gibbs function

$$g = h - Ts \tag{11.21}$$

The Helmholtz and Gibbs functions are properties because each is defined in terms of properties. From inspection of Eqs. 11.20 and 11.21, the units of ψ and g are the same as those of u and h. These two new properties are introduced solely because they contribute to the present discussion, and no physical significance need be attached to them at this point.

Forming the differential $d\psi$

$$d\psi = du - d(Ts) = du - T\,ds - s\,dT$$

Substituting Eq. 11.18 into this gives

$$\boxed{d\psi = -p\,dv - s\,dT} \qquad (11.22)$$

Similarly, forming the differential dg

$$dg = dh - d(Ts) = dh - T\,ds - s\,dT$$

Substituting Eq. 11.19 into this gives

$$\boxed{dg = v\,dp - s\,dT} \qquad (11.23)$$

11.3.2 Property Relations from Exact Differentials

The four differential equations introduced above, Eqs. 11.18, 11.19, 11.22, and 11.23, provide the basis for several important property relations. Since only properties are involved, each is an exact differential exhibiting the general form $dz = M\,dx + N\,dy$ considered in Sec. 11.2. Underlying these exact differentials are, respectively, functions of the form $u(s, v)$, $h(s, p)$, $\psi(v, T)$, and $g(T, p)$. Let us consider these functions in the order given.

The differential of the function $u = u(s, v)$ is

$$du = \left(\frac{\partial u}{\partial s}\right)_v ds + \left(\frac{\partial u}{\partial v}\right)_s dv$$

By comparison with Eq. 11.18, we conclude that

$$\boxed{T = \left(\frac{\partial u}{\partial s}\right)_v} \qquad (11.24)$$

$$\boxed{-p = \left(\frac{\partial u}{\partial v}\right)_s} \qquad (11.25)$$

The differential of the function $h = h(s, p)$ is

$$dh = \left(\frac{\partial h}{\partial s}\right)_p ds + \left(\frac{\partial h}{\partial p}\right)_s dp$$

By comparison with Eq. 11.19, we conclude that

$$\boxed{T = \left(\frac{\partial h}{\partial s}\right)_p} \qquad (11.26)$$

$$\boxed{v = \left(\frac{\partial h}{\partial p}\right)_s} \qquad (11.27)$$

Similarly, the coefficients $-p$ and $-s$ of Eq. 11.22 are partial derivatives of $\psi(v, T)$

$$-p = \left(\frac{\partial \psi}{\partial v}\right)_T \tag{11.28}$$

$$-s = \left(\frac{\partial \psi}{\partial T}\right)_v \tag{11.29}$$

and the coefficients v and $-s$ of Eq. 11.23 are partial derivatives of $g(T, p)$

$$v = \left(\frac{\partial g}{\partial p}\right)_T \tag{11.30}$$

$$-s = \left(\frac{\partial g}{\partial T}\right)_p \tag{11.31}$$

As each of the four differentials introduced above is exact, the second mixed partial derivatives are equal. Thus, in Eq. 11.18, T plays the role of M in Eq. 11.14b and $-p$ plays the role of N in Eq. 11.14b, so

$$\left(\frac{\partial T}{\partial v}\right)_s = -\left(\frac{\partial p}{\partial s}\right)_v \tag{11.32}$$

In Eq. 11.19, T and v play the roles of M and N in Eq. 11.14b, respectively. Thus,

$$\left(\frac{\partial T}{\partial p}\right)_s = \left(\frac{\partial v}{\partial s}\right)_p \tag{11.33}$$

Similarly, from Eqs. 11.22 and 11.23 follow

$$\left(\frac{\partial p}{\partial T}\right)_v = \left(\frac{\partial s}{\partial v}\right)_T \tag{11.34}$$

$$\left(\frac{\partial v}{\partial T}\right)_p = -\left(\frac{\partial s}{\partial p}\right)_T \tag{11.35}$$

Maxwell relations   Equations 11.32 through 11.35 are known as the **Maxwell relations**.

Since each of the properties T, p, v, s appears on the left side of two of the eight equations, Eqs. 11.24 through 11.31, four additional property relations can be obtained by equating such expressions. They are

$$\left(\frac{\partial u}{\partial s}\right)_v = \left(\frac{\partial h}{\partial s}\right)_p, \qquad \left(\frac{\partial u}{\partial v}\right)_s = \left(\frac{\partial \psi}{\partial v}\right)_T$$
$$\left(\frac{\partial h}{\partial p}\right)_s = \left(\frac{\partial g}{\partial p}\right)_T, \qquad \left(\frac{\partial \psi}{\partial T}\right)_v = \left(\frac{\partial g}{\partial T}\right)_p \tag{11.36}$$

Equations 11.24 through 11.36, which are listed in **Table 11.1** for ease of reference, are 16 property relations obtained from Eqs. 11.18, 11.19, 11.22, and 11.23, using the concept of an exact differential. Since Eqs. 11.19, 11.22, and 11.23 can themselves be derived from Eq. 11.18, the important role of the first $T\,dS$ equation in developing property relations is apparent.

TABLE 11.1

Summary of Property Relations from Exact Differentials

Basic relations:

from $u = u(s, \upsilon)$ from $h = h(s, p)$

$$T = \left(\frac{\partial u}{\partial s}\right)_{\upsilon} \quad (11.24) \qquad\qquad T = \left(\frac{\partial h}{\partial s}\right)_{p} \quad (11.26)$$

$$-p = \left(\frac{\partial u}{\partial \upsilon}\right)_{s} \quad (11.25) \qquad\qquad \upsilon = \left(\frac{\partial h}{\partial p}\right)_{s} \quad (11.27)$$

from $\psi = \psi(\upsilon, T)$ from $g = g(T, p)$

$$-p = \left(\frac{\partial \psi}{\partial \upsilon}\right)_{T} \quad (11.28) \qquad\qquad \upsilon = \left(\frac{\partial g}{\partial p}\right)_{T} \quad (11.30)$$

$$-s = \left(\frac{\partial \psi}{\partial T}\right)_{\upsilon} \quad (11.29) \qquad\qquad -s = \left(\frac{\partial g}{\partial T}\right)_{p} \quad (11.31)$$

Maxwell relations:

$$\left(\frac{\partial T}{\partial \upsilon}\right)_{s} = -\left(\frac{\partial p}{\partial s}\right)_{\upsilon} \quad (11.32) \qquad\qquad \left(\frac{\partial p}{\partial T}\right)_{\upsilon} = \left(\frac{\partial s}{\partial \upsilon}\right)_{T} \quad (11.34)$$

$$\left(\frac{\partial T}{\partial p}\right)_{s} = \left(\frac{\partial \upsilon}{\partial s}\right)_{p} \quad (11.33) \qquad\qquad \left(\frac{\partial \upsilon}{\partial T}\right)_{p} = -\left(\frac{\partial s}{\partial p}\right)_{T} \quad (11.35)$$

Additional relations:

$$\left(\frac{\partial u}{\partial s}\right)_{\upsilon} = \left(\frac{\partial h}{\partial s}\right)_{p} \qquad\qquad\qquad \left(\frac{\partial u}{\partial \upsilon}\right)_{s} = \left(\frac{\partial \psi}{\partial \upsilon}\right)_{T} \quad (11.36)$$

$$\left(\frac{\partial h}{\partial p}\right)_{s} = \left(\frac{\partial g}{\partial p}\right)_{T} \qquad\qquad\qquad \left(\frac{\partial \psi}{\partial T}\right)_{\upsilon} = \left(\frac{\partial g}{\partial T}\right)_{p}$$

The utility of these 16 property relations is demonstrated in subsequent sections of this chapter. However, to give a specific illustration at this point, suppose the partial derivative $(\partial s/\partial \upsilon)_T$ involving entropy is required for a certain purpose. The Maxwell relation Eq. 11.34 would allow the derivative to be determined by evaluating the partial derivative $(\partial p/\partial T)_{\upsilon}$, which can be obtained using p–υ–T data only. Further elaboration is provided in Example 11.3.

EXAMPLE 11.3

Applying the Maxwell Relations

Evaluate the partial derivative $(\partial s/\partial \upsilon)_T$ for water vapor at a state fixed by a temperature of 240°C and a specific volume of 0.4646 m³/kg. **(a)** Use the Redlich–Kwong equation of state and an appropriate Maxwell relation. **(b)** Check the value obtained using steam table data.

SOLUTION

Known The system consists of a fixed amount of water vapor at 240°C and 0.4646 m³/kg.

Find Determine the partial derivative $(\partial s/\partial \upsilon)_T$ employing the Redlich–Kwong equation of state, together with a Maxwell relation. Check the value obtained using steam table data.

Engineering Model

1. The system consists of a fixed amount of water at a known equilibrium state.

2. Accurate values for $(\partial s/\partial T)_{\upsilon}$ in the neighborhood of the given state can be determined from the Redlich–Kwong equation of state.

Analysis

a. The Maxwell relation given by Eq. 11.34 allows $(\partial s/\partial \upsilon)_T$ to be determined from the p–υ–T relationship. That is,

$$\left(\frac{\partial s}{\partial \upsilon}\right)_T = \left(\frac{\partial p}{\partial T}\right)_{\upsilon}$$

The partial derivative $(\partial p/\partial T)_{\upsilon}$ obtained from the Redlich–Kwong equation, Eq. 11.7, is

$$\left(\frac{\partial p}{\partial T}\right)_{\upsilon} = \frac{\bar{R}}{\bar{\upsilon} - b} + \frac{a}{2\bar{\upsilon}(\bar{\upsilon} + b)T^{3/2}}$$

At the specified state, the temperature is 513 K and the specific volume on a molar basis is

$$\bar{v} = 0.4646 \frac{m^3}{kg}\left(\frac{18.02 \text{ kg}}{\text{kmol}}\right) = 8.372 \frac{m^3}{\text{kmol}}$$

From Table A-24

$$a = 142.59 \text{ bar}\left(\frac{m^3}{\text{kmol}}\right)^2 (K)^{1/2}, \qquad b = 0.0211\frac{m^3}{\text{kmol}}$$

Substituting values into the expression for $(\partial p/\partial T)_v$

$$\left(\frac{\partial p}{\partial T}\right)_v = \frac{\left(8314\dfrac{N \cdot m}{\text{kmol} \cdot K}\right)}{(8.372 - 0.0211)\dfrac{m^3}{\text{kmol}}}$$

$$+ \frac{142.59 \text{ bar}\left(\dfrac{m^3}{\text{kmol}}\right)^2 (K)^{1/2}}{2\left(8.372\dfrac{m^2}{\text{kmol}}\right)\left(8.3931\dfrac{m^3}{\text{kmol}}\right)(513 \text{ K})^{3/2}}\left|\dfrac{10^5 \text{ N/m}^2}{1 \text{ bar}}\right|$$

$$= \left(1004.3\frac{N \cdot m}{m^3 \cdot K}\right)\left|\frac{1 \text{ kJ}}{10^3 \text{ N} \cdot m}\right|$$

$$= 1.0043\frac{kJ}{m^3 \cdot K}$$

Accordingly

$$\left(\frac{\partial s}{\partial v}\right)_T = 1.0043\frac{kJ}{m^3 \cdot K}$$

b. A value for $(\partial s/\partial v)_T$ can be estimated using a graphical approach with steam table data, as follows: At 240°C, Table A-4 provides the values for specific entropy s and specific volume v tabulated below.

	$T = 240°C$	
p(bar)	s(kJ/kg · K)	v(m³/kg)
1.0	7.9949	2.359
1.5	7.8052	1.570
3.0	7.4774	0.781
5.0	7.2307	0.4646
7.0	7.0641	0.3292
10.0	6.8817	0.2275

With the values for s and v listed in the table, the plot in **Fig. E11.3a** giving s versus v can be prepared. Note that a line representing the tangent to the curve at the given state is shown on the plot. The pressure at this state is 5 bar. The slope of the tangent is $(\partial s/\partial v)_T \approx 1.0 \text{ kJ/m}^3 \cdot K$. Thus, the value of $(\partial s/\partial v)_T$ obtained using the Redlich–Kwong equation agrees closely with the result determined graphically using steam table data.

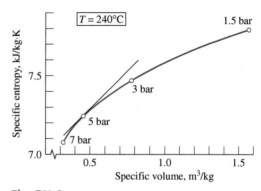

Fig. E11.3a

Alternative Solution Alternatively, the partial derivative $(\partial s/\partial v)_T$ can be estimated using numerical methods and computer-generated data. The following *IT* code illustrates *one way* the partial derivative, denoted dsdv, can be estimated:

```
v = 0.4646 // m³/kg
T = 240 // °C
v2 = v + dv
v1 = v − dv
dv = 0.2
v2 = v_PT ("Water/Steam", p2, T)
v1 = v_PT ("Water/Steam", p1, T)
s2 = s_PT ("Water/Steam", p2, T)
s1 = s_PT ("Water/Steam", p1, T)
dsdv = (s2 − s1)/(v2 − v1)
```

Using the **Explore** button, sweep dv from 0.0001 to 0.2 in steps of 0.001. Then, using the **Graph** button, the following graph can be constructed:

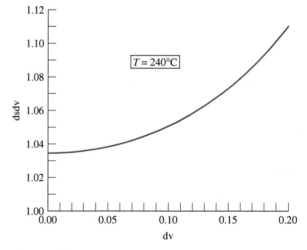

Fig. E11.3b

From the computer data, the y-intercept of the graph is

❶
$$\left(\frac{\partial s}{\partial v}\right)_T = \lim_{\Delta v \to 0}\left(\frac{\Delta s}{\Delta v}\right)_T \approx 1.033\frac{kJ}{m^3 \cdot K}$$

This answer is an estimate because it relies on a numerical approximation of the partial derivative based on the equation of state that underlies the *steam tables*. The values obtained using the Redlich–Kwong equation of state and the graphical method using steam table data agree with this result.

① It is left as an exercise to show that, in accordance with Eq. 11.34, the value of $(\partial p/\partial T)_v$ estimated by a procedure like the one used for $(\partial s/\partial v)_T$ agrees with the value given here.

SKILLS DEVELOPED

Ability to…

- apply a Maxwell relation to evaluate a thermodynamic quantity.
- apply the Redlich–Kwong equation.

- perform a comparison with data from the steam table using graphical and computer-based methods.

Quick Quiz

For steam at $T = 240°C$, $v = 0.4646$ m³/kg, $p = 5$ bar, calculate the value of the compressibility factor Z. Ans. **0.981**.

11.3.3 Fundamental Thermodynamic Functions

A *fundamental thermodynamic function* provides a complete description of the thermodynamic state. In the case of a pure substance with two independent properties, the **fundamental thermodynamic function** can take one of the following four forms:

$$
\begin{aligned}
u &= u(s,\ v)\\
h &= h(s,\ p)\\
\psi &= \psi(T,\ v)\\
g &= g(T,\ p)
\end{aligned}
\qquad (11.37)
$$

fundamental thermodynamic function

Of the four fundamental functions listed in Eqs. 11.37, the Helmholtz function ψ and the Gibbs function g have the greatest importance for subsequent discussions (see Sec. 11.6.2). Accordingly, let us discuss the fundamental function concept with reference to ψ and g.

In principle, all properties of interest can be determined from a fundamental thermodynamic function by differentiation and combination.

FOR EXAMPLE

Consider a fundamental function of the form $\psi(T, v)$. The properties v and T, being the independent variables, are specified to fix the state. The pressure p at this state can be determined from Eq. 11.28 by differentiation of $\psi(T, v)$. Similarly, the specific entropy s at the state can be found from Eq. 11.29 by differentiation. By definition, $\psi = u - Ts$, so the specific internal energy is obtained as

$$u = \psi + Ts$$

With u, p, and v known, the specific enthalpy can be found from the definition $h = u + pv$. Similarly, the specific Gibbs function is found from the definition, $g = h - Ts$. The specific heat c_v can be determined by further differentiation, $c_v = (\partial u/\partial T)_v$. Other properties can be calculated with similar operations.

FOR EXAMPLE

Consider a fundamental function of the form $g(T, p)$. The properties T and p are specified to fix the state. The specific volume and specific entropy at this state can be determined by differentiation from Eqs. 11.30 and 11.31, respectively. By definition, $g = h - Ts$, so the specific enthalpy is obtained as

$$h = g + Ts$$

With h, p, and v known, the specific internal energy can be found from $u = h - pv$. The specific heat c_p can be determined by further differentiation, $c_p = (\partial h/\partial T)_p$. Other properties can be calculated with similar operations.

Like considerations apply for functions of the form $u(s, v)$ and $h(s, p)$, as can readily be verified. Note that a Mollier diagram provides a graphical representation of the fundamental function $h(s, p)$.

11.4 Evaluating Changes in Entropy, Internal Energy, and Enthalpy

With the introduction of the Maxwell relations, we are in a position to develop thermodynamic relations that allow changes in entropy, internal energy, and enthalpy to be evaluated from measured property data. The presentation begins by considering relations applicable to phase changes and then turns to relations for use in single-phase regions.

11.4.1 Considering Phase Change

The objective of this section is to develop relations for evaluating the changes in specific entropy, internal energy, and enthalpy accompanying a change of phase at fixed temperature and pressure. A principal role is played by the *Clapeyron equation*, which allows the change in enthalpy during vaporization, sublimation, or melting at a constant temperature to be evaluated from pressure-specific volume–temperature data pertaining to the phase change. Thus, the present discussion provides important examples of how p–v–T measurements can lead to the determination of other property changes, namely Δs, Δu, and Δh for a change of phase.

Consider a change in phase from saturated liquid to saturated vapor at fixed temperature. For an isothermal phase change, pressure also remains constant, so Eq. 11.19 reduces to

$$dh = T \, ds$$

Integration of this expression gives

$$s_g - s_f = \frac{h_g - h_f}{T} \tag{11.38}$$

Hence, the change in specific entropy accompanying a phase change from saturated liquid to saturated vapor at temperature T can be determined from the temperature and the change in specific enthalpy.

The change in specific internal energy during the phase change can be determined using the definition $h = u + pv$.

$$u_g - u_f = h_g - h_f - p(v_g - v_f) \tag{11.39}$$

Thus, the change in specific internal energy accompanying a phase change at temperature T can be determined from the temperature and the changes in specific volume and enthalpy.

Clapeyron Equation The change in specific enthalpy required by Eqs. 11.38 and 11.39 can be obtained using the Clapeyron equation. To derive the Clapeyron equation, begin with the Maxwell relation

$$\left(\frac{\partial s}{\partial v}\right)_T = \left(\frac{\partial p}{\partial T}\right)_v \tag{11.34}$$

During a phase change at fixed temperature, the pressure is independent of specific volume and is determined by temperature alone. Thus, the quantity $(\partial p / \partial T)_v$ is determined by the temperature and can be represented as

$$\left(\frac{\partial p}{\partial T}\right)_v = \left(\frac{dp}{dT}\right)_{\text{sat}}$$

where "sat" indicates that the derivative is the slope of the saturation pressure–temperature curve at the point determined by the temperature held constant during the phase change (Sec. 11.2). Combining the last two equations gives

$$\left(\frac{\partial s}{\partial v}\right)_T = \left(\frac{dp}{dT}\right)_{sat}$$

Since the right side of this equation is fixed when the temperature is specified, the equation can be integrated to give

$$s_g - s_f = \left(\frac{dp}{dT}\right)_{sat} (v_g - v_f)$$

Introducing Eq. 11.38 into this expression results in the **Clapeyron equation**

Clapeyron equation

$$\boxed{\left(\frac{dp}{dT}\right)_{sat} = \frac{h_g - h_f}{T(v_g - v_f)}} \qquad (11.40)$$

Equation 11.40 allows $(h_g - h_f)$ to be evaluated using only p–v–T data pertaining to the phase change. In instances when the enthalpy change is also measured, the Clapeyron equation can be used to check the consistency of the data. Once the specific enthalpy change is determined, the corresponding changes in specific entropy and specific internal energy can be found from Eqs. 11.38 and 11.39, respectively.

Equations 11.38, 11.39, and 11.40 also can be written for sublimation or melting occurring at constant temperature and pressure. In particular, the Clapeyron equation would take the form

$$\left(\frac{dp}{dT}\right)_{sat} = \frac{h'' - h'}{T(v'' - v')} \qquad (11.41)$$

where $''$ and $'$ denote the respective phases, and $(dp/dT)_{sat}$ is the slope of the relevant saturation pressure–temperature curve.

The Clapeyron equation shows that the slope of a saturation line on a phase diagram depends on the signs of the specific volume and enthalpy changes accompanying the phase change. In most cases, when a phase change takes place with an increase in specific enthalpy, the specific volume also increases, and $(dp/dT)_{sat}$ is positive. However, in the case of the melting of ice and a few other substances, the specific volume decreases on melting. The slope of the saturated solid–liquid curve for these few substances is negative, as pointed out in Sec. 3.2.2 in the discussion of phase diagrams.

An approximate form of Eq. 11.40 can be derived when the following two idealizations are justified: (1) v_f is negligible in comparison to v_g, and (2) the pressure is low enough that v_g can be evaluated from the ideal gas equation of state as $v_g = RT/p$. With these, Eq. 11.40 becomes

$$\left(\frac{dp}{dT}\right)_{sat} = \frac{h_g - h_f}{RT^2/p}$$

which can be rearranged to read

$$\left(\frac{d \ln p}{dT}\right)_{sat} = \frac{h_g - h_f}{RT^2} \qquad (11.42)$$

Equation 11.42 is called the **Clausius–Clapeyron equation**. A similar expression applies for the case of sublimation.

Clausius–Clapeyron equation

The use of the Clapeyron equation in any of the foregoing forms requires an accurate representation for the relevant saturation pressure–temperature curve. This must not only depict the pressure–temperature variation accurately but also enable accurate values of the derivative $(dp/dT)_{sat}$ to be determined. Analytical representations in the form of equations are commonly used. Different equations for different portions of the pressure–temperature curves may be required. These equations can involve several constants. One form that is used for the vapor–pressure curves is the four-constant equation

$$\ln p_{sat} = A + \frac{B}{T} + C \ln T + DT$$

in which the constants A, B, C, D are determined empirically.

The use of the Clapeyron equation for evaluating changes in specific entropy, internal energy, and enthalpy accompanying a phase change at fixed T and p is illustrated in the next example.

▶▶▶ **EXAMPLE 11.4** ▶ ·

Applying the Clapeyron Equation

Using p–υ–T data for saturated water, calculate at 100°C **(a)** $h_g - h_f$, **(b)** $u_g - u_f$, **(c)** $s_g - s_f$. Compare with the respective steam table value.

SOLUTION

Known   The system consists of a unit mass of saturated water at 100°C.

Find   Using saturation data, determine at 100°C the change on vaporization of the specific enthalpy, specific internal energy, and specific entropy, and compare with the respective steam table value.

Analysis   For comparison, Table A-2 gives at 100°C, $h_g - h_f =$ 2257.0 kJ/kg, $u_g - u_f = 2087.6$ kJ/kg, $s_g - s_f = 6.048$ kJ/kg · K.

a. The value of $h_g - h_f$ can be determined from the Clapeyron equation, Eq. 11.40, expressed as

$$h_g - h_f = T(\upsilon_g - \upsilon_f)\left(\frac{dp}{dT}\right)_{sat}$$

This equation requires a value for the slope $(dp/dT)_{sat}$ of the saturation pressure–temperature curve at the specified temperature.

The required value for $(dp/dT)_{sat}$ at 100°C can be estimated graphically as follows. Using saturation pressure–temperature data from the steam tables, the accompanying plot can be prepared. Note that a line drawn tangent to the curve at 100°C is shown on the plot. The slope of this tangent line is about 3570 N/m² · K. Accordingly, at 100°C

$$\left(\frac{dp}{dT}\right)_{sat} \approx 3570 \frac{N}{m^2 \cdot K}$$

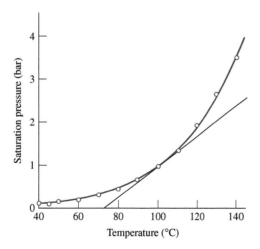

Fig. E11.4

Inserting property data from Table A-2 into the equation for $h_g - h_f$ gives

$$h_g - h_f = (373.15 \text{ K})(1.673 - 1.0435 \times 10^{-3})\left(\frac{m^3}{kg}\right)$$
$$\times \left(3570 \frac{N}{m^2 \cdot K}\right)\left|\frac{1 \text{ kJ}}{10^3 \text{ N} \cdot m}\right|$$
$$= 2227 \text{ kJ/kg}$$

This value is about 1% less than the value read from the steam tables.
① Alternatively, the derivative $(dp/dT)_{sat}$ can be estimated using numerical methods and computer-generated data. The following IT code illustrates *one way* the derivative, denoted dpdT, can be estimated:

```
T = 100 // °C
dT = 0.001
T1 = T - dT
T2 = T + dT
p1 = Psat ("Water/Steam", T1) // bar
p2 = Psat ("Water/Steam", T2) // bar
dpdT = ((p2 - p1) / (T2 - T1)) * 100000
```

Using the **Explore** button, sweep dT from 0.001 to 0.01 in steps of 0.001. Then, reading the limiting value from the computer data

$$\left(\frac{dp}{dT}\right)_{sat} \approx 3616 \frac{N}{m^2 \cdot K}$$

When this value is used in the above expression for $h_g - h_f$, the result is $h_g - h_f = 2256$ kJ/kg, which agrees closely with the value read from the steam tables.

b. With Eq. 11.39

$$u_g - u_f = h_g - h_f - p_{sat}(\upsilon_g - \upsilon_f)$$

Inserting the IT result for $(h_g - h_f)$ from part (a) together with saturation data at 100°C from Table A-2

$$u_g - u_f = 2256 \frac{kJ}{kg} - \left(1.014 \times 10^5 \frac{N}{m^2}\right)\left(1.672 \frac{m^3}{kg}\right)\left|\frac{1 \text{ kJ}}{10^3 \text{ N} \cdot m}\right|$$
$$= 2086.5 \frac{kJ}{kg}$$

which also agrees closely with the value from the steam tables.

c. With Eq. 11.38 and the *IT* result for $(h_g - h_f)$ from part (a)

$$s_g - s_f = \frac{h_g - h_f}{T} = \frac{2256 \text{ kJ/kg}}{373.15 \text{ K}} = 6.046 \frac{\text{kJ}}{\text{kg} \cdot \text{K}}$$

which again agrees closely with the steam table value.

 Also, $(dp/dT)_{\text{sat}}$ might be obtained by differentiating an analytical expression for the vapor pressure curve, as discussed immediately above the introduction of this example.

Quick Quiz

Use the *IT* result $(dp/dT)_{\text{sat}} = 3616 \text{ N/m}^2 \cdot \text{K}$ to extrapolate the saturation pressure, in bar, at 105°C. Ans. 1.195 bar.

11.4.2 Considering Single-Phase Regions

The objective of the present section is to derive expressions for evaluating Δs, Δu, and Δh between states in single-phase regions. These expressions require both p–v–T data and appropriate specific heat data. Since single-phase regions are under present consideration, any two of the properties pressure, specific volume, and temperature can be regarded as the independent properties that fix the state. Two convenient choices are T, v and T, p.

T and v as Independent Properties With temperature and specific volume as the independent properties that fix the state, the specific entropy can be regarded as a function of the form $s = s(T, v)$. The differential of this function is

$$ds = \left(\frac{\partial s}{\partial T}\right)_v dT + \left(\frac{\partial s}{\partial v}\right)_T dv$$

The partial derivative $(\partial s/\partial v)_T$ appearing in this expression can be replaced using the Maxwell relation, Eq. 11.34, giving

$$ds = \left(\frac{\partial s}{\partial T}\right)_v dT + \left(\frac{\partial p}{\partial T}\right)_v dv \tag{11.43}$$

The specific internal energy also can be regarded as a function of T and v: $u = u(T, v)$. The differential of this function is

$$du = \left(\frac{\partial u}{\partial T}\right)_v dT + \left(\frac{\partial u}{\partial v}\right)_T dv$$

With $c_v = (\partial u/\partial T)_v$

$$du = c_v\, dT + \left(\frac{\partial u}{\partial v}\right)_T dv \tag{11.44}$$

Substituting Eqs. 11.43 and 11.44 into $du = T\, ds - p\, dv$ and collecting terms results in

$$\left[\left(\frac{\partial u}{\partial v}\right)_T + p - T\left(\frac{\partial p}{\partial T}\right)_v\right] dv = \left[T\left(\frac{\partial s}{\partial T}\right)_v - c_v\right] dT \tag{11.45}$$

Since specific volume and temperature can be varied independently, let us hold specific volume constant and vary temperature. That is, let $dv = 0$ and $dT \neq 0$. It then follows from Eq. 11.45 that

$$\boxed{\left(\frac{\partial s}{\partial T}\right)_v = \frac{c_v}{T}} \tag{11.46}$$

Similarly, suppose that $dT = 0$ and $dv \neq 0$. It then follows that

$$\left(\frac{\partial u}{\partial v}\right)_T = T\left(\frac{\partial p}{\partial T}\right)_v - p \qquad (11.47)$$

Equations 11.46 and 11.47 are additional examples of useful thermodynamic property relations.

FOR EXAMPLE

Equation 11.47, which expresses the dependence of the specific internal energy on specific volume at fixed temperature, allows us to demonstrate that the internal energy of a gas whose equation of state is $pv = RT$ depends on temperature alone, a result first discussed in Sec. 3.12.2. Equation 11.47 requires the partial derivative $(\partial p/\partial T)_v$. If $p = RT/v$, the derivative is $(\partial p/\partial T)_v = R/v$. Introducing this, Eq. 11.47 gives

$$\left(\frac{\partial u}{\partial v}\right)_T = T\left(\frac{\partial p}{\partial T}\right)_v - p = T\left(\frac{R}{v}\right) - p = p - p = 0$$

This demonstrates that when $pv = RT$, the specific internal energy is independent of specific volume and depends on temperature alone.

Continuing the discussion, when Eq. 11.46 is inserted in Eq. 11.43, the following expression results.

$$ds = \frac{c_v}{T}dT + \left(\frac{\partial p}{\partial T}\right)_v dv \qquad (11.48)$$

Inserting Eq. 11.47 into Eq. 11.44 gives

$$du = c_v\, dT + \left[T\left(\frac{\partial p}{\partial T}\right)_v - p\right]dv \qquad (11.49)$$

Observe that the right sides of Eqs. 11.48 and 11.49 are expressed solely in terms of p, v, T, and c_v.

Changes in specific entropy and internal energy between two states are determined by integration of Eqs. 11.48 and 11.49, respectively.

$$s_2 - s_1 = \int_1^2 \frac{c_v}{T}dT + \int_1^2 \left(\frac{\partial p}{\partial T}\right)_v dv \qquad (11.50)$$

$$u_2 - u_1 = \int_1^2 c_v\, dT + \int_1^2 \left[T\left(\frac{\partial p}{\partial T}\right)_v - p\right]dv \qquad (11.51)$$

To integrate the first term on the right of each of these expressions, the variation of c_v with temperature at one fixed specific volume (isometric) is required. Integration of the second term requires knowledge of the p–v–T relation at the states of interest. An equation of state explicit in pressure would be particularly convenient for evaluating the integrals involving $(\partial p/\partial T)_v$. The accuracy of the resulting specific entropy and internal energy changes would depend on the accuracy of this derivative. In cases where the integrands of Eqs. 11.50 and 11.51 are too complicated to be integrated in closed form they may be evaluated numerically. Whether closed-form or numerical integration is used, attention must be given to the path of integration.

FOR EXAMPLE

Let us consider the evaluation of Eq. 11.51. Referring to Fig. 11.2, if the specific heat c_v is known as a function of temperature along the isometric (constant specific volume) passing through the states x and y, one possible path of integration for determining the change in specific internal energy between states 1 and 2 is 1–x–y–2. The integration would be performed in three steps. Since temperature is constant from state 1 to state x, the first integral of Eq. 11.51 vanishes, so

$$u_x - u_1 = \int_{v_1}^{v_x} \left[T \left(\frac{\partial p}{\partial T} \right)_v - p \right] dv$$

From state x to y, the specific volume is constant and c_v is known as a function of temperature only, so

$$u_y - u_x = \int_{T_x}^{T_y} c_v \, dT$$

where $T_x = T_1$ and $T_y = T_2$. From state y to state 2, the temperature is constant once again, and

$$u_2 - u_y = \int_{v_y = v_x}^{v_2} \left[T \left(\frac{\partial p}{\partial T} \right)_v - p \right] dv$$

When these are added, the result is the change in specific internal energy between states 1 and 2.

T and p as Independent Properties   In this section a presentation parallel to that considered above is provided for the choice of temperature and pressure as the independent properties. With this choice for the independent properties, the specific entropy can be regarded as a function of the form $s = s(T, p)$. The differential of this function is

$$ds = \left(\frac{\partial s}{\partial T} \right)_p dT + \left(\frac{\partial s}{\partial p} \right)_T dp$$

The partial derivative $(\partial s/\partial p)_T$ appearing in this expression can be replaced using the Maxwell relation, Eq. 11.35, giving

$$ds = \left(\frac{\partial s}{\partial T} \right)_p dT - \left(\frac{\partial v}{\partial T} \right)_p dp \tag{11.52}$$

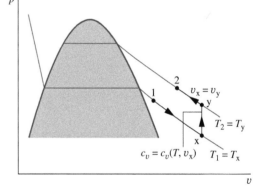

Fig. 11.2 Integration path between two vapor states.

The specific enthalpy also can be regarded as a function of T and p: $h = h(T, p)$. The differential of this function is

$$dh = \left(\frac{\partial h}{\partial T} \right)_p dT + \left(\frac{\partial h}{\partial p} \right)_T dp$$

With $c_p = (\partial h/\partial T)_p$

$$dh = c_p \, dT + \left(\frac{\partial h}{\partial p} \right)_T dp \tag{11.53}$$

Substituting Eqs. 11.52 and 11.53 into $dh = T \, ds + v \, dp$ and collecting terms results in

$$\left[\left(\frac{\partial h}{\partial p} \right)_T + T \left(\frac{\partial v}{\partial T} \right)_p - v \right] dp = \left[T \left(\frac{\partial s}{\partial T} \right)_p - c_p \right] dT \tag{11.54}$$

Since pressure and temperature can be varied independently, let us hold pressure constant and vary temperature. That is, let $dp = 0$ and $dT \neq 0$. It then follows from Eq. 11.54 that

$$\left(\frac{\partial s}{\partial T} \right)_p = \frac{c_p}{T} \qquad (11.55)$$

Similarly, when $dT = 0$ and $dp \neq 0$, Eq. 11.54 gives

$$\left(\frac{\partial h}{\partial p} \right)_T = v - T \left(\frac{\partial v}{\partial T} \right)_p \qquad (11.56)$$

Equations 11.55 and 11.56, like Eqs. 11.46 and 11.47, are useful thermodynamic property relations.

When Eq. 11.55 is inserted in Eq. 11.52, the following equation results:

$$ds = \frac{c_p}{T} dT - \left(\frac{\partial v}{\partial T} \right)_p dp \qquad (11.57)$$

Introducing Eq. 11.56 into Eq. 11.53 gives

$$dh = c_p dT + \left[v - T \left(\frac{\partial v}{\partial T} \right)_p \right] dp \qquad (11.58)$$

Observe that the right sides of Eqs. 11.57 and 11.58 are expressed solely in terms of p, v, T, and c_p.

Changes in specific entropy and enthalpy between two states are found by integrating Eqs. 11.57 and 11.58, respectively,

$$s_2 - s_1 = \int_1^2 \frac{c_p}{T} dT - \int_1^2 \left(\frac{\partial v}{\partial T} \right)_p dp \qquad (11.59)$$

$$h_2 - h_1 = \int_1^2 c_p \, dT + \int_1^2 \left[v - T \left(\frac{\partial v}{\partial T} \right)_p \right] dp \qquad (11.60)$$

To integrate the first term on the right of each of these expressions, the variation of c_p with temperature at one fixed pressure (isobar) is required. Integration of the second term requires knowledge of the p–v–T behavior at the states of interest. An equation of state explicit in v would be particularly convenient for evaluating the integrals involving $(\partial v/\partial T)_p$. The accuracy of the resulting specific entropy and enthalpy changes would depend on the accuracy of this derivative.

Changes in specific enthalpy and internal energy are related through $h = u + pv$ by

$$h_2 - h_1 = (u_2 - u_1) + (p_2 v_2 - p_1 v_1) \qquad (11.61)$$

Hence, only one of Δh and Δu need be found by integration. Then, the other can be evaluated from Eq. 11.61. Which of the two property changes is found by integration depends on the information available. Δh would be found using Eq. 11.60 when an equation of state explicit in v and c_p as a function of temperature at some fixed pressure are known. Δu would be found from Eq. 11.51 when an equation of state explicit in p and c_v as a function of temperature at some specific volume are known. Such issues are considered in Example 11.5.

Evaluating Δs, Δu, and Δh of a Gas

Using the Redlich–Kwong equation of state, develop expressions for the changes in specific entropy, internal energy, and enthalpy of a gas between two states where the temperature is the same, $T_1 = T_2$, and the pressures are p_1 and p_2, respectively.

SOLUTION

Known Two states of a unit mass of a gas as the system are fixed by p_1 and T_1 at state 1 and p_2, T_2 ($= T_1$) at state 2.

Find Determine the changes in specific entropy, internal energy, and enthalpy between these two states.

Schematic and Given Data:

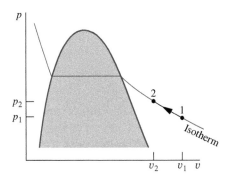

Fig. E11.5

Engineering Model The Redlich–Kwong equation of state represents the p–v–T behavior at these states and yields accurate values for $(\partial p/\partial T)_v$.

Analysis The Redlich–Kwong equation of state is explicit in pressure, so Eqs. 11.50 and 11.51 are selected for determining $s_2 - s_1$ and $u_2 - u_1$. Since $T_1 = T_2$, an isothermal path of integration between the two states is convenient. Thus, these equations reduce to give

$$s_2 - s_1 = \int_1^2 \left(\frac{\partial p}{\partial T}\right)_v dv$$

$$u_2 - u_1 = \int_1^2 \left[T\left(\frac{\partial p}{\partial T}\right)_v - p\right] dv$$

The limits for each of the foregoing integrals are the specific volumes v_1 and v_2 at the two states under consideration. Using p_1, p_2, and the known temperature, these specific volumes would be determined from the Redlich–Kwong equation of state. Since this equation is not explicit in specific volume, the use of an equation solver such as *Interactive Thermodynamics: IT* is recommended.

The above integrals involve the partial derivative $(\partial p/\partial T)_v$, which can be determined from the Redlich–Kwong equation of state as

$$\left(\frac{\partial p}{\partial T}\right)_v = \frac{R}{v - b} + \frac{a}{2v(v + b)T^{3/2}}$$

Inserting this into the expression for $(s_2 - s_1)$ gives

$$s_2 - s_1 = \int_{v_1}^{v_2} \left[\frac{R}{v - b} + \frac{a}{2v(v + b)T^{3/2}}\right] dv$$

$$= \int_{v_1}^{v_2} \left[\frac{R}{v - b} + \frac{a}{2bT^{3/2}}\left(\frac{1}{v} - \frac{1}{v + b}\right)\right] dv$$

$$= R \ln\left(\frac{v_2 - b}{v_1 - b}\right) + \frac{a}{2bT^{3/2}}\left[\ln\left(\frac{v_2}{v_1}\right) - \ln\left(\frac{v_2 + b}{v_1 + b}\right)\right]$$

$$= R \ln\left(\frac{v_2 - b}{v_1 - b}\right) + \frac{a}{2bT^{3/2}} \ln\left[\frac{v_2(v_1 + b)}{v_1(v_2 + b)}\right]$$

With the Redlich–Kwong equation, the integrand of the expression for $(u_2 - u_1)$ becomes

$$\left[T\left(\frac{\partial p}{\partial T}\right)_v - p\right] = T\left[\frac{R}{v - b} + \frac{a}{2v(v + b)T^{3/2}}\right]$$
$$- \left[\frac{RT}{v - b} - \frac{a}{v(v + b)T^{1/2}}\right]$$
$$= \frac{3a}{2v(v + b)T^{1/2}}$$

Accordingly

$$u_2 - u_1 = \int_{v_1}^{v_2} \frac{3a}{2v(v + b)T^{1/2}} dv$$

$$= \frac{3a}{2bT^{1/2}} \int_{v_1}^{v_2} \left(\frac{1}{v} - \frac{1}{v + b}\right) dv$$

$$= \frac{3a}{2bT^{1/2}} \left[\ln\frac{v_2}{v_1} - \ln\left(\frac{v_2 + b}{v_1 + b}\right)\right]$$

$$= \frac{3a}{2bT^{1/2}} \ln\left[\frac{v_2(v_1 + b)}{v_1(v_2 + b)}\right]$$

Finally, $(h_2 - h_1)$ would be determined using Eq. 11.61 together with the known values of $(u_2 - u_1)$, p_1, v_1, p_2, and v_2.

SKILLS DEVELOPED

Ability to...

• perform differentiations and integrations required to evaluate Δu and Δs using the two-constant Redlich–Kwong equation of state.

Quick Quiz

Using results obtained, develop expressions for Δu and Δs of an ideal gas. Ans. $\Delta u = 0$, $\Delta s = R \ln (v_2/v_1)$.

11.5 Other Thermodynamic Relations

The presentation to this point has been directed mainly at developing thermodynamic relations that allow changes in u, h, and s to be evaluated from measured property data. The objective of the present section is to introduce several other thermodynamic relations that are useful for thermodynamic analysis. Each of the properties considered has a common attribute: It is defined in terms of a partial derivative of some other property. The specific heats c_v and c_p are examples of this type of property.

11.5.1 Volume Expansivity, Isothermal and Isentropic Compressibility

In single-phase regions, pressure and temperature are independent, and we can think of the specific volume as being a function of these two, $v = v(T, p)$. The differential of such a function is

$$dv = \left(\frac{\partial v}{\partial T}\right)_p dT + \left(\frac{\partial v}{\partial p}\right)_T dp$$

volume expansivity   Two thermodynamic properties related to the partial derivatives appearing in this differential are the **volume expansivity** β, also called the *coefficient of volume expansion*

$$\boxed{\beta = \frac{1}{v}\left(\frac{\partial v}{\partial T}\right)_p} \tag{11.62}$$

isothermal compressibility   and the **isothermal compressibility** κ

$$\boxed{\kappa = -\frac{1}{v}\left(\frac{\partial v}{\partial p}\right)_T} \tag{11.63}$$

By inspection, the unit for β is seen to be the reciprocal of that for temperature and the unit for κ is the reciprocal of that for pressure. The volume expansivity is an indication of the change in volume that occurs when temperature changes while pressure remains constant. The isothermal compressibility is an indication of the change in volume that takes place when pressure changes while temperature remains constant. The value of κ is positive for all substances in all phases.

The volume expansivity and isothermal compressibility are thermodynamic properties, and like specific volume are functions of T and p. Values for β and κ are provided in compilations of engineering data. **Table 11.2** gives values of these properties for liquid water at a pressure of 1 atm versus temperature. For a pressure of 1 atm, water has a *state of maximum density* at about 4°C. At this state, the value of β is zero.

isentropic compressibility   The **isentropic compressibility** α is an indication of the change in volume that occurs when pressure changes while entropy remains constant:

$$\boxed{\alpha = -\frac{1}{v}\left(\frac{\partial v}{\partial p}\right)_s} \tag{11.64}$$

The unit for α is the reciprocal of that for pressure.

Volume Expansivity β and Isothermal Compressibility κ of Liquid Water at 1 atm Versus Temperature

T (°C)	Density (kg/m³)	$\beta \times 10^6$ (K)$^{-1}$	$\kappa \times 10^6$ (bar)$^{-1}$
0	999.84	-68.14	50.89
10	999.70	87.90	47.81
20	998.21	206.6	45.90
30	995.65	303.1	44.77
40	992.22	385.4	44.24
50	988.04	457.8	44.18

The isentropic compressibility is related to the speed at which sound travels in a substance, and such speed measurements can be used to determine α. In Sec. 9.12.2, the **velocity of sound**, or *sonic velocity*, is introduced as

velocity of sound

$$c = \sqrt{-v^2 \left(\frac{\partial p}{\partial v}\right)_s}$$

(9.36b)

The relationship of the isentropic compressibility and the velocity of sound can be obtained using the relation between partial derivatives expressed by Eq. 11.15. Identifying p with x, v with y, and s with z, we have

$$\left(\frac{\partial p}{\partial v}\right)_s = \frac{1}{(\partial v/\partial p)_s}$$

TAKE NOTE...

Through the ***Mach number***, the sonic velocity c plays an important role in analyzing flow in nozzles and diffusers. See Sec. 9.13.

With this, the previous two equations can be combined to give

$$c = \sqrt{v/\alpha}$$

(11.65)

The details are left as an exercise.

BioConnections

The propagation of elastic waves, such as sound waves, has important implications related to injury in living things. During impact such as a collision in a sporting event (see accompanying figure), elastic waves are created that cause some bodily material to move relative to the rest of the body. The waves can propagate at supersonic, transonic, or subsonic speeds depending on the nature of the impact, and the resulting trauma can cause serious damage. The waves may be focused into a small area, causing localized damage, or they may be reflected at the boundary of organs and cause more widespread damage.

An example of the focusing of waves occurs in some head injuries. An impact to the skull causes flexural and compression waves to move along the curved surface and arrive at the far side of the skull simultaneously. Waves also propagate through the softer brain tissue. Consequently, concussions, skull fractures, and other injuries can appear at locations away from the site of the original impact.

Central to an understanding of traumatic injury is data on speed of sound and other elastic characteristics of organs and tissues. For humans the speed of sound varies widely, from approximately 30–45 m/s in spongy lung tissue to about 1600 m/s in muscle and 3500 m/s in bone. Because the speed of sound in the lungs is relatively low, impacts such as in automobile collisions or even air-bag deployment can set up waves that propagate supersonically. Medical personnel responding to traumas are trained to check for lung injuries.

The study of wave phenomena in the body constitutes an important area in the field of *biomechanics*.

11.5.2 Relations Involving Specific Heats

In this section, general relations are obtained for the difference between specific heats $(c_p - c_v)$ and the ratio of specific heats c_p/c_v.

Evaluating $(c_p - c_v)$ An expression for the difference between c_p and c_v can be obtained by equating the two differentials for entropy given by Eqs. 11.48 and 11.57 and rearranging to obtain

$$(c_p - c_v)dT = T\left(\frac{\partial p}{\partial T}\right)_v dv + T\left(\frac{\partial v}{\partial T}\right)_p dp$$

Considering the equation of state $p = p(T, v)$, the differential dp can be expressed as

$$dp = \frac{\partial p}{\partial T}\Big)_v dT + \frac{\partial p}{\partial v}\Big)_T dv$$

Eliminating dp between the last two equations and collecting terms gives

$$\left[(c_p - c_v) - T\left(\frac{\partial v}{\partial T}\right)_p\left(\frac{\partial p}{\partial T}\right)_v\right]dT = T\left[\left(\frac{\partial v}{\partial T}\right)_p\left(\frac{\partial p}{\partial v}\right)_T + \left(\frac{\partial p}{\partial T}\right)_v\right]dv$$

Since temperature and specific volume can be varied independently, the coefficients of the differentials in this expression must vanish, so

$$c_p - c_v = T\left(\frac{\partial v}{\partial T}\right)_p\left(\frac{\partial p}{\partial T}\right)_v \tag{11.66}$$

$$\left(\frac{\partial p}{\partial T}\right)_v = -\left(\frac{\partial v}{\partial T}\right)_p\left(\frac{\partial p}{\partial v}\right)_T \tag{11.67}$$

Introducing Eq. 11.67 into Eq. 11.66 gives

$$c_p - c_v = -T\left(\frac{\partial v}{\partial T}\right)_p^2\left(\frac{\partial p}{\partial v}\right)_T \tag{11.68}$$

This equation allows c_v to be calculated from observed values of c_p knowing only p–v–T data, or c_p to be calculated from observed values of c_v.

> **FOR EXAMPLE**
>
> For the special case of an ideal gas, Eq. 11.68 reduces to Eq. 3.44: $c_p(T) = c_v(T) + R$, as can readily be shown.

The right side of Eq. 11.68 can be expressed in terms of the volume expansivity β and the isothermal compressibility κ. Introducing Eqs. 11.62 and 11.63, we get

$$c_p - c_v = v\frac{T\beta^2}{\kappa} \tag{11.69}$$

In developing this result, the relationship between partial derivatives expressed by Eq. 11.15 has been used.

Several important conclusions about the specific heats c_p and c_v can be drawn from Eq. 11.69.

FOR EXAMPLE

Since the factor β^2 cannot be negative and κ is positive for all substances in all phases, the value of c_p is always greater than, or equal to, c_v. The specific heats are equal when $\beta = 0$, as occurs in the case of water at 1 atmosphere and 4°C, where water is at its state of maximum density. The two specific heats also become equal as the temperature approaches absolute zero. For some liquids and solids at certain states, c_p and c_v differ only slightly. For this reason, tables often give the specific heat of a liquid or solid without specifying whether it is c_p or c_v. The data reported are normally c_p values, since these are more easily determined for liquids and solids.

Evaluating c_p/c_v   Next, let us obtain expressions for the ratio of specific heats, k. Employing Eq. 11.16, we can rewrite Eqs. 11.46 and 11.55, respectively, as

$$\frac{c_v}{T} = \left(\frac{\partial s}{\partial T}\right)_v = \frac{-1}{(\partial v/\partial s)_T (\partial T/\partial v)_s}$$

$$\frac{c_p}{T} = \left(\frac{\partial s}{\partial T}\right)_p = \frac{-1}{(\partial p/\partial s)_T (\partial T/\partial p)_s}$$

Forming the ratio of these equations gives

$$\frac{c_p}{c_v} = \frac{(\partial v/\partial s)_T (\partial T/\partial v)_s}{(\partial p/\partial s)_T (\partial T/\partial p)_s} \tag{11.70}$$

Since $(\partial s/\partial p)_T = 1/(\partial p/\partial s)_T$ and $(\partial p/\partial T)_s = 1/(\partial T/\partial p)_s$, Eq. 11.70 can be expressed as

$$\frac{c_p}{c_v} = \left[\left(\frac{\partial v}{\partial s}\right)_T \left(\frac{\partial s}{\partial p}\right)_T\right]\left[\left(\frac{\partial p}{\partial T}\right)_s \left(\frac{\partial T}{\partial v}\right)_s\right] \tag{11.71}$$

Finally, the chain rule from calculus allows us to write $(\partial v/\partial p)_T = (\partial v/\partial s)_T (\partial s/\partial p)_T$ and $(\partial p/\partial v)_s = (\partial p/\partial T)_s (\partial T/\partial v)_s$, so Eq. 11.71 becomes

$$k = \frac{c_p}{c_v} = \left(\frac{\partial v}{\partial p}\right)_T \left(\frac{\partial p}{\partial v}\right)_s \tag{11.72}$$

This can be expressed alternatively in terms of the isothermal and isentropic compressibilities as

$$k = \frac{\kappa}{\alpha} \tag{11.73}$$

Solving Eq. 11.72 for $(\partial p/\partial v)_s$ and substituting the resulting expression into Eq. 9.36b gives the following relationship involving the velocity of sound c and the specific heat ratio k

$$c = \sqrt{-kv^2(\partial p/\partial v)_T} \tag{11.74}$$

Equation 11.74 can be used to determine c knowing the specific heat ratio and p–v–T data, or to evaluate k knowing c and $(\partial p/\partial v)_T$.

FOR EXAMPLE

In the special case of an ideal gas, Eq. 11.74 reduces to give Eq. 9.37 (Sec. 9.12.2):

$$c = \sqrt{kRT} \qquad \text{(ideal gas)} \tag{9.37}$$

as can easily be verified.

In the next example we illustrate the use of specific heat relations introduced above.

Using Specific Heat Relations

For liquid water at 1 atm and 20°C, estimate **(a)** the percent error in c_v that would result if it were assumed that $c_p = c_v$, **(b)** the velocity of sound, in m/s.

SOLUTION

Known The system consists of a fixed amount of liquid water at 1 atm and 20°C.

Find Estimate the percent error in c_v that would result if c_v were approximated by c_p, and the velocity of sound, in m/s.

Analysis

a. Equation 11.69 gives the difference between c_p and c_v. Table 11.2 provides the required values for the volume expansivity β, the isothermal compressibility κ, and the specific volume. Thus,

$$c_p - c_v = v\frac{T\beta^2}{\kappa}$$

$$= \left(\frac{1}{998.21 \text{ kg/m}^3}\right)(293 \text{ K})$$

$$\times \left(\frac{206.6 \times 10^{-6}}{\text{K}}\right)^2 \left(\frac{\text{bar}}{45.90 \times 10^{-6}}\right)$$

$$= \left(272.96 \times 10^{-6} \frac{\text{bar} \cdot \text{m}^3}{\text{kg} \cdot \text{K}}\right)$$

$$\times \left|\frac{10^5 \text{ N/m}^2}{1 \text{ bar}}\right|\left|\frac{1 \text{ kJ}}{10^3 \text{ N} \cdot \text{m}}\right|$$

$$= 0.027 \frac{\text{kJ}}{\text{kg} \cdot \text{K}}$$

❶ Interpolating in Table A-19 at 20°C gives $c_p = 4.188$ kJ/kg · K Thus, the value of c_v is

$$c_v = 4.188 - 0.027 = 4.161 \text{ kJ/kg} \cdot \text{K}$$

Using these values, the percent error in approximating c_v by c_p is

❷ $$\left(\frac{c_p - c_v}{c_v}\right)(100) = \left(\frac{0.027}{4.161}\right)(100) = 0.6\%$$

b. The velocity of sound at this state can be determined using Eq. 11.65. The required value for the isentropic compressibility α is calculable in terms of the specific heat ratio k and the isothermal compressibility κ. With Eq. 11.73, $\alpha = \kappa/k$. Inserting this into Eq. 11.65 results in the following expression for the velocity of sound

$$c = \sqrt{\frac{kv}{\kappa}}$$

The values of v and κ required by this expression are the same as used in part (a). Also, with the values of c_p and c_v from part (a), the specific heat ratio is $k = 1.006$. Accordingly

❸ $$c = \sqrt{\frac{(1.006)(10^6) \text{ bar}}{(998.21 \text{ kg/m}^3)(45.90)}\left|\frac{10^5 \text{ N/m}^2}{1 \text{ bar}}\right|\left|\frac{1 \text{ kg} \cdot \text{m/s}^2}{1 \text{ N}}\right|}$$

$$= 1482 \text{ m/s}$$

---

❶ Consistent with the discussion of Sec. 3.10.1, we take c_p at 1 atm and 20°C as the saturated liquid value at 20°C.

❷ The result of part (a) shows that for liquid water at the given state, c_p and c_v are closely equal.

❸ For comparison, the velocity of sound in air at 1 atm, 20°C is about 343 m/s, which can be checked using Eq. 9.37.

SKILLS DEVELOPED

Ability to...

• apply specific heat relations to liquid water.
• evaluate velocity of sound for liquid water.

Quick Quiz

A submarine moves at a speed of 20 knots (1 knot = 1.852 km/h). Using the sonic velocity calculated in part (b), estimate the Mach number of the vessel relative to the water. Ans. 0.0069.

11.5.3 Joule–Thomson Coefficient

Joule–Thomson coefficient

The value of the specific heat c_p can be determined from p–v–T data and the Joule–Thomson coefficient. The **Joule–Thomson coefficient** μ_J is defined as

$$\mu_J = \left(\frac{\partial T}{\partial p}\right)_h \tag{11.75}$$

Like other partial differential coefficients introduced in this section, the Joule–Thomson coefficient is defined in terms of thermodynamic properties only and thus is itself a property. The units of μ_J are those of temperature divided by pressure.

A relationship between the specific heat c_p and the Joule–Thomson coefficient μ_J can be established by using Eq. 11.16 to write

$$\left(\frac{\partial T}{\partial p}\right)_h \left(\frac{\partial p}{\partial h}\right)_T \left(\frac{\partial h}{\partial T}\right)_p = -1$$

The first factor in this expression is the Joule–Thomson coefficient and the third is c_p. Thus,

$$c_p = \frac{-1}{\mu_J (\partial p/\partial h)_T}$$

With $(\partial h/\partial p)_T = 1/(\partial p/\partial h)_T$ from Eq. 11.15, this can be written as

$$c_p = -\frac{1}{\mu_J} \left(\frac{\partial h}{\partial p}\right)_T \tag{11.76}$$

The partial derivative $(\partial h/\partial p)_T$, called the *constant-temperature coefficient*, can be eliminated from Eq. 11.76 by use of Eq. 11.56. The following expression results:

$$c_p = \frac{1}{\mu_J}\left[T\left(\frac{\partial v}{\partial T}\right)_T - v\right] \tag{11.77}$$

Equation 11.77 allows the value of c_p at a state to be determined using p–v–T data and the value of the Joule–Thomson coefficient at that state. Let us consider next how the Joule–Thomson coefficient can be found experimentally.

Experimental Evaluation   The Joule–Thomson coefficient can be evaluated experimentally using an apparatus like that pictured in Fig. 11.3. Consider first Fig. 11.3a, which shows a porous plug through which a gas (or liquid) may pass. During operation at steady state, the gas enters the apparatus at a specified temperature T_1 and pressure p_1 and expands through the plug to a lower pressure p_2, which is controlled by an outlet valve. The temperature T_2 at the exit is measured. The apparatus is designed so that the gas undergoes a *throttling* process (Sec. 4.10) as it expands from 1 to 2. Accordingly, the exit state fixed by p_2 and T_2 has the same value for the specific enthalpy as at the inlet, $h_2 = h_1$. By progressively lowering the

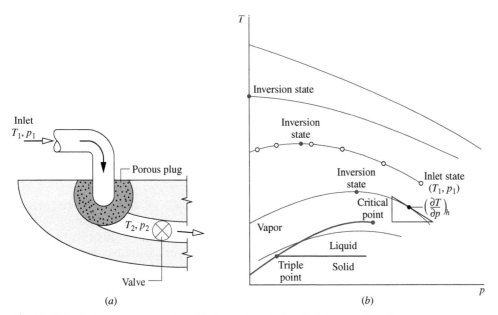

Fig. 11.3 Joule–Thomson expansion. (*a*) Apparatus. (*b*) Isenthalpics on a *T*–*p* diagram.

outlet pressure, a finite sequence of such exit states can be visited, as indicated on Fig. 11.3b. A curve may be drawn through the set of data points. Such a curve is called an isenthalpic (constant enthalpy) curve. An isenthalpic curve is the locus of all points representing equilibrium states of the same specific enthalpy.

inversion states

The slope of an isenthalpic curve at any state is the Joule–Thomson coefficient at that state. The slope may be positive, negative, or zero in value. States where the coefficient has a zero value are called **inversion states**. Notice that not all lines of constant h have an inversion state. The uppermost curve of Fig. 11.3b, for example, always has a negative slope. Throttling a gas from an initial state on this curve would result in an increase in temperature. However, for isenthalpic curves having an inversion state, the temperature at the exit of the apparatus may be greater than, equal to, or less than the initial temperature, depending on the exit pressure specified. For states to the right of an inversion state, the value of the Joule–Thomson coefficient is negative. For these states, the temperature increases as the pressure at the exit of the apparatus decreases. At states to the left of an inversion state, the value of the Joule–Thomson coefficient is positive. For these states, the temperature decreases as the pressure at the exit of the device decreases. This can be used to advantage in systems designed to *liquefy gases.*

Horizons

Small Power Plants Pack Punch

An innovation in power systems moving from concept to reality promises to help keep computer networks humming, hospital operating rooms lit, and shopping centers thriving. Called *distributed generation systems*, compact power plants provide electricity for small loads or are linked for larger applications. With distributed generation, consumers hope to avoid unpredictable price swings and brownouts.

Distributed generation includes a broad range of technologies that provides relatively small levels of power at sites close to users, including but not limited to internal combustion engines, microturbines, fuel cells, and photovoltaic systems.

Although the cost per kilowatt-hour may be higher with distributed generation, some customers are willing to pay more to gain control over their electric supply. Computer networks and hospitals need high reliability, since even short disruptions can be disastrous. Businesses such as shopping centers also must avoid costly service interruptions. With distributed generation, the needed reliability is provided by modular units that can be combined with power management and energy storage systems to ensure quality power is available when needed.

11.6 Constructing Tables of Thermodynamic Properties

The objective of this section is to utilize the thermodynamic relations introduced thus far to describe how tables of thermodynamic properties can be constructed. The characteristics of the tables under consideration are embodied in the tables for water and the refrigerants presented in the Appendix. The methods introduced in this section are extended in Chap. 13 for the analysis of reactive systems, such as gas turbine and vapor power systems involving combustion. The methods of this section also provide the basis for computer retrieval of thermodynamic property data.

Two different approaches for constructing property tables are considered:

- The presentation of Sec. 11.6.1 employs the methods introduced in Sec. 11.4 for assigning specific enthalpy, specific internal energy, and specific entropy to states of pure, simple compressible substances using p–v–T data, together with a limited amount of specific heat data. The principal mathematical operation of this approach is *integration.*

- The approach of Sec. 11.6.2 utilizes the fundamental thermodynamic function concept introduced in Sec. 11.3.3. Once such a function has been constructed, the principal mathematical operation required to determine all other properties is *differentiation.*

11.6.1 Developing Tables by Integration Using p–v–T and Specific Heat Data

In principle, all properties of present interest can be determined using

$$c_p = c_{p0}(T) \qquad (11.78)$$
$$p = p(v, T), \qquad v = v(p, T)$$

In Eqs. 11.78, $c_{p0}(T)$ is the specific heat c_p for the substance under consideration extrapolated to zero pressure. This function might be determined from data obtained calorimetrically or from spectroscopic data, using equations supplied by statistical mechanics. Specific heat expressions for several gases are given in Table A-21. The expressions $p(v, T)$ and $v(p, T)$ represent functions that describe the saturation pressure–temperature curves, as well as the p–v–T relations for the single-phase regions. These functions may be tabular, graphical, or analytical in character. Whatever their forms, however, the functions must not only represent the p–v–T data accurately but also yield accurate values for derivatives such as $(\partial v/\partial T)_p$ and $(dp/dT)_{sat}$.

Figure 11.4 shows eight states of a substance. Let us consider how values can be assigned to specific enthalpy and specific entropy at these states. The same procedures can be used to assign property values at other states of interest. Note that when h has been assigned to a state, the specific internal energy at that state can be found from $u = h - pv$.

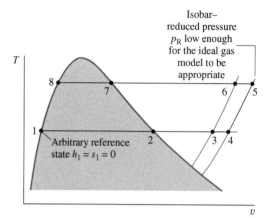

Fig. 11.4 T–v diagram used to discuss how h and s can be assigned to liquid and vapor states.

- Let the state denoted by 1 on Fig. 11.4 be selected as the reference state for enthalpy and entropy. Any value can be assigned to h and s at this state, but a value of zero is usual. It should be noted that the use of an arbitrary reference state and arbitrary reference values for specific enthalpy and specific entropy suffices only for evaluations involving differences in property values between states of the same composition, for then datums cancel.

- Once a value is assigned to enthalpy at state 1, the enthalpy at the saturated vapor state, state 2, can be determined using the Clapeyron equation, Eq. 11.40

TAKE NOTE...

See Sec. 3.6.3 for a discussion of reference states and reference values for Tables A-2 through A-18.

$$h_2 - h_1 = T_1(v_2 - v_1)\left(\frac{dp}{dT}\right)_{sat}$$

where the derivative $(dp/dT)_{sat}$ and the specific volumes v_1 and v_2 are obtained from appropriate representations of the p–v–T data for the substance under consideration. The specific entropy at state 2 is found using Eq. 11.38 in the form

$$s_2 - s_1 = \frac{h_2 - h_1}{T_1}$$

- Proceeding at constant temperature from state 2 to state 3, the entropy and enthalpy are found by means of Eqs. 11.59 and 11.60, respectively. Since temperature is fixed, these equations reduce to give

$$s_3 - s_2 = -\int_{p_2}^{p_3}\left(\frac{\partial v}{\partial T}\right)_p dp \qquad \text{and} \qquad h_3 - h_2 = \int_{p_2}^{p_3}\left[v - T\left(\frac{\partial v}{\partial T}\right)_p\right]dp$$

With the same procedure, s_4 and h_4 can be determined.

- The isobar (constant-pressure line) passing through state 4 is assumed to be at a low enough pressure for the ideal gas model to be appropriate. Accordingly, to evaluate s and h at states such as 5 on this isobar, the only required information would be $c_{p0}(T)$ and the temperatures at these states. Thus, since pressure is fixed, Eqs. 11.59 and 11.60 give, respectively

$$s_5 - s_4 = \int_{T_4}^{T_5} c_{p0} \frac{dT}{T} \quad \text{and} \quad h_5 - h_4 = \int_{T_4}^{T_5} c_{p0}\, dT$$

- Specific entropy and enthalpy values at states 6 and 7 are found from those at state 5 by the same procedure used in assigning values at states 3 and 4 from those at state 2. Finally, s_8 and h_8 are obtained from the values at state 7 using the Clapeyron equation.

11.6.2 Developing Tables by Differentiating a Fundamental Thermodynamic Function

Property tables also can be developed using a fundamental thermodynamic function. It is convenient for this purpose to select the independent variables of the fundamental function from among pressure, specific volume (density), and temperature. This indicates the use of the Helmholtz function $\psi(T, v)$ or the Gibbs function $g(T, p)$. The properties of water tabulated in Tables A-2 through A-6 have been calculated using the Helmholtz function. Fundamental thermodynamic functions also have been employed successfully to evaluate the properties of other substances in the Appendix tables.

The development of a fundamental thermodynamic function requires considerable mathematical manipulation and numerical evaluation. Prior to the advent of high-speed computers, the evaluation of properties by this means was not feasible, and the approach described in Sec. 11.6.1 was used exclusively. The fundamental function approach involves three steps:

1. The first step is the selection of a functional form in terms of the appropriate pair of independent properties and a set of adjustable coefficients, which may number 50 or more. The functional form is specified on the basis of both theoretical and practical considerations.

2. Next, the coefficients in the fundamental function are determined by requiring that a set of carefully selected property values and/or observed conditions be satisfied in a least-squares sense. This generally involves the use of property data requiring the assumed functional form to be differentiated one or more times, such as p–v–T and specific heat data.

3. When all coefficients have been evaluated, the function is carefully tested for accuracy by using it to evaluate properties for which accepted values are known. These may include properties requiring differentiation of the fundamental function two or more times. For example, velocity of sound and Joule–Thomson data might be used.

This procedure for developing a fundamental thermodynamic function is not routine and can be accomplished only with a computer. However, once a suitable fundamental function is established, extreme accuracy in and consistency among the thermodynamic properties is possible.

The form of the Helmholtz function used in constructing the steam tables from which Tables A-2 through A-6 have been extracted is

$$\psi(\rho, T) = \psi_0(T) + RT[\ln \rho + \rho Q(\rho, \tau)] \tag{11.79}$$

where ψ_0 and Q are given as the sums listed in **Table 11.3**. The independent variables are density and temperature. The variable τ denotes $1000/T$. Values for pressure, specific internal energy, and specific entropy can be determined by differentiation of Eq. 11.79. Values for the specific enthalpy and Gibbs function are found from $h = u + pv$ and $g = \psi + pv$, respectively. The specific heat c_v is evaluated by further differentiation, $c_v = (\partial v/\partial T)_v$. With similar operations, other properties can be evaluated. Property values for water calculated from Eq. 11.79 are in excellent agreement with experimental data over a wide range of conditions.

TABLE 11.3

Fundamental Equation Used to Construct the Steam Tables[a,b]

$$\psi = \psi_0(T) + RT[\ln \rho + \rho Q(\rho, \tau)] \tag{1}$$

where

$$\psi_0 = \sum_{i=1}^{6} C_i/\tau^{i-1} + C_7 \ln T + C_8 \ln T/\tau \tag{2}$$

and

$$Q = (\tau - \tau_c)\sum_{j=1}^{7}(\tau - \tau_{aj})^{j-2}\left[\sum_{i=1}^{8} A_{ij}(\rho - \rho_{aj})^{i-1} + e^{-E\rho}\sum_{i=9}^{10} A_{ij}\rho^{i-9}\right] \tag{3}$$

In (1), (2), and (3), T denotes temperature on the Kelvin scale, τ denotes $1000/T$, ρ denotes density in g/cm^3, $R = 4.6151$ bar $\cdot$ $cm^3/g \cdot$ K or 0.46151 J/g $\cdot$ K, $\tau_c = 1000/T_c = 1.544912$, $E = 4.8$, and

$$\tau_{aj} = \tau_c(j = 1) \qquad \rho_{aj} = 0.634(j = 1)$$
$$= 2.5(j > 1) \qquad\quad = 1.0(j > 1)$$

The coefficients for ψ_0 in J/g are given as follows:

$C_1 = 1857.065$	$C_4 = 36.6649$	$C_7 = 46.0$
$C_2 = 3229.12$	$C_5 = -20.5516$	$C_8 = -1011.249$
$C_3 = -419.465$	$C_6 = 4.85233$	

Values for the coefficients A_{ij} are listed in the original source.[a]

[a] J. H. Keenan, F. G. Keyes, P. G. Hill, and J. G. Moore, *Steam Tables,* Wiley, New York, 1969.

[b] Also see L. Haar, J. S. Gallagher, and G. S. Kell, *NBS/NRC Steam Tables,* Hemisphere, Washington, D.C., 1984. The properties of water are determined in this reference using a different functional form for the Helmholtz function than given by Eqs. (1)–(3).

Energy & Environment

Due to the phase-out of refrigerants that contribute significantly to global climate change, new substances and mixtures with lower Global Warming Potential are being developed as alternatives (see Sec. 10.3). This has led to significant research efforts to provide the necessary thermodynamic property data for analysis and design.

The National Institute of Standards and Technology (NIST) has led governmental efforts to provide accurate data. Specifically, data have been developed for high-accuracy p–v–T equations of state from which fundamental thermodynamic functions can be obtained. The equations are carefully validated using data for velocity of sound, Joule–Thomson coefficient, saturation pressure–temperature relations, and specific heats. Such data were used to calculate the property values in Tables A-7 to A-18 in the Appendix. NIST has developed REFPROP, a computer database that is the current standard for refrigerant and refrigerant mixture properties.

Example 11.7 illustrates the use of a fundamental function to determine thermodynamic properties for computer evaluation and to develop tables.

▶ EXAMPLE 11.7 ▶

Determining Properties Using a Fundamental Function

The following expression for the Helmholtz function has been used to determine the properties of water:

$$\psi(\rho, T) = \psi_0(T) + RT[\ln \rho + \rho Q(\rho, \tau)]$$

where ρ denotes density and τ denotes $1000/T$. The functions ψ_0 and Q are sums involving the indicated independent variables and a number of adjustable constants (see Table 11.3). Obtain expressions for **(a)** pressure, **(b)** specific entropy, and **(c)** specific internal energy resulting from this fundamental thermodynamic function.

SOLUTION

Known   An expression for the Helmholtz function ψ is given.

Find   Determine the expressions for pressure, specific entropy, and specific internal energy that result from this fundamental thermodynamic function.

Analysis   The expressions developed below for p, s, and u require only the functions $\psi_0(T)$ and $Q(\rho, \tau)$. Once these functions are determined, p, s, and u can each be determined as a function of density and temperature using elementary mathematical operations.

a. When expressed in terms of density instead of specific volume, Eq. 11.28 becomes

$$p = \rho^2 \left(\frac{\partial \psi}{\partial \rho} \right)_T$$

as can easily be verified. When T is held constant τ is also constant. Accordingly, the following is obtained on differentiation of the given function:

$$\left(\frac{\partial \psi}{\partial \rho} \right)_T = RT \left[\frac{1}{\rho} + Q(\rho, \tau) + \rho \left(\frac{\partial Q}{\partial \rho} \right)_\tau \right]$$

Combining these equations gives an expression for pressure

$$p = \rho RT \left[1 + \rho Q + \rho^2 \left(\frac{\partial Q}{\partial \rho} \right)_\tau \right] \qquad \text{(a)}$$

b. From Eq. 11.29

$$s = -\left(\frac{\partial \psi}{\partial T} \right)_\rho$$

Differentiation of the given expression for ψ yields

$$\left(\frac{\partial \psi}{\partial T} \right)_\rho = \frac{d\psi_0}{dT} + \left[R(\ln \rho + \rho Q) + RT\rho \left(\frac{\partial Q}{\partial \tau} \right)_\rho \frac{d\tau}{dT} \right]$$

$$= \frac{d\psi_0}{dT} + \left[R(\ln \rho + \rho Q) + RT\rho \left(\frac{\partial Q}{\partial \tau} \right)_\rho \left(-\frac{1000}{T^2} \right) \right]$$

$$= \frac{d\psi_0}{dT} + R \left[\ln \rho + \rho Q - \rho \tau \left(\frac{\partial Q}{\partial \tau} \right)_\rho \right]$$

Combining results gives

$$s = -\frac{d\psi_0}{dT} - R \left[\ln \rho + \rho Q - \rho \tau \left(\frac{\partial Q}{\partial \tau} \right)_\rho \right] \qquad \text{(b)}$$

c. By definition, $\psi = u - Ts$. Thus, $u = \psi + Ts$. Introducing the given expression for ψ together with the expression for s from part (b) results in

$$u = [\psi_0 + RT(\ln \rho + \rho Q)] + T \left\{ -\frac{d\psi_0}{dT} \right.$$

$$\left. - R \left[\ln \rho + \rho Q - \rho \tau \left(\frac{\partial Q}{\partial \tau} \right)_\rho \right] \right\}$$

$$= \psi_0 - T \frac{d\psi_0}{dT} + RT\rho\tau \left(\frac{\partial Q}{\partial \tau} \right)_\rho$$

This can be written more compactly by noting that

$$T \frac{d\psi_0}{dT} = T \frac{d\psi_0}{d\tau} \frac{d\tau}{dT} = T \frac{d\psi_0}{d\tau} \left(-\frac{1000}{T^2} \right) = -\tau \frac{d\psi_0}{d\tau}$$

Thus,

$$\psi_0 - T \frac{d\psi_0}{dT} = \psi_0 + \tau \frac{d\psi_0}{d\tau} = \frac{d(\psi_0 \tau)}{d\tau}$$

Finally, the expression for u becomes

$$u = \frac{d(\psi_0 \tau)}{d\tau} + RT\rho\tau \left(\frac{\partial Q}{\partial \tau} \right)_\rho \qquad \text{(c)}$$

SKILLS DEVELOPED

Ability to...

• derive expressions for pressure, specific entropy, and specific internal energy based on a fundamental thermodynamic function.

Quick Quiz

Using results obtained, how can an expression be developed for h? Ans. $h = u + p/\rho$. Substitute Eq. (c) for u and Eq. (a) for p and collect terms.

11.7 Generalized Charts for Enthalpy and Entropy

TAKE NOTE...

Generalized compressibility charts are provided in Appendix figures A-1, A-2, and A-3. See Example 3.7 for an application.

Generalized charts giving the compressibility factor Z in terms of the reduced properties p_R, T_R, and v'_R are introduced in Sec. 3.11. With such charts, estimates of p–v–T data can be obtained rapidly knowing only the critical pressure and critical temperature for the substance of interest. The objective of the present section is to introduce generalized charts that allow changes in enthalpy and entropy to be estimated.

Generalized Enthalpy Departure Chart   The change in specific enthalpy of a gas (or liquid) between two states fixed by temperature and pressure can be evaluated using the identity

$$
h(T_2, p_2) - h(T_1, p_1) = [h^*(T_2) - h^*(T_1)]
$$
$$
+ \{[h(T_2, p_2) - h^*(T_2)] - [h(T_1, p_1) - h^*(T_1)]\} \tag{11.80}
$$

The term $[h(T, p) - h^*(T)]$ denotes the specific enthalpy of the substance relative to that of its ideal gas model when both are at the same temperature. The superscript * is used in this section to identify ideal gas property values. Thus, Eq. 11.80 indicates that the change in specific enthalpy between the two states equals the enthalpy change determined using the ideal gas model plus a correction that accounts for the departure from ideal gas behavior. The correction is shown underlined in Eq. 11.80. The ideal gas term can be evaluated using methods introduced in Chap. 3. Next, we show how the correction term is evaluated in terms of the *enthalpy departure.*

Developing the Enthalpy Departure   The variation of enthalpy with pressure at fixed temperature is given by Eq. 11.56 as

$$
\left(\frac{\partial h}{\partial p}\right)_T = v - T\left(\frac{\partial v}{\partial T}\right)_p
$$

Integrating from pressure p' to pressure p at fixed temperature T

$$
h(T, p) - h(T, p') = \int_{p'}^{p}\left[v - T\left(\frac{\partial v}{\partial T}\right)_p\right]dp
$$

This equation is not altered fundamentally by adding and subtracting $h^*(T)$ on the left side. That is,

$$
[h(T, p) - h^*(T)] - [h(T, p') - h^*(T)] = \int_{p'}^{p}\left[v - T\left(\frac{\partial v}{\partial T}\right)_p\right]dp \tag{11.81}
$$

As pressure tends to zero at fixed temperature, the enthalpy of the substance approaches that of its ideal gas model. Accordingly, as p' tends to zero

$$
\lim_{p'\to 0}[h(T, p') - h^*(T)] = 0
$$

In this limit, the following expression is obtained from Eq. 11.81 for the specific enthalpy of a substance relative to that of its ideal gas model when both are at the same temperature:

$$
h(T, p) - h^*(T) = \int_{0}^{p}\left[v - T\left(\frac{\partial v}{\partial T}\right)_p\right]dp \tag{11.82}
$$

This also can be thought of as the change in enthalpy as the pressure is increased from zero to the given pressure while temperature is held constant. Using p–v–T data only, Eq. 11.82 can be evaluated at states 1 and 2 and thus the correction term of Eq. 11.80 evaluated. Let us consider next how this procedure can be conducted in terms of compressibility factor data and the reduced properties T_R and p_R.

The integral of Eq. 11.82 can be expressed in terms of the compressibility factor Z and the reduced properties T_R and p_R as follows. Solving $Z = pv/RT$ gives

$$
v = \frac{ZRT}{p}
$$

On differentiation

$$\left(\frac{\partial v}{\partial T}\right)_p = \frac{RZ}{p} + \frac{RT}{p}\left(\frac{\partial Z}{\partial T}\right)_p$$

With the previous two expressions, the integrand of Eq. 11.82 becomes

$$v - T\left(\frac{\partial v}{\partial T}\right)_p = \frac{ZRT}{p} - T\left[\frac{RZ}{p} + \frac{RT}{p}\left(\frac{\partial Z}{\partial T}\right)_p\right] = -\frac{RT^2}{p}\left(\frac{\partial Z}{\partial T}\right)_p \tag{11.83}$$

Equation 11.83 can be written in terms of reduced properties as

$$v - T\left(\frac{\partial v}{\partial T}\right)_p = -\frac{RT_c}{p_c}\cdot\frac{T_R^2}{p_R}\left(\frac{\partial Z}{\partial T_R}\right)_{p_R}$$

Introducing this into Eq. 11.82 gives on rearrangement

$$\frac{h^*(T) - h(T, p)}{RT_c} = T_R^2\int_0^{p_R}\left(\frac{\partial Z}{\partial T_R}\right)_{p_R}\frac{dp_R}{p_R}$$

enthalpy departure   Or, on a per mole basis, the **enthalpy departure** is

$$\boxed{\frac{\bar{h}^*(T) - \bar{h}(T, p)}{\bar{R}T_c} = T_R^2\int_0^{p_R}\left(\frac{\partial Z}{\partial T_R}\right)_{p_R}\frac{dp_R}{p_R}} \tag{11.84}$$

The right side of Eq. 11.84 depends only on the reduced temperature T_R and reduced pressure p_R. Accordingly, the quantity $(\bar{h}^* - \bar{h})/\bar{R}T_c$, the enthalpy departure, is a function only of these two reduced properties. Using a generalized equation of state giving Z as a function of T_R and p_R, the enthalpy departure can readily be evaluated with a computer. Tabular representations are also found in the literature. Alternatively, the graphical representation provided in Fig. A-4 can be employed.

Evaluating Enthalpy Change   The change in specific enthalpy between two states can be evaluated by expressing Eq. 11.80 in terms of the enthalpy departure as

$$\boxed{\bar{h}_2 - \bar{h}_1 = \underline{\bar{h}_2^* - \bar{h}_1^*} - \bar{R}T_c\left[\underline{\left(\frac{\bar{h}^* - \bar{h}}{\bar{R}T_c}\right)_2 - \left(\frac{\bar{h}^* - \bar{h}}{\bar{R}T_c}\right)_1}\right]} \tag{11.85}$$

The first underlined term in Eq. 11.85 represents the change in specific enthalpy between the two states assuming ideal gas behavior. The second underlined term is the correction that must be applied to the ideal gas value for the enthalpy change to obtain the actual value for the enthalpy change. Referring to the engineering literature, the quantity $(\bar{h}^* - \bar{h})/\bar{R}T_c$ at states 1 and 2 can be calculated with an equation giving $Z(T_R, p_R)$ or obtained from tables. This quantity also can be evaluated at state 1 from the generalized enthalpy departure chart, Fig. A-4, using the reduced temperature T_{R1} and reduced pressure p_{R1} corresponding to the temperature T_1 and pressure p_1 at the initial state, respectively. Similarly, $(\bar{h}^* - \bar{h})/\bar{R}T_c$ at state 2 can be evaluated from Fig. A-4 using T_{R2} and p_{R2}. The use of Eq. 11.85 is illustrated in the next example.

▸ **EXAMPLE 11.8** ▸

Using the Generalized Enthalpy Departure Chart

Nitrogen enters a turbine operating at steady state at 100 bar and 300 K and exits at 40 bar and 245 K. Using the enthalpy departure chart, determine the work developed, in kJ per kg of nitrogen flowing, if heat transfer with the surroundings can be ignored. Changes in kinetic and potential energy from inlet to exit also can be neglected.

SOLUTION

Known A turbine operating at steady state has nitrogen entering at 100 bar and 300 K and exiting at 40 bar and 245 K.

Find Using the enthalpy departure chart, determine the work developed.

Schematic and Given Data:

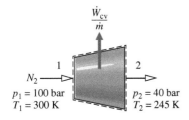

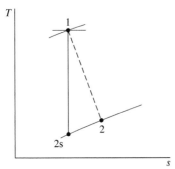

Fig. E11.8

Engineering Model

1. The control volume shown on the accompanying figure operates at steady state.
2. There is no significant heat transfer between the control volume and its surroundings.
3. Changes in kinetic and potential energy between inlet and exit can be neglected.
4. Equilibrium property relations apply at the inlet and exit.

Analysis The mass and energy rate balances reduce at steady state to give

$$0 = \frac{\dot{Q}_{cv}}{\dot{m}} - \frac{\dot{W}_{cv}}{\dot{m}} + \left[h_1 - h_2 + \frac{V_1^2 - V_2^2}{2} + g(z_1 - z_2) \right]$$

where $\dot{m}$ is the mass flow rate. Dropping the heat transfer term by assumption 2 and the kinetic and potential energy terms by assumption 3 gives on rearrangement

$$\frac{\dot{W}_{cv}}{\dot{m}} = h_1 - h_2$$

The term $h_1 - h_2$ can be evaluated as follows:

$$h_1 - h_2 = \frac{1}{M} \left\{ \bar{h}_1^* - \bar{h}_2^* - \bar{R}T_c \left[\left(\frac{\bar{h}^* - \bar{h}}{\bar{R}T_c} \right)_1 - \left(\frac{\bar{h}^* - \bar{h}}{\bar{R}T_c} \right)_2 \right] \right\}$$

In this expression, M is the molecular weight of nitrogen and the other terms have the same significance as in Eq. 11.85.

With specific enthalpy values from Table A-23 at $T_1 = 300$ K and $T_2 = 245$ K, respectively,

$$\bar{h}_1^* - \bar{h}_2^* = 8723 - 7121 = 1602 \text{ kJ/mol}$$

The terms $(\bar{h}^* - \bar{h})/\bar{R}T_c$ at states 1 and 2 required by the above expression for $h_1 - h_2$ can be determined from Fig. A-4. First, the reduced temperature and reduced pressure at the inlet and exit must be determined. From Table A-1, $T_c = 126$ K, $p_c = 33.9$ bar. Thus, at the inlet

$$T_{R1} = \frac{300}{126} = 2.38, \qquad p_{R1} = \frac{100}{33.9} = 2.95$$

At the exit

$$T_{R2} = \frac{245}{126} = 1.94, \qquad p_{R2} = \frac{40}{33.9} = 1.18$$

By inspection of Fig. A-4

❶ $$\left(\frac{\bar{h}^* - \bar{h}}{\bar{R}T_c} \right)_1 \approx 0.5, \qquad \left(\frac{\bar{h}^* - \bar{h}}{\bar{R}T_c} \right)_2 \approx 0.31$$

Substituting values

$$\frac{\dot{W}_{cv}}{\dot{m}} = \frac{1}{28 \frac{\text{kg}}{\text{kmol}}} \left[1602 \frac{\text{kJ}}{\text{kmol}} \right.$$
$$\left. - \left(8.314 \frac{\text{kJ}}{\text{kmol} \cdot \text{K}} \right)(126 \text{ K})(0.5 - 0.31) \right]$$
$$= 50.1 \text{ kJ/kg}$$

---

❶ Due to inaccuracy in reading values from a graph such as Fig. A-4, we cannot expect extreme accuracy in the final calculated result.

SKILLS DEVELOPED

Ability to...

• use data from the generalized enthalpy departure chart to calculate the change in enthalpy of nitrogen.

Quick Quiz

Determine the work developed, in kJ per kg of nitrogen flowing, assuming the ideal gas model. Ans. 57.2 kJ/kg.

Generalized Entropy Departure Chart   A generalized chart that allows changes in specific entropy to be evaluated can be developed in a similar manner to the generalized enthalpy departure chart introduced earlier in this section. The difference in specific entropy between states 1 and 2 of a gas (or liquid) can be expressed as the identity

$$
\begin{aligned}
s(T_2, p_2) - s(T_1, p_1) = {}& s^*(T_2, p_2) - s^*(T_1, p_1) \\
& + \{[s(T_2, p_2) - s^*(T_2, p_2)] - [s(T_1, p_1) - s^*(T_1, p_1)]\}
\end{aligned}
\tag{11.86}
$$

where $[s(T, p) - s^*(T, p)]$ denotes the specific entropy of the substance relative to that of its ideal gas model when both are at the same temperature and pressure. Equation 11.86 indicates that the change in specific entropy between the two states equals the entropy change determined using the ideal gas model plus a correction (shown underlined) that accounts for the departure from ideal gas behavior. The ideal gas term can be evaluated using methods introduced in Sec. 6.5. Let us consider next how the correction term is evaluated in terms of the *entropy departure*.

Developing the Entropy Departure   The following Maxwell relation gives the variation of entropy with pressure at fixed temperature:

$$
\left(\frac{\partial s}{\partial p}\right)_T = -\left(\frac{\partial v}{\partial T}\right)_p
\tag{11.35}
$$

Integrating from pressure p' to pressure p at fixed temperature T gives

$$
s(T, p) - s(T, p') = -\int_{p'}^{p} \left(\frac{\partial v}{\partial T}\right)_p dp
\tag{11.87}
$$

For an ideal gas, $v = RT/p$, so $(\partial v/\partial T)_p = R/p$. Using this in Eq. 11.87, the change in specific entropy assuming ideal gas behavior is

$$
s^*(T, p) - s^*(T, p') = -\int_{p'}^{p} \frac{R}{p} dp
\tag{11.88}
$$

Subtracting Eq. 11.88 from Eq. 11.87 gives

$$
[s(T, p) - s^*(T, p)] - [s(T, p') - s^*(T, p')] = \int_{p'}^{p} \left[\frac{R}{p} - \left(\frac{\partial v}{\partial T}\right)_p\right] dp
\tag{11.89}
$$

Since the properties of a substance tend to merge into those of its ideal gas model as pressure tends to zero at fixed temperature, we have

$$
\lim_{p' \to 0} [s(T, p') - s^*(T, p')] = 0
$$

Thus, in the limit as p' tends to zero, Eq. 11.89 becomes

$$
s(T, p) - s^*(T, p) = \int_{0}^{p} \left[\frac{R}{p} - \left(\frac{\partial v}{\partial T}\right)_p\right] dp
\tag{11.90}
$$

Using p–v–T data only, Eq. 11.90 can be evaluated at states 1 and 2 and thus the correction term of Eq. 11.86 evaluated.

Equation 11.90 can be expressed in terms of the compressibility factor Z and the reduced properties T_R and p_R. The result, on a per mole basis, is the **entropy departure**

entropy departure

$$
\frac{\bar{s}^*(T, p) - \bar{s}(T, p)}{\bar{R}} = \frac{\bar{h}^*(T) - \bar{h}(T, p)}{\bar{R} T_R T_c} + \int_{0}^{p_R} (Z - 1)\frac{dp_R}{p_R}
\tag{11.91}
$$

The right side of Eq. 11.91 depends only on the reduced temperature T_R and reduced pressure p_R. Accordingly, the quantity $(\bar{s}* - \bar{s})/\bar{R}$, the entropy departure, is a function only of these two reduced properties. As for the enthalpy departure, the entropy departure can be evaluated with a computer using a generalized equation of state giving Z as a function of T_R and p_R. Alternatively, tabular data from the literature or the graphical representation provided in Fig. A-5 can be employed.

Evaluating Entropy Change   The change in specific entropy between two states can be evaluated by expressing Eq. 11.86 in terms of the entropy departure as

$$\bar{s}_2 - \bar{s}_1 = \underline{\bar{s}_2^* - \bar{s}_1^*} - \bar{R}\left[\left(\frac{\overline{s* - s}}{\bar{R}}\right)_2 - \left(\frac{\overline{s* - s}}{\bar{R}}\right)_1\right] \qquad (11.92)$$

The first underlined term in Eq. 11.92 represents the change in specific entropy between the two states assuming ideal gas behavior. The second underlined term is the correction that must be applied to the ideal gas value for entropy change to obtain the actual value for the entropy change. The quantity $(\bar{s}* - \bar{s})_1/\bar{R}$ appearing in Eq. 11.92 can be evaluated from the generalized entropy departure chart, Fig. A-5, using the reduced temperature T_{R1} and reduced pressure p_{R1} corresponding to the temperature T_1 and pressure p_1 at the initial state, respectively. Similarly, $(\bar{s}* - \bar{s})_2/\bar{R}$ can be evaluated from Fig. A-5 using T_{R2} and p_{R2}. The use of Eq. 11.92 is illustrated in the next example.

> ▶▶ **EXAMPLE 11.9** ▶ •

Using the Generalized Entropy Departure Chart

For the turbine of Example 11.8, determine **(a)** the rate of entropy production, in kJ/kg · K, and **(b)** the isentropic turbine efficiency.

SOLUTION

Known   A turbine operating at steady state has nitrogen entering at 100 bar and 300 K and exiting at 40 bar and 245 K.

Find   Determine the rate of entropy production, in kJ/kg · K, and the isentropic turbine efficiency.

Schematic and Given Data: See Fig. E11.8.

Engineering Model   See Example 11.8.

Analysis

a. At steady state, the control volume form of the entropy rate equation reduces to give

$$\frac{\dot{\sigma}_{cv}}{\dot{m}} = s_2 - s_1$$

The change in specific entropy required by this expression can be written as

$$s_2 - s_1 = \frac{1}{M}\left\{\bar{s}_2^* - \bar{s}_1^* - \bar{R}\left[\left(\frac{\overline{s* - s}}{\bar{R}}\right)_2 - \left(\frac{\overline{s* - s}}{\bar{R}}\right)_1\right]\right\}$$

where M is the molecular weight of nitrogen and the other terms have the same significance as in Eq. 11.92.
The change in specific entropy $\bar{s}_2^* - \bar{s}_1^*$ can be evaluated using

$$\bar{s}_2^* - \bar{s}_1^* = \bar{s}°(T_2) = \bar{s}°(T_1) - \bar{R}\ln\frac{p_2}{p_1}$$

With values from Table A-23

$$\bar{s}_2^* - \bar{s}_1^* = 185.775 - 191.682 - 8.134\ln\frac{40}{100}$$

$$= 1.711\frac{kJ}{kmol \cdot K}$$

The terms $(\bar{s}* - \bar{s})/\bar{R}$ at the inlet and exit can be determined from Fig. A-5. Using the reduced temperature and reduced pressure values calculated in the solution to Example 11.8, inspection of Fig. A-5 gives

$$\left(\frac{\overline{s* - s}}{\bar{R}}\right)_1 \approx 0.21, \qquad \left(\frac{\overline{s* - s}}{\bar{R}}\right)_2 \approx 0.14$$

Substituting values

$$\frac{\dot{\sigma}_{cv}}{\dot{m}} = \frac{1}{(28\ kg/kmol)}\left[1.711\frac{kJ}{kmol \cdot K}\right.$$

$$\left. - 8.314\frac{kJ}{kmol \cdot K}(0.14 - 0.21)\right]$$

$$= 0.082\frac{kJ}{kg \cdot K}$$

b. The isentropic turbine efficiency is defined in Sec. 6.12 as

$$\eta_t = \frac{(\dot{W}_{cv}/\dot{m})}{(\dot{W}_{cv}/\dot{m})_s}$$

where the denominator is the work that would be developed by the turbine if the nitrogen expanded isentropically from the specified inlet state to the specified exit pressure. Thus, it is necessary to fix the state, call it 2s, at the turbine exit for an expansion in which there is no change in specific entropy from inlet to exit. With $(\bar{s}_{2s} - \bar{s}_1) = 0$ and procedures similar to those used in part (a)

$$0 = \bar{s}_{2s}^* - \bar{s}_1^* - \bar{R}\left[\left(\frac{\bar{s}^* - \bar{s}}{\bar{R}}\right)_{2s} - \left(\frac{\bar{s}^* - \bar{s}}{\bar{R}}\right)_1\right]$$

$$0 = \left[\bar{s}^\circ(T_{2s}) - \bar{s}^\circ(T_1) - \bar{R}\ln\left(\frac{p_2}{p_1}\right)\right]$$
$$\quad - \bar{R}\left[\left(\frac{\bar{s}^* - \bar{s}}{\bar{R}}\right)_{2s} - \left(\frac{\bar{s}^* - \bar{s}}{\bar{R}}\right)_1\right]$$

Using values from part (a), the last equation becomes

$$0 = \bar{s}^\circ(T_{2s}) - 191.682 - 8.314\ln\frac{40}{100} - \bar{R}\left(\frac{\bar{s}^* - \bar{s}}{\bar{R}}\right)_{2s} + 1.746$$

or

$$\bar{s}^\circ(T_{2s}) - \bar{R}\left(\frac{\bar{s}^* - \bar{s}}{\bar{R}}\right)_{2s} = 182.3$$

The temperature T_{2s} can be determined in an iterative procedure using s° data from Table A-23 and $(\bar{s}^* - \bar{s})/\bar{R}$ from Fig. A-5 as follows: First, a value for the temperature T_{2s} is assumed. The corresponding value of s° can then be obtained from Table A-23. The reduced temperature $(T_R)_{2s} = T_{2s}/T_c$, together with $p_{R2} = 1.18$, allows a value for $(\bar{s}^* - \bar{s})/\bar{R}$ to be obtained from Fig. A-5. The procedure continues until agreement with the value on the right side of the above equation is obtained. Using this procedure, T_{2s} is found to be closely 228 K.

With the temperature T_{2s} known, the work that would be developed by the turbine if the nitrogen expanded isentropically from the specified inlet state to the specified exit pressure can be evaluated from

$$\left(\frac{\dot{W}_{cv}}{\dot{m}}\right)_s = h_1 - h_{2s}$$

$$= \frac{1}{M}\left\{(\bar{h}_1^* - \bar{h}_{2s}^*) - \bar{R}T_c\left[\left(\frac{\bar{h}^* - \bar{h}}{\bar{R}T_c}\right)_1 - \left(\frac{\bar{h}^* - \bar{h}}{\bar{R}T_c}\right)_{2s}\right]\right\}$$

From Table A-23, $\bar{h}_{2s}^* = 6654\,\text{kJ/kmol}$. From Fig. A-4 at $p_{R2} = 1.18$ and $(T_R)_{2s} = 228/126 = 1.81$

$$\left(\frac{\bar{h}^* - \bar{h}}{\bar{R}T_c}\right)_{2s} \approx 0.36$$

Values for the other terms in the expression for $(\dot{W}_{cv}/\dot{m})_s$ are obtained in the solution to Example 11.8. Finally,

$$\left(\frac{\dot{W}_{cv}}{\dot{m}}\right)_s = \frac{1}{28}[8723 - 6654 - (8.314)(126)(0.5 - 036)]$$
$$= 68.66\,\text{kJ/kg}$$

With the work value from Example 11.8, the turbine efficiency is

❶
$$\eta_t = \frac{(\dot{W}_{cv}/\dot{m})}{(\dot{W}_{cv}/\dot{m})_s} = \frac{50.1}{68.66} = 0.73(73\%)$$

---

❶ We cannot expect extreme accuracy when reading data from a generalized chart such as Fig. A-5, which affects the final calculated result.

SKILLS DEVELOPED

Ability to…

- use data from the generalized entropy departure chart to calculate the entropy production.
- use data from the generalized enthalpy and entropy departure charts to calculate isentropic turbine efficiency.
- use an iterative procedure to calculate the temperature at the end of an isentropic process using data from the generalized entropy departure chart.

Quick Quiz

Determine the rate of entropy production, in kJ/K per kg of nitrogen flowing, assuming the ideal gas model. Ans. 0.061 kJ/kg · K.

11.8 p–v–T Relations for Gas Mixtures

Many systems of interest involve mixtures of two or more components. The principles of thermodynamics introduced thus far *are* applicable to systems involving mixtures, but to apply such principles requires that mixture properties be evaluated.

Since an unlimited variety of mixtures can be formed from a given set of pure components by varying the relative amounts present, the properties of mixtures are available in tabular, graphical, or equation forms only in particular cases such as air. Generally, special means are required for determining mixture properties.

In this section, methods for evaluating the p–v–T relations for pure components introduced in previous sections of the book are adapted to obtain plausible estimates for gas mixtures. In Sec. 11.9 some general aspects of property evaluation for multicomponent systems are introduced.

To evaluate the properties of a mixture requires knowledge of the composition. The composition can be described by giving the *number of moles* (kmol or lbmol) of each component present. The total number of moles, n, is the sum of the number of moles of each of the components

TAKE NOTE...

Additional content in support of this section is provided in Sec. 12.1. The special case of ideal gas mixtures is detailed in Secs. 12.2–12.4.

$$n = n_1 + n_2 + \cdots + n_j = \sum_{i=1}^{j} n_i \tag{11.93}$$

The *relative* amounts of the components present can be described in terms of *mole fractions.* The mole fraction y_i of component i is defined as

$$y_i = \frac{n_i}{n} \tag{11.94}$$

Dividing each term of Eq. 11.93 by the total number of moles and using Eq. 11.94

$$1 = \sum_{i=1}^{j} y_i \tag{11.95}$$

That is, the sum of the mole fractions of all components present is equal to unity.

Most techniques for estimating mixture properties are empirical in character and are not derived from fundamental principles. The realm of validity of any particular technique can be established only by comparing predicted property values with empirical data. The brief discussion to follow is intended only to show how certain of the procedures for evaluating the *p–v–T* relations of pure components introduced previously can be extended to gas mixtures.

Mixture Equation of State   One way the *p–v–T* relation for a gas mixture can be estimated is by applying to the overall mixture an equation of state such as introduced in Sec. 11.1. The constants appearing in the equation selected would be *mixture values* determined with empirical combining rules developed for the equation. For example, mixture values of the constants a and b for use in the van der Waals and Redlich–Kwong equations would be obtained using relations of the form

$$a = \left(\sum_{i=1}^{j} y_i a_i^{1/2} \right)^2 , \qquad b = \left(\sum_{i=1}^{j} y_i b_i \right) \tag{11.96}$$

where a_i and b_i are the values of the constants for component i and y_i is the mole fraction. Combination rules for obtaining mixture values for the constants in other equations of state also have been suggested.

Kay's Rule   The *principle of corresponding states* method for single components introduced in Sec. 3.11.3 can be extended to mixtures by regarding the mixture as if it were a single pure component having critical properties calculated by one of several mixture rules. Perhaps the simplest of these, requiring only the determination of a mole fraction averaged critical temperature T_c and critical pressure p_c, is **Kay's rule**

Kay's rule

$$\boxed{T_c = \sum_{i=1}^{j} y_i T_{c,i}, \qquad p_c = \sum_{i=1}^{j} y_i p_{c,i}} \tag{11.97}$$

where $T_{c,i}$, $p_{c,i}$, and y_i are the critical temperature, critical pressure, and mole fraction of component i, respectively. Using T_c and p_c, the mixture compressibility factor Z is obtained as for a single pure component. The unknown quantity from among the pressure p, volume V, temperature T, and total number of moles n of the gas mixture can then be obtained by solving

$$Z = \frac{pV}{n\bar{R}T} \tag{11.98}$$

Mixture values for T_c and p_c also can be used to enter the generalized enthalpy departure and entropy departure charts introduced in Sec. 11.7.

Additive Pressure Rule Additional means for estimating p–v–T relations for mixtures are provided by empirical mixture rules, of which several are found in the engineering literature. Among these are the *additive pressure* and *additive volume* rules. According to the **additive pressure rule**, the pressure of a gas mixture occupying volume V at temperature T is expressible as a sum of pressures exerted by the individual components:

additive pressure rule

$$\boxed{p = p_1 + p_2 + p_3 + \cdots]_{T,V}} \tag{11.99a}$$

where the pressures p_1, p_2, and so on are evaluated by considering the respective components to be at the volume and temperature of the mixture. These pressures would be determined using tabular or graphical p–v–T data or a suitable equation of state.

An alternative expression of the additive pressure rule in terms of compressibility factors can be obtained. Since component i is considered to be at the volume and temperature of the mixture, the compressibility factor Z_i for this component is $Z_i = p_i V/n_i \bar{R}T$, so the pressure p_i is

$$p_i = \frac{Z_i n_i \bar{R}T}{V}$$

Similarly, for the mixture

$$p = \frac{Zn\bar{R}T}{V}$$

Substituting these expressions into Eq. 11.99a and reducing gives the following relationship between the compressibility factors for the mixture Z and the mixture components Z_i

$$\boxed{Z = \sum_{i=1}^{j} y_i Z_i]_{T,V}} \tag{11.99b}$$

The compressibility factors Z_i are determined assuming that component i occupies the entire volume of the mixture at the temperature T.

additive volume rule

Additive Volume Rule The underlying assumption of the **additive volume rule** is that the volume V of a gas mixture at temperature T and pressure p is expressible as the sum of volumes occupied by the individual components:

$$\boxed{V = V_1 + V_2 + V_3 + \cdots]_{p,T}} \tag{11.100a}$$

where the volumes V_1, V_2, and so on are evaluated by considering the respective components to be at the pressure and temperature of the mixture. These volumes would be determined from tabular or graphical p–v–T data or a suitable equation of state.

An alternative expression of the additive volume rule in terms of compressibility factors can be obtained. Since component i is considered to be at the pressure and temperature of the mixture, the compressibility factor Z_i for this component is $Z_i = pV_i/n_i\bar{R}T$, so the volume V_i is

$$V_i = \frac{Z_i n_i \bar{R}T}{p}$$

Similarly, for the mixture

$$V = \frac{Zn\bar{R}T}{p}$$

Substituting these expressions into Eq. 11.100a and reducing gives

$$Z = \sum_{i=1}^{j} y_i Z_i]_{p,T}$$ (11.100b)

The compressibility factors Z_i are determined assuming that component i exists at the temperature T and pressure p of the mixture.

The next example illustrates alternative means for estimating the pressure of a gas mixture.

> ▶ **EXAMPLE 11.10** ▶

Estimating Mixture Pressure by Alternative Means

A mixture consisting of 0.18 kmol of methane (CH_4) and 0.274 kmol of butane (C_4H_{10}) occupies a volume of 0.241 m³ at a temperature of 238°C. The experimental value for the pressure is 68.9 bar. Calculate the pressure, in bar, exerted by the mixture by using **(a)** the ideal gas equation of state, **(b)** Kay's rule together with the generalized compressibility chart, **(c)** the van der Waals equation, and **(d)** the rule of additive pressures employing the generalized compressibility chart. Compare the calculated values with the known experimental value.

SOLUTION

Known A mixture of two specified hydrocarbons with known molar amounts occupies a known volume at a specified temperature.

Find Determine the pressure, in bar, using four alternative methods, and compare the results with the experimental value.

Schematic and Given Data:

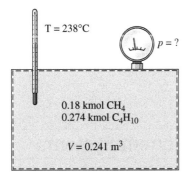

T = 238°C

p = ?

0.18 kmol CH_4
0.274 kmol C_4H_{10}

V = 0.241 m³

Fig. E11.10

Engineering Model As shown in the accompanying figure, the system is the mixture.

Analysis The total number of moles of mixture n is

$$n = 0.18 + 0.274 = 0.454 \text{ kmol}$$

Thus, the mole fractions of the methane and butane are, respectively,

$$y_1 = 0.396 \quad \text{and} \quad y_2 = 0.604$$

The specific volume of the mixture on a molar basis is

$$\bar{v} = \frac{0.241 \text{ m}^3}{(0.18 + 0.274) \text{ kmol}} = 0.531 \frac{\text{m}^3}{\text{kmol}}$$

a. Substituting values into the ideal gas equation of state

$$p = \frac{\bar{R}T}{\bar{v}} = \frac{(8314 \text{ N} \cdot \text{m/kmol} \cdot \text{K})(511 \text{ K})}{(0.531 \text{ m}^3/\text{kmol})} \left| \frac{1 \text{ bar}}{10^5 \text{ N/m}^2} \right|$$

$$= 80.01 \text{ bar}$$

b. To apply Kay's rule, the critical temperature and pressure for each component are required. From Table A-1, for methane

$$T_{c1} = 191 \text{ K}, \qquad p_{c1} = 46.4 \text{ bar}$$

and for butane

$$T_{c2} = 425 \text{ K}, \qquad p_{c2} = 38.0 \text{ bar}$$

Thus, with Eqs. 11.97

$$T_c = y_1 T_{c1} + y_2 T_{c2} = (0.396)(191) + (0.604)(425) = 332.3 \text{ K}$$
$$p_c = y_1 p_{c1} + y_2 p_{c2} = (0.396)(46.4) + (0.604)(38.0) = 41.33 \text{ bar}$$

Treating the mixture as a pure component having the above values for the critical temperature and pressure, the following reduced properties are determined for the mixture:

$$T_R = \frac{T}{T_c} = \frac{511}{332.3} = 1.54$$

$$v_R' = \frac{\bar{v} p_c}{\bar{R} T_c} = \frac{(0.531)(41.33)|10^5|}{(8314)(332.3)}$$

$$= 0.794$$

Turning to Fig. A-2, $Z \approx 0.88$. The mixture pressure is then found from

$$p = \frac{Zn\bar{R}T}{V} = Z\frac{\bar{R}T}{\bar{v}} = 0.88\frac{(8314)(511)}{(0.531)|10^5|}$$

$$= 70.4 \text{ bar}$$

c. Mixture values for the van der Waals constants can be obtained using Eqs. 11.96. This requires values of the van der Waals constants for each of the two mixture components. Table A-24 gives the following values for methane:

$$a_1 = 2.293 \text{ bar} \left(\frac{m^3}{kmol} \right)^2, \qquad b_1 = 0.0428 \frac{m^3}{kmol}$$

Similarly, from Table A-24 for butane

$$a_2 = 13.86 \text{ bar} \left(\frac{m^3}{kmol} \right)^2, \qquad b_2 = 0.1162 \frac{m^3}{kmol}$$

Then, the first of Eqs. 11.96 gives a mixture value for the constant a as

$$a = (y_1 a_1^{1/2} + y_2 a_2^{1/2})^2 = [0.396(2.293)^{1/2} + 0.604(13.86)^{1/2}]^2$$

$$= 8.113 \text{ bar} \left(\frac{m^3}{kmol} \right)^2$$

Substituting into the second of Eqs. 11.96 gives a mixture value for the constant b

$$b = y_1 b_1 + y_2 b_2 = (0.396)(0.0428) + (0.604)(0.1162)$$

$$= 0.087 \frac{m^3}{kmol}$$

Inserting the mixture values for a and b into the van der Waals equation together with known data

$$p = \frac{\bar{R}T}{\bar{v} - b} - \frac{a}{\bar{v}^2}$$

$$= \frac{(8314 \text{ N} \cdot \text{m/kmol} \cdot \text{K})(511 \text{ K})}{(0.531 - 0.087)(\text{m}^3/\text{kmol})} \left| \frac{1 \text{ bar}}{10^5 \text{ N/m}^2} \right|$$

$$- \frac{8.113 \text{ bar (m}^3/\text{kmol})^2}{(0.531 \text{ m}^3/\text{kmol})^2}$$

$$= 66.91 \text{ bar}$$

d. To apply the additive pressure rule with the generalized compressibility chart requires that the compressibility factor for each component be determined assuming that the component occupies the entire volume at the mixture temperature.

With this assumption, the following reduced properties are obtained for methane:

$$T_{R1} = \frac{T}{T_{c1}} = \frac{511}{191} = 2.69$$

$$v'_{R1} = \frac{\bar{v}_1 p_{c1}}{\bar{R} T_{c1}} = \frac{(0.241 \text{ m}^3/0.18 \text{ kmol})(46.4 \text{ bar})}{(8314 \text{ N} \cdot \text{m/kmol} \cdot \text{K})(191 \text{ K})} \left| \frac{10^5 \text{ N/m}^2}{1 \text{ bar}} \right|$$

$$= 3.91$$

With these reduced properties, Fig. A-2 gives $Z_1 \approx 1.0$.

Similarly, for butane

$$T_{R2} = \frac{T}{T_{c2}} = \frac{511}{425} = 1.2$$

$$v'_{R2} = \frac{\bar{v}_2 p_{c2}}{\bar{R} T_{c2}} = \frac{(0.88)(38) \left| 10^5 \right|}{(8314)(425)} = 0.95$$

From Fig. A-2, $Z_2 \approx 0.8$.

The compressibility factor for the mixture determined from Eq. 11.99b is

$$Z = y_1 Z_1 + y_2 Z_2 = (0.396)(1.0) + (0.604)(0.8) = 0.88.$$

Accordingly, the same value for pressure as determined in part (b) using Kay's rule results: $p = 70.4$ bar.

In this particular example, the ideal gas equation of state gives a value for pressure that exceeds the experimental value by nearly 16%. Kay's rule and the rule of additive pressures give pressure values about 3% greater than the experimental value. The van der Waals equation with mixture values for the constants gives a pressure value about 3% less than the experimental value.

SKILLS DEVELOPED

Ability to...

• calculate the pressure of a gas mixture using four alternative methods.

Quick Quiz

Convert the mixture analysis from a molar basis to a mass fraction basis. Ans. Methane: 0.153, Butane: 0.847.

11.9 Analyzing Multicomponent Systems

In the preceding section we considered means for evaluating the p–v–T relation of gas mixtures by extending methods developed for pure components. The current section is devoted to the development of some general aspects of the properties of systems with two or more components. Primary emphasis is on the case of *gas mixtures*, but the methods developed also apply to *solutions*. When liquids and solids are under consideration, the term **solution** is sometimes used in place of mixture. The present discussion is limited to nonreacting mixtures or solutions in a single phase. The effects of chemical reactions and equilibrium between different phases are taken up in Chaps. 13 and 14.

To describe multicomponent systems, composition must be included in our thermodynamic relations. This leads to the definition and development of several new concepts, including the *partial molal property*, the *chemical potential*, and the *fugacity*.

solution

TAKE NOTE...

Section 11.9 may be deferred until Secs. 12.1–12.4 have been studied.

11.9.1 Partial Molal Properties

In the present discussion we introduce the concept of a *partial molal* property and illustrate its use. This concept plays an important role in subsequent discussions of multicomponent systems.

Defining Partial Molal Properties   Any extensive thermodynamic property X of a single-phase, single-component system is a function of two independent intensive properties and the size of the system. Selecting temperature and pressure as the independent properties and the number of moles n as the measure of size, we have $X = X(T, p, n)$. For a single-phase, *multicomponent* system, the extensive property X must then be a function of temperature, pressure, and the number of moles of each component present, $X = X(T, p, n_1, n_2, \ldots, n_j)$.

If each mole number is increased by a factor α, the size of the system increases by the same factor, and so does the value of the extensive property X. That is,

$$\alpha X(T, p, n_1, n_2, \ldots, n_j) = X(T, p, \alpha n_1, \alpha n_2, \ldots, \alpha n_j)$$

Differentiating with respect to α while holding temperature, pressure, and the mole numbers fixed and using the chain rule on the right side gives

$$X = \frac{\partial X}{\partial(\alpha n_1)} n_1 + \frac{\partial X}{\partial(\alpha n_2)} n_2 + \cdots + \frac{\partial X}{\partial(\alpha n_j)} n_j$$

This equation holds for all values of α. In particular, it holds for $\alpha = 1$. Setting $\alpha = 1$

$$X = \sum_{i=1}^{j} n_i \frac{\partial X}{\partial n_i}\bigg)_{T, p, n_l} \tag{11.101}$$

where the subscript n_l denotes that all n's except n_i are held fixed during differentiation.

The **partial molal property** $\overline{X}_i$ is by definition

partial molal property

$$\boxed{\overline{X}_i = \frac{\partial X}{\partial n_i}\bigg)_{T, p, n_l}} \tag{11.102}$$

The partial molal property $\overline{X}_i$ is a property of the mixture and not simply a property of component i, for $\overline{X}_i$ depends in general on temperature, pressure, *and* mixture composition: $\overline{X}_i(T, p, n_1, n_2, \ldots, n_j)$. Partial molal properties are intensive properties of the mixture.

Introducing Eq. 11.102, Eq. 11.101 becomes

$$\boxed{X = \sum_{i=1}^{j} n_i \overline{X}_i} \tag{11.103}$$

This equation shows that the extensive property X can be expressed as a weighted sum of the partial molal properties $\overline{X}_i$.

Selecting the extensive property X in Eq. 11.103 to be volume, internal energy, enthalpy, and entropy, respectively, gives

$$V = \sum_{i=1}^{j} n_i \overline{V}_i, \quad U = \sum_{i=1}^{j} n_i \overline{U}_i, \quad H = \sum_{i=1}^{j} n_i \overline{H}_i, \quad S = \sum_{i=1}^{j} n_i \overline{S}_i \tag{11.104}$$

where $\overline{V}_i, \overline{U}_i, \overline{H}_i, \overline{S}_i$ denote the partial molal volume, internal energy, enthalpy, and entropy. Similar expressions can be written for the Gibbs function G and the Helmholtz function ψ. Moreover, the relations between these extensive properties: $H = U + pV$, $G = H - TS$, $\psi = U - TS$ can be differentiated with respect to n_i while holding temperature, pressure, and the remaining n's constant to produce corresponding relations among partial molal properties: $\overline{H}_i = \overline{U}_i + p\overline{V}_i$, $\overline{G}_i = \overline{H}_i - T\overline{S}_i$, $\overline{\Psi}_i = \overline{U}_i - T\overline{S}_i$, where $\overline{G}_i$ and $\overline{\Psi}_i$ are the partial molal Gibbs function and Helmholtz function, respectively. Several additional relations involving partial molal properties are developed later in this section.

Evaluating Partial Molal Properties Partial molal properties can be evaluated by several methods, including the following:

- If the property X can be measured, $\bar{X}_i$ can be found by extrapolating a plot giving $(\Delta X/\Delta n_i)_{T,p,n_l}$ versus Δn_i. That is,

$$\bar{X}_i = \left(\frac{\partial X}{\partial n_i}\right)_{T,p,n_l} = \lim_{\Delta n_i \to 0}\left(\frac{\Delta X}{\Delta n_i}\right)_{T,p,n_l}$$

- If an expression for X in terms of its independent variables is known, $\bar{X}_i$ can be evaluated by differentiation. The derivative can be determined analytically if the function is expressed analytically or found numerically if the function is in tabular form.

method of intercepts

- When suitable data are available, a simple graphical procedure known as the **method of intercepts** can be used to evaluate partial molal properties. In principle, the method can be applied for any extensive property. To introduce this method, let us consider the volume of a system consisting of two components, A and B. For this system, Eq. 11.103 takes the form

$$V = n_A\bar{V}_A + n_B\bar{V}_B$$

where $\bar{V}_A$ and $\bar{V}_B$ are the partial molal volumes of A and B, respectively. Dividing by the number of moles of mixture n

$$\frac{V}{n} = y_A\bar{V}_A + y_B\bar{V}_B$$

where y_A and y_B denote the mole fractions of A and B, respectively. Since $y_A + y_B = 1$, this becomes

$$\frac{V}{n} = (1 - y_B)\bar{V}_A + y_B\bar{V}_B = \bar{V}_A + y_B(\bar{V}_B - \bar{V}_A)$$

This equation provides the basis for the method of intercepts. For example, refer to Fig. 11.5, in which V/n is plotted as a function of y_B at constant T and p. At a specified value for y_B, a tangent to the curve is shown on the figure. When extrapolated, the tangent line intersects the axis on the left at $\bar{V}_A$ and the axis on the right at $\bar{V}_B$. These values for the partial molal volumes correspond to the particular specifications for T, p, and y_B. At fixed temperature and pressure, $\bar{V}_A$ and $\bar{V}_B$ vary with y_B and are not equal to the molar specific volumes of *pure* A and *pure* B, denoted on the figure as $\bar{v}_A$ and $\bar{v}_B$, respectively. The values of $\bar{v}_A$ and $\bar{v}_B$ are fixed by temperature and pressure only.

Extensive Property Changes on Mixing Let us conclude the present discussion by evaluating the change in volume on mixing of pure components at the same temperature and

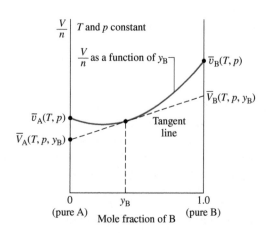

Fig. 11.5 Illustration of the evaluation of partial molal volumes by the method of intercepts.

pressure, a result for which an application is given in the discussion of Eq. 11.135. The total volume of the pure components before mixing is

$$V_{\text{components}} = \sum_{i=1}^{j} n_i \bar{v}_i$$

where $\bar{v}_i$ is the molar specific volume of pure component i. The volume of the mixture is

$$V_{\text{mixture}} = \sum_{i=1}^{j} n_i \bar{V}_i$$

where $\bar{V}_i$ is the partial molal volume of component i in the mixture. The volume change on mixing is

$$\Delta V_{\text{mixing}} = V_{\text{mixture}} - V_{\text{components}} = \sum_{i=1}^{j} n_i \bar{V}_i - \sum_{i=1}^{j} n_i \bar{v}_i$$

or

$$\Delta V_{\text{mixing}} = \sum_{i=1}^{j} n_i (\bar{V}_i - \bar{v}_i) \tag{11.105}$$

Similar results can be obtained for other extensive properties, for example,

$$\Delta U_{\text{mixing}} = \sum_{i=1}^{j} n_i (\bar{U}_i - \bar{u}_i)$$

$$\Delta H_{\text{mixing}} = \sum_{i=1}^{j} n_i (\bar{H}_i - \bar{h}_i) \tag{11.106}$$

$$\Delta S_{\text{mixing}} = \sum_{i=1}^{j} n_i (\bar{S}_i - \bar{s}_i)$$

In Eqs. 11.106, $\bar{u}_i$, $\bar{h}_i$, and $\bar{s}_i$ denote the molar internal energy, enthalpy, and entropy of pure component i, respectively. The symbols $\bar{U}_i$, $\bar{H}_i$, and $\bar{S}_i$ denote the respective partial molal properties.

11.9.2 Chemical Potential

Of the partial molal properties, the partial molal Gibbs function is particularly useful in describing the behavior of mixtures and solutions. This quantity plays a central role in the criteria for both chemical and phase equilibrium (Chap. 14). Because of its importance in the study of multicomponent systems, the partial molal Gibbs function of component i is given a special name and symbol. It is called the **chemical potential** of component i and symbolized by μ_i:

chemical potential

$$\boxed{\mu_i = \bar{G}_i = \frac{\partial G}{\partial n_i}\bigg)_{T, p, n_l}} \tag{11.107}$$

Like temperature and pressure, the chemical potential μ_i is an *intensive* property.

Applying Eq. 11.103 together with Eq. 11.107, the following expression can be written:

$$G = \sum_{i=1}^{j} n_i \mu_i \tag{11.108}$$

Expressions for the internal energy, enthalpy, and Helmholtz function can be obtained from Eq. 11.108, using the definitions $H = U + pV$, $G = H - TS$, and $\psi = U - TS$. They are

$$U = TS - pV + \sum_{i=1}^{j} n_i \mu_i$$

$$H = TS + \sum_{i=1}^{j} n_i \mu_i \qquad (11.109)$$

$$\Psi = -pV + \sum_{i=1}^{j} n_i \mu_i$$

Other useful relations can be obtained as well. Forming the differential of $G(T, p, n_1, n_2, \ldots, n_j)$

$$dG = \frac{\partial G}{\partial p}\bigg)_{T,n} dp + \frac{\partial G}{\partial T}\bigg)_{p,n} dT + \sum_{i=1}^{j}\left(\frac{\partial G}{\partial n_i}\right)_{T,p,n_l} dn_i \qquad (11.110)$$

The subscripts n in the first two terms indicate that all n's are held fixed during differentiation. Since this implies fixed composition, it follows from Eqs. 11.30 and 11.31 (Sec. 11.3.2) that

$$V = \left(\frac{\partial G}{\partial p}\right)_{T,n} \qquad \text{and} \qquad -S = \left(\frac{\partial G}{\partial T}\right)_{p,n} \qquad (11.111)$$

With Eqs. 11.107 and 11.111, Eq. 11.110 becomes

$$\boxed{dG = V\, dp - S\, dT + \sum_{i=1}^{j} \mu_i\, dn_i} \qquad (11.112)$$

which for a multicomponent system is the counterpart of Eq. 11.23.

Another expression for dG is obtained by forming the differential of Eq. 11.108. That is,

$$dG = \sum_{i=1}^{j} n_i\, d\mu_i + \sum_{i=1}^{j} \mu_i\, dn_i$$

Gibbs–Duhem equation   Combining this equation with Eq. 11.112 gives the **Gibbs–Duhem equation**

$$\sum_{i=1}^{j} n_i\, d\mu_i = V\, dp - S\, dT \qquad (11.113)$$

11.9.3 Fundamental Thermodynamic Functions for Multicomponent Systems

A *fundamental thermodynamic function* provides a complete description of the thermodynamic state of a system. In principle, all properties of interest can be determined from such a function by differentiation and/or combination. Reviewing the developments of Sec. 11.9.2, we see that a function $G(T, p, n_1, n_2, \ldots, n_j)$ is a fundamental thermodynamic function for a multicomponent system.

Functions of the form $U(S, V, n_1, n_2, \ldots, n_j)$, $H(S, p, n_1, n_2, \ldots, n_j)$, and $\psi(T, V, n_1, n_2, \ldots, n_j)$ also can serve as fundamental thermodynamic functions for multicomponent systems. To demonstrate this, first form the differential of each of Eqs. 11.109 and use the Gibbs–Duhem equation, Eq. 11.113, to reduce the resultant expressions to obtain

$$dU = T\, dS - p\, dV + \sum_{i=1}^{j} \mu_i\, dn_i \qquad (11.114a)$$

$$dH = T\, dS - V\, dp + \sum_{i=1}^{j} \mu_i\, dn_i \qquad (11.114b)$$

$$d\Psi = -p\, dV - S\, dT + \sum_{i=1}^{j} \mu_i\, dn_i \qquad (11.114c)$$

For multicomponent systems, these are the counterparts of Eqs. 11.18, 11.19, and 11.22, respectively.

The differential of $U(S, V, n_1, n_2, \ldots, n_j)$ is

$$dU = \frac{\partial U}{\partial S}\bigg)_{V,n} dS + \frac{\partial U}{\partial V}\bigg)_{S,n} dV + \sum_{i=1}^{j}\left(\frac{\partial U}{\partial n_i}\bigg)_{S,V,n_l}\right) dn_i$$

Comparing this expression term by term with Eq. 11.114a, we have

$$T = \frac{\partial U}{\partial S}\bigg)_{V,n}, \qquad -p = \frac{\partial U}{\partial V}\bigg)_{S,n}, \qquad \mu_i = \frac{\partial U}{\partial n_i}\bigg)_{S,V,n_l} \qquad (11.115a)$$

That is, the temperature, pressure, and chemical potentials can be obtained by differentiation of $U(S, V, n_1, n_2, \ldots, n_j)$. The first two of Eqs. 11.115a are the counterparts of Eqs. 11.24 and 11.25.

A similar procedure using a function of the form $H(S, p, n_1, n_2, \ldots, n_j)$ together with Eq. 11.114b gives

$$T = \frac{\partial H}{\partial S}\bigg)_{p,n}, \qquad V = \frac{\partial H}{\partial p}\bigg)_{S,n}, \qquad \mu_i = \frac{\partial H}{\partial n_i}\bigg)_{S,p,n_l} \qquad (11.115b)$$

where the first two of these are the counterparts of Eqs. 11.26 and 11.27. Finally, with $\psi(S, V, n_1, n_2, \ldots, n_j)$ and Eq. 11.114c

$$-p = \frac{\partial \Psi}{\partial V}\bigg)_{T,n}, \qquad -S = \frac{\partial \Psi}{\partial T}\bigg)_{V,n}, \qquad \mu_i = \frac{\partial \Psi}{\partial n_i}\bigg)_{T,V,n_l} \qquad (11.115c)$$

The first two of these are the counterparts of Eqs. 11.28 and 11.29. With each choice of fundamental function, the remaining extensive properties can be found by combination using the definitions $H = U + pV$, $G = H - TS$, $\psi = U - TS$.

The foregoing discussion of fundamental thermodynamic functions has led to several property relations for multicomponent systems that correspond to relations obtained previously. In addition, counterparts of the Maxwell relations can be obtained by equating mixed second partial derivatives. For example, the first two terms on the right of Eq. 11.112 give

$$\frac{\partial V}{\partial T}\bigg)_{p,n} = -\frac{\partial S}{\partial p}\bigg)_{T,n} \qquad (11.116)$$

which corresponds to Eq. 11.35. Numerous relationships involving chemical potentials can be derived similarly by equating mixed second partial derivatives. An important example from Eq. 11.112 is

$$\frac{\partial \mu_i}{\partial p}\bigg)_{T,n} = \frac{\partial V}{\partial n_i}\bigg)_{T,p,n_l}$$

Recognizing the right side of this equation as the partial molal volume, we have

$$\boxed{\frac{\partial \mu_i}{\partial p}\bigg)_{T,n} = \overline{V}_i} \qquad (11.117)$$

This relationship is applied in the development of Eqs. 11.126.

The present discussion concludes by listing four different expressions derived above for the chemical potential in terms of other properties. In the order obtained, they are

$$\boxed{\mu_i = \frac{\partial G}{\partial n_i}\bigg)_{T,p,n_l} = \frac{\partial U}{\partial n_i}\bigg)_{S,V,n_i} = \frac{\partial H}{\partial n_i}\bigg)_{S,p,n_i} = \frac{\partial \Psi}{\partial n_i}\bigg)_{T,V,n_i}} \qquad (11.118)$$

Only the first of these partial derivatives is a partial molal property, however, for the term *partial molal* applies only to partial derivatives where the independent variables are temperature, pressure, and the number of moles of each component present.

11.9.4 Fugacity

The chemical potential plays an important role in describing multicomponent systems. In some instances, however, it is more convenient to work in terms of a related property, the fugacity. The fugacity is introduced in the present discussion.

Single-Component Systems Let us begin by taking up the case of a system consisting of a single component. For this case, Eq. 11.108 reduces to give

$$G = n\mu \quad \text{or} \quad \mu = \frac{G}{n} = \bar{g}$$

That is, for a pure component the chemical potential equals the Gibbs function per mole. With this equation, Eq. 11.30 written on a per mole basis becomes

$$\left(\frac{\partial \mu}{\partial p}\right)_T = \bar{v} \tag{11.119}$$

For the special case of an ideal gas, $\bar{v} = \bar{R}T/p$, and Eq. 11.119 assumes the form

$$\left(\frac{\partial \mu^*}{\partial p}\right)_T = \frac{\bar{R}T}{p}$$

where the asterisk denotes ideal gas. Integrating at constant temperature

$$\mu^* = \bar{R}T \ln p + C(T) \tag{11.120}$$

where $C(T)$ is a function of integration. Since the pressure p can take on values from zero to plus infinity, the $\ln p$ term of this expression, and thus the chemical potential, has an inconvenient range of values from minus infinity to plus infinity. Equation 11.120 also shows that the chemical potential can be determined only to within an arbitrary constant.

Introducing Fugacity Because of the above considerations, it is advantageous for many types of thermodynamic analyses to use fugacity in place of the chemical potential, for it is a well-behaved function that can be more conveniently evaluated. We introduce the **fugacity** f by the expression

fugacity

$$\mu = \bar{R}T \ln f + C(T) \tag{11.121}$$

Comparing Eq. 11.121 with Eq. 11.120, the fugacity is seen to play the same role in the general case as pressure plays in the ideal gas case. Fugacity has the same units as pressure.

Substituting Eq. 11.121 into Eq. 11.119 gives

$$\boxed{\bar{R}T\left(\frac{\partial \ln f}{\partial p}\right)_T = \bar{v}} \tag{11.122}$$

Integration of Eq. 11.122 while holding temperature constant can determine the fugacity only to within a constant term. However, since ideal gas behavior is approached as pressure tends to zero, the constant term can be fixed by requiring that the fugacity of a pure component equals the pressure in the limit of zero pressure. That is,

$$\boxed{\lim_{p \to 0} \frac{f}{p} = 1} \tag{11.123}$$

Equations 11.122 and 11.123 then *completely determine* the fugacity function.

Evaluating Fugacity Let us consider next how the fugacity can be evaluated. With $Z = p\bar{v}/\bar{R}T$, Eq. 11.122 becomes

$$\bar{R}T\left(\frac{\partial \ln f}{\partial p}\right)_T = \frac{\bar{R}TZ}{p}$$

or

$$\left(\frac{\partial \ln f}{\partial p}\right)_T = \frac{Z}{p}$$

Subtracting $1/p$ from both sides and integrating from pressure p' to pressure p at fixed temperature T

$$[\ln f - \ln p]_{p'}^{p} = \int_{p'}^{p}(Z - 1)d \ln p$$

or

$$\left[\ln \frac{f}{p}\right]_{p'}^{p} = \int_{p'}^{p}(Z - 1)d \ln p$$

Taking the limit as p' tends to zero and applying Eq. 11.123 results in

$$\ln \frac{f}{p} = \int_{0}^{p}(Z - 1)d \ln p$$

When expressed in terms of the reduced pressure, $p_R = p/p_c$, the above equation is

$$\boxed{\ln \frac{f}{p} = \int_{0}^{p_R}(Z - 1)d \ln p_R} \tag{11.124}$$

Since the compressibility factor Z depends on the reduced temperature T_R and reduced pressure p_R, it follows that the right side of Eq. 11.124 depends on these properties only. Accordingly, the quantity $\ln f/p$ is a function only of these two reduced properties. Using a generalized equation of state giving Z as a function of T_R and p_R, $\ln f/p$ can readily be evaluated with a computer. Tabular representations are also found in the literature. Alternatively, the graphical representation presented in Fig. A-6 can be employed.

FOR EXAMPLE

To illustrate the use of Fig. A-6, consider two states of water vapor at the same temperature, 400°C. At state 1 the pressure is 200 bar, and at state 2 the pressure is 240 bar. The change in the chemical potential between these states can be determined using Eq. 11.121 as

$$\mu_2 - \mu_1 = \bar{R}T \ln \frac{f_2}{f_1} = \bar{R}T \ln\left(\frac{f_2}{p_2}\frac{p_2}{p_1}\frac{p_1}{f_1}\right)$$

Using the critical temperature and pressure of water from Table A-1, at state 1 $p_{R1} = 0.91$, $T_{R1} = 1.04$, and at state 2 $p_{R2} = 1.09$, $T_{R2} = 1.04$. By inspection of Fig. A-6, $f_1/p_1 = 0.755$ and $f_2/p_2 = 0.7$. Inserting values in the above equation

$$\mu_2 - \mu_1 = (8.314)(673.15) \ln\left[(0.7)\left(\frac{240}{200}\right)\left(\frac{1}{0.755}\right)\right] = 597 \text{ kJ/kmol}$$

For a pure component, the chemical potential equals the Gibbs function per mole, $\bar{g} = \bar{h} - T\bar{s}$. Since the temperature is the same at states 1 and 2, the change in the chemical potential can be expressed as $\mu_2 - \mu_1 = \bar{h}_2 - \bar{h}_1 - T(\bar{s}_2 - \bar{s}_1)$. Using steam table data, the value obtained with this expression is 597 kJ/kmol, which agrees with the value determined from the generalized fugacity coefficient chart.

Multicomponent Systems The fugacity of a component i in a mixture can be defined by a procedure that parallels the definition for a pure component. For a pure component, the development begins with Eq. 11.119, and the fugacity is introduced by Eq. 11.121. These are then used to write the pair of equations, Eqs. 11.122 and 11.123, from which the fugacity can be evaluated. For a mixture, the development begins with Eq. 11.117, the counterpart of Eq. 11.119, and the fugacity $\bar{f_i}$ of component i is introduced by

$$\mu_i = \bar{R}T \ln \bar{f_i} + C_i(T) \tag{11.125}$$

fugacity of a mixture component

which parallels Eq. 11.121. The pair of equations that allow the **fugacity of a mixture component**, $\bar{f_i}$, to be evaluated is

$$\bar{R}T \left(\frac{\partial \ln \bar{f_i}}{\partial p} \right)_{T,n} = \bar{V_i} \tag{11.126a}$$

$$\lim_{p \to 0} \left(\frac{\bar{f_i}}{y_i p} \right) = 1 \tag{11.126b}$$

The symbol $\bar{f_i}$ denotes the fugacity of component i in the *mixture* and should be carefully distinguished in the presentation to follow from f_i, which denotes the fugacity of *pure i*.

Discussion Referring to Eq. 11.126b, note that in the ideal gas limit the fugacity $\bar{f_i}$ is not required to equal the pressure p, as for the case of a pure component, but to equal the quantity $y_i p$. To see that this is the appropriate limiting quantity, consider a system consisting of a mixture of gases occupying a volume V at pressure p and temperature T. If the overall mixture behaves as an ideal gas, we can write

$$p = \frac{n\bar{R}T}{V} \tag{11.127}$$

where n is the total number of moles of mixture. Recalling from Sec. 3.12.3 that an ideal gas can be regarded as composed of molecules that exert negligible forces on one another and whose volume is negligible relative to the total volume, we can think of each component i as behaving as if it were an ideal gas alone at the temperature T and volume V. Thus, the pressure exerted by component i would not be the mixture pressure p but the pressure p_i given by

$$p_i = \frac{n_i \bar{R}T}{V} \tag{11.128}$$

where n_i is the number of moles of component i. Dividing Eq. 11.128 by Eq. 11.127

$$\frac{p_i}{p} = \frac{n_i \bar{R}T/V}{n\bar{R}T/V} = \frac{n_i}{n} = y_i$$

On rearrangement

$$p_i = y_i p \tag{11.129}$$

Accordingly, the quantity $y_i p$ appearing in Eq. 11.126b corresponds to the pressure p_i.

Summing both sides of Eq. 11.129, we obtain

$$\sum_{i=1}^{j} p_i = \sum_{i=1}^{j} y_i p = p \sum_{i=1}^{j} y_i$$

Or, since the sum of the mole fractions equals unity

$$p = \sum_{i=1}^{j} p_i \tag{11.130}$$

In words, Eq. 11.130 states that the sum of the pressures p_i equals the mixture pressure. This gives rise to the designation *partial pressure* for p_i. With this background, we now see that

Eq. 11.126b requires the fugacity of component i to approach the partial pressure of component i as pressure p tends to zero. Comparing Eqs. 11.130 and 11.99a, we also see that the *additive pressure rule* is exact for ideal gas mixtures. This special case is considered further in Sec. 12.2 under the heading *Dalton model*.

Evaluating Fugacity in a Mixture   Let us consider next how the fugacity of component i in a mixture can be expressed in terms of quantities that can be evaluated. For a pure component i, Eq. 11.122 gives

$$\bar{R}T\left(\frac{\partial \ln f_i}{\partial p}\right)_T = \bar{v}_i \tag{11.131}$$

where $\bar{v}_i$ is the molar specific volume of pure i. Subtracting Eq. 11.131 from Eq. 11.126a

$$\bar{R}T\left[\frac{\partial \ln (\bar{f}_i/f_i)}{\partial p}\right]_{T,n} = \bar{V}_i - \bar{v}_i \tag{11.132}$$

Integrating from pressure p' to pressure p at fixed temperature and mixture composition

$$\bar{R}T\left[\ln\left(\frac{\bar{f}_i}{f_i}\right)\right]_{p'}^{p} = \int_{p'}^{p}(\bar{V}_i - \bar{v}_i)\, dp$$

In the limit as p' tends to zero, this becomes

$$\bar{R}T\left[\ln\left(\frac{\bar{f}_i}{f_i}\right) - \lim_{p'\to 0}\ln\left(\frac{\bar{f}_i}{f_i}\right)\right] = \int_{0}^{p}(\bar{V}_i - \bar{v}_i)\, dp$$

Since $f_i \to p'$ and $\bar{f}_i \to y_i p'$ as pressure p' tends to zero

$$\lim_{p'\to 0}\ln\left(\frac{\bar{f}_i}{f_i}\right) \to \ln\left(\frac{y_i p'}{p'}\right) = \ln y_i$$

Therefore, we can write

$$\bar{R}T\left[\ln\left(\frac{\bar{f}_i}{f_i}\right) - \ln y_i\right] = \int_{0}^{p}(\bar{V}_i - \bar{v}_i)\, dp$$

or

$$\boxed{\bar{R}T\ln\left(\frac{\bar{f}_i}{y_i f_i}\right) = \int_{0}^{p}(\bar{V}_i - \bar{v}_i)\, dp} \tag{11.133}$$

in which $\bar{f}_i$ is the fugacity of component i at pressure p in a mixture of given composition at a given temperature, and f_i is the fugacity of pure i at the same temperature and pressure. Equation 11.133 expresses the relation between $\bar{f}_i$ and f_i in terms of the difference between $\bar{V}_i$ and $\bar{v}_i$, a measurable quantity.

11.9.5 Ideal Solution

The task of evaluating the fugacities of the components in a mixture is considerably simplified when the mixture can be modeled as an ideal solution. An **ideal solution** is a mixture for which

ideal solution

$$\boxed{\bar{f}_i = y_i f_i \qquad \text{(ideal solution)}} \tag{11.134}$$

Lewis–Randall rule Equation 11.134, known as the **Lewis–Randall rule**, states that the fugacity of each component in an ideal solution is equal to the product of its mole fraction and the fugacity of the pure component at the same temperature, pressure, and state of aggregation (gas, liquid, or solid) as the mixture. Many gaseous mixtures at low to moderate pressures are adequately modeled by the Lewis–Randall rule. The ideal gas mixtures considered in Chap. 12 are an important special class of such mixtures. Some liquid solutions also can be modeled with the Lewis–Randall rule.

As consequences of the definition of an ideal solution, the following characteristics are exhibited:

- Introducing Eq. 11.134 into Eq. 11.132, the left side vanishes, giving $\bar{V}_i - \bar{v}_i = 0$, or

$$\bar{V}_i = \bar{v}_i \tag{11.135}$$

Thus, the partial molal volume of each component in an ideal solution is equal to the molar specific volume of the corresponding pure component at the same temperature and pressure. When Eq. 11.135 is introduced in Eq. 11.105, it can be concluded that there is no volume change on mixing pure components to form an ideal solution.

With Eq. 11.135, the volume of an ideal solution is

$$V = \sum_{i=1}^{j} n_i \bar{V}_i = \sum_{i=1}^{j} n_i \bar{v}_i = \sum_{i=1}^{j} V_i \quad \text{(ideal solution)} \tag{11.136}$$

where V_i is the volume that pure component i would occupy when at the temperature and pressure of the mixture. Comparing Eqs. 11.136 and 11.100a, the *additive volume rule* is seen to be exact for ideal solutions.

- It also can be shown that the partial molal internal energy of each component in an ideal solution is equal to the molar internal energy of the corresponding pure component at the same temperature and pressure. A similar result applies for enthalpy. In symbols

$$\bar{U}_i = \bar{u}_i, \qquad \bar{H}_i = \bar{h}_i \tag{11.137}$$

With these expressions, it can be concluded from Eqs. 11.106 that there is no change in internal energy or enthalpy on mixing pure components to form an ideal solution.

With Eqs. 11.137, the internal energy and enthalpy of an ideal solution are

$$U = \sum_{i=1}^{j} n_i \bar{u}_i \quad \text{and} \quad H = \sum_{i=1}^{j} n_i \bar{h}_i \quad \text{(ideal solution)} \tag{11.138}$$

where $\bar{u}_i$ and $\bar{h}_i$ denote, respectively, the molar internal energy and enthalpy of pure component i at the temperature and pressure of the mixture.

Although there is no change in V, U, or H on mixing pure components to form an ideal solution, we expect an entropy increase to result from the *adiabatic* mixing of different pure components because such a process is irreversible: The separation of the mixture into the pure components would never occur spontaneously. The entropy change on adiabatic mixing is considered further for the special case of ideal gas mixtures in Sec. 12.4.2.

The Lewis–Randall rule requires that the fugacity of mixture component i be evaluated in terms of the fugacity of pure component i at the same temperature and pressure as the mixture and in the *same state of aggregation*. For example, if the mixture were a gas at T, p, then f_i would be determined for pure i at T, p and as a gas. However, at certain temperatures and pressures of interest a component of a gaseous mixture may, as a pure substance, be a liquid or solid. An example is an air–water vapor mixture at 20°C and 1 atm. At this temperature and pressure, water exists not as a vapor but as a liquid. Although not considered here, means have been developed that allow the ideal solution model to be useful in such cases.

11.9.6 Chemical Potential for Ideal Solutions

The discussion of multicomponent systems concludes with the introduction of expressions for evaluating the chemical potential for ideal solutions used in Sec. 14.3.3.

Consider a reference state where component i of a multicomponent system is pure at the temperature T of the system and a reference-state pressure p_{ref}. The difference in the chemical potential of i between a specified state of the multicomponent system and the reference state is obtained with Eq. 11.125 as

$$\mu_i - \mu_i^\circ = \bar{R}T \ln \frac{\bar{f}_i}{f_i^\circ} \tag{11.139}$$

where the superscript $^\circ$ denotes property values at the reference state. The fugacity ratio appearing in the logarithmic term is known as the **activity**, a_i, of component i in the mixture. That is,

activity

$$\boxed{a_i = \frac{\bar{f}_i}{f_i^\circ}} \tag{11.140}$$

For subsequent applications, it suffices to consider the case of gaseous mixtures. For gaseous mixtures, p_{ref} is specified as 1 atm, so μ_i° and f_i° in Eq. 11.140 are, respectively, the chemical potential and fugacity of pure i at temperature T and 1 atm.

Since the chemical potential of a pure component equals the Gibbs function per mole, Eq. 11.139 can be written as

$$\boxed{\mu_i = \bar{g}_i^\circ + \bar{R}T \ln a_i} \tag{11.141}$$

where $\bar{g}_i^\circ$ is the Gibbs function per mole of pure component i evaluated at temperature T and 1 atm: $\bar{g}_i^\circ = \bar{g}_i\,(T, 1\text{ atm})$.

For an ideal solution, the Lewis–Randall rule applies and the activity is

$$a_i = \frac{y_i f_i}{f_i^\circ} \tag{11.142}$$

where f_i is the fugacity of pure component i at temperature T and pressure p. Introducing Eq. 11.142 into Eq. 11.141

$$\mu_i = \bar{g}_i^\circ + \bar{R}T \ln \frac{y_i f_i}{f_i^\circ}$$

or

$$\boxed{\mu_i = \bar{g}_i^\circ + \bar{R}T \ln \left[\left(\frac{f_i}{p} \right) \left(\frac{p_{ref}}{f_i^\circ} \right) \frac{y_i p}{p_{ref}} \right]} \quad \text{(ideal solution)} \tag{11.143}$$

In principle, the ratios of fugacity to pressure shown underlined in this equation can be evaluated from Eq. 11.124 or the generalized fugacity chart, Fig. A-6, developed from it. If component i behaves as an ideal gas at both T, p and T, p_{ref}, we have $f_i/p = f_i^\circ/p_{ref} = 1$; Eq. 11.143 then reduces to

$$\boxed{\mu_i = \bar{g}_i^\circ + \bar{R}T \ln \frac{y_i p}{p_{ref}}} \quad \text{(ideal gas)} \tag{11.144}$$

CHAPTER SUMMARY AND STUDY GUIDE

In this chapter, we introduce thermodynamic relations that allow u, h, and s as well as other properties of simple compressible systems to be evaluated using property data that are more readily measured. The emphasis is on systems involving a single chemical species such as water or a mixture such as air. An introduction to general property relations for mixtures and solutions is also included.

Equations of state relating p, v, and T are considered, including the virial equation and examples of two-constant and multiconstant equations. Several important property relations based on the mathematical characteristics of exact differentials are developed, including the Maxwell relations. The concept of a fundamental thermodynamic function is discussed. Means for evaluating changes in specific internal energy, enthalpy, and entropy are developed and applied to phase change and to single-phase processes. Property relations are introduced involving the volume expansivity, isothermal and isentropic compressibilities, velocity of sound, specific heats and specific heat ratio, and the Joule–Thomson coefficient.

Additionally, we describe how tables of thermodynamic properties are constructed using the property relations and methods developed in this chapter. Such procedures also provide the basis for data retrieval by computer software. Also described are means for using the generalized enthalpy and entropy departure charts and the generalized fugacity coefficient chart to evaluate enthalpy, entropy, and fugacity, respectively.

We also consider p–v–T relations for gas mixtures of known composition, including Kay's rule. The chapter concludes with a discussion of property relations for multicomponent systems, including partial molal properties, chemical potential, fugacity, and activity. Ideal solutions and the Lewis–Randall rule are introduced as a part of that presentation.

The following checklist provides a study guide for this chapter. When your study of the text and end-of-chapter exercises has been completed, you should be able to write out the meanings of the terms listed in the margins throughout the chapter and understand each of the related concepts. The subset of key concepts listed below is particularly important. Additionally, for systems involving a single species you should be able to

- calculate p–v–T data using equations of state such as the Redlich–Kwong and Benedict–Webb–Rubin equations.

- apply the 16 property relations summarized in Table 11.1 and explain how the relations are obtained.

- evaluate Δs, Δu, and Δh, using the Clapeyron equation when considering phase change, and using equations of state and specific heat relations when considering single phases.

- apply the property relations introduced in Sec. 11.5, such as those involving the specific heats, the volume expansivity, and the Joule–Thomson coefficient.

- explain how tables of thermodynamic properties, such as Tables A-2 through A-18, are constructed.

- apply the generalized enthalpy and entropy departure charts, Figs. A-4 and A-5, to evaluate Δh and Δs.

For a *gas mixture* of known composition, you should be able to

- apply the methods introduced in Sec. 11.8 for relating pressure, specific volume, and temperature—Kay's rule, for example.

For *multicomponent systems*, you should be able to

- evaluate extensive properties in terms of the respective partial molal properties.

- evaluate partial molal volumes using the *method of intercepts*.

- evaluate fugacity using data from the generalized fugacity coefficient chart, Fig. A-6.

- apply the ideal solution model.

KEY ENGINEERING CONCEPTS

equation of state	fundamental thermodynamic function	method of intercepts
exact differential	Clapeyron equation	chemical potential
test for exactness	Joule–Thomson coefficient	fugacity
Helmholtz function	enthalpy departure	Lewis–Randall rule
Gibbs function	entropy departure	
Maxwell relations	Kay's rule	

KEY EQUATIONS

Equations of State

$$Z = 1 + \frac{B(T)}{\overline{v}} + \frac{C(T)}{\overline{v}^2} + \frac{D(T)}{\overline{v}^3} + \cdots$$	(11.1)	Virial equation of state
$$p = \frac{\overline{R}T}{\overline{v} - b} - \frac{a}{\overline{v}^2}$$	(11.2)	van der Waals equation of state
$$p = \frac{\overline{R}T}{\overline{v} - b} - \frac{a}{\overline{v}(\overline{v} + b)T^{1/2}}$$	(11.7)	Redlich–Kwong equation of state

Mathematical Relations for Properties

$$\frac{\partial}{\partial y}\left[\left(\frac{\partial z}{\partial x}\right)_y\right]_x = \frac{\partial}{\partial x}\left[\left(\frac{\partial z}{\partial y}\right)_x\right]_y$$	(11.14a)	Test for exactness
$$\left(\frac{\partial M}{\partial y}\right)_x = \left(\frac{\partial N}{\partial x}\right)_y$$	(11.14b)	

$\left(\dfrac{\partial x}{\partial y}\right)_z \left(\dfrac{\partial y}{\partial x}\right)_z = 1$	(11.15)	Important relations among partial derivatives of properties
$\left(\dfrac{\partial y}{\partial z}\right)_x \left(\dfrac{\partial z}{\partial x}\right)_y \left(\dfrac{\partial x}{\partial y}\right)_z = -1$	(11.16)	
Table 11.1	(11.24–11.36)	Summary of property relations from exact differentials

Expressions for Δu, Δh, and Δs

$\left(\dfrac{dp}{dT}\right)_{\text{sat}} = \dfrac{h_{\text{g}} - h_{\text{f}}}{T(v_{\text{g}} - v_{\text{f}})}$	(11.40)	Clapeyron equation
$s_2 - s_1 = \displaystyle\int_1^2 \dfrac{c_v}{T}\, dT + \int_1^2 \left(\dfrac{\partial p}{\partial T}\right)_v dv$	(11.50)	Expressions for changes in s and u with T and v as independent variables
$u_2 - u_1 = \displaystyle\int_1^2 c_v\, dT + \int_1^2 \left[T\left(\dfrac{\partial p}{\partial T}\right)_v - p\right] dv$	(11.51)	
$s_2 - s_1 = \displaystyle\int_1^2 \dfrac{c_p}{T}\, dT - \int_1^2 \left(\dfrac{\partial v}{\partial T}\right)_p dp$	(11.59)	Expressions for changes in s and h with T and p as independent variables
$h_2 - h_1 = \displaystyle\int_1^2 c_p\, dT + \int_1^2 \left[v - T\left(\dfrac{\partial v}{\partial T}\right)_p\right] dp$	(11.60)	
$\bar{h}_2 - \bar{h}_1 = \bar{h}_2^* - \bar{h}_1^* - \bar{R}T_{\text{c}}\left[\left(\dfrac{\bar{h}^* - \bar{h}}{\bar{R}T_{\text{c}}}\right)_2 - \left(\dfrac{\bar{h}^* - \bar{h}}{\bar{R}T_{\text{c}}}\right)_1\right]$	(11.85)	Evaluating enthalpy and entropy changes in terms of generalized enthalpy and entropy departures and data from Figs. A-4 and A-5, respectively
$\bar{s}_2 - \bar{s}_1 = \bar{s}_2^* - \bar{s}_1^* - \bar{R}\left[\left(\dfrac{\bar{s}^* - \bar{s}}{\bar{R}}\right)_2 - \left(\dfrac{\bar{s}^* - \bar{s}}{\bar{R}}\right)_1\right]$	(11.92)	

Additional Thermodynamic Relations

$\psi = u - Ts$	(11.20)	Helmholtz function
$g = h - Ts$	(11.21)	Gibbs function
$c = \sqrt{-v^2\left(\dfrac{\partial p}{\partial v}\right)_s} = \sqrt{-kv^2\left(\dfrac{\partial p}{\partial v}\right)_T}$	(9.36b) (11.74)	Expressions for velocity of sound
$\mu_{\text{J}} = \left(\dfrac{\partial T}{\partial p}\right)_h$	(11.75)	Joule–Thomson coefficient

Properties of Multicomponent Mixtures

$T_{\text{c}} = \displaystyle\sum_{i=1}^{j} y_i T_{\text{c},i}, \quad p_{\text{c}} = \sum_{i=1}^{j} y_i p_{\text{c},i}$	(11.97)	Kay's rule for critical temperature and pressure of mixtures
$\bar{X}_i = \left(\dfrac{\partial X}{\partial n_i}\right)_{T,p,n_l}$	(11.102)	Partial molal property $\bar{X}_i$ and its relation to extensive property X
$X = \displaystyle\sum_{i=1}^{j} n_i \bar{X}_i$	(11.103)	X as a weighted sum of partial molal properties

Properties of Multicomponent Mixtures

$\mu_i = \bar{G}_i = \dfrac{\partial G}{\partial n_i}\bigg)_{T,p,n_l}$	(11.107)	Chemical potential of species i in a mixture
$\bar{R}T\left(\dfrac{\partial \ln f}{\partial p}\right)_T = \bar{v}$	(11.122)	Expressions for evaluating fugacity of a single-component system
$\lim\limits_{p\to 0} \dfrac{f}{p} = 1$	(11.123)	
$\bar{R}T\left(\dfrac{\partial \ln \bar{f}_i}{\partial p}\right)_{T,n} = \bar{V}_i$	(11.126a)	Expressions for evaluating fugacity of mixture component i
$\lim\limits_{p\to 0}\left(\dfrac{\bar{f}_i}{y_i p}\right) = 1$	(11.126b)	
$\bar{f}_i = y_i f_i$	(11.134)	Lewis–Randall rule for ideal solutions
$\mu_i = \bar{g}_i^\circ + \bar{R}T \ln \dfrac{y_i p}{p_{\text{ref}}}$	(11.144)	Chemical potential of component i in an ideal gas mixture

EXERCISES: THINGS ENGINEERS THINK ABOUT

11.1 With so much thermodynamic property data software available today, engineers no longer need be well versed in the fundamentals of property evaluation—or do they?

11.2 To determine the specific volume of superheated water vapor at a known pressure and temperature, when would you use each of the following: the *steam tables*, the generalized compressibility chart, an equation of state, the ideal gas model?

11.3 What is an advantage of using the Redlich–Kwong equation of state in the generalized form given by Eq. 11.9 instead of Eq. 11.7? A disadvantage?

11.4 What is the reference state and accompanying reference property values used in constructing the *steam tables*, Tables A-2 through A-5?

11.5 If the function $p = p(T, v)$ is an equation of state, is $(\partial p/\partial T)_v$ a property? What are the independent variables of $(\partial p/\partial T)_v$?

11.6 In the expression $(\partial u/\partial T)_v$, what does the subscript v signify?

11.7 Explain how a Mollier diagram provides a graphical representation of the fundamental function $h(s, p)$.

11.8 What is the importance of the Clapeyron equation?

11.9 For a gas whose equation of state is $p\bar{v} = \bar{R}T$, are the specific heats $\bar{c}_p$ and $\bar{c}_v$ *necessarily* functions of temperature alone?

11.10 Referring to the p–T diagram for water, explain why ice melts under the blade of an ice skate.

11.11 For an ideal gas, what is the value of the Joule–Thomson coefficient?

11.12 At what states is the entropy departure negligible? The fugacity coefficient, f/p, closely equal to unity?

11.13 In Eq. 11.107, what do the subscripts T, p, and n_l signify? What does i denote?

11.14 How does Eq. 11.108 reduce for a system consisting of a pure substance? Repeat for an ideal gas mixture.

11.15 If two different liquids are mixed, is the final entropy *necessarily* equal to the sum of the original entropy values? Explain.

CHECKING UNDERSTANDING

11.1 For the equation of state,

$$p = \frac{\bar{R}T}{\bar{v} - b} + a$$

demonstrate that the mixed second partial derivatives are equal and explain the mathematical significance of this attribute.

11.2 For liquid water at 1 atm a slight increase in temperature from 0°C results in (a) an increase in specific volume, (b) a decrease in specific volume, (c) no change in specific volume, (d) the specific volume behavior cannot be determined without more information.

11.3 By inspection of the appendix tables for R-22, R-134a, ammonia, and propane, what is the temperature, in °C, at the reference state defined as in Fig. 11.4?

11.4 Extend Example 11.5 by obtaining an expression in terms of known property values for the change in enthalpy, $(h_2 - h_1)$, of the specified process.

11.5 For an ideal gas obtain expressions for the (a) volume expansivity and (b) isothermal compressibility.

11.6 For nitrogen, N_2, at 67.8 bar, −34°C, evaluate the enthalpy departure.

11.7 For the turbine of Examples 11.8 and 11.9 determine the exergetic efficiency. Take $T_0 = 298.15$ K.

11.8 The fugacity of water at 245°C, 133 bar is _____ bar.

11.9 The partial molal Gibbs function of a component i within a mixture or solution is also called the _____.

11.10 A closed, rigid tank holds 20 kg of a gas mixture having the molar analysis 50% nitrogen (N_2), 50% argon. If the mixture is at 180 K, 20 bar, determine the tank volume, in m^3, using Kay's rule.

11.11 Repeat parts (a)–(d) of Example 11.1 if the carbon monoxide is at −120°C, while all other given data remain the same.

11.12 Match the appropriate expression in the right column with each term in the left column.

_____ 1. $dh =$ A. $-p\,dv - s\,dT$
_____ 2. $dg =$ B. $T\,ds + v\,dp$
_____ 3. $du =$ C. $v\,dp - s\,dT$
_____ 4. $d\psi =$ D. $T\,ds - p\,dv$

11.13 Match the appropriate expression in the right column with each term in the left column.

_____ 1. $\left(\dfrac{\partial T}{\partial v}\right)_s =$ A. $\left(\dfrac{\partial \Psi}{\partial v}\right)_T$

_____ 2. $\left(\dfrac{\partial u}{\partial s}\right)_v =$ B. $\left(\dfrac{\partial s}{\partial v}\right)_T$

_____ 3. $\left(\dfrac{\partial p}{\partial T}\right)_v =$ C. $\left(\dfrac{\partial h}{\partial s}\right)_p$

_____ 4. $\left(\dfrac{\partial u}{\partial v}\right)_s =$ D. $-\left(\dfrac{\partial p}{\partial s}\right)_v$

11.14 Match the appropriate expression in the right column with each term in the left column.

_____ 1. $\alpha =$ A. $\left(\dfrac{\partial T}{\partial p}\right)_h$

_____ 2. $c =$ B. $\dfrac{1}{v}\left(\dfrac{\partial v}{\partial T}\right)_p$

_____ 3. $\beta =$ C. $\left(\dfrac{\partial v}{\partial p}\right)_T\left(\dfrac{\partial p}{\partial v}\right)_s$

_____ 4. $\mu_J =$ D. $-\dfrac{1}{v}\left(\dfrac{\partial v}{\partial p}\right)_s$

_____ 5. $\dfrac{c_p}{c_v} =$ E. $\sqrt{-v^2\left(\dfrac{\partial p}{\partial v}\right)_s}$

_____ 6. $\kappa =$ F. $-\dfrac{1}{v}\left(\dfrac{\partial v}{\partial p}\right)_T$

Indicate whether the following statements are true or false. Explain

11.15 The compressibility factor Z for a gas mixture can be evaluated using Kay's rule to determine mixture values for the critical temperature, T_c, and critical pressure, p_c.

11.16 The value of the isothermal compressibility is positive in all phases.

11.17 When pure components are mixed to form an ideal solution, no change in volume, internal energy, enthalpy, or entropy is observed.

11.18 Using only p–v–T data, the change in specific enthalpy for a change in phase from liquid to vapor can be evaluated from

$$h_g - h_f = T(v_g - v_f)\left(\frac{dp}{dT}\right)_{sat}$$

11.19 The following expression is the differential of an equation of state, $p = p(v, T)$:

$$dp = \left(\frac{-R}{v - b}\right)dT + \left[\frac{-RT}{(v-b)^2} + \frac{2a}{v^3}\right]dv$$

11.20 According to the additive pressure rule, the pressure of a gas mixture is expressible as the sum of the pressures exerted by the individual components, assuming each component occupies the entire volume at the mixture temperature.

11.21 Partial molal properties are extensive properties.

11.22 The value of the critical compressibility factor, Z_c, for most substances is in the range of 0.3 to 0.4.

11.23 For a simple compressible system whose pressure increases while temperature remains constant, the specific Gibbs function can only increase.

11.24 In an ideal solution, the *activity* of a component i is a measure of its tendency to react chemically with other components in the solution.

11.25 A Mollier diagram provides a graphical representation of the fundamental function $h(s, p)$.

PROBLEMS: DEVELOPING ENGINEERING SKILLS

C Problem may require use of appropriate computer software in order to complete.

Using Equations of State

11.1 Owing to safety requirements, the pressure within a 0.5 m³ cylinder should not exceed 52 bar. Check the pressure within the cylinder if filled with 45 kg of CO_2 maintained at 100°C using the

  a. van der Waals equation.
  b. compressibility chart.
  c. ideal gas equation of state.

11.2 Five kilograms mass of propane have a volume of 0.06 m³ and a pressure of 4 MPa. Determine the temperature, in K, using the

  a. van der Waals equation.
  b. compressibility chart.

  c. ideal gas equation of state.
  d. propane tables.

11.3 The pressure within a 23.1-m³ tank should not exceed 105 bar. Check the pressure within the tank if filled with 1200 kg of water vapor maintained at 320°C using the

  a. ideal gas equation of state.
  b. van der Waals equation.
  c. Redlich–Kwong equation.
  d. compressibility chart.
  e. steam tables.

11.4 Estimate the pressure of water vapor at a temperature of 480°C and a density of 24.789 kg/m³ using the

   a. steam tables.

   b. compressibility chart.

   c. Redlich–Kwong equation.

   d. van der Waals equation.

   e. ideal gas equation of state.

11.5 **C** Methane gas flows through a pipeline with a volumetric flow rate of 0.3 m³/s at a pressure of 183 bar and a temperature of 13°C. Determine the mass flow rate, in kg/s, using the

   a. ideal gas equation of state.

   b. van der Waals equation.

   c. compressibility chart.

11.6 Determine the specific volume of water vapor at 18 MPa and 440°C, in m³/kg, using the

   a. steam tables.

   b. compressibility chart.

   c. Redlich–Kwong equation.

   d. van der Waals equation.

   e. ideal gas equation of state.

11.7 A vessel whose volume is 1 m³ contains 4 kmol of methane at 100°C. Owing to safety requirements, the pressure of the methane should not exceed 12 MPa. Check the pressure using the

   a. ideal gas equation of state.

   b. Redlich–Kwong equation.

   c. Benedict–Webb–Rubin equation.

11.8 **C** Methane gas at 100 bar and −18°C is stored in a 10-m³ tank. Determine the mass of methane contained in the tank, in kg, using the

   a. ideal gas equation of state.

   b. van der Waals equation.

   c. Benedict–Webb–Rubin equation.

11.9 Using the Benedict–Webb–Rubin equation of state, determine the volume, in m³, occupied by 150 kg of methane at a pressure of 160 atm and temperature of 360 K. Compare with the results obtained using the ideal gas equation of state and the generalized compressibility chart.

11.10 One kilogram mass of air initially occupying a volume of 0.01 m³ at a pressure of 7 MPa expands isothermally and without irreversibilities until the volume is 0.06 m³. Using the Redlich–Kwong equation of state, determine the

   a. temperature, in K.

   b. final pressure, in MPa.

   c. work developed in the process, in kJ.

Using Relations from Exact Differentials

11.11 The differential of pressure obtained from a certain equation of state is given by *one* of the following expressions. Determine the equation of state.

$$dp = \frac{2(v - b)}{RT}dv + \frac{(v - b)^2}{RT^2}dT$$

$$dp = -\frac{RT}{(v - b)^2}dv + \frac{R}{v - b}dT$$

11.12 Show that Eq. 11.16 is satisfied by an equation of state with the form $p = [RT/(v - b)] + a$.

11.13 Using Eq. 11.35, check the consistency of

   a. the steam tables at 0.7 MPa, 316°C.

   b. the Refrigerant 134a tables at 0.3 MPa, 38°C.

11.14 For the functions $x = x(y, w)$, $y = y(z, w)$, $z = z(x, w)$, demonstrate that

$$\left(\frac{\partial x}{\partial y}\right)_w \left(\frac{\partial y}{\partial z}\right)_w \left(\frac{\partial z}{\partial x}\right)_w = 1$$

11.15 At a pressure of 1 bar, liquid water has a state of *maximum* density at about 4°C. What can be concluded about $(\partial s/\partial p)_T$ at

   a. 3°C?

   b. 4°C?

   c. 5°C?

11.16 A gas enters a compressor operating at steady state and is compressed isentropically. Does the specific enthalpy increase or decrease as the gas passes from the inlet to the exit?

11.17 Evaluate p, s, u, h, c_v, and c_p for a substance for which the Helmholtz function has the form

$$\psi = -RT \ln\frac{v}{v'} - cT'\left[1 - \frac{T}{T'} + \frac{T}{T'}\ln\frac{T}{T'}\right]$$

where v' and T' denote specific volume and temperature, respectively, at a reference state, and c is a constant.

11.18 The Mollier diagram provides a graphical representation of the fundamental thermodynamic function $h = h(s, p)$. Show that at any state fixed by s and p the properties T, v, u, ψ, and g can be evaluated using data obtained from the diagram.

Evaluating Δs, Δu, and Δh

11.19 Using $p–v–T$ data for saturated ammonia from Table A-13, calculate at −7°C.

   a. $h_g - h_f$.

   b. $u_g - u_f$.

   c. $s_g - s_f$.

Compare with the values obtained using table data.

11.20 Using $p–v–T$ data for saturated water from the steam tables, calculate at 50°C

   a. $h_g - h_f$.

   b. $u_g - u_f$.

   c. $s_g - s_f$.

Compare with values obtained from the steam tables.

11.21 Using h_{fg}, v_{fg}, and p_{sat} at −12°C from the Refrigerant 134a tables, estimate the saturation pressure at −7°C. Comment on the accuracy of your estimate.

11.22 At 0°C, the specific volumes of saturated solid water (ice) and saturated liquid water are, respectively, $v_i = 1.0911 \times 10^{-3}$ m³/kg and $v_f = 1.0002 \times 10^{-3}$ m³/kg, and the change in specific enthalpy on melting is $h_{if} = 333.4$ kJ/kg. Calculate the melting temperature of ice at (a) 400 bar, (b) 200 bar. Locate your answers on a sketch of the $p–T$ diagram for water.

11.23 Obtain the relationship between c_p and c_v for a gas that obeys the equation of state $p(v - b) = RT$.

11.24 Complete the following exercises dealing with slopes:

  a. At the triple point of water, evaluate the ratio of the slope of the vaporization line to the slope of the sublimation line. Use steam table data to obtain a numerical value for the ratio.

  b. Consider the superheated vapor region of a temperature–entropy diagram. Show that the slope of a constant specific volume line is greater than the slope of a constant pressure line through the same state.

  c. An enthalpy–entropy diagram (Mollier diagram) is often used in analyzing steam turbines. Obtain an expression for the slope of a constant-pressure line on such a diagram in terms of p–v–T data only.

  d. A pressure–enthalpy diagram is often used in the refrigeration industry. Obtain an expression for the slope of an isentropic line on such a diagram in terms of p–v–T data only.

11.25 One kmol of argon at 300 K is initially confined to one side of a rigid, insulated container divided into equal volumes of 0.2 m^3 by a partition. The other side is initially evacuated. The partition is removed and the argon expands to fill the entire container. Using the van der Waals equation of state, determine the final temperature of the argon, in K. Repeat using the ideal gas equation of state.

11.26 The p–v–T relation for a certain gas is represented closely by $v = RT/p + B - A/RT$, where R is the gas constant and A and B are constants. Determine expressions for the changes in specific enthalpy, internal energy, and entropy, $[h(p_2, T) - h(p_1, T)]$, $[u(p_2, T) - u(p_1, T)]$, and $[s(p_2, T) - s(p_1, T)]$, respectively.

11.27 Develop expressions for the specific enthalpy, internal energy, and entropy changes $[h(v_2, T) - h(v_1, T)]$, $[u(v_2, T) - u(v_1, T)]$, $[s(v_2, T) - s(v_1, T)]$, using the

  a. van der Waals equation of state.

  b. Redlich–Kwong equation of state.

11.28 At certain states, the p–v–T data of a gas can be expressed as $Z = 1 - Ap/T^4$, where Z is the compressibility factor and A is a constant.

  a. Obtain an expression for $(\partial p/\partial T)_v$ in terms of p, T, A, and the gas constant R.

  b. Obtain an expression for the change in specific entropy, $[s(p_2, T) - s(p_1, T)]$.

  c. Obtain an expression for the change in specific enthalpy, $[h(p_2, T) - h(p_1, T)]$.

11.29 For a gas whose p–v–T behavior is described by $Z = 1 + Bp/RT$, where B is a function of temperature, derive expressions for the specific enthalpy, internal energy, and entropy changes, $[h(p_2, T) - h(p_1, T)]$, $[u(p_2, T) - u(p_1, T)]$, and $[s(p_2, T) - s(p_1, T)]$.

11.30 For a gas whose p–v–T behavior is described by $Z = 1 + B/v + C/v^2$, where B and C are functions of temperature, derive an expression for the specific entropy change, $[s(v_2, T) - s(v_1, T)]$.

Using Other Thermodynamic Relations

11.31 Develop expressions for the volume expansivity β and the isothermal compressibility κ for

  a. an ideal gas.

  b. a gas whose equation of state is $p(v - b) = RT$.

  c. a gas obeying the van der Waals equation.

11.32 The volume of a 0.5 kg copper sphere is not allowed to vary by more than 0.1%. If the pressure exerted on the sphere is increased from

1 bar while the temperature remains constant at 27°C, determine the maximum allowed pressure, in bar. Average values of ρ, β, and κ are 9300 kg m^3, 4.91×10^{-5} (K)$^{-1}$, and 7.5×10^{-12} m^2/N, respectively.

11.33 Show that the isothermal compressibility κ is nearly always greater than or equal to the isentropic compressibility α.

11.34 Prove that $(\partial \beta/\partial p)_T = -(\partial \kappa/\partial T)_p$.

11.35 For a gas obeying the van der Waals equation of state,

  a. show that $(\partial c_v/\partial v)_T = 0$.

  b. develop an expression for $c_p - c_v$.

  c. develop expressions for $[u(T_2, v_2) - u(T_1, v_1)]$ and $[s(T_2, v_2) - s(T_1, v_1)]$.

  d. complete the Δu and Δs evaluations if $c_v = a + bT$, where a and b are constants.

11.36 At certain states, the p–v–T data for a particular gas can be represented as $Z = 1 - Ap/T^4$, where Z is the compressibility factor and A is a constant. Obtain an expression for the specific heat c_p in terms of the gas constant R, specific heat ratio k, and Z. Verify that your expression reduces to Eq. 3.47a when $Z = 1$.

11.37 For liquid water at 40°C, 1 bar estimate

  a. c_v, in kJ/kg · K.

  b. the velocity of sound, in m/s.

Use data from Table 11.2, as required.

11.38 Using steam table data, estimate the velocity of sound in liquid water at (a) 20°C, 50 bar, (b) 10°C, 10.3 MPa.

11.39 At a certain location in a *wind tunnel*, a stream of air is at 260°C, 1 bar and has a velocity of 635 m/s. Determine the Mach number at this location.

11.40 A gas is described by $v = RT/p - A/T + B$, where A and B are constants. For the gas

  a. obtain an expression for the temperatures at the Joule–Thomson inversion states.

  b. obtain an expression for $c_p - c_v$.

Developing Property Data

11.41 If the specific heat c_v of a gas obeying the van der Waals equation is given at a particular pressure, p', by $c_v = A + BT$, where A and B are constants, develop an expression for the change in specific entropy between any two states 1 and 2: $[s(T_2, p_2) - s(T_1, p_1)]$.

11.42 For air, develop an expression for the change in specific enthalpy from a state where the temperature is 25°C and the pressure is 1 atm to a state where the temperature is T and the pressure is p. Use the van der Waals equation of state and account for the variation of the ideal gas specific heat as in Table A-21.

11.43 Using the Redlich–Kwong equation of state, determine the changes in specific enthalpy, in kJ/kmol, and entropy, in kJ/kmol · K, for ethylene between 400 K, 1 bar and 400 K, 100 bar.

11.44 **C** Using the Redlich–Kwong equation of state together with an appropriate specific heat relation, determine the final temperature for an isentropic expansion of nitrogen from 400 K, 250 bar to 5 bar.

Using Enthalpy and Entropy Departures

11.45 The van der Waals equation is followed by a gas with specific heat capacity at constant volume, $c_v = A + BT$, where A and B are constants. On this basis derive an expression for the change in internal energy of the gas.

11.46 Derive an expression giving

    a. the internal energy of a substance relative to that of its ideal gas model at the same temperature: $[u(T, v) - u^*(T)]$.

    b. the entropy of a substance relative to that of its ideal gas model at the same temperature and specific volume: $[s(T, v) - s^*(T, v)]$.

11.47 Prove that the internal energy is a function of temperature only, that is, $u = u(T)$ for

    a. an incompressible substance.

    b. an ideal gas.

11.48 The following expression for the enthalpy departure is convenient for use with equations of state that are explicit in pressure:

$$\frac{\bar{h}^*(T) - \bar{h}(T, \bar{v})}{\bar{R}T_c} = T_R \left[1 - Z - \frac{1}{\bar{R}T} \int_\infty^{\bar{v}} \left[T \left(\frac{\partial p}{\partial T} \right)_v - p \right] d\bar{v} \right]$$

    a. Derive this expression.

    b. Using the given expression, evaluate the enthalpy departure for a gas obeying the Redlich–Kwong equation of state.

    c. Using the result of part (b), determine the change in specific enthalpy, in kJ/kmol, for CO_2 undergoing an isothermal process at 300 K from 50 to 20 bar.

11.49 Refrigerant-22 at 450 K, 15 bar enters a compressor operating at steady state and is compressed isothermally without internal irreversibilities to 50 bar. Kinetic and potential energy changes are negligible. Evaluate in kJ per kg of R-22 flowing through the compressor

    a. the work required.

    b. the heat transfer.

11.50 Nitrogen (N_2) enters a compressor operating at steady state at 1.5 MPa, 300 K and exits at 8 MPa, 500 K. If the work input is 240 kJ per kg of nitrogen flowing, determine the heat transfer, in kJ per kg of nitrogen flowing. Ignore kinetic and potential energy effects.

11.51 Oxygen (O_2) enters a control volume operating at steady state with a mass flow rate of 9 kg/min at 100 bar, 287 K and is compressed adiabatically to 150 bar, 400 K. Determine the power required, in kW, and the rate of entropy production, in kW/K. Ignore kinetic and potential energy effects.

11.52 A closed, rigid, insulated vessel having a volume of 0.142 m³ contains oxygen (O_2) initially at 100 bar, 7°C. The oxygen is stirred by a paddle wheel until the pressure becomes 150 bar. Determine the

    a. final temperature, in °C

    b. work, in kJ.

    c. amount of exergy destroyed in the process, in kJ.

Let $T_0 = 7°C$.

Evaluating p–v–T for Gas Mixtures

11.53 A preliminary design calls for a 1 kmol mixture of CO_2 and C_2H_2 to occupy a volume of 0.25 m³ at a temperature of 450 K. The mole fraction of CO_2 is 0.4. Owing to safety requirements, the pressure should not exceed 200 bar. Check the pressure using

    a. the ideal gas equation of state.

    b. Kay's rule together with the generalized compressibility chart.

    c. the additive pressure rule together with the generalized compressibility chart.

Compare and discuss these results.

11.54 A gaseous mixture with a molar composition of 60% CO and 40% H_2 enters a turbine operating at steady state at 150°C, 14 MPa and

exits at 100°C, 1 bar with a volumetric flow rate of 540 m³/min. Estimate the volumetric flow rate at the turbine inlet, in m³/min, using Kay's rule. What value would result from using the ideal gas model? Discuss.

11.55 A 0.2-m³ cylinder contains a gaseous mixture with a molar composition of 90% CO and 10% CO_2 initially at 150 bar. Due to a leak, the pressure of the mixture drops to 130 bar while the temperature remains, constant at 30°C. Using Kay's rule, estimate the amount of mixture, in kmol, that leaks from the cylinder.

11.56 A gaseous mixture consisting of 0.8 kmol of hydrogen (H_2) and 0.2 kmol of nitrogen (N_2) occupies 0.08 m³ at 25°C. Estimate the pressure, in bar, using

    a. the ideal gas equation of state.

    b. Kay's rule together with the generalized compressibility chart.

    c. the van der Waals equation together with mixture values for the constants a and b.

    d. the rule of additive pressure together with the generalized compressibility chart.

11.57 **C** A gaseous mixture of 0.25 kmol of methane and 0.25 kmol of propane occupies a volume of 0.21 m³ at a temperature of 90°C. Estimate the pressure using the following procedures and compare each estimate with the measured value of pressure, 50.7 bar:

    a. the ideal gas equation of state.

    b. Kay's rule together with the generalized compressibility chart.

    c. the van der Waals equation together with mixture values for the constants a and b.

    d. the rule of additive pressures together with the van der Waals equation.

    e. the rule of additive pressures together with the generalized compressibility chart.

    f. the rule of additive volumes together with the van der Waals equation.

11.58 Air having an approximate molar composition of 79% N_2 and 21% O_2 fills a 0.36-m³ vessel. The mass of mixture is 100 kg. The measured pressure and temperature are 101 bar and 180 K, respectively. Compare the measured pressure with the pressure predicted using

    a. the ideal gas equation of state.

    b. Kay's rule.

    c. the additive pressure rule with the Redlich–Kwong equation.

    d. the additive volume rule with the Redlich–Kwong equation.

11.59 Two tanks having equal volumes are connected by a valve. One tank contains carbon dioxide gas at 38°C and pressure p. The other tank contains ethylene gas at 38°C and 10.2 MPa. The valve is opened and the gases mix, eventually attaining equilibrium at 38°C and pressure p' with a composition of 20% carbon dioxide and 80% ethylene (molar basis). Using Kay's rule and the generalized compressibility chart, determine in kPa,

    a. the initial pressure of the carbon dioxide, p.

    b. the final pressure of the mixture, p'.

Analyzing Multicomponent Systems

11.60 A binary solution at 25°C consists of 59 kg of ethyl alcohol (C_2H_5OH) and 41 kg of water. The respective partial molal volumes are 0.0573 and 0.0172 m³/kmol. Determine the total volume, in m³. Compare with the volume calculated using the molar specific volumes of the pure components, each a liquid at 25°C, in the place of the partial molal volumes.

11.61 The following data are for a binary mixture of carbon dioxide and methane at a certain temperature and pressure:

mole fraction of methane	0.000	0.204	0.406	0.606	0.847	1.000
volume (in m³) per kmol of mixture		0.090	0.180	0.210	0.230	0.250	0.255

Estimate

a. the specific volumes of pure carbon dioxide and pure methane, each in m³/kmol.

b. the partial molal volumes of carbon dioxide and methane for an equimolar mixture, each in m³/kmol.

11.62 Determine the fugacity, in bar, for

a. butane at 555 K, 150 bar.

b. methane at 49°C, 5.5 MPa.

c. benzene at 495 K, 137 bar.

11.63 Consider a one-inlet, one-exit control volume at steady state through which the flow is internally reversible and isothermal. Show that the work per unit of mass flowing can be expressed in terms of the fugacity f as

$$\left(\frac{\dot{W}_{cv}}{\dot{m}}\right)_{\substack{int \\ rev}} = -RT \ \ln\left(\frac{f_2}{f_1}\right) + \frac{V_1^2 - V_2^2}{2} + g(z_1 - z_2)$$

11.64 Propane (C_3H_8) enters a turbine operating at steady state at 100 bar, 400 K and expands isothermally without irreversibilities to 10 bar. There are no significant changes in kinetic or potential energy. Using data from the generalized fugacity chart, determine the power developed, in kW, for a mass flow rate of 50 kg/min.

11.65 Ethane (C_2H_6) is compressed isothermally without irreversibilities at a temperature of 350 K from 15 to 50 bar. Using data from the generalized fugacity and enthalpy departure charts, determine the work of compression and the heat transfer, each in kJ per kg of ethane

flowing. Assume steady-state operation and neglect kinetic and potential energy effects.

11.66 Methane enters a turbine operating at steady state at 100 bar, 275 K and expands isothermally without irreversibilities to 15 bar. There are no significant changes in kinetic or potential energy. Using data from the generalized fugacity and enthalpy departure charts, determine the power developed and heat transfer, each in kW, for a mass flow rate of 0.5 kg/s.

11.67 Determine the fugacity, in bar, for pure methane at 250 K, 15 bar and as a component with a mole fraction of 0.35 in an ideal solution at the same temperature and pressure.

11.68 A tank contains 100 kg of a gaseous mixture of 80% ethane and 20% nitrogen (molar basis) at 350 K and 150 bar. Determine the volume of the tank, in m³, using data from the generalized compressibility chart together with (a) Kay's rule, (b) the ideal solution model. Compare with the measured tank volume of 1 m³.

11.69 A tank contains a mixture of 75% argon and 25% ethylene on a molar basis at 25°C, 82.5 bar. For 71 kg of mixture, estimate the tank volume, in m³, using

a. the ideal gas equation of state.

b. Kay's rule together with data from the generalized compressibility chart.

c. the ideal solution model together with data from the generalized compressibility chart.

11.70 A tank contains a mixture of 80% ethane and 20% nitrogen (N_2) on a molar basis at 600 K, 100 bar. For 1500 kg of mixture, estimate the tank volume, in m³, using

a. the ideal gas equation of state.

b. Kay's rule together with data from the generalized compressibility chart.

c. the ideal solution model together with data from the generalized compressibility chart.

DESIGN & OPEN-ENDED PROBLEMS: EXPLORING ENGINEERING PRACTICE

11.1D Design a laboratory flask for containing up to 10 kmol of mercury vapor at pressures up to 3 MPa and temperatures from 900 to 1000 K. Consider the health and safety of the technicians who would be working with such a mercury vapor–filled container. Use a p–v–T relation for mercury vapor obtained from the literature, including appropriate property software. Write a report including at least three references.

11.2D Compressed natural gas (CNG) is being used as a fuel to replace gasoline for automobile engines. Aluminum cylinders wrapped in a fibrous composite can provide lightweight, economical, and safe on-board storage. The storage vessels should hold enough CNG for 160 km to 200 km of urban travel, at storage pressures up to 21 MPa, and with a maximum total mass of 68 kg. Adhering to applicable U.S. Department of Transportation standards, specify both the size and number of cylinders that would meet the above design constraints.

11.3D Develop the preliminary design of a thermal storage system that would recover automobile engine *waste heat* for later use in improving the engine cold-start performance. Among the specifications are reliable operation down to an ambient temperature of −30°C, a storage duration of 16 hours, and no more than 15 minutes of urban driving to return the storage medium to its maximum temperature of 200°C. Specify the storage medium and determine whether the medium should be charged by the engine exhaust gases, the engine coolant, or some combination. Explain how the system would be configured and

where it would be located in the automobile.

11.4D Figure P11.4D shows the schematic of a hydraulic accumulator in the form of a cylindrical pressure vessel with a piston separating a hydraulic fluid from a charge of nitrogen gas. The device has been proposed as a means for storing some of the exergy of a decelerating vehicle as it comes to rest. The exergy is stored by compressing the nitrogen. When the vehicle accelerates again, the gas expands and returns some exergy to the hydraulic fluid, which is in communication with the vehicle's drive train, thereby assisting the vehicle to accelerate. In a proposal for one such device, the nitrogen operates in the range 50–150 bar and 200–350 K. Develop a thermodynamic model of the accumulator and use the model to assess its suitability for vehicle deceleration/acceleration.

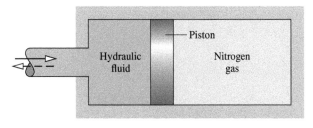

FIGURE P11.4D

11.5D A power plant located at a river's mouth where freshwater river currents meet saltwater ocean tides can generate electricity by exploiting the difference in composition of the freshwater and salt water. The technology for generating power is called *reverse electrodialysis*. While only small-scale demonstration power plants using reverse electrodialysis have been developed thus far, some observers have high expectations for the approach. Investigate the technical readiness and economic feasibility of this renewable power source for providing 3%, or more, of annual U.S. electricity by 2030. Present your conclusions in a report, including a discussion of potential adverse environmental effects of such power plants and at least three references.

11.6D During a phase change from liquid to vapor at fixed pressure, the temperature of a binary *nonazeotropic* solution such as an ammonia–water solution increases rather than remains constant as for a pure substance. This attribute is exploited in both the *Kalina* power cycle and in the *Lorenz* refrigeration cycle. Write a report assessing the status of technologies based on these cycles. Discuss the principal advantages of using binary nonazeotropic solutions. What are some of the main design issues related to their use in power and refrigeration systems?

11.7D The following data are known for a 100-ton ammonia–water absorption system like the one shown in Fig. 10.10. The pump is to handle 259 kg of strong solution per minute. The generator conditions are 1.2 MPa, 104°C. The absorber is at 0.2 MPa with strong solution exiting at 27°C. For the evaporator, the pressure is 0.21 MPa and the exit temperature is −12°C. Specify the type and size, in horsepower, of the pump required. Justify your choices.

11.8D The *Servel* refrigerator works on an absorption principle and requires no moving parts. An energy input by heat transfer is used to drive the cycle, and the refrigerant circulates due to its natural buoyancy. This type of refrigerator is commonly employed in mobile applications, such as recreational vehicles. Liquid propane is burned to provide the required energy input during mobile operation, and electric power is used when the vehicle is parked and can be connected to an electrical outlet. Investigate the principles of operation of commercially available Servel-type systems, and study their feasibility for solar-activated operation. Consider applications in remote locations where electricity or gas is not available. Write a report summarizing your findings.

11.9D In the experiment for the *regelation* of ice, a small-diameter wire weighted at each end is draped over a block of ice. The loaded wire is observed to cut slowly through the ice without leaving a trace. In one such set of experiments, a weighted 1.00-mm-diameter wire is reported to have passed through 0°C ice at a rate of 54 mm/h. Perform the regelation experiment and propose a plausible explanation for this phenomenon.

11.10D An inventor has proposed a new type of marine engine *fueled* by freshwater stored on board and seawater drawn in from the surrounding ocean. At steady state, the engine would develop power from freshwater and seawater streams, each entering the engine at the ambient temperature and pressure. A single mixed stream would be discharged at the ambient temperature and pressure. Critically evaluate this proposal.

CHAPTER 12

Ideal Gas Mixture and Psychrometric Applications

Engineering Context

Many systems of interest involve gas mixtures of two or more components. To apply the principles of thermodynamics introduced thus far to these systems requires that we evaluate properties of the mixtures. Means are available for determining the properties of mixtures from the mixture composition and the properties of the individual pure components from which the mixtures are formed. Methods for this purpose are discussed both in Chap. 11 and in the present chapter.

The **objective** of the present chapter is to study mixtures where the overall mixture and each of its components can be modeled as ideal gases. General ideal gas mixture considerations are provided in the first part of the chapter. Understanding the behavior of ideal gas mixtures of dry air and water vapor is prerequisite to considering air-conditioning processes in the second part of the chapter, which is identified by the heading, *Psychrometric Applications*. In those processes, we sometimes must consider the presence of liquid water as well. We will also need to know how to handle ideal gas mixtures when we study the subjects of combustion and chemical equilibrium in Chaps. 13 and 14, respectively.

LEARNING OUTCOMES

When you complete your study of this chapter, you will be able to...

- Describe ideal gas mixture composition in terms of mass fractions or mole fractions.
- Use the *Dalton model* to relate pressure, volume, and temperature and to calculate changes in *U*, *H*, and *S* for ideal gas mixtures.
- Apply mass, energy, and entropy balances to systems involving ideal gas mixtures, including mixing processes.
- Demonstrate understanding of psychrometric terminology, including humidity ratio, relative humidity, mixture enthalpy, and dew point temperature.
- Use the psychrometric chart to represent common air-conditioning processes and to retrieve data.
- Apply mass, energy, and entropy balances to analyze air-conditioning processes and cooling towers.

Ideal Gas Mixtures: General Considerations

12.1 Describing Mixture Composition

To specify the state of a mixture requires the composition and the values of two independent intensive properties such as temperature and pressure. The object of the present section is to consider ways for describing mixture composition. In subsequent sections, we show how mixture properties other than composition can be evaluated.

Consider a closed system consisting of a gaseous mixture of two or more components. The composition of the mixture can be described by giving the *mass* or the *number of moles* of each component present. With Eq. 1.8, the mass, the number of moles, and the molecular weight of a component i are related by

$$n_i = \frac{m_i}{M_i} \quad (12.1)$$

where m_i is the mass, n_i is the number of moles, and M_i is the molecular weight of component i, respectively. When m_i is expressed in terms of the kilogram, n_i is in kmol. When m_i is in terms of the pound mass, n_i is in lbmol. However, any unit of mass can be used in this relationship.

The total mass of the mixture, m, is the sum of the masses of its components

$$m = m_1 + m_2 + \cdots + m_j = \sum_{i=1}^{j} m_i \quad (12.2)$$

mass fractions

The *relative* amounts of the components present in the mixture can be specified in terms of **mass fractions**. The mass fraction mf_i of component i is defined as

$$mf_i = \frac{m_i}{m} \quad (12.3)$$

gravimetric analysis

A listing of the mass fractions of the components of a mixture is sometimes referred to as a **gravimetric analysis**.

Dividing each term of Eq. 12.2 by the total mass of mixture m and using Eq. 12.3

$$1 = \sum_{i=1}^{j} mf_i \quad (12.4)$$

That is, the sum of the mass fractions of all the components in a mixture is equal to unity.

The total number of moles in a mixture, n, is the sum of the number of moles of each of its components

$$n = n_1 + n_2 + \cdots + n_j = \sum_{i=1}^{i} n_i \quad (12.5)$$

mole fractions

The *relative* amounts of the components present in the mixture also can be described in terms of **mole fractions**. The mole fraction y_i of component i is defined as

$$y_i = \frac{n_i}{n} \quad (12.6)$$

molar analysis
volumetric analysis

A listing of the mole fractions of the components of a mixture may be called a **molar analysis**. An analysis of a mixture in terms of mole fractions is also called a **volumetric analysis**.

Dividing each term of Eq. 12.5 by the total number of moles of mixture n and using Eq. 12.6

$$1 = \sum_{i=1}^{j} y_i \qquad (12.7)$$

That is, the sum of the mole fractions of all the components in a mixture is equal to unity.

The **apparent (or average) molecular weight** of the mixture, M, is defined as the ratio of the total mass of the mixture, m, to the total number of moles of mixture, n

apparent molecular weight

$$M = \frac{m}{n} \qquad (12.8)$$

Equation 12.8 can be expressed in a convenient alternative form. With Eq. 12.2, it becomes

$$M = \frac{m_1 + m_2 + \cdots + m_j}{n}$$

Introducing $m_i = n_i M_i$ from Eq. 12.1

$$M = \frac{n_1 M_1 + n_2 M_2 + \cdots + n_j M_j}{n}$$

Finally, with Eq. 12.6, the apparent molecular weight of the mixture can be calculated as a mole-fraction average of the component molecular weights

$$M = \sum_{i=1}^{j} y_i M_i \qquad (12.9)$$

FOR EXAMPLE

Consider the case of air. A sample of *atmospheric air* contains several gaseous components including water vapor and contaminants such as dust, pollen, and pollutants. The term **dry air** refers only to the gaseous components when all water vapor and contaminants have been removed. The molar analysis of a typical sample of dry air is given in **Table 12.1**. Selecting molecular weights for nitrogen, oxygen, argon, and carbon dioxide from Table A-1, and neglecting the trace substances neon, helium, etc., the apparent molecular weight of dry air obtained from Eq. 12.9 is

dry air

$$M \approx 0.7808(28.02) + 0.2095(32.00) + 0.0093(39.94) + 0.0003(44.01)$$
$$= 28.97 \text{ kg/kmol}$$

This value, which is the entry for air in Table A-1, would not be altered significantly if the trace substances were also included in the calculation.

TABLE 12.1

Approximate Composition of Dry Air

Component	Mole Fraction (%)
Nitrogen	78.08
Oxygen	20.95
Argon	0.93
Carbon dioxide	0.03
Neon, helium, methane, and others	0.01

Next, we consider two examples illustrating, respectively, the conversion from an analysis in terms of mole fractions to an analysis in terms of mass fractions, and conversely.

▶▶▶ **EXAMPLE 12.1** ▶ •

Converting Mole Fractions to Mass Fractions

The molar analysis of the gaseous products of combustion of a certain hydrocarbon fuel is CO_2, 0.08; H_2O, 0.11; O_2, 0.07; N_2, 0.74. **(a)** Determine the apparent molecular weight of the mixture. **(b)** Determine the composition in terms of mass fractions (gravimetric analysis).

SOLUTION

Known The molar analysis of the gaseous products of combustion of a hydrocarbon fuel is given.

Find Determine (a) the apparent molecular weight of the mixture, (b) the composition in terms of mass fractions.

Analysis

a. Using Eq. 12.9 and molecular weights (rounded) from Table A-1

$$M = 0.08(44) + 0.11(18) + 0.07(32) + 0.74(28)$$

$$= 28.46 \text{ kg/kmol}$$

b. Equations 12.1, 12.3, and 12.6 are the key relations required to determine the composition in terms of mass fractions.

❶ Although the actual amount of mixture is not known, the calculations can be based on any convenient amount. Let us base the solution on 1 kmol of mixture. Then, with Eq. 12.6, the amount n_i of each component present in kmol is numerically equal to the mole fraction, as listed in column (ii) of the accompanying table. Column (iii) of the table gives the respective molecular weights of the components.

Column (iv) of the table gives the mass m_i of each component, in kg per kmol of mixture, obtained with Eq. 12.1 in the form $m_i = M_i n_i$. The values of column (iv) are obtained by multiplying each value of column (ii) by the corresponding value of column (iii). The sum of the values in column (iv) is the mass of the mixture: kg of mixture per kmol of mixture. Note that this sum is just the apparent mixture molecular weight determined in part (a). Finally, using Eq. 12.3, column

(v) gives the mass fractions as a percentage. The values of column (v) are obtained by dividing the values of column (iv) by the column (iv) total and multiplying by 100.

(i)	(ii)*		(iii)		(iv)**	(v)
Component	n_i	$\times$	M_i	$=$	m_i	mf_i%
CO_2	0.08	$\times$	44	$=$	3.52	12.37
H_2O	0.11	$\times$	18	$=$	1.98	6.96
O_2	0.07	$\times$	32	$=$	2.24	7.87
N_2	0.74	$\times$	28	$=$	20.72	72.80
	1.00				28.46	100.00

*Entries in this column have units of kmol per kmol of mixture. For example, the first entry is 0.08 kmol of CO_2 per kmol of mixture.

**Entries in this column have units of kg per kmol of mixture. For example, the first entry is 3.52 kg of CO_2 per kmol of mixture. The column sum, 28.46, has units of kg of mixture per kmol of mixture.

---

❶ If the solution to part (b) were conducted on the basis of some other assumed amount of mixture—for example, 100 kmol or 100 lbmol—the same result for the mass fractions would be obtained, as can be verified.

SKILLS DEVELOPED

Ability to…

- calculate the apparent molecular weight with known mole fractions.
- determine the gravimetric analysis given the molar analysis.

Quick Quiz

Determine the mass, in kg, of CO_2 in 0.5 kmol of mixture. **Ans. 1.76 kg.**

▶▶▶ **EXAMPLE 12.2** ▶ •

Converting Mass Fractions to Mole Fractions

A gas mixture has the following composition in terms of mass fractions: H_2, 0.10; N_2, 0.60; CO_2, 0.30. Determine **(a)** the composition in terms of mole fractions and **(b)** the apparent molecular weight of the mixture.

SOLUTION

Known The gravimetric analysis of a gas mixture is known.

Find Determine the analysis of the mixture in terms of mole fractions (molar analysis) and the apparent molecular weight of the mixture.

Analysis

a. Equations 12.1, 12.3, and 12.6 are the key relations required to determine the composition in terms of mole fractions.

❶ Although the actual amount of mixture is not known, the calculation can be based on any convenient amount. Let us base the solution on 100 kg. Then, with Eq. 12.3, the amount m_i of each component present, in kg, is equal to the mass fraction multiplied by 100 kg. The values are listed in column (ii) of the accompanying table. Column (iii) of the table gives the respective molecular weights of the components.

Column (iv) of the table gives the amount n_i of each component, in kmol per 100 kg of mixture, obtained using Eq. 12.1. The values of column (iv) are obtained by dividing each value of column (ii) by the corresponding value of column (iii). The sum of the values of column (iv) is the total amount of mixture, in kmol per 100 kg of mixture. Finally, using Eq. 12.6, column (v) gives the mole fractions as a percentage. The values of column (v) are obtained by dividing the values of column (iv) by the column (iv) total and multiplying by 100.

	(i)	(ii)*		(iii)		(iv)**	(v)
②	Component	m_i	÷	M_i	=	n_i	y_i %
	H_2	10	÷	2	=	5.00	63.9
	N_2	60	÷	28	=	2.14	27.4
	CO_2	30	÷	44	=	0.68	8.7
		100				7.82	100.0

*Entries in this column have units of kg per 100 kg of mixture. For example, the first entry is 10 kg of H_2 per 100 kg of mixture.

**Entries in this column have units of kmol per 100 kg of mixture. For example, the first entry is 5.00 kmol of H_2 per 100 kg of mixture. The column sum, 7.82, has units of kmol of mixture per 100 kg of mixture.

b. The apparent molecular weight of the mixture can be found by using Eq. 12.9 and the calculated mole fractions. The value can be determined alternatively by using the column (iv) total giving the total amount of mixture in kmol per 100 kg of mixture. Thus, with Eq. 12.8

$$M = \frac{m}{n} = \frac{100 \text{ kg}}{7.82 \text{ kmol}} = 12.79 \frac{\text{kg}}{\text{kmol}} = 12.79 \frac{\text{lb}}{\text{lbmol}}$$

❶ If the solution to part (a) were conducted on the basis of some other assumed amount of mixture, the same result for the mass fractions would be obtained, as can be verified.

❷ Although H_2 has the smallest mass fraction, its mole fraction is the largest.

SKILLS DEVELOPED

Ability to...

• determine the molar analysis given the gravimetric analysis.

Quick Quiz

How many kmol of H_2 would be present in 200 kg of mixture? Ans. 10 kmol.

12.2 Relating p, V, and T for Ideal Gas Mixtures[1]

The definitions given in Sec. 12.1 apply generally to mixtures. In the present section we are concerned only with *ideal gas* mixtures and introduce a model commonly used in conjunction with this idealization: the *Dalton model*.

Consider a system consisting of a number of gases contained within a closed vessel of volume V as shown in **Fig. 12.1**. The temperature of the gas mixture is T and the pressure is p. The overall mixture is considered an ideal gas, so p, V, T, and the total number of moles of mixture n are related by the ideal gas equation of state

$$p = n\frac{\bar{R}T}{V} \tag{12.10}$$

With reference to this system let us consider the Dalton model.

The Dalton model is consistent with the concept of an ideal gas as being made up of molecules that exert negligible forces on one another and whose volume is negligible relative to the volume occupied by the gas (Sec. 3.12.3). In the absence of significant intermolecular forces, the behavior of each component is unaffected by the presence of the other components. Moreover, if the volume occupied by the molecules is a very small fraction of the total volume, the molecules of each gas present may be regarded as free to roam throughout the full volume. In keeping with this simple picture, the **Dalton model** assumes that each mixture component behaves as an ideal gas as if it were *alone at the temperature T and volume V of the mixture*.

It follows from the Dalton model that the individual components would not exert the mixture pressure p but rather a *partial pressure*. As shown below, the sum of the partial pressures equals the mixture pressure. The **partial pressure** of component i, p_i, is the pressure that n_i moles of component i would exert if the component were alone in the volume V at the

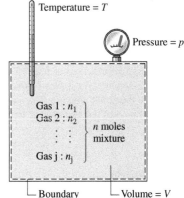

Temperature = T

Pressure = p

Gas 1 : n_1
Gas 2 : n_2
⋮ ⋮
Gas j : n_j
n moles mixture

Boundary

Volume = V

Fig. 12.1 Mixture of several gases.

Dalton model

partial pressure

mixture temperature T. The partial pressure can be evaluated using the ideal gas equation of state

$$p_i = \frac{n_i \bar{R} T}{V} \tag{12.11}$$

Dividing Eq. 12.11 by Eq. 12.10

$$\frac{p_i}{p} = \frac{n_i \bar{R} T / V}{n \bar{R} T / V} = \frac{n_i}{n} = y_i$$

Thus, the partial pressure of component i can be evaluated in terms of its mole fraction y_i and the mixture pressure p

$$\boxed{p_i = y_i p} \tag{12.12}$$

To show that the sum of partial pressures equals the mixture pressure, sum both sides of Eq. 12.12 to obtain

$$\sum_{i=1}^{j} p_i = \sum_{i=1}^{j} y_i p = p \sum_{i=1}^{j} y_i$$

Since the sum of the mole fractions is unity (Eq. 12.7), this becomes

$$\boxed{p = \sum_{i=1}^{j} p_i} \tag{12.13}$$

The Dalton model is a special case of the *additive pressure* rule for relating the pressure, specific volume, and temperature of gas mixtures introduced in Sec. 11.8. Among numerous other mixture rules found in the engineering literature is the *Amagat* model considered in the box that follows.

Introducing the *Amagat Model*

The underlying assumption of the *Amagat model* is that each mixture component behaves as an ideal gas as if it existed separately at the pressure p and temperature T of the mixture.

The volume that n_i moles of component i would occupy if the component existed at p and T is called the *partial volume*, V_i, of component i. As shown below, the sum of the partial volumes equals the total volume. The partial volume can be evaluated using the ideal gas equation of state

$$V_i = \frac{n_i \bar{R} T}{p} \tag{12.14}$$

Dividing Eq. 12.14 by the total volume V

$$\frac{V_i}{V} = \frac{n_i \bar{R} T / p}{n \bar{R} T / p} = \frac{n_i}{n} = y_i$$

Thus, the partial volume of component i also can be evaluated in terms of its mole fraction y_i and the total volume

$$V_i = y_i V \tag{12.15}$$

This relationship between volume fraction and mole fraction underlies the use of the term *volumetric analysis* as signifying an analysis of a mixture in terms of mole fractions.

To show that the sum of partial volumes equals the total volume, sum both sides of Eq. 12.15 to obtain

$$\sum_{i=1}^{j} V_i = \sum_{i=1}^{j} y_i V = V \sum_{i=1}^{j} y_i$$

Since the sum of the mole fractions equals unity, this becomes

$$V = \sum_{i=1}^{j} V_i \tag{12.16}$$

Finally, note that the Amagat model is a special case of the *additive volume* model introduced in Sec. 11.8.

12.3 Evaluating *U, H, S,* and Specific Heats

To apply the conservation of energy principle to a system involving an ideal gas mixture requires evaluation of the internal energy, enthalpy, or specific heats of the mixture at various states. Similarly, to conduct an analysis using the second law normally requires the entropy of the mixture. The objective of the present section is to develop means to evaluate these properties for ideal gas mixtures.

12.3.1 Evaluating *U* and *H*

Consider a closed system consisting of an ideal gas mixture. Extensive properties of the mixture, such as *U, H,* or *S,* can be found by adding the contribution of each component *at the condition at which the component exists in the mixture.* Let us apply this model to internal energy and enthalpy.

Since the internal energy and enthalpy of ideal gases are functions of temperature only, the values of these properties for each component present in the mixture are determined by the mixture temperature alone. Accordingly,

$$U = U_1 + U_2 + \cdots + U_j = \sum_{i=1}^{j} U_i \tag{12.17}$$

$$H = H_1 + H_2 + \cdots + H_j = \sum_{i=1}^{j} H_i \tag{12.18}$$

where U_i and H_i are the internal energy and enthalpy, respectively, of component i evaluated at the mixture temperature.

Equations 12.17 and 12.18 can be rewritten on a molar basis as

$$n\bar{u} = n_1\bar{u}_1 + n_2\bar{u}_2 + \cdots + n_j\bar{u}_j = \sum_{i=1}^{j} n_i\bar{u}_i \tag{12.19}$$

and

$$n\bar{h} = n_1\bar{h}_1 + n_2\bar{h}_2 + \cdots + n_j\bar{h}_j = \sum_{i=1}^{j} n_i\bar{h}_i \tag{12.20}$$

where $\bar{u}$ and $\bar{h}$ are the specific internal energy and enthalpy of the *mixture* per mole of mixture, and $\bar{u}_i$ and $\bar{h}_i$ are the specific internal energy and enthalpy of *component i* per mole of *i.* Dividing by the total number of moles of mixture, n, gives expressions for the specific internal energy and enthalpy of the mixture per mole of mixture, respectively,

$$\bar{u} = \sum_{i=1}^{j} y_i\bar{u}_i \tag{12.21}$$

$$\bar{h} = \sum_{i=1}^{j} y_i\bar{h}_i \tag{12.22}$$

Each of the molar internal energy and enthalpy terms appearing in Eqs. 12.19 through 12.22 is evaluated at the mixture temperature only.

12.3.2 Evaluating c_v and c_p

Differentiation of Eqs. 12.21 and 12.22 with respect to temperature results, respectively, in the following expressions for the specific heats $\bar{c}_v$ and $\bar{c}_p$ of the mixture on a molar basis

$$\bar{c}_v = \sum_{i=1}^{j} y_i\bar{c}_{v,i} \tag{12.23}$$

$$\bar{c}_p = \sum_{i=1}^{j} y_i\bar{c}_{p,i} \tag{12.24}$$

That is, the mixture specific heats $\bar{c}_p$ and $\bar{c}_v$ are mole-fraction averages of the respective component specific heats. The specific heat ratio for the mixture is $k = \bar{c}_p/\bar{c}_v$.

12.3.3 Evaluating *S*

The entropy of a mixture can be found, as for *U* and *H,* by adding the contribution of each component at the condition at which the component exists in the mixture. The entropy of an ideal gas depends on two properties, not on temperature alone as for internal energy and enthalpy. Accordingly, for the mixture

$$S = S_1 + S_2 + \cdots + S_j = \sum_{i=1}^{j} S_i \qquad (12.25)$$

where S_i is the entropy of component i evaluated at the mixture temperature T and partial pressure p_i (or at temperature T and total volume V).

Equation 12.25 can be written on a molar basis as

$$n\bar{s} = n_1\bar{s}_1 + n_2\bar{s}_2 + \cdots + n_j\bar{s}_j = \sum_{i=1}^{j} n_i\bar{s}_i \qquad (12.26)$$

where $\bar{s}$ is the entropy of the *mixture* per mole of mixture and $\bar{s}_i$ is the entropy of *component i* per mole of i. Dividing by the total number of moles of mixture, n, gives an expression for the entropy of the mixture per mole of mixture

$$\boxed{\bar{s} = \sum_{i=1}^{j} y_i\bar{s}_i} \qquad (12.27)$$

In subsequent applications, the specific entropies $\bar{s}_i$ of Eqs. 12.26 and 12.27 are evaluated at the mixture temperature T and the partial pressure p_i.

12.3.4 Working on a Mass Basis

In cases where it is convenient to work on a mass basis, the foregoing expressions are written with the mass of the mixture, m, and the mass of component i in the mixture, m_i, replacing, respectively, the number of moles of mixture, n, and the number of moles of component i, n_i. Similarly, the mass fraction of component i, mf_i, replaces the mole fraction, y_i. All specific internal energies, enthalpies, and entropies are evaluated on a unit mass basis rather than on a per mole basis as above. To illustrate, **Table 12.2** provides property relations on a mass basis for *binary* mixtures. These relations are applicable, in particular, to *moist* air, introduced in Sec. 12.5.

TABLE 12.2

Property Relations on a Mass Basis for Binary Ideal Gas Mixtures

Notation: $m_1 =$ mass of gas 1, $M_1 =$ molecular weight of gas 1
  $m_2 =$ mass of gas 2, $M_2 =$ molecular weight of gas 2
  $m =$ mixture mass $= m_1 + m_2$, $mf_1 = (m_1/m)$, $mf_2 = (m_2/m)$
  $T =$ mixture temperature, $p =$ mixture pressure, $V =$ mixture volume

Equation of state:

$$p = m(\bar{R}/M)T/V \qquad \text{(a)}$$

where $M = (y_1 M_1 + y_2 M_2)$ and the mole fractions y_1 and y_2 are given by

$$y_1 = n_1/(n_1 + n_2), y_2 = n_2/(n_1 + n_2) \qquad \text{(b)}$$

where $n_1 = m_1/M_1$ and $n_2 = m_2/M_2$.

Partial pressures:	$p_1 = y_1 p, p_2 = y_2 p$ (c)

Properties on a mass basis:

Mixture enthalpy:	$H = m_1 h_1(T) + m_2 h_2(T)$	(d)
Mixture internal energy:	$U = m_1 u_1(T) + m_2 u_2(T)$	(e)
Mixture specific heats:	$c_p = (m_1/m)c_{p1}(T) + (m_2/m)c_{p2}(T)$	
	$= (mf_1)c_{p1}(T) + (mf_2)c_{p2}(T)$	(f)
	$c_v = (m_1/m)c_{v1}(T) + (m_2/m)c_{v2}(T)$	
	$= (mf_1)c_{v1}(T) + (mf_2)c_{v2}(T)$	(g)
Mixture entropy:	$S = m_1 s_1(T, p_1) + m_2 s_2(T, p_2)$	(h)

By using the molecular weight of the mixture or of component i, as appropriate, data can be converted from a mass basis to a molar basis or, conversely, with relations of the form

$$\bar{u} = Mu, \quad \bar{h} = Mh, \quad \bar{c}_p = Mc_p, \quad \bar{c}_v = Mc_v, \quad \bar{s} = Ms \quad (12.28)$$

for the mixture and

$$\bar{u}_i = M_i u_i, \quad \bar{h}_i = M_i h_i, \quad \bar{c}_{p,i} = M_i c_{p,i}, \quad \bar{c}_{v,i} = M_i c_{v,i}, \quad \bar{s}_i = M_i s_i \quad (12.29)$$

for component i.

12.4 Analyzing Systems Involving Mixtures

To perform thermodynamic analyses of systems involving *nonreacting* ideal gas mixtures requires no new fundamental principles. The conservation of mass and energy principles and the second law of thermodynamics are applicable in the forms previously introduced. The only new aspect is the proper evaluation of the required property data for the mixtures involved. This is illustrated in the present section, which deals with two classes of problems involving mixtures: In Sec. 12.4.1 the mixture is already formed, and we study processes in which there is no change in composition. Section 12.4.2 considers the formation of mixtures from individual components that are initially separate.

12.4.1 Mixture Processes at Constant Composition

In the present section we are concerned with the case of ideal gas mixtures undergoing processes during which the composition remains constant. The number of moles of each component present and, thus, the total number of moles of mixture remain the same throughout the process. This case is shown schematically in **Fig. 12.2**, which is labeled with expressions for U, H, and S of a mixture at the initial and final states of a process undergone by the mixture. In accordance with the discussion of Sec. 12.3, the specific internal energies and enthalpies of the components are evaluated at the temperature of the mixture. The specific entropy of each component is evaluated at the mixture temperature and the partial pressure of the component in the mixture.

The changes in the internal energy and enthalpy of the mixture during the process are given, respectively, by

$$U_2 - U_1 = \sum_{i=1}^{j} n_i [\bar{u}_i(T_2) - \bar{u}_i(T_1)] \quad (12.30)$$

$$H_2 - H_1 = \sum_{i=1}^{j} n_i [\bar{h}_i(T_2) - \bar{h}_i(T_1)] \quad (12.31)$$

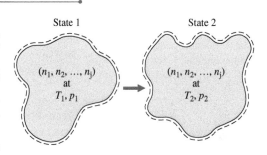

$$U_1 = \sum_{i=1}^{j} n_i \bar{u}_i(T_1) \qquad U_2 = \sum_{i=1}^{j} n_i \bar{u}_i(T_2)$$

$$H_1 = \sum_{i=1}^{j} n_i \bar{h}_i(T_1) \qquad H_2 = \sum_{i=1}^{j} n_i \bar{h}_i(T_2)$$

$$S_1 = \sum_{i=1}^{j} n_i \bar{s}_i(T_1, p_{i1}) \qquad S_2 = \sum_{i=1}^{j} n_i \bar{s}_i(T_2, p_{i2})$$

FIG. 12.2 Process of an ideal gas mixture.

where T_1 and T_2 denote the temperature at the initial and final states. Dividing by the number of moles of mixture, n, expressions for the change in internal energy and enthalpy of the mixture per mole of mixture result

$$\Delta \bar{u} = \sum_{i=1}^{j} y_i [\bar{u}_i(T_2) - \bar{u}_i(T_1)] \quad (12.32)$$

$$\Delta \bar{h} = \sum_{i=1}^{j} y_i [\bar{h}_i(T_2) - \bar{h}_i(T_1)] \quad (12.33)$$

Similarly, the change in entropy for the mixture is

$$S_2 - S_1 = \sum_{i=1}^{j} n_i [\bar{s}_i(T_2, p_{i2}) - \bar{s}_i(T_1, p_{i1})] \quad (12.34)$$

where p_{i1} and p_{i2} denote, respectively, the initial and final partial pressures of component i. Dividing by the total moles of mixture, Eq. 12.34 becomes

$$\Delta\bar{s} = \sum_{i=1}^{j} y_i[\bar{s}_i(T_2, p_{i2}) - \bar{s}_i(T_1, p_{i1})] \tag{12.35}$$

Companion expressions for Eqs. 12.30 through 12.35 on a mass basis also can be written. This is left as an exercise.

The foregoing expressions giving the changes in internal energy, enthalpy, and entropy of the mixture are written in terms of the respective property changes of the components. Accordingly, different datums might be used to assign specific enthalpy values to the various components because the datums would cancel when the component enthalpy changes are calculated. Similar remarks apply to the cases of internal energy and entropy.

Using Ideal Gas Tables

For several common gases modeled as ideal gases, the quantities $\bar{u}_i$ and $\bar{h}_i$ appearing in the foregoing expressions can be evaluated as functions of temperature only from Tables A-22 and A-23. Table A-22 for air gives these quantities on a *mass* basis. Table A-23 gives them on a *molar* basis.

The ideal gas tables also can be used to evaluate the entropy change. The change in specific entropy of component i required by Eqs. 12.34 and 12.35 can be determined with Eq. 6.20b as

$$\Delta\bar{s}_i = \bar{s}_i^\circ(T_2) - \bar{s}_i^\circ(T_1) - \bar{R}\ln\frac{p_{i2}}{p_{i1}}$$

Since the mixture composition remains constant, the ratio of the partial pressures in this expression is the same as the ratio of the mixture pressures, as can be shown by using Eq. 12.12 to write

$$\frac{p_{i2}}{p_{i1}} = \frac{y_i p_2}{y_i p_1} = \frac{p_2}{p_1}$$

Accordingly, when the composition is constant, the change in the specific entropy of component i is simply

$$\Delta\bar{s}_i = \bar{s}_i^\circ(T_2) - \bar{s}_i^\circ(T_1) - \bar{R}\ln\frac{p_2}{p_1} \tag{12.36}$$

where p_1 and p_2 denote, respectively, the initial and final *mixture* pressures. The terms $\bar{s}_i^\circ$ of Eq. 12.36 can be obtained as functions of temperature for several common gases from Table A-23. Table A-22 for air gives s° versus temperature.

Assuming Constant Specific Heats

When the component specific heats $\bar{c}_{v,i}$ and $\bar{c}_{p,i}$ are taken as constants, the specific internal energy, enthalpy, and entropy changes of the mixture and the components of the mixture are given by

$$\Delta\bar{u} = \bar{c}_v(T_2 - T_1), \qquad \Delta\bar{u}_i = \bar{c}_{v,i}(T_2 - T_1) \tag{12.37}$$

$$\Delta\bar{h} = \bar{c}_p(T_2 - T_1), \qquad \Delta\bar{h}_i = \bar{c}_{p,i}(T_2 - T_1) \tag{12.38}$$

$$\Delta\bar{s} = \bar{c}_p\ln\frac{T_2}{T_1} - \bar{R}\ln\frac{p_2}{p_1}, \qquad \Delta\bar{s}_i = \bar{c}_{p,i}\ln\frac{T_2}{T_1} - \bar{R}\ln\frac{p_2}{p_1} \tag{12.39}$$

where the mixture specific heats $\bar{c}_v$ and $\bar{c}_p$ are evaluated from Eqs. 12.23 and 12.24, respectively, using data from Table A-20 or the literature, as required.

The expression for $\Delta\bar{u}$ can be obtained formally by substituting the above expression for $\Delta\bar{u}_i$ into Eq. 12.32 and using Eq. 12.23 to simplify the result. Similarly, the expressions for $\Delta\bar{h}$ and $\Delta\bar{s}$ can be obtained by inserting $\Delta\bar{h}_i$ and $\Delta\bar{s}_i$ into Eqs. 12.33 and 12.35, respectively, and using Eq. 12.24 to simplify. In the equations for entropy change, the ratio of mixture pressures replaces the ratio of partial pressures as discussed above. Similar expressions can be written for the mixture specific internal energy, enthalpy, and entropy changes on a mass basis. This is left as an exercise.

Using Computer Software   The changes in internal energy, enthalpy, and entropy required in Eqs. 12.32, 12.33, and 12.35, respectively, also can be evaluated using computer software. *Interactive Thermodynamics: IT* provides data for a large number of gases modeled as ideal gases, and its use is illustrated in Example 12.4 below.

The next example illustrates the use of ideal gas mixture relations for analyzing a compression process.

> ▶▶ **EXAMPLE 12.3** ▶ ⋅⋅⋅⋅⋅⋅⋅⋅⋅⋅⋅⋅⋅⋅⋅⋅⋅⋅⋅⋅⋅⋅⋅⋅⋅⋅⋅⋅⋅⋅⋅⋅⋅⋅⋅⋅⋅⋅⋅

Compressing an Ideal Gas Mixture

A mixture of 0.3 kg of carbon dioxide and 0.2 kg of nitrogen is compressed from $p_1 = 1$ bar, $T_1 = 300$ K to $p_2 = 3$ bar in a polytropic process for which $n = 1.25$. Determine **(a)** the final temperature, in K, **(b)** the work, in kJ, **(c)** the heat transfer, in kJ, and **(d)** the change in entropy of the mixture, in kJ/K.

SOLUTION

Known   A mixture of 0.3 kg of CO_2 and 0.2 kg of N_2 is compressed in a polytropic process for which $n = 1.25$. At the initial state, $p_1 = 1$ bar, $T_1 = 300°K$. At the final state, $p_2 = 3$ bar.

Find   Determine the final temperature, in K, the work, in kJ, the heat transfer, in kJ, and the entropy change of the mixture in, kJ/K.

Schematic and Given Data:

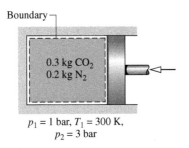

$p_1 = 1$ bar, $T_1 = 300$ K,
$p_2 = 3$ bar

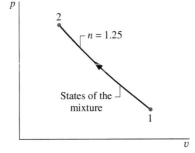

Fig. E12.3

Engineering Model:

1. As shown in the accompanying figure, the system is the mixture of CO and N_2. The mixture composition remains constant during the compression.

2. Each mixture component behaves as if it were an ideal gas occupying the entire system volume at the mixture temperature. The overall mixture acts as an ideal gas.

3. The compression process is a polytropic process for which $n = 1.25$.

4. The changes in kinetic and potential energy between the initial and final states can be ignored.

Analysis

a. For an ideal gas, the temperatures and pressures at the end states of a polytropic process are related by Eq. 3.56

$$T_2 = T_1 \left(\frac{p_2}{p_1} \right)^{(n-1)/n}$$

Inserting values

$$T_2 = 300 \left(\frac{3}{1} \right)^{0.2} = 374 \, \text{K}$$

b. The work for the compression process is given by

$$W = \int_1^2 p \, dV$$

Introducing $pV^n = constant$ and performing the integration

$$W = \frac{p_2 V_2 - p_1 V_1}{1 - n}$$

With the ideal gas equation of state, this reduces to

$$W = \frac{m(\bar{R}/M)(T_2 - T_1)}{1 - n}$$

The mass of the mixture is $m = 0.3 + 0.2 = 0.5$ kg. The apparent molecular weight of the mixture can be calculated using $M = m/n$, where n is the total number of moles of mixture. With Eq. 12.1, the numbers of moles of CO_2 and N_2 are, respectively,

$$n_{CO_2} = \frac{0.3}{44} = 0.0068 \, \text{kmol}, \qquad n_{N_2} = \frac{0.2}{28} = 0.0071 \, \text{kmol}$$

The total number of moles of mixture is then $n = 0.0139$ kmol. The apparent molecular weight of the mixture is $M = 0.5/0.0139 = 35.97$.

Calculating the work

$$W = \frac{(0.5 \, \text{kg}) \left(\dfrac{8.314 \, \text{kJ}}{35.97 \, \text{kg} \cdot °\text{K}} \right)(374 \, \text{K} - 300 \, \text{K})}{1 - 1.25}$$

$$= -34.21 \, \text{kJ}$$

where the minus sign indicates that work is done on the mixture, as expected.

c. With assumption 4, the closed system energy balance can be placed in the form

$$Q = \Delta U + W$$

where ΔU is the change in internal energy of the mixture.

The change in internal energy of the mixture equals the sum of the internal energy changes of the components. With Eq. 12.30

① $\Delta U = n_{CO_2}[\bar{u}_{CO_2}(T_2) - \bar{u}_{CO_2}(T_1)] + n_{N_2}[\bar{u}_{N_2}(T_2) - \bar{u}_{N_2}(T_1)]$

This form is convenient because Table A-23 gives internal energy values for N_2 and CO_2, respectively, on a molar basis. With values from this table

$\Delta U = (0.0068)(9198 - 6939) + (0.0071)(7770 - 6229)$

$= 26.3 \text{ kJ}$

Inserting values for ΔU and W into the expression for Q

$Q = +26.3 - 34.21 = -7.91 \text{ kJ}$

where the minus sign signifies a heat transfer from the system.

d. The change in entropy of the mixture equals the sum of the entropy changes of the components. With Eq. 12.34

$$\Delta S = n_{CO_2}\Delta\bar{s}_{CO_2} + n_{N_2}\Delta\bar{s}_{N_2}$$

where $\Delta\bar{s}_{N_2}$ and $\Delta\bar{s}_{CO_2}$ are evaluated using Eq. 12.36 and values of $\bar{s}°$ for N_2 and CO_2 from Table A-23. That is

$$\Delta S = 0.0068\left(222.475 - 213.915 - 8.314\frac{3}{1}\right)$$

②
$$+ 0.0071\left(198.105 - 191.682 - 8.314\frac{3}{1}\right)$$

$$= -0.0231 \text{ kJ/K}$$

Entropy decreases in the process because entropy is transferred from the system accompanying heat transfer.

① In view of the relatively small temperature change, the changes in the internal energy and entropy of the mixture can be evaluated alternatively using the constant specific heat relations, Eqs. 12.37 and 12.39, respectively. In these equations, $\bar{c}_v$ and $\bar{c}_p$ are specific heats for the mixture determined using Eqs. 12.23 and 12.24 together with appropriate specific heat values for the components chosen from Table A-20.

② Since the composition remains constant, the ratio of mixture pressures equals the ratio of partial pressures, so Eq. 12.36 can be used to evaluate the component specific entropy changes required here.

SKILLS DEVELOPED

Ability to...

- analyze a polytropic process of a closed system for a mixture of ideal gases.
- apply ideal gas mixture principles.
- determine changes in internal energy and entropy for ideal gas mixtures using tabular data.

Quick Quiz

Recalling that polytropic processes are internally reversible, determine for the system the amount of entropy transfer accompanying heat transfer, in kJ/K. Ans. −0.0105 kJ/K.

The next example illustrates the application of ideal gas mixture principles for the analysis of a mixture expanding isentropically through a nozzle. The solution features the use of table data and *IT* as an alternative.

> ▶▶▶ **EXAMPLE 12.4** ▶ ....

Considering an Ideal Gas Mixture Expanding Isentropically Through a Nozzle

A gas mixture consisting of CO_2 and O_2 with mole fractions 0.8 and 0.2, respectively, expands isentropically and at steady state through a nozzle from 700 K, 5 atm, 3 m/s to an exit pressure of 1 atm. Determine **(a)** the temperature at the nozzle exit, in K, **(b)** the entropy changes of the CO_2 and O_2 from inlet to exit, in kJ/kmol · K, **(c)** the exit velocity, in m/s.

SOLUTION

Known A gas mixture consisting of CO_2 and O_2 in specified proportions expands isentropically through a nozzle from specified inlet conditions to a given exit pressure.

Find Determine the temperature at the nozzle exit, in K, the entropy changes of the CO_2 and O_2 from inlet to exit, in kJ/kmol · K, and the exit velocity, in m/s.

Schematic and Given Data:

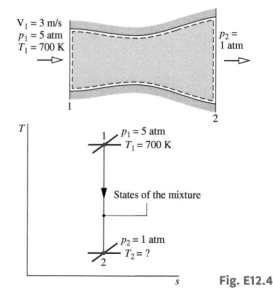

Fig. E12.4

Engineering Model

1. The control volume shown by the dashed line on the accompanying figure operates at steady state.

2. The mixture composition remains constant as the mixture expands isentropically through the nozzle.

3. The Dalton model applies: The overall mixture and each mixture component act as ideal gases. The state of each component is defined by the temperature and the partial pressure of the component.

4. The change in potential energy between inlet and exit can be ignored.

Analysis

a. The temperature at the exit can be determined using the fact that the expansion occurs isentropically: $\bar{s}_2 - \bar{s}_1 = 0$. As there is no change in the specific entropy of the *mixture* between inlet and exit, Eq. 12.35 can be used to write

$$\bar{s}_2 - \bar{s}_1 = y_{O_2} \Delta \bar{s}_{O_2} + y_{CO_2} \Delta \bar{s}_{CO_2} = 0 \qquad (a)$$

Since composition remains constant, the ratio of partial pressures equals the ratio of mixture pressures. Accordingly, the change in specific entropy of each component can be determined using Eq. 12.36. Equation (a) then becomes

$$y_{O_2} \left[\bar{s}^{\circ}_{O_2}(T_2) - \bar{s}^{\circ}_{O_2}(T_1) - \bar{R} \ln \frac{p_2}{p_1} \right]$$

$$+ y_{CO_2} \left[\bar{s}^{\circ}_{CO_2}(T_2) - \bar{s}^{\circ}_{CO_2}(T_1) - \bar{R} \ln \frac{p_2}{p_1} \right] = 0$$

On rearrangement

$$y_{O_2} \bar{s}^{\circ}_{O_2}(T_2) + y_{CO_2} \bar{s}^{\circ}_{CO_2}(T_2)$$

$$= y_{O_2} \bar{s}^{\circ}_{O_2}(T_1) + y_{CO_2} \bar{s}^{\circ}_{CO_2}(T_1) + (y_{O_2} + y_{CO_2}) \bar{R} \ln \frac{p_2}{p_1}$$

The sum of mole fractions equals unity, so the coefficient of the last term on the right side is $(y_{O_2} + y_{CO_2}) = 1$.

Introducing given data and values of $\bar{s}^{\circ}$ for O_2 and CO_2 at $T_1 = 700$ K from Table A-23

$$0.2\bar{s}^{\circ}_{O_2}(T_2) + 0.8\bar{s}^{\circ}_{CO_2}(T_2)$$

$$= 0.2(231.358) + 0.8(250.663) + 8.314 \ln \frac{1}{5}$$

or

$$0.2\bar{s}^{\circ}_{O_2}(T_2) + 0.8\bar{s}^{\circ}_{CO_2}(T_2) = 233.42 \text{ kJ/kmol} \cdot \text{K}$$

To determine the temperature T_2 requires an iterative approach with the above equation: A final temperature T_2 is assumed, and the $\bar{s}^{\circ}$ values for O_2 and CO_2 are found from Table A-23. If these two values do not satisfy the equation, another temperature is assumed. The procedure continues until the desired agreement is attained. In the present case

at $T = 510$ K: $0.2(221.206) + 0.8(235.700) = 232.80$

at $T = 520$ K: $0.2(221.812) + 0.8(236.575) = 233.62$

Linear interpolation between these values gives $T_2 = 517.6$ K.

Alternative Solution

Alternatively, the following *IT* program can be used to evaluate T_2 without resorting to iteration with table data. In the program, yO2 denotes the mole fraction of O_2, p1_O2 denotes the partial pressure of O_2 at state 1, s1_O2 denotes the entropy per mole of O_2 at state 1, and so on.

```
T1 = 700 // K
p1 = 5 // bar
p2 = 1 // bar
yO2 = 0.2
yCO2 = 0.8
p1_O2 = yO2 * p1
p1_CO2 = yCO2 * p1
p2_O2 = yO2 * p2
p2_CO2 = yCO2 * p2
s1_O2 = s_TP ("O2",T1,p1_O2)
s1_CO2 = s_TP ("CO2",T1,p1_CO2)
s2_O2 = s_TP ("O2",T2,p2_O2)
s2_CO2 = s_TP ("CO2",T2,p2_CO2)
// When expressed in terms of these quantities,
Eq. (a) takes the form
yO2 * (s2_O2 - s1_O2) + yCO2 * (s2_CO2 - s1_CO2) = 0
```

Using the **Solve** button, the result is $T_2 = 517.6$ K, which agrees with the value obtained using table data. Note that *IT* provides the value of specific entropy for each component directly and does not return $\bar{s}^{\circ}$ of the ideal gas tables.

❶ b. The change in the specific entropy for each of the components can be determined using Eq. 12.36. For O_2

$$\Delta \bar{s}_{O_2} = \bar{s}^{\circ}_{O_2}(T_2) - \bar{s}^{\circ}_{O_2}(T_1) - \bar{R} \ln \frac{p_2}{p_1}$$

Inserting $\bar{s}^{\circ}$ values for O_2 from Table A-23 at $T_1 = 700$ K and $T_2 = 517.6$ K

$$\Delta \bar{s}_{O_2} = 221.667 - 231.358 - 8.314 \ln(0.2) = 3.69 \text{ kJ/kmol} \cdot \text{K}$$

Similarly, with CO_2 data from Table A-23

$$\Delta \bar{s}_{CO_2} = \bar{s}^{\circ}_{CO_2}(T_2) - \bar{s}^{\circ}_{CO_2}(T_1) - \bar{R} \ln \frac{p_2}{p_1}$$

$$= 236.365 - 250.663 - 8.314 \ln(0.2)$$

❷

$$= -0.92 \text{ kJ/kmol} \cdot \text{K}$$

c. Reducing the energy rate balance for the one-inlet, one-exit control volume at steady state

$$0 = h_1 - h_2 + \frac{V_1^2 - V_2^2}{2}$$

where h_1 and h_2 are the enthalpy *of the mixture,* per unit mass of mixture, at the inlet and exit, respectively. Solving for V_2

$$V_2 = \sqrt{V_1^2 + 2(h_1 - h_2)}$$

The term $(h_1 - h_2)$ in the expression for V_2 can be evaluated as

$$h_1 - h_2 = \frac{\bar{h}_1 - \bar{h}_2}{M} = \frac{1}{M}[y_{O_2}(\bar{h}_1 - \bar{h}_2)_{O_2} + y_{CO_2}(\bar{h}_1 - \bar{h}_2)_{CO_2}]$$

where M is the apparent molecular weight of mixture, and the molar specific enthalpies of O_2 and CO_2 are from Table A-23. With Eq. 12.9, the apparent molecular weight of the mixture is

$$M = 0.8(44) + 0.2(32) = 41.6 \text{ kg/kmol}$$

Then, with enthalpy values at $T_1 = 700$ K and $T_2 = 517.6$ K from Table A-23,

$$h_1 - h_2 = \frac{1}{41.6}[0.2(21{,}184 - 15{,}320) + 0.8(27{,}125 - 18{,}468)]$$

$$= 194.7 \text{ kJ/kg}$$

Finally,

❸ $V_2 = \sqrt{\left(3\frac{m}{s}\right)^2 + 2\left(194.7\frac{kJ}{kg}\right)\left|\frac{1\,kg\cdot m/s^2}{1\,N}\right|\left|\frac{10^3\,N\cdot m}{1\,kJ}\right|} = 624$ m/s

❶ Parts (b) and (c) can be solved alternatively using *IT*. These parts also can be solved using a constant c_p together with Eqs. 12.38 and 12.39. Inspection of Table A-20 shows that the specific heats of CO_2 and O_2 increase only slightly with temperature over the interval from 518 to 700 K, and so suitable constant values of c_p for the components and the overall mixture can be readily determined. These alternative solutions are left as exercises.

❷ Each component experiences an entropy change as it passes from inlet to exit. The increase in entropy of the oxygen and the decrease in entropy of the carbon dioxide are due to entropy transfer accompanying heat transfer from the CO_2 to the O_2 as they expand through the nozzle. However, as indicated by Eq. (a), there is no change in the entropy of the *mixture* as it expands through the nozzle.

❸ Note the use of unit conversion factors in the calculation of V_2.

SKILLS DEVELOPED

Ability to...

• analyze the isentropic expansion of an ideal gas mixture flowing through a nozzle.

• apply ideal gas mixture principles together with mass and energy balances to calculate the exit velocity of a nozzle.

• determine the exit temperature for a given inlet state and a given exit pressure using tabular data and alternatively using *IT*.

Quick Quiz

What would be the exit velocity, in m/s, if the isentropic nozzle efficiency were 90%? Ans. 592 m/s.

12.4.2 Mixing of Ideal Gases

Thus far, we have considered only mixtures that have already been formed. Now let us take up cases where ideal gas mixtures are formed by mixing gases that are initially separate. Such mixing is irreversible because the mixture forms spontaneously, and a work input from the surroundings would be required to separate the gases and return them to their respective initial states. In this section, the irreversibility of mixing is demonstrated through calculations of the entropy production.

Three factors contribute to the production of entropy in mixing processes:

1. The gases are initially at different temperatures.

2. The gases are initially at different pressures.

3. The gases are distinguishable from one another.

Entropy is produced when any of these factors is present during a mixing process. This is illustrated in the next example, where different gases, initially at different temperatures and pressures, are mixed.

▸▸▸ **EXAMPLE 12.5** ▸

Adiabatic Mixing at Constant Total Volume

Two rigid, insulated tanks are interconnected by a valve. Initially 0.79 kmol of nitrogen at 2 bar and 250 K fills one tank. The other tank contains 0.21 kmol of oxygen at 1 bar and 300 K. The valve is opened and the gases are allowed to mix until a final equilibrium state is attained. During this process, there are no heat or work interactions between the tank contents and the surroundings. Determine **(a)** the final temperature of the mixture, in K, **(b)** the final pressure of the mixture, in bar, and **(c)** the amount of entropy produced in the mixing process, in kJ/K.

SOLUTION

Known Nitrogen and oxygen, initially separate at different temperatures and pressures, are allowed to mix without heat or work interactions with the surroundings until a final equilibrium state is attained.

Find Determine the final temperature of the mixture, in K, the final pressure of the mixture, in bar, and the amount of entropy produced in the mixing process, in kJ/K.

Schematic and Given Data:

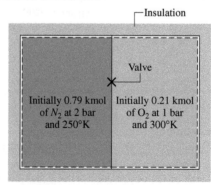

Fig. E12.5

Engineering Model

1. The system is taken to be the nitrogen and the oxygen together.

2. When separate, each of the gases behaves as an ideal gas. The final mixture also acts as an ideal gas. Each mixture component occupies the total volume and exhibits the mixture temperature.

3. No heat or work interactions occur with the surroundings, and there are no changes in kinetic and potential energy.

Analysis

a. The final temperature of the mixture can be determined from an energy balance. With assumption 3, the closed system energy balance reduces to

$$\Delta U = \cancelto{0}{Q} - \cancelto{0}{W} \qquad \text{or} \qquad U_2 - U_1 = 0$$

The initial internal energy of the system, U_1, equals the sum of the internal energies of the two gases when separate

$$U_1 = n_{N_2}\bar{u}_{N_2}(T_{N_2}) + n_{O_2}\bar{u}_{O_2}(T_{O_2})$$

where $T_{N_2} = 250°K$ is the initial temperature of the nitrogen and $T_{O_2} = 300°K$ is the initial temperature of the oxygen. The final internal energy of the system, U_2, equals the sum of the internal energies of the two gases evaluated at the final mixture temperature T_2

$$U_2 = n_{N_2}\bar{u}_{N_2}(T_2) + n_{O_2}\bar{u}_{O_2}(T_2)$$

Collecting the last three equations

$$n_{N_2}[\bar{u}_{N_2}(T_2) - \bar{u}_{N_2}(T_{N_2})] + n_{O_2}[\bar{u}_{O_2}(T_2) - \bar{u}_{O_2}(T_{O_2})] = 0$$

The temperature T_2 can be determined using specific internal energy data from Table A-23 and an iterative procedure like that employed in part (a) of Example 12.4. However, since the specific heats of N_2 and O_2 vary little over the temperature interval from 250 to 300°K, the solution can be conducted accurately on the basis of constant specific heats. Hence, the foregoing equation becomes

$$n_{N_2}\bar{c}_{v,N_2}(T_2 - T_{N_2}) + n_{O_2}\bar{c}_{v,O_2}(T_2 - T_{O_2}) = 0$$

Solving for T_2

$$T_2 = \frac{n_{N_2}\bar{c}_{v,N_2}T_{N_2} + n_{O_2}\bar{c}_{v,O_2}T_{O_2}}{n_{N_2}\bar{c}_{v,N_2} + n_{O_2}\bar{c}_{v,O_2}}$$

Selecting c_v values for N_2 and O_2 from Table A-20 at the average of the initial temperatures of the gases, 275 K, and using the respective molecular weights to convert to a molar basis

$$\bar{c}_{v,N_2} = \left(28.01 \frac{\text{kg}}{\text{mol}}\right)\left(0.743 \frac{\text{kJ}}{\text{kg} \cdot \text{K}}\right) = 20.82 \frac{\text{kJ}}{\text{kmol} \cdot \text{K}}$$

$$\bar{c}_{v,O_2} = \left(32.0 \frac{\text{kg}}{\text{mol}}\right)\left(0.656 \frac{\text{kJ}}{\text{kg} \cdot \text{K}}\right) = 20.99 \frac{\text{kJ}}{\text{kmol} \cdot \text{K}}$$

Substituting values into the expression for T_2

$$T_2 = \frac{\begin{array}{l}(0.79\,\text{kmol})\left(20.82\dfrac{\text{kJ}}{\text{kmol} \cdot \text{K}}\right)(250\,\text{K}) \\[2mm] + (0.21\,\text{kmol})\left(20.99\dfrac{\text{kJ}}{\text{kmol} \cdot \text{K}}\right)(300\,\text{K})\end{array}}{\begin{array}{l}(0.79\,\text{kmol})\left(20.82\dfrac{\text{kJ}}{\text{kmol} \cdot \text{K}}\right) \\[2mm] + (0.21\,\text{kmol})\left(20.99\dfrac{\text{kJ}}{\text{kmol} \cdot \text{K}}\right)\end{array}}$$

$$= 261\,\text{K}$$

b. The final mixture pressure p_2 can be determined using the ideal gas equation of state, $p_2 = n\bar{R}T_2/V$, where n is the total number of moles of mixture and V is the total volume occupied by the mixture. The volume V is the sum of the volumes of the two tanks, obtained with the ideal gas equation of state as follows

$$V = \frac{n_{N_2}\bar{R}T_{N_2}}{p_{N_2}} + \frac{n_{O_2}\bar{R}T_{O_2}}{p_{O_2}}$$

where $p_{N_2} = 2$ atm is the initial pressure of the nitrogen and $p_{O_2} = 1$ atm is the initial pressure of the oxygen. Combining results and reducing

$$p_2 = \frac{(n_{N_2} + n_{O_2})T_2}{\left(\dfrac{n_{N_2}T_{N_2}}{p_{N_2}} + \dfrac{n_{O_2}T_{O_2}}{p_{O_2}}\right)}$$

Substituting values

$$p_2 = \frac{(1.0\,\text{kmol})(261\,\text{K})}{\left[\dfrac{(0.79\,\text{kmol})(250\,\text{K})}{2\,\text{bar}} + \dfrac{(0.21\,\text{kmol})(300\,\text{K})}{1\,\text{bar}}\right]}$$

$$= 1.62\,\text{bar}$$

c. Reducing the closed system form of the entropy balance

$$S_2 - S_1 = \int_1^2 \cancelto{0}{\left(\frac{\delta Q}{T}\right)_b} + \sigma$$

where the entropy transfer term drops out for the adiabatic mixing process. The initial entropy of the system, S_1, is the sum of the entropies of the gases at the respective initial states

$$S_1 = n_{N_2}\bar{s}_{N_2}(T_{N_2}, p_{N_2}) + n_{O_2}\bar{s}_{O_2}(T_{O_2}, p_{O_2})$$

The final entropy of the system, S_2, is the sum of the entropies of the individual components, each evaluated at the final mixture temperature and the partial pressure of the component in the mixture

$$S_2 = n_{N_2}\bar{s}_{N_2}(T_2, y_{N_2}p_2) + n_{O_2}\bar{s}_{O_2}(T_2, y_{O_2}p_2)$$

Collecting the last three equations

$$\sigma = n_{N_2}[\bar{s}_{N_2}(T_2, y_{N_2}p_2) - \bar{s}_{N_2}(T_{N_2}, p_{N_2})]$$
$$+ n_{O_2}[\bar{s}_{O_2}(T_2, y_{O_2}p_2) - \bar{s}_{O_2}(T_{O_2}, p_{O_2})]$$

Evaluating the change in specific entropy of each gas in terms of a constant specific heat $\bar{c}_p$, this becomes

$$\sigma = n_{N_2}\left(\bar{c}_{p,N_2}\ln\frac{T_2}{T_{N_2}} - \bar{R}\ln\frac{y_{N_2}p_2}{p_{N_2}}\right)$$

$$+ n_{O_2}\left(\bar{c}_{p,O_2}\ln\frac{T_2}{T_{O_2}} - \bar{R}\ln\frac{y_{O_2}p_2}{p_{O_2}}\right)$$

The required values for $\bar{c}_p$ can be found by adding $\bar{R}$ to the $\bar{c}_v$ values found previously (Eq. 3.45)

$$\bar{c}_{p,N_2} = 29.13\frac{kJ}{kmol \cdot K} \qquad \bar{c}_{p,O_2} = 29.30\frac{kJ}{kmol \cdot °K}$$

Since the total number of moles of mixture $n = 0.79 + 0.21 = 1.0$, the mole fractions of the two gases are $y_{N_2} = 0.79$ and $y_{O_2} = 0.21$.

Substituting values into the expression for σ gives

$$\sigma = 0.79\,kmol\left[29.13\frac{kJ}{kmol \cdot K}\ln\left(\frac{261°K}{250°K}\right)\right.$$

$$\left. - 8.314\frac{kJ}{kmol \cdot K}\ln\left(\frac{(0.79)(1.62\,bar)}{2\,bar}\right)\right]$$

$$+ 0.21\,kmol\left[29.30\frac{kJ}{kmol \cdot K}\ln\left(\frac{261°K}{300°K}\right)\right.$$

$$\left. - 8.314\frac{kJ}{kmol \cdot K}\ln\left(\frac{(0.21)(1.62\,bar)}{1\,bar}\right)\right]$$

① $= 5.0\,kJ/°K$

---

① Entropy is produced when different gases, initially at different temperatures and pressures, are allowed to mix.

SKILLS DEVELOPED

Ability to...

- analyze the adiabatic mixing of two ideal gases at constant total volume.
- apply energy and entropy balances to the mixing of two gases.
- apply ideal gas mixture principles assuming constant specific heats.

Quick Quiz

Determine the total volume of the final mixture, in m³. Ans. 6 m³.

In the next example, we consider a control volume at steady state where two incoming streams form a mixture. A single stream exits.

► ► **EXAMPLE 12.6** ► ···

Analyzing Adiabatic Mixing of Two Streams

At steady state, 100 m³/min of dry air at 32°C and 1 bar is mixed adiabatically with a stream of oxygen (O_2) at 127°C and 1 bar to form a mixed stream at 47°C and 1 bar. Kinetic and potential energy effects can be ignored. Determine **(a)** the mass flow rates of the dry air and oxygen, in kg/min, **(b)** the mole fractions of the dry air and oxygen in the exiting mixture, and **(c)** the time rate of entropy production, in kJ/K · min.

SOLUTION

Known At steady state, 100 m³/min of dry air at 32°C and 1 bar is mixed adiabatically with an oxygen stream at 127°C and 1 bar to form a mixed stream at 47°C and 1 bar.

Find Determine the mass flow rates of the dry air and oxygen, in kg/min, the mole fractions of the dry air and oxygen in the exiting mixture, and the time rate of entropy production, in kJ/K · min.

Schematic and Given Data:

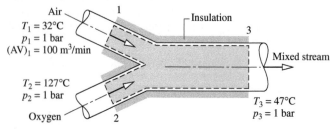

Fig. E12.6

Engineering Model

1. The control volume identified by the dashed line on the accompanying figure operates at steady state.

2. No heat transfer occurs with the surroundings.

3. Kinetic and potential energy effects can be ignored, and $\dot{W}_{cv} = 0$.

4. The entering gases can be regarded as ideal gases. The exiting mixture can be regarded as an ideal gas mixture adhering to the Dalton model.

5. The dry air is treated as a pure component.

Analysis

a. The mass flow rate of the dry air entering the control volume can be determined from the given volumetric flow rate $(AV)_1$

$$\dot{m}_{a1} = \frac{(AV)_1}{v_{a1}}$$

where v_{a1} is the specific volume of the air at 1. Using the ideal gas equation of state

$$v_{a1} = \frac{(\bar{R}/M_a)T_1}{p_1} = \frac{\left(\dfrac{8314\,N \cdot m}{28.97\,kg \cdot K}\right)(305\,K)}{10^5\,N/m^2} = 0.875\frac{m^3}{kg}$$

The mass flow rate of the dry air is then

$$\dot{m}_{a1} = \frac{100 \text{ m}^3/\text{min}}{0.875 \text{ m}^3/\text{kg}} = 114.29 \frac{\text{kg}}{\text{min}}$$

The mass flow rate of the oxygen can be determined using mass and energy rate balances. At steady state, the amounts of dry air and oxygen contained within the control volume do not vary. Thus, for each component individually it is necessary for the incoming and outgoing mass flow rates to be equal. That is,

$$\dot{m}_{a1} = \dot{m}_{a3} \quad \text{(dry air)}$$

$$\dot{m}_{o2} = \dot{m}_{o3} \quad \text{(oxygen)}$$

Using assumptions 1–3 together with the foregoing mass flow rate relations, the energy rate balance reduces to

$$0 = \dot{m}_a h_a(T_1) + \dot{m}_o h_o(T_2) - [\dot{m}_a h_a(T_3) + \dot{m}_o h_o(T_3)]$$

where $\dot{m}_a$ and $\dot{m}_o$ denote the mass flow rates of the dry air and oxygen, respectively. The enthalpy of the mixture at the exit is evaluated by summing the contributions of the air and oxygen, each at the mixture temperature. Solving for $\dot{m}_o$

$$\dot{m}_o = \dot{m}_a \left[\frac{h_a(T_3) - h_a(T_1)}{h_o(T_2) - h_o(T_3)} \right]$$

The specific enthalpies can be obtained from Tables A-22 and A-23. Since Table A-23 gives enthalpy values on a molar basis, the molecular weight of oxygen is introduced into the denominator to convert the molar enthalpy values to a mass basis

$$\dot{m}_o = \frac{(114.29 \text{ kg/min})(320.29 \text{ kJ/kg} - 305.22 \text{ kJ/kg})}{\left(\dfrac{1}{32 \text{ kg/kmol}} \right)(11{,}711 \text{ kJ/kmol} - 9{,}325 \text{ kJ/kmol})}$$

$$= 23.1 \frac{\text{kg}}{\text{min}}$$

b. To obtain the mole fractions of the dry air and oxygen in the exiting mixture, first convert the mass flow rates to molar flow rates using the respective molecular weights

$$\dot{n}_a = \frac{\dot{m}_a}{M_a} = \frac{114.29 \text{ kg/min}}{28.97 \text{ kg/kmol}} = 3.95 \text{ kmol/min}$$

$$\dot{n}_o = \frac{\dot{m}_o}{M_o} = \frac{23.1 \text{ kg/min}}{32 \text{ kg/kmol}} = 0.72 \text{ kmol/min}$$

where $\dot{n}$ denotes molar flow rate. The molar flow rate of the mixture $\dot{n}$ is the sum

$$\dot{n} = \dot{n}_a + \dot{n}_o = 3.95 + 0.72 = 4.67 \text{ kmol/min}$$

The mole fractions of the air and oxygen in the exiting mixture are, respectively,

1 $\quad y_a = \dfrac{\dot{n}_a}{\dot{n}} = \dfrac{3.95}{4.67} = 0.846 \quad$ and $\quad y_o = \dfrac{\dot{n}_o}{\dot{n}} = \dfrac{0.72}{4.67} = 0.154$

c. For the control volume at steady state, the entropy rate balance reduces to

$$0 = \dot{m}_a s_a(T_1, p_1) + \dot{m}_o s_o(T_2, p_2) - [\dot{m}_a s_a(T_3, y_a p_3) + \dot{m}_o s_o(T_3, y_o p_3)] + \dot{\sigma}$$

The specific entropy of each component in the exiting ideal gas mixture is evaluated at its partial pressure in the mixture and at the mixture temperature. Solving for $\dot{\sigma}$

$$\dot{\sigma} = \dot{m}_a[s_a(T_3, y_a p_3) - s_a(T_1, p_1)] + \dot{m}_o[s_o(T_3, y_o p_3) - s_o(T_2, p_2)]$$

Since $p_1 = p_3$, the specific entropy change of the dry air is

$$s_a(T_3, y_a p_3) - s_a(T_1, p_1) = s_a^\circ(T_3) - s_a^\circ(T_1) - \frac{\overline{R}}{M_a} \ln \frac{y_a p_3}{p_1}$$

$$= s_a^\circ(T_3) - s_a^\circ(T_1) - \frac{\overline{R}}{M_a} \ln y_a$$

The s_a° terms are evaluated from Table A-22. Similarly, since $p_2 = p_3$, the specific entropy change of the oxygen is

$$s_o(T_3, y_o p_3) - s_o(T_2, p_2) = \frac{1}{M_o}[\overline{s}_o^\circ(T_3) - \overline{s}_o^\circ(T_2) - \overline{R} \ln y_o]$$

The $\overline{s}_o^\circ$ terms are evaluated from Table A-23. Note the use of the molecular weights M_a and M_o in the last two equations to obtain the respective entropy changes on a mass basis.

The expression for the rate of entropy production becomes

$$\dot{\sigma} = \dot{m}_a \left[s_a^\circ(T_3) - s_a^\circ(T_1) - \frac{\overline{R}}{M_a} \ln y_a \right]$$
$$+ \frac{\dot{m}_o}{M_o}[\overline{s}_o^\circ(T_3) - \overline{s}_o^\circ(T_2) - \overline{R} \ln y_o]$$

Substituting values

$$\dot{\sigma} = \left(114.29 \frac{\text{kg}}{\text{min}} \right) \left[1.7669 \frac{\text{kJ}}{\text{kg} \cdot \text{K}} - 1.71865 \frac{\text{kJ}}{\text{kg} \cdot \text{K}} \right.$$
$$\left. - \left(\frac{8.314}{28.97} \frac{\text{kJ}}{\text{kg} \cdot \text{K}} \right) \ln 0.846 \right]$$
$$+ \left(\frac{23.1 \text{ kg/min}}{32 \text{ kg/kmol}} \right) \left[207.112 \frac{\text{kJ}}{\text{kmol} \cdot \text{K}} - 213.765 \frac{\text{kJ}}{\text{kmol} \cdot \text{K}} \right.$$
$$\left. - \left(8.314 \frac{\text{kJ}}{\text{kmol} \cdot \text{K}} \right) \ln 0.154 \right]$$

2 $\quad = 17.42 \dfrac{\text{kJ}}{\text{K} \cdot \text{min}}$

---

1 This calculation is based on dry air modeled as a pure component (assumption 5). However, since O_2 is a component of dry air (Table 12.1), the *actual* mole fraction of O_2 in the exiting mixture is greater than given here.

② Entropy is produced when different gases, initially at different temperatures, are allowed to mix.

SKILLS DEVELOPED

Ability to...

• analyze the adiabatic mixing of two ideal gas streams at steady state.

• apply ideal mixture principles together with mass, energy, and entropy rate balances.

Quick Quiz

What are the mass fractions of air and oxygen in the exiting mixture? Ans. $mf_{air} = 0.832$, $mf_{O_2} = 0.168$.

BioConnections

Can spending time inside a building cause you to sneeze, cough, or develop a headache? If so, the culprit could be the indoor air. The term *sick building syndrome* (SBS) describes a condition where indoor air quality leads to acute health problems and comfort issues for building occupants. Effects of SBS can be linked to the amount of time an occupant spends within the space; yet while the specific cause and illness are frequently unidentifiable, symptoms typically subside after the occupant leaves the building. If the symptoms persist even after leaving the space and are diagnosed as a specific illness attributed to an airborne contaminant, the term *building-related illness* is a more accurate descriptor.

The U.S. Environmental Protection Agency (EPA) conducted studies from 1994 to 2000 on domestic office buildings to understand the effects of ventilation rates on occupant health. In one study, indoor CO_2 levels were used to approximate levels of other occupant-generated pollutants and as a proxy for ventilation rate per occupant. The results of the EPA investigations agreed with most previous findings that relate

lower ventilation rates per person within office buildings to higher rates of SBS symptom reporting. Building codes and guidelines in the United States generally recommend ventilation rates within office buildings in the range of 0.42–0.57 m^3/min per occupant. Some of the spaces studied had ventilation rates below the guidelines.

Indoor air quality is a significant concern for both building occupants and engineers who design and operate building air delivery systems. Careful design is needed to ensure that air distribution systems deliver acceptable ventilation for each space. Inadequate system installation and improper maintenance also can give rise to indoor air quality problems, even when appropriate standards have been applied in the design. The EPA studies found this to be the case with the systems of many buildings in the study group. Also testing and balancing of the installed systems were never conducted in several of the buildings to ensure that systems were operating according to design intent.

Psychrometric Applications

The remainder of this chapter is concerned with the study of systems involving mixtures of dry air and water vapor. A condensed water phase also may be present. Knowledge of the behavior of such systems is essential for the analysis and design of air-conditioning devices, cooling towers, and industrial processes requiring close control of the vapor content in air. The study of systems involving dry air and water is known as **psychrometrics**.

psychrometrics

12.5 Introducing Psychrometric Principles

The object of the present section is to introduce some important definitions and principles used in the study of systems involving dry air and water.

12.5.1 Moist Air

moist air

The term **moist air** refers to a mixture of dry air and water vapor in which the dry air is treated as if it were a pure component. As can be verified by reference to appropriate property data, the overall mixture and each mixture component behave as ideal gases at the states under present consideration. Accordingly, for the applications to be considered, the ideal gas mixture concepts introduced previously apply directly.

In particular, the Dalton model and the relations provided in Table 12.2 are applicable to moist air mixtures. Simply by identifying gas 1 with dry air, denoted by the subscript a, and gas 2 with water vapor, denoted by the subscript v, the table gives a useful set of moist air property relations. Referring to **Fig. 12.3**, let's verify this point by obtaining a sampling of moist air relations and relating them to entries in Table 12.2.

Shown in Fig. 12.3—a special case of Fig. 12.1—is a closed system consisting of moist air occupying a volume V at mixture pressure p and mixture temperature T. The overall mixture is assumed to obey the ideal gas equation of state. Thus,

$$p = \frac{n\bar{R}T}{V} = \frac{m(\bar{R}/M)T}{V} \qquad (12.40)$$

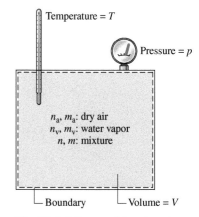

Fig. 12.3 Mixture of dry air and water vapor.

where n, m, and M denote the moles, mass, and molecular weight of the mixture, respectively, and $n = m/M$.

Each mixture component is considered to act as if it existed alone in the volume V at the mixture temperature T while exerting a part of the pressure. The mixture pressure is the sum of the partial pressures of the dry air and the water vapor: $p = p_a + p_v$. That is, the Dalton model applies.

Using the ideal gas equation of state, the partial pressures p_a and p_v of the dry air and water vapor are, respectively,

TAKE NOTE...

Moist air is a *binary* mixture of dry air and water vapor, and the property relations of Table 12.2 apply.

$$p_a = \frac{n_a\bar{R}T}{V} = \frac{m_a(\bar{R}/M_a)T}{V}, \qquad p_v = \frac{n_v\bar{R}T}{V} = \frac{m_v(\bar{R}/M_v)T}{V} \qquad (12.41a)$$

where n_a and n_v denote the moles of dry air and water vapor, respectively; m_a, m_v, M_a, and M_v are the respective masses and molecular weights. The amount of water vapor present is normally much less than the amount of dry air. Accordingly, the values of n_v, m_v, and p_v are small relative to the corresponding values of n_a, m_a, and p_a.

Forming ratios with Eqs. 12.40 and 12.41a, we get the following alternative expressions for p_a and p_v

$$p_a = y_a p, \qquad p_v = y_v p \qquad (12.41b)$$

where y_a and y_v are the mole fractions of the dry air and water vapor, respectively. These moist air expressions conform to Eqs. (c) in Table 12.2.

A typical state of water vapor in moist air is shown in **Fig. 12.4**. At this state, fixed by the partial pressure p_v and the mixture temperature T, the vapor is superheated. When the partial pressure of the water vapor corresponds to the saturation pressure of water at the mixture temperature, p_g of Fig. 12.4, the mixture is said to be *saturated*. **Saturated air** is a mixture of dry air and saturated water vapor. The amount of water vapor in moist air varies from zero in dry air to a maximum, depending on the pressure and temperature, when the mixture is saturated.

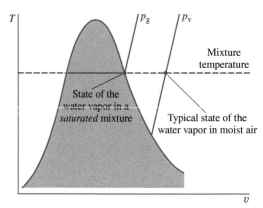

Fig. 12.4 T–v diagram for water vapor in an air–water mixture.

Saturated air

12.5.2 Humidity Ratio, Relative Humidity, Mixture Enthalpy, and Mixture Entropy

A given moist air sample can be described in a number of ways. The mixture can be described in terms of the moles of dry air and water vapor present or in terms of the respective mole fractions. Alternatively, the mass of dry air and water vapor, or the respective mass fractions,

humidity ratio   can be specified. The composition also can be indicated by means of the **humidity ratio** ω, defined as the ratio of the mass of the water vapor to the mass of dry air

$$\boxed{\omega = \frac{m_v}{m_a}} \tag{12.42}$$

The humidity ratio is sometimes referred to as the *specific humidity*.

The humidity ratio can be expressed in terms of partial pressures and molecular weights by solving Eqs. 12.41a for m_a and m_v, respectively, and substituting the resulting expressions into Eq. 12.42 to obtain

$$\omega = \frac{m_v}{m_a} = \frac{M_v p_v V/\overline{R}T}{M_a p_a V/\overline{R}T} = \frac{M_v p_v}{M_a p_a}$$

Introducing $p_a = p - p_v$ and noting that the ratio of the molecular weight of water to that of dry air, M_v/M_a, is approximately 0.622, this expression can be written as

$$\boxed{\omega = 0.622\frac{p_v}{p - p_v}} \tag{12.43}$$

Moist air also can be described in terms of the *relative humidity* ϕ, defined as the ratio of the mole fraction of water vapor y_v in a given moist air sample to the mole fraction $y_{v,\text{sat}}$ in a saturated moist air sample at the same mixture temperature and pressure:

$$\phi = \left.\frac{y_v}{y_{v,\text{sat}}}\right)_{T,\,p}$$

relative humidity   Since $p_v = y_v p$ and $p_g = y_{v,\text{sat}} p$, the **relative humidity** can be expressed as

$$\phi = \left.\frac{p_v}{p_g}\right)_{T,\,p} \tag{12.44}$$

The pressures in this expression for the relative humidity are labeled on Fig. 12.4.

The humidity ratio and relative humidity can be measured. For laboratory measurements of humidity ratio, a *hygrometer* can be used in which a moist air sample is exposed to suitable chemicals until the moisture present is absorbed. The amount of water vapor is determined by weighing the chemicals. Continuous recording of the relative humidity can be accomplished by means of transducers consisting of resistance- or capacitance-type sensors whose electrical characteristics change with relative humidity.

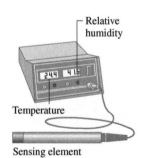

Relative humidity

Temperature

Sensing element

Evaluating H, U, and S for Moist Air

     The values of H, U, and S for moist air modeled as an ideal gas mixture can be found by adding the contribution of each component at the condition at which the component exists in the mixture. For example, the enthalpy H of a given moist air sample is

$$H = H_a + H_v = m_a h_a + m_v h_v \tag{12.45}$$

This moist air expression conforms to Eq. (d) in Table 12.2.

mixture enthalpy   Dividing by m_a and introducing the humidity ratio gives the **mixture enthalpy** *per unit mass of dry air*

$$\boxed{\frac{H}{m_a} = h_a + \frac{m_v}{m_a}h_v = h_a + \omega h_v} \tag{12.46}$$

The enthalpies of the dry air and water vapor appearing in Eq. 12.46 are evaluated at the mixture temperature. An approach similar to that for enthalpy also applies to the evaluation of the internal energy of moist air.

     Reference to steam table data or a Mollier diagram for water shows that the enthalpy of superheated water vapor at *low vapor pressures* is very closely given by the saturated vapor value corresponding to the given temperature. Hence, the enthalpy of the water vapor h_v in Eq. 12.46 can be taken as h_g at the mixture temperature. That is,

$$h_v \approx h_g(T) \qquad (12.47)$$

Equation 12.47 is used in the remainder of the chapter. Enthalpy data for water vapor as an ideal gas from Table A-23 are *not used* for h_v because the enthalpy datum of the ideal gas tables differs from that of the steam tables. These different datums can lead to error when studying systems that contain both water vapor and a liquid or solid phase of water. The enthalpy of dry air, h_a, can be obtained from the appropriate ideal gas table, Table A-22, however, because air is a gas at all states under present consideration and is closely modeled as an ideal gas at these states.

In accord with Eq. (h) in Table 12.2, the moist air **mixture entropy** has two contributions: water vapor and dry air. The contribution of each component is determined at the mixture temperature and the partial pressure of the component in the mixture. Using Eq. 6.18 and referring to Fig. 12.4 for the states, the specific entropy of the water vapor is given by $s_v(T, p_v) = s_g(T) - R \ln p_v/p_g$, where s_g is the specific entropy of saturated vapor at temperature T. Observe that the ratio of pressures, p_v/p_g, can be replaced by the relative humidity ϕ, giving an alternative expression.

mixture entropy

Using Computer Software   Property functions for moist air are listed under the **Properties** menu of *Interactive Thermodynamics: IT*. Functions are included for humidity ratio, relative humidity, specific enthalpy and entropy as well as other psychrometric properties introduced later. The methods used for evaluating these functions correspond to the methods discussed in this chapter, and the values returned by the computer software agree closely with those obtained by hand calculations with table data. The use of *IT* for psychrometric evaluations is illustrated in examples later in the chapter.

BioConnections

Medical practitioners and their patients have long noticed that influenza cases peak during winter. Speculation about the cause ranged widely, including the possibility that people spend more time indoors in winter and thus transmit the flu virus more easily or that the peak might be related to less sunlight exposure during winter, perhaps affecting human immune responses.

Since air is drier in winter, others suspected a link between relative humidity and influenza virus survival and transmission. In a 2007 study, using influenza-infected guinea pigs in climate-controlled habitats, researchers investigated the effects of variable habitat temperature and humidity on the aerosol spread of influenza virus. The study showed there were more infections when it was colder and drier, but relative humidity was a relatively weak variable in explaining findings. This prompted researchers to look for a better rationale.

When data from the 2007 study were reanalyzed, a significant correlation was found between humidity ratio and influenza. Unlike relative humidity, humidity ratio measures the actual amount of moisture present in air. When humidity ratio is low, as in peak winter flu months, the virus survives longer and transmission rates increase, researchers say. These findings strongly point to the value of humidifying indoor air in winter, particularly in high-risk places such as nursing homes.

12.5.3 Modeling Moist Air in Equilibrium with Liquid Water

Thus far, our study of psychrometrics has been conducted as an application of the ideal gas mixture principles introduced in the first part of this chapter. However, many systems of interest are composed of a mixture of dry air and water vapor in contact with a liquid (or solid) water phase. To study these systems requires additional considerations.

Shown in **Fig. 12.5** is a vessel containing liquid water, above which is a mixture of water vapor and dry air. If no interactions with the surroundings are allowed, liquid will evaporate until eventually the gas phase becomes saturated and the system attains an equilibrium state. For many engineering applications, systems consisting of moist

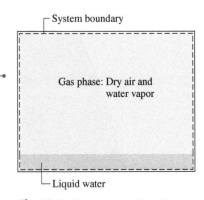

Fig. 12.5 System consisting of moist air in contact with liquid water.

air in *equilibrium* with a liquid water phase can be described simply and accurately with the following idealizations:

- The dry air and water vapor behave as independent ideal gases.
- The equilibrium between the liquid phase and the water vapor is not significantly disturbed by the presence of the air.
- The partial pressure of the water vapor equals the saturation pressure of water corresponding to the temperature of the mixture: $p_v = p_g(T)$.

Similar considerations apply for systems consisting of moist air in equilibrium with a solid water phase. The presence of the air actually alters the partial pressure of the vapor from the saturation pressure by a small amount whose magnitude is calculated in Sec. 14.6.

12.5.4 Evaluating the Dew Point Temperature

A significant aspect of the behavior of moist air is that partial condensation of the water vapor can occur when the temperature is reduced. This type of phenomenon is commonly encountered in the condensation of vapor on windowpanes and on pipes carrying cold water. The formation of dew on grass is another familiar example.

To study such condensation, consider a closed system consisting of a sample of moist air that is cooled at *constant* pressure, as shown in **Fig. 12.6.** The property diagram given on this figure locates states of the water vapor. Initially, the water vapor is superheated at state 1. In the first part of the cooling process, both the system pressure *and* the composition of the moist air remain constant. Accordingly, since $p_v = y_v p$, the *partial pressure* of the water vapor remains constant, and the water vapor cools at constant p_v from state 1 to state d, called the *dew* *point*. The saturation temperature corresponding to p_v is called the **dew point temperature**. This temperature is labeled on Fig. 12.6.

dew point temperature

In the next part of the cooling process, the system cools *below* the dew point temperature and some of the water vapor initially present condenses. At the final state, the system consists of a gas phase of dry air and water vapor in equilibrium with a liquid water phase. In accord with the discussion of Sec. 12.5.3, the vapor that remains is saturated vapor at the final temperature, state 2 of Fig. 12.6, with a partial pressure equal to the saturation pressure p_{g2} corresponding to this temperature. The condensate is saturated liquid at the final temperature: state 3 of Fig. 12.6.

Referring again to Fig. 12.6, note that the partial pressure of the water vapor at the final state, p_{g2}, is less than the initial value, p_{v1}. Owing to condensation, the partial pressure decreases because the amount of water vapor present at the final state is less than at the initial state. Since the amount of dry air is unchanged, the mole fraction of water vapor in the moist air also decreases.

In the next two examples, we illustrate the use of psychrometric properties introduced thus far. The examples consider, respectively, cooling moist air at constant pressure and at constant volume.

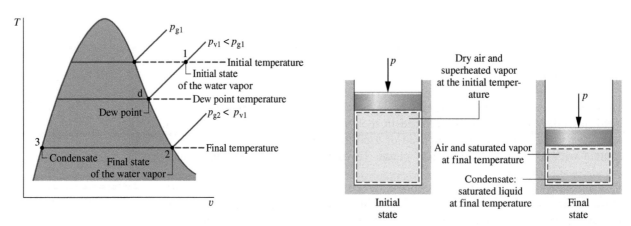

Fig. 12.6 States of water for moist air cooled at constant mixture pressure.

> **EXAMPLE 12.7** ▶

Cooling Moist Air at Constant Pressure

A 1 kg sample of moist air initially at 21°C 1 bar, and 70% relative humidity is cooled to 5°C while keeping the pressure constant. Determine **(a)** the initial humidity ratio, **(b)** the dew point temperature, in °C, and **(c)** the amount of water vapor that condenses, in kg.

SOLUTION

Known   A 1 kg sample of moist air is cooled at a constant mixture pressure of 1 bar from 21 to 5°C. The initial relative humidity is 70%.

Find   Determine the initial humidity ratio, the dew point temperature, in °C, and the amount of water vapor that condenses, in kg.

Schematic and Given Data:

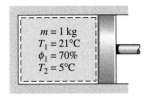

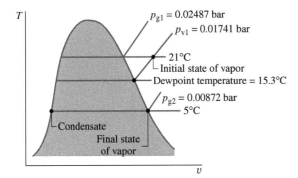

Fig. E12.7

Engineering Model

1. The 1 kg sample of moist air is taken as the closed system. The system pressure remains constant at 1 bar.

2. The gas phase can be treated as an ideal gas mixture. Each mixture component acts as an ideal gas existing alone in the volume occupied by the gas phase at the mixture temperature.

3. When a liquid water phase is present, the water vapor exists as a saturated vapor at the system temperature. The liquid present is a saturated liquid at the system temperature.

Analysis

a. The initial humidity ratio can be evaluated from Eq. 12.43. This requires the partial pressure of the water vapor, p_{v1} which can be found from the given relative humidity and p_g from Table A-2 at 21°C as follows:

$$p_{v1} = \phi p_g = (0.7)(0.02487)\,\text{bar} = 0.01741\,\text{bar}$$

Inserting values in Eq. 12.43

$$\omega_1 = 0.622\left(\frac{0.01741}{1 - 0.01741}\right) = 0.011\,\frac{\text{kg (vapor)}}{\text{kg (dry air)}}$$

b. The dew point temperature is the saturation temperature corresponding to the partial pressure, p_{v1}. Interpolation in Table A-2 gives $T = 15.3°C$. The dew point temperature is labeled on the accompanying property diagram.

c. The amount of condensate, m_w, equals the difference between the initial amount of water vapor in the sample, m_{v1}, and the final amount of water vapor, m_{v2}. That is

$$m_w = m_{v1} - m_{v2}$$

To evaluate m_{v1}, note that the system initially consists of 1 kg of dry air and water vapor, so $1\,\text{kg} = m_a + m_{v1}$, where m_a is the mass of dry air present in the sample. Since $\omega_1 = m_{v1}/m_a$, $m_a = m_{v1}/\omega_1$. With this we get

$$1\,\text{kg} = \frac{m_{v1}}{\omega_1} + m_{v1} = m_{v1}\left(\frac{1}{\omega_1} + 1\right)$$

Solving for m_{v1}

$$m_{v1} = \frac{1\,\text{kg}}{(1/\omega_1) + 1}$$

Inserting the value of determined in part (a)

$$m_{v1} = \frac{1\,\text{kg}}{(1/0.011) + 1} = 0.0109\,\text{kg (vapor)}$$

The mass of dry air present is then $m_a = 1 - 0.0109 = 0.9891$ kg (dry air).

Next, let us evaluate m_{v2}. With assumption 3, the partial pressure of the water vapor remaining in the system at the final state is the saturation pressure corresponding to 5°C: $p_g = 0.00872$ bar. Accordingly, the humidity ratio after cooling is found from Eq. 12.43 as

$$\omega_2 = 0.622\left(\frac{0.00872}{1 - 0.00872}\right) = 0.0054\,\frac{\text{kg (vapor)}}{\text{kg (dry air)}}$$

The mass of the water vapor present at the final state is then

$$m_{v2} = \omega_2 m_a = (0.0054)(0.9891) = 0.0053\,\text{kg (vapor)}$$

Finally, the amount of water vapor that condenses is

$$m_w = m_{v1} - m_{v2} = 0.0109 - 0.0053 = 0.0056\,\text{kg (condensate)}$$

---

❶ The amount of water vapor present in a typical moist air mixture is considerably less than the amount of dry air present.

SKILLS DEVELOPED

Ability to...

- apply psychrometric terminology and principles.

- demonstrate understanding of the dew point temperature and the formation of liquid condensate when pressure is constant.

- retrieve property data for water.

Quick Quiz

Determine the quality of the two-phase, liquid–vapor mixture and the relative humidity of the gas phase at the final state. Ans. 47%, 100%.

EXAMPLE 12.8

Cooling Moist Air at Constant Volume

An air–water vapor mixture is contained in a rigid, closed vessel with a volume of 35 m³ at 1.5 bar, 120°C, and $\phi = 10\%$. The mixture is cooled at constant volume until its temperature is reduced to 22°C. Determine **(a)** the dew point temperature corresponding to the initial state, in °C, **(b)** the temperature at which condensation actually begins, in °C, and **(c)** the amount of water condensed, in kg.

SOLUTION

Known   A rigid, closed tank with a volume of 35 m³ containing moist air initially at 1.5 bar, 120°C, and $\phi = 10\%$ is cooled to 22°C.

Find   Determine the dew point temperature at the initial state, in °C, the temperature at which condensation actually begins, in °C, and the amount of water condensed, in kg.

Schematic and Given Data:

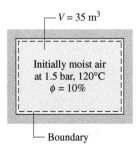

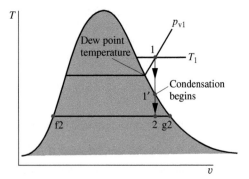

Fig. E12.8

Engineering Model

1. The contents of the tank are taken as a closed system. The system volume remains constant.

2. The gas phase can be treated as an ideal gas mixture. The Dalton model applies: Each mixture component acts as an ideal gas existing alone in the volume occupied by the gas phase at the mixture temperature.

3. When a liquid water phase is present, the water vapor exists as a saturated vapor at the system temperature. The liquid is a saturated liquid at the system temperature.

Analysis

a. The dew point temperature at the initial state is the saturation temperature corresponding to the partial pressure p_{v1}. With the given relative humidity and the saturation pressure at 120°C from Table A-2

$$p_{v1} = \phi_1 p_{g1} = (0.10)(1.985) = 0.1985 \text{ bar}$$

Interpolating in Table A-2 gives the dew point temperature as 60°C, which is the temperature condensation would begin *if* the moist air were cooled at *constant pressure*.

b. Whether the water exists as a vapor only, or as liquid *and* vapor, it occupies the full volume, which remains constant. Accordingly, since the total mass of the water present is also constant, the water undergoes the constant specific volume process illustrated on the accompanying *T–υ* diagram. In the process from state 1 to state 1′, the water exists as a vapor only. For the process from state 1′ to state 2, the water exists as a two-phase liquid–vapor mixture. Note that pressure does not remain constant during the cooling process from state 1 to state 2.

State 1′ on the *T–υ* diagram denotes the state where the water vapor first becomes saturated. The saturation temperature at this state is denoted as *T′*. Cooling to a temperature less than *T′* results in condensation of some of the water vapor present. Since state 1′ is a saturated vapor state, the temperature *T′* can be found by interpolating in Table A-2 with the specific volume of the water at this state. The specific volume of the vapor at state 1′ equals the specific volume of the vapor at state 1, which can be evaluated from the ideal gas equation

$$\upsilon_{v1} = \frac{(\bar{R}/M_v)T_1}{p_{v1}} = \left(\frac{8314}{18} \frac{\text{N} \cdot \text{m}}{\text{kg} \cdot \text{K}} \right) \left(\frac{393 \text{ K}}{0.1985 \times 10^5 \text{ N/m}^2} \right)$$

$$= 9.145 \frac{\text{m}^3}{\text{kg}}$$

① Interpolation in Table A-2 with $\upsilon_{v1} = \upsilon_g$ gives $T' = 56°C$. This is the temperature at which condensation begins.

c. The amount of condensate equals the difference between the initial and final amounts of water vapor present. The mass of the water vapor present initially is

$$m_{v1} = \frac{V}{\upsilon_{v1}} = \frac{35 \text{ m}^3}{9.145 \text{ m}^3/\text{kg}} = 3.827 \text{ kg}$$

The mass of water vapor present finally can be determined from the *quality*. At the final state, the water forms a two-phase liquid–vapor mixture having a specific volume of 9.145 m³/kg. Using this specific volume value, the quality x_2 of the liquid–vapor mixture can be found as

$$x_2 = \frac{\upsilon_{v2} - \upsilon_{f2}}{\upsilon_{g2} - \upsilon_{f2}} = \frac{9.145 - 1.0022 \times 10^{-3}}{51.447 - 1.0022 \times 10^{-3}} = 0.178$$

where υ_{f2} and υ_{g2} are the saturated liquid and saturated vapor specific volumes at $T_2 = 22°C$, respectively.

Using the quality together with the known total amount of water present, 3.827 kg, the mass of the water vapor contained in the system at the final state is

$$m_{v2} = (0.178)(3.827) = 0.681 \text{ kg}$$

The mass of the condensate, m_{w2}, is then

$$m_{w2} = m_{v1} - m_{v2} = 3.827 - 0.681 = 3.146 \text{ kg}$$

1 When a moist air mixture is cooled at constant mixture volume, the temperature at which condensation begins is not the dew point temperature corresponding to the initial state. In this case, condensation begins at 56°C, but the dew point temperature at the initial state, determined in part (a), is 60°C.

SKILLS DEVELOPED

Ability to…

- apply psychrometric terminology and principles.

- demonstrate understanding of the onset of condensation when cooling moist air at constant volume.
- retrieve property data for water.

Quick Quiz

Determine the humidity ratio at the initial state and the amount of dry air present, in kg. Ans. 0.0949, 40.389 kg.

No additional fundamental concepts are required for the study of closed systems involving mixtures of dry air and water vapor. Example 12.9, which builds on Example 12.8, brings out some special features of the use of conservation of mass and conservation of energy in analyzing this kind of system. Similar considerations can be used to study other closed systems involving moist air.

▶▶ **EXAMPLE 12.9** ▶

Evaluating Heat Transfer for Moist Air Cooling at Constant Volume

An air–water vapor mixture is contained in a rigid, closed vessel with a volume of 35 m³ at 1.5 bar, 120°C, and $\phi = 10\%$. The mixture is cooled until its temperature is reduced to 22°C. Determine the heat transfer during the process, in kJ.

SOLUTION

Known   A rigid, closed tank with a volume of 35 m³ containing moist air initially at 1.5 bar, 120°C, and $\phi = 10\%$ is cooled to 22°C.

Find   Determine the heat transfer for the process, in kJ.

Schematic and Given Data: See the figure for Example 12.8.

Engineering Model

1. The contents of the tank are taken as a closed system. The system volume remains constant.

2. The gas phase can be treated as an ideal gas mixture. The Dalton model applies: Each component acts as an ideal gas existing alone in the volume occupied by the gas phase at the mixture temperature.

3. When a liquid water phase is present, the water vapor exists as a saturated vapor and the liquid is a saturated liquid, each at the system temperature.

4. There is no work during the cooling process and no change in kinetic or potential energy.

Analysis   Reduction of the closed system energy balance using assumption 4 results in

$$\Delta U = Q - \cancel{W}^0$$

or

$$Q = U_2 - U_1$$

where

$$U_1 = m_a u_{a1} + m_{v1} u_{v1} = m_a u_{a1} + m_{v1} u_{g1}$$

and

$$U_2 = m_a u_{a2} + m_{v2} u_{v2} + m_{w2} u_{w2} = m_a u_{a2} + m_{v2} u_{g2} + m_{w2} u_{f2}$$

In these equations, the subscripts a, v, and w denote, respectively, dry air, water vapor, and liquid water. The specific internal energy of the water vapor at the initial state can be approximated as the saturated vapor value at T_1. At the final state, the water vapor is assumed to exist as a saturated vapor, so its specific internal energy is u_g at T_2. The liquid water at the final state is saturated, so its specific internal energy is u_f at T_2.

Collecting the last three equations

1 $$Q = \underline{m_a(u_{a2} - u_{a1})} + \underline{m_{v2}u_{g2} + m_{w2}u_{f2} - m_{v1}u_{g1}}$$

The mass of dry air, m_a, can be found using the ideal gas equation of state together with the partial pressure of the dry air at the initial state obtained using $p_{v1} = 0.1985$ bar from the solution to Example 12.8 as follows:

$$m_a = \frac{p_{a1}V}{(\overline{R}/M_a)T_1} = \frac{[(1.5 - 0.1985) \times 10^5 \text{ N/m}^2](35 \text{ m}^3)}{(8314/28.97 \text{ N} \cdot \text{m/kg} \cdot \text{K})(393 \text{ K})}$$

$$= 40.389 \text{ kg}$$

Then, evaluating internal energies of dry air and water from Tables A-22 and A-2, respectively,

$$Q = 40.389(210.49 - 281.1) + 0.681(2405.7) + 3.146(92.32)$$
$$- 3.827(2529.3)$$

$$= -2851.87 + 1638.28 + 290.44 - 9679.63 = -10{,}603 \text{ kJ}$$

The values for m_{v1}, m_{v2}, and m_{w2} are from the solution to Example 12.8.

1   The first underlined term in this equation for Q is evaluated with specific internal energies from the ideal gas table for air, Table A-22. Steam table data are used to evaluate the second underlined term. The different datums for internal energy underlying these tables cancel because each of these two terms involves internal energy *differences*. Since the specific heat c_{va} for dry air varies only slightly over the interval from 120 to 22°C (Table A-20), the specific internal energy change of the dry air could be evaluated alternatively using a constant c_{va} value. See the Quick Quiz that follows.

SKILLS DEVELOPED

Ability to…

• apply psychrometric terminology and principles.

• apply the energy balance to the cooling of moist air at constant volume.

• retrieve property data for water.

Quick Quiz

Calculate the change in internal energy of the *dry air*, in kJ, assuming a constant specific heat c_{va} interpolated from Table A-20 at the average of the initial and final temperatures. Ans. −2854 kJ.

12.5.5 Evaluating Humidity Ratio Using the Adiabatic-Saturation Temperature

adiabatic-saturation temperature

The humidity ratio ω of an air–water vapor mixture can be determined, in principle, knowing the values of three mixture properties: the pressure p, the temperature T, and the **adiabatic-saturation temperature** T_{as} introduced in this section. The relationship among these mixture properties is obtained by applying conservation of mass and conservation of energy to an *adiabatic saturator* (see box).

Equations 12.48 and 12.49 give the humidity ratio ω in terms of the adiabatic-saturation temperature and other quantities:

$$\omega = \frac{h_a(T_{as}) - h_a(T) + \omega'[h_g(T_{as}) - h_f(T_{as})]}{h_g(T) - h_f(T_{as})} \tag{12.48}$$

where h_f and h_g denote the enthalpies of saturated liquid water and saturated water vapor, respectively, obtained from the steam tables at the indicated temperatures. The enthalpies of the dry air h_a can be obtained from the ideal gas table for air. Alternatively, $h_a(T_{as}) - h_a(T) = c_{pa}(T_{as} - T)$, where c_{pa} is an appropriate constant value for the specific heat of dry air. The humidity ratio ω' appearing in Eq. 12.48 is

$$\omega' = 0.622 \frac{p_g(T_{as})}{p - p_g(T_{as})} \tag{12.49}$$

where $p_g(T_{as})$ is the saturation pressure at the adiabatic-saturation temperature and p is the mixture pressure.

Modeling an Adiabatic Saturator

Figure 12.7 shows the schematic and process representations of an adiabatic saturator, which is a two-inlet, single-exit device through which moist air passes. The device is assumed to operate at steady state and without significant heat transfer with its surroundings. An air–water vapor mixture of *unknown* humidity ratio ω enters the adiabatic saturator at a known pressure p and temperature T. As the mixture passes through the device, it comes into contact with a pool of water. If the entering mixture is not saturated ($\phi < 100\%$), some of the water would evaporate. The energy required to evaporate the water would come from the moist air, so the mixture temperature would decrease as the air passes through the duct. For a sufficiently long duct, the mixture would be saturated as it exits ($\phi = 100\%$). Since a saturated mixture would be achieved without heat transfer with the surroundings, the temperature of the exiting mixture is the *adiabatic-saturation temperature*. As indicated on Fig. 12.7, a steady flow of makeup water at temperature T_{as} is added at

the same rate at which water is evaporated. The pressure of the mixture is assumed to remain constant as it passes through the device.

Equation 12.48 giving the humidity ratio ω of the entering moist air in terms of p, T, and T_{as} can be obtained by applying conservation of mass and conservation of energy to the adiabatic saturator, as follows:

At steady state, the mass flow rate of the dry air entering the device, $\dot{m}_a$, must equal the mass flow rate of the dry air exiting. The mass flow rate of the makeup water is the difference between the exiting and entering vapor flow rates denoted by $\dot{m}_v$ and $\dot{m}'_v$, respectively. These flow rates are labeled on Fig. 12.7a. At steady state, the energy rate balance reduces to

$$(\dot{m}_a h_a + \dot{m}_v h_v)_{\substack{\text{moist} \\ \text{air entering}}} + [(\dot{m}'_v - \dot{m}_v)h_w]_{\substack{\text{makeup} \\ \text{water}}} + (\dot{m}_a h_a + \dot{m}'_v h_v)_{\substack{\text{moist air} \\ \text{exiting}}}$$

Several assumptions underlie this expression: Each of the two moist air streams is modeled as an ideal gas mixture of dry air and

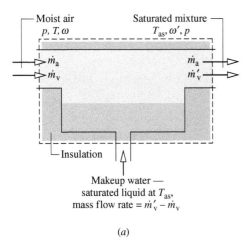
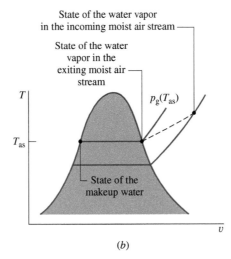

Fig. 12.7 Adiabatic saturator. (*a*) Schematic. (*b*) Process representation.

water vapor. Heat transfer with the surroundings is assumed to be negligible. There is no work $\dot{W}_{cv}$, and changes in kinetic and potential energy are ignored.

Dividing by the mass flow rate of the dry air, $\dot{m}_a$, the energy rate balance can be written on the basis of a unit mass of dry air passing through the device as

$$\underbrace{(h_a + \omega h_g)_{\text{moist air}}}_{\text{entering}} + \underbrace{[(\omega' - \omega)h_f]_{\text{makeup}}}_{\text{water}} = \underbrace{(h_a + \omega' h_g)_{\text{moist air}}}_{\text{exiting}}$$

$$(12.50)$$

where $\omega = \dot{m}_v/\dot{m}_a$ and $\omega' = \dot{m}_v'/\dot{m}_a$.

For the exiting saturated mixture, the partial pressure of the water vapor is the saturation pressure corresponding to the adiabatic-saturation temperature, $p_g(T_{as})$. Accordingly, the humidity ratio ω' can be evaluated knowing T_{as} and the mixture pressure p, as indicated by Eq. 12.49. In writing Eq. 12.50, the specific enthalpy of the entering water vapor has been evaluated as that of saturated water vapor at the temperature of the incoming mixture, in accordance with Eq. 12.47. Since the exiting mixture is saturated, the enthalpy of the water vapor at the exit is given by the saturated vapor value at T_{as}. The enthalpy of the makeup water is evaluated as that of saturated liquid at T_{as}.

When Eq. 12.50 is solved for ω, Eq. 12.48 results. The details of the solution are left as an exercise.

12.6 Psychrometers: Measuring the Wet-Bulb and Dry-Bulb Temperatures

TAKE NOTE...

Although derived with reference to the adiabatic saturator in Fig. 12.7, the relationship provided by Eq. 12.48 applies generally to moist air mixtures and is not restricted to this type of system or even to control volumes. The relationship allows the humidity ratio ω to be determined for *any* moist air mixture for which the pressure p, temperature T, and adiabatic-saturation temperature T_{as} are known.

For moist air mixtures in the normal pressure and temperature ranges of psychrometrics, the readily measured wet-bulb temperature is an important parameter.

The **wet-bulb temperature** is read from a wet-bulb thermometer, which is an ordinary liquid-in-glass thermometer whose bulb is enclosed by a wick moistened with water. The term **dry-bulb temperature** refers simply to the temperature that would be measured by a thermometer placed in the mixture. Often a wet-bulb thermometer is mounted together with a dry-bulb thermometer to form an instrument called a **psychrometer**.

The psychrometer of **Fig. 12.8a** is whirled in the air whose wet- and dry-bulb temperatures are to be determined. This induces air flow over the two thermometers. For the psychrometer of Fig. 12.8b, the air flow is induced by a battery-operated fan. In each type of psychrometer, if the surrounding air is not saturated, water in the wick of the wet-bulb thermometer evaporates and the temperature of the remaining water falls below the dry-bulb temperature. Eventually a steady-state condition is attained by the wet-bulb thermometer. The wet- and dry-bulb temperatures are then read from the respective thermometers. The wet-bulb temperature depends on the rates of heat and mass transfer between the moistened wick and the air. Since these depend in turn on the geometry of the thermometer, air velocity, supply water temperature, and other factors, the wet-bulb temperature is not a mixture property.

wet-bulb temperature

dry-bulb temperature

psychrometer

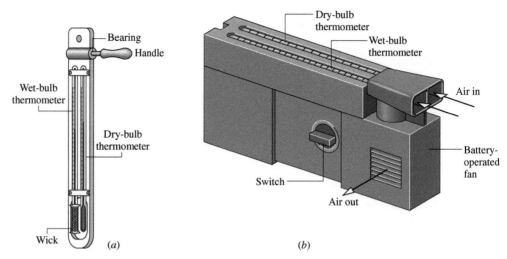

Fig. 12.8 Psychrometers. (*a*) Sling psychrometer. (*b*) Aspirating psychrometer.

For moist air mixtures in the normal pressure and temperature ranges of psychrometric applications, the adiabatic-saturation temperature introduced in Sec. 12.5.5 is closely approximated by the wet-bulb temperature. Accordingly, the humidity ratio for any such mixture can be calculated by using the wet-bulb temperature in Eqs. 12.48 and 12.49 in place of the adiabatic-saturation temperature. Close agreement between the adiabatic-saturation and wet-bulb temperatures is not generally found for moist air departing from normal psychrometric conditions.

BioConnections

The *National Weather Service* is finding better ways to help measure our misery during cold snaps so we can avoid dangerous exposure. The wind chill index, for many years based on a single 1945 study, has been upgraded using new physiological data and computer modeling to better reflect the effects of cold winds and freezing temperatures.

The new wind chill index is a standardized "temperature" that accounts for both the actual air temperature and the wind speed. The formula on which it is based uses measurements of skin tissue thermal resistance and computer models of the wind patterns over the human face, together with principles of heat transfer. Using the new index, an air temperature of 5°F and a

wind speed of 25 miles per hour correspond to a wind chill temperature of −17°F. The old index assigned a wind chill of −36°F to the same conditions. With the new information, people are better armed to avoid exposure that can lead to such serious medical problems as frostbite.

The improved measure was developed by universities, international scientific societies, and government in an effort that led to the new standard being adopted in the United States. Further upgrades may include the amount of cloud cover in the formula, since solar radiation is also an important factor in how cold it feels. In fact under sunny conditions, wind chill temperatures may increase up to 18°F.

12.7 Psychrometric Charts

psychrometric charts

Graphical representations of several important properties of moist air are provided by **psychrometric charts**. The main features of one form of chart are shown in **Fig. 12.9**. Complete charts in SI units are given in Fig. A-9. These charts are constructed for a mixture pressure of 1 atm, but charts for other mixture pressures are also available. When the mixture pressure differs only slightly from 1 atm, Fig. A-9 remain sufficiently accurate for engineering analyses. In this text, such differences are ignored.

Let us consider several features of the psychrometric chart:

- Referring to Fig. 12.9, note that the abscissa gives the dry-bulb temperature and the ordinate provides the humidity ratio. For charts in SI, the temperature is in °C and ω is expressed in kg, or g, of water vapor per kg of dry air.

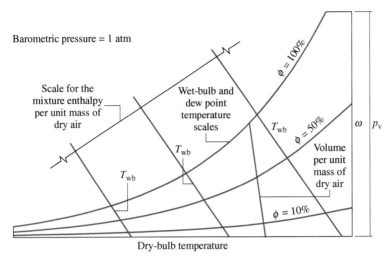

Fig. 12.9 Psychrometric chart.

- Equation 12.43 shows that for fixed mixture pressure there is a direct correspondence between the partial pressure of the water vapor and the humidity ratio. Accordingly, the vapor pressure also can be shown on the ordinate, as illustrated on Fig. 12.9.

- Curves of constant relative humidity are shown on psychrometric charts. On Fig. 12.9, curves labeled $\phi = 100$, 50, and 10% are indicated. Since the dew point is the state where the mixture becomes saturated when cooled at constant vapor pressure, the dew point temperature corresponding to a given moist air state can be determined by following a line of constant ω (constant p_v) to the saturation line, $\phi = 100\%$. The dew point temperature and dry-bulb temperature are identical for states on the saturation curve.

- Psychrometric charts also give values of the mixture enthalpy per unit mass of dry air in the mixture: $h_a + \omega h_v$. In Fig. A-9, the mixture enthalpy has units of kJ per kg of dry air. The numerical values provided on these charts are determined relative to the following *special* reference states and reference values. In Fig. A-9, the enthalpy of the dry air h_a is determined relative to a zero value at 0°C, and not 0 K as in Table A-22. Accordingly, in place of Eq. 3.49 used to develop the enthalpy data of Table A-22, the following expression is employed to evaluate the enthalpy of the dry air for use on the psychrometric chart:

$$h_a = \int_{273.15\,\text{K}}^{T} c_{pa}\, dT = c_{pa} T(°C) \tag{12.51}$$

where c_{pa} is a constant value for the specific heat c_p of dry air and $T(°C)$ denotes the temperature in °C. In the temperature ranges of Fig. A-9, c_{pa} can be taken as 1.005 kJ/kg · K. In Fig. A-9 the enthalpy of the water vapor h_v is evaluated as h_g at the dry-bulb temperature of the mixture from Table A-2, as appropriate.

- Another important parameter on psychrometer charts is the wet-bulb temperature. As illustrated by Fig. A-9, constant T_{wb} lines run from the upper left to the lower right of the chart. The relationship between the wet-bulb temperature and other chart quantities is provided by Eq. 12.48. The wet-bulb temperature can be used in this equation in place of the adiabatic-saturation temperature for the states of moist air located on Fig. A-9.

- Lines of constant wet-bulb temperature are approximately lines of constant mixture enthalpy per unit mass of dry air. This feature can be brought out by study of the energy balance for the adiabatic saturator, Eq. 12.50. Since the contribution of the energy entering the adiabatic saturator with the makeup water is normally much smaller than that of the moist air, the enthalpy of the entering moist air is very nearly equal to the enthalpy of the saturated mixture exiting. Accordingly, all states with the same value of the wet-bulb temperature (adiabatic-saturation temperature) have nearly the same value for the mixture enthalpy per unit mass of dry air. Although Fig. A-9 ignore this slight effect, some psychrometric charts are drawn to show the departure of lines of constant wet-bulb temperature from lines of constant mixture enthalpy.

- As shown on Fig. 12.9, psychrometric charts also provide lines representing volume per unit mass of dry air, V/m_a. Figure A-9 gives this quantity in units of m^3/kg. These specific volume lines can be interpreted as giving the volume of dry air or of water vapor, per unit mass of dry air, since each mixture component is considered to fill the entire volume.

The psychrometric chart is easily used.

FOR EXAMPLE

A psychrometer indicates that in a classroom the dry-bulb temperature is 20°C and the wet-bulb temperature is 15°C. Locating the mixture state on Fig. A-9 corresponding to the intersection of these temperatures, we read $\omega = 0.0092$ kg(vapor)/kg(dry air) and $\phi = 63\%$.

(12.8) Analyzing Air-Conditioning Processes

The purpose of the present section is to study typical air-conditioning processes using the psychrometric principles developed in this chapter. Specific illustrations are provided in the form of solved examples involving control volumes at steady state. In each example, the methodology introduced in Sec. 12.8.1 is employed to arrive at the solution.

To reinforce principles developed in this chapter, the psychrometric parameters required by these examples are determined in most cases using tabular data from Appendix tables. Where a full psychrometric chart solution is not also provided, we recommend the example be solved using the chart, checking results with values from the solution presented.

12.8.1 Applying Mass and Energy Balances to Air-Conditioning Systems

The object of this section is to illustrate the use of the conservation of mass and conservation of energy principles in analyzing systems involving mixtures of dry air and water vapor in which a condensed water phase may be present. The same basic solution approach that has been used in thermodynamic analyses considered thus far is applicable. The only new aspect is the use of the special vocabulary and parameters of psychrometrics.

Systems that accomplish air-conditioning processes such as heating, cooling, humidification, or dehumidification are normally analyzed on a control volume basis. To consider a typical analysis, refer to **Fig. 12.10**, which shows a two-inlet, single-exit control volume at steady state. A moist air stream enters at 1, a moist air stream exits at 2, and a water-only stream enters (or exits) at 3. The water-only stream may be a liquid or a vapor. Heat transfer at the rate $\dot{Q}_{cv}$ can occur between the control volume and its surroundings. Depending on the application, the value of $\dot{Q}_{cv}$ might be positive, negative, or zero.

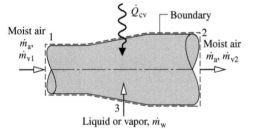

Fig. 12.10 Systems for conditioning moist air.

Mass Balance At steady state, the amounts of dry air and water vapor contained within the control volume cannot vary. Thus, for each component individually it is necessary for the total incoming and outgoing mass flow rates to be equal. That is,

$$\dot{m}_{a1} = \dot{m}_{a2} \qquad \text{(dry air)}$$
$$\dot{m}_{v1} + \dot{m}_w = \dot{m}_{v2} \qquad \text{(water)}$$

For simplicity, the constant mass flow rate of the dry air is denoted by $\dot{m}_a$. The mass flow rates of the water vapor can be expressed conveniently in terms of humidity ratios as $\dot{m}_{v1} = \omega_1 \dot{m}_a$ and $\dot{m}_{v2} = \omega_2 \dot{m}_a$. With these expressions, the mass balance for water becomes

$$\dot{m}_w = \dot{m}_a(\omega_2 - \omega_1) \qquad \text{(water)} \qquad (12.52)$$

When water is added at 3, ω_2 is greater than ω_1.

Energy Balance   Assuming $\dot{W}_{cv} = 0$ and ignoring all kinetic and potential energy effects, the energy rate balance reduces at steady state to

$$0 = \dot{Q}_{cv} + (\dot{m}_a h_{a1} + \dot{m}_{v1} h_{v1}) + \dot{m}_w h_w - (\dot{m}_a h_{a2} + \dot{m}_{v2} h_{v2}) \tag{12.53}$$

In this equation, the entering and exiting moist air streams are regarded as ideal gas mixtures of dry air and water vapor.

Equation 12.53 can be cast into a form that is particularly convenient for the analysis of air-conditioning systems. First, with Eq. 12.47 the enthalpies of the entering and exiting water vapor can be evaluated as the saturated vapor enthalpies corresponding to the temperatures T_1 and T_2, respectively, giving

$$0 = \dot{Q}_{cv} + (\dot{m}_a h_{a1} + \dot{m}_{v1} h_{g1}) + \dot{m}_w h_w - (\dot{m}_a h_{a2} + \dot{m}_{v2} h_{g2})$$

Then, with $\dot{m}_{v1} = \omega_1 \dot{m}_a$ and $\dot{m}_{v2} = \omega_2 \dot{m}_a$, the equation can be expressed as

$$0 = \dot{Q}_{cv} + \dot{m}_a (h_{a1} + \omega_1 h_{g1}) + \dot{m}_w h_w - \dot{m}_a (h_{a2} + \omega_2 h_{g2}) \tag{12.54}$$

Finally, introducing Eq. 12.52, the energy rate balance becomes

$$0 = \dot{Q}_{cv} + \dot{m}_a [(h_{a1} - h_{a2}) + \omega_1 h_{g1} + (\omega_2 - \omega_1) h_w - \omega_2 h_{g2}] \tag{12.55}$$

The first underlined term of Eq. 12.55 can be evaluated from Table A-22 giving the ideal gas properties of air. Alternatively, since relatively small temperature differences are normally encountered in the class of systems under present consideration, this term can be evaluated as $h_{a1} - h_{a2} = c_{pa}(T_1 - T_2)$, where c_{pa} is a constant value for the specific heat of dry air. The second underlined term of Eq. 12.55 can be evaluated using steam table data together with known values for ω_1 and ω_2. As illustrated in discussions to follow, Eq. 12.55 also can be evaluated using the psychrometric chart or *IT*.

TAKE NOTE...

As indicated by the developments of Sec. 12.8.1, several simplifying assumptions are made when analyzing air-conditioning systems considered in Examples 12.10–12.14 to follow. They include:

- The control volume is at steady state.
- Moist air streams are ideal gas mixtures of dry air and water vapor adhering to the Dalton model.
- Flow is one-dimensional where mass crosses the boundary of the control volume, and the effects of kinetic and potential energy at these locations are neglected.
- The only work is *flow work* (Sec. 4.4.2) where mass crosses the boundary of the control volume.

12.8.2 Conditioning Moist Air at Constant Composition

Building air-conditioning systems frequently heat or cool a moist air stream with no change in the amount of water vapor present. In such cases the humidity ratio ω remains constant, while relative humidity and other moist air parameters vary. Example 12.10 gives an elementary illustration using the methodology of Sec. 12.8.1.

▶ ▶ **EXAMPLE 12.10** ▶ ....

Heating Moist Air in a Duct

Moist air enters a duct at 10°C, 80% relative humidity, and a volumetric flow rate of 150 m³/min. The mixture is heated as it flows through the duct and exits at 30°C. No moisture is added or removed, and the mixture pressure remains approximately constant at 1 bar. For steady-state operation, determine **(a)** the rate of heat transfer, in kJ/min, and **(b)** the relative humidity at the exit. Changes in kinetic and potential energy can be ignored.

SOLUTION

Known   Moist air that enters a duct at 10°C and $\phi = 80\%$ with a volumetric flow rate of 150 m³/min is heated at constant pressure and exits at 30°C. No moisture is added or removed.

Find   Determine the rate of heat transfer, in kJ/min, and the relative humidity at the exit.

Schematic and Given Data:

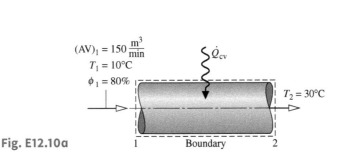
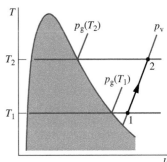

Fig. E12.10a

Engineering Model

1. The control volume shown in the accompanying figure operates at steady state.

2. The changes in kinetic and potential energy between inlet and exit can be ignored and $\dot{W}_{cv} = 0$.

3. The entering and exiting moist air streams are regarded as ideal gas mixtures adhering to the Dalton model.

Analysis

a. The heat transfer rate $\dot{Q}_{cv}$ can be determined from the mass and energy rate balances. At steady state, the amounts of dry air and water vapor contained within the control volume cannot vary. Thus, for each component individually it is necessary for the incoming and outgoing mass flow rates to be equal. That is,

$$\dot{m}_{a1} = \dot{m}_{a2} \qquad \text{(dry air)}$$
$$\dot{m}_{v1} = \dot{m}_{v2} \qquad \text{(water vapor)}$$

For simplicity, the constant mass flow rates of the dry air and water vapor are denoted, respectively, by $\dot{m}_a$ and $\dot{m}_v$. From these considerations, it can be concluded that the humidity ratio is the same at the inlet and exit: $\omega_1 = \omega_2$.

The steady-state form of the energy rate balance reduces with assumption 2 to

$$0 = \dot{Q}_{cv} - \overset{0}{\cancel{\dot{W}_{cv}}} + (\dot{m}_a h_{a1} + \dot{m}_v h_{v1}) - (\dot{m}_a h_{a2} + \dot{m}_v h_{v2})$$

In writing this equation, the incoming and outgoing moist air streams are regarded as ideal gas mixtures of dry air and water vapor.

Solving for $\dot{Q}_{cv}$

$$\dot{Q}_{cv} = \dot{m}_a (h_{a2} - h_{a1}) + \dot{m}_v (h_{v2} - h_{v1})$$

Noting that $\dot{m}_v = \omega \dot{m}_a$, where ω is the humidity ratio, the expression for $\dot{Q}_{cv}$ can be written in the form

❶ $$\dot{Q}_{cv} = \dot{m}_a [(h_{a2} - h_{a1}) + \omega(h_{v2} - h_{v1})] \tag{a}$$

To evaluate $\dot{Q}_{cv}$ from this expression requires the specific enthalpies of the dry air and water vapor at the inlet and exit, the mass flow rate of the dry air, and the humidity ratio.

The specific enthalpies of the dry air are obtained from Table A-22 at the inlet and exit temperatures T_1 and T_2, respectively: $h_{a1} = 283.1$ kJ/kg, $h_{a2} = 303.2$ kJ/kg. The specific enthalpies of the water vapor are found using $h_v \approx h_g$ and data from Table A-2 at T_1 and T_2, respectively: $h_{g1} = 2519.8$ kJ/kg, $h_{g2} = 2556.3$ kJ/kg.

The mass flow rate of the dry air can be determined from the volumetric flow rate at the inlet $(AV)_1$

$$\dot{m}_a = \frac{(AV)_1}{v_{a1}}$$

In this equation, v_{a1} is the specific volume of the dry air evaluated at T_1 and the partial pressure of the dry air p_{a1}. Using the ideal gas equation of state

$$v_{a1} = \frac{(\overline{R}/M)T_1}{p_{a1}}$$

The partial pressure p_{a1} can be determined from the mixture pressure p and the partial pressure of the water vapor p_{v1}: $p_{a1} = p - p_{v1}$. To find p_{v1}, use the given inlet relative humidity and the saturation pressure at 10°C from Table A-2

$$p_{v1} = \phi_1 p_{g1} = (0.8)(0.01228 \text{ bar}) = 0.0098 \text{ bar}$$

Since the mixture pressure is 1 bar, it follows that $p_{a1} = 0.9902$ bar. The specific volume of the dry air is then

$$v_{a1} = \frac{\left(\dfrac{8314 \text{ N} \cdot \text{m}}{28.97 \text{ kg} \cdot \text{K}}\right)(283 \text{ K})}{(0.9902 \times 10^5 \text{ N/m}^2)} = 0.82 \text{ m}^3/\text{kg}$$

Using this value, the mass flow rate of the dry air is

$$\dot{m}_a = \frac{150 \text{ m}^3/\text{min}}{0.82 \text{ m}^3/\text{kg}} = 182.9 \text{ kg/min}$$

The humidity ratio ω can be found from

$$\omega = 0.622\left(\frac{p_{v1}}{p - p_{v1}}\right) = 0.622\left(\frac{0.0098}{1 - 0.0098}\right)$$
$$= 0.00616 \frac{\text{kg (vapor)}}{\text{kg (dry air)}}$$

Finally, substituting values into Eq. (a) we get

$$\dot{Q}_{cv} = 182.9[(303.2 - 283.1) + (0.00616)(2556.3 - 2519.8)]$$
$$= 3717 \text{ kJ/min}$$

b. The states of the water vapor at the duct inlet and exit are located on the accompanying T–v diagram. Since the composition of the moist air and the mixture pressure remain constant, the partial pressure of the water vapor at the exit equals the partial pressure of the water vapor at the inlet: $p_{v2} = p_{v1} = 0.0098$ bar. The relative humidity at the exit is then

❷ $$\phi_2 = \frac{p_{v2}}{p_{g2}} = \frac{0.0098}{0.04246} = 0.231(23.1\%)$$

where p_{g2} is from Table A-2 at 30°C.

Alternative Psychrometric Chart Solution Let us consider an alternative solution using the psychrometric chart. As shown on the sketch of the psychrometric chart, Fig. E12.10b, the state of the moist air at the inlet is defined by $\phi_1 = 80\%$ and a dry-bulb temperature of 10°C. From the solution to part (a), we know that the humidity ratio has the same value at the exit as at the inlet. Accordingly, the state of the moist air at the exit is fixed by $\omega_2 = \omega_1$ and a dry-bulb temperature of 30°C. By inspection of Fig. A-9, the relative humidity at the duct exit is about 23% and, thus, in agreement with the result of part (b).

The rate of heat transfer can be evaluated from the psychrometric chart using the following expression obtained by rearranging Eq. (a) of part (a) to read

$$\dot{Q}_{cv} = \dot{m}_a [(h_a + \omega h_v)_2 - (h_a + \omega h_v)_1] \tag{b}$$

To evaluate $\dot{Q}_{cv}$ from this expression requires values for the mixture enthalpy per unit mass of dry air $(h_a + \omega h_v)$ at the inlet and exit. These can be determined by inspection of the psychrometric chart, Fig. A-9, as $(h_a + \omega h_v)_1 = 25.7$ kJ/kg(dry air), $(h_a + \omega h_v)_2 = 45.9$ kJ/kg(dry air).

Using the specific volume value v_{a1} at the inlet state read from the chart together with the given volumetric flow rate at the inlet, the mass flow rate of the dry air is found as

$$\dot{m}_a = \frac{150 \text{ m}^3/\text{min}}{0.81 \text{ m}^3/\text{kg(dry air)}} = 185 \frac{\text{kg(dry air)}}{\text{min}}$$

Substituting values into the energy rate balance, Eq. (b), we get

$$\dot{Q}_{cv} = 185 \frac{\text{kg(dry air)}}{\text{min}} (45.9 - 25.7) \frac{\text{kJ}}{\text{kg(dry air)}}$$

$$= 3737 \frac{\text{kJ}}{\text{min}}$$

which agrees closely with the result obtained in part (a), as expected.

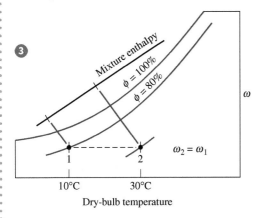

Fig. E12.10b

1 The first underlined term in this equation for $\dot{Q}_{cv}$ is evaluated with specific enthalpies from the ideal gas table for air, Table A-22. Steam table data are used to evaluate the second underlined term. Note that the different datums for enthalpy underlying these tables cancel because each of the two terms involves enthalpy *differences* only. Since the specific heat c_{pa} for dry air varies only slightly over the interval from 10 to 30°C (Table A-20), the specific enthalpy change of the dry air could be evaluated alternatively with $c_{pa} = 1.005$ kJ/kg · K.

2 No water is added or removed as the moist air passes through the duct at constant pressure; accordingly, the humidity ratio ω and the partial pressures p_v and p_a remain constant. However, because the saturation pressure increases as the temperature increases from inlet to exit, the *relative humidity* decreases: $\phi_2 < \phi_1$.

3 The mixture pressure, 1 bar, differs slightly from the pressure used to construct the psychrometric chart, 1 atm. This difference is ignored.

SKILLS DEVELOPED

Ability to...

- apply psychrometric terminology and principles.
- apply mass and energy balances for heating at constant composition in a control volume at steady state.
- retrieve necessary property data.

Quick Quiz

Using the psychrometric chart, what is the dew point temperature, in °C, for the moist air entering? At the exit? Ans. ≈7°C, same.

12.8.3 Dehumidification

When a moist air stream is cooled at constant mixture pressure to a temperature below its dew point temperature, some condensation of the water vapor initially present will occur. **Figure 12.11** shows the schematic of a dehumidifier using this principle. Moist air enters at state 1 and flows across a cooling coil through which a refrigerant or chilled water circulates. Some of the water vapor initially present in the moist air condenses, and a saturated moist air mixture exits the dehumidifier section at state 2. Although water condenses at various temperatures, the condensed water is assumed to be cooled to T_2 before it exits the dehumidifier. Since the moist air leaving the humidifier is saturated at a temperature lower than the temperature of the moist air entering, the moist air stream at state 2 might be uncomfortable for direct use in occupied spaces. However, by passing the stream through a following heating section, it can be brought to a condition—state 3—most occupants would regard as comfortable.

Let us sketch the procedure for evaluating the rates at which condensate exits and refrigerant circulates. This requires the use of mass and energy rate balances for the dehumidifier section. They are developed next.

TAKE NOTE...

A dashed line on a property diagram signals only that a process has occurred between initial and final equilibrium states and does not define a *path* for the process.

Mass Balance The mass flow rate of the condensate $\dot{m}_w$ can be related to the mass flow rate of the dry air $\dot{m}_a$ by applying conservation of mass separately for the dry air and water passing through the dehumidifier section. At steady state

$$\dot{m}_{a1} = \dot{m}_{a2} \qquad \text{(dry air)}$$
$$\dot{m}_{v1} = \dot{m}_w + \dot{m}_{v2} \qquad \text{(water)}$$

The common mass flow rate of the dry air is denoted as $\dot{m}_a$. Solving for the mass flow rate of the condensate

$$\dot{m}_w = \dot{m}_{v1} - \dot{m}_{v2}$$

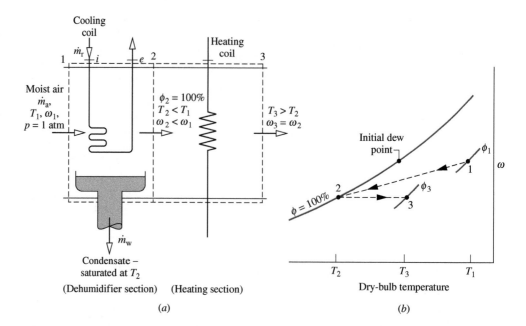

Fig. 12.11 Dehumidification.
(*a*) Equipment schematic.
(*b*) Psychrometric chart
representation.

Introducing $\dot{m}_{v1} = \omega_1 \dot{m}_a$ and $\dot{m}_{v2} = \omega_2 \dot{m}_a$, the amount of water condensed per unit mass of dry air passing through the device is

$$\frac{\dot{m}_w}{\dot{m}_a} = \omega_1 - \omega_2$$

This expression requires the humidity ratios ω_1 and ω_2. Because no moisture is added or removed in the heating section, it can be concluded from conservation of mass that $\omega_2 = \omega_3$, so ω_3 can be used in the above equation in place of ω_2.

Energy Balance   The mass flow rate of the refrigerant through the cooling coil $\dot{m}_r$ can be related to the mass flow rate of the dry air $\dot{m}_a$ by means of an energy rate balance applied to the dehumidifier section. With $\dot{W}_{cv} = 0$, negligible heat transfer with the surroundings, and no significant kinetic and potential energy changes, the energy rate balance reduces at steady state to

$$0 = \dot{m}_r(h_i - h_e) + (\dot{m}_a h_{a1} + \dot{m}_{v1} h_{v1}) - \dot{m}_w h_w - (\dot{m}_a h_{a2} + \dot{m}_{v2} h_{v2})$$

where h_i and h_e denote the specific enthalpy values of the refrigerant entering and exiting the dehumidifier section, respectively. Introducing $\dot{m}_{v1} = \omega_1 \dot{m}_a$, $\dot{m}_{v2} = \omega_2 \dot{m}_a$, and $\dot{m}_w = (\omega_1 - \omega_2)\dot{m}_a$

$$0 = \dot{m}_r(h_i - h_e) + \dot{m}_a[(h_{a1} - h_{a2}) + \omega_1 h_{g1} - \omega_2 h_{g2} - (\omega_1 - \omega_2)h_{f2}]$$

where the specific enthalpies of the water vapor at 1 and 2 are evaluated at the saturated vapor values corresponding to T_1 and T_2, respectively. Since the condensate is assumed to exit as a saturated liquid at T_2, $h_w = h_{f2}$. Solving for the refrigerant mass flow rate per unit mass of dry air flowing through the device

$$\frac{\dot{m}_r}{\dot{m}_a} = \frac{(h_{a1} - h_{a2}) + \omega_1 h_{g1} - \omega_2 h_{g2} - (\omega_1 - \omega_2)h_{f2}}{h_e - h_i}$$

The accompanying psychrometric chart, Fig. 12.11*b*, illustrates important features of the processes involved. As indicated by the chart, the moist air first cools from state 1, where the temperature is T_1 and the humidity ratio is ω_1, to state 2, where the mixture is saturated ($\phi_2 = 100\%$), the temperature is $T_2 < T_1$, and the humidity ratio is $\omega_2 < \omega_1$. During the subsequent heating process, the humidity ratio remains constant, $\omega_2 = \omega_3$, and the temperature increases to T_3. Since all states visited are not equilibrium states, these processes are indicated on the psychrometric chart by dashed lines.

The example that follows provides an illustration involving dehumidification where one of the objectives is the refrigerating capacity of the cooling coil.

EXAMPLE 12.11 ►

Assessing Dehumidifier Performance

Moist air at 30°C and 50% relative humidity enters a dehumidifier operating at steady state with a volumetric flow rate of 280 m³/min. The moist air passes over a cooling coil and water vapor condenses. Condensate exits the dehumidifier saturated at 10°C. Saturated moist air exits in a separate stream at the same temperature. There is no significant loss of energy by heat transfer to the surroundings and pressure remains constant at 1.013 bar. Determine **(a)** the mass flow rate of the dry air, in kg/min, **(b)** the rate at which water is condensed, in kg per kg of dry air flowing through the control volume, and **(c)** the required refrigerating capacity, in tons.

SOLUTION

Known Moist air enters a dehumidifier at 30°C and 50% relative humidity with a volumetric flow rate of 280 m³/min. Condensate and moist air exit in separate streams at 10°C.

Determine Find the mass flow rate of the dry air, in kg/min, the rate at which water is condensed, in kg per kg of dry air, and the required refrigerating capacity, in tons.

Schematic and Given Data:

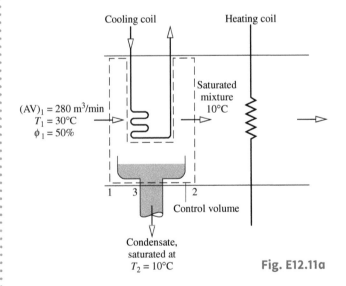

Fig. E12.11a

Engineering Model

1. The control volume shown in the accompanying figure operates at steady state. Changes in kinetic and potential energy can be neglected, and $\dot{W}_{cv} = 0$.

2. There is no significant heat transfer to the surroundings.

3. The pressure remains constant throughout at 1.013 bar.

4. At location 2, the moist air is saturated. The condensate exits at location 3 as a saturated liquid at temperature T_2.

5. The moist air streams are regarded as ideal gas mixtures adhering to the Dalton model.

Analysis

a. At steady state, the mass flow rates of the dry air entering and exiting are equal. The common mass flow rate of the

dry air can be determined from the volumetric flow rate at the inlet

$$\dot{m}_a = \frac{(AV)_1}{v_{a1}}$$

The specific volume of the dry air at inlet 1, v_{a1}, can be evaluated using the ideal gas equation of state, so

$$\dot{m}_a = \frac{(AV)_1}{(\overline{R}/M_a)(T_1/p_{a1})}$$

The partial pressure of the dry air p_{a1} can be determined from $p_{a1} = p_1 - p_{v1}$. Using the relative humidity at the inlet ϕ_1 and the saturation pressure at 30°C from Table A-2

$$p_{v1} = \phi_1 p_{g1} = (05)(0.04246) = 0.02123 \text{ bar}$$

Thus, $p_{a1} = 1.013 - 0.02123 = 0.99177$ bar. Inserting values into the expression for $\dot{m}_a$ gives

$$\dot{m}_a = \frac{(280 \text{ m}^3/\text{min})(0.99177 \times 10^5 \text{ N/m}^2)}{(8314/28.97 \text{ N} \cdot \text{m/kg} \cdot \text{K})(303 \text{ K})} = 319.35 \text{ kg/min}$$

b. Conservation of mass for the water requires $\dot{m}_{v1} = \dot{m}_{v2} + \dot{m}_w$. With $\dot{m}_{v1} = \omega_1 \dot{m}_a$ and $\dot{m}_{v2} = \omega_2 \dot{m}_a$, the rate at which water is condensed per unit mass of dry air is

$$\frac{\dot{m}_w}{\dot{m}_a} = \omega_1 - \omega_2$$

The humidity ratios ω_1 and ω_2 can be evaluated using Eq. 12.43. Thus, ω_1 is

$$\omega_1 = 0.662 \left(\frac{p_{v1}}{p_1 - p_{v1}} \right) = 0.622 \left(\frac{0.02123}{0.99177} \right) = 0.0133 \frac{\text{kg(vapor)}}{\text{kg(dry air)}}$$

Since the moist air is saturated at 10°C, p_{v2} equals the saturation pressure at 10°C: $p_g = 0.01228$ bar from Table A-2. Equation 12.43 then gives $\omega_2 = 0.0076$ kg(vapor)/kg(dry air). With these values for ω_1 and ω_2

$$\frac{\dot{m}_w}{\dot{m}_a} = 0.0133 - 0.0076 = 0.0057 \frac{\text{kg(condensate)}}{\text{kg(dry air)}}$$

c. The rate of heat transfer $\dot{Q}_{cv}$ between the moist air stream and the refrigerant coil can be determined using an energy rate balance. With assumptions 1 and 2, the steady-state form of the energy rate balance reduces to

$$0 = \dot{Q}_{cv} + (\dot{m}_a h_{a1} + \dot{m}_{v1} h_{v1}) - \dot{m}_w h_w - (\dot{m}_a h_{a2} + \dot{m}_{v2} h_{v2}) \quad \text{(a)}$$

With $\dot{m}_{v1} = \omega_1 \dot{m}_a$, $\dot{m}_{v2} = \omega_2 \dot{m}_a$, and $\dot{m}_w = (\omega_1 - \omega_2)\dot{m}_a$, this becomes

$$\dot{Q}_{cv} = \dot{m}_a [(h_{a2} - h_{a1}) - \omega_1 h_{g1} + \omega_2 h_{g2} + (\omega_1 - \omega_2)h_{f2}] \quad \text{(b)}$$

which agrees with Eq. 12.55. In Eq. (b), the specific enthalpies of the water vapor at 1 and 2 are evaluated at the saturated vapor values corresponding to T_1 and T_2, respectively, and the specific enthalpy of the exiting condensate is evaluated as h_f at T_2.

Selecting enthalpies from Tables A-2 and A-22, as appropriate, Eq. (b) reads

$$\dot{Q}_{cv} = (319.35)[(283.1 - 303.2) - 0.0133(2556.3)$$
$$+ 0.0076(2519.8) + 0.0057(42.01)]$$

$$= -11,084 \text{ kJ/min}$$

Since 1 ton of refrigeration equals a heat transfer rate of 211 kJ/min (Sec. 10.2.1), the required refrigerating capacity is 52.5 tons.

Alternative Psychrometric Chart Solution Let us consider an alternative solution using the psychrometric chart. As shown on the sketch of the psychrometric chart, **Fig. E12.11b**, the state of the moist air at the

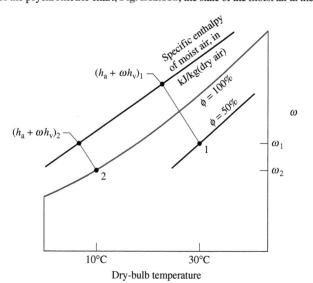

Fig. E12.11b

inlet 1 is defined by $\phi = 50\%$ and a dry-bulb temperature of 30°C. At 2, the moist air is saturated at 10°C. Rearranging Eq. (a), we get

$$\dot{Q}_{cv} = \dot{m}_a [\underline{(h_a + \omega h_v)_2} - \underline{(h_a + \omega h_v)_1} + (\omega_1 - \omega_2)h_w] \quad \text{(c)}$$

The underlined terms and humidity ratios, ω_1 and ω_2, can be read directly from the chart. The mass flow rate of the dry air can be determined using the volumetric flow rate at the inlet and v_{a1} read from the chart. The specific enthalpy h_w is obtained (as above) from Table A-2: h_f at T_2. The details are left as an exercise.

SKILLS DEVELOPED

Ability to...

- apply psychrometric terminology and principles.
- apply mass and energy balances for a dehumidification process in a control volume at steady state.
- retrieve property data for dry air and water.
- apply the psychrometric chart.

Quick Quiz

Using the psychrometric chart, determine the wet-bulb temperature of the moist air entering the dehumidifier, in °C. Ans. 22°C.

12.8.4 Humidification

It is often necessary to increase the moisture content of the air circulated through occupied spaces. One way to accomplish this is to inject steam. Alternatively, liquid water can be sprayed into the air. Both cases are shown schematically in **Fig. 12.12a**. The temperature of the moist air as it exits the humidifier depends on the condition of the water introduced. When relatively high-temperature steam is injected, both the humidity ratio and the dry-bulb temperature are increased. This is illustrated by the accompanying psychrometric chart of Fig. 12.12b. If liquid water is injected instead of steam, the moist air may exit the humidifier with a *lower* temperature than at the inlet. This is illustrated in Fig. 12.12c. The example to follow illustrates the case of steam injection. The case of liquid water injection is considered further in the next section.

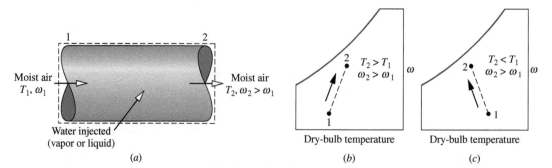

Fig. 12.12 Humidification. (*a*) Control volume. (*b*) Steam injected. (*c*) Liquid injected.

> ► ► ► **EXAMPLE 12.12** ► ●

Analyzing a Steam-Spray Humidifier

Moist air with a temperature of 22°C and a wet-bulb temperature of 9°C enters a steam-spray humidifier. The mass flow rate of the dry air is 90 kg/min. Saturated water vapor at 110°C is injected into the mixture at a rate of 52 kg/h. There is no heat transfer with the surroundings, and the pressure is constant throughout at 1 bar. Using the psychrometric chart, determine at the exit **(a)** the humidity ratio and **(b)** the temperature, in °C.

SOLUTION

Known Moist air enters a humidifier at a temperature of 22°C and a wet-bulb temperature of 9°C. The mass flow rate of the dry air is 90 kg/min. Saturated water vapor at 110°C is injected into the mixture at a rate of 52 kg/h.

Find Using the psychrometric chart, determine at the exit the humidity ratio and the temperature, in °C.

Schematic and Given Data:

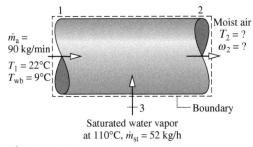

Moist air
$T_2 = ?$
$\omega_2 = ?$

$\dot{m}_a =$ 90 kg/min
$T_1 = 22°C$
$T_{wb} = 9°C$

3 — Boundary

Saturated water vapor at 110°C, $\dot{m}_{st} = 52$ kg/h

Fig. E12.12a

Engineering Model

1. The control volume shown in the accompanying figure operates at steady state. Changes in kinetic and potential energy can be neglected and $\dot{W}_{cv} = 0$.

2. There is no heat transfer with the surroundings.

3. The pressure remains constant throughout at 1 bar. Figure A-9 remains valid at this pressure.

4. The moist air streams are regarded as ideal gas mixtures adhering to the Dalton model.

Analysis

a. The humidity ratio at the exit ω_2 can be found from mass rate balances on the dry air and water individually. Thus,

$$\dot{m}_{a1} = \dot{m}_{a2} \quad \text{(dry air)}$$
$$\dot{m}_{v1} + \dot{m}_{st} = \dot{m}_{v2} \quad \text{(water)}$$

With $\dot{m}_{v1} = \omega_1 \dot{m}_a$, and $\dot{m}_{v2} = \omega_2 \dot{m}_a$, where $\dot{m}_a$ is the mass flow rate of the air, the second of these becomes

$$\omega_2 = \omega_1 + \frac{\dot{m}_{st}}{\dot{m}_a}$$

Using the inlet dry-bulb temperature, 22°C, and the inlet wet-bulb temperature, 9°C, the value of

the humidity ratio ω_1 can be found by inspection of the psychrometric chart, Fig. A-9. The result is $\omega_1 = 0.002$ kg (vapor)/kg(dry air). This value should be verified as an exercise. Inserting values into the expression for ω_2

$$\omega_2 = 0.002 + \frac{(52 \text{ kg/h})|1 \text{ h/60 min}|}{90 \text{ kg/min}} = 0.0116 \frac{\text{kg(vapor)}}{\text{kg(dry air)}}$$

b. The temperature at the exit can be determined using an energy rate balance. With assumptions 1 and 2, the steady-state form of the energy rate balance reduces to a special case of Eq. 12.55. Namely,

$$0 = h_{a1} - h_{a2} + \omega_1 h_{g1} + (\omega_2 - \omega_1)h_{g3} - \omega_2 h_{g2} \quad \text{(a)}$$

In writing this, the specific enthalpies of the water vapor at 1 and 2 are evaluated as the respective saturated vapor values, and h_{g3} denotes the enthalpy of the saturated vapor injected into the moist air.

Equation (a) can be rearranged in the following form suitable for use with the psychrometric chart.

❶ $$(h_a + \omega h_g)_2 = (h_a + \omega h_g)_1 + (\omega_2 - \omega_1)h_{g3} \quad \text{(b)}$$

As shown on the sketch of the psychrometric chart, Fig. E12.12b, the first term on the right of Eq. (b) can be obtained from Fig. A-9 at the inlet state, defined by the intersection of the inlet dry-bulb temperature, 22°C, and the inlet wet-bulb temperature, 9°C; the value is 27.2 kJ/kg(dry air). The second term on the right can be evaluated using the known humidity ratios ω_1 and ω_2 and h_{g3} from Table A-2: 2691.5 kJ/kg(vapor). The value of the second term of Eq. (b) is 25.8 kJ/kg (dry air). The state at the exit is then fixed by ω_2 and $(h_a + \omega h_g)_2 = 53$ kJ/kg (dry air), calculated from the two values just determined. Finally, the temperature at the exit can be read directly from the chart. The result is $T_2 \approx 23.5°C$.

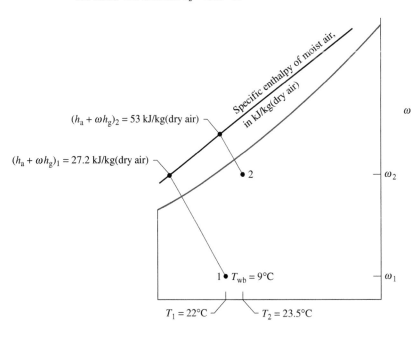

Specific enthalpy of moist air, in kJ/kg(dry air)

$(h_a + \omega h_g)_2 = 53$ kJ/kg(dry air)

$(h_a + \omega h_g)_1 = 27.2$ kJ/kg(dry air)

ω

ω_2

2

ω_1

1 • $T_{wb} = 9°C$

$T_1 = 22°C$ $T_2 = 23.5°C$

Fig. E12.12b

Alternative *IT* Solution

❷ The following program allows T_2 to be determined using *IT*, where $\dot{m}_a$ is denoted as mdota, $\dot{m}_{st}$ is denoted as mdotst, w1 and w2 denote ω_1 and ω_2, respectively, and so on.

```
// Given data
T1 = 22 // °C
Twb1 = 9 // °C
mdota = 90 // kg/min
p = 1 // bar
Tst = 110 // °C
mdotst = (52 / 60) // converting kg/h to kg/min

// Evaluate humidity ratios
w1 = w_TTwb (T1,Twb1,p)
w2 = w1 + (mdotst / mdota)

// Denoting the enthalpy of moist air at state
1 by
// h1, etc., the energy balance, Eq. (a),
becomes
0 = h1 – h2 + (w2 – w1)*hst

// Evaluate enthalpies
h1 = ha_Tw(T1,w1)
h2 = ha_Tw(T2,w2)
hst = hsat_Px("Water/Steam",psat,1)
psat = Psat_T("Water/Steam ",Tst)
```

Using the **Solve** button, the result is $T_2 = 23.4°C$, which agrees closely with the values obtained above, as expected.

---

❶ A solution of Eq. (b) using data from Tables A-2 and A-22 requires an iterative (trial) procedure. The result is $T_2 = 24°C$, as can be verified.

❷ Note the use of special *Moist Air* functions listed in the **Properties** menu of *IT*.

SKILLS DEVELOPED

Ability to...

• apply psychrometric terminology and principles.

• apply mass and energy balances for a spray humidification process in a control volume at steady state.

• retrieve necessary property data using the psychrometric chart.

• apply *IT* for psychrometric analysis.

Quick Quiz

Using the psychrometric chart, what is the relative humidity at the exit? Ans. $\approx 63\%$.

12.8.5 Evaporative Cooling

Cooling in hot, relatively dry climates can be accomplished by *evaporative cooling*. This involves either spraying liquid water into air or forcing air through a soaked pad that is kept replenished with water, as shown in **Fig. 12.13**. Owing to the low humidity of the moist air entering at state 1, part of the injected water evaporates. The energy for evaporation is provided by the air stream, which is reduced in temperature and exits at state 2 with a lower temperature than the entering stream. Because the incoming air is relatively dry, the additional moisture carried by the exiting moist air stream is normally beneficial.

For negligible heat transfer with the surroundings, no work $\dot{W}_{cv}$, and no significant changes in kinetic and potential energy, the steady-state forms of the mass and energy rate balances reduce for the control volume of Fig. 12.13a to this special case of Eq. 12.55:

$$(h_{a2} + \omega_2 h_{g2}) = \underline{(\omega_2 - \omega_1)h_f} + (h_{a1} + \omega_1 h_{g1})$$

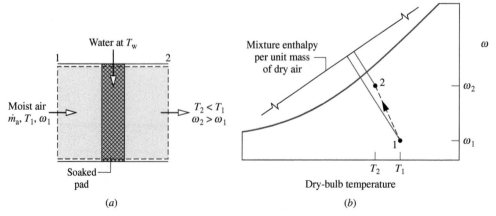

Fig. 12.13 Evaporative cooling. (*a*) Equipment schematic. (*b*) Psychrometric chart representation.

where h_f denotes the specific enthalpy of the liquid stream entering the control volume. All the injected water is assumed to evaporate into the moist air stream. The underlined term accounts for the energy carried in with the injected liquid water. This term is normally much smaller in magnitude than either of the two moist air enthalpy terms. Accordingly, the enthalpy of the moist air varies only slightly, as illustrated on the psychrometric chart of Fig. 12.13b. Recalling that lines of constant mixture enthalpy are closely lines of constant wet-bulb temperature (Sec. 12.7), it follows that evaporative cooling takes place at a nearly constant wet-bulb temperature.

In the next example, we consider the analysis of an evaporative cooler.

TAKE NOTE...

Evaporative cooling takes place at a nearly constant wet-bulb temperature.

▶ EXAMPLE 12.13 ▶

Evaporative Cooler

Air at 38°C and 10% relative humidity enters an evaporative cooler with a volumetric flow rate of 140 m³/min. Moist air exits the cooler at 21°C. Water is added to the soaked pad of the cooler as a liquid at 21°C and evaporates fully into the moist air. There is no heat transfer with the surroundings and the pressure is constant throughout at 1 bar. Determine **(a)** the mass flow rate of the water to the soaked pad, in kg/h, and **(b)** the relative humidity of the moist air at the exit to the evaporative cooler.

SOLUTION

Known Air at 38°C and $\phi = 10\%$ enters an evaporative cooler with a volumetric flow rate of 140 m³/min. Moist air exits the cooler at 21°C. Water is added to the soaked pad of the cooler at 21°C.

Find Determine the mass flow rate of the water to the soaked pad, in kg/h, and the relative humidity of the moist air at the exit of the cooler.

Schematic and Given Data:

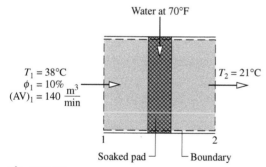

Water at 70°F

$T_1 = 38°C$
$\phi_1 = 10\%$
$(AV)_1 = 140 \frac{m^3}{min}$

$T_2 = 21°C$

1   2

Soaked pad   Boundary

Fig. E12.13

Engineering Model

1. The control volume shown in the accompanying figure operates at steady state. Changes in kinetic and potential energy can be neglected and $\dot{W}_{cv} = 0$.

2. There is no heat transfer with the surroundings.

3. The water added to the soaked pad enters as a liquid and evaporates fully into the moist air.

4. The pressure remains constant throughout at 1 atm.

Analysis

a. Applying conservation of mass to the dry air and water individually as in previous examples gives

$$\dot{m}_w = \dot{m}_a(\omega_2 - \omega_1)$$

where $\dot{m}_w$ is the mass flow rate of the water to the soaked pad. To find $\dot{m}_w$ requires ω_1, $\dot{m}_a$, and ω_2. These will now be evaluated in turn.

The humidity ratio ω_1 can be found from Eq. 12.43, which requires p_{v1}, the partial pressure of the moist air entering the control volume. Using the given relative humidity ϕ_1 and p_g at T_1 from Table A-2, we have $p_{v1} = \phi_1 p_{g1} = 0.0066$ bar. With this, $\omega_1 = 0.00408$/kg(vapor) kg(dry air).

The mass flow rate of the dry air $\dot{m}_a$ can be found as in previous examples using the volumetric flow rate and specific volume of the dry air. Thus

$$\dot{m}_a = \frac{(AV)_1}{v_{a1}}$$

The specific volume of the dry air can be evaluated from the ideal gas equation of state. The result is $v_{a1} = 0.887$ m³/kg(dry air). Inserting values, the mass flow rate of the dry air is

$$\dot{m}_a = \frac{140 \text{ m}^3/\text{min}}{0.887 \text{ m}^3/\text{kg(dry air)}} = 157.8 \frac{\text{kg(dry air)}}{\text{min}}$$

To find the humidity ratio ω_2, reduce the steady-state forms of the mass and energy rate balances using assumption 1 to obtain

$$0 = (\dot{m}_a h_{a1} + \dot{m}_{v1} h_{v1}) + \dot{m}_w h_w - (\dot{m}_a h_{a2} + \dot{m}_{v2} h_{v2})$$

With the same reasoning as in previous examples, this can be expressed as

❶ $$0 = (h_a + \omega h_g)_1 + \underline{(\omega_2 - \omega_1)h_f} - (h_a + \omega h_g)_2$$

where h_f denotes the specific enthalpy of the water entering the control volume at 21°C. Solving for ω_2

$$\omega_2 = \frac{h_{a1} - h_{a2} + \omega_1(h_{g1} - h_f)}{h_{g2} - h_f} = \frac{c_{pa}(T_1 - T_2) + \omega_1(h_{g1} - h_f)}{h_{g2} - h_f}$$

❷ where $c_{pa} = 1.005$ kJ/kg · K. With h_f, h_{g1}, and h_{g2} from Table A-2

$$\omega_2 = \frac{(1.005)(38 - 21) + 0.00408(2570.7 - 88.14)}{2451.8}$$

$$= 0.011 \frac{\text{kg(vapor)}}{\text{kg(dry air)}}$$

Substituting values for $\dot{m}_a$, ω_1, and ω_2 into the expression for $\dot{m}_w$

$$\dot{m}_w = \left[157.8 \frac{\text{kg(dry air)}}{\text{min}} \left| \frac{60 \text{ min}}{1 \text{ h}} \right| \right] (0.011 - 0.00408) \frac{\text{kg(water)}}{\text{kg(dry air)}}$$

$$= 65.5 \frac{\text{kg(water)}}{\text{h}}$$

b. The relative humidity of the moist air at the exit can be determined using Eq. 12.44. The partial pressure of the water vapor required by this expression can be found by solving Eq. 12.43 to obtain

$$p_{v2} = \frac{\omega_2 p}{\omega_2 + 0.622}$$

Inserting values

$$p_{v2} = \frac{(0.011)(1.01325 \text{ bar})}{(0.011 + 0.622)} = 0.0176 \text{ bar}$$

At 21°C, the saturation pressure is 0.02487 bar. Thus, the relative humidity at the exit is

$$\phi_2 = \frac{0.0176}{0.02478} = 0.708(70.8\%)$$

---

1 Since the underlined term in this equation is much smaller than either of the moist air enthalpies, the enthalpy of the moist air remains nearly constant, and thus evaporative cooling takes place at nearly constant wet-bulb temperature. This can be verified by locating the incoming and outgoing moist air states on the psychrometric chart.

2 A constant value of the specific heat c_{pa} has been used here to evaluate the term $(h_{a1} - h_{a2})$. As shown in previous examples, this term can be evaluated alternatively using the ideal gas table for air.

SKILLS DEVELOPED

Ability to...

- apply psychrometric terminology and principles.
- apply mass and energy balances for an evaporative cooling process in a control volume at steady state.
- retrieve property data for dry air and water

Quick Quiz

Using steam table data, what is the dew point temperature at the exit, in °C? Ans. 15.3°C.

12.8.6 Adiabatic Mixing of Two Moist Air Streams

A common process in air-conditioning systems is the mixing of moist air streams, as shown in **Fig. 12.14**. The objective of the thermodynamic analysis of such a process is normally to fix the flow rate and state of the exiting stream for specified flow rates and states of each of the two inlet streams. The case of adiabatic mixing is governed by Eqs. 12.56 to follow.

The mass rate balances for the dry air and water vapor at steady state are, respectively,

$$\dot{m}_{a1} + \dot{m}_{a2} = \dot{m}_{a3} \quad \text{(dry air)} \tag{12.56a}$$

$$\dot{m}_{v1} + \dot{m}_{v2} = \dot{m}_{v3} \quad \text{(water vapor)}$$

With $\dot{m}_v = \omega \dot{m}_a$, the water vapor mass balance becomes

$$\omega_1 \dot{m}_{a1} + \omega_2 \dot{m}_{a2} = \omega_3 \dot{m}_{a3} \quad \text{(water vapor)} \tag{12.56b}$$

Assuming $\dot{Q}_{cv} = \dot{W}_{cv} = 0$ and ignoring the effects of kinetic and potential energy, the energy rate balance reduces at steady state to

$$\dot{m}_{a1}(h_{a1} + \omega_1 h_{g1}) + \dot{m}_{a2}(h_{a2} + \omega_2 h_{g2}) = \dot{m}_{a3}(h_{a3} + \omega_3 h_{g3}) \tag{12.56c}$$

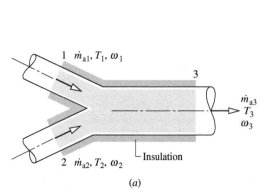

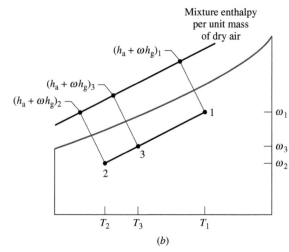

Fig. 12.14 Adiabatic mixing of two moist air streams. (*a*) Equipment representation. (*b*) Psychrometric chart representation.

where the enthalpies of the entering and exiting water vapor are evaluated as the saturated vapor values at the respective dry-bulb temperatures.

If the inlet flow rates and states are known, Eqs. 12.56 are three equations in three unknowns: $\dot{m}_{a3}$, ω_3, and $(h_{a3} + \omega_3 h_{g3})$. The solution of these equations is illustrated by Example 12.14.

Let us also consider how Eqs. 12.56 can be solved *geometrically* with the psychrometric chart: Using Eq. 12.56a to eliminate $\dot{m}_{a3}$, the mass flow rate of dry air at 3, from Eqs. 12.56b and 12.56c, we get

$$\frac{\dot{m}_{a1}}{\dot{m}_{a2}} = \frac{\omega_3 - \omega_2}{\omega_1 - \omega_3} = \frac{(h_{a3} + \omega_3 h_{g3}) - (h_{a2} + \omega_2 h_{g2})}{(h_{a1} + \omega_1 h_{g1}) - (h_{a3} + \omega_3 h_{g3})} \tag{12.57}$$

From the relations of Eqs. 12.56, we conclude that on a psychrometric chart state 3 of the mixture lies on a straight line connecting states 1 and 2 of the two streams before mixing. This is shown in Fig. 12.14b.

Analyzing Adiabatic Mixing of Two Moist Air Streams

A stream consisting of 142 m³/min of moist air at a temperature of 5°C and a humidity ratio of 0.002 kg(vapor)/kg(dry air) is mixed adiabatically with a second stream consisting of 425 m³/min of moist air at 24°C and 50% relative humidity. The pressure is constant throughout at 1 bar. Determine **(a)** the humidity ratio and **(b)** the temperature of the exiting mixed stream, in °C.

SOLUTION

Known A moist air stream at 5°C, $\omega = 0.002$ kg(vapor)/kg(dry air), and a volumetric flow rate of 142 m³/min is mixed adiabatically with a stream consisting of 425 m³/min of moist air at 24°C and $\phi = 50\%$.

Find Determine the humidity ratio and the temperature, in °C, of the mixed stream exiting the control volume.

Schematic and Given Data:

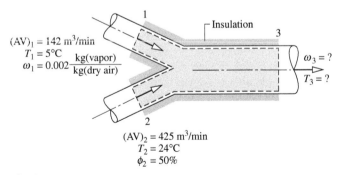

$(AV)_1 = 142$ m³/min
$T_1 = 5°C$
$\omega_1 = 0.002 \frac{\text{kg(vapor)}}{\text{kg(dry air)}}$

$\omega_3 = ?$
$T_3 = ?$

$(AV)_2 = 425$ m³/min
$T_2 = 24°C$
$\phi_2 = 50\%$

Insulation

Fig. E12.14

Engineering Model

1. The control volume shown in the accompanying figure operates at steady state. Changes in kinetic and potential energy can be neglected and $\dot{W}_{cv} = 0$.

2. There is no heat transfer with the surroundings.

3. The pressure remains constant throughout at 1 bar.

4. The moist air streams are regarded as ideal gas mixtures adhering to the Dalton model.

Analysis

a. The humidity ratio ω_3 can be found by means of mass rate balances for the dry air and water vapor, respectively,

$$\dot{m}_{a1} + \dot{m}_{a2} = \dot{m}_{a3} \quad \text{(dry air)}$$
$$\dot{m}_{v1} + \dot{m}_{v2} = \dot{m}_{v3} \quad \text{(water vapor)}$$

With $\dot{m}_{v1} = \omega_1 \dot{m}_{a1}$, $\dot{m}_{v2} = \omega_2 \dot{m}_{a2}$, and $\dot{m}_{v3} = \omega_3 \dot{m}_{a3}$, the second of these balances becomes (Eq. 12.56b)

$$\omega_1 \dot{m}_{a1} + \omega_2 \dot{m}_{a2} = \omega_3 \dot{m}_{a3}$$

Solving

$$\omega_3 = \frac{\omega_1 \dot{m}_{a1} + \omega_2 \dot{m}_{a2}}{\dot{m}_{a3}}$$

Since $\dot{m}_{a3} = \dot{m}_{a1} + \dot{m}_{a2}$, this can be expressed as

$$\omega_3 = \frac{\omega_1 \dot{m}_{a1} + \omega_2 \dot{m}_{a2}}{\dot{m}_{a1} + \dot{m}_{a2}}$$

To determine ω_3 requires values for ω_3, $\dot{m}_{a1}$, and $\dot{m}_{a2}$. The mass flow rates of the dry air, $\dot{m}_{a1}$ and $\dot{m}_{a2}$, can be found as in previous examples using the given volumetric flow rates

$$\dot{m}_{a1} = \frac{(AV)_1}{v_{a1}}, \qquad \dot{m}_{a2} = \frac{(AV)_2}{v_{a2}}$$

The values of v_{a1}, v_{a2}, and ω_2 are readily found from the psychrometric chart, Fig. A-9. Thus, at $\omega_1 = 0.002$ and $T_1 = 5°C$, $v_{a1} = 0.79$ m³/kg(dry air). At $\phi_2 = 50\%$ and $T_2 = 24°C$, $v_{a2} = 0.855$ m³/kg(dry air) and $\omega_2 = 0.0094$. The mass flow rates of the dry air are then $\dot{m}_{a1} = 180$ kg(dry air)/min and $\dot{m}_{a2} = 497$ kg(dry air)/min. Inserting values into the expression for ω_3

$$\omega_3 = \frac{(0.002)(180) + (0.0094)(497)}{180 + 497} = 0.0074 \frac{\text{kg(vapor)}}{\text{kg(dry air)}}$$

b. The temperature T_3 of the exiting mixed stream can be found from an energy rate balance. Reduction of the energy rate balance using assumptions 1 and 2 gives (Eq. 12.56c)

$$\dot{m}_{a1}(h_a + \omega h_g)_1 + \dot{m}_{a2}(h_a + \omega h_g)_2 = \dot{m}_{a3}(h_a + \omega h_g)_3 \tag{a}$$

Solving

$$(h_a + \omega h_g)_3 = \frac{\dot{m}_{a1}(h_a + \omega h_g)_1 + \dot{m}_{a2}(h_a + \omega h_g)_2}{\dot{m}_{a1} + \dot{m}_{a2}} \tag{b}$$

With $(h_a + \omega h_g)_1 = 10$ kJ/kg(dry air) and $(h_a + \omega h_g)_2 = 47.8$ kJ/kg (dry air) from Fig. A-9 and other known values

$$(h_a + \omega h_g)_3 = \frac{180(10) + 497(47.8)}{180 + 497} = 37.7 \frac{kJ}{kg(dry\ air)}$$

❶ This value for the enthalpy of the moist air at the exit, together with the previously determined value for ω_3, fixes the state of the exiting moist air. From inspection of Fig. A-9, $T_3 = 19°C$.

Alternative Solutions The use of the psychrometric chart facilitates the solution for T_3. Without the chart, an iterative solution of Eq. (b) using data from Tables A-2 and A-22 could be used. Alternatively, T_3 can be determined using the following *IT* program, where ϕ_2 is denoted as phi2, the volumetric flow rates at 1 and 2 are denoted as AV1 and AV2, respectively, and so on.

```
// Given data
T1 = 5 // °C
w1 = 0.002 // kg(vapor) / kg(dry air)
AV1 = 142 // m³/min
T2 = 24 // °C
phi2 = 0.5
AV2 = 425 // m³/min
p = 1 // bar
// Mass balances for water vapor and dry air:
w1 * mdota1 + w2 * mdota2 = w3 * mdota3
mdota1 + mdota2 = mdota3
// Evaluate mass flow rates of dry air
mdota1 = AV1 / va1
```

```
② va1 = va_Tw(T1, w1, p)
   mdota2 = AV2 / va2
   va2 = va_Tphi(T2, phi2, p)

   // Determine w2
   w2 = w_Tphi(T2, phi2, p)
   // The energy balance, Eq. (a), reads
   mdota1 * h1 + mdota2 * h2 = mdota3 * h3
   h1 = ha_Tw(T1, w1)
   h2 = ha_Tphi(T2, phi2, p)
   h3 = ha_Tw(T3, w3)
```

Using the **Solve** button, the result is $T_3 = 19.01°C$ and $\omega_3 = 0.00745$ kg(vapor)/kg(dry air), which agree with the psychrometric chart solution.

❶ A solution using the geometric approach based on Eq. 12.57 is left as an exercise.

❷ Note the use here of special *Moist Air* functions listed in the **Properties** menu of *IT*.

SKILLS DEVELOPED

Ability to...

- apply psychrometric terminology and principles.
- apply mass and energy balances for an adiabatic mixing process of two moist air streams in a control volume at steady state.
- retrieve property data for moist air using the psychrometric chart.
- apply *IT* for psychrometric analysis.

Quick Quiz

Using the psychrometric chart, what is the relative humidity at the exit? Ans. ≈ 53%.

⓬.9 Cooling Towers

Power plants invariably discharge considerable energy to their surroundings by heat transfer (Chap. 8). Although water drawn from a nearby river or lake can be employed to carry away this energy, cooling towers provide an alternative in locations where sufficient cooling water cannot be obtained from natural sources or where concerns for the environment place a limit on the temperature at which cooling water can be returned to the surroundings. Cooling towers also are frequently employed to provide chilled water for applications other than those involving power plants.

Cooling towers can operate by *natural* or *forced* convection. Also they may be *counterflow*, *cross-flow*, or a combination of these. A schematic diagram of a forced-convection, counterflow cooling tower is shown in **Fig. 12.15**. The warm water to be cooled enters at 1 and is sprayed from the top of the tower. The falling water usually passes through a series of baffles intended to keep it broken up into fine drops to promote evaporation. Atmospheric air drawn in at 3 by the fan flows upward, counter to the direction of the falling water droplets. As the two streams interact, a fraction of the entering liquid water stream evaporates into the moist air, which exits at 4 with a greater humidity ratio than the incoming moist air at 3, while liquid water exits at 2 with a lower temperature than the water entering at 1. Since some of the incoming water is evaporated into the moist air stream, an equivalent amount of makeup water

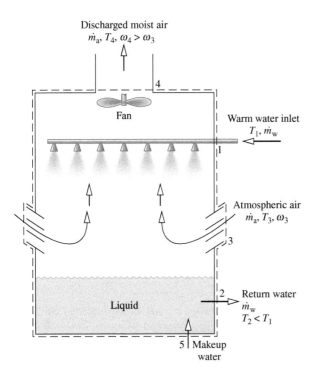

Discharged moist air
$\dot{m}_a, T_4, \omega_4 > \omega_3$

Fig. 12.15 Schematic of a cooling tower.

is added at 5 so that the return mass flow rate of the cool water equals the mass flow rate of the warm water entering at 1.

For operation at steady state, mass balances for the dry air and water and an energy balance on the overall cooling tower provide information about cooling tower performance. In applying the energy balance, heat transfer with the surroundings is usually neglected. The power input to the fan of forced-convection towers also may be negligible relative to other energy rates involved. The example to follow illustrates the analysis of a cooling tower using conservation of mass and energy together with property data for the dry air and water.

►►► EXAMPLE 12.15 ►

Determining Mass Flow Rates for a Power Plant Cooling Tower

Water exiting the condenser of a power plant at 38°C enters a cooling tower with a mass flow rate of 4.5×10^7 kg/h. A stream of cooled water is returned to the condenser from a cooling tower with a temperature of 30°C and the same flow rate. Makeup water is added in a separate stream at 20°C. Atmospheric air enters the cooling tower at 25°C and 35% relative humidity. Moist air exits the tower at 35°C and 90% relative humidity. Determine the mass flow rates of the dry air and the makeup water, in kg/h. The cooling tower operates at steady state. Heat transfer with the surroundings and the fan power can each be neglected, as can changes in kinetic and potential energy. The pressure remains constant throughout at 1 atm.

Schematic and Given Data:

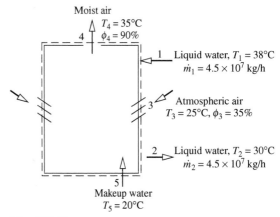

Moist air
$T_4 = 35°C$
$\phi_4 = 90\%$

Liquid water, $T_1 = 38°C$
$\dot{m}_1 = 4.5 \times 10^7$ kg/h

Atmospheric air
$T_3 = 25°C, \phi_3 = 35\%$

Liquid water, $T_2 = 30°C$
$\dot{m}_2 = 4.5 \times 10^7$ kg/h

Makeup water
$T_5 = 20°C$

Fig. E12.15

SOLUTION

Known A liquid water stream enters a cooling tower from a condenser at 38°C with a known mass flow rate. A stream of cooled water is returned to the condenser at 30°C and the same mass flow rate. Makeup water is added at 20°C. Atmospheric air enters the tower at 25°C and $\phi = 35\%$. Moist air exits the tower at 35°C and $\phi = 90\%$.

Find Determine the mass flow rates of the dry air and the makeup water, in kg/h.

Engineering Model

1. The control volume shown in the accompanying figure operates at steady state. Heat transfer with the surroundings can be neglected, as can changes in kinetic and potential energy; also $\dot{W}_{cv} = 0$.

2. To evaluate specific enthalpies, each liquid stream is regarded as a saturated liquid at the corresponding specified temperature.

3. The moist air streams are regarded as ideal gas mixtures adhering to the Dalton model.

4. The pressure is constant throughout at 1 atm.

Analysis The required mass flow rates can be found from mass and energy rate balances. Mass balances for the dry air and water individually reduce at steady state to

$$\dot{m}_{a3} = \dot{m}_{a4} \qquad \text{(dry air)}$$

$$\dot{m}_1 + \dot{m}_5 + \dot{m}_{v3} = \dot{m}_2 + \dot{m}_{v4} \qquad \text{(water)}$$

The common mass flow rate of the dry air is denoted as $\dot{m}_a$. Since $\dot{m}_1 = \dot{m}_2$, the second of these equations becomes

$$\dot{m}_5 = \dot{m}_{v4} - \dot{m}_{v3}$$

With $\dot{m}_{v3} = \omega_3 \dot{m}_a$ and $\dot{m}_{v4} = \omega_4 \dot{m}_a$

$$\dot{m}_5 = \dot{m}_a(\omega_4 - \omega_3)$$

Accordingly, the two required mass flow rates, $\dot{m}_a$ and $\dot{m}_5$, are related by this equation. Another equation relating the flow rates is provided by the energy rate balance.

Reducing the energy rate balance with assumption 1 results in

$$0 = \dot{m}_1 h_{w1} + (\dot{m}_a h_{a3} + \dot{m}_{v3} h_{v3}) + \dot{m}_5 h_{w5} - \dot{m}_2 h_{w2} - (\dot{m}_a h_{a4} + \dot{m}_{v4} h_{v4})$$

Evaluating the enthalpies of the water vapor as the saturated vapor values at the respective temperatures and the enthalpy of each liquid stream as the saturated liquid enthalpy at each respective temperature, the energy rate equation becomes

$$0 = \dot{m}_1 h_{f1} + (\dot{m}_a h_{a3} + \dot{m}_{v3} h_{g3}) + \dot{m}_5 h_{f5} - \dot{m}_2 h_{f2} - (\dot{m}_a h_{a4} + \dot{m}_{v4} h_{g4})$$

Introducing $\dot{m}_1 = \dot{m}_2$, $\dot{m}_5 = \dot{m}_a(\omega_4 - \omega_3)$, $\dot{m}_{v3} = \omega_3 \dot{m}_a$, and $\dot{m}_{v4} = \omega_4 \dot{m}_a$ and solving for $\dot{m}_a$

$$\dot{m}_a = \frac{\dot{m}_1(h_{f1} - h_{f2})}{h_{a4} - h_{a3} + \omega_4 h_{g4} - \omega_3 h_{g3} - (\omega_4 - \omega_3)h_{f5}} \qquad \text{(a)}$$

The humidity ratios ω_3 and ω_4 required by this expression can be determined from Eq. 12.43, using the partial pressure of the water vapor obtained with the respective relative humidity. Thus,

$\omega_3 = 0.00688$ kg(vapor)/kg(dry air) and $\omega_4 = 0.0327$ kg(vapor)/kg(dry air).

With enthalpies from Tables A-2 and A-22, as appropriate, and the known values for ω_3, ω_4, and $\dot{m}_1$, the expression for $\dot{m}_a$ becomes

$$\dot{m}_a = \frac{(4.5 \times 10^7)(159.21 - 125.79)}{(308.2 - 298.2) + (0.0327)(2565.3)}$$
$$\frac{}{\qquad - (0.00688)(2547.2) - (0.0258)(83.96)}$$

$$= 2.03 \times 10^7 \text{ kg/h}$$

Finally, inserting known values into the expression for $\dot{m}_5$ results in

$$\dot{m}_5 = (2.03 \times 10^7)(0.0327 - 0.00688) = 5.24 \times 10^5 \text{ kg/h}$$

Alternative Psychrometric Chart Solution Equation (a) can be rearranged to read

$$\dot{m} = \frac{\dot{m}_1(h_{f1} - h_{f2})}{(\underline{h_{a4} + \omega_4 h_{g4}}) - (\underline{h_{a3} + \omega_3 h_{g3}}) - (\omega_4 - \omega_3)h_{f5}}$$

The specific enthalpy terms h_{f1}, h_{f2}, and h_{f5} are obtained from Table A-2, as above. The underlined terms and ω_3 and ω_4 can be obtained by inspection of a psychrometric chart from the engineering literature providing data at states 3 *and* 4. Figure A-9 does not suffice in this application at state 4. The details are left as an exercise.

SKILLS DEVELOPED

Ability to...

• apply psychrometric terminology and principles.

• apply mass and energy balances for a cooling tower process in a control volume at steady state.

• retrieve property data for dry air and water.

Quick Quiz

Using steam table data, determine the partial pressure of the water vapor in the entering moist air stream, p_{v3}, in bar. Ans. 0.0111 bar.

CHAPTER SUMMARY AND STUDY GUIDE

In this chapter we have applied the principles of thermodynamics to systems involving ideal gas mixtures, including the special case of *psychrometric* applications involving air–water vapor mixtures, possibly in the presence of liquid water. Both closed system and control volume applications are presented.

The first part of the chapter deals with general ideal gas mixture considerations and begins by describing mixture composition in terms of the mass fractions or mole fractions. The Dalton model, which brings in the partial pressure concept, is then introduced for the p–v–T relation of ideal gas mixtures.

Means are also introduced for evaluating the enthalpy, internal energy, and entropy of a mixture by adding the contribution of each component at its condition in the mixture. Applications are considered where ideal gas mixtures undergo processes at constant composition as well as where ideal gas mixtures are formed from their component gases.

In the second part of the chapter, we study *psychrometrics*. Special terms commonly used in psychrometrics are introduced,

including moist air, humidity ratio, relative humidity, mixture enthalpy, and the dew point, dry-bulb, and wet-bulb temperatures. The *psychrometric chart*, which gives a graphical representation of important moist air properties, is introduced. The principles of conservation of mass and energy are formulated in terms of psychrometric quantities, and typical air-conditioning applications are considered, including dehumidification and humidification, evaporative cooling, and mixing of moist air streams. A discussion of cooling towers is also provided.

The following list provides a study guide for this chapter. When your study of the text and end-of-chapter exercises has been completed, you should be able to

• write out the meanings of the terms listed in the margin throughout the chapter and understand each of the related concepts. The subset of key concepts listed below is particularly important.

- describe mixture composition in terms of mass fractions or mole fractions.
- relate pressure, volume, and temperature of ideal gas mixtures using the Dalton model, and evaluate U, H, c_v, c_p, and S of ideal gas mixtures in terms of the mixture composition and the respective contribution of each component.
- apply the conservation of mass and energy principles and the second law of thermodynamics to systems involving ideal gas mixtures.

For psychrometric applications, you should be able to

- evaluate the humidity ratio, relative humidity, mixture enthalpy, and dew point temperature.
- use the psychrometric chart.
- apply the conservation of mass and energy principles and the second law of thermodynamics to analyze air-conditioning processes and cooling towers.

KEY ENGINEERING CONCEPTS

mass fraction
gravimetric analysis
mole fraction
molar (volumetric) analysis
apparent molecular weight
Dalton model

partial pressure
psychrometrics
moist air
humidity ratio
relative humidity
mixture enthalpy

dew point temperature
dry-bulb temperature
wet-bulb temperature
psychrometric chart

KEY EQUATIONS

Ideal Gas Mixtures: General Considerations

$mf_i = m_i/m$	(12.3)	Analysis in terms of mass fractions
$1 = \sum_{i=1}^{j} mf_i$	(12.4)	
$y_i = n_i/n$	(12.6)	Analysis in terms of mole fractions
$1 = \sum_{i=1}^{j} y_i$	(12.7)	
$M = \sum_{i=1}^{j} y_i M_i$	(12.9)	Apparent molecular weight
$p_i = y_i p$	(12.12)	Partial pressure of component i and relation to mixture pressure p
$p = \sum_{i=1}^{j} p_i$	(12.13)	
$\bar{u} = \sum_{i=1}^{j} y_i \bar{u}_i$	(12.21)	Internal energy, enthalpy, and entropy per mole of mixture. $\bar{u}_i$ and $\bar{h}_i$ evaluated at mixture temperature T. $\bar{s}_i$ evaluated at T and partial pressure p_i.
$\bar{h} = \sum_{i=1}^{j} y_i \bar{h}_i$	(12.22)	
$\bar{s} = \sum_{i=1}^{j} y_i \bar{s}_i$	(12.27)	
$\bar{c}_v = \sum_{i=1}^{j} y_i \bar{c}_{v,i}$	(12.23)	Mixture specific heats on a molar basis
$\bar{c}_p = \sum_{i=1}^{j} y_i \bar{c}_{p,i}$	(12.24)	

Psychrometric Applications

$\omega = \dfrac{m_v}{m_a} = 0.622\dfrac{p_v}{p - p_v}$	(12.42, 12.43)	Humidity ratio
$\phi = \dfrac{p_v}{p_g}\bigg)_{T, p}$	(12.44)	Relative humidity
$\dfrac{H}{m_a} = h_a + \omega h_v$	(12.46)	Mixture enthalpy per unit mass of dry air

EXERCISES: THINGS ENGINEERS THINK ABOUT

12.1 How do you calculate the specific heat ratio, k, at 300 K for a mixture of H_2, O_2, and CO if you know the molar analysis?

12.2 In an *equimolar* mixture of O_2 and N_2, are the mass fractions equal?

12.3 During winter months in cold climate zones, people often feel as if the outdoor air is dry. Are outdoor relative humidity levels typically low in these regions? Explain.

12.4 The molar analysis of an ideal gas mixture is $\{y_{CO_2} = 0.4, y_{N_2} = 0.25, y_{O_2}\}$. How many kmol of oxygen are present in 5 kmol of mixture?

12.5 Which do you think is more closely related to human comfort, the humidity ratio or relative humidity? Explain.

12.6 A rigid, insulated container is divided into two compartments by a partition, and each compartment contains air at the same temperature and pressure. If the partition is removed, is entropy produced within the container? Explain.

12.7 Can the dry-bulb and wet-bulb temperatures be equal? Explain.

12.8 Which component of the fuel–air mixture in a cylinder of an automobile engine would have the greater mass fraction?

12.9 Can cooling towers operate in cold regions when the winter temperatures drop below freezing? Explain.

12.10 During winter, why do eyeglasses fog up when the wearer enters a warm building?

12.11 Does operating a car's air-conditioning system affect its fuel economy? Explain.

12.12 Why does your bathroom mirror often fog up when you shower?

12.13 Although water vapor in air is typically a superheated vapor, why can we use the saturated vapor value, $h_g(T)$, to represent its enthalpy?

12.14 What is the difference between a *steam sauna* and a *steam room*?

12.15 How do you explain the water dripping from the tailpipe of an automobile on a cold morning?

CHECKING UNDERSTANDING

For Problems 12.1–12.6, match the appropriate definition in the right column with each term in the left column.

12.1 Mass Fractions

12.2 Mole Fractions

12.3 Gravimetric Analysis

12.4 Humidity Ratio

12.5 Psychrometrics

12.6 Apparent Molecular Weight

A. The study of systems involving dry air and water

B. The relative amounts of the components present in a mixture on a mass basis

C. A listing of the mass fractions of the components of a mixture

D. The ratio of the total mass of the mixture to the total number of moles of the mixture

E. The ratio of the mass of the water vapor to the mass of the dry air

F. The relative amounts of the components in a mixture on a molar basis

12.7 Component i within a mixture consists of 5 kmol with a mass of 8.8 kg. What is the molecular weight of the substance?

　a. 1.76 kg/kmol

　b. 0.57 kg/kmol

　c. 44.00 kg/kmol

　d. 42.08 kg/kmol

12.8 For the steady-state humidification processes shown in Fig. 12.12, which of the following is a false statement?

　a. The dry-bulb temperature increases.

　b. The dry-bulb temperature decreases.

　c. The humidity ratio decreases.

　d. Water is injected to the stream.

12.9 The *Dalton model* assumes that each mixture component behaves as an ideal gas as if it were alone at the temperature and _____ of the mixture.

　a. pressure

　b. volume

　c. mass

　d. humidity ratio

12.10 Which of the following can be associated with cooling tower operation?

    a. Natural convection heat transfer

    b. Counterflow heat exchange

    c. Cross-flow heat exchange

    d. All of the above

12.11 For the steady-state dehumidification process shown in Fig. 12.11, which of the following does not occur?

    a. The mixture pressure remains constant.

    b. The temperature drops below the dew point temperature.

    c. Water condenses.

    d. Water evaporates.

12.12 A mixture of dry air and saturated water vapor is called _____.

12.13 During an evaporative cooling process as illustrated in Fig. 12.13, which of the following occurs?

    a. The wet-bulb temperature changes significantly.

    b. The dry-bulb temperature decreases.

    c. The humidity ratio decreases.

    d. The relative humidity decreases.

Indicate whether the following statements are true or false. Explain.

12.14 In a gravimetric analysis, the summation of all the components' mass fractions must be equal to one.

12.15 Dehumidification is a process that involves condensation.

12.16 While the dry air component of a moist air stream can be treated as an ideal gas, the water vapor component cannot be treated this way.

12.17 The wet-bulb temperature is the temperature measured by a thermometer placed in the condensed liquid from a moist air stream.

12.18 There is no difference between volumetric analysis and molar analysis for an ideal gas mixture.

12.19 The wet-bulb temperature and the dry-bulb temperature can be measured using a *psychrometer*.

12.20 The sum of the mass fractions of all the components in a mixture must be greater than unity.

12.21 In a mixture, the Dalton model assumes that the summation of each component's volume equals the mixture volume.

12.22 The humidity ratio of moist air increases when it is heated in a steady-state flow process.

12.23 It is possible to cool moist air without changing its corresponding humidity ratio.

12.24 In moist air, when the partial pressure of water vapor is greater than the saturation pressure corresponding to the mixture's temperature, the mixture is said to be saturated.

12.25 On a psychrometric chart, lines of constant wet-bulb temperature are approximately lines of constant mixture enthalpy per unit mass of dry air.

PROBLEMS: DEVELOPING ENGINEERING SKILLS

C Problem may require use of appropriate computer software in order to complete.

Determining Mixture Composition

12.1 The analysis on a molar basis of a gas mixture at 5°C, 1 bar is 20% Ar, 35% CO_2, 45% O_2. Determine

    a. the analysis in terms of mass fractions.

    b. the partial pressure of each component, in kPa.

    c. the volume occupied by 5 kg of mixture, in m^3.

12.2 The analysis on a mass basis of a gas mixture at 4°C, 140 kPa is 60% CO_2, 25% CO, 15% O_2. Determine

    a. the analysis in terms of mole fractions.

    b. the partial pressure of each component, in kPa.

    c. the volume occupied by 10 kg of the mixture, in m^3.

12.3 The molar analysis of a gas mixture at 30°C, 2 bar is 40% N_2, 50% CO_2, 10% CH_4. Determine

    a. the analysis in terms of mass fractions.

    b. the partial pressure of each component, in bar.

    c. the volume occupied by 10 kg of mixture, in m^3.

12.4 Natural gas at 23°C, 1 bar enters a furnace with the following molar analysis: 40% propane (C_3H_8), 40% ethane (C_2H_6), 20% methane (CH_4). Determine

    a. the analysis in terms of mass fractions.

    b. the partial pressure of each component, in bar.

    c. the mass flow rate, in kg/s, for a volumetric flow rate of 20 m^3/s.

12.5 A rigid vessel having a volume of 3 m^3 initially contains a mixture at 21°C, 1 bar consisting of 79% N_2 and 21% O_2 on a molar basis. Helium is allowed to flow into the vessel until the pressure is 2 bar. If the final temperature of the mixture within the vessel is 27°C, determine the mass, in kg, of each component present.

12.6 A flue gas in which the mole fraction of H_2S is 0.002 enters a *scrubber* operating at steady state at 93°C, 1 bar and a volumetric flow rate of 540 m^3/h. If the scrubber removes 92% (molar basis) of the entering H_2S, determine the rate at which H_2S is removed, in kg/h. Comment on why H_2S should be removed from the gas stream.

12.7 A control volume operating at steady state has two entering streams and a single exiting stream. A mixture with a mass flow rate of 11.67 kg/min and a molar analysis 9% CH_4, 91% air enters at one location and is diluted by a separate stream of air entering at another location. The molar analysis of the air is 21% O_2, 79% N_2. If the mole fraction of CH_4 in the exiting stream is required to be 5%, determine

    a. the molar flow rate of the entering air, in kmol/min.

    b. the mass flow rate of oxygen in the exiting stream, in kg/min.

Considering Constant-Composition Processes

12.8 A gas mixture consists of 2 kg of N_2 and 3 kg of He. Determine

    a. the composition in terms of mass fractions.

    b. the composition in terms of mole fractions.

c. the heat transfer, in kJ, required to increase the mixture temperature from 21 to 66°C, while keeping the pressure constant.

d. the change in entropy of the mixture for the process of part (c), in kJ/K.

For parts (c) and (d), use the ideal gas model with constant specific heats.

12.9 A closed, rigid tank having a volume of 0.1 m³ contains 0.7 kg of N_2 and 1.1 kg of CO_2 at 27°C. Determine

a. the analysis of the mixture in terms of mass fractions.

b. the analysis of the mixture in terms of mole fractions.

c. the partial pressure of each component, in bar.

d. the mixture pressure, in bar.

e. the heat transfer, in kJ, required to bring the mixture to 127°C.

f. the entropy change of the mixture for the process of part (e) in kJ/K.

12.10 A mixture consisting of 2.8 kg of N_2 and 3.2 kg of O_2 is compressed from 1 bar, 300 K to 2 bar, 600 K. During the process there is heat transfer from the mixture to the surroundings, which are at 27°C. The work done on the mixture is claimed to be 2300 kJ. Can this value be correct?

12.11 A mixture of 2 kg of H_2 and 4 kg of N_2 is compressed in a piston–cylinder assembly in a polytropic process for which $n = 1.2$. The temperature increases from 22 to 150°C. Using constant values for the specific heats, determine

a. the heat transfer, in kJ.

b. the entropy change, in kJ/K.

12.12 A gas mixture at 1500 K with the molar analysis 10% CO_2, 20% H_2O, 70% N_2 enters a waste-heat boiler operating at steady state, and exits the boiler at 600 K. A separate stream of saturated liquid water enters at 25 bar and exits as saturated vapor with a negligible pressure drop. Ignoring stray heat transfer and kinetic and potential energy changes, determine the mass flow rate of the exiting saturated vapor, in kg per kmol of gas mixture.

12.13 A gas turbine receives a mixture having the following molar analysis: 20% CO_2, 30% H_2O, 50% N_2 at 740 K, 0.40 MPa and a volumetric flow rate of 3.5 m³/s. Products exit the turbine at 400 K, 0.15 MPa. For adiabatic operation with negligible kinetic and potential energy effects, determine the power developed at steady state, in kW.

12.14 An equimolar mixture of helium and carbon dioxide enters an insulated nozzle at 127°C, 5 bar, 30 m/s and expands isentropically to a pressure of 3.24 bar. Determine the temperature, in °C, and the velocity, in m/s, at the nozzle exit. Neglect potential energy effects.

12.15 A mixture having an analysis on a mass basis of 80% N_2, 20% CO_2 enters a nozzle operating at steady state at 1000 K with a velocity of 5 m/s and expands adiabatically through a 7.5:1 pressure ratio, exiting with a velocity of 900 m/s. Determine the isentropic nozzle efficiency.

12.16 A gas mixture having a molar analysis of 60% O_2 and 40% N_2 enters an insulated compressor operating at steady state at 1 bar, 20°C with a mass flow rate of 0.5 kg/s and is compressed to 5.4 bar. Kinetic and potential energy effects are negligible. For an isentropic compressor efficiency of 78%, determine

a. the temperature at the exit, in °C.

b. the power required, in kW.

c. the rate of entropy production, in kW/K.

12.17 Natural gas having a molar analysis of 60% methane (CH_4) and 40% ethane (C_2H_6) enters a compressor at 340 K, 6 bar and is compressed isothermally without internal irreversibilities to 20 bar. The compressor operates at steady state, and kinetic and potential energy effects are negligible.

a. Assuming ideal gas behavior, determine for the compressor the work and heat transfer, each in kJ per kmol of mixture flowing.

b. Compare with the values for work and heat transfer, respectively, determined assuming ideal solution behavior (Sec. 11.9.5). For the pure components at 340 K:

	h (kJ/kg)		s (kJ/kg · K)	
	6 bar	20 bar	6 bar	20 bar
Methane	715.33	704.40	10.9763	10.3275
Ethane	462.39	439.13	7.3493	6.9680

Forming Mixtures

12.18 An insulated tank has two compartments connected by a valve. Initially, one compartment contains 0.7 kg of CO_2 at 500 K, 6.0 bar and the other contains 0.3 kg of N_2 at 300 K, 6.0 bar. The valve is opened and the gases are allowed to mix until equilibrium is achieved. Determine

a. the final temperature, in K.

b. the final pressure, in bar.

c. the amount of entropy produced, in kJ/K.

12.19 Carbon dioxide (CO_2) at 197°C, 2 bar enters a chamber at steady state with a molar flow rate of 2 kmol/s and mixes with nitrogen (N_2) entering at 27°C, 2 bar with a molar flow rate of 1 kmol/s. Heat transfer from the mixing chamber occurs at an average surface temperature of 127°C. A single stream exits the mixing chamber at 127°C, 2 bar and passes through a duct, where it cools at constant pressure to 42°C through heat transfer with the surroundings at 27°C. Kinetic and potential energy effects can be ignored. Determine the rates of heat transfer and exergy destruction, each in kW, for control volumes enclosing

a. the mixing chamber only.

b. the mixing chamber and enough of the nearby surroundings that heat transfer occurs at 27°C.

c. the duct and enough of the nearby surroundings that heat transfer occurs at 27°C.

Let $T_0 = 27°C$.

12.20 One kilogram of argon at 27°C, 1 bar is contained in a rigid tank connected by a valve to another rigid tank containing 0.8 kg of O_2 at 127°C, 5 bar. The valve is opened, and the gases are allowed to mix, achieving an equilibrium state at 87°C. Determine

a. the volume of each tank, in m³.

b. the final pressure, in bar.

c. the heat transfer to or from the gases during the process, in kJ.

d. the entropy change of each gas, in kJ/K.

12.21 A rigid insulated tank has two compartments. Initially one contains 0.5 kmol of carbon dioxide (CO_2) at 27°C, 2 bar and the other contains 1 kmol of oxygen (O_2) at 152°C, 5 bar. The gases are allowed to mix while 500 kJ of energy are added by electrical work. Determine

a. the final temperature, in °C.

b. the final pressure, in bar.

c. the change in exergy, in kJ, for $T_0 = 20°C$.

d. the exergy destruction, in kJ.

12.22 A stream of air (21% O_2 and 79% N_2 on a molar basis) at 300 K and 0.1 MPa is to be *separated* into pure oxygen and nitrogen streams, each at 300 K and 0.1 MPa. A device to achieve the separation is claimed to require a work input at steady state of 1200 kJ per kmol of air. Heat transfer between the device and its surroundings occurs at 300 K. Ignoring kinetic and potential energy effects, evaluate whether the work value can be as claimed.

12.23 Air at 77°C, 1 bar, and a molar flow rate of 0.1 kmol/s enters an insulated mixing chamber operating at steady state and mixes with water vapor entering at 277°C, 1 bar, and a molar flow rate of 0.3 kmol/s. The mixture exits at 1 bar. Kinetic and potential energy effects can be ignored. For the chamber, determine

a. the temperature of the exiting mixture, in °C.

b. the rate of entropy production, in kW/K.

12.24 Argon (Ar), at 300 K, 1 bar with a mass flow rate of 1 kg/s enters the insulated mixing chamber shown in **Fig. P12.24** and mixes with carbon dioxide (CO_2) entering as a separate stream at 575 K, 1 bar with a mass flow rate of 0.5 kg/s. The mixture exits at 1 bar. Assume ideal gas behavior with $k = 1.67$ for Ar and $k = 1.25$ for CO_2. For steady-state operation, determine

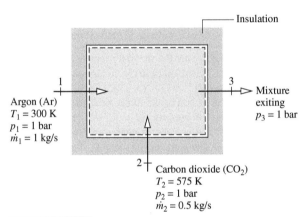

FIGURE P12.24

a. the molar analysis of the exiting mixture.

b. the temperature of the exiting mixture, in K.

c. the rate of entropy production, in kW/K.

12.25 **C** Helium at 400 K, 1 bar enters an insulated mixing chamber operating at steady state, where it mixes with argon entering at 300 K, 1 bar. The mixture exits at a pressure of 1 bar. If the argon mass flow rate is x times that of helium, plot versus x

a. the exit temperature, in K.

b. the rate of exergy destruction within the chamber, in kJ per kg of helium entering.

Kinetic and potential energy effects can be ignored. Let $T_0 = 300$ K.

12.26 A gas mixture required in an industrial process is prepared by first allowing carbon monoxide (CO) at 27°C, 125 kPa to enter an insulated mixing chamber operating at steady state and mix with argon (Ar) entering at 190°C, 125 kPa. The mixture exits the chamber at 60°C, 110 kPa and is then allowed to expand in a throttling process through a valve to 101.3 kPa. Determine

a. the mass and molar analyses of the mixture.

b. the temperature of the mixture at the exit of the valve, in °C.

c. the rates of exergy destruction for the mixing chamber and the valve, each in kJ per kg of mixture, for $T_0 = 4°C$.

Kinetic and potential energy effects can be ignored.

12.27 Hydrogen (H_2) at 77°C, 4 bar enters an insulated chamber at steady state, where it mixes with nitrogen (N_2) entering as a separate stream at 277°C, 4 bar. The mixture exits at 3.8 bar with the molar analysis 75% H_2, 25% N_2. Kinetic and potential energy effects can be ignored. Determine

a. the temperature of the exiting mixture, in °C.

b. the rate at which entropy is produced, in kJ/K per kmol of mixture exiting.

12.28 **C** An insulated, rigid tank initially contains 1 kmol of argon (Ar) at 300 K, 1 bar. The tank is connected by a valve to a large vessel containing nitrogen (N_2) at 500 K, 4 bar. A quantity of nitrogen flows into the tank, forming an argon–nitrogen mixture at temperature T and pressure p. Plot T, in K, and p, in bar, versus the amount of N_2 within the tank, in kmol.

12.29 A 27 m^3 tank initially filled with N_2 at 21°C, 35 kPa is connected by a valve to a large vessel containing O_2 at 21°C, 138 kPa. Oxygen is allowed to flow into the tank until the pressure in the tank becomes 103 kPa. If heat transfer with the surroundings maintains the tank contents at a constant temperature, determine

a. the mass of oxygen that enters the tank, in kg.

b. the heat transfer, in kJ.

Exploring Psychrometric Principles

12.30 A water pipe at 5°C runs above ground between two buildings. The surrounding air is at 35°C. What is the maximum relative humidity the air can have before condensation occurs on the pipe?

12.31 A large room contains moist air at 30°C, 102 kPa. The partial pressure of water vapor is 1.5 kPa. Determine

a. the relative humidity.

b. the humidity ratio.

c. the dew point temperature, in °C.

d. the mass of dry air, in kg, if the mass of water vapor is 10 kg.

12.32 A fixed amount of moist air initially at 1 bar and a relative humidity of 60% is compressed isothermally until condensation of water begins. Determine the pressure of the mixture at the onset of condensation, in bar. Repeat if the initial relative humidity is 90%.

12.33 As shown in **Fig. P12.33**, moist air at 30°C, 2 bar, and 50% relative humidity enters a heat exchanger operating at steady state with a mass flow rate of 600 kg/h and is cooled at constant pressure to 20°C. Ignoring kinetic and potential energy effects, determine the rate of heat transfer from the moist air stream, in kJ/h.

FIGURE P12.33

12.34 One kg of moist air initially at 27°C, 1 bar, 50% relative humidity is compressed isothermally to 3 bar. If condensation occurs, determine the amount of water condensed, in kg. If there is no condensation, determine the final relative humidity.

12.35 A closed, rigid tank initially contains 0.5 m³ of moist air in equilibrium with 0.1 m³ of liquid water at 80°C and 0.1 MPa. If the tank contents are heated to 200°C, determine

   a. the final pressure, in MPa.

   b. the heat transfer, in kJ.

12.36 Air at 12°C, 1 atm, and 40% relative humidity enters a heat exchanger with a volumetric flow rate of 1 m³/s. A separate stream of dry air enters at 280°C, 1 atm with a mass flow rate of 0.875 kg/s and exits at 220°C. Neglecting heat transfer between the heat exchanger and its surroundings, pressure drops of each stream, and kinetic and potential energy effects, determine

   a. the temperature of the exiting moist air, in °C.

   b. the rate of exergy destruction, in kW, for $T_0 = 12$°C.

12.37 To what temperature, in °C, must moist air with a humidity ratio of 5×10^{-3} be cooled at a constant pressure of 2 bar to become saturated moist air?

12.38 Moist air initially at 125°C, 4 bar, and 50% relative humidity is contained in a 2.5-m³ closed, rigid tank. The tank contents are cooled. Determine the heat transfer, in kJ, if the final temperature in the tank is (a) 110°C, (b) 30°C.

12.39 Air at 30°C, 1.05 bar, and 80% relative humidity enters a dehumidifier operating at steady state. Moist air exits at 15°C, 1 bar, and 95% relative humidity. Condensate exits in a separate stream at 15°C. A refrigerant flows through the cooling coil of the dehumidifier with an increase in its specific enthalpy of 100 kJ per kg of refrigerant flowing. Heat transfer between the humidifier and its surroundings and kinetic and potential energy effects can be ignored. Determine the refrigerant flow rate, in kg per kg of dry air.

12.40 Air at 35°C, 3 bar, 30% relative humidity, and a velocity of 50 m/s expands isentropically through a nozzle. Determine the lowest exit pressure, in bar, that can be attained without condensation. For this exit pressure, determine the exit velocity, in m/s. The nozzle operates at steady state and without significant potential energy effects.

12.41 Dry air enters a device operating at steady state at 27°C, 2 bar with a volumetric flow rate of 300 m³/min. Liquid water is injected and a moist air stream exits at 15°C, 2 bar, and 91% relative humidity. Determine

   a. the mass flow rate at the exit, in kg/min.

   b. the temperature, in °C, of the liquid water injected into the air stream.

Ignore heat transfer between the device and its surroundings and neglect kinetic and potential energy effects.

12.42 At steady state, moist air at 29°C, 1 bar, and 50% relative humidity enters a device with a volumetric flow rate of 13 m³/s. Liquid water at 40°C is sprayed into the moist air with a mass flow rate of 22 kg/s. The liquid water that does not evaporate into the moist air stream is drained and flows to another device at 26°C with a mass flow rate of 21.55 kg/s. A single moist air stream exits at 1 bar. Determine the temperature and relative humidity of the moist air stream exiting. Ignore heat transfer between the device and its surroundings and kinetic and potential energy effects.

12.43 Moist air at 20°C, 1.05 bar, 85% relative humidity and a volumetric flow rate of 0.3 m³/s enters a well-insulated compressor operating at steady state. If moist air exits at 100°C, 2.0 bar, determine

   a. the relative humidity at the exit.

   b. the power input, in kW.

   c. the rate of entropy production, in kW/K.

12.44 Moist air at 20°C, 1 atm, 43% relative humidity and a volumetric flow rate of 900 m³/h enters a control volume at steady state and flows along a surface maintained at 65°C, through which heat transfer occurs. Liquid water at 20°C is injected at a rate of 5 kg/h and evaporates into the flowing stream. For the control volume, $\dot{W}_{cv} = 0$, and kinetic and potential energy effects are negligible. Moist air exits at 32°C, 1 atm. Determine

   a. the rate of heat transfer, in kW.

   b. the rate of entropy production, in kW/K.

12.45 A fan within an insulated duct delivers moist air at the duct exit at 35°C, 50% relative humidity, and a volumetric flow rate of 0.4 m³/s. At steady state, the power input to the fan is 1.7 kW. The pressure in the duct is nearly 1 atm throughout. Using the psychrometric chart, determine the temperature, in °C, and relative humidity at the duct inlet.

12.46 Air enters a compressor operating at steady state at 50°C, 0.9 bar, 70% relative humidity with a volumetric flow rate of 0.8 m³/s. The moist air exits the compressor at 195°C, 1.5 bar. Assuming the compressor is well insulated, determine

   a. the relative humidity at the exit.

   b. the power input, in kW.

   c. the rate of entropy production, in kW/K.

Considering Air-Conditioning Applications

12.47 Each case listed gives the dry-bulb temperature and relative humidity of the moist air stream entering an air-conditioning system: **(a)** 30°C, 40%, **(b)** 17°C, 60%, **(c)** 25°C, 70%, **(d)** 15°C, 40%, **(e)** 27°C, 30%. The condition of the moist air stream exiting the system must satisfy these *constraints:* $22 \leq T_{db} \leq 27$°C, $40 \leq \phi \leq 60$%. In each case, develop a schematic of equipment and processes from Sec. 12.9 that would achieve the desired result. Sketch the processes on a psychrometric chart.

12.48 Moist air enters a device operating at steady state at 1 atm with a dry-bulb temperature of 55°C and a wet-bulb temperature of 25°C. Liquid water at 20°C is sprayed into the air stream, bringing it to 40°C, 1 atm at the exit. Determine

   a. the relative humidities at the inlet and exit.

   b. the rate that liquid water is sprayed into the air stream, in kg per kg of dry air.

12.49 Moist air at 33°C and 60% relative humidity enters a dehumidifier operating at steady state with a volumetric flow of rate of 230 m³/min. The moist air passes over a cooling coil and water vapor condenses. Condensate exits the dehumidifier saturated at 12°C. Saturated moist air exits in a separate stream at the same temperature. There is no significant loss of energy by heat transfer to the surroundings and pressure remains constant at 1 bar. Determine

   a. the mass flow rate of the dry air, in kg/min.

   b. the rate at which water is condensed, in kg per kg of dry air flowing through the control volume.

   c. the required refrigerating capacity, in tons.

12.50 An air conditioner operating at steady state takes in moist air at 28°C, 1 bar, and 70% relative humidity. The moist air first passes over a cooling coil in the dehumidifier unit and some water vapor is condensed. The rate of heat transfer between the moist air and the cooling coil is 11 tons. Saturated moist air and condensate streams exit the dehumidifier unit at the same temperature. The moist air then passes through a heating unit, exiting at 24°C, 1 bar, and 40%

relative humidity. Neglecting kinetic and potential energy effects, determine

a. the temperature of the moist air exiting the dehumidifier unit, in °C.

b. the volumetric flow rate of the air entering the air conditioner, in m³/min.

c. the rate water is condensed, in kg/min.

d. the rate of heat transfer to the air passing through the heating unit, in kW.

12.51 **Figure P12.51** shows a compressor followed by an aftercooler. Atmospheric air at 100 kPa, 30°C, and 75% relative humidity enters the compressor with a volumetric flow rate of 0.05 m³/s. The compressor power input is 10 kW. The moist air exiting the compressor at 670 kPa, 200°C flows through the aftercooler, where it is cooled at constant pressure, exiting saturated at 40°C. Condensate also exits the aftercooler at 40°C. For steady-state operation and negligible kinetic and potential energy effects, determine

a. the rate of heat transfer from the compressor to its surroundings, in kJ/s.

b. the mass flow rate of the condensate, in kg/s.

c. the rate of heat transfer from the moist air to the refrigerant circulating in the cooling coil, in tons of refrigeration.

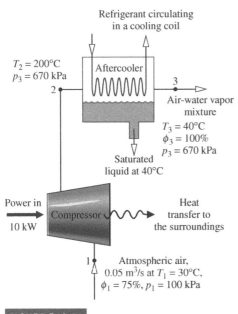

FIGURE P12.51

12.52 Outside air at 10°C, 1 bar, and 40% relative humidity enters an air-conditioning device operating at steady state. Liquid water is injected at 5°C and a moist air stream exits with a volumetric flow rate of 0.5 m³/s at 30°C, 1 bar and a relative humidity of 40%. Neglecting kinetic and potential energy effects, determine

a. the rate water is injected, in kg/s.

b. the rate of heat transfer to the moist air, in kJ/s.

12.53 An air-conditioner system is designed to operate under the following conditions:

Required indoor conditions—22°C, 70% relative humidity. Outdoor conditions—30°C, 75% relative humidity. Dew point temperature for the coil—14°C. Mass flow rate of moist air circulated—3.5 m³/s. The above conditions are achieved by initially cooling which is followed by dehumidification and heating. Determine

a. the capacity of the cooling coil in kW.

b. the capacity of the heating coil in kW.

c. the amount of water vapour removed in kg/s.

Use psychrometric chart Fig. A-9 to solve this problem.

12.54 Atmospheric air having dry-bulb and wet-bulb temperatures of 33 and 29°C, respectively, enters a well-insulated chamber operating at steady state and mixes with air entering with dry-bulb and wet-bulb temperatures of 16 and 12°C, respectively. The volumetric flow rate of the lower temperature stream is twice that of the other stream. A single mixed stream exits. Determine for the exiting stream

a. the relative humidity.

b. the temperature, in °C.

Pressure is uniform throughout at 1 atm. Neglect kinetic and potential energy effects.

12.55 Moist air at 25°C, 1 atm, and 40% relative humidity enters an evaporative cooling unit operating at steady state consisting of a heating section followed by a soaked pad evaporative cooler operating adiabatically. The air passing through the heating section is heated to 43°C. Next, the air passes through a soaked pad exiting with 40% relative humidity. Using data from the psychrometric chart, determine

a. the humidity ratio of the entering moist air mixture.

b. the rate of heat transfer to the moist air passing through the heating section, in kJ per kg of mixture.

c. the humidity ratio and temperature, in °C, at the exit of the evaporative cooling section.

12.56 At steady state, a moist air stream (stream 1) is mixed adiabatically with another stream (stream 2). Stream 1 is at 13°C, 1 bar, and 20% relative humidity, with a volumetric flow rate of 0.3 m³/s. A single stream exits the mixing chamber at 19°C, 1 bar, and 60% relative humidity, with a volumetric flow rate of 0.7 m³/s. Determine for stream 2

a. the relative humidity.

b. the temperature, in °C.

c. the mass flow rate, in kg/s.

12.57 At steady state, a device for heating and humidifying air has 0.12 m³/s of air at 4°C, 1 bar, and 80% relative humidity entering at one location, 0.5 m³/s of air at 15°C, 1 bar, and 80% relative humidity entering at another location, and liquid water injected at 12°C. A single moist air stream exits at 30°C, 1 bar, and 35% relative humidity. Determine

a. the rate of heat transfer to the device, in kJ/s.

b. the rate at which liquid water is injected, in kg/s.

Neglect kinetic and potential energy effects.

12.58 Air at 30°C, 1 bar, 50% relative humidity enters an insulated chamber operating at steady state with a mass flow rate of 3 kg/min and mixes with a saturated moist air stream entering at 5°C, 1 bar with a mass flow rate of 5 kg/min. A single mixed stream exits at 1 bar. Determine

a. the relative humidity and temperature, in °C, of the exiting stream.

b. the rate of exergy destruction, in kW, for $T_0 = 20$°C.

Neglect kinetic and potential energy effects.

12.59 At steady state, moist air at 42°C, 1 atm, 30% relative humidity is mixed adiabatically with a second moist air stream entering at 1 atm. The mass flow rates of the two streams are the same. A single mixed stream exits at 29°C, 1 atm, 40% relative humidity with a mass flow rate of 2 kg/s. Kinetic and potential energy effects are negligible.

For the second entering moist air stream, determine, using data from the psychrometric chart,

  a. the relative humidity.

  b. the temperature, in °C.

12.60 [C] At steady state, moist air is to be supplied to a classroom at a specified volumetric flow rate and temperature T. Air is removed from the classroom in a separate stream at a temperature of 27°C and 50% relative humidity. Moisture is added to the air in the room from the occupants at a rate of 4.5 kg/h. The moisture can be regarded as saturated vapor at 33°C. Heat transfer into the occupied space from all sources is estimated to occur at a rate of 34,000 kJ/h. The pressure remains uniform at 1 atm.

  a. For a supply air volumetric flow rate of 40 m³/min, determine the supply air temperature T, in °C, and the relative humidity.

  b. Plot the supply air temperature, in °C, and relative humidity, each versus the supply air volumetric flow rate ranging from 35 to 90 m³/min.

12.61 Air at 30°C, 1 bar, 50% relative humidity enters an insulated chamber operating at steady state with a mass flow rate of 3 kg/min and mixes with a saturated moist air stream entering at 5°C, 1 bar with a mass flow rate of 5 kg/min. A single mixed stream exits at 1 bar. Determine

  a. the relative humidity and temperature, in °C, of the exiting stream.

  b. the rate of exergy destruction, in kW, for $T_0 = 20$°C.

Neglect kinetic and potential energy effects.

12.62 Figure P12.62 shows a device for conditioning moist air entering at 5°C, 1 atm, 90% relative humidity, and a volumetric flow rate of 60 m³/min. The incoming air is first heated at essentially constant pressure to 24°C. Superheated steam at 1 atm is then injected, bringing the moist air stream to 25°C, 1 atm, and 45% relative humidity. Determine for steady-state operation

  a. the rate of heat transfer to the air passing through the heating section, in kJ/min.

  b. the mass flow rate of the injected steam, in kg/min.

  c. If the injected steam expands through a valve from a saturated vapor condition at the valve inlet, determine the inlet pressure, in bar.

Neglect kinetic and potential energy effects.

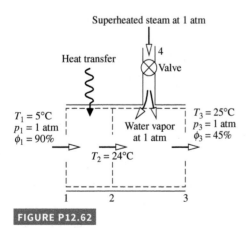

FIGURE P12.62

12.63 Air at 35°C, 1 bar, and 10% relative humidity enters an evaporative cooler operating at steady state. The volumetric flow rate of the incoming air is 50 m³/min. Liquid water at 20°C enters

the cooler and fully evaporates. Moist air exits the cooler at 25°C, 1 bar. If there is no significant heat transfer between the device and its surroundings, determine

  a. the rate at which liquid enters, in kg/min.

  b. the relative humidity at the exit.

  c. the rate of exergy destruction, in kJ/min, for $T_0 = 20$°C.

Neglect kinetic and potential energy effects.

12.64 Figure P12.64 shows the adiabatic mixing of two moist air streams at steady state. Kinetic and potential energy effects are negligible. Determine the rate of exergy destruction, in kJ/s, for $T_0 = 35$°C.

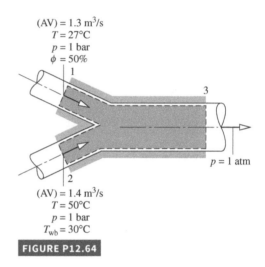

FIGURE P12.64

Analyzing Cooling Towers

12.65 In the condenser of a power plant, energy is discharged by heat transfer at a rate of 836 MW to cooling water that exits the condenser at 40°C into a cooling tower. Cooled water at 20°C is returned to the condenser. Atmospheric air enters the tower at 25°C, 1 bar, 35% relative humidity. Moist air exits at 35°C, 1 bar, 90% relative humidity. Makeup water is supplied at 20°C. For operation at steady state, determine the mass flow rate, in kg/s, of

  a. the entering atmospheric air.

  b. the makeup water.

Ignore kinetic and potential energy effects.

12.66 [C] Liquid water at 50°C enters a cooling tower operating at steady state with a mass flow rate of 64 kg/s. Atmospheric air enters at 27°C, 1 bar, 30% relative humidity. Saturated air exits at 38°C, 1 bar. No makeup water is provided. Plot the mass flow rate of dry air required, in kg/s, versus the temperature at which cooled water exits the tower. Consider temperatures ranging from 15 to 32°C. Ignore kinetic and potential energy effects.

12.67 Liquid water at 43°C and a volumetric flow rate of 0.1 m³/s enters a cooling tower operating at steady state. Cooled water exits the cooling tower at 30°C. Atmospheric air enters the tower at 27°C, 1 bar, 40% relative humidity, and saturated moist air at 40°C, 1 bar exits the cooling tower. Determine

  a. the mass flow rates of the dry air and the cooled water, each in kg/s.

  b. the rate of exergy destruction within the cooling tower, in kJ/s, for $T_0 = 25$°C.

Ignore kinetic and potential energy effects.

12.68 Liquid water at 50°C enters a forced draft cooling tower operating at steady state. Cooled water exits the tower with a mass flow rate of 80 kg/min. No makeup water is provided. A fan located within the tower draws in atmospheric air at 17°C, 0.098 MPa, 60% relative humidity with a volumetric flow rate of 110 m³/min. Saturated air exits the tower at 30°C, 0.098 MPa. The power input to the fan is 8 kW. Ignoring kinetic and potential energy effects, determine

    a. the mass flow rate of the liquid stream entering, in kg/min.

    b. the temperature of the cooled liquid stream exiting, in °C.

DESIGN & OPEN-ENDED PROBLEMS: EXPLORING ENGINEERING PRACTICE

12.1D About half the air we breathe on some airplanes is fresh air, and the rest is recirculated. Investigate typical equipment schematics for providing a blend of fresh and recirculated filtered air to the passenger cabins of commercial airplanes. What types of filters are used and how do they work? Write a report including at least three references.

12.2D **Figure P12.2D** shows a spray cooler included on the schematic of a plant layout. Arguing that it can only be the *addition* of makeup water at location 3 that is intended, an engineer orders the direction of the arrow at this location on the schematic to be reversed. Evaluate the engineer's order and write a memorandum explaining your evaluation.

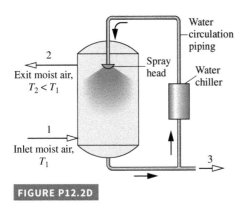

FIGURE P12.2D

12.3D Study the air-conditioning system for one of the classrooms you frequent where occupant comfort is unsatisfactory and describe the system in detail, including the control strategy used. Propose modifications aimed at improving occupant satisfaction, including a new heating, ventilation, and air conditioning (HVAC) system for the room, if warranted. Compare the proposed and existing systems in terms of occupant comfort, potential impact on productivity, and energy requirements. Detail your findings in an executive summary and PowerPoint presentation.

12.4D Identify a campus, commercial, or other building in your locale with an air-conditioning system installed 20 or more years ago. Critically evaluate the efficacy of the system in terms of comfort level provided, operating costs, maintenance costs, global warming potential of the refrigerant used, and other pertinent issues. On this basis, recommend specific system upgrades or a full system replacement, as warranted. Present your findings in a PowerPoint presentation.

12.5D **Figure P12.5D** shows a system for supplying a space with 2100 m³/min of conditioned air at a dry-bulb temperature of 22°C and a relative humidity of 60% when the outside air is at a dry-bulb temperature of 35°C and a relative humidity of 55%. Dampers A and B can be set to give three alternative operating modes: (1) Both dampers closed (no use of recirculated air). (2) Damper A open and damper B closed. One-third of the conditioned air comes from outside air. (3) Both dampers open. One-third of the conditioned air comes from outside air. One-third of the recirculated air bypasses the dehumidifier

via open damper B, and the rest flows through the damper A. Which of the three operating modes should be used? Discuss.

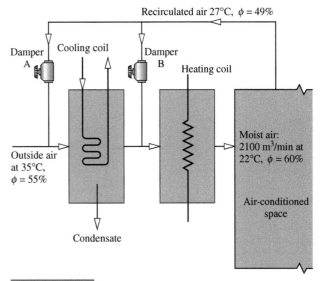

FIGURE P12.5D

12.6D Tunnel-type spray coolers are used to cool vegetables. In one application, tomatoes on a 1.2 m wide conveyor belt pass under a 1°C water spray and cool from 20 to 4°C. The water is collected in a reservoir below and recirculated through the evaporator of an ammonia refrigeration unit. Recommend the length of the tunnel and the conveyor speed. Estimate the number of tomatoes that can be cooled in an hour with your design. Write a report including calculations and at least three references.

12.7D Phosphorous compounds and zinc are used as additives in larger cooling tower systems to control corrosion and deposition of solids. Regulations are emerging that will limit the use of phosphorous compounds in these systems, especially those discharging process water directly to public waterways. Write a report explaining how cooling towers are maintained. Include relevant chemistry and explore how corrosion and deposition are currently controlled. Examine emerging regulations and describe ways by which new and existing designs will need to be modified to comply with new regulations. Include in your report at least three references.

12.8D An air-handling system is being designed for a 12 m × 12 m × 2.5 m biological research facility that houses 3000 laboratory mice. The indoor conditions must be maintained at 24°C, 60% relative humidity when the outdoor air conditions are 32°C, 70% relative humidity. Develop a preliminary design of an air-conditioning and distribution system that satisfies National Institute of Health (NIH) standards for animal facilities. Assume a *biological safety level of one* (BSL-1), and that two thirds of the floor space is devoted to animal care. Since an interruption in ventilation or air conditioning could place the laboratory animals under stress and compromise the research under way in the facility, account for *redundancy* in your design.

12.9D Figure P12.9D shows a steam-injected gas-turbine cogeneration system that produces power and process steam: a simplified *Cheng* cycle. The steam is generated by a heat-recovery steam generator (HRSG) in which the hot gas exiting the turbine at state 4 is cooled and discharged at state 5. A separate stream of return water enters the HRSG at state 6 and steam exits at states 7 and 8. The superheated steam exiting at 7 is injected into the combustor. In what ways does the steam-injected system shown in the figure differ from the humid air turbine (HAT) cycle? Draw the schematic of a HAT cycle. Discuss appropriate applications of each cycle.

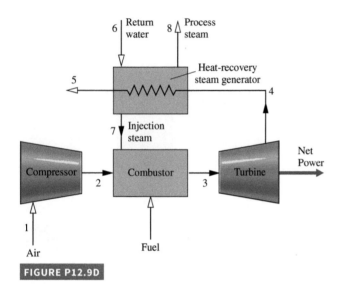

FIGURE P12.9D

12.10D Adequate levels of ventilation reduce the likelihood of *sick building syndrome*. (See BIOCONNECTIONS in section 12.4.2.) The outdoor air used for ventilation must be conditioned, and this requires energy. Consider the air-handling system for the commercial building illustrated in Fig. P12.10D, consisting of ducting, two dampers labeled A and B, a vapor-compression dehumidifier, and a heater. The system supplies 25 m³/s of conditioned air at 20°C and a relative humidity of 55% to maintain the interior space at 25°C and a relative humidity of 50%. The recirculated air has the same conditions as the air in the interior space. A minimum of 5 m³/s of outdoor air is required to provide adequate ventilation. Dampers A and B can be set to provide alternative operating modes to maintain required ventilation rates. On a given summer day when the outside air dry-bulb temperature and relative humidity are 25°C and 60%, respectively, which of the following three operating modes is best from the standpoint of minimizing the total heat transfer of energy *from* the conditioned air *to* the cooling coil and *to* the conditioned air *from* the heating coil?

1. Dampers A and B closed.
2. Damper A open and damper B closed with outside air contributing one-quarter of the total supply air.
3. Dampers A and B open. One-quarter of the conditioned air comes from outside air and one-third of the recirculated air bypasses the dehumidifier via open damper B; the rest flows through damper A.

Present your recommendation together with your reasoning in a PowerPoint presentation suitable for your class. Additionally, in an accompanying memorandum, provide well-documented sample calculations in support of your recommendation.

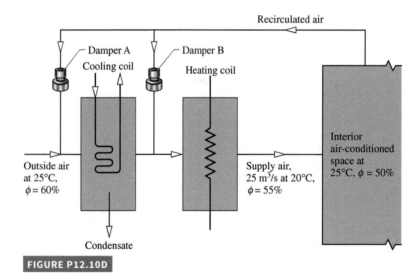

FIGURE P12.10D

CHAPTER 13

Reacting Mixtures and Combustion

Engineering Context

The **objective** of this chapter is to study systems involving chemical reactions. Since the *combustion* of hydrocarbon fuels occurs in most power-producing devices (Chaps. 8 and 9), combustion is emphasized.

The thermodynamic analysis of reacting systems is primarily an extension of principles introduced thus far. The concepts applied in the *first part* of the chapter dealing with combustion fundamentals remain the same: conservation of mass, conservation of energy, and the second law. It is necessary, though, to modify the methods used to evaluate specific enthalpy, internal energy, and entropy, by accounting for changing chemical composition. Only the manner in which these properties are evaluated represents a departure from previous practice, for once appropriate values are determined they are used as in earlier chapters in the energy and entropy balances for the system under consideration. In the *second part* of the chapter, the exergy concept of Chap. 7 is extended by introducing chemical exergy.

The principles developed in this chapter allow the equilibrium composition of a mixture of chemical substances to be determined. This topic is studied in Chap. 14. The subject of *dissociation* is also deferred until then. Prediction of *reaction rates* is not within the scope of classical thermodynamics, so the topic of chemical kinetics, which deals with reaction rates, is not discussed in this text.

LEARNING OUTCOMES

When you complete your study of this chapter, you will be able to...

- Define complete combustion, theoretical air, enthalpy of formation, and adiabatic flame temperature, and compute values associated with each term.
- Develop balanced reaction equations for combustion of hydrocarbon fuels.
- Apply mass, energy, and entropy balances to closed systems and control volumes involving chemical reactions.
- Conduct exergy analyses, including chemical exergy and the evaluation of exergetic efficiencies.

Combustion Fundamentals

13.1 Introducing Combustion

reactants

products

When a chemical reaction occurs, the bonds within molecules of the **reactants** are broken, and atoms and electrons rearrange to form **products**. In combustion reactions, rapid oxidation of combustible elements of the fuel results in energy release as combustion products are formed. The three major combustible chemical elements in most common fuels are carbon, hydrogen, and sulfur. Sulfur is usually a relatively unimportant contributor to the energy released, but it can be a significant cause of pollution and corrosion problems.

complete combustion

Combustion is **complete** when all the carbon present in the fuel is burned to carbon dioxide, all the hydrogen is burned to water, all the sulfur is burned to sulfur dioxide, and all other combustible elements are fully oxidized. When these conditions are not fulfilled, combustion is *incomplete*.

In this chapter, we deal with combustion reactions expressed by chemical equations of the form

$$\text{reactants} \rightarrow \text{products}$$

or

$$\text{fuel} + \text{oxidizer} \rightarrow \text{products}$$

When dealing with chemical reactions, it is necessary to remember that mass is conserved, so the mass of the products equals the mass of the reactants. The total mass of each chemical *element* must be the same on both sides of the equation, even though the elements exist in different chemical compounds in the reactants and products. However, the number of moles of products may differ from the number of moles of reactants.

FOR EXAMPLE

Consider the complete combustion of hydrogen with oxygen

$$1H_2 + \tfrac{1}{2}O_2 \rightarrow 1H_2O \tag{13.1}$$

In this case, the reactants are hydrogen and oxygen. Hydrogen is the fuel and oxygen is the oxidizer. Water is the only product of the reaction. The numerical coefficients in the equation, which precede the chemical symbols to give equal amounts of each chemical element on both sides of the equation, are called **stoichiometric coefficients**. In words, Eq. 13.1 states

stoichiometric coefficients

$$1 \text{ kmol } H_2 + \tfrac{1}{2} \text{ kmol } O_2 \rightarrow 1 \text{ kmol } H_2O$$

Note that the total numbers of moles on the left and right sides of Eq. 13.1 are not equal. However, because mass is conserved, the total mass of reactants must equal the total mass of products. Since 1 kmol of H_2 equals 2 kg, $\tfrac{1}{2}$ kmol of O_2 equals 16 kg, and 1 kmol of H_2O equals 18 kg, Eq. 13.1 can be interpreted as stating

$$2 \text{ kg } H_2 + 16 \text{ kg } O_2 \rightarrow 18 \text{ kg } H_2O$$

In the remainder of this section, consideration is given to the makeup of the fuel, oxidizer, and combustion products typically involved in engineering combustion applications.

13.1.1 Fuels

A *fuel* is simply a combustible substance. In this chapter emphasis is on hydrocarbon fuels, which contain hydrogen and carbon. Sulfur and other chemical substances also may be present. Hydrocarbon fuels can exist as liquids, gases, and solids.

Liquid hydrocarbon fuels are commonly derived from crude oil through distillation and cracking processes. Examples are gasoline, diesel fuel, kerosene, and other types of fuel oils. Most liquid fuels are mixtures of hydrocarbons for which compositions are usually given in terms of mass fractions. For simplicity in combustion calculations, gasoline is often modeled as octane, C_8H_{18}, and diesel fuel as dodecane, $C_{12}H_{26}$.

Gaseous hydrocarbon fuels are obtained from natural gas wells or are produced in certain chemical processes. Natural gas normally consists of several different hydrocarbons, with the major constituent being methane, CH_4. The compositions of gaseous fuels are usually given in terms of mole fractions. Both gaseous and liquid hydrocarbon fuels can be synthesized from coal, oil shale, and tar sands.

Coal is a familiar solid fuel. Its composition varies considerably with the location from which it is mined. For combustion calculations, the composition of coal is usually expressed as an ultimate analysis. The **ultimate analysis** gives the composition on a *mass basis* in terms of the relative amounts of chemical elements (carbon, sulfur, hydrogen, nitrogen, oxygen) and ash.

ultimate analysis

13.1.2 Modeling Combustion Air

Oxygen is required in every combustion reaction. Pure oxygen is used only in special applications such as cutting and welding. In most combustion applications, air provides the needed oxygen. The composition of a typical sample of dry air is given in Table 12.1. For the combustion calculations of this book, however, the following model is used for simplicity:

- All components of dry air other than oxygen are lumped together with nitrogen. Accordingly, air is considered to be 21% oxygen and 79% nitrogen on a molar basis. With this idealization the molar ratio of the nitrogen to the oxygen is $0.79/0.21 = 3.76$. When air supplies the oxygen in a combustion reaction, therefore, every mole of oxygen is accompanied by 3.76 moles of nitrogen.

- We also assume that the nitrogen present in the combustion air does *not* undergo chemical reaction. That is, nitrogen is regarded as inert. The nitrogen in the products is at the same temperature as the other products, however. Accordingly, nitrogen undergoes a change of state if the products are at a temperature other than the reactant air temperature. If a high enough product temperature is attained, nitrogen can form compounds such as nitric oxide and nitrogen dioxide. Even trace amounts of oxides of nitrogen appearing in the exhaust of internal combustion engines can be a source of air pollution.

TAKE NOTE...

In this model, air is assumed to contain no water vapor. When *moist* air is involved in combustion, the water vapor present is usually considered in writing the combustion equation.

Air–Fuel Ratio   Two parameters that are frequently used to quantify the amounts of fuel and air in a particular combustion process are the air–fuel ratio and its reciprocal, the fuel–air ratio. The **air–fuel ratio** is simply the ratio of the amount of air in a reaction to the amount of fuel. The ratio can be written on a molar basis (moles of air divided by moles of fuel) or on a mass basis (mass of air divided by mass of fuel). Conversion between these values is accomplished using the molecular weights of the air, M_{air}, and fuel, M_{fuel},

air–fuel ratio

$$\frac{\text{mass of air}}{\text{mass of fuel}} = \frac{\text{moles of air} \times M_{air}}{\text{moles of fuel} \times M_{fuel}}$$

$$= \frac{\text{moles of air}}{\text{moles of fuel}} \left(\frac{M_{air}}{M_{fuel}} \right)$$

or

$$\boxed{AF = \overline{AF} \left(\frac{M_{air}}{M_{fuel}} \right)} \tag{13.2}$$

where $\overline{AF}$ is the air–fuel ratio on a molar basis and AF is the ratio on a mass basis. For the combustion calculations of this book the molecular weight of air is taken as 28.97. Table A-1 provide the molecular weights of several important hydrocarbons. Since AF is a ratio, it has the same value whether the quantities of air and fuel are expressed in SI units.

Theoretical Air The minimum amount of air that supplies sufficient oxygen for the complete combustion of all the carbon, hydrogen, and sulfur present in the fuel is called theoretical air   the **theoretical amount of air**. For complete combustion with the theoretical amount of air, the products consist of carbon dioxide, water, sulfur dioxide, the nitrogen accompanying the oxygen in the air, and any nitrogen contained in the fuel. No free oxygen appears in the products.

FOR EXAMPLE

Let us determine the theoretical amount of air for the complete combustion of methane. For this reaction, the products contain only carbon dioxide, water, and nitrogen. The reaction is

$$CH_4 + a(O_2 + 3.76N_2) \rightarrow bCO_2 + cH_2O + dN_2 \tag{13.3}$$

where a, b, c, and d represent the numbers of moles of oxygen, carbon dioxide, water, and nitrogen. In writing the left side of Eq. 13.3, 3.76 moles of nitrogen are considered to accompany each mole of oxygen. Applying the conservation of mass principle to the carbon, hydrogen, oxygen, and nitrogen, respectively, results in four equations among the four unknowns:

$$
\begin{aligned}
\text{C:} &\quad b = 1 \\
\text{H:} &\quad 2c = 4 \\
\text{O:} &\quad 2b + c = 2a \\
\text{N:} &\quad d = 3.76a
\end{aligned}
$$

Solving these equations, the *balanced* chemical equation is

$$CH_4 + 2(O_2 + 3.76N_2) \rightarrow CO_2 + 2H_2O + 7.52N_2 \tag{13.4}$$

The coefficient 2 before the term $(O_2 + 3.76N_2)$ in Eq. 13.4 is the number of moles of *oxygen* in the combustion air, per mole of fuel, and *not* the amount of air. The amount of combustion air is 2 moles of oxygen *plus* 2×3.76 moles of nitrogen, giving a total of 9.52 moles of air per mole of fuel. Thus, for the reaction given by Eq. 13.4 the air–fuel ratio on a molar basis is 9.52. To calculate the air–fuel ratio on a mass basis, use Eq. 13.2 to write

$$AF = \overline{AF}\left(\frac{M_{air}}{M_{fuel}}\right) = 9.52\left(\frac{28.97}{16.04}\right) = 17.19$$

---

percent of theoretical air

percent excess air

Normally the amount of air supplied is either greater or less than the theoretical amount. The amount of air actually supplied is commonly expressed in terms of the **percent of theoretical air**. For example, 150% of theoretical air means that the air actually supplied is 1.5 times the theoretical amount of air. The amount of air supplied can be expressed alternatively as a **percent excess air** or a *percent deficiency* of air. Thus, 150% of theoretical air is equivalent to 50% excess air, and 80% of theoretical air is the same as a 20% deficiency of air.

FOR EXAMPLE

Consider the *complete* combustion of methane with 150% theoretical air (50% excess air). The balanced chemical reaction equation is

$$CH_4 + (1.5)(2)(O_2 + 3.76N_2) \rightarrow CO_2 + 2H_2O + O_2 + 11.28N_2 \tag{13.5}$$

In this equation, the amount of air per mole of fuel is 1.5 times the theoretical amount determined by Eq. 13.4. Accordingly, the air–fuel ratio is 1.5 times the air–fuel ratio determined for Eq. 13.4. Since complete combustion is assumed, the products contain only carbon dioxide, water, nitrogen, and oxygen. The excess air supplied appears in the products as uncombined oxygen and a greater amount of nitrogen than in Eq. 13.4, based on the theoretical amount of air.

The **equivalence ratio** is the ratio of the actual fuel–air ratio to the fuel–air ratio for complete combustion with the theoretical amount of air. The reactants are said to form a *lean* mixture when the equivalence ratio is less than unity. When the ratio is greater than unity, the reactants are said to form a *rich* mixture.

In Example 13.1, we use conservation of mass to obtain balanced chemical reactions. The air–fuel ratio for each of the reactions is also calculated.

equivalence ratio

▶ ▶ ▶ **EXAMPLE 13.1** ▶ ·················

Determining the Air–Fuel Ratio for Complete Combustion of Octane

Determine the air–fuel ratio on both a molar and mass basis for the complete combustion of octane, C_8H_{18}, with **(a)** the theoretical amount of air, **(b)** 150% theoretical air (50% excess air).

SOLUTION

Known Octane, C_8H_{18}, is burned completely with (a) the theoretical amount of air, (b) 150% theoretical air.

Find Determine the air–fuel ratio on a molar and a mass basis.

Engineering Model

1. Each mole of oxygen in the combustion air is accompanied by 3.76 moles of nitrogen.

2. The nitrogen is inert.

3. Combustion is complete.

Analysis

a. For complete combustion of C_8H_{18} with the theoretical amount of air, the products contain carbon dioxide, water, and nitrogen only. That is,

$$C_8H_{18} + a(O_2 + 3.76N_2) \rightarrow bCO_2 + cH_2O + dN_2$$

Applying the conservation of mass principle to the carbon, hydrogen, oxygen, and nitrogen, respectively, gives

C:	$b = 8$
H:	$2c = 18$
O:	$2b + c = 2a$
N:	$d = 3.76a$

Solving these equations, $a = 12.5$, $b = 8$, $c = 9$, $d = 47$. The balanced chemical equation is

$$C_8H_{18} + 12.5(O_2 + 3.76N_2) \rightarrow 8CO_2 + 9H_2O + 47N_2$$

The air–fuel ratio on a molar basis is

$$\overline{AF} = \frac{12.5 + 12.5(3.76)}{1} = \frac{12.5(4.76)}{1} = 59.5 \frac{\text{kmol (air)}}{\text{kmol (fuel)}}$$

The air–fuel ratio expressed on a mass basis is

$$AF = \left[59.5 \frac{\text{kmol (air)}}{\text{kmol (fuel)}} \right] \left[\frac{28.97 \dfrac{\text{kg (air)}}{\text{kmol (air)}}}{114.22 \dfrac{\text{kg (fuel)}}{\text{kmol (fuel)}}} \right] = 15.1 \frac{\text{kg (air)}}{\text{kg (fuel)}}$$

b. For 150% theoretical air, the chemical equation for complete combustion takes the form

❶ $C_8H_{18} + 1.5(12.5)(O_2 + 3.76N_2) \rightarrow bCO_2 + cH_2O + dN_2 + eO_2$

Applying conservation of mass

C:	$b = 8$
H:	$2c = 18$
O:	$2b + c + 2e = (1.5)(12.5)(2)$
N:	$d = (1.5)(12.5)(3.76)$

Solving this set of equations, $b = 8$, $c = 9$, $d = 70.5$, $e = 6.25$, giving a balanced chemical equation

$$C_8H_{18} + 18.75(O_2 + 3.76N_2) \rightarrow 8CO_2 + 9H_2O + 70.5N_2 + 6.25O_2$$

The air–fuel ratio on a molar basis is

$$\overline{AF} = \frac{18.75(4.76)}{1} = 89.25 \frac{\text{kmol (air)}}{\text{kmol (fuel)}}$$

On a mass basis, the air–fuel ratio is 22.6 kg (air)/kg (fuel), as can be verified.

---

❶ When complete combustion occurs with *excess air*, oxygen appears in the products, in addition to carbon dioxide, water, and nitrogen.

SKILLS DEVELOPED

Ability to…

• balance a chemical reaction equation for complete combustion with theoretical air and with excess air.

• apply definitions of air–fuel ratio on mass and molar bases.

Quick Quiz

For the condition in part (b), determine the *equivalence ratio*. Ans. 0.67.

13.1.3 Determining Products of Combustion

In each of the illustrations given above, complete combustion is assumed. For a hydrocarbon fuel, this means that the only allowed products are CO_2, H_2O, and N_2, with O_2 also present when excess air is supplied. If the fuel is specified and combustion is complete, the respective amounts of the products can be determined by applying the conservation of mass principle to the chemical equation. The procedure for obtaining the balanced reaction equation of an *actual* reaction where combustion is incomplete is not always so straightforward.

Combustion is the result of a series of very complicated and rapid chemical reactions, and the products formed depend on many factors. When fuel is burned in the cylinder of an internal combustion engine, the products of the reaction vary with the temperature and pressure in the cylinder. In combustion equipment of all kinds, the degree of mixing of the fuel and air is a controlling factor in the reactions that occur once the fuel and air mixture is ignited. Although the amount of air supplied in an actual combustion process may exceed the theoretical amount, it is not uncommon for some carbon monoxide and unburned oxygen to appear in the products. This can be due to incomplete mixing, insufficient time for complete combustion, and other factors. When the amount of air supplied is less than the theoretical amount of air, the products may include both CO_2 and CO, and there also may be unburned fuel in the products. Unlike the complete combustion cases considered above, the products of combustion of an actual combustion process and their relative amounts can be determined only by *measurement*.

Among several devices for measuring the composition of products of combustion are the *Orsat analyzer, gas chromatograph, infrared analyzer,* and *flame ionization detector.* Data from these devices can be used to determine the mole fractions of the gaseous products of combustion. The analyses are often reported on a "dry" basis. In a **dry product analysis**, the mole fractions are given for all gaseous products *except* the water vapor. In Examples 13.2 and 13.3, we show how analyses of the products of combustion on a dry basis can be used to determine the balanced chemical reaction equations.

Since water is formed when hydrocarbon fuels are burned, the mole fraction of water vapor in the gaseous products of combustion can be significant. If the gaseous products of combustion are cooled at constant mixture pressure, the *dew point temperature* is reached when water vapor begins to condense. Since water deposited on duct work, mufflers, and other metal parts can cause corrosion, knowledge of the dew point temperature is important. Determination of the dew point temperature is illustrated in Example 13.2, which also features a dry product analysis.

TAKE NOTE...

In *actual* combustion processes, the products of combustion and their relative amounts can be determined only through measurement.

dry product analysis

TAKE NOTE...

For cooling of combustion products at constant pressure, the *dew point temperature* marks the onset of condensation of water vapor present in the products. See Sec. 12.5.4 to review this concept.

▶▶▶ EXAMPLE 13.2 ▶

Using a Dry Product Analysis for Combustion of Methane

Methane, CH_4, is burned with dry air. The molar analysis of the products on a dry basis is CO_2, 9.7%; CO, 0.5%; O_2, 2.95%; and N_2, 86.85%. Determine **(a)** the air–fuel ratio on both a molar and a mass basis, **(b)** the percent theoretical air, **(c)** the dew point temperature of the products, in °C, if the mixture were cooled at 1 bar.

SOLUTION

Known Methane is burned with dry air. The molar analysis of the products on a dry basis is provided.

Find Determine (a) the air–fuel ratio on both a molar and a mass basis, (b) the percent theoretical air, and (c) the dew point temperature of the products, in °C, if cooled at 1 atm.

Engineering Model

1. Each mole of oxygen in the combustion air is accompanied by 3.76 moles of nitrogen, which is inert.

2. The products form an ideal gas mixture and the dew point temperature of the mixture is conceptualized as in Sec. 12.5.4.

Analysis

① a. The solution is conveniently conducted on the basis of 100 kmol of dry products. The chemical equation then reads

$$aCH_4 + b(O_2 + 3.76N_2) \rightarrow$$
$$9.7CO_2 + 0.5CO + 2.95O_2 + 86.85N_2 + cH_2O$$

In addition to the assumed 100 kmol of dry products, water must be included as a product.

Applying conservation of mass to carbon, hydrogen, and oxygen, respectively,

C:	$9.7 + 0.5 = a$
H:	$2c = 4a$
O:	$(9.7)(2) + 0.5 + 2(2.95) + c = 2b$

2 Solving this set of equations gives $a = 10.2$, $b = 23.1$, $c = 20.4$. The balanced chemical equation is

$$10.2CH_4 + 23.1(O_2 + 3.76N_2) \rightarrow$$
$$9.7CO_2 + 0.5CO + 2.95O_2 + 86.85N_2 + 20.4H_2O$$

On a molar basis, the air–fuel ratio is

$$\overline{AF} = \frac{23.1(4.76)}{10.2} = 10.78 \frac{\text{kmol (air)}}{\text{kmol (fuel)}}$$

On a mass basis

$$AF = (10.78)\left(\frac{28.97}{16.04}\right) = 19.47 \frac{\text{kg (air)}}{\text{kg (fuel)}}$$

b. The balanced chemical equation for the *complete combustion* of methane with the *theoretical amount* of air is

$$CH_4 + 2(O_2 + 3.76N_2) \rightarrow CO_2 + 2H_2O + 7.52N_2$$

The theoretical air–fuel ratio on a molar basis is

$$(\overline{AF})_{\text{theo}} = \frac{2(4.76)}{1} = 9.52 \frac{\text{kmol (air)}}{\text{kmol (fuel)}}$$

The percent theoretical air is then found from

$$\% \text{ theoretical air} = \frac{(\overline{AF})}{(\overline{AF})_{\text{theo}}}$$
$$= \frac{10.78 \text{ kmol (air)/kmol (fuel)}}{9.52 \text{ kmol (air)/kmol (fuel)}}$$
$$= 1.13 \ (113\%)$$

c. To determine the dew point temperature requires the partial pressure of the water vapor p_v. The partial pressure p_v is found from $p_v = y_v p$, where y_v is the mole fraction of the water vapor in the combustion products and p is 1 atm.

Referring to the balanced chemical equation of part (a), the mole fraction of the water vapor is

$$y_v = \frac{20.4}{100 + 20.4} = 0.169$$

3 Thus, $p_v = 0.169$ atm $= 0.1712$ bar. Interpolating in Table A-2, $T = 57°C$.

---

1 The solution could be obtained on the basis of any assumed amount of dry products—for example, 1 kmol. With some other assumed amount, the values of the coefficients of the balanced chemical equation would differ from those obtained in the solution, but the air–fuel ratio, the value for the percent of theoretical air, and the dew point temperature would be unchanged.

2 The three unknown coefficients, a, b, and c, are evaluated here by application of conservation of mass to carbon, hydrogen, and oxygen. As a check, note that the nitrogen also balances

$$\text{N:} \qquad b(3.76) = 86.85$$

This confirms the accuracy of both the given product analysis and the calculations conducted to determine the unknown coefficients.

3 If the products of combustion were cooled at constant pressure below the dew point temperature of 57°C, some condensation of the water vapor would occur.

SKILLS DEVELOPED

Ability to...

- balance a chemical reaction equation for incomplete combustion given the analysis of dry products of combustion.
- apply definitions of air–fuel ratio on mass and molar bases as well as percent theoretical air.
- determine the dew point temperature of combustion products.

Quick Quiz

Recalculate the dew point temperature as in part (c) if the air supply were *moist* air, including 3.53 kmol of additional water vapor. Ans. 59°C.

In Example 13.3, a fuel mixture having a known molar analysis is burned with air, giving products with a known dry analysis.

> ▶ ▶ **EXAMPLE 13.3** ▶

Burning Natural Gas with Excess Air

A natural gas has the following molar analysis: CH_4, 80.62%; C_2H_6, 5.41%; C_3H_8, 1.87%; C_4H_{10}, 1.60%; N_2, 10.50%. The gas is burned with dry air, giving products having a molar analysis on a dry basis: CO_2, 7.8%; CO, 0.2%; O_2, 7%; N_2, 85%. **(a)** Determine the air–fuel ratio on a molar basis. **(b)** Assuming ideal gas behavior for the fuel mixture, determine the amount of products in kmol that would be formed from 100 m^3 of fuel mixture at 300 K and 1 bar. **(c)** Determine the percent of theoretical air.

SOLUTION

Known A natural gas with a specified molar analysis burns with dry air, giving products having a known molar analysis on a dry basis.

Find Determine the air–fuel ratio on a molar basis, the amount of products in kmol that would be formed from 100 m³ of natural gas at 300 K and 1 bar, and the percent of theoretical air.

Engineering Model

1. Each mole of oxygen in the combustion air is accompanied by 3.76 moles of nitrogen, which is inert.

2. The fuel mixture can be modeled as an ideal gas.

Analysis

a. The solution can be conducted on the basis of an assumed amount of fuel mixture or on the basis of an assumed amount of dry products. Let us illustrate the first procedure, basing the solution on 1 kmol of fuel mixture. The chemical equation then takes the form

$$(0.8062CH_4 + 0.0541C_2H_6 + 0.0187C_3H_8 + 0.0160C_4H_{10}$$
$$+ 0.1050N_2) + a(O_2 + 3.76N_2) \rightarrow b(0.078CO_2$$
$$+ 0.002CO + 0.07O_2 + 0.85N_2) + cH_2O$$

The products consist of b kmol of dry products and c kmol of water vapor, each per kmol of fuel mixture.

Applying conservation of mass to carbon

$$b(0.078 + 0.002) = 0.8062 + 2(0.0541) + 3(0.0187) + 4(0.0160)$$

Solving gives $b = 12.931$. Conservation of mass for hydrogen results in

$$2c = 4(0.8062) + 6(0.0541) + 8(0.0187) + 10(0.0160)$$

which gives $c = 1.93$. The unknown coefficient a can be found from either an oxygen balance or a nitrogen balance. Applying conservation of mass to oxygen

$$12.931[2(0.078) + 0.002 + 2(0.07)] + 1.93 = 2a,$$

❶ giving $a = 2.892$.

The balanced chemical equation is then

$$(0.8062CH_4 + 0.0541C_2H_6 + 0.0187C_3H_8 + 0.0160C_4H_{10}$$
$$+ 0.1050N_2) + 2.892(O_2 + 3.76N_2) \rightarrow 12.931(0.078CO_2$$
$$+ 0.002CO + 0.07O_2 + 0.85N_2) + 1.93H_2O$$

The air–fuel ratio on a molar basis is

$$\overline{AF} = \frac{(2.892)(4.76)}{1} = 13.77 \frac{\text{kmol (air)}}{\text{kmol (fuel)}}$$

b. By inspection of the chemical reaction equation, the total amount of products is $b + c = 12.931 + 1.93 = 14.861$ kmol of products per kmol of fuel. The amount of fuel in kmol, n_F,

present in 100 m³ of fuel mixture at 300 K and 1 bar can be determined from the ideal gas equation of state as

$$n_F = \frac{pV}{\overline{R}T}$$
$$= \frac{(10^5 \text{ N/m}^2)(100 \text{ m}^3)}{(8314 \text{ N} \cdot \text{m/kmol} \cdot \text{K})(300 \text{ K})} = 4.01 \text{ kmol (fuel)}$$

Accordingly, the amount of product mixture that would be formed from 100 m³ of fuel mixture is $(14.861)(4.01) = 59.59$ kmol of product gas.

c. The balanced chemical equation for the complete combustion of the fuel mixture with the *theoretical amount* of air is

$$(0.8062CH_4 + 0.0541C_2H_6 + 0.0187C_3H_8 + 0.0160C_4H_{10} + 0.1050N_2)$$
$$+ 2(O_2 + 3.76N_2) \rightarrow 1.0345CO_2 + 1.93H_2O + 7.625N_2$$

The theoretical air–fuel ratio on a molar basis is

$$(\overline{AF})_{\text{theo}} = \frac{2(4.76)}{1} = 9.52 \frac{\text{kmol (air)}}{\text{kmol (fuel)}}$$

The percent theoretical air is then

$$\% \text{ theoretical air} = \frac{13.77 \text{ kmol (air)/kmol (fuel)}}{9.52 \text{ kmol (air)/kmol (fuel)}} = 1.45 \text{ (145\%)}$$

---

❶ A check on both the accuracy of the given molar analyses and the calculations conducted to determine the unknown coefficients is obtained by applying conservation of mass to nitrogen. The amount of nitrogen in the reactants is

$$0.105 + (3.76)(2.892) = 10.98 \text{ kmol/kmol of fuel}$$

The amount of nitrogen in the products is $(0.85)(12.931) = 10.99$ kmol/kmol of fuel. The difference can be attributed to round-off.

SKILLS DEVELOPED

Ability to…

- balance a chemical reaction equation for incomplete combustion of a fuel mixture given the analysis of dry products of combustion.

- apply the definition of air–fuel ratio on a molar basis as well as percent theoretical air.

Quick Quiz

Determine the mole fractions of the products of combustion. Ans. $y_{CO_2} = 0.0679, y_{CO} = 0.0017, y_{O_2} = 0.0609, y_{N_2} = 0.7396, y_{H_2O} = 0.1299.$

13.1.4 Energy and Entropy Balances for Reacting Systems

Thus far our study of reacting systems has involved only the conservation of mass principle. A more complete understanding of reacting systems requires application of the first and second laws of thermodynamics. For these applications, energy and entropy balances play important roles, respectively. Energy balances for reacting systems are developed and applied in Secs. 13.2 and 13.3; entropy balances for reacting systems are the subject of Sec. 13.5. To apply such balances, it is necessary to take special care in how internal energy, enthalpy, and entropy are evaluated.

For the energy and entropy balances of this chapter, combustion air and (normally) products of combustion are modeled as ideal gas mixtures. Accordingly, ideal gas mixture

TABLE 13.1

Internal Energy, Enthalpy, and Entropy for Ideal Gas Mixtures

Notation: n_i = moles of gas i, y_i = mole fraction of gas i

T = mixture temperature, p = mixture pressure

$p_i = y_i p$ = partial pressure of gas i

$\bar{u}_i$ = specific internal energy of gas i, per mole of i

$\bar{h}_i$ = specific enthalpy of gas i, per mole of i

$\bar{s}_i$ = specific entropy of gas i, per mole of i

Mixture internal energy:

$$U = n_1\bar{u}_1 + n_2\bar{u}_2 + \cdots + n_j\bar{u}_j = \sum_{i=1}^{j} n_i\bar{u}_i(T) \tag{12.19}$$

Mixture enthalpy:

$$H = n_1\bar{h}_1 + n_2\bar{h}_2 + \cdots + n_j\bar{h}_j = \sum_{i=1}^{j} n_i\bar{h}_i(T) \tag{12.20}$$

Mixture entropy:

$$S = n_1\bar{s}_1 + n_2\bar{s}_2 + \cdots + n_j\bar{s}_j = \sum_{i=1}^{j} n_i\bar{s}_i(T, p_i) \tag{12.26}$$

▶ With Eq. 6.18:

$$\bar{s}_i(T, p_i) = \bar{s}_i(T, p) - \bar{R} \ln \frac{p_i}{p}$$

$$= \bar{s}_i(T, p) - \bar{R} \ln y_i \tag{a}$$

▶ With Eq. 6.18 and $p_{\text{ref}} = 1$ atm:

$$\bar{s}_i(T, p_i) = \bar{s}_i(T, p_{\text{ref}}) - \bar{R} \ln \frac{p_i}{p_{\text{ref}}}$$

$$= \bar{s}_i^{\circ}(T) - \bar{R} \ln \frac{y_i p}{p_{\text{ref}}} \tag{b}[1]$$

where $\bar{s}_i^{\circ}$ is obtained from Table A-23.

[1]Equation (b) corresponds to Eq. 13.23.

principles introduced in the first part of Chap. 12 play a role. For ease of reference, Table 13.1 summarizes ideal gas mixture relations introduced in Chap. 12 that are used in this chapter.

13.2 Conservation of Energy—Reacting Systems

The objective of the present section is to illustrate the application of the conservation of energy principle to reacting systems. The forms of the conservation of energy principle introduced previously remain valid whether or not a chemical reaction occurs within the system. However, the methods used for evaluating the properties of reacting systems differ somewhat from the practices used to this point.

13.2.1 Evaluating Enthalpy for Reacting Systems

In most tables of thermodynamic properties used thus far, values for the specific internal energy, enthalpy, and entropy are given relative to some arbitrary datum state where the enthalpy (or alternatively the internal energy) and entropy are set to zero. This approach is satisfactory for evaluations involving *differences* in property values between states of the same composition, for then arbitrary datums cancel. However, when a chemical reaction occurs, reactants disappear and products are formed, so differences cannot be calculated for all substances involved. For reacting systems, it is necessary to evaluate h, u, and s in such a way that there are no subsequent ambiguities or inconsistencies in evaluating properties. In this section, we will

standard reference state

enthalpy of formation

consider how this is accomplished for h and u. The case of entropy is handled differently and is taken up in Sec. 13.5.

An enthalpy datum for the study of reacting systems can be established by assigning arbitrarily a value of zero to the enthalpy of the *stable elements* at a state called the **standard reference state** and defined by $T_{ref} = 298.15$ K (25°C) and $p_{ref} = 1$ atm. Note that only *stable* elements are assigned a value of zero enthalpy at the standard state. The term *stable* simply means that the particular element is in a chemically stable form. For example, at the standard state the stable forms of hydrogen, oxygen, and nitrogen are H_2, O_2, and N_2 and not the monatomic H, O, and N. No ambiguities or conflicts result with this choice of datum.

Enthalpy of Formation   Using the datum introduced above, enthalpy values can be assigned to *compounds* for use in the study of reacting systems. The enthalpy of a compound at the standard state equals its *enthalpy of formation*, symbolized $\bar{h}_f^\circ$. The **enthalpy of formation** is the energy released or absorbed when the compound is formed from its elements, the compound and elements all being at T_{ref} and p_{ref}. The enthalpy of formation is usually determined by application of procedures from statistical thermodynamics using observed spectroscopic data.

The enthalpy of formation also can be found in principle by measuring the heat transfer in a reaction in which the compound is formed from the elements.

> **FOR EXAMPLE**
>
> Consider the simple reactor shown in **Fig. 13.1**, in which carbon and oxygen each enter at T_{ref} and p_{ref} and react completely at steady state to form carbon dioxide at the same temperature and pressure. Carbon dioxide is *formed* from carbon and oxygen according to
>
> $$C + O_2 \rightarrow CO_2 \tag{13.6}$$
>
> This reaction is *exothermic*, so for the carbon dioxide to exit at the same temperature as the entering elements, there would be a heat transfer from the reactor to its surroundings. The rate of heat transfer and the enthalpies of the incoming and exiting streams are related by the energy rate balance
>
> $$0 = \dot{Q}_{cv} + \dot{m}_C h_C + \dot{m}_{O_2} h_{O_2} - \dot{m}_{CO_2} h_{CO_2}$$
>
> where $\dot{m}$ and h denote, respectively, mass flow rate and specific enthalpy. In writing this equation, we have assumed no work $\dot{W}_{cv}$ and negligible effects of kinetic and potential energy. For enthalpies on a molar basis, the energy rate balance appears as
>
> $$0 = \dot{Q}_{cv} + \dot{n}_C \bar{h}_C + \dot{n}_{O_2} \bar{h}_{O_2} - \dot{n}_{CO_2} \bar{h}_{CO_2}$$
>
> where $\dot{n}$ and $\bar{h}$ denote, respectively, the molar flow rate and specific enthalpy per mole. Solving for the specific enthalpy of carbon dioxide and noting from Eq. 13.6 that all molar flow rates are equal,
>
> $$\bar{h}_{CO_2} = \frac{\dot{Q}_{cv}}{\dot{n}_{CO_2}} + \frac{\dot{n}_C}{\dot{n}_{CO_2}} \bar{h}_C + \frac{\dot{n}_{O_2}}{\dot{n}_{CO_2}} \bar{h}_{O_2} = \frac{\dot{Q}_{cv}}{\dot{n}_{CO_2}} + \bar{h}_C + \bar{h}_{O_2} \tag{13.7}$$
>
> Since carbon and oxygen are stable elements at the standard state, $\bar{h}_C = \bar{h}_{O_2} = 0$, and Eq. 13.7 becomes
>
> $$\bar{h}_{CO_2} = \frac{\dot{Q}_{cv}}{\dot{n}_{CO_2}} \tag{13.8}$$
>
> Accordingly, the value assigned to the specific enthalpy of carbon dioxide at the standard state, the enthalpy of formation, equals the heat transfer, per mole of CO_2, between the reactor and its surroundings. If the heat transfer could be measured accurately, it would be found to equal $-393,520$ kJ per kmol of carbon dioxide formed.

Fig. 13.1 Reactor used to discuss the enthalpy of formation concept.

Table A-25 gives values of the enthalpy of formation for several compounds in units of kJ/kmol. In this text, the superscript $\circ$ is used to denote properties at 1 atm. For the case of the enthalpy of formation, the reference temperature T_{ref} is also intended by this symbol. The values of $\bar{h}_f^\circ$ listed in Table A-25 for CO_2 correspond to those given in the previous example.

The sign associated with the enthalpy of formation values appearing in Table A-25 corresponds to the sign convention for heat transfer. If there is heat transfer *from* a reactor in which a compound is formed from its elements (an *exothermic* reaction as in the previous example), the enthalpy of formation has a negative sign. If a heat transfer *to* the reactor is required (an *endothermic* reaction), the enthalpy of formation is positive.

Evaluating Enthalpy   The specific enthalpy of a compound at a state other than the standard state is found by adding the specific enthalpy change $\Delta \overline{h}$ between the standard state and the state of interest to the enthalpy of formation

$$\overline{h}(T, p) = \overline{h}_f^\circ + [\overline{h}(T, p) - \overline{h}(T_{\text{ref}}, p_{\text{ref}})] = \overline{h}_f^\circ + \Delta \overline{h} \qquad (13.9)$$

That is, the enthalpy of a compound is composed of $\overline{h}_f^\circ$, associated with the formation of the compound from its elements, and $\Delta \overline{h}$, associated with a change of state at constant composition. An arbitrary choice of datum can be used to determine $\Delta \overline{h}$, since it is a *difference* at constant composition. Accordingly, $\Delta \overline{h}$ can be evaluated from tabular sources such as the steam tables, the ideal gas tables when appropriate, and so on. Note that as a consequence of the enthalpy datum adopted for the stable elements, the specific enthalpy determined from Eq. 13.9 is often negative.

Table A-25 provide two values of the enthalpy of formation of water: $\overline{h}_f^\circ (1), \overline{h}_f^\circ (g)$. The first is for liquid water and the second is for water vapor. Under equilibrium conditions, water exists only as a liquid at 25°C and 1 atm. The vapor value listed is for a *hypothetical* ideal gas state in which water is a vapor at 25°C and 1 atm. The difference between the two enthalpy of formation values is given closely by the enthalpy of vaporization $\overline{h}_{\text{fg}}$ at T_{ref}. That is,

$$\overline{h}_f^\circ(g) - \overline{h}_f^\circ(1) \approx \overline{h}_{\text{fg}}(25°C) \qquad (13.10)$$

Similar considerations apply to other substances for which liquid and vapor values for $\overline{h}_f^\circ$ are listed in Table A-25.

> **TAKE NOTE...**
>
> When applying Eq. 13.9 to water vapor, we use the *vapor* value of the enthalpy of formation of water, $\overline{h}_f^\circ(g)$, from Table A-25 together with $\Delta \overline{h}$ for water vapor from Table A-23.

13.2.2 Energy Balances for Reacting Systems

Several considerations enter when writing energy balances for systems involving combustion. Some of these apply generally, without regard for whether combustion takes place. For example, it is necessary to consider if significant work and heat transfers take place and if the respective values are known or unknown. Also, the effects of kinetic and potential energy must be assessed. Other considerations are related directly to the occurrence of combustion. For example, it is important to know the state of the fuel before combustion occurs. Whether the fuel is a liquid, a gas, or a solid is important. It is also necessary to consider whether the fuel is premixed with the combustion air or the fuel and air enter a reactor separately.

The state of the combustion products also must be assessed. For instance, the presence of water vapor should be noted, for some of the water present will condense if the products are cooled sufficiently. The energy balance must then be written to account for the presence of water in the products as both a liquid and a vapor. For cooling of combustion products at constant pressure, the dew point temperature method of Example 13.2 is used to determine the temperature at the onset of condensation.

Analyzing Control Volumes at Steady State   To illustrate the many considerations involved when writing energy balances for reacting systems, we consider special cases of broad interest, highlighting the underlying assumptions. Let us begin by considering the steady-state reactor shown in **Fig. 13.2**, in which a hydrocarbon fuel C_aH_b burns completely with the theoretical amount of air according to

$$C_aH_b + \left(a + \frac{b}{4}\right)(O_2 + 3.76N_2) \rightarrow aCO_2 + \frac{b}{2}H_2O + \left(a + \frac{b}{4}\right)3.76N_2 \qquad (13.11)$$

The fuel enters the reactor in a stream separate from the combustion air, which is regarded as an ideal gas mixture. The products of combustion also are assumed to form an ideal gas mixture. Kinetic and potential energy effects are ignored.

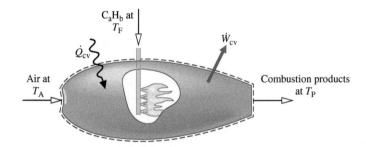

Fig. 13.2 Reactor at steady state.

With the foregoing idealizations, the mass and energy rate balances for the two-inlet, single-exit reactor can be used to obtain the following equation on a *per mole of fuel basis:*

$$\frac{\dot{Q}_{cv}}{\dot{n}_F} - \frac{\dot{W}_{cv}}{\dot{n}_F} = \underline{\left[a\bar{h}_{CO_2} + \frac{b}{2}\bar{h}_{H_2O} + \left(a + \frac{b}{4} \right)3.76\bar{h}_{N_2} \right]}$$
$$- \bar{h}_F - \underline{\left[\left(a + \frac{b}{4} \right)\bar{h}_{O_2} + \left(a + \frac{b}{4} \right)3.76\bar{h}_{N_2} \right]} \tag{13.12a}$$

where $\dot{n}_F$ denotes the molar flow rate of the fuel. Note that each coefficient on the right side of this equation is the same as the coefficient of the corresponding substance in the reaction equation.

The first underlined term on the right side of Eq. 13.12a is the enthalpy of the exiting gaseous products of combustion *per mole of fuel*. The second underlined term on the right side is the enthalpy of the combustion air *per mole of fuel*. In accord with Table 13.1, the enthalpies of the combustion products and the air have been evaluated by adding the contribution of each component present in the respective ideal gas mixtures. The symbol $\bar{h}_F$ denotes the molar enthalpy of the fuel.

Equation 13.12a can be expressed more concisely as

$$\frac{\dot{Q}_{cv}}{\dot{n}_F} - \frac{\dot{W}_{cv}}{\dot{n}_F} = \bar{h}_P - \bar{h}_R \tag{13.12b}$$

where $\bar{h}_P$ and $\bar{h}_R$ denote, respectively, the enthalpies of the products and reactants per mole of fuel.

Evaluating Enthalpy Terms   Once the energy balance has been written, the next step is to evaluate the individual enthalpy terms. Since each component of the combustion products is assumed to behave as an ideal gas, its contribution to the enthalpy of the products depends solely on the temperature of the products, T_P. Accordingly, for each component of the products, Eq. 13.9 takes the form

$$\bar{h} = \bar{h}_f^\circ + [\bar{h}(T_P) - \bar{h}(T_{ref})] \tag{13.13}$$

In Eq. 13.13, $\bar{h}_f^\circ$ is the enthalpy of formation from Table A-25. The second term accounts for the change in enthalpy from the temperature T_{ref} to the temperature T_P. For several common gases, this term can be evaluated from tabulated values of enthalpy versus temperature in Table A-23. Alternatively, the term can be obtained by integration of the ideal gas specific heat $\bar{c}_p$ obtained from Table A-21 or some other source of data.

A similar approach is employed to evaluate the enthalpies of the oxygen and nitrogen in the combustion air. For these

$$\bar{h} = \bar{h}_f^{\circ\,0} + [\bar{h}(T_A) - \bar{h}(T_{ref})] \tag{13.14}$$

where T_A is the temperature of the air entering the reactor. Note that the enthalpy of formation for oxygen and nitrogen is zero by definition and thus drops out of Eq. 13.14 as indicated.

The evaluation of the enthalpy of the fuel is also based on Eq. 13.9. If the fuel can be modeled as an ideal gas, the fuel enthalpy is obtained using an expression of the same form as Eq. 13.13, with the temperature of the incoming fuel replacing T_P.

With the foregoing considerations, Eq. 13.12a takes the form

$$\frac{\dot{Q}_{cv}}{\dot{n}_F} - \frac{\dot{W}_{cv}}{\dot{n}_F} = a(\bar{h}_f^\circ + \Delta\bar{h})_{CO_2} + \frac{b}{2}(\bar{h}_f^\circ + \Delta\bar{h})_{H_2O} + \left(a + \frac{b}{4}\right)3.76(\bar{h}_f^{\not{\circ}0} + \Delta\bar{h})_{N_2}$$

$$- (\bar{h}_f^\circ + \Delta\bar{h})_F - \left(a + \frac{b}{4}\right)(\bar{h}_f^{\not{\circ}0} + \Delta\bar{h})_{O_2} - \left(a + \frac{b}{4}\right)3.76(\bar{h}_f^{\not{\circ}0} + \Delta\bar{h})_{N_2} \qquad \text{(13.15a)}$$

The terms set to zero in this expression are the enthalpies of formation of oxygen and nitrogen.

Equation 13.15a can be written more concisely as

$$\boxed{\frac{\dot{Q}_{cv}}{\dot{n}_F} - \frac{\dot{W}_{cv}}{\dot{n}_F} = \sum_P n_e(\bar{h}_f^\circ + \Delta\bar{h})_e - \sum_R n_i(\bar{h}_f^\circ + \Delta\bar{h})_i} \qquad \text{(13.15b)}$$

where i denotes the incoming fuel and air streams and e the exiting combustion products.

Although Eqs. 13.15 have been developed with reference to the reaction of Eq. 13.11, equations having the same general forms would be obtained for other combustion reactions.

In Examples 13.4 and 13.5, the energy balance is applied together with tabular property data to analyze control volumes at steady state involving combustion. Example 13.4 involves a reciprocating internal combustion engine while Example 13.5 involves a simple gas turbine power plant.

TAKE NOTE...

The coefficients n_i and n_e of Eq. 13.15b correspond to the respective coefficients of the reaction equation giving the moles of reactants and products *per mole of fuel*, respectively.

> > > **EXAMPLE 13.4** ▶ ·

Analyzing an Internal Combustion Engine

Liquid octane enters an internal combustion engine operating at steady state with a mass flow rate of 1.8×10^{-3} kg and is mixed with the theoretical amount of air. The fuel and air enter the engine at 25°C and 1 atm. The mixture burns completely and combustion products leave the engine at 890 K. The engine develops a power output of 37 kW horsepower. Determine the rate of heat transfer from the engine, in kW, neglecting kinetic and potential energy effects.

SOLUTION

Known Liquid octane and the theoretical amount of air enter an internal combustion engine operating at steady state in separate streams at 25°C, 1 atm. Combustion is complete and the products exit at 890 K. The power developed by the engine and fuel mass flow rate are specified.

Find Determine the rate of heat transfer from the engine, in kW.

Schematic and Given Data:

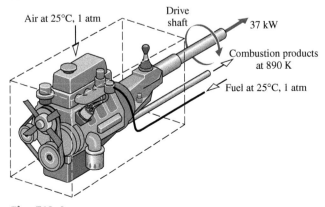

Fig. E13.4

Engineering Model

1. The control volume identified by a dashed line on the accompanying figure operates at steady state.

2. Kinetic and potential energy effects can be ignored.

3. The combustion air and the products of combustion each form ideal gas mixtures.

4. Each mole of oxygen in the combustion air is accompanied by 3.76 moles of nitrogen. The nitrogen is inert and combustion is complete.

Analysis The balanced chemical equation for complete combustion with the theoretical amount of air is obtained from the solution to Example 13.1 as

$$C_8H_{18} + 12.5O_2 + 47N_2 \rightarrow 8CO_2 + 9H_2O + 47N_2$$

The energy rate balance reduces, with assumptions 1–3, to give

$$\frac{\dot{Q}_{cv}}{\dot{n}_F} = \frac{\dot{W}_{cv}}{\dot{n}_F} + \bar{h}_P - \bar{h}_R$$

$$= \frac{\dot{W}_{cv}}{\dot{n}_F} + \{8[\bar{h}_f^\circ + \Delta\bar{h}]_{CO_2} + 9[\bar{h}_f^\circ + \Delta\bar{h}]_{H_2O(g)} + 47[\bar{h}_f^{\not{\circ}0} + \Delta\bar{h}]_{N_2}\}$$

$$- \{[\bar{h}_f^\circ + \Delta\bar{h}^{\not{0}}]_{C_8H_{18}(l)} + 12.5[\bar{h}_f^{\not{\circ}0} + \Delta\bar{h}^{\not{0}}]_{O_2} + 47[\bar{h}_f^{\not{\circ}0} + \Delta\bar{h}^{\not{0}}]_{N_2}\}$$

where each coefficient is the same as the corresponding term of the balanced chemical equation and Eq. 13.9 has been used to evaluate enthalpy terms. The enthalpy of formation terms for oxygen and nitrogen are zero, and $\Delta\bar{h} = 0$ for each of the reactants, because the fuel and combustion air enter at 25°C.

With the enthalpy of formation for $C_8H_{18}(l)$ from Table A-25

$$\bar{h}_R = (\bar{h}_f^\circ)_{C_8H_{18}(l)} = -249,910 \text{ kJ/kmol (fuel)}$$

With enthalpy of formation values for CO_2 and $H_2O(g)$ from Table A-25, and enthalpy values for N_2, H_2O, and CO_2 from Table A-23

$$\overline{h}_P = 8[-393,520 + (36,876 - 9364)] + 9[-241,820$$
$$+ (31,429 - 9,904)] + 47[26,568 - 8669]$$
$$= -4,069,466 \text{ kJ/kmol}$$

Using the molecular weight of the fuel from Table A-1, the molar flow rate of the fuel is

$$\dot{n}_F = \frac{1.8 \times 10^{-3} \text{ kg(fuel)/s}}{114.22 \text{ kg(fuel)/kmol(fuel)}} = 1.58 \times 10^{-5} \text{ kmol(fuel)/s}$$

Inserting values into the expression for the rate of heat transfer

$$\dot{Q}_{cv} = \dot{W}_{cv} + \dot{n}_F(\overline{h}_P - \overline{h}_R)$$

$$= 37 \text{ kW} + \left[1.58 \times 10^{-5} \frac{\text{kmol(fuel)}}{\text{s}}\right]$$

$$[-4,069,466 - (-249,910)]\left(\frac{\text{kJ}}{\text{kmol(fuel)}}\right)\left(\frac{1 \text{ kW}}{1 \text{ kg/s}}\right)$$

$$= -23.3 \text{ kW}$$

SKILLS DEVELOPED

Ability to...

• balance a chemical reaction equation for complete combustion of octane with theoretical air.

• apply the control volume energy balance to a reacting system.

• evaluate enthalpy values appropriately.

Quick Quiz

If the density of octane is 0.7 kg/L, how many liters of fuel would be used in 2 h of continuous operation of the engine? Ans. 18.5 L.

> > > ▶ **EXAMPLE 13.5** ▶

Analyzing a Gas Turbine Fueled with Methane

Methane (CH_4) at 25°C enters the combustor of a simple open gas turbine power plant and burns completely with 400% of theoretical air entering the compressor at 25°C, 1 atm. Products of combustion exit the turbine at 730 K, 1 atm. The rate of heat transfer from the power plant is estimated as 3% of the net power developed. Determine the net power developed, in MW, if the fuel mass flow rate is 20 kg/min. For the entering air and exiting combustion products, kinetic and potential energy effects are negligible.

SOLUTION

Known Steady-state operating data are provided for a simple gas turbine power plant.

Find The net power developed, in MW, for a given fuel mass flow rate.

Schematic and Given Data:

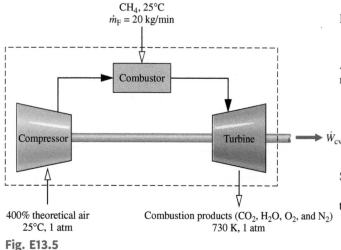

Fig. E13.5

Engineering Model

1. The control volume identified by a dashed line on the accompanying figure operates at steady state.

2. Kinetic and potential energy effects can be ignored where mass enters and exits the control volume.

3. The ideal gas model is applicable to the fuel; the combustion air and the products of combustion each form ideal gas mixtures.

4. Each mole of oxygen in the combustion air is accompanied by 3.76 moles of nitrogen, which is inert. Combustion is complete.

Analysis The balanced chemical equation for complete combustion of methane with the theoretical amount of air is given by Eq. 13.4:

$$CH_4 + 2(O_2 + 3.76N_2) \rightarrow CO_2 + 2H_2O + 7.52N_2$$

For combustion of fuel with 400% of theoretical air

$$CH_4 + (4.0)2(O_2 + 3.76N_2) \rightarrow aCO_2 + bH_2O + cO_2 + dN_2$$

Applying conservation of mass to carbon, hydrogen, oxygen, and nitrogen, respectively,

$$\begin{array}{lll} \text{C:} & & 1 = a \\ \text{H:} & & 4 = 2b \\ \text{O:} & (4.0)(2)(2) = 2a + b + 2c \\ \text{N:} & (4.0)(2)(3.76)(2) = 2d \end{array}$$

Solving these equations, $a = 1$, $b = 2$, $c = 6$, $d = 30.08$.

The balanced chemical equation for complete combustion of the fuel with 400% of theoretical air is

$$CH_4 + 8(O_2 + 3.76N_2) \rightarrow CO_2 + 2H_2O(g) + 6O_2 + 30.08N_2$$

Schematic labels:

CH_4, 25°C
$\dot{m}_F = 20$ kg/min

Combustor

Compressor

Turbine

$\dot{W}_{cv}$

400% theoretical air
25°C, 1 atm

Combustion products (CO_2, H_2O, O_2, and N_2)
730 K, 1 atm

The energy rate balance reduces, with assumptions 1–3, to give

❶

$$0 = \frac{\dot{Q}_{cv}}{\dot{n}_F} - \frac{\dot{W}_{cv}}{\dot{n}_F} + \bar{h}_R - \bar{h}_P$$

Since the rate of heat transfer from the power plant is 3% of the net power developed, we have $\dot{Q}_{cv} = -0.03\dot{W}_{cv}$. Accordingly, the energy rate balance becomes

$$\frac{1.03\dot{W}_{cv}}{\dot{n}_F} = \bar{h}_R - \bar{h}_P$$

Evaluating terms, we get

$$\frac{1.03\dot{W}_{cv}}{\dot{n}_F} = \{[\bar{h}_f^\circ + \Delta\bar{h}]^0_{CH_4} + 8[\bar{h}_f^{\circ 0} + \Delta\bar{h}]^0_{O_2} + 30.08[\bar{h}_f^{\circ 0} + \Delta\bar{h}]^0_{N_2}\}$$
$$- \{[\bar{h}_f^\circ + \Delta\bar{h}]_{CO_2} + 2[\bar{h}_f^\circ + \Delta\bar{h}]_{H_2O(g)} + 6[\bar{h}_f^{\circ 0} + \Delta\bar{h}]_{O_2}$$
$$+ 30.08[\bar{h}_f^{\circ 0} + \Delta\bar{h}]_{N_2}\}$$

where each coefficient is the same as the corresponding term of the balanced chemical equation and Eq. 13.9 has been used to evaluate enthalpy terms. The enthalpy of formation terms for oxygen and nitrogen are zero, and $\Delta\bar{h} = 0$ for each of the reactants because the fuel and combustion air enter at 25°C.

With the enthalpy of formation for $CH_4(g)$ from Table A-25

❷

$$\bar{h}_R = (\bar{h}_f^\circ)_{CH_4(g)} = -74,850 \text{ kJ/kmol(fuel)}$$

With enthalpy of formation values for CO_2 and $H_2O(g)$ from Table A-25, and enthalpy values for CO_2, H_2O, O_2, and N_2 at 730 K and 298 K from Table A-23

$$\bar{h}_P = [-393,520 + 28,622 - 9,364] + 2[-241,820 + 25,218$$
$$- 9,904] + 6[22,177 - 8,682] + 30.08[21,529 - 8,669]$$
$$\bar{h}_P = -359,475 \text{ kJ/kmol(fuel)}$$

Using the molecular weight of methane from Table A-1, the molar flow rate of the fuel is

$$\dot{n}_F = \frac{\dot{m}_F}{M_F} = \frac{20 \text{ kg(fuel)/min}}{16.04 \text{ kg(fuel)/kmol(fuel)}} \left| \frac{1 \text{ min}}{60 \text{ s}} \right|$$
$$= 0.02078 \text{ kmol(fuel)/s}$$

Inserting values into the expression for the power

$$\dot{W}_{cv} = \frac{\dot{n}_F(\bar{h}_R - \bar{h}_P)}{1.03}$$

$$\dot{W}_{cv} = \frac{\left(0.02078 \dfrac{\text{kmol(fuel)}}{\text{s}}\right)[-74,850 - (-359,475)]\dfrac{\text{kJ}}{\text{kmol(fuel)}}}{1.03}$$

$$\times \left| \frac{1 \text{ MW}}{10^3 \dfrac{\text{kJ}}{\text{s}}} \right| = 5.74 \text{ MW}$$

The positive sign indicates power is *from* the control volume.

---

❶ This expression corresponds to Eq. 13.12b.

❷ In the combustor, fuel is injected into air at a pressure greater than 1 atm because combustion air pressure has been increased in passing through the compressor. Still, since ideal gas behavior is assumed for the fuel, the fuel enthalpy is determined only by its temperature, 25°C.

SKILLS DEVELOPED

Ability to...

- balance a chemical reaction equation for complete combustion of methane with 400% of theoretical air.

- apply the control volume energy balance to a reacting system.

- evaluate enthalpy values appropriately.

Quick Quiz

Determine the net power developed, in MW, if the rate of heat transfer from the power plant is 10% of the net power developed. Ans. 5.38 MW.

···

Analyzing Closed Systems Let us consider next a closed system involving a combustion process. In the absence of kinetic and potential energy effects, the appropriate form of the energy balance is

$$U_P - U_R = Q - W$$

where U_R denotes the internal energy of the reactants and U_P denotes the internal energy of the products.

If the reactants and products form ideal gas mixtures, the energy balance can be expressed as

$$\sum_P n\bar{u} - \sum_R n\bar{u} = Q - W \tag{13.16}$$

where the coefficients n on the left side are the coefficients of the reaction equation giving the moles of each reactant or product.

Since each component of the reactants and products behaves as an ideal gas, the respective specific internal energies of Eq. 13.16 can be evaluated as $\bar{u} = \bar{h} - \bar{R}T$, so the equation becomes

$$Q - W = \sum_P n(\bar{h} - \bar{R}T_P) - \sum_R n(\bar{h} - \bar{R}T_R) \tag{13.17a}$$

where T_P and T_R denote the temperature of the products and reactants, respectively. With expressions of the form of Eq. 13.13 for each of the reactants and products, Eq. 13.17a can be written alternatively as

$$
\begin{aligned}
Q - W &= \sum_P n(\bar{h}_f^\circ + \Delta\bar{h} - \bar{R}T_P) - \sum_R n(\bar{h}_f^\circ + \Delta\bar{h} - \bar{R}T_R) \\
&= \sum_P n(\bar{h}_f^\circ + \Delta\bar{h}) - \sum_R n(\bar{h}_f^\circ + \Delta\bar{h}) - \bar{R}T_P\sum_P n + \bar{R}T_R\sum_R n \quad (13.17b)
\end{aligned}
$$

The enthalpy of formation terms are obtained from Table A-25. The $\Delta\bar{h}$ terms are evaluated from Table A-23.

The foregoing concepts are illustrated in Example 13.6, where a gaseous mixture burns in a closed, rigid container.

▶ **EXAMPLE 13.6** ▶ . . . .

Analyzing Combustion of Methane with Oxygen at Constant Volume

A mixture of 1 kmol of gaseous methane and 2 kmol of oxygen initially at 25°C and 1 atm burns completely in a closed, rigid container. Heat transfer occurs until the products are cooled to 900 K. If the reactants and products each form ideal gas mixtures, determine **(a)** the amount of heat transfer, in kJ, and **(b)** the final pressure, in atm.

SOLUTION

Known A mixture of gaseous methane and oxygen, initially at 25°C and 1 atm, burns completely within a closed rigid container. The products are cooled to 900 K.

Find Determine the amount of heat transfer, in kJ, and the final pressure of the combustion products, in atm.

Schematic and Given Data:

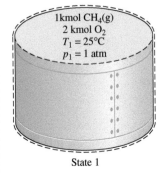
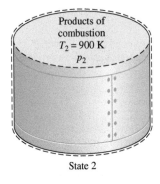

State 1   State 2

Fig. E13.6

Engineering Model

1. The contents of the closed, rigid container are taken as the system.

2. Kinetic and potential energy effects are absent, and $W = 0$.

3. Combustion is complete.

4. The initial mixture and the products of combustion each form ideal gas mixtures.

5. The initial and final states are equilibrium states.

Analysis The chemical reaction equation for the complete combustion of methane with oxygen is

$$CH_4 + 2O_2 \rightarrow CO_2 + 2H_2O(g)$$

a. With assumptions 2 and 3, the closed system energy balance takes the form

$$U_P - U_R = Q - W^0$$

or

$$Q = U_P - U_R = (1\bar{u}_{CO_2} + 2\bar{u}_{H_2O(g)}) - (1\bar{u}_{CH_4(g)} + 2\bar{u}_{O_2})$$

Each coefficient in this equation is the same as the corresponding term of the balanced chemical equation.

Since each reactant and product behaves as an ideal gas, the respective specific internal energies can be evaluated as $\bar{u} = \bar{h} - \bar{R}T$. The energy balance then becomes

$$
\begin{aligned}
Q = &[1(\bar{h}_{CO_2} - \bar{R}T_2) + 2(\bar{h}_{H_2O(g)} - \bar{R}T_2)] \\
&- [1(\bar{h}_{CH_4(g)} - \bar{R}T_1) + 2(\bar{h}_{O_2} - \bar{R}T_1)]
\end{aligned}
$$

where T_1 and T_2 denote, respectively, the initial and final temperatures. Collecting like terms

$$Q = (\bar{h}_{CO_2} + 2\bar{h}_{H_2O(g)} - \bar{h}_{CH_4(g)} - 2\bar{h}_{O_2}) + 3\bar{R}(T_1 - T_2)$$

The specific enthalpies are evaluated in terms of the respective enthalpies of formation to give

❶ $$
\begin{aligned}
Q = &[(\bar{h}_f^\circ + \Delta\bar{h})_{CO_2} + 2(\bar{h}_f^\circ + \Delta\bar{h})_{H_2O(g)} \\
&- (\bar{h}_f^\circ + \Delta\bar{h}^{\,0})_{CH_4(g)} - 2(\bar{h}_f^{\circ\,0} + \Delta\bar{h}^{\,0})_{O_2}] + 3\bar{R}(T_1 - T_2)
\end{aligned}
$$

Since the methane and oxygen are initially at 25°C, $\Delta\bar{h} = 0$ for each of these reactants. Also, $\bar{h}_f^\circ = 0$ for oxygen.

With enthalpy of formation values for CO_2, $H_2O(g)$ and $CH_4(g)$ from Table A-25 and enthalpy values for H_2O and CO_2 from Table A-23

$$
\begin{aligned}
Q = &[-393{,}520 + (37{,}405 - 9364)] + 2[-241{,}820 \\
&+ (31{,}828 - 9904)] - (-74{,}850) \\
&+ 3(8.314)(298 - 900) \\
= &-745{,}436 \text{ kJ}
\end{aligned}
$$

b. By assumption 4, the initial mixture and the products of combustion each form ideal gas mixtures. Thus, for the reactants

$$p_1 V = n_R \bar{R} T_1$$

where n_R is the total number of moles of reactants and p_1 is the initial pressure. Similarly, for the products

$$p_2 V = n_P \bar{R} T_2$$

where n_P is the total number of moles of products and p_2 is the final pressure.

Since $n_R = n_P = 3$ and volume is constant, these equations combine to give

$$p_2 = \frac{T_2}{T_1} p_1 = \left(\frac{900 \text{ K}}{298 \text{ K}} \right) (1 \text{ atm}) = 3.02 \text{ atm}$$

❶ This expression corresponds to Eq. 13.17b.

SKILLS DEVELOPED

Ability to...

- apply the closed system energy balance to a reacting system.
- evaluate property data appropriately.
- apply the ideal gas equation of state.

Quick Quiz

Calculate the volume of the system, in m³. Ans. 73.36 m³.

13.2.3 Enthalpy of Combustion and Heating Values

Although the enthalpy of formation concept underlies the formulations of the energy balances for reactive systems presented thus far, the enthalpy of formation of fuels is not always tabulated.

FOR EXAMPLE

Fuel oil and coal are normally composed of several individual chemical substances, the relative amounts of which may vary considerably, depending on the source. Owing to the wide variation in composition that these fuels can exhibit, we do not find their enthalpies of formation listed in Table A-25 or similar compilations of thermophysical data.

In many cases of practical interest, however, the *enthalpy of combustion,* which is accessible experimentally, can be used to conduct an energy analysis when enthalpy of formation data are lacking.

The **enthalpy of combustion** $\bar{h}_{RP}$ is defined as the difference between the enthalpy of the products and the enthalpy of the reactants when *complete* combustion occurs at a *given temperature and pressure.* That is,

enthalpy of combustion

$$\bar{h}_{RP} = \sum_P n_e \bar{h}_e - \sum_R n_i \bar{h}_i \qquad (13.18)$$

where the n's correspond to the respective coefficients of the reaction equation giving the moles of reactants and products per mole of fuel. When the enthalpy of combustion is expressed on a unit mass of fuel basis, it is designated h_{RP}. Tabulated values are usually given at the standard temperature T_{ref} and pressure p_{ref} introduced in Sec. 13.2.1. The symbol $\bar{h}_{RP}^{\circ}$ or h_{RP}° is used for data at this temperature and pressure.

The *heating value* of a fuel is a positive number equal to the magnitude of the enthalpy of combustion. Two heating values are recognized by name: the **higher heating value** (HHV) and the **lower heating value** (LHV). The higher heating value is obtained when all the water formed by combustion is a liquid; the lower heating value is obtained when all the water formed by combustion is a vapor. The higher heating value exceeds the lower heating value by the energy that would be released were all water in the products condensed to liquid. Values for the HHV and LHV also depend on whether the fuel is a liquid or a gas. Heating value data for several hydrocarbons are provided in Table A-25.

higher and lower heating values

The calculation of the enthalpy of combustion, and the associated heating value, using table data is illustrated in the next example.

Calculating Enthalpy of Combustion of Methane

Calculate the enthalpy of combustion of gaseous methane, in kJ per kg of fuel, **(a)** at 25°C, 1 atm with liquid water in the products, **(b)** at 25°C, 1 atm with water vapor in the products. **(c)** Repeat part (b) at 1000 K, 1 atm.

SOLUTION

Known   The fuel is gaseous methane.

Find   Determine the enthalpy of combustion, in kJ per kg of fuel, (a) at 25°C, 1 atm with liquid water in the products, (b) at 25°C, 1 atm with water vapor in the products, (c) at 1000 K, 1 atm with water vapor in the products.

Engineering Model

1. Each mole of oxygen in the combustion air is accompanied by 3.76 moles of nitrogen, which is inert.

2. Combustion is complete, and both reactants and products are at the same temperature and pressure.

3. The ideal gas model applies for methane, the combustion air, and the gaseous products of combustion.

Analysis   The combustion equation is obtained from Eq. 13.4

$$CH_4 + 2O_2 + 7.52N_2 \rightarrow CO_2 + 2H_2O + 7.52N_2$$

Using Eq. 13.9 in Eq. 13.18, the enthalpy of combustion is

$$\bar{h}_{RP} = \sum_P n_e(\bar{h}_f^\circ + \Delta\bar{h})_e - \sum_R n_i(\bar{h}_f^\circ + \Delta\bar{h})_i$$

Introducing the coefficients of the combustion equation and evaluating the specific enthalpies in terms of the respective enthalpies of formation

$$\bar{h}_{RP} = \bar{h}_{CO_2} + 2\bar{h}_{H_2O} - \bar{h}_{CH_4(g)} - 2\bar{h}_{O_2}$$
$$= (\bar{h}_f^\circ + \Delta\bar{h})_{CO_2} + 2(\bar{h}_f^\circ + \Delta\bar{h})_{H_2O}$$
$$- (\bar{h}_f^\circ + \Delta\bar{h})_{CH_4(g)} - 2(\overset{0}{\cancel{\bar{h}_f^\circ}} + \Delta\bar{h})_{O_2}$$

For nitrogen, the enthalpy terms of the reactants and products cancel. Also, the enthalpy of formation of oxygen is zero by definition. On rearrangement, the enthalpy of combustion expression becomes

$$\bar{h}_{RP} = (\bar{h}_f^\circ)_{CO_2} + 2(\bar{h}_f^\circ)_{H_2O} - (\bar{h}_f^\circ)_{CH_4(g)}$$
$$+ [(\Delta\bar{h})_{CO_2} + 2(\Delta\bar{h})_{H_2O} - (\Delta\bar{h})_{CH_4(g)} - 2(\Delta\bar{h})_{O_2}]$$
$$= \bar{h}_{RP}^\circ + [(\Delta\bar{h})_{CO_2} + 2(\Delta\bar{h})_{H_2O} - (\Delta\bar{h})_{CH_4(g)} - 2(\Delta\bar{h})_{O_2}] \quad (1)$$

The values for $\bar{h}_{RP}^\circ$ and $(\Delta\bar{h})_{H_2O}$ depend on whether the water in the products is a liquid or a vapor.

a. Since the reactants and products are at 25°C, 1 atm in this case, the $\Delta\bar{h}$ terms drop out of Eq. (1) giving the expression for $\bar{h}_{RP}$. Thus, for liquid water in the products, the enthalpy of combustion is

$$\bar{h}_{RP}^\circ = (\bar{h}_f^\circ)_{CO_2} + 2(\bar{h}_f^\circ)_{H_2O(l)} - (\bar{h}_f^\circ)_{CH_4(g)}$$

With enthalpy of formation values from Table A-25

$$\bar{h}_{RP}^\circ = -393,520 + 2(-285,830) - (-74,850)$$
$$= -890,330 \text{ kJ/kmol (fuel)}$$

Dividing by the molecular weight of methane places this result on a unit mass of fuel basis

$$h_{RP}^\circ = \frac{-890,330 \text{ kJ/kmol (fuel)}}{16.04 \text{ kg (fuel)/kmol (fuel)}} = -55,507 \text{ kJ/kg (fuel)}$$

The magnitude of this value agrees with the higher heating value of methane given in Table A-25.

b. As in part (a), the $\Delta\bar{h}$ terms drop out of the expression for $\bar{h}_{RP}$, Eq. (1), which for water vapor in the products reduces to $\bar{h}_{RP}^\circ$, where

$$\bar{h}_{RP}^\circ = (\bar{h}_f^\circ)_{CO_2} + 2(\bar{h}_f^\circ)_{H_2O(g)} - (\bar{h}_f^\circ)_{CH_4(g)}$$

With enthalpy of formation values from Table A-25

$$\bar{h}_{RP}^\circ = -393,520 + 2(-241,820) - (-74,850)$$
$$= -802,310 \text{ kJ/kmol (fuel)}$$

On a unit of mass of fuel basis, the enthalpy of combustion for this case is

$$h_{RP}^\circ = \frac{-802,310}{16.04} = -50,019 \text{ kJ/kg (fuel)}$$

The magnitude of this value agrees with the lower heating value of methane given in Table A-25.

c. For the case where the reactants and products are at 1000 K, 1 atm, the term $\bar{h}_{RP}^\circ$ in Eq. (1) giving the expression for $\bar{h}_{RP}$ has the value determined in part (b): $\bar{h}_{RP}^\circ = -802,310$ kJ/kmol (fuel), and the $\Delta\bar{h}$ terms for O_2, $H_2O(g)$, and CO_2 can be evaluated using specific enthalpies at 298 and 1000 K from Table A-23. The results are

$$(\Delta\bar{h})_{O_2} = 31,389 - 8682 = 22,707 \text{ kJ/kmol}$$
$$(\Delta\bar{h})_{H_2O(g)} = 35,882 - 9904 = 25,978 \text{ kJ/kmol}$$
$$(\Delta\bar{h})_{CO_2} = 42,769 - 9364 = 33,405 \text{ kJ/kmol}$$

For methane, the $\bar{c}_p$ expression of Table A-21 can be used to obtain

$$(\Delta\bar{h})_{CH_4(g)} = \int_{298}^{1000} \bar{c}_p \, dT$$
$$= \bar{R} \left(3.826T - \frac{3.979}{10^3}\frac{T^2}{2} + \frac{24.558}{10^6}\frac{T^3}{3} \right.$$
$$\left. - \frac{22.733}{10^9}\frac{T^4}{4} + \frac{6.963}{10^{12}}\frac{T^5}{5} \right)\Big|_{298}^{1000}$$

①
$$= 38,189 \text{ kJ/kmol (fuel)}$$

Substituting values into the expression for the enthalpy of combustion, Eq. (1), we get

$$\bar{h}_{RP} = -802,310 + [33,405 + 2(25,978) - 38,189 - 2(22,707)]$$
$$= -800,522 \text{ kJ/kmol (fuel)}$$

On a unit mass basis

② $\quad h_{RP} = \dfrac{-800,552}{16.04} = -49,910 \text{ kJ/kg (fuel)}$

❶ Using *Interactive Thermodynamics: IT,* we get 38,180 kJ/kmol (fuel).

❷ Comparing the values of parts (b) and (c), the enthalpy of combustion of methane is seen to vary little with temperature. The same is true for many hydrocarbon fuels. This fact is sometimes used to simplify combustion calculations.

SKILLS DEVELOPED

Ability to...

- calculate the enthalpy of combustion at standard temperature and pressure.
- calculate the enthalpy of combustion at an elevated temperature and standard pressure.

Quick Quiz

What is the lower heating value of methane, in kJ/kg (fuel) at 25°C, 1 atm? Ans. 50,020 kJ/kg (Table A-25).

Evaluating Enthalpy of Combustion by Calorimetry When enthalpy of formation data are available for *all* the reactants and products, the enthalpy of combustion can be calculated directly from Eq. 13.18, as illustrated in Example 13.7. Otherwise, it must be obtained experimentally using devices known as *calorimeters*. Both constant-volume (bomb calorimeters) and flow-through devices are employed for this purpose. Consider as an illustration a reactor operating at steady state in which the fuel is burned completely with air. For the products to be returned to the same temperature as the reactants, a heat transfer from the reactor would be required. From an energy rate balance, the required heat transfer is

$$\frac{\dot{Q}_{cv}}{\dot{n}_F} = \sum_P n_e \overline{h}_e - \sum_R n_i \overline{h}_i \tag{13.19}$$

where the symbols have the same significance as in previous discussions. The heat transfer per mole of fuel, $\dot{Q}_{cv}/\dot{n}_F$, would be determined from measured data. Comparing Eq. 13.19 with the defining equation, Eq. 13.18, we have $\overline{h}_{RP} = \dot{Q}_{cv}/\dot{n}_F$. In accord with the usual sign convention for heat transfer, the enthalpy of combustion would be negative.

As noted previously, the enthalpy of combustion can be used for energy analyses of reacting systems.

FOR EXAMPLE

Consider a control volume at steady state in which a fuel oil reacts completely with air. The energy rate balance is given by Eq. 13.15b

$$\frac{\dot{Q}_{cv}}{\dot{n}_F} - \frac{\dot{W}_{cv}}{\dot{n}_F} = \sum_P n_e (\overline{h}_f^\circ + \Delta \overline{h})_e - \sum_R n_i (\overline{h}_f^\circ + \Delta \overline{h})_i$$

All symbols have the same significance as in previous discussions. This equation can be rearranged to read

$$\frac{\dot{Q}_{cv}}{\dot{n}_F} - \frac{\dot{W}_{cv}}{\dot{n}_F} = \underline{\sum_P n_e (\overline{h}_f^\circ)_e - \sum_R n_i (\overline{h}_f^\circ)_i} + \sum_P n_e (\Delta \overline{h})_e - \sum_R n_i (\Delta \overline{h})_i$$

For a complete reaction, the underlined term is just the enthalpy of combustion $\overline{h}_{RP}^\circ$, at T_{ref} and p_{ref}. Thus, the equation becomes

$$\frac{\dot{Q}_{cv}}{\dot{n}_F} - \frac{\dot{W}_{cv}}{\dot{n}_F} = \overline{h}_{RP}^\circ + \sum_P n_e (\Delta \overline{h})_e - \sum_R n_i (\Delta \overline{h})_i \tag{13.20}$$

The right side of Eq. 13.20 can be evaluated with an experimentally determined value for $\overline{h}_{RP}^\circ$ and $\Delta \overline{h}$ values for the reactants and products determined as discussed previously.

adiabatic flame temperature

13.3 Determining the Adiabatic Flame Temperature

Let us reconsider the reactor at steady state pictured in Fig. 13.2. In the absence of work $\dot{W}_{cv}$ and appreciable kinetic and potential energy effects, the energy liberated on combustion is transferred from the reactor in two ways only: by energy accompanying the exiting combustion products and by heat transfer to the surroundings. The smaller the heat transfer, the greater the energy carried out with the combustion products and thus the greater the temperature of the products. The temperature that would be achieved by the products in the limit of adiabatic operation of the reactor is called the **adiabatic flame temperature** or **adiabatic combustion** temperature.

The adiabatic flame temperature can be determined by use of the conservation of mass and conservation of energy principles. To illustrate the procedure, let us suppose that the combustion air and the combustion products each form ideal gas mixtures. Then, with the other assumptions stated previously, the energy rate balance on a per mole of fuel basis, Eq. 13.12b, reduces to the form $\bar{h}_P = \bar{h}_R$—that is,

$$\sum_P n_e \bar{h}_e = \sum_R n_i \bar{h}_i \tag{13.21a}$$

where i denotes the incoming fuel and air streams and e the exiting combustion products. With this expression, the adiabatic flame temperature can be determined using table data or computer software, as follows.

13.3.1 Using Table Data

When using Eq. 13.9 with table data to evaluate enthalpy terms, Eq. 13.21a takes the form

$$\sum_P n_e (\bar{h}_f^\circ + \Delta\bar{h})_e = \sum_R n_i (\bar{h}_f^\circ + \Delta\bar{h})_i$$

or

$$\sum_P n_e (\Delta\bar{h})_e = \sum_R n_i (\Delta\bar{h})_i + \sum_R n_i \bar{h}_{fi}^\circ - \sum_P n_e \bar{h}_{fe}^\circ \tag{13.21b}$$

The n's are obtained on a per mole of fuel basis from the balanced chemical reaction equation. The enthalpies of formation of the reactants and products are obtained from Table A-25. Enthalpy of combustion data might be employed in situations where the enthalpy of formation for the fuel is not available. Knowing the states of the reactants as they enter the reactor, the $\Delta\bar{h}$ terms for the reactants can be evaluated as discussed previously. Thus, all terms on the right side of Eq. 13.21b can be evaluated. The terms $(\Delta\bar{h})_e$ on the left side account for the changes in enthalpy of the products from T_{ref} to the unknown adiabatic flame temperature. Since the unknown temperature appears in each term of the sum on the left side of the equation, determination of the adiabatic flame temperature requires *iteration:* A temperature for the products is assumed and used to evaluate the left side of Eq. 13.21b. The value obtained is compared with the previously determined value for the right side of the equation. The procedure continues until satisfactory agreement is attained. Example 13.8 gives an illustration.

13.3.2 Using Computer Software

TAKE NOTE...

The adiabatic flame temperature can be determined iteratively using table data or *IT*. See Example 13.8.

Thus far we have emphasized the use of Eq. 13.9 together with table data when evaluating the specific enthalpies required by energy balances for reacting systems. Such enthalpy values also can be retrieved using *Interactive Thermodynamics: IT*. With *IT*, the quantities on the right side of Eq. 13.9 are evaluated by software, and $\bar{h}$ data are returned *directly*.

> **FOR EXAMPLE**
>
> Consider CO_2 at 500 K modeled as an ideal gas. The specific enthalpy is obtained from *IT* as follows:
>
> ```
> T = 500 // K
> h = h_T("CO₂", T)
> ```

Choosing K for the temperature unit and moles for the amount under the Units menu, *IT* returns $h = -3.852 \times 10^5$ kJ/kmol.

This value agrees with the value calculated from Eq. 13.9 using enthalpy data for CO_2 from Table A-23, as follows

$$\bar{h} = \bar{h}_f^\circ + [\bar{h}(500 \text{ K}) - \bar{h}(298 \text{ K})]$$
$$= -393,520 + [17,678 - 9364]$$
$$= -3.852 \times 10^5 \text{ kJ/kmol}$$

As suggested by this discussion, *IT* is also useful for analyzing reacting systems. In particular, the equation solver and property retrieval features of *IT* allow the adiabatic flame temperature to be determined without the iteration required when using table data.

In Example 13.8, we show how the adiabatic flame temperature can be determined iteratively using table data or *Interactive Thermodynamics: IT*.

EXAMPLE 13.8

Determining the Adiabatic Flame Temperature for Complete Combustion of Liquid Octane

Liquid octane at 25°C, 1 atm enters a well-insulated reactor and reacts with air entering at the same temperature and pressure. For steady-state operation and negligible effects of kinetic and potential energy, determine the temperature of the combustion products for complete combustion with **(a)** the theoretical amount of air, **(b)** 400% theoretical air.

SOLUTION

Known Liquid octane and air, each at 25°C and 1 atm, burn completely within a well-insulated reactor operating at steady state.

Find Determine the temperature of the combustion products for (a) the theoretical amount of air and (b) 400% theoretical air.

Schematic and Given Data:

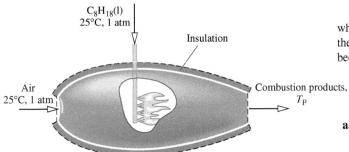

Fig. E13.8

Engineering Model

1. The control volume indicated on the accompanying figure by a dashed line operates at steady state.

2. For the control volume, $\dot{Q}_{cv} = 0$, $\dot{W}_{cv} = 0$, and kinetic and potential effects are negligible.

3. The combustion air and the products of combustion each form ideal gas mixtures.

4. Combustion is complete.

5. Each mole of oxygen in the combustion air is accompanied by 3.76 moles of nitrogen, which is inert.

Analysis At steady state, the control volume energy rate balance Eq. 13.12b reduces with assumptions 2 and 3 to give Eq. 13.21a:

$$\sum_P n_e \bar{h}_e = \sum_R n_i \bar{h}_i \tag{1}$$

When Eq. 13.9 and table data are used to evaluate the enthalpy terms, Eq. (1) is written as

$$\sum_P n_e (\bar{h}_f^\circ + \Delta\bar{h})_e = \sum_R n_i (\bar{h}_f^\circ + \Delta\bar{h})_i$$

On rearrangement, this becomes

$$\sum_P n_e (\Delta\bar{h})_e = \sum_R n_i (\Delta\bar{h})_i + \sum_R n_i \bar{h}_{fi}^\circ - \sum_P n_e \bar{h}_{fe}^\circ$$

which corresponds to Eq. 13.21b. Since the reactants enter at 25°C, the $(\Delta\bar{h})_i$ terms on the right side vanish, and the energy rate equation becomes

$$\sum_P n_e (\Delta\bar{h})_e = \sum_R n_i \bar{h}_{fi}^\circ - \sum_P n_e \bar{h}_{fe}^\circ \tag{2}$$

a. For combustion of liquid octane with the theoretical amount of air, the chemical equation is

$$C_8H_{18}(l) + 12.5O_2 + 47N_2 \rightarrow 8CO_2 + 9H_2O(g) + 47N_2$$

Introducing the coefficients of this equation, Eq. (2) takes the form

$$8(\Delta\bar{h})_{CO_2} + 9(\Delta\bar{h})_{H_2O(g)} + 47(\Delta\bar{h})_{N_2}$$
$$= [(\bar{h}_f^\circ)_{C_8H_{18}(l)} + 12.5(\bar{h}_f^\circ)_{O_2}^0 + 47(\bar{h}_f^\circ)_{N_2}^0]$$
$$- [8(\bar{h}_f^\circ)_{CO_2} + 9(\bar{h}_f^\circ)_{H_2O(g)} + 47(\bar{h}_f^\circ)_{N_2}^0]$$

The right side of the above equation can be evaluated with enthalpy of formation data from Table A-25, giving

$$8(\Delta\bar{h})_{CO_2} + 9(\Delta\bar{h})_{H_2O(g)} + 47(\Delta\bar{h})_{N_2} = 5,074,630 \text{ kJ/kmol(fuel)}$$

Each $\Delta\bar{h}$ term on the left side of this equation depends on the temperature of the products, T_P. This temperature can be determined by an iterative procedure.

The following table gives a summary of the iterative procedure for three trial values of T_P. Since the summation of the enthalpies of the products equals 5,074,630 kJ/kmol, the actual value of T_P is in the interval from 2350 to 2400 K. Interpolation between these temperatures gives $T_P = 2395$ K.

T_P	2500 K	2400 K	2350 K
$8(\Delta\bar{h})_{CO_2}$	975,408	926,304	901,816
$9(\Delta\bar{h})_{H_2O(g)}$	890,676	842,436	818,478
$47(\Delta\bar{h})_{N_2}$	3,492,664	3,320,597	3,234,869
$\sum\limits_{P} n_e(\Delta\bar{h})_e$	5,358,748	5,089,337	4,955,163

Alternative Solution

The following *IT* code can be used as an alternative to iteration with table data, where hN2_R and hN2_P denote the enthalpy of N_2 in the reactants and products, respectively, and so on. In the **Units** menu, select temperature in K and amount of substance in moles.

```
TR = 25 + 273.15 // K
// Evaluate reactant and product enthalpies,
hR and hP, respectively
hR = hC8H18 + 12.5 * hO2_R + 47 * hN2_R
hP = 8 * hCO2_P + 9 * hH2O_P + 47 * hN2_P
hC8H18 = −249910 // kj/kmol (value from Table
A-25)
hO2_R = h_T("O2",TR)
hN2_R = h_T("N2",TR)
hCO2_P = h_T("CO2",TP)
hH2O_P = h_T("H2O",TP)
hN2_P = h_T("N2",TP)
// Energy balance
hP = hR
```

Using the **Solve** button, the result is TP = 2394 K, which agrees closely with the result obtained above.

b. For complete combustion of liquid octane with 400% theoretical air, the chemical equation is

$$C_8H_{18}(l) + 50O_2 + 188N_2 \rightarrow$$
$$8CO_2 + 9H_2O(g) + 37.5O_2 + 188N_2$$

The energy rate balance, Eq. (2), reduces for this case to

$$8(\Delta\bar{h})_{CO_2} + 9(\Delta\bar{h})_{H_2O(g)} + 37.5(\Delta\bar{h})_{O_2} + 188(\Delta\bar{h})_{N_2}$$
$$= 5,074,630 \text{ kJ/kmol (fuel)}$$

① Observe that the right side has the same value as in part (a). Proceeding iteratively as above, the temperature of the products is $T_P = 962$ K. The use of *IT* to solve part (b) is left as an exercise.

① The temperature determined in part (b) is considerably lower than the value found in part (a). This shows that once enough oxygen has been provided for complete combustion, bringing in more air dilutes the combustion products, lowering their temperature.

SKILLS DEVELOPED

Ability to...

- apply the control volume energy balance to calculate the adiabatic flame temperature.
- evaluate enthalpy values appropriately.

Quick Quiz

If octane gas entered instead of liquid octane, would the adiabatic flame temperature increase, decrease, or stay the same? Ans. Increase.

13.3.3 Closing Comments

For a specified fuel and specified temperature and pressure of the reactants, the *maximum* adiabatic flame temperature is for complete combustion with the theoretical amount of air. The measured value of the temperature of the combustion products may be several hundred degrees below the maximum adiabatic flame temperature, however, for several reasons:

- Once adequate oxygen has been provided to permit complete combustion, bringing in more air dilutes the combustion products, lowering their temperature.
- Incomplete combustion also tends to reduce the temperature of the products, and combustion is seldom complete (see Sec. 14.4).
- Heat losses can be reduced but not altogether eliminated.
- As a result of the high temperatures achieved, some of the combustion products may dissociate. Endothermic dissociation reactions lower the product temperature. The effect of dissociation on the adiabatic flame temperature is considered in Sec. 14.4.

13.4 Fuel Cells

A **fuel cell** is an *electrochemical* device in which fuel and an oxidizer (normally oxygen from air) undergo a chemical reaction, providing electrical current to an external circuit and producing products. The fuel and oxidizer react catalytically in stages on separate electrodes: the anode and the cathode. An electrolyte separating the two electrodes allows passage of ions formed by reaction. Depending on the type of fuel cell, the ions may be positively or negatively charged. Individual fuel cells are connected in parallel or series to form fuel cell *stacks* to provide the desired level of power output.

Fuel cell

With today's technology, the preferred fuel for oxidation at the fuel cell anode is hydrogen because of its exceptional ability to produce electrons when suitable catalysts are used, while producing no harmful emissions from the fuel cell itself. Depending on the type of fuel cell, methanol (CH_3OH) and carbon monoxide (CO) can be oxidized at the anode in some applications but often with performance penalties.

Since hydrogen is not naturally occurring, it must be produced. Production methods include electrolysis of water (see Sec. 2.7) and chemically reforming hydrogen-bearing fuels, predominantly hydrocarbons. See the following box.

Hydrogen Production by Reforming of Hydrocarbons

Steam reforming of natural gas is currently the most common method of producing hydrogen. To illustrate this simply, consider steam reforming of methane, typically the largest component of natural gas.

Methane undergoes an *endothermic* reaction with steam to yield syngas (synthesis gas) consisting of H_2 and CO:

$$CH_4 + H_2O(g) \rightarrow 3\,H_2 + CO$$

In a second step, additional hydrogen is produced using the *exothermic* water–gas shift reaction

$$CO + H_2O(g) \rightarrow H_2 + CO_2$$

The shift reaction also eliminates carbon monoxide, which poisons platinum catalysts used in some fuel cells to promote reaction rates.

Hydrocarbon reforming can occur either separately or within the fuel cell (depending on type). Hydrogen produced by reforming fuel separately from the fuel cell itself is known as *external reforming*. If not fed directly from the reformer to a fuel cell, hydrogen can be stored as a compressed gas, a cryogenic liquid, or atoms absorbed within metallic structures and then provided to fuel cells from storage, when required. *Internal reforming* refers to applications where hydrogen production by reforming fuel is integrated within the fuel cell. Owing to limitations of current technology, internal reforming is feasible only in fuel cells operating at temperatures above about 600°C.

Rates of reaction in fuel cells are limited by the time it takes for diffusion of chemical species through the electrodes and the electrolyte and by the speed of the chemical reactions themselves. The reaction in a fuel cell is *not* a combustion process. These features result in fuel cell internal irreversibilities that are inherently less significant than those encountered in power systems employing combustion. Thus, fuel cells have the *potential* of providing more power from a given supply of fuel and oxidizer than conventional internal combustion engines and gas turbines.

Fuel cells do not operate as thermodynamic power cycles, and thus the notion of a limiting thermal efficiency imposed by the second law is not applicable. However, as for all power systems, the power provided by fuel cell *systems* is eroded by inefficiencies in auxiliary equipment. For fuel cells this includes heat exchangers, compressors, and humidifiers. Irreversibilities and losses inherent in hydrogen production also can be greater than those seen in production of more conventional fuels.

In comparison to reciprocating internal combustion engines and gas turbines that incorporate combustion, fuel cells typically produce relatively few damaging emissions as they develop power. Still, such emissions accompany production of fuels consumed by fuel cells as well as the manufacture of fuel cells and their supporting components.

Despite potential thermodynamic advantages, widespread fuel cell use has not occurred thus far owing primarily to cost. Table 13.2 summarizes the most promising fuel cell

TAKE NOTE...

Hydrogen production by electrolysis of water and by reforming of hydrocarbons are each burdened by the second law. Significant exergy destruction is observed with each method of production.

Characteristics of Major Fuel Cell Types

	Proton Exchange Membrane Fuel Cell (PEMFC)	Phosphoric Acid Fuel Cell (PAFC)	Molten Carbonate Fuel Cell (MCFC)	Solid Oxide Fuel Cell (SOFC)
Transportation application	Automotive power	Large vehicle power	None	Vehicle auxiliary power Heavy vehicle propulsion
Other applications	Portable power Small-scale stationary power	On-site cogeneration Electric power generation	On-site cogeneration Electric power generation	On-site cogeneration Electric power generation
Electrolyte	Ion exchange membrane	Liquid phosphoric acid	Liquid molten carbonate	Solid oxide ceramic
Charge carrier	H^+	H^+	$CO_3^=$	$O^=$
Operating temperature	60–80°C	150–220°C	600–700°C	600–1000°C
Fuel oxidized at anode	H_2 or methanol	H_2	H_2	H_2 or CO
Fuel reforming	External	External	Internal or external	Internal or external
Fuel typically used for internal reforming	None	None	CO Light hydrocarbons (e.g., methane, propane) Methanol	Light hydrocarbons (e.g., methane, propane) Synthetic diesel and gasoline

Sources: *Fuel Cell Handbook*, Seventh Edition, 2004, EG&G Technical Services, Inc., DOE Contract No. DE-AM26-99FT40575. Larminie, J., and Dicks, A., 2000, *Fuel Cell Systems Explained*, John Wiley & Sons, Ltd., Chichester, West Sussex, England.

technologies currently under consideration. Included are potential applications and other characteristics.

Cooperative efforts by government and industry have fostered advances in both proton exchange membrane fuel cells and solid oxide fuel cells, which appear to provide the greatest range of potential applications in transportation, portable power, and stationary power. The proton exchange membrane fuel cell and the solid oxide fuel cell are discussed next.

13.4.1 Proton Exchange Membrane Fuel Cell

The fuel cells shown in Fig. 13.3 are *proton exchange membrane* fuel cells (PEMFCs). At the anode, hydrogen ions (H^+) and electrons (e^-) are produced. At the cathode, oxygen, hydrogen ions, and electrons react to produce water.

- The fuel cell shown schematically in Fig. 13.3a operates with hydrogen (H_2) as the fuel and oxygen (O_2) as the oxidizer. The reactions at these electrodes and the *overall* cell reaction are labeled on the figure. The only products of this fuel cell are water, power generated, and waste heat.
- The fuel cell shown schematically in Fig. 13.3b operates with humidified methanol ($CH_3OH + H_2O$) as the fuel and oxygen (O_2) as the oxidizer. This type of PEMFC is a *direct-methanol fuel cell*. The reactions at these electrodes and the *overall* cell reaction are labeled on the figure. The only products of this fuel cell are water, carbon dioxide, power generated, and waste heat.

For PEMFCs, charge-carrying hydrogen ions are conducted through the electrolytic membrane. For acceptable ion conductivity, high membrane water content is required. This requirement restricts the fuel cell to operating below the boiling point of water, so PEMFCs typically operate at temperatures in the range 60–80°C. Cooling is generally needed to maintain the fuel cell at the operating temperature.

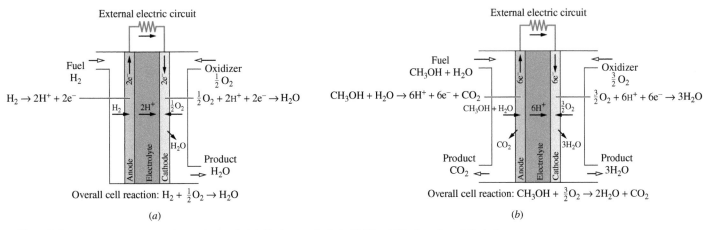

Fig. 13.3 Proton exchange membrane fuel cell. (*a*) Hydrogen fueled. (*b*) Humidified-methanol fueled.

A Bright Future for Fuel Cells?

Fuel cells pushed into public awareness during the first decades of the U.S. space program. Observers learned then that fuel cells were being used in orbit, on the moon, and in outer space, producing power for various tasks and drinking water for crew members. More recently, in the early years of the twenty-first century, fuel cells once again enjoyed broad popular appeal owing to their anticipated deployment in vehicles for personal transportation. Fuel cells seemed ideal for powering cars because at the point of use they exhibit high second law efficiencies and low emissions. But the lack of infrastructure to fuel them and other issues have greatly slowed entry of fuel cell vehicles into the marketplace. Interest among the public also subsided as automakers began to favor hybrid and all-electric vehicles. With such a history, many are wondering what the future

holds for fuel cells. Some say fuel cells for portable electronics will be the first fuel cells most of us will actually encounter because of strong consumer demand for cost-competitive, long-lasting, instantly rechargeable power. Yet without much fanfare commercial-scale fuel cells currently provide emergency power in the event of outages. They are also deployed increasingly for cogeneration duty, including industrial and commercial applications. Fuel cells still may play an important role in transportation, powering buses, fleet vehicles, and ships, for example.

Looking more to the future, the U.S. Department of Energy actively encourages fuel cell innovations through the Energy Efficiency and Renewable Energy (EERE) program and has identified specific goals for fuel cell technology through 2020.

Owing to the relatively low-temperature operation of proton exchange membrane fuel cells, hydrogen derived from hydrocarbon feedstock must be produced using external reforming, while costly platinum catalysts are required at both the anode and cathode to increase ionization reaction rates. Due to an extremely slow reaction rate at the anode, the direct-methanol fuel cell requires several times as much platinum catalyst as the hydrogen-fueled PEMFC to improve the anode reaction rate. Catalytic activity is more important in lower-temperature fuel cells because rates of reaction at the anode and cathode tend to decrease with decreasing temperature.

Automakers continue to strive toward the production of market-ready hydrogen-fueled proton exchange membrane fuel cell vehicles. Still, just two such fuel cell vehicles are currently available in the United States. To promote further fuel cell development major automakers have formed partnerships among themselves. Thus far, several major brands have formed three distinct partnerships, each having two or three automakers as members. Both hydrogen-fueled and direct-methanol PEMFCs have potential to replace batteries in portable devices such as cellular phones, laptop computers, and video players. Hurdles to wider deployment of PEMFCs include extending stack life, simplifying system integration, and reducing costs.

13.4.2 Solid Oxide Fuel Cell

For scale, **Fig. 13.4a** shows a *solid oxide* fuel cell (SOFC) module. The fuel cell schematic shown in Fig. 13.4b operates with hydrogen (H_2) as the fuel and oxygen (O_2) as the oxidizer. At the anode, water (H_2O) and electrons (e^-) are produced. At the cathode, oxygen reacts with electrons (e^-) to produce oxygen ions ($O^=$) that migrate through the electrolyte to the anode.

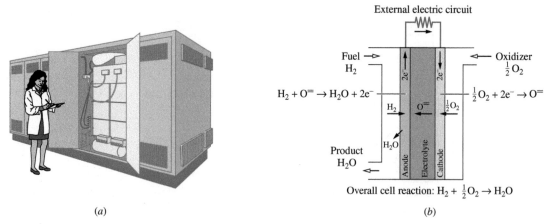

Fig. 13.4 Solid oxide fuel cell. (*a*) Module. (*b*) Schematic.

The reactions at these electrodes and the *overall* cell reaction are labeled on the figure. The only products of this fuel cell are water, power generated, and waste heat.

For SOFCs, an alternative fuel to hydrogen is carbon monoxide (CO) that produces carbon dioxide (CO_2) and electrons (e^-) during oxidation at the anode. The cathode reaction is the same as that in Fig. 13.4*b*. Due to their high-temperature operation, solid oxide fuel cells can incorporate internal reforming of various hydrocarbon fuels to produce hydrogen and/or carbon monoxide at the anode.

Since waste heat is produced at relatively high temperature, solid oxide fuel cells can be used for cogeneration of power and process heat or steam. SOFCs also can be used for distributed (decentralized) power generation and for fuel cell–microturbine hybrids. Such applications are very attractive because they achieve objectives without using highly irreversible combustion.

For instance, a solid oxide fuel cell replaces the combustor in the gas turbine shown in the fuel cell–microturbine schematic in **Fig. 13.5**. The fuel cell produces electric power while its high-temperature exhaust expands through the microturbine, producing shaft power $\dot{W}_{net}$. By producing power electrically *and* mechanically without combustion, fuel cell–microturbine hybrids have the potential of significantly improving effectiveness of fuel utilization over that achievable with comparable conventional gas turbine technology *and* with fewer harmful emissions.

13.5 Absolute Entropy and the Third Law of Thermodynamics

Thus far, our analyses of reacting systems have been conducted using the conservation of mass and conservation of energy principles. In the present section some of the implications of the second law of thermodynamics for reacting systems are considered. The discussion continues in

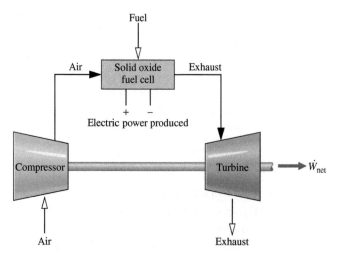

Fig. 13.5 Solid oxide fuel cell–microturbine hybrid.

the second part of this chapter dealing with the exergy concept and in the next chapter where the subject of chemical equilibrium is taken up.

13.5.1 Evaluating Entropy for Reacting Systems

The property entropy plays an important part in quantitative evaluations using the second law of thermodynamics. When reacting systems are under consideration, the same problem arises for entropy as for enthalpy and internal energy: A common datum must be used to assign entropy values for each substance involved in the reaction. This is accomplished using the *third law* of thermodynamics and the *absolute entropy* concept.

The **third law** deals with the entropy of substances at the absolute zero of temperature. Based on empirical evidence, this law states that the entropy of a pure crystalline substance is zero at the absolute zero of temperature, 0 K. Substances not having a pure crystalline structure at absolute zero have a nonzero value of entropy at absolute zero. The experimental evidence on which the third law is based is obtained primarily from studies of chemical reactions at low temperatures and specific heat measurements at temperatures approaching absolute zero.

third law of thermodynamics

Absolute Entropy For present considerations, the importance of the third law is that it provides a datum relative to which the entropy of each substance participating in a reaction can be evaluated so that no ambiguities or conflicts arise. The entropy relative to this datum is called the **absolute entropy**. The change in entropy of a substance between absolute zero and any given state can be determined from precise measurements of energy transfers and specific heat data or from procedures based on statistical thermodynamics and observed molecular data.

absolute entropy

Table A-25 gives the value of the absolute entropy for selected substances at the standard reference state, $T_{ref} = 298.15$ K, $p_{ref} = 1$ atm, in units of kJ/kmol · K. Two values of absolute entropy for water are provided. One is for liquid water and the other is for water vapor. As for the case of the enthalpy of formation of water considered in Sec. 13.2.1, the vapor value listed is for a *hypothetical* ideal gas state in which water is a vapor at 25°C and $p_{ref} = 1$ atm.

TAKE NOTE...

Entropy data retrieved from Tables A-2 through A-18 are not absolute values.

Table A-23 gives tabulations of absolute entropy versus temperature at a pressure of 1 atm for selected gases. In these tables, the absolute entropy at 1 atm and temperature T is designated $\bar{s}°(T)$, and ideal gas behavior is assumed for the gases.

Using Absolute Entropy When the absolute entropy is known at the standard state, the specific entropy at any other state can be found by adding the specific entropy change between the two states to the absolute entropy at the standard state. Similarly, when the absolute entropy is known at the pressure p_{ref} and temperature T, the absolute entropy at the same temperature and any pressure p can be found from

$$\bar{s}(T, p) = \bar{s}(T, p_{ref}) + [\bar{s}(T, p) - \bar{s}(T, p_{ref})]$$

For the ideal gases listed in Table A-23, the first term on the right side of this equation is $\bar{s}°(T)$, and the second term on the right can be evaluated using Eq. 6.18. Collecting results, we get

$$\boxed{\bar{s}(T, p) = \bar{s}°(T) - \bar{R} \ln \frac{p}{p_{ref}} \qquad \text{(ideal gas)}} \qquad (13.22)$$

To reiterate, $\bar{s}°(T)$ is the absolute entropy at temperature T and pressure $p_{ref} = 1$ atm.

The entropy of the ith component of an ideal gas mixture is evaluated at the mixture temperature T and the *partial* pressure $p_i : \bar{s}_i(T, p_i)$. The partial pressure is given by $p_i = y_i p$, where y_i is the mole fraction of component i and p is the mixture pressure. Thus, Eq. 13.22 takes the form

$$\bar{s}_i(T, p_i) = \bar{s}_i°(T) - \bar{R} \ln \frac{p_i}{p_{ref}}$$

or

$$\boxed{\bar{s}_i(T, p_i) = \bar{s}_i°(T) - \bar{R} \ln \frac{y_i p}{p_{ref}} \qquad \left(\begin{array}{c} \text{component } i \text{ of an} \\ \text{ideal gas mixture} \end{array} \right)} \qquad (13.23)$$

where $\bar{s}_i^\circ(T)$ is the absolute entropy of component i at temperature T and $p_{\text{ref}} = 1$ atm. Equation 13.23 corresponds to Eq. (b) of Table 13.1.

Finally, note that *Interactive Thermodynamics (IT)* returns absolute entropy directly and does not use the special function $\bar{s}^\circ$.

13.5.2 Entropy Balances for Reacting Systems

Many of the considerations that enter when energy balances are written for reacting systems also apply to entropy balances. The writing of entropy balances for reacting systems will be illustrated by referring to special cases of broad interest.

Control Volumes at Steady State   Let us begin by reconsidering the steady-state reactor of Fig. 13.2, for which the combustion reaction is given by Eq. 13.11. The combustion air and the products of combustion are each assumed to form ideal gas mixtures, and thus Eq. 12.26 from Table 13.1 for mixture entropy is applicable to them. The entropy rate balance for the two-inlet, single-exit reactor can be expressed on a *per mole of fuel* basis as

$$0 = \sum_i \frac{\dot{Q}_j/T_j}{\dot{n}_F} + \bar{s}_F + \left[\left(a + \frac{b}{4}\right)\bar{s}_{O_2} + \left(a + \frac{b}{4}\right)3.76\,\bar{s}_{N_2}\right]$$
$$- \left[a\bar{s}_{CO_2} + \frac{b}{2}\bar{s}_{H_2O} + \left(a + \frac{b}{4}\right)3.76\,\bar{s}_{N_2}\right] + \frac{\dot{\sigma}_{cv}}{\dot{n}_F} \tag{13.24}$$

where $\dot{n}_F$ is the molar flow rate of the fuel and the coefficients appearing in the underlined terms are the same as those for the corresponding substances in the reaction equation.

All entropy terms of Eq. 13.24 are absolute entropies. The first underlined term on the right side of Eq. 13.24 is the entropy of the combustion air *per mole of fuel*. The second underlined term is the entropy of the exiting combustion products *per mole of fuel*. In accord with Table 13.1, the entropies of the air and combustion products are evaluated by adding the contribution of each component present in the respective gas mixtures. For instance, the specific entropy of a substance in the combustion products is evaluated from Eq. 13.23 using the temperature of the combustion products and the partial pressure of the substance in the combustion product mixture. Such considerations are illustrated in Example 13.9.

▶ ▶ ▶ **EXAMPLE 13.9** ▶ ·

Evaluating Entropy Production for a Reactor Fueled by Liquid Octane

Liquid octane at 25°C, 1 atm enters a well-insulated reactor and reacts with air entering at the same temperature and pressure. The products of combustion exit at 1 atm pressure. For steady-state operation and negligible effects of kinetic and potential energy, determine the rate of entropy production, in kJ/K per kmol of fuel, for complete combustion with **(a)** the theoretical amount of air, **(b)** 400% theoretical air.

Schematic and Given Data:

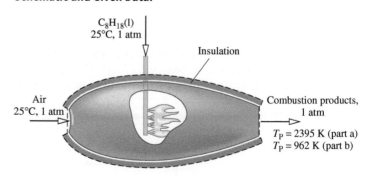

Fig. E13.9

SOLUTION

Known   Liquid octane and air, each at 25°C and 1 atm, burn completely within a well-insulated reactor operating at steady state. The products of combustion exit at 1 atm pressure.

Find   Determine the rate of entropy production, in kJ/K per kmol of fuel, for combustion with (a) the theoretical amount of air, (b) 400% theoretical air.

Engineering Model

1. The control volume shown on the accompanying figure by a dashed line operates at steady state and without heat transfer with its surroundings.

2. Combustion is complete. Each mole of oxygen in the combustion air is accompanied by 3.76 moles of nitrogen, which is inert.

3. The combustion air can be modeled as an ideal gas mixture, as can the products of combustion.

4. The reactants enter at 25°C, 1 atm. The products exit at a pressure of 1 atm.

Analysis The temperature of the exiting products of combustion T_P was evaluated in Example 13.8 for each of the two cases. For combustion with the theoretical amount of air, $T_P = 2395$ K. For complete combustion with 400% theoretical air, $T_P = 962$ K.

a. For combustion of liquid octane with the theoretical amount of air, the chemical equation is

$$C_8H_{18}(l) + 12.5O_2 + 47N_2 \rightarrow 8CO_2 + 9H_2O(g) + 47N_2$$

With assumptions 1 and 3, the entropy rate balance on a per mole of fuel basis, Eq. 13.24, takes the form

$$0 = \sum_j \frac{\overset{0}{\dot{Q}_j/T_j}}{\dot{n}_F} + \overline{s}_F + (12.5\overline{s}_{O_2} + 47\overline{s}_{N_2})$$
$$- (8\overline{s}_{CO_2} + 9\overline{s}_{H_2O(g)} + 47\overline{s}_{N_2}) + \frac{\dot{\sigma}_{cv}}{\dot{n}_F}$$

or on rearrangement

$$\frac{\dot{\sigma}_{cv}}{\dot{n}_F} = (8\overline{s}_{CO_2} + 9\overline{s}_{H_2O(g)} + 47\overline{s}_{N_2}) - \overline{s}_F - (12.5\overline{s}_{O_2} + 47\overline{s}_{N_2}) \quad (1)$$

Each coefficient of this equation is the same as for the corresponding term of the balanced chemical equation.

The fuel enters the reactor separately at T_{ref}, p_{ref}. The absolute entropy of liquid octane required by the entropy balance is obtained from Table A-25 as 360.79 kJ/kmol · K.

The oxygen and nitrogen in the combustion air enter the reactor as components of an ideal gas mixture at T_{ref}, p_{ref}. With Eq. 13.23 and absolute entropy data from Table A-23

$$\overline{s}_{O_2} = \overline{s}^\circ_{O_2}(T_{ref}) - \overline{R} \ln \frac{y_{O_2} p_{ref}}{p_{ref}}$$
$$= 205.03 - 8.314 \ln 0.21 = 218.01 \text{ kJ/kmol} \cdot \text{K}$$

$$\overline{s}_{N_2} = \overline{s}^\circ_{N_2}(T_{ref}) - \overline{R} \ln \frac{y_{N_2} p_{ref}}{p_{ref}}$$
$$= 191.5 - 8.314 \ln 0.79 = 193.46 \text{ kJ/kmol} \cdot \text{K}$$

The product gas exits as an ideal gas mixture at 1 atm, 2395 K with the following composition: $y_{CO_2} = 8/64 = 0.125$, $y_{H_2O(g)} = 9/64 = 0.1406$, $y_{N_2} = 47/64 = 0.7344$. With Eq. 13.23 and absolute entropy data at 2395 K from Table A-23

$$\overline{s}_{CO_2} = \overline{s}^\circ_{CO_2} - \overline{R} \ln y_{CO_2}$$
$$= 320.173 - 8.314 \ln 0.125 = 337.46 \text{ kJ/kmol} \cdot \text{K}$$

$$\overline{s}_{H_2O} = 273.986 - 8.314 \ln 0.1406 = 290.30 \text{ kJ/kmol} \cdot \text{K}$$

$$\overline{s}_{N_2} = 258.503 - 8.314 \ln 0.7344 = 261.07 \text{ kJ/kmol} \cdot \text{K}$$

Inserting values into Eq. (1), the expression for the rate of entropy production, we get

$$\frac{\dot{\sigma}_{cv}}{\dot{n}_F} = 8(337.46) + 9(290.30) + 47(261.07)$$
$$- 360.79 - 12.5(218.01) - 47(193.46)$$
$$= 5404 \text{ kJ/kmol (octane)} \cdot \text{K}$$

Alternative Solution

As an alternative, the following *IT* code can be used to determine the entropy production per mole of fuel entering, where sigma denotes $\dot{\sigma}_{cv}/\dot{n}_F$, and sN2_R and sN2_P denote the entropy of N_2 in the reactants and products, respectively, and so on. In the **Units** menu, select temperature in K, pressure in bar, and amount of substance in moles.

```
TR = 25 + 273.15 // K
p = 1.01325 // bar
TP = 2394 // K (Value from the IT alternative
solution of Example 13.8)

// Determine the partial pressures
pO2_R = 0.21 * p
pN2_R = 0.79 * p
pCO2_P = (8/64) * p
pH2O_P = (9/64) * p
pN2_P = (47/64) * p
// Evaluate the absolute entropies
sC8H18 = 360.79 // kJ/kmol·K (from Table A-25)
sO2_R = s_TP("O2", TR, pO2_R)
sN2_R = s_TP("N2", TR, pN2_R)
sCO2_P = s_TP("CO2", TP, pCO2_P)
sH2O_P = s_TP("H2O", TP, pH2O_P)
sN2_P = s_TP("N2", TP, pN2_P)

// Evaluate the reactant and product
entropies, sR and sP, respectively
sR = sC8H18 + 12.5 * sO2_R + 47 * sN2_R
sR = 8 * sCO2_P + 9 * sH2O_P + 47 * sN2_P

// Entropy balance, Eq. (1)
sigma = sP - sR
```

❶

Using the **Solve** button, the result is sigma = 5404 kJ/kmol (octane) · K, which agrees with the result obtained above.

b. The complete combustion of liquid octane with 400% theoretical air is described by the following chemical equation:

$$C_8H_{18}(l) + 50O_2 + 188N_2 \rightarrow$$
$$8CO_2 + 9H_2O(g) + 37.5O_2 + 188N_2$$

The entropy rate balance on a per mole of fuel basis takes the form

$$\frac{\dot{\sigma}_{cv}}{\dot{n}_F} = (8\overline{s}_{CO_2} + 9\overline{s}_{H_2O(g)} + 37.5\overline{s}_{O_2} + 188\overline{s}_{N_2})$$
$$- \overline{s}_F - (50\overline{s}_{O_2} + 188\overline{s}_{N_2})$$

The specific entropies of the reactants have the same values as in part (a). The product gas exits as an ideal gas mixture at 1 atm, 962 K with the following composition: $y_{CO_2} = 8/242.5 = 0.033$, $y_{H_2O(g)} = 9/242.5 = 0.0371$, $y_{O_2} = 37.5/242.5 = 0.1546$, $y_{N_2} = 0.7753$. With the same approach as in part (a)

$$\overline{s}_{CO_2} = 267.12 - 8.314 \ln 0.033 = 295.481 \text{ kJ/kmol} \cdot \text{K}$$
$$\overline{s}_{H_2O} = 231.01 - 8.314 \ln 0.0371 = 258.397 \text{ kJ/kmol} \cdot \text{K}$$
$$\overline{s}_{O_2} = 242.12 - 8.314 \ln 0.1546 = 257.642 \text{ kJ/kmol} \cdot \text{K}$$
$$\overline{s}_{N_2} = 226.795 - 8.314 \ln 0.7753 = 228.911 \text{ kJ/kmol} \cdot \text{K}$$

Inserting values into the expression for the rate of entropy production

$$\frac{\dot{\sigma}_{cv}}{\dot{n}_F} = 8(295.481) + 9(258.397) + 37.5(257.642)$$

$$+ 188(228.911) - 360.79 - 50(218.01) - 188(193.46)$$

② $= 9754$ kJ/kmol (octane) · K

The use of *IT* to solve part (b) is left as an exercise.

---

① For several gases modeled as ideal gases, *IT* directly returns the absolute entropies required by entropy balances for reacting systems. The entropy data obtained from *IT* agree with values calculated from Eq. 13.23 using table data.

② Although the rates of entropy production calculated in this example are positive, as required by the second law, this does not mean that the proposed reactions necessarily would occur, for the results are based on the assumption of *complete* combustion.

The possibility of achieving complete combustion with specified reactants at a given temperature and pressure can be investigated with the methods of Chap. 14, dealing with chemical equilibrium. For further discussion, see Sec. 14.4.1.

SKILLS DEVELOPED

Ability to…

- apply the control volume entropy balance to a reacting system.
- evaluate entropy values appropriately based on absolute entropies.

Quick Quiz

How do combustion product temperature and rate of entropy production vary, respectively, as percent excess air increases? Assume complete combustion. Ans. Decrease, increase.

Closed Systems   Next consider an entropy balance for a process of a closed system during which a chemical reaction occurs

$$S_P - S_R = \int \left(\frac{\delta Q}{T} \right)_b + \sigma \tag{13.25}$$

S_R and S_P denote, respectively, the entropy of the reactants and the entropy of the products.

When the reactants and products form ideal gas mixtures, the entropy balance can be expressed on a *per mole of fuel* basis as

$$\sum_P n\bar{s} - \sum_R n\bar{s} = \frac{1}{n_F} \int \left(\frac{\delta Q}{T} \right)_b + \frac{\sigma}{n_F} \tag{13.26}$$

where the coefficients *n* on the left are the coefficients of the reaction equation giving the moles of each reactant or product *per mole of fuel*. The entropy terms of Eq. 13.26 are evaluated from Eq. 13.23 using the temperature and partial pressures of the reactants or products, as appropriate. In any such application, the fuel is mixed with the oxidizer, so this must be taken into account when determining the partial pressures of the reactants.

Example 13.10 provides an illustration of the evaluation of entropy change for combustion at constant volume.

▶ ▶ ▶ **EXAMPLE 13.10** ▶ · · · · · · · · · · · · · · · · · · ·

Determine Entropy Change for Combustion of Gaseous Methane with Oxygen at Constant Volume

Determine the change in entropy of the system of Example 13.6 in kJ/K.

SOLUTION

Known   A mixture of gaseous methane and oxygen, initially at 25°C and 1 atm, burns completely within a closed rigid container. The products are cooled to 900 K, 3.02 atm.

Find   Determine the change entropy for the process in kJ/K.

Schematic and Given Data: See Fig. E13.6.

Engineering Model

1. The contents of the container are taken as the system.

2. The initial mixture can be modeled as an ideal gas mixture, as can the products of combustion.

3. Combustion is complete.

Analysis   The chemical equation for the complete combustion of methane with oxygen is

$$CH_4 + 2O_2 \rightarrow CO_2 + 2H_2O(g)$$

The change in entropy for the process of the closed system is $\Delta S = S_P - S_R$, where S_R and S_P denote, respectively, the initial and final entropies of the system. Since the initial mixture forms an ideal gas mixture (assumption 2), the entropy of the reactants can be expressed as the sum of the contributions of the components, each evaluated at the mixture temperature and the partial pressure of the component. That is,

$$S_R = 1\bar{s}_{CH_4}(T_1, y_{CH_4} p_1) + 2\bar{s}_{O_2}(T_1, y_{O_2} p_1)$$

where $y_{CH_4} = 1/3$ and $y_{O_2} = 2/3$ denote, respectively, the mole fractions of the methane and oxygen in the initial mixture. Similarly, since the products of combustion form an ideal gas mixture (assumption 2)

$$S_P = 1\bar{s}_{CO_2}(T_2, y_{CO_2} p_2) + 2\bar{s}_{H_2O}(T_2, y_{H_2O} p_2)$$

where $y_{CO_2} = 1/3$ and $y_{H_2O} = 2/3$ denote, respectively, the mole fractions of the carbon dioxide and water vapor in the products of combustion. In these equations, p_1 and p_2 denote the pressure at the initial and final states, respectively.

The specific entropies required to determine S_R can be calculated from Eq. 13.23. Since $T_1 = T_{ref}$ and $p_1 = p_{ref}$, absolute entropy data from Table A-25 can be used as follows

$$\bar{s}_{CH_4}(T_1, y_{CH_4} p_1) = \bar{s}_{CH_4}^\circ(T_{ref}) - \bar{R}\ln\frac{y_{CH_4} p_{ref}}{p_{ref}}$$

$$= 186.16 - 8.314 \ \ln\frac{1}{3} = 195.294 \text{ kJ/kmol} \cdot \text{K}$$

Similarly

$$\bar{s}_{O_2}(T_1, y_{O_2} p_1) = \bar{s}_{O_2}^\circ(T_{ref}) - \bar{R}\ln\frac{y_{O_2} p_{ref}}{p_{ref}}$$

$$= 205.03 - 8.314\ln\frac{2}{3} = 208.401 \text{ kJ/kmol} \cdot \text{K}$$

At the final state, the products are at $T_2 = 900$ K and $p_2 = 3.02$ atm. With Eq. 13.23 and absolute entropy data from Table A-23

$$\bar{s}_{CO_2}(T_2, y_{CO_2} p_2) = \bar{s}_{CO_2}^\circ(T_2) - \bar{R}\ln\frac{y_{CO_2} p_2}{p_{ref}}$$

$$= 263.559 - 8.314\ln\frac{(1/3)(3.02)}{1}$$

$$= 263.504 \text{ kJ/kmol} \cdot \text{K}$$

$$\bar{s}_{H_2O}(T_2, y_{H_2O} p_2) = \bar{s}_{H_2O}^\circ(T_2) - \bar{R}\ln\frac{y_{H_2O} p_2}{p_{ref}}$$

$$= 228.321 - 8.314\ln\frac{(2/3)(3.02)}{1}$$

$$= 222.503 \text{ kJ/kmol} \cdot \text{K}$$

Finally, the entropy change for the process is

$$\Delta S = S_P - S_R$$

$$= [263.504 + 2(222.503)] - [195.294 + 2(208.401)]$$

$$= 96.414 \text{ kJ/K}$$

SKILLS DEVELOPED

Ability to…

- apply the closed system entropy balance to a reacting system.
- evaluate entropy values appropriately based on absolute entropies.

Quick Quiz

Applying the entropy balance, Eq. 13.25, is σ greater than, less than, or equal to ΔS? Ans. Greater than.

13.5.3 Evaluating Gibbs Function for Reacting Systems

The thermodynamic property known as the Gibbs function plays a role in the second part of this chapter dealing with exergy analysis. The *specific Gibbs function* $\bar{g}$, introduced in Sec. 11.3, is

$$\bar{g} = \bar{h} - T\bar{s} \tag{13.27}$$

The procedure followed in setting a datum for the Gibbs function closely parallels that used in defining the enthalpy of formation: To each stable element at the standard state is assigned a zero value of the Gibbs function. The **Gibbs function of formation** of a compound, $\bar{g}_f^\circ$, equals the change in the Gibbs function for the reaction in which the compound is formed from its elements, the compound and the elements all being at $T_{ref} = 25°C$ and $p_{ref} = 1$ atm. Table A-25 gives the Gibbs function of formation, $\bar{g}_f^\circ$, for selected substances.

The Gibbs function at a state other than the standard state is found by adding to the Gibbs function of formation the change in the specific Gibbs function $\Delta\bar{g}$ between the standard state and the state of interest:

$$\boxed{\bar{g}(T, p) = \bar{g}_f^\circ + [\bar{g}(T, p) - \bar{g}(T_{ref}, p_{ref})] = \bar{g}_f^\circ + \Delta\bar{g}} \tag{13.28a}$$

With Eq. 13.27, $\Delta\bar{g}$ can be written as

$$\boxed{\Delta\bar{g} = [\bar{h}(T, p) - \bar{h}(T_{ref}, p_{ref})] - [T\bar{s}(T, p) - T_{ref}\bar{s}(T_{ref}, p_{ref})]} \tag{13.28b}$$

The Gibbs function of component i in an ideal gas mixture is evaluated at the *partial pressure* of component i and the mixture temperature.

The procedure for determining the Gibbs function of formation is illustrated in the next example.

Gibbs function of formation

TAKE NOTE...

Gibbs function is introduced here because it contributes to subsequent developments of this chapter.

Gibbs function is a property because it is defined in terms of properties. Like enthalpy, introduced as a combination of properties in Sec. 3.6.1, Gibbs function has no physical significance—in general.

Determining the Gibbs Function of Formation for Methane

Determine the Gibbs function of formation of methane at the standard state, 25°C and 1 atm, in kJ/kmol, and compare with the value given in Table A-25.

SOLUTION

Known The compound is methane.

Find Determine the Gibbs function of formation at the standard state, in kJ/kmol, and compare with the Table A-25 value.

Assumptions In the formation of methane from carbon and hydrogen (H_2), the carbon and hydrogen are each initially at 25°C and 1 atm. The methane formed is also at 25°C and 1 atm.

Analysis Methane is formed from carbon and hydrogen according to $C + 2H_2 \rightarrow CH_4$. The change in the Gibbs function for this reaction is

$$\bar{g}_P - \bar{g}_R = (\bar{h} - T\bar{s})_{CH_4} - (\bar{h} - T\bar{s})_C - 2(\bar{h} - T\bar{s})_{H_2}$$
$$= (\bar{h}_{CH_4} - \bar{h}_C - 2\bar{h}_{H_2}) - T(\bar{s}_{CH_4} - \bar{s}_C - 2\bar{s}_{H_2}) \quad (1)$$

where $\bar{g}_P$ and $\bar{g}_R$ denote, respectively, the Gibbs functions of the products and reactants, each per kmol of methane.

In the present case, all substances are at the same temperature and pressure, 25°C and 1 atm, which correspond to the standard reference state values. At the standard reference state, the enthalpies and

Gibbs functions for carbon and hydrogen are zero by definition. Thus, in Eq. (1), $\bar{g}_R = \bar{h}_C = \bar{h}_{H_2} = 0$. Also, $\bar{g}_P = (\bar{g}_f^\circ)_{CH_4}$. Eq. (1) then reads

$$(\bar{g}_f^\circ)_{CH_4} = (\bar{h}_f^\circ)_{CH_4} - T_{ref}(\bar{s}_{CH_4}^\circ - \bar{s}_C^\circ - 2\bar{s}_{H_2}^\circ) \quad (2)$$

where all properties are at T_{ref}, p_{ref}. With enthalpy of formation and absolute entropy data from Table A-25, Eq. (2) gives

$$(\bar{g}_f^\circ)_{CH_4} = -74{,}850 - 298.15[186.16 - 5.74 - 2(130.57)]$$
$$= -50{,}783 \text{ kJ/kmol}$$

The slight difference between the calculated value for the Gibbs function of formation of methane and the value from Table A-25 can be attributed to round-off.

SKILLS DEVELOPED

Ability to...

• apply the definition of Gibbs function of formation to calculate $\bar{g}_f^\circ$.

Quick Quiz

Using the method applied in the example, calculate $\bar{g}_f^\circ$ for monatomic oxygen at the standard state, in kJ/kmol. Begin by writing $\frac{1}{2}O_2 \rightarrow O$. Ans. 231,750 kJ/kmol, which agrees with Table A-25.

Chemical Exergy

The objective of this part of the chapter is to extend the exergy concept introduced in Chap. 7 to include chemical exergy. Several important exergy aspects are listed in Sec. 7.3.1. We suggest you review that material before continuing the current discussion.

A key aspect of the Chap. 7 presentation is that *thermomechanical* exergy is a measure of the departure of the temperature and pressure of a system from those of a thermodynamic model of Earth and its atmosphere called the *exergy reference environment* or, simply, the *environment*. In the current discussion, the departure of the system state from the environment centers on composition, for *chemical* exergy is a measure of the departure of the composition of a system from that of the exergy reference environment.

Exergy is the *maximum* theoretical work obtainable from an overall system of system plus environment as the system passes from a specified state to equilibrium with the environment. Alternatively, exergy is the *minimum* theoretical work *input* required to form the system from the environment and bring it to the specified state.

For conceptual and computational ease, we think of the system passing to equilibrium with the environment in two steps. With this approach, exergy is the sum of two contributions: *thermomechanical*, developed in Chap. 7, and *chemical*, developed in this chapter.

13.6 Conceptualizing Chemical Exergy

In this section, we consider a thought experiment to bring out important aspects of chemical exergy. This involves

• a set of substances represented by $C_aH_bO_c$ (see **Table 13.3**),

• an *environment* modeling Earth's atmosphere (see **Table 13.4**), and

• an *overall* system including a control volume (see **Fig. 13.6**).

TABLE 13.3

Set of Substances Represented by $C_aH_bO_c$

	C	H_2	C_aH_b	CO	CO_2	H_2O(liq.)
a	1	0	a	1	1	0
b	0	2	b	0	0	2
c	0	0	0	1	2	1

TABLE 13.4

Exergy Reference Environment Used in Sec. 13.6

Gas phase at $T_0 = 298.15$ K (25°C), $p_0 = 1$ atm

Component	y^e(%)
N_2	75.67
O_2	20.35
H_2O(g)	3.12
CO_2	0.03
Other	0.83

Referring to Table 13.4, the exergy reference environment considered in the present discussion is an ideal gas mixture modeling Earth's atmosphere. T_0 and p_0 denote the temperature and pressure of the environment, respectively. The composition of the environment is given in terms of mole fractions denoted by y^e, where superscript e is used to signal the mole fraction of an environmental component. The values of these mole fractions, and the values of T_0 and p_0, are specified and remain unchanged throughout the development to follow. The gas mixture modeling the atmosphere adheres to the Dalton model (Sec. 12.2).

Considering Fig. 13.6, a substance represented by $C_aH_bO_c$ enters the control volume at T_0, p_0. Depending on the particular substance, compounds present in the environment enter (O_2) and exit (CO_2 and H_2O(g)) at T_0 and their respective partial pressures in the environment. All substances enter and exit with negligible effects of motion and gravity. Heat transfer between the control volume and environment occurs only at temperature T_0. The control volume operates at steady state, and the ideal gas model applies to all gases. Finally, for the overall system whose boundary is denoted by the dotted line, total volume is constant and there is no heat transfer across the boundary.

Next, we apply conservation of mass, an energy balance, and an entropy balance to the control volume of Fig. 13.6 with the objective of determining the maximum theoretical work per mole of substance $C_aH_bO_c$ entering—namely, the maximum theoretical value of $\dot{W}_{cv}/\dot{n}_F$. This value is the molar chemical exergy of the substance. The chemical exergy is given by

$$\overline{e}^{ch} = \left[\overline{h}_F + \left(a + \frac{b}{4} - \frac{c}{2}\right)\overline{h}_{O_2} - a\overline{h}_{CO_2} - \frac{b}{2}\overline{h}_{H_2O}\right]$$

$$- T_0\left[\overline{s}_F + \left(a + \frac{b}{4} - \frac{c}{2}\right)\overline{s}_{O_2} - a\overline{s}_{CO_2} - \frac{b}{2}\overline{s}_{H_2O}\right] \quad (13.29)$$

where the superscript ch is used to distinguish this contribution to the exergy magnitude from the thermomechanical exergy introduced in Chap. 7. The subscript F denotes the substance represented by $C_aH_bO_c$. The other molar enthalpies and entropies appearing in Eq. 13.29 refer to the substances entering and exiting the control volume, each evaluated at the state at which it enters or exits. See the following box for the derivation of Eq. 13.29.

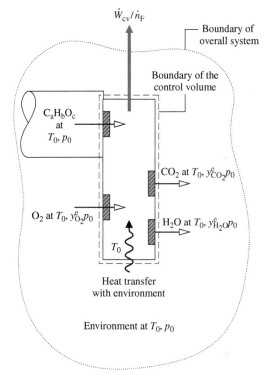

$$C_aH_bO_c + [a + b/4 - c/2]O_2 \rightarrow aCO_2 + b/2\ H_2O(g)$$

Fig. 13.6 Illustration used to conceptualize chemical exergy.

Evaluating Chemical Exergy

Although chemical reaction does not occur in every case we will be considering, conservation of mass is accounted for generally by the following expression

$$C_aH_bO_c + [a + b/4 - c/2]O_2 \rightarrow aCO_2 + b/2\ H_2O(g) \quad (13.30)$$

which assumes that when reaction occurs, the reaction is complete.

For steady-state operation, the energy rate balance for the control volume of Fig. 13.6 reduces to give

$$\frac{\dot{W}_{cv}}{\dot{n}_F} = \frac{\dot{Q}_{cv}}{\dot{n}_F} + \overline{h}_F + \left(a + \frac{b}{4} - \frac{c}{2}\right)\overline{h}_{O_2} - a\overline{h}_{CO_2} - \frac{b}{2}\overline{h}_{H_2O} \quad (13.31)$$

where the subscript F denotes a substance represented by $C_aH_bO_c$ (Table 13.3). Since the control volume is at steady state, its volume does not change with time, so no portion of $\dot{W}_{cv}/\dot{n}_F$ is required to displace the environment. Thus, in keeping with all specified idealizations, Eq. 13.31 also gives the work developed by the overall system of control volume plus environment whose boundary is denoted on

Fig. 13.6 by a dotted line. The potential for such work is the difference in composition between substance $C_aH_bO_c$ and the environment.

Heat transfer is assumed to occur with the environment only at temperature T_0. An entropy balance for the control volume takes the form

$$0 = \frac{\dot{Q}_{cv}/\dot{n}_F}{T_0} + \bar{s}_F + \left(a + \frac{b}{4} - \frac{c}{2}\right)\bar{s}_{O_2} - a\bar{s}_{CO_2} - \frac{b}{2}\bar{s}_{H_2O} + \frac{\dot{\sigma}_{cv}}{\dot{n}_F} \tag{13.32}$$

Eliminating the heat transfer rate between Eqs. 13.31 and 13.32 gives

$$\frac{\dot{W}_{cv}}{\dot{n}_F} = \left[\bar{h}_F + \left(a + \frac{b}{4} - \frac{c}{2}\right)\bar{h}_{O_2} - a\bar{h}_{CO_2} - \frac{b}{2}\bar{h}_{H_2O}\right]$$
$$- T_0\left[\bar{s}_F + \left(a + \frac{b}{4} - \frac{c}{2}\right)\bar{s}_{O_2} - a\bar{s}_{CO_2} - \frac{b}{2}\bar{s}_{H_2O}\right] - T_0\frac{\dot{\sigma}_{cv}}{\dot{n}_F} \tag{13.33}$$

In Eq. 13.33, the specific enthalpy h_F and specific entropy s_F are evaluated at T_0 and p_0. Since the ideal gas model applies to the environment (Table 13.4), the specific enthalpies of the first underlined term of Eq. 13.33 are determined knowing only the temperature T_0. Further, the specific entropy of each substance of the second underlined term is determined by temperature T_0 and the partial pressure in the environment of that substance. Accordingly, since the environment is specified, all enthalpy and entropy terms of Eq. 13.33 are known and independent of the nature of the processes occurring within the control volume.

The term $T_0\dot{\sigma}_{cv}$ depends explicitly on the nature of such processes, however. In accordance with the second law, $T_0\dot{\sigma}_{cv}$ is positive whenever irreversibilities are present, vanishes in the limiting case of no irreversibilities, and is never negative. The *maximum theoretical value* for the work developed is obtained when no irreversibilities are present. Setting $T_0\dot{\sigma}_{cv}$ to zero in Eq. 13.33 yields the expression for *chemical exergy* given by Eq. 13.29.

13.6.1 Working Equations for Chemical Exergy

TAKE NOTE...

Observe that the approach used here to evaluate chemical exergy parallels those used in Secs. 7.3 and 7.5 to evaluate exergy of a system and flow exergy. In each case, energy and entropy balances are applied to evaluate maximum theoretical work in the limit as entropy production tends to zero.

For computational convenience, the chemical exergy given by Eq. 13.29 is written as Eqs. 13.35 and 13.36. The first of these is obtained by recasting the specific entropies of O_2, CO_2, and H_2O using the following equation obtained by application of Eq. (a) of Table 13.1:

$$\bar{s}_i(T_0, y_i^e p_0) = \bar{s}_i(T_0, p_0) - \bar{R}\ln y_i^e \tag{13.34}$$

The first term on the right is the absolute entropy at T_0 and p_0, and y_i^e is the mole fraction of component i in the environment.

Applying Eq. 13.34, Eq. 13.29 becomes

$$\bar{e}^{ch} = \left[\bar{h}_F + \left(a + \frac{b}{4} - \frac{c}{2}\right)\bar{h}_{O_2} - a\bar{h}_{CO_2} - \frac{b}{2}\bar{h}_{H_2O(g)}\right](T_0, p_0)$$
$$- T_0\left[\bar{s}_F + \left(a + \frac{b}{4} - \frac{c}{2}\right)\bar{s}_{O_2} - a\bar{s}_{CO_2} - \frac{b}{2}\bar{s}_{H_2O(g)}\right](T_0, p_0)$$
$$+ \bar{R}T_0\ln\left[\frac{(y_{O_2}^e)^{a+b/4-c/2}}{(y_{CO_2}^e)^a(y_{H_2O}^e)^{b/2}}\right] \tag{13.35}$$

where the notation (T_0, p_0) signals that the specific enthalpy and entropy terms of Eq. 13.35 are each evaluated at T_0 and p_0, although T_0 suffices for the enthalpy of substances modeled as ideal gases.

Recognizing the Gibbs function in Eq. 13.35—$\bar{g}_F = \bar{h}_F - T_0\bar{s}_F$, for instance—Eq. 13.35 can be expressed alternatively in terms of the Gibbs functions of the several substances as

$$\bar{e}^{ch} = \left[\bar{g}_F + \left(a + \frac{b}{4} - \frac{c}{2}\right)\bar{g}_{O_2} - a\bar{g}_{CO_2} - \frac{b}{2}\bar{g}_{H_2O(g)}\right](T_0, p_0)$$
$$+ \bar{R}T_0\ln\left[\frac{(y_{O_2}^e)^{a+b/4-c/2}}{(y_{CO_2}^e)^a(y_{H_2O}^e)^{b/2}}\right] \tag{13.36}$$

The logarithmic term common to Eqs. 13.35 and 13.36 typically contributes only a few percent to the chemical exergy magnitude. Other observations follow:

- The specific Gibbs functions of Eq. 13.36 are evaluated at the temperature T_0 and pressure p_0 of the environment. These terms can be determined with Eq. 13.28a as

$$\bar{g}(T_0, p_0) = \bar{g}_f^\circ + [\bar{g}(T_0, p_0) - \bar{g}(T_{ref}, p_{ref})] \qquad (13.37)$$

  where $\bar{g}_f^\circ$ is the Gibbs function of formation and $T_{ref} = 25°C$, $p_{ref} = 1$ atm.

- For the *special case* where T_0 and p_0 are the same as T_{ref} and p_{ref}, respectively, the second term on the right of Eq. 13.37 vanishes and the specific Gibbs function is just the Gibbs function of formation. That is, the Gibbs function values of Eq. 13.36 can be simply read from Table A-25 or similar compilations.

- Finally, note that the underlined term of Eq. 13.36 can be written more compactly as $-\Delta G$: the negative of the change in Gibbs function for the reaction, Eq. 13.30, regarding each substance as separate at temperature T_0 and pressure p_0.

13.6.2 Evaluating Chemical Exergy for Several Cases

Cases of practical interest corresponding to selected values of a, b, and c in the representation $C_aH_bO_c$ can be obtained from Eq. 13.36. For example, a = 8, b = 18, c = 0 corresponds to octane, C_8H_{18}. An application of Eq. 13.36 to evaluate the chemical exergy of octane is provided in Example 13.12. Further special cases follow:

- Consider the case of pure carbon monoxide at T_0, p_0. For CO we have a = 1, b = 0, c = 1. Accordingly, Eq. 13.30 reads $CO + \frac{1}{2}O_2 \rightarrow CO_2$, and the chemical exergy obtained from Eq. 13.36 is

$$\bar{e}_{CO}^{ch} = [\bar{g}_{CO} + \tfrac{1}{2}\bar{g}_{O_2} - \bar{g}_{CO_2}](T_0, p_0) + \bar{R}T_0 \ln\left[\frac{(y_{O_2}^e)^{1/2}}{y_{CO_2}^e}\right] \qquad (13.38)$$

  If carbon monoxide is not pure but a component of an ideal gas mixture at T_0, p_0, each component i of the mixture enters the control volume of Fig. 13.6 at temperature T_0 and the respective partial pressure $y_i p_0$. The contribution of carbon monoxide to the chemical exergy of the mixture, per mole of CO, is then given by Eq. 13.38, but with the mole fraction of carbon monoxide in the mixture, y_{CO}, appearing in the numerator of the logarithmic term that then reads $\ln[y_{CO}(y_{O_2}^e)^{1/2}/y_{CO_2}^e]$. This becomes important when evaluating the exergy of combustion products involving carbon monoxide.

TAKE NOTE...

For liquid water, we think only of the work that could be developed as water expands through a turbine, or comparable device, from pressure p_0 to the partial pressure of the water vapor in the environment:

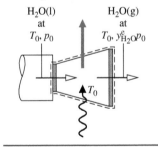

- Consider the case of pure water at T_0 and p_0. Water is a liquid when at T_0, p_0, but is a vapor within the environment of Table 13.4. Thus, water enters the control volume of Fig. 13.6 as a liquid and exits as a vapor at T_0, $y_{H_2O}^e p_0$, with *no chemical reaction required*. In this case, a = 0, b = 2, and c = 1. Equation 13.36 gives the chemical exergy as

$$\bar{e}_{H_2O}^{ch} = [\bar{g}_{H_2O(l)} - \bar{g}_{H_2O(g)}](T_0, p_0) + \bar{R}T_0 \ln\left(\frac{1}{y_{H_2O}^e}\right) \qquad (13.39)$$

- Consider the case of pure carbon dioxide at T_0, p_0. Like water, carbon dioxide is present within the environment and thus requires no chemical reaction to evaluate its chemical exergy. With a = 1, b = 0, c = 2, Eq. 13.36 gives the chemical exergy simply in terms of a logarithmic expression of the form

$$\bar{e}^{ch} = \bar{R}T_0 \ln\left(\frac{1}{y_{CO_2}^e}\right) \qquad (13.40)$$

Provided the appropriate mole fraction y^e is used, Eq. 13.40 also applies to other substances that are gases in the environment, in particular to O_2 and N_2. Moreover, Eqs. 13.39 and 13.40 reveal that a chemical reaction does not always play a part when conceptualizing chemical exergy. In the cases of liquid water, CO_2, O_2, N_2, and other gases present in the environment, we think of the work that could be done as the particular substance passes by *diffusion* from the dead state, where its pressure is p_0, to the environment, where its pressure is the partial pressure, $y^e p_0$.

- Finally, for an ideal gas mixture at T_0, p_0 consisting *only* of substances present as gases in the environment, the chemical exergy is obtained by summing the contributions of each of the components. The result, per mole of mixture, is

$$\overline{e}^{ch} = \overline{R}T_0 \sum_{i=1}^{j} y_i \ln\left(\frac{y_i}{y_i^e}\right) \tag{13.41a}$$

where y_i and y_i^e denote, respectively, the mole fraction of component i in the mixture at T_0, p_0 and in the environment.

Expressing the logarithmic term as $(\ln(1/y_i^e) + \ln y_i)$ and introducing a relation like Eq. 13.40 for each gas i, Eq. 13.41a can be written alternatively as

$$\overline{e}^{ch} = \sum_{i=1}^{j} y_i \overline{e}_i^{ch} + \overline{R}T_0 \sum_{i=1}^{j} y_i \ln y_i \tag{13.41b}$$

The development of Eqs. 13.41a and 13.41b is left as an exercise.

TAKE NOTE...

Equation 13.41b is also applicable for mixtures containing gases other than those present in the reference environment, for example, gaseous fuels. Moreover, this equation can be applied to mixtures that do not adhere to the ideal gas model. In all such applications, the terms $\overline{e}_i^{ch}$ may be selected from a table of standard chemical exergies, introduced in Sec. 13.7 to follow.

13.6.3 Closing Comments

The approach introduced in this section for conceptualizing the chemical exergy of the set of substances represented by $C_aH_bO_c$ can also be applied, in principle, for other substances. In any such application, the chemical exergy is the maximum theoretical work that could be developed by a control volume like that considered in Fig. 13.6 where the substance of interest enters the control volume at T_0, p_0 and reacts completely with environmental components to produce environmental components. All participating environmental components enter and exit the control volume at their conditions within the environment. By describing the environment appropriately, this approach can be applied to many substances of practical interest.[1]

13.7 Standard Chemical Exergy

While the approach used in Sec. 13.6 for conceptualizing chemical exergy can be applied to many substances of practical interest, complications are soon encountered. For one thing, the environment generally must be extended; the simple environment of Table 13.4 no longer suffices. In applications involving coal, for example, sulfur dioxide or some other sulfur-bearing compound must appear among the environmental components. Furthermore, once the environment is determined, a series of calculations is required to obtain exergy values for the substances of interest. These complexities can be sidestepped by using a table of *standard chemical exergies*.

standard chemical exergy

Standard chemical exergy values are based on a standard exergy reference environment exhibiting standard values of the environmental temperature T_0 and pressure p_0 such as 298.15 K and 1 atm, respectively. The exergy reference environment also consists of a set of reference substances with standard concentrations reflecting as closely as possible the chemical makeup of the natural environment. To exclude the possibility of developing work from

---

[1]For further discussion see M. J. Moran, *Availability Analysis: A Guide to Efficient Energy Use*, ASME Press, New York, 1989, pp. 169–170.

interactions among parts of the environment, these reference substances must be in equilibrium mutually.

The reference substances generally fall into three groups: gaseous components of the atmosphere, solid substances from Earth's crust, and ionic and nonionic substances from the oceans. A common feature of standard exergy reference environments is a gas phase, intended to represent air, that includes N_2, O_2, CO_2, $H_2O(g)$, and other gases. The ith gas present in this gas phase is assumed to be at temperature T_0 and the partial pressure $p_i^e = y_i^e p_0$.

Two standard exergy reference environments are considered in this book, called *Model I* and *Model II*. For each of these models, Table A-26 gives values of the standard chemical exergy for several substances, in units of kJ/kmol, together with a brief description of the underlying rationale. The methods employed to determine the tabulated standard chemical exergy values are detailed in the references accompanying the tables. Only one of the two models should be used in a particular analysis.

The use of a table of standard chemical exergies often simplifies the application of exergy principles. However, the term *standard* is somewhat misleading, for there is no one specification of the environment that suffices for *all* applications. Still, chemical exergies calculated relative to alternative specifications of the environment are generally in good agreement. For a broad range of engineering applications, the convenience of using standard values generally outweighs the slight lack of accuracy that might result. In particular, the effect of slight variations in the values of T_0 and p_0 about their standard values normally can be neglected.

TAKE NOTE...

Standard exergy Model II is commonly used today. Model I is provided to show that other standard reference environments can at least be imagined.

13.7.1 Standard Chemical Exergy of a Hydrocarbon: C_aH_b

In principle, the standard chemical exergy of a substance *not* present in the environment can be evaluated by considering a reaction of the substance with other substances for which the chemical exergies *are known*.

To illustrate this for the case of a pure hydrocarbon fuel C_aH_b at T_0, p_0, refer to the control volume at steady state shown in **Fig. 13.7** where the fuel reacts completely with oxygen to form carbon dioxide and *liquid water*. All substances are assumed to enter and exit at T_0, p_0 and heat transfer occurs only at temperature T_0.

Assuming no irreversibilities, an exergy rate balance for the control volume reads

$$0 = \sum_j \left[1 - \frac{\cancel{T_0}}{T_j} \right]^{0} \left(\frac{\dot{Q}_j}{\dot{n}_F} \right) - \left(\frac{\dot{W}_{cv}}{\dot{n}_F} \right)_{\substack{int \\ rev}} + \overline{e}_F^{ch} + \left(a + \frac{b}{4} \right) \overline{e}_{O_2}^{ch}$$

$$- a\overline{e}_{CO_2}^{ch} - \left(\frac{b}{2} \right) \overline{e}_{H_2O(l)}^{ch} - \cancel{\dot{E}_d^{0}}$$

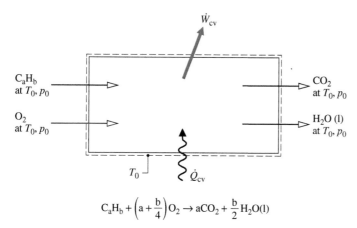

$$C_aH_b + \left(a + \frac{b}{4} \right) O_2 \rightarrow aCO_2 + \frac{b}{2} H_2O(l)$$

Fig. 13.7 Reactor used to introduce the standard chemical exergy of C_aH_b.

where the subscript F denotes C_aH_b. Solving for the chemical exergy $\bar{e}_F^{ch}$, we get

$$\bar{e}_F^{ch} = \left(\frac{\dot{W}_{cv}}{\dot{n}_F}\right)_{\substack{int \\ rev}} + a\bar{e}_{CO_2}^{ch} + \left(\frac{b}{2}\right)\bar{e}_{H_2O(l)}^{ch} - \left(a + \frac{b}{4}\right)\bar{e}_{O_2}^{ch} \qquad (13.42)$$

Applying energy and entropy balances to the control volume, as in the development in the preceding box giving the derivation of Eq. 13.29, we get

$$\left(\frac{\dot{W}_{cv}}{\dot{n}_F}\right)_{\substack{int \\ rev}} = \left[\underline{\bar{h}_F + \left(a + \frac{b}{4}\right)\bar{h}_{O_2} - a\bar{h}_{CO_2} - \frac{b}{2}\bar{h}_{H_2O(l)}}\right](T_0, p_0)$$

$$- T_0\left[\bar{s}_F + \left(a + \frac{b}{4}\right)\bar{s}_{O_2} - a\bar{s}_{CO_2} - \frac{b}{2}\bar{s}_{H_2O(l)}\right](T_0, p_0) \qquad (13.43)$$

The underlined term in Eq. 13.43 is recognized from Sec. 13.2.3 as the molar higher heating value $\overline{HHV}(T_0, p_0)$. Substituting Eq. 13.43 into Eq. 13.42, we obtain

$$\bar{e}_F^{ch} = \overline{HHV}(T_0, p_0) - T_0\left[\bar{s}_F + \left(a + \frac{b}{4}\right)\bar{s}_{O_2} - a\bar{s}_{CO_2} - \frac{b}{2}\bar{s}_{H_2O(l)}\right](T_0, p_0)$$

$$+ a\underline{\bar{e}_{CO_2}^{ch}} + \left(\frac{b}{2}\right)\underline{\bar{e}_{H_2O(l)}^{ch}} - \left(a + \frac{b}{4}\right)\underline{\bar{e}_{O_2}^{ch}} \qquad (13.44a)$$

Equations 13.42 and 13.43 can be expressed alternatively in terms of molar Gibbs functions as follows

$$\bar{e}_F^{ch} = \left[\bar{g}_F + \left(a + \frac{b}{4}\right)\bar{g}_{O_2} - a\bar{g}_{CO_2} - \frac{b}{2}\bar{g}_{H_2O(l)}\right](T_0, p_0)$$

$$+ a\underline{\bar{e}_{CO_2}^{ch}} + \left(\frac{b}{2}\right)\underline{\bar{e}_{H_2O(l)}^{ch}} - \left(a + \frac{b}{4}\right)\underline{\bar{e}_{O_2}^{ch}} \qquad (13.44b)$$

With Eqs. 13.44, the standard chemical exergy of the hydrocarbon C_aH_b can be calculated using the standard chemical exergies of O_2, CO_2, and $H_2O(l)$, together with selected property data: the higher heating value and absolute entropies, or Gibbs functions.

> **FOR EXAMPLE**
>
> Consider the case of methane, CH_4, and $T_0 = 298.15$ K (25°C), $p_0 = 1$ atm. For this application we can use Gibbs function data directly from Table A-25, and standard chemical exergies for CO_2, $H_2O(l)$, and O_2 from Table A-26 (Model II), since each source corresponds to 298.15 K, 1 atm. With a = 1, b = 4, Eq. 13.44b gives 831,680 kJ/kmol. This agrees with the value listed for methane in Table A-26 for Model II.

We conclude the present discussion by noting special aspects of Eqs. 13.44:

- First, Eq. 13.44a requires the higher heating value and the absolute entropy $\bar{s}_F$. When data from property compilations are lacking for these quantities, as in the cases of coal, char, and fuel oil, the approach of Eq. 13.44a can be invoked using a *measured* or *estimated* heating value and an *estimated* value of the absolute entropy $\bar{s}_F$ determined with procedures discussed in the literature.[2]

- Next, note that the first term of Eq. 13.44b can be written more compactly as $-\Delta G$: the negative of the change in Gibbs function for the reaction.

- Finally, note that only the underlined terms of Eqs. 13.44 require chemical exergy data relative to the model selected for the exergy reference environment.

In Example 13.12 we compare the use of Eq. 13.36 and Eq. 13.44b for evaluating the chemical exergy of a pure hydrocarbon fuel.

---

[2]See, for example, A. Bejan, G. Tsatsaronis, and M. J. Moran, *Thermal Design and Optimization,* Wiley, New York, 1996, Secs. 3.4.3 and 3.5.4.

EXAMPLE 13.12

Evaluating the Chemical Exergy of Liquid Octane

Determine the chemical exergy of liquid octane at 25°C, 1 atm, in kJ/kg. **(a)** Using Eq. 13.36, evaluate the chemical exergy for an environment corresponding to Table 13.4—namely, a gas phase at 25°C, 1 atm obeying the ideal gas model with the following composition on a molar basis: N_2, 75.67%; O_2, 20.35%; H_2O, 3.12%; CO_2, 0.03%; other, 0.83%. **(b)** Evaluate the chemical exergy using Eq. 13.44b and standard chemical exergies from Table A-26 (Model II). Compare each calculated exergy value with the standard chemical exergy for liquid octane reported in Table A-26 (Model II).

SOLUTION

Known The fuel is liquid octane.

Find Determine the chemical exergy (a) using Eq. 13.36 relative to an environment consisting of a gas phase at 25°C, 1 atm with a specified composition, (b) using Eq. 13.44b and standard chemical exergies. Compare calculated values with the value reported in Table A-26 (Model II).

Schematic and Given Data:

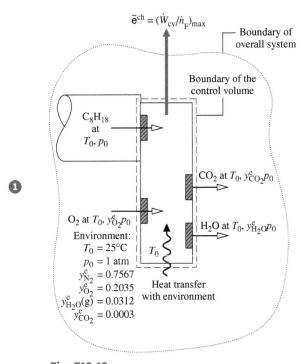

Fig. E13.12

Engineering Model For part (a), as shown in Fig. E13.12, the environment consists of an ideal gas mixture with the molar analysis: N_2, 75.67%; O_2, 20.35%; H_2O, 3.12%; CO_2, 0.03%; other, 0.83%. For part (b), Model II of Table A-26 applies.

Analysis

a. Since a = 8, b = 18, c = 0, Eq. 13.30 gives the following expression for the complete combustion of liquid octane with O_2:

$$C_8H_{18}(l) + 12.5O_2 \rightarrow 8CO_2 + 9H_2O(g)$$

Furthermore, Eq. 13.36 takes the form

$$\overline{e}^{ch} = [\overline{g}_{C_8H_{18}(l)} + 12.5\,\overline{g}_{O_2} - 8\overline{g}_{CO_2} - 9\overline{g}_{H_2O(g)}](T_0, p_0)$$

$$+ \overline{R}T_0 \ln\left[\frac{(y_{O_2}^e)^{12.5}}{(y_{CO_2}^e)^8 (y_{H_2O(g)}^e)^9}\right]$$

Since $T_0 = T_{ref}$ and $p_0 = p_{ref}$, the required specific Gibbs functions are just the Gibbs functions of formation from Table A-25. With the given composition of the environment and data from Table A-25, the above equation gives

$$\overline{e}^{ch} = [6610 + 12.5(0) - 8(-394,380) - 9(-228,590)]$$

$$+ 8.314(298.15) \ln\left[\frac{(0.2035)^{12.5}}{(0.0003)^8 (0.0312)^9}\right]$$

② $= 5,218,960 + 188,883 = 5,407,843$ kJ/kmol

This value agrees closely with the standard chemical exergy for liquid octane reported in Table A-26 (Model II): 5,413,100 kJ/kmol.

Dividing by the molecular weight, the chemical exergy is obtained on a unit mass basis:

$$e^{ch} = \frac{5,407,843}{114.22} = 47,346 \text{ kJ/kg}$$

b. Using coefficients from the reaction equation above, Eq. 13.44b reads

$$\overline{e}^{ch} = [\overline{g}_{C_8H_{18}(l)} + 12.5\overline{g}_{O_2} - 8\overline{g}_{CO_2} - 9\overline{g}_{H_2O(l)}](T_0, p_0)$$

$$+ 8\overline{e}_{CO_2}^{ch} + 9\overline{e}_{H_2O(l)}^{ch} - 12.5\overline{e}_{CO_2}^{ch}$$

With data from Table A-25 and Model II of Table A-26, the above equation gives

$$\overline{e}^{ch} = [6610 + 12.5(0) - 8(-394,380) - 9(-237,180)]$$

$$+ 8(19,870) + 9(900) - 12.5(3970)$$

$$= 5,296,270 + 117,435 = 5,413,705 \text{ kJ/kmol}$$

As expected, this agrees closely with the value listed for octane in Table A-26 (Model II): 5,413,100 kJ/kmol. Dividing by the molecular weight, the chemical exergy is obtained on a unit mass basis:

$$e^{ch} = \frac{5,413,705}{114.22} = 47,397 \text{ kJ/kg}$$

The chemical exergies determined with the two approaches used in parts (a) and (b) also closely agree.

---

❶ A molar analysis of this environment on a *dry* basis reads: O_2: 21%, N_2, CO_2 and the other dry components: 79%. This is consistent with the dry air analysis used throughout the chapter. The water vapor present in the assumed environment corresponds to the amount of vapor that would be present were the gas phase saturated with water at the specified temperature and pressure.

❷ The value of the logarithmic term of Eq. 13.36 depends on the composition of the environment. In the present case, this term contributes about 3% to the magnitude of the chemical exergy. The contribution of the logarithmic term is usually small. In such instances, a satisfactory approximation to the chemical exergy can be obtained by omitting the term.

Quick Quiz

Would the higher heating value (HHV) of liquid octane provide a plausible estimate of the chemical exergy in this case? Ans. Yes, Table A-25 gives 47,900 kJ/kg, which is approximately 1% greater than values obtained in parts (a) and (b).

13.7.2 Standard Chemical Exergy of Other Substances

By paralleling the development given in Sec. 13.7.1 leading to Eq. 13.44b, we can in principle determine the standard chemical exergy of any substance not present in the environment. With such a substance playing the role of C_aH_b in the previous development, we consider a reaction of the substance with other substances for which the standard chemical exergies *are known,* and write

$$\overline{e}^{ch} = -\Delta G + \underline{\sum_P n\overline{e}^{ch}} - \sum_R n\overline{e}^{ch} \tag{13.45}$$

where ΔG is the change in Gibbs function for the reaction, regarding each substance as separate at temperature T_0 and pressure p_0. The underlined term corresponds to the underlined term of Eq. 13.44b and is evaluated using the *known* standard chemical exergies, together with the n's giving the moles of these reactants and products per mole of the substance whose chemical exergy is being evaluated.

> **FOR EXAMPLE**
>
> Consider the case of ammonia, NH_3, and $T_0 = 298.15$ K (25°C), $p_0 = 1$ atm. Letting NH_3 play the role of C_aH_b in the development leading to Eq. 13.44b, we can consider any reaction of NH_3 with other substances for which the standard chemical exergies are known. For the reaction
>
> $$NH_3 + \tfrac{3}{4}O_2 \rightarrow \tfrac{1}{2}N_2 + \tfrac{3}{2}H_2O(l)$$
>
> Eq. 13.45 takes the form
>
> $$\overline{e}^{ch}_{NH_3} = \left[\overline{g}_{NH_3} + \tfrac{3}{4}\overline{g}_{O_2} - \tfrac{1}{2}\overline{g}_{N_2} - \tfrac{3}{2}\overline{g}_{H_2O(l)}\right](T_0, p_0)$$
> $$+ \tfrac{1}{2}\overline{e}^{ch}_{N_2} + \tfrac{3}{2}\overline{e}^{ch}_{H_2O(l)} - \tfrac{3}{4}\overline{e}^{ch}_{O_2}$$
>
> Using Gibbs function data from Table A-25, and standard chemical exergies for O_2, N_2, and $H_2O(l)$ from Table A-26 (Model II), $\overline{e}^{ch}_{NH_3} = 337,910$ kJ/kmol. This agrees closely with the value for ammonia listed in Table A-26 for Model II.

13.8 Applying Total Exergy

The exergy associated with a specified state of a substance is the sum of two contributions: the thermomechanical contribution introduced in Chap. 7 and the chemical contribution introduced in this chapter. On a unit mass basis, the **total exergy** is

total exergy

$$e = \underline{(u - u_0) + p_0(v - v_0) - T_0(s - s_0) + \frac{V^2}{2} + gz} + e^{ch} \tag{13.46}$$

where the underlined term is the thermomechanical contribution (Eq. 7.2) and e^{ch} is the chemical contribution evaluated as in Sec. 13.6 or 13.7. Similarly, the **total flow exergy** associated with a specified state is the sum

total flow exergy

$$e_f = \underline{h - h_0 - T_0(s - s_0) + \frac{V^2}{2} + gz} + e^{ch} \tag{13.47}$$

where the underlined term is the thermomechanical contribution (Eq. 7.14) and e^{ch} is the chemical contribution.

13.8.1 Calculating Total Exergy

Exergy evaluations considered in previous chapters of this book have been alike in this respect: Differences in exergy or flow exergy between states of the same composition have been evaluated. In such cases, the chemical exergy contribution cancels, leaving just the difference in thermomechanical contributions to exergy. However, for many evaluations it is necessary to account explicitly for chemical exergy—for instance, chemical exergy is important when evaluating processes involving combustion.

When using Eqs. 13.46 and 13.47 to evaluate total exergy at a state, we first think of bringing the substance from that state to the state where it is in thermal and mechanical equilibrium with the environment—that is, to the **dead state** where temperature is T_0 and pressure is p_0. In applications dealing with gas mixtures involving water vapor, such as combustion products of hydrocarbons, some condensation of water vapor to liquid may occur in such a process. If so, at the dead state the initial gas mixture consists of a gas phase containing water vapor in equilibrium with liquid water. See the solution of Example 13.15 for an illustration. Yet total exergy evaluations *are* simplified by assuming at the dead state that all water present in the mixture exists only as a vapor, and this *hypothetical* dead state condition often suffices. Also see Example 13.15 for an illustration.

dead state

In Examples 13.13–13.15, we evaluate the total flow exergy for an application involving a cogeneration system.

> **EXAMPLE 13.13** ▸ •

Evaluating the Total Flow Exergy of Process Steam

The cogeneration system shown in Fig. E13.13 develops both power and process steam. At steady state, process steam exits the heat-recovery steam generator at state 9 as saturated vapor at 20 bar with a mass flow rate 14 kg/s. Evaluate the total flow exergy rate, in MW, of the process steam at this state relative to the exergy reference environment of Table A-26 (Model II). Neglect the effects of motion and gravity.

SOLUTION

Known Process steam exits a heat-recovery steam generator at a specified state and a specified mass flow rate.

Find Determine the total flow exergy rate of the process steam, in MW.

Schematic and Given Data:

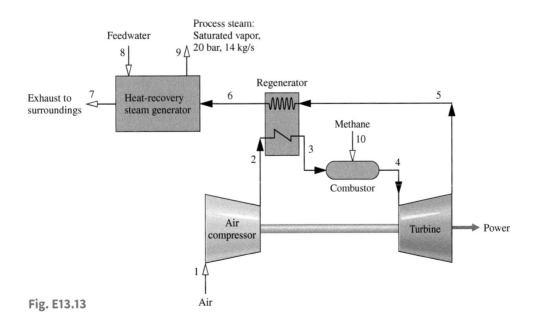

Fig. E13.13

Engineering Model

1. The heat-recovery steam generator is analyzed as a control volume at steady state.

2. The effects of motion and gravity are ignored.

3. For liquid water at T_0, p_0, $h(T_0, p_0) \approx h_f(T_0)$, $s(T_0, p_0) \approx s_f(T_0)$.

4. The exergy reference environment of Table A-26 (Model II) applies.

Analysis   The total flow exergy on a unit of mass basis is given by Eq. 13.47 as

$$e_f = \underline{(h_9 - h_0) - T_0(s_9 - s_0)} + e^{ch}$$

where the underlined term is the thermomechanical contribution to the flow exergy, Eq. 7.14, subject to assumption 2. With steam table data and assumption 3

$$(h_9 - h_0) - T_0(s_9 - s_0) = (2799.5 - 104.9)\frac{kJ}{kg}$$

❶

$$- 298.15 \text{ K}(6.3409 - 0.3674)\frac{kJ}{kg \cdot K}$$

$$= 913.6\frac{kJ}{kg}$$

Water is a liquid at the dead state. Thus, the chemical exergy contribution to the total flow exergy is read from Table A-26 (Model II) as 900 kJ/kmol. Converting to a mass basis, e^{ch} = 49.9 kJ/kg.

Adding the two contributions

❷

$$\dot{E}_9 = 14\frac{kg}{s}(913.6 + 49.9)\frac{kJ}{kg}\left|\frac{1 \text{ MW}}{10^3 \text{ kJ/s}}\right|$$

$$= 13.49 \text{ MW}$$

❶ When evaluating thermomechanical exergy at a state, we think of bringing the substance from the given state to the state where it is in thermal and mechanical equilibrium with the exergy reference environment—that is, to the *dead state* where temperature is T_0 and pressure is p_0. In this application water is a liquid at the dead state.

❷ In keeping with expectations for high-pressure process steam, the thermomechanical contribution dominates; in this case it accounts for 95% of the total.

SKILLS DEVELOPED

Ability to…

- determine the flow exergy, including the chemical exergy contribution of steam.

Quick Quiz

Evaluating the process steam at 8 cents per kW · h of exergy, determine its value, in $/year, for 4000 hours of operation annually. Ans. $4.3 million.

EXAMPLE 13.14

Evaluating the Total Flow Exergy of Fuel Entering a Combustor

Reconsider the cogeneration system of Example 13.13. The combustor is fueled by methane, which enters the combustor at state 10 at 25°C, 12 bar and a mass flow rate of 1.64 kg/s. Evaluate the total flow exergy rate, in MW, of the methane relative to the exergy reference environment of Table A-26 (Model II). Assume the ideal gas model and neglect the effects of motion and gravity.

SOLUTION

Known   Methane enters a combustor at a specified state and a specified mass flow rate.

Find   Determine the total flow exergy rate of the methane, in MW.

Schematic and Given Data: See Fig. E13.13.

Engineering Model

1. The combustor is analyzed as a control volume at steady state.

2. The effects of motion and gravity can be ignored.

3. The methane adheres to the ideal gas model.

4. The exergy reference environment of Table A-26 (Model II) applies.

Analysis   The total flow exergy on a unit of mass basis is given by Eq. 13.47 as

$$e_f = \underline{(h_{10} - h_0) - T_0(s_{10} - s_0)} + e^{ch}$$

where the underlined term is the thermomechanical contribution to flow exergy, Eq. 7.14, subject to assumption 2.

Since the methane is modeled as an ideal gas and enters the combustor at the dead state temperature, 298.15 K (25°C), the thermomechanical contribution reduces with Eqs. 3.43 and 6.18 to give

❶

$$(h_{10} - h_0^0) - T_0(s_{10} - s_0) = RT_0 \ln\left(\frac{p_{10}}{p_0}\right)$$

$$= \left(\frac{8.314}{16.04}\frac{kJ}{kg \cdot K}\right)(298.15 \text{ K})$$

$$\times \ln\left(\frac{12 \text{ bar}}{1.01325 \text{ bar}}\right)$$

$$= 382\frac{kJ}{kg}$$

The chemical exergy contribution is read from Table A-26 (Model II) as 831,650 kJ/kmol. Converting to a mass basis, the chemical exergy is 51,849 kJ/kg.

Adding the two exergy contributions, we get on a time-rate basis

② $\dot{E}_{10} = 1.64 \dfrac{kg}{s}(382 + 51{,}849) \dfrac{kJ}{kg}\left|\dfrac{1\ MW}{10^3\ kJ/s}\right|$

$\quad = 85.7\ MW$

---

① When evaluating thermomechanical exergy at a state, we think of bringing the substance from the given state to the dead state, where temperature is T_0 and pressure is p_0.

② In keeping with expectations for a fuel, the chemical exergy contribution dominates; in this case it accounts for 99% of the total.

SKILLS DEVELOPED

Ability to...

- determine the flow exergy, including the chemical exergy contribution, of methane modeled as an ideal gas.

Quick Quiz

If hydrogen (H_2) enters the combustor at the same state as for the methane, determine the hydrogen mass flow rate, in kg/s, required to give the total flow exergy rate determined above: 85.7 MW. Ans. 0.71.

EXAMPLE 13.15

Evaluating the Total Flow Exergy of Combustion Products

Evaluate the total flow exergy, in MW, of the combustion products exiting the combustor considered in Example 13.14 relative to the exergy reference environment of Table A-26 (Model II). The molar analysis of the combustion products is

$$N_2, 75.07\%;\ O_2, 13.72\%;\ CO_2, 3.14\%;\ H_2O(g), 8.07\%$$

and the mixture molecular weight is 28.25. The combustion products form an ideal gas mixture and the effects of motion and gravity can be ignored. Steady-state data at the combustor exit, state 4, are provided in the accompanying table.

State	$\dot{m}$(kg/s)	T(K)	p(bar)	h(kJ/kg)	s(kJ/kg · K)
4	92.92	1520	9.14	322	8.32

SOLUTION

Known Combustion products exit a combustor at a specified state and a specified mass flow rate.

Find Determine the total flow exergy rate of the combustion products, in MW.

Schematic and Given Data:

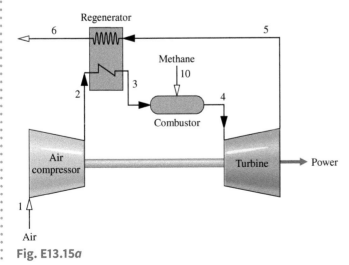

Fig. E13.15a

Engineering Model

1. The combustor is analyzed as a control volume at steady state.

2. The effects of motion and gravity are ignored.

3. The combustion products form an ideal gas mixture.

4. The exergy reference environment of Table A-26 (Model II) applies.

Analysis The total flow exergy on a unit of mass basis is given by Eq. 13.47 as

$$e_f = \underline{(h_4 - h_0) - T_0(s_4 - s_0)} + e^{ch} \qquad (a)$$

where the underlined term is the thermomechanical contribution to flow exergy, Eq. 7.14, subject to assumption 2, and e^{ch} is the chemical contribution, each on a unit of mass basis. These contributions will now be considered in turn.

Thermomechanical Contribution The thermomechanical contribution to Eq. (a) requires values for h_4, s_4, h_0, and s_0. The table provided with the problem statement gives h_4 and s_4. They are mixture values determined using Eqs. 13.9 and 13.23, respectively. While the evaluation of h_4 and s_4 is left as an exercise (Problem 13.107), the evaluation of h_0 and s_0 is detailed next.

When determining thermomechanical exergy at a state, we think of bringing the substance from the given state to the dead state where temperature is T_0 and pressure is p_0. In applications dealing with gas mixtures involving water vapor, some condensation of water vapor to liquid may occur in such a process, which is the case in the present application. See note **①**.

As shown in **Fig. E13.15b**, a 1-kmol sample of the combustion products at the dead state where $T_0 = 298.15$ K, $p_0 = 1$ atm consists of a gas phase, including the "dry" products N_2, O_2, and CO_2 together with water vapor in equilibrium with the condensed water. In kmol, the analysis of the gas phase is N_2, 0.7507; O_2, 0.1372; CO_2, 0.0314; $H_2O(g)$, 0.0297, while the liquid water phase contains the rest of the water formed on combustion: 0.0510 kmol. Within the gas phase N_2, O_2, CO_2, and $H_2O(g)$ have the following mole fractions, denoted by y':

② $y'_{N_2} = 0.7910,\ y'_{O_2} = 0.1446,\ y'_{CO_2} = 0.0331,\ y'_{H_2O(g)} = 0.0313$ \qquad (b)

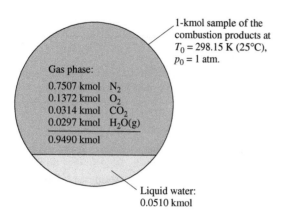

1-kmol sample of the combustion products at $T_0 = 298.15$ K (25°C), $p_0 = 1$ atm.

Gas phase:

0.7507 kmol N_2
0.1372 kmol O_2
0.0314 kmol CO_2
0.0297 kmol $H_2O(g)$
_____
0.9490 kmol

Liquid water:
0.0510 kmol

Fig. E13.15b

The mixture enthalpy h_0 is determined using Eq. 13.9 to obtain the specific enthalpy of each mixture component and then summing those values using the known molar analysis of the mixture. For each component, the $\Delta\bar{h}$ term of Eq. 13.9 vanishes, leaving just the enthalpy of formation. Accordingly, with data from Table A-25,

$$\bar{h}_0 = 0.7507(0) + 0.1372(0) + 0.0314(-393,520)$$
$$+ 0.0297(-241,820) + 0.0510(-285,830)$$
$$= -34,116 \text{ kJ/kmol}$$

The first four terms of this calculation correspond to the gas phase whereas the last term is the contribution of the liquid water phase. Using the mixture molecular weight, we get the mixture enthalpy on a unit mass basis:

$$h_0 = \frac{-34,116 \text{ kJ/kmol}}{28.25 \text{ kg/kmol}} = -1208 \frac{\text{kJ}}{\text{kg}}$$

The mixture entropy s_0 is also determined by summing the entropies of the gas and liquid water phases. For each substance in the gas phase the following special form of Eq. 13.23 is used:

③ $$\bar{s}_i = \bar{s}°(T_0) - \bar{R} \ln y' \qquad \text{(c)}$$

where y' is the mole fraction of that substance in the gas phase, as given by Eqs. (b).

With absolute entropy data from Table A-23, we get the following values, each in kJ/kmol, for the gas phase

$$\bar{s}_{N_2} = 191.502 - 8.314 \ln(0.7910) = 193.45$$
$$\bar{s}_{O_2} = 205.033 - 8.314 \ln(0.1446) = 221.11$$
$$\bar{s}_{CO_2} = 213.685 - 8.314 \ln(0.0331) = 242.02$$
$$\bar{s}_{H_2O(g)} = 188.720 - 8.314 \ln(0.0313) = 217.52$$

The liquid water value is obtained from Table A-25 as

$$\bar{s}_{H_2O(l)} = 69.95 \text{ kJ/kmol}$$

Summing as in the calculation of h_0 above

$$\bar{s}_0 = 0.7507(193.45) + 0.1372(221.11)$$
$$+ 0.0314(242.02) + 0.0297(217.52) + 0.0510(69.95)$$
$$= 193.19 \text{ kJ/kmol} \cdot \text{K}$$

When expressed on a unit mass basis

$$s_0 = \frac{193.19 \text{ kJ/kmol}}{28.25 \text{ kg/kmol}} = 6.84 \text{ kJ/kg} \cdot \text{K}$$

The thermomechanical contribution to the total flow exergy is then

$$(h_4 - h_0) - T_0(s_4 - s_0) = (322 - (-1208)) \text{kJ/kg}$$
$$- 298.15 \text{ K}(8.32 - 6.84) \text{ kJ/kg} \cdot \text{K}$$
$$= 1089 \text{ kJ/kg}$$

Chemical Contribution

At the dead state a sample of the combustion products consists of a gas phase and a liquid water phase. The chemical exergy is determined by adding the chemical exergies of these phases.

For the gas phase, Eq. 13.41b is applied in the form

$$\bar{e}_{gas}^{ch} = \sum_{i=1}^{j} y_i' \bar{e}_i^{ch} + \bar{R}T_0 \sum_{i=1}^{j} y_i' \ln y_i' \qquad \text{(d)}$$

where y' is the mole fraction of component i of the gas phase, as given by Eqs. (b). Accordingly, with chemical exergy values from Table A-26 (Model II),

$$\bar{e}_{gas}^{ch} = [0.7910(720) + 0.1446(3970)$$
$$+ 0.0331(19,870) + 0.0313(9500)]$$
$$+ (8.314)(298.15)[0.7910 \ln(0.7910)$$
$$+ 0.1446 \ln(0.1446) + 0.0331 \ln(0.0331)$$
$$+ 0.0313 \ln(0.0313)]$$
$$= 397 \text{ kJ per kmol of gas}$$

For the liquid phase, Table A-26 (Model II) gives 900 kJ per kmol of liquid.

On the basis of 1 kmol of combustion products at T_0, p_0, the gas phase accounts for 0.949 kmol and the liquid phase accounts for 0.0510 kmol. The chemical exergy is then

$$\bar{e}^{ch} = (0.949 \text{ kmol})(397 \text{ kJ/kmol})$$
$$+ (0.051 \text{ kmol})(900 \text{ kJ/kmol})$$
$$= 423 \text{ kJ/kmol}$$

When expressed on a unit mass basis,

$$e^{ch} = \frac{423 \text{ kJ/kmol}}{28.25 \text{ kg/kmol}} = 15 \frac{\text{kJ}}{\text{kg}}$$

Collecting results, Eq. (a) gives the total flow exergy as the sum

$$e_f = (1089 + 15) \frac{\text{kJ}}{\text{kg}} = 1104 \frac{\text{kJ}}{\text{kg}}$$

On a time-rate basis

④ ⑤ $$\dot{E}_4 = \left(92.92 \frac{\text{kg}}{\text{s}}\right)\left(1104 \frac{\text{kJ}}{\text{kg}}\right)\left|\frac{1 \text{ MW}}{10^3 \text{ kJ/s}}\right|$$
$$= 102.6 \text{ MW}$$

---

① At the dead state where $T_0 = 25$°C, $p_0 = 1$ atm, a 1-kmol sample of the combustion products consists of 0.9193 kmol of "dry products" (N_2, O_2, CO_2) plus 0.0807 kmol of water. Of the water, n kmol is water vapor and the rest is liquid. Considering the gas phase, the partial pressure of the water vapor is the saturation pressure at 25°C: 0.0317 bar. The partial pressure also is the product of the water vapor mole fraction and the mixture pressure, 1.01325 bar. Collecting results,

$$0.0317 = \left(\frac{n}{n + 0.9193}\right)1.01325$$

Solving, the amount of water vapor is $n = 0.0297$ kmol. The amount of liquid is $(0.0807 - n) = 0.0510$ kmol. These values appear in Fig. E13.15b. See Example 13.2(d) for a closely similar analysis.

❷ On the basis of 1 kmol of combustion products the gas phase accounts for 0.949 kmol. Then, for N_2, $y' = 0.7507/0.949 = 0.7910$; for O_2, $y' = 0.1372/0.949 = 0.1446$; and so on.

❸ In the present application, $T = T_0$, $p_i = y_i' p_0$, where $p_0 = p_{ref} = 1$ atm.

❹ In keeping with expectations for high-temperature combustion products, the thermomechanical exergy contribution dominates; chemical exergy contributes only 1.4% to the total.

❺ In this application, evaluation of the total flow exergy at state 4 can be simplified by assuming a *hypothetical* dead state where all water formed by combustion is in vapor form only. With this assumption, the thermomechanical and chemical exergy contributions to the total flow exergy are 1086 kJ/kg and 17 kJ/kg, respectively, as can be verified (Problem 13.107). The total flow exergy is then 1103 kJ/kg, which differs negligibly from the value determined in the solution: 1104 kJ/kg.

SKILLS DEVELOPED

Ability to...

- determine the flow exergy, including the chemical exergy contribution, of combustion products modeled as an ideal gas mixture.

Quick Quiz

If the total flow exergy of the preheated compressed air entering the combustor at state 3 is 41.9 MW and heat transfer from the combustor is ignored, evaluate the rate exergy is destroyed within the combustor, in MW. Ans. 25.

13.8.2 Calculating Exergetic Efficiencies of Reacting Systems

Devices designed to do work by utilization of a combustion process, such as vapor and gas power plants and reciprocating internal combustion engines, invariably have irreversibilities and losses associated with their operation. Accordingly, actual devices produce work equal to only a fraction of the maximum theoretical value that might be obtained. The vapor power plant exergy analysis of Sec. 8.6 and the combined cycle exergy analysis of Example 9.12 provide illustrations.

The performance of devices whose primary function is to do work can be evaluated as the ratio of the actual work developed to the exergy of the fuel consumed in producing that work. This ratio is an *exergetic efficiency*. The relatively low exergetic efficiency exhibited by many common power-producing devices suggests that thermodynamically more thrifty ways of utilizing the fuel to develop power might be possible. However, efforts in this direction must be tempered by the economic imperatives that govern the practical application of all devices. The trade-off between fuel savings and the additional costs required to achieve those savings must be carefully weighed.

The fuel cell provides an illustration of a relatively fuel-efficient device. We noted previously (Sec. 13.4) that the chemical reactions in fuel cells are more controlled than the rapidly occurring, highly irreversible combustion reactions taking place in conventional power-producing systems. In principle, fuel cells can achieve greater exergetic efficiencies than many such devices. Still, relative to conventional power systems, fuel cell systems typically cost more per unit of power generated, and this has limited their deployment.

The examples to follow illustrate the evaluation of an exergetic efficiency for an internal combustion engine and a reactor. In each case, standard chemical exergies are used in the solution.

▶▶ **EXAMPLE 13.16** ▶

Exergetic Efficiency of an Internal Combustion Engine

Devise and evaluate an exergetic efficiency for the internal combustion engine of Example 13.4. For the fuel, use the chemical exergy value determined in Example 13.12(a).

SOLUTION

Known Liquid octane and the theoretical amount of air enter an internal combustion engine operating at steady state in separate streams at 25°C, 1 atm, and burn completely. The combustion products exit at 890°C. The power developed by the engine is 37 kW, and the fuel mass flow rate is 1.8×10^{-3} kg/s.

Find Devise and evaluate an exergetic efficiency for the engine using the fuel chemical exergy value determined in Example 13.12(a).

Schematic and Given Data: See Fig. E13.4.

Engineering Model

1. See the assumptions listed in the solution to Example 13.4.
2. The environment is the same as used in Example 13.12(a).
3. The combustion air enters at the condition of the environment.
4. Kinetic and potential energy effects are negligible.

Analysis An exergy balance can be used in formulating an exergetic efficiency for the engine: At steady state, the rate at which exergy enters the engine equals the rate at which exergy exits plus the rate at which exergy is destroyed within the engine. As the combustion air enters at the condition of the environment, and thus with zero exergy, exergy enters the engine only with the fuel. Exergy exits the engine accompanying heat and work, and with the products of combustion.

If the power developed is taken to be the *product* of the engine, and the heat transfer and exiting product gas are regarded as losses, an exergetic efficiency expression that gauges the extent to which the exergy entering the engine with the fuel is converted to the product is

❶
$$\varepsilon = \frac{\dot{W}_{cv}}{\dot{E}_F}$$

where $\dot{E}_F$ denotes the rate at which exergy enters with the fuel.

Since the fuel enters the engine at 25°C and 1 atm, which correspond to the values of T_0 and p_0 of the environment, and kinetic and potential energy effects are negligible, the exergy of the fuel is just the chemical exergy evaluated in Example 13.12(a). There is no thermomechanical contribution. Thus

$$\dot{E}_F = \dot{m}_F e^{ch} = \left(1.8 \times 10^{-3}\,\frac{kg}{s}\right)\left(47{,}346\,\frac{kJ}{kg}\right)\left|\frac{1\,kW}{1\,kJ/s}\right| = 85.2\,kW$$

The exergetic efficiency is then

$$\varepsilon = \frac{37\,kW}{85.2\,kW} = 0.434\ (43.4\%)$$

❶ The "waste heat" from large engines may be utilizable by some other device—for example, an absorption heat pump. In such cases, some of the exergy accompanying heat transfer and the exiting product gas might be included in the numerator of the exergetic efficiency expression. Since a greater portion of the entering fuel exergy would be utilized in such arrangements, the value of ε would be greater than that evaluated in the solution.

SKILLS DEVELOPED

Ability to…

- devise and evaluate an exergetic efficiency for an internal combustion engine.

Quick Quiz

For the internal combustion engine, list sources of exergy loss. Ans. Flow exergy of the exiting combustion products, exergy loss associated with heat transfer to the surroundings.

In the next example, we evaluate an exergetic efficiency for a reactor. In this case, the exergy of the combustion products, not power developed, is the valuable output.

▶ **EXAMPLE 13.17** ▶

Evaluating Exergetic Efficiency of a Reactor Fueled by Liquid Octane

For the reactor of Examples 13.8 and 13.9, determine the exergy destruction, in kJ per kmol of fuel, and devise and evaluate an exergetic efficiency. Consider two cases: complete combustion with the theoretical amount of air, and complete combustion with 400% theoretical air. For the fuel, use the standard chemical exergy value from Table A-26 (Model II).

SOLUTION

Known Liquid octane and air, each at 25°C and 1 atm, burn completely in a well-insulated reactor operating at steady state. The products of combustion exit at 1 atm pressure.

Find Determine the exergy destruction, in kJ per kmol of fuel, and evaluate an exergetic efficiency for complete combustion with the theoretical amount of air and 400% theoretical air.

Schematic and Given Data: See Fig. E13.9.

Engineering Model

1. See assumptions listed in Examples 13.8 and 13.9.

2. The environment corresponds to Model II of Table A-26.
3. Air entering the reactor at 25°C, 1 atm with the composition 21% O_2, 79% N_2 has negligible exergy.

❶

Analysis An exergy rate balance can be used in formulating an exergetic efficiency: At steady state, the rate at which exergy enters the reactor equals the rate at which exergy exits plus the rate at which exergy is destroyed within the reactor. With assumption 3 exergy enters the reactor only with the fuel. The reactor is well insulated, so there is no exergy transfer accompanying heat transfer. There is also no work $\dot{W}_{cv}$. Accordingly, exergy exits only with the combustion products, which is the valuable output in this case. The exergy rate balance then reads

$$\dot{E}_F = \dot{E}_{products} + \dot{E}_d \qquad (a)$$

where $\dot{E}_F$ is the rate at which exergy enters with the fuel, $\dot{E}_{products}$ is the rate at which exergy exits with the combustion products, and $\dot{E}_d$ is the rate of exergy destruction within the reactor.

An exergetic efficiency then takes the form

$$\varepsilon = \frac{\dot{E}_{products}}{\dot{E}_F} \qquad \text{(b)}$$

The rate at which exergy exits with the products can be evaluated with the approach used in the solution to Example 13.15. But in the present case effort is saved with the following approach: Using the exergy balance for the reactor, Eq. (a), the exergetic efficiency expression, Eq. (b), can be written alternatively as

$$\varepsilon = \frac{\dot{E}_F - \dot{E}_d}{\dot{E}_F} = 1 - \frac{\dot{E}_d}{\dot{E}_F} \qquad \text{(c)}$$

The exergy destruction term appearing in Eq. (b) can be found from the relation

$$\frac{\dot{E}_d}{\dot{n}_F} = T_0 \frac{\dot{\sigma}_{cv}}{\dot{n}_F}$$

where T_0 is the temperature of the environment and $\dot{\sigma}_{cv}$ is the rate of entropy production. The rate of entropy production is evaluated in the solution to Example 13.9 for each of the two cases. For the case of complete combustion with the theoretical amount of air

$$\frac{\dot{E}_d}{\dot{n}_F} = (298\ \text{K})\left(5404\ \frac{\text{kJ}}{\text{kmol} \cdot \text{K}}\right) = 1,610,392\ \frac{\text{kJ}}{\text{kmol}}$$

Similarly, for the case of complete combustion with 400% of the theoretical amount of air

$$\frac{\dot{E}_d}{\dot{n}_F} = (298)(9754) = 2,906,692\ \frac{\text{kJ}}{\text{kmol}}$$

Since the fuel enters the reactor at 25°C, 1 atm, which correspond to the values of T_0 and p_0 of the environment, and kinetic and potential effects are negligible, the exergy of the fuel is just the standard chemical exergy from Table A-26 (Model II): 5,413,100 kJ/kmol.

There is no thermomechanical contribution. Thus, for the case of complete combustion with the theoretical amount of air, Eq. (c) gives

$$\varepsilon = 1 - \frac{1,610,392}{5,413,100} = 0.703\ (70.3\%)$$

Similarly, for the case of complete combustion with 400% of the theoretical amount of air, we get

② $$\varepsilon = 1 - \frac{2,906,692}{5,413,100} = 0.463\ (46.3\%)$$

① The entering air has chemical exergy that can be calculated from Eq. 13.41b using the known oxygen and nitrogen mole fractions together with their chemical exergies from Table A-26. The result is 129 kJ per kmol of air. Compared to the chemical exergy of the fuel, such a value is negligible.

② The calculated efficiency values show that a substantial portion of the fuel exergy is destroyed in the combustion process. In the case of combustion with the theoretical amount of air, about 30% of the fuel exergy is destroyed. In the excess air case, over 50% of the fuel exergy is destroyed. Further exergy destructions would take place as the hot gases are utilized. It might be evident, therefore, that the overall conversion from fuel input to end use would have a relatively low exergetic efficiency. The vapor power plant exergy analysis of Sec. 8.6 illustrates this point.

SKILLS DEVELOPED

Ability to...

- determine exergy destruction for a reactor.
- devise and evaluate an appropriate exergetic efficiency.

Quick Quiz

For complete combustion with 300% of theoretical air, would the exergetic efficiency be greater than, or less than, the exergetic efficiency determined for the case of 400% of theoretical air? Ans. Greater than.

CHAPTER SUMMARY AND STUDY GUIDE

In this chapter we have applied the principles of thermodynamics to systems involving chemical reactions, with emphasis on systems involving the combustion of hydrocarbon fuels. We also have extended the notion of exergy to include chemical exergy.

The first part of the chapter begins with a discussion of concepts and terminology related to fuels, combustion air, and products of combustion. The application of energy balances to reacting systems is then considered, including control volumes at steady state and closed systems. To evaluate the specific enthalpies required in such applications, the enthalpy of formation concept is introduced and illustrated. The determination of the adiabatic flame temperature is considered as an application.

The use of the second law of thermodynamics is also discussed. The absolute entropy concept is developed to provide the specific entropies required by entropy balances for systems involving chemical reactions. The related Gibbs function of formation concept is introduced. The first part of the chapter also includes a discussion of fuel cells.

In the second part of the chapter, we extend the exergy concept of Chap. 7 by introducing chemical exergy. The *standard* chemical exergy concept is also discussed. Means are developed and illustrated

for evaluating the chemical exergies of hydrocarbon fuels and other substances. The presentation concludes with a discussion of exergetic efficiencies of reacting systems.

The following list provides a study guide for this chapter. When your study of the text and end-of-chapter exercises has been completed, you should be able to

- write out the meaning of the terms listed in the margin throughout the chapter and understand each of the related concepts. The subset of key concepts listed below is particularly important.

- determine balanced reaction equations for the combustion of hydrocarbon fuels, including complete and incomplete combustion with various percentages of theoretical air.

- apply energy balances to systems involving chemical reactions, including the evaluation of enthalpy using Eq. 13.9 and the evaluation of the adiabatic flame temperature.

- apply entropy balances to systems involving chemical reactions, including the evaluation of the entropy produced.

- evaluate the chemical exergy of hydrocarbon fuels and other substances using Eqs. 13.35 and 13.36, as well as the standard chemical exergy using Eqs. 13.44 and 13.45.

- evaluate total exergy using Eqs. 3.46 and 3.47.
- apply exergy analysis, including chemical exergy and the evaluation of exergetic efficiencies.

KEY ENGINEERING CONCEPTS

Complete combustion	enthalpy of formation	chemical exergy
air–fuel ratio	heating values	standard chemical exergy
theoretical air	adiabatic flame temperature	dead state
percent of theoretical air	fuel cell	
dry product analysis	absolute entropy	

KEY EQUATIONS

$$AF = \overline{AF}\left(\frac{M_{air}}{M_{fuel}}\right)$$	(13.2)	Relation between air–fuel ratios on mass and molar bases
$$\overline{h}(T, p) = \overline{h}_f^\circ + [\overline{h}(T, p) - \overline{h}(T_{ref}, p_{ref})] = \overline{h}_f^\circ + \Delta\overline{h}$$	(13.9)	Evaluating enthalpy at T, p in terms of enthalpy of formation
$$\frac{\dot{Q}_{cv}}{\dot{n}_F} - \frac{\dot{W}_{cv}}{\dot{n}_F} = \sum_P n_e(\overline{h}_f^\circ + \Delta\overline{h})_e - \sum_R n_i(\overline{h}_f^\circ + \Delta\overline{h})_i$$	(13.15b)	Energy rate balance for a control volume at steady state per mole of fuel entering
$$Q - W = \sum_P n(\overline{h}_f^\circ + \Delta\overline{h}) - \sum_R n(\overline{h}_f^\circ + \Delta\overline{h}) - \overline{R}T_p\sum_P n + \overline{R}T_R\sum_R n$$	(13.17b)	Closed system energy balance, where reactants and products are each ideal gas mixtures
$$\overline{s}(T, p) = \overline{s}^\circ(T) - \overline{R}\ln\frac{p}{p_{ref}}$$	(13.22)	Absolute entropy of an ideal gas (molar basis) at T, p, where $\overline{s}^\circ(T)$ is from Table A-23
$$\overline{s}_i(T, p_i) = \overline{s}_i^\circ(T) - \overline{R}\ln\frac{y_i p}{p_{ref}}$$	(13.23)	Absolute entropy for component i of an ideal gas mixture (molar basis) at T, p, where $\overline{s}_i^\circ(T)$ is from Table A-23
$$\overline{g}(T, p) = \overline{g}_f^\circ + [g(T, p) - \overline{g}(T_{ref}, p_{ref})] = \overline{g}_f^\circ + \Delta\overline{g}$$	(13.28a)	Evaluating Gibbs function at T, p in terms of Gibbs function of formation
where $$\Delta\overline{g} = [\overline{h}(T, p) - \overline{h}(T_{ref}, p_{ref})] - [T\overline{s}(T, p) - T_{ref}\overline{s}(T_{ref}, p_{ref})]$$	(13.28b)	(see Table A-25 for $\overline{g}_f^\circ$ values)
$$\mathbf{e}_f = h - h_0 - T_0(s - s_0) + \frac{V^2}{2} + gz + \mathbf{e}^{ch}$$	(13.47)	Total specific flow exergy including thermomechanical and chemical contributions (see Secs. 13.6 and 13.7 for chemical exergy expressions)

EXERCISES: THINGS ENGINEERS THINK ABOUT

13.1 Is combustion an inherently irreversible process? Why or why not?

13.2 If an engine burns *rich,* is the percent of theoretical air greater than 100% or less than 100%?

13.3 What steps, both indoor and outdoor, should homeowners take to protect their dwellings from fire?

13.4 When I burn wood in my fireplace, do I contribute to *global warming*? Explain.

13.5 Can coal be converted to a liquid diesel-like fuel? Explain.

13.6 What is the difference between *octane rating* and *octane*?

13.7 How is the desired air–fuel ratio maintained in automotive internal combustion engines?

13.8 If the enthalpy of formation of carbon dioxide (CO_2) were assigned a value of zero, would the enthalpy of formation of pure carbon be positive or negative in value?

13.9 Why do oil companies still *flare* natural gas? What are the alternatives?

13.10 How might I define an adiabatic flame temperature for constant volume combustion?

13.11 How do those instant hot and cold packs used by athletes to treat injuries work? For what kinds of injury is each type of pack best suited?

13.12 What is an advantage of using standard chemical exergies? A disadvantage?

13.13 How might an exergetic efficiency be defined for the hybrid power system of Fig. 13.5?

CHECKING UNDERSTANDING

13.1 The *lower* heating value of a hydrocarbon corresponds to the case where all of the water formed by combustion is

a. a liquid

b. a solid

c. a vapor

d. a two-phase mixture of liquid and vapor

13.2 Carbon dioxide gas at 400 K, 1 atm exits a combustor, which has carbon and oxygen (O_2) entering in separate streams, each at 25°C, 1 atm. These are the only entering and exiting streams. To apply an energy balance to the combustor, the specific enthalpies of the carbon and carbon dioxide on a molar basis are _____ kJ/kmol and _____ kJ/kmol, respectively. Assume the ideal gas model.

13.3 Referring to Question 13.2, the *absolute entropy* of the carbon and carbon dioxide on a molar basis are _____ kJ/kmol · K and _____ kJ/kmol · K, respectively.

13.4 The higher heating value of liquid octane at the standard state is _____ kJ/kg.

13.5 In a *dry* product analysis, the mole fractions are given for all gaseous products except

a. N_2, which is inert

b. any unburned fuel

c. water

d. all of the above

13.6 Referring to Example 13.15, the *chemical exergy* at state 5, per kmol of mixture, is _____ kJ/kmol.

13.7 For each of the following reactions:

(i) $H_2 + \frac{1}{2}O_2 \rightarrow H_2O(g)$

(ii) $H_2 + \frac{1}{2}(O_2 + 3.76N_2) \rightarrow H_2O(g) + 1.88N_2$

determine the temperature, in °C, at which water begins to condense when products are cooled at a pressure of 1 bar.

13.8 Pulverized carbon at 25°C, 1 atm enters an insulated reactor operating at steady state and burns completely with 200% of theoretical air entering at 25°C, 1 atm. The adiabatic flame temperature, in K, is closely

a. 1470

b. 1490

c. 1510

d. 1530

13.9 At 25°C, 1 atm, water exists only as a liquid. Yet two values of the enthalpy of formation are given for water in Table A-25. Explain.

13.10 At 500 K and 1 atm, an ideal gas mixture consists of 1 kmol of O_2 and 1 kmol of N_2. The absolute entropy per kmol of mixture is _____ kJ/kmol · K.

13.11 An ideal gas mixture of 1 kmol of hydrogen (H_2) and 2 kmol of oxygen (O_2), initially at 25°C and 1 atm, burns completely in a closed,

rigid cylinder. Finally, the cylinder contains an ideal gas mixture at 1516°C. The pressure of the mixture at its final state is _____ atm.

13.12 If the reactants form a *rich* mixture, then the percent of theoretical air for the combustion reaction is

a. greater than 100%

b. less than 100%

c. cannot be determined without more information. Explain.

13.13 For complete combustion of hydrogen sulfide, H_2S, with the theoretical amount of air, the products consist of _____.

13.14 When combustion is with 400% of theoretical air and $T_0 = 298.15$ K, $p_0 = 1$ atm, the rate of exergy destruction within the reactor of Example 13.9, in kJ per kmol of octane consumed, is _____.

13.15 In words, stoichiometric coefficients are _____.

13.16 In symbols, the *Gibbs function* is

a. $\bar{u} + p\bar{v}$

b. $\bar{u} - T\bar{s}$

c. $\bar{h} + T\bar{s}$

d. $\bar{h} - T\bar{s}$

13.17 Which substance, H_2 or CH_4, will store the greater total exergy, in kJ, in a closed tank of volume V when each substance is at 25°C, 1 atm? _____ Explain.

Indicate whether the following statements are true or false. Explain.

13.18 Methane at 25°C, 1 atm enters a reactor operating at steady state and reacts with greater than the theoretical amount of air entering at the same temperature and pressure. Compared to the maximum adiabatic flame temperature, the measured temperature of the exiting combustion products is greater.

13.19 Even very small amounts of oxides of nitrogen in the exhaust of an internal combustion engine can be a source of air pollution.

13.20 The conservation of mass principle requires that the total number of moles on each side of a chemical equation is equal.

13.21 The *enthalpy of formation* is the energy released or absorbed when a compound is formed from its elements, the compound and elements all being at the standard reference state.

13.22 The third law of thermodynamics states that at a temperature of absolute zero the entropy of a pure crystalline substance cannot be negative.

13.23 A limiting thermal efficiency for fuel cells is imposed by the second law in the form of the *Carnot efficiency*.

13.24 A fuel whose *ultimate analysis* is 85% C, 15% H is represented closely as C_8H_{17}.

13.25 Specific entropy values retrieved from the *steam tables* are absolute entropy values.

PROBLEMS: DEVELOPING ENGINEERING SKILLS

C Problem may require use of appropriate computer software in order to complete.

Working with Reaction Equations

13.1 Twenty grams of propane (C_3H_8) burns with just enough oxygen (O_2) for complete combustion. Determine the amount of oxygen required and the amount of combustion products formed, each in grams.

13.2 Ethane (C_2H_6) burns completely with the theoretical amount of air. Determine the air–fuel ratio on a (a) molar basis, (b) mass basis.

13.3 A gas turbine burns octane (C_8H_{18}) completely with 350% of theoretical air. Determine the amount of N_2 in the products, in kmol per kmol of fuel.

13.4 A closed, rigid vessel initially contains a mixture of 40% CO and 60% O_2 on a mass basis. These substances react giving a final mixture of CO_2 and O_2. Determine the balanced reaction equation.

13.5 Propane (C_3H_8) is burned with air. For each case, obtain the balanced reaction equation for complete combustion

a. with the theoretical amount of air.

b. with 30% excess air.

c. with 30% excess air, but only 80% of the propane being consumed in the reaction.

13.6 Butane (C_4H_{10}) burns completely with air. The equivalence ratio is 0.9. Determine

a. the balanced reaction equation.

b. the percent excess air.

13.7 A fuel mixture with the molar analysis 40% CH_3OH, 50% C_2H_5OH, and 10% N_2 burns completely with 33% excess air. Determine

a. the balanced reaction equation.

b. the air–fuel ratio, both on a molar and a mass basis.

13.8 A gas mixture with the molar analysis 0.50 H_2, 0.50 CO, 1.0 O_2 reacts to form products consisting of CO_2, H_2O, and O_2 only. Determine the amount of each product, in kg per kg of mixture.

13.9 A natural gas with the molar analysis 78% CH_4, 13% C_2H_6, 6% C_3H_8, 1.7% C_4H_{10}, 1.3% N_2 burns completely with 40% excess air in a reactor operating at steady state. If the molar flow rate of the fuel is 0.5 kmol/h, determine the molar flow rate of the air, in kmol/h.

13.10 A natural gas fuel mixture has the molar analysis shown below. Determine the molar analysis of the products for complete combustion with 70% excess dry air.

Fuel	CH_4	H_2	NH_3
y_i	25%	30%	45%

13.11 Coal with the mass analysis 77.54% C, 4.28% H, 1.46% S, 7.72% O, 1.34% N, 7.66% noncombustible ash burns completely with 120% of theoretical air. Determine

a. the balanced reaction equation.

b. the amount of SO_2 produced, in kg per kg of coal.

13.12 A sample of dried Appanoose County coal has a mass analysis of 68.8% carbon, 6.1% hydrogen (H), 8.8% oxygen (O), 1.7% nitrogen (N_2), 6.6% sulfur, and the rest noncombustible ash. For complete combustion with the theoretical amount of air, determine

a. the amount of SO_2 produced, in kg per kg of coal.

b. the air–fuel ratio on a mass basis.

13.13 Octane (C_8H_{18}) burns completely with 120% of theoretical air. Determine

a. the air–fuel ratio on a molar and mass basis.

b. the dew point temperature of the combustion products, in °C, when cooled at 1 atm.

13.14 A gaseous fuel mixture with a specified molar analysis burns completely with moist air to form gaseous products as shown in Fig. P13.14. Determine the dew point temperature of the products, in °C.

13.15 Acetylene (C_2H_2) enters a torch and burns completely with 110% of theoretical air entering at 25°C, 1 atm, 50% relative humidity. Obtain the balanced reaction equation, and determine the dew point temperature of the products, in °C, at 1 atm.

13.16 Butane (C_4H_{10}) burns completely with 150% of theoretical air at 30°C, 1 atm, and 80% relative humidity. Determine

a. the balanced reaction equation.

b. the dew point temperature, in °C, of the products, when cooled at 1 atm.

13.17 Ethane (C_2H_6) enters a furnace and burns completely with 130% of theoretical air entering at 25°C, 85 kPa, 50% relative humidity. Determine

a. the balanced reaction equation.

b. the dew point temperature of the combustion products, in °C, at 85 kPa.

13.18 Methyl alcohol (CH_3OH) burns with 200% theoretical air, yielding CO_2, H_2O, O_2, and N_2. Determine the

a. balanced reaction equation.

b. air–fuel ratio on a mass basis.

c. molar analysis of the products.

13.19 Propane (C_3H_8) reacts with 80% of theoretical air to form products including CO_2, CO, H_2O, and N_2 only. Determine

a. the balanced reaction equation.

b. the air–fuel ratio on a mass basis.

c. the analysis of the products on a dry molar basis.

13.20 Hexane (C_6H_{14}) burns with dry air to give products with the dry molar analysis 8.0% CO_2, 5.5% CO, 3.1% O_2, 83.4% N_2. Determine

a. the balanced reaction equation.

b. the percent of theoretical air.

c. the dew point temperature, in °C, of the products at 1 atm.

13.21 The combustion of a hydrocarbon fuel, represented as C_aH_b, results in products with the dry molar analysis 11% CO_2, 0.5% CO, 2% CH_4, 1.5% H_2, 6% O_2, and 79% N_2. Determine the air–fuel ratio on (a) a molar basis, (b) a mass basis.

13.22 Butane (C_4H_{10}) burns with air, giving products having the dry molar analysis 11.0% CO_2, 1.0% CO, 3.5% O_2, 84.5% N_2. Determine

a. the percent theoretical air.

b. the dew point temperature of the combustion products, in °C, at 1 bar.

Fuel
70% CH_4, 10% H_2,
3% O_2, 5% CO_2, 12% N_2

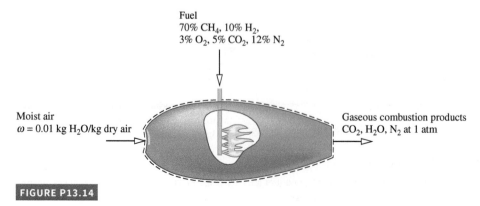

Moist air
$\omega = 0.01$ kg H_2O/kg dry air

Gaseous combustion products
CO_2, H_2O, N_2 at 1 atm

FIGURE P13.14

13.23 Ethyl alcohol (C_2H_5OH) burns with air. The product gas is analyzed and the laboratory report gives only the following percentages on a dry molar basis: 6.9% CO_2, 1.4% CO, 0.5% C_2H_5OH. Assuming the remaining components consist of O_2 and N_2 determine

    a. the percentages of O_2 and N_2 in the dry molar analysis.

    b. the percent excess air.

13.24 A fuel oil with the mass analysis 85% C, 13% H, 1.4% S, 0.6% inert matter burns with 150% of theoretical air. The hydrogen and sulfur are completely oxidized, but 95% of the carbon is oxidized to CO_2 and the remainder to CO.

    a. Determine the balanced reaction equation.

    b. For the CO and SO_2, determine the amount, in kmol per 10^6 kmol of combustion products (that is, the amount in *parts per million*).

13.25 A natural gas with the volumetric analysis 97.3% CH_4, 2.3% CO_2, 0.4% N_2 is burned with air in a furnace to give products having a dry molar analysis of 9.20% CO_2, 3.84% O_2, 0.64% CO, and the remainder N_2. Determine

    a. the percent theoretical air.

    b. the dew point temperature, in °C, of the combustion products at 1 atm.

13.26 Methyl alcohol (CH_3OH) burns in dry air according to the reaction

$$CH_3OH + 3.3(O_2 + 3.76N_2) \rightarrow CO_2 + 2H_2O$$
$$+ 1.8O_2 + 12.408N_2$$

Determine the

    a. air–fuel ratio on a mass basis.

    b. equivalence ratio.

    c. percent excess air.

13.27 For each of the following mixtures, determine the equivalence ratio and indicate if the mixture is lean or rich:

    a. 1 kmol of butane (C_4H_{10}) and 32 kmol of air.

    b. 1 kg of propane (C_3H_8) and 6.5 kg of air.

13.28 Octane (C_8H_{18}) enters an engine and burns with air to give products with the dry molar analysis of CO_2, 10.5%; CO, 5.8%; CH_4, 0.9%; H_2, 2.6%; O_2, 0.3%; N_2, 79.9%. Determine the equivalence ratio.

Applying the First Law to Reacting Systems

13.29 Butane (C_4H_{10}) at 25°C, 1 atm enters a combustion chamber operating at steady state and burns completely with the theoretical amount of air entering at the same conditions. If the products exit at 93°C, 1 atm, determine the rate of heat transfer from the combustion chamber, in kJ per kmol of fuel. Kinetic and potential energy effects are negligible.

13.30 Propane (C_3H_8) at 298 K, 1 atm, enters a combustion chamber operating at steady state with a molar flow rate of 0.7 kmol/s and burns completely with 200% of theoretical air entering at 298 K, 1 atm. Kinetic and potential energy effects are negligible. If the combustion products exit at 560 K, 1 atm, determine the rate of heat transfer for the combustion chamber, in kW. Repeat for an exit temperature of 298 K.

13.31 Methane (CH_4) at 25°C, 1 atm enters a furnace operating at steady state and burns completely with 140% of theoretical air entering at 400 K, 1 atm. The products of combustion exit at 700 K, 1 atm. Kinetic and potential energy effects are negligible. If the rate of heat transfer *from* the furnace to the surroundings is 400 kW, determine the mass flow rate of methane, in kg/s.

13.32 Methane gas (CH_4) at 25°C, 1 atm enters a steam generator operating at steady state. The methane burns completely with 140% of theoretical air entering at 127°C, 1 atm. Products of combustion exit at 427°C, 1 atm. In a separate stream, saturated liquid water enters at 8 MPa and exits as superheated vapor at 480°C with a negligible pressure drop. If the vapor mass flow rate is 3.7×10^5 kg/h, determine the volumetric flow rate of the methane, in m^3/h.

13.33 Liquid ethanol (C_2H_5OH) at 25°C, 1 atm enters a combustion chamber operating at steady state and burns with air entering at 227°C, 1 atm. The fuel flow rate is 25 kg/s and the equivalence ratio is 1.2. Heat transfer from the combustion chamber to the surroundings is at a rate of 3.75×10^5 kJ/s. Products of combustion, consisting of CO_2, CO, $H_2O(g)$, and N_2, exit. Ignoring kinetic and potential energy effects, determine

    a. the exit temperature, in K.

    b. the air–fuel ratio on a mass basis.

13.34 Benzene gas (C_6H_6) at 25°C, 1 atm enters a combustion chamber operating at steady state and burns with 95% theoretical air entering at 25°C, 1 atm. The combustion products exit at 1000 K and include only CO_2, CO, H_2O, and N_2. Determine the mass flow rate of the fuel, in kg/s, to provide heat transfer at a rate of 1000 kW.

13.35 Liquid propane (C_3H_8) at 25°C, 1 atm, enters a well-insulated reactor operating at steady state. Air enters at the same temperature and pressure. For liquid propane, $\bar{h}_f^\circ = -118,900$ KJ/kmol. Determine the temperature of the combustion products, in K, for complete combustion with

    a. the theoretical amount of air.

    b. 300% of theoretical air.

13.36 Methane (CH_4) at 25°C, enters the combustor of a simple open gas turbine power plant and burns completely with 400% of theoretical air entering the compressor at 25°C, 1 atm. Products of combustion exit the turbine at 577°C, 1 atm. The rate of heat transfer from the gas turbine is estimated as 10% of the net power developed. Determine the net power output, in MW, if the fuel mass flow rate is 1200 kg/h. Kinetic and potential energy effects are negligible.

13.37 **Figure P13.37** provides data for a boiler and air preheater operating at steady state. Methane (CH_4) entering the boiler at 25°C, 1 atm is burned completely with 170% of theoretical air. Ignoring stray heat transfer and kinetic and potential energy effects, determine the temperature, in °C, of the combustion air entering the boiler from the preheater.

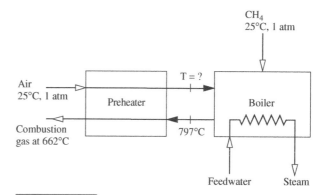

FIGURE P13.37

13.38 A rigid tank initially contains 7.27 kg of CH_4 and 44 kg of O_2 at 25°C, 1 atm. After complete combustion, the pressure in the tank is 3.352 atm. Determine the heat transfer, in kJ.

13.39 A closed rigid vessel initially contains a gaseous mixture at 25°C, 1 atm with the molar analysis of 25% ethylene (C_2H_4), 75% oxygen (O_2). The mixture burns completely and the products are cooled to 500 K. Determine the heat transfer between the vessel and its surroundings, in kJ per kmol of fuel present initially, and the final pressure, in atm.

13.40 A mixture of 1 kmol of hydrogen (H_2) and n kmol of oxygen (O_2), initially at 25°C and 1 atm, burns completely in a closed, rigid, insulated container. The container finally holds a mixture of water vapor and O_2 at 3000 K. The ideal gas model applies to each mixture and there is no change in kinetic or potential energy between the initial and final states. Determine

  a. the value of n.

  b. the final pressure, in atm.

13.41 Calculate the enthalpy of combustion of gaseous pentane (C_5H_{12}), in kJ per kmol of fuel, at 25°C with water vapor in the products.

13.42 Determine the higher heating value, in kJ per kmol of fuel and in kJ per kg of fuel, at 25°C, 1 atm for

  a. liquid octane (C_8H_{18}).

  b. gaseous hydrogen (H_2).

  c. liquid methanol (CH_3OH).

  d. gaseous butane (C_4H_{10}).

Compare with the values listed in Table A-25.

13.43 Liquid octane (C_8H_{18}) at 25°C, 1 atm enters an insulated reactor operating at steady state and burns with 90% of theoretical air at 25°C, 1 atm to form products consisting of CO_2, CO, H_2O, and N_2 only. Determine the temperature of the exiting products, in K. Compare with the results of Example 13.8 and comment.

13.44 Methane (CH_4) at 25°C, 1 atm enters an insulated reactor operating at steady state and burns with the theoretical amount of air entering at 25°C, 1 atm. Determine the temperature of the exiting combustion products, in K, if 90% of the carbon in the fuel burns to CO_2 and the rest to CO. Neglect kinetic and potential energy effects.

13.45 Methane (CH_4) at 25°C, 1 atm enters an insulated reactor operating at steady state and burns with the theoretical amount of air entering at 25°C, 1 atm. The products contain CO_2, CO, H_2O, O_2, and N_2, and exit at 2260 K. Determine the fractions of the entering carbon in the fuel that burn to CO_2 and CO, respectively.

13.46 Liquid methanol (CH_3OH) at 25°C, 1 atm enters an insulated reactor operating at steady state and burns completely with air entering at 100°C, 1 atm. If the combustion products exit at 1256°C, determine the percent excess air used. Neglect kinetic and potential energy effects.

13.47 Air enters the compressor of a simple gas turbine power plant at 21°C, 1 atm, is compressed adiabatically to 275 kPa, and then enters the combustion chamber where it burns completely with propane gas (C_3H_8) entering at 25°C, 275 kPa and a molar flow rate of 0.77 kmol/h. The combustion products at 727°C, 275 kPa enter the turbine and expand adiabatically to a pressure of 1 atm. The isentropic compressor efficiency is 83.3% and the isentropic turbine efficiency is 90%. Determine at steady state

  a. the percent of theoretical air required.

  b. the net power developed, in kW.

13.48 Propane gas (C_3H_8) at 25°C, 1 atm enters an insulated reactor operating at steady state and burns completely with air entering at 25°C, 1 atm. Determine the percent of theoretical air if the combustion products exit at

  a. 616°C.

  b. 1227°C.

Neglect kinetic and potential energy effects.

13.49 A mixture of gaseous octane (C_8H_{18}) and 200% of theoretical air, initially at 25°C, 1 atm, reacts completely in a rigid vessel.

  a. If the vessel were well-insulated, determine the temperature, in °C, and the pressure, in atm, of the combustion products.

  b. If the combustion products were cooled at constant volume to 25°C, determine the final pressure, in atm, and the heat transfer, in kJ per kmol of fuel.

13.50 A mixture of methane (CH_4) and 200% of theoretical air, initially at 25°C, 1 atm, reacts completely in an insulated vessel. Determine the temperature, in °C, of the combustion products if the reaction occurs

  a. at constant volume.

  b. at constant pressure in a piston–cylinder assembly.

Applying the Second Law to Reacting Systems

13.51 Carbon monoxide (CO) at 25°C, 1 atm enters an insulated reactor operating at steady state and reacts completely with the theoretical amount of air entering in a separate stream at 25°C, 1 atm. The products of combustion exit as a mixture at 1 atm. For the reactor, determine the rate of entropy production, in kJ/K per kmol of CO entering. Neglect kinetic and potential energy effects.

13.52 Pentane (C_5H_{12}) gas enters a well-insulated reactor at 25°C, 1.5 atm and reacts completely with excess air entering at 500 K, 1.5 atm. The products exit at 1800 K, 1.5 atm. For operation at steady state and ignoring kinetic and potential energy effects, determine (a) the percent excess air, (b) the rate of entropy production, in kJ/K per kmol of pentane.

13.53 Ethylene (C_2H_4) gas enters a well-insulated reactor and reacts completely with 400% of theoretical air, each at 25°C, 2 atm. The products exit the reactor at 2 atm. For operation at steady state and ignoring kinetic and potential energy effects, determine (a) the balanced reaction equation, (b) the temperature, in K, at which the products exit, (c) the rate of entropy production, in kJ/K per kmol of ethylene.

13.54 Methane (CH_4) at 25°C, 1 atm enters an insulated reactor operating at steady state and burns completely with air entering in a separate stream at 25°C, 1 atm. The products of combustion exit as a mixture at 1 atm. For the reactor, determine the rate of entropy production, in kJ/K per kmol of methane entering, for combustion with

  a. the theoretical amount of air.

  b. 200% of theoretical air.

Neglect kinetic and potential energy effects.

13.55 A gaseous mixture of butane (C_4H_{10}) and 80% excess air at 25°C, 3 atm enters a reactor operating at steady state. Complete combustion occurs and the products exit as a mixture at 1200 K, 3 atm. Refrigerant 134a with a mass flow rate of 5 kg/s enters an outer cooling jacket as saturated liquid and exits the jacket as saturated vapor, each at 25°C. No stray heat transfer occurs from the outside of the jacket, and kinetic and potential energy effects are negligible. Determine for the jacketed reactor

  a. the molar flow rate of the fuel, in kmol/s.

  b. the rate of entropy production, in kW/K.

  c. the rate of exergy destruction, in kW, for $T_0 = 25°C$.

13.56 Liquid ethanol (C_2H_5OH) at 25°C, 1 atm enters a reactor operating at steady state and burns completely with 130% of theoretical air entering in a separate stream at 25°C, 1 atm. Combustion products exit at 227°C, 1 atm. Heat transfer from the reactor takes place at an average surface temperature of 127°C. Determine

  a. the rate of entropy production within the reactor, in kJ/K per kmol of fuel,

  b. the rate of exergy destruction within the reactor, in kJ per kmol of fuel. Kinetic and potential energy effects are negligible. Let $T_0 = 25°C$.

13.57 [C] A gaseous mixture of ethane (C_2H_6) and the theoretical amount of air at 25°C, 1 atm enters a reactor operating at steady state and burns completely. Combustion products exit at 627°C, 1 atm. Heat transfer from the reactor takes place at an average surface temperature T_b. For T_b ranging from 25 to 600°C, determine the rate of exergy destruction within the reactor, in kJ per kmol of fuel. Kinetic and potential energy effects are negligible. Let $T_0 = 25$°C.

13.58 If the water vapor is present in the final products then determine the enthalpy of combustion for gaseous butane (C_4H_{10}), in kJ per kmol of fuel and kJ per kg of fuel at 25°C, 1 atm.

13.59 For (a) carbon, (b) hydrogen (H_2), (c) methane (CH_4), (d) carbon monoxide, (e) liquid methanol (CH_3OH), (f) nitrogen (N_2), (g) oxygen (O_2), (h) carbon dioxide, and (i) water, determine the chemical exergy, in kJ/kg, relative to the following environment in which the gas phase obeys the ideal gas model:

Environment $T_0 = 298.15$ K (25°C), $p_0 = 1$ atm		
Gas Phase:	Component	y^e (%)
	N_2	75.67
	O_2	20.35
	$H_2O(g)$	3.12
	CO_2	0.03
	Other	0.83

13.60 Using data from Tables A-25 and A-26, together with Eq. 13.45, determine the standard molar chemical exergy, in kJ/kmol, of

    a. ammonia (NH_3).

    b. propane (C_3H_8).

13.61 Nitrogen (N_2) flows through a duct. At a particular location the temperature is 400 K, the pressure is 4 atm, and the velocity is 350 m/s. Assuming the ideal gas model and ignoring the effect of gravity, determine the total specific flow exergy, in kJ/kmol. Perform calculations relative to the environment of Table A-26 (Model II).

13.62 Evaluate the total specific flow exergy of an equimolar mixture of oxygen (O_2) and nitrogen (N_2), in kJ/kg, at 227°C, 1 atm. Neglect the effects of motion and gravity. Perform calculations

    a. relative to the environment of Table 13.4.

    b. using data from Table A-26 (Model II).

13.63 A mixture having an analysis on a molar basis of 85% dry air, 15% CO enters a device at 125°C, 2.1 atm, and a velocity of 250 m/s. If the mass flow rate is 1.0 kg/s, determine the rate exergy enters, in kW. Neglect the effect of gravity. Perform calculations

    a. relative to the environment of Problem 13.59.

    b. using data from Table A-26.

13.64 A mixture of methane gas (CH_4) and 150% of theoretical air enters a combustion chamber at 25°C, 1 atm. Determine the specific flow exergy of the entering mixture, in kJ per kmol of methane. Ignore the effects of motion and gravity. Perform calculations

    a. relative to the environment of Problem 13.59.

    b. using data from Table A-26.

13.65 The following flow rates in kg/h are reported for the exiting SNG (substitute natural gas) stream in a certain process for producing SNG from bituminous coal:

CH_4	194,905 kg/h
CO_2	4,125 kg/h
N_2	1,697 kg/h
H_2	261 kg/h
CO	92 kg/h
H_2O	27 kg/h

If the SNG stream is at 25°C, 1 atm, determine the rate at which exergy exits. Perform calculations relative to the environment of Problem 13.59. Neglect the effects of motion and gravity.

Exergy Analysis of Reacting and Psychrometric Systems

13.66 Propane (C_3H_8) gas at 25°C, 1 atm and a mass flow rate of 0.67 kg/min enters an internal combustion engine operating at steady state. The fuel burns with air entering at 25°C, 1 atm according to

$$C_3H_8 + 4.5[O_2 + 3.76N_2] \rightarrow 2.7CO_2 + 0.3CO + 3.3H_2O + 0.7H_2 + 16.92N_2$$

The combustion products exit at 1000 K, 1 atm and the rate of energy transfer by heat from the engine is 100 kW. For the hydrogen, $\bar{c}_p = 29.5$ KJ/kmol · K. The effects of motion and gravity can be ignored. Using the environment of Table A-26 (Model II), evaluate an exergetic efficiency for the engine.

13.67 [C] Figure P13.67 shows a coal gasification reactor making use of the *carbon–steam* process. The energy required for the endothermic reaction is supplied by an electrical resistor. The reactor operates at steady state, with no stray heat transfers and negligible kinetic and potential energy effects. Evaluate in kJ per kmol of carbon entering

    a. the required electrical input.

    b. the exergy entering with the carbon.

    c. the exergy entering with the steam.

    d. the exergy exiting with the product gas.

    e. the exergy destruction within the reactor.

Perform calculations relative to the environment of Problem 13.59. Ignore kinetic and potential energy effects.

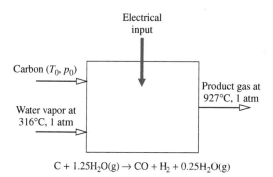

$$C + 1.25H_2O(g) \rightarrow CO + H_2 + 0.25H_2O(g)$$

FIGURE P13.67

13.68 Carbon monoxide at 25°C, 1 atm enters an insulated reactor operating at steady state and reacts completely with the theoretical amount of air entering in a separate stream at 25°C, 1 atm. The products exit as a mixture at 1 atm. Determine in kJ per kmol of CO

    a. the exergy entering with the carbon monoxide.

    b. the exergy exiting with the products.

    c. the rate of exergy destruction.

Also, evaluate an exergetic efficiency for the reactor. Perform calculations relative to the environment of Problem 13.59. Ignore kinetic and potential energy effects.

13.69 Carbon monoxide (CO) at 25°C, 1 atm enters an insulated reactor operating at steady state and reacts completely with the theoretical amount of air entering in a separate stream at 25°C, 1 atm. The products exit as a mixture at 2665 K, 1 atm. Determine in kJ per kmol of CO

a. the exergy entering with the carbon monoxide.

b. the exergy exiting with the products.

c. the rate of exergy destruction.

Also evaluate an exergetic efficiency for the reactor. Perform calculations relative to the environment of Table A-26 (Model II). Neglect the effects of motion and gravity.

13.70 Propane gas (C_3H_8) at 25°C, 1 atm and a volumetric flow rate of 0.03 m³/min enters a furnace operating at steady state and burns completely with 200% of theoretical air entering at 25°C, 1 atm. The furnace provides energy by heat transfer at 227°C for an industrial process and combustion products at 227°C, 1 atm for cogeneration of hot water. For the furnace, determine

a. the rate of heat transfer, in kJ/min, and

b. the rate of entropy production, in kJ/K · min.

c. Also devise and evaluate an exergetic efficiency for the furnace relative to the environment of Table A-26 (Model II). Ignore the effects of motion and gravity.

13.71 Liquid octane (C_8H_{18}) at 25°C, 1 atm and a mass flow rate of 0.57 kg/h enters a small internal combustion engine operating at steady state. The fuel burns with air entering the engine in a separate stream at 25°C, 1 atm. Combustion products exit at 670 K, 1 atm with a dry molar analysis of 11.4% CO_2, 2.9% CO, 1.6% O_2, and 84.1% N_2. If the engine develops power at the rate of 1 kW, determine

a. the rate of heat transfer from the engine, in kW.

b. an exergetic efficiency for the engine.

Use the environment of Problem 13.59 and neglect kinetic and potential energy effects.

13.72 Propane gas (C_3H_8) at 25°C, 1 atm and a volumetric flow rate of 0.03 m³/min enters a furnace operating at steady state and burns completely with 200% of theoretical air entering at 25°C, 1 atm. Combustion

products exit at 227°C, 1 atm. The furnace provides energy by heat transfer at 227°C, for an industrial process. For the furnace, compare the rate of exergy transfer accompanying heat transfer with the rate of exergy destruction, each in kJ/min. Let $T_0 = 25$°C and ignore kinetic and potential energy effects.

13.73 Consider a furnace operating at steady state idealized as shown in **Fig. P13.73**. The fuel is methane, which enters at 25°C, 1 atm and burns completely with 200% theoretical air entering at the same temperature and pressure. The furnace delivers energy by heat transfer at 600 K. Combustion products at 600 K, 1 atm are provided to the surroundings for cogeneration of steam. There are no stray heat transfers, and the effects of motion and gravity can be ignored. Assuming all water present in the combustion products is a vapor at the dead state, determine in kJ per kmol of fuel

a. the exergy entering the furnace with the fuel.

b. the exergy exiting with the products.

c. the rate of exergy destruction.

Also devise and evaluate an exergetic efficiency for the furnace and comment. Perform calculations relative to the environment of Table A-26 (Model II).

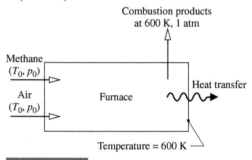

FIGURE P13.73

DESIGN & OPEN-ENDED PROBLEMS: EXPLORING ENGINEERING PRACTICE

13.1D The term *acid rain* is frequently used today. Define what is meant by the term. Write a report in which you discuss the origin and consequences of acid rain as well as options for its control. Include at least three references.

13.2D Municipal solid waste (MSW), often called garbage or trash, consists of the combined solid waste produced by homes and workplaces. In the United States, a portion of the annual MSW accumulation is burned to generate steam for producing electricity, to heat buildings and water, and for other uses, while several times as much MSW is buried in *sanitary* landfills. Investigate these two MSW disposal approaches. For each approach, prepare a list of at least three advantages and three disadvantages, together with brief discussions of each advantage and disadvantage. Report your findings in a PowerPoint presentation suitable for a community planning group.

13.3D Figure P13.3D shows a natural gas–fired boiler for steam generation integrated with a *direct-contact condensing* heat exchanger that discharges warm water for tasks such as space heating, water heating, and combustion air preheating. For 7200 h of operation annually, estimate the annual savings for fuel, in dollars, by integrating these functions. What other cost considerations would enter into the decision to install such a condensing heat exchanger? Discuss.

13.4D A proposed simple gas turbine will produce power at a rate of 0.5 MW by burning fuel with 200% theoretical air in the combustor. Air temperature and pressure at the compressor inlet are 298 K and 100 kPa, respectively. Fuel enters the combustor at 298 K, while products of combustion consisting of CO_2, H_2O, O_2, and N_2 exit the combustor with no significant change in pressure. Metallurgical

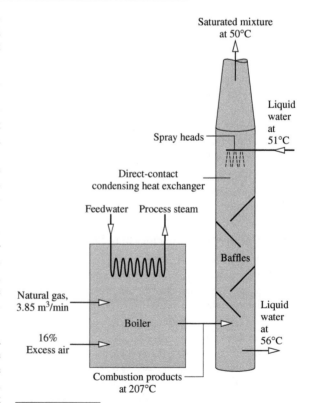

FIGURE P13.3D

considerations require the turbine inlet temperature to be no greater than 1500 K. Products of combustion exit the turbine at 100 kPa. The compressor has an isentropic efficiency of 85%, while the turbine isentropic efficiency is 90%. Three fuels are being considered: methane (CH_4), ethylene (C_2H_4), and ethane (C_2H_6). With the aim of minimizing fuel use, recommend a fuel, turbine inlet temperature, and compressor pressure ratio for the gas turbine. Summarize your findings in a report supported by well-documented sample calculations and a full discussion of the thermodynamic modeling used.

13.5D Figure P13.5D illustrates the schematic of an air preheater for the boiler of a coal-fired power plant. An alternative design would eliminate the air preheater, supplying air to the furnace directly at 30°C and discharging flue gas at 260°C. All other features would remain unchanged. On the basis of thermodynamic and economic analyses, recommend one of the two options for further consideration. The analysis of the coal on a mass basis is

{76% C, 5% H, 8% O, 11% noncombustible ash}

The coal higher heating value is 26,000 kJ/kg. The molar analysis of the flue gas on a dry basis is

{7.8% CO_2, 1.2% CO, 11.4% O_2, 79.6% N_2}

13.6D Identify and research a fuel cell system for combined heat and power integrated with a building in your locale. Describe each component in the fuel cell system and create a schematic of the system to include fuel cell stack, its auxiliary components, and its integration with the building to provide electricity and heating. Contact the building supervisor to identify any installation, operational, and/or maintenance issues. Estimate total costs (components, installation, and annual fuel and operating costs) for the fuel cell system and compare to the prior system's cost, assuming the same annual electricity and heating requirements. Summarize your findings in a PowerPoint presentation.

13.7D Fuel or chemical leaks and spills can have catastrophic ramifications; thus the hazards associated with such events must be well understood. Prepare a memorandum for one of the following:

a. Experience with interstate pipelines shows that propane leaks are usually much more hazardous than leaks of natural gas or liquids such as gasoline. Why is this so?

b. The most important parameter in determining the accidental rate of release from a fuel or chemical storage vessel is generally the size of the opening. Roughly how much faster would such a substance be released from a 1-cm hole than from a 1-mm hole? What are the implications of this?

13.8D The chemical exergies of common hydrocarbons C_aH_b can be represented in terms of their respective lower heating value, $\overline{LHV}$, by an expression of the form

$$\frac{\overline{e}^{ch}}{\overline{LHV}} = c_1 + c_2(b/a) - c_3/a$$

where c_1, c_2, and c_3 are constants. Evaluate the constants to obtain an expression applicable to the gaseous hydrocarbons of Table A-26 (Model II).

13.9D Mounting evidence suggests that more frequent severe weather events, melting of glaciers and polar icecaps, and other observed changes in nature are linked to emissions from fossil fuel combustion. Researchers have developed detailed computer simulations to predict global weather patterns and investigate the effects of CO_2 and other gaseous emissions on the weather. Investigate the underlying modeling assumptions and characteristics of computer simulations used to predict global weather patterns. Write a report, including at least three references, of your findings.

13.10D Coal mining operations in certain regions of the United States have created vast amounts of waste coal known as *culm*. Some power plants have been built near culm banks to generate electricity from this waste resource. A particular culm sample has the following ultimate analysis: 44.1% C, 2.9% H, 16% O, 0.5% N, and 0.5% S. The higher heating value including moisture and ash is 15,600 kJ/kg; the higher heating value on a dry and ash-free basis is 32,600 kJ/kg. Estimate the chemical exergy of this sample, in kJ/kg. For comparison, also determine the chemical exergy value of anthracite coal. Investigate the advantages and disadvantages of using culm instead of coal in power plants. Write a report summarizing your findings, including a comparison of the chemical exergy values obtained, sample calculations, and at least three references.

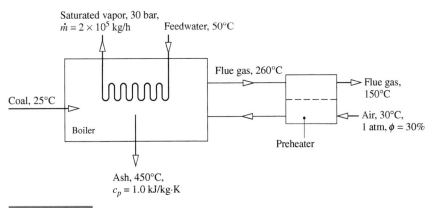

Saturated vapor, 30 bar, $\dot{m} = 2 \times 10^5$ kg/h
Feedwater, 50°C
Flue gas, 260°C
Coal, 25°C
Boiler
Flue gas, 150°C
Air, 30°C, 1 atm, $\phi = 30\%$
Preheater
Ash, 450°C, $c_p = 1.0$ kJ/kg·K

FIGURE P13.5D

CHAPTER 14

Chemical and Phase Equilibrium

Engineering Context

The **objective** of the present chapter is to consider the concept of equilibrium in greater depth than has been done thus far. In the first part of the chapter, we develop the fundamental concepts used to study chemical and phase equilibrium. In the second part of the chapter, the study of reacting systems initiated in Chap. 13 is continued with a discussion of *chemical* equilibrium in a single phase. Particular emphasis is placed on the case of reacting ideal gas mixtures. The third part of the chapter concerns *phase* equilibrium. The equilibrium of multicomponent, multiphase, nonreacting systems is considered and the *phase rule* is introduced.

LEARNING OUTCOMES

When you complete your study of this chapter, you will be able to...

- Explain key concepts related to chemical and phase equilibrium, including criteria for equilibrium, the equilibrium constant, and the Gibbs phase rule.
- Apply the equilibrium constant relationship, Eq. 14.35, to relate pressure, temperature, and equilibrium constant for ideal gas mixtures involving individual and multiple reactions.
- Apply chemical equilibrium concepts with the energy balance.
- Determine the equilibrium flame temperature.
- Apply the Gibbs phase rule, Eq. 14.68.

Equilibrium Fundamentals

In this part of the chapter, fundamental concepts are developed that are useful in the study of chemical and phase equilibrium. Among these are equilibrium criteria and the chemical potential concept.

14.1 Introducing Equilibrium Criteria

thermodynamic equilibrium

A system is said to be in **thermodynamic equilibrium** if, when it is isolated from its surroundings, there would be no macroscopically observable changes. An important requirement for equilibrium is that the temperature be uniform throughout the system or each part of the system in thermal contact. If this condition were not met, spontaneous heat transfer from one location to another could occur when the system was isolated. There must also be no unbalanced forces between parts of the system. These conditions ensure that the system is in thermal and mechanical equilibrium, but there is still the possibility that complete equilibrium does not exist. A process might occur involving a chemical reaction, a transfer of mass between phases, or both. The objective of this section is to introduce criteria that can be applied to decide whether a system in a particular state is in equilibrium. These criteria are developed using the conservation of energy principle and the second law of thermodynamics as discussed next.

Consider the case of a simple compressible system of fixed mass for which temperature and pressure are uniform with position throughout. In the absence of overall system motion and ignoring the influence of gravity, the energy balance in differential form (Eq. 2.36) is

$$dU = \delta Q - \delta W$$

If volume change is the only work mode and pressure is uniform with position throughout the system, $\delta W = p \, dV$. Introducing this in the energy balance and solving for δQ gives

$$\delta Q = dU + p \, dV$$

Since temperature is uniform with position throughout the system, the entropy balance in differential form (Eq. 6.25) is

$$dS = \frac{\delta Q}{T} + \delta \sigma$$

Eliminating δQ between the last two equations

$$T \, dS - dU - p \, dV = T \, \delta \sigma \tag{14.1}$$

Entropy is produced in all actual processes and conserved only in the absence of irreversibilities. Hence, Eq. 14.1 provides a constraint on the direction of processes. The only processes allowed are those for which $\delta \sigma \geq 0$. Thus,

$$\boxed{T \, dS - dU - p \, dV \geq 0} \tag{14.2}$$

Equation 14.2 can be used to study equilibrium under various conditions.

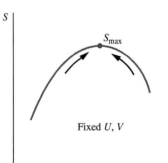

S

S_{max}

Fixed U, V

FOR EXAMPLE

A process taking place in an insulated, constant-volume vessel, where $dU = 0$ and $dV = 0$, must be such that

$$dS]_{U,V} \geq 0 \tag{14.3}$$

Equation 14.3 suggests that changes of state of a closed system at constant internal energy and volume can occur only in the direction of *increasing entropy*. The expression also implies that entropy approaches a *maximum* as a state of equilibrium is approached. This is a special case of the increase of entropy principle introduced in Sec. 6.8.1.

An important case for the study of chemical and phase equilibria is one in which temperature and pressure are fixed. For this, it is convenient to employ the **Gibbs function** in extensive form

Gibbs function

$$G = H - TS = U + pV - TS$$

Forming the differential

$$dG = dU + p\,dV + V\,dp - T\,dS - S\,dT$$

or on rearrangement

$$dG - V\,dp + S\,dT = -(T\,dS - dU - p\,dV)$$

Except for the minus sign, the right side of this equation is the same as the expression appearing in Eq. 14.2. Accordingly, Eq. 14.2 can be written as

$$dG - V\,dp + S\,dT \leq 0 \qquad (14.4)$$

where the inequality reverses direction because of the minus sign noted above.

It can be concluded from Eq. 14.4 that any process taking place at a specified temperature and pressure ($dT = 0$ and $dp = 0$) must be such that

$$\boxed{dG]_{T,\,p} \leq 0} \qquad (14.5)$$

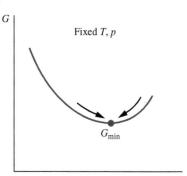

This inequality indicates that the Gibbs function of a system at fixed T and p *decreases* during an irreversible process. Each step of such a process results in a decrease in the Gibbs function of the system and brings the system closer to equilibrium. The equilibrium state is the one having the *minimum* value of the Gibbs function. Therefore, when

$$\boxed{dG]_{T,\,p} = 0} \qquad (14.6)$$

we have equilibrium. In subsequent discussions, we refer to Eq. 14.6 as the **equilibrium criterion**.

equilibrium criterion

Equation 14.6 provides a relationship among the properties of a system when it is *at* an equilibrium state. The manner in which the equilibrium state is reached is unimportant, however, for once an equilibrium state is obtained, a system exists at a particular T and p and no further spontaneous changes can take place. When applying Eq. 14.6, therefore, we may specify the temperature T and pressure p, but it is unnecessary to require additionally that the system actually achieves equilibrium at fixed T and fixed p.

14.1.1 Chemical Potential and Equilibrium

In the present discussion, the Gibbs function is considered further as a prerequisite for application of the equilibrium criterion $dG]_{T,\,p} = 0$ introduced above. Let us begin by noting that any *extensive* property of a single-phase, single-component system is a function of two independent intensive properties and the size of the system. Selecting temperature and pressure as the independent properties and the number of moles n as the measure of size, the Gibbs function can be expressed in the form $G = G(T, p, n)$. For a single-phase, **multicomponent system**, G may then be considered a function of temperature, pressure, and the number of moles of each component present, written $G = G(T, p, n_1, n_2, \ldots, n_j)$.

multicomponent system

If each mole number is multiplied by α, the size of the system changes by the same factor and so does the value of every extensive property. Thus, for the Gibbs function we may write

$$\alpha G(T, p, n_1, n_2, \ldots, n_j) = G(T, p, \alpha n_1, \alpha n_2, \ldots, \alpha n_j)$$

Differentiating with respect to α while holding temperature, pressure, and the mole numbers fixed and using the chain rule on the right side gives

$$G = \frac{\partial G}{\partial(\alpha n_1)}n_1 + \frac{\partial G}{\partial(\alpha n_2)}n_2 + \cdots + \frac{\partial G}{\partial(\alpha n_j)}n_j$$

This equation holds for all values of α. In particular, it holds for $\alpha = 1$. Setting $\alpha = 1$, the following expression results:

$$G = \sum_{i=1}^{j} n_i \left(\frac{\partial G}{\partial n_i} \right)_{T, p, n_l} \tag{14.7}$$

where the subscript n_l denotes that all n's except n_i are held fixed during differentiation.

chemical potential

The partial derivatives appearing in Eq. 14.7 have such importance for our study of chemical and phase equilibrium that they are given a special name and symbol. The **chemical potential** of component i, symbolized by μ_i, is defined as

$$\mu_i = \left(\frac{\partial G}{\partial n_i} \right)_{T, p, n_l} \tag{14.8}$$

The chemical potential is an *intensive property*. With Eq. 14.8, Eq. 14.7 becomes

$$G = \sum_{i=1}^{j} n_i \mu_i \tag{14.9}$$

The equilibrium criterion given by Eq. 14.6 can be written in terms of chemical potentials, providing an expression of fundamental importance for subsequent discussions of equilibrium. Forming the differential of $G(T, p, n_1, \ldots, n_j)$ while holding temperature and pressure fixed results in

$$dG]_{T, p} = \sum_{i=1}^{j} \left(\frac{\partial G}{\partial n_i} \right)_{T, p, n_l} dn_i$$

The partial derivatives are recognized from Eq. 14.8 as the chemical potentials, so

$$dG]_{T, p} = \sum_{i=1}^{j} \mu_i \, dn_i \tag{14.10}$$

With Eq. 14.10, the equilibrium criterion $dG]_{T, p} = 0$ can be placed in the form

$$\sum_{i=1}^{j} \mu_i \, dn_i = 0 \tag{14.11}$$

Like Eq. 14.6, from which it is obtained, this equation provides a relationship among properties of a system when the system is *at* an equilibrium state where the temperature is T and the pressure is p. Like Eq. 14.6, this equation applies to a particular state, and the manner in which that state is attained is not important.

14.1.2 Evaluating Chemical Potentials

Means for evaluating the chemical potentials for two cases of interest are introduced in this section: a single phase of a pure substance and an ideal gas mixture.

Single Phase of a Pure Substance   An elementary case considered later in this chapter is that of equilibrium between two phases involving a pure substance. For a single phase of a pure substance, Eq. 14.9 becomes simply

$$G = n\mu$$

or

$$\mu = \frac{G}{n} = \bar{g} \tag{14.12}$$

That is, the chemical potential is just the Gibbs function per mole.

Ideal Gas Mixture An important case for the study of chemical equilibrium is that of an ideal gas mixture. The enthalpy and entropy of an ideal gas mixture are given by

$$H = \sum_{i=1}^{j} n_i \, \bar{h}_i(T) \quad \text{and} \quad S = \sum_{i=1}^{j} n_i \bar{s}_i(T, p_i)$$

where $p_i = y_i p$ is the partial pressure of component i. Accordingly, the Gibbs function takes the form

$$G = H - TS = \sum_{i=1}^{j} n_i \, \bar{h}_i(T) - T \sum_{i=1}^{j} n_i \, \bar{s}_i(T, p_i)$$

$$= \sum_{i=1}^{j} n_i [\bar{h}_i(T) - T \, \bar{s}_i(T, p_i)] \quad \text{(ideal gas)} \tag{14.13}$$

Introducing the molar Gibbs function of component i

$$\bar{g}_i(T, p_i) = \bar{h}_i(T) - T \, \bar{s}_i(T, p_i) \quad \text{(ideal gas)} \tag{14.14}$$

Equation 14.13 can be expressed as

$$\boxed{G = \sum_{i=1}^{j} n_i \bar{g}_i(T, p_i) \quad \text{(ideal gas)}} \tag{14.15}$$

Comparing Eq. 14.15 to Eq. 14.9 suggests that

$$\boxed{\mu_i = \bar{g}_i(T, p_i) \quad \text{(ideal gas)}} \tag{14.16}$$

That is, the chemical potential of component i in an ideal gas mixture is equal to its Gibbs function per mole of i, evaluated at the mixture temperature and the partial pressure of i in the mixture. Equation 14.16 can be obtained formally by taking the partial derivative of Eq. 14.15 with respect to n_i, holding temperature, pressure, and the remaining n's constant, and then applying the definition of chemical potential, Eq. 14.8.

The chemical potential of component i in an ideal gas mixture can be expressed in an alternative form that is somewhat more convenient for subsequent applications. Using Eq. 13.23, Eq. 14.14 becomes

$$\mu_i = \bar{h}_i(T) - T \bar{s}_i(T, p_i)$$

$$= \bar{h}_i(T) - T \left(\bar{s}_i^\circ(T) - \bar{R} \ln \frac{y_i p}{p_{\text{ref}}} \right)$$

$$= \bar{h}_i(T) - T \bar{s}_i^\circ(T) + \bar{R} T \ln \frac{y_i p}{p_{\text{ref}}}$$

where p_{ref} is 1 atm and y_i is the mole fraction of component i in a mixture at temperature T and pressure p. The last equation can be written compactly as

$$\boxed{\mu_i = \bar{g}_i^\circ + \bar{R} T \ln \frac{y_i p}{p_{\text{ref}}} \quad \text{(ideal gas)}} \tag{14.17}$$

where $\bar{g}_i^\circ$ is the Gibbs function of component i evaluated at temperature T and a pressure of 1 atm. Additional details concerning the chemical potential concept are provided in Sec. 11.9. Equation 14.17 is the same as Eq. 11.144 developed there.

BioConnections

The human body, as well as all living things, exists in a type of equilibrium that is considered in biology to be a *dynamic* equilibrium called *homeostasis*. The term refers to the ability of the body to regulate its internal state, such as body temperature, within specific limits necessary for life despite the conditions of the external environment, or the level of exertion. Within the body, feedback mechanisms regulate numerous variables that must be kept under control. If this dynamic equilibrium cannot be maintained because of the severe influence of the surroundings, living things can become sick and die.

The numerous control mechanisms in the body are varied, but they share some common elements. Generally, they involve a feedback loop that includes a way to sense a fluctuation of a variable from its desired condition and generate a corrective action to bring the variable back into the desired range. Feedback loops exist within the body at the levels of cells, body fluid systems (such as the circulatory system), tissues, and organs. These feedback loops

also interact to maintain homeostasis much as a home thermostat regulates the heating and cooling system in a house. The mechanisms are highly stable as long as the external conditions are not too severe.

An example of the complexity of homeostasis is the way in which the body maintains the levels of glucose in the blood within desired limits. Glucose is an essential "fuel" for the processes within cells. Glucose absorbed by the digestive system from food, or stored as glycogen in the liver, is distributed by the bloodstream to the cells. The body senses the level of glucose in the blood and various glands produce hormones to *stimulate* the conversion of glucose from stored glycogen, if needed to supplement food intake, or to *inhibit* the release of glucose, as necessary to maintain the desired levels. Several different hormones and numerous organs within the body are involved. The result is a balance that is highly stable within the limits needed by the body for homeostatic equilibrium.

Chemical Equilibrium

In this part of the chapter, the equilibrium criterion $dG]_{T,p} = 0$ introduced in Sec. 14.1 is used to study the equilibrium of reacting mixtures. The objective is to establish the composition present at equilibrium for a specified temperature and pressure. An important parameter for determining the equilibrium composition is the *equilibrium constant*. The equilibrium constant is introduced and its use illustrated by several solved examples. The discussion is concerned only with equilibrium states of reacting systems, and no information can be deduced about the *rates of reaction*. Whether an equilibrium mixture would form quickly or slowly can be determined only by considering the *chemical kinetics*, a topic that is not treated in this text.

14.2 Equation of Reaction Equilibrium

In Chap. 13 the conservation of mass and conservation of energy principles are applied to reacting systems by assuming that the reactions can occur as written. However, the extent to which a chemical reaction proceeds is limited by many factors. In general, the composition of the products actually formed from a given set of reactants, and the relative amounts of the products, can be determined only from experiment. Knowledge of the composition that would be present were a reaction to proceed to equilibrium is frequently useful, however. The *equation of reaction equilibrium* introduced in the present section provides the basis for determining the equilibrium composition of a reacting mixture.

14.2.1 Introductory Case

Consider a closed system consisting initially of a gaseous mixture of hydrogen and oxygen. A number of reactions might take place, including

$$1H_2 + \tfrac{1}{2}O_2 \rightleftarrows 1H_2O \qquad (14.18)$$

$$1H_2 \rightleftarrows 2H \qquad (14.19)$$

$$1O_2 \rightleftarrows 2O \qquad (14.20)$$

Let us consider for illustration purposes only the first of the reactions given above, in which hydrogen and oxygen combine to form water. At equilibrium, the system will consist in general of three components: H_2, O_2, and H_2O, for not all of the hydrogen and oxygen initially present need be reacted. *Changes* in the amounts of these components during each differential step of the reaction leading to the formation of an equilibrium mixture are governed by Eq. 14.18. That is

$$dn_{\text{H}_2} = -dn_{\text{H}_2\text{O}}, \qquad dn_{\text{O}_2} = -\tfrac{1}{2}dn_{\text{H}_2\text{O}} \tag{14.21a}$$

where dn denotes a differential change in the respective component. The minus signs signal that the amounts of hydrogen and oxygen present decrease when the reaction proceeds toward the right. Equations 14.21a can be expressed alternatively as

$$\frac{-dn_{\text{H}_2}}{1} = \frac{-dn_{\text{O}_2}}{\tfrac{1}{2}} = \frac{dn_{\text{H}_2\text{O}}}{1} \tag{14.21b}$$

which emphasizes that increases and decreases in the components are proportional to the stoichiometric coefficients of Eq. 14.18.

Equilibrium is a condition of *balance*. Accordingly, as suggested by the direction of the arrows in Eq. 14.18, when the system is at equilibrium, the tendency of the hydrogen and oxygen to form water is just balanced by the tendency of water to dissociate into oxygen and hydrogen. The equilibrium criterion $dG]_{T,p} = 0$ can be used to determine the composition at an equilibrium state where the temperature is T and the pressure is p. This requires evaluation of the differential $dG]_{T,p}$ in terms of system properties.

For the present case, Eq. 14.10 giving the difference in the Gibbs function of the mixture between two states having the same temperature and pressure, but compositions that differ infinitesimally, takes the following form:

$$dG]_{T,p} = \mu_{\text{H}_2}\, dn_{\text{H}_2} + \mu_{\text{O}_2}\, dn_{\text{O}_2} + \mu_{\text{H}_2\text{O}}\, dn_{\text{H}_2\text{O}} \tag{14.22}$$

The changes in the mole numbers are related by Eqs. 14.21. Hence,

$$dG]_{T,p} = (-1\mu_{\text{H}_2} - \tfrac{1}{2}\mu_{\text{O}_2} + 1\mu_{\text{H}_2\text{O}})\, dn_{\text{H}_2\text{O}}$$

At equilibrium, $dG]_{T,p} = 0$, so the term in parentheses must be zero. That is,

$$-1\mu_{\text{H}_2} - \tfrac{1}{2}\mu_{\text{O}_2} + 1\mu_{\text{H}_2\text{O}} = 0$$

When expressed in a form that resembles Eq. 14.18, this becomes

$$1\mu_{\text{H}_2} + \tfrac{1}{2}\mu_{\text{O}_2} = 1\mu_{\text{H}_2\text{O}} \tag{14.23}$$

Equation 14.23 is the *equation of reaction equilibrium* for the case under consideration. The chemical potentials are functions of temperature, pressure, and composition. Thus, the composition that would be present at equilibrium for a given temperature and pressure can be determined, in principle, by solving this equation. The solution procedure is described in Sec. 14.3.

14.2.2 General Case

The foregoing development can be repeated for reactions involving any number of components. Consider a closed system containing *five* components, A, B, C, D, and E, at a given temperature and pressure, subject to a single chemical reaction of the form

$$\boxed{\nu_A A + \nu_B B \rightleftarrows \nu_C C + \nu_D D} \tag{14.24}$$

where the ν's are stoichiometric coefficients. Component E is assumed to be inert and thus does not appear in the reaction equation. As we will see, component E does influence the equilibrium composition even though it is not involved in the chemical reaction. The form of Eq. 14.24 suggests that at equilibrium the tendency of A and B to form C and D is just balanced by the tendency of C and D to form A and B.

The stoichiometric coefficients ν_A, ν_B, ν_C and ν_D do not correspond to the respective number of moles of the components present. The amounts of the components present are designated n_A, n_B, n_C, n_D, and n_E. However, *changes* in the amounts of the components present do bear a definite relationship to the values of the stoichiometric coefficients. That is,

$$\boxed{\frac{-dn_A}{\nu_A} = \frac{-dn_B}{\nu_B} = \frac{dn_C}{\nu_C} = \frac{dn_D}{\nu_D}} \tag{14.25a}$$

where the minus signs indicate that A and B would be consumed when C and D are produced. Since E is inert, the amount of this component remains constant, so $dn_E = 0$.

Introducing a proportionality factor $d\varepsilon$, Eqs. 14.25a take the form

$$\frac{-dn_A}{\nu_A} = \frac{-dn_B}{\nu_B} = \frac{dn_C}{\nu_C} = \frac{dn_D}{\nu_D} = d\varepsilon$$

from which the following expressions are obtained:

$$\boxed{\begin{aligned} dn_A &= -\nu_A\, d\varepsilon, & dn_B &= -\nu_B\, d\varepsilon \\ dn_C &= \nu_C\, d\varepsilon, & dn_D &= \nu_D\, d\varepsilon \end{aligned}}$$

(14.25b)

extent of reaction

The parameter ε is sometimes referred to as the **extent of reaction**.

For the system under present consideration, Eq. 14.10 takes the form

$$dG]_{T,p} = \mu_A\, dn_A + \mu_B\, dn_B + \mu_C\, dn_C + \mu_D\, dn_D + \mu_E\, dn_E$$

Introducing Eqs. 14.25b and noting that $dn_E = 0$, this becomes

$$dG]_{T,p} = (-\nu_A\mu_A - \nu_B\mu_B + \nu_C\mu_C + \nu_D\mu_D)\, d\varepsilon$$

At equilibrium, $dG]_{T,p} = 0$, so the term in parentheses must be zero. That is,

$$-\nu_A\mu_A - \nu_B\mu_B + \nu_C\mu_C + \nu_D\mu_D = 0$$

or when written in a form resembling Eq. 14.24

$$\boxed{\nu_A\mu_A + \nu_B\mu_B = \nu_C\mu_C + \nu_D\mu_D}$$

(14.26)

equation of reaction equilibrium

For the present case, Eq. 14.26 is the **equation of reaction equilibrium**. In principle, the composition that would be present at equilibrium for a given temperature and pressure can be determined by solving this equation. The solution procedure is simplified through the *equilibrium constant* concept introduced in the next section.

14.3 Calculating Equilibrium Compositions

The objective of the present section is to show how the equilibrium composition of a system at a specified temperature and pressure can be determined by solving the equation of reaction equilibrium. An important part is played in this by the *equilibrium constant*.

14.3.1 Equilibrium Constant for Ideal Gas Mixtures

The first step in solving the equation of reaction equilibrium, Eq. 14.26, for the equilibrium composition is to introduce expressions for the chemical potentials in terms of temperature, pressure, and composition. For an ideal gas mixture, Eq. 14.17 can be used for this purpose. When this expression is introduced for each of the components A, B, C, and D, Eq. 14.26 becomes

$$\nu_A\left(\bar{g}_A^\circ + \bar{R}T\ln\frac{y_A p}{p_{ref}}\right) + \nu_B\left(\bar{g}_B^\circ + \bar{R}T\ln\frac{y_B p}{p_{ref}}\right)$$
$$= \nu_C\left(\bar{g}_C^\circ + \bar{R}T\ln\frac{y_C p}{p_{ref}}\right) + \nu_D\left(\bar{g}_D^\circ + \bar{R}T\ln\frac{y_D p}{p_{ref}}\right)$$

(14.27)

where $\bar{g}_i^\circ$ is the Gibbs function of component i evaluated at temperature T and the pressure $p_{ref} = 1$ atm. Equation 14.27 is the basic working relation for chemical equilibrium in a mixture of ideal gases. However, subsequent calculations are facilitated if it is written in an alternative form, as follows.

Collect like terms and rearrange Eq. 14.27 as

$$(v_C \bar{g}_C^\circ + v_D \bar{g}_D^\circ - v_A \bar{g}_A^\circ - v_B \bar{g}_B^\circ)$$

$$= -\bar{R}T \left(v_C \ln \frac{y_C p}{p_{ref}} + v_D \ln \frac{y_D p}{p_{ref}} - v_A \ln \frac{y_A p}{p_{ref}} - v_B \ln \frac{y_B p}{p_{ref}} \right) \qquad (14.28)$$

The term on the left side of Eq. 14.28 can be expressed concisely as ΔG°. That is,

$$\boxed{\Delta G^\circ = v_C \bar{g}_C^\circ + v_D \bar{g}_D^\circ - v_A \bar{g}_A^\circ - v_B \bar{g}_B^\circ} \qquad (14.29a)$$

which is the change in the Gibbs function for the reaction given by Eq. 14.24 if each reactant and product were separate at temperature T and a pressure of 1 atm. This expression can be written alternatively in terms of specific enthalpies and entropies as

$$\Delta G^\circ = v_C(\bar{h}_C - T\bar{s}_C^\circ) + v_D(\bar{h}_D - T\bar{s}_D^\circ) - v_A(\bar{h}_A - T\bar{s}_A^\circ) - v_B(\bar{h}_B - T\bar{s}_B^\circ)$$

$$= (v_C\bar{h}_C + v_D\bar{h}_D - v_A\bar{h}_A - v_B\bar{h}_B) - T(v_C\bar{s}_C^\circ + v_D\bar{s}_D^\circ - v_A\bar{s}_A^\circ - v_B\bar{s}_B^\circ) \qquad (14.29b)$$

Since the enthalpy of an ideal gas depends on temperature only, the $\bar{h}$'s of Eq. 14.29b are evaluated at temperature T. As indicated by the superscript $^\circ$, each of the entropies is evaluated at temperature T and a pressure of 1 atm.

Introducing Eq. 14.29a into Eq. 14.28 and combining the terms involving logarithms into a single expression gives

$$\boxed{-\frac{\Delta G^\circ}{\bar{R}T} = \ln \left[\frac{y_C^{v_C} y_D^{v_D}}{y_A^{v_A} y_B^{v_B}} \left(\frac{p}{p_{ref}} \right)^{v_C + v_D - v_A - v_B} \right]} \qquad (14.30)$$

Equation 14.30 is simply the form taken by the equation of reaction equilibrium, Eq. 14.26, for an ideal gas mixture subject to the reaction Eq. 14.24. As illustrated by subsequent examples, similar expressions can be written for other reactions.

Equation 14.30 can be expressed concisely as

$$\boxed{-\frac{\Delta G^\circ}{\bar{R}T} = \ln K(T)} \qquad (14.31)$$

where K is the **equilibrium constant** defined by

equilibrium constant

$$\boxed{K(T) = \frac{y_C^{v_C} y_D^{v_D}}{y_A^{v_A} y_B^{v_B}} \left(\frac{p}{p_{ref}} \right)^{v_C + v_D - v_A - v_B}} \qquad (14.32)$$

Given the values of the stoichiometric coefficients, v_A, v_B, v_C, and v_D and the temperature T, the left side of Eq. 14.31 can be evaluated using either of Eqs. 14.29 together with the appropriate property data. The equation can then be solved for the value of the equilibrium constant K. Accordingly, for selected reactions K can be evaluated and tabulated against temperature. It is common to tabulate $\log_{10}K$ or $\ln K$ versus temperature, however. A tabulation of $\log_{10}K$ values over a range of temperatures for several reactions is provided in Table A-27, which is extracted from a more extensive compilation.

The terms in the numerator and denominator of Eq. 14.32 correspond, respectively, to the products and reactants of the reaction given by Eq. 14.24 as it proceeds from left to right as written. For the *inverse* reaction $v_C C + v_D D \rightleftarrows v_A A + v_B B$, the equilibrium constant takes the form

$$K^* = \frac{y_A^{v_A} y_B^{v_B}}{y_C^{v_C} y_D^{v_D}} \left(\frac{p}{p_{ref}} \right)^{v_A + v_B - v_C - v_D} \qquad (14.33)$$

Comparing Eqs. 14.32 and 14.33, it follows that the value of K^* is just the reciprocal of K: $K^* = 1/K$. Accordingly,

$$\log_{10} K^* = \log_{10} K \qquad (14.34)$$

Hence, Table A-27 can be used both to evaluate K for the reactions listed proceeding in the direction left to right and to evaluate K^* for the inverse reactions proceeding in the direction right to left.

Example 14.1 illustrates how the $\log_{10} K$ values of Table A-27 are determined. Subsequent examples show how the $\log_{10} K$ values can be used to evaluate equilibrium compositions.

► EXAMPLE 14.1 ►

Evaluating the Equilibrium Constant at a Specified Temperature

Evaluate the equilibrium constant, expressed as $\log_{10} K$, for the reaction $CO + \frac{1}{2} O_2 \rightleftarrows CO_2$ at **(a)** 298 K and **(b)** 2000 K. Compare with the value obtained from Table A-27.

SOLUTION

Known The reaction is $CO + \frac{1}{2} O_2 \rightleftarrows CO_2$.

Find Determine the equilibrium constant for $T = 298$ K (25°C) and $T = 2000$ K.

Engineering Model The ideal gas model applies.

Analysis The equilibrium constant requires the evaluation of $\Delta G°$ for the reaction. Invoking Eq. 14.29b for this purpose, we have

$$\Delta G° = (\bar{h}_{CO_2} - \bar{h}_{CO} - \tfrac{1}{2}\bar{h}_{O_2}) - T(\bar{s}°_{CO_2} - \bar{s}°_{CO} - \tfrac{1}{2}\bar{s}°_{O_2})$$

where the enthalpies are evaluated at temperature T and the absolute entropies are evaluated at temperature T and a pressure of 1 atm. Using Eq. 13.9, the enthalpies are evaluated in terms of the respective enthalpies of formation, giving

$$\Delta G° = [(\bar{h}°_f)_{CO_2} - (\bar{h}°_f)_{CO} - \tfrac{1}{2}\overset{0}{(\bar{h}°_f)_{O_2}}] + [(\Delta\bar{h})_{CO_2} - (\Delta\bar{h})_{CO}$$
$$- \tfrac{1}{2}(\Delta\bar{h})_{O_2}] - T(\bar{s}°_{CO_2} - \bar{s}°_{CO} - \tfrac{1}{2}\bar{s}°_{O_2})$$

where the $\Delta\bar{h}$ terms account for the change in specific enthalpy from $T_{ref} = 298$ K to the specified temperature T. The enthalpy of formation of oxygen is zero by definition.

a. When $T = 298$ K, the $\Delta\bar{h}$ terms of the above expression for $\Delta G°$ vanish. The required enthalpy of formation and absolute entropy values can be read from Table A-25, giving

$$\Delta G° = [(-393,520) - (-110,530) - \tfrac{1}{2}(0)]$$
$$- 298[213.69 - 197.54 - \tfrac{1}{2}(205.03)]$$
$$= -257,253 \text{ kJ/kmol}$$

With this value for $\Delta G°$, Eq. 14.31 gives

$$\ln K = -\frac{(-257,253 \text{ kJ/kmol})}{(8.314 \text{ kJ/kmol} \cdot \text{K})(298 \text{ K})} = 103.83$$

which corresponds to $\log_{10} K = 45.093$.

Table A-27 gives the logarithm to the base 10 of the equilibrium constant for the inverse reaction: $CO_2 \rightleftarrows CO + \frac{1}{2} O_2$. That is, $\log_{10} K^* = -45.066$. Thus, with Eq. 14.34, $\log_{10} K = 45.066$, which agrees closely with the calculated value.

b. When $T = 2000$ K, the $\Delta\bar{h}$ and $\bar{s}°$ terms for O_2, CO, and CO_2 required by the above expression for $\Delta G°$ are evaluated from Table A-23. The enthalpy of formation values are the same as in part (a). Thus,

$$\Delta G° = [(-393,520) - (-110,530) - \tfrac{1}{2}(0)] + [(100,804 - 9364)$$
$$- (65408 - 8669) - \tfrac{1}{2}(67,881 - 8682)] - 2000[309.210$$
$$- 258.600 - \tfrac{1}{2}(268.655)]$$
$$= -282,990 + 5102 + 167,435 = -110,453 \text{ kJ/kmol}$$

With this value, Eq. 14.31 gives

$$\ln K = -\frac{(-110,453)}{(8.314)(2000)} = 6.643$$

which corresponds to $\log_{10} K = 2.885$.

At 2000 K, Table A-27 gives $\log_{10} K^* = -2.884$. With Eq. 14.34, $\log_{10} K = 2.884$, which is in agreement with the calculated value.

Using the procedures described above, it is straightforward to determine $\log_{10} K$ versus temperature for each of several specified reactions and tabulate the results as in Table A-27.

SKILLS DEVELOPED

Ability to...

• evaluate $\log_{10} K$ based on Eq. 14.31 and data from Tables A-23 and A-25.

• use the relation of Eq. 14.34 for inverse reactions.

Quick Quiz

If $\ln K = 23.535$ for the given reaction, use Table A-27 to determine T, in K. **Ans. 1000 K.**

14.3.2 Illustrations of the Calculation of Equilibrium Compositions for Reacting Ideal Gas Mixtures

It is often convenient to express Eq. 14.32 explicitly in terms of the number of moles that would be present at equilibrium. Each mole fraction appearing in the equation has the form $y_i = n_i/n$, where n_i is the amount of component i in the equilibrium mixture and n is the total number of moles of mixture. Hence, Eq. 14.32 can be rewritten as

$$K = \frac{n_C^{\nu_C}\, n_D^{\nu_D}}{n_A^{\nu_A}\, n_B^{\nu_B}} \left(\frac{p/p_{\text{ref}}}{n} \right)^{\nu_C + \nu_D - \nu_A - \nu_B} \qquad (14.35)$$

The value of n must include not only the reacting components A, B, C, and D but also all inert components present. Since inert component E has been assumed present, we would write $n = n_A + n_B + n_C + n_D + n_E$.

Equation 14.35 provides a relationship among the temperature, pressure, and composition of an ideal gas mixture at equilibrium. Accordingly, if any two of temperature, pressure, and composition are known, the third can be found by solving this equation.

FOR EXAMPLE

Suppose that the temperature T and pressure p are known and the object is the equilibrium composition. With temperature known, the value of K can be obtained from Table A-27. The n's of the reacting components A, B, C, and D can be expressed in terms of a single unknown variable through application of the conservation of mass principle to the various chemical species present. Then, since the pressure is known, Eq. 14.35 constitutes a single equation in a single unknown, which can be solved using an *equation solver* or iteratively with a hand calculator.

In Example 14.2, we apply Eq. 14.35 to study the effect of pressure on the equilibrium composition of a mixture of CO_2, CO, and O_2.

▶▶ **EXAMPLE 14.2** ▶ ···

Determining Equilibrium Composition Given Temperature and Pressure

One kilomole of carbon monoxide, CO, reacts with $\frac{1}{2}$ kmol of oxygen, O_2, to form an equilibrium mixture of CO_2, CO, and O_2 at 2500 K and **(a)** 1 atm, **(b)** 10 atm. Determine the equilibrium composition in terms of mole fractions.

SOLUTION

Known A system initially consisting of 1 kmol of CO and $\frac{1}{2}$ kmol of O_2 reacts to form an equilibrium mixture of CO_2, CO, and O_2. The temperature of the mixture is 2500 K and the pressure is (a) 1 atm, (b) 10 atm.

Find Determine the equilibrium composition in terms of mole fractions.

Engineering Model The equilibrium mixture is modeled as an ideal gas mixture.

Analysis Equation 14.35 relates temperature, pressure, and composition for an ideal gas mixture at equilibrium. If any two are known, the third can be determined using this equation.

In the present case, T and p are known, and the composition is unknown.

Applying conservation of mass, the overall balanced chemical reaction equation is

$$1CO + \frac{1}{2}O_2 \rightarrow zCO + \frac{z}{2}O_2 + (1-z)CO_2$$

where z is the amount of CO, in kmol, present in the equilibrium mixture. Note that $0 \le z \le 1$.

The total number of moles n in the equilibrium mixture is

$$n = z + \frac{z}{2} + (1-z) = \frac{2+z}{2}$$

Accordingly, the molar analysis of the equilibrium mixture is

$$y_{CO} = \frac{2z}{2+z}, \qquad y_{O_2} = \frac{z}{2+z}, \qquad y_{CO_2} = \frac{2(1-z)}{2+z}$$

At equilibrium, the tendency of CO and O_2 to form CO_2 is just balanced by the tendency of CO_2 to form CO and O_2, so we have $CO_2 \rightleftarrows CO + \frac{1}{2}O_2$. Accordingly, Eq. 14.35 takes the form

$$K = \frac{z(z/2)^{1/2}}{(1-z)}\left[\frac{p/p_{\text{ref}}}{(2+z)/2}\right]^{1+1/2-1} = \frac{z}{1-z}\left(\frac{z}{2+z}\right)^{1/2}\left(\frac{p}{p_{\text{ref}}}\right)^{1/2}$$

At 2500 K, Table A-27 gives $\log_{10}K = -1.440$. Thus, $K = 0.0363$. Inserting this value into the last expression

$$0.0363 = \frac{z}{1-z}\left(\frac{z}{2+z}\right)^{1/2}\left(\frac{p}{p_{\text{ref}}}\right)^{1/2} \qquad \text{(a)}$$

a. When $p = 1$ atm, Eq. (a) becomes

$$0.0363 = \frac{z}{1-z}\left(\frac{z}{2+z}\right)^{1/2}$$

Using an equation solver or iteration with a hand calculator, $z = 0.129$. The equilibrium composition in terms of mole fractions is then

$$y_{CO} = \frac{2(0.129)}{2.129} = 0.121, \qquad y_{O_2} = \frac{0.129}{2.129} = 0.061,$$

$$y_{CO_2} = \frac{2(1-0.129)}{2.129} = 0.818$$

b. When $p = 10$ atm, Eq. (a) becomes

$$0.0363 = \frac{z}{1-z}\left(\frac{z}{2+z}\right)^{1/2}(10)^{1/2}$$

❶ Solving, $z = 0.062$. The corresponding equilibrium composition in terms of mole fractions is $y_{CO} = 0.06$, $y_{O_2} = 0.03$, $y_{CO_2} = 0.91$.

❶ Comparing the results of parts (a) and (b) we conclude that the extent to which the reaction proceeds toward completion (the extent to which CO_2 is formed) is increased by increasing the pressure.

SKILLS DEVELOPED

Ability to...

- apply Eq. 14.35 to determine equilibrium composition given temperature and pressure.
- retrieve and use data from Table A-27.

Quick Quiz

If $z = 0.0478$ (corresponding to $p = 22.4$ atm, $T = 2500$ K), what would be the mole fraction of each constituent of the equilibrium mixture? Ans. $y_{CO} = 0.0467$, $y_{O_2} = 0.0233$, $y_{CO_2} = 0.9300$.

In Example 14.3, we determine the temperature of an equilibrium mixture when the pressure and composition are known.

EXAMPLE 14.3

Determining Equilibrium Temperature Given Pressure and Composition

Measurements show that at a temperature T and a pressure of 1 atm, the equilibrium mixture for the system of Example 14.2 has the composition $y_{CO} = 0.298$, $y_{O_2} = 0.149$, $y_{CO_2} = 0.553$. Determine the temperature T of the mixture, in K.

SOLUTION

Known The pressure and composition of an equilibrium mixture of CO, O_2, and CO_2 are specified.

Find Determine the temperature of the mixture, in K.

Engineering Model The mixture can be modeled as an ideal gas mixture.

Analysis Equation 14.35 relates temperature, pressure, and composition for an ideal gas mixture at equilibrium. If any two are known, the third can be found using this equation. In the present case, composition and pressure are known, and the temperature is the unknown.

Equation 14.35 takes the same form here as in Example 14.2. Thus, when $p = 1$ atm, we have

$$K(T) = \frac{z}{1-z}\left(\frac{z}{2+z}\right)^{1/2}$$

where z is the amount of CO, in kmol, present in the equilibrium mixture and T is the temperature of the mixture.

The solution to Example 14.2 gives the following expression for the mole fraction of the CO in the mixture: $y_{CO} = 2z/(2+z)$. Since $y_{CO} = 0.298$, $z = 0.35$.

❶ Inserting this value for z into the expression for the equilibrium constant gives $K = 0.2078$. Thus, $\log_{10}K = -0.6824$. Interpolation in Table A-27 then gives $T = 2881$ K.

❶ Comparing this example with part (a) of Example 14.2, we conclude that the extent to which the reaction proceeds to completion (the extent to which CO_2 is formed) is decreased by increasing the temperature.

Quick Quiz

Determine the temperature, in K, for a pressure of 2 atm if the equilibrium composition were unchanged. Ans. ≈ 2970 K.

In Example 14.4, we consider the effect of an inert component on the equilibrium composition.

► EXAMPLE 14.4 ►

Considering the Effect on Equilibrium of an Inert Component

One kilomole of carbon monoxide reacts with the theoretical amount of air to form an equilibrium mixture of CO_2, CO, O_2, and N_2 at 2500 K and 1 atm. Determine the equilibrium composition in terms of mole fractions, and compare with the result of Example 14.2.

SOLUTION

Known A system initially consisting of 1 kmol of CO and the theoretical amount of air reacts to form an equilibrium mixture of CO_2, CO, O_2, and N_2. The temperature and pressure of the mixture are 2500 K and 1 atm.

Find Determine the equilibrium composition in terms of mole fractions, and compare with the result of Example 14.2.

Engineering Model The equilibrium mixture can be modeled as an ideal gas mixture wherein N_2 is inert.

Analysis For a *complete reaction* of CO with the theoretical amount of air

$$CO + \tfrac{1}{2}O_2 + 1.88N_2 \rightarrow CO_2 + 1.88N_2$$

Accordingly, the reaction of CO with the theoretical amount of air to form CO_2, CO, O_2, and N_2 is

$$CO + \tfrac{1}{2}O_2 + 1.88N_2 \rightarrow zCO + \frac{z}{2}O_2 + (1-z)CO_2 + 1.88N_2$$

where z is the amount of CO, in kmol, present in the equilibrium mixture.

The total number of moles n in the equilibrium mixture is

$$n = z + \frac{z}{2} + (1-z) + 1.88 = \frac{5.76 + z}{2}$$

The composition of the equilibrium mixture in terms of mole fractions is

$$y_{CO} = \frac{2z}{5.76 + z}, \qquad y_{O_2} = \frac{z}{5.76 + z},$$

$$y_{CO_2} = \frac{2(1-z)}{5.76 + z}, \qquad y_{N_2} = \frac{3.76}{5.76 + z}$$

At equilibrium we have $CO_2 \rightleftarrows CO + \tfrac{1}{2}O_2$. So Eq. 14.35 takes the form

$$K = \frac{z(z/2)^{1/2}}{(1-z)}\left[\frac{p/p_{\text{ref}}}{(5.76 + z)/2}\right]^{1/2}$$

The value of K is the same as in the solution to Example 14.2, $K = 0.0363$. Thus, since $p = 1$ atm, we have

$$0.0363 = \frac{z}{1-z}\left(\frac{z}{5.76 + z}\right)^{1/2}$$

Solving, $z = 0.175$. The corresponding equilibrium composition is $y_{CO} = 0.059$, $y_{CO_2} = 0.278$, $y_{O_2} = 0.029$, $y_{N_2} = 0.634$.

Comparing this example with Example 14.2, we conclude that the presence of the inert component nitrogen reduces the extent to which the reaction proceeds toward completion at the specified temperature and pressure (reduces the extent to which CO_2 is formed).

Quick Quiz

Determine the amounts, in kmol, of each component of the equilibrium mixture. Ans. $n_{CO} = 0.175$, $n_{O_2} = 0.0875$, $n_{CO_2} = 0.8250$, $n_{N_2} = 1.88$.

In the next example, the equilibrium concepts of this chapter are applied together with the energy balance for reacting systems developed in Chap. 13.

> ▶ ▶ **EXAMPLE 14.5** ▶

Using Equilibrium Concepts with the Energy Balance

Carbon dioxide at 25°C, 1 atm enters a reactor operating at steady state and dissociates, giving an equilibrium mixture of CO_2, CO, and O_2 that exits at 3200 K, 1 atm. Determine the heat transfer to the reactor, in kJ per kmol of CO_2 entering. The effects of kinetic and potential energy can be ignored and $\dot{W}_{cv} = 0$.

SOLUTION

Known Carbon dioxide at 25°C, 1 atm enters a reactor at steady state. An equilibrium mixture of CO_2, CO, and O_2 exits at 3200 K, 1 atm.

Find Determine the heat transfer to the reactor, in kJ per kmol of CO_2 entering.

Schematic and Given Data:

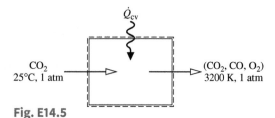

Fig. E14.5

Engineering Model

1. The control volume shown on the accompanying sketch by a dashed line operates at steady state with $\dot{W}_{cv} = 0$. Kinetic energy and potential energy effects can be ignored.

2. The entering CO_2 is modeled as an ideal gas.

3. The exiting mixture of CO_2, CO, and O_2 is an equilibrium ideal gas mixture.

Analysis The required heat transfer can be determined from an energy rate balance for the control volume, but first the composition of the exiting equilibrium mixture must be determined.

Applying the conservation of mass principle, the overall dissociation reaction is described by

$$CO_2 \rightarrow z\,CO_2 + (1-z)CO + \left(\frac{1-z}{2}\right)O_2$$

where z is the amount of CO_2, in kmol, present in the mixture exiting the control volume, per kmol of CO_2 entering. The total number of moles n in the mixture is then

$$n = z + (1-z) + \left(\frac{1-z}{2}\right) = \frac{3-z}{2}$$

The exiting mixture is assumed to be an equilibrium mixture (assumption 3). Thus, for the mixture we have $CO_2 \rightleftarrows CO + \frac{1}{2}O_2$. Equation 14.35 takes the form

$$K = \frac{(1-z)[(1-z)/2]^{1/2}}{z}\left[\frac{p/p_{ref}}{(3-z)/2}\right]^{1+1/2-1}$$

Rearranging and noting that $p = 1$ atm

$$K = \left(\frac{1-z}{z}\right)\left(\frac{1-z}{3-z}\right)^{1/2}$$

At 3200 K, Table A-27 gives $\log_{10}K = -0.189$. Thus, $K = 0.647$, and the equilibrium constant expression becomes

$$0.647 = \left(\frac{1-z}{z}\right)\left(\frac{1-z}{3-z}\right)^{1/2}$$

Solving, $z = 0.422$. The composition of the exiting equilibrium mixture, in kmol per kmol of CO_2 entering, is then 0.422CO_2, 0.578CO, 0.289O_2.

When expressed per kmol of CO_2 entering the control volume, the energy rate balance reduces by assumption 1 to

$$0 = \frac{\dot{Q}_{cv}}{\dot{n}_{CO_2}} - \frac{\dot{W}_{cv}^{\,0}}{\dot{n}_{CO_2}} + \overline{h}_{CO_2} - (0.422\overline{h}_{CO_2} + 0.578\overline{h}_{CO} + 0.289\overline{h}_{O_2})$$

Solving for the heat transfer per kmol of CO_2 entering, and evaluating each enthalpy in terms of the respective enthalpy of formation

$$\frac{\dot{Q}_{cv}}{\dot{n}_{CO_2}} = 0.422(\overline{h}_f^\circ + \Delta\overline{h})_{CO_2} + 0.578(\overline{h}_f^\circ + \Delta\overline{h})_{CO}$$
$$+ 0.289(\overline{h}_f^{\circ\,0} + \Delta\overline{h})_{O_2} - (\overline{h}_f^\circ + \Delta\overline{h}^{\,0})_{CO_2}$$

The enthalpy of formation of O_2 is zero by definition; $\Delta\overline{h}$ for the CO_2 at the inlet vanishes because CO_2 enters at 25°C.

With enthalpy of formation values from Table A-25 and $\Delta\overline{h}$ values for O_2, CO, and CO_2 from Table A-23

$$\frac{\dot{Q}_{cv}}{\dot{n}_{CO_2}} = 0.422[-393{,}520 + (174{,}695 - 9364)] + 0.578[-110{,}530$$
$$+ (109{,}667 - 8669)] + 0.289(114{,}809 - 8682) - (-393{,}520)$$

❶ $$= 322{,}385 \text{ kJ/kmol}(CO_2)$$

---

❶ For comparison, let us determine the heat transfer if we assume no dissociation—namely, when CO_2 alone exits the reactor. With data from Table A-23, the heat transfer is

$$\frac{\dot{Q}_{cv}}{\dot{n}_{CO_2}} = \overline{h}_{CO_2}(3200 \text{ K}) - \overline{h}_{CO_2}(298 \text{ K})$$

$$= 174{,}695 - 9364 = 165{,}331 \text{ kJ/kmol}(CO_2)$$

The value is much less than the value obtained in the solution above because the dissociation of CO_2 requires more energy input (an endothermic reaction).

SKILLS DEVELOPED

Ability to...

- apply Eq. 14.35 together with the energy balance for reacting systems to determine heat transfer for a reactor.

- retrieve and use data from Tables A-23, A-25, and A-27.

Quick Quiz

Determine the heat transfer rate, in kW, and the molar flow rate of mixture exiting, in kmol/s, for a flow rate of 3.1×10^{-5} kmol/s of CO_2 entering. Ans. 10 kW, 4×10^{-5} kmol/s.

14.3.3 Equilibrium Constant for Mixtures and Solutions

TAKE NOTE...

Study of Sec. 14.3.3 requires content from Sec. 11.9.

The procedures that led to the equilibrium constant for reacting ideal gas mixtures can be followed for the general case of reacting mixtures by using the fugacity and activity concepts introduced in Sec. 11.9. In principle, equilibrium compositions of such mixtures can be determined with an approach paralleling the one for ideal gas mixtures.

Equation 11.141 can be used to evaluate the chemical potentials appearing in the equation of reaction equilibrium (Eq. 14.26). The result is

$$\nu_A(\bar{g}_A^\circ + \bar{R}T \ln a_A) + \nu_B(\bar{g}_B^\circ + \bar{R}T \ln a_B) = \nu_C(\bar{g}_C^\circ + \bar{R}T \ln a_C) + \nu_D(\bar{g}_D^\circ + \bar{R}T \ln a_D)$$

(14.36)

where $\bar{g}_i^\circ$ is the Gibbs function of pure component i at temperature T and the pressure $p_{ref} = 1$ atm, and a_i is the *activity* of that component.

Collecting terms and employing Eq. 14.29a, Eq. 14.36 becomes

$$-\frac{\Delta G^\circ}{\bar{R}T} = \ln\left(\frac{a_C^{\nu_C} a_D^{\nu_D}}{a_A^{\nu_A} a_B^{\nu_B}}\right)$$

(14.37)

This equation can be expressed in the same form as Eq. 14.31 by defining the equilibrium constant as

$$K = \frac{a_C^{\nu_C} a_D^{\nu_D}}{a_A^{\nu_A} a_B^{\nu_B}}$$

(14.38)

Since Table A-27 and similar compilations are constructed simply by evaluating $-\Delta G^\circ/\bar{R}T$ for specified reactions at several temperatures, such tables can be employed to evaluate the more general equilibrium constant given by Eq. 14.38. However, before Eq. 14.38 can be used to determine the equilibrium composition for a known value of K, it is necessary to evaluate the activity of the various mixture components. Let us illustrate this for the case of mixtures that can be modeled as *ideal solutions*.

Ideal Solutions   For an ideal solution, the activity of component i is given by

$$a_i = \frac{y_i f_i}{f_i^\circ}$$

(11.142)

where f_i is the fugacity of pure i at the temperature T and pressure p of the mixture, and f_i° is the fugacity of pure i at temperature T and the pressure p_{ref}. Using this expression to evaluate a_A, a_B, a_C, and a_D, Eq. 14.38 becomes

$$K = \frac{(y_C f_C/f_C^\circ)^{\nu_C}(y_D f_D/f_D^\circ)^{\nu_D}}{(y_A f_A/f_A^\circ)^{\nu_A}(y_B f_B/f_B^\circ)^{\nu_B}}$$

(14.39a)

which can be expressed alternatively as

$$K = \left[\frac{(f_C/p)^{\nu_C}(f_D/p)^{\nu_D}}{(f_A/p)^{\nu_A}(f_B/p)^{\nu_B}}\right]\left[\frac{(f_A^\circ/p_{ref})^{\nu_A}(f_B^\circ/p_{ref})^{\nu_B}}{(f_C^\circ/p_{ref})^{\nu_C}(f_D^\circ/p_{ref})^{\nu_D}}\right]\underline{\left[\frac{y_C^{\nu_C} y_D^{\nu_D}}{y_A^{\nu_A} y_B^{\nu_B}}\left(\frac{p}{p_{ref}}\right)^{\nu_C+\nu_D-\nu_A-\nu_B}\right]}$$

(14.39b)

The ratios of fugacity to pressure in this equation can be evaluated, in principle, from Eq. 11.124 or the generalized fugacity chart, Fig. A-6, developed from it. In the special case when each component behaves as an ideal gas at both T, p and T, p_{ref}, these ratios equal unity and Eq. 14.39b reduces to the underlined term, which is just Eq. 14.32.

Horizons

Methane, *Another* Greenhouse Gas

While carbon dioxide is often mentioned by the media because of its effect on global climate change, and rightly so, other gases released to the atmosphere also contribute to climate change but get less publicity. In particular, methane, CH_4, which receives little notice as a greenhouse gas, has a Global Warming Potential of 25, compared to carbon dioxide's GWP of 1 (see Table 10.1).

Sources of methane related to human activity include fossil-fuel (coal, natural gas, and petroleum) production, distribution, combustion, and other uses. Wastewater treatment, landfills, and agriculture, including ruminant animals raised for food, are also human-related sources of methane. Natural sources of methane include wetlands and methane *hydrate* deposits in seafloor sediments.

For decades, the concentration of methane in the atmosphere has increased significantly. But some observers report that the increase has slowed recently and may be ceasing. While this could be only a temporary pause, reasons have been advanced to explain the development. Some say governmental actions aimed at reducing release of methane have begun to show results. Changes in agricultural practices, such as the way rice is produced, also

may be a factor in the reported reduction of methane in the atmosphere.

Another view is that the plateau in atmospheric methane may at least in part be due to chemical equilibrium: Methane released to the atmosphere is balanced by its consumption in the atmosphere. Methane is consumed in the atmosphere principally by its reaction with the hydroxyl radical (OH), which is produced through decomposition of atmospheric ozone by action of solar radiation. For instance, OH reacts with methane to yield water and CH_3, a methyl radical, according to $CH_4 + OH \rightarrow H_2O + CH_3$. Other reactions follow this, leading eventually to water-soluble products that are *washed out* of the atmosphere by rain and snow.

Understanding the reasons for the apparent slowing rate of growth of methane in the atmosphere will take effort, including quantifying changes in the various sources of methane and pinpointing natural mechanisms by which it is removed from the atmosphere. Better understanding will enable us to craft measures aimed at curbing release of methane, allowing the atmosphere's natural ability to cleanse itself to assist in maintaining a healthier balance.

14.4 Further Examples of the Use of the Equilibrium Constant

Some additional aspects of the use of the equilibrium constant are introduced in this section: the equilibrium flame temperature, the van't Hoff equation, and chemical equilibrium for ionization reactions and simultaneous reactions. To keep the presentation at an introductory level, only the case of ideal gas mixtures is considered.

14.4.1 Determining Equilibrium Flame Temperature

In this section, the effect of incomplete combustion on the adiabatic flame temperature, introduced in Sec. 13.3, is considered using concepts developed in the present chapter. We begin with a review of some ideas related to the adiabatic flame temperature by considering a reactor operating at steady state for which no significant heat transfer with the surroundings takes place.

Let carbon monoxide gas entering at one location react *completely* with the theoretical amount of air entering at another location as follows:

$$CO + \tfrac{1}{2}O_2 + 1.88N_2 \rightarrow CO_2 + 1.88N_2$$

As discussed in Sec. 13.3, the products would exit the reactor at a temperature we have designated the *maximum* adiabatic flame temperature. This temperature can be determined by solving a *single* equation, the energy equation. At such an elevated temperature, however, there would be a tendency for CO_2 to dissociate

$$CO_2 \rightarrow CO + \tfrac{1}{2}O_2$$

Since dissociation requires energy (an endothermic reaction), the temperature of the products would be *less than* the maximum adiabatic temperature found under the assumption of complete combustion.

When dissociation takes place, the gaseous products exiting the reactor would not be CO_2 and N_2 but a mixture of CO_2, CO, O_2, and N_2. The balanced chemical reaction equation would read

$$CO + \tfrac{1}{2}O_2 + 1.88N_2 \rightarrow zCO + (1 - z)CO_2 + \frac{z}{2}O_2 + 1.88N_2 \qquad (14.40)$$

where z is the amount of CO, in kmol, present in the exiting mixture for each kmol of CO entering the reactor.

Accordingly, there are *two* unknowns: z and the temperature of the exiting stream. To solve a problem with two unknowns requires two equations. One is provided by an energy equation. If the exiting gas mixture is in equilibrium, the other equation is provided by the equilibrium constant, Eq. 14.35. The temperature of the products may then be called the **equilibrium flame temperature**. The equilibrium constant used to evaluate the equilibrium flame temperature would be determined with respect to $CO_2 \rightleftharpoons CO + \tfrac{1}{2}O_2$.

Although only the dissociation of CO_2 has been discussed, other products of combustion may dissociate, for example,

$$H_2O \rightleftharpoons H_2 + \tfrac{1}{2}O_2$$
$$H_2O \rightleftharpoons OH + \tfrac{1}{2}H_2$$
$$O_2 \rightleftharpoons 2O$$
$$H_2 \rightleftharpoons 2H$$
$$N_2 \rightleftharpoons 2N$$

When there are many dissociation reactions, the study of chemical equilibrium is facilitated by the use of computers to solve the *simultaneous* equations that result. Simultaneous reactions are considered in Sec. 14.4.4. The following example illustrates how the equilibrium flame temperature is determined when one dissociation reaction occurs.

equilibrium flame temperature

▶▶ EXAMPLE 14.6 ▶

Determining the Equilibrium Flame Temperature

Carbon monoxide at 25°C, 1 atm enters a well-insulated reactor and reacts with the theoretical amount of air entering at the same temperature and pressure. An equilibrium mixture of CO_2, CO, O_2, and N_2 exits the reactor at a pressure of 1 atm. For steady-state operation and negligible effects of kinetic and potential energy, determine the composition and temperature of the exiting mixture, in K.

SOLUTION

Known Carbon monoxide at 25°C, 1 atm reacts with the theoretical amount of air at 25°C, 1 atm to form an equilibrium mixture of CO_2, CO, O_2, and N_2 at temperature T and a pressure of 1 atm.

Find Determine the composition and temperature of the exiting mixture.

Schematic and Given Data:

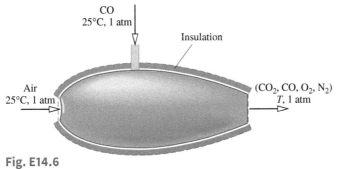

Fig. E14.6

Engineering Model

1. The control volume shown on the accompanying sketch by a dashed line operates at steady state with $\dot{Q}_{cv} = 0, \dot{W}_{cv} = 0$, and negligible effects of kinetic and potential energy.

2. The entering gases are modeled as ideal gases.

3. The exiting mixture is an ideal gas mixture at equilibrium wherein N_2 is inert.

Analysis The overall reaction is the same as in the solution to Example 14.4:

$$CO + \tfrac{1}{2}O_2 + 1.88N_2 \rightarrow zCO + \frac{z}{2}O_2 + (1 - z)CO_2 + 1.88N_2$$

By assumption 3, the exiting mixture is an equilibrium mixture. The equilibrium constant expression developed in the solution to Example 14.4 is

$$K(T) = \frac{z(z/2)^{1/2}}{(1 - z)} \left(\frac{p/p_{ref}}{(5.76 + z)/2} \right)^{1/2} \qquad (a)$$

Since $p = 1$ atm, Eq. (a) reduces to

$$K(T) = \frac{z}{(1 - z)} \left(\frac{z}{5.76 + z} \right)^{1/2} \qquad (b)$$

This equation involves two unknowns: z and the temperature T of the exiting equilibrium mixture.

Another equation involving the two unknowns is obtained from an energy rate balance of the form Eq. 13.12b, which reduces with assumption 1 to

$$\bar{h}_R = \bar{h}_P \qquad (c)$$

where

$$\bar{h}_R = (\bar{h}_f^\circ + \Delta\bar{h}^0)_{CO} + \frac{1}{2}(\bar{h}_f^{\circ\,0} + \Delta\bar{h}^0)_{O_2} + 1.88(\bar{h}_f^{\circ\,0} + \Delta\bar{h}^0)_{N_2}$$

and

$$\bar{h}_P = z(\bar{h}_f^\circ + \Delta\bar{h})_{CO} + \frac{z}{2}(\bar{h}_f^{\circ\,0} + \Delta\bar{h})_{O_2} + (1-z)(\bar{h}_f^\circ + \Delta\bar{h})_{CO_2}$$
$$+ 1.88(\bar{h}_f^{\circ\,0} + \Delta\bar{h})_{N_2}$$

The enthalpy of formation terms set to zero are those for oxygen and nitrogen. Since the reactants enter at 25°C, the corresponding $\Delta\bar{h}$ terms also vanish. Collecting and rearranging, we get

$$z(\Delta\bar{h})_{CO} + \frac{z}{2}(\Delta\bar{h})_{O_2} + (1-z)(\Delta\bar{h})_{CO_2} + 1.88(\Delta\bar{h})_{N_2}$$
$$+ (1-z)[(\bar{h}_f^\circ)_{CO_2} - (\bar{h}_f^\circ)_{CO}] = 0 \qquad (d)$$

Equations (b) and (d) are simultaneous equations involving the unknowns z and T. When solved *iteratively* using *tabular data*, the results are $z = 0.125$ and $T = 2399$ K, as can be verified. The composition of the equilibrium mixture, in kmol per kmol of CO entering the reactor, is then $0.125CO$, $0.0625O_2$, $0.875CO_2$, $1.88N_2$.

SKILLS DEVELOPED

Ability to...

• apply Eq. 14.35 together with the energy balance for reacting systems to determine equilibrium flame temperature.

• retrieve and use data from Tables A-23, A-25, and A-27.

Quick Quiz

If the CO and air each entered at 500°C, would the equilibrium flame temperature increase, decrease, or stay constant? Ans. Increase.

As illustrated by Example 14.7, the equation solver and property retrieval features of *Interactive Thermodynamics: IT* allow the equilibrium flame temperature and composition to be determined without the iteration required when using table data.

> ▷ ▷ **EXAMPLE 14.7** ▷ •

Determining the Equilibrium Flame Temperature Using Software

Solve Example 14.6 using *Interactive Thermodynamics: IT* and plot equilibrium flame temperature and z, the amount of CO present in the exiting mixture, each versus pressure ranging from 1 to 10 atm.

SOLUTION

Known See Example 14.6.

Find Using *IT*, plot the equilibrium flame temperature and the amount of CO present in the exiting mixture of Example 14.6, each versus pressure ranging from 1 to 10 atm.

Engineering Model See Example 14.6.

Analysis Equation (a) of Example 14.6 provides the point of departure for the *IT* solution

$$K(T) = \frac{z(z/2)^{1/2}}{(1-z)}\left[\frac{p/p_{ref}}{(5.76+z)/2}\right]^{1/2} \qquad (a)$$

For a given pressure, this expression involves two unknowns: z and T.

Also, from Example 14.6, we use the energy balance, Eq. (c)

$$\bar{h}_R = \bar{h}_P \qquad (c)$$

where

$$\bar{h}_R = (\bar{h}_{CO})_R + \frac{1}{2}(\bar{h}_{O_2})_R + 1.88(\bar{h}_{N_2})_R$$

and

$$\bar{h}_P = z(\bar{h}_{CO})_P + (z/2)(\bar{h}_{O_2})_P + (1-z)(\bar{h}_{CO_2})_P + 1.88(\bar{h}_{N_2})_P$$

where the subscripts R and P denote reactants and products, respectively, and z denotes the amount of CO in the products, in kmol per kmol of CO entering.

With pressure known, Eqs. (a) and (c) can be solved for T and z using the following *IT* code. Choosing SI from the **Units** menu and amount of substance in moles, and letting hCO_R denote the specific enthalpy of CO in the reactants, and so on, we have

```
// Given data
TR = 25 + 273.15 // K
p = 1 // atm
pref = 1 // atm

// Evaluating the equilibrium constant using
Eq. (a)
K = ((z * (z/2)^0.5) / (1 - z)) * ((p / pref) /
((5.76 + z) / 2))^0.5

// Energy balance: Eq. (c)
hR = hP
hR = hCO_R + (1/2) * hO2_R + 1.88 * hN2_R
hP = z * hCO_P + (z /2) * hO2_P + (1 - z) * hCO2_P +
1.88 * hN2_P

hCO_R = h_T("CO",TR)
hO2_R = h_T("O2",TR)
hN2_R = h_T("N2",TR)

hCO_P = h_T("CO",T)
hO2_P = h_T("O2",T)
```

```
hCO2_P = h_T ("CO2",T)
hN2_P = h_T ("N2",T)
```

```
    /* To obtain data for the equilibrium constant
    use the Look-up Table option under the Edit
    menu. Load the file "eqco2.lut". Data for CO₂ ⇌
    CO + 1/2 O2 from Table A-27 are stored in the
    look-up table as T in column 1 and log10(K) in
    column 2. To retrieve the data use */
```

```
log(K) = lookupvall(eqco2, 1, T,2)
```

Obtain a solution for $p = 1$ using the **Solve** button. To ensure rapid convergence, restrict T and K to positive values, and set a lower limit of 0.001 and an upper limit of 0.999 for z. The results are $T = 2399$ K and $z = 0.1249$, which agree with the values obtained in Example 14.6.

Now, use the **Explore** button and sweep p from 1 to 10 atm in steps of 0.01. Using the **Graph** button, construct the following plots:

From **Fig. E14.7**, we see that as pressure increases more CO is oxidized to CO_2 (z decreases) and temperature increases.

❶ Similar files are included in *IT* for each of the reactions in Table A-27.

SKILLS DEVELOPED

Ability to...

• apply Eq. 14.35 together with the energy balance for reacting systems to determine equilibrium flame temperature.

• perform equilibrium calculations using *Interactive Thermodynamics: IT*.

Quick Quiz

If the CO and air each entered at 500°C, determine the equilibrium flame temperature in K using *Interactive Thermodynamics: IT*. Ans. 2575.

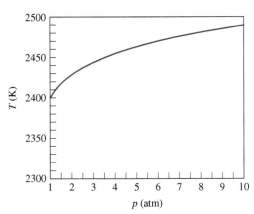
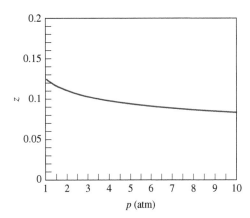

Fig. E14.7

14.4.2 Van't Hoff Equation

The dependence of the equilibrium constant on temperature exhibited by the values of Table A-27 follows from Eq. 14.31. An alternative way to express this dependence is given by the van't Hoff equation, Eq. 14.43b.

The development of this equation begins by introducing Eq. 14.29b into Eq. 14.31 to obtain on rearrangement

$$\bar{R}T \ln K = -[(\nu_C \bar{h}_C + \nu_D \bar{h}_D - \nu_A \bar{h}_A - \nu_B \bar{h}_B) - T(\nu_C \bar{s}^{\circ}_C + \nu_D \bar{s}^{\circ}_D - \nu_A \bar{s}^{\circ}_A - \nu_B \bar{s}^{\circ}_B)] \quad (14.41)$$

Each of the specific enthalpies and entropies in this equation depends on temperature alone. Differentiating with respect to temperature

$$\bar{R}T \frac{d \ln K}{dT} + \bar{R} \ln K = -\left[\nu_C \left(\frac{d\bar{h}_C}{dT} - T \frac{d\bar{s}^{\circ}_C}{dT} \right) + \nu_D \left(\frac{d\bar{h}_D}{dT} - T \frac{d\bar{s}^{\circ}_D}{dT} \right) \right.$$
$$\left. - \nu_A \left(\frac{d\bar{h}_A}{dT} - T \frac{d\bar{s}^{\circ}_A}{dT} \right) - \nu_B \left(\frac{d\bar{h}_B}{dT} - T \frac{d\bar{s}^{\circ}_B}{dT} \right) \right]$$
$$+ (\nu_C \bar{s}^{\circ}_C + \nu_D \bar{s}^{\circ}_D - \nu_A \bar{s}^{\circ}_A - \nu_B \bar{s}^{\circ}_B)$$

From the definition of $\bar{s}^\circ(T)$ (Eq. 6.19), we have $d\bar{s}^\circ/dT = \bar{c}_p/T$. Moreover, $d\bar{h}/dT = \bar{c}_p$. Accordingly, each of the underlined terms in the above equation vanishes identically, leaving

$$\bar{R}T \frac{d \ln K}{dT} + \bar{R} \ln K = (\nu_C \bar{s}_C^\circ + \nu_D \bar{s}_D^\circ - \nu_A \bar{s}_A^\circ - \nu_B \bar{s}_B^\circ) \tag{14.42}$$

Using Eq. 14.41 to evaluate the second term on the left and simplifying the resulting expression, Eq. 14.42 becomes

$$\frac{d \ln K}{dT} = \frac{(\nu_C \bar{h}_C + \nu_D \bar{h}_D - \nu_A \bar{h}_A - \nu_B \bar{h}_B)}{\bar{R}T^2} \tag{14.43a}$$

or, expressed more concisely,

$$\boxed{\frac{d \ln K}{dT} = \frac{\Delta H}{\bar{R}T^2}} \tag{14.43b}$$

van't Hoff equation

which is the **van't Hoff equation**.

In Eq. 14.43b, ΔH is the *enthalpy of reaction* at temperature T. The van't Hoff equation shows that when ΔH is negative (exothermic reaction), K decreases with temperature, whereas for ΔH positive (endothermic reaction), K increases with temperature.

The enthalpy of reaction ΔH is often very nearly constant over a rather wide interval of temperature. In such cases, Eq. 14.43b can be integrated to yield

$$\ln \frac{K_2}{K_1} = -\frac{\Delta H}{\bar{R}} \left(\frac{1}{T_2} - \frac{1}{T_1} \right) \tag{14.44}$$

where K_1 and K_2 denote the equilibrium constants at temperatures T_1 and T_2, respectively. This equation shows that $\ln K$ is linear in $1/T$. Accordingly, plots of $\ln K$ versus $1/T$ can be used to determine ΔH from experimental equilibrium composition data. Alternatively, the equilibrium constant can be determined using enthalpy data.

14.4.3 Ionization

The methods developed for determining the equilibrium composition of a reactive ideal gas mixture can be applied to systems involving ionized gases, also known as *plasmas*. In previous sections we considered the chemical equilibrium of systems where dissociation is a factor. For example, the dissociation reaction of diatomic nitrogen

$$N_2 \rightleftarrows 2N$$

can occur at elevated temperatures. At still higher temperatures, ionization may take place according to

$$N \rightleftarrows N^+ + e^- \tag{14.45}$$

That is, a nitrogen atom loses an electron, yielding a singly ionized nitrogen atom N^+ and a free electron e^-. Further heating can result in the loss of additional electrons until all electrons have been removed from the atom.

For some cases of practical interest, it is reasonable to think of the neutral atoms, positive ions, and electrons as forming an ideal gas mixture. With this idealization, ionization equilibrium can be treated in the same manner as the chemical equilibrium of reacting ideal gas mixtures. The change in the Gibbs function for the equilibrium ionization reaction required to evaluate the ionization-equilibrium constant can be calculated as a function of temperature by using the procedures of statistical thermodynamics. In general, the extent of ionization increases as the temperature is raised and the pressure is lowered.

Example 14.8 illustrates the analysis of ionization equilibrium.

> ▶ ▶ **EXAMPLE 14.8** ▶

Considering Ionization Equilibrium

Consider an equilibrium mixture at 2000 K consisting of Cs, Cs^+, and e^-, where Cs denotes neutral cesium atoms, Cs^+ singly ionized cesium ions, and e^- free electrons. The ionization-equilibrium constant at this temperature for

$$Cs \rightleftarrows Cs^+ + e^-$$

is $K = 15.63$. Determine the pressure, in atmospheres, if the ionization of Cs is 95% complete, and plot percent completion of ionization versus pressure ranging from 0 to 10 atm.

SOLUTION

Known   An equilibrium mixture of Cs, Cs^+, e^- is at 2000 K. The value of the equilibrium constant at this temperature is known.

Find   Determine the pressure of the mixture if the ionization of Cs is 95% complete. Plot percent completion versus pressure.

Engineering Model   Equilibrium can be treated in this case using ideal gas mixture equilibrium considerations.

Analysis   The ionization of cesium to form a mixture of Cs, Cs^+, and e^- is described by

$$Cs \rightarrow (1 - z)Cs + z\ Cs^+ + ze^-$$

where z denotes the extent of ionization, ranging from 0 to 1. The total number of moles of mixture n is

$$n = (1 - z) + z + z = 1 + z$$

At equilibrium, we have $Cs \rightleftarrows Cs^+ + e^-$, so Eq. 14.35 takes the form

$$K = \frac{(z)(z)}{(1-z)} \left[\frac{p/p_{ref}}{(1+z)} \right]^{1+1-1} = \left(\frac{z^2}{1-z^2} \right) \left(\frac{p}{p_{ref}} \right) \quad \text{(a)}$$

Solving for the ratio p/p_{ref} and introducing the known value of K

$$\frac{p}{p_{ref}} = (15.63)\left(\frac{1-z^2}{z^2} \right)$$

For $p_{ref} = 1$ atm and $z = 0.95$ (95%), $p = 1.69$ atm. Using an equation-solver and plotting package, the following plot can be constructed:

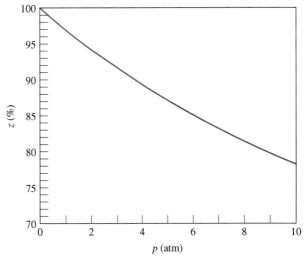

Fig. E14.8

Figure E14.8 shows that ionization tends to occur to a lesser extent as pressure is raised. Ionization also tends to occur to a greater extent as temperature is raised at fixed pressure.

SKILLS DEVELOPED

Ability to...

- apply Eq. 14.35 to determine the extent of ionization of cesium given temperature and pressure.

Quick Quiz

Solving Eq. (a) for z, determine the percent of ionization of Cs at $T = 1600$ K ($K = 0.78$) and $p = 1$ atm. Ans. 66.2%.

14.4.4 Simultaneous Reactions

Let us return to the discussion of Sec. 14.2 and consider the possibility of more than one reaction among the substances present within a system. For the present application, the closed system is assumed to contain a mixture of *eight* components A, B, C, D, E, L, M, and N, subject to *two* independent reactions:

$$(1) \qquad \nu_A A + \nu_B B \rightleftarrows \nu_C C + \nu_D D \qquad (14.24)$$

$$(2) \qquad \nu_{A'} A + \nu_L L \rightleftarrows \nu_M M + \nu_N N \qquad (14.46)$$

As in Sec. 14.2, component E is inert. Also, note that component A has been taken as common to both reactions but with a possibly different stoichiometric coefficient ($\nu_{A'}$ is not necessarily equal to ν_A).

The stoichiometric coefficients of the above equations do not correspond to the numbers of moles of the respective components present within the system, but *changes* in the amounts of the components are related to the stoichiometric coefficients by

$$\frac{-dn_A}{\nu_A} = \frac{-dn_B}{\nu_B} = \frac{dn_C}{\nu_C} = \frac{dn_D}{\nu_D} \tag{14.25a}$$

following from Eq. 14.24, and

$$\frac{-dn_A}{\nu_{A'}} = \frac{-dn_L}{\nu_L} = \frac{dn_M}{\nu_M} = \frac{dn_N}{\nu_N} \tag{14.47a}$$

following from Eq. 14.46. Introducing a proportionality factor $d\varepsilon_1$, Eqs. 14.25a may be represented by

$$
\begin{array}{ll}
dn_A = -\nu_A d\varepsilon_1, & dn_B = -\nu_B d\varepsilon_1 \\
dn_C = \nu_C d\varepsilon_1, & dn_D = \nu_D d\varepsilon_1
\end{array} \tag{14.25b}
$$

Similarly, with the proportionality factor $d\varepsilon_2$, Eqs. 14.47a may be represented by

$$
\begin{array}{ll}
dn_A = -\nu_{A'} d\varepsilon_2, & dn_L = -\nu_L d\varepsilon_2 \\
dn_M = \nu_M d\varepsilon_2, & dn_N = \nu_N d\varepsilon_2
\end{array} \tag{14.47b}
$$

Component A is involved in both reactions, so the total change in A is given by

$$dn_A = -\nu_A d\varepsilon_1 - \nu_{A'} d\varepsilon_2 \tag{14.48}$$

Also, we have $dn_E = 0$ because component E is inert.

For the system under present consideration, Eq. 14.10 is

$$
\begin{aligned}
dG]_{T,p} = {}& \mu_A dn_A + \mu_B dn_B + \mu_C dn_C + \mu_D dn_D \\
& + \mu_E dn_E + \mu_L dn_L + \mu_M dn_M + \mu_N dn_N
\end{aligned} \tag{14.49}
$$

Introducing the above expressions giving the changes in the n's, this becomes

$$
\begin{aligned}
dG]_{T,p} = {}& (-\nu_A \mu_A - \nu_B \mu_B + \nu_C \mu_C + \nu_D \mu_D)\, d\varepsilon_1 \\
& + (-\nu_{A'} \mu_A - \nu_L \mu_L + \nu_M \mu_M + \nu_N \mu_N)\, d\varepsilon_2
\end{aligned} \tag{14.50}
$$

Since the two reactions are independent, $d\varepsilon_1$ and $d\varepsilon_2$ can be independently varied. Accordingly, when $dG]_{T,p} = 0$, the terms in parentheses must be zero and *two* equations of reaction equilibrium result, one corresponding to each of the foregoing reactions:

$$\nu_A \mu_A + \nu_B \mu_B = \nu_C \mu_C + \nu_D \mu_D \tag{14.26b}$$

$$\nu_{A'} \mu_A + \nu_L \mu_L = \nu_M \mu_M + \nu_N \mu_N \tag{14.51}$$

The first of these equations is exactly the same as that obtained in Sec. 14.2. For the case of reacting ideal gas mixtures, this equation can be expressed as

$$-\left(\frac{\Delta G^\circ}{\bar{R}T}\right)_1 = \ln\left[\frac{y_C^{\nu_C} y_D^{\nu_D}}{y_A^{\nu_A} y_B^{\nu_B}} \left(\frac{p}{p_{ref}}\right)^{\nu_C + \nu_D - \nu_A - \nu_B}\right] \tag{14.52}$$

Similarly, Eq. 14.51 can be expressed as

$$-\left(\frac{\Delta G^\circ}{\bar{R}T}\right)_2 = \ln\left[\frac{y_M^{\nu_M} y_N^{\nu_N}}{y_{A'}^{\nu_{A'}} y_L^{\nu_L}} \left(\frac{p}{p_{ref}}\right)^{\nu_M + \nu_N - \nu_{A'} - \nu_L}\right] \tag{14.53}$$

In each of these equations, the $\Delta G°$ term is evaluated as the change in Gibbs function for the respective reaction, regarding each reactant and product as separate at temperature T and a pressure of 1 atm.

From Eq. 14.52 follows the equilibrium constant

$$K_1 = \frac{y_C^{\nu_C} y_D^{\nu_D}}{y_A^{\nu_A} y_B^{\nu_B}} \left(\frac{p}{p_{ref}}\right)^{\nu_C + \nu_D - \nu_A - \nu_B} \tag{14.54}$$

and from Eq. 14.53 follows

$$K_2 = \frac{y_M^{\nu_M} y_N^{\nu_N}}{y_{A'}^{\nu_{A'}} y_L^{\nu_L}} \left(\frac{p}{p_{ref}}\right)^{\nu_M + \nu_N - \nu_{A'} - \nu_L} \tag{14.55}$$

The equilibrium constants K_1 and K_2 can be determined from Table A-27 or a similar compilation. The mole fractions appearing in these expressions must be evaluated by considering *all* the substances present within the system, including the inert substance E. Each mole fraction has the form $y_i = n_i/n$, where n_i is the amount of component i in the equilibrium mixture and

$$n = n_A + n_B + n_C + n_D + n_E + n_L + n_M + n_N \tag{14.56}$$

The n's appearing in Eq. 14.56 can be expressed in terms of *two* unknown variables through application of the conservation of mass principle to the various chemical species present. Accordingly, for a specified temperature and pressure, Eqs. 14.54 and 14.55 give *two* equations in *two* unknowns. The composition of the system at equilibrium can be determined by solving these equations simultaneously. This procedure is illustrated by Example 14.9.

The procedure discussed in this section can be extended to systems involving several simultaneous independent reactions. The number of simultaneous equilibrium constant expressions that results equals the number of independent reactions. As these equations are nonlinear and require simultaneous solution, the use of a computer is usually required.

▶ ▶ EXAMPLE 14.9 ▶

Considering Equilibrium with Simultaneous Reactions

As a result of heating, a system consisting initially of 1 kmol of CO_2, $\frac{1}{2}$ kmol of O_2, and $\frac{1}{2}$ kmol of N_2 forms an equilibrium mixture of CO_2, CO, O_2, N_2, and NO at 3000 K, 1 atm. Determine the composition of the equilibrium mixture.

SOLUTION

Known   A system consisting of specified amounts of CO_2, O_2, and N_2 is heated to 3000 K, 1 atm, forming an equilibrium mixture of CO_2, CO, O_2, N_2, and NO.

Find   Determine the equilibrium composition.

Engineering Model   The final mixture is an equilibrium mixture of ideal gases.

Analysis   The overall reaction has the form

❶ $1CO_2 + \frac{1}{2}O_2 + \frac{1}{2}N_2 \rightarrow aCO + bNO + cCO_2 + dO_2 + eN_2$

Applying conservation of mass to carbon, oxygen, and nitrogen, the five unknown coefficients can be expressed in terms of any two of

the coefficients. Selecting a and b as the unknowns, the following balanced equation results:

$$1CO_2 + \tfrac{1}{2}O_2 + \tfrac{1}{2}N_2 \rightarrow aCO + bNO + (1-a)CO_2$$
$$+ \tfrac{1}{2}(1 + a - b)O_2 + \tfrac{1}{2}(1-b)N_2$$

The total number of moles n in the mixture formed by the products is

$$n = a + b + (1-a) + \tfrac{1}{2}(1+a-b) + \tfrac{1}{2}(1-b) = \frac{4+a}{2}$$

At equilibrium, two independent reactions relate the components of the product mixture:

1. $CO_2 \rightleftarrows CO + \tfrac{1}{2}O_2$

2. $\tfrac{1}{2}O_2 + \tfrac{1}{2}N_2 \rightleftarrows NO$

For the first of these reactions, the form taken by the equilibrium constant when $p = 1$ atm is

$$K_1 = \frac{a[\frac{1}{2}(1 + a - b)]^{1/2}}{(1 - a)}\left[\frac{1}{(4 + a)/2}\right]^{1 + 1/2 - 1} = \frac{a}{1 - a}\left(\frac{1 + a - b}{4 + a}\right)^{1/2}$$

Similarly, the equilibrium constant for the second of the reactions is

$$K_2 = \frac{b}{[\frac{1}{2}(1 + a - b)]^{1/2}[\frac{1}{2}(1 - b)]^{1/2}}\left[\frac{1}{(4 + a)/2}\right]^{1 - 1/2 - 1/2}$$

$$= \frac{2b}{[(1 + a - b)(1 - b)]^{1/2}}$$

At 3000 K, Table A-27 provides $\log_{10} K_1 = -0.485$ and $\log_{10} K_2 = -0.913$, giving $K_1 = 0.3273$ and $K_2 = 0.1222$. Accordingly, the two equations that must be solved simultaneously for the two unknowns a and b are

$$0.3273 = \frac{a}{1 - a}\left(\frac{1 + a - b}{4 + a}\right)^{1/2}, \quad 0.1222 = \frac{2b}{[(1 + a - b)(1 - b)]^{1/2}}$$

The solution is $a = 0.3745$, $b = 0.0675$, as can be verified. The composition of the equilibrium mixture, in kmol per kmol of CO_2 present initially, is then $0.3745CO$, $0.0675NO$, $0.6255CO_2$, $0.6535O_2$, $0.4663N_2$.

───

❶ If high enough temperatures are attained, nitrogen can combine with oxygen to form components such as nitric oxide. Even trace amounts of oxides of nitrogen in products of combustion can be a source of air pollution.

SKILLS DEVELOPED

Ability to...

- apply Eqs. 14.54 and 14.55 to determine equilibrium composition given temperature and pressure for two simultaneous equilibrium reactions.

- retrieve and use data from Table A-27.

───

Quick Quiz

Determine the mole fractions of the components of the equilibrium mixture. Ans. $y_{CO} = 0.171$, $y_{NO} = 0.031$, $y_{CO2} = 0.286$, $y_{O2} = 0.299$, $y_{N2} = 0.213$.

Phase Equilibrium

In this part of the chapter the equilibrium condition $dG]_{T,p} = 0$ introduced in Sec. 14.1 is used to study the equilibrium of multicomponent, multiphase, nonreacting systems. The discussion begins with the elementary case of equilibrium between two phases of a pure substance and then turns to the general case of several components present in several phases.

14.5 Equilibrium between Two Phases of a Pure Substance

Consider the case of a system consisting of two phases of a pure substance at equilibrium. Since the system is at equilibrium, each phase is at the same temperature and pressure. The Gibbs function for the system is

$$G = n'\bar{g}'(T, p) + n''\bar{g}''(T, p) \tag{14.57}$$

where the primes ′ and ″ denote phases 1 and 2, respectively.

Forming the differential of G at fixed T and p

$$dG]_{T,p} = \bar{g}' dn' + \bar{g}'' dn'' \tag{14.58}$$

Since the total amount of the pure substance remains constant, an increase in the amount present in one of the phases must be compensated by an equivalent decrease in the amount present in the other phase. Thus, we have $dn'' = -dn'$, and Eq. 14.58 becomes

$$dG]_{T,p} = (\bar{g}' - \bar{g}'') dn'$$

At equilibrium, $dG]_{T,p} = 0$, so

$$\boxed{\bar{g}' = \bar{g}''} \tag{14.59}$$

At equilibrium, the molar Gibbs functions of the phases are equal.

Clapeyron Equation   Equation 14.59 can be used to derive the *Clapeyron* equation, obtained by other means in Sec. 11.4. For two phases at equilibrium, variations in pressure are uniquely related to variations in temperature: $p = p_{sat}(T)$; thus, differentiation of Eq. 14.59 with respect to temperature gives

$$\left.\frac{\partial \bar{g}'}{\partial T}\right)_p + \left.\frac{\partial \bar{g}'}{\partial p}\right)_T \frac{dp_{sat}}{dT} = \left.\frac{\partial \bar{g}''}{\partial T}\right)_p + \left.\frac{\partial \bar{g}''}{\partial p}\right)_T \frac{dp_{sat}}{dT}$$

With Eqs. 11.30 and 11.31, this becomes

$$-\bar{s}' + \bar{v}' \frac{dp_{sat}}{dT} = -\bar{s}'' + \bar{v}'' \frac{dp_{sat}}{dT}$$

Or on rearrangement

$$\frac{dp_{sat}}{dT} = \frac{\bar{s}'' - \bar{s}'}{\bar{v}'' - \bar{v}'}$$

This can be expressed alternatively by noting that, with $\bar{g} = \bar{h} - T\bar{s}$, Eq. 14.59 becomes

$$\bar{h}' - T\bar{s}' = \bar{h}'' - T\bar{s}''$$

or

$$\bar{s}'' - \bar{s}' = \frac{\bar{h}'' - \bar{h}'}{T} \tag{14.60}$$

Combining results, the **Clapeyron equation** is obtained

Clapeyron equation

$$\boxed{\frac{dp_{sat}}{dT} = \frac{1}{T}\left(\frac{\bar{h}'' - \bar{h}'}{\bar{v}'' - \bar{v}'}\right)} \tag{14.61}$$

An application of the Clapeyron equation is provided in Example 11.4.

   A special form of Eq. 14.61 for a system at equilibrium consisting of a liquid or solid phase and a vapor phase can be obtained simply. If the specific volume of the liquid or solid, $\bar{v}'$, is negligible compared with the specific volume of the vapor, $\bar{v}''$, and the vapor can be treated as an ideal gas, $\bar{v}'' = \bar{R}T/p_{sat}$, Eq. 14.61 becomes

$$\frac{dp_{sat}}{dT} = \frac{\bar{h}'' - \bar{h}'}{\bar{R}T^2/p_{sat}}$$

or

$$\boxed{\frac{d \ln p_{sat}}{dT} = \frac{\bar{h}'' - \bar{h}'}{\bar{R}T^2}} \tag{14.62}$$

TAKE NOTE...

Equations 11.40 and 11.42 are special cases of Eqs. 14.61 and 14.62, respectively.

which is the **Clausius–Clapeyron equation**. The similarity in form of Eq. 14.62 and the van't Hoff equation, Eq. 14.43b, may be noted. The van't Hoff equation for chemical equilibrium is the counterpart of the Clausius–Clapeyron equation for phase equilibrium.

Clausius–Clapeyron equation

14.6 Equilibrium of Multicomponent, Multiphase Systems

The equilibrium of systems that may involve several phases, each having a number of components present, is considered in this section. The principal result is the Gibbs phase rule, which summarizes important limitations on multicomponent, multiphase systems at equilibrium.

14.6.1 Chemical Potential and Phase Equilibrium

Phase 1
Component A, n'_A, μ'_A
Component B, n'_B, μ'_B

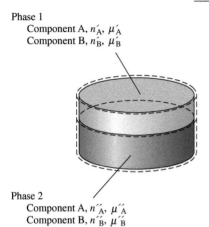

Phase 2
Component A, n''_A, μ''_A
Component B, n''_B, μ''_B

Fig. 14.1 System consisting of two components in two phases.

Figure 14.1 shows a system consisting of *two* components A and B in *two* phases 1 and 2 that are at the same temperature and pressure. Applying Eq. 14.10 to each of the phases

$$dG']_{T,p} = \mu'_A \, dn'_A + \mu'_B \, dn'_B$$
$$dG'']_{T,p} = \mu''_A \, dn''_A + \mu''_B \, dn''_B$$

(14.63)

where as before the primes identify the two phases.

When matter is transferred between the two phases in the absence of chemical reaction, the total amounts of A and B must remain constant. Thus, the increase in the amount present in one of the phases must be compensated by an equivalent decrease in the amount present in the other phase. That is,

$$dn''_A = -dn'_A, \qquad dn''_B = -dn'_B$$

(14.64)

With Eqs. 14.63 and 14.64, the change in the Gibbs function for the system is

$$dG]_{T,p} = dG']_{T,p} + dG'']_{T,p}$$
$$= (\mu'_A - \mu''_A) \, dn'_A + (\mu'_B - \mu''_B) \, dn'_B$$

(14.65)

Since n'_A and n'_B can be varied independently, it follows that when $dG]_{T,p} = 0$, the terms in parentheses are zero, resulting in

$$\boxed{\mu'_A = \mu''_A \qquad \text{and} \qquad \mu'_B = \mu''_B}$$

(14.66)

At equilibrium, the chemical potential of each component is the same in each phase.

The significance of the chemical potential for phase equilibrium can be brought out simply by reconsidering the system of Fig. 14.1 in the special case when the chemical potential of component B is the same in both phases: $\mu'_B = \mu''_B$. With this constraint, Eq. 14.65 reduces to

$$dG]_{T,p} = (\mu'_A - \mu''_A) \, dn'_A$$

Any spontaneous process of the system taking place at a fixed temperature and pressure must be such that the Gibbs function decreases: $dG]_{T,p} < 0$. Thus, with the above expression we have

$$(\mu'_A - \mu''_A) dn'_A < 0$$

Accordingly,

- when the chemical potential of A is greater in phase 1 than in phase 2 ($\mu'_A > \mu''_A$), it follows that $dn'_A < 0$. That is, substance A passes from phase 1 to phase 2.
- when the chemical potential of A is greater in phase 2 than in phase 1 ($\mu''_A > \mu'_A$), it follows that $dn'_A > 0$. That is, substance A passes from phase 2 to phase 1.

At equilibrium, the chemical potentials are equal ($\mu'_A = \mu''_A$), and there is no net transfer of A between the phases.

With this reasoning, we see that the chemical potential can be regarded as a measure of the *escaping tendency* of a component. If the chemical potential of a component is not the same in each phase, there will be a tendency for that component to pass from the phase having the higher chemical potential for that component to the phase having the lower chemical potential. When the chemical potential is the same in both phases, there is no tendency for a net transfer to occur from one phase to the other.

In Example 14.10, we apply phase equilibrium principles to provide a rationale for the model introduced in Sec. 12.5.3 for moist air in contact with liquid water.

▶ ▶ EXAMPLE 14.10 ▶ · · · · · · · · · · · · · · · · · · ·

Equilibrium of Moist Air in Contact with Liquid Water

A closed system at a temperature of 20°C and a pressure of 1 atm consists of a pure liquid water phase in equilibrium with a vapor phase composed of water vapor and dry air. Determine the departure, in percent, of the partial pressure of the water vapor from the saturation pressure of water at 20°C.

SOLUTION

Known A phase of liquid water only is in equilibrium with *moist air* at 20°C and 1 bar.

Find Determine the percentage departure of the partial pressure of the water vapor in the moist air from the saturation pressure of water at 20°C.

Schematic and Given Data:

Gas phase of water vapor and dry air

Liquid water

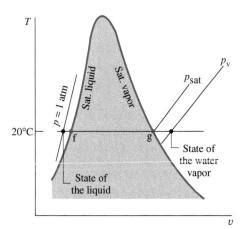

Fig. E14.10

Engineering Model

1. The gas phase can be modeled as an ideal gas mixture.

2. The liquid phase is pure water only.

Analysis For phase equilibrium, the chemical potential of the water must have the same value in both phases: $\mu_1 = \mu_v$, where μ_1 and μ_v denote, respectively, the chemical potentials of the pure liquid water in the liquid phase and the water vapor in the vapor phase.

The chemical potential μ_1 is the Gibbs function per mole of pure liquid water (Eq. 14.12)

$$\mu_1 = \bar{g}(T, p)$$

Since the vapor phase is assumed to form an ideal gas mixture, the chemical potential μ_v equals the Gibbs function per mole evaluated at temperature T and the partial pressure p_v of the water vapor (Eq. 14.16)

$$\mu_v = \bar{g}(T, p_v)$$

For phase equilibrium, $\mu_1 = \mu_v$, or

$$\bar{g}(T, p_v) = \bar{g}(T, p)$$

With $\bar{g} = \bar{h} - T\bar{s}$, this can be expressed alternatively as

$$\bar{h}(T, p_v) - T\bar{s}(T, p_v) = \bar{h}(T, p) - T\bar{s}(T, p)$$

The water vapor is modeled as an ideal gas. Thus, the enthalpy is given closely by the saturated vapor value at temperature T

$$\bar{h}(T, p_v) \approx \bar{h}_g$$

Furthermore, with Eq. 6.20b, the difference between the specific entropy of the water vapor and the specific entropy at the corresponding saturated vapor state is

$$\bar{s}(T, p_v) - \bar{s}_g(T) = -\bar{R} \ln \frac{p_v}{p_{sat}}$$

or

$$\bar{s}(T, p_v) = \bar{s}_g(T) - \bar{R} \ln \frac{p_v}{p_{sat}}$$

where p_{sat} is the saturation pressure at temperature T.
With Eq. 3.13, the enthalpy of the liquid is closely

$$\bar{h}(T, p) \approx \bar{h}_f + \bar{v}_f(p - p_{sat})$$

where $\bar{v}_f$ and $\bar{h}_f$ are the saturated liquid specific volume and enthalpy at temperature T. Furthermore, with Eq. 6.5

$$\bar{s}(T, p) \approx \bar{s}_f(T)$$

where $\bar{s}_f$ is the saturated liquid specific entropy at temperature T.
Collecting the foregoing expressions, we have

$$\bar{h}_g - T\left(\bar{s}_g - \bar{R} \ln \frac{p_v}{p_{sat}}\right) = \bar{h}_f + \bar{v}_f(p - p_{sat}) - T\bar{s}_f$$

or

$$\bar{R}T \ln \frac{p_v}{p_{sat}} = \bar{v}_f(p - p_{sat}) - \overline{[(h_g - h_f) - T(s_g - s_f)]}$$

The underlined term vanishes by Eq. 14.60, leaving

$$\ln \frac{p_v}{p_{sat}} = \frac{\bar{v}_f(p - p_{sat})}{\bar{R}T} \quad \text{or} \quad \frac{p_v}{p_{sat}} = \exp\frac{\bar{v}_f(p - p_{sat})}{\bar{R}T}$$

With data from Table A-2 at 20°C, $v_f = 1.0018 \times 10^{-3}$ m³/kg and $p_{sat} = 0.0239$ bar, we have

$$\frac{v_f(p - p_{sat})}{RT} = \frac{1.0018 \times 10^{-3} \text{ m}^3/\text{kg}(1 - 0.0239) \times 10^5 \text{ N/m}^2}{\left(\dfrac{8314 \text{ N} \cdot \text{m}}{18.02 \text{ kg} \cdot \text{K}}\right)(293.15 \text{ K})}$$

$$= 7.23 \times 10^{-4}$$

Finally

$$\frac{p_v}{p_{sat}} = \exp(7.23 \times 10^{-4}) = 1.00072$$

When expressed as a percentage, the departure of p_v from p_{sat} is

② $\left(\dfrac{p_v - p_{sat}}{p_{sat}}\right)(100) = (1.00072 - 1)(100) = 0.072\%$

---

① For phase equilibrium, there would be a small, but finite, concentration of air in the liquid water phase. However, this small amount of dissolved air is ignored in the present development.

② The departure of p_v from p_{sat} is negligible at the specified conditions. This suggests that at normal temperatures and pressures the equilibrium between the liquid water phase and the water vapor is not significantly disturbed by the presence of the dry air. Accordingly, the partial pressure of the water vapor can be taken as equal to the saturation pressure of the water at the system temperature. This model, introduced in Sec. 12.5.3, is used extensively in Chap. 12.

SKILLS DEVELOPED

Ability to…

- apply the concept of phase equilibrium expressed in Eq. 14.66 to an air–water–vapor mixture in equilibrium with liquid water.

Quick Quiz

Using the methods of Sec. 12.5.2, determine the specific humidity, ω, of the air–water–vapor mixture. Ans. 0.01577.

14.6.2 Gibbs Phase Rule

The requirement for equilibrium of a system consisting of two components and two phases, given by Eqs. 14.66, can be extended with similar reasoning to nonreacting multicomponent, multiphase systems. At equilibrium, the chemical potential of each component must be the same in all phases. For the case of N components that are present in P phases we have, therefore, the following set of $N(P - 1)$ equations:

$$N\begin{cases} \overbrace{\mu_1^1 = \mu_1^2 = \mu_1^3 = \cdots = \mu_1^P}^{P-1} \\ \mu_2^1 = \mu_2^2 = \mu_2^3 = \cdots = \mu_2^P \\ \vdots \\ \mu_N^1 = \mu_N^2 = \mu_N^3 = \cdots = \mu_N^P \end{cases} \tag{14.67}$$

where μ_i^j denotes the chemical potential of the ith component in the jth phase. This set of equations provides the basis for the *Gibbs phase rule,* which allows the determination of the number of *independent intensive* properties that may be arbitrarily specified in order to fix the *intensive* state of the system. The number of independent intensive properties is called the

degrees of freedom **degrees of freedom** (or the *variance*).

Since the chemical potential is an intensive property, its value depends on the relative proportions of the components present and not on the amounts of the components. In other words, in a given phase involving N components at temperature T and pressure p, the chemical potential is determined by the *mole fractions* of the components present and not the respective n's. However, as the mole fractions add to unity, at most $N - 1$ of the mole fractions can be independent. Thus, for a system involving N components, there are at most $N - 1$ independently variable mole fractions for each phase. For P phases, therefore, there are at most $P(N - 1)$ independently variable mole fractions. In addition, the temperature and pressure, which are the same in each phase, are two further intensive properties, giving a maximum of $P(N - 1) + 2$ independently variable intensive properties for the system. But because of the $N(P - 1)$ equilibrium conditions represented by Eqs. 14.67 among these properties, the number of intensive properties that are freely variable, the degrees of freedom F, is

$$\boxed{F = [P(N - 1) + 2] - N(P - 1) = 2 + N - P} \tag{14.68}$$

Gibbs phase rule which is the **Gibbs phase rule.**

In Eq. 14.68, F is the number of intensive properties that may be arbitrarily specified and that must be specified to fix the intensive state of a nonreacting system at equilibrium.

> **FOR EXAMPLE**
>
> Let us apply the Gibbs phase rule to a liquid solution consisting of water and ammonia such as considered in the discussion of absorption refrigeration (Sec. 10.5). This solution involves two components and a single phase: $N = 2$ and $P = 1$. Equation 14.68 then gives $F = 3$, so the intensive state is fixed by giving the values of *three* intensive properties, such as temperature, pressure, and the ammonia (or water) mole fraction.

The phase rule summarizes important limitations on various types of systems. For example, for a system involving a single component such as water, $N = 1$ and Eq. 14.68 becomes

$$F = 3 - P \qquad (14.69)$$

- The minimum number of phases is one, corresponding to $P = 1$. For this case, Eq. 14.69 gives $F = 2$. That is, *two* intensive properties must be specified to fix the intensive state of the system. This requirement is familiar from our use of the steam tables and similar property tables. To obtain properties of superheated vapor, say, from such tables requires that we give values for *any two* of the tabulated properties, for example, T and p.

- When two phases are present in a system involving a single component, $N = 1$ and $P = 2$. Equation 14.69 then gives $F = 1$. That is, the intensive state is determined by a single intensive property value. For example, the intensive states of the separate phases of an equilibrium mixture of liquid water and water vapor are completely determined by specifying the temperature.

- The minimum allowable value for the degrees of freedom is zero: $F = 0$. For a single-component system, Eq. 14.69 shows that this corresponds to $P = 3$, a three-phase system. Thus, *three* is the maximum number of different phases of a pure component that can coexist in equilibrium. Since there are no degrees of freedom, both temperature and pressure are fixed at equilibrium. For example, there is only a single temperature 0.01°C and a single pressure 0.6113 kPa (0.006 atm) for which ice, liquid water, and water vapor are in equilibrium.

The phase rule given here must be modified for application to systems in which chemical reactions occur. Furthermore, the system of equations, Eqs. 14.67, giving the requirements for phase equilibrium at a specified temperature and pressure can be expressed alternatively in terms of partial molal Gibbs functions, fugacities, and activities, all of which are introduced in Sec. 11.9. To use any such expression to determine the equilibrium composition of the different phases present within a system at equilibrium requires a model for each phase that allows the relevant quantities—the chemical potentials, fugacities, and so on—to be evaluated for the components present in terms of system properties that can be determined. For example, a gas phase might be modeled as an ideal gas mixture or, at higher pressures, as an ideal solution.

CHAPTER SUMMARY AND STUDY GUIDE

In this chapter, we have studied chemical equilibrium and phase equilibrium. The chapter opens by developing criteria for equilibrium and introducing the chemical potential. In the second part of the chapter, we study the chemical equilibrium of ideal gas mixtures using the equilibrium constant concept. We also utilize the energy balance and determine the equilibrium flame temperature as an application. The final part of the chapter concerns phase equilibrium, including multicomponent, multiphase systems and the Gibbs phase rule.

The following list provides a study guide for this chapter. When your study of the text and end-of-chapter exercises has been completed, you should be able to

- write out the meaning of the terms listed in the margin throughout the chapter and understand each of the related

concepts. The subset of key concepts listed below is particularly important.

- apply the equilibrium constant relationship, Eq. 14.35, to determine the third quantity when *any two* of temperature, pressure, and equilibrium composition of an ideal gas mixture are known. Special cases include applications with simultaneous reactions and systems involving ionized gases.

- apply chemical equilibrium concepts with the energy balance, including determination of the equilibrium flame temperature.

- apply Eq. 14.43b, the van't Hoff equation, to determine the enthalpy of reaction when the equilibrium constant is known, and conversely.

- apply the Gibbs phase rule, Eq. 14.68.

Gibbs function   equation of reaction equilibrium   Gibbs phase rule
equilibrium criterion   equilibrium constant
chemical potential   equilibrium flame temperature

KEY EQUATIONS

$dG]_{T,P} = 0$	(14.6)	Equilibrium criterion.
$\mu_i = \left(\dfrac{\partial G}{\partial n_i} \right)_{T,P,n_l}$	(14.8)	Chemical potential of component i in a mixture.
$G = \sum\limits_{i=1}^{j} n_i \bar{g}_i(T, p_i)$	(14.15)	Gibbs function and chemical potential relations for ideal gas mixtures.
$\mu_i = \bar{g}_i(T, p_i)$	(14.16)	
$\mu_i = \bar{g}_i^{\circ} + \bar{R}T \ln \dfrac{y_i p}{p_{\text{ref}}}$	(14.17)	
$K(T) = \dfrac{y_C^{\nu_C} y_D^{\nu_D}}{y_A^{\nu_A} y_B^{\nu_B}} \left(\dfrac{p}{p_{\text{ref}}} \right)^{\nu_C + \nu_D - \nu_A - \nu_B}$	(14.32)	Equilibrium constant expressions for an equilibrium mixture of ideal gases.
$K = \dfrac{n_C^{\nu_C} n_D^{\nu_D}}{n_A^{\nu_A} n_B^{\nu_B}} \left(\dfrac{p/p_{\text{ref}}}{n} \right)^{\nu_C + \nu_D - \nu_A - \nu_B}$	(14.35)	
$\dfrac{d \ln K}{dT} = \dfrac{\Delta H}{\bar{R}T^2}$	(14.43b)	van't Hoff equation.
$\bar{g}' = \bar{g}''$	(14.59)	Phase equilibrium criterion for a pure substance.
$\mu_A' = \mu_A'' \qquad \mu_B' = \mu_B''$	(14.66)	Phase equilibrium criteria for two-component, two-phase systems.
$F = 2 + N - P$	(14.68)	Gibbs phase rule.

EXERCISES: THINGS ENGINEERS THINK ABOUT

14.1 Why is using the Gibbs function advantageous when studying chemical and phase equilibrium?

14.2 For Eq. 14.6 to apply at equilibrium, must a system attain equilibrium at fixed T and p?

14.3 Show that $(dA)_{T,V} = 0$ is a valid equilibrium criterion, where $A = U - TS$ is the *Helmholtz function*.

14.4 A mixture of 1 kmol of CO and $\frac{1}{2}$ kmol of O_2 is held at ambient temperature and pressure. After 100 hours only an insignificant amount of CO_2 has formed. Why?

14.5 Why might oxygen contained in an iron tank be treated as *inert* in a thermodynamic analysis even though iron *oxidizes* in the presence of oxygen?

14.6 For $CO_2 + H_2 \rightleftarrows CO + H_2O$, how does pressure affect the equilibrium composition?

14.7 For each of the reactions listed in Table A-27, the value of $\log_{10}K$ increases with increasing temperature. What does this imply?

14.8 For each of the reactions listed in Table A-27, the value of the equilibrium constant K at 298 K is relatively small. What does this imply?

14.9 If a system initially containing CO_2 and H_2O were held at fixed T, p, list chemical species that *might* be present at equilibrium.

14.10 Using Eq. 14.12 together with phase equilibrium considerations, suggest how the chemical potential of a mixture component could be evaluated.

14.11 Note 1 of Example 14.10 refers to the small amount of air that would be dissolved in the liquid phase. For equilibrium, what must be true of the chemical potentials of air in the liquid and gas phases?

14.12 Can liquid water, water vapor, and two different phases of ice exist in equilibrium? Explain.

CHECKING UNDERSTANDING

14.1 With reference to equilibrium criteria, express each of the following as an inequality:

   a. $dG]_{T,p}$

   b. $dS]_{U,V}$

   c. $T\,ds - dU - p\,dV$

14.2 In words, what do the symbols T and p in the expression of question 1(a) signal?

14.3 The ionization of argon, Ar, to form a mixture of Ar, Ar^+, and e^- is described by

$$Ar \rightarrow (1 - z)\,Ar + z\,Ar^+ + z e^-$$

where z denotes the extent of ionization. Derive an expression for z in terms of the ionization-equilibrium constant $K(T)$ and pressure ratio p/p_{ref}.

14.4 For the case of Question 3, if pressure increases while temperature remains constant, the value of z (a) increases, (b) decreases, (c) remains unchanged, (d) may increase, decrease, or remain constant depending on the value of temperature.

14.5 Referring to Example 14.1, if $\ln K = 6.641$ for the given reaction, the temperature is _____ K.

14.6 If $T = 2500$ K and $p = 22.4$ atm, the amount of CO present in the equilibrium mixture of Example 14.2 is _____ kmol.

14.7 For the following reaction, $\log_{10}K$ at 25°C is _____.

$$2H_2O(g) \rightleftarrows 2H_2 + O_2$$

14.8 Referring to Example 14.3, if the composition is unchanged but the temperature is 3000 K, the pressure of the equilibrium mixture is _____ atm.

14.9 If the pressure of the equilibrium mixture of Example 14.5 is 1.3 atm, while the temperature remains 3200 K, the heat transfer to the reactor, in kJ per kmol of CO_2 entering, is _____.

14.10 Referring to Example 14.7, if the amount of CO present in the equilibrium mixture is 0.10 kmol, the equilibrium flame temperature is closely (a) 2400 K, (b) 2425 K, (c) 2450 K, (d) 2475 K.

14.11 At a specified temperature, the equilibrium constant for the following reaction is 0.2:

$$\tfrac{1}{2}O_2 + \tfrac{1}{2}N_2 \rightleftarrows NO$$

The equilibrium constant at this temperature for the following reaction is then _____.

$$O_2 + N_2 \rightleftarrows 2NO$$

For each of the specified systems, determine N, P, and F in accordance with the Gibbs phase rule.

System	N	P	F
14.12 Liquid water	—	—	—
14.13 Liquid solution of water and lithium bromide	—	—	—
14.14 Liquid water and liquid mercury together with a water vapor–mercury vapor mixture	—	—	—
14.15 Liquid water and water vapor	—	—	—

Indicate whether the following statements are true or false. Explain.

14.16 Processes taking place in an insulated constant-volume tank can occur only in the direction of increasing entropy.

14.17 As a result of the high temperatures achieved during combustion some combustion products may dissociate, thereby reducing the product temperature.

14.18 In Eq. 14.54, the mole fractions are evaluated considering just the reacting substances A, B, C, D, and the inert substance E.

14.19 The chemical potential of a component can be regarded as a measure of its *escaping tendency*.

14.20 When two phases of a *pure substance* are in equilibrium, the chemical potential of the substance is the same in each phase.

14.21 When the value for the degrees of freedom is zero, the number of phases must be greater than two.

14.22 The chemical potential is an *intensive* property.

14.23 According to the van't Hoff equation, when the enthalpy of reaction, ΔH, is positive the equilibrium constant increases with temperature.

14.24 In general, the composition of the products actually formed from a given set of reactants, and their relative amounts, can be determined only from experiment.

14.25 The composition that would be present in an equilibrium mixture at T and p for a given set of reactants can be determined by solving the Clapeyron equation.

PROBLEMS: DEVELOPING ENGINEERING SKILLS

C Problem may require use of appropriate computer software in order to complete.

Working with the Equilibrium Constant

14.1 Calculate the equilibrium constant, expressed as $\log_{10} K$, for $CO_2 \rightleftarrows CO + \tfrac{1}{2}O_2$ at (a) 500 K, (b) 1000 K. Compare with values from Table A-27.

14.2 Calculate the equilibrium constant, expressed as $\log_{10} K$, for the *water–gas* reaction $CO + H_2O(g) \rightleftarrows CO_2 + H_2$ at (a) 298 K, (b) 1200 K. Compare with values from Table A-27.

14.3 Calculate the equilibrium constant, expressed as $\log_{10} K$, for $H_2O \rightleftarrows H_2 + \tfrac{1}{2}O_2$ at (a) 298 K, (b) 2000 K. Compare with values from Table A-27.

14.4 Using data from Table A-27, determine $\log_{10} K$ at 1000 K for

   a. $H_2O \rightleftarrows H_2 + \tfrac{1}{2}O_2$.

   b. $H_2 + \tfrac{1}{2}O_2 \rightleftarrows H_2O$.

   c. $2H_2O \rightleftarrows 2H_2 + O_2$.

14.5 Determine the relationship between the ideal gas equilibrium constants K_1 and K_2 for the following two alternative ways of expressing the ammonia synthesis reaction:

   1. $\tfrac{1}{2}N_2 + \tfrac{3}{2}H_2 \rightleftarrows NH_3$

   2. $N_2 + 3H_2 \rightleftarrows 2NH_3$

14.6 Consider the reactions

1. $CO + H_2O \rightleftarrows H_2 + CO_2$
2. $2CO_2 \rightleftarrows 2CO + O_2$
3. $2H_2O \rightleftarrows 2H_2 + O_2$

Show that $K_1 = (K_3/K_2)^{1/2}$.

14.7 Consider the reactions

1. $CO_2 + H_2 \rightleftarrows CO + H_2O$
2. $CO_2 \rightleftarrows CO + \frac{1}{2}O_2$
3. $H_2O \rightleftarrows H_2 + \frac{1}{2}O_2$

a. Show that $K_1 = K_2/K_3$.

b. Evaluate $\log_{10}K_1$ at 298 K, 1 atm using the expression from part (a), together with $\log_{10}K$ data from Table A-27.

c. Check the value for $\log_{10}K_1$ obtained in part (b) by applying Eq. 14.31 to reaction 1.

14.8 One kmol of carbon reacts with 2 kmol of oxygen (O_2) to form an equilibrium mixture of CO_2, CO, and O_2 at 2727°C, 1 atm. Determine the equilibrium composition.

14.9 **C** One kmol of CO_2 dissociates to form an equilibrium ideal gas mixture of CO_2, CO, and O_2 at temperature T and pressure p.

a. For $T = 3000$ K, plot the amount of CO present, in kmol, versus pressure for $1 \leq p \leq 10$ atm.

b. For $p = 1$ atm, plot the amount of CO present, in kmol, versus temperature for $2000 \leq T \leq 3500$ K.

14.10 **C** One kmol of H_2O dissociates to form an equilibrium ideal gas mixture of H_2O, H_2, and O_2 at temperature T and pressure p.

a. For $T = 3000°$K, plot the amount of H_2 present, in kmol, versus pressure ranging from 1 to 10 atm.

b. For $p = 1$ atm, plot the amount of H_2 present, in kmol, versus temperature ranging from 2000 to 3500 K.

14.11 **C** One kmol of H_2O together with x kmol of N_2 (inert) forms an equilibrium mixture at 3000 K, 1 atm consisting of H_2O, H_2, O_2, and N_2. Plot the amount of H_2 present in the equilibrium mixture, in kmol, versus x ranging from 0 to 2.

14.12 **C** An equimolar mixture of CO and O_2 reacts to form an equilibrium mixture of CO_2, CO, and O_2 at 3000 K. Determine the effect of pressure on the composition of the equilibrium mixture. Will lowering the pressure while keeping the temperature fixed increase or decrease the amount of CO_2 present? Explain.

14.13 An equimolar mixture of CO and $H_2O(g)$ reacts to form an equilibrium mixture of CO_2, CO, H_2O, and H_2 at 1727°C, 1 atm.

a. Will lowering the temperature increase or decrease the amount of H_2 present? Explain.

b. Will decreasing the pressure while keeping the temperature constant increase or decrease the amount of H_2 present? Explain.

14.14 **C** One kmol of CO_2 dissociates to form an equilibrium mixture of CO_2, CO, and O_2 at 2500 K. Determine the equilibrium composition of the mixture if the pressure of the mixture is 1.75 atm. Plot the amount of CO present in the mixture versus pressure for $1 \leq p \leq 10$ atm at 2500 K.

14.15 A vessel initially contains 1 kmol of O_2 and 3 kmol of He. An equilibrium mixture of O_2, O, and He forms at 1.25 atm. Determine the temperature of the mixture when 0.04 kmol of O is present.

14.16 **C** A vessel initially containing 1 kmol of $H_2O(g)$ and x kmol of N_2 forms an equilibrium mixture at 1 atm consisting of $H_2O(g)$, H_2,

O_2, and N_2 in which 0.5 kmol of $H_2O(g)$ is present. Plot x versus the temperature T for $3000 \leq T \leq 3600$ K.

14.17 **C** A vessel initially containing 2 kmol of N_2 and 1 kmol of O_2 forms an equilibrium mixture at 1 atm consisting of N_2, O_2, and NO. Plot the amount of NO formed versus temperature T for $2000 \leq T \leq 3500$ K.

14.18 A vessel initially containing 2 kmol of CO and 9.52 kmol of dry air forms an equilibrium mixture of CO_2, CO, O_2, and N_2 at 2000 K, 1 atm. Determine the equilibrium composition.

14.19 A vessel initially containing 1 kmol of O_2, 2 kmol of N_2, and 1 kmol of Ar forms an equilibrium mixture of O_2, N_2, NO, and Ar at 2727°C, 1 atm. Determine the equilibrium composition.

14.20 An ideal gas mixture with the molar analysis 30% CO, 10% CO_2, 40% H_2O, 20% inert gas enters a reactor operating at steady state. An equilibrium mixture of CO, CO_2, H_2O, H_2, and the inert gas exits at 1 atm.

a. If the equilibrium mixture exits at 1200 K, determine on a molar basis the ratio of the H_2 in the equilibrium mixture to the H_2O in the entering mixture.

b. If the mole fraction of CO present in the equilibrium mixture is 7.5%, determine the temperature of the equilibrium mixture, in K.

14.21 Methane burns with 90% of theoretical air to form an equilibrium mixture of CO_2, CO, $H_2O(g)$, H_2, and N_2 at 1000 K, 1 atm. Determine the composition of the equilibrium mixture, per kmol of mixture.

Chemical Equilibrium and the Energy Balance

14.22 **C** Carbon dioxide gas at 27°C, 4.1 atm enters a heat exchanger operating at steady state. An equilibrium mixture of CO_2, CO, and O_2 exits at 2227°C, 4 atm. Determine, per kmol of CO_2 entering,

a. the composition of the exiting mixture.

b. the heat transfer to the gas stream, in kJ.

Neglect kinetic and potential energy effects.

14.23 **C** Saturated water vapor at 103 kPa enters a heat exchanger operating at steady state. An equilibrium mixture of $H_2O(g)$, H_2, and O_2 exits at 2627°C, 1 atm. Determine, per kmol of steam entering,

a. the composition of the exiting mixture.

b. the heat transfer to the flowing stream, in kJ.

Neglect kinetic and potential energy effects.

14.24 **C** Carbon at 25°C, 1 atm enters a reactor operating at steady state and burns with oxygen entering at 117°C, 1 atm. The entering streams have equal molar flow rates. An equilibrium mixture of CO_2, CO, and O_2 exits at 2627°C, 1 atm. Determine, per kmol of carbon,

a. the composition of the exiting mixture.

b. the heat transfer between the reactor and its surroundings, in kJ.

Neglect kinetic and potential energy effects.

14.25 **C** Methane gas at 25°C, 1 atm enters a reactor operating at steady state and burns with 80% of theoretical air entering at 227°C, 1 atm. An equilibrium mixture of CO_2, CO, $H_2O(g)$, H_2, and N_2 exits at 1427°C, 1 atm. Determine, per kmol of methane entering,

a. the composition of the exiting mixture.

b. the heat transfer between the reactor and its surroundings, in kJ.

Neglect kinetic and potential energy effects.

14.26 **C** Gaseous propane (C_3H_8) at 25°C, 1 atm enters a reactor operating at steady state and burns with 80% of theoretical air entering separately at 25°C, 1 atm. An equilibrium mixture of CO_2, CO, $H_2O(g)$, H_2, and N_2 exits at 1227°C, 1 atm. Determine the heat transfer between the reactor and its surroundings, in kJ per kmol of propane entering. Neglect kinetic and potential energy effects.

14.27 **C** Gaseous propane (C_3H_8) at 25°C, 1 atm enters a reactor operating at steady state and burns with the theoretical amount of air entering separately at 115°C, 1 atm. An equilibrium mixture of CO_2, CO, $H_2O(g)$ O_2, and N_2 exits at 1730°C, 1 atm. Determine the heat transfer between the reactor and its surroundings, in kJ per kmol of propane entering. Neglect kinetic and potential energy effects.

14.28 One kmol of CO_2 in a piston–cylinder assembly, initially at temperature T and 1 atm, is heated at constant pressure until a final state is attained consisting of an equilibrium mixture of CO_2, CO, and O_2 in which the amount of CO_2 present is 0.422 kmol. Determine the heat transfer and the work, each in kJ, if T is (a) 298 K, (b) 400 K.

14.29 **C** Hydrogen gas (H_2) at 25°C, 1 atm enters an insulated reactor operating at steady state and reacts with 250% excess oxygen entering at 227°C, 1 atm. The products of combustion exit at 1 atm. Determine the temperature of the products, in K, if

    a. combustion is complete.

    b. an equilibrium mixture of H_2O, H_2, and O_2 exits.

Kinetic and potential energy effects are negligible.

14.30 **C** Methane at 25°C, 1 atm enters an insulated reactor operating at steady state and burns with 90% of theoretical air entering separately at 25°C, 1 atm. The products exit at 1 atm as an equilibrium mixture of CO_2, CO, $H_2O(g)$ H_2, and N_2. Determine the temperature of the exiting products, in K. Kinetic and potential energy effects are negligible.

14.31 **C** Carbon monoxide at 25°C, 1 atm enters an insulated reactor operating at steady state and burns with excess oxygen (O_2) entering at 25°C, 1 atm. The products exit at 2950 K, 1 atm as an equilibrium mixture of CO_2, CO, and O_2. Determine the percent excess oxygen. Kinetic and potential energy effects are negligible.

14.32 **C** A gaseous mixture of carbon monoxide and the theoretical amount of air at 130°C, 1.5 atm enters an insulated reactor operating at steady state. An equilibrium mixture of CO_2, CO, O_2, and N_2 exits at 1.5 atm. Determine the temperature of the exiting mixture, in K. Kinetic and potential energy effects are negligible.

14.33 **C** Methane at 25°C, 1 atm enters an insulated reactor operating at steady state and burns with oxygen entering at 127°C, 1 atm. An equilibrium mixture of CO_2, CO, O_2, and $H_2O(g)$ exits at 3250 K, 1 atm. Determine the rate at which oxygen enters the reactor, in kmol per kmol of methane. Kinetic and potential energy effects are negligible.

14.34 A mixture consisting of 1 kmol of CO and the theoretical amount of air, initially at 60°C, 1 atm, reacts in a closed, rigid, insulated vessel to form an equilibrium mixture. An analysis of the products shows that there are 0.808 kmol of CO_2, 0.192 kmol of CO, and 0.096 kmol of O_2 present. The temperature of the final mixture is measured as 2465°C. Check the consistency of these data.

Using the van't Hoff Equation, Ionization

14.35 Estimate the enthalpy of reaction at 2000 K, in kJ/kmol, for $CO_2 \rightleftarrows CO + \frac{1}{2}O_2$ using the van't Hoff equation and equilibrium constant data. Compare with the value obtained for the enthalpy of reaction using enthalpy data.

14.36 Estimate the equilibrium constant at 2700 K for the reaction $H_2O \rightleftarrows H_2 + \frac{1}{2}O_2$ using the equilibrium constant at 2100 K from Table A-27, together with the van't Hoff equation and enthalpy data.

Compare with the value for the equilibrium constant obtained from Table A-27.

14.37 If the ionization-equilibrium constants for $Cs \rightleftarrows Cs^+ + e^-$ at 1800 and 2200 K are $K = 8.21$ and $K = 23.06$, respectively, estimate the enthalpy of ionization, in kJ/kmol, at 2000 K using the van't Hoff equation.

14.38 An equilibrium mixture at 2000 K, 1 atm consists of Cs, Cs^+, and e^-. Based on 1 kmol of Cs present initially, determine the percent ionization of cesium. At 2000 K, the ionization-equilibrium constant for $Cs \rightleftarrows Cs^+ + e^-$ is $K = 15.63$.

14.39 **C** An equilibrium mixture at 10,000 K and pressure p consists of Ar, Ar^+, and e^-. Based on 1 kmol of neutral argon present initially, plot the percent ionization of argon versus pressure for $0.01 \le p \le 0.05$ atm. At 10,000 K, the ionization-equilibrium constant for $Ar \rightleftarrows Ar^+ + e^-$ is $K = 4.2 \times 10^{-4}$.

14.40 At 10,000 K and 5 atm, 1 kmol of N ionizes to form an equilibrium mixture of N, N^+, and e^- in which the amount of N present is 0.9 kmol. Determine the ionization-equilibrium constant at this temperature for $N \rightleftarrows N^+ + e^-$.

Considering Simultaneous Reactions

14.41 **C** Carbon dioxide (CO_2), oxygen (O_2), and nitrogen (N_2) enter a reactor operating at steady state with equal molar flow rates. An equilibrium mixture of CO_2, O_2, N_2, CO, and NO exits at 2800 K, 4 atm. Determine the molar analysis of the equilibrium mixture.

14.42 **C** An equimolar mixture of carbon monoxide and water vapor enters a heat exchanger operating at steady state. An equilibrium mixture of CO, CO_2, O_2, $H_2O(g)$, and H_2 exits at 2227°C, 1 atm. Determine the molar analysis of the exiting equilibrium mixture.

14.43 **C** A closed vessel initially contains a gaseous mixture consisting of 1.5 kmol of CO_2, 3 kmol of CO, and 0.5 kmol of H_2. An equilibrium mixture at 2325°C, 1 atm is formed containing CO_2, CO, H_2O, H_2, and O_2. Determine the composition of the equilibrium mixture.

14.44 **C** One kmol of $H_2O(g)$ dissociates to form an equilibrium mixture at 2778 K, 1 atm consisting of $H_2O(g)$, H_2, O_2, and OH. Determine the equilibrium composition.

Considering Phase Equilibrium

14.45 A closed system at a temperature of 25°C and a pressure of 1 bar consists of a pure liquid water phase in equilibrium with a vapor phase composed of water vapor and dry air. Determine the departure, in percent, of the partial pressure of the water vapor from the saturation pressure of water at 25°C.

14.46 An isolated system has two phases, denoted by A and B, each of which consists of the same two substances, denoted by 1 and 2. Show that necessary conditions for equilibrium are

    1. the temperature of each phase is the same, $T_A = T_B$.

    2. the pressure of each phase is the same, $p_A = p_B$.

    3. the chemical potential of each component has the same value in each phase, $\mu_1^A = \mu_1^B$, $\mu_2^A = \mu_2^B$.

14.47 An isolated system has two phases, denoted by A and B, each of which consists of the same two substances, denoted by 1 and 2. The phases are separated by a freely moving *thin* wall permeable only by substance 2. Determine the necessary conditions for equilibrium.

14.48 Derive an expression for estimating the pressure at which graphite and diamond exist in equilibrium at 25°C in terms of the specific volume, specific Gibbs function, and isothermal compressibility of each phase at 25°C, 1 atm. Discuss.

14.49 What is the maximum number of homogeneous phases that can exist at equilibrium for a system involving

    a. one component?

    b. two components?

    c. three components?

14.50 Determine the number of degrees of freedom for systems composed of

    a. liquid water and water vapor.

    b. ice, liquid water, and water vapor.

    c. ice, water vapor, and dry air.

    d. N_2 and O_2 at 20°C, 1 atm.

    e. a liquid phase and a vapor phase, each of which contains ammonia and water.

    f. liquid mercury, liquid water, and a vapor phase of mercury and water.

14.51 Develop the *phase rule* for chemically reacting systems.

14.52 For a gas–liquid system in equilibrium at temperature T and pressure p, *Raoult's law* models the relation between the partial pressure of substance i in the gas phase, p_i, and the mole fraction of substance i in the liquid phase, y_i, as follows:

$$p_i = y_i p_{\text{sat},i}(T)$$

where $p_{\text{sat},i}(T)$ is the saturation pressure of pure i at temperature T. The gas phase is assumed to form an ideal gas mixture; thus, $p_i = x_i p$ where x_i is the mole fraction of i in the gas phase. Apply Raoult's law to the following cases, which are representative of conditions that might be encountered in ammonia–water absorption systems (Sec. 10.5):

    a. Consider a two-phase, liquid–vapor ammonia–water system in equilibrium at 20°C. The mole fraction of ammonia in the liquid phase is 80%. Determine the pressure, in bar, and the mole fraction of ammonia in the vapor phase.

    b. Determine the mole fractions of ammonia in the liquid and vapor phases of a two-phase ammonia–water system in equilibrium at 40°C, 12 bar.

DESIGN & OPEN-ENDED PROBLEMS: EXPLORING ENGINEERING PRACTICE

14.1D Spark-ignition engine exhaust gases contain several air pollutants including the oxides of nitrogen, NO, and NO_2, collectively known as NO_x. Additionally, the exhaust gases may contain carbon monoxide (CO) and unburned or partially burned hydrocarbons (HC).

    a. The pollutant amounts actually present depend on engine design and operating conditions, and typically differ significantly from values calculated on the basis of chemical equilibrium. Discuss both the reasons for these discrepancies and the possible mechanisms by which such pollutants are formed in an actual engine.

    b. For spark-ignition engines, the average production of pollutants upstream of the catalyst, in g per mile of vehicle travel, are nitric oxides, 1.5; hydrocarbons, 2; and carbon monoxide, 20. For a city in your locale having a population of 100,000 or more, estimate the annual amount, in kg, of each pollutant that would be discharged if automobiles had no emission control devices. Repeat if the vehicles adhere to current U.S. government emissions standards.

14.2D Using appropriate software, develop plots giving the variation with equivalence ratio of the equilibrium products of octane–air mixtures at 30 atm and selected temperatures ranging from 1700 to 2800 K. Consider equivalence ratios in the interval from 0.2 to 1.4 and equilibrium products including, but not necessarily limited to, CO_2, CO, H_2O, O_2, O, H_2, N_2, NO, OH. Under what conditions is the formation of nitric oxide (NO) and carbon monoxide (CO) most significant? Discuss.

14.3D The Federal Clean Air Act of 1970 and succeeding Clean Air Act Amendments target the oxides of nitrogen NO and NO_2, collectively known as NO_x, as significant air pollutants. NO_x is formed in combustion via three primary mechanisms: *thermal* NO_x formation, *prompt* NO_x formation, and *fuel* NO_x formation. Discuss these formation mechanisms, including a discussion of thermal NO_x formation by the *Zeldovich mechanism*. What is the role of NO_x in the formation of ozone? What are some NO_x reduction strategies?

14.4D The amount of sulfur dioxide (SO_2) present in *off gases* from industrial processes can be reduced by oxidizing the SO_2 to SO_3 at an elevated temperature in a catalytic reactor. The SO_3 can be reacted in turn with water to form sulfuric acid that has economic value. For an off gas at 1 atm having the molar analysis of 12% SO_2, 8% O_2, 80% N_2, estimate the range of temperatures at which a *substantial* conversion of SO_2 to SO_3 might be realized. Report your findings in a PowerPoint presentation suitable for your class. Additionally, in an accompanying memorandum, discuss your modeling assumptions and provide sample calculations.

14.5D A gaseous mixture of hydrogen (H_2) and carbon monoxide (CO) enters a catalytic reactor and a gaseous mixture of methanol (CH_3OH), hydrogen, and carbon monoxide exits. At the preliminary process design stage, plausible estimates are required of the inlet hydrogen mole fraction, y_{H_2}, the temperature of the exiting mixture, T_e, and the pressure of the exiting mixture, p_e, subject to the following four constraints: (1) $0.5 \leq y_{H_2} \leq 0.75$, (2) $300 \leq T_e \leq 400$ K, (3) $1 \leq p_e \leq 10$ atm, and (4) the exiting mixture contains at least 75% methanol on a molar basis. In a memorandum, provide your estimates together with a discussion of the modeling employed and sample calculations.

14.6D When systems in thermal, mechanical, and chemical equilibrium are *perturbed*, changes within the systems can occur, leading to a new equilibrium state. The effects of perturbing the system considered in developing Eqs. 14.32 and 14.33 can be determined by study of these equations. For example, at fixed pressure and temperature it can be concluded that an increase in the amount of the inert component E would lead to increases in n_C and n_D when $\Delta \nu = (\nu_C + \nu_D - \nu_A - \nu_B)$ is positive, to decreases in n_C and n_D when $\Delta \nu$ is negative, and no change when $\Delta \nu = 0$.

    a. For a system consisting of NH_3, N_2, and H_2 at fixed pressure and temperature, subject to the reaction

$$2NH_3(g) \rightleftarrows N_2(g) + 3H_2(g)$$

investigate the effects, in turn, of additions in the amounts present of NH_3, H_2, and N_2.

b. For the *general case* of Eqs. 14.32 and 14.33, investigate the effects, in turn, of additions of A, B, C, and D.

Present your findings, together with the modeling assumptions used, in a PowerPoint presentation suitable for your class.

14.7D With reference to the equilibrium constant data of Table A-27:

a. For each of the tabulated reactions plot $\log_{10}K$ versus $1/T$ and determine the slope of the line of best fit. What is the thermodynamic significance of the slope? Check your conclusion about the slope using data from the *JANAF* tables.[1]

b. A textbook states that the magnitude of the equilibrium constant often signals the importance of a reaction, and offers this *rule of thumb*: When $K < 10^{-3}$, the extent of the reaction is usually not significant, whereas when $K > 10^3$ the reaction generally proceeds closely to equilibrium. Confirm or deny this rule.

Present your findings and conclusions in a report including at least three references.

14.8D **a.** For an equilibrium ideal gas mixture of N_2, H_2, and NH_3, evaluate the equilibrium constant from an expression you derive from the *van't Hoff equation* that requires only standard state enthalpy of formation and Gibbs function of formation data together with suitable analytical expressions in terms of temperature for the ideal gas specific heats of N_2, H_2, NH_3.

b. For the synthesis of ammonia by

$$\tfrac{1}{2}N_2 + \tfrac{3}{2}H_2 \rightarrow NH_3$$

provide a recommendation for the ranges of temperature and pressure for which the mole fraction of ammonia in the mixture is at least 0.5.

Write a report including your derivation, recommendations for the ranges of temperature and pressure, sample calculations, and at least three references.

14.9D U.S. Patent 5,298,233 describes a means for converting industrial wastes to carbon dioxide and water vapor. Hydrogen- and carbon-containing feed, such as organic or inorganic sludge, low-grade fuel oil, or municipal garbage, is introduced into a molten bath consisting of two immiscible molten metal phases. The carbon and hydrogen of the feed are converted, respectively, to dissolved carbon and dissolved hydrogen. The dissolved carbon is oxidized in the first molten metal phase to carbon dioxide, which is released from the bath. The dissolved hydrogen migrates to the second molten metal phase, where it is oxidized to form water vapor, which is also released from the bath. Critically evaluate this technology for waste disposal. Is the technology promising commercially? Compare with alternative waste management practices such as pyrolysis and incineration. Write a report including at least three references.

14.10D **Figure P14.10D** gives a table of data for a lithium bromide–water absorption refrigeration cycle together with the sketch of a property diagram showing the cycle. The property diagram plots the vapor pressure versus the lithium bromide concentration. Apply the *phase rule* to verify that the numbered states are fixed by the property values provided. What does the *crystallization* line on the equilibrium diagram represent, and what is its significance for absorption cycle operation? Locate the numbered states on an enthalpy-concentration diagram for lithium bromide–water solutions obtained from the literature. Finally, develop a sketch of the equipment schematic for this refrigeration cycle. Present your findings in a report including at least three references.

State	Temperature (°F)	Pressure (in. Hg)	$(mf)_{LiBr}$ (%)
1	115	0.27	63.3
2	104	0.27	59.5
3	167	1.65	59.5
4	192	3.00	59.5
5	215	3.00	64.0
6	135	0.45	64.0
7	120	0.32	63.3

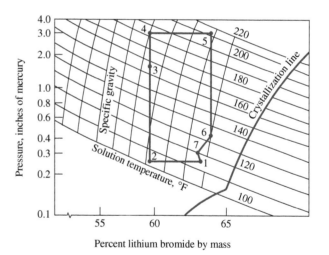

Percent lithium bromide by mass

FIGURE P14.10D

---

[1]Stull, D. R., and H. Prophet, *JANAF Thermochemical Tables*, 2nd ed., NSRDS-NBS 37, National Bureau of Standards, Washington, DC, June 1971.

Index to Tables in SI Units

TABLE A-1 Atomic or Molecular Weights and Critical Properties of Selected Elements and Compounds

Substance	Chemical Formula	M (kg/kmol)	T_c (K)	p_c (bar)	$Z_c = \dfrac{p_c v_c}{RT_c}$
Acetylene	C_2H_2	26.04	309	62.8	0.274
Air (equivalent)	—	28.97	133	37.7	0.284
Ammonia	NH_3	17.03	406	112.8	0.242
Argon	Ar	39.94	151	48.6	0.290
Benzene	C_6H_6	78.11	563	49.3	0.274
Butane	C_4H_{10}	58.12	425	38.0	0.274
Carbon	C	12.01	—	—	—
Carbon dioxide	CO_2	44.01	304	73.9	0.276
Carbon monoxide	CO	28.01	133	35.0	0.294
Copper	Cu	63.54	—	—	—
Ethane	C_2H_6	30.07	305	48.8	0.285
Ethanol	C_2H_5OH	46.07	516	63.8	0.249
Ethylene	C_2H_4	28.05	283	51.2	0.270
Helium	He	4.003	5.2	2.3	0.300
Hydrogen	H_2	2.016	33.2	13.0	0.304
Methane	CH_4	16.04	191	46.4	0.290
Methanol	CH_3OH	32.04	513	79.5	0.220
Nitrogen	N_2	28.01	126	33.9	0.291
Octane	C_8H_{18}	114.22	569	24.9	0.258
Oxygen	O_2	32.00	154	50.5	0.290
Propane	C_3H_8	44.09	370	42.7	0.276
Propylene	C_3H_6	42.08	365	46.2	0.276
Refrigerant 12	CCl_2F_2	120.92	385	41.2	0.278
Refrigerant 22	$CHClF_2$	86.48	369	49.8	0.267
Refrigerant 134a	CF_3CH_2F	102.03	374	40.7	0.260
Sulfur dioxide	SO_2	64.06	431	78.7	0.268
Water	H_2O	18.02	647.3	220.9	0.233

Sources: Adapted from *International Critical Tables* and L. C. Nelson and E. F. Obert, Generalized Compressibility Charts, *Chem. Eng., 61:* 203 (1954).

Sources for Tables A-2 through A-18.

Tables A-2 through A-6 are extracted from J. H. Keenan, F. G. Keyes, P. G. Hill, and J. G. Moore, *Steam Tables,* Wiley, New York, 1969.

Tables A-7 through A-9 are calculated based on equations from A. Kamei and S. W. Beyerlein, "A Fundamental Equation for Chlorodifluoromethane (R-22)," *Fluid Phase Equilibria,* Vol. 80, No. 11, 1992, pp. 71–86.

Tables A-10 through A-12 are calculated based on equations from D. P. Wilson and R. S. Basu, "Thermodynamic Properties of a New Stratospherically Safe Working Fluid — Refrigerant 134a," *ASHRAE Trans.,* Vol. 94, Pt. 2, 1988, pp. 2095–2118.

Tables A-13 through A-15 are calculated based on equations from L. Haar and J. S. Gallagher, "Thermodynamic Properties of Ammonia," *J. Phys. Chem. Reference Data,* Vol. 7, 1978, pp. 635–792.

Tables A-16 through A-18 are calculated based on B. A. Younglove and J. F. Ely, "Thermophysical Properties of Fluids. II. Methane, Ethane, Propane, Isobutane and Normal Butane," *J. Phys. Chem. Ref. Data,* Vol. 16, No. 4, 1987, pp. 577–598.

TABLE A-2 Properties of Saturated Water (Liquid–Vapor): Temperature Table

Pressure Conversions: 1 bar = 0.1 MPa = 10^2 kPa		Specific Volume m^3/kg		Internal Energy kJ/kg		Enthalpy kJ/kg			Entropy kJ/kg · K		
Temp. °C	Press. bar	Sat. Liquid $v_f \times 10^3$	Sat. Vapor v_g	Sat. Liquid u_f	Sat. Vapor u_g	Sat. Liquid h_f	Evap. h_{fg}	Sat. Vapor h_g	Sat. Liquid s_f	Sat. Vapor s_g	Temp. °C
.01	0.00611	1.0002	206.136	0.00	2375.3	0.01	2501.3	2501.4	0.0000	9.1562	.01
4	0.00813	1.0001	157.232	16.77	2380.9	16.78	2491.9	2508.7	0.0610	9.0514	4
5	0.00872	1.0001	147.120	20.97	2382.3	20.98	2489.6	2510.6	0.0761	9.0257	5
6	0.00935	1.0001	137.734	25.19	2383.6	25.20	2487.2	2512.4	0.0912	9.0003	6
8	0.01072	1.0002	120.917	33.59	2386.4	33.60	2482.5	2516.1	0.1212	8.9501	8
10	0.01228	1.0004	106.379	42.00	2389.2	42.01	2477.7	2519.8	0.1510	8.9008	10
11	0.01312	1.0004	99.857	46.20	2390.5	46.20	2475.4	2521.6	0.1658	8.8765	11
12	0.01402	1.0005	93.784	50.41	2391.9	50.41	2473.0	2523.4	0.1806	8.8524	12
13	0.01497	1.0007	88.124	54.60	2393.3	54.60	2470.7	2525.3	0.1953	8.8285	13
14	0.01598	1.0008	82.848	58.79	2394.7	58.80	2468.3	2527.1	0.2099	8.8048	14
15	0.01705	1.0009	77.926	62.99	2396.1	62.99	2465.9	2528.9	0.2245	8.7814	15
16	0.01818	1.0011	73.333	67.18	2397.4	67.19	2463.6	2530.8	0.2390	8.7582	16
17	0.01938	1.0012	69.044	71.38	2398.8	71.38	2461.2	2532.6	0.2535	8.7351	17
18	0.02064	1.0014	65.038	75.57	2400.2	75.58	2458.8	2534.4	0.2679	8.7123	18
19	0.02198	1.0016	61.293	79.76	2401.6	79.77	2456.5	2536.2	0.2823	8.6897	19
20	0.02339	1.0018	57.791	83.95	2402.9	83.96	2454.1	2538.1	0.2966	8.6672	20
21	0.02487	1.0020	54.514	88.14	2404.3	88.14	2451.8	2539.9	0.3109	8.6450	21
22	0.02645	1.0022	51.447	92.32	2405.7	92.33	2449.4	2541.7	0.3251	8.6229	22
23	0.02810	1.0024	48.574	96.51	2407.0	96.52	2447.0	2543.5	0.3393	8.6011	23
24	0.02985	1.0027	45.883	100.70	2408.4	100.70	2444.7	2545.4	0.3534	8.5794	24
25	0.03169	1.0029	43.360	104.88	2409.8	104.89	2442.3	2547.2	0.3674	8.5580	25
26	0.03363	1.0032	40.994	109.06	2411.1	109.07	2439.9	2549.0	0.3814	8.5367	26
27	0.03567	1.0035	38.774	113.25	2412.5	113.25	2437.6	2550.8	0.3954	8.5156	27
28	0.03782	1.0037	36.690	117.42	2413.9	117.43	2435.2	2552.6	0.4093	8.4946	28
29	0.04008	1.0040	34.733	121.60	2415.2	121.61	2432.8	2554.5	0.4231	8.4739	29
30	0.04246	1.0043	32.894	125.78	2416.6	125.79	2430.5	2556.3	0.4369	8.4533	30
31	0.04496	1.0046	31.165	129.96	2418.0	129.97	2428.1	2558.1	0.4507	8.4329	31
32	0.04759	1.0050	29.540	134.14	2419.3	134.15	2425.7	2559.9	0.4644	8.4127	32
33	0.05034	1.0053	28.011	138.32	2420.7	138.33	2423.4	2561.7	0.4781	8.3927	33
34	0.05324	1.0056	26.571	142.50	2422.0	142.50	2421.0	2563.5	0.4917	8.3728	34
35	0.05628	1.0060	25.216	146.67	2423.4	146.68	2418.6	2565.3	0.5053	8.3531	35
36	0.05947	1.0063	23.940	150.85	2424.7	150.86	2416.2	2567.1	0.5188	8.3336	36
38	0.06632	1.0071	21.602	159.20	2427.4	159.21	2411.5	2570.7	0.5458	8.2950	38
40	0.07384	1.0078	19.523	167.56	2430.1	167.57	2406.7	2574.3	0.5725	8.2570	40
45	0.09593	1.0099	15.258	188.44	2436.8	188.45	2394.8	2583.2	0.6387	8.1648	45

v_f = (table value)/1000

H₂O

H₂O

TABLE A-2 Properties of Saturated Water (Liquid–Vapor): Temperature Table (*Continued*)

Temp. °C	Press. bar	Specific Volume m³/kg Sat. Liquid $v_f \times 10^3$	Sat. Vapor v_g	Internal Energy kJ/kg Sat. Liquid u_f	Sat. Vapor u_g	Enthalpy kJ/kg Sat. Liquid h_f	Evap. h_{fg}	Sat. Vapor h_g	Entropy kJ/kg·K Sat. Liquid s_f	Sat. Vapor s_g	Temp. °C
50	.1235	1.0121	12.032	209.32	2443.5	209.33	2382.7	2592.1	.7038	8.0763	50
55	.1576	1.0146	9.568	230.21	2450.1	230.23	2370.7	2600.9	.7679	7.9913	55
60	.1994	1.0172	7.671	251.11	2456.6	251.13	2358.5	2609.6	.8312	7.9096	60
65	.2503	1.0199	6.197	272.02	2463.1	272.06	2346.2	2618.3	.8935	7.8310	65
70	.3119	1.0228	5.042	292.95	2469.6	292.98	2333.8	2626.8	.9549	7.7553	70
75	.3858	1.0259	4.131	313.90	2475.9	313.93	2321.4	2635.3	1.0155	7.6824	75
80	.4739	1.0291	3.407	334.86	2482.2	334.91	2308.8	2643.7	1.0753	7.6122	80
85	.5783	1.0325	2.828	355.84	2488.4	355.90	2296.0	2651.9	1.1343	7.5445	85
90	.7014	1.0360	2.361	376.85	2494.5	376.92	2283.2	2660.1	1.1925	7.4791	90
95	.8455	1.0397	1.982	397.88	2500.6	397.96	2270.2	2668.1	1.2500	7.4159	95
100	1.014	1.0435	1.673	418.94	2506.5	419.04	2257.0	2676.1	1.3069	7.3549	100
110	1.433	1.0516	1.210	461.14	2518.1	461.30	2230.2	2691.5	1.4185	7.2387	110
120	1.985	1.0603	0.8919	503.50	2529.3	503.71	2202.6	2706.3	1.5276	7.1296	120
130	2.701	1.0697	0.6685	546.02	2539.9	546.31	2174.2	2720.5	1.6344	7.0269	130
140	3.613	1.0797	0.5089	588.74	2550.0	589.13	2144.7	2733.9	1.7391	6.9299	140
150	4.758	1.0905	0.3928	631.68	2559.5	632.20	2114.3	2746.5	1.8418	6.8379	150
160	6.178	1.1020	0.3071	674.86	2568.4	675.55	2082.6	2758.1	1.9427	6.7502	160
170	7.917	1.1143	0.2428	718.33	2576.5	719.21	2049.5	2768.7	2.0419	6.6663	170
180	10.02	1.1274	0.1941	762.09	2583.7	763.22	2015.0	2778.2	2.1396	6.5857	180
190	12.54	1.1414	0.1565	806.19	2590.0	807.62	1978.8	2786.4	2.2359	6.5079	190
200	15.54	1.1565	0.1274	850.65	2595.3	852.45	1940.7	2793.2	2.3309	6.4323	200
210	19.06	1.1726	0.1044	895.53	2599.5	897.76	1900.7	2798.5	2.4248	6.3585	210
220	23.18	1.1900	0.08619	940.87	2602.4	943.62	1858.5	2802.1	2.5178	6.2861	220
230	27.95	1.2088	0.07158	986.74	2603.9	990.12	1813.8	2804.0	2.6099	6.2146	230
240	33.44	1.2291	0.05976	1033.2	2604.0	1037.3	1766.5	2803.8	2.7015	6.1437	240
250	39.73	1.2512	0.05013	1080.4	2602.4	1085.4	1716.2	2801.5	2.7927	6.0730	250
260	46.88	1.2755	0.04221	1128.4	2599.0	1134.4	1662.5	2796.6	2.8838	6.0019	260
270	54.99	1.3023	0.03564	1177.4	2593.7	1184.5	1605.2	2789.7	2.9751	5.9301	270
280	64.12	1.3321	0.03017	1227.5	2586.1	1236.0	1543.6	2779.6	3.0668	5.8571	280
290	74.36	1.3656	0.02557	1278.9	2576.0	1289.1	1477.1	2766.2	3.1594	5.7821	290
300	85.81	1.4036	0.02167	1332.0	2563.0	1344.0	1404.9	2749.0	3.2534	5.7045	300
320	112.7	1.4988	0.01549	1444.6	2525.5	1461.5	1238.6	2700.1	3.4480	5.5362	320
340	145.9	1.6379	0.01080	1570.3	2464.6	1594.2	1027.9	2622.0	3.6594	5.3357	340
360	186.5	1.8925	0.006945	1725.2	2351.5	1760.5	720.5	2481.0	3.9147	5.0526	360
374.14	220.9	3.155	0.003155	2029.6	2029.6	2099.3	0	2099.3	4.4298	4.4298	374.14

Pressure Conversions:
1 bar = 0.1 MPa
= 10² kPa

v_f = (table value)/1000

TABLE A-3   **Properties of Saturated Water (Liquid–Vapor): Pressure Table**

Pressure Conversions: 1 bar = 0.1 MPa = 10² kPa		Specific Volume m³/kg		Internal Energy kJ/kg		Enthalpy kJ/kg			Entropy kJ/kg · K		
Press. bar	Temp. °C	Sat. Liquid $v_f \times 10^3$	Sat. Vapor v_g	Sat. Liquid u_f	Sat. Vapor u_g	Sat. Liquid h_f	Evap. h_{fg}	Sat. Vapor h_g	Sat. Liquid s_f	Sat. Vapor s_g	Press. bar
0.04	28.96	1.0040	34.800	121.45	2415.2	121.46	2432.9	2554.4	0.4226	8.4746	0.04
0.06	36.16	1.0064	23.739	151.53	2425.0	151.53	2415.9	2567.4	0.5210	8.3304	0.06
0.08	41.51	1.0084	18.103	173.87	2432.2	173.88	2403.1	2577.0	0.5926	8.2287	0.08
0.10	45.81	1.0102	14.674	191.82	2437.9	191.83	2392.8	2584.7	0.6493	8.1502	0.10
0.20	60.06	1.0172	7.649	251.38	2456.7	251.40	2358.3	2609.7	0.8320	7.9085	0.20
0.30	69.10	1.0223	5.229	289.20	2468.4	289.23	2336.1	2625.3	0.9439	7.7686	0.30
0.40	75.87	1.0265	3.993	317.53	2477.0	317.58	2319.2	2636.8	1.0259	7.6700	0.40
0.50	81.33	1.0300	3.240	340.44	2483.9	340.49	2305.4	2645.9	1.0910	7.5939	0.50
0.60	85.94	1.0331	2.732	359.79	2489.6	359.86	2293.6	2653.5	1.1453	7.5320	0.60
0.70	89.95	1.0360	2.365	376.63	2494.5	376.70	2283.3	2660.0	1.1919	7.4797	0.70
0.80	93.50	1.0380	2.087	391.58	2498.8	391.66	2274.1	2665.8	1.2329	7.4346	0.80
0.90	96.71	1.0410	1.869	405.06	2502.6	405.15	2265.7	2670.9	1.2695	7.3949	0.90
1.00	99.63	1.0432	1.694	417.36	2506.1	417.46	2258.0	2675.5	1.3026	7.3594	1.00
1.50	111.4	1.0528	1.159	466.94	2519.7	467.11	2226.5	2693.6	1.4336	7.2233	1.50
2.00	120.2	1.0605	0.8857	504.49	2529.5	504.70	2201.9	2706.7	1.5301	7.1271	2.00
2.50	127.4	1.0672	0.7187	535.10	2537.2	535.37	2181.5	2716.9	1.6072	7.0527	2.50
3.00	133.6	1.0732	0.6058	561.15	2543.6	561.47	2163.8	2725.3	1.6718	6.9919	3.00
3.50	138.9	1.0786	0.5243	583.95	2546.9	584.33	2148.1	2732.4	1.7275	6.9405	3.50
4.00	143.6	1.0836	0.4625	604.31	2553.6	604.74	2133.8	2738.6	1.7766	6.8959	4.00
4.50	147.9	1.0882	0.4140	622.25	2557.6	623.25	2120.7	2743.9	1.8207	6.8565	4.50
5.00	151.9	1.0926	0.3749	639.68	2561.2	640.23	2108.5	2748.7	1.8607	6.8212	5.00
6.00	158.9	1.1006	0.3157	669.90	2567.4	670.56	2086.3	2756.8	1.9312	6.7600	6.00
7.00	165.0	1.1080	0.2729	696.44	2572.5	697.22	2066.3	2763.5	1.9922	6.7080	7.00
8.00	170.4	1.1148	0.2404	720.22	2576.8	721.11	2048.0	2769.1	2.0462	6.6628	8.00
9.00	175.4	1.1212	0.2150	741.83	2580.5	742.83	2031.1	2773.9	2.0946	6.6226	9.00
10.0	179.9	1.1273	0.1944	761.68	2583.6	762.81	2015.3	2778.1	2.1387	6.5863	10.0
15.0	198.3	1.1539	0.1318	843.16	2594.5	844.84	1947.3	2792.2	2.3150	6.4448	15.0
20.0	212.4	1.1767	0.09963	906.44	2600.3	908.79	1890.7	2799.5	2.4474	6.3409	20.0
25.0	224.0	1.1973	0.07998	959.11	2603.1	962.11	1841.0	2803.1	2.5547	6.2575	25.0
30.0	233.9	1.2165	0.06668	1004.8	2604.1	1008.4	1795.7	2804.2	2.6457	6.1869	30.0
35.0	242.6	1.2347	0.05707	1045.4	2603.7	1049.8	1753.7	2803.4	2.7253	6.1253	35.0
40.0	250.4	1.2522	0.04978	1082.3	2602.3	1087.3	1714.1	2801.4	2.7964	6.0701	40.0
45.0	257.5	1.2692	0.04406	1116.2	2600.1	1121.9	1676.4	2798.3	2.8610	6.0199	45.0
50.0	264.0	1.2859	0.03944	1147.8	2597.1	1154.2	1640.1	2794.3	2.9202	5.9734	50.0
60.0	275.6	1.3187	0.03244	1205.4	2589.7	1213.4	1571.0	2784.3	3.0267	5.8892	60.0
70.0	285.9	1.3513	0.02737	1257.6	2580.5	1267.0	1505.1	2772.1	3.1211	5.8133	70.0
80.0	295.1	1.3842	0.02352	1305.6	2569.8	1316.6	1441.3	2758.0	3.2068	5.7432	80.0
90.0	303.4	1.4178	0.02048	1350.5	2557.8	1363.3	1378.9	2742.1	3.2858	5.6772	90.0
100.	311.1	1.4524	0.01803	1393.0	2544.4	1407.6	1317.1	2724.7	3.3596	5.6141	100.
110.	318.2	1.4886	0.01599	1433.7	2529.8	1450.1	1255.5	2705.6	3.4295	5.5527	110.

$v_f = $ (table value)/1000

H₂O

TABLE A-3 **Properties of Saturated Water (Liquid–Vapor): Pressure Table** (*Continued*)

Pressure Conversions: 1 bar = 0.1 MPa = 10² kPa		Specific Volume m³/kg		Internal Energy kJ/kg		Enthalpy kJ/kg			Entropy kJ/kg · K		
Press. bar	Temp. °C	Sat. Liquid $v_f \times 10^3$	Sat. Vapor v_g	Sat. Liquid u_f	Sat. Vapor u_g	Sat. Liquid h_f	Evap. h_{fg}	Sat. Vapor h_g	Sat. Liquid s_f	Sat. Vapor s_g	Press. bar
120.	324.8	1.5267	0.01426	1473.0	2513.7	1491.3	1193.6	2684.9	3.4962	5.4924	120.
130.	330.9	1.5671	0.01278	1511.1	2496.1	1531.5	1130.7	2662.2	3.5606	5.4323	130.
140.	336.8	1.6107	0.01149	1548.6	2476.8	1571.1	1066.5	2637.6	3.6232	5.3717	140.
150.	342.2	1.6581	0.01034	1585.6	2455.5	1610.5	1000.0	2610.5	3.6848	5.3098	150.
160.	347.4	1.7107	0.009306	1622.7	2431.7	1650.1	930.6	2580.6	3.7461	5.2455	160.
170.	352.4	1.7702	0.008364	1660.2	2405.0	1690.3	856.9	2547.2	3.8079	5.1777	170.
180.	357.1	1.8397	0.007489	1698.9	2374.3	1732.0	777.1	2509.1	3.8715	5.1044	180.
190.	361.5	1.9243	0.006657	1739.9	2338.1	1776.5	688.0	2464.5	3.9388	5.0228	190.
200.	365.8	2.036	0.005834	1785.6	2293.0	1826.3	583.4	2409.7	4.0139	4.9269	200.
220.9	374.1	3.155	0.003155	2029.6	2029.6	2099.3	0	2099.3	4.4298	4.4298	220.9

v_f = (table value)/1000

TABLE A-4   **Properties of Superheated Water Vapor**

H₂O

T °C	v m³/kg	u kJ/kg	h kJ/kg	s kJ/kg·K	v m³/kg	u kJ/kg	h kJ/kg	s kJ/kg·K
	p = 0.06 bar = 0.006 MPa (T_{sat} = 36.16°C)				p = 0.35 bar = 0.035 MPa (T_{sat} = 72.69°C)			
Sat.	23.739	2425.0	2567.4	8.3304	4.526	2473.0	2631.4	7.7158
80	27.132	2487.3	2650.1	8.5804	4.625	2483.7	2645.6	7.7564
120	30.219	2544.7	2726.0	8.7840	5.163	2542.4	2723.1	7.9644
160	33.302	2602.7	2802.5	8.9693	5.696	2601.2	2800.6	8.1519
200	36.383	2661.4	2879.7	9.1398	6.228	2660.4	2878.4	8.3237
240	39.462	2721.0	2957.8	9.2982	6.758	2720.3	2956.8	8.4828
280	42.540	2781.5	3036.8	9.4464	7.287	2780.9	3036.0	8.6314
320	45.618	2843.0	3116.7	9.5859	7.815	2842.5	3116.1	8.7712
360	48.696	2905.5	3197.7	9.7180	8.344	2905.1	3197.1	8.9034
400	51.774	2969.0	3279.6	9.8435	8.872	2968.6	3279.2	9.0291
440	54.851	3033.5	3362.6	9.9633	9.400	3033.2	3362.2	9.1490
500	59.467	3132.3	3489.1	10.1336	10.192	3132.1	3488.8	9.3194
	p = 0.70 bar = 0.07 MPa (T_{sat} = 89.95°C)				p = 1.0 bar = 0.10 MPa (T_{sat} = 99.63°C)			
Sat.	2.365	2494.5	2660.0	7.4797	1.694	2506.1	2675.5	7.3594
100	2.434	2509.7	2680.0	7.5341	1.696	2506.7	2676.2	7.3614
120	2.571	2539.7	2719.6	7.6375	1.793	2537.3	2716.6	7.4668
160	2.841	2599.4	2798.2	7.8279	1.984	2597.8	2796.2	7.6597
200	3.108	2659.1	2876.7	8.0012	2.172	2658.1	2875.3	7.8343
240	3.374	2719.3	2955.5	8.1611	2.359	2718.5	2954.5	7.9949
280	3.640	2780.2	3035.0	8.3162	2.546	2779.6	3034.2	8.1445
320	3.905	2842.0	3115.3	8.4504	2.732	2841.5	3114.6	8.2849
360	4.170	2904.6	3196.5	8.5828	2.917	2904.2	3195.9	8.4175
400	4.434	2968.2	3278.6	8.7086	3.103	2967.9	3278.2	8.5435
440	4.698	3032.9	3361.8	8.8286	3.288	3032.6	3361.4	8.6636
500	5.095	3131.8	3488.5	8.9991	3.565	3131.6	3488.1	8.8342
	p = 1.5 bar = 0.15 MPa (T_{sat} = 111.37°C)				p = 3.0 bar = 0.30 MPa (T_{sat} = 133.55°C)			
Sat.	1.159	2519.7	2693.6	7.2233	0.606	2543.6	2725.3	6.9919
120	1.188	2533.3	2711.4	7.2693				
160	1.317	2595.2	2792.8	7.4665	0.651	2587.1	2782.3	7.1276
200	1.444	2656.2	2872.9	7.6433	0.716	2650.7	2865.5	7.3115
240	1.570	2717.2	2952.7	7.8052	0.781	2713.1	2947.3	7.4774
280	1.695	2778.6	3032.8	7.9555	0.844	2775.4	3028.6	7.6299
320	1.819	2840.6	3113.5	8.0964	0.907	2838.1	3110.1	7.7722
360	1.943	2903.5	3195.0	8.2293	0.969	2901.4	3192.2	7.9061
400	2.067	2967.3	3277.4	8.3555	1.032	2965.6	3275.0	8.0330
440	2.191	3032.1	3360.7	8.4757	1.094	3030.6	3358.7	8.1538
500	2.376	3131.2	3487.6	8.6466	1.187	3130.0	3486.0	8.3251
600	2.685	3301.7	3704.3	8.9101	1.341	3300.8	3703.2	8.5892

H₂O

Pressure Conversions:
1 bar = 0.1 MPa
= 10² kPa

TABLE A-4 **Properties of Superheated Water Vapor (*Continued*)**

T °C	v m³/kg	u kJ/kg	h kJ/kg	s kJ/kg·K	v m³/kg	u kJ/kg	h kJ/kg	s kJ/kg·K
	$p = 5.0$ bar $= 0.50$ MPa ($T_{sat} = 151.86$°C)				$p = 7.0$ bar $= 0.70$ MPa ($T_{sat} = 164.97$°C)			
Sat.	0.3749	2561.2	2748.7	6.8213	0.2729	2572.5	2763.5	6.7080
180	0.4045	2609.7	2812.0	6.9656	0.2847	2599.8	2799.1	6.7880
200	0.4249	2642.9	2855.4	7.0592	0.2999	2634.8	2844.8	6.8865
240	0.4646	2707.6	2939.9	7.2307	0.3292	2701.8	2932.2	7.0641
280	0.5034	2771.2	3022.9	7.3865	0.3574	2766.9	3017.1	7.2233
320	0.5416	2834.7	3105.6	7.5308	0.3852	2831.3	3100.9	7.3697
360	0.5796	2898.7	3188.4	7.6660	0.4126	2895.8	3184.7	7.5063
400	0.6173	2963.2	3271.9	7.7938	0.4397	2960.9	3268.7	7.6350
440	0.6548	3028.6	3356.0	7.9152	0.4667	3026.6	3353.3	7.7571
500	0.7109	3128.4	3483.9	8.0873	0.5070	3126.8	3481.7	7.9299
600	0.8041	3299.6	3701.7	8.3522	0.5738	3298.5	3700.2	8.1956
700	0.8969	3477.5	3925.9	8.5952	0.6403	3476.6	3924.8	8.4391
	$p = 10.0$ bar $= 1.0$ MPa ($T_{sat} = 179.91$°C)				$p = 15.0$ bar $= 1.5$ MPa ($T_{sat} = 198.32$°C)			
Sat.	0.1944	2583.6	2778.1	6.5865	0.1318	2594.5	2792.2	6.4448
200	0.2060	2621.9	2827.9	6.6940	0.1325	2598.1	2796.8	6.4546
240	0.2275	2692.9	2920.4	6.8817	0.1483	2676.9	2899.3	6.6628
280	0.2480	2760.2	3008.2	7.0465	0.1627	2748.6	2992.7	6.8381
320	0.2678	2826.1	3093.9	7.1962	0.1765	2817.1	3081.9	6.9938
360	0.2873	2891.6	3178.9	7.3349	0.1899	2884.4	3169.2	7.1363
400	0.3066	2957.3	3263.9	7.4651	0.2030	2951.3	3255.8	7.2690
440	0.3257	3023.6	3349.3	7.5883	0.2160	3018.5	3342.5	7.3940
500	0.3541	3124.4	3478.5	7.7622	0.2352	3120.3	3473.1	7.5698
540	0.3729	3192.6	3565.6	7.8720	0.2478	3189.1	3560.9	7.6805
600	0.4011	3296.8	3697.9	8.0290	0.2668	3293.9	3694.0	7.8385
640	0.4198	3367.4	3787.2	8.1290	0.2793	3364.8	3783.8	7.9391
	$p = 20.0$ bar $= 2.0$ MPa ($T_{sat} = 212.42$°C)				$p = 30.0$ bar $= 3.0$ MPa ($T_{sat} = 233.90$°C)			
Sat.	0.0996	2600.3	2799.5	6.3409	0.0667	2604.1	2804.2	6.1869
240	0.1085	2659.6	2876.5	6.4952	0.0682	2619.7	2824.3	6.2265
280	0.1200	2736.4	2976.4	6.6828	0.0771	2709.9	2941.3	6.4462
320	0.1308	2807.9	3069.5	6.8452	0.0850	2788.4	3043.4	6.6245
360	0.1411	2877.0	3159.3	6.9917	0.0923	2861.7	3138.7	6.7801
400	0.1512	2945.2	3247.6	7.1271	0.0994	2932.8	3230.9	6.9212
440	0.1611	3013.4	3335.5	7.2540	0.1062	3002.9	3321.5	7.0520
500	0.1757	3116.2	3467.6	7.4317	0.1162	3108.0	3456.5	7.2338
540	0.1853	3185.6	3556.1	7.5434	0.1227	3178.4	3546.6	7.3474
600	0.1996	3290.9	3690.1	7.7024	0.1324	3285.0	3682.3	7.5085
640	0.2091	3362.2	3780.4	7.8035	0.1388	3357.0	3773.5	7.6106
700	0.2232	3470.9	3917.4	7.9487	0.1484	3466.5	3911.7	7.7571

TABLE A-4   **Properties of Superheated Water Vapor** (*Continued*)

H₂O

Pressure Conversions:
1 bar = 0.1 MPa
= 10² kPa

T °C	v m³/kg	u kJ/kg	h kJ/kg	s kJ/kg·K	v m³/kg	u kJ/kg	h kJ/kg	s kJ/kg·K
	p = 40 bar = 4.0 MPa (T_{sat} = 250.4°C)				p = 60 bar = 6.0 MPa (T_{sat} = 275.64°C)			
Sat.	0.04978	2602.3	2801.4	6.0701	0.03244	2589.7	2784.3	5.8892
280	0.05546	2680.0	2901.8	6.2568	0.03317	2605.2	2804.2	5.9252
320	0.06199	2767.4	3015.4	6.4553	0.03876	2720.0	2952.6	6.1846
360	0.06788	2845.7	3117.2	6.6215	0.04331	2811.2	3071.1	6.3782
400	0.07341	2919.9	3213.6	6.7690	0.04739	2892.9	3177.2	6.5408
440	0.07872	2992.2	3307.1	6.9041	0.05122	2970.0	3277.3	6.6853
500	0.08643	3099.5	3445.3	7.0901	0.05665	3082.2	3422.2	6.8803
540	0.09145	3171.1	3536.9	7.2056	0.06015	3156.1	3517.0	6.9999
600	0.09885	3279.1	3674.4	7.3688	0.06525	3266.9	3658.4	7.1677
640	0.1037	3351.8	3766.6	7.4720	0.06859	3341.0	3752.6	7.2731
700	0.1110	3462.1	3905.9	7.6198	0.07352	3453.1	3894.1	7.4234
740	0.1157	3536.6	3999.6	7.7141	0.07677	3528.3	3989.2	7.5190
	p = 80 bar = 8.0 MPa (T_{sat} = 295.06°C)				p = 100 bar = 10.0 MPa (T_{sat} = 311.06°C)			
Sat.	0.02352	2569.8	2758.0	5.7432	0.01803	2544.4	2724.7	5.6141
320	0.02682	2662.7	2877.2	5.9489	0.01925	2588.8	2781.3	5.7103
360	0.03089	2772.7	3019.8	6.1819	0.02331	2729.1	2962.1	6.0060
400	0.03432	2863.8	3138.3	6.3634	0.02641	2832.4	3096.5	6.2120
440	0.03742	2946.7	3246.1	6.5190	0.02911	2922.1	3213.2	6.3805
480	0.04034	3025.7	3348.4	6.6586	0.03160	3005.4	3321.4	6.5282
520	0.04313	3102.7	3447.7	6.7871	0.03394	3085.6	3425.1	6.6622
560	0.04582	3178.7	3545.3	6.9072	0.03619	3164.1	3526.0	6.7864
600	0.04845	3254.4	3642.0	7.0206	0.03837	3241.7	3625.3	6.9029
640	0.05102	3330.1	3738.3	7.1283	0.04048	3318.9	3723.7	7.0131
700	0.05481	3443.9	3882.4	7.2812	0.04358	3434.7	3870.5	7.1687
740	0.05729	3520.4	3978.7	7.3782	0.04560	3512.1	3968.1	7.2670
	p = 120 bar = 12.0 MPa (T_{sat} = 324.75°C)				p = 140 bar = 14.0 MPa (T_{sat} = 336.75°C)			
Sat.	0.01426	2513.7	2684.9	5.4924	0.01149	2476.8	2637.6	5.3717
360	0.01811	2678.4	2895.7	5.8361	0.01422	2617.4	2816.5	5.6602
400	0.02108	2798.3	3051.3	6.0747	0.01722	2760.9	3001.9	5.9448
440	0.02355	2896.1	3178.7	6.2586	0.01954	2868.6	3142.2	6.1474
480	0.02576	2984.4	3293.5	6.4154	0.02157	2962.5	3264.5	6.3143
520	0.02781	3068.0	3401.8	6.5555	0.02343	3049.8	3377.8	6.4610
560	0.02977	3149.0	3506.2	6.6840	0.02517	3133.6	3486.0	6.5941
600	0.03164	3228.7	3608.3	6.8037	0.02683	3215.4	3591.1	6.7172
640	0.03345	3307.5	3709.0	6.9164	0.02843	3296.0	3694.1	6.8326
700	0.03610	3425.2	3858.4	7.0749	0.03075	3415.7	3846.2	6.9939
740	0.03781	3503.7	3957.4	7.1746	0.03225	3495.2	3946.7	7.0952

H₂O

TABLE A-4 **Properties of Superheated Water Vapor** (*Continued*)

T °C	υ m³/kg	u kJ/kg	h kJ/kg	s kJ/kg · K	υ m³/kg	u kJ/kg	h kJ/kg	s kJ/kg · K
	p = 160 bar = 16.0 MPa (T_{sat} = 347.44°C)				*p* = 180 bar = 18.0 MPa (T_{sat} = 357.06°C)			
Sat.	0.00931	2431.7	2580.6	5.2455	0.00749	2374.3	2509.1	5.1044
360	0.01105	2539.0	2715.8	5.4614	0.00809	2418.9	2564.5	5.1922
400	0.01426	2719.4	2947.6	5.8175	0.01190	2672.8	2887.0	5.6887
440	0.01652	2839.4	3103.7	6.0429	0.01414	2808.2	3062.8	5.9428
480	0.01842	2939.7	3234.4	6.2215	0.01596	2915.9	3203.2	6.1345
520	0.02013	3031.1	3353.3	6.3752	0.01757	3011.8	3378.0	6.2960
560	0.02172	3117.8	3465.4	6.5132	0.01904	3101.7	3444.4	6.4392
600	0.02323	3201.8	3573.5	6.6399	0.02042	3188.0	3555.6	6.5696
640	0.02467	3284.2	3678.9	6.7580	0.02174	3272.3	3663.6	6.6905
700	0.02674	3406.0	3833.9	6.9224	0.02362	3396.3	3821.5	6.8580
740	0.02808	3486.7	3935.9	7.0251	0.02483	3478.0	3925.0	6.9623
	p = 200 bar = 20.0 MPa (T_{sat} = 365.81°C)				*p* = 240 bar = 24.0 MPa			
Sat.	0.00583	2293.0	2409.7	4.9269				
400	0.00994	2619.3	2818.1	5.5540	0.00673	2477.8	2639.4	5.2393
440	0.01222	2774.9	3019.4	5.8450	0.00929	2700.6	2923.4	5.6506
480	0.01399	2891.2	3170.8	6.0518	0.01100	2838.3	3102.3	5.8950
520	0.01551	2992.0	3302.2	6.2218	0.01241	2950.5	3248.5	6.0842
560	0.01689	3085.2	3423.0	6.3705	0.01366	3051.1	3379.0	6.2448
600	0.01818	3174.0	3537.6	6.5048	0.01481	3145.2	3500.7	6.3875
640	0.01940	3260.2	3648.1	6.6286	0.01588	3235.5	3616.7	6.5174
700	0.02113	3386.4	3809.0	6.7993	0.01739	3366.4	3783.8	6.6947
740	0.02224	3469.3	3914.1	6.9052	0.01835	3451.7	3892.1	6.8038
800	0.02385	3592.7	4069.7	7.0544	0.01974	3578.0	4051.6	6.9567
	p = 280 bar = 28.0 MPa				*p* = 320 bar = 32.0 MPa			
400	0.00383	2223.5	2330.7	4.7494	0.00236	1980.4	2055.9	4.3239
440	0.00712	2613.2	2812.6	5.4494	0.00544	2509.0	2683.0	5.2327
480	0.00885	2780.8	3028.5	5.7446	0.00722	2718.1	2949.2	5.5968
520	0.01020	2906.8	3192.3	5.9566	0.00853	2860.7	3133.7	5.8357
560	0.01136	3015.7	3333.7	6.1307	0.00963	2979.0	3287.2	6.0246
600	0.01241	3115.6	3463.0	6.2823	0.01061	3085.3	3424.6	6.1858
640	0.01338	3210.3	3584.8	6.4187	0.01150	3184.5	3552.5	6.3290
700	0.01473	3346.1	3758.4	6.6029	0.01273	3325.4	3732.8	6.5203
740	0.01558	3433.9	3870.0	6.7153	0.01350	3415.9	3847.8	6.6361
800	0.01680	3563.1	4033.4	6.8720	0.01460	3548.0	4015.1	6.7966
900	0.01873	3774.3	4298.8	7.1084	0.01633	3762.7	4285.1	7.0372

TABLE A-5 **Properties of Compressed Liquid Water**

T °C	$v \times 10^3$ m³/kg	u kJ/kg	h kJ/kg	s kJ/kg·K	$v \times 10^3$ m³/kg	u kJ/kg	h kJ/kg	s kJ/kg·K

H₂O

Pressure Conversions:
1 bar = 0.1 MPa
= 10² kPa

T °C	$v \times 10^3$ m³/kg	u kJ/kg	h kJ/kg	s kJ/kg·K	$v \times 10^3$ m³/kg	u kJ/kg	h kJ/kg	s kJ/kg·K
	p = 25 bar = 2.5 MPa (T_{sat} = 223.99°C)				**p = 50 bar = 5.0 MPa** (T_{sat} = 263.99°C)			
20	1.0006	83.80	86.30	.2961	.9995	83.65	88.65	.2956
40	1.0067	167.25	169.77	.5715	1.0056	166.95	171.97	.5705
80	1.0280	334.29	336.86	1.0737	1.0268	333.72	338.85	1.0720
100	1.0423	418.24	420.85	1.3050	1.0410	417.52	422.72	1.3030
140	1.0784	587.82	590.52	1.7369	1.0768	586.76	592.15	1.7343
180	1.1261	761.16	763.97	2.1375	1.1240	759.63	765.25	2.1341
200	1.1555	849.9	852.8	2.3294	1.1530	848.1	853.9	2.3255
220	1.1898	940.7	943.7	2.5174	1.1866	938.4	944.4	2.5128
Sat.	1.1973	959.1	962.1	2.5546	1.2859	1147.8	1154.2	2.9202
	p = 75 bar = 7.5 MPa (T_{sat} = 290.59°C)				**p = 100 bar = 10.0 MPa** (T_{sat} = 311.06°C)			
20	.9984	83.50	90.99	.2950	.9972	83.36	93.33	.2945
40	1.0045	166.64	174.18	.5696	1.0034	166.35	176.38	.5686
80	1.0256	333.15	340.84	1.0704	1.0245	332.59	342.83	1.0688
100	1.0397	416.81	424.62	1.3011	1.0385	416.12	426.50	1.2992
140	1.0752	585.72	593.78	1.7317	1.0737	584.68	595.42	1.7292
180	1.1219	758.13	766.55	2.1308	1.1199	756.65	767.84	2.1275
220	1.1835	936.2	945.1	2.5083	1.1805	934.1	945.9	2.5039
260	1.2696	1124.4	1134.0	2.8763	1.2645	1121.1	1133.7	2.8699
Sat.	1.3677	1282.0	1292.2	3.1649	1.4524	1393.0	1407.6	3.3596
	p = 150 bar = 15.0 MPa (T_{sat} = 342.24°C)				**p = 200 bar = 20.0 MPa** (T_{sat} = 365.81°C)			
20	.9950	83.06	97.99	.2934	.9928	82.77	102.62	.2923
40	1.0013	165.76	180.78	.5666	.9992	165.17	185.16	.5646
80	1.0222	331.48	346.81	1.0656	1.0199	330.40	350.80	1.0624
100	1.0361	414.74	430.28	1.2955	1.0337	413.39	434.06	1.2917
140	1.0707	582.66	598.72	1.7242	1.0678	580.69	602.04	1.7193
180	1.1159	753.76	770.50	2.1210	1.1120	750.95	773.20	2.1147
220	1.1748	929.9	947.5	2.4953	1.1693	925.9	949.3	2.4870
260	1.2550	1114.6	1133.4	2.8576	1.2462	1108.6	1133.5	2.8459
300	1.3770	1316.6	1337.3	3.2260	1.3596	1306.1	1333.3	3.2071
Sat.	1.6581	1585.6	1610.5	3.6848	2.036	1785.6	1826.3	4.0139
	p = 250 bar = 25 MPa				**p = 300 bar = 30.0 MPa**			
20	.9907	82.47	107.24	.2911	.9886	82.17	111.84	.2899
40	.9971	164.60	189.52	.5626	.9951	164.04	193.89	.5607
100	1.0313	412.08	437.85	1.2881	1.0290	410.78	441.66	1.2844
200	1.1344	834.5	862.8	2.2961	1.1302	831.4	865.3	2.2893
300	1.3442	1296.6	1330.2	3.1900	1.3304	1287.9	1327.8	3.1741

v = (table value)/1000 v = (table value)/1000

H₂O

TABLE A-6 Properties of Saturated Water (Solid–Vapor): Temperature Table

Pressure Conversions:
1 bar = 0.1 MPa = 10² kPa

Temp. °C	Pressure kPa	Specific Volume m³/kg		Internal Energy kJ/kg			Enthalpy kJ/kg			Entropy kJ/kg·K		
		Sat. Solid $v_i \times 10^3$	Sat. Vapor v_g	Sat. Solid u_i	Subl. u_{ig}	Sat. Vapor u_g	Sat. Solid h_i	Subl. h_{ig}	Sat. Vapor h_g	Sat. Solid s_i	Subl. s_{ig}	Sat. Vapor s_g
.01	.6113	1.0908	206.1	−333.40	2708.7	2375.3	−333.40	2834.8	2501.4	−1.221	10.378	9.156
0	.6108	1.0908	206.3	−333.43	2708.8	2375.3	−333.43	2834.8	2501.3	−1.221	10.378	9.157
−2	.5176	1.0904	241.7	−337.62	2710.2	2372.6	−337.62	2835.3	2497.7	−1.237	10.456	9.219
−4	.4375	1.0901	283.8	−341.78	2711.6	2369.8	−341.78	2835.7	2494.0	−1.253	10.536	9.283
−6	.3689	1.0898	334.2	−345.91	2712.9	2367.0	−345.91	2836.2	2490.3	−1.268	10.616	9.348
−8	.3102	1.0894	394.4	−350.02	2714.2	2364.2	−350.02	2836.6	2486.6	−1.284	10.698	9.414
−10	.2602	1.0891	466.7	−354.09	2715.5	2361.4	−354.09	2837.0	2482.9	−1.299	10.781	9.481
−12	.2176	1.0888	553.7	−358.14	2716.8	2358.7	−358.14	2837.3	2479.2	−1.315	10.865	9.550
−14	.1815	1.0884	658.8	−362.15	2718.0	2355.9	−362.15	2837.6	2475.5	−1.331	10.950	9.619
−16	.1510	1.0881	786.0	−366.14	2719.2	2353.1	−366.14	2837.9	2471.8	−1.346	11.036	9.690
−18	.1252	1.0878	940.5	−370.10	2720.4	2350.3	−370.10	2838.2	2468.1	−1.362	11.123	9.762
−20	.1035	1.0874	1128.6	−374.03	2721.6	2347.5	−374.03	2838.4	2464.3	−1.377	11.212	9.835
−22	.0853	1.0871	1358.4	−377.93	2722.7	2344.7	−377.93	2838.6	2460.6	−1.393	11.302	9.909
−24	.0701	1.0868	1640.1	−381.80	2723.7	2342.0	−381.80	2838.7	2456.9	−1.408	11.394	9.985
−26	.0574	1.0864	1986.4	−385.64	2724.8	2339.2	−385.64	2838.9	2453.2	−1.424	11.486	10.062
−28	.0469	1.0861	2413.7	−389.45	2725.8	2336.4	−389.45	2839.0	2449.5	−1.439	11.580	10.141
−30	.0381	1.0858	2943	−393.23	2726.8	2333.6	−393.23	2839.0	2445.8	−1.455	11.676	10.221
−32	.0309	1.0854	3600	−396.98	2727.8	2330.8	−396.98	2839.1	2442.1	−1.471	11.773	10.303
−34	.0250	1.0851	4419	−400.71	2728.7	2328.0	−400.71	2839.1	2438.4	−1.486	11.872	10.386
−36	.0201	1.0848	5444	−404.40	2729.6	2325.2	−404.40	2839.1	2434.7	−1.501	11.972	10.470
−38	.0161	1.0844	6731	−408.06	2730.5	2322.4	−408.06	2839.0	2430.9	−1.517	12.073	10.556
−40	.0129	1.0841	8354	−411.70	2731.3	2319.6	−411.70	2838.9	2427.2	−1.532	12.176	10.644

$v = $ (table value)/1000

TABLE A-7   **Properties of Saturated Refrigerant 22 (Liquid–Vapor): Temperature Table**

Temp. °C	Press. bar	Specific Volume m³/kg		Internal Energy kJ/kg		Enthalpy kJ/kg			Entropy kJ/kg · K		Temp. °C
		Sat. Liquid $v_f \times 10^3$	Sat. Vapor v_g	Sat. Liquid u_f	Sat. Vapor u_g	Sat. Liquid h_f	Evap. h_{fg}	Sat. Vapor h_g	Sat. Liquid s_f	Sat. Vapor s_g	
−60	0.3749	0.6833	0.5370	−21.57	203.67	−21.55	245.35	223.81	−0.0964	1.0547	−60
−50	0.6451	0.6966	0.3239	−10.89	207.70	−10.85	239.44	228.60	−0.0474	1.0256	−50
−45	0.8290	0.7037	0.2564	−5.50	209.70	−5.44	236.39	230.95	−0.0235	1.0126	−45
−40	1.0522	0.7109	0.2052	−0.07	211.68	0.00	233.27	233.27	0.0000	1.0005	−40
−36	1.2627	0.7169	0.1730	4.29	213.25	4.38	230.71	235.09	0.0186	0.9914	−36
−32	1.5049	0.7231	0.1468	8.68	214.80	8.79	228.10	236.89	0.0369	0.9828	−32
−30	1.6389	0.7262	0.1355	10.88	215.58	11.00	226.77	237.78	0.0460	0.9787	−30
−28	1.7819	0.7294	0.1252	13.09	216.34	13.22	225.43	238.66	0.0551	0.9746	−28
−26	1.9345	0.7327	0.1159	15.31	217.11	15.45	224.08	239.53	0.0641	0.9707	−26
−22	2.2698	0.7393	0.0997	19.76	218.62	19.92	221.32	241.24	0.0819	0.9631	−22
−20	2.4534	0.7427	0.0926	21.99	219.37	22.17	219.91	242.09	0.0908	0.9595	−20
−18	2.6482	0.7462	0.0861	24.23	220.11	24.43	218.49	242.92	0.0996	0.9559	−18
−16	2.8547	0.7497	0.0802	26.48	220.85	26.69	217.05	243.74	0.1084	0.9525	−16
−14	3.0733	0.7533	0.0748	28.73	221.58	28.97	215.59	244.56	0.1171	0.9490	−14
−12	3.3044	0.7569	0.0698	31.00	222.30	31.25	214.11	245.36	0.1258	0.9457	−12
−10	3.5485	0.7606	0.0652	33.27	223.02	33.54	212.62	246.15	0.1345	0.9424	−10
−8	3.8062	0.7644	0.0610	35.54	223.73	35.83	211.10	246.93	0.1431	0.9392	−8
−6	4.0777	0.7683	0.0571	37.83	224.43	38.14	209.56	247.70	0.1517	0.9361	−6
−4	4.3638	0.7722	0.0535	40.12	225.13	40.46	208.00	248.45	0.1602	0.9330	−4
−2	4.6647	0.7762	0.0501	42.42	225.82	42.78	206.41	249.20	0.1688	0.9300	−2
0	4.9811	0.7803	0.0470	44.73	226.50	45.12	204.81	249.92	0.1773	0.9271	0
2	5.3133	0.7844	0.0442	47.04	227.17	47.46	203.18	250.64	0.1857	0.9241	2
4	5.6619	0.7887	0.0415	49.37	227.83	49.82	201.52	251.34	0.1941	0.9213	4
6	6.0275	0.7930	0.0391	51.71	228.48	52.18	199.84	252.03	0.2025	0.9184	6
8	6.4105	0.7974	0.0368	54.05	229.13	54.56	198.14	252.70	0.2109	0.9157	8
10	6.8113	0.8020	0.0346	56.40	229.76	56.95	196.40	253.35	0.2193	0.9129	10
12	7.2307	0.8066	0.0326	58.77	230.38	59.35	194.64	253.99	0.2276	0.9102	12
16	8.1268	0.8162	0.0291	63.53	231.59	64.19	191.02	255.21	0.2442	0.9048	16
20	9.1030	0.8263	0.0259	68.33	232.76	69.09	187.28	256.37	0.2607	0.8996	20
24	10.164	0.8369	0.0232	73.19	233.87	74.04	183.40	257.44	0.2772	0.8944	24
28	11.313	0.8480	0.0208	78.09	234.92	79.05	179.37	258.43	0.2936	0.8893	28
32	12.556	0.8599	0.0186	83.06	235.91	84.14	175.18	259.32	0.3101	0.8842	32
36	13.897	0.8724	0.0168	88.08	236.83	89.29	170.82	260.11	0.3265	0.8790	36
40	15.341	0.8858	0.0151	93.18	237.66	94.53	166.25	260.79	0.3429	0.8738	40
45	17.298	0.9039	0.0132	99.65	238.59	101.21	160.24	261.46	0.3635	0.8672	45
50	19.433	0.9238	0.0116	106.26	239.34	108.06	153.84	261.90	0.3842	0.8603	50
60	24.281	0.9705	0.0089	120.00	240.24	122.35	139.61	261.96	0.4264	0.8455	60

Pressure Conversions: 1 bar = 0.1 MPa = 10^2 kPa

R-22

v_f = (table value)/1000

TABLE A-8   **Properties of Saturated Refrigerant 22 (Liquid–Vapor): Pressure Table**

Pressure Conversions: 1 bar = 0.1 MPa = 10² kPa		Specific Volume m³/kg		Internal Energy kJ/kg		Enthalpy kJ/kg			Entropy kJ/kg · K		
Press. bar	Temp. °C	Sat. Liquid $v_f \times 10^3$	Sat. Vapor v_g	Sat. Liquid u_f	Sat. Vapor u_g	Sat. Liquid h_f	Evap. h_{fg}	Sat. Vapor h_g	Sat. Liquid s_f	Sat. Vapor s_g	Press. bar
0.40	−58.86	0.6847	0.5056	−20.36	204.13	−20.34	244.69	224.36	−0.0907	1.0512	0.40
0.50	−54.83	0.6901	0.4107	−16.07	205.76	−16.03	242.33	226.30	−0.0709	1.0391	0.50
0.60	−51.40	0.6947	0.3466	−12.39	207.14	−12.35	240.28	227.93	−0.0542	1.0294	0.60
0.70	−48.40	0.6989	0.3002	−9.17	208.34	−9.12	238.47	229.35	−0.0397	1.0213	0.70
0.80	−45.73	0.7026	0.2650	−6.28	209.41	−6.23	236.84	230.61	−0.0270	1.0144	0.80
0.90	−43.30	0.7061	0.2374	−3.66	210.37	−3.60	235.34	231.74	−0.0155	1.0084	0.90
1.00	−41.09	0.7093	0.2152	−1.26	211.25	−1.19	233.95	232.77	−0.0051	1.0031	1.00
1.25	−36.23	0.7166	0.1746	4.04	213.16	4.13	230.86	234.99	0.0175	0.9919	1.25
1.50	−32.08	0.7230	0.1472	8.60	214.77	8.70	228.15	236.86	0.0366	0.9830	1.50
1.75	−28.44	0.7287	0.1274	12.61	216.18	12.74	225.73	238.47	0.0531	0.9755	1.75
2.00	−25.18	0.7340	0.1123	16.22	217.42	16.37	223.52	239.88	0.0678	0.9691	2.00
2.25	−22.22	0.7389	0.1005	19.51	218.53	19.67	221.47	241.15	0.0809	0.9636	2.25
2.50	−19.51	0.7436	0.0910	22.54	219.55	22.72	219.57	242.29	0.0930	0.9586	2.50
2.75	−17.00	0.7479	0.0831	25.36	220.48	25.56	217.77	243.33	0.1040	0.9542	2.75
3.00	−14.66	0.7521	0.0765	27.99	221.34	28.22	216.07	244.29	0.1143	0.9502	3.00
3.25	−12.46	0.7561	0.0709	30.47	222.13	30.72	214.46	245.18	0.1238	0.9465	3.25
3.50	−10.39	0.7599	0.0661	32.82	222.88	33.09	212.91	246.00	0.1328	0.9431	3.50
3.75	−8.43	0.7636	0.0618	35.06	223.58	35.34	211.42	246.77	0.1413	0.9399	3.75
4.00	−6.56	0.7672	0.0581	37.18	224.24	37.49	209.99	247.48	0.1493	0.9370	4.00
4.25	−4.78	0.7706	0.0548	39.22	224.86	39.55	208.61	248.16	0.1569	0.9342	4.25
4.50	−3.08	0.7740	0.0519	41.17	225.45	41.52	207.27	248.80	0.1642	0.9316	4.50
4.75	−1.45	0.7773	0.0492	43.05	226.00	43.42	205.98	249.40	0.1711	0.9292	4.75
5.00	0.12	0.7805	0.0469	44.86	226.54	45.25	204.71	249.97	0.1777	0.9269	5.00
5.25	1.63	0.7836	0.0447	46.61	227.04	47.02	203.48	250.51	0.1841	0.9247	5.25
5.50	3.08	0.7867	0.0427	48.30	227.53	48.74	202.28	251.02	0.1903	0.9226	5.50
5.75	4.49	0.7897	0.0409	49.94	227.99	50.40	201.11	251.51	0.1962	0.9206	5.75
6.00	5.85	0.7927	0.0392	51.53	228.44	52.01	199.97	251.98	0.2019	0.9186	6.00
7.00	10.91	0.8041	0.0337	57.48	230.04	58.04	195.60	253.64	0.2231	0.9117	7.00
8.00	15.45	0.8149	0.0295	62.88	231.43	63.53	191.52	255.05	0.2419	0.9056	8.00
9.00	19.59	0.8252	0.0262	67.84	232.64	68.59	187.67	256.25	0.2591	0.9001	9.00
10.00	23.40	0.8352	0.0236	72.46	233.71	73.30	183.99	257.28	0.2748	0.8952	10.00
12.00	30.25	0.8546	0.0195	80.87	235.48	81.90	177.04	258.94	0.3029	0.8864	12.00
14.00	36.29	0.8734	0.0166	88.45	236.89	89.68	170.49	260.16	0.3277	0.8786	14.00
16.00	41.73	0.8919	0.0144	95.41	238.00	96.83	164.21	261.04	0.3500	0.8715	16.00
18.00	46.69	0.9104	0.0127	101.87	238.86	103.51	158.13	261.64	0.3705	0.8649	18.00
20.00	51.26	0.9291	0.0112	107.95	239.51	109.81	152.17	261.98	0.3895	0.8586	20.00
24.00	59.46	0.9677	0.0091	119.24	240.22	121.56	140.43	261.99	0.4241	0.8463	24.00

$v_f =$ (table value)/1000

R-22

TABLE A-9   **Properties of Superheated Refrigerant 22 Vapor**

Pressure Conversions:
1 bar = 0.1 MPa
= 10^2 kPa

T °C	v m³/kg	u kJ/kg	h kJ/kg	s kJ/kg·K	v m³/kg	u kJ/kg	h kJ/kg	s kJ/kg·K
	_colspan4							

T °C	v m³/kg	u kJ/kg	h kJ/kg	s kJ/kg·K	v m³/kg	u kJ/kg	h kJ/kg	s kJ/kg·K
	p = 0.4 bar = 0.04 MPa (T_{sat} = −58.86°C)				p = 0.6 bar = 0.06 MPa (T_{sat} = −51.40°C)			
Sat.	0.50559	204.13	224.36	1.0512	0.34656	207.14	227.93	1.0294
−55	0.51532	205.92	226.53	1.0612				
−50	0.52787	208.26	229.38	1.0741	0.34895	207.80	228.74	1.0330
−45	0.54037	210.63	232.24	1.0868	0.35747	210.20	231.65	1.0459
−40	0.55284	213.02	235.13	1.0993	0.36594	212.62	234.58	1.0586
−35	0.56526	215.43	238.05	1.1117	0.37437	215.06	237.52	1.0711
−30	0.57766	217.88	240.99	1.1239	0.38277	217.53	240.49	1.0835
−25	0.59002	220.35	243.95	1.1360	0.39114	220.02	243.49	1.0956
−20	0.60236	222.85	246.95	1.1479	0.39948	222.54	246.51	1.1077
−15	0.61468	225.38	249.97	1.1597	0.40779	225.08	249.55	1.1196
−10	0.62697	227.93	253.01	1.1714	0.41608	227.65	252.62	1.1314
−5	0.63925	230.52	256.09	1.1830	0.42436	230.25	255.71	1.1430
0	0.65151	233.13	259.19	1.1944	0.43261	232.88	258.83	1.1545
	p = 0.8 bar = 0.08 MPa (T_{sat} −45.73°C)				p = 1.0 bar = 0.10 MPa (T_{sat} = −41.09°C)			
Sat.	0.26503	209.41	230.61	1.0144	0.21518	211.25	232.77	1.0031
−45	0.26597	209.76	231.04	1.0163				
−40	0.27245	212.21	234.01	1.0292	0.21633	211.79	233.42	1.0059
−35	0.27890	214.68	236.99	1.0418	0.22158	214.29	236.44	1.0187
−30	0.28530	217.17	239.99	1.0543	0.22679	216.80	239.48	1.0313
−25	0.29167	219.68	243.02	1.0666	0.23197	219.34	242.54	1.0438
−20	0.29801	222.22	246.06	1.0788	0.23712	221.90	245.61	1.0560
−15	0.30433	224.78	249.13	1.0908	0.24224	224.48	248.70	1.0681
−10	0.31062	227.37	252.22	1.1026	0.24734	227.08	251.82	1.0801
−5	0.31690	229.98	255.34	1.1143	0.25241	229.71	254.95	1.0919
0	0.32315	232.62	258.47	1.1259	0.25747	232.36	258.11	1.1035
5	0.32939	235.29	261.64	1.1374	0.26251	235.04	261.29	1.1151
10	0.33561	237.98	264.83	1.1488	0.26753	237.74	264.50	1.1265
	p = 1.5 bar = 0.15 MPa (T_{sat} = −32.08°C)				p = 2.0 bar = 0.20 MPa (T_{sat} = −25.18°C)			
Sat.	0.14721	214.77	236.86	0.9830	0.11232	217.42	239.88	0.9691
−30	0.14872	215.85	238.16	0.9883				
−25	0.15232	218.45	241.30	1.0011	0.11242	217.51	240.00	0.9696
−20	0.15588	221.07	244.45	1.0137	0.11520	220.19	243.23	0.9825
−15	0.15941	223.70	247.61	1.0260	0.11795	222.88	246.47	0.9952
−10	0.16292	226.35	250.78	1.0382	0.12067	225.58	249.72	1.0076
−5	0.16640	229.02	253.98	1.0502	0.12336	228.30	252.97	1.0199
0	0.16987	231.70	257.18	1.0621	0.12603	231.03	256.23	1.0310
5	0.17331	234.42	260.41	1.0738	0.12868	233.78	259.51	1.0438
10	0.17674	237.15	263.66	1.0854	0.13132	236.54	262.81	1.0555
15	0.18015	239.91	266.93	1.0968	0.13393	239.33	266.12	1.0671
20	0.18355	242.69	270.22	1.1081	0.13653	242.14	269.44	1.0786
25	0.18693	245.49	273.53	1.1193	0.13912	244.97	272.79	1.0899

R-22

Pressure Conversions:
1 bar = 0.1 MPa
= 10^2 kPa

TABLE A-9   **Properties of Superheated Refrigerant 22 Vapor (*Continued*)**

T °C	v m³/kg	u kJ/kg	h kJ/kg	s kJ/kg·K	v m³/kg	u kJ/kg	h kJ/kg	s kJ/kg·K
	p = 2.5 bar = 0.25 MPa (T_{sat} = −19.51°C)				p = 3.0 bar = 0.30 MPa (T_{sat} = −14.66°C)			
Sat.	0.09097	219.55	242.29	0.9586	0.07651	221.34	244.29	0.9502
−15	0.09303	222.03	245.29	0.9703				
−10	0.09528	224.79	248.61	0.9831	0.07833	223.96	247.46	0.9623
−5	0.09751	227.55	251.93	0.9956	0.08025	226.78	250.86	0.9751
0	0.09971	230.33	255.26	1.0078	0.08214	229.61	254.25	0.9876
5	0.10189	233.12	258.59	1.0199	0.08400	232.44	257.64	0.9999
10	0.10405	235.92	261.93	1.0318	0.08585	235.28	261.04	1.0120
15	0.10619	238.74	265.29	1.0436	0.08767	238.14	264.44	1.0239
20	0.10831	241.58	268.66	1.0552	0.08949	241.01	267.85	1.0357
25	0.11043	244.44	272.04	1.0666	0.09128	243.89	271.28	1.0472
30	0.11253	247.31	275.44	1.0779	0.09307	246.80	274.72	1.0587
35	0.11461	250.21	278.86	1.0891	0.09484	249.72	278.17	1.0700
40	0.11669	253.13	282.30	1.1002	0.09660	252.66	281.64	1.0811
	p = 3.5 bar = 0.35 MPa (T_{sat} = −10.39°C)				p = 4.0 bar = 0.40 MPa (T_{sat} = −6.56°C)			
Sat.	0.06605	222.88	246.00	0.9431	0.05812	224.24	247.48	0.9370
−10	0.06619	223.10	246.27	0.9441				
−5	0.06789	225.99	249.75	0.9572	0.05860	225.16	248.60	0.9411
0	0.06956	228.86	253.21	0.9700	0.06011	228.09	252.14	0.9542
5	0.07121	231.74	256.67	0.9825	0.06160	231.02	225.66	0.9670
10	0.07284	234.63	260.12	0.9948	0.06306	233.95	259.18	0.9795
15	0.07444	237.52	263.57	1.0069	0.06450	236.89	262.69	0.9918
20	0.07603	240.42	267.03	1.0188	0.06592	239.83	266.19	1.0039
25	0.07760	243.34	270.50	1.0305	0.06733	242.77	269.71	1.0158
30	0.07916	246.27	273.97	1.0421	0.06872	245.73	273.22	1.0274
35	0.08070	249.22	227.46	1.0535	0.07010	248.71	276.75	1.0390
40	0.08224	252.18	280.97	1.0648	0.07146	251.70	280.28	1.0504
45	0.08376	255.17	284.48	1.0759	0.07282	254.70	283.83	1.0616
	p = 4.5 bar = 0.45 MPa (T_{sat} = −3.08°C)				p = 5.0 bar = 0.50 MPa (T_{sat} = 0.12°C)			
Sat.	0.05189	225.45	248.80	0.9316	0.04686	226.54	249.97	0.9269
0	0.05275	227.29	251.03	0.9399				
5	0.05411	230.28	254.63	0.9529	0.04810	229.52	253.57	0.9399
10	0.05545	233.26	258.21	0.9657	0.04934	232.55	257.22	0.9530
15	0.05676	236.24	261.78	0.9782	0.05056	235.57	260.85	0.9657
20	0.05805	239.22	265.34	0.9904	0.05175	238.59	264.47	0.9781
25	0.05933	242.20	268.90	1.0025	0.05293	241.61	268.07	0.9903
30	0.06059	245.19	272.46	1.0143	0.05409	244.63	271.68	1.0023
35	0.06184	248.19	276.02	1.0259	0.05523	247.66	275.28	1.0141
40	0.06308	251.20	279.59	1.0374	0.05636	250.70	278.89	1.0257
45	0.06430	254.23	283.17	1.0488	0.05748	253.76	282.50	1.0371
50	0.06552	257.28	286.76	1.0600	0.05859	256.82	286.12	1.0484
55	0.06672	260.34	290.36	1.0710	0.05969	259.90	289.75	1.0595

TABLE A-9   **Properties of Superheated Refrigerant 22 Vapor (*Continued*)**

T °C	v m³/kg	u kJ/kg	h kJ/kg	s kJ/kg·K	v m³/kg	u kJ/kg	h kJ/kg	s kJ/kg·K
	p = 5.5 bar = 0.55 MPa (T_{sat} = 3.08°C)				p = 6.0 bar = 0.60 MPa (T_{sat} = 5.85°C)			
Sat.	0.04271	227.53	251.02	0.9226	0.03923	228.44	251.98	0.9186
5	0.04317	228.72	252.46	0.9278				
10	0.04433	231.81	256.20	0.9411	0.04015	231.05	255.14	0.9299
15	0.04547	234.89	259.90	0.9540	0.04122	234.18	258.91	0.9431
20	0.04658	237.95	263.57	0.9667	0.04227	237.29	262.65	0.9560
25	0.04768	241.01	267.23	0.9790	0.04330	240.39	266.37	0.9685
30	0.04875	244.07	270.88	0.9912	0.04431	243.49	270.07	0.9808
35	0.04982	247.13	274.53	1.0031	0.04530	246.58	273.76	0.9929
40	0.05086	250.20	278.17	1.0148	0.04628	249.68	277.45	1.0048
45	0.05190	253.27	281.82	1.0264	0.04724	252.78	281.13	1.0164
50	0.05293	256.36	285.47	1.0378	0.04820	255.90	284.82	1.0279
55	0.05394	259.46	289.13	1.0490	0.04914	259.02	288.51	1.0393
60	0.05495	262.58	292.80	1.0601	0.05008	262.15	292.20	1.0504
	p = 7.0 bar = 0.70 MPa (T_{sat} = 10.91°C)				p = 8.0 bar = 0.80 MPa (T_{sat} = 15.45°C)			
Sat.	0.03371	230.04	253.64	0.9117	0.02953	231.43	255.05	0.9056
15	0.03451	232.70	256.86	0.9229				
20	0.03547	235.92	260.75	0.9363	0.03033	234.47	258.74	0.9182
25	0.03639	239.12	264.59	0.9493	0.03118	237.76	262.70	0.9315
30	0.03730	242.29	268.40	0.9619	0.03202	241.04	266.66	0.9448
35	0.03819	245.46	272.19	0.9743	0.03283	244.28	270.54	0.9574
40	0.03906	248.62	275.96	0.9865	0.03363	247.52	274.42	0.9700
45	0.03992	251.78	279.72	0.9984	0.03440	250.74	278.26	0.9821
50	0.04076	254.94	283.48	1.0101	0.03517	253.96	282.10	0.9941
55	0.04160	258.11	287.23	1.0216	0.03592	257.18	285.92	1.0058
60	0.04242	261.29	290.99	1.0330	0.03667	260.40	289.74	1.0174
65	0.04324	264.48	294.75	1.0442	0.03741	263.64	293.56	1.0287
70	0.04405	267.68	298.51	1.0552	0.03814	266.87	297.38	1.0400
	p = 9.0 bar = 0.90 MPa (T_{sat} = 19.59°C)				p = 10.0 bar = 1.00 MPa (T_{sat} = 23.40°C)			
Sat.	0.02623	232.64	256.25	0.9001	0.02358	233.71	257.28	0.8952
20	0.02630	232.92	256.59	0.9013				
30	0.02789	239.73	264.83	0.9289	0.02457	238.34	262.91	0.9139
40	0.02939	246.37	272.82	0.9549	0.02598	245.18	271.17	0.9407
50	0.03082	252.95	280.68	0.9795	0.02732	251.90	279.22	0.9660
60	0.03219	259.49	288.46	1.0033	0.02860	258.56	287.15	0.9902
70	0.03353	266.04	296.21	1.0262	0.02984	265.19	295.03	1.0135
80	0.03483	272.62	303.96	1.0484	0.03104	271.84	302.88	1.0361
90	0.03611	279.23	311.73	1.0701	0.03221	278.52	310.74	1.0580
100	0.03736	285.90	319.53	1.0913	0.03337	285.24	318.61	1.0794
110	0.03860	292.63	327.37	1.1120	0.03450	292.02	326.52	1.1003
120	0.03982	299.42	335.26	1.1323	0.03562	298.85	334.46	1.1207
130	0.04103	306.28	343.21	1.1523	0.03672	305.74	342.46	1.1408
140	0.04223	313.21	351.22	1.1719	0.03781	312.70	350.51	1.1605
150	0.04342	320.21	359.29	1.1912	0.03889	319.74	358.63	1.1790

Pressure Conversions:
1 bar = 0.1 MPa
= 10^2 kPa

TABLE A-9 **Properties of Superheated Refrigerant 22 Vapor (*Continued*)**

T °C	v m³/kg	u kJ/kg	h kJ/kg	s kJ/kg·K	v m³/kg	u kJ/kg	h kJ/kg	s kJ/kg·K
	p = 12.0 bar = 1.20 MPa (T_{sat} = 30.25°C)				p = 14.0 bar = 1.40 MPa (T_{sat} = 36.29°C)			
Sat.	0.01955	235.48	258.94	0.8864	0.01662	236.89	260.16	0.8786
40	0.02083	242.63	267.62	0.9146	0.01708	239.78	263.70	0.8900
50	0.02204	249.69	276.14	0.9413	0.01823	247.29	272.81	0.9186
60	0.02319	256.60	284.43	0.9666	0.01929	254.52	281.53	0.9452
70	0.02428	263.44	292.58	0.9907	0.02029	261.60	290.01	0.9703
80	0.02534	270.25	300.66	1.0139	0.02125	268.60	298.34	0.9942
90	0.02636	277.07	308.70	1.0363	0.02217	275.56	306.60	1.0172
100	0.02736	283.90	316.73	1.0582	0.02306	282.52	314.80	1.0395
110	0.02834	290.77	324.78	1.0794	0.02393	289.49	323.00	1.0612
120	0.02930	297.69	332.85	1.1002	0.02478	296.50	331.19	1.0823
130	0.03024	304.65	340.95	1.1205	0.02562	303.55	339.41	1.1029
140	0.03118	311.68	349.09	1.1405	0.02644	310.64	347.65	1.1231
150	0.03210	318.77	357.29	1.1601	0.02725	317.79	355.94	1.1429
160	0.03301	325.92	365.54	1.1793	0.02805	324.99	364.26	1.1624
170	0.03392	333.14	373.84	1.1983	0.02884	332.26	372.64	1.1815
	p = 16.0 bar = 1.60 MPa (T_{sat} = 41.73°C)				p = 18.0 bar = 1.80 MPa (T_{sat} = 46.69°C)			
Sat.	0.01440	238.00	261.04	0.8715	0.01265	238.86	261.64	0.8649
50	0.01533	244.66	269.18	0.8971	0.01301	241.72	265.14	0.8758
60	0.01634	252.29	278.43	0.9252	0.01401	249.86	275.09	0.9061
70	0.01728	259.65	287.30	0.9515	0.01492	257.57	284.43	0.9337
80	0.01817	266.86	295.93	0.9762	0.01576	265.04	293.40	0.9595
90	0.01901	274.00	304.42	0.9999	0.01655	272.37	302.16	0.9839
100	0.01983	281.09	312.82	1.0228	0.01731	279.62	310.77	1.0073
110	0.02062	288.18	321.17	1.0448	0.01804	286.83	319.30	1.0299
120	0.02139	295.28	329.51	1.0663	0.01874	294.04	327.78	1.0517
130	0.02214	302.41	337.84	1.0872	0.01943	301.26	336.24	1.0730
140	0.02288	309.58	346.19	1.1077	0.02011	308.50	344.70	1.0937
150	0.02361	316.79	354.56	1.1277	0.02077	315.78	353.17	1.1139
160	0.02432	324.05	362.97	1.1473	0.02142	323.10	361.66	1.1338
170	0.02503	331.37	371.42	1.1666	0.02207	330.47	370.19	1.1532
	p = 20.0 bar = 2.00 MPa (T_{sat} = 51.26°C)				p = 24.0 bar = 2.4 MPa (T_{sat} = 59.46°C)			
Sat.	0.01124	239.51	261.98	0.8586	0.00907	240.22	261.99	0.8463
60	0.01212	247.20	271.43	0.8873	0.00913	240.78	262.68	0.8484
70	0.01300	255.35	281.36	0.9167	0.01006	250.30	274.43	0.8831
80	0.01381	263.12	290.74	0.9436	0.01085	258.89	284.93	0.9133
90	0.01457	270.67	299.80	0.9689	0.01156	267.01	294.75	0.9407
100	0.01528	278.09	308.65	0.9929	0.01222	274.85	304.18	0.9663
110	0.01596	285.44	317.37	1.0160	0.01284	282.53	313.35	0.9906
120	0.01663	292.76	326.01	1.0383	0.01343	290.11	322.35	1.0137
130	0.01727	300.08	334.61	1.0598	0.01400	297.64	331.25	1.0361
140	0.01789	307.40	343.19	1.0808	0.01456	305.14	340.08	1.0577
150	0.01850	314.75	351.76	1.1013	0.01509	312.64	348.87	1.0787
160	0.01910	322.14	360.34	1.1214	0.01562	320.16	357.64	1.0992
170	0.01969	329.56	368.95	1.1410	0.01613	327.70	366.41	1.1192
180	0.02027	337.03	377.58	1.1603	0.01663	335.27	375.20	1.1388

TABLE A-10   **Properties of Saturated Refrigerant 134a (Liquid–Vapor): Temperature Table**

Pressure Conversions: 1 bar = 0.1 MPa = 10² kPa		Specific Volume m³/kg		Internal Energy kJ/kg		Enthalpy kJ/kg			Entropy kJ/kg · K		
Temp. °C	Press. bar	Sat. Liquid $v_f \times 10^3$	Sat. Vapor v_g	Sat. Liquid u_f	Sat. Vapor u_g	Sat. Liquid h_f	Evap. h_{fg}	Sat. Vapor h_g	Sat. Liquid s_f	Sat. Vapor s_g	Temp. °C
−40	0.5164	0.7055	0.3569	−0.04	204.45	0.00	222.88	222.88	0.0000	0.9560	−40
−36	0.6332	0.7113	0.2947	4.68	206.73	4.73	220.67	225.40	0.0201	0.9506	−36
−32	0.7704	0.7172	0.2451	9.47	209.01	9.52	218.37	227.90	0.0401	0.9456	−32
−28	0.9305	0.7233	0.2052	14.31	211.29	14.37	216.01	230.38	0.0600	0.9411	−28
−26	1.0199	0.7265	0.1882	16.75	212.43	16.82	214.80	231.62	0.0699	0.9390	−26
−24	1.1160	0.7296	0.1728	19.21	213.57	19.29	213.57	232.85	0.0798	0.9370	−24
−22	1.2192	0.7328	0.1590	21.68	214.70	21.77	212.32	234.08	0.0897	0.9351	−22
−20	1.3299	0.7361	0.1464	24.17	215.84	24.26	211.05	235.31	0.0996	0.9332	−20
−18	1.4483	0.7395	0.1350	26.67	216.97	26.77	209.76	236.53	0.1094	0.9315	−18
−16	1.5748	0.7428	0.1247	29.18	218.10	29.30	208.45	237.74	0.1192	0.9298	−16
−12	1.8540	0.7498	0.1068	34.25	220.36	34.39	205.77	240.15	0.1388	0.9267	−12
−8	2.1704	0.7569	0.0919	39.38	222.60	39.54	203.00	242.54	0.1583	0.9239	−8
−4	2.5274	0.7644	0.0794	44.56	224.84	44.75	200.15	244.90	0.1777	0.9213	−4
0	2.9282	0.7721	0.0689	49.79	227.06	50.02	197.21	247.23	0.1970	0.9190	0
4	3.3765	0.7801	0.0600	55.08	229.27	55.35	194.19	249.53	0.2162	0.9169	4
8	3.8756	0.7884	0.0525	60.43	231.46	60.73	191.07	251.80	0.2354	0.9150	8
12	4.4294	0.7971	0.0460	65.83	233.63	66.18	187.85	254.03	0.2545	0.9132	12
16	5.0416	0.8062	0.0405	71.29	235.78	71.69	184.52	256.22	0.2735	0.9116	16
20	5.7160	0.8157	0.0358	76.80	237.91	77.26	181.09	258.36	0.2924	0.9102	20
24	6.4566	0.8257	0.0317	82.37	240.01	82.90	177.55	260.45	0.3113	0.9089	24
26	6.8530	0.8309	0.0298	85.18	241.05	85.75	175.73	261.48	0.3208	0.9082	26
28	7.2675	0.8362	0.0281	88.00	242.08	88.61	173.89	262.50	0.3302	0.9076	28
30	7.7006	0.8417	0.0265	90.84	243.10	91.49	172.00	263.50	0.3396	0.9070	30
32	8.1528	0.8473	0.0250	93.70	244.12	94.39	170.09	264.48	0.3490	0.9064	32
34	8.6247	0.8530	0.0236	96.58	245.12	97.31	168.14	265.45	0.3584	0.9058	34
36	9.1168	0.8590	0.0223	99.47	246.11	100.25	166.15	266.40	0.3678	0.9053	36
38	9.6298	0.8651	0.0210	102.38	247.09	103.21	164.12	267.33	0.3772	0.9047	38
40	10.164	0.8714	0.0199	105.30	248.06	106.19	162.05	268.24	0.3866	0.9041	40
42	10.720	0.8780	0.0188	108.25	249.02	109.19	159.94	269.14	0.3960	0.9035	42
44	11.299	0.8847	0.0177	111.22	249.96	112.22	157.79	270.01	0.4054	0.9030	44
48	12.526	0.8989	0.0159	117.22	251.79	118.35	153.33	271.68	0.4243	0.9017	48
52	13.851	0.9142	0.0142	123.31	253.55	124.58	148.66	273.24	0.4432	0.9004	52
56	15.278	0.9308	0.0127	129.51	255.23	130.93	143.75	274.68	0.4622	0.8990	56
60	16.813	0.9488	0.0114	135.82	256.81	137.42	138.57	275.99	0.4814	0.8973	60
70	21.162	1.0027	0.0086	152.22	260.15	154.34	124.08	278.43	0.5302	0.8918	70
80	26.324	1.0766	0.0064	169.88	262.14	172.71	106.41	279.12	0.5814	0.8827	80
90	32.435	1.1949	0.0046	189.82	261.34	193.69	82.63	276.32	0.6380	0.8655	90
100	39.742	1.5443	0.0027	218.60	248.49	224.74	34.40	259.13	0.7196	0.8117	100

v_f = (table value)/1000

R-134a

TABLE A-11   **Properties of Saturated Refrigerant 134a (Liquid–Vapor): Pressure Table**

Pressure Conversions: 1 bar = 0.1 MPa = 10² kPa		Specific Volume m³/kg		Internal Energy kJ/kg		Enthalpy kJ/kg			Entropy kJ/kg · K		
Press. bar	Temp. °C	Sat. Liquid $v_f \times 10^3$	Sat. Vapor v_g	Sat. Liquid u_f	Sat. Vapor u_g	Sat. Liquid h_f	Evap. h_{fg}	Sat. Vapor h_g	Sat. Liquid s_f	Sat. Vapor s_g	Press. bar
0.6	−37.07	0.7097	0.3100	3.41	206.12	3.46	221.27	224.72	0.0147	0.9520	0.6
0.8	−31.21	0.7184	0.2366	10.41	209.46	10.47	217.92	228.39	0.0440	0.9447	0.8
1.0	−26.43	0.7258	0.1917	16.22	212.18	16.29	215.06	231.35	0.0678	0.9395	1.0
1.2	−22.36	0.7323	0.1614	21.23	214.50	21.32	212.54	233.86	0.0879	0.9354	1.2
1.4	−18.80	0.7381	0.1395	25.66	216.52	25.77	210.27	236.04	0.1055	0.9322	1.4
1.6	−15.62	0.7435	0.1229	29.66	218.32	29.78	208.19	237.97	0.1211	0.9295	1.6
1.8	−12.73	0.7485	0.1098	33.31	219.94	33.45	206.26	239.71	0.1352	0.9273	1.8
2.0	−10.09	0.7532	0.0993	36.69	221.43	36.84	204.46	241.30	0.1481	0.9253	2.0
2.4	−5.37	0.7618	0.0834	42.77	224.07	42.95	201.14	244.09	0.1710	0.9222	2.4
2.8	−1.23	0.7697	0.0719	48.18	226.38	48.39	198.13	246.52	0.1911	0.9197	2.8
3.2	2.48	0.7770	0.0632	53.06	228.43	53.31	195.35	248.66	0.2089	0.9177	3.2
3.6	5.84	0.7839	0.0564	57.54	230.28	57.82	192.76	250.58	0.2251	0.9160	3.6
4.0	8.93	0.7904	0.0509	61.69	231.97	62.00	190.32	252.32	0.2399	0.9145	4.0
5.0	15.74	0.8056	0.0409	70.93	235.64	71.33	184.74	256.07	0.2723	0.9117	5.0
6.0	21.58	0.8196	0.0341	78.99	238.74	79.48	179.71	259.19	0.2999	0.9097	6.0
7.0	26.72	0.8328	0.0292	86.19	241.42	86.78	175.07	261.85	0.3242	0.9080	7.0
8.0	31.33	0.8454	0.0255	92.75	243.78	93.42	170.73	264.15	0.3459	0.9066	8.0
9.0	35.53	0.8576	0.0226	98.79	245.88	99.56	166.62	266.18	0.3656	0.9054	9.0
10.0	39.39	0.8695	0.0202	104.42	247.77	105.29	162.68	267.97	0.3838	0.9043	10.0
12.0	46.32	0.8928	0.0166	114.69	251.03	115.76	155.23	270.99	0.4164	0.9023	12.0
14.0	52.43	0.9159	0.0140	123.98	253.74	125.26	148.14	273.40	0.4453	0.9003	14.0
16.0	57.92	0.9392	0.0121	132.52	256.00	134.02	141.31	275.33	0.4714	0.8982	16.0
18.0	62.91	0.9631	0.0105	140.49	257.88	142.22	134.60	276.83	0.4954	0.8959	18.0
20.0	67.49	0.9878	0.0093	148.02	259.41	149.99	127.95	277.94	0.5178	0.8934	20.0
25.0	77.59	1.0562	0.0069	165.48	261.84	168.12	111.06	279.17	0.5687	0.8854	25.0
30.0	86.22	1.1416	0.0053	181.88	262.16	185.30	92.71	278.01	0.6156	0.8735	30.0

v_f = (table value)/1000

TABLE A-12   **Properties of Superheated Refrigerant 134a Vapor**

Pressure Conversions:
1 bar = 0.1 MPa
= 10^2 kPa

T °C	v m³/kg	u kJ/kg	h kJ/kg	s kJ/kg·K	v m³/kg	u kJ/kg	h kJ/kg	s kJ/kg·K
	\multicolumn{4}{c}{$p = 0.6$ bar = 0.06 MPa ($T_{sat} = 37.07°C$)}				
Sat.	0.31003	206.12	224.72	0.9520	0.19170	212.18	231.35	0.9395
−20	0.33536	217.86	237.98	1.0062	0.19770	216.77	236.54	0.9602
−10	0.34992	224.97	245.96	1.0371	0.20686	224.01	244.70	0.9918
0	0.36433	232.24	254.10	1.0675	0.21587	231.41	252.99	1.0227
10	0.37861	239.69	262.41	1.0973	0.22473	238.96	261.43	1.0531
20	0.39279	247.32	270.89	1.1267	0.23349	246.67	270.02	1.0829
30	0.40688	255.12	279.53	1.1557	0.24216	254.54	278.76	1.1122
40	0.42091	263.10	288.35	1.1844	0.25076	262.58	287.66	1.1411
50	0.43487	271.25	297.34	1.2126	0.25930	270.79	296.72	1.1696
60	0.44879	279.58	306.51	1.2405	0.26779	279.16	305.94	1.1977
70	0.46266	288.08	315.84	1.2681	0.27623	287.70	315.32	1.2254
80	0.47650	296.75	325.34	1.2954	0.28464	296.40	324.87	1.2528
90	0.49031	305.58	335.00	1.3224	0.29302	305.27	334.57	1.2799

The headers for the two pressure blocks above: left block $p = 0.6$ bar = 0.06 MPa ($T_{sat} = 37.07°C$); right block $p = 1.0$ bar = 0.10 MPa ($T_{sat} = 26.43°C$).

T °C	v m³/kg	u kJ/kg	h kJ/kg	s kJ/kg·K	v m³/kg	u kJ/kg	h kJ/kg	s kJ/kg·K
	\multicolumn{8}{c}{$p = 1.4$ bar = 0.14 MPa ($T_{sat} = −18.80°C$) / $p = 1.8$ bar = 0.18 MPa ($T_{sat} = −12.73°C$)}						
Sat.	0.13945	216.52	236.04	0.9322	0.10983	219.94	239.71	0.9273
−10	0.14549	223.03	243.40	0.9606	0.11135	222.02	242.06	0.9362
0	0.15219	230.55	251.86	0.9922	0.11678	229.67	250.69	0.9684
10	0.15875	238.21	260.43	1.0230	0.12207	237.44	259.41	0.9998
20	0.16520	246.01	269.13	1.0532	0.12723	245.33	268.23	1.0304
30	0.17155	253.96	277.97	1.0828	0.13230	253.36	277.17	1.0604
40	0.17783	262.06	286.96	1.1120	0.13730	261.53	286.24	1.0898
50	0.18404	270.32	296.09	1.1407	0.14222	269.85	295.45	1.1187
60	0.19020	278.74	305.37	1.1690	0.14710	278.31	304.79	1.1472
70	0.19633	287.32	314.80	1.1969	0.15193	286.93	314.28	1.1753
80	0.20241	296.06	324.39	1.2244	0.15672	295.71	323.92	1.2030
90	0.20846	304.95	334.14	1.2516	0.16148	304.63	333.70	1.2303
100	0.21449	314.01	344.04	1.2785	0.16622	313.72	343.63	1.2573

T °C	v m³/kg	u kJ/kg	h kJ/kg	s kJ/kg·K	v m³/kg	u kJ/kg	h kJ/kg	s kJ/kg·K
	\multicolumn{8}{c}{$p = 2.0$ bar = 0.20 MPa ($T_{sat} = −10.09°C$) / $p = 2.4$ bar = 0.24 MPa ($T_{sat} = −5.37°C$)}						
Sat.	0.09933	221.43	241.30	0.9253	0.08343	224.07	244.09	0.9222
−10	0.09938	221.50	241.38	0.9256				
0	0.10438	229.23	250.10	0.9582	0.08574	228.31	248.89	0.9399
10	0.10922	237.05	258.89	0.9898	0.08993	236.26	257.84	0.9721
20	0.11394	244.99	267.78	1.0206	0.09399	244.30	266.85	1.0034
30	0.11856	253.06	276.77	1.0508	0.09794	252.45	275.95	1.0339
40	0.12311	261.26	285.88	1.0804	0.10181	260.72	285.16	1.0637
50	0.12758	269.61	295.12	1.1094	0.10562	269.12	294.47	1.0930
60	0.13201	278.10	304.50	1.1380	0.10937	277.67	303.91	1.1218
70	0.13639	286.74	314.02	1.1661	0.11307	286.35	313.49	1.1501
80	0.14073	295.53	323.68	1.1939	0.11674	295.18	323.19	1.1780
90	0.14504	304.47	333.48	1.2212	0.12037	304.15	333.04	1.2055
100	0.14932	313.57	343.43	1.2483	0.12398	313.27	343.03	1.2326

R-134a

Pressure Conversions:
1 bar = 0.1 MPa
= 10² kPa

TABLE A-12   **Properties of Superheated Refrigerant 134a Vapor (*Continued*)**

T °C	υ m³/kg	u kJ/kg	h kJ/kg	s kJ/kg·K	υ m³/kg	u kJ/kg	h kJ/kg	s kJ/kg·K
	p = 2.8 bar = 0.28 MPa (*T*ₛₐₜ = −1.23°C)				*p* = 3.2 bar = 0.32 MPa (*T*ₛₐₜ = 2.48°C)			
Sat.	0.07193	226.38	246.52	0.9197	0.06322	228.43	248.66	0.9177
0	0.07240	227.37	247.64	0.9238				
10	0.07613	235.44	256.76	0.9566	0.06576	234.61	255.65	0.9427
20	0.07972	243.59	265.91	0.9883	0.06901	242.87	264.95	0.9749
30	0.08320	251.83	275.12	1.0192	0.07214	251.19	274.28	1.0062
40	0.08660	260.17	284.42	1.0494	0.07518	259.61	283.67	1.0367
50	0.08992	268.64	293.81	1.0789	0.07815	268.14	293.15	1.0665
60	0.09319	277.23	303.32	1.1079	0.08106	276.79	302.72	1.0957
70	0.09641	285.96	312.95	1.1364	0.08392	285.56	312.41	1.1243
80	0.09960	294.82	322.71	1.1644	0.08674	294.46	322.22	1.1525
90	0.10275	303.83	332.60	1.1920	0.08953	303.50	332.15	1.1802
100	0.10587	312.98	342.62	1.2193	0.09229	312.68	342.21	1.2076
110	0.10897	322.27	352.78	1.2461	0.09503	322.00	352.40	1.2345
120	0.11205	331.71	363.08	1.2727	0.09774	331.45	362.73	1.2611
	p = 4.0 bar = 0.40 MPa (*T*ₛₐₜ = 8.93°C)				*p* = 5.0 bar = 0.50 MPa (*T*ₛₐₜ = 15.74°C)			
Sat.	0.05089	231.97	252.32	0.9145	0.04086	235.64	256.07	0.9117
10	0.05119	232.87	253.35	0.9182				
20	0.05397	241.37	262.96	0.9515	0.04188	239.40	260.34	0.9264
30	0.05662	249.89	272.54	0.9837	0.04416	248.20	270.28	0.9597
40	0.05917	258.47	282.14	1.0148	0.04633	256.99	280.16	0.9918
50	0.06164	267.13	291.79	1.0452	0.04842	265.83	290.04	1.0229
60	0.06405	275.89	301.51	1.0748	0.05043	274.73	299.95	1.0531
70	0.06641	284.75	311.32	1.1038	0.05240	283.72	309.92	1.0825
80	0.06873	293.73	321.23	1.1322	0.05432	292.80	319.96	1.1114
90	0.07102	302.84	331.25	1.1602	0.05620	302.00	330.10	1.1397
100	0.07327	312.07	341.38	1.1878	0.05805	311.31	340.33	1.1675
110	0.07550	321.44	351.64	1.2149	0.05988	320.74	350.68	1.1949
120	0.07771	330.94	362.03	1.2417	0.06168	330.30	361.14	1.2218
130	0.07991	340.58	372.54	1.2681	0.06347	339.98	371.72	1.2484
140	0.08208	350.35	383.18	1.2941	0.06524	349.79	382.42	1.2746
	p = 6.0 bar = 0.60 MPa (*T*ₛₐₜ = 21.58°C)				*p* = 7.0 bar = 0.70 MPa (*T*ₛₐₜ = 26.72°C)			
Sat.	0.03408	238.74	259.19	0.9097	0.02918	241.42	261.85	0.9080
30	0.03581	246.41	267.89	0.9388	0.02979	244.51	265.37	0.9197
40	0.03774	255.45	278.09	0.9719	0.03157	253.83	275.93	0.9539
50	0.03958	264.48	288.23	1.0037	0.03324	263.08	286.35	0.9867
60	0.04134	273.54	298.35	1.0346	0.03482	272.31	296.69	1.0182
70	0.04304	282.66	308.48	1.0645	0.03634	281.57	307.01	1.0487
80	0.04469	291.86	318.67	1.0938	0.03781	290.88	317.35	1.0784
90	0.04631	301.14	328.93	1.1225	0.03924	300.27	327.74	1.1074
100	0.04790	310.53	339.27	1.1505	0.04064	309.74	338.19	1.1358
110	0.04946	320.03	349.70	1.1781	0.04201	319.31	348.71	1.1637
120	0.05099	329.64	360.24	1.2053	0.04335	328.98	359.33	1.1910
130	0.05251	339.38	370.88	1.2320	0.04468	338.76	370.04	1.2179
140	0.05402	349.23	381.64	1.2584	0.04599	348.66	380.86	1.2444
150	0.05550	359.21	392.52	1.2844	0.04729	358.68	391.79	1.2706
160	0.05698	369.32	403.51	1.3100	0.04857	368.82	402.82	1.2963

TABLE A-12   Properties of Superheated Refrigerant 134a Vapor (*Continued*)

Pressure Conversions:
1 bar = 0.1 MPa
= 10^2 kPa

T °C	v m³/kg	u kJ/kg	h kJ/kg	s kJ/kg·K	v m³/kg	u kJ/kg	h kJ/kg	s kJ/kg·K
	— p = 8.0 bar = 0.80 MPa (T_{sat} = 31.33°C)				p = 9.0 bar = 0.90 MPa (T_{sat} = 35.53°C)			
Sat.	0.02547	243.78	264.15	0.9066	0.02255	245.88	266.18	0.9054
40	0.02691	252.13	273.66	0.9374	0.02325	250.32	271.25	0.9217
50	0.02846	261.62	284.39	0.9711	0.02472	260.09	282.34	0.9566
60	0.02992	271.04	294.98	1.0034	0.02609	269.72	293.21	0.9897
70	0.03131	280.45	305.50	1.0345	0.02738	279.30	303.94	1.0214
80	0.03264	289.89	316.00	1.0647	0.02861	288.87	314.62	1.0521
90	0.03393	299.37	326.52	1.0940	0.02980	298.46	325.28	1.0819
100	0.03519	308.93	337.08	1.1227	0.03095	308.11	335.96	1.1109
110	0.03642	318.57	347.71	1.1508	0.03207	317.82	346.68	1.1392
120	0.03762	328.31	358.40	1.1784	0.03316	327.62	357.47	1.1670
130	0.03881	338.14	369.19	1.2055	0.03423	337.52	368.33	1.1943
140	0.03997	348.09	380.07	1.2321	0.03529	347.51	379.27	1.2211
150	0.04113	358.15	391.05	1.2584	0.03633	357.61	390.31	1.2475
160	0.04227	368.32	402.14	1.2843	0.03736	367.82	401.44	1.2735
170	0.04340	378.61	413.33	1.3098	0.03838	378.14	412.68	1.2992
180	0.04452	389.02	424.63	1.3351	0.03939	388.57	424.02	1.3245
	p = 10.0 bar = 1.00 MPa (T_{sat} = 39.39°C)				p = 12.0 bar = 1.20 MPa (T_{sat} = 46.32°C)			
Sat.	0.02020	247.77	267.97	0.9043	0.01663	251.03	270.99	0.9023
40	0.02029	248.39	268.68	0.9066				
50	0.02171	258.48	280.19	0.9428	0.01712	254.98	275.52	0.9164
60	0.02301	268.35	291.36	0.9768	0.01835	265.42	287.44	0.9527
70	0.02423	278.11	302.34	1.0093	0.01947	275.59	298.96	0.9868
80	0.02538	287.82	313.20	1.0405	0.02051	285.62	310.24	1.0192
90	0.02649	297.53	324.01	1.0707	0.02150	295.59	321.39	1.0503
100	0.02755	307.27	334.82	1.1000	0.02244	305.54	332.47	1.0804
110	0.02858	317.06	345.65	1.1286	0.02335	315.50	343.52	1.1096
120	0.02959	326.93	356.52	1.1567	0.02423	325.51	354.58	1.1381
130	0.03058	336.88	367.46	1.1841	0.02508	335.58	365.68	1.1660
140	0.03154	346.92	378.46	1.2111	0.02592	345.73	376.83	1.1933
150	0.03250	357.06	389.56	1.2376	0.02674	355.95	388.04	1.2201
160	0.03344	367.31	400.74	1.2638	0.02754	366.27	399.33	1.2465
170	0.03436	377.66	412.02	1.2895	0.02834	376.69	410.70	1.2724
180	0.03528	388.12	423.40	1.3149	0.02912	387.21	422.16	1.2980
	p = 14.0 bar = 1.40 MPa (T_{sat} = 52.43°C)				p = 16.0 bar = 1.60 MPa (T_{sat} = 57.92°C)			
Sat.	0.01405	253.74	273.40	0.9003	0.01208	256.00	275.33	0.8982
60	0.01495	262.17	283.10	0.9297	0.01233	258.48	278.20	0.9069
70	0.01603	272.87	295.31	0.9658	0.01340	269.89	291.33	0.9457
80	0.01701	283.29	307.10	0.9997	0.01435	280.78	303.74	0.9813
90	0.01792	293.55	318.63	1.0319	0.01521	291.39	315.72	1.0148
100	0.01878	303.73	330.02	1.0628	0.01601	301.84	327.46	1.0467
110	0.01960	313.88	341.32	1.0927	0.01677	312.20	339.04	1.0773
120	0.02039	324.05	352.59	1.1218	0.01750	322.53	350.53	1.1069
130	0.02115	334.25	363.86	1.1501	0.01820	332.87	361.99	1.1357
140	0.02189	344.50	375.15	1.1777	0.01887	343.24	373.44	1.1638
150	0.02262	354.82	386.49	1.2048	0.01953	353.66	384.91	1.1912
160	0.02333	365.22	397.89	1.2315	0.02017	364.15	396.43	1.2181
170	0.02403	375.71	409.36	1.2576	0.02080	374.71	407.99	1.2445
180	0.02472	386.29	420.90	1.2834	0.02142	385.35	419.62	1.2704
190	0.02541	396.96	432.53	1.3088	0.02203	396.08	431.33	1.2960
200	0.02608	407.73	444.24	1.3338	0.02263	406.90	443.11	1.3212

TABLE A-13 **Properties of Saturated Ammonia (Liquid–Vapor): Temperature Table**

Temp. °C	Press. bar	Specific Volume m³/kg Sat. Liquid $v_f \times 10^3$	Specific Volume m³/kg Sat. Vapor v_g	Internal Energy kJ/kg Sat. Liquid u_f	Internal Energy kJ/kg Sat. Vapor u_g	Enthalpy kJ/kg Sat. Liquid h_f	Enthalpy kJ/kg Evap. h_{fg}	Enthalpy kJ/kg Sat. Vapor h_g	Entropy kJ/kg·K Sat. Liquid s_f	Entropy kJ/kg·K Sat. Vapor s_g	Temp. °C
−50	0.4086	1.4245	2.6265	−43.94	1264.99	−43.88	1416.20	1372.32	−0.1922	6.1543	−50
−45	0.5453	1.4367	2.0060	−22.03	1271.19	−21.95	1402.52	1380.57	−0.0951	6.0523	−45
−40	0.7174	1.4493	1.5524	−0.10	1277.20	0.00	1388.56	1388.56	0.0000	5.9557	−40
−36	0.8850	1.4597	1.2757	17.47	1281.87	17.60	1377.17	1394.77	0.0747	5.8819	−36
−32	1.0832	1.4703	1.0561	35.09	1286.41	35.25	1365.55	1400.81	0.1484	5.8111	−32
−30	1.1950	1.4757	0.9634	43.93	1288.63	44.10	1359.65	1403.75	0.1849	5.7767	−30
−28	1.3159	1.4812	0.8803	52.78	1290.82	52.97	1353.68	1406.66	0.2212	5.7430	−28
−26	1.4465	1.4867	0.8056	61.65	1292.97	61.86	1347.65	1409.51	0.2572	5.7100	−26
−22	1.7390	1.4980	0.6780	79.46	1297.18	79.72	1335.36	1415.08	0.3287	5.6457	−22
−20	1.9019	1.5038	0.6233	88.40	1299.23	88.68	1329.10	1417.79	0.3642	5.6144	−20
−18	2.0769	1.5096	0.5739	97.36	1301.25	97.68	1322.77	1420.45	0.3994	5.5837	−18
−16	2.2644	1.5155	0.5291	106.36	1303.23	106.70	1316.35	1423.05	0.4346	5.5536	−16
−14	2.4652	1.5215	0.4885	115.37	1305.17	115.75	1309.86	1425.61	0.4695	5.5239	−14
−12	2.6798	1.5276	0.4516	124.42	1307.08	124.83	1303.28	1428.11	0.5043	5.4948	−12
−10	2.9089	1.5338	0.4180	133.50	1308.95	133.94	1296.61	1430.55	0.5389	5.4662	−10
−8	3.1532	1.5400	0.3874	142.60	1310.78	143.09	1289.86	1432.95	0.5734	5.4380	−8
−6	3.4134	1.5464	0.3595	151.74	1312.57	152.26	1283.02	1435.28	0.6077	5.4103	−6
−4	3.6901	1.5528	0.3340	160.88	1314.32	161.46	1276.10	1437.56	0.6418	5.3831	−4
−2	3.9842	1.5594	0.3106	170.07	1316.04	170.69	1269.08	1439.78	0.6759	5.3562	−2
0	4.2962	1.5660	0.2892	179.29	1317.71	179.96	1261.97	1441.94	0.7097	5.3298	0
2	4.6270	1.5727	0.2695	188.53	1319.34	189.26	1254.77	1444.03	0.7435	5.3038	2
4	4.9773	1.5796	0.2514	197.80	1320.92	198.59	1247.48	1446.07	0.7770	5.2781	4
6	5.3479	1.5866	0.2348	207.10	1322.47	207.95	1240.09	1448.04	0.8105	5.2529	6
8	5.7395	1.5936	0.2195	216.42	1323.96	217.34	1232.61	1449.94	0.8438	5.2279	8
10	6.1529	1.6008	0.2054	225.77	1325.42	226.75	1225.03	1451.78	0.8769	5.2033	10
12	6.5890	1.6081	0.1923	235.14	1326.82	236.20	1217.35	1453.55	0.9099	5.1791	12
16	7.5324	1.6231	0.1691	253.95	1329.48	255.18	1201.70	1456.87	0.9755	5.1314	16
20	8.5762	1.6386	0.1492	272.86	1331.94	274.26	1185.64	1459.90	1.0404	5.0849	20
24	9.7274	1.6547	0.1320	291.84	1334.19	293.45	1169.16	1462.61	1.1048	5.0394	24
28	10.993	1.6714	0.1172	310.92	1336.20	312.75	1152.24	1465.00	1.1686	4.9948	28
32	12.380	1.6887	0.1043	330.07	1337.97	332.17	1134.87	1467.03	1.2319	4.9509	32
36	13.896	1.7068	0.0930	349.32	1339.47	351.69	1117.00	1468.70	1.2946	4.9078	36
40	15.549	1.7256	0.0831	368.67	1340.70	371.35	1098.62	1469.97	1.3569	4.8652	40
45	17.819	1.7503	0.0725	393.01	1341.81	396.13	1074.84	1470.96	1.4341	4.8125	45
50	20.331	1.7765	0.0634	417.56	1342.42	421.17	1050.09	1471.26	1.5109	4.7604	50

Pressure Conversions:
1 bar = 0.1 MPa
= 10^2 kPa

v_f = (table value)/1000

Ammonia

TABLE A-14   **Properties of Saturated Ammonia (Liquid–Vapor): Pressure Table**

Press. bar	Temp. °C	Specific Volume m³/kg Sat. Liquid $v_f \times 10^3$	Sat. Vapor v_g	Internal Energy kJ/kg Sat. Liquid u_f	Sat. Vapor u_g	Enthalpy kJ/kg Sat. Liquid h_f	Evap. h_{fg}	Sat. Vapor h_g	Entropy kJ/kg·K Sat. Liquid s_f	Sat. Vapor s_g	Press. bar
0.40	−50.36	1.4236	2.6795	−45.52	1264.54	−45.46	1417.18	1371.72	−0.1992	6.1618	0.40
0.50	−46.53	1.4330	2.1752	−28.73	1269.31	−28.66	1406.73	1378.07	−0.1245	6.0829	0.50
0.60	−43.28	1.4410	1.8345	−14.51	1273.27	−14.42	1397.76	1383.34	−0.0622	6.0186	0.60
0.70	−40.46	1.4482	1.5884	−2.11	1276.66	−2.01	1389.85	1387.84	−0.0086	5.9643	0.70
0.80	−37.94	1.4546	1.4020	8.93	1279.61	9.04	1382.73	1391.78	0.0386	5.9174	0.80
0.90	−35.67	1.4605	1.2559	18.91	1282.24	19.04	1376.23	1395.27	0.0808	5.8760	0.90
1.00	−33.60	1.4660	1.1381	28.03	1284.61	28.18	1370.23	1398.41	0.1191	5.8391	1.00
1.25	−29.07	1.4782	0.9237	48.03	1289.65	48.22	1356.89	1405.11	0.2018	5.7610	1.25
1.50	−25.22	1.4889	0.7787	65.10	1293.80	65.32	1345.28	1410.61	0.2712	5.6973	1.50
1.75	−21.86	1.4984	0.6740	80.08	1297.33	80.35	1334.92	1415.27	0.3312	5.6435	1.75
2.00	−18.86	1.5071	0.5946	93.50	1300.39	93.80	1325.51	1419.31	0.3843	5.5969	2.00
2.25	−16.15	1.5151	0.5323	105.68	1303.08	106.03	1316.83	1422.86	0.4319	5.5558	2.25
2.50	−13.67	1.5225	0.4821	116.88	1305.49	117.26	1308.76	1426.03	0.4753	5.5190	2.50
2.75	−11.37	1.5295	0.4408	127.26	1307.67	127.68	1301.20	1428.88	0.5152	5.4858	2.75
3.00	−9.24	1.5361	0.4061	136.96	1309.65	137.42	1294.05	1431.47	0.5520	5.4554	3.00
3.25	−7.24	1.5424	0.3765	146.06	1311.46	146.57	1287.27	1433.84	0.5864	5.4275	3.25
3.50	−5.36	1.5484	0.3511	154.66	1313.14	155.20	1280.81	1436.01	0.6186	5.4016	3.50
3.75	−3.58	1.5542	0.3289	162.80	1314.68	163.38	1274.64	1438.03	0.6489	5.3774	3.75
4.00	−1.90	1.5597	0.3094	170.55	1316.12	171.18	1268.71	1439.89	0.6776	5.3548	4.00
4.25	−0.29	1.5650	0.2921	177.96	1317.47	178.62	1263.01	1441.63	0.7048	5.3336	4.25
4.50	1.25	1.5702	0.2767	185.04	1318.73	185.75	1257.50	1443.25	0.7308	5.3135	4.50
4.75	2.72	1.5752	0.2629	191.84	1319.91	192.59	1252.18	1444.77	0.7555	5.2946	4.75
5.00	4.13	1.5800	0.2503	198.39	1321.02	199.18	1247.02	1446.19	0.7791	5.2765	5.00
5.25	5.48	1.5847	0.2390	204.69	1322.07	205.52	1242.01	1447.53	0.8018	5.2594	5.25
5.50	6.79	1.5893	0.2286	210.78	1323.06	211.65	1237.15	1448.80	0.8236	5.2430	5.50
5.75	8.05	1.5938	0.2191	216.66	1324.00	217.58	1232.41	1449.99	0.8446	5.2273	5.75
6.00	9.27	1.5982	0.2104	222.37	1324.89	223.32	1227.79	1451.12	0.8649	5.2122	6.00
7.00	13.79	1.6148	0.1815	243.56	1328.04	244.69	1210.38	1455.07	0.9394	5.1576	7.00
8.00	17.84	1.6302	0.1596	262.64	1330.64	263.95	1194.36	1458.30	1.0054	5.1099	8.00
9.00	21.52	1.6446	0.1424	280.05	1332.82	281.53	1179.44	1460.97	1.0649	5.0675	9.00
10.00	24.89	1.6584	0.1285	296.10	1334.66	297.76	1165.42	1463.18	1.1191	5.0294	10.00
12.00	30.94	1.6841	0.1075	324.99	1337.52	327.01	1139.52	1466.53	1.2152	4.9625	12.00
14.00	36.26	1.7080	0.0923	350.58	1339.56	352.97	1115.82	1468.79	1.2987	4.9050	14.00
16.00	41.03	1.7306	0.0808	373.69	1340.97	376.46	1093.77	1470.23	1.3729	4.8542	16.00
18.00	45.38	1.7522	0.0717	394.85	1341.88	398.00	1073.01	1471.01	1.4399	4.8086	18.00
20.00	49.37	1.7731	0.0644	414.44	1342.37	417.99	1053.27	1471.26	1.5012	4.7670	20.00

Pressure Conversions:
1 bar = 0.1 MPa
= 10² kPa

v_f = (table value)/1000

Ammonia

TABLE A-15   **Properties of Superheated Ammonia Vapor**

T °C	v m³/kg	u kJ/kg	h kJ/kg	s kJ/kg·K	v m³/kg	u kJ/kg	h kJ/kg	s kJ/kg·K
	p = 0.4 bar = 0.04 MPa (T_{sat} = −50.36°C)				**p = 0.6 bar = 0.06 MPa** (T_{sat} = −43.28°C)			
Sat.	2.6795	1264.54	1371.72	6.1618	1.8345	1273.27	1383.34	6.0186
−50	2.6841	1265.11	1372.48	6.1652				
−45	2.7481	1273.05	1382.98	6.2118				
−40	2.8118	1281.01	1393.48	6.2573	1.8630	1278.62	1390.40	6.0490
−35	2.8753	1288.96	1403.98	6.3018	1.9061	1286.75	1401.12	6.0946
−30	2.9385	1296.93	1414.47	6.3455	1.9491	1294.88	1411.83	6.1390
−25	3.0015	1304.90	1424.96	6.3882	1.9918	1303.01	1422.52	6.1826
−20	3.0644	1312.88	1435.46	6.4300	2.0343	1311.13	1433.19	6.2251
−15	3.1271	1320.87	1445.95	6.4711	2.0766	1319.25	1443.85	6.2668
−10	3.1896	1328.87	1456.45	6.5114	2.1188	1327.37	1454.50	6.3077
−5	3.2520	1336.88	1466.95	6.5509	2.1609	1335.49	1465.14	6.3478
0	3.3142	1344.90	1477.47	6.5898	2.2028	1343.61	1475.78	6.3871
5	3.3764	1352.95	1488.00	6.6280	2.2446	1351.75	1486.43	6.4257
	p = 0.8 bar = 0.08 MPa (T_{sat} = −37.94°C)				**p = 1.0 bar = 0.10 MPa** (T_{sat} = −33.60°C)			
Sat.	1.4021	1279.61	1391.78	5.9174	1.1381	1284.61	1398.41	5.8391
−35	1.4215	1284.51	1398.23	5.9446				
−30	1.4543	1292.81	1409.15	5.9900	1.1573	1290.71	1406.44	5.8723
−25	1.4868	1301.09	1420.04	6.0343	1.1838	1299.15	1417.53	5.9175
−20	1.5192	1309.36	1430.90	6.0777	1.2101	1307.57	1428.58	5.9616
−15	1.5514	1317.61	1441.72	6.1200	1.2362	1315.96	1439.58	6.0046
−10	1.5834	1325.85	1452.53	6.1615	1.2621	1324.33	1450.54	6.0467
−5	1.6153	1334.09	1463.31	6.2021	1.2880	1332.67	1461.47	6.0878
0	1.6471	1342.31	1474.08	6.2419	1.3136	1341.00	1472.37	6.1281
5	1.6788	1350.54	1484.84	6.2809	1.3392	1349.33	1483.25	6.1676
10	1.7103	1358.77	1495.60	6.3192	1.3647	1357.64	1494.11	6.2063
15	1.7418	1367.01	1506.35	6.3568	1.3900	1365.95	1504.96	6.2442
20	1.7732	1375.25	1517.10	6.3939	1.4153	1374.27	1515.80	6.2816
	p = 1.5 bar = 0.15 MPa (T_{sat} = −25.22°C)				**p = 2.0 bar = 0.20 MPa** (T_{sat} = −18.86°C)			
Sat.	0.7787	1293.80	1410.61	5.6973	0.59460	1300.39	1419.31	5.5969
−25	0.7795	1294.20	1411.13	5.6994				
−20	0.7978	1303.00	1422.67	5.7454				
−15	0.8158	1311.75	1434.12	5.7902	0.60542	1307.43	1428.51	5.6328
−10	0.8336	1320.44	1445.49	5.8338	0.61926	1316.46	1440.31	5.6781
−5	0.8514	1329.08	1456.79	5.8764	0.63294	1325.41	1452.00	5.7221
0	0.8689	1337.68	1468.02	5.9179	0.64648	1334.29	1463.59	5.7649
5	0.8864	1346.25	1479.20	5.9585	0.65989	1343.11	1475.09	5.8066
10	0.9037	1354.78	1490.34	5.9981	0.67320	1351.87	1486.51	5.8473
15	0.9210	1363.29	1501.44	6.0370	0.68640	1360.59	1497.87	5.8871
20	0.9382	1371.79	1512.51	6.0751	0.69952	1369.28	1509.18	5.9260
25	0.9553	1380.28	1523.56	6.1125	0.71256	1377.93	1520.44	5.9641
30	0.9723	1388.76	1534.60	6.1492	0.72553	1386.56	1531.67	6.0014

TABLE A-15   **Properties of Superheated Ammonia Vapor (*Continued*)**

Pressure Conversions:
1 bar = 0.1 MPa
= 10^2 kPa

T °C	v m³/kg	u kJ/kg	h kJ/kg	s kJ/kg·K	v m³/kg	u kJ/kg	h kJ/kg	s kJ/kg·K
	\multicolumn{4}{c}{p = 2.5 bar = 0.25 MPa (T_{sat} = −13.67°C)}	\multicolumn{4}{c}{p = 3.0 bar = 0.30 MPa (T_{sat} = −9.24°C)}
Sat.	0.48213	1305.49	1426.03	5.5190	0.40607	1309.65	1431.47	5.4554
−10	0.49051	1312.37	1435.00	5.5534				
−5	0.50180	1321.65	1447.10	5.5989	0.41428	1317.80	1442.08	5.4953
0	0.51293	1330.83	1459.06	5.6431	0.42382	1327.28	1454.43	5.5409
5	0.52393	1339.91	1470.89	5.6860	0.43323	1336.64	1466.61	5.5851
10	0.53482	1348.91	1482.61	5.7278	0.44251	1345.89	1478.65	5.6280
15	0.54560	1357.84	1494.25	5.7685	0.45169	1355.05	1490.56	5.6697
20	0.55630	1366.72	1505.80	5.8083	0.46078	1364.13	1502.36	5.7103
25	0.56691	1375.55	1517.28	5.8471	0.46978	1373.14	1514.07	5.7499
30	0.57745	1384.34	1528.70	5.8851	0.47870	1382.09	1525.70	5.7886
35	0.58793	1393.10	1540.08	5.9223	0.48756	1391.00	1537.26	5.8264
40	0.59835	1401.84	1551.42	5.9589	0.49637	1399.86	1548.77	5.8635
45	0.60872	1410.56	1562.74	5.9947	0.50512	1408.70	1560.24	5.8998
	\multicolumn{4}{c}{p = 3.5 bar = 0.35 MPa (T_{sat} = −5.36°C)}	\multicolumn{4}{c}{p = 4.0 bar = 0.40 MPa (T_{sat} = −1.90°C)}
Sat.	0.35108	1313.14	1436.01	5.4016	0.30942	1316.12	1439.89	5.3548
0	0.36011	1323.66	1449.70	5.4522	0.31227	1319.95	1444.86	5.3731
10	0.37654	1342.82	1474.61	5.5417	0.32701	1339.68	1470.49	5.4652
20	0.39251	1361.49	1498.87	5.6259	0.34129	1358.81	1495.33	5.5515
30	0.40814	1379.81	1522.66	5.7057	0.35520	1377.49	1519.57	5.6328
40	0.42350	1397.87	1546.09	5.7818	0.36884	1395.85	1543.38	5.7101
60	0.45363	1433.55	1592.32	5.9249	0.39550	1431.97	1590.17	5.8549
80	0.48320	1469.06	1638.18	6.0586	0.42160	1467.77	1636.41	5.9897
100	0.51240	1504.73	1684.07	6.1850	0.44733	1503.64	1682.58	6.1169
120	0.54136	1540.79	1730.26	6.3056	0.47280	1539.85	1728.97	6.2380
140	0.57013	1577.38	1776.92	6.4213	0.49808	1576.55	1775.79	6.3541
160	0.59876	1614.60	1824.16	6.5330	0.52323	1613.86	1823.16	6.4661
180	0.62728	1652.51	1872.06	6.6411	0.54827	1651.85	1871.16	6.5744
200	0.65572	1691.15	1920.65	6.7460	0.57322	1690.56	1919.85	6.6796
	\multicolumn{4}{c}{p = 4.5 bar = 0.45 MPa (T_{sat} = 1.25°C)}	\multicolumn{4}{c}{p = 5.0 bar = 0.50 MPa (T_{sat} = 4.13°C)}
Sat.	0.27671	1318.73	1443.25	5.3135	0.25034	1321.02	1446.19	5.2765
10	0.28846	1336.48	1466.29	5.3962	0.25757	1333.22	1462.00	5.3330
20	0.30142	1356.09	1491.72	5.4845	0.26949	1353.32	1488.06	5.4234
30	0.31401	1375.15	1516.45	5.5674	0.28103	1372.76	1513.28	5.5080
40	0.32631	1393.80	1540.64	5.6460	0.29227	1391.74	1537.87	5.5878
60	0.35029	1430.37	1588.00	5.7926	0.31410	1428.76	1585.81	5.7362
80	0.37369	1466.47	1634.63	5.9285	0.33535	1465.16	1632.84	5.8733
100	0.39671	1502.55	1681.07	6.0564	0.35621	1501.46	1679.56	6.0020
120	0.41947	1538.91	1727.67	6.1781	0.37681	1537.97	1726.37	6.1242
140	0.44205	1575.73	1774.65	6.2946	0.39722	1574.90	1773.51	6.2412
160	0.46448	1613.13	1822.15	6.4069	0.41749	1612.40	1821.14	6.3537
180	0.48681	1651.20	1870.26	6.5155	0.43765	1650.54	1869.36	6.4626
200	0.50905	1689.97	1919.04	6.6208	0.45771	1689.38	1918.24	6.5681

Ammonia

Ammonia

TABLE A-15 Properties of Superheated Ammonia Vapor (*Continued*)

T °C	v m³/kg	u kJ/kg	h kJ/kg	s kJ/kg·K	v m³/kg	u kJ/kg	h kJ/kg	s kJ/kg·K
	p = 5.5 bar = 0.55 MPa (T_{sat} = 6.79°C)				*p* = 6.0 bar = 0.60 MPa (T_{sat} = 9.27°C)			
Sat.	0.22861	1323.06	1448.80	5.2430	0.21038	1324.89	1451.12	5.2122
10	0.23227	1329.88	1457.63	5.2743	0.21115	1326.47	1453.16	5.2195
20	0.24335	1350.50	1484.34	5.3671	0.22155	1347.62	1480.55	5.3145
30	0.25403	1370.35	1510.07	5.4534	0.23152	1367.90	1506.81	5.4026
40	0.26441	1389.64	1535.07	5.5345	0.24118	1387.52	1532.23	5.4851
50	0.27454	1408.53	1559.53	5.6114	0.25059	1406.67	1557.03	5.5631
60	0.28449	1427.13	1583.60	5.6848	0.25981	1425.49	1581.38	5.6373
80	0.30398	1463.85	1631.04	5.8230	0.27783	1462.52	1629.22	5.7768
100	0.32307	1500.36	1678.05	5.9525	0.29546	1499.25	1676.52	5.9071
120	0.34190	1537.02	1725.07	6.0753	0.31281	1536.07	1723.76	6.0304
140	0.36054	1574.07	1772.37	6.1926	0.32997	1573.24	1771.22	6.1481
160	0.37903	1611.66	1820.13	6.3055	0.34699	1610.92	1819.12	6.2613
180	0.39742	1649.88	1868.46	6.4146	0.36390	1649.22	1867.56	6.3707
200	0.41571	1688.79	1917.43	6.5203	0.38071	1688.20	1916.63	6.4766
	p = 7.0 bar = 0.70 MPa (T_{sat} = 13.79°C)				*p* = 8.0 bar = 0.80 MPa (T_{sat} = 17.84°C)			
Sat.	0.18148	1328.04	1455.07	5.1576	0.15958	1330.64	1458.30	5.1099
20	0.18721	1341.72	1472.77	5.2186	0.16138	1335.59	1464.70	5.1318
30	0.19610	1362.88	1500.15	5.3104	0.16948	1357.71	1493.29	5.2277
40	0.20464	1383.20	1526.45	5.3958	0.17720	1378.77	1520.53	5.3161
50	0.21293	1402.90	1551.95	5.4760	0.18465	1399.05	1546.77	5.3986
60	0.22101	1422.16	1576.87	5.5519	0.19189	1418.77	1572.28	5.4763
80	0.23674	1459.85	1625.56	5.6939	0.20590	1457.14	1621.86	5.6209
100	0.25205	1497.02	1673.46	5.8258	0.21949	1494.77	1670.37	5.7545
120	0.26709	1534.16	1721.12	5.9502	0.23280	1532.24	1718.48	5.8801
140	0.28193	1571.57	1768.92	6.0688	0.24590	1569.89	1766.61	5.9995
160	0.29663	1609.44	1817.08	6.1826	0.25886	1607.96	1815.04	6.1140
180	0.31121	1647.90	1865.75	6.2925	0.27170	1646.57	1863.94	6.2243
200	0.32571	1687.02	1915.01	6.3988	0.28445	1685.83	1913.39	6.3311
	p = 9.0 bar = 0.90 MPa (T_{sat} = 21.52°C)				*p* = 10.0 bar = 1.00 MPa (T_{sat} = 24.89°C)			
Sat.	0.14239	1332.82	1460.97	5.0675	0.12852	1334.66	1463.18	5.0294
30	0.14872	1352.36	1486.20	5.1520	0.13206	1346.82	1478.88	5.0816
40	0.15582	1374.21	1514.45	5.2436	0.13868	1369.52	1508.20	5.1768
50	0.16263	1395.11	1541.47	5.3286	0.14499	1391.07	1536.06	5.2644
60	0.16922	1415.32	1567.61	5.4083	0.15106	1411.79	1562.86	5.3460
80	0.18191	1454.39	1618.11	5.5555	0.16270	1451.60	1614.31	5.4960
100	0.19416	1492.50	1667.24	5.6908	0.17389	1490.20	1664.10	5.6332
120	0.20612	1530.30	1715.81	5.8176	0.18478	1528.35	1713.13	5.7612
140	0.21788	1568.20	1764.29	5.9379	0.19545	1566.51	1761.96	5.8823
160	0.22948	1606.46	1813.00	6.0530	0.20598	1604.97	1810.94	5.9981
180	0.24097	1645.24	1862.12	6.1639	0.21638	1643.91	1860.29	6.1095
200	0.25237	1684.64	1911.77	6.2711	0.22670	1683.44	1910.14	6.2171

| TABLE A-15 | Properties of Superheated Ammonia Vapor (*Continued*) |

T °C	v m³/kg	u kJ/kg	h kJ/kg	s kJ/kg·K	v m³/kg	u kJ/kg	h kJ/kg	s kJ/kg·K
	$p = 12.0$ bar $= 1.20$ MPa ($T_{sat} = 30.94°C$)				$p = 14.0$ bar $= 1.40$ MPa ($T_{sat} = 36.26°C$)			
Sat.	0.10751	1337.52	1466.53	4.9625	0.09231	1339.56	1468.79	4.9050
40	0.11287	1359.73	1495.18	5.0553	0.09432	1349.29	1481.33	4.9453
60	0.12378	1404.54	1553.07	5.2347	0.10423	1396.97	1542.89	5.1360
80	0.13387	1445.91	1606.56	5.3906	0.11324	1440.06	1598.59	5.2984
100	0.14347	1485.55	1657.71	5.5315	0.12172	1480.79	1651.20	5.4433
120	0.15275	1524.41	1707.71	5.6620	0.12986	1520.41	1702.21	5.5765
140	0.16181	1563.09	1757.26	5.7850	0.13777	1559.63	1752.52	5.7013
160	0.17072	1601.95	1806.81	5.9021	0.14552	1598.92	1802.65	5.8198
180	0.17950	1641.23	1856.63	6.0145	0.15315	1638.53	1852.94	5.9333
200	0.18819	1681.05	1906.87	6.1230	0.16068	1678.64	1903.59	6.0427
220	0.19680	1721.50	1957.66	6.2282	0.16813	1719.35	1954.73	6.1485
240	0.20534	1762.63	2009.04	6.3303	0.17551	1760.72	2006.43	6.2513
260	0.21382	1804.48	2061.06	6.4297	0.18283	1802.78	2058.75	6.3513
280	0.22225	1847.04	2113.74	6.5267	0.19010	1845.55	2111.69	6.4488
	$p = 16.0$ bar $= 1.60$ MPa ($T_{sat} = 41.03°C$)				$p = 18.0$ bar $= 1.80$ MPa ($T_{sat} = 45.38°C$)			
Sat.	0.08079	1340.97	1470.23	4.8542	0.07174	1341.88	1471.01	4.8086
60	0.08951	1389.06	1532.28	5.0461	0.07801	1380.77	1521.19	4.9627
80	0.09774	1434.02	1590.40	5.2156	0.08565	1427.79	1581.97	5.1399
100	0.10539	1475.93	1644.56	5.3648	0.09267	1470.97	1637.78	5.2937
120	0.11268	1516.34	1696.64	5.5008	0.09931	1512.22	1690.98	5.4326
140	0.11974	1556.14	1747.72	5.6276	0.10570	1552.61	1742.88	5.5614
160	0.12663	1595.85	1798.45	5.7475	0.11192	1592.76	1794.23	5.6828
180	0.13339	1635.81	1849.23	5.8621	0.11801	1633.08	1845.50	5.7985
200	0.14005	1676.21	1900.29	5.9723	0.12400	1673.78	1896.98	5.9096
220	0.14663	1717.18	1951.79	6.0789	0.12991	1715.00	1948.83	6.0170
240	0.15314	1758.79	2003.81	6.1823	0.13574	1756.85	2001.18	6.1210
260	0.15959	1801.07	2056.42	6.2829	0.14152	1799.35	2054.08	6.2222
280	0.16599	1844.05	2109.64	6.3809	0.14724	1842.55	2107.58	6.3207
	$p = 20.0$ bar $= 2.00$ MPa ($T_{sat} = 49.37°C$)							
Sat.	0.06445	1342.37	1471.26	4.7670				
60	0.06875	1372.05	1509.54	4.8838				
80	0.07596	1421.36	1573.27	5.0696				
100	0.08248	1465.89	1630.86	5.2283				
120	0.08861	1508.03	1685.24	5.3703				
140	0.09447	1549.03	1737.98	5.5012				
160	0.10016	1589.65	1789.97	5.6241				
180	0.10571	1630.32	1841.74	5.7409				
200	0.11116	1671.33	1893.64	5.8530				
220	0.11652	1712.82	1945.87	5.9611				
240	0.12182	1754.90	1998.54	6.0658				
260	0.12706	1797.63	2051.74	6.1675				
280	0.13224	1841.03	2105.50	6.2665				

Pressure Conversions:
1 bar = 0.1 MPa
= 10^2 kPa

Ammonia

TABLE A-16 **Properties of Saturated Propane (Liquid–Vapor): Temperature Table**

Pressure Conversions: 1 bar = 0.1 MPa = 10^2 kPa		Specific Volume m³/kg		Internal Energy kJ/kg		Enthalpy kJ/kg			Entropy kJ/kg · K		
Temp. °C	Press. bar	Sat. Liquid $v_f \times 10^3$	Sat. Vapor v_g	Sat. Liquid u_f	Sat. Vapor u_g	Sat. Liquid h_f	Evap. h_{fg}	Sat. Vapor h_g	Sat. Liquid s_f	Sat. Vapor s_g	Temp. °C
−100	0.02888	1.553	11.27	−128.4	319.5	−128.4	480.4	352.0	−0.634	2.140	−100
−90	0.06426	1.578	5.345	−107.8	329.3	−107.8	471.4	363.6	−0.519	2.055	−90
−80	0.1301	1.605	2.774	−87.0	339.3	−87.0	462.4	375.4	−0.408	1.986	−80
−70	0.2434	1.633	1.551	−65.8	349.5	−65.8	453.1	387.3	−0.301	1.929	−70
−60	0.4261	1.663	0.9234	−44.4	359.9	−44.3	443.5	399.2	−0.198	1.883	−60
−50	0.7046	1.694	0.5793	−22.5	370.4	−22.4	433.6	411.2	−0.098	1.845	−50
−40	1.110	1.728	0.3798	−0.2	381.0	0.0	423.2	423.2	0.000	1.815	−40
−30	1.677	1.763	0.2585	22.6	391.6	22.9	412.1	435.0	0.096	1.791	−30
−20	2.444	1.802	0.1815	45.9	402.4	46.3	400.5	446.8	0.190	1.772	−20
−10	3.451	1.844	0.1309	69.8	413.2	70.4	388.0	458.4	0.282	1.757	−10
0	4.743	1.890	0.09653	94.2	423.8	95.1	374.5	469.6	0.374	1.745	0
4	5.349	1.910	0.08591	104.2	428.1	105.3	368.8	474.1	0.410	1.741	4
8	6.011	1.931	0.07666	114.3	432.3	115.5	362.9	478.4	0.446	1.737	8
12	6.732	1.952	0.06858	124.6	436.5	125.9	356.8	482.7	0.482	1.734	12
16	7.515	1.975	0.06149	135.0	440.7	136.4	350.5	486.9	0.519	1.731	16
20	8.362	1.999	0.05525	145.4	444.8	147.1	343.9	491.0	0.555	1.728	20
24	9.278	2.024	0.04973	156.1	448.9	158.0	337.0	495.0	0.591	1.725	24
28	10.27	2.050	0.04483	166.9	452.9	169.0	329.9	498.9	0.627	1.722	28
32	11.33	2.078	0.04048	177.8	456.7	180.2	322.4	502.6	0.663	1.720	32
36	12.47	2.108	0.03659	188.9	460.6	191.6	314.6	506.2	0.699	1.717	36
40	13.69	2.140	0.03310	200.2	464.3	203.1	306.5	509.6	0.736	1.715	40
44	15.00	2.174	0.02997	211.7	467.9	214.9	298.0	512.9	0.772	1.712	44
48	16.40	2.211	0.02714	223.4	471.4	227.0	288.9	515.9	0.809	1.709	48
52	17.89	2.250	0.02459	235.3	474.6	239.3	279.3	518.6	0.846	1.705	52
56	19.47	2.293	0.02227	247.4	477.7	251.9	269.2	521.1	0.884	1.701	56
60	21.16	2.340	0.02015	259.8	480.6	264.8	258.4	523.2	0.921	1.697	60
65	23.42	2.406	0.01776	275.7	483.6	281.4	243.8	525.2	0.969	1.690	65
70	25.86	2.483	0.01560	292.3	486.1	298.7	227.7	526.4	1.018	1.682	70
75	28.49	2.573	0.01363	309.5	487.8	316.8	209.8	526.6	1.069	1.671	75
80	31.31	2.683	0.01182	327.6	488.2	336.0	189.2	525.2	1.122	1.657	80
85	34.36	2.827	0.01011	347.2	486.9	356.9	164.7	521.6	1.178	1.638	85
90	37.64	3.038	0.008415	369.4	482.2	380.8	133.1	513.9	1.242	1.608	90
95	41.19	3.488	0.006395	399.8	467.4	414.2	79.5	493.7	1.330	1.546	95
96.7	42.48	4.535	0.004535	434.9	434.9	454.2	0.0	454.2	1.437	1.437	96.7

v_f = (table value)/1000

Propane

TABLE A-17   **Properties of Saturated Propane (Liquid–Vapor): Pressure Table**

Press. bar	Temp. °C	Specific Volume m³/kg		Internal Energy kJ/kg		Enthalpy kJ/kg			Entropy kJ/kg · K		Press. bar
Pressure Conversions: 1 bar = 0.1 MPa = 10² kPa		Sat. Liquid $v_f \times 10^3$	Sat. Vapor v_g	Sat. Liquid u_f	Sat. Vapor u_g	Sat. Liquid h_f	Evap. h_{fg}	Sat. Vapor h_g	Sat. Liquid s_f	Sat. Vapor s_g	
0.05	−93.28	1.570	6.752	−114.6	326.0	−114.6	474.4	359.8	−0.556	2.081	0.05
0.10	−83.87	1.594	3.542	−95.1	335.4	−95.1	465.9	370.8	−0.450	2.011	0.10
0.25	−69.55	1.634	1.513	−64.9	350.0	−64.9	452.7	387.8	−0.297	1.927	0.25
0.50	−56.93	1.672	0.7962	−37.7	363.1	−37.6	440.5	402.9	−0.167	1.871	0.50
0.75	−48.68	1.698	0.5467	−19.6	371.8	−19.5	432.3	412.8	−0.085	1.841	0.75
1.00	−42.38	1.719	0.4185	−5.6	378.5	−5.4	425.7	420.3	−0.023	1.822	1.00
2.00	−25.43	1.781	0.2192	33.1	396.6	33.5	406.9	440.4	0.139	1.782	2.00
3.00	−14.16	1.826	0.1496	59.8	408.7	60.3	393.3	453.6	0.244	1.762	3.00
4.00	−5.46	1.865	0.1137	80.8	418.0	81.5	382.0	463.5	0.324	1.751	4.00
5.00	1.74	1.899	0.09172	98.6	425.7	99.5	372.1	471.6	0.389	1.743	5.00
6.00	7.93	1.931	0.07680	114.2	432.2	115.3	363.0	478.3	0.446	1.737	6.00
7.00	13.41	1.960	0.06598	128.2	438.0	129.6	354.6	484.2	0.495	1.733	7.00
8.00	18.33	1.989	0.05776	141.0	443.1	142.6	346.7	489.3	0.540	1.729	8.00
9.00	22.82	2.016	0.05129	152.9	447.6	154.7	339.1	493.8	0.580	1.726	9.00
10.00	26.95	2.043	0.04606	164.0	451.8	166.1	331.8	497.9	0.618	1.723	10.00
11.00	30.80	2.070	0.04174	174.5	455.6	176.8	324.7	501.5	0.652	1.721	11.00
12.00	34.39	2.096	0.03810	184.4	459.1	187.0	317.8	504.8	0.685	1.718	12.00
13.00	37.77	2.122	0.03499	193.9	462.2	196.7	311.0	507.7	0.716	1.716	13.00
14.00	40.97	2.148	0.03231	203.0	465.2	206.0	304.4	510.4	0.745	1.714	14.00
15.00	44.01	2.174	0.02997	211.7	467.9	215.0	297.9	512.9	0.772	1.712	15.00
16.00	46.89	2.200	0.02790	220.1	470.4	223.6	291.4	515.0	0.799	1.710	16.00
17.00	49.65	2.227	0.02606	228.3	472.7	232.0	285.0	517.0	0.824	1.707	17.00
18.00	52.30	2.253	0.02441	236.2	474.9	240.2	278.6	518.8	0.849	1.705	18.00
19.00	54.83	2.280	0.02292	243.8	476.9	248.2	272.2	520.4	0.873	1.703	19.00
20.00	57.27	2.308	0.02157	251.3	478.7	255.9	265.9	521.8	0.896	1.700	20.00
22.00	61.90	2.364	0.01921	265.8	481.7	271.0	253.0	524.0	0.939	1.695	22.00
24.00	66.21	2.424	0.01721	279.7	484.3	285.5	240.1	525.6	0.981	1.688	24.00
26.00	70.27	2.487	0.01549	293.1	486.2	299.6	226.9	526.5	1.021	1.681	26.00
28.00	74.10	2.555	0.01398	306.2	487.5	313.4	213.2	526.6	1.060	1.673	28.00
30.00	77.72	2.630	0.01263	319.2	488.1	327.1	198.9	526.0	1.097	1.664	30.00
35.00	86.01	2.862	0.009771	351.4	486.3	361.4	159.1	520.5	1.190	1.633	35.00
40.00	93.38	3.279	0.007151	387.9	474.7	401.0	102.3	503.3	1.295	1.574	40.00
42.48	96.70	4.535	0.004535	434.9	434.9	454.2	0.0	454.2	1.437	1.437	42.48

v_f = (table value)/1000

Propane

Propane

Pressure Conversions:
1 bar = 0.1 MPa
= 10^2 kPa

TABLE A-18   Properties of Superheated Propane Vapor

T °C	v m³/kg	u kJ/kg	h kJ/kg	s kJ/kg·K	v m³/kg	u kJ/kg	h kJ/kg	s kJ/kg·K
	p = 0.05 bar = 0.005 MPa (T_{sat} = −93.28°C)				p = 0.1 bar = 0.01 MPa (T_{sat} = −83.87°C)			
Sat.	6.752	326.0	359.8	2.081	3.542	367.3	370.8	2.011
−90	6.877	329.4	363.8	2.103				
−80	7.258	339.8	376.1	2.169	3.617	339.5	375.7	2.037
−70	7.639	350.6	388.8	2.233	3.808	350.3	388.4	2.101
−60	8.018	361.8	401.9	2.296	3.999	361.5	401.5	2.164
−50	8.397	373.3	415.3	2.357	4.190	373.1	415.0	2.226
−40	8.776	385.1	429.0	2.418	4.380	385.0	428.8	2.286
−30	9.155	397.4	443.2	2.477	4.570	397.3	443.0	2.346
−20	9.533	410.1	457.8	2.536	4.760	410.0	457.6	2.405
−10	9.911	423.2	472.8	2.594	4.950	423.1	472.6	2.463
0	10.29	436.8	488.2	2.652	5.139	436.7	488.1	2.520
10	10.67	450.8	504.1	2.709	5.329	450.6	503.9	2.578
20	11.05	270.6	520.4	2.765	5.518	465.1	520.3	2.634
	p = 0.5 bar = 0.05 MPa (T_{sat} = −56.93°C)				p = 1.0 bar = 0.1 MPa (T_{sat} = −42.38°C)			
Sat.	0.796	363.1	402.9	1.871	0.4185	378.5	420.3	1.822
−50	0.824	371.3	412.5	1.914				
−40	0.863	383.4	426.6	1.976	0.4234	381.5	423.8	1.837
−30	0.903	396.0	441.1	2.037	0.4439	394.2	438.6	1.899
−20	0.942	408.8	455.9	2.096	0.4641	407.3	453.7	1.960
−10	0.981	422.1	471.1	2.155	0.4842	420.7	469.1	2.019
0	1.019	435.8	486.7	2.213	0.5040	434.4	484.8	2.078
10	1.058	449.8	502.7	2.271	0.5238	448.6	501.0	2.136
20	1.096	464.3	519.1	2.328	0.5434	463.3	517.6	2.194
30	1.135	479.2	535.9	2.384	0.5629	478.2	534.5	2.251
40	1.173	494.6	553.2	2.440	0.5824	493.7	551.9	2.307
50	1.211	510.4	570.9	2.496	0.6018	509.5	569.7	2.363
60	1.249	526.7	589.1	2.551	0.6211	525.8	587.9	2.419
	p = 2.0 bar = 0.2 MPa (T_{sat} = −25.43°C)				p = 3.0 bar = 0.3 MPa (T_{sat} = −14.16°C)			
Sat.	0.2192	396.6	440.4	1.782	0.1496	408.7	453.6	1.762
−20	0.2251	404.0	449.0	1.816				
−10	0.2358	417.7	464.9	1.877	0.1527	414.7	460.5	1.789
0	0.2463	431.8	481.1	1.938	0.1602	429.0	477.1	1.851
10	0.2566	446.3	497.6	1.997	0.1674	443.8	494.0	1.912
20	0.2669	461.1	514.5	2.056	0.1746	458.8	511.2	1.971
30	0.2770	476.3	531.7	2.113	0.1816	474.2	528.7	2.030
40	0.2871	491.9	549.3	2.170	0.1885	490.1	546.6	2.088
50	0.2970	507.9	567.3	2.227	0.1954	506.2	564.8	2.145
60	0.3070	524.3	585.7	2.283	0.2022	522.7	583.4	2.202
70	0.3169	541.1	604.5	2.339	0.2090	539.6	602.3	2.258
80	0.3267	558.4	623.7	2.394	0.2157	557.0	621.7	2.314
90	0.3365	576.1	643.4	2.449	0.2223	574.8	641.5	2.369

TABLE A-18   **Properties of Superheated Propane Vapor (*Continued*)**

T °C	v m³/kg	u kJ/kg	h kJ/kg	s kJ/kg·K	v m³/kg	u kJ/kg	h kJ/kg	s kJ/kg·K
	$p = 4.0$ bar $= 0.4$ MPa ($T_{sat} = -5.46$°C)				$p = 5.0$ bar $= 0.5$ MPa ($T_{sat} = 17.4$°C)			
Sat.	0.1137	418.0	463.5	1.751	0.09172	425.7	471.6	1.743
0	0.1169	426.1	472.9	1.786				
10	0.1227	441.2	490.3	1.848	0.09577	438.4	486.3	1.796
20	0.1283	456.6	507.9	1.909	0.1005	454.1	504.3	1.858
30	0.1338	472.2	525.7	1.969	0.1051	470.0	522.5	1.919
40	0.1392	488.1	543.8	2.027	0.1096	486.1	540.9	1.979
50	0.1445	504.4	562.2	2.085	0.1140	502.5	559.5	2.038
60	0.1498	521.1	581.0	2.143	0.1183	519.4	578.5	2.095
70	0.1550	538.1	600.1	2.199	0.1226	536.6	597.9	2.153
80	0.1601	555.7	619.7	2.255	0.1268	554.1	617.5	2.209
90	0.1652	573.5	639.6	2.311	0.1310	572.1	637.6	2.265
100	0.1703	591.8	659.9	2.366	0.1351	590.5	658.0	2.321
110	0.1754	610.4	680.6	2.421	0.1392	609.3	678.9	2.376
	$p = 6.0$ bar $= 0.6$ MPa ($T_{sat} = 7.93$°C)				$p = 7.0$ bar $= 0.7$ MPa ($T_{sat} = 13.41$°C)			
Sat.	0.07680	432.2	478.3	1.737	0.06598	438.0	484.2	1.733
10	0.07769	435.6	482.2	1.751				
20	0.08187	451.5	500.6	1.815	0.06847	448.8	496.7	1.776
30	0.08588	467.7	519.2	1.877	0.07210	465.2	515.7	1.840
40	0.08978	484.0	537.9	1.938	0.07558	481.9	534.8	1.901
50	0.09357	500.7	556.8	1.997	0.07896	498.7	554.0	1.962
60	0.09729	517.6	576.0	2.056	0.08225	515.9	573.5	2.021
70	0.1009	535.0	595.5	2.113	0.08547	533.4	593.2	2.079
80	0.1045	552.7	615.4	2.170	0.08863	551.2	613.2	2.137
90	0.1081	570.7	635.6	2.227	0.09175	569.4	633.6	2.194
100	0.1116	589.2	656.2	2.283	0.09482	587.9	654.3	2.250
110	0.1151	608.0	677.1	2.338	0.09786	606.8	675.3	2.306
120	0.1185	627.3	698.4	2.393	0.1009	626.2	696.8	2.361
	$p = 8.0$ bar $= 0.8$ MPa ($T_{sat} = 18.33$°C)				$p = 9.0$ bar $= 0.9$ MPa ($T_{sat} = 22.82$°C)			
Sat.	0.05776	443.1	489.3	1.729	0.05129	447.2	493.8	1.726
20	0.05834	445.9	492.6	1.740				
30	0.06170	462.7	512.1	1.806	0.05355	460.0	508.2	1.774
40	0.06489	479.6	531.5	1.869	0.05653	477.2	528.1	1.839
50	0.06796	496.7	551.1	1.930	0.05938	494.7	548.1	1.901
60	0.07094	514.0	570.8	1.990	0.06213	512.2	568.1	1.962
70	0.07385	531.6	590.7	2.049	0.06479	530.0	588.3	2.022
80	0.07669	549.6	611.0	2.107	0.06738	548.1	608.7	2.081
90	0.07948	567.9	631.5	2.165	0.06992	566.5	629.4	2.138
100	0.08222	586.5	652.3	2.221	0.07241	585.2	650.4	2.195
110	0.08493	605.6	673.5	2.277	0.07487	604.3	671.7	2.252
120	0.08761	625.0	695.1	2.333	0.07729	623.7	693.3	2.307
130	0.09026	644.8	717.0	2.388	0.07969	643.6	715.3	2.363
140	0.09289	665.0	739.3	2.442	0.08206	663.8	737.7	2.418

Pressure Conversions:
1 bar $= 0.1$ MPa
$= 10^2$ kPa

Propane

Propane

Pressure Conversions:
1 bar = 0.1 MPa
= 10² kPa

TABLE A-18 **Properties of Superheated Propane Vapor (Continued)**

T °C	υ m³/kg	u kJ/kg	h kJ/kg	s kJ/kg·K	υ m³/kg	u kJ/kg	h kJ/kg	s kJ/kg·K
	$p = 10.0$ bar = 1.0 MPa ($T_{sat} = 26.95°C$)				$p = 12.0$ bar = 1.2 MPa ($T_{sat} = 34.39°C$)			
Sat.	0.04606	451.8	497.9	1.723	0.03810	459.1	504.8	1.718
30	0.04696	457.1	504.1	1.744				
40	0.04980	474.8	524.6	1.810	0.03957	469.4	516.9	1.757
50	0.05248	492.4	544.9	1.874	0.04204	487.8	538.2	1.824
60	0.05505	510.2	565.2	1.936	0.04436	506.1	559.3	1.889
70	0.05752	528.2	585.7	1.997	0.04657	524.4	580.3	1.951
80	0.05992	546.4	606.3	2.056	0.04869	543.1	601.5	2.012
90	0.06226	564.9	627.2	2.114	0.05075	561.8	622.7	2.071
100	0.06456	583.7	648.3	2.172	0.05275	580.9	644.2	2.129
110	0.06681	603.0	669.8	2.228	0.05470	600.4	666.0	2.187
120	0.06903	622.6	691.6	2.284	0.05662	620.1	688.0	2.244
130	0.07122	642.5	713.7	2.340	0.05851	640.1	710.3	2.300
140	0.07338	662.8	736.2	2.395	0.06037	660.6	733.0	2.355
	$p = 14.0$ bar = 1.4 MPa ($T_{sat} = 40.97°C$)				$p = 16.0$ bar = 1.6 MPa ($T_{sat} = 46.89°C$)			
Sat.	0.03231	465.2	510.4	1.714	0.02790	470.4	515.0	1.710
50	0.03446	482.6	530.8	1.778	0.02861	476.7	522.5	1.733
60	0.03664	501.6	552.9	1.845	0.03075	496.6	545.8	1.804
70	0.03869	520.4	574.6	1.909	0.03270	516.2	568.5	1.871
80	0.04063	539.4	596.3	1.972	0.03453	535.7	590.9	1.935
90	0.04249	558.6	618.1	2.033	0.03626	555.2	613.2	1.997
100	0.04429	577.9	639.9	2.092	0.03792	574.8	635.5	2.058
110	0.04604	597.5	662.0	2.150	0.03952	594.7	657.9	2.117
120	0.04774	617.5	684.3	2.208	0.04107	614.8	680.5	2.176
130	0.04942	637.7	706.9	2.265	0.04259	635.3	703.4	2.233
140	0.05106	658.3	729.8	2.321	0.04407	656.0	726.5	2.290
150	0.05268	679.2	753.0	2.376	0.04553	677.1	749.9	2.346
160	0.05428	700.5	776.5	2.431	0.04696	698.5	773.6	2.401
	$p = 18.0$ bar = 1.8 MPa ($T_{sat} = 52.30°C$)				$p = 20.0$ bar = 2.0 MPa ($T_{sat} = 57.27°C$)			
Sat.	0.02441	474.9	518.8	1.705	0.02157	478.7	521.8	1.700
60	0.02606	491.1	538.0	1.763	0.02216	484.8	529.1	1.722
70	0.02798	511.4	561.8	1.834	0.02412	506.3	554.5	1.797
80	0.02974	531.6	585.1	1.901	0.02585	527.1	578.8	1.867
90	0.03138	551.5	608.0	1.965	0.02744	547.6	602.5	1.933
100	0.03293	571.5	630.8	2.027	0.02892	568.1	625.9	1.997
110	0.03443	591.7	653.7	2.087	0.03033	588.5	649.2	2.059
120	0.03586	612.1	676.6	2.146	0.03169	609.2	672.6	2.119
130	0.03726	632.7	699.8	2.204	0.03299	630.0	696.0	2.178
140	0.03863	653.6	723.1	2.262	0.03426	651.2	719.7	2.236
150	0.03996	674.8	746.7	2.318	0.03550	672.5	743.5	2.293
160	0.04127	696.3	770.6	2.374	0.03671	694.2	767.6	2.349
170	0.04256	718.2	794.8	2.429	0.03790	716.2	792.0	2.404
180	0.04383	740.4	819.3	2.484	0.03907	738.5	816.6	2.459

TABLE A-18   **Properties of Superheated Propane Vapor** (*Continued*)

T °C	v m³/kg	u kJ/kg	h kJ/kg	s kJ/kg·K	v m³/kg	u kJ/kg	h kJ/kg	s kJ/kg·K
	$p = 22.0$ bar $= 2.2$ MPa ($T_{sat} = 61.90°C$)				$p = 24.0$ bar $= 2.4$ MPa ($T_{sat} = 66.21°C$)			
Sat.	0.01921	481.8	524.0	1.695	0.01721	484.3	525.6	1.688
70	0.02086	500.5	546.4	1.761	0.01802	493.7	536.9	1.722
80	0.02261	522.4	572.1	1.834	0.01984	517.0	564.6	1.801
90	0.02417	543.5	596.7	1.903	0.02141	539.0	590.4	1.873
100	0.02561	564.5	620.8	1.969	0.02283	560.6	615.4	1.941
110	0.02697	585.3	644.6	2.032	0.02414	581.9	639.8	2.006
120	0.02826	606.2	668.4	2.093	0.02538	603.2	664.1	2.068
130	0.02949	627.3	692.2	2.153	0.02656	624.6	688.3	2.129
140	0.03069	648.6	716.1	2.211	0.02770	646.0	712.5	2.188
150	0.03185	670.1	740.2	2.269	0.02880	667.8	736.9	2.247
160	0.03298	691.9	764.5	2.326	0.02986	689.7	761.4	2.304
170	0.03409	714.1	789.1	2.382	0.03091	711.9	786.1	2.360
180	0.03517	736.5	813.9	2.437	0.03193	734.5	811.1	2.416
	$p = 26.0$ bar $= 2.6$ MPa ($T_{sat} = 70.27°C$)				$p = 30.0$ bar $= 3.0$ MPa ($T_{sat} = 77.72°C$)			
Sat.	0.01549	486.2	526.5	1.681	0.01263	488.2	526.0	1.664
80	0.01742	511.0	556.3	1.767	0.01318	495.4	534.9	1.689
90	0.01903	534.2	583.7	1.844	0.01506	522.8	568.0	1.782
100	0.02045	556.4	609.6	1.914	0.01654	547.2	596.8	1.860
110	0.02174	578.3	634.8	1.981	0.01783	570.4	623.9	1.932
120	0.02294	600.0	659.6	2.045	0.01899	593.0	650.0	1.999
130	0.02408	621.6	684.2	2.106	0.02007	615.4	675.6	2.063
140	0.02516	643.4	708.8	2.167	0.02109	637.7	701.0	2.126
150	0.02621	665.3	733.4	2.226	0.02206	660.1	726.3	2.186
160	0.02723	687.4	758.2	2.283	0.02300	682.6	751.6	2.245
170	0.02821	709.9	783.2	2.340	0.02390	705.4	777.1	2.303
180	0.02918	732.5	808.4	2.397	0.02478	728.3	802.6	2.360
190	0.03012	755.5	833.8	2.452	0.02563	751.5	828.4	2.417
	$p = 35.0$ bar $= 3.5$ MPa ($T_{sat} = 86.01°C$)				$p = 40.0$ bar $= 4.0$ MPa ($T_{sat} = 93.38°C$)			
Sat.	0.00977	486.3	520.5	1.633	0.00715	474.7	503.3	1.574
90	0.01086	502.4	540.5	1.688				
100	0.01270	532.9	577.3	1.788	0.00940	512.1	549.7	1.700
110	0.01408	558.9	608.2	1.870	0.01110	544.7	589.1	1.804
120	0.01526	583.4	636.8	1.944	0.01237	572.1	621.6	1.887
130	0.01631	607.0	664.1	2.012	0.01344	597.4	651.2	1.962
140	0.01728	630.2	690.7	2.077	0.01439	621.9	679.5	2.031
150	0.01819	653.3	717.0	2.140	0.01527	645.9	707.0	2.097
160	0.01906	676.4	743.1	2.201	0.01609	669.7	734.1	2.160
170	0.01989	699.6	769.2	2.261	0.01687	693.4	760.9	2.222
180	0.02068	722.9	795.3	2.319	0.01761	717.3	787.7	2.281
190	0.02146	746.5	821.6	2.376	0.01833	741.2	814.5	2.340
200	0.02221	770.3	848.0	2.433	0.01902	765.3	841.4	2.397

Pressure Conversions:
1 bar $= 0.1$ MPa
$= 10^2$ kPa

Propane

TABLE A-19   **Properties of Selected Solids and Liquids: c_p, ρ, and κ**

Substance	Specific Heat, c_p (kJ/kg $\cdot$ K)	Density, ρ (kg/m³)	Thermal Conductivity, κ (W/m $\cdot$ K)
Selected Solids, **300K**			
Aluminum	0.903	2700	237
Coal, anthracite	1.260	1350	0.26
Copper	0.385	8930	401
Granite	0.775	2630	2.79
Iron	0.447	7870	80.2
Lead	0.129	11300	35.3
Sand	0.800	1520	0.27
Silver	0.235	10500	429
Soil	1.840	2050	0.52
Steel (AISI 302)	0.480	8060	15.1
Tin	0.227	7310	66.6
Building Materials, **300K**			
Brick, common	0.835	1920	0.72
Concrete (stone mix)	0.880	2300	1.4
Glass, plate	0.750	2500	1.4
Hardboard, siding	1.170	640	0.094
Limestone	0.810	2320	2.15
Plywood	1.220	545	0.12
Softwoods (fir, pine)	1.380	510	0.12
Insulating Materials, **300K**			
Blanket (glass fiber)	—	16	0.046
Cork	1.800	120	0.039
Duct liner (glass fiber, coated)	0.835	32	0.038
Polystyrene (extruded)	1.210	55	0.027
Vermiculite fill (flakes)	0.835	80	0.068
Saturated Liquids			
Ammonia, 300K	4.818	599.8	0.465
Mercury, 300K	0.139	13529	8.540
Refrigerant 22, 300K	1.267	1183.1	0.085
Refrigerant 134a, 300K	1.434	1199.7	0.081
Unused Engine Oil, 300K	1.909	884.1	0.145
Water, 275K	4.211	999.9	0.574
300K	4.179	996.5	0.613
325K	4.182	987.1	0.645
350K	4.195	973.5	0.668
375K	4.220	956.8	0.681
400K	4.256	937.4	0.688

Sources: Drawn from several sources, these data are only representative. Values can vary depending on temperature, purity, moisture content, and other factors.

TABLE A-20 **Ideal Gas Specific Heats of Some Common Gases (kJ/kg · K)**

Temp. K	c_p	c_v	k	c_p	c_v	k	c_p	c_v	k	Temp. K
	Air			Nitrogen, N_2			Oxygen, O_2			
250	1.003	0.716	1.401	1.039	0.742	1.400	0.913	0.653	1.398	250
300	1.005	0.718	1.400	1.039	0.743	1.400	0.918	0.658	1.395	300
350	1.008	0.721	1.398	1.041	0.744	1.399	0.928	0.668	1.389	350
400	1.013	0.726	1.395	1.044	0.747	1.397	0.941	0.681	1.382	400
450	1.020	0.733	1.391	1.049	0.752	1.395	0.956	0.696	1.373	450
500	1.029	0.742	1.387	1.056	0.759	1.391	0.972	0.712	1.365	500
550	1.040	0.753	1.381	1.065	0.768	1.387	0.988	0.728	1.358	550
600	1.051	0.764	1.376	1.075	0.778	1.382	1.003	0.743	1.350	600
650	1.063	0.776	1.370	1.086	0.789	1.376	1.017	0.758	1.343	650
700	1.075	0.788	1.364	1.098	0.801	1.371	1.031	0.771	1.337	700
750	1.087	0.800	1.359	1.110	0.813	1.365	1.043	0.783	1.332	750
800	1.099	0.812	1.354	1.121	0.825	1.360	1.054	0.794	1.327	800
900	1.121	0.834	1.344	1.145	0.849	1.349	1.074	0.814	1.319	900
1000	1.142	0.855	1.336	1.167	0.870	1.341	1.090	0.830	1.313	1000
Temp. K	Carbon Dioxide, CO_2			Carbon Monoxide, CO			Hydrogen, H_2			Temp. K
250	0.791	0.602	1.314	1.039	0.743	1.400	14.051	9.927	1.416	250
300	0.846	0.657	1.288	1.040	0.744	1.399	14.307	10.183	1.405	300
350	0.895	0.706	1.268	1.043	0.746	1.398	14.427	10.302	1.400	350
400	0.939	0.750	1.252	1.047	0.751	1.395	14.476	10.352	1.398	400
450	0.978	0.790	1.239	1.054	0.757	1.392	14.501	10.377	1.398	450
500	1.014	0.825	1.229	1.063	0.767	1.387	14.513	10.389	1.397	500
550	1.046	0.857	1.220	1.075	0.778	1.382	14.530	10.405	1.396	550
600	1.075	0.886	1.213	1.087	0.790	1.376	14.546	10.422	1.396	600
650	1.102	0.913	1.207	1.100	0.803	1.370	14.571	10.447	1.395	650
700	1.126	0.937	1.202	1.113	0.816	1.364	14.604	10.480	1.394	700
750	1.148	0.959	1.197	1.126	0.829	1.358	14.645	10.521	1.392	750
800	1.169	0.980	1.193	1.139	0.842	1.353	14.695	10.570	1.390	800
900	1.204	1.015	1.186	1.163	0.866	1.343	14.822	10.698	1.385	900
1000	1.234	1.045	1.181	1.185	0.888	1.335	14.983	10.859	1.380	1000

Source: Adapted from K. Wark, *Thermodynamics*, 4th ed., McGraw-Hill, New York, 1983, as based on "Tables of Thermal Properties of Gases," NBS Circular 564, 1955.

Table A-20

TABLE A-21   **Variation of $\bar{c}_p$ with Temperature for Selected Ideal Gases**

$$\frac{\bar{c}_p}{R} = \alpha + \beta T + \gamma T^2 + \delta T^3 + \varepsilon T^4$$

T is in K, equations valid from 300 to 1000 K

Gas	α	$\beta \times 10^3$	$\gamma \times 10^6$	$\delta \times 10^9$	$\varepsilon \times 10^{12}$
CO	3.710	−1.619	3.692	−2.032	0.240
CO_2	2.401	8.735	−6.607	2.002	0
H_2	3.057	2.677	−5.810	5.521	−1.812
H_2O	4.070	−1.108	4.152	−2.964	0.807
O_2	3.626	−1.878	7.055	−6.764	2.156
N_2	3.675	−1.208	2.324	−0.632	−0.226
Air	3.653	−1.337	3.294	−1.913	0.2763
SO_2	3.267	5.324	0.684	−5.281	2.559
CH_4	3.826	−3.979	24.558	−22.733	6.963
C_2H_2	1.410	19.057	−24.501	16.391	−4.135
C_2H_4	1.426	11.383	7.989	−16.254	6.749
Monatomic gases[a]	2.5	0	0	0	0

[a]For monatomic gases, such as He, Ne, and Ar, $\bar{c}_p$ is constant over a wide temperature range and is very nearly equal to $5/2\,\bar{R}$.

Source: Adapted from K. Wark, *Thermodynamics*, 4th ed., McGraw-Hill, New York, 1983, as based on NASA SP-273, U.S. Government Printing Office, Washington, DC, 1971.

TABLE A-22 **Ideal Gas Properties of Air**

				T(K), h and u (kJ/kg), $s°$ (kJ/kg · K)							
				when $\Delta s = 0$[1]						when $\Delta s = 0$	
T	h	u	$s°$	p_r	v_r	T	h	u	$s°$	p_r	v_r
200	199.97	142.56	1.29559	0.3363	1707.	450	451.80	322.62	2.11161	5.775	223.6
210	209.97	149.69	1.34444	0.3987	1512.	460	462.02	329.97	2.13407	6.245	211.4
220	219.97	156.82	1.39105	0.4690	1346.	470	472.24	337.32	2.15604	6.742	200.1
230	230.02	164.00	1.43557	0.5477	1205.	480	482.49	344.70	2.17760	7.268	189.5
240	240.02	171.13	1.47824	0.6355	1084.	490	492.74	352.08	2.19876	7.824	179.7
250	250.05	178.28	1.51917	0.7329	979.	500	503.02	359.49	2.21952	8.411	170.6
260	260.09	185.45	1.55848	0.8405	887.8	510	513.32	366.92	2.23993	9.031	162.1
270	270.11	192.60	1.59634	0.9590	808.0	520	523.63	374.36	2.25997	9.684	154.1
280	280.13	199.75	1.63279	1.0889	738.0	530	533.98	381.84	2.27967	10.37	146.7
285	285.14	203.33	1.65055	1.1584	706.1	540	544.35	389.34	2.29906	11.10	139.7
290	290.16	206.91	1.66802	1.2311	676.1	550	554.74	396.86	2.31809	11.86	133.1
295	295.17	210.49	1.68515	1.3068	647.9	560	565.17	404.42	2.33685	12.66	127.0
300	300.19	214.07	1.70203	1.3860	621.2	570	575.59	411.97	2.35531	13.50	121.2
305	305.22	217.67	1.71865	1.4686	596.0	580	586.04	419.55	2.37348	14.38	115.7
310	310.24	221.25	1.73498	1.5546	572.3	590	596.52	427.15	2.39140	15.31	110.6
315	315.27	224.85	1.75106	1.6442	549.8	600	607.02	434.78	2.40902	16.28	105.8
320	320.29	228.42	1.76690	1.7375	528.6	610	617.53	442.42	2.42644	17.30	101.2
325	325.31	232.02	1.78249	1.8345	508.4	620	628.07	450.09	2.44356	18.36	96.92
330	330.34	235.61	1.79783	1.9352	489.4	630	638.63	457.78	2.46048	19.84	92.84
340	340.42	242.82	1.82790	2.149	454.1	640	649.22	465.50	2.47716	20.64	88.99
350	350.49	250.02	1.85708	2.379	422.2	650	659.84	473.25	2.49364	21.86	85.34
360	360.58	257.24	1.88543	2.626	393.4	660	670.47	481.01	2.50985	23.13	81.89
370	370.67	264.46	1.91313	2.892	367.2	670	681.14	488.81	2.52589	24.46	78.61
380	380.77	271.69	1.94001	3.176	343.4	680	691.82	496.62	2.54175	25.85	75.50
390	390.88	278.93	1.96633	3.481	321.5	690	702.52	504.45	2.55731	27.29	72.56
400	400.98	286.16	1.99194	3.806	301.6	700	713.27	512.33	2.57277	28.80	69.76
410	411.12	293.43	2.01699	4.153	283.3	710	724.04	520.23	2.58810	30.38	67.07
420	421.26	300.69	2.04142	4.522	266.6	720	734.82	528.14	2.60319	32.02	64.53
430	431.43	307.99	2.06533	4.915	251.1	730	745.62	536.07	2.61803	33.72	62.13
440	441.61	315.30	2.08870	5.332	236.8	740	756.44	544.02	2.63280	35.50	59.82

1. p_r and v_r data for use with Eqs. 6.41 and 6.42, respectively.

Table A-22

TABLE A-22 Ideal Gas Properties of Air (*Continued*)

				$T(\text{K})$, h and u (kJ/kg), s° (kJ/kg $\cdot$ K)							
				when $\Delta s = 0$[1]						when $\Delta s = 0$	
T	h	u	s°	p_r	v_r	T	h	u	s°	p_r	v_r
750	767.29	551.99	2.64737	37.35	57.63	1300	1395.97	1022.82	3.27345	330.9	11.275
760	778.18	560.01	2.66176	39.27	55.54	1320	1419.76	1040.88	3.29160	352.5	10.747
770	789.11	568.07	2.67595	41.31	53.39	1340	1443.60	1058.94	3.30959	375.3	10.247
780	800.03	576.12	2.69013	43.35	51.64	1360	1467.49	1077.10	3.32724	399.1	9.780
790	810.99	584.21	2.70400	45.55	49.86	1380	1491.44	1095.26	3.34474	424.2	9.337
800	821.95	592.30	2.71787	47.75	48.08	1400	1515.42	1113.52	3.36200	450.5	8.919
820	843.98	608.59	2.74504	52.59	44.84	1420	1539.44	1131.77	3.37901	478.0	8.526
840	866.08	624.95	2.77170	57.60	41.85	1440	1563.51	1150.13	3.39586	506.9	8.153
860	888.27	641.40	2.79783	63.09	39.12	1460	1587.63	1168.49	3.41247	537.1	7.801
880	910.56	657.95	2.82344	68.98	36.61	1480	1611.79	1186.95	3.42892	568.8	7.468
900	932.93	674.58	2.84856	75.29	34.31	1500	1635.97	1205.41	3.44516	601.9	7.152
920	955.38	691.28	2.87324	82.05	32.18	1520	1660.23	1223.87	3.46120	636.5	6.854
940	977.92	708.08	2.89748	89.28	30.22	1540	1684.51	1242.43	3.47712	672.8	6.569
960	1000.55	725.02	2.92128	97.00	28.40	1560	1708.82	1260.99	3.49276	710.5	6.301
980	1023.25	741.98	2.94468	105.2	26.73	1580	1733.17	1279.65	3.50829	750.0	6.046
1000	1046.04	758.94	2.96770	114.0	25.17	1600	1757.57	1298.30	3.52364	791.2	5.804
1020	1068.89	776.10	2.99034	123.4	23.72	1620	1782.00	1316.96	3.53879	834.1	5.574
1040	1091.85	793.36	3.01260	133.3	22.39	1640	1806.46	1335.72	3.55381	878.9	5.355
1060	1114.86	810.62	3.03449	143.9	21.14	1660	1830.96	1354.48	3.56867	925.6	5.147
1080	1137.89	827.88	3.05608	155.2	19.98	1680	1855.50	1373.24	3.58335	974.2	4.949
1100	1161.07	845.33	3.07732	167.1	18.896	1700	1880.1	1392.7	3.5979	1025	4.761
1120	1184.28	862.79	3.09825	179.7	17.886	1750	1941.6	1439.8	3.6336	1161	4.328
1140	1207.57	880.35	3.11883	193.1	16.946	1800	2003.3	1487.2	3.6684	1310	3.944
1160	1230.92	897.91	3.13916	207.2	16.064	1850	2065.3	1534.9	3.7023	1475	3.601
1180	1254.34	915.57	3.15916	222.2	15.241	1900	2127.4	1582.6	3.7354	1655	3.295
1200	1277.79	933.33	3.17888	238.0	14.470	1950	2189.7	1630.6	3.7677	1852	3.022
1220	1301.31	951.09	3.19834	254.7	13.747	2000	2252.1	1678.7	3.7994	2068	2.776
1240	1324.93	968.95	3.21751	272.3	13.069	2050	2314.6	1726.8	3.8303	2303	2.555
1260	1348.55	986.90	3.23638	290.8	12.435	2100	2377.4	1775.3	3.8605	2559	2.356
1280	1372.24	1004.76	3.25510	310.4	11.835	2150	2440.3	1823.8	3.8901	2837	2.175
						2200	2503.2	1872.4	3.9191	3138	2.012
						2250	2566.4	1921.3	3.9474	3464	1.864

Source: Table A-22 is based on J. H. Keenan and J. Kaye, *Gas Tables*, Wiley, New York, 1945.

TABLE A-23   Ideal Gas Properties of Selected Gases

Enthalpy $\bar{h}(T)$ and internal energy $\bar{u}(T)$, in kJ/kmol. Absolute entropy at 1 atm $\bar{s}°(T)$, in kJ/kmol · K.

T(K)	Carbon Dioxide, CO_2 ($\bar{h}_f° = -393{,}520$ kJ/kmol)			Carbon Monoxide, CO ($\bar{h}_f° = -110{,}530$ kJ/kmol)			Water Vapor, H_2O ($\bar{h}_f° = -241{,}820$ kJ/kmol)			Oxygen, O_2 ($\bar{h}_f° = 0$ kJ/kmol)			Nitrogen, N_2 ($\bar{h}_f° = 0$ kJ/kmol)			T(K)
	$\bar{h}$	$\bar{u}$	$\bar{s}°$	$\bar{h}$	$\bar{u}$	$\bar{s}°$	$\bar{h}$	$\bar{u}$	$\bar{s}°$	$\bar{h}$	$\bar{u}$	$\bar{s}°$	$\bar{h}$	$\bar{u}$	$\bar{s}°$	
0	0	0	0	0	0	0	0	0	0	0	0	0	0	0	0	0
220	6,601	4,772	202.966	6,391	4,562	188.683	7,295	5,466	178.576	6,404	4,575	196.171	6,391	4,562	182.638	220
230	6,938	5,026	204.464	6,683	4,771	189.980	7,628	5,715	180.054	6,694	4,782	197.461	6,683	4,770	183.938	230
240	7,280	5,285	205.920	6,975	4,979	191.221	7,961	5,965	181.471	6,984	4,989	198.696	6,975	4,979	185.180	240
250	7,627	5,548	207.337	7,266	5,188	192.411	8,294	6,215	182.831	7,275	5,197	199.885	7,266	5,188	186.370	250
260	7,979	5,817	208.717	7,558	5,396	193.554	8,627	6,466	184.139	7,566	5,405	201.027	7,558	5,396	187.514	260
270	8,335	6,091	210.062	7,849	5,604	194.654	8,961	6,716	185.399	7,858	5,613	202.128	7,849	5,604	188.614	270
280	8,697	6,369	211.376	8,140	5,812	195.173	9,296	6,968	186.616	8,150	5,822	203.191	8,141	5,813	189.673	280
290	9,063	6,651	212.660	8,432	6,020	196.735	9,631	7,219	187.791	8,443	6,032	204.218	8,432	6,021	190.695	290
298	9,364	6,885	213.685	8,669	6,190	197.543	9,904	7,425	188.720	8,682	6,203	205.033	8,669	6,190	191.502	298
300	9,431	6,939	213.915	8,723	6,229	197.723	9,966	7,472	188.928	8,736	6,242	205.213	8,723	6,229	191.682	300
310	9,807	7,230	215.146	9,014	6,437	198.678	10,302	7,725	190.030	9,030	6,453	206.177	9,014	6,437	192.638	310
320	10,186	7,526	216.351	9,306	6,645	199.603	10,639	7,978	191.098	9,325	6,664	207.112	9,306	6,645	193.562	320
330	10,570	7,826	217.534	9,597	6,854	200.500	10,976	8,232	192.136	9,620	6,877	208.020	9,597	6,853	194.459	330
340	10,959	8,131	218.694	9,889	7,062	201.371	11,314	8,487	193.144	9,916	7,090	208.904	9,888	7,061	195.328	340
350	11,351	8,439	219.831	10,181	7,271	202.217	11,652	8,742	194.125	10,213	7,303	209.765	10,180	7,270	196.173	350
360	11,748	8,752	220.948	10,473	7,480	203.040	11,992	8,998	195.081	10,511	7,518	210.604	10,471	7,478	196.995	360
370	12,148	9,068	222.044	10,765	7,689	203.842	12,331	9,255	196.012	10,809	7,733	211.423	10,763	7,687	197.794	370
380	12,552	9,392	223.122	11,058	7,899	204.622	12,672	9,513	196.920	11,109	7,949	212.222	11,055	7,895	198.572	380
390	12,960	9,718	224.182	11,351	8,108	205.383	13,014	9,771	197.807	11,409	8,166	213.002	11,347	8,104	199.331	390
400	13,372	10,046	225.225	11,644	8,319	206.125	13,356	10,030	198.673	11,711	8,384	213.765	11,640	8,314	200.071	400
410	13,787	10,378	226.250	11,938	8,529	206.850	13,699	10,290	199.521	12,012	8,603	214.510	11,932	8,523	200.794	410
420	14,206	10,714	227.258	12,232	8,740	207.549	14,043	10,551	200.350	12,314	8,822	215.241	12,225	8,733	201.499	420
430	14,628	11,053	228.252	12,526	8,951	208.252	14,388	10,813	201.160	12,618	9,043	215.955	12,518	8,943	202.189	430
440	15,054	11,393	229.230	12,821	9,163	208.929	14,734	11,075	201.955	12,923	9,264	216.656	12,811	9,153	202.863	440
450	15,483	11,742	230.194	13,116	9,375	209.593	15,080	11,339	202.734	13,228	9,487	217.342	13,105	9,363	203.523	450
460	15,916	12,091	231.144	13,412	9,587	210.243	15,428	11,603	203.497	13,535	9,710	218.016	13,399	9,574	204.170	460
470	16,351	12,444	232.080	13,708	9,800	210.880	15,777	11,869	204.247	13,842	9,935	218.676	13,693	9,786	204.803	470
480	16,791	12,800	233.004	14,005	10,014	211.504	16,126	12,135	204.982	14,151	10,160	219.326	13,988	9,997	205.424	480
490	17,232	13,158	233.916	14,302	10,228	212.117	16,477	12,403	205.705	14,460	10,386	219.963	14,285	10,210	206.033	490
500	17,678	13,521	234.814	14,600	10,443	212.719	16,828	12,671	206.413	14,770	10,614	220.589	14,581	10,423	206.630	500
510	18,126	13,885	235.700	14,898	10,658	213.310	17,181	12,940	207.112	15,082	10,842	221.206	14,876	10,635	207.216	510
520	18,576	14,253	236.575	15,197	10,874	213.890	17,534	13,211	207.799	15,395	11,071	221.812	15,172	10,848	207.792	520
530	19,029	14,622	237.439	15,497	11,090	214.460	17,889	13,482	208.475	15,708	11,301	222.409	15,469	11,062	208.358	530
540	19,485	14,996	238.292	15,797	11,307	215.020	18,245	13,755	209.139	16,022	11,533	222.997	15,766	11,277	208.914	540
550	19,945	15,372	239.135	16,097	11,524	215.572	18,601	14,028	209.795	16,338	11,765	223.576	16,064	11,492	209.461	550
560	20,407	15,751	239.962	16,399	11,743	216.115	18,959	14,303	210.440	16,654	11,998	224.146	16,363	11,707	209.999	560
570	20,870	16,131	240.789	16,701	11,961	216.649	19,318	14,579	211.075	16,971	12,232	224.708	16,662	11,923	210.528	570
580	21,337	16,515	241.602	17,003	12,181	217.175	19,678	14,856	211.702	17,290	12,467	225.262	16,962	12,139	211.049	580
590	21,807	16,902	242.405	17,307	12,401	217.693	20,039	15,134	212.320	17,609	12,703	225.808	17,262	12,356	211.562	590

Table A-23

Table A-23

TABLE A-23   **Ideal Gas Properties of Selected Gases (Continued)**

$\bar{h}$ and $\bar{u}$ in kJ/kmol. $\bar{s}°$ in kJ/kmol · K

T(K)	Carbon Dioxide, CO_2 ($\bar{h}_f° = -393{,}520$ kJ/kmol) $\bar{h}$	$\bar{u}$	$\bar{s}°$	Carbon Monoxide, CO ($\bar{h}_f° = -110{,}530$ kJ/kmol) $\bar{h}$	$\bar{u}$	$\bar{s}°$	Water Vapor, H_2O ($\bar{h}_f° = -241{,}820$ kJ/kmol) $\bar{h}$	$\bar{u}$	$\bar{s}°$	Oxygen, O_2 ($\bar{h}_f° = 0$ kJ/kmol) $\bar{h}$	$\bar{u}$	$\bar{s}°$	Nitrogen, N_2 ($\bar{h}_f° = 0$ kJ/kmol) $\bar{h}$	$\bar{u}$	$\bar{s}°$	T(K)
600	22,280	17,291	243.199	17,611	12,622	218.204	20,402	15,413	212.920	17,929	12,940	226.346	17,563	12,574	212.066	600
610	22,754	17,683	243.983	17,915	12,843	218.708	20,765	15,693	213.529	18,250	13,178	226.877	17,864	12,792	212.564	610
620	23,231	18,076	244.758	18,221	13,066	219.205	21,130	15,975	214.122	18,572	13,417	227.400	18,166	13,011	213.055	620
630	23,709	18,471	245.524	18,527	13,289	219.695	21,495	16,257	214.707	18,895	13,657	227.918	18,468	13,230	213.541	630
640	24,190	18,869	246.282	18,833	13,512	220.179	21,862	16,541	215.285	19,219	13,898	228.429	18,772	13,450	214.018	640
650	24,674	19,270	247.032	19,141	13,736	220.656	22,230	16,826	215.856	19,544	14,140	228.932	19,075	13,671	214.489	650
660	25,160	19,672	247.773	19,449	13,962	221.127	22,600	17,112	216.419	19,870	14,383	229.430	19,380	13,892	214.954	660
670	25,648	20,078	248.507	19,758	14,187	221.592	22,970	17,399	216.976	20,197	14,626	229.920	19,685	14,114	215.413	670
680	26,138	20,484	249.233	20,068	14,414	222.052	23,342	17,688	217.527	20,524	14,871	230.405	19,991	14,337	215.866	680
690	26,631	20,894	249.952	20,378	14,641	222.505	23,714	17,978	218.071	20,854	15,116	230.885	20,297	14,560	216.314	690
700	27,125	21,305	250.663	20,690	14,870	222.953	24,088	18,268	218.610	21,184	15,364	231.358	20,604	14,784	216.756	700
710	27,622	21,719	251.368	21,002	15,099	223.396	24,464	18,561	219.142	21,514	15,611	231.827	20,912	15,008	217.192	710
720	28,121	22,134	252.065	21,315	15,328	223.833	24,840	18,854	219.668	21,845	15,859	232.291	21,220	15,234	217.624	720
730	28,622	22,552	252.755	21,628	15,558	224.265	25,218	19,148	220.189	22,177	16,107	232.748	21,529	15,460	218.059	730
740	29,124	22,972	253.439	21,943	15,789	224.692	25,597	19,444	220.707	22,510	16,357	233.201	21,839	15,686	218.472	740
750	29,629	23,393	254.117	22,258	16,022	225.115	25,977	19,741	221.215	22,844	16,607	233.649	22,149	15,913	218.889	750
760	30,135	23,817	254.787	22,573	16,255	225.533	26,358	20,039	221.720	23,178	16,859	234.091	22,460	16,141	219.301	760
770	30,644	24,242	255.452	22,890	16,488	225.947	26,741	20,339	222.221	23,513	17,111	234.528	22,772	16,370	219.709	770
780	31,154	24,669	256.110	23,208	16,723	226.357	27,125	20,639	222.717	23,850	17,364	234.960	23,085	16,599	220.113	780
790	31,665	25,097	256.762	23,526	16,957	226.762	27,510	20,941	223.207	24,186	17,618	235.387	23,398	16,830	220.512	790
800	32,179	25,527	257.408	23,844	17,193	227.162	27,896	21,245	223.693	24,523	17,872	235.810	23,714	17,061	220.907	800
810	32,694	25,959	258.048	24,164	17,429	227.559	28,284	21,549	224.174	24,861	18,126	236.230	24,027	17,292	221.298	810
820	33,212	26,394	258.682	24,483	17,665	227.952	28,672	21,855	224.651	25,199	18,382	236.644	24,342	17,524	221.684	820
830	33,730	26,829	259.311	24,803	17,902	228.339	29,062	22,162	225.123	25,537	18,637	237.055	24,658	17,757	222.067	830
840	34,251	27,267	259.934	25,124	18,140	228.724	29,454	22,470	225.592	25,877	18,893	237.462	24,974	17,990	222.447	840
850	34,773	27,706	260.551	25,446	18,379	229.106	29,846	22,779	226.057	26,218	19,150	237.864	25,292	18,224	222.822	850
860	35,296	28,125	261.164	25,768	18,617	229.482	30,240	23,090	226.517	26,559	19,408	238.264	25,610	18,459	223.194	860
870	35,821	28,588	261.770	26,091	18,858	229.856	30,635	23,402	226.973	26,899	19,666	238.660	25,928	18,695	223.562	870
880	36,347	29,031	262.371	26,415	19,099	230.227	31,032	23,715	227.426	27,242	19,925	239.051	26,248	18,931	223.927	880
890	36,876	29,476	262.968	26,740	19,341	230.593	31,429	24,029	227.875	27,584	20,185	239.439	26,568	19,168	224.288	890
900	37,405	29,922	263.559	27,066	19,583	230.957	31,828	24,345	228.321	27,928	20,445	239.823	26,890	19,407	224.647	900
910	37,935	30,369	264.146	27,392	19,826	231.317	32,228	24,662	228.763	28,272	20,706	240.203	27,210	19,644	225.002	910
920	38,467	30,818	264.728	27,719	20,070	231.674	32,629	24,980	229.202	28,616	20,967	240.580	27,532	19,883	225.353	920
930	39,000	31,268	265.304	28,046	20,314	232.028	33,032	25,300	229.637	28,960	21,228	240.953	27,854	20,122	225.701	930
940	39,535	31,719	265.877	28,375	20,559	232.379	33,436	25,621	230.070	29,306	21,491	241.323	28,178	20,362	226.047	940
950	40,070	32,171	266.444	28,703	20,805	232.727	33,841	25,943	230.499	29,652	21,754	241.689	28,501	20,603	226.389	950
960	40,607	32,625	267.007	29,033	21,051	233.072	34,247	26,265	230.924	29,999	22,017	242.052	28,826	20,844	226.728	960
970	41,145	33,081	267.566	29,362	21,298	233.413	34,653	26,588	231.347	30,345	22,280	242.411	29,151	21,086	227.064	970
980	41,685	33,537	268.119	29,693	21,545	233.752	35,061	26,913	231.767	30,692	22,544	242.768	29,476	21,328	227.398	980
990	42,226	33,995	268.670	30,024	21,793	234.088	35,472	27,240	232.184	31,041	22,809	243.120	29,803	21,571	227.728	990

TABLE A-23   Ideal Gas Properties of Selected Gases *(Continued)*

$\bar{h}$ and $\bar{u}$ in kJ/kmol. $\bar{s}°$ in kJ/kmol·K

T(K)	Carbon Dioxide, CO_2 ($\bar{h}_f° = -393{,}520$ kJ/kmol) $\bar{h}$	$\bar{u}$	$\bar{s}°$	Carbon Monoxide, CO ($\bar{h}_f° = -110{,}530$ kJ/kmol) $\bar{h}$	$\bar{u}$	$\bar{s}°$	Water Vapor, H_2O ($\bar{h}_f° = -241{,}820$ kJ/kmol) $\bar{h}$	$\bar{u}$	$\bar{s}°$	Oxygen, O_2 ($\bar{h}_f° = 0$ kJ/kmol) $\bar{h}$	$\bar{u}$	$\bar{s}°$	Nitrogen, N_2 ($\bar{h}_f° = 0$ kJ/kmol) $\bar{h}$	$\bar{u}$	$\bar{s}°$	T(K)
1000	42,769	34,455	269.215	30,355	22,041	234.421	35,882	27,568	232.597	31,389	23,075	243.471	30,129	21,815	228.057	1000
1020	43,859	35,378	270.293	31,020	22,540	235.079	36,709	28,228	233.415	32,088	23,607	244.164	30,784	22,304	228.706	1020
1040	44,953	36,306	271.354	31,688	23,041	235.728	37,542	28,895	234.223	32,789	24,142	244.844	31,442	22,795	229.344	1040
1060	46,051	37,238	272.400	32,357	23,544	236.364	38,380	29,567	235.020	33,490	24,677	245.513	32,101	23,288	229.973	1060
1080	47,153	38,174	273.430	33,029	24,049	236.992	39,223	30,243	235.806	34,194	25,214	246.171	32,762	23,782	230.591	1080
1100	48,258	39,112	274.445	33,702	24,557	237.609	40,071	30,925	236.584	34,899	25,753	246.818	33,426	24,280	231.199	1100
1120	49,369	40,057	275.444	34,377	25,065	238.217	40,923	31,611	237.352	35,606	26,294	247.454	34,092	24,780	231.799	1120
1140	50,484	41,006	276.430	35,054	25,575	238.817	41,780	32,301	238.110	36,314	26,836	248.081	34,760	25,282	232.391	1140
1160	51,602	41,957	277.403	35,733	26,088	239.407	42,642	32,997	238.859	37,023	27,379	248.698	35,430	25,786	232.973	1160
1180	52,724	42,913	278.362	36,406	26,602	239.989	43,509	33,698	239.600	37,734	27,923	249.307	36,104	26,291	233.549	1180
1200	53,848	43,871	279.307	37,095	27,118	240.663	44,380	34,403	240.333	38,447	28,469	249.906	36,777	26,799	234.115	1200
1220	54,977	44,834	280.238	37,780	27,637	241.128	45,256	35,112	241.057	39,162	29,018	250.497	37,452	27,308	234.673	1220
1240	56,108	45,799	281.158	38,466	28,155	241.686	46,137	35,827	241.773	39,877	29,568	251.079	38,129	27,819	235.223	1240
1260	57,244	46,768	282.066	39,154	28,678	242.236	47,022	36,546	242.482	40,594	30,118	251.653	38,807	28,331	235.766	1260
1280	58,381	47,739	282.962	39,884	29,201	242.780	47,912	37,270	243.183	41,312	30,670	252.219	39,488	28,845	236.302	1280
1300	59,522	48,713	283.847	40,534	29,725	243.316	48,807	38,000	243.877	42,033	31,224	252.776	40,170	29,361	236.831	1300
1320	60,666	49,691	284.722	41,266	30,251	243.844	49,707	38,732	244.564	42,753	31,778	253.325	40,853	29,878	237.353	1320
1340	61,813	50,672	285.586	41,919	30,778	244.366	50,612	39,470	245.243	43,475	32,334	253.868	41,539	30,398	237.867	1340
1360	62,963	51,656	286.439	42,613	31,306	244.880	51,521	40,213	245.915	44,198	32,891	254.404	42,227	30,919	238.376	1360
1380	64,116	52,643	287.283	43,309	31,836	245.388	52,434	40,960	246.582	44,923	33,449	254.932	42,915	31,441	238.878	1380
1400	65,271	53,631	288.106	44,007	32,367	245.889	53,351	41,711	247.241	45,648	34,008	255.454	43,605	31,964	239.375	1400
1420	66,427	54,621	288.934	44,707	32,900	246.385	54,273	42,466	247.895	46,374	34,567	255.968	44,295	32,489	239.865	1420
1440	67,586	55,614	289.743	45,408	33,434	246.876	55,198	43,226	248.543	47,102	35,129	256.475	44,988	33,014	240.350	1440
1460	68,748	56,609	290.542	46,110	33,971	247.360	56,128	43,989	249.185	47,831	35,692	256.978	45,682	33,543	240.827	1460
1480	69,911	57,606	291.333	46,813	34,508	247.839	57,062	44,756	249.820	48,561	36,256	257.474	46,377	34,071	241.301	1480
1500	71,078	58,606	292.114	47,517	35,046	248.312	57,999	45,528	250.450	49,292	36,821	257.965	47,073	34,601	241.768	1500
1520	72,246	59,609	292.888	48,222	35,584	248.778	58,942	46,304	251.074	50,024	37,387	258.450	47,771	35,133	242.228	1520
1540	73,417	60,613	293.654	48,928	36,124	249.240	59,888	47,084	251.693	50,756	37,952	258.928	48,470	35,665	242.685	1540
1560	74,590	61,620	294.411	49,635	36,665	249.695	60,838	47,868	252.305	51,490	38,520	259.402	49,168	36,197	243.137	1560
1580	76,767	62,630	295.161	50,344	37,207	250.147	61,792	48,655	252.912	52,224	39,088	259.870	49,869	36,732	243.585	1580
1600	76,944	63,741	295.901	51,053	37,750	250.592	62,748	49,445	253.513	52,961	39,658	260.333	50,571	37,268	244.028	1600
1620	78,123	64,653	296.632	51,763	38,293	251.033	63,709	50,240	254.111	53,696	40,227	260.791	51,275	37,806	244.464	1620
1640	79,303	65,668	297.356	52,472	38,837	251.470	64,675	51,039	254.703	54,434	40,799	261.242	51,980	38,344	244.896	1640
1660	80,486	66,592	298.072	53,184	39,382	251.901	65,643	51,841	255.290	55,172	41,370	261.690	52,686	38,884	245.324	1660
1680	81,670	67,702	298.781	53,895	39,927	252.329	66,614	52,646	255.873	55,912	41,944	262.132	53,393	39,424	245.747	1680
1700	82,856	68,721	299.482	54,609	40,474	252.751	67,589	53,455	256.450	56,652	42,517	262.571	54,099	39,965	246.166	1700
1720	84,043	69,742	300.177	55,323	41,023	253.169	68,567	54,267	257.022	57,394	43,093	263.005	54,807	40,507	246.580	1720
1740	85,231	70,764	300.863	56,039	41,572	253.582	69,550	55,083	257.589	58,136	43,669	263.435	55,516	41,049	246.990	1740

Table A-23

Table A-23

TABLE A-23   **Ideal Gas Properties of Selected Gases (Continued)**

$\bar{h}$ and $\bar{u}$ in kJ/kmol, $\bar{s}°$ in kJ/kmol·K

T(K)	Carbon Dioxide, CO_2 ($\bar{h}_f° = -393{,}520$ kJ/kmol) $\bar{h}$	$\bar{u}$	$\bar{s}°$	Carbon Monoxide, CO ($\bar{h}_f° = -110{,}530$ kJ/kmol) $\bar{h}$	$\bar{u}$	$\bar{s}°$	Water Vapor, H_2O ($\bar{h}_f° = -241{,}820$ kJ/kmol) $\bar{h}$	$\bar{u}$	$\bar{s}°$	Oxygen, O_2 ($\bar{h}_f° = 0$ kJ/kmol) $\bar{h}$	$\bar{u}$	$\bar{s}°$	Nitrogen, N_2 ($\bar{h}_f° = 0$ kJ/kmol) $\bar{h}$	$\bar{u}$	$\bar{s}°$	T(K)
1760	86,420	71,787	301.543	56,756	42,123	253.991	70,535	55,902	258.151	58,800	44,247	263.861	56,227	41,594	247.396	1760
1780	87,612	72,812	302.271	57,473	42,673	254.398	71,523	56,723	258.708	59,624	44,825	264.283	56,938	42,139	247.798	1780
1800	88,806	73,840	302.884	58,191	43,225	254.797	72,513	57,547	259.262	60,371	45,405	264.701	57,651	42,685	248.195	1800
1820	90,000	74,868	303.544	58,910	43,778	255.194	73,507	58,375	259.811	61,118	45,986	265.113	58,363	43,231	248.589	1820
1840	91,196	75,897	304.198	59,629	44,331	255.587	74,506	59,207	260.357	61,866	46,568	265.521	59,075	43,777	248.979	1840
1860	92,394	76,929	304.845	60,351	44,886	255.976	75,506	60,042	260.898	62,616	47,151	265.925	59,790	44,324	249.365	1860
1880	93,593	77,962	305.487	61,072	45,441	256.361	76,511	60,880	261.436	63,365	47,734	266.326	60,504	44,873	249.748	1880
1900	94,793	78,996	306.122	61,794	45,997	256.743	77,517	61,720	261.969	64,116	48,319	266.722	61,220	45,423	250.128	1900
1920	95,995	80,031	306.751	62,516	46,552	257.122	78,527	62,564	262.497	64,868	48,904	267.115	61,936	45,973	250.502	1920
1940	97,197	81,067	307.374	63,238	47,108	257.497	79,540	63,411	263.022	65,620	49,490	267.505	62,654	46,524	250.874	1940
1960	98,401	82,105	307.992	63,961	47,665	257.868	80,555	64,259	263.542	66,374	50,078	267.891	63,381	47,075	251.242	1960
1980	99,606	83,144	308.604	64,684	48,221	258.236	81,573	65,111	264.059	67,127	50,665	268.275	64,090	47,627	251.607	1980
2000	100,804	84,185	309.210	65,408	48,780	258.600	82,593	65,965	264.571	67,881	51,253	268.655	64,810	48,181	251.969	2000
2050	103,835	86,791	310.701	67,224	50,179	259.494	85,156	68,111	265.838	69,772	52,727	269.588	66,612	49,567	252.858	2050
2100	106,864	89,404	312.160	69,044	51,584	260.370	87,735	70,275	267.081	71,668	54,208	270.504	68,417	50,957	253.726	2100
2150	109,898	92,023	313.589	70,864	52,988	261.226	90,330	72,454	268.301	73,573	55,697	271.399	70,226	52,351	254.578	2150
2200	112,939	94,648	314.988	72,688	54,396	262.065	92,940	74,649	269.500	75,484	57,192	272.278	72,040	53,749	255.412	2200
2250	115,984	97,277	316.356	74,516	55,809	262.887	95,562	76,855	270.679	77,397	58,690	273.136	73,856	55,149	256.227	2250
2300	119,035	99,912	317.695	76,345	57,222	263.692	98,199	79,076	271.839	79,316	60,193	273.981	75,676	56,553	257.027	2300
2350	122,091	102,552	319.011	78,178	58,640	264.480	100,846	81,308	272.978	81,243	61,704	274.809	77,496	57,958	257.810	2350
2400	125,152	105,197	320.302	80,015	60,060	265.253	103,508	83,553	274.098	83,174	63,219	275.625	79,320	59,366	258.580	2400
2450	128,219	107,849	321.566	81,852	61,482	266.012	106,183	85,811	275.201	85,112	64,742	276.424	81,149	60,779	259.332	2450
2500	131,290	110,504	322.808	83,692	62,906	266.755	108,868	88,082	276.286	87,057	66,271	277.207	82,981	62,195	260.073	2500
2550	134,368	113,166	324.026	85,537	64,335	267.485	111,565	90,364	277.354	89,004	67,802	277.979	84,814	63,613	260.799	2550
2600	137,449	115,832	325.222	87,383	65,766	268.202	114,273	92,656	278.407	90,956	69,339	278.738	86,650	65,033	261.512	2600
2650	140,533	118,500	326.396	89,230	67,197	268.905	116,991	94,958	279.441	92,916	70,883	279.485	88,488	66,455	262.213	2650
2700	143,620	121,172	327.549	91,077	68,628	269.596	119,717	97,269	280.462	94,881	72,433	280.219	90,328	67,880	262.902	2700
2750	146,713	123,849	328.684	92,930	70,066	270.285	122,453	99,588	281.464	96,852	73,987	280.942	92,171	69,306	263.577	2750
2800	149,808	126,528	329.800	94,784	71,504	270.943	125,198	101,917	282.453	98,826	75,546	281.654	94,014	70,734	264.241	2800
2850	152,908	129,212	330.896	96,639	72,945	271.602	127,952	104,256	283.429	100,808	77,112	282.357	95,859	72,163	264.895	2850
2900	156,009	131,898	331.975	98,495	74,383	272.249	130,717	106,605	284.390	102,793	78,682	283.048	97,705	73,593	265.538	2900
2950	159,117	134,589	333.037	100,352	75,825	272.884	133,486	108,959	285.338	104,785	80,258	283.728	99,556	75,028	266.170	2950
3000	162,226	137,283	334.084	102,210	77,267	273.508	136,264	111,321	286.273	106,780	81,837	284.399	101,407	76,464	266.793	3000
3050	165,341	139,982	335.114	104,073	78,715	274.123	139,051	113,692	287.194	108,778	83,419	285.060	103,260	77,902	267.404	3050
3100	168,456	142,681	336.126	105,939	80,164	274.730	141,846	116,072	288.102	110,784	85,009	285.713	105,115	79,341	268.007	3100
3150	171,576	145,385	337.124	107,802	81,612	275.326	144,648	118,458	288.999	112,795	86,601	286.355	106,972	80,782	268.601	3150
3200	174,695	148,089	338.109	109,667	83,061	275.914	147,457	120,851	289.884	114,809	88,203	286.989	108,830	82,224	269.186	3200
3250	177,822	150,801	339.069	111,534	84,513	276.494	150,272	123,250	290.756	116,827	89,804	287.614	110,690	83,668	269.763	3250

Source: Table A-23 is based on the JANAF Thermochemical Tables, NSRDS-NBS-37, 1971.

TABLE A-24   **Constants for the van der Waals, Redlich–Kwong, and Benedict–Webb–Rubin Equations of State**

1. van der Waals and Redlich–Kwong: Constants for pressure in bar, specific volume in $m^3/kmol$, and temperature in K

| Substance | van der Waals | | Redlich–Kwong | |
	a $\mathrm{bar}\left(\dfrac{m^3}{kmol}\right)^2$	b $\dfrac{m^3}{kmol}$	a $\mathrm{bar}\left(\dfrac{m^3}{kmol}\right)^2 K^{1/2}$	b $\dfrac{m^3}{kmol}$
Air	1.368	0.0367	15.989	0.02541
Butane (C_4H_{10})	13.86	0.1162	289.55	0.08060
Carbon dioxide (CO_2)	3.647	0.0428	64.43	0.02963
Carbon monoxide (CO)	1.474	0.0395	17.22	0.02737
Methane (CH_4)	2.293	0.0428	32.11	0.02965
Nitrogen (N_2)	1.366	0.0386	15.53	0.02677
Oxygen (O_2)	1.369	0.0317	17.22	0.02197
Propane (C_3H_8)	9.349	0.0901	182.23	0.06242
Refrigerant 12	10.49	0.0971	208.59	0.06731
Sulfur dioxide (SO_2)	6.883	0.0569	144.80	0.03945
Water (H_2O)	5.531	0.0305	142.59	0.02111

Source: Calculated from critical data.

2. Benedict–Webb–Rubin: Constants for pressure in bar, specific volume in $m^3/kmol$, and temperature in K

Substance	a	A	b	B	c	C	α	γ
C_4H_{10}	1.9073	10.218	0.039998	0.12436	3.206×10^5	1.006×10^6	1.101×10^{-3}	0.0340
CO_2	0.1386	2.7737	0.007210	0.04991	1.512×10^4	1.404×10^5	8.47×10^{-5}	0.00539
CO	0.0371	1.3590	0.002632	0.05454	1.054×10^3	8.676×10^3	1.350×10^{-4}	0.0060
CH_4	0.0501	1.8796	0.003380	0.04260	2.579×10^3	2.287×10^4	1.244×10^{-4}	0.0060
N_2	0.0254	1.0676	0.002328	0.04074	7.381×10^2	8.166×10^3	1.272×10^{-4}	0.0053

Source: H. W. Cooper and J. C. Goldfrank, *Hydrocarbon Processing, 46* (12): 141 (1967).

Table A-24

| | TABLE A-25 | Thermochemical Properties of Selected Substances at 298K and 1 atm | | | | | |

TABLE A-25 Thermochemical Properties of Selected Substances at 298K and 1 atm

Substance	Formula	Molar Mass, M (kg/kmol)	Enthalpy of Formation, $\bar{h}_f^\circ$ (kJ/kmol)	Gibbs Function of Formation, $\bar{g}_f^\circ$ (kJ/kmol)	Absolute Entropy, $\bar{s}^\circ$ (kJ/kmol · K)	Heating Values Higher, HHV (kJ/kg)	Lower, LHV (kJ/kg)
Carbon	C(s)	12.01	0	0	5.74	32,770	32,770
Hydrogen	H$_2$(g)	2.016	0	0	130.57	141,780	119,950
Nitrogen	N$_2$(g)	28.01	0	0	191.50	—	—
Oxygen	O$_2$(g)	32.00	0	0	205.03	—	—
Carbon monoxide	CO(g)	28.01	−110,530	−137,150	197.54	—	—
Carbon dioxide	CO$_2$(g)	44.01	−393,520	−394,380	213.69	—	—
Water	H$_2$O(g)	18.02	−241,820	−228,590	188.72	—	—
Water	H$_2$O(l)	18.02	−285,830	−237,180	69.95	—	—
Hydrogen peroxide	H$_2$O$_2$(g)	34.02	−136,310	−105,600	232.63	—	—
Ammonia	NH$_3$(g)	17.03	−46,190	−16,590	192.33	—	—
Oxygen	O(g)	16.00	249,170	231,770	160.95	—	—
Hydrogen	H(g)	1.008	218,000	203,290	114.61	—	—
Nitrogen	N(g)	14.01	472,680	455,510	153.19	—	—
Hydroxyl	OH(g)	17.01	39,460	34,280	183.75	—	—
Methane	CH$_4$(g)	16.04	−74,850	−50,790	186.16	55,510	50,020
Acetylene	C$_2$H$_2$(g)	26.04	226,730	209,170	200.85	49,910	48,220
Ethylene	C$_2$H$_4$(g)	28.05	52,280	68,120	219.83	50,300	47,160
Ethane	C$_2$H$_6$(g)	30.07	−84,680	−32,890	229.49	51,870	47,480
Propylene	C$_3$H$_6$(g)	42.08	20,410	62,720	266.94	48,920	45,780
Propane	C$_3$H$_8$(g)	44.09	−103,850	−23,490	269.91	50,350	46,360
Butane	C$_4$H$_{10}$(g)	58.12	−126,150	−15,710	310.03	49,500	45,720
Pentane	C$_5$H$_{12}$(g)	72.15	−146,440	−8,200	348.40	49,010	45,350
Octane	C$_8$H$_{18}$(g)	114.22	−208,450	17,320	463.67	48,260	44,790
Octane	C$_8$H$_{18}$(l)	114.22	−249,910	6,610	360.79	47,900	44,430
Benzene	C$_6$H$_6$(g)	78.11	82,930	129,660	269.20	42,270	40,580
Methanol	CH$_3$OH(g)	32.04	−200,890	−162,140	239.70	23,850	21,110
Methanol	CH$_3$OH(l)	32.04	−238,810	−166,290	126.80	22,670	19,920
Ethanol	C$_2$H$_5$OH(g)	46.07	−235,310	−168,570	282.59	30,590	27,720
Ethanol	C$_2$H$_5$OH(l)	46.07	−277,690	−174,890	160.70	29,670	26,800

Source: Based on JANAF Thermochemical Tables, NSRDS-NBS-37, 1971; *Selected Values of Chemical Thermodynamic Properties,* NBS Tech. Note 270-3, 1968; and *API Research Project 44,* Carnegie Press, 1953. Heating values calculated.

TABLE A-26	Standard Molar Chemical Exergy, $\bar{e}^{ch}$ (kJ/kmol), of Selected Substances at 298 K and p_0		
Substance	**Formula**	**Model I**[a]	**Model II**[b]
Nitrogen	$N_2(g)$	640	720
Oxygen	$O_2(g)$	3,950	3,970
Carbon dioxide	$CO_2(g)$	14,175	19,870
Water	$H_2O(g)$	8,635	9,500
Water	$H_2O(l)$	45	900
Carbon (graphite)	$C(s)$	404,590	410,260
Hydrogen	$H_2(g)$	235,250	236,100
Sulfur	$S(s)$	598,160	609,600
Carbon monoxide	$CO(g)$	269,410	275,100
Sulfur dioxide	$SO_2(g)$	301,940	313,400
Nitrogen monoxide	$NO(g)$	88,850	88,900
Nitrogen dioxide	$NO_2(g)$	55,565	55,600
Hydrogen sulfide	$H_2S(g)$	799,890	812,000
Ammonia	$NH_3(g)$	336,685	337,900
Methane	$CH_4(g)$	824,350	831,650
Acetylene	$C_2H_2(g)$	—	1,265,800
Ethylene	$C_2H_4(g)$	—	1,361,100
Ethane	$C_2H_6(g)$	1,482,035	1,495,840
Propylene	$C_3H_6(g)$	—	2,003,900
Propane	$C_3H_8(g)$	—	2,154,000
Butane	$C_4H_{10}(g)$	—	2,805,800
Pentane	$C_5H_{12}(g)$	—	3,463,300
Benzene	$C_6H_6(g)$	—	3,303,600
Octane	$C_8H_{18}(l)$	—	5,413,100
Methanol	$CH_3OH(g)$	715,070	722,300
Methanol	$CH_3OH(l)$	710,745	718,000
Ethanol	$C_2H_5OH(g)$	1,348,330	1,363,900
Ethanol	$C_2H_5OH(l)$	1,342,085	1,357,700

[a]J. Ahrendts, "Die Exergie Chemisch Reaktionsfähiger Systeme," *VDI-Forschungsheft*, VDI-Verlag, Dusseldorf, 579, 1977. Also see "Reference States," *Energy—The International Journal, 5:* 667–677, 1980. In Model I, $p_0 = 1.019$ atm. This model attempts to impose a criterion that the reference environment be in equilibrium. The reference substances are determined assuming restricted chemical equilibrium for nitric acid and nitrates and unrestricted thermodynamic equilibrium for all other chemical components of the atmosphere, the oceans, and a portion of the Earth's crust. The chemical composition of the gas phase of this model approximates the composition of the natural atmosphere.

[b]J. Szargut, D. R. Morris, and F. R. Steward, *Exergy Analysis of Thermal,* Chemical, and *Metallurgical Processes,* Hemisphere, New York, 1988. In Model II, $p_0 = 1.0$ atm. In developing this model a reference substance is selected for each chemical element from among substances that contain the element being considered and that are abundantly present in the natural environment, even though the substances are not in completely mutual stable equilibrium. An underlying rationale for this approach is that substances found abundantly in nature have little economic value. On an overall basis, the chemical composition of the exergy reference environment of Model II is closer than Model I to the composition of the natural environment, but the equilibrium criterion is not always satisfied.

TABLE A-27 Logarithms to the Base 10 of the Equilibrium Constant K

Temp. K	$H_2 \rightleftharpoons 2H$	$O_2 \rightleftharpoons 2O$	$N_2 \rightleftharpoons 2N$	$\frac{1}{2}O_2 + \frac{1}{2}N_2 \rightleftharpoons NO$	$H_2O \rightleftharpoons H_2 + \frac{1}{2}O_2$	$H_2O \rightleftharpoons OH + \frac{1}{2}H_2$	$CO_2 \rightleftharpoons CO + \frac{1}{2}O_2$	$CO_2 + H_2 \rightleftharpoons CO + H_2O$	Temp. °R
298	−71.224	−81.208	−159.600	−15.171	−40.048	−46.054	−45.066	−5.018	537
500	−40.316	−45.880	−92.672	−8.783	−22.886	−26.130	−25.025	−2.139	900
1000	−17.292	−19.614	−43.056	−4.062	−10.062	−11.280	−10.221	−0.159	1800
1200	−13.414	−15.208	−34.754	−3.275	−7.899	−8.811	−7.764	+0.135	2160
1400	−10.630	−12.054	−28.812	−2.712	−6.347	−7.021	−6.014	+0.333	2520
1600	−8.532	−9.684	−24.350	−2.290	−5.180	−5.677	−4.706	+0.474	2880
1700	−7.666	−8.706	−22.512	−2.116	−4.699	−5.124	−4.169	+0.530	3060
1800	−6.896	−7.836	−20.874	−1.962	−4.270	−4.613	−3.693	+0.577	3240
1900	−6.204	−7.058	−19.410	−1.823	−3.886	−4.190	−3.267	+0.619	3420
2000	−5.580	−6.356	−18.092	−1.699	−3.540	−3.776	−2.884	+0.656	3600
2100	−5.016	−5.720	−16.898	−1.586	−3.227	−3.434	−2.539	+0.688	3780
2200	−4.502	−5.142	−15.810	−1.484	−2.942	−3.091	−2.226	+0.716	3960
2300	−4.032	−4.614	−14.818	−1.391	−2.682	−2.809	−1.940	+0.742	4140
2400	−3.600	−4.130	−13.908	−1.305	−2.443	−2.520	−1.679	+0.764	4320
2500	−3.202	−3.684	−13.070	−1.227	−2.224	−2.270	−1.440	+0.784	4500
2600	−2.836	−3.272	−12.298	−1.154	−2.021	−2.038	−1.219	+0.802	4680
2700	−2.494	−2.892	−11.580	−1.087	−1.833	−1.823	−1.015	+0.818	4860
2800	−2.178	−2.536	−10.914	−1.025	−1.658	−1.624	−0.825	+0.833	5040
2900	−1.882	−2.206	−10.294	−0.967	−1.495	−1.438	−0.649	+0.846	5220
3000	−1.606	−1.898	−9.716	−0.913	−1.343	−1.265	−0.485	+0.858	5400
3100	−1.348	−1.610	−9.174	−0.863	−1.201	−1.103	−0.332	+0.869	5580
3200	−1.106	−1.340	−8.664	−0.815	−1.067	−0.951	−0.189	+0.878	5760
3300	−0.878	−1.086	−8.186	−0.771	−0.942	−0.809	−0.054	+0.888	5940
3400	−0.664	−0.846	−7.736	−0.729	−0.824	−0.674	+0.071	+0.895	6120
3500	−0.462	−0.620	−7.312	−0.690	−0.712	−0.547	+0.190	+0.902	6300

Source: Based on data from the JANAF Thermochemical Tables, NSRDS-NBS-37, 1971.

Index to Figures and Charts

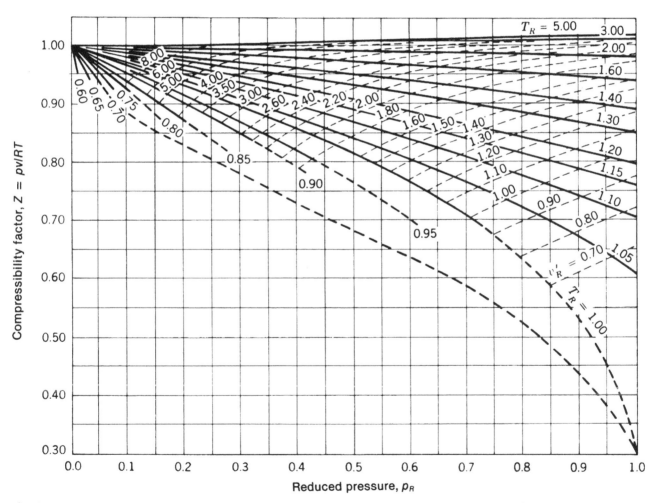

Fig. A.1 Generalized compressibility chart, $p_R \leq 1.0$. *Source*: E. F. Obert, *Concepts of Thermodynamics*, McGraw-Hill, New York, 1960.

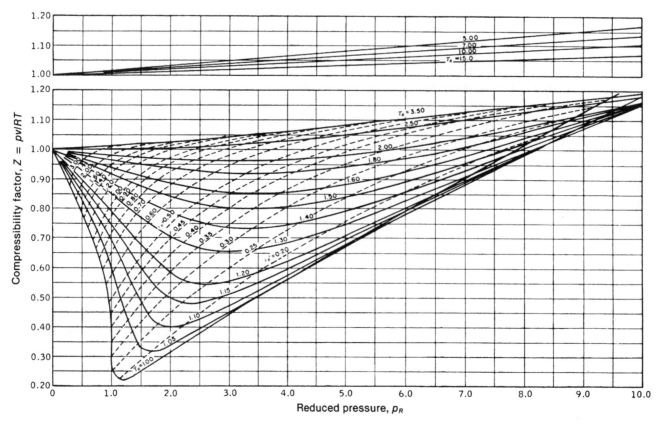

Fig. A.2 Generalized compressibility chart, $p_R \leq 10.0$. *Source*: E. F. Obert, *Concepts of Thermodynamics*, McGraw-Hill, New York, 1960.

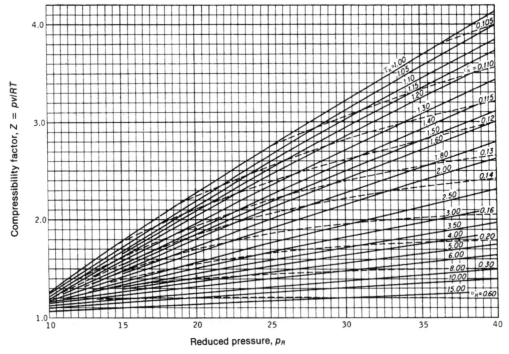

Fig. A.3 Generalized compressibility chart, $10 \leq p_R \leq 40$. *Source*: E. F. Obert. *Concepts of Thermodynamics*, McGraw-Hill, New York, 1960.

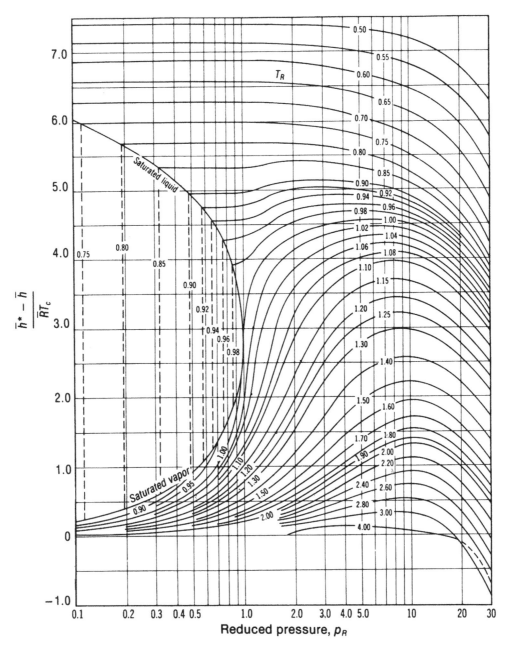

Fig. A.4 Generalized enthalpy correction chart. *Source:* Adapted from G. J. Van Wylen and R. E. Sonntag, *Fundamentals of Classical Thermodynamics*, 3rd. ed., English/SI, Wiley, New York, 1986.

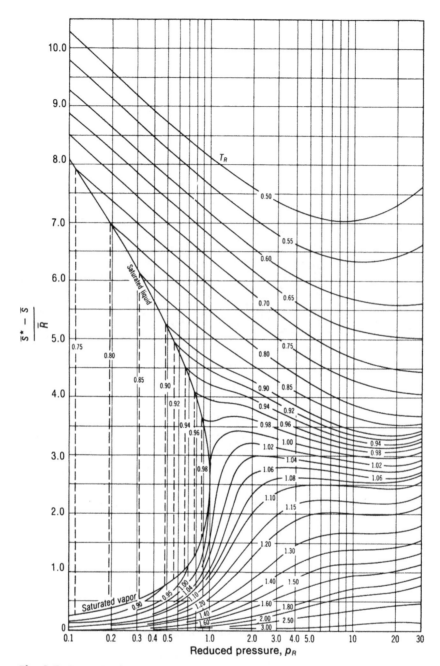

Fig. A.5 Generalized entropy correction chart. *Source*: Adapted from G. J. Van Wylen and R. E. Sonntag, *Fundamentals of Classical Thermodynamics*, 3rd. ed., English/SI, Wiley, New York, 1986.

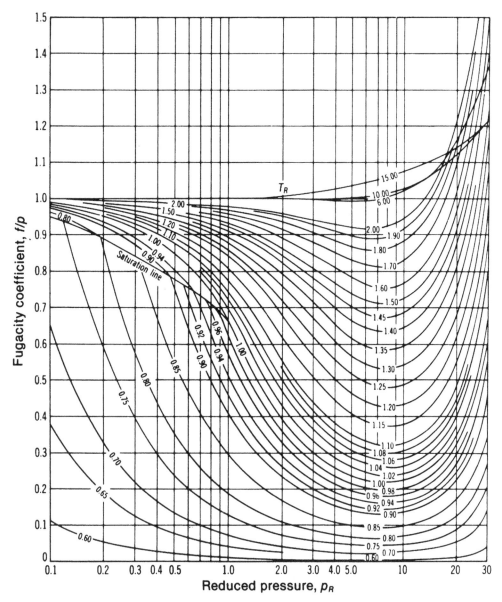

Fig. A.6 Generalized fugacity coefficient chart. *Source*: G. J. Van Wylen and R. E. Sonntag, *Fundamentals of Classical Thermodynamics*, 3rd. ed., English/SI, Wiley, New York, 1986.

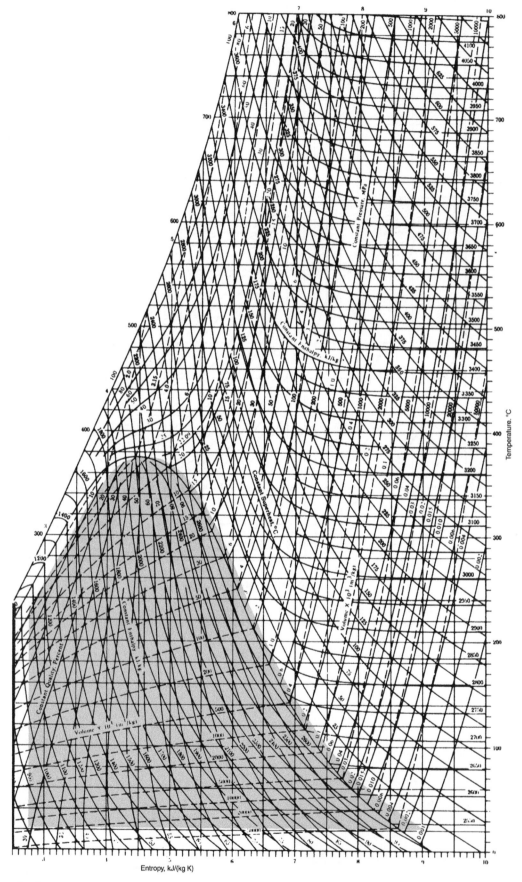

Fig. A.7 Temperature–entropy diagram for water (SI units). *Source*: J. H. Keenan, F. G. Keyes, P. G. Hill, and J. G. Moore, *Steam Tables*, Wiley, New York, 1978.

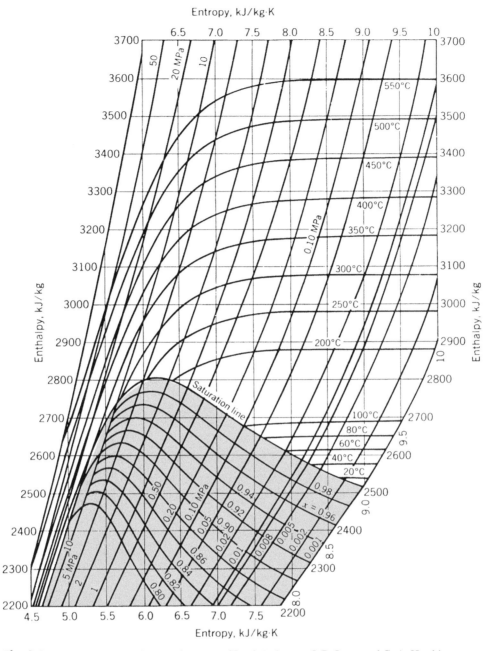

Fig. A.8 Enthalpy–entropy diagram for water (SI units). *Source*: J. B. Jones and G. A. Hawkins, *Engineering Thermodynamics*, 2nd ed., Wiley, New York, 1986.

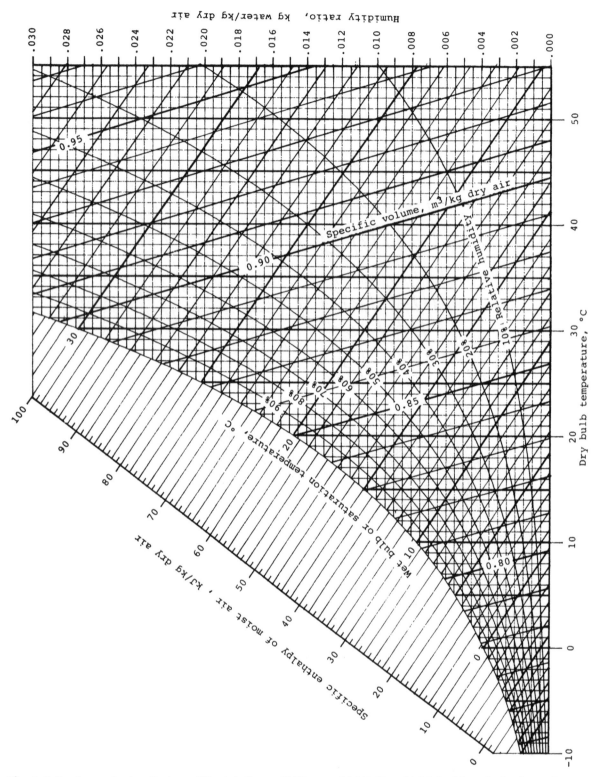

Fig. A.9 Psychrometric chart for 1 atm (SI units). *Source*: Z. Zhang and M. B. Pate, "A Methodology for Implementing a Psychrometric Chart in a Computer Graphics System," *ASHRAE Transactions*, Vol. 94, Pt. 1, 1988.

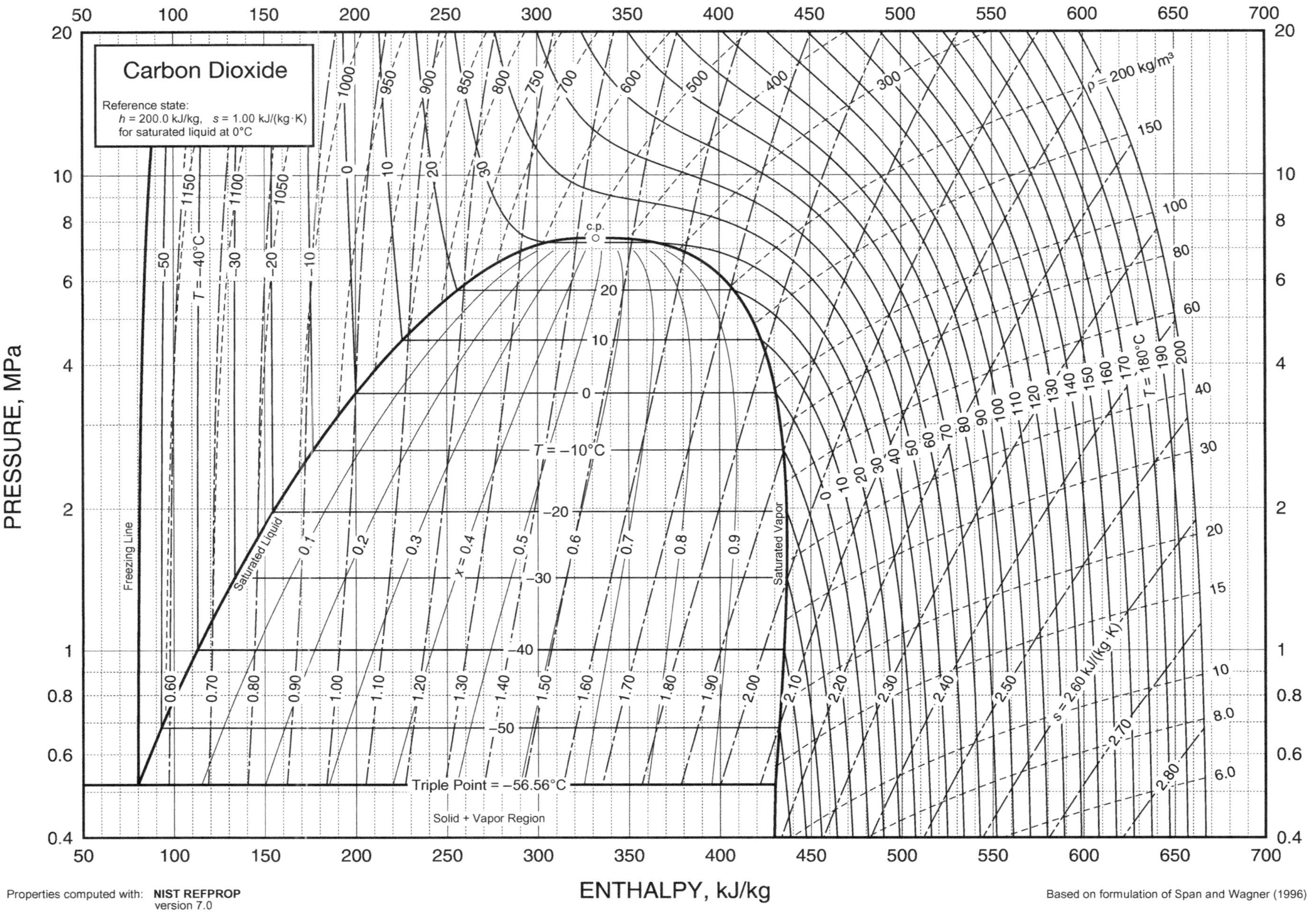

Properties computed with: **NIST REFPROP** version 7.0

Based on formulation of Span and Wagner (1996)

Fig. A.10 Pressure-enthalpy diagram for carbon dioxide (SI units). *Source*: ©ASHRAE, www.ashrae.org. 2009 ASHRAE Handbook of Fundamentals—Fundamentals.

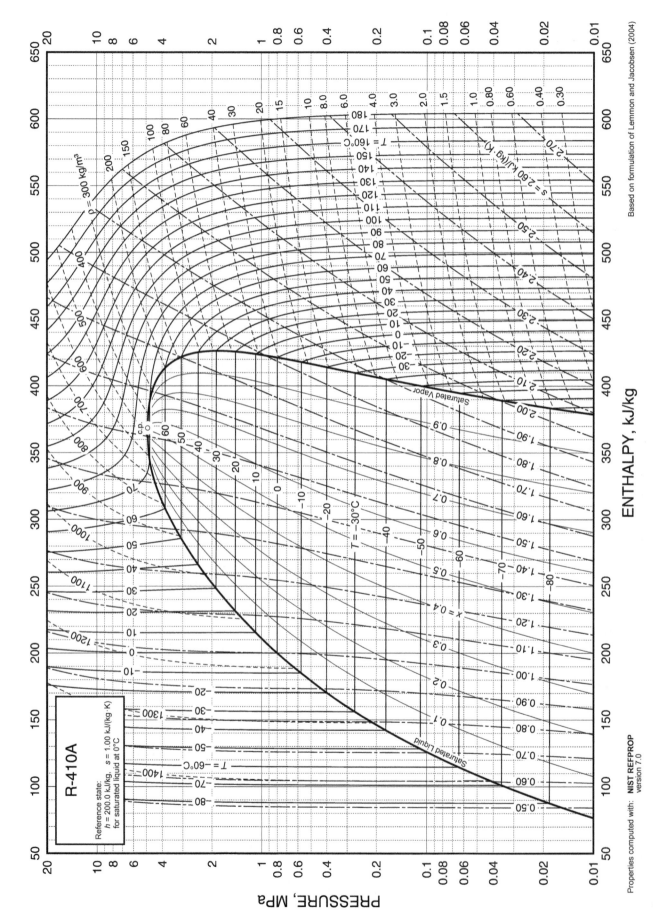

Fig. A.11 Pressure-enthalpy diagram for Refrigerant 410A (SI units). *Source:* ©ASHRAE, www.ashrae.org. 2009 ASHRAE Handbook of Fundamentals—Fundamentals.

Properties computed with: **NIST REFPROP** version 7.0

Based on formulation of Lemmon and Jacobsen (2004)

Index

Symbols

a	acceleration, activity	mep	mean effective pressure
A	area	mf	mass fraction
AF	air–fuel ratio	n	number of moles, polytropic exponent
bwr	back work ratio	N	number of components in the phase rule
c	specific heat of an incompressible substance, velocity of sound	p	pressure
c	unit cost	p_{atm}	atmospheric pressure
$\dot{C}$	cost rate	p_i	pressure associated with mixture component i, partial pressure of i
C_aH_b	hydrocarbon fuel	p_r	relative pressure as used in Tables A-22
c_p	specific heat at constant pressure, $\partial h/\partial T)_p$	p_R	reduced pressure: p/p_c
c_v	specific heat at constant volume, $\partial u/\partial T)_v$	P	number of phases in the phase rule
c_{p0}	specific heat c_p at zero pressure	**P**	electric dipole moment per unit volume
e, E	energy per unit of mass, energy	pe, PE	potential energy per unit of mass, potential energy
e, E	exergy per unit of mass, exergy		
$e_f, \dot{E}_f$	specific flow exergy, flow exergy rate	$\dot{q}$	heat flux
$E_d, \dot{E}_d$	exergy destruction, exergy destruction rate	Q	heat transfer
$E_q, \dot{E}_q$	exergy transfer accompanying heat transfer, rate of exergy transfer accompanying heat transfer	$\dot{Q}$	heat transfer rate
		$\dot{Q}_x$	conduction rate
E_w	*exergy* transfer accompanying work	$\dot{Q}_c, \dot{Q}_e$	convection rate, thermal radiation rate
E	electric field strength	r	compression ratio
$\mathscr{E}$	electrical potential, electromotive force (emf)	r_c	cutoff ratio
f	fugacity	R	gas constant: $\bar{R}/M$, resultant force, electric resistance
$\bar{f}_i$	fugacity of component i in a mixture		
F	degrees of freedom in the phase rule	$\bar{R}$	universal gas constant
F, F	force vector, force magnitude	s, S	entropy per unit of mass, entropy
FA	fuel–air ratio	s°	entropy function as used in Tables A-22, absolute entropy at the standard reference pressure as used in Table A-23
g	acceleration of gravity		
g, G	Gibbs function per unit of mass, Gibbs function		
$\bar{g}_f^\circ$	Gibbs function of formation per mole at standard state	t	time
		T	temperature
h, H	enthalpy per unit of mass, enthalpy	T_R	reduced temperature: T/T_c
h	heat transfer coefficient	$\mathscr{T}$	torque
H	magnetic field strength	u, U	internal energy per unit of mass, internal energy
$\bar{h}_f^\circ$	enthalpy of formation per mole at standard state	v, V	specific volume, volume
		V, V	velocity vector, velocity magnitude
$\bar{h}_{RP}$	enthalpy of combustion per mole	v_r	relative volume as used in Tables A-22
HHV	higher heating value	v_R'	pseudoreduced specific volume: $\bar{v}/(\bar{R}T_c/p_c)$
i	electric current	V_i	volume associated with mixture component i, partial volume of i
k	specific heat ratio: c_p/c_v		
k	Boltzmann constant	W	work
K	equilibrium constant	$\dot{W}$	rate of work, or power
ke, KE	kinetic energy per unit of mass, kinetic energy	x	quality, position
l, L	length	X	extensive property
LHV	lower heating value	y	mole fraction, mass flow rate ratio
m	mass	z	elevation, position
$\dot{m}$	mass flow rate	Z	compressibility factor, electric charge
M	molecular weight, Mach number	$\dot{Z}$	cost rate of owning/operating
M	magnetic dipole moment per unit volume		

Greek Letters

α	isentropic compressibility
β	coefficient of performance for a refrigerator, volume expansivity
γ	coefficient of performance for a heat pump, activity coefficient
Δ	change = final minus initial
ε	exergetic (second law) efficiency, emissivity, extent of reaction
η	efficiency, effectiveness
θ	temperature
κ	thermal conductivity, isothermal compressibility
μ	chemical potential
μ_J	Joule–Thomson coefficient
ν	stoichiometric coefficient
ρ	density
$\sigma, \dot{\sigma}$	entropy production, rate of entropy production
σ	normal stress, Stefan–Boltzmann constant
Σ	summation
τ	surface tension
ϕ	relative humidity
ψ, Ψ	Helmholtz function per unit of mass, Helmholtz function
ω	humidity ratio (specific humidity), angular velocity

Subscripts

a	dry air
ad	adiabatic
as	adiabatic saturation
ave	average
b	boundary
c	property at the critical point, compressor, overall system
cv	control volume
cw	cooling water
C	cold reservoir, low temperature
db	dry bulb
e	state of a substance exiting a control volume
e	exergy reference environment
f	property of saturated liquid, temperature of surroundings, final value
F	fuel
fg	difference in property for saturated vapor and saturated liquid
g	property of saturated vapor
H	hot reservoir, high temperature
i	state of a substance entering a control volume, mixture component
i	initial value, property of saturated solid
I	irreversible
ig, if	difference in property for saturated vapor (saturated liquid) and saturated solid
isol	isolated
int rev	internally reversible
j	portion of the boundary, number of components present in a mixture
n	normal component
p	pump
ref	reference value or state
reg	regenerator
res	reservoir
P	products
R	reversible, reactants
s	isentropic
sat	saturated
surr	surroundings
t	turbine
tp	triple point
0	property at the dead state, property of the surroundings
o	stagnation property
v	vapor
w	water
wb	wet bulb
x	upstream of a normal shock
y	downstream of a normal shock
1,2,3	different states of a system, different locations in space

Superscripts

ch	chemical exergy
e	component of the exergy reference environment
–	bar over symbol denotes property on a molar basis (over X, V, H, S, U, Ψ, G, the bar denotes partial molal property)
·	dot over symbol denotes time rate
°	property at standard state or standard pressure
*	ideal gas, quantity corresponding to sonic velocity